ELSEVIER'S DICTIONARY OF PLANT NAMES OF NORTH AMERICA INCLUDING MEXICO

ELSEVIER'S DICTIONARY OF PLANT NAMES OF NORTH AMERICA INCLUDING MEXICO

in

Latin, Spanish (Mexican and European) and English (American)

compiled by

RICHARD C. WHITE
Hollister, CA, U.S.A.

2003
ELSEVIER
Amsterdam – Boston – Heidelberg – London – New York – Oxford
Paris – San Diego – San Francisco – Singapore – Sydney – Tokyo

ELSEVIER SCIENCE B.V.
Sara Burgerhartstraat 25
P.O. Box 211, 1000 AE Amsterdam, The Netherlands

First edition 2003

Library of Congress Cataloging in Publication Data
A catalog record from the Library of Congress has been applied for.

British Library Cataloguing in Publication Data
A catalogue record from the British Library has been applied for.

ISBN 0-444-51272-1

∞ The paper used in this publication meets the requirements of ANSI/NISO Z39.48-1992 (Permanence of Paper).
Printed in The Netherlands.

PREFACE

The border between the United States of America and Mexico is the busiest in the world. This area is also the meeting place of the two great cultures of the Western Hemisphere, Spanish-speaking Latin America and English-speaking North America. Recent demographic migrations coupled with increasing globalization have necessitated closer cooperation and communication between these groups.

As a botanist, the area of communication I am interested in expanding upon centers around the vernacular or common names of plants. Many recent immigrants from Mexico and further south have gained employment in areas of agriculture, landscaping, and commercial plant nurseries. The new residents also bring with them a rich history of herbal medicines that are becoming increasingly popular in the United States. Both groups share interests in each other's cuisines with respect to food plants and spices.

This dictionary contains the Latin, English, and Spanish names for over 7,000 species of the most important plants found mainly in the southwestern United States and northern Mexico. Included are native and naturalized plants as well as plants of economic value. The main section of the book is organized alphabetically by the scientific Latin genus and species of each plant. The book is also fully indexed by common names primarily in American-English and Mexican-Spanish.

Richard C. White
Hollister, California
March, 2003

PREFACIO

La frontera entre los Estados Unidos de América y México es la más activa en el mundo. Esta área es asimismo lugar de encuentro de dos de las grandes culturas del hemisferio occidental: la cultura hispanoparlante latinoamericana y la cultura angloparlante norteamericana. Los recientes flujos migratorios combinados con el incremento del fenómeno de globalización hacen necesaria una mayor cooperación y comunicación entre estos grupos.

El área de comunicación que me interesa como botánico se centra alrededor de los nombres vernáculos o comunes de las plantas. Muchos de los migrantes recientes de México y Centroamérica se han empleado en las áreas de agricultura, jardinería y viveros comerciales. Los nuevos residentes traen consigo una rica historia de plantas medicinales que se han popularizado en los Estados Unidos. Ambos grupos comparten un interés común en sus respectivas cocinas, en sus plantas y sus especias.

Este diccionario contiene los nombres en latín, inglés y español de más de siete mil especies de las plantas más populares que se encuentran principalmente en el suroeste de los Estados Unidos y el norte de México. Se incluyen plantas nativas e importantes asi como plantas de valor comercial. La sección principal de este libro está organizada alfabéticamente por la denominación científica en latín de cada planta por su género y especie. El libro cuenta también con un índice completo, arreglado por nombres communes en inglés americano y en español mexicano.

Richard C. White
Hollister, California
marzo, 2003

BIBLIOGRAPHY

Ajilvsgi, G. 1984. *Wildflowers of Texas*. Shearer Publishing. Fredericksburg, TX

Aleman, B.C. 1968. Flores de México. *Artes de México* 104. Nobilis-Lees. México, DF

Anderson, E.F. 2001. *The Cactus Family*. Timber Press. Portland, OR

Andrews, J. 1995. *Peppers, the Domesticated Capsicums*. University of Texas Press. Austin, TX

Anonymous. 2002. *La Guía de Plantas*. http://www.supercable.es/~criatura/plantasweb/

Anonymous. 2002. *Planthogar*. http://www.planthogar.net

Arzeni, C.B. & T.M. Simon. 1974. *Plants of Mexico*. Eastern Illinois University. Charleston, IL

Balick, M.J., M.H. Nee, & D.E. Atha. 2000. *Checklist of the Vascular Plants of Belize*. The New York Botanical Garden Press. Bronx, NY

Balls, E.K. 1972. *Early Uses of California Plants*. University of California Press. Berkeley, CA

Bayless, R. 2000. *Mexico One Plate at a Time*. Scribner. New York, NY

Benson, L. & R.A. Darrow. 1954. *The Trees and Shrubs of the Southwestern Deserts*. University of Arizona Press. Tucson, AZ

Bianchini, F. & A.C. Pantano. 1975. *Guía de Plantas y Flores*. Grijalbo. Barcelona, Spain

Bourke, D.O., L. Fanjul, & A.J. Rendell-Dunn. 1987. *Spanish-English Horticultural Dictionary*. C.A.B. International. Wallingford, UK

Brenzel, K.N., ed. 2001. *Sunset Western Garden Book*. Sunset Publishing Corp. Menlo Park, CA

Bright, W. 1998. *1500 California Place Names*. University of California Press. Berkeley, CA

Bye, R.A. 1986. Medicinal Plants of the Sierra Madre: Comparative Study of Tarahumara and Mexican Market Plants. *Economic Botany,* 40: 103-124

Cecchini, T. 1983. *Enciclopedia de las Hierbas y de las Plantas Medicinales*. Editorial De Vecchi. Barcelona, Spain

Cendroso Curiel, O. 1976. *Botánica*. Libreria de Porrua Hnos. México, DF

Chase, J.S. 1919. *California Desert Trails*. Houghton Mifflin. New York, NY

Clark, P. 1964. *A Guide to Mexican Flora*. Editorial Minutiae Mexicana. Mexico, DF

Clark, P. 1972. *A Flower Lover's Guide to Mexico*. Editorial Minutiae Mexicana. Mexico, DF

Clason, W.E. 1989. *Elsevier's Dictionary of Wild and Cultivated Plants*. Elsevier Science. Amsterdam, The Netherlands

Coyle, J. & N.C. Roberts. 1975. *A Field Guide to the Common and Interesting Plants of Baja California*. Natural History Publishing. La Jolla, CA

Creasy, R. 2000. *The Edible Mexican Garden*. Periplus. Boston, MA

Creasy, R. 2000. *The Edible Pepper Garden*. Periplus. Boston, MA

Cunningham, S. 1999. *Enciclopedia de las Hierbas Mágicas*. Llewellyn Español. St. Paul, MN

De la Rosa, F. 1992. *Yerbas y Plantas Medicinales de México*. Editores Mexicanos Unidos. México, DF

Del Barco, M. 1980. *The Natural History of Baja California*. Dawson's Book Shop. Los Angeles, CA

DeStefano, A.M. 2001. *Hierbas Buenas*. Random House Español. New York, NY

DeStefano, A.M. 2001. *Latino Folk Medicine*. Ballantine Books. New York, NY

Dewitt, D. & **P.W. Bosland.** 1996. *Peppers of the World*. Ten Speed Press. Berkeley, CA

Dimayuga, R.E. & **J. Agundez.** 1986. Traditional Medicine of Baja California Sur (México) I. *Journal of Ethnopharmacology,* 17: 183-193

Dimayuga, R.E., R.F. Murillo, & **M.L. Pantoja.** 1987. Traditional Medicine of Baja California Sur (México) II. *Journal of Ethnopharmacology,* 20: 209-222

Dodge, N.N. 1961. *Flowers of the Southwest Deserts*. Southwestern Monuments Association. Globe, AZ

Dressler, R.L. 1953. The Pre-Columbian Cultivated Plants of Mexico. *Botanical Museum Leaflets, Harvard University,* 16 (6)

Duke, J.A. 1998. *La Farmacia Natural*. Rodale Press. Emmaus, PA

Duke, J.A. 2001. *Tico Ethnobotanical Dictionary*. http://www.ars-grin.gov/duke/dictionary/tico/b.html

Elmore, F.H. 1976. *Shrubs and Trees of the Southwest Uplands*. Southwest Parks and Monuments. Globe, AZ

Emmart, E.W. 1940. *The Badianus Manuscript*. The Johns Hopkins Press. Baltimore, MD

Faucon, P. 2002. *Desert Tropicals*. http://www.desert-tropicals.com

Felger, R.S. 2000. *Flora of the Gran Desierto and Río Colorado of Northwestern Mexico*. The University of Arizona Press. Tucson, AZ

Felger, R.S., M.B. Johnson, & **M.F. Wilson.** 2001. *The Trees of Sonora, Mexico*. Oxford University Press. New York, NY

Ferguson, C.E. 1968. *Pequeño Glosario de Plantas y Árboles Útiles*. Agencia para el Desarrollo Internacional. México, DF

Ferris, R.S. 1927. *Preliminary Report on the Flora of the Tres Marías Islands*. Stanford University Press. Stanford, CA

Flores, G.J.A., et al. 1996. *Flora Representativa de Chianque*. Universidad Autónoma de Nuevo León. Monterrey, México

Flores, M. 2001. *Plant Common Names in English & Spanish*. http://www.floresflowers.com/EnglishAndSpanish.html

Ford, K.C. 1975. Las Yerbas de la Gente: A Study of Hispano-American Medicinal Plants. *University of Michigan, Anthropological Papers,* 60

Fultz, F.M. 1928. *Lily, Iris, and Orchid of Southern California*. Spanish American Institute Press. Gardena, CA

Geddes, B. 2000. *World Food Mexico*. Lonely Planet Publications. Victoria, Australia

Gentry, H.S. 1942. Rio Mayo Plants. *Carnegie Institution of Washington Publication* 527. Washington, DC

Gentry, H.S. 1982. *Agaves of Continental North America*. The University of Arizona Press. Tucson, AZ

Gerhard, P. & H.E. Gulick. 1962. *Lower California Guidebook*. Arthur H. Clark Company. Glendale, CA

Gomez-Pompa, A. 1985. *Los Recursos Bióticos de México: Reflexiones*. Editorial Alhambra Mexicana. México, DF

Gonzales, C. (n.d.). *Beliefs, Herbs, and Health*. No publisher

Gould, F.W. & R. Moran. 1981. The Grasses of Baja California, Mexico. *San Diego Society of Natural History,* 12

Grandtner, M.M. 2001. *World Dictionary of Trees*. Laval University, Quebec, Canada. http://www.wdt.qc.ca/

Griffin, J. 2000. *Hierbas de la Madre Naturaleza*. Llewellyn Español. St. Paul, MN

Hargreaves, D. & B. Hargreaves. 1960. *Tropical Blossoms of the Caribbean*. Hargreaves Company. Kailua, HI

Hargreaves, D. & B. Hargreaves. 1965. *Tropical Trees Found in the Caribbean, South America, Central America, Mexico*. Hargreaves Company. Kailua, HI

Haughton, C. 2001. *Purple Sage*. http://www.purplesage.org.uk/profiles

Heinerman, J. 1997. *Enciclopedia de Jugos Curativos*. Prentice Hall Press. Paramus, NJ

Heinerman, J. 1999. *Milagrosas Hierbas Curativas*. Prentice Hall Press. Paramus, NJ

Heinsen, V. (n.d.). *Mission San Antonio de Padua Herbs*. No publisher

Henderson, A., G. Galeano, & R. Bernal. 1995. *Field Guide to the Palms of the Americas*. Princeton University Press. Princeton, NJ

Herbario de Universidad de Sonora. 2001. *Catálago de Plantas Silvestres de Sonora*. http://aguaje.dictus.uson.mx

Herwig, R. 1981. *Enciclopedia de Plantas de Jardín*. Ediciones Omega. Barcelona, Spain

Hessayon, D. G. 1999. *Flores de Jardín*. Leopold Blume. Barcelona, Spain

Hessayon, D. G. 2001. *Árboles y Arbustos de Jardín*. Leopold Blume. Barcelona, Spain

Hessayon, D. G. 2001. *Plantas de Interior*. Leopold Blume. Barcelona, Spain

Heywood, V.H. 1978. *Flowering Plants of the World*. Mayflower Books. New York, NY

Hickman, J.C. 1993. *The Jepson Manual, Higher Plants of California*. University of California Press. Berkeley, CA

Hilton-Taylor, C. 2000. *IUCN Red List of Threatened Species*. http://www.redlist.org

Hitchcock, A.S. 1950. *Manual of the Grasses of the United States*. U.S. Government Printing Office. Washington, DC

Hoyos de Martens, V. & M.L.N. Palma de Villarreal. 1995. *Diccionario de Floricultura Ball/Ball Floriculture Dictionary*. Ball Publishing. Batavia, IL

Huxtable, R.J. 1983. Herbs along the Western Mexican-American Border. *Proceedings of the Western Pharmacological Society,* 26: 185-191

International Seed Testing Association. 1982. *A Multilingual Glossary of Common Plant Names*

Jaeger, E.C. 1940. *Desert Wildflowers*. Stanford University Press. Stanford, CA

Kasbeer, T. 1971. *Flora of Baja Norte*. La Siesta Press. Glendale, CA

Kay, M.A. 1978. The Florilegio Medicinal: Source of Southwest Ethnomedicine. *Ethnohistory,* 24: 251-259

Kay, M.A. 1996. *Healing with Plants in the American and Mexican West*. The University of Arizona Press. Tucson, AZ

Kays, S. & J.C. Silva Dias. 1998. Common Names of Commercially Cultivated Vegetables of the World in 15 Languages. *Economic Botany,* 49 (2): 115-152

Kearney, T.H. and R.H. Peebles. 1969. *Arizona Flora*. University of California Press. Berkeley, CA

Kelsey, H.P. & W.A. Dayton. 1942. *Standard Plant Names*. J. Horace McFarland Company. Harrisburg, PA

Kennedy, P.B. 1917. *Annotated List of the Wild Flowers of California*. Leviston Printing Co. San Francisco, CA

Lamb, S.H. 1989. *Woody Plants of the Southwest*. Sunstone Press. Santa Fe, NM

Langman, I.K. 1956. Botanical Gardens in Ancient Mexico. *Missouri Botanical Garden Bulletin,* Feb.

Leavenworth, W.C. 1946. A Preliminary Study of the Vegetation of the Region Between Cerro Tancitaro and the Rio Tepalcatepec, Michoacan, Mexico. *The American Midland Naturalist,* 36: 137-206

Lenz, L.W. 1992. *Plants of the Cape Region*. The Cape Press. Claremont, CA

Little, E.L. 1968. Southwestern Trees. *Agriculture Handbook,* no. 9. U.S. Dept. of Agriculture. Washington, DC

Lozoya, X. 1998. *La Herbolaria en México*. Consejo Nacional para la Cultura y las Artes. México, DF

Martínez, M. 1936. *Plantas Útiles de México*. Ediciones Botas. México, DF

Martínez, M. 1937. *Catálogo de Nombres Vulgares y Científicos de Plantas Mexicanas*. Ediciones Botas. México, DF

Martínez, M. 1947. *Baja California, Reseña Histórica del Territorio y de su Flora*. Ediciones Botas. México, DF

Martínez, M. 1959. *Las Plantas Medicinales de México*. Ediciones Botas. México, DF

Martínez, M. 1963. *Las Pináceas Mexicanas*. Universidad Nacional Autónoma de México. México, DF

Martínez, M. 1979. *Catálogo de Nombres Vulgares y Científicos de Plantas Mexicanas*. Fondo de Cultura Económica, México, DF

Mason Jr., C.T. & P.B. Mason. 1987. *A Handbook of Mexican Roadside Flora*. The University of Arizona Press. Tucson, AZ

Matschat, C.H. 1935. *Mexican Plants for American Gardens*. Houghton Mifflin. New York, NY

McKechnie, J.L., ed. 1983. *Webster's New Twentieth Century Dictionary of the English Language*. Second edition, unabridged. Simon and Schuster. New York, NY

Mejía M., M. 1984. *Nombres Científicos y Vulgares de Especies Forrajeras Tropicales*. Centro Internacional de Agricultura Tropical. Cali, Colombia

Mejía-Saulés, M.T. & P.D. Aranda. 1992. *Gramíneas Útiles de México.* Universidad Nacional Autónoma de México. México, DF

Merrill, R. & J. Ortiz. 2000. *The Gardener's Table.* Ten Speed Press. Berkeley, CA

Mielke, J. 1993. *Native Plants for Southwestern Landscapes.* University of Texas Press. Austin, TX

Miller, G.O. 1991. *Landscaping with Native Plants of Texas and the Southwest.* Voyageur Press. Stillwater, MN

Minnich, R.A. & E.F. Vizcaíno. 1998. *Land of Chamise and Pines.* University of California Press. Berkeley, CA

Mitchell, A. 1979. *Guía de Campo de los Árboles de Europa.* Ediciones Omega. Barcelona, Spain

Molino, R.T. 2001. *Flora y Vegetación de España.* http://www.unex.es/botanica/flora/index.htm

Mölzer, V. 1989. *Plantas de Jardín.* Susaeta. Madrid, Spain

Moore, M. 1979. *Medical Plants of the Mountain West.* Museum of New Mexico Press. Santa Fe, NM

Moore, M. 1989. *Medicinal Plants of the Desert and Canyon West.* Museum of New Mexico Press. Santa Fe, NM

Moore, M. 1990. *Los Remedios.* Red Crane Books. Santa Fe, NM

Moore, M. 1997. *Southwest School of Botanical Medicine.* Bisbee, Arizona. http://chili.rt66.com/hrbmoore/HOMEPAGE/HomePage.html

Moser, M.B. & S.A. Marlett. 1999. *Seri Dictionary: Plants.* Summer Institute of Linguistics, University of North Dakota. http://www.und.nodak.edu/dept/linguistics/wp/1999Moser-Marlett-Plants.PDF

Nash, J.R. 1997. *Spanish-English-Latin Plant Dictionary.* http://www.echonet.org/tropicalag/spanish.engligh.dictionary.htm

Niehaus, T.F., C.L. Ripper, & V.Savage. 1984. *Southwestern and Texas Wildflowers.* Houghton Mifflin Company. Boston, MA

O'Gorman, H. 1961. *Mexican Flowering Trees and Plants.* Ammex Associados. Mexico, DF

Parsons, M.E. 1912. *The Wild Flowers of California.* Cunningham, Curtiss & Welch. San Francisco, CA

Patterson, J. 2001. *Spanish-English Fruit, Herb, and Vegetable Commodity Reference with Botanical Names.* http://www.nappo.org/span-dic_e.htm

Pennington, C.W. 1973. Plantas Medicinales Utilizadas por el Pima Montañés de Chihuahua. *Américana Indigena,* 33: 213-232

Pesman, M.W. 1962. *Meet Flora Mexicana.* Dale Stuart King. Globe, AZ

Phillips, J. 1987. *Southwestern Landscaping with Native Plants.* The Museum of New Mexico Press. Santa Fe, NM

Phillips, R. & M. Rix. 1993. *Vegetables.* Random House. New York, NY

Phillips, S.J. & P.W. Comus. 2000. *A Natural History of the Sonoran Desert.* Arizona-Sonora Desert Museum Press. Tucson, AZ

Polunin, O. & **B.E. Smythies.** 1981. *Guía de Campo de las Flores de España*. Ediciones Omega. Barcelona, Spain

Porcher, M.H., et al. 2001. *Searchable World Wide Web Multilingual Multiscript Plant Name Database*. The Institute of Land and Food Resources, University of Melbourne, Australia.
http://gmr.landfood.unimelb.edu.au/Plantnames/

Ramamoorthy, T.P., R. Bye, J. Fa, & **A. Lot.** 1993. *Biological Diversity of Mexico: Origins and Distribution*. Oxford University Press. New York, NY

Real Academia Española. 1970. *Diccionario de la Lengua Española*. Decimonovena Edición. Madrid, Spain

Record, S.J. & **R.W. Hess.** 1943. *Timbers of the New World*. Yale University Press. New Haven, CT

Rhode, D. 2002. *Native Plants of Southern Nevada*. The University of Utah Press. Salt Lake City, UT

Ritchason, J. 1984. *La Pequeña Enciclopedia de Hierbas*. Woodland Books

Rivera Morales, I. & **F. Miranda.** 1942. Nombres Vulgares de Plantas en el S.O. del Estado de Puebla. *An. Inst. Biol. Mex.*, 13: 493-498

Roberts, N.C. 1989. *Baja California Plant Field Guide*. Natural History Publishing. La Jolla, CA

Rohwer, J.G. 2002. *Tropical Plants of the World*. Sterling Publishing. New York, NY

Rose, J.N. 1899. Notes on Useful Plants of Mexico. *U.S. National Herbarium,* vol. V, no. 4

Rzedowski, J. 1986. *Vegetación de México*. Editorial Limusa. México, DF

Rzedowski, J. & **G.C. de Rzedowski.** 1979. *Flora Fanerogámica del Valle de México*. Vol. I. Compañia Editorial Continental. México, DF

Rzedowski, J. & **G.C. de Rzedowski.** 1985. *Flora Fanerogámica del Valle de México*. Vol. II. Instituto de Ecología. México, DF

Rzedowski, J. & **M. Equihua.** 1987. *Atlas Cultural de México: Flora*. Grupo Editorial Planeta. México, DF

Sánchez de Lorenzo Cáceres, J.M. 2000. *Buscaplantas*.
http://www.floraguide.es/buscaplantas/index.htm

Sanchez Sanchez, O. 1984. *La Flor del Valle de México*. Editorial Herrero. México, DF

Saunders, C.F. 1919. *With the Flowers and Trees in California*. Robert M. McBride & Co. New York, NY

Saunders, C.F. 1927. *The Wild Gardens of Old California*. Wallace Hebberd. Santa Barbara, CA

Saunders, C.F. 1933. *Western Wild Flowers*. Doubleday, Doran. Garden City, NJ

Schlesinger, V. 2001. *Animals & Plants of the Ancient Maya*. University of Texas Press. Austin, TX

Schoenhals, L.C. 1988. *A Spanish-English Glossary of Mexican Flora and Fauna*. Summer Institute of Linguistics

Schultes, R.E. & **A. Hofmann.** 1992. *Plants of the Gods*. Healing Arts Press. Rochester, VT

Septien, C.V. 1994. *De México al Mundo – Plantas*. Grupo Azabache. México, DF

Shreve, F. & I.L. Wiggins. 1964. *Vegetation and Flora of the Sonoran Desert.* 2 Vol. Stanford University Press. Stanford, CA

Standley, P.C. 1920-1926. Trees and Shrubs of Mexico. *Contributions from the U.S. National Herbarium,* vol. 23 (1982, Strauss & Cramer)

Thurston, C. 1936. *Wildflowers of Southern California.* Esto Publishing. Pasadena, CA

Tierney, G.D. 1983. *Roadside Plants of Northern New Mexico.* The Lightning Tree. Santa Fe, NM

Turner, R.M., J.E. Bowers, & T.L. Burgess. 1995. *Sonoran Desert Plants.* The University of Arizona Press. Tucson, AZ

Universidad Michoacana. 1998. *Museo Virtual.* http://www.ccu.umich.mx/museo/hist-natural/botanica/botanica.html

University of Bonn, Germany. Department of Botany. *Gymnosperm Database.* http://www.botanik.uni-bonn.de/conifers

Vázquez-Yanes, C.A., et al. 1999. *Árboles y Arbustos Potencialmente Valiosos para la Restauración Ecológica y la Reforestación.* Reporte Técnico del Proyecto J084. CONABO – Instituto de Ecología, UNAM. http://www.conabio.gob.mx/arboles/indice_especies.html

Weber, Msgr. F.J. 1978. *Andrew Garriga's Compilation of Herbs & Remedies Used by the Indians and Spanish Californians.* The Plantain Press. Los Angeles, CA

Weniger, D. 1991. *Cacti of Texas and Neighboring States.* University of Texas Press. Austin, TX

West, S. 2000. *Northern Chihuahuan Desert Wildflowers.* Falcon Publishing. Helena, MT

White, S.S. 1948. The Vegetation and Flora of the Region of the Rio Bavispe in Northeastern Sonora, Mexico. *Lloydia,* 11 (4): 229-302

Wiggins, I.L. 1980. *Flora of Baja California.* Stanford University Press. Stanford, CA

Winkelman, M. 1986. Frequently Used Medicinal Plants in Baja California Norte. *Journal of Ethnopharmacology,* 18: 109-131

Wooton, E.O. & P.C. Standley. 1915. Flora of New Mexico. *Contributions from the United States National Herbarium,* vol. 19. U.S. Government Printing Office. Washington, DC

Wright, N.P. 1958. *Orquídeas de México.* Editorial Fournier. México, DF

Zingg, R.M. 1932. Mexican Folk Remedies of Chihuahua. *Journal of the Washington Academy of Sciences,* 22 (7): 174-181

Zwinger, A. 1961. *A Desert Country Near the Sea.* Harper & Row. New York, NY

Basic Table

A

Abelia floribunda – abelia – Mexican abelia
Abelia* x *grandiflora – abelia – glossy abelia
Abelmoschus (Hibiscus) manihot – pajiza, viuda
Abelmoschus moschatus (Hibiscus abelmoschus) – abelmosco, algalia, almizclillo, antiviperina, hierba de la culebra, semilla de culebra, viborana – ambrette, musk mallow, silk flower
Abelmoschus = Hibiscus
Abies amabilis – abeto amábilis, abeto del Pacífico, abeto rojo americano – cascade fir, Pacific silver fir, silver fir
Abies balsamea – abeto balsámico, bálsamo del Canadá – balsam fir
Abies bracteata (venusta) – abeto bracteado, abeto cedroso, abeto de Santa Lucía, árbol de incienso – bristlecone fir, Santa Lucia fir
Abies cephalonica – abeto griego
Abies concolor – abeto, abeto del Colorado, pinabete, pino blanco, pino real blanco – white fir
Abies durangensis – guayamé, oyamel, pinabete – Durango fir
Abies excelsa – abeto
Abies grandis – abeto de Vancuver, abeto gigante, abeto grande americano – giant fir, grand fir
Abies guatemalensis – abeto, abeto guatemalteco, pinabete, plumajatzin – Guatemalan fir
Abies hickeli – ocopetla, pinabete, plumajillo, plumajillo de montaña
Abies koreana – abeto de Corea – Korean fir
Abies lasiocarpa – abeto blanco americano, abeto blanco de las rocosas, abeto blanco de las sierras, abeto lasiocarpa americano, blanco de las sierras, pino real, pino real blanco, pino real blanco de las sierras, pino real de las sierras – alpine fir, Rocky Mountain fir, subalpine fir
Abies magnifica – abeto, abeto magnífico – California red fir
Abies magnifica* var *shastensis – abeto rojo shasta, abeto shasta – Shasta red fir
Abies mexicana – guayamé blanco, pinabete
Abies nordmanniana – abeto del Cáucaso, abeto de Nordmann – Nordmann fir
Abies pectinata (alba) – abeto, abeto blanco, abeto común, abeto plateado, pinabete – European silver fir
Abies pinsapo – abeto de España, abeto español, pinsapo – Spanish fir
Abies procera (nobilis) – abeto azul de Oregón, abeto noble, abeto prócer – noble fir
Abies religiosa – abeto, acxoyatl, bansú, cahuite, guayamé, hoja de petate, jalocote, ocopetla, oyamel, oyametl, picea común, pinabete, pino, pino oyamel, tucumu, xalócotl – sacred fir
Abies sibirica – abeto ruso, abeto siberiano – Siberian fir
Abies vejari – abeto, guayamé blanco, hallarín, oyamel – vejar fir
Abronia fragrans – alfambrilla, lechuguilla – heart's-delight, Lasater's pride, sweet sand verbena
Abronia gracilis – alfambrilla, alfombrilla – sand verbena
Abronia maritima – alfambrilla, alfombrilla – beach sand verbena
Abronia umbellata – alfambrilla – sand verbena
Abronia villosa – alfambrilla – desert ground-cover, desert sand verbena, sand verbena
Abrus precatorius – jequirity, oxoak, peonía, semilla de culebra, xocoac – prayer bean, rosary pea
Abutilon abutiloides – pintapan
Abutilon aurantiacum – pelotazo
Abutilon avicennae – hierba de la China, malva índica, malvavisco de la India – button weed
Abutilon fruticosa – pelotazo – Indian mallow
Abutilon hypoleucum – monacillo, tzacotxójol – Rio Grande abutilon
Abutilon incanum – escoba, malva, malva peluda, pelotazo, pelotazo chico, tronador – Indian mallow
Abutilon lignosum – colotahue, cotolahue, malva, mizbil, pelotazo, pelotazo bronco, yax-holché, zacxiu
Abutilon megapotamicum – abutilon trepador – flowering maple, trailing abutilon

Abutilon notolophium – canastilla, mojagua – abutilon
Abutilon reventum – malva colorada
Abutilon sonorae – pintapan cimarrón
Abutilon umbellatum – malva, pelotazo peludo, sacxiu
Abutilon vitifolium – abutilon, malvavisco de las Indias
Acacia angustifolia – cantemó, timbre
Acacia angustissima – barba de chivo, cantemó, cheramusco, day, gavia, guajillo, palo de pulque, timbre, timbrillo, uaxim, xaax, xisich – fern acacia, prairie acacia, shame-weed, white-ball acacia
Acacia armata – acacia de espinas, espino de canguro – kangaroo thorn
Acacia berlandieri – espino, guajillo, huajillo, huispancle, matorral, mimbre, thóbem – guajillo, thornless catclaw
Acacia brandegeana – agigandú, teso, vinorama – acacia
Acacia constricta – chaparro prieto, gigantillo, huisache, largancillo, largoncillo, mezquitillo, vara prieta, vinorama – Mescat acacia, whitethorn, whitethorn acacia
Acacia coulteri – guajillo, huajillo, maguó, palo blanco, palo de arco, tepeguaje
Acacia crinita – gato, nudita
Acacia cyclops – acacia de la costa, acacia majorera
Acacia cymbispina (cochlicantha) – cherahuí, chirahui, chiroui, corteno, cubata, cucharilla, cucharita, cucharo, encinilla, espino, huinol, huinole, huisache, huisache costeño, huisache tepame, palo de cucharitas, sínala, vinolo, vinolo prieto, vinorama – boat-thorn acacia
Acacia dealbata – acacia mimosa, mimosa, mimosa plateada – silver wattle
Acacia decurrens – mimosa común
Acacia farnesiana – aromo, aromo espinoso, bihi, cantiriz, cascalote, chapote, cuca, espinillo, espino, espino blanco, fiñisachi, flor de niño, gavia, huichaxin, huisache, huisache de semilla, huisache yóndiro, huitzaxin, kankilizche, kantiriz, kuká, matita, mimosa, subinché, uña de cabra, vinorama, zubin, zubinache – cassie, huisache, sweet acacia
Acacia gladiata – ishcanal negro – flat-thorn acacia
Acacia goldmanii – frijolillo
Acacia greggii – algarrobo, cimmarón, gatuña, palo chino, patito, tepame, tésota, tis, uña de gato – catclaw, catclaw acacia, devil's claw
Acacia hindsii – carretadero, cornezuelo, huisache costeño, ishcanal, ixiconal, jarretadera – hat-thorn acacia
Acacia longifolia – acacia falsa, mimosa dorada de Sidney – golden wattle, Sydney golden wattle
Acacia (Poponax) macracantha – algarrobo, chimay, cornezuelo de la playa, cují negro, espino de tinta, faique, garrobo, güinote blanco, hediondo, úveda, vinolo blanco
Acacia mcmurphyi – frijolillo
Acacia melanoxylon – acacia de madera negra, mimosa – blackwood acacia
Acacia millefolia – culuncuahui, mauto, tepemezquite blanco, vara prieta – Santa Rita acacia
Acacia occidentalis – teso, tésota, tesoto, uasiva, uña de gato – tree catclaw
Acacia peninsularis – palo chino – acacia
Acacia pennatula – algarrobo, algarrobo espino, cajui, chírahui, coquette, cucabit, cuquet, espino, espino blanco, garrobo, huisache, huisache tepame, palo garrobo, tepame, xabxib, yepouecha – feather acacia, fernleaf acacia, Sierra Madre acacia
Acacia pringlei – cherahuí, gavia, guamúchil, guamuchilillo, teso, vinorama
Acacia pringlei (california)* ssp *california – cherahuí, chicoria, guamuchilillo, teso, vinorama
Acacia retinodes (floribunda) – acacia plateada, mimosa – everblooming acacia
Acacia rigidula – chaparro prieto, gavia – blackbrush acacia, catclaw
Acacia roemeriana – palo de arco – Roemer acacia
Acacia saligna – acacia de hojas azules – blue-leaf wattle
Acacia schaffneri – huisache, huisache chino
Acacia senegal – acacia gomosa, goma arábiga – gum Arabic
Acacia seyal – acacia – shittah tree
Acacia spadicigera (cornigera) – árbol de cuerno, chixcanal, cornezuelo, cuernito, cuernos del toro, espino blanco,

hoitzmamazali, huitzmamaxalli, ishcanal, tepame, torito, vinorama, zubin, zubinché – bullhorn acacia

Acacia villosa – dais

Acacia willardiana – palo blanco, palo liso, vara blanca – palo blanco, Willard's acacia

Acacia wrightii – negra, uña de gato, uña de gato negro – Texas catclaw

Acaena elongata – cadillo, cardillo, pegarropa, tlalsometl

Acaena novae-zelandiae – planta erizo – biddy-biddy

Acalypha aliena – hierba del cáncer

Acalypha arvensis – cola de gato, espinosilla, gatito, hierba del cáncer, hierba del gusano – cattail

Acalypha californica – hierba de la gangrena, hierba del cáncer – California copperleaf

Acalypha comonduana – hierba del cáncer

Acalypha hispida – acalifa, califa, cola de gato, cola de zorro roja, nemiz, rabo de gato, rosa de mono – chenille plant, red cats-tail

Acalypha lindheimeri – duraznillo, hierba del cáncer – copper leaf, three-seeded mercury

Acalypha wilkesiana (tricolor) – acalifa, capa roja, hoja de cobre, payasito, planta bistec – beefsteak plant, copper leaf, fire dragon, Jacob's coat, match-me-if-you-can

Acantherum brachychaetum (Stipa brachychaeta) – pasto puna – spear grass

Acanthocereus occidentalis – tasajo

Acanthocereus subinermis – pitahaya

Acanthocereus (Cereus) tetragonus (pentagonus) – chacha, chaco, jacube, nuntzutzuy, órgano, pitahaya, pitahaya marítima, pitahaya morada, pitahaya naranjada, pitaya, tuna silvestre – barbed-wire cactus

Acanthopanax = Eleutherococcus

Acanthus balcanicus – acanto balcánico – Balkan acanthus

Acanthus mollis – acanto, acanto mullido, branca ursina, carnerona, hierba gigante – bear's breech

Acanthus spinosus – acanto espinoso – spiny acanthus

Acca = Feijoa

Acer brachypterum – arce, encino haya, haya, palo de azúcar, palo santo

Acer buergeranum – arce trífido, arce trinerve – trident maple

Acer campestre – arce menor, arce silvestre – hedge maple

Acer capillipes – arce rojo japonés de corteza serpentina

Acer cappadocicum – arce de capadocia – coliseum maple

Acer davidii – arce chino del padre David – David's maple

Acer griseum – arce chino gris – paperbark maple

Acer japonicum – arce afelpado japonés, arce afelpado del Japón, arce japonés – fullmoon maple

Acer macrophyllum – arce de Oregón, arce macrofilo – bigleaf maple

Acer negundo – acecinte, acezintle, arce, arce de hoja ceniza, arce de hojas de fresno, arce negundo, negundo – box elder

Acer palmatum – arce de Japón, arce japonés, arce palmado japonés, arce palmado, arce polimorfo – Japanese maple

Acer pensylvanicum – arce de Pensilvania

Acer platanoides – arce noruego, arce platanoide, arce real, plátano – Norway maple

Acer pseudoplatanus – arce blanco, arce plátano falso, falso plátano, plátano común, plátano falso, sicomoro – sycamore maple

Acer pseudosieboldianum – arce coreano

Acer rubrum – arce rojo americano – red maple, scarlet maple

Acer saccharinum – acero blanco, acero blanco americano, arce plateado, arce sacarino, maple – silver maple

Acer saccharum – arce azucarero, arce de azúcar, arce sacarino, meple – sugar maple

Acer saccharum (grandidentatum)* ssp *grandidentatum – palo de azúcar – bigtooth maple, Rocky Mountain sugar maple

Acer tartaricum – arce tártaro – Tartarian maple

Acer tartaricum (ginnala)* var *ginnala – arce del Amur, arce de Manchuria, arce ruso – Amur maple

Aceras anthropophorum – flor del hombre ahorcado – man orchid

Achillea ageratifolia – agérato, altarreina, hierba julia – Greek yarrow

Achillea clavennae – milenrama blanca – silvery yarrow
Achillea filipendulina – aquilea amarilla, aquilea de hojas de filipéndula, milefolio, milenrama – fernleaf yarrow
Achillea lanulosa – plumajillo, real de oro – western yarrow
Achillea millefolium (borealis) – achilea, alcaina, alcanfor, alcanforero, altarreina, aquilea, artemisa bastarda, cientoenrama, flor de pluma, hierba de Aquiles, hierba de la cortadura, hierba de la muela, hierba del carpintero, hierba del cortes, hierba del soldado, milefolio, milenrama, milhojas, perla, plumajillo, plumajo, real de oro, tlalquequetzal – milfoil, yarrow
Achillea ptarmica – botón de plata, hierba estonutatoria, ptármica – sneezewort, yarrow
Achillea tomentosa – milefolio alpino, milenrama alpina, milenrama amarilla – woolly yarrow
Achimenes grandiflora – patito
Achnatherum (Oryzopsis) hymenoides – arroz indio – Indian ricegrass
Achras (Manilkara) bidentata – balata – balata tree, bullet wood
Achras chicle – chicozapote de hoja ancha, zapotillo – crown gum
Achras zapota (Manilkara achras, M. zapotilla) – balata, caucho, chapote, chicle, chicle zapote, chico, chicozapote, chicozapotl, chiku, guendaxiña, muyozapot, níspero, palo María, peruétano, tzaput, xeink, xicotzápotl, ya', zapote, zapote blanco, zapote chico, zapote colorado, zapote de abejas, zapote de chicle, zapote montés, zapote morado, zapote-níspero, zapotillo, zaya – chicle, naseberry, sapodilla
Achyranthes aspera – rabo de gato, zorrillo blanco, zorro – prickly chaff-flower
Acineta barkeri – boca de león
Acleisanthes longiflora – hierba de la rabia, hierba santa, plateada, trompetilla – angel's trumpet, angel trumpets
Acoelorraphe wrightii – chi-it, palma, palmetto, papta, pim-int, prementa, primenta, taciste, tasiste – Everglades palm, hairy Tom palmetto, paurotis palm, pimento palm
Aconitum napellus – acónito, acónito común, alva lupina, anapelo, capucha de monje, casco de Júpiter, cogulla de fraile, matalobos, napelo, pardal, raíz del diablo, uva verga, vedegambre azul – garden monkshood, monkshood, wolfsbane, wolf's-bane
Acorus calamus – ácoro, ácoro verdadero, cálamo aromático, cálanis, iris amarillo, lirio de mirto, lirio dulce – flagroot, sweet flag
Acorus gramineus – ácoro – Japanese sweet flag
Acourtia (Perezia) adnata – cola de zorra, espantavaqueros, hojarasca, pipitzáhuac
Acourtia cordata (Perezia hebeclada) – coapatli, pipitzáhuac, tairago, xararo, zazanaca
Acourtia (Perezia) microcephala – sacapeyote
Acourtia (Perezia) montana – pipichowa
Acourtia (Perezia) nana – pichichagua – deseret holly
Acourtia (Perezia) runcinata – peonía
Acourtia (Perezia) thurberi – árnica, cola de zorra, hierba de la tarantuala, matagusanos, matarique morada, pipichowa
Acourtia wrightii (Perezia arizonica) – pipitzáhuac – brownfoot, desert holly, perezia
Acroclinium = Helipterum
Acrocomia aculeata – cocoyol, coyol, sipa – suppa palm
Acrocomia crispa – corojo – corojo palm
Acrocomia mexicana – cocoyol, colconab, coquito baboso, corojo, corozo, coyol, coyol baboso, coyol redondo, cuau-coyotli, guacoyul, palma, palma coca, palma de coyol, palma redonda, sago, tuk, xocoyol – coyal palm, feather palm
Acrocomia sclerocarpa – mocaya – macaw palm
Acrocomia totai – palmera totai – totai palm
Acrocomia vinifera – coyol, palma coyol
Acrostichum aureum – helecho, helecho de manglar – mangrove fern, tiger bush
Actaea arguta – hierba del peco – baneberry
Actaea rubra – hierba del peco – baneberry
Actaea spicata – actea, hierba de San Cristóbal – baneberry, herb Christopher
Actinea gaillardia – manta india
Actinea = Hymenoxys
Actinidia deliciosa (chinensis) – grosellero chino, kiwi – kiwi

Adansonia digitata – baobab – baobab tree
Adelia virgata – ona újugo, pimentilla
Adenanthera pavonina – condori, coralito, cresta de pavo – redhead, sandal beadtree
Adenium obesum – rosa del desierto – desert rose, impala lily, Japanese frangipani, mock azalea, sabi star
Adenophora liliifolia – adenófora olorosa, vela escamada – lady bell
Adenophyllum porophylloides – cambray, cempasúchil chiquito
Adenostoma fasciculatum – chamizo, chamizo de vara prieta, vara prieta – chamise, greasewood
Adenostoma sparsifolium – chamizo, chamizo colorado, chamizo de vara colorada, hierba del pasmo, palo amarillo – red shank, ribbon bush
Adiantum aleuticum (pedatum) – adianto del Canadá, culantrillo – five-finger fern, maidenhair fern
Adiantum capillus-veneris – adianto, arañuela, avenca, cabellos de Venus, capilaria, culantrillo, culantrillo de ojo de agua, culantrillo de pozo, helecho, helecho de culantrillo – maidenhair fern, southern maidenhair, Venus-hair fern
Adiantum concinnum – culantrillo de México – maidenhair fern
Adiantum hispidulum – culantrillo – rosy maidenhair
Adiantum raddianum (cuneatum) – culantrillo, helecho – maidenhair fern
Adiantum tenerum – culantrillo, culantrillo de México, culantrillo de pozo, palo negro, pich, rok-chik-kuán – black stick maidenhair
Adiantum trapeziforma – culantrillo, culantrillo de pozo, tlalquequetzal
Adiantum trapezoides – helecho
Adiantum wilsonii – rah-li-ch'och, Santa María
Adolphia californica – junco – California adolphia, spinebush, spineshrub
Adolphia infesta – abrojo, junco
Adonis aestivalis – adonis de verano – summer pheasant's eye
Adonis annua – adonis de otoño – pheasant's eye
Adonis flamea – adonis escarlato, ojo de perdiz – large pheasant's eye
Adonis vernalis – adonis, adonis amarillo, adonis de primavera, adonis vernal, ojo de perdiz – yellow pheasant's eye
Aechmea bracteata – axs'que, chacana, chaccanahzihi, flor de tecolón, sobrepalo, tobe
Aechmea chantinii – acmea
Aechmea fasciata – acmea, billbergia – silver vase, urn plant
Aechmea fulgens – acmea
Aechmea lueddemanniana – gallibasco – wild pine
Aechmea mexicana – axs'que
Aechmea = Ananas
Aegilops cylindrica – egílope cilindrica – jointed goatgrass
Aegilops ovata – egílope geniculada – goatgrass
Aegilops triuncialis – egílope de tres uncias – barbed goatgrass
Aegle marmelos – bael, bela, membrillo de Bengala – bael fruit
Aegopodium podagraria – dauco, hierba de San Gerardo – bishop's weed, gout weed, ground elder
Aeonium arboreum – eonio de las Canarias
Aeonium canariense – eonio de las Canarias
Aeschynomene americana – tamarindillo – American jointvetch
Aeschynomene nivea – mata de Dios, popotillo, vara prieta – jointvetch
Aesculus californica – berraco, hierba del pescado, trompo – California buckeye
Aesculus × carnea – castaño de Indias rojo, castaño rubio – red horse chestnut
Aesculus hippocastanum – castaño caballuno, castaño de Indias, castaño de Indias común, castaño de la India, castaño falso, falso castaño – horse chestnut
Aesculus parryi – berraco, trompo – buckeye
Aesculus pavia – castaño de frutos lisos, falso castaño de flores rojas, pavía roja – red buckeye
Aesculus urbinata – castaño japonés – Japanese horse chestnut tree
Aethionema schistosum – etiónema – stonecress

Aethusa cynapium – apio de perro, cicuta menor, etusa, perejil de perro – fool's parsley
Aframomum melegueta – alguita, grano del paraíso, malagueta – grains of paradise, Malaguetta pepper
Agapanthus africanus – agapanto, azucena africana, lirio africano – agapanthus, lily-of-the-Nile
Agaricus arvensis – champiñón grande, hongo blanco – horse mushroom
Agaricus augustus – champiñón grande, mazayel
Agaricus bisporus* var *hortensis – champiñón, champiñón de París – cultivated mushroom
Agaricus bitorquis – agárico bianillado, llanero – white mushroom
Agaricus (Psaliota) campestris – champiñón, champiñón de campo, corralito, hongo, hongo común, hongo corriente, hongo de campo, hongo de San Juan, hongo de sombrero, llanero, seta, tarecua, yotito – field mushroom, meadow mushroom
Agaricus silvicola – champiñón anisado, champiñón grande – wood mushroom
Agastache mexicana – tlalámatl, toronjil, toronjil del país, toronjil morado – giant Mexican lemon hyssop
Agastache micrantha – húpachi, té de menta – giant hyssop, horsemint
Agastache urticifolia – toronjil – horsemint, hyssop
Agathis alba – copal de Manila – white Dammar pine
Agathosma = Barosma
Agave americana – agave, cabuya, henequén blanco, maguey, maguey chichimeco, maguey de mezcal, maguey de pulque, maguey meco, maguey pinto, mecoztli, mezcal, pita, pita común, saqui, teometl – agave, century plant
Agave angustifolia – babki, cántula, maguey de mezcal, maguey lechuguilla, mezcal de bacanora, mescal de maguey, zapupe
Agave applanata – maguey de ixtle, maguey de la casa, socolume
Agave asperrima – maguey, maguey cenizo, maguey de cerro – century plant
Agave atrovirens – amole, maguey, maguey del cumbre, maguey de pulque, maguey javalin, maguey manso, maguey manso fino, maguey pulquero, maguey tobala, maguey verde grande, metl, mezcal, raíz de maguey, teometl, tepeme, tlacametl, tobala, tuah – pulque agave
Agave atrovirens* var *mirabilis – maguey blanco, maguey cenizo
Agave attenuata – cola de león – drooping agave, foxtail agave
Agave aurea – lechuguilla, lechuguilla mezcal, maguey, mezcal – mescal, agave
Agave bovicornuta – lechuguilla, lechuguilla sabalí, lechuguilla verde, mezcal lechuguilla, noriba, sapari, sapulí, zabila – cow horn agave
Agave cantala – cantala, maguey, maguey de Manila – cantala maguey, Manila maguey
Agave capensis – mezcal
Agave cocui – cocuí
Agave colorata – cenizo, mezcal, mezcal cenizo
Agave datylio – datilillo, datillo, mezcal, mezcalillo – agave
Agave dentiens – mezcal – agave
Agave deserti – amul, agave, maguey, mezcal – desert agave
Agave deserti* ssp *simplex – lechuguilla, mezcal – desert agave
Agave deweyana – maguey mezatillo, zapupe, zapupe de Huastusco, zapupe de Tantoyuca, zapupe verde
Agave falcata – espadín, guapilla, ixtle, palmita, sotolillo, sotolito, soyate, soyatillo
Agave felgeri – mezcalito
Agave filifera – lechuguilla mansa, maguey de maceta, noriba
Agave fortiflora – lechuguilla
Agave fourcroydes – cahum, chelem, henequén, henequén blanco, ixtle, pita grande, saqui, sosquil – gray agave, henequen agave, Mexican sisal, sisal
Agave funkiana – ixtle de jaumave, juamave, juamave ixtle, juamave lechuguilla
Agave guadalajarana – maguey chato
Agave hartmanii – amole, magueyito, taiehcholi
Agave heteracantha – ixtle de Tula – Tula istle
Agave huachucensis – huachuca
Agave jaiboli – jaiboli, temeshi

Agave kerchovei – cacaya, maguey pichomel, rabo de león

Agave lechuguilla – amole, ixtle, ixtle de Tampico, lechuguilla, maguey, maguey lechuguilla, mezcal lechuguilla, metometl, Tampico, tula istle, tzuta – lechuguilla agave

Agave macroculmis – maguey, maguey verde

Agave mapisaga – maguey, maguey manso, maguey mapisaga

Agave margaritae – mezcal – agave

Agave marmorata – huiscole, maguey curandero, maguey tepestate, papalometl, pitzometl

Agave multifilifera – chahuiqui

Agave murpheyi – maguey – Hohokam agave

Agave ocahui – amolillo, majahui, ocahui

Agave palmeri – amole, lechuguilla, mezcal – century plant, Palmer agave

Agave parrasana – noa

Agave parryi – mezcal – cabbage agave, Parry agave

Agave parviflora* ssp *flexiflora – sóbari, tauta, tautilla

Agave peacockii – capulixtl, tlalometl

Agave pelona – mezcal pelón

Agave polianthiflora – taiehcholi

Agave potatorum – acamba, papalometl

Agave promontorii – mezcal – agave

Agave pseudotequilana – mezcal blanco, mezcal cucharo, mezcal maguey – mescal fiber agave

Agave quiotifera – maguey cenizo, quiote

Agave salmiana – maguey chalqueño, maguey de pulque, maguey manso, maguey verde, metl, tlacametl, tlacoamatl – pulque agave

Agave schidigera – maguey

Agave schottii – amole, amolillo, maguey – shindagger

Agave shawii – maguey, mezcal – coastal agave, Shaw's agave

Agave shrevei – cenizo, lechuguilla, lechuguilla ceniza, mezcal ceniza, totosá, totosali

Agave shrevei* ssp *matapensis – mezcal blanco

Agave sisalana (rigida) – cáñamo sisal, henequén, henequén verde, ixtle, maguey de Yucatán, maguey tuxtleco, sisal, sosquil, yaxqui – sisal, sisal agave, sisal hemp

Agave sobria – lechuguilla, mezcal pardo, pardito – agave

Agave striata – espadillo, espadín, estoquillo, guapilla, peinecillo, sotolito, soyate

Agave striata* ssp *falcata – espadín

Agave stricta – rabo de león

Agave sullivanii – mezcal – agave

Agave tequilana – agave azul, azul, bermejo, chato, chino azul, chino bermejo, maguey azul, maguey de tequila, maguey mezcal, maguey tequilero, mano larga, mezcal, mezcal azul, mezcal blanco, pata de mula, pie de mula, tequila, zapalote, zapote, zapotl, zapupe, zeguín, zopilote – mezcal, tequila agave, Weber blue agave

Agave utahensis – mezcal, yant – Utah agave

Agave victoriae-reginae – agave de la Reina Victoria, noa, pintilla

Agave vilmoriniana – amole – octopus agave

Agave vivipara – tapemete

Agave wocomahi – ojcome, wocomahi

Agave zapupe – zapupe, zapupe azul, zapupe de estopier, zapupe de San Bernardo

Agave zebra – mezcal lechuguilla

Agave = Manfreda

Agdestis clematides – hierba del indio, tripa baja, tripa de Judas, tush, velo de novia – sweet-and-vile vine

Ageratina adenophora (Eupatorium adenophorum) – eupatorio adenóforo

Ageratina (Eupatorium) herbacea – mata, tabardillo – western thoroughwort

Ageratina (Eupatorium) wrightii – orégano – Wright thoroughwort

Ageratum conyzoides – bak-elus, mejorana, sereno – ageratum

Ageratum corymbosum – cielito, hierba del ángel, mendrasto, mota morada – ageratum

Ageratum houstonianum – agérato, hierba de la sarna – floss flower

Aglaonema modestum – aglaonema, cuna de Moisés cimarrona, siempreviva china – Chinese evergreen

Agonandra racemosa – chilillo, limoncillo, margarita, matachamacos, murió, palo del golpe, palo verde, pega hueso, peincillo, suelda, suelda consuelda, usimyuera

Agrimonia eupatoria – agrimonia, agrimonia común, hierba de San Guillermo – agrimony

Agrimonia gryposepala – agrimonia – common agrimony
Agropyron = Elytrigia, Pascopyrum, Pseudoroegneria
Agrostemma githago – candelaria, candilejo, neguilla, neguillón – corn-cockle
Agrostis canina – agróstide canina, agróstide de las turberas, agróstide de perro – brown bent
Agrostis capillaris (tenuis) – agróstide común, chepica alemana, hierba fina – brown top, colonial bent, common bent
Agrostis gigantea (alba) – agróstide blanca, agróstide mayor, castillito, zacate de piedras castillitos – red-top bent grass
Agrostis palustris – agróstide de los pantanos, agróstide palustre, zacate de piedras quillas – creeping bent, marsh bent
Agrostis scabra – agróstide escabra – bent grass
Agrostis stolonifera – agróstide blanca, agróstide de renuevos, agróstide estolonífera, agróstide rastrera, nombre de Dios – creeping bent
Agrostis viridis (semiverticillata) – castillito de agua, castillo, castillo de agua, cola de ardilla, cola de zorrillo – water bent grass
Ailanthus altissima (glandulosa) – ailanto, árbol del cielo, árbol de los dioses, falso barniz, falso zumaque, maque – tree of heaven
Aiphanes acanthophylla – coyor, palma de coyor – coyor ruffle-palm
Aiphanes caryotifolia – cariota, corozo, corozo del Orinoco, mararay
Aiphanes minima – coyor
Aira caryophyllea – heno, heno mayor – silver European hairgrass
Aira elegantissima – heno elegante – elegant European hairgrass
Aira praecox – heno precoz – early hairgrass
Aizoon hispanicum – gazul
Ajuga chamaepitys – hierba artética, hierba de las junturas, pinillo oloroso – ground pine
Ajuga genevensis – pinillo azul, pinillo de ginebra – blue bugle
Ajuga pyramidalis – búgula piramidal – carpet bugle
Ajuga reptans – ajuga, búgula menor, búgula reptante, consuelda media, consuelda media rastrera, pinillo búgula, pinillo rastrero – carpet bugle
Albizzia julibrissin – acacia de Constantinopla, acacia de Persia, árbol de seda – mimosa, Persian acacia, silk tree
Albizzia lebbeck – acacia amarilla, cabellos de ángel, casia amarilla, ébano de oriente, tibet – East Indian walnut, siris tree
Albizzia occidentalis – algarrobo de olor, árbol de la lengua de mujer, arellano, bolillo, cico, palo escopeta, palo fierro, trucha
Albizzia saman – samán – cow tamarind, monkey pod tree, rain tree
Albizzia sinaloënsis – cayábajo, joso, palo joso
Alcea (Althaea) rosea – alcea, altea, amapola grande, escarapela, flor de San José, malva, malva arbórea, malva loca, malva real, malva rosa, vara de San José – hollyhock
Alchemilla alpina – alquimila alpina – lady's-mantle
Alchemilla glaucescens – alquimila híbrida – blue-green lady's-mantle
Alchemilla mollis – alquimila, pie de león – lady's-mantle
Alchemilla vulgaris – alquimila, alquimila común, manto de la dama, pie de león, pie de oso – lady's mantle
Aletris farinosa – raíz de ague, sasafrás – ague root, star grass
Aleurites cordata – abrasín del Japón – Japanese tung
Aleurites fordii – árbol del tung, tungo – tung
Aleurites moluccana – aleurita, árbol candil, árbol de la cera, lumbán, nogal de la India, nuez de la India – candlenut tree, kukui, tungoil tree
Alhagi pseudalhagi (camelorum) – maná – camel thorn, manna, manna hebraica, Sinai manna
Alibertia edulis – canilla de venado, cascarita, costarrica, guayabo de monte, guayabo de venado, guayabito, malaquito, naranjillo, palo de jarro – gibnut fruit
Alisma gramineum – llantén de hojas estrechas – water plantain
Alisma lanceolatum – llantén lanceolado – water plantain
Alisma plantago-aquatica – alisma, azúmbar,

llantén acuático, llantén de agua, oreja de liebre, pan de ranas – water plantain

Alkanna tinctoria – alcana, onoquiles, orcaneta, orcaneta azul, palomilla, palomilla de tintes, pie de paloma – alkanet, dyer's alkanet

Allamanda cathartica – alamanda, amanda, campana, campanita amarilla, canaria, copa de oro, flor de barbero, jazmín amarillo, llamarada, San José, trompeta de oro – buttercups, golden trumpet, yellow allamanda, yellow bell

Allenrolfea occidentalis – chamizo, hierba del burro – iodine bush, pickleweed

Allionia incarnata – guapile, hierba de la hormiga, hierba de la mosca, hierba del golpe – trailing four-o'clock, umbrella-wort, windmills

Allium ampeloprasum – cebollino común, puerro silvestre – Levant garlic, wild leek

Allium carinatum* ssp *pulchellum – ajo carinado – keeled allium

Allium cepa – 1. cebolla, cucut, tumos, xonácatl – onion; 2. cebolla de rabo – peasant onion, Texas scallion; 3. cebolla larga, cebollita de Cambray, puerro – green onion, scallion

Allium cepa* var *aggregatum – cebolla patata, cebolla subterránea, chalote, escaloña – multiplier onion, potato onion

Allium cepa* var *ascalonicum – ajo chalote, ascalonia, cebolla, cebolla escalonia, chalote, escalona, escalonia, escaloña – shallot

Allium cepa* var *viviparum – cebolla bulbífera, cebolla de Egipto – tree onion

Allium cernuum – ajo cebolla, cebolleta del campo – nodding onion

Allium christophii (albopilosum) – ajo floreciente – star of Persia

Allium fistulosum – ajo fistuloso, cebolla de hoja, cebolla de invierno, cebollaja, cebolla junca, cebolla verde, cebolleta, cebollín, cebollino inglés, cebollino de verdeo, puerro de piedra – bunching onion, Welsh onion

Allium giganteum – puerro gigante – giant allium

Allium glandulosum – cebolleta

Allium haematochiton – cebolla nativa, cebolleta, ceborrincha – onion, wild onion

Allium karataviense – ajo del Turquestán – Turkestan allium

Allium moly – ajo dorado, ajo silvestre – golden garlic, moly, yellow allium

Allium neapolitanum – ajo napolitano – Naples garlic, onion of Naples

Allium oreophilum – ajo – ornamental allium

Allium paniculatum – ajo paniculado

Allium porrum (ampeloprasum* var *porrum) – ajo porro, ajo puerro, puerro – leek

Allium sativum – ajo, ajo sativo – garlic

Allium sativum* var *ophioscorodon – ajo francés, ajo italiano, rocambola – rocambole

Allium scaposum – cebolleta, cebollín, cebollina – wild onion

Allium schoenoprasum – ajo cebollino, ajo moruno, cebolino, cebolla, cebolla de rama, cebollana, cebolleja, cebolleta, cebollín, cebollina, cebollino, cebollino común, cebollino francés – chive, chives

Allium sphaerocephalum – ajo de cabeza redonda – drumsticks, round-headed garlic

Allium triquetrum – ajo triángulo, lágrimas de la Virgen

Allium tuberosum – cebollino chino – Chinese chives, garlic chives

Allium vineale – ajo de las viñas – crow garlic, wild onion

Alnus acuminata – abedul, aile, aliso, ilite, ilite verde, palo de águila, yaga-bizie

Alnus arguta – abedul, aile, aliso, hilite, ilite, olmo del país, palo de águila

Alnus cordata – aliso corso, aliso de Córcega, aliso de Italia, aliso italiano, aliso napolitano – Italian alder

Alnus glutinosa – aliso, aliso común, aliso negro, alno, humero – black alder

Alnus incana (tenuifolia) – aliso, aliso blanco, aliso cano, aliso gris, baraña – mountain alder, thin-leaf alder

Alnus jorullensis – abedul, aile, aliso, ilite verde, ilitl – wax alder

Alnus oblongifolia – alamillo, aliso, júcaro – Arizona alder, New Mexican alder

Alnus rhombifolia – aliso blanco – white alder

Alnus rubra (oregona) – aile, ailitl, aliso, aliso de Oregón, aliso rojo, aliso rojo americano, ilite, olmo del país, saúco – Oregon alder, red alder

Alnus viridis – aliso verde – Sitka alder, thinleaf alder
Alocasia cuprea – chactzotzmacal – elephant's ear
Alocasia macrorrhiza – chactzotzmacal – upright elephant's ear
Aloe ferox – áloe – bitter aloe
Aloe socotrina – acíbar, áloe, zabila – aloe
Aloe variegata – batoba-xha – partridge-breast aloe, tiger aloe
Aloe vera (barbadensis) – acíbar, áloe, áloe de Barbados, áloe de Zanzíbar, áloe vera, atimorreal, batoba-xha, hunpetskinci, tlailochtía, zabida, zabila – aloe vera, Barbados aloe, medicinal aloe, Zanzibar aloe
Alopecurus aequalis – alopecuro rojizo – short-awn foxtail
Alopecurus geniculatus – alopecuro geniculado, cola de zorra – water foxtail
Alopecurus myosuroides – alopecuro de los campos – black twitch, foxtail
Alopecurus pratensis – alopecuro, alopecuro de los prados, alopecuro pratense, cola de zorra, rabo de zorro – meadow foxtail
Aloysia gratissima – cedrón, hierba dulce, oreganillo, troncoso, vara dulce – bee brush, white brush
Aloysia sonorensis – mariola
Aloysia triphylla (Lippia citriodora) – cedrón, cedrón de Castilla, hierba de la princesa, hierba Luisa, María Luisa, Reina Luisa, té cedrón – lemon verbena
Aloysia (Lippia) wrightii – altamisa, oreganillo, vara dulce – bee brush, Mexican oregano, oreganillo, whitebrush
Alpinia officinarum – galanga, galanga de la China – galingale, lesser galangal
Alpinia purpurata – jenjibre, jenjibre ornamental, jenjibre rojo – red ginger
Alpinia zerumbet (nutans, speciosa) – alpina, colonia, lirio de Colón – pink porcelain lily, shell flower, shell ginger
Alsophila aculeata – helecho arborescente – tree fern
Alsophila salvinii – helecho arborescente, helecho de árbol, ocopetate, pelma, pesma, pesto, rabo de mico, rabo de mono – tree fern
Alstonia scholaris – ditá – dita tree
Alstroemeria aurea (aurantiaca) – amancay, azucena peruana, lirio del Perú – Peruvian lily
Alstroemeria pelegrina – alstroemeria, peregrina de Lima – Peruvian lily
Alternanthera achyrantha – verdolaga de puerco
Alternanthera caracasana (repens) – alternantera repente, kabal-xtes, tececeque, témix-tapextli, tiánguis, tianguispepetla, tlalpetate, tumina, verdolaga de puerco – alligator weed, chaff-flower
Alternanthera sessilis – sanguinaria
Althaea officinalis – altea, bismalva, hierba cañamera, malva real, malvavisco – marshmallow
Althaea = Alcea
Alvaradoa amorphoides – bezinic-ché, camarón, charagallo, cuachipile, cuetzé, flor de piojo, guachipil, guaje, gueguetzí, huezequí, huichipil, palo de hormiga, palo piojo, palo torcido, peinecillo, pie de gallo, plumajillo, ruda cimarrona, suetsinic-ché – Mexican alvaradoa, pinkwing
Alyssum alyssoides – mastuerzo anual – small alison
Alyssum minus – mastuerzo de los campos – field alison
Alyssum montanum – broqueleto, cestillo de oro, mastuerzo montano
Alyssum strigosum – mastuerzo estrigoso
Alyssum = Aurinia, Lobularia
Amanita caesarea – ahuevado, chichimán, jicarita, oronja, sochi, tecomate, xochinanácatl, yema, yema de huevo, yemita – Caesar's mushroom
Amanita muscaria – cuacicitlal, falsa oronja, kaquijá, mosquero, seta del bonete rojo, seta mágica, seta sagrada, tarecua cahuica, terecua-varirapeni, tzontecomananácatl – fly agaric
Amanita vaginatum – pollita, venado, yema ceniza
Amaracus = Origanum
Amaranthus albus – amaranto blanco, bledo, quelite – tumbleweed
Amaranthus blitoides (graecizans) – chile puerco, huee'e, quelite, quelite de agua, quelite rodador – prostrate amaranth
Amaranthus caudatus (leucocarpus) –

abanico, alegría, amaranto, amaranto cinco de mayo, amaranto rojo, bledo, cinco de mayo, cordón de obispo, flor de seda, huautli, moco de pavo – love-lies-bleeding, quilete, tassel flower

Amaranthus cruentus – alegría, amaranto, amaranto pátulo, bledo, huautli, quelite morado – spreading amaranth

Amaranthus deflexus – amaranto prostrado – prostrate amaranth

Amaranthus fimbriatus – bledo, quelite, quelitillo – fringed amaranth

Amaranthus gangeticus – papagayo, tampala – Joseph's coat

Amaranthus hybridus (hypochondriacus) – achita, amaranto de espigas verdes, bledo, qual, quelite, quelite blanco, quelite de cochino, quelite de espiga, quelite morado, quílitl, quintonil, quintonile, quixxtez – green amaranth, prince's feather

Amaranthus palmeri – bledo, quelite, quelite manso, quelite rojo – amaranth, careless weed, pigweed, red root

Amaranthus paniculatus – alegría, amaranto, quelite morado

Amaranthus powellii – quelite colorado yus, quelite yus – amaranth

Amaranthus retroflexus – amaranto común, bledo

Amaranthus spinosus – amaranto espinoso, bledo, bledo cimarrón, bledo espinoso, bledo macho, bledo rojo, epazote de mula, pira brava, quelite espinoso, quixxtez – amaranth, pigweed

Amaranthus tricolor – ala de oro, ala de perico, amaranto, moco de pavo – Joseph's coat

Amaranthus watsonii – bledo, quelite – amaranth, careless weed, pigweed

Amaryllis belladonna (Brunsvigia rosea) – amarilis, azucena rosa, lirio de San Jaime – belladonna lily, naked lady

Amaryllis formossisima – flor de lis

Amaryllis longifolia – lirio listado

Amaryllis reginae – azucena roja, lirio rojo

Amberboa = Centaurea

Amblyopappus pusillus – coquimbo

Ambrosia (Franseria) acanthicarpa – estafiate, hierba del sapo, rosetilla – annual bur-sage, bur-weed, sandbur

Ambrosia (Franseria) ambrosioides – chicoria, chicura, chicurilla, estafiate, hierba amargosa, hierba de la muela, hierba del sapo, jiogo, tincl – bur-sage, canyon ragweed

Ambrosia aptera – chuchuni – ragweed

Ambrosia artemisiifolia – altamisa, amargosa, ambrosía de hojas de ajenjo, artemisa, cambeba, estafiate, hierba amargosa – common ragweed, low ragweed

Ambrosia (Fransera) bryantii – chicura – ragweed

Ambrosia (Franseria) camphorata – estafiate

Ambrosia (Franseria) chenopodiifolia – huizapol, huizapolito – burbush, San Diego bur-sage

Ambrosia (Franseria) confertiflora – altamisa del campo, chichibo, chichuro, estafiate, hierba del sapo – bur-sage

Ambrosia (Franseria) cordifolia – chicurilla

Ambrosia cordifolia – chicura, chicurilla – bur-sage

Ambrosia (Franseria) deltoidea – chamizo, chamizo forrajero, estafiate, franseria – triangle-leaf bur-sage

Ambrosia (Franseria) dumosa – chamizo, chicurilla, hierba del burro, huizapol – burrobush, burro-weed, white bur-sage

Ambrosia psilostachya – altamisa, amargosa, chipuna, estafiate – western ragweed

Amelanchier alnifolia – cornijuelo, cornillo – service-berry, shadbush

Amelanchier canadensis – guillomo del Canadá, níspero del Canadá – June berry, service-berry, shad bush

Amelanchier denticulata – duraznillo, madronillo, membrillito, membrillo, membrillo cimarrón, tlaxistle, tomistlacati

Amelanchier × grandiflora – guillomo grandifloro – large-flowered June berry

Amelanchier laevis – amelanchier arbóreo, guillomo arbóreo – Allegheny service-berry

Amelanchier ovalis – bullonera, carrasquilla, cornijuelo, curroñé, guillomo, guillomo común – service-berry, snowy mespilus

Amelanchier vulgaris – bullonera, cornillo, curroñé – service-berry

Ammi majus – ameos mayor, ami, ami

vulgare, encaji, espuma de mar, khella – bishop's weed

Ammi visnaga – ameos común, biznaga, escuradentis, fistra, hierba de la gota, hierba de San Gerardo – lesser bishop's weed, khella

Ammobroma = Pholisma

Ammophila arenaria – arenaria, barrón, caña de las arenas, grama de las dunas – European beachgrass, marram grass

Amomum = Elettaria

Amoreuxia gonzalezii – saya, saya mome

Amoreuxia palmatifida – saya, sayasaga, témaqui – Arizona yellow show, saya

Amorpha fruticosa – añil bastardo, índigo bastardo – false indigo

Amorphophalus rivieri – corona imperial, imperial, paraguas chino – corpse flower

Ampelopsis cordata – labrusca – ivy vine

Amphipterygium (Juliania) adstringens – cuachalalate, quetchalalatl

Amphipterygium glaucum – cuachalalate

Amsinckia douglasiana – amsinckia de Douglas, zacate gordo – fiddleneck

Amsinckia menziesii (intermedia, lycopsoides)* var *intermedia – amsinckia intermedia – rancher's fireweed

Amsinckia spectabilis – zacate gordo – fiddleneck

Amsinckia tessellata – amsinckia teselada – bristly fiddleneck, devil's lettuce

Amygdalus = Prunus

Amyris balsamifera – candil, cuaba blanca, cuaba de la costa, limoncillo, ocotillo blanco, palo de gas, palo de rosa, palo limón, quigua, tojtanyuc – torchwood

Anacardium occidentale – acajú, acajuiba, anacardo, cajú, cajueiro, cajuil, castaño de cajú, jocote marañón, marañón, merey, nuez de la India, pajul – cashew

Anacardium rhinocarpus – caracolí, nariz

Anacyclus depressus (pyrethrum* var *depressus) – margarita de Atlas, pelitre, pelitre extranjero – Mount Atlas daisy

Anagallis arvensis – anagálide, bocado de gallina, coralillo, hierba coral, hierba del pájaro, hierba pajarera, ixcuicuil, murajes, murajes arvenses, saponaria, toboso – poor man's weatherglass, scarlet pimpernel

Anagallis monelli – centaurea real – pimpernel

Anagyris foetida – altramuz del diablo, altramuz hediondo – bean trefoil

Anamirta paniculata – coca de Levante, morga – India berry tree

Ananas comosus – ananás, corojo, muatzate, noai, piña, piña de Indias, pita de corojo – pineapple

Ananas (Aechmea) magdalenae – pita, pita floja – pingwing

Ananas sativus – abacachi, ananás, matsatli, muatzate, noai, piña, piña de América – pineapple

Anaphalis margaritacea – gordolobo, pie de gato, sempiterna – pearly everlasting

Anchusa arvensis – buglosa de los campos, licópside, miel de avispas – bugloss

Anchusa azurea (italica) – alcana, anchusa, buglosa, chupamiel, lengua de buey, lenguaza, raíz de fuego – alkanet, bugloss

Anchusa capensis – buglosa del Cabo – Cape forget-me-not, summer forget-me-not

Anchusa officinalis – albarines, ancusa, buglosa común, lengua de buey, melera – alkanet

Andira inermis – caca de caballo, cuaxtololote, cuchucua, cuilimbuca, huijol, huixe, iximche, juanjunesco, llaba, lombricero, maca colorada, macayo, macayo chico, moca, moca colorada, pacay, pangelín, tinco, tololote, uxe, yabo – angelin tree, cabbage angelin, partridge wood

Andromeda polifolia – andrómeda – bog rosemary

Andromeda = Pieris

Andropogon glomeratus* var *scabriglumus – cola de zorra, popotillo matorralero, tallo azul matorralero – southwestern bushy bluestem

Andropogon hallii – popotillo arenoso – sand bluestem, turkey-foot

Andropogon schaenanthus – té de limón – lemon grass

Andropogon virginicus – popotillo pajón – broomsedge bluestem

Andropogon = Bothriochloa, Dichanthium, Schizachyrium, Sorghum

Androsace elongata – androsace alargada – rock-jasmine

Androsace maxima – androsace de los

campos, androsace anual, cantarillo, picacuello – rock-jasmine
Androsace sarmentosa – jazmín rocoso – rock-jasmine
Androsace septentrionalis – androsace septentrional – rock-jasmine
Anemone blanda – anémona – anemone, windflower
Anemone coronaria – anémona, anémona coronaria, coronaria – lilies of the field, poppy-flowered anemone
Anemone* x *fulgens – anémona ojo de pavo – scarlet windflower
Anemone* x *hybrida (japonica) – anémona del Japón, anémona japonesa – Japanese anemone
Anemone nemorosa – ahogaviejas, anémona de los bosques, anémona silvia, nemorosa, pulsatila – wood anemone
Anemone sylvestris – anémona, angélica borde – snowdrop anemone
Anemone = Pulsatilla
Anemopsis californica – babiza, hierba del manso, hierba mansa, manso, raíz del manso, verissima hierba – lizard's tail, swamp root, yerba mansa
Anethum graveolens – eneldo, falso anís, hinojo fétido, hinojo hediondo, neldo – dill, dillweed
Angelica archangelica (officinalis) – angélica, hierba angélica, hierba del Espíritu Santo, hierba del Santo Espíritu, oshá del campo – angelica, archangel
Angelica arguta – oshá del campo – angelica
Angelica sinensis – angélica china, dong quai – Chinese angelica, dong quai
Anisacanthus quadrifidus (wrightii)* ssp *wrightii – muicle – desert honeysuckle, flame acanthus
Anisacanthus thurberi – chuparrosa, cola de gallo, hierba del cáncer, lustiej, muicle – chuparosa, desert honeysuckle, hummingbird bush
Anisostichus capreolatus (Bignonia capreolata) – bignonia, hibuero – cross-vine
Annona cherimola – anona, chirimorriñón, chirimoyo, cuauhtzapotl, lamatzapotl, matzapotl, pox, tsulipox, uruata – cherimoya, soursop
Annona diversifolia – anona blanca, anonilla, ilama, ilamatzapotl, papausa – cherimoya-of-the-lowlands, custard apple, ilama
Annona glabra – anona lisa, árbol del corcho, corazón cimarrón, corcho, coyur, mag, palo de corcho – alligator apple, pond apple
Annona muricata – anona amarilla, anona de broquel, cabeza de negro, catuche, chincuya, corosol, corosol espinoso, graviola, guanaba, guanábano, guendabixhu, ilama de Tehuántepec, ilamatzapotl, polvox, takob, zapote de viejas – custard apple, soursop
Annona purpurea – ahate, anona morada, cabeza de ilama, cabeza de negro, chacoop, chincuya, ilama, ilama de Tehuántepic, papanaca, polvox, pox, soncoya – anona, custard apple
Annona reticulata – anona, anona colorada, anona de tierra caliente, anona morada, anonilla, chirimoyo, cocax, cocox, corazón, corazón de buey, cuauhtzapotl, ilama, mamón, op, pox, tsulipox, zulipox – bullock's heart, cachimant
Annona squamosa – ahate, anona, anona blanca, anona tis, chirimoyo, chirimoyo del Senegal, cuauhtzapotl, fruta de condesa, riñón, saramuyo, texaltzápotl, tzalmuy – sugar-apple, sweetsop
Anoda acerifolia – violeta, violeta del campo
Anoda crenatiflora – amapolita, malva
Anoda cristata – aguatosa, alachi, altea, amapolita, amapolita del campo, amapolita morada, halanche, malva con espuelas, malva de Castilla, pintapan, quesito, tzalyaltzai, violeta, violeta del campo, violeta del país – violettas
Anoda thurberi – San Miguelito
Antennaria dioica – pie de gato – pussy-toes
Anthemis arvensis – manzanilla bastarda – corn chamomile
Anthemis cotula – magarzuela, manzanilla, manzanilla fétida, manzanilla hedionda – chamomile, dog-fennel, mayweed, stinkweed
Anthemis tinctoria – manzanilla amarilla, manzanilla de los tintoreros, manzanilla loca – dyer's chamomile, golden marguerite
Anthemis = Chamaemelum
Anthericum torreyi – coyamol – amber-lily, Torrey's crag-lily

Anthoxanthum aristatum – alesta aristosa – annual vernal grass

Anthoxanthum odoratum – alesta olorosa, grama de los prados, grama de olor, zacate primavera – sweet vernal grass

Anthriscus caucalis – ahogagatos – bur-chervil

Anthriscus cerefolium – cerafolio, perifollo, salsa de pastor – chervil

Anthriscus sylvestris – cerafolio agreste – cow parsley

Anthurium andraeanum – anturio, cresta de gallo, lengua del diablo – anthurium, flamingo lily, spathe flower, tail flower

Anthurium crassinervium – cola de ratón, hoja de cuero, hoja de viento

Anthurium crystallinum – anturio – spathe flower

Anthurium scandens – maicillo, maíz del monte, tepecentli – spathe flower

Anthurium scherzerianum – anturio, lengua del diablo, planta flamingo, rabito cochino – flamingo flower, pigtail, spathe flower

Anthurium schlechtendalii – cola de faisán – pheasant's tail

Anthurium xanthosomifolium – anturio, capa de pobre, conte cinco dedos, flor porcelana, macal – anthurium, waxflower

Anthyllis barba-jovis – barba de Júpiter, vulneraria barba de Júpiter – Jupiter's beard

Anthyllis vulneraria – trébol amarillo, vulneraria – kidney vetch

Antigonon leptopus – bejuco de colación, bellísima, cadena de amor, chaclomacal, colación, confite, coral, coralillo, coralito, corona, corona de la reina, coronilla, cuamecate, cuamécatl, enredadera coralina, enredadera de San Diego, flor de San Diego, flor de San Miguel, flor de San Miguelito, fulmina, hierba de Santa Rosa, jololito, rosa de mayo, rosa de montaña, San Diego, San Miguel, San Miguelito, tunuc – bride's tears, chain of love, Confederate vine, coral-vine, Honolulu creeper, Mexican creeper, pink vine, queen's wreath, rosa de montana

Antirrhinum majus – antirrino, becerra, boca de dragón, boca de león, cabeza de dragón, conejito, dragón, dragoncillo, hierba becerra, linaria, perrito, tarasca de jardín – common snapdragon, snapdragon

Antirrhinum orontium – boca de dragón campestre, hierba becerra – weasel's snout

Anulocaulis eriosolenus – pegajosa – ringstem

Apera interrupta – apera interrumpida

Apera spica-venti – apera de los campos – windlestraw

Aphananthe monoica – guasimilla, mohuarí

Aphanes occidentalis – alquimila – piert

Aphelandra deppeana – añililo

Aphelandra squarrosa – afelandra, afelandra rayadita – saffron-spike, zebra plant

Aphyllon = Orobanche

Apium graveolens – apio, apio borde, apio común, panul – celery

Apium graveolens* var *rapaceum – apio, apio nabo, apirrábano, nabo – celeriac

Apium nodiflorum – apio de hojas nudosas – fool's water cress, marshwort

Apium = Ciclospermum

Aplopappus = Ericameria, Isocoma

Apocynum androsaemifolium – apocino, lechuguilla – bitter dogbane

Apocynum cannabinum – cáñamo del Canadá, lechuguilla, madregosa – Indian hemp

Apodanthera undulata – meloncillo, melón de coyote, melón loco – melon-loco

Apoplanesia paniculata – arco negro, cacanaguaste, chulul, ébano, guie-biche, matagallinas, palo de arco, palo de arco negro, palo matagallinas

Aporocactus = Disocactus

Aptenia cordifolia – hierba cristalina, hierba cristalina de hojas cordiformes – baby sun rose

Aquilaria agallocha – áloe – aloes, eagle wood, lign aloe, paradise wood

Aquilegia alpina – aguileña alpina – alpine columbine

Aquilegia caerulea – aguileña azul – Rocky Mountain columbine

Aquilegia hybrida – ancolia – columbine

Aquilegia vulgaris – aguileña, aguileña común, clérigo, clérigo boca abajo, farolillo, guileña, manto real, muela de San Cristobál, pajarilla, pajarito, paloma, palomilla, palomita, pelícano – European columbine

Arabidopsis thaliana – arabeta, arabeta de Thalius – mouse-ear cress, thale cress

Arabis alpina – arábide alpina, arábide de los Alpes, pelusilla – mountain rock cress
Arabis blepharophylla – arábide – California rock cress, coast rock cress, rose cress
Arabis caucasica (albida) – arábide caucásica, arábide de los Alpes, berro de ganso – wall rock cress
Arabis glabra (perfoliata) – arábide glabrescente, hierba del sapo – rock cress, tower mustard
Arabis hirsuta – arábide hirsuta – hairy rock cress
Arabis mexicana – lentejuelilla – rock cress
Arabis sparsiflora – arábide ciliada – rock cress
Arachis hypogaea – alfóncigo de tierra, aráquida, cacahuate, fruta del maní, maní, maní español, tlalcacahuatl – peanut
Aralia californica – aralia – spikenard
Aralia cordata – aralia
Aralia elata – angélica japonesa – Japanese angelica tree
Aralia humilis (pubescens) – aralia, cuajilotillo, curguatón, palo santo, tepetate
Aralia racemosa – aralia, nardo americano – American spikenard
Aralia = Fatsia
Araucaria araucana (imbricata) – araucaria, araucaria de Chile, araucaria de Neuquén, araucaria espinuda, cola de macaco, pehuén, pino chileno, pino de Chile, pino de pisos, rompecabeza de mono – monkey puzzle tree
Araucaria bidwillii – araucaria, araucaria de Bidwill, bungya – bunya-bunya
Araucaria heterophylla (excelsa) – araucaria, pino de pisos, pino de Norfolk – Norfolk Island pine
Arbutus andrachne – madroño de Grecia, madroño oriental
Arbutus arizonica – madroño – Arizona madrone, Arizona madroño
Arbutus canariensis – madroño canario
Arbutus glandulosa – aile, jucay, madroño, nuzu-ndu, panán-gsuni – madrone
Arbutus menziesii – madroño, madroño de las montañas rocosas – madrona, madrone, madroño, Pacific madrone
Arbutus peninsularis – madroño
Arbutus tessellata – madroño, madroño chino, madroño negro – checkerbark madrone
Arbutus texana – madroño, manzanita, nuzu-ndu – Texas madrone
Arbutus unedo – alborto, borrubiete, borto, madroñero, madroño, madroño común, tomazquitl – strawberry tree
Arbutus xalapensis – amazaquitl, madroño, madroño blanco, madroño borracho, madroño colorado, manzanita, nuzu-ndu, urúbishi, xoxocote – Mexican madrone, smooth madrone
Arceuthobium campylopodum – toji – western dwarf mistletoe
Arceuthobium divaricatum – toji – pinyon dwarf mistletoe
Arceuthobium globosum – flor de ocote, flor de pino, retoño de ocote – dwarf mistletoe, pine mistletoe
Arceuthobium vaginatum – flor de ocote, flor de pino, injerto – dwarf mistletoe, pine mistletoe
Archibaccharis serratifolia – hierba del carbonero
Archontophoenix alexandrae – palma alejandra – Alexandra palm
Archontophoenix cunninghamia (Seaforthia elegans) – palma de Cunningham, seaforthia – king palm
Arctium lappa (Lappa major) – bardana, bardana mayor, cadillo, cardo ajonjero, hierba del amor, hierba del tiñoso, lampazo, lapa, pegadillo – burdock, cocklebur
Arctium minus – bardana común, bardana menor, lampazo menor – lesser burdock
Arctostaphylos alpinus – gayuba alpina – alpine bearberry
Arctostaphylos arguta (discolor) – garambullo, madroño, madroño borracho, pingüica
Arctostaphylos glauca – madroño, manzanita – bigberry manzanita
Arctostaphylos manzanita – manzanita – bearberry, common manzanita, Parry manzanita
Arctostaphylos nevadensis – coralillo – pinemat manzanita
Arctostaphylos patula – madroño, manzanita – greenleaf manzanita
Arctostaphylos peninsularis – manzanita

Arctostaphylos polifolia – chino, cucuato, madroño, madroño chino, manzanilla, pingüica, tnutque

Arctostaphylos pringlei – manzanita – bracted manzanita

Arctostaphylos pringlei* ssp *drupacea – manzanita – pink-bracted manzanita

Arctostaphylos pungens – gayuba del país, gayuba de México, leño colorado, madroño, manzanilla, manzanita, manzano, palo de pingüica, pepioita, pingüica, tepesquisúchil, tepezquite, tepezquitl, tnundido, tomazquitl, ují, yori – Mexican manzanita, point-leaf manzanita

Arctostaphylos tomentosa – madroño, madroño de árbol, manzanita, tepetomate – marganita, shaggy-barked manzanita

Arctostaphylos uva-ursi – aguavilla, coralillo, gayuba, gorrinchia, madroño rastrero, manzanita, pingüica, uva de oso, uva de zorro, uvaduz – bearberry, kinnikinnik, sandberry

Arctostaphylos viscida – manzanita – whiteleaf manzanita

Arctostaphylos xalapensis – madroño borracho

Arctotis grandis – oreja de oso

Arctotis stoechadifolia (venusta) – margarita africana – African daisy, blue-eyed African daisy

Ardisia revoluta – aguapepe, arrayán, camaca, capulín, capulincillo, capulín manso, laurel, laurel de la sierra, negrito, pimientillo, sirasil – marlberry

Areca catechu – areca, areca catecu, palma catecu, palmera de betel – arecanut palm, betelnut

Areca = Chrysalidocarpus

Arecaceae – palma, palmera – palm

Arecastrum = Syagrus

Arenaria balearica – arenaria, arenaria de las Baleares – Corsican sandwort

Arenaria bourgaei – paletaria

Arenaria montana – arenaria montana, césped espinoso, pamplina falsa espinosa – mountain sandwort

Arenaria serpyllifolia – arenaria, arenaria de hojas de tomillo – sandwort

Arenaria = Minuartia, Sagina

Arenga pinnata – barú, palma de azúcar, palmera azucarera – sugar palm

Arenga saccharifera – palmera azucarera – sugar palm

Arethusa grandiflora – vara de San Miguel

Argania sideroxylon – erguén – argan tree

Argemone alba – cardo – prickly poppy

Argemone corymbosa (intermedia) – chicalote – Mojave prickly poppy

Argemone gracilenta – cardo, chicalote – cowboy's fried egg, white-flowered prickly poppy

Argemone intermedia – chicalote – prickle poppy

Argemone mexicana – adormidera espinosa, amapola blanca, amapola mexicana, cardo, cardo amarillo, cardo blanco, cardo santo, cardo santo campestre, chicalote, chicalotl, chichillotl, chillazotl, kanlal, quixkanlol, quixzaclol, tachino, tlamexcaltzin, xaté, xicólotl – devil's fig, Mexican poppy, prickly poppy

Argemone munita (platyceras) – amapola blanca, cardo, cardo santo, chicalote, huitzquilitl, nardo – flatbud prickly poppy

Argemone ochroleuca – amapola amarilla, cardo, cardo santo, chicalote, chicalotl, hierba loca, San Carlos, tachino – yellow prickly poppy

Argemone platyceras – chicalote – prickle poppy

Argemone subintegrifolia – cardo

Argemone superba – chicalote

Ariocarpus (Roseocactus) fissuratus (lloydia) – chautle, hikora sunama, peyote, peyote cimarrón, peyotl, roca viviente, sunami, wanamé – crack star, false peyote, living-rock cactus

Ariocarpus (Roseocactus) kotschoubeyanus – pata de venado, pezuña de venado – living rock

Ariocarpus retusus – chautle, peyote cimarrón, tsuwíri – living rock cactus, seven stars

Arisaema dracontium – culebrilla, hierba de la culebra – green dragon, jack-in-the-pulpit

Aristida adscensionis – tres aristas de agua, tres barbas, tres barbas anual, tres barbas de semillas, zacate cola de zorra, zacate de agua, zacate de agua tres barbas, zacate de semilla, zacate tres barbas – six-weeks three-awn

Aristida arizonica – tres aristas arizónica – Arizona three-awn

Aristida barbata – tres aristas barbado – Havard three-awn

Aristida californica – tres barbas de California – California three-awn

Aristida divaricata – tres aristas abierto, tres aristas barbado, tres barbas abierto – poverty three-awn

Aristida gentilis – tres barbas gentil

Aristida orcuttiana – tres barbas volador – beggar-tick grass

Aristida pansa – tres aristas perenne – Wooton three-awn

Aristida purpurea – tres aristas morado – purple three-awn

Aristida purpurea (fendleriana)* var *fendleriana – tres aristas largo – Fendler three-awn

Aristida purpurea (longiseta)* var *longiseta – tres aristas rojo – red three-awn

Aristida purpurea (glauca)* var *nealleyi – tres barbas, tres barbas liso – Nealley three-awn

Aristida purpurea (wrightii)* var *wrightii – tres barbas perenne

Aristida roemeriana – cola de zorra, tres aristas mexicano

Aristida scabra – aceitilla

Aristida ternipes – chak-suuk, tres barbas arqueado, zacate araña – spidergrass

Aristida ternipes* var *gentilis – kan-zuuk, zacate araña, zacate araña de tres barbas, zacate volador – poverty three-awn

Aristida ternipes* var *hamulosa – tres aristas arqueado – hook three-awn

Aristida tuberculosa – zacate torcido

Aristolochia brevipes – hierba del indio, Santa María – birthwort, Indian root

Aristolochia californica – guaco, hierba del indio, indio, pipa de mico, tacopate – birthwort, California Dutchman's pipe, pipevine

Aristolochia clematitis – aristoloquia, aristoloquia clematítide, clematítide, guaco – birthwort

Aristolochia durior (macrophylla) – aristoloquia de hojas grandes, aristoloquia macrofila, pipa de los Holandeses – Dutchman's pipe

Aristolochia grandiflora – alcatraz, bonete, bonete del diablo, chanuahko, contribo, flor de pato, flor de pelícano, guaco, güegüecho, hierba del indio, itamo real, pato, sombrero de don Zope, tacopaste, tacopate, tacopatl, tacopaxtle, tlacopatl, tlacopatle, tlacopatli, tlacopaxtle, uahko – calico flower, large-flowered pelican flower

Aristolochia longa – aristoloquia larga – long birthwort

Aristolochia mexicana – guaco, guaco del sur, raíz del flato, tacopaxtle, tlacopatle

Aristolochia monticola – hierba del indio – Dutchman's pipe

Aristolochia pentandra – guaco de Merida, guaco de San Cristobal, hierba del indio

Aristolochia porphyrophylla – hierba del indio – Dutchman's pipe

Aristolochia quercetorum – hierba del indio

Aristolochia rotunda – aristoloquia redonda – round-leaved birthwort

Aristolochia serpentaria – aristoloquia de Virginia, hierba de la víbora, viperina de Virginia – Virginia snakeroot

Aristolochia taliscana – canastilla, guaco, matanene de mar, patito, zapatito

Aristolochia watsonii – guaco, hierba del indio, indio, raíz del indio – Indian root, southwestern pipevine, Watson's Dutchman's pipe

Armeniaca = Prunus

Armeria alliacea – armeria de Jersey – Jersey thrift

Armeria girardii (juncea) – armeria falso junco – rush-like thrift

Armeria juniperifolia (caespitosa) – armeria, armeria cespitosa – sea-pink, thrift

Armeria maritima – armeria, armeria marítima, césped de Olimpo, césped marino, clavel silvestre, estátice – sea-pink, common thrift

Armillaria mellea – babosito, cazahuate, clavo, hongo, hongo comestible, sopitza – honey agaric, honey mushroom, oak fungus

Armoracia rusticana – rábano picante, rábano rusticano, rábano silvestre, tamarrago, vagiseo – horseradish

Arnica montana – árnica, árnica montana, estornudadera, hierba de caída, tabaco borde, tabaco de montaña, talpica – arnica
Arpophyllum spicatum – tzauxilotl
Arrabidaea spp – güirote negro
Arracacia atropurpurea – acocote, acocotli tepecuacuilcense, comino, comino rústico, hierba del oso
Arracacia brandegeei – acocote, chuchupate
Arracacia edulis – hierba del oso
Arracacia xanthorrhiza (esculenta) – apio, arracacha – Peruvian carrot
Arrhenatherum elatius – avena desarrollada, avena elevada, avena falsa alta, fromental, gramari, mazorra, porrilla, tortero – tall oatgrass
Artemisia abrotanum – abrótano, abrótano macho, ajenjo arbóreo, ajenjo moruno, boja, hierba lombriguera, toronjil – herb royal, old man, southernwood
Artemisia absinthium – absintio, ajenjo, ajenjo extranjero, ajenjo mayor, alonsa, asenjo, encienso, estafiate, hierba santa – absinth, absinthe, common wormwood
Artemisia annua – ajenjo, ajenjo anual, ajenjo dulce, estafiate – sweet Annie, sweet wormwood
Artemisia arborescens – ajenjo arborescente, ajenjo moruno – tree-like mugwort
Artemisia biennis – ajenjo bisanuo – biannual mugwort
Artemisia californica – romerillo, romerito – California sagebrush
Artemisia campestris – ajenjo de los campos, escobilla parda – field southernwood
Artemisia caucasica – ajenjo del Cáucaso – silver spreader
Artemisia douglasiana – estafiate – mugwort
Artemisia dracunculoides – estafiate, hierba niso – absinth, false tarragon
Artemisia dracunculus – artemisa, dragoncillo, estragón, hierba niso, hierba nora – French tarragon, tarragon, wormwood
Artemisia filifolia – estafiate, romerillo – sand sage, sand sagebrush
Artemisia franserioides – altamisa, altamisa de la sierra – mountain mugwort
Artemisia frigida – altamisa de la sierra, estafiate, plumajillo de la sierra, romerillo – fringed wormwood, silver sage
Artemisia judaica – amargo – wormwood
Artemisia klotzschiana – estafiate
Artemisia lactiflora – ajenjo lactescente – white mugwort
Artemisia ludoviciana – ajenjo, ajenjo del país, altamisa de la casa, estafiate, estafiate del bosque, iztauhyatl, mostafiete del campo, musha, romerillo, rosabari, rosáwari – silver wormwood, western mugwort
Artemisia ludoviciana* ssp *albula – chamizo cenizo, estafiate – western mugweed – silver king artemisia, white sage
Artemisia ludoviciana* ssp *incompta – estafiate – mountain sagewort
Artemisia mexicana – ajenjo, ajenjo del país, altamisa, artemisa, azumate de Puebla, estafiate, hierba maestra, iztauhyatl, nexmitzi – sage, wormwood
Artemisia norvegica – ajenjo norvego – boreal sagewort
Artemisia pauciflora – santónico, semencontra
Artemisia pontica – ajenjo póntico – Roman wormwood
Artemisia redolens – anisote
Artemisia stellerana – ajenjo de las playas – beach wormwood, dusty miller, old woman
Artemisia tridentata – artemisa, chamizo, chamizo blanco, chamizo ganado, chamizo hediondo, estafiate – basin sagebrush, big sagebrush
Artemisia vulgaris – ajenjo, ajenjo común, anastasia, artemisa, corona de San Juan, estafiate, hierba artemis, hierba de San Juan, planta de San Juan, romerillo, tzitzim – mugwort
Arthrostylidium racemiflora – caña brava, guiyaa, otate – wild cane
Artocarpus altilis (communis) – árbol de pan, castaño, fruta de pan, palo de pan, pana, pana cimarrona, pana de pepita, panapén, pan de pepita – breadfruit, breadnut
Artocarpus heterophyllus – árbol de jack, artocárpeo, jaca – jackfruit, poor man's bread
Artocarpus incisa – árbol de pan, palo de pan – breadfruit
Artocarpus integrifolia – árbol de pan, buen

pan, castaño, jaca, jaqueira, pana cimarrona, pan de fruta – jackfruit, jacktree

Arum italicum – aro, aro italiano, flor de primavera, jaro, maxcalzots – Italian arum

Arum maculatum – aro, aro manchado, flor de primavera, hoja de culebrón, jaro, tragontina – arum

Arum palestinum – aro de Levante – black calla

Aruncus dioicus (sylvester) – barba cabruna, barba de carbon, barba de chivo – goat's beard

Arundinaria longifolia – caña brava, otate – bamboo, cane, wild cane

Arundo donax – acatl, baca, caña, caña castilla, caña común, caña de Castilla, caña gigante, caña hueca, cañaveral, cañutillo, carricillo, carrizo, carrizo de la selva, carrizo de la sierra, gubaguih, haca, halal, jara, ocatl, pakaab, patamu, tarrotekhalal, tekhalal – giant reed

Asarina (Maurandya) antirrhiniflora – mipil – chickabiddy, snapdragon vine

Asarina (Maurandya) barclaiana – cundeamor – climbing snapdragon

Asarina (Maurandya) scandens – tlaltecomaxóchitl

Asarum europaeum – asarabácara, ásaro, ásaro común, aserina, oreja de fraile – asarabacca, wild ginger

Asclepias albicans – jumete, matacandelillas, yamete, yumete – wax milkweed, white-stemmed milkweed

Asclepias asperula (capricornu* ssp *occidentalis) – inmortal – antelope horns, spider milkweed

Asclepias atropurpurea – oreja de venado

Asclepias curassavica – algodoncillo, analkak, burladora, cancerillo, cancerina, chacanalkak, chilillo, chilillosxiu, chontalpa, cochinita, cocholloxin, cojón de gato, corcalito, cuchilxiu, flor de culebra, flor de la calentura, flor de sangre, hierba de leche, hierba María, hoja de veneno, maleza de mariposa, matacaballos, matacoyotes, pedo de culebra, pelo de gato, plato y taza, polkuch, ponchilhuits, punchuhuiche, quema cosas, quiebra muela, quiebra ojo, rejalgar, revienta, revienta muelas, señorita, sintescochit, solidadito, veintiunilla, venenillo, veneno rojo, víbora, viborona, xpolcútzil, yuquilla, zacacanzelxiu – blood flower, butterfly-weed, Curaçao milkweed, Indian root, milkweed, orange milkweed

Asclepias erosa – hierba del cuervo, lechona, yumete – desert milkweed

Asclepias glaucescens – guievada, najcala, oreja de mula, palohuisi, señorita – milkweed

Asclepias incarnata – algodón rojo, señorita, soldadillo – butterfly plant, swamp milkweed

Asclepias latifolia – lechón – broad-leaved milkweed

Asclepias lemmonii – inmortal – Lemmon's milkweed

Asclepias linaria – algodoncillo, hierba de la punzada, hierba del cuervo, inmortal, lechestrenza, mapipitza, pinillo, plumerillo, romerillo, solimán, teperomero, tezonpatli, tlalacxoyatl, tlalayote, tlalnochtli, torbisco, torito, venenillo – pine-needle milkweed

Asclepias masonii – yamete – milkweed

Asclepias oenotheroides – hierba de zizotes – green milkweed

Asclepias speciosa – lechero, lechón – Greek milkweed, showy milkweed

Asclepias subulata – ajamete, cadenilla bronca, candelilla, candelilla bronca, jumete, matacandelillas, yamete, yumete – ajamete, desert milkweed, rush milkweed

Asclepias syriaca – algodoncillo, algodón cornudo, asclepias, mata de la seda – milkweed, swallow-wort

Asclepias tuberosa – algodón bulboso, hierba de la mariposa, inmortal, raíz de pleuresia, tlalacxóyatl – butterfly weed, orange milkweed, pleurisy root

Asclepias vestita – hierba lechera – woolly milkweed

Asimina triloba – papayo – American pawpaw

Asparagus asparagoides – esmilax – smilax asparagus

Asparagus densiflorus (sprengeri)* var *sprengeri – espárrago, espárrago densifloro, espárrago de Sprenger, esparraguera – Japanese asparagus, Sprenger asparagus

Asparagus medeoloides – zarzaparrilla – florist's smilax

Asparagus officinalis – espárrago, espárrago

común, esparraguera – asparagus, garden asparagus
Asparagus setaceus (plumosus) – espárrago, espárrago plumoso, esparraguera, esparraguera de adorno, helecho de esparraguera – asparagus fern, fern asparagus
Asperugo procumbens – azotalengua, hierba de la rabia, miosota, raspilla procumbente – madwort
Asperula = Galium
Asphodeline lutea – asfodelina, bastón de Jacob – king's spear, yellow asphodel
Asphodelus albus – asfódelo blanco, gamón – king's spear
Asphodelus fistulosus – asfódelo, asfódelo fistuloso, cebolleta, estrella del norte, gamoncillo – asphodel
Aspicarpa hirtella – carrechuela, hierba del zorro
Aspidistra elatior (lurida) – aspidistra, hoja de maíz, hoja de salón, oreja de burro – cast-iron plant
Aspidium = Dryopteris
Aspidosperma quebracho – quebracho, quebracho blanco, volador – malady, my lady, white quebracho
Asplenium (Athyrium) filix-femina – helecho hembra – lady fern
Asplenium fontanum – asplenio de roca – rock spleenwort
Asplenium monanthes – calaguala, negrito – spleenwort
Asplenium nidus (nidus-avis) – asplenio, helecho nido, helecho nido de pájaro, nido de ave – bird's nest fern
Asplenium trichomanes (vespertinum) – culantrillo bastardo – maidenhair spleenwort
Asplenium trichomanes-ramosum (viride) – asplenio verde – green spleenwort
Asplenium = Phyllitis
Asplundia chiapensis – confetillo – mantle flower
Aster alpinus – aster, aster alpino, aster de montaña – alpine aster
Aster amellus – aster de montaña, aster de otoño, aster de Virgilio – Italian aster
Aster dumosus – aster de otoño
Aster laevis – ramillete – Geyer's aster, smooth aster
Aster lanceolatus – aster lanceolado – aster
Aster moranensis – flor de María
Aster novae-angliae – aster de la Nueva-Inglaterra, margarita – New England aster
Aster novi-belgii – aster brumal, cielo estrellado, margarita michaelmas – Michaelmas daisy, New York aster
Aster pauciflorus – aster, escobilla – aster
Aster sibiricus – aster de los Pirineos – Pyrenean aster
Aster subulatus – aster, escobilla, estrellita, hierba del marrano, lucerillo – annual aster, awlleaf aster
Aster = Chloracantha, Machaeranthera
Asteraceae (Compositae) – compuesta – comp, composite
Asteriscus maritimus – asterisco del mar – marine asteriscus
Astianthus viminalis – ahuejote, axóchitl, azúchil, flor de agua, palo de agua, palo de San Nicolás, sabina de arroyos, tronador – water blossom, yellow grass tree
Astilbe arendsii – astilbe, falsa spiraea – false spiraea, meadow sweet
Astragalus amphioxys – hierba loca
Astragalus boeticus – astrágalo español, café de la China, café mexicano – Andalusian milk-vetch, Swedish coffee
Astragalus francisquitensis – cascabelito – locoweed
Astragalus glycyphyllos – astrágalo de hojas de regaliz, orozuz falso – wild licorice
Astragalus gummifer – alquitira, goma adragante, tragacanto – tragacanth
Astragalus humboldtii – chinche, chinchín, garbancillo, hierba loca, moradilla, pavito cimarrón, sonadora
Astragalus lusitanicus – garbancillo
Astragalus mollissimus – astrágalo, cascabelillo, cascabelito, chinchín, garbancillo, hierba loca, hierba plata, loco, sonadora – woolly locoweed
Astragalus nuttallianus – loco – milk-vetch, Nuttall's locoweed, turkey-pea
Astragalus potosensis – garbancillo, patito, sonadora
Astragalus tragacantha – perfume – thorny astragal, wild tragacanth

Astrantia major – astrancia, astrancia mayor, sanícula hembra – masterwort
Astrocaryum mexicanum – ak-té, chapaya, chicalito de Túxtepec, chichón, chipi, chocho, chocón, güiscoyol, lancetilla, pacaya, palma, palma chapay – cohune, warree cohune
Astrocaryum murumuru – palma murumuru – murumuru palm
Astrocaryum vulgare – chambira – tucuma palm
Astrolepis cochisensis (Cheilanthes sinuata) – helecho – scaly cloak fern
Astronium graveolens – chaperla, copaiba, culinché, gateado, gateado galán, jobillo, jocotillo, palo de cera, palo de culebra, ronrón, usipon, vaga-biche, zongolica
Astrophytum asterias – peyote, peyote queso – sand dollar cactus, sea urchin cactus
Astrophytum capricorne – biznaga de estropajos – bishop's cap, bishop's hat, goat's horn
Astrophytum myriostigma – birreta de obispo, bonete, gorro de cardenal, mitra, peyote cimarrón – bishop's cap, cardinal's hat
Astrophytum ornatum – mitra – monk's hood
Atamisquea emarginata – jubaibena, palo hediondo, palo zorrillo
Atamosco carinata – atzcalxóchitl, azcasúchil, cebolla de mayo, flor de mayo, mayito, mayo
Atamosco citrina – cucutchom, mayito
Athyrium = Asplenium
Atractylis cancellata – cardo cancelado – distaff thistle
Atractylis gummifera – cameleón blanco, cardo de liga – gum-bearing carline thistle
Atractylodes lancea – atractylis
Atriplex acanthocarpa – quelite
Atriplex barclayana – chamizo, saladilla – coast saltbush
Atriplex californica – chamizo – California saltbush
Atriplex canescens – cenizo, chamizo, chamizo blanco, chamizo cenizo, chico, costilla de vaca, saladilla – fourwing saltbush, sagebrush, shad-scale, wingscale
Atriplex elegans – cenizo, chamizo cenizo – wheelscale
Atriplex halimus – álimo, armuelle glauco, orzaga, saladilla, salado, salgada, salgadera – mallows, Mediterranean saltbush, sea purslane
Atriplex heterosperma – armuelle heterosperma
Atriplex hortensis – armuelle, armuelle de huerta, armuelle molle, bledo molle, orzaga, quenopodio de los jardines – garden orache, garden oracle
Atriplex julacea – chamizo – saltbush
Atriplex lentiformis – cenizo, chamizo – big saltbush, lens scale, quail brush, quail bush, white thistle
Atriplex linearis – chamizo – fourwing saltbush, narrow-wing saltbush
Atriplex linifolia – quelite de puerco, romerillo
Atriplex patula – armuelle, armuelle pátulo, armuelle patente – common orache
Atriplex polycarpa – cenizo, chamizo, chamizo cenizo – all-scale, cattle spinach, desert saltbush, sage, sagebrush
Atriplex rosea – armuelle róseo – tumbling oracle
Atriplex semibaccata – romerillo, saladilla – Australian saltbush
Atropa belladona – atropa, baya de la bruja, belladama, belladona, botón negro, guindo de la costa, solano furioso, tabaco borde – belladonna, deadly nightshade
Attalea butyracea – corozo, coyol real
Attalea (Scheelia) gomphococca – palma real, verdadera palma real
Attalea (Orbignya) speciosa – babasú – babassu palm
Attalea = Orbignya
Aubrieta deltoidea – aubrecia – common aubretia, rock cress
Aucuba japonica – aucuba, laurel manchado – Japanese aucuba, spotted laurel
Aulonemia fulgor – relámpago
Auricularia auricula-judae – oreja de Judas – cloud ear fungus
Auricularia polytricha – oreja, oreja de Judas, oreja de Judío – wood ear fungus
Aurinia saxatilis (Alyssum saxatile) – alisón, canastilla, cesta de oro, mastuerzo saxátil – basket-of-gold, golden alison
Austrocedrus chilensis – ciprés de la cordillera – Chilean cedar

Avena barbata – avena barbada – slender wild oat
Avena fatua – avena, avena cimarrona, avena guacha, avena loca, avena silvestre, ballueca, triguillo loco – wild oat
Avena sativa – avena, avena común – cultivated oat
Avena sterilis – avena estéril, avena de invierno, avena mayor – winter wild oat
Avena strigosa – avena estrigosa – black oat
Averrhoa bilimbi – bilimbi, tiriguro – bilimbi tree
Averrhoa carambola – caimito, carambola, tamarindo chino, tamarindo culí – carambola, star apple, starfruit
Avicennia germinans (nitida) – árbol de sal, chifle de vaca, ciali, culumate, istaten, madre de sal, mangle, mangle blanco, mangle bobo, manglecito, mangle negro, mangle prieto, mangle salado, palo de sal, paseo, puyeque, saladilla – black mangrove, honey mangrove
Avicennia marina – mangle rojo – red mangrove
Axonopus affinis – cañutillo, carpeta, cuchipen, grama tapete, hierba prado, pasto carpeta de hojas angostas, pasto chato común, zacate alfombra, zacate amargo – carpetgrass
Axonopus compressus – alfombra, amargo, caña dulce, grama argentina, grama trenza, hierba césped, hierba de Luisiana, pasto alfombra, pasto carpeta de hojas angostas, pasto chato, pasto chato de alfombra, pasto natural, trencilla, zacate amargo – carpet grass, savanna grass
Axonopus scoparius – carpeta, gramalote, imperial, guaratara, hierba imperial, micay, pasto carpeta, pasto chato, pasto imperial, zacate imperial – imperial grass
Ayenia pusilla – hierba del cáncer, piston-xiu, pixtonchich
Azadirachta indica – árbol del neem, nim, sicomoro – neem, neem tree
Azalea = Rhododendron
Azolla filiculoides – chilacaxtli – mosquito fern

B

Baccharis conferta – escoba, escoba del monte, escobilla, escobilla del carbonero, hierba del carbonero, jarilla, limpiatunas, popotillo, quauhizquitli, tepopote, tepopotl, tepopotli – charcoal shrub

Baccharis neglecta – jarilla, jarilla del río – poverty weed, Roosevelt weed

Baccharis pteronioides (ramulosa) – boshi, carátacua, escobilla, hierba del carbonero, hierba del pasmo, jaral blanco, popotillo, tepopote, tepopotl

Baccharis salicifolia (glutinosa) – bachomo, batamote, chamizo, chilca, escobilla, guatamote, hierba del carbonero, hierba del pasmo, huatamote, jara, jara dulce, jaral, jarilla, jarilla común, jarilla del río, vara dulce – mule fat, seep-willow, water-motie, water-wally, water willow

Baccharis sarothroides – bácaris, batamote, escoba amarga, guatamote, hierba del pasmo, huatamote, romerillo, romerillón – broom baccharis, desert broom, rosin brush

Baccharis sergiloides – escoba amarga – desert baccharis, squaw waterweed

Baccharis thesioides – batamote del monte, hierba del aire, hierba del pasmo, uachamo – Arizona baccharis

Bacopa monnieri – baraima, ya'ax-kach – water hyssop

Bactris gasipaes – chontaduro, chonta, palmito, pejibaye – peach palm

Bactris major – biscoyol, huiscoyol, jahuacté

Bactris mexicana – chapaya, chischi, coyolillo, guacoyul, huiscoyol, jahuacté, palma de espina, palma garrocha, quauhcoyolli – spiny club palm

Baeria = Lasthenia

Bahia absinthifolia – caliche, hierba ratón – hairy-seed bahia

Baileya multiradiata – baileya del desierto, calancapatle, flor de papel, hierba amarilla, palancapatli, tecomblate – desert-marigold, paper daisy

Baileya pauciradiata – tecomblate – drooping marigold, few-flowered desert marigold

Baileya pleniradiata – margarita silvestre, tecomblate – woolly desert marigold, woolly marigold

Balanites aegyptiaca – bálsamo – balm

Ballota nigra – balota, balota negra, marrubio fétido, marrubio negro – black horehound

Ballota pseudodictamnus – balota falso-tarraquillo – false dittany

Balsamorhiza sagittata – balsamillo, chuchupate – balsam-root

Bambusa aculeata – caña brava, otate

Bambusa arundinacea – bambú, bambú japonés, caña brava, otate – reed bamboo

Bambusa spinosa – bambú, caña espina – spiny bamboo

Bambusa vulgaris – bacáu, bambú, bambú común, bambú patamba, carrizo, cupamu, otate, sacaú – common bamboo

Bambusa = Guadua

Banisteriopsis caapi – ayahuasca, caapi, yaje

Barbarea praecox – berrillo, berro de jardín, hierba de Santa Bárbara precoz – winter cress

Barbarea verna – berro de tierra, hierba de Santa Bárbara, hierba de Santa Bárbara primaveral, mastuerzo de jardín – early winter cress, scurvy grass

Barbarea vulgaris – berro terrestre, hierba del carpintero, hierba de Santa Bárbara, hierba de Santa Bárbara común – common winter cress, yellow rocket

Barosma (Agathosma) betulina – buchú – buchu

Basella alba – bretaña, espinaca blanca, espinaca de Ceilán – Ceylon spinach

Basidiomycetes – champiñón, hongo, mízcalo, seta – mushroom

Basistelma angustifolium – caballito, cabellito

Bassia hyssopifolia – albahaca larga, barriola, bassia de hojas de hisopo, mirabel, sisallo – mock cypress, smother weed

Bastardiastrum cinctum – malva blanca

Batis maritima – barilla, chamis, dedito, saladilla, vidrillo – beachwort, saltwort

Bauhinia divaricata (peninsularis) – barba de mantel, calzoncillo, huamúchil, papalocuáhuitl, pata de cabra, pata de chivo, pata de res, pata de vaca, pie de cabra, pie de vaca, pie de venado, tsomeltok, tsulubtok – orchid tree

Bauhinia forficata – árbol orquídea, pata de venado – Brazilian butterfly tree
Bauhinia monandra – casco de buey, palo de orquídeas, urape – butterfly-flower tree, orchid tree, ox hoof
Bauhinia pringlei – palo piojo blanco
Bauhinia variegata – árbol orquídea, calzoncillo, orquídea, palo de orquídeas, pata de cabra, pata de vaca, salacasúchil, urape, urape morado, urape orquídea – buddhist bauhinia, bull hoof, mountain ebony, purple orchid tree
Beaucarnea gracilis – izote, sotolín
Beaucarnea inermis – coyolito, palma barrigona, palma culona, palmitón, soyate – soyate
Beaucarnea = Nolina
Beaumontia grandiflora – beaumontia – herald's trumpet, easter lily vine
Bebbia atriplicifolia – apá, apán
Bebbia juncea – apá, chuparrosa, junco – chuckwalla's delight, sweetbush
Begonia fuchsioides – corazón de ángel, corazón de Jesús
Begonia gracilis – ala de ángel, begonia, cañaigre, carne de doncella, coyol, hierba de doncella, jocoyol color de rosa, sangre de doncella, totoncaxoxocoyolin – tuberous begonia
Begonia martiana – cañaigre, hierba ácida, hupari
Begonia nelumbiifolia – cañita agria, mano de león, xocoyol cimarrón
Begonia rex – begonia de hojas ornamentals, begonia regia, begonia rex – rex begonia
Begonia semperflorens – hermosa – bedding begonia, semperflorens begonia, wax-leaf begonia
Begonia solananthera – begonia colgante – trailing begonia
Begonia tuberosa – begonia, begonia bulbosa, begonia tuberosa, hierba de doncella – tuberous begonia
Belamcanda chinensis – belamcanda – blackberry lily
Bellardia trixago – gallocresta
Bellis perennis – bellorita, chiribita, dormilona, mancerina, margarita, margarita de los prados, margarita enana, margarita menor, margarita pequeña, maya, maya perenne, vellorita – English daisy, English lawn daisy, herb Margaret
Beloperone = Justicia
Belotia mexicana – capulín, capulín blanco, capulincillo, cerezo, corcho colorado, cotomate, coyotillo, cuapetate, guacimilla, jonote coyotillo, yaco de venado
Benincasa hispida – calabaza blanca, melón blanco – wax gourd
Berberis (Mahonia) aquifolium (repens) – hierba de sangre, mahonia, sangre de Cristo, sangre de drago, uva de Oregón, uva silvestre de Oregón, vid de Oregón – creeping barberry, creeping mahonia, holly-grape, Oregon-grape
Berberis buxifolia – berberís de hojas de palma, calafate – Magellan barberry
Berberis darwinii – agracejo de Darwin – Darwin barberry
Berberis fendleri – palo amarillo – western barberry
Berberis (Mahonia) fremontii (higginsae) – acebo, agracejo, algerita, berberís, bérbero, mahonia, palo amarillo – barberry, desert mahonia
Berberis (Mahonia) haematocarpa – algerita, palo amarillo – barberry, red barberry, red mahonia
Berberis julianae – bérbero – wintergreen barberry
Berberis longipes – palo amarillo – barberry
Berberis* x *stenophylla – berberís de hojas cortas – rosemary barberry
Berberis thunbergii – berberís japonés – Japanese barberry
Berberis (Mahonia, Odostemon) trifoliolata – agarito, agrillo, algerita, hierba de sangre, palo amarillo – agarita, Texas algerita
Berberis vulgaris – agracejina, agracejo, agracillo, agrazón, arlo, berberís común, berbero – English barberry, European barberry
Bergenia cordifolia – bergenia, saxífraga de hojas grandes – heartleaf bergenia
Bergenia crassifolia – hortensia de invierno – winter-blooming bergenia
Bergerocactus (Cereus, Echinocereus) emoryi

– cacto aterciopelado – golden-spined cereus, velvet cactus

Berlandiera lyrata – cocolmeca, coronilla, hierba de la víbora, peonía – chocolate flower, lyreleaf green eyes

Bernardia myricifolia (incana) – palo de tarugo, oreja de ratón

Bernoullia flammea – ala de cucaracha, amapola, cuesa, marquesote, palo de cuesa, palo de perdiz, palo de tortilla, platanillo, pojote – locust tree

Berthelotia = Pluchea

Bertholletia excelsa – castaño, castaño de Brasil, castaño de maranhoa, castaño de pará, coquito de Brasil, juvia, nuez de Brasil, nuez de pará – Brazil nut

Berula erecta (Sium erectum) – arsáfraga, barrera, berro, berula acuática de hojas estrechas, queza-pijchi – cutleaf water-parsnip

Beschorneria yuccoides – yuca – yucca

Bessera elegans – arete, aretillo, coralito, gota de coral mexicano, zarcillo – Mexican coral drops

Beta vulgaris – 1. betabel, betarraga, nabo colorado de jaraua, remolacha, remolacha comestible, remolacha de mesa, remolacha de huerto, remolacha roja – beet, garden beet; 2. betarraga forrajera, remolacha forrajera – fodder beet

Beta vulgaris* ssp *cicla – acelga, acelga suiza, alcachola, armuelle, bleda – chard, Swiss chard

Beta vulgaris* ssp *saccharifera – betarraga azucarera, remolacha azucarera – sugar beet

Betonica = Stachys

Betula alba (pendula) – abedul, abedul blanco, abedul común europeo, árbol de la sabiduría, bétula, madera blanca – European white birch

Betula albo-sinensis – abedul chino de corteza roja

Betula alleghaniensis – abedul amarillo americano, abedul amarillo canadiense, abedul de los Alleghanys

Betula jacquemontii (utilis) – abedul de Jacquemont, abedul de la India, abedul del Himalaya, abedul del Himalaya de corteza blanca – Himalayan birch

Betula lenta – abedul, abedul-cerezo americano, abedul de hojas de carpe, abedul dulce americano – cherry birch, sweet birch

Betula maximowicziana – abedul japonés de Maximowicz – monarch birch

Betula nigra (lenta) – abedul, abedul de agua, abedul fluvial, abedul negro, abedul negro americano – red birch, river birch

Betula occidentalis – abedul – water birch

Betula papyrifera – abedul de los canoes, abedul papirífero – canoe birch, paper birch

Betula pendula – abedul común, abedul de Europa, abedul péndulo, abedul verrugoso – European white birch

Betula pendula* var *youngii – abedul llorón plateado – Young's weeping birch

Betula pubescens – abedul, abedul pubescente, bedoll, bétula pubescente, bidveiro, violo – common white birch

Betula verrucosa – abedul, abedul blanco, abedul plateado, aliso blanco, bétula verrugosa – silver birch

Bidens aurea – acahual, acahualillo, aceitilla, bidente de hojas desiguales, capitaneja, lampotillo, té de corral, té de milpa, té de milpa amarillo, xiuh-elo-quílitl – Spanish-needles

Bidens bigelovii – aceitilla, jara – Spanish-needles

Bidens cernua – bidente penché – nodding bur-marigold

Bidens ferulaefolia – juve – Spanish-needles

Bidens frondosa – aceitilla, bidente pedunculado – sticktight

Bidens grandiflora (serrulata) – acahualillo, aceitilla, té de milpa – bur-marigold

Bidens laevis – girasol de agua, jara, té de milpa – bur-marigold, wild goldenglow

Bidens odorata – acahual blanco, aceitilla, cisiquelite, rosatilla, rosetilla, té de milpa blanco

Bidens pilosa – acahual, acahual blanco, acahualillo de flor blanca, aceitilla, acocotli, bidente piloso, cadillo negro, canmul, eloquiltic, escobilla de calendula, margarita, mozote, mozote blanco, mulito, ortiga española, rosilla, rosilla del valle, saetilla, té de corral, té de milpa – common beggar-ticks, Spanish-needles

Bidens pilosa* var *leucantha – acocotli, canmul, mulito, té de milpa
Bidens pilosa (leucantha)* var *radiata – acahual blanco, aceitilla, aceitilla blanca, té de milpa blanco
Bidens refracta – rama pioja
Bidens tetragosa – acahual, acahualillo, aceitilla, mozote, té de milpa, té de milpa amarillo
Bidens tripartita – bidente tripartito, cáñamo acuático
Bigelovia diffusa – hierba del pasmo
Bigelovia = Isocoma
Bignonia = Anisostichus, Campsis, Distictis, Macfadyena, Pyrostegia
Bismarckia nobilis – palmera de Bismarck
Bituminaria (Psoralea) bituminosa – hierba cabruna, higueruela, psoralea bituminosa, trébol hediondo, triguerula
Bixa orellana – acanguarica, achiote, achiotillo, achiotl, arnato, bija, changuarica, chaya, cuajachote, cuxub, pumacua, tlapalachiyotl, urucú – annatto, lipstick tree
Blepharoneuron tricholepis – pastillo del pinar, popotillo del pinar – hairy dropseed, pine dropseed
Blechnum gibbum – helecho cola de pavo – blechnum fern, dwarf tree fern
Bletia campanulata – chautle, flor de muerto, tzacutli, tzacuxóchitl, tzautli
Bletia coccinea – orquídea del campo, tonaloxóchitl, tzacuxóchitl – bletia
Bletia reflexa – flor de muerto
Bletia stricta – orquídea
Blighia sapida – akí, seso vegetál – ackee, akee
Bocconia arborea – ahuacachili, árbol de Judas, boconia, capulín, chicalote, chicalote de árbol, chicalote grande, cocoxíhuitl, engüemba, guachili, hediondilla, inguande, jauque, llora sangre, mano de león, palmilla, palo amarillo, palo de Judas, palo del diablo, saúco, tlacoxihuitl – tree celandine
Bocconia frutescens – boconia, cachisda, calderón, chapalhuacán, chicalote de árbol, clacojiguite, coacuahuitl, cocojegüite, cocoxihuitl, cuatlatlaya, cuauchichili, engüemba, gordolobo, guachichil, guachichile, guachili, inguande, llora sangre, manalote, mano de león, palo amargo, palo amarillo, palo santo, tlacoxihuitl – tears of blood
Bocconia = Macleaya
Boehmeria nivea – ortiga blanca, ortiga de la China, pasto de la China, ramio, seda vegetal – false nettle, ramie plant
Boerhavia coccinea (caribaea) – chacilxiu, coanenepili, juanamipili, mata de pavo, mochi, sambesarambe, sambesarambe macho – scarlet spiderling, wine-flower
Boerhavia coulteri – juanamipili, juantilipin – spiderling
Boerhavia erecta – golondrina, hierba blanca, juantilipin, mochi, saranda, xacil, xacilxacxiu, xacxiu – spiderling
Boerhavia hirsuta – chual, hierba tostada, pega pinto – hog weed, pig weed
Boerhavia intermedia (erecta* var *intermedia) – hierba blanca – spiderling
Boerhavia spicata – juanamipili, sambesarambe hembra – spiderling
Boerhavia xanti – juanamipili
Boldo = Peumus
Boletus edulis – aubarell, boleto, cemita, cepa, cigró, ciunery, corralito, loyo, mazayel, pambazo, panadero de encino, pancita, pancita blanca, poposito, rodellón – cèpe, edible bolete, penny bun
Boletus impolitus – boleto amarillento
Boletus luridus – cemita, galambo, galambo bueno, mataparen, pancita azul – lurid bolete
Boletus luteus – higado – yellow-brown bolete
Boletus pinicola – cemita, cepa, mazayel, pambazo, pancita blanca
Boletus regius – guarín, hongorado, panza roja, penadero de encino, penadero de oyamel
Boletus reticulatus – aubarell, cemita, cepa, cigró, ciunery, marceyalito, pambazo, panadero, rodellón – summer king bolete
Bomarea acutifolia (edulis) – coyolxóchitl, flor de sonaja, granadita, yatzí
Bomarea hirtella – arete de India, coyolxóchitl, granadita, jícama montés
Bombax ceiba – malabarico – red silk cotton tree
Bombax ellipticum – amapola, amapola blanca, amapola colorada, bailador, bailarina, bote, cabellitos de ángel, cabellos de ángel,

carolina, ceiba, chakuyche, chilochuchi, chospó, chucté, chulté, chusté, clavellina, clavellina de la barranca, clavellina roja, coquito, coquito blanco, coquito de puerto ángel, cuajilote, disciplina, guajilote, güi-güi, itztamatl, itzmal, jiquique, jumallo, lana, lele, luya, mapola, mocoque, ococ, palo de lana, pochote, pochotl, pongolote, sospó, tambor, tetique, tiati, tindusa, titilamatl, xanacol, xihuixcán, xiloxóchitl, yaco de la costa, zackuyché – cotton tree, shaving brush tree, wool tree

Bombax* = *Pseudobombax

Bonplandia geminiflora – hierba del toro, pegajosa, sciatica, tepzotzo, tetzotzs

Borago officinalis – becoquino, borraja, borraja común, borraja fina, borrajina, borrajita – borage, common bugloss

Borassus flabellifer – palma de palmira, palmera de palmira – lontar palm, palmyra palm, toddy palm, wine palm

Boswellia carteri – incienso – frankincense

Boswellia thurifera – incienso, olíbano – frankincense

Bothriochloa barbinodis – popotillo algodonoso, popotillo cola de coyote, popotillo perforado, popotillo plateado, zacate popotillo – cane bluestem

Bothriochloa ischaemum* var *songarica – tallo azul de Kingranch

Bothriochloa pertusa – sac suuk

Bothriochloa (Andropogon) saccharoides – pasto blanco, popotillo, popotillo cola de zorra, popotillo plateado, zacate aceite – silver beard grass

Bothriochloa (Andropogon) wrightii – popotillo cola de coyote, popotillo hoyuelo

Botrychium lunaria – lunaria – moonwort

Bougainvillea spectabilis (glabra) – azalea de guía, bombilla, buganvilla, camelina, clarín, flor de papel, flor de verano, lustrosa, manto de Jesús, Napoleón, pompilla, tres Marías, trinitaria, veranera – bougainvillea

Bourreria huanita – bacalché, behek, esquisúchil, flor de palomita, huanita, guiexoba, izquixóchitl, jazmín de la India, jazmín del Istmo, jazmín de Oaxaca, jazmín de palo, jazmín de Tehuántepec, kakalché, munisté, opay, sacbayeck, teoyzquixóchitl, tlapalixquixóchitl, vanita, yaga guiexoba – huanita tree

Bourreria sonorae – chocolatillo, Juanita, vanita

Bouteloua aristidoides – aceitilla, navajita aguja, pasto de cabra, tochite, zacate aceitillo – needle grama

Bouteloua barbata – navajita, navajita anual, navajita de agua, navajita liebrera, navajita racimosa, pata de cuervo, zacate liebre, zacate liebrero – six-weeks grama

Bouteloua breviseta – chino grama, navajita china

Bouteloua chondrosioides – navajita morada, navajita peinada

Bouteloua curtipendula – banderilla, gramilla, navajita banderilla, pasto banderilla – side-oats grama

Bouteloua eludens – navajita Santa Rita

Bouteloua eriopoda – grama negra, navajita negra – black grama

Bouteloua filiformis (repens) – navajita pelillo, navajita rastrera, pelillo – slender grama

Bouteloua gracilis – artiguilla, grama, navajita, navajita azul, zacate cepillo – blue grama, mosquito grass

Bouteloua hirsuta – grama, navajita, navajita velluda – hairy grama

Bouteloua pectinata – navajita

Bouteloua radicosa – navajita morada – purple grama

Bouteloua reflexa – zacate mateado – grama

Bouteloua simplex – navajita simple – mat grama

Bouteloua trifida – navajita china, navajita roja – red grama

Bouteloua uniflora – banderilla simple

Bouvardia glaberrima – chicoria, chuparrosa, mirto

Bouvardia longiflora – buvardia, flor de San Juan, huele de noche, jazmín de San Juan, rosa de San Juan, tlacoxóchitl – St. John's flower, white bouvardia

Bouvardia multiflora – clavelito

Bouvardia ternifolia – cadillo, contrahierba, contrahierba colorada, doncellita, expatli, hierba del indio, hierba del pasmo, hierba del zorrillo, indita, mirto, mirto del campo, rosa

trompetilla, tabaquillo, tlacosúchil, tlacoxóchitl, tonati-shochit, trompetillo – scarlet bouvardia, trompetilla
Brachiaria arizonica – zacate guía
Brachiaria fasciculata – kanchin, paja del camino, panizo fasciculado, zacate, zacate panizo café – birdseed grass, browntop
Brachiaria mutica (Panicum purpurascens) – admirable, camalote, egipto, grama de para, hierba del parral, hierba de pará, malojillo, nilo, paraná, pasto admirable, pasto de laguna, pasto malojillo, piojillo para quixicoba, zacate camalote, zacate egipto, zacate pará – para grass
Brachiaria texana (Panicum texanum) – panizo texano – Texas millet
Brachychiton acerifolius – árbol del fuego, braquiquito rojo – Australian flame tree, flame tree
Brachychiton discolor – árbol sombrero – pink flame tree, Queensland lacebark
Brachychiton populneus – braquiquito blanco – bottle tree
Brachyglottis = Senecio
Brachypodium distachyon – lastón anual – annual false-brome
Brachypodium pinnatum – lastón pinado – heath false-brome
Brahea = Erythea
Brassaia = Schefflera
Brassavola = Laelia
Brassica japonica – mostaza japonesa – mustard greens
Brassica juncea – mostaza china, mostaza de Abisinia, mostaza de hojas, mostaza de la China, mostaza de la India, mostaza hindu – brown mustard, Indian mustard, wild mustard
Brassica juncea* var *sareptana – mostaza de Sarepta – Sarepta mustard
Brassica napus* var *arvensis – colza – colza
Brassica napus* var *napobrassica – colinabo, flor de nabo, nabo, nabo comestible, nabo forrajero, nabo sueco, rutabaga – rapeseed, rutabaga, Swede, Swede rape
Brassica napus* var *oleifera – colza, naba, nabo, nabo colza – canola, oilseed rape, rape
Brassica napus* var *pabularia – col de la Sibéria, siberiana – Siberian kale
Brassica nigra – ajenabo, jenabe, mostaza, mostaza marrón, mostaza negra, pamita, quelite, verduras de mostaza, xenelba – black mustard, mustard greens
Brassica oleracea* var *acephala – berza, berza col, berza común, col, col caballar, col de hoja, col forragera, col gallega, col rizada, llanta – collard greens, kale
Brassica oleracea* var *alboglabra – bróculi chino – Chinese broccoli, Chinese kale
Brassica oleracea* var *botrytis – coliflor, colinabo – cauliflower
Brassica oleracea* var *capitata – 1. asa de cántaro, berza, col, col común, col repollo, gogolo, repollo – cabbage; 2. berza roja, col lombarda, col roja, lombarda, repollo colorado, repollo rojo – red cabbage
Brassica oleracea* var *conica – col picuda – oxheart cabbage
Brassica oleracea* var *costata – col portuguesa – Portuguese cabbage
Brassica oleracea* var *gemmifera – col de Bruselas, repollito de Bruselas, repollo de Bruselas – Brussels sprouts
Brassica oleracea (cauolorapa)* var *gongylodes – colinabo, colirrábano, col rábano, naba, nabicol – kohlrabi
Brassica oleracea* var *italica – 1. brécol, bróculi – broccoli; 2. bróculi de yemas – Calabrese broccoli
Brassica oleracea* var *laciniata – col crespa, col frizada – borecole
Brassica oleracea* var *medullosa – col caballar, col de meollo, col forrajera, col medular, col meollosa – marrow-stemmed kale
Brassica oleracea* var *ramosa – col caballar – Siberian kale
Brassica oleracea* var *sabauda – berza de Saboya, col de Milán, col de Saboya – Savoy cabbage
Brassica oleracea* var *sabellica – berza col, col crespa – curly kale
Brassica oleracea* var *tronchuda – col gallega, culash – Portuguese kale
Brassica oleracea* var *viridis – berza, col forrajera – collards
Brassica rapa (campestris) – 1. flor de nabo, flor de pájaro, mostacilla, mostaza, mostaza amarilla, pata de cuervo, semilla para los

pájaros, vaina, verduras de nabo – birdsrape mustard, field mustard, turnip greens; 2. naba, nabiza, nabo, nabo hortelano, rapo – turnip; 3. naba, nabo sueco, rutabaga – rutabaga

Brassica rapa* var *campestris – nabo de verano – annual turnip rape, bird rape, canola

Brassica rapa* var *chinensis – acelga, col china, col de China, repollo de China – bok choy

Brassica rapa* var *oleifera – nabina, nabo de invierno – biennial turnip rape, bird rape, canola

Brassica rapa* var *pekinensis – col china, col de China, col petsai, repollo chino – Chinese cabbage, Napa cabbage

Brassica rapa* var *perviridis – mostaza espinaca – spinach mustard

Brassica rapa* var *praecox – nabo de verano – summer turnip rape

Brassica tournefortii – mostaza – Sahara mustard, wild turnip

Brassica = Eruca, Sinapis

Brickellia californica – hamula, hierba de la vaca, pachaba, prodigiosa – brickellbush

Brickellia cavanillesi – atanasia, atanasia amarga, gobernadora de Puebla, hierba de la gobernadora de Puebla, hierba del becerro, pextón, prodigiosa

Brickellia coulteri – bacotomi, hierba de la cruz, miona blanca, pachaba – desert brickellia

Brickellia desertorum – pachaba – brickellbush

Brickellia grandiflora – atanasia amarga, hamula, prodigiosa – brickellbush, tassel flower

Brickellia greggii – San Juan del monte

Brickellia hastata – apan

Brickellia veronicaefolia (Coleosanthus veronicaefolius) – gobernadora, gobernadora de Puebla, hierba de la gobernadora de Puebla, hierba del perro, mejorana, orégano del campo, orégano del cerro, orégano del monte, pextón – brickellia

Brittonastrum mexicanum – toronjil – Mexican balm

Briza minor – briza corta, briza menor – quaking grass

Briza maxima – cedacillo, tembladera – quaking grass

Brodiaea coronaria – brodiea – harvest brodiaea

Brodiaea = Dichelostemma

Bromelia balansae – timbiriche – heart of flame

Bromelia karatas – aguama, aguava, bichicol, cardo, ch'om, chuqui, guapilla, jocuiste, jocuiste de majahua, jocuistle, piñuela, tasuchi, timbiriche

Bromelia lasiantha – caraguatá, chaguar, maya, teco

Bromelia magdalenae – maya, pita, pita floja – silk grass

Bromelia pinguin – aguama, aguava, baguela, cardo, cardón, chihue-chihue, ch'om, guámara, hman, ixtle de jocuiztle, ix-tot, jocuistle, piña ratón, piña silvestre, pinguin, piñuela, piñuelilla, pita-puerco, timbiriche, tsalbay – wild pine, wild pineapple

Bromus alopecurus – bromo falso alopecuro – foxtail-brome

Bromus anomalus (porteri) – bromo dormilón, bromo frondoso – brome

Bromus carinatus (compresa, laciniatus, pendulinus) – basicuáare, bromo de California, masiyague, pipilo – California brome, mountain brome

Bromus catharticus (unioloides, willdenowii) – bromo cebadilla, cebadilla, cebadilla criolla, guilmo, zacate de rescate – rescue grass

Bromus diandrus – bromo grande – ripgut grass

Bromus hordeaceus (mollis) – bromo blando, bromo mediterráneo, bromo peloso, bromo suelto – soft brome

Bromus inermis – bromo cebadilla, bromo de Hungría, bromo inerme, bromo liso, bromo mocho, bromo perenne, bromo suave, cebadilla de Hungría, cebadilla perenne – smooth brome

Bromus japonicus – bromo pátulo, bromo pratense – Japanese brome

Bromus madritensis – bromo de Madrid, bromo español – foxtail chess

Bromus madritensis (rubens)* ssp *rubens – bromo rojezo, bromo rojo – red brome

Bromus secalinus – bromo falso centeno – rye-brome
Bromus stamineus – bromo, bromo estamíneo
Bromus sterilis – bromo, bromo estéril – poverty brome
Bromus tectorum – bromo de los techos – cheat grass, downy brome
Brongniartia alamosana – cacachila, palo piojo, vara prieta
Brongniartia palmeri – palo piojo
Brosimum alicastrum – aja, apomo, ash, capomo, huje, Juan Diego, mojito, mojo, motsoque, mujú, nazareno, ojite, ojoche, ojoche blanco, ojochillo, ox, ox dulce, oxitl, oxocoxóchitl, oxotzin, pan de nuez, ramón, ramón blanco, ramón colorado, ramón de mico, ramón rosa, samaritano, talcoite, tsotz, ují – breadnut, ramon breadnut
Brosimum utile – árbol de la leche, árbol de la vaca, palo de vaca – cow tree, milk tree
Broussonetia papyrifera – moral de la China, moral de papel, morera de papel, papelero – paper mulberry
Brownea grandiceps – rosa de montaña – rose of Venezuela
Brugmansia (Datura) arborea – borrachero, cambanda, campanilla, floripondio, palpanichim, trómbita, trompeta, trompeta de juicio – angel's trumpet
Brugmansia aurea – borrachero – yellow angel's trumpet
Brugmansia (Datura)* x *candida – almizclillo, campanilla, campanilla blanca, estramonio, floripondio, palpanichim, trómbita, trompeta – angel's trumpet
Brugmansia (Datura) sanguinea – campanilla encarnada, floripondio, floripondio encarnado, floripondio rojo – mountain angel's trumpet
Brugmansia (Datura) suaveolens – bomba, borrachero, campana, campanilla, campanilla blanca, chamico, flor de campana, floripondio, floripondio blanco, galán de noche, huacacachu, huanto – angel's trumpet, tree datura
Brunfelsia pauciflora – galán de noche – yesterday, today and tomorrow
Brunnera macrophylla – nomeolvides del Cáucaso, nomeolvides perenne
Brunsvigia* = *Amaryllis
Brya ebenus – ébano americano – ebony cocuswood
Bryonia alba – brionia, brionia blanca, colubrine, nueza – bryony, wild hop
Bryonia cretica – nueza
Bryonia dioica – alfesira, anorza, brionia, ceñidero, nabo del diablo, nueza – white bryony
Bryophyllum calycinum – hoja bruja, hoja de Columbia, prodigiosa, siempreviva
Bryophyllum* = *Kalanchoe
***Bryum* spp** – musgo – moss
Buchloë dactyloides – zacate búfalo, zacate chino – buffalo grass
Buddleja americana – axtlacapali, cayolizán, hierba de la mosca, layolizán, salvia real, tepozá, tepozán, zayolizcán, zompantle
Buddleja cordata – tepozá, tepozán, tepozán blanco – rusty buddleja
Buddleja crotonoides – lengua de buey
Buddleja davidii – budleya, lila de verano, mariposa de David – butterfly bush, summer lilac
Buddleja marrubifolia – azafrán, azafrancillo, azafrán del campo, azafrán de los pobres, azafranillo – woolly butterfly bush
Buddleja parviflora – tepozán cimarrón, tepozán de cerro, zayolizco
Buddleja perfoliata – orejón, salvia aromatica, salvia de bolita, salvia india, salvia real, salvia real de bolita
Buddleja scordioides – escobilla, escobilla savilla, golondrilla, hierba de la escoba, salvia
Buddleja sessiliflora – hierba del tepozán, lengua de buey, lengua de vaca, mispatle, pahtlaxoxoctic, popotzo, quimixpatli, salvia, salvia común, tepozán, tepozancillo, tepozán verde
Buddleja tomentella – palo cenizo, tepozán
Buddleja wrightii – lengua de buey, salvia, tapiro, tepozán
Buglossoides* = *Lithospermum
Bumelia* = *Sideroxylon
Bunchosia sonorensis – garbancillo, garbanzo de coyote, palo cenizo, saspuchi, tóroco
Bunium bulbocastanum – castaña de tierra – earth nut

Bupleurum chinense – bupléurum, perfoliada china – bupleurum
Bupleurum fruticosum – adelfilla, amarguera, amarguera fruticosa – shrubby hare's ear
Bupleurum lancifolium – amarguera de hojas lanceoladas – narrow thorow wax
Bursera aptera – cuajiote
Bursera (Elaphrium) cerasifolia – copal, copalquin, torote
Bursera confusa – copal, torote, torote blanco, torote chutama, torote papelillo – elephant tree
Bursera cuneata – copal
Bursera epinnata – torote
Bursera (Elaphrium) fagarioides – chutama, copal, cuajiote, cuajiote amarillo, cuajiote blanco, cuajiote colorado, cuajiote verde, palo mulato, to'oro, toró chutama, torote, torote amarillo, torote blanco, torote de vaca, torote de venado, torote papelillo, torote prieto, ugedri – fragrant elephant tree
Bursera (Elaphrium) filicifolia – torote, torote blanco
Bursera (Elaphrium) grandifolia – chicopun, chutama, jiote blanco, napitoro, palo mulato, to'oro mulato
Bursera gummifera – almácigo, almácigo amarillo, caratero, tacamahaca – West Indian elemi
Bursera (Elaphrium) hindsiana (epinnata) – copal, copalquín, copal rojo, torote, torote prieto – red elephant tree
Bursera inopinnata – torote copal, torote prieto
Bursera jorullensis (Elaphrium jorullense) – copal, copal blanco, copal de penca, copal de santo, copalli, copalquáhuitl, ngedni
Bursera lancifolia (fragilis) – cuajiote colorado, gomosilla, incienso, palo mulato, tacamahaca, to'oro chutama, toró, torote, torote amarillo, torote blanco, torote colorado, torote de incienso, torote jolopete, torote papelillo, torote prieto
Bursera laxiflora (Elaphrium laxiflorum) – chutama, copal, palo mulato, to'oro chucuri, to'oro ochónome, toró, torote, torote papelillo, torote prieto
Bursera (Elaphrium) macdougalii – copal – elephant tree
Bursera microphylla (Elaphrium microphyllum) – árbol elefante, chukuri to'oro, copal, copalquín, cuajiote blanco, cuajiote colorado, palo colorado, pata de elefante, to'oro, toró chukuri, torote, torote blanco, torote colorado, torote prieto, torote rojo, xoop – elephant tree
Bursera odorata (Elaphrium odoratum) – chutama, copalcuáhuitl, copali, copalquín, cuajiote amarillo, cuajiote blanco, cuajiote verde, jiotillo, papelillo, patlahoac, tecopalcuáhuitl, torote, torote blanco, torote colorado, xochicopalli – elephant tree
Bursera palmeri – copal, palo de cuchara
Bursera penicillata (Elaphrium penicillatum) – copal, copal de santo, copalillo, to'oro, topocá, toró, torote, torote colorado, torote copal, torote de puntagruesa, torote incienso, torote prieto
Bursera (Elaphrium) simaruba – almácigo, caratero, chaca, chocohuite, cohuite, copalillo, desnudo, gumbo limbo blanco, hukup, indio desnudo, indio peludo, jiote, jiote colorado, jobo, marare, mulato, palo chino, palo colorado, palo jiote, palo mulato, palo mulato de Chiapas, palo retinto, papelillo, percha, piocha, quiote, sac-chacah, torote colorado, zongolica – gumbolimbo, gum elemi, mastic tree, tropical birch
Bursera simplicifolia – torote prieto
Bursera stenophylla (Elaphrium stenophyllum) – jaboncillo, palo copal, to'oro sajo, torote blanco, torote camforado, torote copal
Bursera (Elaphrium) submoniliformis – copal grueso, copalillo
Butia capitata – palmera de la jalea, pindo – jelly palm, pindo palm
Butia eriospatha – butia lanosa
Butia yatay – yatay
Butomus umbellatus – junco florido
Butyrospermum parkii – árbol de la manteca, árbol de shea, árbol mantequero, butirospermo – shea-nut tree
Buxus sempervirens – alarquez, boj, boj común – common boxwood, English boxwood
Byrsonima crassifolia – changungo, chaparro de chinche, chaparro manteco, chengua, chi,

nanche, nanche agrio, nanche amarillo, nanche de perro, nanche dulce, nantzinquáhuitl, nantzinxócotl, zacpah – golden spoon, pickle tree

Byrsonima coriacea – chaparro – shoemaker's bark

Byttneria aculeata – almohada de culebra

C

Cacalia = Psacalium
Cactaceae – cacto – cactus
Caesalpinia arenosa – barba de gallo
Caesalpinia cacalaco – cacalaco, cacalaxtli, cascalote, chalalá, cuachalate, dividivi, guachalalá, huisache, huisache bola, huisache marismeño, nacazcul, nacazcólotl, oreja retorcida, patlahoachoixachin – bird-of-paradise, cascalote
Caesalpinia caladenia – gueto pochi, jícamuchi, palo piojo blanco
Caesalpinia californica – tabachín, vara prieta
Caesalpinia coriaria – cascalote, dividivi, ébano del país, nacazcul, nacazcólotl – divi-divi
Caesalpinia crista – brasil, brasilete colorado, cojón de gato, guacolote, guacolote prieto, haba de San Antonio, huitzquahuitl, jabillo, jabillo del mar, ojo de águila, ojo de venado, piedra del águila, taray – nicker nut
Caesalpinia echinata – palo de Brasil – Brazil wood
Caesalpinia (Poinciana) gilliesii – espuela de Santiago, guacamaya, huisache, mal de ojo, poinciana, poinciana amarilla, tabachín – bird-of-paradise, yellow bird-of-paradise
Caesalpinia mexicana – biche, ébano, guajillo, guayabillo, gunabia prieto, haya, hierba del potro, retamilla, tabachín del monte – Mexican bird-of-paradise
Caesalpinia palmeri – jícamuchi, palo piojo, piojito, piojo, polilla – poinciana
Caesalpinia pannosa – palo estaca
Caesalpinia platyloba – arellano, barba de viejo, brasil, brasilete, chacte, hueilaqui, palo colorado, quebracho, veylaqui – poinciana
Caesalpinia (Poinciana) pulcherrima – bandarigu de Juchitán, barba del sol, barbasúchil, cacalaco, cacaloxóchitl, camarón, chacamalxóchitl, chacasúchitl, chachalcahuite, chactsinkin, chalcasúchil, chamalxóchitl, chamol, chamolxóchitl, chinchemalinche, clavellina, clavellina colorada, flor de camarón, flor de guacamaya, flor de San Francisco, framboyán francés, guacamaya, Guatemala, hojasén, kanzinkin, maravilla costeña, maravilla moreña, matachín, mechuda, nochebuena, pequeño flamboyant, poinciana, poinciana enana, sirundanicua, tabachín, tabachín amarillo, tabachín de la sierra, talcapache, talpakapache, tetezo, xiloxóchitl, zinkin, zurandanicua – Barbados flower, Barbados pride, dwarf poinciana, peacock flower, pride of Barbados, red bird-of-paradise
Caesalpinia sappan – brasilete de la India, palo sapán – sappan wood tree
Caesalpinia sclerocarpa – ébano, huisache bola, palo freno, tubchi – Mexican ebony
Caesalpinia spinosa – acacia amarilla, dividivi de los Andes, tara – tara
Caesalpinia standleyi – huetapoche, margarita
Cajanus cajun (indicus) – cachito, cajan, chícharo, chícharo de árbol, chícharo gandul, frijol congo, frijol de árbol, frijol de la India, frijol del monte, frijol de palo, frijol gandul, gandul, gandul verde, guandú, guisante de Angola, guisante de paloma, guisante pajarero, poroto granado – gongo pea, pigeon pea
Cakile maritima – oruga marítima, roqueta marítima, rucamar – sea rocket
Caladium bicolor – caladio, capa de rey, capote de payaso, capotillo, corazón de cabrito, corazón de Jesús, corazón sangriento, hocha de leche, hoja de adorno, manto de la reina, manto de la Virgen, papagayo, papagayo del Brasil, turiara – caladium, fancy-leaved caladium, heart of Jesus
Caladium = Colocasia
Calamagrostis stricta – calamagrostis rígida, carrizo – reed grass
Calamintha grandiflora – calamento de flores grandes – calamint
Calamintha nepeta – anéota, calamento, calamento de flores pequeñas, hedéota – lesser calamint
Calamus draco – caña de las Indias, sangre de venado – rattan palm
Calamus optimus – palasán, ratán – rattan palm
Calamus rotang – cálomo rota, cálomo roten, caña de Bengala – rattan palm, rotang palm
Calandrinia ciliata (micrantha) – lengua de pájaro – red maids

Calathea allouia – lairén, macure, macús – sweet corn root, topee tambu
Calathea lancifolia (insignis) – hoja de zope, quequeste de puerco
Calathea lutea – caussú, galatea, hoja blanca, hoja de chombo, hoja de piedra, hoja de sal, hoja de zopilote, huacanalá, joltuluc, popal, quelite de puerco, quequeste de puerco, sacjavén – cachibou, maranta, water chestnut of the West Indies
Calathea makoyana – calatea, galatea sapo – peacock plant
Calathea zebrina – galatea, papel de música – zebra plant
Calatola mollis – aguacate de mono, calatola, calatolazno, duraznillo, durazno de mono, matacaballos, nogal, nuez de calatola, nuez de montaña, omanche, onmachinté, palo de tinto, zapote de mono – pecan
Calceolaria herbeohybrida – calceolaria, monedero
Calceolaria mexicana – bolsa de clerigo, bolsa del pastor, portamoneda, pulicua
Calea zacatechichi – ahuapatli, hierba de la puebla, juralillo, Oaxaqueña, prodigiosa, simonillo, tzicin, xaxachichitl, zacachichi, zacachichic, zacate amargo, zacatechi, zacatechichi, zacate de perro – bitter's calea
Calendula arvensis – caléndula campestre, caléndula silvestre, hierba lavamanos, maravilla de los campos, maravilla silvestre – field-marigold
Calendula officinalis – caléndula, caléndula de jardín, caléndula officinal, china, flamenquilla, flor de muerto, flor de todos los meses, maravilla, maravilla común, mercadela, reinita, virreina, virreinita – calendula, pot-marigold
Calla = Zantedeschia
Callaeum = Mascagnia
Calliandra brandegei – tabachín, tabardillo – fairy duster
Calliandra californica – cabellos de ángel, chuparrosa, hierba del tabardillo, tabardillo, zapotillo – Baja fairyduster
Calliandra emarginata – pie de venado – dwarf powder puff
Calliandra eriophylla – brasilillo, ángel, cabellitos de ángel, cabellos de ángel, cabeza de ángel, charresquillo, cósahui, cósahui del norte, huajillo, mezquitilla, pelo de ángel, plumita – fairyduster, false mesquite, mock mesquite
Calliandra gentryi – clavellina
Calliandra gracilis – tosapolo
Calliandra grandiflora – cabellos de ángel, tzonxóchitl
Calliandra haematocephala – bellota, calliandra, flor de la cruz – pink powder puff, red powder puff
Calliandra houstoniana – barba de chivato, barba de chivo, cabellos de ángel, charamusco, chivato grande, day, flor de sangre, flor mechudo, hierba del burro, huajillo prieto, huitote, pambotano, tabardillo, tzotzocoli
Calliandra peninsularis – tabardillo, tabardillo de la sierra, zapotillo – fairy duster
Calliandra tweedii – arbusto de la llama – Brazilian flame bush, Trinidad flame bush
Calliandra = Zapoteca
Callicarpa americana – uvilla – beautyberry
Callistemon citrinus – calistemon – crimson bottlebrush, lemon bottlebrush
Callistemon viminalis – calistemon, callistemo, limpia tubos llorón – bottlebrush, weeping bottlebrush
Callistephus chinensis (sinensis) – aster, aster de China, aster de septiembre, aster de verano, astras, coronado, estrella, flor de estrella, maravilla, margarita, margarita de San Miguel, margarita reina, maya, Reina Margarita – China aster
Callitriche hermaphroditica – estrella de agua de otoño – water-starwort
Callitriche stagnalis – estrella de agua de los terrenos paludosos – water-starwort
Callitriche verna – estrella de agua común, estrella de agua de los pantanos – water-starwort
Calluna vulgaris – biércol, brecina, brecina común, brezo, brezo común, brezo escocés, querihuela – Scotch heather
Calocarpum = Lucuma
Calocedrus (Libocedrus) decurrens – cedro de incienso, cedro de incienso de California, ciprés, libocedro, libocedro de California – incense cedar

Calochortus barbatus – ayatito, flor de María, gallito
Calochortus bruneaunis (nuttallii* var *bruneaunis) – lirio sego blanco, sego – sego lily
Calochortus flavus – ayatito, gallito, lirio mariposa – mariposa lily
Calochortus kennedyi – mariposa – desert mariposa
Calochortus panamintensis (nuttallii* var *panamintensis) – lirio sego blanco – Panamint mariposa lily
Calocybe = Tricholoma
Calonyction = Ipomoea
Calophyllum braziliense – barí, barillo, barí oscuro, cedro cimarrón, cimarrón, guaya, leche amarilla, leche María, limoncillo de Córdoba, María, ocú, palo María, Santa María
Calophyllum calaba – aceite, árbol de María, bálsamo de María, barrillo, María, ocuje, palo de María – calaba
Calotropis procera – akund, algodón de seda, árbol de seda, bomba, calotropis – rooster tree, Sodom apple
Caltha leptosepala – hierba centella – marsh marigold
Caltha palustris – botón de oro, calta, calta palustre, hierba centella, hierba del rosario, hierba velluda, maravilla palustre – marsh marigold
***Calumus* spp** – bejuco, rota – rattan
Calvatia gigantea – cuesco grande de lobo, pedo de venado – giant puffball
Calycanthus floridus – calicanto de Florida – Carolina allspice
Calycanthus occidentalis – calicanto, calicanto de occidente – spicebush, sweet-shrub
Calycophyllum candidissimum – camarón, canela, chacalí, colorado, dagame, madrón, palo calabaza, palo colorado – lemon wood
Calycophyllum multiflorum – palo blanco
Calypso bulbosa – calipso bullosa – bulbous calypso, fairy slipper
Calyptrogyne ghiesbreghtiana – capuche, guanito talis, pumúk – San Miguel palm
Calyptronoma rivalis – palma de manaca
Calystegia sepium – berza común, correhuela mayor – hedge bindweed
Calystegia soldanella – berza amarga, berza marina, col marina, soldanela mayor – morning-glory, sea bindweed
Camassia quamash (esculenta) – camasia, quamash – camas, camass, Indian hyacinth
Camelina microcarpa – camelina de fruto pequeño – false flax
Camelina sativa – camelina, camelina pilosa – gold of pleasure
Camellia asamica – té – Assam tea
Camellia japonica – camelia – camellia
Camellia (Thea) sinensis – árbol de té, té, té negro, té verde – tea
Campanula carpatica – farolillo de los Cárpatos – tussock bellflower
Campanula cochlearifolia – campánula, campánula en ramilletes – fairy thimbles
Campanula garganica – campánula
Campanula isophylla – campanilla, estrella de Belén, estrella de Italia – Italian bellflower, star of Bethlehem
Campanula lactiflora – farolillo blanco
Campanula medium – besico, campanilla, campánula, campánula media, farolillo, hierba media – Canterbury bell, cup-and-saucer
Campanula persicifolia – campanilla de hojas de melocotonero – peach-leafed bluebell
Campanula portenschlagiana – campánula de pared – Dalmatian bellflower
Campanula rapunculoides – aguinaldo, rapónchigo – bellflower
Campanula rapunculus – rapónchigo, ruiponce – rampion
Campanula rotundifolia – campánula azul, campánula de hojas circulares – bluebell, harebell
Campanula trachelium – campánula de hojas de ortiga, hermosilla – throatwort
Camphora = Cinnamomum
Campsis grandiflora (Bignonia chinensis) – trompetilla – Chinese trumpet creeper
Campsis (Bignonia, Tecoma) radicans – bignonia, flor de trompeta, jazmín de Virginia, jazmín trompeta, monapesto – common trumpet creeper
Campsis* x *tagliabuana – trompeta trepadora
Cananga odorata – cananga, ilang-ilang – ylang-ylang

Canarium album – pisa – Chinese olive
Canarium commune – almendro de Java, canari – Java almond
Canarium ovatum – árbol de la nuez pili, nuez pili – pili nut tree
Canavalia ensiformis – canavalia, canavalia gigante, costilla de caballo, frijol blanco, frijol de playa, frijol de sable, frijol espada, frijolón, guisante sable, haba blanca, haba criolla, haba de burro, haba de caballo, judía de caballo, judía de puerco, judía espía, judía sable, poroto sable, teijoco de puerco – jackbean
Canavalia gladiata – carabanz, caraota grande, frijol de machete, frijol de sable, frijol espada, frijol grande, haba de burro, poroto sable, teijoco de sable – swordbean
Canavalia maritima – frijol de playa
Canavalia plagiosperma – poroto gigante – giant bean
Canavalia villosa – frijol de playa, frijolillo, frijolito, gallinita, judía de puerco, judía sable, patito, talayote de costilla – jackbean
Candida albicans – cándida – candida
Canna coccinea – achira, bandera española – Indian shot
Canna edulis – achira, caña comestible, chancle, lengua de dragón, platanillo – Queensland arrowroot
Canna generalis – lengua de dragón, platanillo – canna
Canna indica – achira, bandera, bandera española, caña, caña comestible, caña coro, caña de cuentas, caña de las Indias, caña india, chancalá, chancle, changara, chilalaga, coshu, coyol, flor de cangrejo, frutilla, hierba del rosario, junco de Indias, platanillo, yuquilla – canna, Indian shot
Cannabis ruderalis – cáñamo silvestre – wild hemp
Cannabis sativa (indica) – cáñamo, cáñamo común, cáñamo cultivado, consentida, doña Juanita, grifa, hachís, hapis-coil, haxix, jashis, Juanita, macusi, marihuana, marijuana, mota, rosa María, Santa Rosa, tronador – cannabis, grass, hemp, marijuana, pot
Canotia holacantha – corona de Cristo, junco, palo verde – crucifixion thorn
Cantharellus cibarius – amarillo, duraznillo, embudo, fuchila, hongo amarillo, membrillo, mízcalo, rebozuelo, rossinyol – chanterelle
Cantharellus edulis – enchilado – chanterelle, chantrelle
Cantua buxifolia – cantúa, lluvia – magic flower, sacred flower of the Incas
Capparis atamisquea – jito, juvavena, palo hediondo, palo zorrillo – caper bush
Capparis flexuosa – clavellina, jos, mimbre del monte, mosto, palo de burro, tabarjeca, tebelojeca – caper tree
Capparis incana – matagallinas, palo paloma, vara blanca
Capparis indica – alcaparra, arete, colorín, falsa alcaparra, mangle de la sierra, palo sapo, prieta, taiché, vara prieta
Capparis odoratissima – carbonero, naranjillo, olivo, paniagua
Capparis spinosa – alcaparra, alcaparra común, alcaparra parrera, alcaparra parro, alcaparrera, alcaparrón, capote, zapote de perro, taparera – caper bush, desire
Capparis verrucosa – ajito, coquito, limoncillo, naranjillo, palo sapo, vara prieta
Capriola = Cynodon
Capsella bursa-pastoris – bolsa del pastor, comida de pajaritos, pan y quesillo, zurrón de pastor – shepherd's purse
Capsicum annuum – 1. ají, chacic, chile, chilli, guindilla, pimiento picante – chile, hot pepper; 2. chile colorado, pimentón, pimentón paprika, pimiento, pimientón dulce – paprika; 3. pimiento – pimento, pimiento; 4. chile ancho, chile jarál, chile joto, chile mulato, chile para rellenar, chile pasilla, chile poblano, pasilla, pasilla roja – ancho chile, poblano chile; 5. chile chilchote, chile cuaresmeño, chile gordo, chile jalapeño, jalapeño, tepochilli, texochilli, tornachile, tzonchilli – jalapeno pepper; 6. chile verde, chile verde del norte – New Mexico chile, Anaheim chile; 7. chilaca seca, chile achocolatado, chile chilaca, chile cuernillo, chile negro, chile para deshebrar, chile pasilla – black chile; 8. chile cascabel, bola, bolita, boludo; 9. chile serrano, chile tampiqueño, chile verde, serrano – serrano chile; 10. chile de árbol, cola de rata, milchilli, pico de pájaro, pimiento alfilerillo – tree chile, tree

pepper; 11. chile de agua – water chile; 12. carricillo, chile caricillo, chile cera, chile cristal, chile güero, chile largo, chile xcatic, tonalchilli – Fresno chile, wax chile, wax pepper; 13. chile chilacate; 14. cascabel, chilcoztli, chile guajillo, chile travieso, chile trompa, guajillo, mirasol, puya – mirasol pepper; 15. chilhuacle, chilhuacle amarillo, chilhuacle negro, chilhuacle rojo; 16. chilaile, mora, morita; 17. chile rubio, chile roja – Turkish pepper; 18. chile blanco, chile inayague, costeño amarillo, mirasol, viejo arruga dulce; 19. chile de Chiapas, pico de paloma; 20. chile mulato, chile achocolatado, ancho mulato, ancho negro – chocolate ancho pepper, mulato pepper

Capsicum annuum **var** ***grossum*** – ají, ají dulce, chile campana, chile de agua, chile dulce, chile morrón, pimentón, pimiento dulce, pimiento morrón – bell pepper

Capsicum annuum **var** ***minimum (aviculare, glabrisculum)*** – chile amomo, chile chipilquitín, chile del monte, chile de pájaro, chile de perro, chile enano, chile max, chile mosquito, chile piquín, chile silvestre, chililillo pequeño, chilipiquín, chiltepín, chiltepiquín, estiércol de pájaro, piquín, quauhchilli, tepín, tlilchilli, totocuitlatl, tzanalchilli – bird-pepper, wild pepper

Capsicum baccatum – chiltepiquín, chiltepín, cocorit – bird-pepper

Capsicum chinense (frutescens) – 1. chile habañero, guindilla, habañero, onza – habanero chile, Scotch bonnet; 2. ají dátil, chile dátil – datil pepper; 3. chile criollo – creole pepper; 4. chile rocotillo – rocotillo pepper

Capsicum fastigiatum – ají de Cayena, chile cayena, pimentón, pimiento cayena inglesa, pimiento de Cayena, pimiento rojo – cayenne

Capsicum frutescens – 1. chile japonés, maxak, Tabasco – tabasco; 2. ají, chile, guindilla, pimiento picante – bird pepper

Capsicum microcarpum – chicnigui, chile amash

Capsicum pubescens – 1. ají, canario, chile caballo, chile canario, chile manzano, chile perón, pimiento bravo, pimiento caribe, rocoto, rócot uchu – rocoto pepper; 2. chile caballo, chile canario – yellow-fruited rocoto pepper; 3. chile jutiapa, llata, manzano – black-seeded pepper

Capsicum violaceum – tzinquauhyo

Caragana arborescens – acacia de Rusia, guisante arbóreo – Siberian peashrub

Cardamine bellidifolia – mastuerzo alpino, mastuerzo de hojas de margaritilla – bitter-cress

Cardamine hirsuta – mastuerzo, mastuerzo hirsuto – hairy bitter-cress

Cardamine pratensis – berro de prado, berza de prado, dentaria, mastuerzo de prado – bitter-cress, lady's smock, toothwort

Cardaria draba – draba, lepidio draba – heart-podded hoary cress, whitetop

Cardiocrinum (Lilium) giganteum – azucena gigante, lirio gigantesco

Cardiospermum halicacabum (corindum) – besico, bolsilla, bombilla, farolito, farolito de enredadera, farolito de jardín, farolito de la Virgen, güirote cachora, hierba de chivato, huayunac, huevito, huevo de gato, huevo de toro, Juanita, mundito, ocotillo, rayo, tolicarochá, tomatillo, tlazoltomatl, tronador, vaval – baloon vine, love-in-a-puff

Carduus acanthoides – cardito, cardo de burro – plumeless thistle

Carduus cripus – cardo borriquero, cardo corriente, cardo de burro – curly welted thistle

Carduus nutans – cardo almizclero, cardo rojo – musk thistle

Carduus pycnocephalus – cardo cabezudo, cardo de calvero – Italian thistle

Carduus tenuiflorus – cardo menor – slender thistle

Carex aquatilis – carrizo acuático – water sedge

Carex arenaria – cañuela, carrizo arenario, grama roja, zarzaparrilla de Alemania – sedge

Carex atherodes – carrizo aristado – sedge

Carex buchananii – carrizo – leather leaf sedge

Carex buxbaumii – carrizo de Buxbaum – brownish sedge, club sedge

Carex canescens – carrizo canescente – whitish sedge

Carex capitata – carrizo capitado – sedge

Carex diandra – carrizo diándrico – lesser tussock sedge
Carex disperma – carrizo dispermo – sedge
Carex echinata – carrizo hirsuto – sedge
Carex elata – carrizo rígido – tufted sedge
Carex flacca – carrizo glauco – blue sedge
Carex lasiocarpa – carrizo lasiocarpo – slender sedge
Carex limosa – carrizo limoso – bog sedge, mud sedge
Carex morrowii* var *expallida (variegata) – junco – variegated Japanese sedge
Carex nigra – cárex común, carrizo negro – common sedge
Carex norvegica – carrizo noruego – Scandanavian sedge
Carex ovalis – carrizo pie de liebre – hare's foot sedge
Carex utriculata – carrizo rostrado – bottle sedge
Carex vesicaria – carrizo vesicular – bladder sedge
Carex viridula – carrizo serótino – late-flowering sedge
Carex vulpinoidea – carrizo falso de zorro – fox sedge
Carica candamarcensis – papayo – mountain pawpaw
Carica cauliflora – papayo oreja de mico, papayito – wild papaya
Carica mexicana (Pileus mexicanus) – bonete, cuahuayote, guaguayote, kunché, papayo montés, papayo orejón – wild papaya
Carica papaya – árbol de melon, chichput, fruta bomba, lechoso, melón camote, melón chapote, melón zapote, papagayo, papaya de árbol, papayero, papayo, papayo del pájaro, papayo real, put – papaw, papaya, pawpaw
Carica pentagona – babaco – babaco
Carica pubescens – melón chapote, melón zapote, pitzahuac – papaya
Carissa grandiflora – amatúngula, cerezo de Natal – Natal plum
Carlina vulgaris – cardo ajonjero, carlina, carlina de monte, carlina silvestre, cepa caballo – carline thistle
Carlowrightia arizonica – chuparrosa, lemilla – Arizona carlowrightia
Carlowrightia californica – aretito, rama de toro
Carlowrightia lindauiana – lemilla
Carludovica palmata (insignis) – bombonaje, chidra, coyacal, guano, iraca, jipi, jipijapa, jipijapa de Panamá, palma de jipijapa, palma de sombrero, palmilla, soyacal, toquilla – Panama hat palm
Carnegiea gigantea (Cereus giganteus) – mojepe, pitahaya, pitaya, saguaro, sahuaro – giant cactus, saguaro, saguaro cactus
Carpenteria californica – carpenteria – tree-anemone
Carphochaete pringlei – hierba de la muela
Carpinus betulus – abedulillo, árbol de hiero, carpe, carpe blanco, carpe común, carpe europeo, charmilla, hojaranzo, lechillo, ojaranzo – European hornbeam
Carpinus caroliniana – lechillo, moral de la sierra, palo de barranco, palo liso, palo silo, palo sino – American hornbeam
Carpinus japonica – carpe japonés
Carpobrotus chilensis (Mesembryanthemum chilense) – algazul, cabellera de la reina, escarchada, flor de cuchillo, flor de sol, margarita de Livingstone, mesembriantemo, molina – ice plant, sea fig
Carpobrotus edulis (Mesembryanthemum edule) – higo de Hotentote comestible, higo del Cabo, higo marino – Hottentot fig, ice plant
Carthamus lanatus – azotacristos, cardo cabrero – woolly distaff thistle
Carthamus tinctorius – alazor, alazor bastardo, azafrán, azafrán bastardo, azafrancillo, azafrán de los pobres, azafranillo, azafrán rumi, cártamo, simiente de papagayos – American saffron, false saffron, safflower
Carum carvi – alcaravea, alcaravea de prado, alcosoria, carvi, comino de los prados, hinojo de prado – caraway
Carya amara – caria amarga – bitternut
Carya cordiformis – jicoria amarga – bitternut
Carya glabra – jicoria glabra – pignut hickory
Carya (Hicoria) illinoensis – damzá, jicoria pecán, nogal, nogal americano, nogal de Cuilapán, nogal de nuez chica, nogal de nuez lisa, nogal encarcelado, nogal liso, nogal morado, nogal pecanero, nuez, nuez de

Castilla, nuez de pacana, nuez encarcelada, nuez pacana, nuez pecan, pacana – hickory, pecan
Carya laciniosa – gran jicoria de corteza conchosa – big shell-bark hickory
Carya (Hicoria) mexicana – nogalillo, nogal motudo, nogal rayado
Carya ovata – caria blanca, jicoria ovada, nogalillo, nogal motudo, nogal rayado – shagbark hickory
Carya (Hicoria) pecan – napacoma, nogal de Cuilapán, nogal de nuez chica, nogal morado, nuez chiquita, nuez de Cuilapán, nuez encarcelada, nuez lisa, pacanero, pecanero, quauhcacoatl
Carya tomentosa – jicoria tomentosa – mockernut
Caryocar amygdaliferum – achiotillo, almendrón – butter nut, souari nut
Caryophyllus (Syzygium) aromaticus (Eugenia caryophyllata) – árbol del clavo, clavero, clavero oloroso, clavo, clavo de olor, eugenia aromatica – clove
Caryopteris incana (mastacanthus) – cariopteris – blue spiraea, common bluebeard
Caryota mitis – palma de cola de pescado, palmera de cola de pescado – clustered fishtail palm
Caryota urens – palma cola de pescado, palma de sagú de Malabar, palmera cola de pescado, palmera de sagú de Malabar – fishtail wine palm, toddy palm
Casearia aculeata – cedrón, limoncillo, ramoncillo – wild lime
Casearia dolichophylla – chilillo, copalillo, garabatillo, palo güemuchi, tepezapote
Casearia nitida – botoncillo, café cimarrón, cafetillo, obatel – Indian boxwood
Casearia sylvestris – aguedita blanca, aguedita dulce, cafetillo, chalcahuite, guayabillo, mierda de loro, naranjillo, palo blanco – wild coffee
Casimiroa edulis – chapote, chochizapote, coaxmuttza, cochil sapota, cochitzápotl, hyuy, iztactzápotl, matasano, pera mexicana, tzápotl, yuy, zapote, zapote blanco, zapote de rata, zapote dormilón, zapote dulce – Mexican apple, white sapote
Casimiroa sapota – ajté, matasano, tzucui
Cassia fistula (Senna excelsa) – cañafistula, casia, casia purgante, casia purgativa, chacara, hojasén, laburno de la India, lluvia de oro, lluvia dorada – casse, crown-of-gold-tree, golden shower, Indian laburnum, purging cassia, yellow shower
Cassia (Senna) grandis – cañafistula, cañafistula grande, quauhuayo – coral shower, pink shower
Cassia marilandica – senna – senna
Cassia = Senna
Cassytha filiformis – zacatlaxcalli
Castanea dentata – castaño americano – American chestnut tree
Castanea pumila – chincapino – chinquapin, dwarf chestnut
Castanea sativa – castaño, castaño común, castaño regoldo, marrón, regoldo – Spanish chestnut
Castanopsis = Chrysolepis
Castanospermum australe – castaño de Australia – Moreton Bay chestnut tree
Castela (Holacantha) emoryi – amargosa, chaparro amargoso, corona de Cristo – crucifixion thorn
Castela peninsularis – amargosa
Castela polyandra – amargosa
Castela texana – amargosa, bisbirinda, chaparro, chaparro amargoso – allthorn castela
Castela tortuosa – amargosa, bisbirinda, chaparro, chaparro amargoso, hierba del perro, palo amargoso
Castilla elastica – árbol del hule, batey, caucho, caucho de Castilla, cuauchile, goma, holli, hule, kiche, olicuahuitl, olli, ollin, palo de hule, quiche, tarantacua, yaxha – castilloa rubber, gumtree, rubber tree
Castilla fallax – caucho macho, hule macho – Panama rubber
Castilleja angustifolia (chromosa) – garoñona, mano de muerto – desert Indian paintbrush
Castilleja arvensis – bella Inés, cabeza de gorrión, castilleja, catoxitzi, chupirin, cola de borrego, copete de faisán, cresta de gallo, enchilada, enchiladita, flor de milpa, garañona, lupita, mano de muerto, rosilla, sagalejo, saumayate, uitzacua, zagalejo

Castilleja bryantii – castilleja, hierba del cáncer – Indian paint brush
Castilleja canescens – bella Inés, chupirin vitzaqua, cola de borrego, enchilada, enchiladita, garañona, mano de muerto, mirto cimarrón – castilleja
Castilleja densiflora (Orthocarpus densiflorus) – escobita – owl's-clover
Castilleja exserta (Orthocarpus purpurascens) – escobita – purple owl's-clover
Castilleja gracilis – bella Inés, brocha de indio, chupón, cola de borrego, mirto cimarrón – Indian paintbrush
Castilleja indivisa – brocha india, brocha tejana – scarlet paintbrush
Castilleja integra – brocha india – southwestern paintbrush
Castilleja linariifolia – flor de Santa Rita – Wyoming paintbrush
Castilleja tenuiflora – calzón de indio, cola de borrego, coneja, hierba de la piedra, hierba del cáncer, saca miel
Casuarina cunninghamiana – casuarina, pino, pino australiano, pino de Australia – river she-oak
Casuarina equisetifolia – casuarina, ciprés, palo fierro, pino, pino australiano, pino de Australia, pino marítimo, sauce – Australian pine, beefwood, horsetail tree
Catalpa bignonioides – árbol de las trompetas, catalpa, catalpa común americana – common catalpa, Indian bean
Catalpa* x *erubescens – catalpa híbrida americana
Catalpa speciosa – catalpa especiosa
Catananche caerulea – flecha de Cupido, hierba cupido, hierba cupido azul – Cupid's dart
Catasetum maculatum – chinela, chit-cuuc, monje, palma de ardilla
Catasetum punctatum – chit-cuuc
Catha edulis – kat, té de Arabia – Arabian tea, khat
Catharanthus roseus (Vinca rosea) – chata, chula, clavellina, flor de príncipe, hierba doncella, indita, maravilla de España, pervinca, pervinca de Madagascar, pervinca rosada, vicaria, vinca, vincapervinca, vincapervinca de Madagascar, vinca rosea – Madagascar periwinkle, old maid, periwinkle, rosy periwinkle
Cathestecum erectum – grama, grama china, pasto chino, zacate borreguero, zacate liebrero
Cathestecum prostratum – grama china
Cattleya citrina – aróracua, azucena amarilla, azucena del monte, costicoatenoxóchitl, costicoatxóchitl, cozticoatzontecoxóchitl, coztictepetlacuxóchitl, flor de coco, flor de huevo, tatsingueni
Cattleya deckeri – flor de San Sebastián
Cattleya labiata – flor de mayo
Cattleya skinneri – candelaria, guaria morada, San Sebestián – Skinner's cattleya
Cattleya skinneri* var *alba – flor de la candelaria, flor de San Sebastián, lirio de San Sebastián
Caulophyllum thalictroides – arándano azul, cimifuga azul – blue cohosh, blue ginseng
Ceanothus americanus – raíz roja – redroot
Ceanothus coeruleus – chaquira, chaquirilla, cuaicuastle, huacalillo, palo colorado, sayolistle, tlalixtli, tlaxistle, tnuyucú
Ceanothus cordulatus – bracillo, lilac – mountain whitethorn
Ceanothus cuneatus – bracillo, lila, lilac, pelampacate – buck brush
Ceanothus greggii – lila – California-lilac, desert ceanothus
Ceanothus huichagorae – huichagorare, junco
Ceanothus leucodermis – bracillo, damiana, lila, lilac – chaparral whitethorn
Ceanothus oliganthus – lila – California-lilac
Ceanothus spinosus – chaquira, lila – greenbark ceanothus, red-heart
Ceanothus thyrsiflorus – lila de California – blue blossom, blue brush
Ceanothus verrucosus – lila – wart-stemmed ceanothus
Cecropia obtusifolia – chancarro, guarina, guarumo, hormiguillo, jarilla, palo de violin, sarumo, trompeta, trompetilla – pumpwood, trumpet tree
Cecropia peltata – árbol atrompetado, chancarro, chancarro blanco, coilotápalo, guarina, guarumo, imbauha, narciso atrompetado, sarumo, serracenia, trompeta,

yagrumo, yagrumo hembra – pumpwood, trumpet tree

Cedrela mexicana – calicedra, cedro, cedro amargo, cedro chino, cedro colorado, cedro de la Habana, cedro fino, cedro hembro, cedro liso, cedro macho, cedro oloroso, cedro real, cedro rojo, cóbano, cuiche, culche, teocuahuitl – cigarbox tree, Mexican cedrella, royal cedar

Cedrela odorata – calicedra, cedrela americana, cedro, cedro colorado, cedro de España, cedro de la Habana, cedro de México, cedro español, cedro mexicano, cedro oloroso, cedro rojo – cigarbox wood, red cedar, Spanish cedar

Cedrela poblensis – cuachichile

Cedrela sinensis – cedrela de China, cedro de China

Cedrela toona – cedro australiano, cedro índico – red cedar

Cedronella mexicana – noritén, toronjil, toronjil del país, toronjillo, tzompili-huitzpatli – balm

Cedrus atlantica (libani* ssp *atlantica) – cedro atlántico, cedro azul, cedro del atlante, cedro del Atlas – Atlas cedar

Cedrus brevifolia (libani* ssp *brevifolia) – cedro de Chipre, cedro de hojas cortas – Cyprian cedar, Cyprus cedar

Cedrus deodara – cedro, cedro asiático, cedro de la India, cedro del Himalaya, cedro deodar, deodar – deodar cedar

Cedrus libani – cedro, cedro del Líbano, cedro de Salomón – cedar of Lebanon

Cedrus lusitanica – cedro blanco – Portuguese cedar

Ceiba acuminata – ceiba, guacapí, lantá, mosmote, pochote – kapok

Ceiba aesculifolia – ceiba, lantá de cerro, mosmote de cerro, pochote, pochote de aguas

Ceiba pentandra – algodón de árbol, árbol de algodón, cabellitos de ángel, cabellos de ángel, capoca, capoquero, ceiba, cho, kinim, mosmote, peem, pitón, pochote, pochotillo, pochotle, xiloxóchitl, yaxché – kapok tree, silkcotton tree

Ceiba speciosa – ceiba del Brasil – floss silk tree

Celaenodendron mexicanum – palo prieto

Celastrus scandens – celastro, dulcamara – American bittersweet

Celosia argentea – abancillo, abanico, amorseco, boria, celosia, cresta de gallo, moco de pavo, moño, penacho, terciopelo – Chinese woolflower, cockscomb, fireweed, quailgrass, red fox

Celosia cristata – celosia, cinco de mayo, cresta de gallo, mano de león, moco, terciopelo – crested cockscomb

Celosia floribunda – bledo, cresta de gallo

Celosia plumosa – amaranto plumoso – plume cockscomb

Celtis australis – aligonero, almecino, almez, almez del sur, latonero – European hackberry, nettle-tree

Celtis douglassii – palo blanco – Douglas hackberry

Celtis iguanea – bainoro, cagalera, chaparro blanco, cumbro, garabato, garabato blanco, granjeno, guaraguao, muc, naranjillo cimarrón, palo de arco, uña de gato, vainor, zitsmuc – hackberry

Celtis occidentalis – almez americano, almez occidental, latonero – common hackberry

Celtis pallida – acebuche, bainoro, capulín, chindule, cumbro, cúmero, cungo, garabato, garambullo, granjeno, granjeno amarillo, granjeno huasteco, guichibezia, huasteco, palo de águila, rompecapa, yagabecie – desert hackberry, granjeno

Celtis reticulata – acebuche, aceituna, bainoro, cúmaro, cumbro, garabato blanco, machaquí, membrillo, palo blanco, palo cumbro, palo duro, palo mulato, vainor – net-leaf hackberry, sugar berry, western hackberry

Celtis trinervia – anasillo, palo amargo, ramón de costa, tala

Cenchrus brownii – cadillo aglomerado, huizapol

Cenchrus ciliaris – buffel africano, buffel azul, buffel pretoria, cadillo, carricera africana, guizazo, pasto buffel, zacate buffel – buffelgrass, sandbur

Cenchrus echinatus – abrojo, cadillo, cadillo austral, guachapori, guisaso, huitzapole, huizapol, huizapotl, mozote, mul, ojo de arriera, ojo de hormiga, t'oyol, zacate toboso – southern sandbur

Cenchrus incertus – abrojo roseta, cadillo austral, cadillo de playa, guachapori, huizapori, roseta, zacate cochinillo – coast sandbur
Cenchrus longispinus (pauciflorus) – abrojo roseta, cadillo, chancaquilla, huitzapole, huizapol, huizapotl, ladilla, roseta, rosetilla, zacate cochinillo – mat sandbur
Cenchrus palmeri – guachapori, huizapol, huizapol gordo, huizapori – giant sandbur
Cenchrus pilosus – cadillo, muul-suuk
Centaurea americana – cardo, cardo del valle, cardo santo, centaura americana, flor de cesta, tacalote – basket flower, star-thistle, thornless thistle
Centaurea calcitrapa – calcitrapa, cardo estrellado, trepacaballo – calcitrapa, purple star-thistle
Centaurea cineraria – centaura alba – dusty miller, white knapweed
Centaurea cyanus – aciano, aldiza, azuelo, azulejo, azulete, azulina, botoncillo, botoncillo azul, cabezuela, centaurea, ciano, coronita, liebrecilla, pincel – bachelor's button, bluebottle, cornflower
Centaurea diffusa – centaura difusa – diffuse knapweed, wide-spreading knapweed
Centaurea diluta – centaura fragil – weak knapweed
Centaurea macrocephala – centaura macrocéfala – knapweed
Centaurea maculosa – centaura maculada – spotted knapweed
Centaurea melitensis – centaura de Malta, tocalote – Malta star-thistle, tocalote
Centaurea montana – aciano de montaña, cabezuela, centaura montana, centaurea – perennial cornflower, mountain bluet, mountain knapweed
Centaurea (Amberboa) moschata – centaura olor de almizcle, centaurea – sweet sultan
Centuarea nigra – centaura negra, centaura obscura – black knapweed, hardheads, iron weed
Centaurea* x *pratensis – centaura de los prados – meadow knapweed
Centaurea salmantica – escobilla – star-thistle
Centaurea scabiosa – centaura mayor, liebrecilla – greater knapweed
Centaurea solstitialis – abrepuño, centaura del solsticio – yellow star-thistle
Centaurea sulphurea – centaura azufrada – Sicilian star-thistle
Centaurium calycosum – rosita – marsh centaury
Centaurium erythraea – centaura menor, centaurea umbelada, cintoria, hiel de tierra, hierba de Quirón – bitter herb, centaury, gale of the east
Centaurium (Erythraea) muehlenbergii – canchalagua – centaury
Centaurium nudicaule – canchalagua, conchita – centaury
Centaurium (Erythraea) venustum – canchalagua – canchalagua
Centella = Hydrocotyle
Centranthus (Valeriana) ruber – hierba de San Jorge, milamores, milamores roja, valeriana roja – Jupiter's beard, red valerian
Centrolobium platinense – palisandro – canary wood
Cephaëlis ipecacuanha – ipeca, ipecacuana, raicilla, raíz brasileña – ipecac
Cephaëlis tomentosa – cresta de gallo, doradilla, hierba gallina, oropelo – red twincup
Cephalanthera austiniae – cefalantera – phantom orchid
Cephalanthus occidentalis – cahuapate, hierba de la Virgen, jazmín, mimbre, rosa de San Juan, uveral, uvero – button bush, button willow
Cephalanthus salicifolius – atepule, cuaje, guatájori, jazmín blanco, mimbre – button bush
Cephalaria leucantha – escabiosa blanca – white-headed giant scabious
Cephalocereus (Pachycereus) columna-trajani – cardón blanco, higo de tetetzo, tetetzo
Cephalocereus senilis – cabeza de viejo, viejito, viejo – old man cactus
Cephalocereus = Neobuxbaumia, Pilosocereus
Cephalotaxus fortunei – cefalotaxo de fortune, cefalotejo de fortune, tejo de fortune
Cephalotaxus harringtonia – cefalotaxo

drupáceo, cefalotejo drupáceo, cefalotejo oriental, tejo de Hokkaido – plum yew

Cerastium arvense – cerastio de los campos – field chickweed

Cerastium fontanum – cerastio de la fuente – mouse-ear chickweed

Cerastium glomeratum – cerastio aglomerado – mouse-ear chickweed

Cerastium tomentosum – cerastio, cerastio tomentoso – dusty miller, snow-in-summer

Cerasus capollin – capolín

Ceratonia siliqua – algarrobo, algarrobo de España, arveja, arvejera, caroba, garrofero – carob tree, locust, St. John's bread

Ceratophyllum demersum – ceratófila común – hornwort

Ceratopteris thalictroides – helecho de agua – water fern

Ceratostigma plumbaginoides – falso plumbago – dwarf plumbago

Ceratozamia latifolia – costilla de león, palma imperial, piña del monte – Mexican horncone

Ceratozamia mexicana – amenduai, costilla de león, palma, palma imperial, piña del monte, tapacapón – cycad, Mexican horncone

Cercidiphyllum japonicum – árbol de Judas, árbol de Judea, árbol de Katsura, katsura – katsura tree

Cercidium floridum (Parkinsonia torreyanum) – árbol de la brea, baco muteca, brea, dipúa, lluvia de oro, medesá, palo brea, palo de pua, palo verde, palo verde azul, palo verde chino, retama – blue palo verde

Cercidium (Parkinsonia) microphyllum – dipúa, junco, lebón, lebón retama, medesá, palo brea, palo verde, retama – foothill palo verde, little-leaf palo verde

Cercidium peninsulare – dipúa, palo de púa, palo verde

Cercidium praecox (plurifoliolatum) – brea, mantecoso, palo brea, palo de berria, palo mantecoso, palo verde, retama, tsole – palo brea, Sonoran palo verde

Cercidium (Parkinsonia)* x *sonorae – brea, palo estribo, palo verde – Sonoran palo verde

Cercidium texanum – retama china – Texas paloverde

Cercis canadensis – duraznillo, pata de vaca – eastern redbud

Cercis occidentalis – árbol de Judas, botón encarnado – Judas tree, western redbud

Cercis siliquastrum – algarrobo loco, árbol de Judas, árbol de Judea, árbol del amor, arjorán, cercis, ciclamor, sicamor – Asiatic redbud, Judas tree

Cercocarpus betuloides – caoba de montaña, lentisco, palo duro, palo ludo, poñil, ramón, zunuiña – birch-leaf mountain-mahogany, hardtack, mountain ironwood, sweet brush

Cercocarpus ledifolius – caoba – curl-leaf mountain-mahogany

Cercocarpus montanus – lentisco, palo duro – alder-leaf mountain-mahogany, deer browse, mountain-mahogany

Cereus fernambucensis (variabilis) – pitahaya

Cereus hexagonus – cardón, cendrero, reina de la noche – blue cereus, queen-of-the-night

Cereus repandus (peruvianus) – cardón – Peruvian apple cactus

Cereus = Acanthocereus, Bergerocactus, Carnegiea, Echinopsis, Myrtillocactus, Pachycereus, Peniocereus

Cerinthe major – becoquino, ceriflor, palomera – honeywort

Ceroxylon andicola – palma de cera

Cestrum anagyris – agua bendita, hierba del zopilote, paloma

Cestrum dumentorum – chahuacho, galán, galán de tarde, hediondilla, huele de día, palo hediondo, potonxihuite, tepozan

Cestrum elegans – flor del soldado, palquis – red cestrum

Cestrum fasciculatum – hierba del perro, tlanchinol – early cestrum

Cestrum nocturnum – akabyon, dama de noche, galán de noche, galán de tarde, hediondilla, hierba hedionda, huele de noche, parqui, pipilojihuite, pipiloxíhuitl – night jessamine

Ceterach officinarum – culantrillo escamoso, doradilla – scale fern

Cetraria islandica – liquen de Islandia, musgo de Islandia – Iceland moss

Chaenomeles sinensis – membrillero chino – Chinese quince

Chaenomeles speciosa (lagenaria) – membrillero del Japón, membrillero japonés,

membrillero ornamental, membrillo elegante, membrillo japonés – flowering quince
Chaerophyllium tuberosum – perifollo de raíz de chirivía – chervil root
Chaetoptelea = Ulmus
Chamaebatia foliolosa – hierba del oso – bear's clover, mountain misery
Chamaecyparis obtusa – camecíparis japonés hinoki, falso ciprés japonés hinoki – Hinoki false cypress
Chamaecyparis pisifera – camecíparis sawara, ciprés de guisantes, ciprés de hondo, ciprés pisífero, falso ciprés sawara – sawara false cypress
Chamaecyparis thyoides – camecíparis blanco americano, cedro blanco americano, falso ciprés blanco americano – white cedar
Chamaecyparis = Cupressus
Chamaecytisus proliferus – cítiso enano – dwarf broom
Chamaedorea arenbergiana – chib-kuúk – squirrel palm
Chamaedorea cataractarum – guayita, guayita de los arroyos – cascade palm, cat palm
Chamaedorea costaricana – chamadorea de Costa Rica
Chamaedorea elatior – guaya de cerro, junco de bejuco, palma trepadora, tepejilote
Chamaedorea elegans – camedor de cambray, chamadorea, neanthe bella, pacaya, palma cambray, palma camedora, palmilla, palmita, xate hembra – parlor palm
Chamaedorea ernesti-augustii – camedor, guaya, guaya de abajo, guayita, pacaya, palma camedora, palmilla, tepejilote, xoantepxilotl – chamaedor palm
Chamaedorea glaucifolia – kiba
Chamaedorea graminifolia – chapai, chiat, pacaya, palma, quiba
Chamaedorea humilis – bomcabalsáh, palma enana
Chamaedorea ibarrae – cib
Chamaedorea klotzschiana – tepejilote
Chamaedorea metallica – metálica
Chamaedorea microspadix – palmilla
Chamaedorea oblongata – litampa, tepejilote, xate macho
Chamaedorea oreophila – tepejilote
Chamaedorea pinnatifrons – chem-chem, pacaya, tepejilotillo, xate – monkey-tail palm
Chamaedorea pochutlensis – canelilla
Chamaedorea sartorii – tepejilote
Chamaedorea schiedeana – pacaya
Chamaedorea (Edanthe) tepejilote – guaya, khíb-shul, pacaya, tepejilote, tepexilotl – pacaya
Chamaedorea tuerckheimii – guonay
Chamaedorea wendlandiana – pacaya, tepejilote
Chamaedorea woodsoniana – tepejilote
Chamaemelum nobile (Anthemis nobilis) – camomila, camomila de jardín, camomila romana, manzanilla, manzanilla buena, manzanilla fina, manzanilla noble, manzanilla olorosa, manzanilla romana, manzanilla verdadera – chamomile, Roman chamomile, Russian chamomile
Chamaerops humilis – margallón, palma enana, palmera enana, palmera de abanico, palmitera, palmito, palmito blandito, palmito europeo – Mediterranean fan palm
Chamaesyce (Euphorbia) abramsiana – golondrina – prostrate spurge
Chamaesyce (Euphorbia) albomarginata – golondrina, hierba de la víbora – rattlesnake weed, whitemargin sandmat
Chamaesyce (Euphorbia) arizonica – golondrina – Arizona spurge
Chamaesyce (Euphorbia) fendleri – golondrina
Chamaesyce (Euphorbia) glyptosperma – golondrina
Chamaesyce (Euphorbia) indivisa – golondrina
Chamaesyce (Euphorbia) leucophylla – golondrina, liga – spurge
Chamaesyce (Euphorbia) maculata – celidonia, corape, euforbio manchado, hierba de la golondrina, golondrina, memeya-xíhuitl, tlalmemejan – spotted spurge, swallow herb
Chamaesyce (Euphorbia) melanadenia – golondrina – prostrate spurge
Chamaesyce (Euphorbia) micromera – golondrina
Chamaesyce (Euphorbia) nutans – euforbio nutante, golondrina – drooping spurge

Chamaesyce (Euphorbia) ocellata – hierba de la golondrina – valley spurge
Chamaesyce (Euphorbia) pediculifera – golondrina, liga – prostrate spurge
Chamaesyce (Euphorbia) polycarpa – golondrina, liga – prostrate spurge, rattlesnake weed
Chamaesyce (Euphorbia) prostrata – celidonia, euforbio prostrado, golondrina, hierba de la golondrina, xanabmucuy – prostrate spurge
Chamaesyce (Euphorbia) revoluta – golondrina
Chamaesyce (Euphorbia) serpens – euforbio serpenteado, golondrina – prostrate spurge, twisting spurge
Chamaesyce (Euphorbia) serpyllifolia – hierba de la golondrina – thyme-leaved spurge
Chamaesyce (Euphorbia) serrula – golondrina
Chamaesyce (Euphorbia) setiloba – golondrina – fringed spurge
Chamaesyce (Euphorbia) stictospora – golondrina
Chamomilla (Matricaria) courrantiana – manzanilla, manzanilla de Castilla – false chamomile
Chamomilla suaveolens (Matricaria matricarioides) – camomila, camomila alemana, manzanilla, manzanilla alemana, manzanilla común, manzanilla de Alemania, manzanilla de Aragón, manzanilla del barrio – German chamomile, pineapple weed, rayless chamomile
Cheilanthes alabamensis – helecho de Alabama
Cheilanthes leucopoda – peludita
Cheilanthes lindheimeri – helecho, hierba de la peña – lip-fern
Cheilanthes = Astrolepis
Cheiranthus = Erysimum
Chelidonium majus – amapola amarilla, celidonia, celidonia mayor, golondrinera, hierba de la golondrina, hierba de la golondrina extranjera, hierba golondrinera, hierba verruguera, hirundinaria – celandine
Chenopodium album – anserina, ashecani, bledo, ceñiglo, cenizo, cenizo blanco, chual, chual blanco, chual morado, cotasula, pie de anade, quelite, quelite cenizo, quelite de trigo, quelite salado, quilitl – fat hen, lamb's quarters, pigweed
Chenopodium (Teloxys) ambrosioides – ambrosía de México, amush, chichiquelite, epazote, epazote de comer, epazote de zorrillo, epazotl, hierba de santa, huacatay, lucumxiu, matruz, paico, paico mocho, patusai, quelite, santónico americano, té borde, té de España, té de los Jesuits, té de México, té español, té mexicana – epazote, Mexican tea, Spanish tea, wormseed
Chenopodium arizonicum – chual
Chenopodium berlandieri – cenizo de Berlandier, guauzoncles, huaquilitl, huauzontle, huauzontlí – pitseed goosefoot
Chenopodium bonus-henricus – anserina, armuelle bueno, buen Enrique, zurrón – all-good, good King Henry
Chenopodium botrys – biengranada, hierba de chivatito, hierba de zorrillo cimarrón, milgranada – feather geranium, Jerusalem oak
Chenopodium chenopodioides – cenizo botrioide – small red goosefoot
Chenopodium foetidum – cenizo fétido, quaich
Chenopodium foliosum – bledera, cenizo folioso – red strawberry spinach, strawberry blite
Chenopodium glaucum – cenizo glauco – oak-leaved goosefoot
Chenopodium graveolens – epazote, epazote de coyote, epazote de toro, epazote de zorrillo, epazotl, hediondilla, hierba de chivato, hierba del zorrillo, quelite apestoso, quelite de zorrillo, quelite hediondo, sadugua – goosefoot
Chenopodium incisum – epazote cimarrón, epazote de toro, epazote de zorrillo, hediondilla, hierba del zorrillo, quelite de zorrillo
Chenopodium leptophyllum – cenizo leptofilo – narrow-leaved goosefoot
Chenopodium multifidum – cenizo multifido – goosefoot
Chenopodium murale – cenizo mural, chual, chual de cochi, chual morado, hediondilla,

hierba del perro, pie de ganso, quelite cenizo, quilitl, tzoalli – nettle-leaf goosefoot
Chenopodium neomexicanum – chual, hierba del pescado, quelite
Chenopodium nuttaliae – cuatztli, cuauhzontli, epazote, guauzoncles, huauquilitl, huauzontle, michiuauhtli, quautzontetl – pitseed goosefoot
Chenopodium quinoa – quinoa – quinoa
Chenopodium rubrum – cenizo rojo, renegada, mala mujer – red goosefoot
Chenopodium vulvaria – hediondilla, vulvaria – goosefoot, pigweed
Chilopsis linearis – acuáhuitl, jano, mimbre, peludita, sauce – desert catalpa, desert-willow, flowering-willow, willow-leaved catalpa
Chimaphila dasystephana – hierba del hígado – prince's pine
Chimaphila umbellata – peralito umbelado, pirola umbelada – prince's pine
Chimonanthus praecox (fragrans) – calicanto del Japón – wintersweet
Chiococca alba – aceitilla, cahinca, cainia, canchacche, canica, caninara, lágrimas de San Pedro, lágrimas de María, madreselva, oreja de ratón, pegajosa, perlilla, suelda consuelda – David's root, milkberry, snowberry
Chiococca petrina – cacachila blanca, granadillo blanco
Chiranthodendron pentadactylon – árbol de las manitas, camxóchitl, canaco, canague, flor de las manitas, huiahuonahua, macpalxochicuahuitl, macpalxóchitl, manita de león, mano de dragón, mano de león, mapasúchil, palo de mecate, palo de tayuyo, teyacua – hand-flower tree
Chloracantha spinosa (Aster spinosus) – buena mujer, mo'oso, scoba, tulillo – Mexican devilweed, spiny aster
Chloris andropogonoides – verdillo esbelto
Chloris chloridea – verdillo cacahuatoide – buryseed chloris
Chloris ciliata – verdillo de fleco
Chloris (Trichloris) crinita – triverdin de tres
Chloris cucullata – verdillo papalote
Chloris dandyana – zacate araña
Chloris elegans – meex-mascual, zacate
Chloris gayana – grama de Rhodes, hierba rhodes, pasto rhodes, rhodes, rhodes masaba, rhodes pokot, zacate pata de gallo, zacate rhodes – Rhodes grass
Chloris (Trichloris) pluriflora – triverdin de cuatro
Chloris subdolichostachya (latisquamea) – verdillo norteño
Chloris submutica (polystachya) – pata de gallo mexicano
Chloris virgata – barba de indio, cebadilla, verdillo plumerito, zacate, zacate cola de zorra, zacate lagunero, zacate mota – feather finger grass, finger grass
Chlorogalum pomeridianum – amole, amolli, escobeta, jamatai – amole, soap plant, soaproot
Chloroleucon mangense* var *leucospermum* *(Pithecellobium leucospermum) – ébano, palo éban, palo fierro, palo pinto, pinto uus
Chlorophora = Maclura
Chlorophytum comosum – cinta, clorofito, mala madre, pasto listón – spider plant
Choisya arizonica – zorrillo – Arizona star-leaf, star-leaf, zorillo
Choisya dumosa – zorrillo – Mexican orange, star-leaf
Choisya ternata – clavillo, clavo de olor, flor de clavo, hierba de clavo, naranjo mexicano – Mexican orange
Chondodendron tomentosum – curare, urari – curare
Chondrilla juncea – achicoria dulce, achicoria juncal, ajonjera, condrila, husillo, lechuguilla, nusillo, paja – skeleton weed
Chondrus crispus – musgo de Irlanda, musgo de perla – carrageen
Chorisia insignis – palo borracho, samuhú, yuchán
Chorizanthe fimbriata – bachata, hierba del empacho – fringed spineflower
Chrysactinia mexicana – calanca, damiana, damianita, falsa damiana, garañona, hierba de San Nicolás, mariola, romerillo, San Nicolás, yeyepaxtle – damianita
Chrysalidocarpus lucubensis – palma de Madagascar
Chrysalidocarpus (Areca) lutescens – areca, palma amarilla, palma areca – butterfly palm, Madagascar palm, yellow palm

Chrysanthemum (Pyrethrum, Tanacetum) balsamita – balsamita, costo hortense, hierba amarga, hierba balsámica, hierba de la tos, hierba de Nuestra Señora, hierba de San Pedro – costmary

Chrysanthemum carinatum (tricolor) – crisantemo annual, crisantemo carinado, margarita – summer chrysanthemum, tricolor chrysanthemum

Chrysanthemum cinerariifolium – pelitre, piretro – pyrethrum

Chrysanthemum (Tanacetum) coccineum (Pyrethrum roseum) – margarita roja, matricaria, piretro – painted daisy, pyrethrum

Chrysanthemum coronarium – antimaño, crisantemo, crisantemo coronado, crisantemo silvestre, filipéndula, flor de muerto, flor de oro, margarita, magarza, mirabel, moya, pajito, pelitre de Dalmacia, santimonia – crown daisy, chop suey greens, garland chrysanthemum, garland daisy, summer chrysanthemum

Chrysanthemum frutescens – margarita – marguerite, Paris daisy

Chrysanthemum indicum – artemia, bella Luisa, crisantemo, filipéndula, flor de oro, margarita

Chrysanthemum maximum – crisantemo, crisantemo gigantesco, crisantemo máximo, margarita gigante – Shasta daisy

Chrysanthemum morifolium (Dendranthema grandiflorum) – flor de oro – florist's chrysanthemum

Chrysanthemum segetum – corona de rey, crisantemo de las mieses, crisantemo segetal, margarita, ojo del sembrado – corn chrysanthemum

Chrysanthemum = Leucanthemum, Tanacetum

Chrysobalanus icaco – caco, capollaxipehualli, ciruelo de América, ciruelo de paloma, icaco, icaquero, jicaco, xicaco – coco plum, pigeon-plum

Chrysolepis (Castanopsis) chrysophylla – chincapín dorado – giant chinquapin

Chrysophyllum bicolor – caimitillo – wild cainit

Chrysophyllum cainito – caimito, caimito verde, cainito, canela, carambola, cayumito, palo de canela, zapote caimito – star apple

Chrysophyllum mexicanum – caimito, caimito silvestre, canela, chiceh, palo de canela, zapote caimito – wild star apple

Chrysophyllum oliviforme – caimitillo de perro – satin leaf

Chrysosplenium glechomifolium – saxífraga dorada – golden saxifrage

Chrysothamnus nauseosus – chamizo, chamizo amarillo, chamizo blanco, hierba liebrera – chamisa, rabbitbrush

Chusquea bilimekii – carrizo, otate

Chusquea circinata – carrizo, otate, otate chino

Chusquea nelsoni – carrizo, ojotate

Chusquea = Cortaderia

Chysis aurea – acamayo, chisis, flor de un día – smooth-lipped chysis

Chysis bractescens – orquídea blanca, flor de cera

Cibotium regale – helecho real – tree fern

Cicer arietinum – chícharo, garbanzo – chick-pea, garbanzo

Cichorium endivia – 1. endibia, escarola crespa – endive; 2. escarola – Batavian endive, escarole

Cichorium intybus – 1. achicoria, achicoria amarga, achicoria de hojas verdes, achicoria de raíz, achicoria silvestre, ahuiron, almirón, amargón, camarroya, chicoria – chicory, witloof; 2. achicoria de Bruselas, achicoria de hojas blancas, endibia, endibia de Belga – Belgian endive; 3. achicoria de hojas rojas, endibia acogollada roja – radicchio

Ciclospermum (Apium) leptophyllum – apio, culantro de zopilote – marsh parsley

Cigarrilla mexicana – cigarrilla, flor de San Pedro

Cimicifuga racemosa – actea, cimicífuga, cimicífuga negra, espantachinches, raíz de cascabel, raíz de culebra, raíz de serpiente negra – black cohosh, black snakeroot, bugbane

Cinchona calisaya – calisaya, cascarillo, quino, quino amarillo, quino calisaya – quinine, yellow bark

Cinchona ledgeriana – chinchona, corteza

peruana, polvo de los Jesuitas, quino – cinchona, ledger bark

Cinchona officinalis – árbol de la fiebre, árbol de la quina, cascarillo verde, cinchona, corteza de cinchona, corteza de quina, membrillo bronco, quino, quino gris, quino negro, quino oficinal, quino rojo – crown bark, fever tree, loxa, quinine

Cinchona pitayensis – quino de Java, quino rojo – red cinchona

Cinchona succirubra – copalquín, quino, quino rojo – quinine, redbark cinchona

Cinnamomum camphora (Camphora officinalis) – alcánfor, alcanforero, árbol del alcanfor, canfor – camphor tree

Cinnamomum cassia – canela de la China, cassia, cassia acanelada – bastard cinnamon, cassia, Chinese cassia tree

Cinnamomum hartmanii – chihuana cara, haya

Cinnamomum parthenoxylon – alcanforero amarillo

Cinnamomum zeylanicum – árbol de la canela, canela, canelero, canela de Ceilán – cinnamon

Circaea alpina – hierba de San Simón alpina – enchanter's nightshade

Circaea lutetiana – hierba de San Simón – enchanter's nightshade

Cirsium arizonicum – cardo santo – Arizona thistle

Cirsium arvense –cardo cundidor, cardo negro – Canada thistle, creeping thistle

Cirsium bipinnatum – cardo santo – thistle

Cirsium ehrenbergii – cardo, cardo colorado, cardo de montaña, cardo santo, cuautlahuizquílitl, mala mujer, rosa de las nieves

Cirsium mexicanum – cardo, cardo bronco, cardo emplumado, cardo santo, cardo santo cimarrón, chacámecua, cherémecua, chinaca, espino de sol, huitzquílitl, omil, quauhtla huitzquílitl – plumed thistle

Cirsium mohavense – cardo santo – Mojave thistle

Cirsium texanum – cardo – Texas thistle

Cirsium undulatum – cardo santo, hierba del sapo, lleva del sapo – pasture thistle, wavyleaf thistle

Cirsium vulgare – cardo lanceolado – bull thistle

Cissampelos pareira – bejuco, butua, curarina, hierba de la víbora, ixtagua, iztaccoanenepilli, iztactepetzacuxóchitl, oreja de ratón, pareira brava, pegamo, peteltun, tlaxcalxíhuitl, tortilla de los sapos

Cissus cucurbitina – bejuco de coral, temecatl, tripa de Judas, vid coral, xtab-canil – coral vine, possum grape

Cissus gossypifolia – Juan mecate

Cissus rhombifolia – come mano, come mano de llano, palo huaco, palo hueco, tripa de zopilote, yaxtabcanil – grape ivy

Cissus sicyoides – bejuco loco, bejuco loco rojo, come mano, hierba del buey, molonquí, parra de monte, parra silvestre, sanalotodo, temecate, tepemecatl, tripa de Judas, tripa de vaca, tripa de zopilote, tumba vaquero, uva, uva silvestre, vid silvestre, yaxtabcanil, yocohuiro – treebine, water withe

Cissus trifoliata – bolontobi, hierba del buey, quereque del monte – grape ivy

Cissus trilobata – molonqui

Cistus albidus – estepa blanca, jaguarzo blanco, jara blanca – white-leaved rock-rose

Cistus creticus – estepa de Creta – Cretan sun-rose, lot plant, myrrh, rock-rose

Cistus crispus – jaguarzo, jaguarzo prieto, jara rizada – rock-rose

Cistus incanus (villosus) – estepa grisácea, jara de flores rosas – gray sun-rose, rock-rose

Cistus ladanifer – cisto, jara, jara de cádamo, jara del ládano, jara pringosa – crimson-spot rock-rose, gum cistus, onycha

Cistus laurifolius – estepa, estepa de hojas de laurel – laurel-leaved sun-rose, rock-rose

Cistus monspeliensis – estepa de Montpellier, estepa negra, jaguarzo, jaguarzo negro – Montpellier sun-rose, rock-rose

Cistus salvifolius – estepa negra, jaguarzo morisco, jara – sageleaf rock-rose

Citharexylum berlandieri – cacachila, negrito, orejuela, panochillo, revienta cabra, salacate, saúco hediondo

Citharexylum scabrum – chaparacoba, salacate – fiddlewood

Citrullus colocynthis (vulgaris) – alkandal, calabaza silvestre, cidra cayote, coloquíntida, gualeya, gueto xiña, hualeya, itsi, manzano amargo, melón de agua, naranjo del diablo, patilla, pepón, sácobari, sandía, tuera, xantea, zis coil – bitter apple, bitter gourd, colocynth, globe cucumber, prophet's cucumber, watermelon

Citrullus lanatus – melón de agua, sandía – watermelon

***Citrus* spp** – agrios, cítricos – citrus fruit

Citrus aurantifolia – lima, lima ácida, lima agria, lima chica, lima chichona, lima reina, limero, limón, limón agrio, limón chiquito, limón criollo, limón dulce, limonero, limón liso, limón mexicano, limón verde – Key lime, lime, Mexican lime, West Indian lime

Citrus aurantium (vulgaris) – azahar, naranjo, naranjo ácido, naranjo agrio, naranjo amargo, naranjo amateco, naranjo cajel, naranjo de fruta agria, naranjo hojas de mirto, naranjo lima, zamboa, zutspacal – bitter orange, Seville orange, sour orange

Citrus bergamotia (bergamia) – bergamoto, toronjo – bergamot orange

Citrus clementine – clementina – clementine tree

Citrus deliciosa – mandarinero, mandarino, naranjo mandarino – mandarine tree

Citrus grandis (maxima) – cimboa, pamplemusa, pomelo, toronjo – pummelo, shaddock

Citrus hystrix – combava – Mauritius papeda

Citrus jambhiri – limón rugoso – citronelle, rough lemon

Citrus japonica – kunkuat, naranjito chino

Citrus latifolia – lima común de Persia, lima persa, limero de Tahiti, limón persa, limón pérsico – Persian lime, Tahiti lime

Citrus limetta – 1. chichona, lima chichona, lima dulce, limero, limón dulce, limonero dulce, limón real – Adam's apple, limetta, Mediterranean sweet lemon, millsweet, sweet lemon; 2. lima, lima agria, limero – bitter lime

Citrus limettioides – fruta verada, lima dulce, limón dulce – Indian sweet lime, sweet lime of India

Citrus limon (limonia) – árbol de limón, limero, limón, limón agrio, limón amarillo, limón chichón, limón dulce, limonero, limón francés, limón real, nimu, tasposh, tzaposh – lemon

Citrus limon* var *ponderosa – limón ponderosa – ponderosa lemon

Citrus limonia – árbol de limón, limón real, rangpur – Chinese lemon, mandarin lime, rangpur lime

Citrus madurensis – calamondín – calamondin

Citrus medica – azamboero, cedro limón, cidral, cidra limón, cidrelo, cidrello, cidrera, cidro, citrón, limón chivo, limón real, poncidre, poncil, poncilero, toronjo – Assyrian apple, cedrat, citron, ethrog, goodly fruit

Citrus mitis – naranjo enano – dwarf orange

Citrus nobilis – mandarino, naranjo king, toronjo de semilla – king orange, mandarin orange

Citrus paradisi – naranjeto, pamplemusa, toronjo, toronjo agrio, toronjero, toronjo – grapefruit

Citrus reticulata – mandarino, naranjo mandarino, satsuma, tangerina, xokoklavoxochit – clementine, mandarin orange, tangerine

Citrus reticulata* x *C. paradisi – tangelo – tangelo

Citrus reticulata* x *C. sinensis – tangor – tangor

Citrus sinensis – azahar, chuhucpakal, náaraso, nancha, naranjo, naranjo chino, naranjo de la China, naranjo dulce, pacal, sahmées, tsapquinc, tsaptsuic, tuzán, xidni, yaganaraxo – orange

Citrus sinensis* var *melitense – naranjo sanguino – blood orange

Citrus sinensis* x *C. paradisi – chironja – chironja

Citrus = Poncirus

Cladium californicum – galega, masiega – saw-grass

Cladrastis kentukea – cladrastis amarillo – yellow wood

Clarkia amoena (Godetia grandiflora) – adiós, adiós primavera, godecia, última flor de primavera, vara de San José – farewell-to-spring, godetia

Clarkia unguiculata (elegans) – clarkia – clarkia, mountain garland

Clavaria flava – arbolito, coral, escobeta, manita, manita amarilla – coral fungus

Claviceps purpurea – carbón del centeno, cornezuelo, cornezuelo de centeno, cuernecillo de centeno, esclerocio, hongo de la espiga del centeno – ergot

Clavulina rugosa – escobetilla

Claytonia (Montia) perfoliata – claitonia, verdolaga de Cuba, verdolaga de invierno, verdolaga perfoliada – Indian lettuce, miner's lettuce

Clematis alpina – clemátide alpina – alpine clematis

Clematis armandii – clemátide – evergreen clematis

Clematis dioica – barba de chivato, barba de chivo, barba de gato, barba de viejo, borla española, cabeza de viejo, chilillo, chilillo del cerro, ojo de tecolote – virgin's bower

Clematis drummondii – barba de chivo, barba de chivato, chilillo, hierba del avero – old-man's-beard, Texas virgin's bower

Clematis flammula – clemátide flámula – fragrant clematis

Clematis integrifolia – clemátide de hojas enteras

Clematis* x *jackmanii – clemátide jackmanni

Clematis lasiantha – barba de chivo, clemátide – pipestems, virgin's bower

Clematis ligusticifolia – barba de chivato, barba de chivo, barba de viejo, clemátide, diosa, hierba de chivo, hierba de chivato, muermera – virgin's bower, yerba de chiva

Clematis montana – clemátide montana – anemone clematis

Clematis recta – clemátide recta – upright clematis

Clematis vitalba – clemátide, clemátide vitalba, hierba de pordiosero, vidalba – old man's beard, traveller's joy

Clematis viticella – clemátide de Virginia, clemátide italiana – vine bower

Cleome guianensis – barbona – bee plant

Cleome hasslerana (houtteana, spinosa) – barbona rosada, flor de araña, garcita, pata de vaca, volantín – spider flower

Cleome serulata – guaco – Rocky Mountain bee plant

Cleome = Isomeris

Cleomella longipes – guaco

Clerodendrum buchananii (speciosissimum)* var *fallax – amor de Pedro, clerodendro, metroceder – Java glorybower, pagoda flower

Clerodendrum bungei – clerodendro chino – Cashmere boquet

Clerodendrum chinense (fragrans)* var *pleniflorum – bocamelia, jazmín de España, jazmín de Italia, metroceder, volcameria

Clerodendrum ligustrinum – árbol sagrado, cógetelo todo, guardia civil, musté, señora de noche

Clerodendrum thomsoniae (balfouri) – arete de India, clerodendro, paulla – bagflower, bleeding heart glorybower, broken hearts

Clerodendrum trichotomum – clerodendro japonés – harlequin glorybower

Clethra lanata – aguacatillo, jicarillo, madroño, mamalhuaztli, mameyito blanco, mameyito negro, paguilla

Clethra mexicana – jaboncillo, madroño, mameyito, mamojuaxtle, paguilla, palo batea, palo cuchara, tlacoguilla – pepperbush, sweet pepperbush

Clianthus formosus – guisante del paraíso – glory pea

Clinopodium laevigatum – cuencuentz, hierba del borracho, nurite, poleo, tabaquillo, tabaquillo grande, té del monte, té de nurite – wild-basil

Clinopodium vulgare – clinopodio, clinopodio común – wild basil

Clitocybe squamulosa – corneta, hongo, oreja, señorita, tejamanil, tejamanilero

Clitoria terneata – azulejo, bejuco de conchitas, clitoria, clorique, conchita azul, cuña, papito, zapatillo de la reina – blue pea, butterfly pea, cordofan pea

Clivia miniata – clivia – Kaffir lily

Clusia major – copey, tampico – autograph tree

Clusia rosea – copey

Cneorum tricoccon – cnero tricoco – spurge olive

Cnicus benedictus – bendito, cardo bendito, cardo santo, centaurea bendita – blessed thistle

Cnidoscolus aconitifolius – chaya, chaya mansa, chaya pica, mala mujer, tza

Cnidoscolus angustidens – caribe, mala mujer, ortiga, ortiguilla, quemador – treadsoftly
Cnidoscolus chayamansa – chaya, chaya de Castilla, chaya mansa, mala mujer
Cnidoscolus elasticus – chaya, chilte, chilte colorado, chilte rojo, chiquité, cucharacho, tepechicle – chilte rubber
Cnidoscolus maculatus – caribe – tread softly
Cnidoscolus multilobus – chichicastle, mala mujer, mala mujer lisa, ortiga, tepum, tzitzicaitl
Cnidoscolus palmeri – mala mujer, ortiguilla – tread softly
Cnidoscolus texanus – mala mujer – bull-nettle, tread-softly
Cnidoscolus (Jatropha) urens – chaya, guaritoto, mala mujer, ortiga, ortiguilla – spurge nettle, treadsoftly
Cobaea scandens – campanilla, cobea, cundeamor, flor de campana, hiedra, hiedra morada, hiedra violeta, hierba morada, mango de niña, vid de taza y platillo – cathedral bell, cup-and-saucer vine, Mexican ivy, violet ivy
Coccoloba = Cocoloba
Coccothrinax argentea – guamito, nakaz, palmera plateada – silver thatch palm
Coccothrinax miraguama – miraguama – miraguama palm
Cocculus diversifolius – correhuela, uva trepadora, uva tullidora, uvita tullidora – snailseed vine
Cocculus laurifolius – cóculo
Cochemia halei – biznaguita
Cochemia maritima – biznaguita
Cochemia (Mammillaria) pondii – biznaguita
Cochemia (Mammillaria) poselgeri – biznaguita
Cochemia setispina – biznaguita
Cochlearia officinalis – coclearia común, cuchareta, hierba de la cuchara, hierba del escorbuto – scurvygrass
Cochlospermum religiosum – capoquero blanco – white silk cotton plant
Cochlospermum (Maximilianea) vitifolium – algodón silvestre, apánico, apompo, bototo, carne de perro, carnestolenda, chimu, chum, cianori, cocito, cojón de oro, cojón de toro, coquito, flor izquierda, guate, huarumbo, madera de pasta, mirasol, palo amarillo, palo barriga, palo barril, palo bobo, palo cuchara, palo de rosa, palo de rosa amarilla, panaco, pangolote, panigua, panuco, pochote, pochotle, pomposúchitl, poro poro, potija, pumpunsuchi, quie-quega, quieriga, rosa amarilla, rosa china, rosa imperial, tecomasuche, tecomasúchil, tecomate, tecomaxóchitl – buttercup tree, golden eggs, rose imperial, shell seed, wild cotton tree
Cocoloba (Coccoloba) barbadensis – bobché, carnero, carnero de la costa, hoja dura, palo de carnero, roble de la costa, tamulero, tepalcahuite, uvero
Cocoloba (Coccoloba) uvifera – carnero, cocoloba, kiiché, manzano, niche, papaturro, roble de la costa, uva, uva de la costa, uva de la playa, uva del mar, uverillo, uvero, uvero de la playa, uvero del mar – sea grape
Cocos nucifera – coco, coco de agua, coco de Castilla, cocotero, nuez de coco, palma de coco, palma indiana, palmera de coco – coconut palm
Cocos = Syagrus
Codiaeum variegatum – cintillo, croto, crotón, tanche, xabalam – croton, variegated laurel
Coffea arabica – café, café arábica, café criollo, cafeto, cafeto arábica, haba del Yemen – arabica coffee, coffee
Coffea canephora – café robusto, cafeto robusto – Congo coffee tree, robusta coffee
Coffea liberica – café, café de Liberia, café liberia, cafeto, cafeto de Liberia – Liberian coffee tree, liberica coffee
Coffea racemosa – cafeto del Perú – inhambane coffee
Coix lacryma-jobi – acayacoyotl, adlay, arrocillo, batagá, camándula, lágrimas, lágrimas de David, lágrimas de Job, lágrimas de San Pedro, zacate de cuentas – Job's tear's
Cola = Kola
Colchicum autumnale – azafrán, azafrán bastardo, azafrán otoñal, azafrán silvestre, cólquico, cólquico autumnal, cólquico de otoño, flor de otoño, hermodátiles, mataperros, narciso de otoño, quitameriendas – autumn crocus, meadow saffron
Colchicum speciosum – azafrán de los prados, cólquico, quitameriendas – meadow saffron
Coldenia = Tiquilia

Coleosanthus* = *Brickellia
Coleus ambolnicos (aromaticus) – orégano, orégano brujo, orégano de España – country borage, Spanish thyme
Coleus blumei – ahijado, mantel, manto regio, niño, orégano de España – coleus
Coleus hybridus – coleo – coleus
Coleus pumilus – macho – coleus
Coleus rotundifolius – cantang, patata de los Huasas – Hausa potato
Colocasia antiquorum – col caribe, colocasia, malanga, ñame, taro – cocoyam, taro
Colocasia esculenta (Caladium esculentum) – alcocaz, bore, caramicua, cascamote, chamol, chonque, cocoyán, col caribe, colocasia, makal, malanga, ñame, ñampi, ocumo chino, oreja de elefante, quequeste, yame de Canarias, yuatia – cocoyam, dasheen, elephant ears, elephant's ear, taro, yam
Cologania obovata – frijolillo
Colpothrynax* = *Pritchardia
Colubrina californica (texensis* var *californica) – palo colorado – California colubrina, Las Animas colubrina, snakewood
Colubrina ferruginosa – cascalote, cascarillo, churumai, guayul, manzanita, pimiento-che – bitters, snakewood
Colubrina glomerata – amolillo, guacimilla, palillo, quino
Colubrina triflora – cuta guícori, manzanita del monte, palo cachorro
Colubrina viridis (glabra) – espino, granadita, palo colorado, palo santo, pelo rojo – nakedwood
Colutea arborescens – espantalobos, espantalobos arborescente, sen indígena – bladder senna
Combretum farinosum (fruticosum) – abacamiel, angarilla, bejuco angarilla, bejuco de peine, bejuco de piedra, bejuco de toro, carape, cepillo, cepillo del diablo, chupamiel, compio, cuaumecate, empanada, papamiel, peinecillo, peine del diablo, peine de mico, peineta, peinetillo, quietzine, tamborillo – hairbrush vine, orange caterpillar
Commelina coelestis – coapatlí, hierba del pollo, lágrimas de María, matlaliztic, quesadilla, quesadillita, rosilla, texcocana, tripa de pollo – day flower, sky-flower
Commelina dianthifolia – barba de buey, flor de perrito, hierba del pollo, kasalá, quesadilla – birdbill dayflower
Commelina erecta – espuelita, hierba del pollo, mataliste, mataliz, rosilla – whitemouth dayflower, widow's tears
Commelina nudiflora – sinvergüenza, sinvergünza silvestre
Commelina pallida – comelina, hierba del pollo, matlaliztic, quesadilla, rosilla
Commelina scabra – hierba del mal de ojos, hierba del pollo, quesadilla
Commelina tuberosa – coapatli, hierba del pollo, matlaliztic, quesadillita, rosilla, texcocana, yoyolxóchitl, zoyalxóchitl
Commicarpus brandegeei – guananiquil
Commicarpus scandens – bejuco de purgación, miona, mochi, sonorita, xicomini
Commiphora abyssinica – mirra – Abyssinian myrrh, myrrh
Commiphora africana – resina – bdellium
Commiphora molmol – mirra – myrrh
Commiphora myrrha – mirra – common myrrh, myrrh
Commiphora opobalsamum – árbol balsámico, bálsamo de Gilead, bálsamo de la Meca, opobálsamo – balm of Gilead, Mecca myrrh
Compositae* = *Asteraceae
Comptonia aspleniifolia – perifollo oloroso – sweet fern
Condalia brandegeei – ho'i, junco
Condalia ericoides – tecomblate – javelina bush
Condalia globosa – bachata, barachata, crucerilla, crucillo, jutuqui, mezquite sarnoso, papache, sarampión, tecomblate – bitter condalia, crucillo
Condalia hookeri – brasil, capulín, capulín negro, chaparro blanco – bluewood
Condalia mexicana – bindó, bizcolote, membrillo cimarrón – Mexican bluewood
Condalia obovata – brasil, capulín, capulín negro, chaparral, cuahuitl, huitzquahuitl, mezquitillo – bluewood condalia, logwood
Condalia* = *Ziziphus
Conium maculatum – cañaheja, cañerla, cicuta, cicuta maculada, cicuta mayor, hierba del venado, perejil de lobo, perejil venenoso – poison hemlock

Conocarpus erecta – botoncahui, botoncillo, canche, estacahuite, iztac-cuahuitl, madre de sal, mangle, mangle blanco, mangle botoncillo, mangle cenizo, mangle de botón, mangle negro, mangle prieto, tabché, yana – buttonwood mangrove

***Conocybe* spp** – hongo, teonanacatl, teunamacatlth, teyhuintli – psychedelic mushroom

Conopholis americana – elotillo, elotito, mazorca de cuervo, mazorquilla – squaw root

Conopholis mexicana (alpina) – elotillo – Mexican cancer-root

Conostegia xalapensis – capulín, capulincillo, capulín de cotorra, chicab, cinco negritos, frutilla, hojalatillo blanco, locito, lolito, mora, nigua, pasita, peonía, popu, sedita, serita, tecapulín, tesguate colorado, xococquáuitl – tropical blueberry

Conringia orientalis – collegón oriental – hare's ear

Consolida ambigua (Delphinium ajacis) – conejito, consuelda ambigua, espuela de caballero, miramelinda – annual delphinium, larkspur

Convallaria majalis – convalaria, escalera de Jacob, lágrimas de Salomón, lirio convalario, lirio del valle, muguete, muguete de los valles – lily-of-the-valley

Convolvulus althaeoides – campanilla rosa, correhuela itálica – Italian bindweed, morning-glory

Convolvulus arvensis – campanilla menor, campanita, campánula, chiqueo, convulvo campestre, correhuela, correhuela de los prados, correhuela loca, hiedra, manto blanco – field bindweed, orchard morning-glory

Convolvulus cneorum – correhuela turca – bush morning-glory

Convolvulus sabatius – correhuela de la Riviera – ground morning-glory

Convolvulus scammonia – escamonea – scammony

Convolvulus tricolor – campanilla, correhuela tricolor, dondiego de día, dondiego de día enano, enredadera, maravilla – dwarf morning-glory

Conyza bonariensis – coniza de Buenos Aires – crisp fleabane

Conyza canadensis – cola de caballo, coniza del Canadá – Canadian fleabane, horseweed

Conyza confussa – simonillo

Conyza filaginoides – simonillo, zacachichi, zacatechichic

Conyza floribunda – coniza pálida

Conyza lyrata – hierba del histérico

Conyza sophiaefolia – manzanilla del río

Conzattia multiflora – árbol de águila, huautzin, joso, joso de la sierra, navío, palo blanco, palo de totole, palo joso, zopilote

Conzattia sericea – navío, palo joso

Cooperia drummondii – cebolleta, cebollina, cebollita, huele de noche, mayito – Drummond's rain lily, evening-star rail lily

Cooperia pedunculata – flor de mayo

Copaifera copallifera – copal – gum copal

Copaifera coriacea – bálsamo de copaifera – copaiba balsam

Copaifera officinalis – bálsamo, bálsamo copaiba, bálsamo del Jesuita, copaiba, copal, copayero, palo de aceite, tecopalcuáhuitl – copaiba, Jesuit's balsam

Copaifera reticulata – copaiba de pará – para copaiba

Copernicia australis – caranday – caranday palm

Copernicia cerifera – carnauba – wax palm

Copernicia hospita – palma hospita – hospita palm

Copernicia macroglossa – palma jata, palmera de abrigo – petticoat palm

Copernicia tectorum – palma llanera

Copernicia vespertilionum – yuraguana

***Coprinus* spp** – coprino – ink-cap

Coptis groenlandica – coptis – cankerroot, goldthread

Corallorhiza trifida – coraloriza trífida – northern coralroot

Corchorus olitorius – cáñamo de las Indias, yute – jute

Corchorus siliquosus – chichibe, hierba té, malva, malva cubierta, malva té, malvavisco grande, té de hiera, té de la tierra, té de malva – kalallo bush

Cordia alba – caujaro, gomo blanco, gulabere, uva gomosa, zazamil

Cordia alliodora – aguardientillo, amapa, amapa boba, amapa prieta, anacahuite del

Istmo, bojón, bojón blanco, bojón prieto, cueramo, hormiguero, hormiguillo blanco, laurel, laurel blanco, leche María, nopo, nopota peste, pajarito, pajarito prieto, palo de hormiga, palo de rosa, palo María, palo viga, pomarrosa, popocotle, rosadillo, solería, suchicahue, tambor, tepesúchil – onion cordia, samwood, Spanish elm

Cordia boissieri – anacahuite, anacuáhuitl, cicua de cueramo, cueramo, cupape, macahuite, micacuáhuitl, nacagua, nacahuite, nacahuitl, palo del muerto, rasca viejo, rosea vieja, siricote, trompillo – Texas olive

Cordia brevispicata (curassavica) – azota caballo, bolita prieta, chiricote, confiturilla, oreja de ratón, trompillo

Cordia dodecandra – anacahuite, copito, copté, cupapé, guayabillo, siricote, zac-copté, zericote – geiger tree, starbell

Cordia eleagnoides – anacahuite de Tehuántepec, bocote, cueramo, gretaña, grisiño, ocotillo, ocotillo meco

Cordia greggii – palo prieto, San Juanito, vara prieta

Cordia myxa – sebestén – jujube tree, sebesten plum

Cordia parvifolia – chiricote, dodecandra, palo prieto, San Juanito, vara prieta – little-leaf cordia

Cordia salicifolia – chá de burge, café de burge, café de los bosques, café de mato

Cordia sebestena – áloe, amapa asta, amapa boba, anacahuite, asta, baría, bocote, bojón, capá, copté, cueramo, cutiperí, grisiño, habeem, nomeolvides, ocotillo meco, palo de asta, San Bartolemé, siricote, siricote blanco, vomitel, vomitel colorado, zac-copté – aloe wood, cordia, geiger tree, scarlet cordia, zerecote

Cordia sonorae – amapa blanca, amapa boba, asta, chirare, comahuí, palo de asta, palo mirto, palo prieto, pómajo

Cordyline (Dracaena) australis – cordiline de Nueva Zelanda, dracena, falsa palma de Nueva Zelanda, palmera real

Cordyline fruticosa (terminalis) – caña de indio, cordiline, drago rojo, pablona, palmita roja – dracena, flaming dragon tree, good luck plant, ti plant, tree-of-kings

Cordyline indivisa – cordiline – blue dracena

Cordyline stricta – cordiline crasa

Coreopsis grandiflora – coreopsis – coreopsis, golden-wave, tickseed

Coreopsis maritima – dalia marina, dalia marítima – sea-dahlia

Coreopsis tinctoria – berro, coreopsis – annual coreopsis, calliopsis, goldenwaves

Coriandrum sativum – cilantro, coriandro, coriandro cultivado, culantro, perejil, tantsanacua – Chinese parsley, cilantro, coriander

Coriaria myrtifolia – emborrachacabras, garapalo, hierba zapatera, roldón, zumaque de hojas de mirto – Mediterranean coriaria

Coriaria ruscifolia – tlalocopetate

Cornus alba – cornejo albo – Tatarian dogwood

Cornus capitata – cornejo de Bentham – evergreen dogwood

Cornus controversa – cornejo del Japón – giant dogwood

Cornus disciflora – abiodo, asintla, asisincle, canelo, isimac, mimbre pasilla, palo canelo, pasilla blanca, xochilcorona

Cornus excelsa – acaciste, aceitunillo, canelo, corona de Moctezuma, corona de San Pedro, jazmín cimarrón, membrillo, mimbre, mimbre blanco, palo canelo, palo de membrillo, tepecuilo, tepecuilote, tepecuilotl, tepozá – dogwood

Cornus florida – cornejo americano, cornejo de Florida, cornejo floreciente, corona de San Pedro – eastern dogwood, flowering dogwood

Cornus kousa – cornejo del Japón – Japanese dogwood, Kousa dogwood

Cornus mas – cerezo silvestre, cornejo macho – Cornelian cherry

Cornus nuttallii – cornejo blanco americano, cornejo de Nuttall – mountain dogwood, Pacific dogwood, western dogwood

Cornus sanguinea – cornejo, cornejo hembra, cornejo sanguino, durillo encardado, sanguiñuelo – bloodtwig dogwood, red cornel

Cornus stolonifera – cornejo estolonífero – red-osier dogwood, redtwig dogwood

Coronilla varia – coletuy, coletuy multicolor,

cornejón variado, coronilla, ruda inglesa – crown vetch
Coronopus didymus – corónopo dídimo – swine cress
Coronopus squamatus – corónopo procumbente, mastuerzo silvestre – swine cress, wart-cress
Correa speciosa – correa – Australian fuchsia
Cortaderia selloana (Chusquea quila) – carrizo, carrizo de las pampas, cortadera, ginero, hierba de las pampas, hierba de Santa Teresa, pluma de las pampas – pampas grass
Cortaderia = Gynerium
Corydalis cava – corídalo hueco, fumaria, tijerilla bulbosa – bulbous corydalis, bulbous fumitory
Corydalis lutea – corídalo amarillo, fumaria amarilla, tijerilla amarilla – yellow corydalis
Corydalis solida – corídalo digitado, corídalo primaveral, fumaria, tijerilla primaveral – spring fumitory
Corylopsis spicata – avellano de invierno – spike winter hazel
Corylus avellana – avellanera, avellano, avellano común, nochizo – European filbert, hazel tree
Corylus colurna – avellano arbóreo, avellano de Levante, avellano mediterráneo, avellano turco – Turkish hazel
Corylus cornuta* var *californica – ablano, avellano – western hazelnut
Corylus maxima – avellano, avellano de Lambert, avellano turco – giant filbert, hazelnut
Corynanthe yohimbe – yohimbina – yohimbe
Corynocarpus laevigata – laurel de Nueva Zelanda – New Zealand laurel
Corypha taliera – palma taliera
Corypha umbraculifera – palmera de sombrilla, talipot, talipote – talipot palm
Coryphantha elephantidens – biznaga – elephant's tooth, starry ball
Coryphantha longicornis – biznaga de piña
Coryphantha macromeris – borrachito, doña Ana, mulato – big needle cactus, pineapple
Coryphantha pallida – biznaga
Coryphantha (Mammillaria) potosiana – chilito de biznaga – pincushion cactus
Coryphantha pycnacantha (connivens) – chichita de burro, chilito
Coryphantha radians – huevo de coyote – pincushion cactus
Coryphantha retusa – chilito
Coryphantha = Escobaria
Cosmos bipinnatus – cosmea, cosmos, girasol, girasol morado, mirasol, mirasol xocoatole, xaricamata, xocoatole – garden cosmos
Cosmos crithmifolius – bavisa
Cosmos diversifolius – gallito
Cosmos parviflorus – acetilla, amor, tostón
Cosmos pringlei – bavisa, matagusanos
Cosmos sulphureus – axalxóchitl, cempoal, chochopali, ehcaxtli, flor que pinta, girasol amarillo, mirasol, mirasol amarillo, San Miguel, sumpual, ximula, xochipali, xochipalli, zempual – yellow cosmos
Costus mexicanus – apagafuego, caña agria, caña brava, caña cristo, caña de jabalí, caña de tigre, caña de venado, cañuela, hierba del jabalí, huilamole, pahtsab, tirabuzón – spiral ginger
Costus pulverulentus – caña de Cristo
Costus speciosus – caña mexicana – crape ginger, Malay ginger, spiral ginger, white costus
Costus spicatus – caña agria, caña amarga, caña de jabalí, caña de puerco, cañita agria, coyame-catl, tirabuzón – sour cane
Cotinus coggygria (Rhus cotinus) – árbol de las pelucas, cotinus, fustete – smoke tree, Venetian sumac
Cotoneaster adpressus – membrillo falso procumbente – cotoneaster
Cotoneaster franchetii – membrillo falso de Franchet
Cotoneaster horizontalis – cotoneaster, cotoneaster en espina de pescado, membrillo falso horizontal – rock cotoneaster
Cotoneaster microphyllus – cotoneaster de hojas pequeñas, membrillo falso microfilo – rockspray cotoneaster
Cotoneaster salicifolius – cotoneaster, membrillo falso de hojas de sauce – willowleaf cotoneaster
Cottea pappophoroides – cotea, pelusilla – cotta grass

Cotula coronopifolia – cótula sudafricana – brass-buttons
Cotyledon orbiculata – conchita
Cotyledon = Oliveranthus
Couepia polyandra – carnero, fraile, frailecillo, guayabillo de tinta, guayabito de tinta, guayo, olosapo, palo de fraile, zapote amarillo, zapotillo – monkey cup
Coumarouna = Dipteryx
Couroupita guianensis – abahuma, bala de cañón, coco de mono, granadillo, guira, higuera, maraco, muco, taparo de chuco, taparón, totumo, zapote de mono – cannonball tree
Coursetia glandulosa – causamo, chino, chipile, chipilillo, cousano, gomea, gomilla, palo dulce, samán, samo, samo prieto, samota, tepechipile
Coutarea hexandra – chichipate, copalchi, copalchi de Colima, copalchi de Michoacán, palo de quina, quino
Coutarea pterosperma – copalchi, copalquín, coparche, hutatiyo, palo amargo, palo muela, quino
Coutarea = Hintonia
Covillea = Larrea
Cowania = Purshia
Crambe maritima – col de mar, col marina, crambe marítima – seakale
Cranichis speciosa – acatlzauhtli, atzahutli, chautle, flor de corpus, saucle, tzacuxóchitl, tzautli
Cranichis tubularis – acatlzauhtli, tzautli
Crassula aquatica – crásula acuática – water tillaea
Crassula lycopodioides – crásula
Crassula tillaea – crásula musgosa – mossy stonecrop
Crataegus azarolus – acerolo, manzanita de dama – azarole tree, Neapolitan medlar tree
Crataegus crus-galli – espino americano, espino pata de gallo, espolón de gallo – cockspur thorn
Crataegus erythropoda – manzano de puya larga – cerro hawthorn
Crataegus laevigata (oxyacantha, oxycanthoides) – cratego, espino, espino albar, espino blanco, espino de dos estilos, espino majolar, espino navarro, majuelo, majuelo navarro, manuelo, marjoleto, oxiacanto – English hawthorn, Mayflower, midland hawthorn
Crataegus* x *lavallei (carrierei) – espino híbrido, espolón de gallo – carriere hawthorn
Crataegus mexicana – carasu, manzanilla, manzanita, manzanita tejocotera, tejocote, texócotl – Mexican hawthorn
Crataegus mollis – acerolo rojo – downy hawthorn
Crataegus monogyna – cratego, espinablo, espino, espino albar, espino blanco, espino europeo, espino majuelo, espino monógino, majuelo, marjoleto – hawthorn
Crataegus pubescens – chisté, manzanilla, manzanita, tejocote, texócotl
Crataegus wootoniana – tejocote, texócotl – Wooton hawthorn
Craterellus cornucopioides – corneta prieta, trompeta de los muertos – cornucopia, horn-of-plenty
Crateva palmeri – capúsari, cascarón, perihuite, tres Marías, trompo, zapotillo amarillo – garlic pear
Crepis capillaris – crepis capilar – hawksbeard
Crepis setosa – crepis setoso – hawksbeard
Crepis vesicaria – crepis vesiculoso – hawksbeard
Crescentia alata – ayale, cirián, cuautecomate, cuautecomatl, cuiro, guaje cirial, guaje cirián, güiro, jayacaste, jayacastle, jicarita, jícaro, luch, morro, tecomate, tecomate ayele, tecomate de sonajas, tecomatl, urani – calabash tree, gourd tree
Crescentia cujete – árbol de las calabazas, ayale, boch, calabacero, calabazo, chayote, cirián, couis, cuautecomate, cujete, guaje, guaje cirial, guaje cirián, guacal, güiro, güiro de Yucatán, güiro totumo, higuero, huas, huingo, jacaparí, jícaro, jícaro de Yucatán, luch, mate, morro, morro guacalero, palo de huacal, pog, taparo, tecomate, tecomatl, totumo, zacual – calabash tree
Crinum bulbispermum (longifolium) – amarilis de hojas largas
Crinum zeylanicum – corona del marques, lirio, lirio de cinta, lirio listado, reina – spider lily
Crithmum maritimum – cresta marina, hinojo

marino, hinojo marítima, perejil de mar, perejil marino – samphire
Crocosmia (Montbretia, Tritonia)* x *crocosmiiflora – montbretia, palmira – montbretia
Crocus angustifolius (susianus) – azafrán dorado, paño de oro – cloth of gold crocus
Crocus biflorus – azafrán bifloro – Scottish crocus
Crocus chrysanthus – azafrán amarillo – garden crocus
Crocus imperati – precioso
Crocus sativus – azafrán, azafrán amarillo, azafrán cultivado, azafrán oficinal, azafrán verdadero, croco – saffron
Crocus speciosus – azafrán
Crocus vernus – azafrán albo, azafrán primaveral, azafrán silvestre – Dutch crocus
Crotalaria eriocarpa – cascabelillo, cascabelito, tronador
Crotalaria incana – ajonjolí, cascabelillo, chipil, chipilín cimarrón, sacpet, sonaja, tenégara, tronador
Crotalaria juncea – cáñamo de Bengala, crotalaria – sunn hemp
Crotalaria longirostrata – cascabel, cascabel de víbora, chipil, chipilín, chipilín cimarrón, garbancillo, tronador, vichi – rattlepod
Crotalaria lupulina – hierba loca, tronador
Crotalaria mucronata – crotalaria – striped crotalaria
Crotalaria pumila – cascabel, chipil, chipil-max, hierba del cuervo, sonadora, tronador – low rattlebox
Crotalaria sagittalis – ajonjolí, ajonjolí de monte, sonaja – rattlebox
Croton alamosanum – ocotillo, vara prieta
Croton boregensis – hierba de la mula, malva lanosa
Croton californicus – barbasco, copalquín, hierba del pescado, vara blanca – croton
Croton ciliato-glandulosus – canelilla, chirca, ciega vista, cuanaxunaxe, dominguilla, enchiladora, hierba de la cruz, matagusanos, palillo, picosa, rama del caballo, Solimán, Solimán blanco, tatio, trucha, uruquenia, xonaxe, xunalixase – scented croton
Croton corymbulosus – encinilla
Croton dioicus – encembla, encinilla, hierba del gato, hierba del zorrillo, izpaxíhuitl, robaldo, rosval, rubaldo, vara blanca – New Mexico croton
Croton draco – chorro de sangre, cuate, dominguilla, drago, etzcuáhuitl, grado, llora sangre, palo muela, sangre de drago, sangre de grado, sangre de perro, sangregado
Croton eluteria – eleuteria, quino morado – cascarilla bark
Croton fantzianus – vara blanca
Croton flavescens – vara prieta
Croton fragilis – tanché, vara blanca
Croton fruticulosus – encinilla, hierba loca – bush croton
Croton guatemalensis – algodoncillo, cascarillo, chulché, copalchi, copalquín cortez amarilla, huilote, juilocuáhuitl, palo blanco, quino blanco, vara blanca, vidrioso
Croton lecheri – becerra, boca de dragón, drago, sangre de drago, sangre de dragón, sangre de grado – dragon's blood
Croton magdalenae – duraznillo, hormiguillo, José gris
Croton monanthogynus – encinilla
Croton niveus – cascarillo, chulché, copalchi, copalchi blanco, palo almizcle, quino, quino blanco, vara blanca
Croton sanguifluus – ezpatli, sangre de drago
Croton sonorae – hierba del pescado, vara prieta – Sonora croton
Croton texensis – barbasco, hierba de la tórtola, hierba del pescado, tortolita – dove weed, Texas croton
Croton tiglium – crotón – purging croton
Crucianella angustifolia – crucianella angustifolia, crucianella de hojas estrechas, espigadilla – cross-wort
Crucianella maritima – rubia de mar
Crupina vulgaris – crupina común – bearded creeper
Crusea calocephala – azulejo, golondrina, rosa morada, zorrillo
Cryosophila nana – escoba, palma de abanico, palma de coateco, palma de escoba, palmilla, soyamiche, soyaviche
Cryosophila stauracantha – akuum, escoba – give-and-take palm
Crypsis alopecuroides – cripsis alopecuro – prickle grass

Crypsis schoenoides – cripsis falso-junco – swamp grass
Cryptantha angustifolia – peluda
Cryptantha barbigera – nievitas
Cryptantha costata – nievitas – ribbed cryptantha
Cryptantha grayi – espumila, nievitas
Cryptantha holoptera (inaequata) – nievitas – winged cryptantha
Cryptantha maritima – peluda, peludita
Cryptantha muricata – nievitas
Cryptantha nevadensis – nievitas
Cryptantha racemosa – nievitas
Cryptanthus zonatus – bromelia, estrella de la tierra
Cryptomeria japonica – canalete, cedro del Japón, cedro japonés, criptomeria japonesa, sugi – Japanese cryptomeria
Cryptostegia grandiflora – bejuco, Belén, chicote, chirrión del diablo, chirrionera, clavel alemán, clavel de España, cuaumecate chayote, cuerno, hiedra, hierba de Judas, hule, pichuco, quiebraplato
Cucumeropsis edulis – calabaza pamúe – egusi melon
Cucumis anguria – cohombro, cohombro espinoso, halu, melón de coyote, pepinillo, pepinito, pepino, pepino de sabana, pepino espinoso, pepino silvestre, sandía chom, sandía del diablo, sandía de zopilote – West Indian gherkin
Cucumis melo – calabaza melón, gueto bitoo, melón, minari – melon
Cucumis melo var cantalupensis – cantalupo, melón, melón cantalupo, melón chino, melón de cantalú, melón de Castilla – cantaloupe
Cucumis melo var chito – melón chito, melón mango – chate of Egypt, cucumber melon, mango melon
Cucumis melo var conomon – melón manzana, pepino limón – pickling melon
Cucumis melo var dudaim – meloncillo de coyote, melón de coyote – dudaim melon
Cucumis melo var flexuosus – aficoz, cohombro largo, cohombro serpentino, melón flexuoso, melón pepino, melón serpentino – snake cucumber, snake melon
Cucumis melo var indorus – melón chino, melón de Indias – casava
Cucumis melo var reticulatus – melón bordado – Persian melon
Cucumis metuliferus – kiwano – African horned melon
Cucumis sativus – cohombro, pepinillo, pepino – cucumber
Cucurbita digitata – calabacilla, calabaza amarga, chichicoyote, meloncillo, melón de coyote – finger-leaved gourd
Cucurbita ficifolia – alcayota común, chilacayote, chilacayotl, chirigaita, chiverre, cidra cayote, lacayote, mail, ticati, ticatsa, tzilacayotli – fig-leaved gourd, Malabar gourd
Cucurbita foetidissima – alficaz, calabacilla, calabicilla amarga, calabacilla loca, calabaza amargosa, calabaza búfalo, chichic-amole, chichicoyote, cohombrillo, cohombro, coloquinto, hierba de la víbora, jamacha, oreja de burro, pepino del diablo, zaracoche – buffalo gourd, calabazilla, coyote melon, wild gourd
Cucurbita maxima – 1. ahuyama, ayotli, ayoxochquílitl, calabaza, calabaza confitera, calabaza gigante, calabaza grande, calabaza tamala, calabaza tamalayota, gueto, guu, tamaloyota, tecats, totanera, xigaba – gourd, pumpkin; 2. ayote, calabaza amarilla, calabaza de invierno, calabazona – winter squash; 3. bellota – acorn squash
Cucurbita mixta – ayote – green-striped cushaw
Cucurbita moschata – 1. ayotli, ayoxochquílitl, calabaza, calabaza almisclada, calabaza amarilla, calabaza melón, calabaza moscada, cum, melocotón, sehualca – butternut, cushaw, pumpkin; 2. calabaza de cuello torcido – crookneck squash
Cucurbita palmata – calabacilla, calabaza amarga, gueto bichi, meloncillo, melón de coyote – coyote melon
Cucurbita palmeri – calabacilla, chichicayota, haya
Cucurbita pepo – 1. ahuyama, calabaza, calabaza biche, calabaza chompa, calabaza común, calabaza india, calabaza turbante, cum, puru, tzol, zapallo – pumpkin; 2. bonetillo, cabeza de Turco, calabacín, chicayota, zapallito – squash; 3. ayote,

calabaza de verano – summer squash; 4. calabacín, calabacita, calabacita italiana, calabacita verde – marrow, zucchini
Cuitlauzinia pendula – chorrito, pozolillo
Cuminum cyminum – comino, comino cultivado – cumin
Cunninghamia lanceolata – abeto chino, cuningamia, cuningamia de la China, pino de Cunningham, san-shu – China fir
Cupania americana (Paullinia cupana) – chichón colorado, cupana, guara, guaraná, paulina – guarana
Cuphea aequipetala – alcáncer, apanchola, atlanchana, atlanchane, calavera, cáncer, hierba del cáncer, pegamosca
Cuphea ignea (platycentra) – cufea, moradilla – cigar flower, cigar plant
Cuphea jorullensis (micropetalia) (Parsonsia tricolor) – alcáncer, clamería, hierba del cáncer – beetlesage cuphea
Cuphea (Parsonsia) llavea – chupamiel, flor roja, hierba de San Pedro, xoniguilli – bat-faced cuphea
Cuphea wrightii – hierba del cáncer
Cupressocyparis leyandii – ciprés de Leyland, ciprés híbrido de Leyland
Cupressus arizonica – arizónica, cedro, cedro blanco, cedro de la sierra, ciprés, ciprés de Arizona, ciprés de Arizona de corteza áspera, ciprés de México, pinabete, sabina, táscate – Arizona cypress, Cuyamaca cypress
Cupressus arizonica (glabra)* ssp *glabra – ciprés glabro de Arizona – smooth Arizona cypress
Cupressus forbesii – cedro, cedro blanco, madroño – Tecate cypress
Cupressus funebris – ciprés fúnebre, ciprés llorón chino
Cupressus goveniana – ciprés de California, ciprés de Gowen – Gowen cypress
Cupressus guadalupensis – cedro blanco – Guadalupe cypress
Cupressus (Chamaecyparis) lawsoniana – camecíparis de Lawson, cedro de Oregón, cedro de Puerto Oxford, ciprés de Lawson, falso ciprés, falso ciprés de Lawson – Lawson cedar, Lawson cypress, Port Orford cedar
Cupressus lusitanica (benthamii, lindleyi) – cedro, cedro blanco, cedro de busaco, ciprés, ciprés de México, ciprés de Portugal, gretado amarillo, gretado galán, nucalpat, sabina, táscate, teatlale, teotlate, tlatzcan, tláxcal – Mexican cypress, Portuguese cypress
Cupressus macrocarpa – ciprés americano, ciprés de Californica, ciprés de frutos grandes, ciprés de Lambert, ciprés de Monterrey – Monterey cypress
Cupressus montana – cedro de la sierra
Cupressus (Chamaecyparis) nootkatensis – camecíparis de Nootka, ciprés americano, ciprés de Nootka, falso ciprés, falso ciprés de Alaska, falso ciprés de Nootka – Alaska cedar, Nootka cypress, yellow cypress
Cupressus sargentii – ciprés – Sargent cypress
Cupressus sempervirens – ciprés, ciprés común, ciprés de los cementerios, ciprés italiano, ciprés pirámidal – Italian cypress
Curatella americana – cacaíto, caticón, chaparro, chumico, jamán, ojaman, rasca viejo, raspa viejo, sahá, tachicón, tlachicón, vacabuey, yahá – sandpaper tree
Curcuma angustifolia – arrurruz de las Indias Orientales – East Indian arrowroot
Curcuma domestica (longa) – azafrán de la India, azafrán de las Indias, camotillo, cúrcuma, cúrcuma larga, raicilla, yuquilla – turmeric
Curcuma xanthorrhiza – cúrcuma javanesa
Curcuma zedoaria – cedoaria, cúrcuma – zedoary
Cuscuta americana – azafrancillo, azafrán zacatlaxcale, canlecay, cuerda de violin, cúscuta, fideo, hilo de oro, tiripu, tripa de Judas, zacatlaxcale
Cuscuta approximata – cuscuta de cáliz dorado – dodder
Cuscuta denticulata (veatchii) – cochear, manto de la Virgen – witch's hair
Cuscuta gracillima – hierba sin raíz
Cuscuta pentagona (campestris) – cochear, cuscuta del campo, cuscuta glandulosa – dodder
Cuscuta subinclusa – cochear – dodder
Cuscuta tinctoria – zacatlaxcale
Cuscuta umbellata – cuscuta, tiripu, zacapal, zacatlaxcale

Cyamopsis tetragonoloba – goma de guar, guar – guar gum
Cyathea delicatula – helecho arborescente, helecho de árbol, ocopetate, pelma, pesma, pesto, rabo de mico, rabo de mono – tree fern
Cyathea mexicana – chino pezmo, cola de mono, malque, ocopetate, palma helecho, rabo de mico, rabo de mono – tree fern
Cyathea princeps – cola de chango, cola de machín, cola de mico, rabo de chango, rabo de mico
Cybistax = Tabebuia
Cycas revoluta – cica, palmera de sagú – cycad palm, sago palm
Cyclamen cilicum – ciclamen de Chile – Chilean cyclamen
Cyclamen coum – ciclamino de hojas circulares – cyclamen
Cyclamen hederifolium (neapolitanum) – ciclamen de Nápoles, ciclamino de hojas de hiedra – hardy cyclamen, sowbread
Cyclamen persicum – ciclamen, ciclamino, ciclamino de Persia, pan de puerco, violeta de los Alpes – florist's cyclamen
Cyclamen purpurascens (europaeum) – ciclamino, ciclamino de Europa, ciclamino europeano, mitra, pamporcino, pan de puerco, violeta de los Alpes – cyclamen
Cyclamen repandum – ciclamino primaveral – cyclamen
Cyclanthera dissecta – raíz de fregar
Cyclanthera pedata – achojcha, caigua
Cydonia oblonga – membrillero, membrillo, membrillo común – quince
Cylindropuntia (Opuntia) acanthocarpa **(var** ***coloradensis)*** – cholla, hannam, tasajo – buckhorn cholla
Cylindropuntia alcahes (Opuntia brevispina, O. burragena) – clavellina, nopal, tuna
Cylindropuntia (Opuntia) arbuscula – cholla, chollita, chumbera, siviri, tasajo – pencil cholla
Cylindropuntia (Opuntia) bigelovii – cholla, cholla del oso, cholla güera, ciribe, cumbera, guerra, vela de coyote – ball cholla, jumping cholla, teddy-bear cholla
Cylindropuntia bigelovii **var** ***ciribe (Opuntia ciribe)*** – ciribe
Cylindropuntia californica (Opuntia parryi) – cholla – California cholla, cane cholla, snake cholla
Cylindropuntia (Opuntia) cholla – cholla, cholla pelona – chain-link cholla
Cylindropuntia (Opuntia) fulgida – brincadora, cholla, cholla brincadora, tootoco, vela de coyote – chain-fruit cholla, jumping cholla
Cylindropuntia fulgida **var** ***mamillata (Opuntia mammillata)*** – cholla – Sonoran jumping cholla
Cylindropuntia imbricata (Opuntia arborescens, O. lloydia) – abrojo, cardencha, cardón, cholla, coyonostle, coyonostli, entraña, estrara, joconoxtle, tasajo, tasajo macho, tuna huell, tuna joconoxtli, vela de coyote, xoconóchitl, xoconochtli, xoconoxtle – tree cholla, walkingstick cholla
Cylindropuntia (Opuntia) kleiniae – cardoncillo, tasajillo, tasajillo rojo, tasajo – Klein's pencil cholla
Cylindropuntia leptocaulis (Opuntia brittoni, leptocaulis) – agujilla, alfilerillo, catalinaria, catalinia, híxul, tasajillo, tasajo – desert Christmas cactus, pencil cactus, pipestem cactus, rat-tail cactus
Cylindropuntia molesta – agujilla, clavellina – cholla
Cylindropuntia (Opuntia) ramosissima – tasajillo, tasajo – diamond cholla, pencil cactus
Cylindropuntia rosea – cardencha, cardo
Cylindropuntia (Opuntia) spinosior – tasajo – cane cholla, handlegrip cholla
Cylindropuntia tesajo – tasajo – pencil cholla
Cylindropuntia (Opuntia) thurberi (alamosensis) – cholla, siviri
Cylindropuntia (Opuntia) tunicata – abrojo, clavellina, coyonoxtle, huichacame, tencholote – sheathed cholla, thistle cholla
Cylindropuntia (Opuntia) versicolor – cholla, cholla morada, siviri, tasajo – deerhorn cholla, staghorn cholla
Cymbalaria aequitriloba – cimbalaria de tres lobos iguales – toadflax
Cymbalaria muralis (Linaria cymbalaria) – barba de chivo, cimbalaria, cimbalaria de hojas de hiedra, linaria cimbalaria, mosca – kenilworth ivy, ivy-leaved toadflax

Cymbopetalum mayanum – guanabano, marta boni
Cymbopetalum penduliflorum – eriojuelo, flor de oreja, guineillo, guineillo prieto, guineo, hueynacaztli, orejuela, orejuela para chocolate, orifela, teunacaztli, xochinacastle
Cymbopogon citratus – citronela, grama oriental, hierba de limón, hierba limón, hierba Luisa, limoncillo, pasto de limón, pasto limón, té de limón, té limón, zacate de limón, zacate limón – citronella grass, lemon grass
Cymbopogon martinii – palma rosa – palmarosa
Cymbopogon nardus – citronela, hierba de limón, té de limón, té limón, zacate citronella, zacate de limón, zacate limón – citronella grass, nardo grass
Cymopterus fendleri – chimaha, chimaya – biscuit root
Cymopterus purpurascens – chimaha, chimaya – biscuit root, wafer parsnip
Cynanchum acutum – correhuela lechosa, escamonea falsa, falsa escamonea, matacán – stranglewort, vine milkweed
Cynanchum utahense – correhuela lechosa, matacán – Utah cynanchum, vine milkweed
Cynanchum vincetoxicum – berza de perro, berza perruna, vencetósigo – white swallow-wort
Cynara cardunculus – alcaucil, alcaucil silvestre, cardo, cardo comestible, cardo de arrecife, cardo de comer, cardo de huerta – artichoke thistle, cardoon
Cynara scolymus – alcachofa, alcachofera, alcaucil, artichaut, cardo, cardo alcachofero, alcaulera – artichoke, globe artichoke
Cynodon (Capriola) dactylon – acaxacahuitztli, agrarista, asqui, Bermuda, Bermuda de la costa, brena, canzuuc, chepica, diente de perro, estrella gigante, ferrer, gallito, grama, grama común, grama de Bermuda, grama de la costa, grama de marajó, grama de pará, grama de seda, grama menor, grama pie de pollo, grama zastrera, gramilla brava, gramilla colorada, gramilla forestall, gramilla italiana, gramillón, gramón, hierba de Bermuda, hierba de la Virgen, paja bermuda, paja de la Virgen, pasto alicia, pasto de Bermuda, pasto estrella, pasto ferrer, pata de gallo, pata de perdíz, pata de pollo, tejedora, uña de gato, zacate agrarista, zacate agujilla, zacate alicia, zacate bermuda, zacate borrego, zacate chino, zacate de gallina, zacate del conejo, zacate de luna, zacate ferrer, zacate inglés, zacate pelillo, zarzuue – Bermuda grass
Cynodon oleufuensis – Santo Domingo
Cynodon plectostachyus – estrella africana, pasto estrella, pasto estrella gigante, zacate bermuda estrella, zacate estrella – star grass
Cynoglossum officinale – cardillo, cinoglosa, lengua de perro, oreja de liebre, viniebla – hound's tongue
Cynosurus cristatus – cinosuro crestado, cinura, cola de perro – crested dogtail
Cynosurus echinatus – cinosuro espinoso, cinosuro gigantesco – hedgehog dogtail
Cyperus acuminatus – coquillo azul – nutsedge
Cyperus alternifolius – palmilla, paraguas – umbrella plant
Cyperus articulatus – apoyomatli, camelote, chintule, chintule grande, junco, tule chico
Cyperus canus – tule, tule ahuapetla, zacate de tule
Cyperus diandrus – galanga, guasca – galingale
Cyperus difformis – juncia de flores pequeñas, juncia diforme – galingale, small-flowered galingale
Cyperus digitatus – tule
Cyperus eragrostis – juncia comestible – American galingale, edible galingale, umbrella nutsedge
Cyperus esculentus – cebollín, chufa, coquillo, coquillo amarillo, corocito, cotufa, cuca, juncia avellanada, juncia redonda, peonía, peonía de botica, peonía del país, tolpatle, tolpatli, tule, tulillo, zacate de tule – chufa, earth almond, rush-nut, yellow nut-grass, yellow nutsedge
Cyperus flavus – tulillo
Cyperus fuscus – juncia morena – brown galingale
Cyperus hermaphroditus – tule, tulillo – flat sedge
Cyperus involucratus (alternifolius) – juncia, palmilla – nutsedge, umbrella plant

Cyperus laevigatus – coquito, juncia distica, juncia levigada, palla – flat sedge, umbrella sedge
Cyperus laxus – sueño
Cyperus ligularis – tule – nutsedge
Cyperus longus – juncia común, juncia larga, juncia loca, juncia olorosa – galingale
Cyperus niger – tule – nutsedge
Cyperus odoratus (ferax) – cuentas de Santa Elena, tule, tuxpata – nutsedge
Cyperus papyrus – cyperus, papiro, papiro de Egipto, papyrus de Egipto – bulrush, Egyptian paper plant, papyrus
Cyperus perennis – conisaca, coquillo, tule
Cyperus rotundus – cebollín, cípero, conisaca, coquillo, coquillo morado, coquito, coyolillo, juncia de tallo circular, juncia redonda, tuk'uch – nut-grass, purple galingale, purple nut-grass, purple nutsedge
Cyperus sesleroides – tulillo, zacate de toche
Cyperus squarrosus (aristatus) – apoyamate, tule, tulillo – march sedge, nutsedge
Cyperus subambiguus – tule
Cyperus surinamensis – tule, zacate cabezón
Cyperus tenuis – tule
Cyphomandra betacea (crassicaulis) – árbol del tomate, berenjena, grosello, palo de tomate, tamarilla, tomate arbóreo, tomate de árbol, tomate de la paz, tomate de palo, tomatero – tamarillo, tree tomato
Cypripedium calceolus – zapato de Venus, zueco
Cypripedium irapeanum – bolsita, cipripedio, flor de calavera, flor de mocasín, flor de pelícano, orquídea zapatilla, pichohuastle, zapatillo, zapatillo de mujer, zapatillo de Venus, zapatito de dama – lady's-slipper
Cypripedium turgidum – pipixiuhatstli
Cyrtocarpa edulis – ciruelo, ciruelo cimarrón, ciruelo de monte – wild plum
Cyrtocarpa procera – baricoca, berraco, chupandía, coco de cerro, copal, copal cocote, copalcoxócotl, copal jocote, copalxocotl, cupandilla, machocote, maxacocote, palo de chupandía, pompoqua, popoacua
Cytisus multiflorus – escoba blanca, hiniesta genista, marihuela – Portuguese broom, Spanish broom
Cytisus purgans – calabón, piorno, piorno serrano, retama serrana – Provence broom
Cytisus scoparius – cítiso, escoba, escobón, hiniesta, hiniesta de escobas, hiniesta escocesa, retama, retama de escobas, retama escocesa, retama negra – Scotch broom
Cytisus striatus – escobón estriado
Cytisus = Genista, Laburnum

D

Daboecia cantabrica – brezo, brezo de Irlanda, brezo irlandés, brezo polifolio – Irish heath
Dacryodes edulis – uva del país – bush butter tree
Dactylis glomerata – alpistillo, azul orchoro, dáctilo, dáctilo aglomerado, dáctilo apelotonado, dáctilo ramoso, gallo, grama en jopillos, gramilla de huerto, hierba del huerto, orcho, pasto olivo, pasto ovillo, pata de liebre, pata de pollo, yopillo, zacate de la huerta – cocksfoot, orchard grass
Dactyloctenium aegyptium – pasto dáctilo, pata de cuervo, pata de gallo, pata de grulla, pata de pollo, zacate pata de cuervo – crowfoot grass, Durban crowfoot grass
Daemonorops draco – sangre de dragón, sangre de venado – dragon's blood
Dahlia coccinea – acocotli, acocoxóchitl, charahuesca, dalia, dalia del campo, dalia encarnada, flor de invierno, galusa, jícama del colera, jicamite xicamaxóchitl – dahlia
Dahlia excelsa – dalia de árbol, dalia morada, dalia plana, gigantón, invierno, palo de Oaxaca, Santa Catarina, tecotéhue – flat tree dahlia
Dahlia imperialis – dalia, dalia de árbol, dalia plana, gigantón, ocotehue, palo de Oaxaca, tecotéhue – bell tree dahlia, tree dahlia
Dahlia rosea – dalia – wild dahlia
Dahlia variabilis – amapola silvestre, charahuesca, coanenepilli, cohuanenepili, dalia, dalia morada, jícama del colera, jicamaxóchitl, jicamite
Dalbergia glabra – bejuco de panune, camac, muc
Dalbergia granadillo – cocobolo, granadillo, palo de granadillo morado, tapintzirán – cocobolo
Dalbergia nigra – madera de palisandro – Brazilian rosewood
Dalbergia retusa – cocobolo, cocolobo – cocobolo
Dalbergia sissoo – bálsamo, granadillo – rosewood
Dalea alamosana – ramita colorada
Dalea bicolor* var *orcuttiana – damiana, hierba del borrego – indigo bush
Dalea (Parosela) citriodora – engordacabra, hierba de la pulga, limoncillo, ruda cabruna, ruda capraria, terciopelillo – indigo bush
Dalea (Parosela) diffusa – cadillo, escoba, escoba colorada, escoba larga, hipechila, ratón, xolteco
Dalea formosa – hierba de Alonso García – feather plume
Dalea (Parosela) frutescens – terciopelillo – black dalea, shrubby dalea
Dalea grayi – popote
Dalea greggii – orégano cimarrón – trailing indigo bush
Dalea hospes – dalea
Dalea (Parosela) lagopus – terciopelillo
Dalea lumholtzii – limoncillo, té limón – pea bush
Dalea parryi – popotillo
Dalea pogonanthera – hierba del corazón – bearded dalea
Dalea pringlei* var *multijuga – popote, popotillo – indigo bush
Dalea revoluta – popote chiquito
Dalea (Parosela) seemannii – engordacabra
Dalea tomentosa – mota – hairy prairie clover
Dalea (Parosela) wislizeni – mariola
Dalea = Psorothamnus
Damasonium alisma – cola de golondrina, saeta de agua – star fruit, thrumwort
Danae racemosa – laurel alejandrino
Daphne cneorum – adelfilla, torvisco de los Pirineos – garland daphne
Daphne laureola – adelfilla, lauréola, lauréola macho, torvisco macho – daphne, spurge laurel
Daphne mezereum – hoja de San Pedro, lauréola, lauréola hembra, leño gentil, mezereo, mezéreon, olivaseba, torvisco – February daphne, mezereon
Dasylirion acrotriche – cucharilla, palmilla
Dasylirion leiophyllum – sotol – desert candle, spoonplant
Dasylirion longissimum (quadrangulatum) – junquillo, padillo, vara de cohete, vara de cuete – Mexican grass tree
Dasylirion serratifolium – cucharita, tehuizote
Dasylirion texanum – sotol – desert-candle

Dasylirion wheeleri – cucharilla, sano, sotol, tehuizote – desert spoon, sotol
Datura ceratocaula – atlinán, nexehuac, toloache, tornaloco
Datura discolor – belladona, toloache – desert thorn-apple, jimsonweed, trumpet flower
Datura ferox – chamico, datura feroce – jimson weed, thorn-apple
Datura lanosa – toloache
Datura metel – burladora, taguaro, toloache – sacred datura
Datura quercifolia – toloache
Datura stramonium – azcapán, amizolillo, berenjena del diablo, biguero del infierno, borrachero, campanilla silvestre, chamico, datura estramonia, estramonio, flor de trompeta, flor de muerto, galán de noche, hierba del diablo, hierba del topo, hierba hedionda, hierba loca, higuera del infierno, higuera loca, manzano del diablo, manzano espinoso, mehenxtohku, nacazcul, ñongue, piñón de Tlapa, quiebra plato, tapate, tecuyani, tlapa, tlápatl, tlazol-patli, toaloatzin, tohku, toloache, torescua, trompetilla, tzitzintlaptl – jimson weed, thorn-apple
Datura wrightii (meteloides, inoxia) – belladona, cacaíto, chamico, datura inocua, estramonio, hierba ponzoñosa, nacazcul, tecuyani, tlapatl, tohku, toloache, toloache grande, tolohuaxíhuitl – Indian apple, jimson weed, sacred datura, thorn apple
Datura = Brugmansia
Daucus carota – dauco, zanahoria, zanahoria silvestre – carrot, Queen Anne's lace, wild carrot
Daucus carota* ssp *sativus – zanahoria – carrot, garden carrot
Daucus pusillus – hierba de la víbora, zanahoria silvestre – rattlesnake weed, seed-ticks, wild carrot
Davidia involucrata – árbol de las palomas, árbol de los pañuelos, davidia – dove tree
Decaisnea fargesii – habichuela azul
Delairea odorata (Senecio mikanioides) – hiedra alemana, senecio escandente – Cape ivy, German ivy
Delilia biflora – hermana de albaca
Delonix (Poinciana) regia – acacia, árbol del fuego, barambuyan de nejapa, espuela de cabellero, flamboyán, flamboyán rojo, flor de camarón, framboyán, framboyán rojo, guacamaya, pata de león, poinciana real, tabachín, tabachín grande – flamboyant tree, flame tree, poinciana, royal poinciana
Delphinium carolinianum – espuela de caballero – blue larkspur
Delphinium consolida – conejito, consólida real, consuelda, espuela de caballero – forking larkspur
Delphinium elatum – espuela de caballero – candle delphinium, candle larkspur
Delphinium madrense – espuela de caballero
Delphinium nudicaule – espuela, espuela de caballero, espuela de caballero nudicaule, espuelita – orange larkspur, red larkspur
Delphinium staphisagria – albarraz, estafisagria, hierba piojera – stavesacre
Delphinium = Consolida
Dendranthema = Chrysanthemum
Dendrocalamus giganteus – bambú gigante – giant bamboo
Dendropanax = Gilibertia
Dendropogon = Tillandsia
Deschampsia cespitosa – aira de césped – tufted hairgrass
Deschampsia flexuosa – aira flexuosa – crinkled hairgrass
Descuriana pinnata – lepidio, pamita, pamitón, sinapismo, suavoli – tansy mustard
Descuriana sophia – hierba Sofia – flixweed
Desmanthus covillei – batillo, escoba, frijolillo, frutillo, sitiporo
Desmanthus fruticosus – daí, fresa, frijolillo, frijolito, frutillo – bundleflower
Desmodium barbatum – barbadino
Desmodium canum – amor de campo, amorseco, corteza fuerte, pega pega, pegarropa, tortilla de ratón, vid de hierro – iron vine, strong bark
Desmodium cinerascens – alma – tick clover
Desmodium scorpiurus – frijol de gordoniz, trencilla
Desmodium styracifolium – pega pega – beggar lice
Desmoncus chinantlensis – ballí, bayal, bejuco, bejuco de canasto, junco, junco de jaguay, matambilla

Desmoncus orthacanthos – ballí, bayal, matambilla – basket tie-tie, basket whisk
Desmoncus oxyacanthus – bas, bayal
Dianthus alpinus – clavel alpino, clavel de los Alpes – alpine pink
Dianthus arenarius – clavel arenario – sand pink
Dianthus armeria – clavel rugoso – Deptford pink, grass pink
Dianthus barbatus – clavel barbado, clavel de nobleza, clavel de poeta, clavel de ramillete, clavel de San Isidro, clavellina, clavel minutisa, macetilla, minutisa, ramillete de Constantinopla – sweet William
Dianthus caryophyllus – alhelí, clavel, clavel común, clavel de jardín, clavel hortense, clavellina, clavel reventón, clavillo – carnation, clove pink
Dianthus chinensis (sinensis) – clavel, clavel anual, clavel de China, clavel encarnado, clavel imperial, clavellina – Chinese pink, rainbow pink
Dianthus deltoides – clavel de los prados, clavel deltoideo, clavellina rocosa – maiden pink, meadow pink
Dianthus gratianopolitanus (caesius) – clavel de cheddar – cheddar pink
Dianthus plumarius – clavel coronado, clavellina – cottage pink, garden pink, grass pink
Dicentra formosa – corazón, corazón de María, corazón sangrante, dicentra – bleeding heart, western bleeding heart
Dicentra spectabilis – corazón, corazón de María, corazón sangrante, dicentra – common bleeding heart
Dichaea muricata – panza de culebra
Dichanthelium (Panicum) oligosanthes* var *scribnerianum– panizo de hojas anchas, zacate hoja ancha
Dichanthelium (Panicum) sphaerocarpon – panizo redondo
Dichanthium annulatum (Andropogon annulatus) – zacate angleton pretoria
Dichanthium aristatum (Andropogon nodosum) – popotillo robusto, zacate angleton aristado
Dichelostemma (Brodiaea) capitatum (pulchella) – brodiea, cacomite, coquito de tierra, covena, saita – blue dicks, desert hyacinth, grass-nuts, wild hyacinth
Dichelostemma (Brodiaea) capitatum* ssp *pauciflorum – coveria – blue dicks
Dichondra argentea – arejuela de ratón, golondrina, oreja de ratón, orejuela de ratón – dichondra
Dicliptera resupinata – alfalfilla, rama de toro – false mint
Dicliptera (Jacobinia, Justicia) suberecta – jacobina
Dicranopteris pectinata – helecho
Dictamnus albus – díctamo albo, díctamo blanco, díctamo real, fraxinella, fresnillo – dittany, fraxinella, gas plant
Didymaea mexicana – ocosúchil, ocoxóchitl, trébol del monte
Dieffenbachia amoena – amoena, caña muda, difenbaquia, galatea, rabano – dumb cane
Dieffenbachia maculata (picta) – difenbaquia manchada – dumb cane
Digitalis ferruginea – dedalera, digital ferrogínea – rusty foxglove
Digitalis grandiflora (ambigua) – digital amarilla grande – yellow foxglove
Digitalis lanata – dedalera, digital lanosa – Grecian foxglove
Digitalis lutea – digital amarilla – small yellow foxglove
Digitalis obscura – corrigia, crujía, digital negra, estepilla – dark foxglove, narrow-leaf foxglove
Digitalis purpurea – chupamiel, dedalera, digital, digital morada, digital purpúrea, giloria, gualdaperra, guante de Nuestra Señora, guantelete, hierba de jarazo, viluria – digitalis, common foxglove, foxglove
Digitalis thapsi – giloria – foxglove
Digitaria (Trichachne) californica – plumero blanco, zacate punta blanca – California cottontop
Digitaria ciliaris – digitaria ciliada, pasto de gallina, pata de gallo, zacate cangrejo velludo – fringed crabgrass
Digitaria decumbens – hierba pangola, pangola, pasto pangola, zacate pangola – pangola grass
Digitaria (Trichachne) insularis – camalote, plumerillo café, zacate bobo, zacate mano

punta café, zacate taiwan – cotton grass, sourgrass
Digitaria ischaemum – digitaria glabra, garrachuelo pequeño – smooth finger grass
Digitaria (Trichachne) patens – zacate mano texano
Digitaria pentzii* var *minor – zacate mano pangola
Digitaria sanguinalis – digitaria común, garrachuelo, gramilla zacate, granadilla, granadilla de cuaresma, hierba de la vaca macho, hierba del cangrejo, paja colchón, pasto blanco, pata de gallina, plumero áspero, zacate agrio, zacate cangrejo velludo – crabgrass
Digitaria villosa – zacate cangrejo piloso
Dillenia indica (speciosa) – dilenia, hondapara – elephant apple, Indian dillenia, Indian roseapple
Dimorphotheca sinuata (aurantiaca) – caléndula del Cabo, dimorfoteca – African daisy, Cape-marigold
Dion = Dioon
Dionaea muscipula – atrapamoscas, dionea – Venus fly trap
Dioon edule – amenduai, cabeza de chamal, chamal, jangó, palma de dolores, palma de jango, palma de la Virgen, palma de maceta, sotol, tzamal – chestnut dion
Dioon purpusii – palma de la Virgen, palmita
Dioscorea alata – aquilmacal, bejuco de culebra, cabeza de negro, camote real, iñame, maaxcal, macal, ñame, ñame blanco, ñame común, ñame de agua, ñame de diez meses, ñame grande, ñangate – winged yam
Dioscorea bulbifera – masoco, ñame, papa caribe, papa del aire, papa voladora, volador – aerial yam, air potato
Dioscorea composita – barbasco, barbasco de camote, cabeza de negro, camote de cerro, camote gigante, camote matapecado, corrimiento, tepecamote
Dioscorea convolvulacea (galeottiana) – camote blanco, cuachalalate, madre del maíz
Dioscorea esculenta – diente de perro, tongo – lesser yam
Dioscorea laxiflora – bejuco de biznaga, bejuco de chilillo, falsa cocolmeca, popo, rabo de iguana
Dioscorea mexicana – barbasco, cabeza de negro, cocolmeca – yam
Dioscorea macrostachya – bejuco de coraza, cocolmeca, cocolmecatl
Dioscorea sativa – ñame común – yam
Dioscorea triphylla – mapuey, ñame morado
Dioscorea villosa – aje, hueso del diablo, ñame mexicano silvestre, ñame silvestre – wild Mexican yam
Diospyros californica – chapote, guayparín, zapote, zapote prieto, zapotillo – wild persimmon
Diospyros digyna – mabolo, sahual, tilzápotl, totolcuitlatzápotl, zapote, zapote negro, zapote prieto – black sapote
Diospyros ebenaster – biaqui, ébano, ébano agrio, sirunda urveta, tauchya, tliltzápotl, totolcuitlatzápotl, zapote negro, zapote prieto – black sapote, Indian ebony persimmon
Diospyros ebenum – ébano asiático – macassar ebony
Diospyros ebenus – ébano – ebony
Diospyros (Maba) intricata – zapotillo
Diospyros kaki – caqui, caqui de China, caquilero, guayacana, palo santo, parsimonio, persimo, placaminero – Japanese persimmon, oriental persimmon
Diospyros lotus – diospiro loto, ébano loto, mostellar – date-plum
Diospyros mespiliformis – ébano de Mozambique – west African ebony
Diospyros sonorae – caguo rara, guayparín, guayule – Sonoran persimmon
Diospyros texana – chapote, chapote manzano, chapote prieto, zapote negro, zapote prieto – black persimmon, Mexican persimmon, Texas persimmon
Diospyros virginiana – caqui, caqui americano, caqui de Virginia, diospiro, ébano, persimo, placaminero – American persimmon
Dipholis salicifolia – sak-chum, zapotillo – bustic, jungle plum
Diphysa occidentalis – bolsa de coyote, guicobo, huichuri, huiloche, jusiro, palo nahuila
Diphysa racemosa – flor de iguana, huiloche
Diphysa suberosa – colchol, corcho, juyapori, palo santo

Diplotaxis erucoides – oruga silvestre, tabaniza blanca – wall rocket
Diplotaxis muralis – jaramago, oruga de los muros – wall rocket
Diplotaxis tenuifolia – oruga de hojas tenues – wall rocket
Dipsacus fullonum – cardencha, cardencha común, cardo de batanero, cardo de cardador, cardo de cardar, dipsaco, escobilla, oruga silvestre, raspasayo – wild teasel
Dipsacus sativus – cardencha cultivada, cardo de cardar – fuller's teasel
Dipteryx (Coumarouna) odorata – cumarú, haba tonka, judía tonquin, sarapia, tonka – tonka bean
Disella = Sida
Disocactus (Aporocactus) flagelliformis – cacto cola de rata, canchoh, cola de iguana, cuerno, flor de cuerno, flor de la alferecía, flor de látigo, floricuerno, hierba de la alferecía, huitzocuitlapilli, junco, junquillo, nopalillo – rat-tail cactus
Disocactus martianus (Aporocactus conzattii) – junco
Disocactus (Nopalxochia) phyllanthoides – marta, nopalillo, nopalillo color de rosa, nopalillo de flor, nopalxochicuezaltic, sol de agua
Disocactus schrankii (Heliocereus elegantissimus) – junco, nopalillo, pitahaya de roca, texcaxóchitl, tuna de agua, tuna verde
Disocactus (Heliocereus) speciosus – ahuahuasúchil, ahuahuaxóchitl, cola de gato, flor espinosa, junco rojo, nopalillo, órgano, pitahaya del cerro, pitaya de agua, Santa María, Santa Marta, xoalácatl – sun cereus
Distichlis palmeri – arroz del mar, zacate salado – Palmer grass
Distichlis spicata – hierba del burro, huizapol, tequixquiçacatl, zacate salado, zacahuistle – saltgrass
Distictis buccinatoria (Bignonia cherere, Phaedranthus buccinatorius) – bocina, cinatoria, clarín, hiedra, hiedra bocina, palomita, tonacaxóchitl, trompetilla grande, vid de trompeta mexicana – blood-red trumpet vine, blood trumpet, Mexican trumpet vine
Distictis laxiflora – jeroliz morado, manto de la Virgen – vanilla trumpet vine
Ditaxis heterantha – azafrancillo, azafrán de bolita
Ditaxis tinctoria – azafrán, tinta roja
Dizygotheca = Schefflera
Dodecatheon clevelandii – sarapico – shooting star
Dodecatheon meadia – flor de los dioses, sarapico – eastern shooting star
Dodonaea viscosa (angustifolia) – airía, alamillo, chapuliz, chupuliztle, cucaracha, cuerno de cabra, gitarán, granadina, guachomó, guayabillo, hierba de la cucaracha, huesito, jarilla, jarilla de loma, jirimú, mundito, ocotillo, palomilto, pirimu, tapachile, tarachique, tonalcotl-xihuitl, varal – hopbush, hopseed bush, switch-sorrel
Dolicholus phaseoloides – atecuixtle, colorín chiquito, colorincito, frijol de chintlatlahua, frijolillo, negrito, ojito de pichón, ojo de cangrejo, ojo de chanate, ojo de pajarito, ojo de zanate, peonía, peonía de niños, pipilzintli, pulguita, purensapichu, xanecuilche
Dolichos = Lablab
Dolichothele = Mammillaria
Dondia palmeri – saladilla
Dondia (Suadea) suffrutescens – quelite salado, romerillo, romerito, sosa – sea-blite
Dondia = Suadea
Doronicum cordatum – matalobos de colonia, matalobos de hojas cordiformes – leopard's bane
Doronicum pardalianches – matalobos de flores compuestas – leopard's bane
Doronicum plantagineum – dorónico, matalobos de hojas de llantén – plantain leopard's bane
Dorstenia contrayerba – barbudillo, cabalahauche, cohuanenepili, contrahierba, cresta de gallo, hierba del sapo, hierba santa, lombrecina, matadolor, pata de gallo, tuzpatli – torus herb
Dorstenia drakeana – baiburilla, baiburina, barbasco, barbudillo, contrahierba, hierba del pasmo, olotillo, pionilla, tuzpatli
Dovyalis caffra – kei – kei apple
Dovyalis hebecarpa – kitembilla – Ceylon gooseberry

Doxantha = Macfadyena
Draba aizoides – draba de verdura perenne – yellow whitlow-grass
Draba cuneifolia (platycarpa) – gasa, sanguinaria menor – whitlow-grass
Draba nemorosa – draba de los bosques – wood whitlow-grass
Draba verna – draba primaveral – spring whitlow-grass
Dracaena americana – campanilla, izote del monte – candlewood
Dracaena deremensis – dracena
Dracaena draco – drago, drago de Canarias, dragonal de Canarias, palma del desierto, sangre de dragón – dragon tree
Dracaena fragrans – árbol de la felicidad, dracena verde, palo de Brasil, palo de la felicidad – corn plant
Dracaena marginata – dracena tricolor, drago de Madagascar – Madagascar dragon tree, Madasgascar plant
Dracaena reflexa – pleomele – song of India
Dracaena sanderana – dracena sander, muñeco, sanderiana – ribbon plant, song of India
Dracaena = Cordyline
Dracunculus vulgaris – culebrera, culebrillo, dragoncillo, dragoneta, serpentaria, zaragutia – dragonwort, green dragon
Drejella = Justicia
Drimys winteri – aktarcin, canelo, canelo de invierno, canelo de paramo, casca de anta, chachaca, chilillo, corteza de pimienta, falso canelo de Sudamérica, palo de chile, palo picante, paramo – winter's bark
Drosera anglica – drosera mayor – English sundew, long-leaved sundew
Drosera rotundifolia – atrapamoscas, drosera, drosera de hojas circulares, hierba de la gota, rocío del sol, rosolí – dew plant, round-leaved sundew
Dryas octopetala – cariofilada de montaña, dríada de ocho pétalos, dríada octopétala
Drymaria arenarioides – alfombrilla – lightningweed
Drymaria glandulosa – encaje
Drymaria gracilis – cadenilla, hueypali
Drymaria leptophylla* var *nadosa – flor de peña
Dryopteris arguta (Aspidium rigidum* ssp *argutum) – helecho, hierba del golpe – California wood fern, coastal wood fern, wood fern
Dryopteris (Aspidium, Polypodium, Polystichum) filix-mas – dentabrón, helecho macho – male fern, shield fern
Dryopteris karwinskyana – chinita
Drypetes gentryi – cortopico, joyarí, palo verde
Duchesnea indica – fresa silvestre, frutilla – Indian mock strawberry, mock strawberry
Dudleya acuminata – siempreviva
Dudleya albiflora – siempreviva
Dudleya anthonyi – gallina con pollos – stonecrop
Dudleya attenuata – siempreviva – live-forever
Dudleya attenuata* ssp *orcuttii – siempreviva – Orcutt's dudleya
Dudleya lanceolata – echeveria – dudleya
Dudleya nubigena – siempreviva
Dudleya pulverulenta – siempreviva – chalk dudleya, chalk-lettuce, live-forever
Dugaldia (Helenium) hoopesii – hierba del lobo, raíz del lobo – orange sneezeweed, owl claws, sneezeweed
Dugesia mexicana – alchicolia, hierba del puerco, raíz del suelo
Duranta erecta (repens) – adonis blanco, adonis morada, azota caballo, celosa, celosa cimmarona, cólera de novio, cuenta de oro, duranta, espino blanco, espino de paloma, espino negro, flor de cielo, fruta de iguana, garbancillo, hombocoche, kanppocoche, lluvia, pukil, velo de novia, velo de viuda, zarza – forget-me-not tree, golden dewdrop, pigeonberry, sky flower, Yucatan sky flower
Durio zibethinus – durián, durio, durión – doorian, durian
Dyschoriste decumbens – hierba de la víbora – snake herb, spreading silverleaf
Dyschoriste hirsutissima – sciatica
Dyssodia (Thymophylla) acerosa – contrahierba, hierba del arriero, hierba de San Nicolás
Dyssodia anomala – mirasol chiquito, turasali
Dyssodia chrysanthemoides – anisillo, flor de muerto
Dyssodia concinna – manzanita de cogote

Dyssodia papposa – flor de muerto, pagué – cinchweed, fetid marigold

Dyssodia = Thymophylla

E

Ecballium elaterium – calabacilla, calabacilla amarga, chombrillo amargo, elaterio, espantamuchachos, pepinillo del diablo, pepino de lagarto – squirting cucumber
Echeandia brevifolia – azucena del monte
Echeveria agavoides – echeveria
Echeveria elegans – conchita, flor de piedra, tememetlatl – hen-and-chicks
Echeveria gibbiflora – oreja de burro, tememetlatl – lucky plant
Echeveria glauca – conchita
Echeveria mucronata – siempreviva
Echeveria pulverulenta – conchita, orejita, rueda de la fortuna
Echeveria rusbyi – siempreviva
Echeveria secunda – conchita, flor de peña, oreja de ratón, siempreviva – hen and chicks
Echeveria simulans – hiedra del monte, hiedra grande, siempreviva
Echeveria subrigida – oreja de burro
Echinacea angustifolia – cabeza de negro, equinacia, equiseto, flor de cono, rudbekia, Sansón negro – echinacea, handleleaf coneflower, purple coneflower
Echinacea (Rudbeckia) purpurea – cabezuela purpúrea, equinacia, equiseto – black eyed Susan, black Sampson, echinacea, purple coneflower
Echinocactus grusonii – asiento de la suegra, peyote – golden barrel cactus
Echinocactus horizonthalonius – biznaga, biznaga de dulce, biznaga meloncillo, manca caballo, manca mula – blue barrel cactus, devil's head, eaglesclaw cactus, Turk's head cactus
Echinocactus platyacanthus (grandis, ingens, palmeri, visnaga) – acitrón, biznaga, biznaga burra, biznaga de acitrón, biznaga de dulce, biznaga de lana, biznaga gigante – giant barrel cactus
Echinocactus polycephalus – biznaga de chilitos – clustered barrel cactus, cottontop cactus
Echinocactus (Homalocephala) texensis – asiento de la suegra, biznaga, biznaga burra, manca, manca caballo – devil's head, devil's pincushion, horse-crippler cactus
Echinocactus = Ferocactus
Echinocereus barthelowanus – pitayita – hedgehog cactus
Echinocereus berlandieri (blanckii) – alicoche – pickle cactus
Echinocereus brandegeei – casa de rata, pitahayita, pitayita – strawberry cactus
Echinocereus cinerascens – alicoche, cocuá, organito – lily cactus
Echinocereus dasyacanthus – pitaya, pitaya amarilla – Texas rainbow cactus, Texas rainbow hedgehog, yellow-flowered pitaya
Echinocereus engelmannii – cacto fresa, pitayita, sinita barbona – hedgehog cactus, strawberry cactus
Echinocereus enneacanthus (merkeri) – alicoche, pitaya – purple pitaya, strawberry hedgehog cactus
Echinocereus fasciculatus – pitahayita – cockscomb hedgehog, strawberry hedgehog
Echinocereus ferreirianus – pitayita – Lindsay's hedgehog cactus
Echinocereus grandis – pitayita – hedgehog cactus
Echinocereus knippelianus – peyote verde
Echinocereus maritimus – pitayita
Echinocereus papillosus (texensis) – alicoche – alicoche, yellow alicoche
Echinocereus pectinatus – pitaya – rainbow cactus, rainbow hedgehog
Echinocereus (Morangaya) pensilis – clavellina, pitayita – hanging cactus, snake cactus
Echinocereus pentalophus – alicoche, cardoncillo – lady finger cactus
Echinocereus polyacanthus* ssp *pacificus (E. pacificus) – huevo de toro, pitahaya, pitaya, pitayita – claret-cup cactus, spiny hedgehog
Echinocereus (Wilcoxia) poselgeri – sacasil, zacoxóchitl – dahlia cactus, lead pencil cactus
Echinocereus pulchellus – pitaya
Echinocereus rigidissimus (pectinatus* var *rigidissimus) – cabeza de viejo, cacto arcoiris – Arizona rainbow hedgehog cactus
Echinocereus rosei – pitahaya
Echinocereus stoloniferus – agí
Echinocereus stramineus (conglomeratus) –

agosteña, alicoche, órgano, pitahaya, pitahaya de agosto, pitaya, sanjuanero – spiny strawberry hedgehog, strawberry cactus
Echinocereus subinermis (luteus) – agí, agui
Echinocereus triglochidiatus – hikuri, pitayita – claret cup, spineless hedgehog
Echinocereus viridiflorus – pitaya, pitaya verde – green hedgehog, New Mexico rainbow cactus, nylon hedgehog cactus
Echinocereus = Bergerocactus
Echinochloa colona – arrocillo, arroz del monte, paja americana, pierna de gallo meridional, zacate de agua, zacate pinto, zacate rayado, zacate tigre – jungle ricegrass
Echinochloa crus-galli – grama morada, hualcacho, mijo de los arrozales, pasto alemán, pasto mijillo, pata de gallo, pie de gallina, pierna de gallo, zacate de agua, zacate de corral, zacate de manati – barnyard grass
Echinochloa crus-pavonis – teozintle, zacate camelote
Echinochloa frumentacea – barba de indio, mijo de cuadra, mijo de los arrozales, mijo japonés, pierna de gallo, zacate de agua – billion dollar grass, Japanese millet
Echinochloa oryzoides – mijo de los arrozales, pasto del arroz, pierna de gallo mayor – Asian paddy weed
Echinochloa polystachya – alemán, arrocera, hierba de cayena, hierba de corso, janiero, pasto alemán, pasto de Angola, zacate alemán, zacate arrocillo, zacate de Angola
Echinocystis = Marah
Echinofossulocactus = Stenocactus
Echinomastus erectocentrus – acuña, biznaguita – acuña cactus, red pineapple cactus, purple-spined viznagita
Echinomastus intertextus – biznaguita – white-flowered viznagita, woven-spine pineapple cactus
Echinomastus johnsonii – biznaguita – beehive cactus, Johnson's pineapple cactus, pink viznagita
Echinops exaltatus (ritro) – bola de añil, cardo erizo, cardo globoso, cardo multicolor, cardo yesquero – globe thistle, southern ball thistle
Echinops sphaerocephalus – cardiduca, cardimea, cardo erizo, planta de la abeja – globe thistle
Echinops strigosus – cardo yesquero
Echinopsis (Cereus) chiloensis – cacto, cirio de Chile, quisco – rain-stick cactus
Echinopsis (Trichocereus) pachanoi – achuma, aguacolla, gigantón, huachuma, San Pedro – San Pedro cactus
Echium lycopsis – chupamiel, viperina falso-llantén – bugloss
Echium plantagineum – lengua de buey – viper's bugloss
Echium vulgare – hierba de la víbora, viborera, viborera común, viperina, viperina común – viper's bugloss
Eclipta prostrata (alba) – chile de agua, epazotillo, hierba de tajo, soguilla, zarzaparrilla – false daisy
Ehretia anacuna – anacua, anama, yara – knackaway, sugarberry
Ehretia elliptica – anacahuite, anacua, cacala, coposo, manzanilla, manzanita, raspa sombrero, tlalahuacate – sugarberry
Ehretia latifolia – tepoyán
Ehretia tinifolia – beek, capulín cimarrón, frutillo, malinche, manzano, manzanita, nandimbo, pingüico, quilalinacate, roble, saúco
Eichhornia crassipes – buchón de agua, camalote, carolina, choreja, cucharilla, flor de huachinango, huauchinango, jacinto, jacinto acuático, jacinto de agua, lampazo, lechuga de agua, lirio, lirio acuático, lirio de agua, lirio japonés, papalacate, pico de pato, reina, reina del río, tamborcillo, violeta de agua – water hyacinth
Elaeis guineensis – corojo de Guinea, palma africana, palma de aceite, palmera de aceite – oil palm
Elaeis melanococca – coquito de aceite, corozo oleifera, nolí – American oil palm
Elaphrium = Bursera
Eleagnus angustifolius – árbol del paraíso, eleagno de Bohemia, olivo de Bohemia, paraíso – Jerusalem willow, oil tree, oleaster, Russian olive
Eleagnus philippensis – alingaro
Eleagnus pungens – cinamomo – silverberry

Eleocharis (Heleocharis) acicularis – eleocharis acicular, tule, tulillo – spikerush
Eleocharis (Heleocharis) atropurpurea – basalaca, eleocharis roja subida, zacate de agua, zacate de burro, zacatito del agua – dark blood-red spikerush
Eleocharis dulcis – cabeza de negrito, castaña de agua – water chestnut
Eleocharis (Heleocharis) geniculata – chintule, junco, tulillo – spikerush
Eleocharis (Heleocharis) macrostachya (palustris) – tule, tulillo – spikerush
Eleocharis (Heleocharis) montevidensis – capiro – spikerush
Eleocharis parvula – eleocharis enana – small spikerush
Eleocharis rostellata – tulillo – traveling spikerush
Elettaria (Amomum) cardamomum – cardamomo, grana del paraíso – cardamom
Eleusine coracana – coracán, eleusina, mijo coracano – coracan, finger millet
Eleusine coracana* ssp *africana – mijo africano – African millet
Eleusine indica – cola de gallo, grama de caballo, grama de camino, grama dulce, grama pata de gallina, guaratara, hierba del paisano, hierba dulce, matojo dulce, pasto dulce, pata de gallina, pata de gallo, pata de ganso, zacate de ganso, zacate guácima – goose grass, yard grass
Eleutherococcus (Acanthopanax) senticosus – ginseng siberiano, ginseng ruso – Russian ginseng, Siberian ginseng
Elodea canadensis – elodea, elodea del Canadá – common waterweed
Eloeis guineensis – palma avoira
Elymus canadensis – zacate silvestre del Canadá – Canadian wildrye
Elymus elymoides (Sitanion hystrix) – triguillo desértico, zacate cebadilla – squirreltail
Elymus longifolius (Sitanium longifolium) – cola de ardilla, triguillo desértico – longleaf squirreltail
Elymus (Agropyron) trachycaulus – agropiro – slender wheatgrass
Elyonurus tripsacoides – colita
Elytraria imbricata – cola de alacrán, cordoncillo, hierba del toro
Elytraria squamosa – anisillo, cabalxaan, cordoncillo, nachachícoli, papachilla, pie de gallo, soguilla
Elytraria tridentata – coquillo
Elytrigia arizonicus (Agropyron arizonicum) – agropiro azul, triguillo de Arizona – Arizona wheatgrass
Elytrigia elongata (Agropyron elongatum) – grama alargada – tall wheatgrass
Elytrigia intermedia (Agropyron intermedium) – grama intermedia – intermediate wheatgrass
Elytrigia juncea (Agropyron junceum) – grama de las playas – sand couch-grass
Elytrigia (Agropyron) repens – acazacahuiztli, agropiro, canaria, canzuuc, cerrillo, grama, grama canina, grama del norte, grama de prados, grama repente, hierba estañera, hierba rastrera, pie de gallina – quackgrass
Elytrigia stebbinsii (Agropyron parishii) – agropiro de Parish
Elytrigia trachycaulus (Agropyron trachycaulum) – agropiro delgado, triguillo largo – slender wheatgrass
Embothrium coccineum – arbusto de fuego chileno, embotrio – Chilean fire bush
Emmenanthe penduliflora – campanita de suspiro – whispering bell
Empetrum nigrum – camarino negro – black crowberry
Encelia californica – incienso
Encelia farinosa – chamizo, hierba ceniza, hierba de incienso, hierba de la ánima, hierba del bazo, hierba del vaso, incienso, palo blanco, rama blanca – brittlebush, incienso, white brittlebush
Encelia mexicana – acahual, girasol, pegarropa
Encelia palmeri – mirasol
Encyclia citrina – azucena amarilla, limoncito, olóricua
Encyclia cochleata – am, flor de concha, pulpito
Encyclia = Epidendrum
Endymion = Hyacinthoides
Enkianthus campanulatus – enovianto, enquianto

Enneapogon desvauxii – zacate cola de zorra, zacate ladera, zacate lobero – nine-awned pappus grass
Ensete ventricosum (Musa ensete) – guineo, plátano, plátano de Abisinia, plátano gigante – Abyssinian banana, plantain
Entada phaseoloides – alampepe, barbasco, bejuco de jaba, bejuco de tacalote, chupapaya, cupayaya, haba, haba de la costa, tacalote – sea-bean
Entada polystachya – alampepe, bejuco de agua, bejuco de amole, bejuco de estribo, bejuco de mondongo, bejuco de panune, bejuco prieto, cepillo, haba de la costa, haba grande, tacalote – snuffbox bean
Entada scandens – alampepe, haba, haba de la costa, peseta del diablo, tacalote – match-box bean
Enterolobium cyclocarpum – aguacasle, caracaro, caro, cascabel, cascabel sonaja, conacaste, cuaunacaxtle, guanacaste, guanacastle, huanacaxtle, juana costa, nacaste, nacastillo, nacaxtle, oreja de Judío, oreja de mono, oreja de mulato, oreja de vaina, orejera, orejón, parota, parote, perota, pich, timbó – eardrop tree, elephant's ear
Ephedra antisyphilitica – cañutillo, popote, tepopote – clap-weed
Ephedra aspera – cañutillo, grangene, hintimorreal, itamo real, pitamo real, popotillo, sanguinaria, té mormona, tepopote – boundary ephedra
Ephedra californica – cañutillo, popotillo, té mormona, teposote – desert tea
Ephedra compacta – cañutillo, comida de víbora, itamo real, pitamo real, popotillo, retama real, sanguinaria, tepopote
Ephedra nevadensis – cañutillo, té brigham, té mormona – Mormon tea, Nevada joint fir
Ephedra pedunculata – cañutillo, itamo real, popotillo, retama real, sanguinaria, tepopote
Ephedra sinica – belcho, efedra, efedrina, mahuang – Chinese ephedra, mahuang
Ephedra torreyana – cañutillo, popotillo, tepopote – Brigham tea, Mexican tea, Mormon tea
Ephedra trifurca – cañutillo, cañutillo del campo, hierba de la coyuntura, itamo real, popotillo, tepopote – joint fir, Mexican tea
Ephedra viridis – cañutillo, efedra verde, popotillo, té brigham, té mormona – green ephedra, green Mormon tea, Indian tea, joint fir
Epicampes = Muhlenbergia
Epidendrum atropurpureum – flor de candelaria, flor de San Sebastián
Epidendrum (Encyclia) erubescens – güitzl – blushing epidendrum
Epidendrum ibaguense – boca de fuego, rancho viejo – fiery reed orchid, reed stem epidendrum
Epidendrum mariae – María
Epidendrum nemorale – trompillo
Epidendrum parkinsonianum (falcatum) – bayoneta, flor de bayoneta, garza real, lirio pata de gallo, lirio pata de paloma, pata de gallo, pata de paloma – aloe-leaved epidendrum
Epidendrum pastoris – amatzauhtli, papiro, saucle, tzacouhxóchitl, tzacutli
Epidendrum verrucosum – chautle
Epidendrum vitellinum (Encyclia vitellina) – cabellera, Manuelito, Miguelito, etoalxóchitl
Epilobium anagallidifolium – adelfilla de hojas de anagálida – alpine willow herb, pimpernel willow herb
Epilobium angustifolium – adelfilla, adelfilla de hojas estrechas, laurel de San Antonio – blooming sally, red fireweed
Epilobium brachycarpum (paniculatum) – campamocha – parched fireweed
Epilobium canum (Zauschneria californica) – balsamea, bálsamo – California fuchsia, zauschneria
Epilobium ciliatum – adelfilla ciliada, adelfilla glandulosa – American willow herb
Epilobium hirsutum – adelfilla hirsuta, adelfilla pelosa, hierba de San Antonio, laurel de San Antonio – great willow herb
Epilobium lactiflorum – adelfilla de flores blancas – white-flowered willow herb
Epilobium palustre – adelfilla de los pantanos – marsh willow herb
Epimedium alpinum – epimedio alpino – alpine barrenwort, bishop's hat
Epimedium grandiflorum – epimedio – bishop's hat, longspur epimedium

Epipactis helleborine – epipactis común – broad-leaved helleborine
Epiphyllum anguliger – jara de pitayita, jarana de pitahayita, nopalillo, nopalillo blanco, nopalillo de olor, pitahaya, pitahayita del cerro, pitayita del cerro – moon cactus, queen-of-the-night
Epiphyllum hookeri (strictum) – Santa Rita
Epiphyllum oxypetalum – dama de noche, junco oloroso, nopal de lengüita, pitayita de agua, reina de la noche – Dutchman's pipe cactus, orchid cactus
Epiphyllum truncatum – nopalillo – orchid
Epiphyllum x _Disocactus_ – junco, nopalillo, pitahaya del cerro, xoalacatl, Santa María – orchid cactus
Epipogium aphyllum – epipógio sin hojas – ghost orchid
Epipremnum pinnatum (aureum) – pothos, teléfono – devil's ivy, pothos
Epithelantha bokei – boquillo – Boke's button cactus
Epithelantha micromeris – mulato, tapone – button cactus
Equisetum arvense – cañutillo, cañutillo del llano, cola de caballo, cola de caballo de los campos, cola de rata de los campos, equiseto, equiseto menor – common horsetail
Equisetum hyemale – cañutillo del llano, cola de caballo, cola de caballo de invierno, equiseto, junco, limpia platas, platero – common scouring rush, horsetail
Equisetum laevigatum – cañuela, cola de caballo – horsetail, smooth scouring rush
Equisetum palustre – cola de caballo de los pantanos – marsh horsetail
Equisetum ramosissimum – cañuela, cola de caballo mucho ramoso – Boston horsetail
Equisetum robustum – bejuquillo, cañuela, carricillo, cola de caballo
Equisetum telmateia – cola de caballo, equiseto mayor, lepra caballo, rabo de asno – giant horsetail
Eragrostis barrelieri – eragrostis de Barrelier – lovegrass
Eragrostis chloromelas – Boer – Boer lovegrass
Eragrostis cilianensis – amorseco, amorseco ciliado, eragrostis mayor, pasto hediondo, zacate apestoso – stinkgrass
Eragrostis curvula – amor, amorseco curvado, amorseco llorón, eragrostis llorona, falsa barba de node, llorona, milena, pasto llorón – weeping lovegrass
Eragrostis curvula var _conferta_ – pasto boer – Boer grass
Eragrostis hypnoides – pan caliente – lovegrass
Eragrostis intermedia – zacate llanero – plains lovegrass
Eragrostis lehmanniana – amorseco africano, pasto africano, zacate africano – African grass, Lehmann lovegrass
Eragrostis lugens – amorseco llorón, zacate llorón cimarrón
Eragrostis mexicana – bayal, bayal-suuk, eragrostis mexicana, gramilla de huerto, zacate de agua – Mexican lovegrass
Eragrostis minor – eragrostis menor – lesser lovegrass
Eragrostis obtusiflora – pasto jigüite, zacate jigüite – alkali lovegrass
Eragrostis palmeri – amorseco, amorseco norteño
Eragrostis pilosa – amorseco piloso, barba de indio, eragrostis peluda, jaraguá gris, zacate de amor – India lovegrass
Eragrostis reptans – pan caliente – lovegrass
Eragrostis superba – pasto avena – sawtooth lovegrass
Eranthis hyemalis – acónito invernal, eléboro de invierno – winter aconite
Erblichia odorata – axóchitl, azuche, chamizo, copa de oro, jarro de oro, sanjuanero, suelda consuelda – butterfly tree
Eremalche (Malvastrum) scoparium – alatle, escoba blanca, escobilla, huinare prieto, malva – false mallow
Eremocarpus setigerus – hierba del pescado, hierba del venado – dove weed, turkey mullein
Eremocitrus glauca – lima del desierto
Eremurus robustus – azucena de rabo de zorro, rueca de Cleopatra – desert candle, foxtail lily
Erica arborea – brezo arbóreo, brezo blanco, ud, urce – tree heath, tree heather

Erica australis – brezo colorado, brezo español – southern heath, Spanish heath
Erica carnea (herbacea) – brezo de invierno, brezo encarnado – spring heath
Erica ciliaris – brezo, brezo ciliado, carroncha – Dorset heath
Erica cinerea – argoña, brezo nazareno – bell heather, Scotch heath, twisted heath
Erica gracilis – brezo – heath
Erica lusitanica – brezo portugués – Portuguese heath
Erica mediterranea – brezo mediterráneo – Biscay heath
Erica scoparia – brezo de escobas – broom heath, green heather
Erica tetralix – brezo de hojas en cruz, breza de turbera, carroncha – cross-leaved heath
Erica vagans – brezo, brezo de Cornualles, brezo vagabundo, bruco, lizarra, rugón – Cornish heath
Ericameria (Haplopappus) arborescens – rama hechicera – goldenbush, golden-fleece
Ericameria brachylepis (Haplopappus propinquus) – hierba del pasmo, romerillo amargo – boundary goldenbush
Ericameria (Haplopappus) heterophyllus – hierba de la víbora – jimmyweed, rayless goldenrod
Ericameria laricifolia (Haplopappus laricifolius) – hierba del pasmo – larchleaf goldenweed, turpentine-brush, turpentine bush
Ericameria (Haplopappus) palmeri – pasmore – Palmer's goldenbush
Ericameria (Aplopappus, Haplopappus) sonorensis – hierba de la víbora, hierba del pasmo, paroqui, romerillo amargo
Ericameria (Aplopappus, Haplopappus) spinulosus – árnica flor, hierba de la quintana, hierba de la víbora, romerillo – spiny aster
Erigeron affinis – chalchuán, chilcuán, chilmecatl, pelitre, pelitre del país – fleabane daisy
Erigeron annuus – erigerón anual – fleabane daisy
Erigeron canadensis – hierba del burro – horseweed
Erigeron glaucus – erigerón de color verdemar, erigerón glauco – seaside daisy
Erigeron karvinskianus (mucronatus) – erigerón mucronado, hierba pulguera – Mexican daisy, Santa Barbara daisy
Erigeron longipes – chalchuán
Erigeron modestus – erigerón – plains fleabane, prairie fleabane
Erigeron nudiflorus – zarzilla – naked stem fleabane, running fleabane
Erigeron pubescens – chalchuán, hierba del naciente, manzanilla cimarrona
Erigeron speciosus – hierba pulguera – fleabane, southern daisy
Erigeron strigosus – erigerón ramoso – fleabane daisy
Eriobotrya japonica – nisperero del Japón, níspero, níspero del Japón, níspero japonés – loquat
Eriochloa acuminata (gracilis) – zacate salado, zacate taza – southwestern cupgrass
Eriochloa aristata – zacate taza – bearded cupgrass
Eriochloa contracta – zacate taza – prairie cupgrass
Eriochloa polystachya – hierba caribe, janiero, malojillo, pasto caribe, pasto palojilla, zacate caribe – carib grass
Eriochloa punctata – piojillo peludo
Eriodictyon angustifolium – hierbabuena, hierba santa – mountain balm, narrow-leaved yerba santa
Eriodictyon californicum – hierba santa – yerba santa
Eriodictyon glutinosum – hierba santa
Eriodictyon sessilifolium – hierba santa – yerba santa
Eriodictyon tomentosum – hierba blanca, hierba santa – yerba santa
Eriodictyon trichocalyx (lanatum) – hierba santa – mountain balm, yerba santa
Eriogonum ainsliei – escobita de la Virgen
Eriogonum annuum – hierba colorado – wild buckwheat
Eriogonum fasciculatum – colita de ratón, gordolobo, maderista, matanga, valeriana – California buckwheat, flat-top buckwheat-brush

Eriogonum inflatum – guinagua, té, tibinagua – desert spoon, desert trumpet
Eriogonum nudum – tibinagua – buckwheat
Eriogonum rotundifolium – chachaca – false buckwheat, roundleaf wild buckwheat
Eriogonum rupestre – hierba colorada
Eriogonum tenellum – chacate, chuchaca, flor de novia
Erioneuron (Tridens) pulchellum – falso tridente borreguero, zacate borreguero – fluff grass
Eriosema grandiflorum – guayabillo, hierba del duende, igualamilla, jarilla de la sierra
Erodium botrys – pico de garza arracimado – storksbill
Erodium brachycarpum – pico de garza graquicarpo – storksbill
Erodium cicutarium – aguja del pastor, agujita, alfilerillo, alfilerillo de pastor, hierba de chuparrosa, hierba de la coralina, peine de bruja, pico de ciqüeña – filaree, pin-clover, storksbill
Erodium malacoides – cigüeña malva – soft storksbill
Erodium moschatum – aguja, aguja del pastor, agujita, alfilerillo, almizcleña, almizclera, hierba musgo – filaree, musk clover, storksbill
Erodium reichardii – pico de garza de Reichard – cranesbill
Erodium texanum – alfilerillo – heronbill, Texas filaree, Texas storksbill
Erophila verna – draba primaveral, erofila temprana – whitlow grass
Eruca vesicaria (sativa)* ssp *sativa (Brassica eruca) – chipiquelite, cruz de pasto, eruca, eruca roqueta, jaramago, mostacilla, nabo, oruga, oruga común, oruga roqueta, roqueta, roqueta común, saramago – arugula, garden-rocket, rocket-salad, roquette, salad-rocket
Erucastrum gallicum – jaramago francés – French bastard rocket
Eryngium alpinum – eringe alpinum – alpine sea holly, donard
Eryngium amethystinum – cardo corredor, eringe de color de amatista – amethyst eryngium, sea holly
Eryngium campestre – barba cabruna, cardo corredor, corredor, eringe, eringe campestre, setero – field eryngo
Eryngium carlinae – hierba del sapo, mosquita, saibari – button-celery
Eryngium comosum – escorzonera, hierba del sapo, ocopiaxtli, piñata, raíz del sapo
Eryngium foetidum – cilantro, culantro, hierba del sapo, hoitzcolotli, tlipoton – culantro, eryngo, wild coriander
Eryngium giganteum – cardo corredor, eringe marfilado – Miss Willmott's ghost
Eryngium hemsleyanum – babadai, hierba del sapo
Eryngium heterophyllum – hierba del sapo – Mexican thistle
Eryngium lemmoni – hierba del sapo – button snakeroot, eryngo
Eryngium longifolium – chinaca
Eryngium maritimum – cardo corredor marino, cardo de mar, cardo marítimo, eringe marítimo – sea holly
Eryngium nasturtiifolium – hierba del sapo, perejililo
Eryngium planum – eringe de hojas aplanadas – false sea holly
Eryngium rosei – espíritu desilachado
Eryngium wrightii – hierbo del sapo
Erysimum capitatum – hierba de Apache – western wallflower
Erysimum cheiranthoides – erisimo común, erisimo de los campos – wormseed mustard
Erysimum (Cheiranthus) cheiri – alhelí, alhelí amarillo – English wallflower, wallflower
Erysimum hieraciifolium (alpinum) – alhelí alpino, alhelí de Siberia – Siberian wallflower
Erysimum repandum – erisimo repando – wallflower
Erythea (Brahea) aculeata – cohuehui, palma, palma blanca, palmilla – Sonoran hesper palm
Erythea (Brahea) armata – palma azul, palma blanca, palma ceniza, palmera azul – blue palm, blue fan palm, Mexican blue palm
Erythea (Brahea) brandegeei – babisco, coguegue, palma, palma colorada, palma de taco, palma de Tlaco, palma negra, palma palmilla – San Jose hesper palm
Erythea (Brahea) edulis (Brahelia dulcis) – cacaiste, capulín, cocaiste, guano de

sombrero, ixhuhuatl, izote, izhuate, palma, palma apache, palma ceniza, palma de abanico, palma de Guadalupe, palma de matón, palma popotla, palma de sombrero, palma dulce, palma soyal, palmera de Guadalupe, palmito, pima, pumu, soyal, soyal capulín, soyale, soyate, yagaxiña, yucuteyeye, yutnuñun – Guadalupe island palm

Erythea (Brahea) nitida – babisco, jabehui

Erythraea = Centaurium

Erythrina americana – alcaparra, chacmolché, chilocate, chocolín, colorín, coral, cozquelite, equimite, jiquimite, madre brava, madre chontal, madre de cacao, papal, patol, pichoco, pito, purenchecua, pureque, quemite, tzinacancuáhuitl, tzompancuáhuitl, tzompomitl, xoyo, zompantle – coral tree

Erythrina caffra – árbol coral – coral tree

Erythrina coralloides – colorín, pito, purenchecua, pureque – naked coral tree

Erythrina crista-galli – árbol del coral – cockspur coral tree, common coral tree, cry baby, fireman's helmet

Erythrina flabelliformis – aphosi, chijol, chilicote, colcho, colorín, corcho, guaposi, jévero, peonía, pionilla, tzinacancuáhuitl, xololco, zompantle – Indian bean, southwest coral bean, western coral bean,

Erythrina glauca – chamburo, madre alcaparra, madre blanca, madre chontal, madre de agua, madre mansa, madre prieta, moté – coral tree

Erythrina leptorhiza – cochizquilitl, cococha, colorín negro, patol, patolli

Erythrina occidentalis – chilicote, colorín, peonía

Erythronium americanum – lengua de víbora – adder's tongue

Erythronium dens-canis – diente de perro, violeta de diente de perro – dog-tooth violet

Erythroxylon = Erythroxylum

Erythroxylum coca – coca, cocaína, haya – cocaine, coca plant, huanuco cocaine tree

Erythroxylum mexicanum – acusá, agusá, coca de México, escobilla, mamoa, matanene hembra, palillo, palo chino, pata de pájaro, pie de pájaro, zapotillo

Erythroxylum novagranatense – coca, cocaína – cocaine, coca plant, Peruvian cocaine tree

Erythroxylum truxillense – coca, cocaína – cocaine, coca plant, Trujillo cocaine tree

Escallonia illinita – escalonia – white escallonia

Eschscholzia californica – ama-flores, amapola, amapola amarilla, amapola de California, calce de oro, copa de oro, dormidera, toronja, torosa – California poppy

Eschscholzia glyptosperma – copa de oro – goldpoppy

Eschscholzia mexicana – amapola, amapola amarilla, amapola del campo – Mexican poppy

Eschscholzia minutiflora – amapola – pygmy poppy

Eschscholzia palmeri – amapola – poppy

Eschscholzia parishii – amapola – poppy

Eschscholzia ramosa (elegans) – amapola – island poppy

Escobaria (Coryphanta) vivipara – estrella de la tarde – beehive cactus

Escontria chiotilla – chiotilla, jiotillo, xixuega

Esenbeckia flava – palo amarillo

Esenbeckia hartmanii – jócono, mamoa, palo amarillo, panalillo, samota, toshcura – wild orange

Esenbeckia minutiflora – amapolita

Etlingera elatior – boca de dragón – Philippine wax flower, torch ginger

Eucalyptus camaldulensis – eucalipto, eucalipto rojo – red gum, river red gum

Eucalyptus fibrosa – árbol de la fiebre, eucalipto, gigante, ocalo – eucalyptus

Eucalyptus ficifolia – eucalipto, eucalipto rojo – red-flowering gum

Eucalyptus globulus – alcanfor, árbol contra la fiebre, árbol de la fiebre australiano, eucalipto, eucalipto azul, eucalipto blanco, eucalipto común, eucalipto de hojas de salce, eucalipto gigante, gigante, ocalo – blue gum

Eucalyptus gunnii – eucalipto, eucalipto de Gunn, eucalipto de la sidra – cider gum

Eucalyptus pauciflora (niphophila)* ssp *niphophila – eucalipto nival – snow gum

Eucalyptus robusta – eucalipto del alcanfor, eucalipto de pantano – eucalyptus, swamp mahogany

Eucalyptus rostratus – eucalipto – eucalyptus, red gum

Eucalyptus salmonophloia – eucalipto salmón – salmon gum
Eucalyptus sideroxylon – eucalipto – red iron bark
Eucalyptus viminalis – eucalipto del manná – mana gum
Euchlaena (Zea) mexicana – acecé, atzitzintle, huaxcatol, maicillo, maíz correcaminos, maíz de coyote, maíz de diente, maíz primitivo, maíz silvestre, tecointli, teosinte, teozintle – teosinte
Eucnide (Sympetaleia) aurea – pega pega
Eucnide cordata – pega pega
Eucnide (Sympetaleia) rupestris – flor de piedra – rock nettle
Eucommia ulmoides – árbol de la gutapercha – hardy rubber tree
Eucrypta chrysanthemifolia – palo levender
Eugenia (Pimenta) acris – bayrum – bay rum
Eugenia capuli – arrayán, cacho de venado, calarni, capulín, chi-lun-che, chrimis-embra, escobilla, granada cimarrona, guayabillo cimarrón, indio desnudo, rayanillo, vaina de espada, yagalán – blossom berry, walk naked
Eugenia dombeyi – grumichama
Eugenia fragrans – arrayán, arrayán prieto, capulín de hueso, guayabillo, guayabo agrio, pimientillo
Eugenia (Syzygium) jambolana – guayabo pesgua, jambolán, jambul – Java plum
Eugenia (Syzygium) jambos (Jambos vulgaris) – jambo, jamrosade, manzano rosa, manzanito, pomarrosa – plum rose, rose-apple
Eugenia javanica – makopa – Java apple
Eugenia klotzschiana – perado campo
Eugenia (Jambos, Syzygium) malaccensis – jambo, manzano, manzano de agua, manzano de Malaya, marañón japonés, pera de agua, pomagás, pomalaca, pomarrosa americana, pomarrosa del Brasil, pomarrosa de Malaca – Malay apple, Malay rose-apple, mountain apple, otaheite apple, pomerac
Eugenia pimenta (Pimenta dioica, P. officinalis) – malagueta, patalolote, patolote, pimentero, pimentero de Jamaica, pimentero de Tabasco, pimentero de tierra, pimentero dulce, pimentero gordo, pimentero inglés, pimentón, xocoxóchitl – allspice, pimento
Eugenia (Pimenta) racemosa – bayrum, malagueta – bay tree, West Indian bay
Eugenia tomentosa – cabelluda
Eugenia uniflora – cerezo de Cayena, pitanga – Brazil cherry, Cayenne cherry, Florida cherry, pitanga, Surinam cherry
Eugenia = Caryophyllus
Euonymus alatus – bonetero alato – winged euonymus
Euonymus europaeus – bonetero, bonetero común, bonetero de Europa, evónimo, evónimo europeo – spindle tree
Euonymus fortunei – bonetero radicante – spindle-tree
Euonymus japonicus – evónimo, evónimo del Japón – evergreen euonymus
Euonymus japonicus* var *aureo-variegatus – evónimo manchado – variegated euonymus
Euonymus occidentalis – evónimo – evergreen euonymus, western burning bush
Eupatorium collinum – corazón de perro, hierba de chucho, hierba del ángel, yolochichitl, yoloxóchitl – angel herb
Eupatorium deltoideum – árbol María, hierba del ángel, yolochichitl
Eupatorium glabratum – hierba del golpe, hilo, mixtlacotl
Eupatorium hyssopium – hierba de la muela, tabardillo – thoroughwort
Eupatorium maculatum – eupatorio maculado – Joe Pye weed
Eupatorium monanthes – zacate minero
Eupatorium odoratum – cihuapatli, crucetillo, crucito, tokabán
Eupatorium perfoliatum – eupatorio, planta de la fiebre – boneset
Eupatorium quadrangulare – barba de chivato
Eupatorium = Ageratina
Euphorbia adenoptera – golondrina, golondrina serrana
Euphorbia amygdaloides – euforbio falsa-almendra – wood spurge
Euphorbia antisyphilitica – candelilla – candelilla, wax-plant
Euphorbia apicata – golondrina – spurge
Euphorbia brandegeei – golondrina – spurge
Euphorbia californica – jumetón, liga, sipehui – cliff spurge
Euphorbia capitellata – liga – spurge

Euphorbia cedroderma – candelilla
Euphorbia characias – euforbio caracioide, tártago mayor
Euphorbia colletioides – bacachari, candelilla, candelilla bronca, flor blanca, jumete
Euphorbia (Poinsettia) colorata – flor de leche, quanta hierba, tainorej
Euphorbia cotinifolia – mala mujer, matagallinas – Caribbean copper plant
Euphorbia cyparissias – euforbio ciprés – cypress spurge
Euphorbia dentata – hierba de la araña, lechillo, liga – green poinsettia, toothed spurge
Euphorbia dentata (cuphosperma)* var *cuphosperma – congocopac, picachlich
Euphorbia dentonsa – liga – spurge
Euphorbia dulcis – euforbio dulce – sweet spurge
Euphorbia epithymoides (polychroma) – euforbio, euforbio de cabellos de tomillo, euforbio policromo
Euphorbia eriantha – liga – beetle spurge, threaded spurge
Euphorbia esula – euforbio de hojas no numerosas, lechetrenza – leafy spurge
Euphorbia fulgens – chapizlillo, corona de Cristo, euforbio brillante, euforbio fulgente, lechetrenza – scarlet plume
Euphorbia griffithii – euforbio del Himalaya
Euphorbia helioscopia – euforbio helioscopo, lecherina, lecheruela, lechetrezna, solar – sun spurge
Euphorbia (Poinsettia) heterophylla – catalina, catarina, chulex-xiu, contrahierba, flor de navidad, hobon-k'aax, liga, nela, picachlich, tianguis – false poinsettia, Mexican fire plant, painted spurge, redhead
Euphorbia hirta – golondrina, golondrina grande, hierba del pollo, sacateta
Euphorbia hyssopifolia – liga – spurge
Euphorbia incerta – liga – spurge
Euphorbia incisa – liga – Mohave spurge
Euphorbia indivisa – hierba de la golondrina
Euphorbia lactea – candelabro, cardón
Euphorbia lathyris – catapucia menor, hierba topera, lechetrenza, piñoncillo, tártago – caper spurge, gopher plant, mole plant
Euphorbia magdalenae – liga – spurge
Euphorbia milii (splendens) – corona de Cristo, corona de espinas, espino de Cristo, euforbio brillante – Christ thorn, crown-of-thorns
Euphorbia misera – jumetón, liga, sangrengrado, tacora – cliff spurge
Euphorbia myrsinites – euforbio de font quer – spurge
Euphorbia palustris – euforbio de los pantanos, lechetrenza palustre – marsh spurge
Euphorbia peninsularis – liga – spurge
Euphorbia peplus – euforbio menor, hierba del coyote – petty spurge
Euphorbia petrina – golondrina
Euphorbia plicata – candelaria china, candelilla china, candelilla de palo, jumete
Euphorbia polychroma – euforbio de cabellos de tomillo, euforbio policromo – polychrome spurge
Euphorbia pulcherrima – bandera, bebeta, catalina, catarina, cuitlaxóchitl, flor de fuego, flor de navidad, flor de nochebuena, flor de pascua, flor de Santa Catalina, flor de Santa Catarina, guacamaya, listoncillo, nochebuena, paño de Holanda, paño holandés, papagayo, pascua, pastora, poinsetia, rosa del pastor, Santa Catarina, uanipeni – poinsettia
Euphorbia rigida – euforbio rígido – spurge
Euphorbia seguierana – euforbio de seguier – spurge
Euphorbia serrata – euforbio serrado – saw-toothed spurge
Euphorbia tomentulosa – liga – spurge
Euphorbia xantii – jumetón, liga – cliff spurge
Euphorbia = Chamaesyce
Euphrasia officinalis – eufrasia, eufrasia común – eyebright
Eurotia = Krascheninnikovia
Eustoma exaltatum – cimarrón – catchfly gentian
Eustoma grandiflorum – lira de San Pedro – bluebell, showy prairie gentian
Euterpe precatoria – jal-au-té – mountain cabbage palm
Evolvulus alsinoides – fulgencia, ojitos azules, ojo de víbora – dwarf morning-glory
Evolvulus alsinoides* var *angustifolia – oreja de ratón
Evolvulus prostrates – huevo de rana

Exacum affine – violeta persa – Persian violet
Exogonium bracteatum (Ipomoea bracteata) – azalea de la barranca, bejuco blanco, bejuco de la candelaria, camote de Tehuántepec, carnestolenda, cozcanantzi, empanada, empanadilla, empanadita, empanochilla, flor de la candelaria, gallinita del cerro, jícama, jícama del monte, macazpipilloli, papada de gallo
Exogonium = Ipomoea
Exostema caribaeum – copalchi, copalchi de Jojutla, copalchi de Michoacán, falsa quina, huesillo, jocotillo de cerro, quino de Michoacán, sabakché – princewood
Eysenhardtia orthocarpa – koksigam, matariqui, palo dulce – kidneywood
Eysenhardtia polystachya – coate, coatillo, coatli, cuate, cuatle, leño nefrítico, palo azul, palo cuate, palo dulce, palo dulce blanco, rosilla, sipariqui, taray, taray de México, tlapahuaxpatli, tlapalezpatli, urza, vara dulce, varaduz – kidneywood
Eysenhardtia spinosa – chapanillo
Eysenhardtia texana – palo dulce, vara dulce – bee-brush, Texas kidneywood

F

Fagonia cretica – risa de la Virgen
Fagopyrum esculentum – alforfón, alforfón común, trigo mariso, trigo negro, trigo sarraceno – buckwheat
Fagus grandifolia – haya, haya americana, totolcal – American beech
Fagus sylvatica – haya, haya común, haya roja – European beech
Fagus sylvatica* var *purpurea – haya cobriza, haya roja – copper beech
Fallopia baldschuanica (Polygonum aubertii) – polígono de Aubert, polígono de Bokhara – lace vine, silver lace vine
Fallopia = Polygonum
Fallugia paradoxa – poñil – Apache plume, ponil
Fatsia japonica (Aralia japonica, A. sieboldii) – aralia, falso ricino, fatsia, planta del aceite de castor – Japanese aralia
Feijoa (Acca) sellowiana – feijoa, guayabo del Brasil, guayabo-piña – feijoa, pineapple guava
Fendlerella utahensis – hierba desierto – yerba desierto
Ferocactus alamosanus – biznaga, teocomite – barrel cactus
Ferocactus chrysacanthus – biznaga – barrel cactus
Ferocactus cylindraceus (Echinocactus acanthodes) – biznaga – barrel cactus, California barrel cactus, compass barrel cactus, red barrel cactus
Ferocactus cylindraceus (lecontei)* ssp *lecontei – biznaga – California barrel cactus
Ferocactus diguetii – biznaga – barrel cactus
Ferocactus echidne – biznaga
Ferocactus emoryi (covillei) – biznaga – barrel cactus
Ferocactus emoryi (rectispinus)* ssp *rectispinus – biznaga – barrel cactus
Ferocactus fordii – biznaga – barrel cactus
Ferocactus gracilis – biznaga, biznaga colorada – fire barrel cactus
Ferocactus gracilis (coloratus)* ssp *coloratus – biznaga colorada
Ferocactus gracilis (gatesii)* ssp *gatesii – biznaga
Ferocactus hamatacanthus – biznaga costillona, biznaga de tuna, biznaga espinosa, biznaga ganchuda, biznaga limilla, limilla – Turk's head, viznaga barrel cactus, whiskered barrel cactus
Ferocactus histrix (melocactiformis) – biznaga costillona
Ferocactus johnstonianus – biznaga – barrel cactus
Ferocactus latispinus – biznaga de chilitos, biznaga ganchuda, pocha, uña de águila, uña de gato
Ferocactus macrodiscus – biznaga caballona, biznaga de dulce
Ferocactus peninsulae – biznaga – barrel cactus
Ferocactus pilosus (pringlei, stainesii) – biznaga colorada, biznaga de lima – Mexican lime cactus
Ferocactus robustus – piñita
Ferocactus townsendianus – biznaga – barrel cactus
Ferocactus viridescens – biznaga, huitznáhuac, sisal – coast barrel cactus, keg cactus
Ferocactus (Echinocactus) wislizeni – biznaga, biznaga de agua, biznaga gigantesca, biznaga hembra – Arizona barrel cactus, fishhook barrel cactus
Ferocactus = Thelocactus
Ferula asafoetida – asafétida – asafetida
Ferula communis – cañaheja, cañareja, ferula, ferula común – giant fennel
Ferula galbaniflua – gálbano – galbanum
Festuca amethystina – festuca de color amatista – tufted fescue
Festuca arizonica – cañuela arizónica – Arizona fescue
Festuca arundinacea – cañuela alta, cañuela de los prados, cañuela del Rhin, cañuela descollada, cañuela gigante, cañuela noble, festuca alta, festuca arundinácea, festuca canosa, festuca descollada, festuca elevada, festuca gigante – tall fescue
Festuca glauca – festuca glauca, festuca incurvada – common blue fescue
Festuca longifolia – cañuela durilla, festuca de hojas grandes – hard fescue

Festuca ovina – cañuela borreguerra, cañuela de oveja, festuca de oveja – alpine fescue, sheep fescue
Festuca pratensis – cañuela, cañuela común, cañuela de los prados, festuca de los prados – meadow fescue
Festuca rubra – cañuela roja, festuca roja – red fescue
Festuca trachyphylla – festuca de hojas ásperas – hard fescue, sheep fescue
Festuca valesiaca – festuca galesa – Valais fescue, Wallis fescue
Ficus aspera – higuera abigarrada
Ficus benghalensis – amate, banyan, higuera arqueada, higuera de Bengala, higuera de las pagodas, higuera india, laurel de India – banyan tree, East India fig, Indian banyan
Ficus benjamina – ficus llorón, higuera llorona, laurel benjamín – Benjamin fig, Benjamin tree, weeping Chinese banyan, weeping fig
Ficus carica – breva, chuna, higuera, higuera común, yaga-yaxo-Castilla – edible fig
Ficus cotinifolia – álamo, amate prieto, camuchín, capulín, capulina, ceiba, chipil, chuna, cobó, copoy, guaorócochi, guojtoli, higuera, higuera americana, higuerón, matapalo, nacapul, nacapuli, tescalama, texcalamate, tlilamatl – strangler fig
Ficus elastica – amate, árbol del caucho, árbol del hule, caucho, caucho de la India, ficus, goma elástica, gomero, higuera cauchera, higuera del caucho, higuera del hule, higuera del látex, higuera de Sumatra, hule, hule de la India, palo de goma, palo de hule, planta del caucho – Assam rubber tree, Indian rubber plant, rubber tree
Ficus glabrata – álamo, amate, amate blanco, chimón, macahuite, siranda
Ficus glaucescens – amate prieto, higoloxe grande, nacapuli, siranda
Ficus goldmanii – (trigonata) amate, chala, chalate, higuera, popatza, xalama, xalamat, zalate
Ficus indica – goma laca
Ficus insipida – chalate, chuna, higuera, huilocochi
Ficus involuta – amate, amate blanco, chalate, matapalo, tecomasuchiamatl, zalate
Ficus lyrata – ficus, pandureca, pandurita – fiddleleaf fig
Ficus maxima – chalate de burro, chuna
Ficus mexicana – copoy, higuera, higuera blanca, jopoy, sabalí, saccabah
Ficus microcarpa (nitida, retusa) – álamo extranjero, amate, fícus, laurel, laurel de India, yucateco – banyan, Chinese banyan, Indian laurel, Indian laurel fig
Ficus microchlamys – zalate, zalate blanco, zalate bronco
Ficus padifolia (pertusa) – amatillo, amezquite, cabra-higo, camichín, capulín, capulín grande, chalate, chinito, chuna, chusnito, coamichin, cozahuique, higuera amate, higuera negra, higuito, jalamata, jalamate, matapalo, nacapuli, palo de coco, samatito, tabla, tucuchí, xalama limón – plumleaf fig, rock fig, strangler fig
Ficus palmeri (brandegeei) – anaba, higuera, higuera silvestre, tescalama, zalate – cliff fig, desert rock fig, wild fig
Ficus petiolaris – amacoztic, amate, amate amarillo, chichic-texcal-amatl, chilamate, higuera, higuera silvestre, higuerón, higuerote, palo amarillo, palo chilamate, palo María, tepeamate, tepeamatl, tepecamichín, tescalama, tescalama lechosa, tescalamate, texcalamacoztli, texcalamatl, texcalamatlchichic, texcomaxochiamatl – cliff fig, rock fig
Ficus pumila (repens) – amate, ficus rastrero, hiedra, higuera serpenteante, higuera trepadora – creeping fig
Ficus radulina – chacalate, chalate, higuera, nacapul, zalate
Ficus religiosa – álamo, álamo cubano, árbol de goma laca, banyan, higuera de agua, laurel de India – bo-tree, peepul, sacred fig
Ficus superba – higuera marítima
Ficus sycomorus – higuera – sycamine, sycamore-fig
Ficus tecolutensis – abrecapalo, amate, amate prieto, higuera, macahuite, matapalo, matapalo liso
Ficus vulgaris – higuera silvestre, sicomoro – goat fig, wild fig
Fidicina mannifera – chicharra de la hierba mate, maté cicada

Filago gallica – filago de hojas tenues, filago francés – herba impia, narrow cudweed
Filago germanica – filago común, flor de la vida – common cudweed, cotton rose
Filago pyramidata – filago espatulado – broad-leaved cudweed, herba impia
Filipendula ulmaria – ewina de los prados, filipéndula ulmaria, reina de los prados, ulmaria – meadow sweet
Filipendula vulgaris (hexapetala) – filipéndula, filipéndula de seis pétalos, hierba del pobre hombre – dropwort
Fimbristylis thermalis (spadicea) – camalote, esparto, esparto chino, esparto mulato – hot-springs fimbristylis
Firmiana simplex – esterculia – Chinese parasol tree
Fistulina hepatica – higado de buey, lengua de buey – beefsteak fungus
Fitzroya cupressoides – alerce, falso ciprés de Patagonia – Patagonian cypress
Flacourtia indica – ciruelo gobernador – governor's plum
Flacourtia inermis – lovi-lovi – governor's plum
Flaveria angustifolia – contrahierba, escoba amarga, hierba de la barranca, tlaco-popotl
Flaveria contrayerba – chicata, contrahierba, flaveria
Flaveria oppositifolia – baiquillo
Flaveria trinervia – contrahierba
Flourensia cernua – hojasén – blackbrush, tarbush, varnishbrush
***Foeniculum vulgare* (var *azoricum*)** – anís, anís hinojo, eneldo, fenícula, hierbanís, hinojo, hinojo común, hinojo de Castilla, hinojo tuberoso – fennel, finocchio, sweet fennel
Fomes pinicola – hongo del oyamel, hongo del pino, hongo de palo – red belt fungus
Forchammeria watsonii – jito, palo de San Juan, palo jito, tres Marías – lollipop tree
Forestiera angustifolia – acebuche, chaparral blanco, panalero, pico de pájaro – narrow-leaf forestiera
Forestiera neomexicana (pubescens) – adelia, palo blanco – adelia, desert olive, New Mexican privet
Forsythia* x *intermedia – forsitia – golden bell
Forsythia suspensa – campanita china, forsitia – weeping forsythia
Fortunella margarita – cumquat, naranjo chino, naranjito – kumquat, nagami kumquat
Fouquieria burragei – ocotillo blanco, ocotillo de flor blanca, palo de Adán, pichilinge
Fouquieria campanulata – alabarda, albarda, barda, chiquiña, gavia, ocotillo, teocotillo
Fouquieria (Idria) columnaris – cirio, cotootaj, milapa – boojum tree
Fouquieria diguetii (peninsularis) – cirio, mureo, ocotillo, palo de Adán, palvadán – Adam's tree
Fouquieria fasciculata – alabarda, albarda, árbol de barril, barda, chiquiña, ocotillo, teocotillo, torote, torote verde
Fouquieria formosa – colorín cimmarón, ocotillo, palo santo, rabo de iguana, rosadillo, rosalillo, tlapacón – ocotillo
Fouquieria macdougalii – chumari, jaboncillo, mureo, ocotillo, ocotillo macho, palo verde, torote espinoso, torote verde, torotillo – Mexican tree ocotillo
Fouquieria splendens – albarda,barda, cirio, colorín cimmarón, mureo, ocotillo, ocotillo de corral, palo de Adán, xeshish, xong – candlewood, coach-whip, ocotillo, slimwood
Fouquieria = Idria
Fourcraea = Fourcroya
Fourcroya (Fourcraea) humboldtiana – cocuiza
Fragaria ananassa – fresa, fresa ananás, fresera, frutilla – garden strawberry, strawberry
Fragaria chiloensis – fresa chilena, fresón, frutilla – beach strawberry, sand strawberry
Fragaria mexicana – fresa, fresa silvestre
Fragaria vesca – fresa, fresa común, fresa silvestre, fresera, frutilla, mayueta – alpine strawberry, fraise du bois, wood strawberry
Fragaria virginiana – fresa de Virginia – mountain strawberry
Francoa sonchifolia – llaupangue – pink francoa
Frangula = Rhamnus
Frankenia laevis – brezo de mar
Frankenia palmeri – hierba reuma, saladito – Palmer's frankenia

Frankenia salina (grandifolia) – hierba reuma – alkali heath
Franseria = Ambrosia
Frasera = Swertia
Fraxinus americana – fresno, fresno blanco, fresno blanco americano – white ash
Fraxinus angustifolia (oxycarpa) – fresno de hojas estrechas, fresno de hojas pequeñas, fresno de la tierra, fresno del sur – ash
Fraxinus anomala – fresno – single-leaf ash
Fraxinus berlandieriana – fresno, plumero – Mexican ash
Fraxinus cuspidata* var *macropetala – fresno – flowering ash, fragrant ash
Fraxinus excelsior – fresno, fresno común, fresno de Viscaya, fresno grande – European ash
Fraxinus gooddingii – fresnillo – Goodding ash, littleleaf ash
Fraxinus greggii – barreta china, barreta de cochino, escobilla, fresno, manzanilla, palo roñoso – Gregg ash, little leaf ash
Fraxinus ornus – árbol del maná, fresno de flor, fresno del maná, fresno florido, fresno florifero, maná de Calabria, maná de Oriente, orno – flowering ash
Fraxinus papillosa – fresno – Chihuahua ash
Fraxinus pennsylvanica (lanceolata) – fresno de Pensilvania, fresno rojo americano, fresno verde – green ash, red ash
Fraxinus trifoliata – crucecillo, fresno, palo blanco, varejón – chaparral ash
Fraxinus uhdei – fresno, madre de agua – evergreen ash, shamel ash
Fraxinus velutina – botavara, fresno, fresno terciopelo – Arizona ash, velvet ash
Freesia refracta – Africanas, fresia, fresia de exterior – freesia
Fremontodendron californicum – fremontia – California slippery elm, flannelbush, fremontia
Fritillaria biflora – campanario, fritilaria – chocolate lily, mission bell
Fritillaria imperialis – corona imperial, fritilaria imperial – crown imperial
Fritillaria meleagris – fritilaria de cabeza de serpiente, maleagrida, tablero de damas – checkered lily, Guinea hen flowers, snakeshead
Fritillaria persica – fritilaria pérsica – Persian bell, Persian fritillary
Fructus colocyathidis – coloquintida
Fuchsia arborescens – adelaida, arete, aretillo, atesúchil, atexóchitl, chorro, dondiego de día, flor de arete – fuchsia
Fuchsia fulgens – adelaida, aretillo, flor de arete, fucsia
Fuchsia* x *hybrida – fucsia – hybrid fuchsia
Fuchsia magellanica – fucsia – fuchsia
Fuchsia microphylla – aretillo del monte – mountain fuchsia
Fucus vesiculosus – alga marina, caloca, encino de mar, encino marino, fuco, fuco marino, quelpo, sargazo vejigoso – bladderwrac, kelp, seaweed
Fumaria capreolata – conejillo, fumaria, palomilla, palomilla pintada – ramping fumitory
Fumaria officinalis – fumaria, fumaria común, hierba del conejo, palomilla, palomilla común, sangre de Cristo – fumitory
Fumaria parviflora – fumaria, fumaria de flores pequeñas, pajarilla, palomilla – fumitory
Funastrum clausum – bejuco de leche, mata-torsalo, petaquilla, quichi-nixe
Funastrum heterophyllum – hortensia de guía, sandía de la pasión – climbing milkweed
Furcraea gigantea – cáñamo de Mauricio, pita – Mauritius hemp
Furcraea hexapetala – cáñamo cubano, pitre – pitre
Furcraea humboldtiana – cocuiza
Fusarium solani – cenicilla del tomate, podredumbre – tomato wilt

G

Gaillardia aristata (grandiflora) – gaillardia, gallardía – Indian-blanket
Gaillardia mexicana – gallardía
Gaillardia nervosa – árnica
Gaillardia pinnatifida – coronilla, hierba del sol – blanket flower
Gaillardia pulchella – escarpela, gaillardia annual, gallardía – blanket flower, firewheel, Indian-blanket
Galanthus elwesii – campanilla de invierno gigante – giant snowdrop
Galanthus nivalis – campanilla blanca, campanilla de febrero, campanilla de invierno, campanilla de invierno común, doncella de febrero, galanto de nieve, narciso de las nieves silvestre, rompenieve – common snowdrop, white snowdrops
Galega officinalis – caprago, galega, índigo falso, ruda cabruna – goat's rue
Galinsoga parviflora – acetilla, estrellita, galinsoga, galinsoga de flores pequeñas, guasca, mercurial, tacote – gallant soldier, quickweed
Galinsoga quadriradiata – galinsoga ciliada – shaggy soldier
Galium aparine – amor de hortelano, aspérgula, azotalengua, cuajaleche, lapa, lárgalo, presera – beggar's lice, catchweed bedstraw, cleavers, cleaverwort, common bedstraw, goose grass
Galium boreale – galio boreal – northern bedstraw
Galium divaricatum – galio divaricado – Lamarck's bedstraw
Galium mexicanum – amor de hortelano, cuajaleche, hierba de la pulga, hierba del coyote, pegajosa, pegarropa, presera – bedstraw
Galium microphyllum – hierba del coyote – bedstraw
Galium mollugo – galio de los prados – false baby's breath
Galium murale – galio mural – tiny bedstraw
Galium nuttallii – pegajosa, pega pega – San Diego bedstraw
Galium odoratum (Asperula odorata) – asperilla dulce, asperilla olorosa, galio, galio oloroso, hepatica estrellada, reina de los bosques, rubilla – sweet woodruff
Galium parisiense – galio parisino – wall bedstraw
Galium saxatile – galio de las rocas – heath bedstraw
Galium tricornatum – galio de tres cuernos – corn cleavers, rough corn bedstraw
Galium trifidum – cuajaleche, galio trífido, pegarropa
Galium triflorum – amor de hortelano, galio de tres flores – sweet-scented bedstraw
Galium verum – amor de hortelano, cuajaleche, galio, galio primaveral, hierba sanjuanera, presera – ladies' bedstraw, lady's bedstraw, yellow bedstraw
Galphimia (Thryallis) glauca – calderona amarilla, consulita, corpionchi, cortinchi, flor de chinche, flor de diciembre, flor de mazorca, garbancillo, hierba de la hormiga, hierba del piojo, hierba del venado, huachacata, huitlatexihuitl, lluvia de oro, mazorquita, nachácata, palo del muerto, palo de San Vicente, rama de oro, San Vicente, vachácata, xaxaxacótic – rain of gold
Galtonia candicans – jacinto del Cabo, jacinto de Sudáfrica, jacinto de verano, jacinto estival – summer hyacinth
Gandoderma lucidum – hongo reishi, reishi – ling chih, reishi mushroom
Garcinia cambogia – citrí, tamarindo malabar
Garcinia hanburyi – gomaguta – gamboge
Garcinia mangostana – mangostán – mangosteen
Gardenia augusta (jasminoides) – gardenia, jazmín del Cabo – cape jasmine, common gardenia, gardenia
Gardenia florida – gardenia, jazmín del Cabo – Cape jasmine
Garrya eliptica – garrya – coast silk tassel, coast silk tassel bush
Garrya flavescens – cuauchichic, guachichi – yellow silk tassel
Garrya fremontii – quinana – quinine bush
Garrya laurifolia – chichicahuite, chichicuáhuitl, cuauchichic, guachichi, guachichi ovitano, hediondillón, ovitano, palo negro, zapotillo – silk tassel bush

Garrya longifolia – palo amargoso – longleaf silk tassel
Garrya ovata – chichicuáhuitl, cuauchichic
Garrya salicifolia – hierba prieta, rama prieta
Gastridium ventricosum – gastridia ventricosa – nit grass
Gaudichaudia mucronata – chilillo-ak, correhuela, hierba del zorro
Gaultheria acuminata – ajocopaque, ajopatla, arrayán, axocopaque, trampa de perro, xiopatla – wintergreen
Gaultheria odorata – arrayán, axocopaque, pesjúa, té del Canadá
Gaultheria procumbens – gaulteria, té del Canadá – checkerberry, teaberry, wintergreen
Gaultheria shallon – gaulteria – salal
Gaultheria* = *Pernettya
Gaura coccinea – aretito, hierba de la Virgen, hierba del golpe, linda tarde – linda tarde, scarlet bee blossom, wild honeysuckle
Gaura parviflora – hierba salada, linda tarde, punche ratón – lizard-tail, midget bee blossom, velvet weed, wild honeysuckle
Gaussia maya – llume, palma, palmacito – Maya palm
Gazania nivea – gazania – gazania
Gelidium cartilagineum – agar-agar, gelatina vegetal
Gelsemium sempervirens – gelsemio, jazmín amarillo, jazmín de Carolina, jazmín silvestre, madreselva – Carolina jessamine, logania
Genipa americana – genipa, illuale, jagua azul, jagua blanca, maluco, tejoroso, yoale – genipap
Genista aetnensis – retama del Monte Etna – Mt. Etna broom
Genista anglica – aulaga, espino, gatiña, retama, retama inglesa – needle broom, needle whim
Genista (Cytisus) canariensis – genista, retama de Canarias – Canary Island broom
Genista hispanica – aulaga, aulaga occidental, retama española – Spanish broom
Genista linifolia – retama de hojas de lino – flax-leaved broom
Genista lydia – retama – broom
Genista monspessulana (Cytisus monspessulanus) – escobón – French broom
Genista pilosa – retama pelosa – hairy greenweed
Genista sagittalis – carquexia fina
Genista tinctoria – retama de los montes, retama de tintoreros – dyer's greenweed, woadwaxen
Gentiana acaulis – genciana acaule, genciana alpina – stemless gentian
Gentiana adsurgens – flor de los hielos, genciana, hierba del aire – gentian
Gentiana asclepiadea – genciana, genciana asclepiadea, genciana de hojas de algodoncillo – willow gentian
Gentiana calyculata – flor de nieve, flor de Santo Domingo
Gentiana clusii – genciana acaule, genciana de Clausio, genciana de flores grandes, gencianilla – large-flowered gentian
Gentiana cruciata – cruciata, genciana cruciata – cross gentian
Gentiana lutea – genciana, genciana amarilla, genciana mayor – yellow gentian
Gentiana prostrata – genciana prostrada – pygmy gentian
Gentiana spathacea – flor de los hielos, genciana, hielera
Gentiana superba – clavel
Gentiana verna – genciana de primavera, genciana primaveral – spring gentian
Gentiana villosa – genciana, raíz de Sansón – gentian
Gentianella amarella – genciana axilar – autumn gentian
Gentianella tenella – genciana pedunculada – slender gentian
Geoffroea decorticans – chanar – bastard cabbage tree
Geonoma congesta – cortadera
Geonoma interrupta – cambuc, chocho blanco, pacaya – monkey-tail palm
Geonoma magnifica – pojai, talife, talis
***Gerando* spp** – gabat
Geranium argenteum – cordillera cantábrica, geranio argénteo – silver cranesbill
Geranium caespitosum – patita de león – purple geranium, storksbill
Geranium carolinianum – nesesehuilone, tlatlauhcapatli – cranesbill, Carolina geranium

Geranium cinereum – geranio cinéreo – ash-coloured cranesbill
Geranium dissectum – geranio disecado – cut-leaved cranesbill
Geranium endressii – geranio de Endress – French cranesbill
Geranium flaccidum – alquimila – geranium
Geranium macrorrhizum – geranio de raíces grandes – rock cranesbill
Geranium maculatum – geranio, geranio silvestre – cranesbill, geranium, spotted cranesbill, wild cranesbill
Geranium mexicanum – alquimila, alquimila del país, geranio silvestre, mano de león, pata de león – Mexican geranium
Geranium molle – geranio de los caminos, geranio muelle – dove's foot cranesbill
Geranium phaeum – geranio moreno – dusky cranesbill, mourning widow
Geranium platypetalum – geranio platipétalo – large-flowered cranesbill
Geranium potentillifolium – pata de león, soldadiente
Geranium potentilloides – pata de león, soldadiente – cranesbill, geranium
Geranium pratense – geranio de los prados – meadow cranesbill
Geranium pusillum – geranio menor – small-flowered cranesbill
Geranium robertianum – cicuta roja, geranio, hierba de San Roberto, pico de grulla – herb Robert
Geranium sanguineum – geranio sanguineo – bloody cranesbill
Geranium sylvaticum – geranio selvático – wood cranesbill
Gerbera jamesonii – gerbera – barberton daisy, Transvaal daisy
Geum aleppicum – cariofilada rígida – avens
Geum chiloense – cariofilada, geum de Chile, hierba de San Benito – avens
Geum japonicum – avens japonés
Geum macrophyllum – cariofilada de hojas grandes – bigleaf avens
Geum montanum – cariofilada de montaña, geum alpino – alpine avens
Geum urbanum – avens, cariofilada oficinal, hierba de San Benito, sanamunda – cloveroot, herb Bennet, wood avens
Gilia capitata – alhelilla – blue thimble flower, gilia
Gilia gloriosa – ubi
Gilia (Ipomopsis) tenuiflora – agua bonita – gilia
Gilibertia arborea (Dendropanax arboreus) – cajeta, hoja fresca, mano de danta, mano de león, mano de oso, mano de sapo, nixtamalcuáhuitl, nixtamalillo, palo blanco, palo de agua, palo de danta, palo santo, pingüico, vidrioso – white gumbolimbo
Ginkgo biloba – árbol de los cuarenta escudos, culantrillo, ginko – ginkgo, maidenhair tree
Gladiolus* x *colvillei – gladiolo, gladiolo menor – baby gladiolus
Gladiolus communis (byzantinus)* ssp *byzantinum – gladiolo bizantino, gladiolo oriental – Byzantine gladiolus
Gladiolus dalenii – gladiolo primulinus – butterfly gladiolus
Gladiolus hortulanus – cresta de gallo, espadilla, estoque, gladiolo, palma gladiola, pluma – gladiola
Gladiolus italicus – gladiolo de los campos – corn flag
Glaucium corniculatum – glaucio roso – red horned poppy
Glaucium flavum – adormidera marina, amapola de playa, glaucio, glaucio amarillo – sea poppy
Glaux maritima – glaux marítima – sea-milkwort
Glechoma = Glecoma
Glecoma (Glechoma) hederacea – hiedra de tierra, hiedra sueca, hiedra terrestre, hierba de San Juan – gill-over-the-ground, ground ivy, robin-run-in-the-hedge
Gleditsia amorphoides – espino corona – honey locust
Gleditsia triacanthos – acacia de tres espinas, acacia negra, falsa acacia de tres espinas, gleditsia de tres espinas – Christ's thorn, honey locust
Gliricidia sepium (maculata) – cacahuananche, cacahuanantl, cacahuano, chanté, cocohuite, cocuite, cuchunuc, flor de San José, frijolillo, iati, lengua de perico, madera negra, madre de cacao, matarrata, matarratón, muitle, palo de corral, palo de

sombra, piñón violento, rabo de ratón, ratón, sacyab, yaguaguyt, yaité – mouse killer
Globularia alypium – álipo, coronilla de fraile, coronilla de rey, coronilla real – cespitose globe daisy
Globularia cordifolia – globularia cordifoliada, globularia de hojas cordiformes, globularia menor – heart-leaved globe daisy
Globularia vulgaris – cebollada, globularia, globularia común, globularia mayor – globe daisy
Gloriosa superba (rothschildiana) – gloriosa, gorro de Turco, lirio de Malabar, lirio gloriosa – climbing lily, flame lily, gloriosa lily, glory lily, Malabar lily, superb lily
Gloxinia = Sinningia
Glyceria fluitans – hierba del maná – mannagrass
Glyceria striata – gliceria estriada – mannagrass
Glycine (Soya) max – caraota soya, frijol soya, haba de soja, habichuela soya, poroto soya, soja, soja hispida, soja max, soya – soy bean
Glycyrrhiza glabra – alcazuz, licórice, orozuz, palo dulce, paloduz, regaliera, regaliz, regaliz glabro – licorice, liquorice
Glycyrrhiza lepidota – amolillo, orozuz, orozuz americano, raíz del desierto, regaliz – licorice, licorice root, wild licorice
Gnaphalium bicolor – gordolobo – cudweed, everlasting
Gnaphalium bourgovii – gordolobo, lobito, manzanilla, manzanilla del río
Gnaphalium canescens – gordolobo, papaconi, tlacochichic, tzompotonic – cudweed, everlasting
Gnaphalium conoideum – chichictzompotonic, gordolobo, gordolobo del país – cudweed
Gnaphalium crenatum – gordolobo
Gnaphalium leucocephalum – gordolobo, manzanilla, manzanilla del campo – white cudweed
Gnaphalium leucophyllum – manzanilla del río, talamapate, talampacate – cudweed
Gnaphalium luteo-album – borrosa amarilla pálida – Jersey cudweed
Gnaphalium macounii – gordolobo – cudweed
Gnaphalium oxyphyllum – gordolobo, manzanilla del río – cudweed
Gnaphalium palustre – gordolobo – cudweed, everlasting
Gnaphalium purpureum – gordolobo – cudweed
Gnaphalium sonorae – gordolobo
Gnaphalium wrightii – gordolobo, manzanilla del río – cudweed
Gochnatia arborescens – ocote
Godetia = Clarkia
Goldmania foetida – chicoche, chile cimarrón, cochilla, coyacate, cusa, guamú, guarade, huisache, huisache conchilla, pinzanguarimbo, sicoche, yóndiro
Gomphrena decumbens – amorseco, cabezona, confitillo, cordón de obispo, flor de San Francisco, inmortal, Juana milpilla, lupita, madroño del campo, madroño rastrero, moradilla, sanguinaria, sempiterna, siempreviva, siempreviva silvestre
Gomphrena globosa – amarantina, amorseco, chacmol, cordoncillo, flor de San Francisco, guirnalda, hierba del cáncer, inmortal, madroño, permanente, sempiterna, siempreviva, tmuul – bachelor's button, globe amaranth
Gomphrena nitida – amorseco, cordón de San Francisco
Gomphrena sonorae – cordón del obispo, siempreviva cimarrona
Gomphus floccosus – trompeta, hongo – edible mushroom
Gonolobus erianthus – cachayumbo, chicuipo, flor de muerto, pongolote, talayote, tlalayote – angle-pod
Goodyera oblongifolia – godiera – rattlesnake plantain
Gossypiospermum praecox – zapatero – West Indian boxwood
Gossypium barbadense – algodón, algodón común, algodón del país, algodonero, tamán, ts'iin, xurata – cotton, Egyptian cotton, pima cotton, sea island cotton
Gossypium davidsonii – algodón, algodón cimarrón, algodón silvestre – cotton
Gossypium harknessii – algodón, algodón cimarrón, algodón silvestre – cotton
Gossypium herbaceum – algodón,

algodoncillo, algodonero, algodonero herbáceo, chiní, chinique – cotton
Gossypium hirsutum – algodón, algodón coyuche, algodonero, algodonero americano, íxcatl, mooj, musá, suruata, taman, xurata – upland cotton
Gossypium mexicanum – algodón, algodón de árbol, íxcatl, ixcaxíhuitl – cotton
Gossypium thurberi (Thurberia thespesioides) – algodoncillo, algodoncillo del campo, algodoncillo del monte – desert cotton
Gouania rosei – güirote de violín
Gouinia virgata – carricillo, gramilla perdiz, itza-zuuk, zacate colorado
Govenia liliacea – azucena del monte, icahueyo, iztactepetzacuxóchitl, petzanxóchitl
Govenia superba – azucena amarilla, azucena del monte, cozticzacatzacuxóchitl
Gramineae = Poaceae
Gratiola officinalis – graciola, graciola común, hierba de la calentura, hierba del pobre – hedge-hyssop
Grevillea robusta – grevillo, pino australiano – silk oak
Grindelia aphanactis – grindelia, hierba del buey, palancapatli de puebla – gumweed, mountain gumplant
Grindelia hirsutula – hierba de la tos, hierba del buey – gumplant
Grindelia oxylepis – árnica flor – matchweed
Grusonia bradtiana (Opuntia cereiformis) – organillo, viejo
Grusonia (Opuntia) invicta – casa rata – dagger cholla
Grusonia (Opuntia) schottii – clavellina, perrito – devil cactus, dog cholla, Schott's dwarf cholla
Guadua (Bambusa) aculeata – caña brava, caña otate, carrizo, cupamu, jimba, ojtotl, otate, otlatl, tarro, tzajib
Guadua amplexifolia – caña mansa, caña otate, cauro, jimba, ocotate, otate
Guadua (Bambusa) angustifolia – guadua, otate, tarro – American narrow-leaved bamboo
Guadua (Bambusa) inermis – otate, tarro
Guadua (Bambusa) latifolia – guadua, juájuas, tacuara, yaripas – American broad-leaved bamboo
Guadua (Bambusa) longifolia – bambú espinudo, caña brava, cañizo, jimba, jimbilla, otate – American long-leaved bamboo
Guadua paniculata – otate amargo
Guadua velutina – otate
Guaiacum (Porlieria) angustifolium – guayacán – soapbush
Guaiacum coulteri – árbol santo, chumchintoc, guayacán, huaxaxán, juyago, matlalcuahuitl, palo santo, yaga-na, yutnu-tandaa – lignum vitae
Guaiacum officinale – guayacán, guayacán amarillo, guayaco, matlalcuahuitl, palo de rosa, palo santo – lignum vitae
Guaiacum sanctum – guayacán, guayacán real, hoaxacan, huesito, ken, lignum guaiaci, lignum sanctum, palo guayacán, palo santo, zon – hollywood lignum vitae, holywood lignum vitae
Guardiola platyphylla – papalotilla
Guassia maya – cambo, palma cimarrona
Guazuma grandiflora – cupuassú
Guazuma ulmifolia (tomentosa) – ajillá, aquiche, bolita, bulines, cabal-pixoy, cuaulote, cuauolotl, guácimo, guácimo blanco, guácimo cimarrón, guácimo de ternero, guácimo dulce, majagua de toro, palote negro, parandesicua, pixoy, quauhalagua, quauhólotl, tablote, tapaculo, tzuny, yaco de granadillo, yaco de venado – bastard cedar, pricklenut
Guilielma gasipaes – chontaduro, palma melocotonera, pejibaye – peach palm
Guilielma insignis – palmera timbe
Guilleminea densa – rosa de Castilla cimarrona
Guizotia abyssinica – ramtil, ramatilla – ramtilla
Gunnera insignis – capa de pobre – gunnera
Gunnera mexicana – capa de pobre
Gustavia augusta – guatoso, matamatá – gustavia
Gutierrezia (Xanthocephalum) sarothrae – coyaye, escoba de la víbora, hierba de la víbora, hierba de San Nicolás, rosita – broom snakeweed, matchweed, snake broom
Gymnocladus dioica – cafeto de Kentucky, gimnoclado del Canadá, gimnoclado dioico,

gimnoclado norteamericano, raigón del Canadá – Kentucky coffee tree

Gymnosperma = Selloa

Gynerium (Cortaderia) argenteum – cortaderia

Gynerium sagittatum – caña amarga, caña blanca, caña brava, caña de Castilla, carrizo, carrizo real, lata, otatillo – uva grass, wild cane

Gypsophila elegans – gipsófila, gipsófila elegante, nube – baby's-breath

Gypsophila paniculata – gipsófila, gipsófila paniculada, jabonera, nebulosa, nube, nube de novia, saponaria – baby's breath

Gypsophila repens – gipsófila repente – creeping gypsophila

Gyromitra esculenta – murgula – false morel, poisonous brain mushroom

H

Hackelochloa granularis – zacate, sit-suuk
Haemanthus = Scadoxus
Haematoxylum brasiletto – azulillo, Brasil, Brasil azulillo, brasilete, brasilillo, brasilillo marismeño, Campeche, curaqua, huachaco, huichachago, huichilaco, huitzquahuitl, marismeño, palo de Brasil, palo de Campeche, palo de tinta, palo rojo, quamóchitl, sitagape – Brazil bloodwood tree, Brazilwood
Haematoxylum campechianum – Brasil, Campeche, ek, huitzquahuitl, palo atolito, palo azul, palo de Brasil, palo de Campeche, palo de sangre, palo de tinta, tinta, tinte, tinto, yatolito – campeachy wood, camwood, logwood
Hainardia cylindrica – cizaña cilíndrica – hard-grass
Halesia monticola – alesia, árbol de las campanillas de invierno – mountain silver bell
Halimium glomeratum – hierba de la gallina, Juanita, masiola, nexticinatl, sanguinaria
Halimium lasianthum – jaguarzo occidental
Halimium ocymoides – alcayuela
Halimium umbellatum – ardivieja, jaguarcillo, jarilla
Halogeton glomeratus – aral barilla, barilla – halogeton
Hamamelis japonica – nogal de las brujas japonés – Japanese witch hazel
Hamamelis mollis – nogal de la brujería, nogal de las brujas chino – Chinese witch hazel
Hamamelis virginiana – agua maravilla, aliso manchado, avellano chasqueante, avellano de bruja, avellano de la hechicera, flor de invierno, hamamelina, hamamelis, hamélide de Virginia, nogal de las brujas, olmo escocés, vara mosqueada – common witch hazel
Hamatocactus = Thelocactus
Hamelia erecta (patens) – arbusto escarlata, bálsamo, cacahuapastle, cacahuaxóchitl, cacanapazue, cacapuate, canan, canela montés, cañutillo, chacloco, chichipín, chocho, coralillo, cuasipinsé, hierba del toro, madura plátano, maravilla, pañete, pie de pájaro, sangre de toro, suspinché, tochimitillo, valletilla, vara prieta, zipate – scarlet-bush
Hancornia speciosa – mangabera – mangabeira
Haplopappus = Ericameria, Isocoma
Haplophyton cimicidum (crooksii) – actimpatli, hierba de la cucaracha, hierba del piojo, palo bolero, raíz de la cucaracha – cockroach plant
Harfordia macroptera – bolsa de conejo – rabbit's purse
Harpagophytum procumbens – uña del diablo – devil's claw, grapple tree
Hartmannia = Oenothera
Havardia mexicana (Pithecellobium mexicanum) – chino, guamúchil, huahuichó, joso, manto, palo chino, rama de chivato, tepeguaje – Mexican ebony
Havardia (Pithecellobium) pallens – carbonero, guajillo, huajillo, huisache, tenaza – ape's earring, mimosa-bush, tenaza
Havardia (Pithecellobium) sonorae – gato, jócono, palo gato, palo jócono, uña de gato – Sonoran ebony
Hebe buxifolia – verónica arbustiva – boxleaf hebe
Hechtia montana – chikana, datillo, hichiconi, magueycillo, mezcalito – false-agave
Hechtia podantha – guapilla
Hechtia texensis – aguapié
Hedeoma drummondii – poleo – mock pennyroyal
Hedeoma floribunda – mapá, orégano
Hedeoma nanum – poleo, poleo chino – dwarf pennyroyal, mock pennyroyal
Hedeoma oblongifolia – poleo chino – mock pennyroyal, pennyroyal
Hedeoma patens – mapá, menta, orégano
Hedeoma piperita – hierbabuena, piperita de México, poleo chino, Santo Domingo, tabaquillo, tabaquillo chico, tabaquillo oloroso, té de rancheros – mock pennyroyal
Hedeoma pulegioides – poleo, poleo americano, poleo chino – American pennyroyal, mock pennyroyal, pennyroyal
Hedera canariensis – hiedra de las Canarias, hiedra persa – Algerian ivy

Hedera colchica – hiedra de Persia – Persian ivy
Hedera helix – hiedra, hiedra común, hiedra extranjera, hiedra hélice, hiedra inglesa – English ivy
Hedera ibernica – hiedra irlandesa
Hedychium coronarium – lirio del arroyo, mariposa, palomita – butterfly lily, garland flower, white ginger, white ginger lily
Hedychium gardnerianum – edichio – kahili ginger
Hedyotis alexanderae – trompetilla
Hedyotis wrightii – anisillo – bluets
Heimia salicifolia – escoba, escoba colorada, escobilla del río, flor de San Francisco, garañona, granadillo, hauchinol, hauchinoli, hierba jonequil, jarilla, jonequil, sinicuilche
Helenium autumnale (latifolium) – cabezona, chapuz, helenio, manzanilla, manzanilla de Yucatán – sneezeweed
Helenium hybridum – helenio – sneezeweed
Helenium mexicanum – amargosa, árnica del país, cabezona, chapuz, exoxóchitl, hierba cabezona, hierba de la ánima, hierba de la víbora, manzanilla montés, rosilla, rosilla de Puebla – sneezeweed
Helenium puberulum – rosilla – sneezeweed
Helenium thurberi – rosilla
Helenium = Dugaldia
Heleocharis = Eleocharis
Helianthella madrensis – cachana
Helianthemum apenninum* var *roseum – heliantemo – Wisley pink
Helianthemum glomeratum – cenicilla, hierba de la gallina, janajuana, Juanita, sanguinaria – rock-rose, rush-rose, sun-rose
Helianthemum lavandulifolium – heliantemo de hojas de lavándula, hierba sana, romerillo – sun-rose
Helianthemum nummularium – estepa, heliantemo, heliantemo amarillo, heliantemo mayor, jarilla, tamarilla – rock-rose, sun-rose
Helianthus annuus – acahual, acahualli, añil, chimalácatl, chimalate, copa de Júpiter, corona real, flor de gigantón, flor de sol, gigantón, girasol, girasol anual, girasol común, hierba del sol, lampote, maíz de teja, maíz de Texas, maíz meco, maravilla, mirabel, mirasol, mirasol anaranjado, quisnaniquitonale, sol de las Indias, tlacopatli, tornasol, xaricamata – common sunflower, sunflower
Helianthus ciliaris – amargosa, hierba parda – blueweed, Texas blueweed
Helianthus gracilentus – girasol – sunflower
Helianthus niveus – girasol – dune sunflower
Helianthus similis – girasol – sunflower
Helianthus tuberosus – aguaturma, alcachofa de Jerusalén, cotufa, girasol, Jerusalén, papa real, pataca, patata de caña, topinambur, tupinambo – Jerusalem artichoke, sunchoke
Helichrysum bracteatum – perpetua, siempreviva, siempreviva bracteada – strawflower
Helichrysum italicum (angustifolium) – siempreviva de hojas estrechas, siempreviva itálica – curry plant
Helichrysum orientale – azaya, cantueso, inmortal, sempiterna – yellow everlasting flower
Heliconia bihai – biahi, cachete colorado, papacla, platanillo, plátano silvestre, tanay, tanay colorado – wild plantain
Heliconia caribaea – platanillo – wild plantain
Heliconia latispatha – bihagua, cachete, guacamaya, pico de gorrión, platanillo, podomatz, tanay – golden heliconia, lobster plant
Heliconia psitiacorum – platanillo enano – gold torch, parrot heliconia
Heliconia schiedeana – papatla, papatlilla, platanillo – heliconia
Helictotrichon sempervirens – avena perenne – blue oatgrass
Heliocarpus attenuatus – samo, samo baboso, samo blanco, sanarí, yamo baboso – sun fruit
Heliocarpus donnell-smithii – balsa, cajeta, chintule, corcho, jolotzin, jonote, jonote blanco, majagua jolocín, mozote – corkwood tree
Heliocarpus palmeri – samo ochóname, samo prieto
Heliocarpus polyandrus – palo chino, sasaró
Heliocarpus tomentosus – aguajpo, alagüé, cahualagua, copal de Veracruz, fruta sol, jonote, jonote blanco, majagua, xolotsin – sunfruit
Heliocereus = Disocactus

Heliopsis helianthoides – manzanilla loca, ojo de buey – oxeye, ringflower
Heliopsis rubra – mirasol – ox-eye
Heliotropium angiospermum – cola de alacrán, cotsnemax, heliotropo, macapura, nemaax, rabo de mico, soguilla – scorpion tail
Heliotropium arborescens (peruvianum) – heliotropo, heliotropo peruano, hierba de la vainilla – common heliotrope, garden heliotrope
Heliotropium corymbosum – heliotropo del cerro – heliotrope
Heliotropium curassavicum – alacrancillo, alacrancillo de playa, cola de alacrán, cola de mico, cotorrera de playa, hediondilla, heliotropo, heliotropo cimarrón, heliotropo de Curazao, hierba del sapo, hierba de torojo, jumete, rabo de mico – Chinese pusley, quail plant, salt heliotrope
Heliotropium europaeum – heliotropo europeo, hierba de la verruga, verrucaria – heliotrope
Heliotropium fallax – heliotropo
Heliotropium fruticosum – heliotropo, hierba de la mula, nemaax
Heliotropium hintoni – heliotropo
Heliotropium hybrida – heliotropo
Heliotropium indicum – alacrán, alacrancillo, bigotito, cola de alacrán, hierba de la mula, lengua de sapo, nemaax, rabo de mico – turnsole
Heliotropium macrostachyum – pintapan, tetetahuicha, violincillo
Heliotropium procumbens – heliotropo, hierba del gusano – creeping heliotrope
Helipterum (Acroclinium) roseum – acroclíneo – everlasting, pink and white everlasting, pink paper daisy, sunray
Helleborus foetidus – eléboro, eléboro fétido, hierba ballestera – bear's-foot hellebore
Helleborus lividus – eléboro pálido – pale hellebore
Helleborus niger – eléboro, eléboro negro, melampodo, rosa de navidad – Christmas rose
Helleborus officinalis – rosa de cuaresma
Helleborus orientalis – eléboro, rosa de cuaresma – lenten rose
Helleborus viridis – eléboro, eléboro verde – bear's foot, green hellebore
Helosis mexicana – mazorca de culebra
Helvella lacunosa – gachupín, gachupín negro, hongo, negrito
Hemerocallis fulva – azucena, azucena amarilla, azucena azafranada, azucena tabacal, flor de un día, hererocálide, lirio de San Juan – common orange daylily, tawny daylily
Hemerocallis lilioasphodelus (flava) – azucena amarilla, lirio de San Juan – lemon daylily
Henrya insularis – rama de toro
Hepatica nobilis – hepática, hepática de tres lóbulos – liverleaf
Hepaticae – hepáticas – liverworts
Heracleum lanatum – hierba del oso, raíz del oso – cow parsnip
Heracleum mantegazzianum – acanto de Cáucaso, branca ursina falsa, espondilio – giant hogweed
Herissantia crispa – malva, malva negra
Hernandia sonora – mago, palo de chicalpexte
Herniaria glabra – herniaria, herniaria glabra, hierba turca, milengrana, milgranos, quebrantapiedras – burstwort, green carpet, rupture wort
Herniaria hirsuta – herniaria hirsuta – rupturewort
Hesperaloe funifera – ixtle, pita samandoque, samandoque, yuca – yucca
Hesperaloe parviflora – ixtle – tampico fiber, red yucca
Hesperis matronalis – arbelcorán, juliana matronal, violeta damasquina – dame's rocket, sweet rocket
Hesperocallis undulata – ajo, ajo silvestre – desert lily
Hesperoyucca = Yucca
Heteranthera inuloides – chía – mud plantain
Heteranthera limosa – ha'kolel, patito – mud plantain
Heteromeles (Photinia) arbutifolia – fusique, toyón – California holly, Christmas berry, hollywood, toyon
Heteropogon contortus – barba negra, heteropogon de allioni, hierba negra de los prados, rabo de asno, retorcido moreno,

zacate aceitillo, zacate colorado, zacate retorcido – tanglehead
Heteropogon melanocarpus – barba negra dulce, retorcido moreno, retorcido negro, zacatillo, zacatoncillo – sweet tanglehead
Heteropteris laurifolia – bejuco de caballo, escobilla, matapiojos, pinzanillo
Heteropteris palmeri – compio
Heterospermum pinnatum – jarilla
Heterotheca grandiflora – manzanita de cerro, miona – telegraph weed
Heterotheca inuloides – acahual, acáhuatl, árnica, árnica del país, cuauteteco, falsa árnica, lampotillo – goldenaster, telegraph weed
Heterotheca psammophila (subaxillaris) – árnica mexicana, gordolobo – camphor weed, telegraph plant
Heuchera mexicana – falsa begonia
Heuchera sanguinea – campanilla de coral – alum root, coral bell
Hevea brasiliensis – árbol del caucho, árbol del jebe, cauchera, caucho, hule, jacia, jebe, pará, seringa, siringa – para rubber tree
Hibiscus bifurcatus – alalatz, flor de paisto, mala mujer, malva rosa del monte, tulipán del monte – mountain rose mallow
Hibiscus brasiliensis – tosapolo
Hibiscus cannabinus – cáñamo de gombó, clavellina, kenaf, pavona encendida – kenaf, thai jute
Hibiscus cardiophyllus – malva rosa del monte, tulipán del monte – heartleaf hibiscus, mountain rose-mallow
Hibiscus coulteri – hibisco, tulipán – desert hibiscus, desert rose mallow
Hibiscus elatus – majagua – blue mahue
Hibiscus (Abelmoschus) esculentus – abelmosco, ají turco, algalia, angú, café del duque, café de los pobres, café del país, café extranjero, café mareño, calalú, chaucha turca, gombó, nafe, ñajú, ocra, quingombó – okra
Hibiscus mutabilis – amistad, amistad del día, amor al uso, amor de estos tiempos, cortejo, flor de un día – Confederate rose
Hibiscus rosa-sinensis – bis, cayena, clavel japonés, cucarda, gallarda, gallardete, hibisco, hibisco chino, Jamaica, lamparilla, mar pacífico, pabona, obelisco, rosa china, rosa de China, súchil, tulipán, tulipán rojo – Chinese hibiscus, Chinese rose, rose-mallow, tropical hibiscus
Hibiscus sabdariffa – acedera de Guinea, acedera roja de Guinea, cáñamo de Guinea, flor de Jamaica, Jamaica, rosa de Jamaica, rosa jamaica, rosella, vinuela – Jamaica flower, Jamaica sorrel, red sorrel, roselle, tropical cranberry
Hibiscus tilliaceus – holol, majagua, majagua blanca, majagua de Cuba, majagua de mar, mazahua, pox – coast hibiscus, mahoe, tree hibiscus, wild cotton tree
Hibiscus trionum (syriacus) – altea, aurora común, flor de una hora, granado, hibisco, rosa de Siria – flower-of-an-hour, rose of Sharon, shrub althea
Hibiscus = Abelmoschus
Hicoria = Carya
Hidalgoa ternata (Childsia werklei) – bejuco roñoso, dalia – climbing dahlia, treasure vine
Hieracium aurantiacum – vellosilla de color anaranjado – orange hawkweed
Hieracium fendleri – lechuguilla, tihuinahua – Fendler's hawkweed
Hieracium mexicanum – lechuguilla, oreja de gato – hawkweed
Hieracium wrightii – lechuguilla
Hierochloë odorata – alestra olorosa, hierba dulce – aromatic holy grass, vanilla grass
Hilaria = Pleuraphis
Himantoglossum longibracteatum – orquídea gigante – giant orchid
Hintonia (Coutarea) latiflora – árbol amargo, amargo, campanilla, clusia, copalchi, copalchi de Jojutla, copalquín, corteza de Jojutla, falsa quina, jutitió, palo amargo, quino, quino de Michoacán, tapichoqua – clusia
Hippeastrum equestre – adonis, amarilis escarlata, azucena encarnada, azucena roja
Hippeastrum hybrida – amarilis – amaryllis
Hippeastrum puniceum – adonis, azucena, azucena de México, azucena roja, tararaco
Hippeastrum vittatum – amarilillo, amarilis, amarilis del Perú, azucena – amaryllis
Hippocratea utilis (volubilis) – bejuco colorado
Hippomane mancinella – árbol de la muerte,

hincha huevos, manzanilla, manzanilla de Indias, manzanilla de playa – manchineel

Hippophae rhamnoides – espino amarillo, espino falso – sea buckthorn

Hippuris vulgaris – correhuela hembra – mare's tail

Hirschfeldia incana – hirschfeldia incana – hoary mustard

Hirtella americana – cabeza de mico, cajetillo, icaquillo, limoncillo – pigeon plum, wild coco-plum

Hirtella racemosa – cajetillo, icaquillo, matalí – pigeon plum

Hoffmannseggia glauca (densiflora) – camote de ratón, coquito, tlamizquitl, wisachito – hog potato, Indian rush pea, pig-nut

Hoffmannseggia multijuga – tabachín chico

Holacantha = Castela

Holcus lanatus – heno blanco, holco lanoso, pasto miel – common velvet grass, Yorkshire fog

Holcus mollis – heno muelle – creeping velvet grass

Holcus = Sorghum

Holosteum umbellatum – holostea estrellada, holostea umbelada – jagged chickweed

Homalocephala = Echinocactus

Hordeum arizonicum – cebada – barley

Hordeum jubatum – cebada cimarrona, cebada crinita, cola de zorrillo – foxtail barley, squirreltail

Hordeum marinum – cebada marítima – Mediterranean barley

Hordeum murinum (leporinum) – cebada de los muros, cebadilla silvestre – wild barley

Hordeum vulgare – castilla, cebada, cebada común, cebada cultivada, cebada perla, hordio, ndexu, tago-mani, xooba-yati – barley

Horsfordia alata – malva blanca, mariola

Hosta plantaginea (grandiflora, subcordata) – hermosa de día – fragrant plantain lily

Howea (Kentia) belmoreana – kentia – sentry palm

Howea forsteriana – kentia, palma kentia – kentia palm, paradise palm

Hoya bella – flor de cera

Hoya carnosa – flor de cera, hoya, hoya de cera, hule cera, planta de cera – honey plant, porcelain flower, wax flower, waxplant

Huernia schneideriana – uernia

Hulsea vestita* ssp *callicarpha – caparossa

Humulus lupulus – espárrago falso, hombrecillo, lupo, lúpulo, zarsa, zarza del campo, zarzaparilla – common hop, European hop, hops

Humulus lupulus (americanus)* ssp *neomexicanus – lúpulo, zarsa – American hops

Hunnemannia fumariifolia – amapola, amapola amarilla – golden cup, Mexican tulip poppy

Huperzia serrata – licopodio chino

Hura polyandra (crepitans) – árbol del diablo, chico muselo, cuatatachi, cuatlatlatzin, haba, haba de Guatemala, haba del indio, haba de San Ignacio, habilla, habilla de Guatemala, habilla de San Ignacio, molinillo, ovillo, palo villa, pepino de San Ignacio, pepita de San Ignacio, quauhayohuatli, quauhtlat-latzin, salvadera, solimanché, tescalama, tronador – cannon-ball tree, monkey dinner bell, possumwood, sandbox tree

Hyacinthoides hispanica (Scilla campanulata, S. hispanica) – campánula gigante, escila española, escila hispánica – Spanish bluebell, wood hyacinth

Hyacinthoides non-scripta (Endymion non-scriptus, Scilla nutans) – campánula, jacinto silvestre – English bluebell, wood hyacinth

Hyacinthus orientalis – jacinto, jacinto común, jacinto holandés, jacinto legítimo – common hyacinth

Hyacinthus orientalis* var *albulus – jacinto romano – French Roman hyacinth, Roman hyacinth

Hybanthus mexicanus* ssp *occidentalis – jarilla

Hydrangea anomala (petiolaris)* ssp *petiolaris – hortensia trepadora – climbing hydrangea

Hydrangea arborescens – hortensia arborea, hortensia leñosa – smooth hydrangea

Hydrangea macrophylla (hortensia) – hortensia – bigleaf hydrangea, common hydrangea, garden hydrangea

Hydrangea quercifolia – hortensia de hojas de roble – oakleaf hydrangea

Hydrastis canadensis – acónito americano,

botón de oro, hidraste, hidrastis, raíz de oro, sello de oro, sello dorado – golden seal
Hydrocotyle (Centella) asiatica – gotu kola – gotu kola
Hydrocotyle bonariensis – sombrerillo – salt pennywort
Hydrocotyle mexicana – sombrerillo de agua
Hydrocotyle ranunculoides – mamalacote, ombligo de Venus, sombrerito de agua – floating pennywort
Hydrocotyle umbellata – ombligo de Venus, sombrerillo de agua – marsh pennywort, pennywort
Hydrocotyle verticillata – hidrocotule común, ombligo de Venus, pacanle – pennywort, water pennywort
Hydrocotyle vulgaris – hierba del sapo, organillo, sombrerito de agua – pennywort
Hylocereus ocamponis – pitahaya, pitaya
Hylocereus purpusii – reina de la noche
Hylocereus trigonus – pitahaya – night-blooming cereus
Hylocereus undatus – chacam, flor de cáliz, jacube, junco, junco tapatío, pitahaya, pitahaya blanca, pitahaya orejona, pitahaya roja, pitahaya de Yucatán, pitaya, pitaya orejona, reina de la noche, tasajo, zacoub – night-blooming cereus, queen-of-the-night
Hymenaea courbaril – algarrobo, anime, azúcar huayo, courbaril, cuapinol, cuapinole, cuauhpinoli, dedo maloliente, incienso de la tierra, incienso de petapa, jatoba, langosta antillana, nere, pacuy, quauhpinoli – courbaril, sausage tree, stinking tree, West Indian locust
Hymenocallis amancaes – amancay – Peruvian daffodil
Hymenocallis americana – anaite, lirio – spider lily
Hymenocallis harrisiana – amole, araña, arañuela, barba de gato, cebolla albarrana, estrella de San Nicolás, flor de araña, flor de estrella
Hymenocallis narcissiflora (Ismene calathina) – himenocalis – basket flower, Peruvian daffodil
Hymenocallis rotata – amole, amolli, araña, azucena, barba de gato, cebolla albarrana, escila del país, estrella de mar, estrella de San Nicolás, flor de estrella, lirio blanco – spider lily
Hymenoclea monogyra – jeco, jécota, romerillo, romerillo dulce – burrobrush, winged ragweed
Hymenoclea salsola – jécota – burrobrush, burrowbush
Hymenoxys (Actinea) odorata – amargosa – bitterweed actinea, poison bitterweed
Hymenoxys richardsonii – pingué – colorado rubber-weed, owl's foot, pingwing, rubber bush
Hymenoxys (Actinea, Tetraneuris) scaposa – pingué – clustered goldflower, slender-stem bitterweed, yellow daisy
Hyophorbe lagenicaulis – palmera botella – bottle palm
Hyoscyamus niger – beleño, beleño negro, casílago, dulcamara negra, hierba de la punzada, hierba loca, ojo del diablo – black henbane
Hyparrhenia hirta – cimbopogon mediterráneo, jaraguá, jaraguá gris – haired brush grass
Hyparrhenia rufa – bermejo, hierba bermeja, hierba del Brasil, jaraguá, jaraguá bermeja, pasto jaraguá, pasto puntero, provisor, sape gigante, yagale, zacate bermejo, zacate jaraguá, zacate puntero – jaragua grass
Hypericum androsaemum – androsemo, hipérico androsemo, hipérico arbustivo, todabuena, todasana – St. John's wort, tutsan
Hypericum calycinum – granado blanco, hierba de San Juan, hipérico, hipérico calicino, hipericón, rosa de Siria, rosal de Siria – Aaron's beard, creeping St. John's wort, rose of Sharon
Hypericum coris – corazoncito, hipérico de hojas de conífera – needle-leaved St. John's wort
Hypericum formosum – grano de oro, hipericón – St. John's wort
Hypericum perforatum – corazoncillo, cori, espantadiablos, hierba de la herida, hierba de San Juan, hipérico, hipericón, hipérico perforado, perforada – Klamathweed, St. John's wort
Hypericum pratense – sanguinaria, tenchalita
Hyphaene thebaica – duma – doum plum

Hypochaeris glabra – hierba de halcón lisa – smooth cat's-ear
Hypochaeris radicata – hierba de halcón radicada – rough cat's-ear
Hypopitys = Monotropa
Hyptis albida – babino, orégano, salvia, salvia real – bushmint
Hyptis collina – salvia para la misa
Hyptis emoryi – chía, hierba del becero, lavanda, salvia, salvia del desierto, salvia real – bee sage, desert-lavender
Hyptis laniflora – salvia, salvia real – desert lavender
Hyptis seemannii – confitura, jeco, joco, pestón, salvia
Hyptis suaveolens – chan, chía, chía de Colima, chía gorda, chía grande, confitura, confiturilla, conibari, mastranzo, orégano cimarrón
Hyssopus officinalis – hisopo, hisopo común – blue hyssop, hyssop

I

Iberis amara – carraspique amara, zarapinto – hyacinth-flowered candytuft, rocket candytuft
Iberis gibraltarica – carraspique de Gibraltar – Gibraltar candytuft
Iberis intermedia – carraspique intermedia – candytuft
Iberis sempervirens – carraspique, carraspique siempreverde, cesta de plata, zarapinto – evergreen candytuft
Iberis umbellata – carraspique, carraspique umbelada, pinito de flor, zarapinto – globe candytuft
Ibervillea insularis – melón de coyote – coyote melon
Ibervillea sonorae – calabaza amarga, guareque, güierequi, melón de coyote – coyote melon
Ichnanthus palleus – alpiste
Idria = Fouquieria
Ilex* x *altaclerensis – acebo híbrido de highclere – highclere holly, Wilson holly
Ilex aquifolium – acebo, acebo común, agrifolio, aquifolio – English holly, Christmas holly
Ilex brandegeana – junco serrano, manzanita
Ilex crenata – acebo – Japanese holly
Ilex decidua – yaupon – possum haw
Ilex dugesii – naranjillo
Ilex nitida – briqueta, cuero de sapo, hueso prieto
Ilex opaca – acebo – American holly
Ilex paraguariensis – árbol de la hierba, hierba del Paraguay, hierba mate, mate, té de Paraguay – mate, matte
Ilex pernyi – acebo chino de Perny
Ilex rubra – chilillo – holly
Ilex tolucana – aceituna, aceitunillo, limoncillo, orilla de arroyo, palo extraño, tepezapote – holly
Ilex tolucana* var *californica – palo extraño
Ilex vomitoria – yaupon – evergreen holly, yaupon
Illicium anisatum (verum) – anís de la China, anís estrella, anís estrellado, badiana – anise shrub, Chinese anise, Japanese anise, star anise
Illicium floridanum – ixcapatl, matacaballos – Florida anise tree, poison bay, sweet laurel
Impatiens balsamina – balsamina, balsamina cultivada, Belén, chino, coneja, flor de China, madama, miramelindo, Nicaragua – balsam, garden balsam, impatiens
Impatiens capensis – balsamina del Cabo, balsamina del monte, hierba de Santa Catalina – jewelweed, orange balsam
Impatiens glandulifera – balsamina glandulífera, hierba de Santa Catalina
Impatiens pallida – balsamina, balsamina manchada silvestre, mírame-y-no-me-toques – balsam, jewelweed, spotted touch-me-not
Impatiens walleriana (sultani) – balsamina nicaragua, Belen, chinito, coqueta, gliceria acuática, alegría de la casa, miramelindo – busy Lizzie, sweet sultan
Imperata brasiliensis – zacate clavo, zacate cortador, zacate de venado
Imperata cylindrica – choza de paja, cisca, cogón, imperata, yabona, zacate de akalcheses – lalang, spear grass
Incarvillea delavayi – gloxínea de intemperie, incarvillea – incarvillea, trumpet flower
Indigofera fruticosa – añil montés – anil
Indigofera gerardiana (heterantha) – índigo – indigo
Indigofera nelsonii – rama prieta – anil
Indigofera suffruticosa – añil, añil colorado, añil índigo, añil jiquelite, añil montés, azuelo, ch'oh, cornezuelo, hierba turqueza, índigo, jiquilite, mahuitli, pitzahuac, tlaceoitli, xiquelite, xiuquilitl – indigo
Indigofera tinctoria – añil blanco, xiuquilitl
Inga edulis – bitze, chalahuite, cuil de agua, guamo, huitz, pacay – food inga
Inga spuria – agopote, bitze, chalahuite, chelele, coctzán, coyuacate, cuajinicuil, cuajinicuil silvestre, cuil de agua, guatope, jinicuile, san, tiltikchalauij, timbre, vainilla – inga, star tree
Inocarpus edulis – nuez de Tahití – Polynesian chestnut
Inodes = Sabal
Inula helenium – ala, astabaca, énula, énula campana, énula griega, helenio, hierba del ala, hierba del moro, olivarda – elecampane, elf-dock, horse-elder

Ionidium polygalaefolium – hierba de San Nicolás, ipecacuana del país, ipecacuana de México, quimixcuitlaxcoli, tripa de ratón
Iostephane heterophylla – cuatolotlanenzi, escorzonera, hierba contra los venenos, hierba del manso, huichocachacua, liga, quauhtolotlanenci, raíz de liga, raíz del manso, tlacopatli, tlalpopolote
Iostephane madrensis – cachana
Ipomoea alba (Calonyction aculeatum) – amule, bejuco de cuajar hule, bejuco de puerco, bejuco de vaca, cuajaleche, flor de luna, gamuza, guamol, hedrón, mash, nata, nicua, oración, zutub – moonflower
Ipomoea ancisa – baburia, romerillo de la sierra – morning-glory
Ipomoea aquatica – batata acuática, espinaca acuática, espinaca de agua, tangkong – Chinese water spinach
Ipomoea arborescens – cazahuate, cazahuate blanco, itate, ozote, pájaro bobo, palo blanco, palo bobo, palo cabra, palo del muerto, palo santo, palo santo amarillo, patancán, quauhzahuatl, xegua – morning-glory tree, tree morning-glory
Ipomoea batatas – axi, batata, batata dulce, boniato, buniato, camajtli, camote, camote amarillo, camote blanco, camote morado, camotic, camotli, guniró, iz, manta, moniato, papa dulce, uarasu – sweet potato, yam
Ipomoea bona-nox – dama de noche – moonflower
Ipomoea cairica – aurora – Cairo morning glory, five-leaf morning glory, mile-a-minute, railway creeper
Ipomoea capillacea – coquito
Ipomoea cristulata – higuerilla
Ipomoea desertorum – trompillo
Ipomoea fistulosa – amapola, flor de la mañana, palo santo de Castilla
Ipomoea hederacea – enredadera de campanilla, flor de verano, manto de la Virgen, trompillo morado – ivyleaf morning-glory
Ipomoea heterophylla – trompillo
Ipomoea hirsutula – manto mexicano, trompillo
Ipomoea lindheimeri – manto de la Virgen – blue morning-glory
Ipomoea longifolia – alcaparra, cebolleta, huirote, mataliste – longleaf morning-glory
Ipomoea macrorrhiza – batatilla de playa, zacualtipan
Ipomoea muricata (Calonyction muricatum) – maxh, nata, San Miguelito, trompillo
Ipomoea murucoides – árbol de la muerte, árbol del venado, cazahuate, cazahuate prieto, cazahuatl, micacuáhuitl, ozote, pájaro bobo, palo bobo, palo del muerto, palo de ozote, palo santo – morning-glory tree
Ipomoea nil – enredadera de campanilla – ivy morning-glory, morning-glory
Ipomoea orizabensis – escamonea– Mexican scammony
Ipomoea pedicellaris – mantel de María
Ipomoea pes-caprae – bejuco de playa, cambule, hierba de la raya, pata de vaca, riñonina, tripa de aura – beach morning glory, goat-foot creeper, morning-glory, railroad vine, sea-grape
Ipomoea plummerae – trompillo – plummer's morning-glory
Ipomoea purga (Exogonium jalapa) – camotic, jalap, jalapa, jalapa de Veracruz, Juan el conquistador, limoncillo, purga, purga de Jalapa, raíz de Jalapa, tlanoquiloni, tolompatl – jalap
Ipomoea purpurea – aurora, batata pavonada, campanilla, campanita, correhuela, dondiego de día, enredadera de campanilla, gloria de la mañana, hiedra morada, manto, manto de la Virgen, mecapatli, metlacasis, ololiuquí, quiebra plato, trompillo, xhail – common morning-glory
Ipomoea quamoclit (Quamoclit pinnata) – bejuco estrella, cabellos de ángel, cambustera, chisme, chismito, clarín, clarincillo, cundeamor, estrella, estrella del sol, hiedra roja, lágrimas de la Virgen, regadero, trompillo – cardinal climber, cupid flower, cypress vine, Indian pink, red jasmine, star glory
Ipomoea scopulorum – aurora, hiedra morada, jalapa limoncilla – morning-glory
Ipomoea sinuata (Merremia dissecta) – correhuela de las doce – alamo vine, noon flower
Ipomoea stans – cacastlapa, castlapa,

espantalobos, espantavaqueros, limpiatunas, manto, maromero, pegajosa, quiebra plato, tanibata, tlaxcapán, tumba vaquero – morning-glory

Ipomoea tricolor – badoh negro, badungás, campanilla, dondiego de día, enredadera de campanilla, ipomea, manto, manto de la Virgen, quiebra plato, rompeplato, yaxcelil – heavenly-blue morning-glory, morning-glory

Ipomoea triloba – trompillo, trompillo blanco, trompillo lila

Ipomoea valida – manto

Ipomoea violacea (rubrocaerulea) – badoh negro, badungás, hiedra terrestre, hierba María, machacuana, manto de la Virgen, piule, semilla de la Virgen, tlitliltzin – heavenly blues, morning-glory, pearly gates

Ipomoea = Exogonium, Merremia

Ipomopsis = Gilia

Iresine calea – acapatle, amarasillo, amargosillo, atlatlancuaya, carricillo, cecepatli, cecetzin, cerote de coche, hierba de la calentura, hierba del frío, hierba del tabardillo, nemaxrla, pelusita, pie de paloma, salvia, siempre me veras asi, tascuaya, tepozán, tlatlancuaye – bloodleaf

Iresine celosioides – gusanera, gusanillo, pie de paloma, tlatón, toncnaya

Iresine grandis – pie de paloma, tepozán

Iresine herbstii – quelite morado, siempre me veras asi – bloodleaf

Iresine paniculata – hierba de la hinchazón

***Iriartea* spp** – jira – stilt palm

Iris barbata – lirio – bearded iris

Iris ensata (kaempferi) – lirio japonés – Japanese iris

Iris florentina – flor de lis, lirio blanco, lirio de Florencia, lirio florentino – Florentine iris

Iris foetidissima – efémero, espadaña hedionda, espátula fétida, íride, jíride, lirio hediondo – coral iris, gladwin iris

Iris germanica – arco iris, flor de lis, lirio, lirio alemán, lirio azul, lirio cárdeno, lirio común, lirio de Alemania, lirio germánico, lirio iris, lirio morado, lirio pascual – bearded iris, blue flag, fleur-de-lis, garden iris, German iris, iris, white flag

Iris* x *hollandica – iris, lirio de Holanda, lirio holandés – Dutch iris

Iris innominata – lirio rocoso – Del Norte county iris

Iris latifolia – lirio de Inglaterra – English iris

Iris missouriensis – lirio – Rocky Mountain iris, western blue flag

Iris pseudacorus – ácoro bastardo, espadaña fina, falso ácoro, lirio amarillo, lirio espadañal, rabiacán – fleur de lis, yellow flag, yellowflag iris

Iris pumila – lirio enano – dwarf iris

Iris reticulata – lirio reticulado – reticulata iris

Iris siberica – lirio siberiano, lirio sibirica – Siberian iris

Iris sisyrinchium – mazuáa, patita de burro – Barbary nut

Iris unguicularis – lirio unguiculado – winter iris

Iris versicolor – lirio azul, lirio multicolor – blue flag

Iris xiphium – lirio de España, lirio español – Spanish iris

Irvingia gabonensis – árbol chocolate – dika, wild mango

Isatis tinctoria – glasto, hierba pastel, hierba pastel de los tintoreros – dyer's woad, woad

Ismene = Hymenocallis

Isocoma menziesii (Aplopappus, Bigelovia, Haplopappus venetus) – boxosdá, damiana, falsa damiana, jarilla chura – goldenbush

Isolatocereus (Lemaireocereus, Stenocereus) dumortieri – candelabro, órgano, pitaya, pitaya cimarrona – candelabra cactus

Isomeris arborea (Cleome isomeris) – ejotillo – bladderpod, burro fat

Isomeris (Cleome) spinosa – barbona rosada – common spider flower

Isonandra = Palaquium

Iva ambrosiaefolia – amargosa – marsh elder, poverty weed, wormwood

Ixia maculata – ixia – African corn-lily, ixia

Ixora coccinea – cruz de Malta, isoque, ixora roja, lantán, Santa Rita – burning love, flame of the woods, ixora, jungle flame

J

Jacaranda mimosifolia (acutifolia) – flamboyán azul, flor azul, framboyán azul, gualanday, guarupa, jacarandá, jacarandá de hojas de mimosa, palisandro, palo de buba, tabachín, tarco, yarabisco – jacaranda
Jacaratia = Leucopremna
Jacobinia umbrosa (aurea) – aloloé amarillo, cola de ardilla, cresta de gallo, mohuitli, monte de oro, pluma de oro, vara de San José
Jacobinia = Dicliptera, Justicia
Jacquemontia abutiloides – campanilla – morning-glory
Jacquemontia pringlei – San Miguelito
Jacquinia aurantiaca – chacsis, churutana, flor de ensartar, flor de nacimiento, flor de niño, guie-zee, hoitxóchitl, mata de San José, muyché, neucxóchitl, palo de las ánimas, profeta, rosadillo, rosalía, sacramento, San Juan, San Juanito, sicaján, sicaté, sicche, sicjá, siqueté, xochipatli
Jacquinia macrocarpa (pungens)* ssp *pungens – barbasco, corpus, limoncillo, palo de San Juan, pinicua, rosadillo, San Juanito, siqueté, tásiro – cudjoe wood
Jaltomata = Saracha
Jambos = Eugenia
Janusia californica – quechesohui
Janusia gracilis – fermina, gallinita, matanena
Janusia linearis – gallinita, matanena
Jarilla caudata – bonete, jarrilla
Jarilla chocola – capiaj, chócola
Jarilla heterophylla – granadilla, jarrilla
Jasminum grandiflorum (officinale) – granduque, guirnalda, ilanilan, jazmín, jazmín blanco, jazmín común, jazmín de Arabia, jazmín de España, jazmín de flores grandes, jazmín de la India, jazmín de olor, jazmín italiano, jazmín real – common white jasmine, poet's jasmine, Spanish jasmine
Jasminum humile – gelsemio – Italian jasmine, yellow jasmine
Jasminum mesnyi (primulinum) – jazmín amarillo – primrose jasmine
Jasminum multiflorum – guirnalda – downy jasmine
Jasminum nudiflorum – jazmín de invierno, jazmín de San José, jazmín nudifloro – winter jasmine
Jasminum polyanthum – jazmín de China, jazmín rosa
Jasminum sambac – diamela, jazmín, jazmín de Arabia, sampaguita – Arabian jasmine, pikake
Jatropha angustidens – caribe, mala mujer, mala mujer china, quemador, tachinole
Jatropha cardiophylla – chichioaquáhuitl, ensangregado, matacora, sangre de Cristo, sangre de drago, sangregado, sangregado azul, sangregado chaparro, sapo, tacote prieto, torote, torote común – dragon's blood, limberbush
Jatropha cinerea – boroquim, ensangregado, lomboy, lomboy blanco, nombó, sangre de drago, sangre de grado, sangregado, sapo, torotillo, torotito – ashy limberbush, lomboy
Jatropha cordata – copalillo, jay sapo, jiotillo, matamala, matamuchachos, Miguelito, papelillo, sangregado, sapo, to'oro mo'mo, torote, torote blanco, torote papelillo, torote prieto, tzontecpatli – leatherplant, Sonoran tree-limberbush
Jatropha curcas – achcuauit, avellano purgante, axté, coquillo, cuipuy, frailejón, matacora, matamuchachos, nacuala, niin, piñón, piñoncillo, piñón de Indias, piñón purgante, quahaychuachilli, quauhayohuachtli, ricino de América, sangre de Cristo, sangre de drago, sangregado, tempate, vico, xcacalche, zicilté, ziquiltle – Barbados nut, physic-nut
Jatropha gossypiifolia – frailecillo, mala mujer, triatua
Jatropha macrorhiza – jalapa, jicamilla, purga macha, sangre de drago
Jatropha malacophylla – jeyó, ratowa, sangregado, sapo – sycamore-leaf jatropha
Jatropha platanifolia – ensangregado
Jatropha podagrica – capa de rey, ruibardo – gout plant, Guatemala rhubarb, tartago
Jatropha spathulata (cuneata, dioica) – batácora, coatli, drago, gualulo, matacora, piñón de cerro, sangre de drago, sangre de grado, sangregado, sapo, telondillo, tlapalezpatli, torote, torote amarillo, torote prieto – leatherplant, limberbush, rubber plant

Jatropha vellanowethii – lomboy vida
Jatropha vernicosa – lomboy brilloso, lomboy colorado, lomboy negro, lomboy rojo, sangre de drago, sangregado
Jatropha = Cnidoscolus
Jessenia bataua – patana, seje – bataua palm
Jessenia polycarpa – ungurahui
Juania australis – chonta
Jubaea chilensis – palmera del vino – Chilean wine palm
Jubaea spectabilis – coco de Chile, coquito – Chile wine palm
Juglans californica – nogal – California black walnut
Juglans cinera – nogal blanco, nogal blanco americano, nogal ceniciento, nogal de Cuba, nogal gris americano – butternut
Juglans honorei – tocte – Ecuador walnut
Juglans major – nogal, nogal de cerro, nogal silvestre – Arizona walnut, nogal
Juglans mexicana – namboca
Juglans microcarpa – namboca, nogal, nogalillo, nogalito – Mexican walnut
Juglans mollis – nogal, nogal encarcelado, nogal nuez meca, nuez meca
Juglans nigra – nogal americano, nogal negro, nogal negro americano, nuez americana, nuez de nogal, nuez negra, tocte – black walnut
Juglans regia – nogal, nogal común, nogal de Castilla, nogal español, nogal europeo, nuez, nuez de Castilla, nuez de Europa, nuez de nogal, nuez grande, quaucacoatl – English walnut, Persian walnut
Juglans rupestris – nogal
Juncus acutus – junco, junco agudo – rush, spiny rush
Juncus ambiguus – junco de las ranas – frog rush
Juncus articulatus – junco articulado – jointed rush
Juncus balticus – junco báltico, tulillo – Baltic rush, wiregrass
Juncus bufonius – junco de los escuerzos – toad rush
Juncus capitatus – junco capitado – rush
Juncus cooperi – junco – rush
Juncus dudleyi – junco de Dudley – Dudley's rush
Juncus ebracteatus – tulillo
Juncus effuses – junco, junco de esteras, junquera, tulillo – bog rush, rush, wire-grass
Juncus ensifolius – junco de hojas ensiformes – sword-leaved rush
Juncus mexicanus – junco, tulillo – Mexican rush, rush
Juncus tenuis – junco tenue – slender rush
Juncus torreyi – tulito – rush
Juniperus ashei – cedro, cedro blanco, enebro, sabina, táscate – Ashe juniper
Juniperus blancoi – cedro, táscate – blanco juniper
Juniperus californica – cedro, ciprés, huata, táscate, tláxcal – California juniper
Juniperus cedrus – cedro canario, cedro de Canarias – Canary Island juniper
Juniperus chinensis – enebro de la China
Juniperus coahuilensis – huata, táscate – rose-fruited juniper
Juniperus comitana – cedro, ciprés, enebro, táscate – comitan juniper
Juniperus communis – arabota, archenas, brujo, cedro, enebro, enebro achaparrado, enebro albar, enebro común, enebro espinoso, enebro morisquillo, enebro rastrero, enebro real, ginebra, gorojo, ijinebre, jabino, jinebro, junípero, nebrina, nebro, orrea, pinchorrero, rastrero, rosafin, sabina macho, sabina morisca, táscate – common juniper
Juniperus communis* ssp *nana – enebro enano – dwarf juniper
Juniperus deppeana – aorí, cedro, cedro chino, cedro de incienso, huata, pino, sabina, táscate, táscate blanco, tláxcal – alligator juniper
Juniperus drupacea – enebro de Grecia – Syrian juniper
Juniperus durangensis – táscate, táscate espinosa – Durango juniper
Juniperus excelsa – sándalo – algum tree, Grecian juniper, savin
Juniperus flaccida – cedrillo, cedro, cedro blanco, cedro colorado, cedro liso, cedro tasco, enebro, sabina, táscate, tláxcal – Mexican drooping juniper, weeping juniper
Juniperus gamboana – cedro, ciprés, táscate – Gamboa juniper
Juniperus horizontalis – enebro enano
Juniperus jaliscana – cedro

Juniperus mexicana – cedro, cedro colorado, ciprés silvestre, enebro, sabina, táscate – Mexican juniper
Juniperus monosperma (erythrocarpa) – cedro, sabina, sabina de la sierra, táscate – one-seed juniper
Juniperus monticola – cedro, cedro blanco, cedro colorado, ciprés, enebro, junípero, sabina, táscate, tláxcal – juniper
Juniperus monticola* var *compacta – tláxcal
Juniperus osteosperma – cedro, cedrón, sabina, sabina morena, táscate – Utah juniper, western juniper
Juniperus oxycedrus – enebro, enebro albar, enebro de bayas rojas, enebro de la miera – cedar, prickly juniper
Juniperus patoniana – táscate
Juniperus phoenicea – enebro fenicio, sabina, sabina negra
Juniperus pinchotii – cedro, táscate – redberry juniper
Juniperus recurva – enebro llorón
Juniperus sabina – enebro español, enebro sabina, sabina, sabina común – savin juniper
Juniperus scopulorum – cedro, cedro rojo, sabina, táscate – Rocky Mountain juniper, western redcedar
Juniperus sibirica – pino macho, sabina macho – dwarf juniper
Juniperus squamata – enebro escamoso
Juniperus standleyi – táscate
Juniperus thurifera – cedro de España, junípero turífero, sabina albar, sabina española, sabina turífera
Juniperus virginiana – cedro de Virginia, cedro rojo, enebro, enebro americano, enebro de Virginia, enebro virginiano, sabina de Virginia – eastern red cedar
Jussiaea = Ludwigia
Justicia aurea – cola de ardilla, monte de oro, pluma de oro – water willow
Justicia brandegeana (Beloperone (Drejella) guttata) – carpintero, cola de camarón, corazón, flor de camarón, moco de guajolote – false hop, shrimp plant
Justicia (Beloperone) californica – beloperone, chuparrosa, koomeme, semaluca – beloperone, California beloperone, chuparosa, honeysuckle, hummingbird flower
Justicia (Jacobinia) candicans – chuparrosa, espuela de caballero, rama de venado
Justicia (Jacobinia) carnea – isopo rojizo, jacobina, sacatinta, tango rojizo – Brazilian plume flower, flamingo plant
Justicia (Beloperone) purpusii – chuparrosa – hummingbird flower
Justicia (Jacobinia) spicigera – charaitzicua, hierba añil, hierba azul, hierba tinta, mirto del cerro, mohuitle, moyote, moyotli, muicle, siitz, trompetilla, yichcaan – Mexican honeysuckle
Justicia = Dicliptera

K

Kalanchoe blossfeldiana – calancoe – flaming Katy, kalanchoe

Kalanchoe pinnata (Bryophyllum pinnatum) – admirable, amor tras de la puerta, baiburín, bruja, flor de aire, hierba bruja, hoja bruja, hoja del aire, hoja fresca, maravillosa, prodigiosa, sanalotodo, siempreviva, sinvergüenza, sisalxiu, tzitzalxiu – airplant, life plant

Kallstroemia californica – contrahierba, golondrina, jimuri, mal de ojo – California caltrop, California carpetweed

Kallstroemia grandiflora (Tribulus grandiflorus) – abrojo de flor amarilla, baiburín, mal de ojo, San Miguelito – Arizona poppy, desert-poppy, Mexican poppy, orange caltrop, summer poppy

Kallstroemia maxima – bola de hilo, geonduna, tripa de pollo

Kallstroemia parviflora – baiburín, mal de ojo – warty carpetweed

Kallstroemia peninsularis – palagallina, pelagallina

Kalmia latifolia – calico, calmia, laurel de montaña – calico bush, mountain laurel

Kalopanax septemlobus (pictus) – espino del aceite de castor, sen – castor aralia, tree aralia

Karwinskia humboldtiana – aahuatin, árbol que tulle, aroyoguo, cacachila, cacachila china, cacachila silvestre, capulín, capulincillo, capulincillo cimarrón, capulín cimarrón, capulín de zorra, carabullo, chanchanote, cholchonote, coyotillo, diente de molino, espinosilla, frutillo, guasalaco, himoli, itzil, jimuari, margarita, margarita del cerro, negrito, palo apestoso, palo negrito, pimientillo, piojillo, tlalcapolín, tullidor, yagalán, zazacatzin – coffeeberry, dot and dash plant, Humboldt coyotillo

Keckiella (Penstemon) antirrhinoides – romerillo – yellow bush penstemon

Kentia = Howea

Kerria japonica – malva de los Judios, querria, querria japonesa – Jew's mallow

Khaya senegalensis – caoba – African mahogany

Kickxia elatine – desparramadora, hierba del perro, kickxia elatina – fluellin

Kickxia spuria – kickxia de hojas circulares – fluellin

Kigelia africana – árbol de las salchichas – African sausage tree

Knautia arvensis – escabiosa arvense, viuda silvestre – blue buttons

Kniphofia (Tritoma) uvaria – atizador al rojo vivo, bandera española, tritomo, tritomo rojo – red-hot poker, torch-lily

Kobresia bellardii – cobresia falsa-cola de ratón, elina de bellard

Kochia scoparia – ciprés de verano escobiforme, mirabel, morenita – belvedere, broom cypress, mock cypress, summer cypress

Koeberlinia spinosa – abrojo, corona de Cristo, junco – allthorn, crown-of-thorns, crucifixion thorn

Koeleria macrantha (cristata, pyramidata) – koeleria grácil, zacate de cresta – junegrass

Koeleria phleoides – lofochloa crestada – annual hairgrass, crested hairgrass

Koelreuteria bipinnata – árbol de la lluvia chino – Chinese flame tree

Koelreuteria paniculata – árbol chino del barniz, árbol de los farolitos, jabonera de la China, orgullo de la India – goldenrain tree, pride of India

Kohleria deppeana – caracolillo, lágrimas de San Diego, tatachinole, tlachichinole, tochimitillo, tochómitl – tree gloxinia

Kola (Cola) acuminata – cola, colatero, nuez de cola, nuez de gouro – kola nut, kola tree

Kosteletzkya thurberi – limpiatunas

Krameria cytisoides – donapé

Krameria erecta (parvifolia) – chacate, cósahui, cósahui del sur, mezquitillo, pima, tamichil – pima rhatany, purple heather, range rhatany, rhatany

Krameria grayi (canescens) – chacate, cósahui, cósahui del sur, crameria, mamelique, mezquitillo, ratania, temitzo – crimson beak, white rhatany

Krameria lanceolata – chacate, crameria – range rhatany, rhatany

Krameria paucifolia – mezquitillo

Krameria ramosissima – calderona

Krameria secundiflora – crameria, mezquitillo, raíz de cuculillo, zarzaparrilla
Krameria trianda – mapato, pumacuchu, ratania, ratania del Perú, ratania para los dientes, ratania peruana – Peruvian rhatany
Krascheninnikovia (Eurotia) lanata – romerillo – winter fat

L

Lablab purpureus (Dolichos lablab, Rhynchosia lablab) – carmelita, chícharo, dólico, dólico gigante, dólicho, dólicho gigante, dólicho rongai, frijol caballero, frijol de adorno, frijol de enredadera, frijol jacinto, frijol punzada, frijol trepador, gallinazo blanco, gallinazo morado, gallinita, garbanzo indio, indianela, judía de Egipto, lablab – hyacinth bean

Laburnum alpinum – codeso alpino, codeso de los Alpes, laburno escocés – Scotch laburnum

Laburnum anagyroides (Cytisus laburnum) – cítiso, codeso, ébano falso, falso ébano, laburno, laburno común, lluvia de oro – common goldenchain, golden chain tree

Laburnum* x *watereri – laburno híbrido de Voss

Laccaria laccata – hongo manzana, manzanilla, manzanita, tejamanilero – the deceiver

Lachenalia aloides – prímula del Cabo – Cape cowslip

Lactarius deliciosus – chilpán, enchilado, enchilado de ocote, mízcalo, oreja de puerco, rovelló, rubellón – delicious milk mushroom, saffron milk cap

Lactarius piperatus – trompa – peppery milk cap

Lactarius vellereus – borrego – velvet milk cap

Lactuca canadensis – lechuga silvestre – lettuce

Lactuca saligna – lechuga de hojas de sauce – least lettuce

Lactuca sativa* var *angustana – lechuga espárrago

Lactuca sativa* var *capitata – lechuga, lechuga acogollada, lechuga arrepollada, lechuga de bola, lechuga orejona común – head lettuce, iceberg lettuce

Lactuca sativa* var *crispa – lechuga de hojas sueltas, lechuga de repollo abierto – butter lettuce

Lactuca sativa* var *longifolia – lechuga, lechuga cos, lechuga de hojas, lechuga orejona, lechuga romana, lechuga romanita, lechuguino – leaf lettuce, romaine lettuce

Lactuca serriola – lechuga espinaca, lechuga silvestre, lechuguilla – prickly lettuce

Lactuca tatarica – lechuga tártara – Tartaric lettuce

Lactuca virosa – lechuga silvestre, lechuga venenosa, lechuga virosa, serrallón – bitter lettuce

Laelia albida – flor de San Francisco, monjita, tzicxóchitl

Laelia anceps – flor de muerto, flor de San Diego, flor de San Miguel, güichila, lirio, lirio del monte, lirio de todos santos, San Dieguito, tzicxóchitl, vara de San Diego

Laelia autumnalis – ahuasúchil, ahuaxóchitl, chichiltepetzacuxóchitl, Diego, flor de calavera, flor de catarinas, flor de las ánimas, flor de los santos, flor de muerto, flor de todos santos, kikí, lirio, lirio de San Francisco, San Diego, tzacuxóchitl, tzautli – autumn-flowering laelia

Laelia (Brassavola) digbyana – flor de calaverita, monjita

Laelia furfuracea – flor de San Francisco, lirio, monjita

Laelia (Brassavola) glauca – candelaria blanca, flor blanca de la candelaria, lirio blanco – glaucous-stemmed brassavola

Laelia grandiflora – chichiltepetzacuxóchitl, flor de corpus, flor de mayo, hierba de corpus, itzmaqua

Laelia (Brassavola) nodosa – bailarina, barba de viejo, oreja de burro, sah-yak

Laelia speciosa (majalis) – corpus, flor de corpus, flor de mayo, itzmaqua, lirio de corpus – Corpus Christi flower, May-flowering laelia

Laelia superbiens – flor de la candelaria, flor de la candelaria de vara larga, flor de la candelaria grande, flor de niño, palo de águila, tanal, vara de San José – gorgeous laelia

Laelia (Schomburgkia) tibicinis – cacho de toro, confesionario, dac kisin, flor de cacho, fotuto, homicin – trumpeter's schomburgkia

Lagascea decipiens – confiturilla, confiturilla amarilla, confiturilla grande

Lagenaria siceraria (vulgaris) – acocote,

acócotl, alacate, buli, bush, calabacero, calabaza, calabaza de peregrino, calabaza de San Roque, calabaza vinatera, camasa, chinchin, cogorda, cuyabra, estropajo silvestre, guaje, guaje de peregrino, guajito, güiro, huaje, huaje de los niños, huajito, huajito de niños, itzu, leque, mux, pepino chino, pumpo, sahualca, tecomate, uili, xigaba – bottle gourd, white-flowered gourd

Lagerstroemia indica – árbol de Júpiter, astromelia, astromero, astronómica, atmósfera, atmosférica, crespón, crespón de China, Júpiter, reina de las flores – crape myrtle, grape flower, June rose, pride of India

Lagerstroemia speciosa – astromelia, astromero, flor de la reina, reina de las flores – giant crape myrtle, queen flower, queen of flowers, queen's crepe myrtle

Lagunaria patersonii – árbol pica-pica, lagunaria – cow itch tree, primrose tree

Laguncularia racemosa – cincahuite, mangle, mangle amarillo, mangle blanco, mangle bobo, mangle chico, mangle chino, mangle prieto, moyet, patabán, posillo tosa, zacolcom – white buttonwood, white mangrove

Lagurus ovatus – cola de conejo, cola de liebre, lágrimas de la Virgen – hare's tail

Lallemantia iberica – melisa de Moldavia

Lamarckia aurea – cepillito, grama dorada, lamarckia dorada – goldentop

Lamiastrum = Lamium

***Laminaria* spp** – alga marina, quelpo – kelp, seaweed

Lamium amplexicaule – lamio, ortiga muerta – dead nettle, henbit

Lamium (Lamiastrum) galeobdolon – ortiga muerta amarilla, ortiga muerta menor – dead nettle, yellow archangel

Lamium maculatum – lamio maculado, ortiga muerta maculada – dead nettle, spotted nettle

Lamium purpureum – lamio púrpura, ortiga muerta, ortiga muerta purpúrea – dead nettle, red henbit

Lamourouxia multifida – jarrito

Lamourouxia rhinanthifolia – jarrito

Lampranthus deltoides – lampranthus – ice plant

Landolphia florida – mangaba – gum vine

Lansium domesticum – ayer ayer, lansa, lanzón – langsat

Lantana brasilensis – hierba sagrada

Lantana camara – alantana, albahaca cimarrona, albahaca de caballo, alfombrilla hedionda, bandera española, camará, cariaquito colorado, chancaquilla, chancara xiutl, chichiquelite, cinco negritos, coahuitzxóchit, confite, confite negro, confitura, confitura blanca, confiturilla, confiturilla blanca, corona de sol, coronita del sol, filigrana, flor de San Cayetano, frutilla, hierba de Cristo, ikilhaxiu, lantana, margarita del campo, matizadilla, maxte, mocseté, mora, moxete, ojo de pescado, orozuz, orozuz del país, palabra de caballero, palabra de mujer, peonía negra, petelkín, piltzintecouhxóchitl, quita pesar, siete colores, sonora, sonora roja, tarepe, tomatillo, tres colores, uña de gato, venturosa, xobarroba, xohexnuc, zapatillo, zapotillo, zarzamora – cherry pie, lantana, shrub verbena, wild sage

Lantana camara* var *nana – bandera española

Lantana glandulosissima – confiturilla amarilla, salvia real

Lantana hispida – ojo de ratón

Lantana horrida – confiturilla, confiturilla negra – calico bush, Texas lantana

Lantana involucrata (velutina) – caca de mono, chapul, cinco negritos, confite, confiturilla, confiturilla blanca, duraznillo, manzanita, mocseté, orégano, peonía, peonía colorada, sonora, tarepe, tilihuete, uvalama, zicilhaxiu

Lantana macropoda – majorana, mayorana, mejorana – desert lantana, Mexican marjoram

Lapageria rosea – copihue – Chilean bellflower

Lappa = Arctium

Lappula squarrosa – lapolla común – stickseed

Lapsana communis – lámpsana, lámpsana común – nipplewort

Larix decidua (europaea) – alerce, alerce común, alerce deciduo, alerce de Europa, alerce europeo, alerce melis, lárice, pino negral – European larch

Larix gmelinii – alerce dahuriano, alerce de Gmelin – Dahurian larch
Larix kaempferi – alerce del Japón, alerce japonés – Japanese larch
Larix laricina – alerce americano oriental, tamarack – eastern larch, hackmatack, tamarack
Larix occidentalis – alerce americano occidental – tamarack, western larch
Larix sibirica (russica) – alerce siberiano – Siberian larch
Larrea divericata – chaparro, gobernadora, jarrilla – chaparral
Larrea (Covillea) tridentata (divaricata) – chaparral, falsa alcaparra, gobernadora, guamis, háaxat, hediondilla, hediondo, hierba hedionda, huamis, jarilla, wame gobernadora – creosote bush, greasewood
Lasiacis divaricata – carricillo, carrizo, carrizo de ratón, cebadilla, levadura, mehenzit, pichut, pitut, sit – tibisee
Lasianthaea podocephala – pionilla
Lasthenia (Baeria) californica (chrysostoma) – brillo de sol, si me quieres no me quieres – goldfields
Lastrea = Thelypteris
Latania loddigesii – latania azul
Latania lontaroides – latania roja
Lathraea squamaria – escuamaria común, hierba de la madre – toothwort
Lathyrus angulatus – cicércula angulosa, pluma de ángel
Lathyrus aphaca – afaca, alverja silvestre – yellow vetchling
Lathyrus cicera – almorta de monte, cicércula, cicércula roja, galgana, galanor, titarro – flatpod peavine, red vetchling
Lathyrus clymenum – cicércula española, cicércula de hojas tenues, cicércula purpurina – Spanish vetchling
Lathyrus hirsutus – cicércula hirsuta, guija velluda – caley pea, singletary pea
Lathyrus japonicus – cicércula marítima – maritime vetchling, sea pea
Lathyrus latifolius – choreque, cicércula de hojas anchas, cicércula latifolia, guija de hojas anchas – everlasting pea, perennial sweet pea
Lathyrus linifolius – guija de monte – wood pea
Lathyrus ochrus – alverjana, cicércula ocre, tapisote – Cyprus vetch
Lathyrus odoratus – arvejilla de olor, caracolillo de olor, chícharo, chícharo de olor, choreque, cicércula olorosa, clarín, frijol de olor, guisante de olor, guisante oloroso, haba de las Indias, perrito – common sweet pea, sweet pea
Lathyrus palustris – cicércula de los pantanos – marsh pea
Lathyrus sativus – almorta, alverja, arvejo, chícharo, guija, muela, tito – chickpea, sweet pea, wild pea
Lathyrus sphaericus – cicércula de frutos esféricos
Lathyrus splendens – chícharo – campo pea, pride of California
Lathyrus sylvestris – cicércula silvestre, guija silvestre – flat peavine
Lathyrus tingitanus – cicércula de Tánger, guisante tangerino – Tangier pea
Lathyrus tuberosus – arveja tuberosa, cicércula tuberosa – earthnut pea, tuberous vetch
Lathyrus vernus – cicércula primaveral, haba, oreja de gato, orobo vernal – spring pea, vetchling
Lathyrus vestitus* var *alefeldii (laetiflorus* ssp *alefeldii) – chícharo – wild pea
Laurus camphora – alcanforero, cardamomo, galanga, sasafrás
Laurus cerassus – laurel-cerezo
Laurus nobilis – laurel, laurel de Apolo, laurel del poeta, laurel noble, lauro – bay, bay laurel, Grecian laurel, sweet bay
Lavandula angustifolia (officinalis, vera) – alhucema, espliego, espliego común, lavanda, lavanda verdadera, lavándula, lavándula hembra, lavándula holandesa, lavándula inglesa, lavándula vera – English lavender
Lavandula dentata – cantueso dentado, cantueso rizado – French lavender, toothed lavender
Lavandula lanata – alhucema, lavándula lanada – woolly lavender
Lavandula latifolia (spica) – alhucema,

lavándula, lavándula de hojas anchas – broadleaf lavender, spike lavender

Lavandula multifida – espliego de hojas divididas, lavándula multifida – fernleaf lavender

Lavandula stoechas – cantueso, espliego, hierba de San Juan, lavándula francesa, tomillo borriquero – French lavender, Spanish lavender

Lavandula viridis – lavándula vividina – green lavender, yellow lavender

Lavatera arborea – lavatera arbórea – tree-mallow

Lavatera assurgentiflora – malva, malva real, malva rosa, malvón – island mallow, malva rosa

Lavatera cretica – malva de Creta – smaller tree-mallow

Lavatera maritima – malva marítima, malvavisco marino, paloma de mar – sea mallow, tree-mallow

Lavatera trimestris – malva anual, malva de flores grandes, malva real – annual mallow

Lawsonia = Reseda

Lecanora esculenta – maná – manna

Leccinum aurantiacum – muñeco, pambazo, panza

Leccinum corsicum – faisán de la jara

Leccinum lepidum – faisán

Lecythis elliptica – sapucaia – sapucaia nut

Lecythis ollaria – coco de mono

Lecythis zabucajo – nuez de paraíso, olla de mono, zabucajo – monkey pot, paradise nut, sapucaia nut

Ledum palustre – romero de pantano

Leersia hexandra – arrocillo, arroz bravo, barit, hierba de arroz, lambedor, pastito de agua, pasto de agua, pelillo – cutgrass, swamp ricegrass

Leersia monandra – arrocillo cimarrón

Leersia oryzoides – arrocillo áspero – rice cutgrass

Legenaria vulgaris – calabaza, guiro – calabash

Lemairocereus = Isolatocereus, Pachycereus, Polaskia, Stenocereus

Lemboglossum maculatum – mariposa amarilla

Lemna gibba – chichicastle, lenteja de agua con gibas, lentejilla, lentejilla de agua – duckweed, fat duckweed

Lemna minor – chichicastle, chilacastle, iximha, lampazo, lenteja, lenteja de agua, lenteja de agua menor, lentejilla, lentejilla de agua – common duckweed, duckweed

Lemna minuscula (minima) – lenteja de agua minúscola, lentejilla – duckweed, lesser duckweed

Lemna trisulca – lenteja de agua trisulca – duckweed, ivy-leaved duckweed

Lennoa coerulea – angiolote, flor de tierra

Lens culinaris (esculenta) – lenteja – lentil

Lentinus edodes – champiñón chinesco, hongo shiitake, seta japonesa, shiitake – shiitake mushroom

Leonotis leonurus – oreja de león – lion's tail

Leonotis nepetaefolia – bastón de San Francisco, cordón de fraile, jara, molinillo, vara de San José

Leontodon taraxacoides – diente de león hirto – hawkbit

Leontopodium alpinum – edelweis, edelweis alpino, flor de nieve, leontopodio, pie de león – edelweiss

Leonurus cardiaca – agripalma, cardíaca, cola de león, mano de Santa María, marihuanilla – motherwort

Lepachys mexicana – howinowa

Lepechinia caulescens – salvia

Lepechinia hastata – chicura de la sierra, chuchupate, hierba de jarazo – pitcher sage

Lepidium campestre – candelero de Salomón, lepidio campestre, mastuerzo, mostaza silvestre – field pepperwort

Lepidium densiflorum – lentejilla, lepidio densifloro – peppergrass

Lepidium eastwoodiae – pamita cimarrón

Lepidium heterophyllum – lepidio heterofilo – Smith's pepperwort

Lepidium intermedium – mixixi

Lepidium lasiocarpum – lentejilla, epazote – hairy peppergrass

Lepidium latifolium – coclearia del país, lepidio, lepidio de hojas anchas, mastuerzo mayor, mostacilla, piperisa – broad-leaved pepperwort, dittander, peppergrass

Lepidium perfoliatum – lepidio perfoliado

Lepidium ramosissimum – lepidio ramificación compacta
Lepidium sativum – cresón, lepidio, lepidio hortense, mastuerzo, mastuerzo hortense – garden cress
Lepidium sordidum – lentejilla
Lepidium virginicum – coclearia del país, comida de pajaritos, huisique, isohuanquil, lentejilla, lepidio, lepidio de Virginia, mastuerzo, mexix-quílitl, panalillo, pata de cuervo, putcan, putxiu, tapona, xixinda – poor man's pepperwort, Virginia peppergrass
Lepiota = Macrolepiota
Lepista nuda – morada, pie violeta, pimpinela – amethyst agaric, wood blewit
Lepista saeva – pie azul, seta de vaca – field blewit
Leptochloa dubia – desparramado dubiano, hierba de hilo, pasto gigante, zacate gigante – green sprangletop
Leptochloa fascicularis – zacate gigante anual – bearded sprangletop
Leptochloa filiformis – cola de zorra, desparramado rojo, puh-suuk, scoco-sacat, zacate gigante, zacate salado – red spangletop
Leptochloa fusca ssp uninervia – zacate salado mexicano – Mexican spangletop
Leptochloa mucronata – zacate araña, zacate liendrilla, zacate salado
Leptochloa panicea ssp brachiata – desparramado rojo, zacate salado – red spangletop
Leptodactylon spp – hierba del coyote – prickly phlox
Leptoloma cognatum – zacate escobilla – fall witchgrass
Lespedeza striata – trébol del Japón – bush-clover, common lespedeza
Leucaena brandegeei – guaje
Leucaena esculenta – árbol huaje, chajkal, guaje, guaxi, huachi blanco, huaje, huasi, huaxin, tepahuaje, tlapahuaxin
Leucaena glauca – acacia pálida, aroma blanca, campeche, chajkal, granotino, guaje, guaje de Castilla, hediondilla, huasi, huatsin, leleques, leucena, nacaste, tamarindillo – jumpy bean, white popinac lead-tree, wild tamarind
Leucaena involucrata – vara blanca
Leucaena lanceolata (microcarpa) – batachín, batillo, bolillo, guaje, guajillo, huajillo, nasiva
Leucaena leucocephala – acacia, acacia bella rosa, acacia frondosa, aroma boba, aroma mansa, guaje, huaxin, leucena, mimosa, salsa, tamarindo bastardo, tamarindo silvestre – white lead-tree
Leucanthemum maximum – margarita shasta
Leucanthemum vulgare (Chrysanthemum leucanthemum) – hierba amarilla, margarita, margarita mayor, margarita silvestre, margaritón, maya, pajito – ox-eye daisy
Leuchtenbergia principis – cacto-agave – agave cactus
Leucojum aestivum – campanilla de primavera, campanilla de verano, nevadilla de verano – summer snowflake
Leucojum autumnale – campanilla de otoño, nevadilla de otoño – autumn snowflake
Leucojum trichophyllum – campanilla de hojas gramíneas, campanilla de verano
Leucojum vernum – campanilla, campanilla de primavera, campanilla primaveral, nevadilla de primavera – spring snowflake
Leucophyllum ambiguum – cenizo, palo cenizo, raguño
Leucophyllum candidum – cenizo
Leucophyllum frutescens (texanum) – cenicilla, cenizo, chamizo, hierba del cenizo, palo cenizo – cenizo, purple sage, Texas ranger, Texas sage, Texas silverleaf, wild lilac
Leucophyllum laevigatum – cenizo
Leucophyllum minus – cenizo – Big Bend silverleaf, cenizo
Leucophyllum violaceum – cenizo – cenizo
Leucopremna (Jacaratia) mexicana – árbol del bonete, bonete, cuaguayote, cuayote,
Leucothoe mexicana – hierba de chivo, hierba de la sarna, hierba del empeine
Levisticum officinale (Ligusticum officinalis) – apio de montaña, apio de monte, apio montano, levístico, ligústico, zazalicpatli – lovage
Lewisia cotyledon – lewisia – cliff maids
Liabum glabrum – palo de agua, palo de San Nicolás, quelite
Liatris punctata – cachana – blazing star, button snakeroot, gayfeather

Liatris spicata – cachana, liátride, liátride en espiga, serrátula – blazing star, button snakeroot, gayfeather
Libocedrus = Calocedrus
Licania arborea – cacahuananche, cacahuanantzin, cacao volador, caña dulce, carnero, carnero blanco, conduce, frailecillo, nacaco, palo de carnero, palo de fraile, quirindol, quirindol cacahuananche, tlinete, toposte, topostle
Licania (Moquilea) platypus – cabeza de mico, caca de niño, cacahuananche, huicume, icaquito, itzampi, mezonzapote, pío, tzacots, zapote amarillo, zapote borracho, zapote cabello, zapote de mico, zapote de mono, zonzapote – Indian mango, monkey apple
Licania rigida – oiticica
Licuala grandis – licuala – fan palm
Licuala spinosa – licuala de manglar
Ligusticum chuanxiong – lovaja sichiana
Ligusticum grayi – chuchupate, oshá, xoxouhca-pahtli – Indian parsley
Ligusticum porteri – chuchupate, guariaca, hierba de cochino, levístico, oshá, oshá de la sierra, perejil del campo – lovage
Ligusticum = Levisticum
Ligustrum japonicum – aligustre, aligustrón, ligustro japonés, prim, trueno – Japanese privet, wax-leaf privet
Ligustrum lucidum – aligustre de China, aligustre lustroso de la China, trueno – glossy privet
Ligustrum ovalifolium – aligustre japonés – California privet
Ligustrum vulgare – alheña, aligustre, capicuerno, cavicuerna, ligustro, prim, malmadurillo, matahombres, truanilla, trueno – common privet
Lilium auratum – azucena áurea, azucena de rayas dorados, azucena dorada, lirio dorado – fragrant oriental lily, gold-band lily, golden-rayed lily
Lilium bulbiferum – azucena anaranjada, azucena roja, lirio naranja – orange lily
Lilium canadense – azucena canadiense
Lilium candidum – azucena, azucena blanca, azucena común, azucena de la sierra, azucena de la Virgen, azucena serrana, lirio blanco, lirio de San Antonio, lirio de San Luis, sacha huasca blanca – easter lily, madonna lily
Lilium chalcedonicum – azucena – lily, scarlet martagon
Lilium hansonii – azucena de gorro de Turco dorada
Lilium henryi – azucena de Henry – Turk's cap lily
Lilium lancifolium (tigrinum) – azucena atigrada, azucena tigrina, flor de lazo, lirio atrigado, lirio del Japón, lirio de tigre, lirio tigrado – tiger lily
Lilium longiflorum – azucena, azucena de pascua, azucena de trompeta, lirio, lirio de flores largas, lirio de pascua – easter lily
Lilium martagon – azucena de gorro de Turco, azucena martagón, azucena silvestre, lirio silvestre, martagón – Turk's cap lily
Lilium pardalinum – azucena leopardo – leopold lily, panther lily
Lilium pumilum – azucena de coral – coral lily
Lilium regale – azucena real, azucena regia, lirio real – regal lily
Lilium speciosum – azucena, lirio espléndido, lirio soberbio – splendid lily
Lilium = Cardiocrinum
Limnanthemum = Nymphoides
Limnanthes douglasii – limnanta – meadowfoam, poached eggs
Limnocharis flava – cebolla de chucho, hoja de buitre – sawah flowering rush, sawah lettuce, yellow burhead, yellow velvetleaf
Limonium californicum – lavanda del mar, siempreviva azul – sea-lavender, statice, western marsh-rosemary
Limonium sinuatum (Statice sinuata) – espliego de mar, espliego marino, estátice, estaticia, estátil, limonio, limonio sinuado, siempreviva, siempreviva azul – sea lavender
Limosella aquatica – limosella acuática – mudwort
Linaria bipartita – linaria bipartida – toadflax
Linaria genistifolia (dalmatica) – linaria dalmática, linaria de hojas de retama – Dalmatian toadflax
Linaria maroccana – linaria, linaria marroquí – baby snapdragon, toadflax
Linaria purpurea – linaria purpúrea – purple toadflax

Linaria vulgaris – linaria, linaria común, lino montesino, pajarito – butter-and-eggs
Linaria = Cymbalaria
Lindera benzoin – lindera – spicebush
Lindernia dubia – lindernia dudosa – false pimpernel
Linnaea borealis – linnea boralis, té de Suecia – twin flower
Linum bienne – lino bienal – pale flax
Linum flavum – lino amarillo – golden flax
Linum grandiflorum – lino, lino de flores grandes – flowering flax, scarlet flax
Linum lewisii – linaza – blue flax, prairie flax
Linum narbonense – lino azul, lino bravo – blue flax
Linum perenne – lino perenne – perennial blue flax
Linum usitatissimum – linaza, linera, lino, lino cultivado – common flax, cultivated flax, linseed
Lippia alba – té de Castilla, valeriana
Lippia barbata – orégano
Lippia berlandieri – canelilla, epazote, epazotl, hahuiya-xíhuitl, hierba dulce, orégano, salvia – Mexican oregano
Lippia callicarpaefolia – mastranzo, rosa de Castilla, rosa de Castilla cimarrona, salvia poblana, salvia real, salvia real poblana – frog-fruit
Lippia dulcis – hierbabuena, hierba dulce, neuctixihuitl, orozuz, tzopelicxihuitl
Lippia formosa – orégano, oreganón
Lippia graveolens – hierba dulce, neuctixihuitl, orégano, oregano castillo, orégano del burro, orozuz del país, romerillo del monte, sacmumutz, té del país, xakilche – Mexican oregano, Spanish thyme
Lippia ligustrina – agrito, cabradora, hierba de la princesa, hierba dulce, huele de noche, jaboncillo, jazmincillo, vara blanca, vara dulce – bee blossom, bee brush, white brush
Lippia origanoides – orégano
Lippia palmeri – orégano – Mexican oregano
Lippia satureja – orégano – Oaxacan oregano
Lippia umbellata (pringlei) – bacatón, batayáqui, colmillo de víbora, hierba de la mula, macuay, nacare, salvia poblana, tabaquillo, talacáo, tepozán
Lippia = Aloysia
Liquidambar formosana – liquidámbar chino, liquidámbar de Formosa, thau – Chinese sweet gum
Liquidambar orientalis – estoraque, liquidámbar oriental – oriental sweet gum
Liquidambar styraciflua – abeto, árbol del ambar, bálsamo copalme, bito, copalillo, copalme, estoraque, liquidámbar, liquidámbar americano, maripenda, nabá, nite-biito, nogal satinado, ococote, ocotzoquiauitl, ocótzotl, ocozote, ocozotl, pinabete, somerio, suchete, trementina, vito pijte, xochiocotzocuahuitl, xochicuahuitl, xochiocótzotl, yagabicigui, yagabito, zacapoaxtla – American sweet gum, liquidambar
Liriodendron tulipifera – árbol del tulipán, liriodentro, tulipanero, tulipero, tulipero de Virginia, tulipífero americano, tulipífero de Virginia – tulip tree
Liriope muscari – liriope – big blue lily turf
Listera cordata – listera de hojas cordiformes – heart-leaved twayblade
Litchi chinensis (Nephelium litchi) – lichi, mamoncillo chino – litchi nut, lychee
Lithodora diffusa (Lithospermum diffusum) – granos de amor, hierba de la perla, perlita – stoneseed
***Lithops* spp** – piedra viva – living stone, pebble plant, stoneface
Lithospermum (Buglossoides) arvensis – cornicabra, mijo de sol agreste – gromwell
Lithospermum calycosum – hierba de la perlita
Lithospermum cobrense – manzanilla del río – mountain gromwell, puccoon, stoneseed
Lithospermum officinale – aljófar, grano de amor, hierba de la perla, mijo de sol, mijo de sol común – gromwell
Lithospermum = Lithodora
Litsea glaucescens – laurel, laurel del país, laurel de la sierra, sufricago, sufricalla, zizuch – bush laurel, Mexican bay leaf
Livistonia australis – livistona, palmera de abanico australiana
Livistonia chinensis – livistona, palmera de abanico china – Chinese fan palm, Chinese fountain palm
Llavea cordifolia – helecho sombrilla
Lobelia cardinalis – acaxóchitl, cardenal,

cardenal de la laguna, cardenal de maceta, lobelia, lobelia cardinal, lobelia roja – cardinal flower, scarlet lobelia

Lobelia cardinalis (fulgens)* var *pseudosplendens* (ssp *graminea) – cardenal de la laguna, cubanita, lobelia, pajarito del norte

Lobelia erinus – lobelia, lobelia azulada, lobelia del jardín, mosquita – blue lobelia

Lobelia fenestralis – cola de zorra, gusanillo, lobelia morada, talcote – Indian tobacco

Lobelia gruina – flor de María, lobelia – lobelia

Lobelia inflata – hierba del asma, lobelia, tabaco indio – Indian tobacco

Lobelia laxiflora – acaxóchitl, aretito, chilpantlazolli, chilpanxóchitl, cúralo todo, diente de chucho, Guadalupe, hierba del zopilote, hoitzitziltentli, jarrito, lobelia, panxóchitl, pastorcillo, pipilosúchil, pipiloxóchitl, tozcuitlapilxóchitl, xochipile, zarcilillo, zarcillo – giant orange lobelia

Lobelia siphilitica – lobelia glandulosa, lobelia sifilítica

Lobelia urens – matacaballos – heath lobelia

Lobularia maritima (Alyssum maritimum) – alhelicillo, alisón, bola de hilo, mastuerzo marítimo, panal, panalillo, panalito – sweet alyssum

Loeselia ciliata – guachichila, hierba de la Virgen

Loeselia mexicana – chuparrosa, cuachile, cuauchichil, espinonsilla, flor de chupamirto, flor de colibrí, hierba de la Virgen, hierba de San Antonio, hinseseli, huichichil, huichichile, huicicillo, huitzilzilxóchitl, huitzilzitzin, mirto, mirto silvestre, panalillo – hummingbird flower

Loeselia scariosa – huichichile

Loeselia tenuifolia – ubi

Lolium multiflorum – ballico anual, ballico doméstico, ballico italiano, cebolillo, centeno italiano, pasto italiano, raygrás, raygrás annual, raygrás criollo, raygrás italiano, zacate italiano – Italian ryegrass

Lolium perenne – ballico, ballico inglés, ballico perenne, ballico vivaz, césped inglés, cizaña, inglés, pasto inglés, raygrás inglés, zacate inglés – perrenial ryegrass

Lolium temulentum – borrachuela, cizaña, cominillo, joyo, rabillo – bearded darnel, darnel, tares

Lomatium dissectum – oshá, tohza – biscuit root, carrot-leaf, Indian root, wild carrot

Lomatium foeniculaceum – semilla del amor – lomatium

Lonchocarpus hermannii (Willardia mexicana) – japiró, nesco, palo nesco, palo piojo, samo, taliste – Venus tree

Lonchocarpus longistylus – balche, guama de México, palo de patlaches – lance pod

Lonchocarpus utilis – barbasco, cubé

Lonicera arizonica – madreselva – Arizona honeysuckle

Lonicera caerulea – madreselva azul – blue honeysuckle

Lonicera etrusca – madreselva etrusca – Etruscan honeysuckle

Lonicera involucrata – madreselva – bearberry honeysuckle, ink-berry, pigeon bush, twinberry

Lonicera japonica – madreselva, madreselva japonesa – Japanese honeysuckle

Lonicera nitida – madreselva – box honeysuckle

Lonicera periclymenum – madreselva, madreselva atlántica, madreselva de jardín, madreselva inglesa, madreselva selvática – woodbine

Lonicera sempervirens – madreselva, madreselva siempreverde – coral honeysuckle, trumpet honeysuckle

Lonicera subspicata – moronel – honeysuckle

Lonicera tatarica – madreselva tártara – Tartarian honeysuckle

Lopezia mexicana – alfilerillo, guayabillo, hierba del golpe, mitlalxihuitl, perlilla, zancudo

Lopezia racemosa – alfilerillo, aretillo, guayabillo, hierba del golpe, mitlalxihuitl, perlilla, zancudo

Lophocereus = Pachycereus

Lophophora diffusa – peyote

Lophophora williamsii – ahuitzyó, botón de mezcal, camaba, hikora wanami, hikuli, hikuri, jicurite, kamaba, mezcal, peyote, peyote liso, peyotl, raíz del diablo, raíz diabolica, señif, teocomitl, teonanácatl,

tlalcoyote, tuna de tierra, wokowi – dry whisky, mescal buttons, peyote, whisky cactus
Lophosoria quadripinnata – palma, pelma, pesma
***Loranthus* spp** – cepa caballero, pajarito, tiña
Loranthus = Psitticanthus
Lotus angustissimus – cuernecillo pequeño – slender bird's-foot trefoil
Lotus corniculatus – cornezuelo, corona de rey, cuernecillo, loto, loto corniculado, loto de cuernecillo, loto de hojas anchas, loto de los prados, pata de pájaro, trébol de cuernecillo, trébol de cuernos, trébol español – birdfoot trefoil, bird's-foot trefoil
Lotus repens – alfalfa del campo – deer vetch
Lotus scoparius – casa de indio, jiguata, pata del pájaro – California broom, deerweed
Lotus strigosus (tomentellus) – verdolaga
Lotus uliginosus – cuernecillo de los pantanos, cuernecillo grande, loto de los pantanos – large bird's-foot trefoil
Lucuma mammosa (Calocarpum mammosum) – árbol de mamey, atzapotlquahuitl, chachalhaas, haas, kaukpahk, mamey, mamey colorado, mamey zapote, pizle, potkak, tetzapotl, tezonzapote, tsapassabani, tzapotl, zapote, zapote colorado, zapote mamey – mamee sapote, mamey sapote, marmalade fruit, sapote
Lucuma rivicoa – canistel, jacana, lúcuma, zapotillo – canistel
Lucuma (Calocarpum) salicifolia – acamallotetl, acamayo, atzapotlquahuitl, atzápotl, cozticzápotl, huicumo, nochi, zapote amarillo, zapote borracho – yellow sapote
Lucuma (Calocarpum) viride – chulul, injerto, jaca, raxtul, tezonzapote, zapote verde – green sapote
Lucuma = Pouteria
Ludwigia octovalvis – clavillo blanco, clavito, clavo, fresno – water primrose, wild clove
Ludwigia palustris – ludwigia de los pantanos – Hampshire purslane
Ludwigia peploides (Jussiaea repens* var *peploides) – clavillo, flor de camarón, hoja de clavo, verdolaga acuática, verdolaga de agua – false loosestrife, water primrose, yellow waterweed
Ludwigia (Jussiaea) repens – verdolaga de agua
Luffa acutangula – calabaza, calabaza de aristas, estropajo, esponja vegetal, lufa – angled loofah
Luffa aegyptiaca – calabaza de aristas, esponja, esponja vegetal, esponjilla, estropajo, lufa, servilleta del pobre, tzonayotli – loofah
Luffa cylindrica – estropajo, lufa
Luma apiculata (Myrtus luma) – temu – temu
Lunaria annua – hierba de la plata, lunaria, lunaria anual, moneda del papa – honesty, money plant, moonwort
Lunaria biennis – hierba de la plata, moneda del papa – honesty
Lupinus albus – altramuz, altramuz blanco, chocho, lupino blanco – white lupine
Lupinus angustifolius – altramuz azul, altramuz de hojas estrechas, lupino azul – annual lupin, blue lupine
Lupinus arboreus – altramuz arbórea, altramuz arborescente, lupino arbórea – tree lupine, yellow bush lupine
Lupinus arizonicus – lupino – Arizona lupine, bluebonnet
Lupinus aschenbornii – amaxíhuitl
Lupinus bilineatus – amaxíhuitl, hierba del caballo
Lupinus elegans – altramuz, cola de zorra, garbancillo, hierba del pescado, hierba loca, ilamaxóchitl, lupino – lupine
Lupinus leptocarpus – piojillo
Lupinus luteus – altramuz amarillo, lupino amarillo – sweet lupin, yellow lupine
Lupinus polyphyllus – altramuz, altramuz perenne – garden lupin
Lupinus russeli – altramuz – Russell hybrid lupine
Lupinus succulentus – garbancillo – arroyo lupine
Lupinus texensis – altramuz de Texas, garbancillo, lupino – bluebonnet
Luzula parviflora – lúzula de flores pequeñas – hairy wood rush
Luzula spicata – lúzula espigada – hairy wood rush
Lycaste aromatica – canela, canelita

Lycaste virginalis (skinneri) – búcaro, monja blanca, monjita – Skinner's maxillaria
Lychnis chalcedonica – coronaria escarlata, cruz de Jerusalén, cruz de Malta – Maltese cross
Lychnis coronaria – clavellina, coronaria, coronaria coriácea, flor de cuclillo, sangre de Cristo – crown-pink, dusty miller, mullein pink, rose campion
Lychnis flos-cuculi – flor de cuclillo – ragged robin
Lychnis flos-jovis – coronaria de Júpiter, flor de Júpiter – Jupiter's campion
Lychnis viscaria – coronaria viscosa, cruz de Malta alpina – sticky catchfly
Lycium andersonii – frutilla, hierba del monte, jejiri, manzanita, salicieso, sigreropo, vara prieta – box thorn, desert thorn, desert wolfberry, tomato berry, waterjacket
Lycium barbarum – cambronera de hojas de jarilla – Duke of Argyll's tea plant, matrimony vine
Lycium berlandieri – barchata, cilindrillo, frutilla, garambullo, joso, tomatillo, salicieso – wolfberry
Lycium brevipes (richii) – cilindrillo, frutilla – box thorn
Lycium brevipes* var *hassei – frutilla – Santa Catalina Island desert-thorn
Lycium californicum – frutilla – box thorn, coast desert-thorn
Lycium carinatum – salicieso
Lycium carolinianum – frutilla
Lycium chinense – alquitira, cambronera, cambronera de la China, licio de China – Chinese box thorn, Duke of Argyll's tea-tree, matrimony vine
Lycium fremontii – frutilla – box thorn, desert wolfberry
Lycium pallidum – cambronera, chico, frutilla, tomatillo – box thorn, pale wolfberry
Lycium parishii – salicieso – Parish's desert-thorn
Lycium parviflorum – tomatillo
Lycium torreyi – garambullo, tomatillo – box thorn
Lycoperdon candidum – bolita de conejo, ternerita de llano
Lycoperdon mixtecorum – hongo, hongo de huevo, kalamoto, quesito – puffball
Lycoperdon perlatum – bejín, bola, bolita de conejo, bolita de hongo, bolita de San Juan, bomba reventadora, cuesco de lobo, panza, pedo de coyote, pedo del diablo, pedo de lobo, ternerita, ternerita de bosque, ternerita de monte – common puffball
Lycoperdon pratense – cuesco de lobo – puffball
Lycoperdon rimulatum – ternerita de bosque
Lycoperdon umbrinum – bolita de conejo, pedo de lobo, ternerita, ternerita de bosque, ternerita de monte
Lycopersicon (Lycopersicum) esculentum – jítomate, jítomate redondo, manzano del amor, manzano de oro, pak, tomate, tomate bola, tomate de bola, tomatera, tomatl, tsulubppac, tumatle, xictomatl, xitotomatl, xucupara – tomato
Lycopersicon esculentum* var *cerasiforme – cuatomate, jitomate, menudo, mini jitomate dulce, ojo de venado, tinguaraque, tomate, tomate cereza, tomate chitalino, tomate cimarrón, tomate del monte, zitlali – cherry tomato
Lycopersicon esculentum* var *pyriforme – guaje, guajillo, jitomate guaje, jitomate guajito, tomate roma – roma tomato
Lycopersicon pimpinellifolium – tomate cimarrón – currant tomato
Lycopersicum = Lycopersicon
Lycopodium clavatum – azufre vegetal, licopodio, polvo de palo – club-moss, running-pine
Lycopus americanus – licipio, menta de lobo, pinillo – bugleweed, water-horehound
Lycopus asper (lucidus) – licipio, menta de lobo, pinillo – bugleweed
Lycopus europaeus – marrubio acuático, menta de lobo, pie de lobo – bugle-weed, gipsywort, water horehound
Lycoris radiata – licóride – magic lily, red spider lily
Lycurus phleoides – zacate cola de zorra, zacate lobero – Texas timothy, wolftail
Lygeum spartum – albardín, esparto, esparto basto – esparto grass

Lygodesmia juncea – chicote embarañada, chiquete de embarañada – skeleton plant
Lygodium venustum – alambre, bejuco de alambre, corremiento, xix-el-bá – wire whisk
Lyophyllum aggregatum – girbola de bruch – clustered tricholoma
Lyophyllum decastes – amontonado, clavito, clavito grande, jolete, liofilio agregado, moloche, tejamanilero, tzenso
Lysiloma acapulcensis – ébano, guamúchil, laaguia, matze, tepeguaje, tepehuaje, tepehoaxin, tepemizquitl, tripal, yagayace
Lysiloma bahamensis – tzalam, tzucté, xiazek, zalam – wild tamarind
Lysiloma candida – palo blanco
Lysiloma divaricata – jeseri, mauto, mauto colorado, mayo, mezquite prieto, palo de arco, quiebracha, quiebrahacha, rajador, tepeguaje, tepemezquite, tepemezquite blanco, tepemezquite colorado – borderpod acacia, feather bush
Lysiloma microphylla – jeseri, manta, mauto, quiebracha, tepeguaje, tlahuilote – feather bush, fern-of-the-desert
Lysiloma watsonii – machaquí, tepeguaje – feather tree
Lysimachia ciliata – lisimaquia ciliada – fringe-leaved loosestrife
Lysimachia clethroides – lisimaquia china, lisimaquia de hojas de clethra – gooseneck loosestrife
Lysimachia ephemerum – lisimaquia roja
Lysimachia nummularia – hierba de la moneda, lisimaquia nummular, lisimaquia rastrera – creeping jenny, moneywort
Lysimachia punctata – lisimaquia amarilla, lisimaquia punteada – dotted loosestrife
Lysimachia salicaria – lisimaquia roja – purple loosestrife
Lysimachia thyrsiflora – lisimaquia de los pantanos – tufted loosestrife
Lythrum alatum – atlanchane, fraile, hierba colorada, hierba del cáncer, salicaria – loosestrife
Lythrum hyssopifolium – arroyuela de hojas de hisopo, hierba del espanto, hierba del toro – grass poly, hyssop loosestrife
Lythrum portula – arroyuela portula – water purslane
Lythrum salicaria – arroyuela, frailecillo, fraile, lisimaquia purpúrea, lisimaquia roja, litro, salicaria – purple loosestrife
Lythrum tribracteatum – arroyuela de tres brácteas – willow herb
Lythrum virgatum – arroyuela menor, lisimaquia purpúrea – purple loosestrife, willow herb
Lytocaryum weddelianum – palmerita de Petrópolis

M

Maba latifolia – estrellita, guayparín, jegito, sabor, zapotillo
Maba salicifolia – coacoyutillo, ébano
Maba = Diospyros
Macadamia integrifolia – macadamia, nuez de macadamia – macademia nut, smooth-shell macademia
Macadamia ternifolia – macadamia, nogal de Australia, nuez de Australia – macadamia
Macfadyena (Bignonia, Doxantha) unguis-cati – bejuco, bejuco cachorro, bejuco de chinaca, bejuco de ratón, bignonia, ek-kixilak, gasa, hiedra, mano de cachorro, San Pedro de Guía – cat's claw, yellow trumpet vine
Machaeranthera carnosa (Aster intricatus) – hierba del pasmo – shrubby alkali aster
Machaerocereus = Stenocereus
Macleaya (Bocconia) cordata – amapola plumosa – plume poppy
Maclura pomifera – maclura, madera de arco, naranjo chino, naranjo del Osage – mock orange, Osage orange
Maclura (Chlorophora) tinctoria – barossa, chijchijtli, fustete, fustete amarillo, moradilla, moral, moral amarillo, moral de clavo, moral liso, palo amarillo, palo de Cuba, palo de mora, palo moral, tzitzil, yaga-huil – drug fustic tree, drug Osageorange, dyer's mulberry, fustic
Macrochloa = Stipa
Macrocystis pyrifera – alga marina, encino del mar, sargazo gigante – kelp, seaweed
Macrolepiota procera – apagallum, gallinita, parasol – field parasol
Macroscepis obovata – talayote
Macrosiphonia brachysiphon – cacaloxóchitl
Macrosiphonia hesperia – jazmín de la sierra
Macrosiphonia hypoleuca – flor de San Juan, güirambo, hierba de la cucaracha, hierba de San Juan, maravilla silvestre, rosa de San Juan, rosa de San Pedro, San Juan, San Juanito
Macrosiphonia macrosiphoni – flor de San Juan, hierba de San Juan – rocktrumpet
Madia sativa – madia, melosa, yarcas – Chilean tarweed, madia oil plant
Magnolia acuminata – magnolia acuminada norteamericana – cucumber tree
Magnolia campbellii – magnolia de Campbell
Magnolia dealbata – elosúchil, guía-lachi
Magnolia delavayi – magnolia perenne de Delavay, magnolia perenne de la China
Magnolia denudata – yulan – yulan magnolia
Magnolia grandiflora – anonillo, flor del corazón, magnolia, magnolia común, magnolia grandiflora, palo de cacique, semiramis – bull bay, southern magnolia
Magnolia hypoleuca – magnolia japonesa de hojas grandes
Magnolia kobus – magnolia kobus del norte del Japón – kobus magnolia
Magnolia* x *loebneri – magnolia de Loebner
Magnolia macrophylla – magnolia americana de hojas grandes – bigleaf magnolia
Magnolia salicifolia – magnolia japonesa de hojas de sauce – anise magnolia
Magnolia schiedeana – corpus, corpus de cacique, elexóchitl, palo de cacique, yolosúchil
Magnolia sieboldii – magnolia de flores pequeñas – oyama magnolia
Magnolia* x *soulangiana – árbol lirio, magnolia china – saucer magnolia, tulip tree
Magnolia stellata – magnolia estrellada – star magnolia
Magnolia* x *veitchii – magnolia híbrida de Veitch – Veitch magnolia
Mahonia = Berberis
Maianthemum dilatatum – gramigna del Parnaso – false lily-of-the-valley
Majorana = Origanum
Malacothamnus fasciculatus – malva – chaparral mallow
Malacothrix xantii – clavelito, clavelito del monte
Malcolmia africana – malcolmia africana – African stock
Malcolmia maritima – alhelí, alhelí de Virginia, mahonesa, malcomia marítima – Virginian stock
Malope trifida – falsa malva
Malosma (Rhus) laurina – lentil, lentisco – laurel sumac

Malpighia diversifolia – manzanita, manzano, margarita
Malpighia emarginata (umbellata) – acerolo, escobilla, granadilla, grosellero, mora del campo, semeruco, sidra, sire – Barbados cherry, West Indian cherry
Malpighia glabra – acerolo, ahualxócotl, arrayancito, boxhuayalté, capulín, canibinche, cerezo, cerezo de Castilla, cerezo de Jamaica, cerezo tropical, chi, chía, escobilla, escoba blanca, guindo de Indias, huaxacote, kanibinche, manzanita, nanche del monte, palo de gallina, palo de lumbre, tomatillo, xocat, xochtatl – acerola cherry, Barbados cherry, West Indian cherry
Malus floribunda – manzano floreciente, manzano floribundo del Japón, manzano japonés, manzano multifloro, manzano rosa – Japanese flowering crabapple
Malus fusca – manzano silvestre de Oregón – Oregon crabapple
Malus hupehensis (theifera) – manzano silvestre chino de Hupeh – tea carbapple
Malus pumila* var *paradisiacal – paradís – paradise apple
Malus pumila* var *sylvestris – doucin – rootstock apple
Malus* x *purpurea – manzano híbrido purpúreo, manzano rojo – Aldenham crabapple, purple crab
Malus spectabilis – manzano silvestre de la China – Chinese crabapple
Malus sylvestris (Pyrus malus) – 1. belehui, manzano, manzano camueso, manzano chato, perón, tnutinumi – apple, common apple; 2. maguillo, maíllo, manzano silvestre, manzano silvestre europeo – crabapple
Malva alcea – malvavisco salvaje – Vervain's mallow
Malva moschata – ambarina, malva de olor de almizcle – musk mallow
Malva neglecta (rotundifolia) – malva, malva del campo, malva de Castilla, malva de hojas redondas, malva de quesitos, malva enana, malva medicinal – cheeses, common mallow, umbrella mallow
Malva nicaeensis – malva, malva de niza – bull mallow
Malva parviflora – malva, malva de Castilla, malva de flores pequeñas, malva del campo, malva de quesitos, quesito, salicieso – cheeseweed, little mallow
Malva sylvestris – alboheza, belehui, malva, malva blanca, malva común, malva selvática, quelite – high mallow
Malva = Sida
Malvastrum bicuspidatum – malva, tuchi
Malvastrum coromandelianum – chichichbe, escoba blanca, malva, totopzots
Malvastrum sylvestris – malva, malva grande, malva real – mallow
Malvastrum = Eremalche, Modiolastrum
Malvaviscus arboreus – amapola, bizil, chanita, chilillo, chupamirto, civil, malva, manzanita, mazapán, monacillo, obelisco de la sierra, papito de monte, quesillo, quesito, tamanchich, tulipán, tulipancillo, tulipán del monte, tzopelchichilxóchitl – bootblack flowers, sleeping hibiscus, Texas mallow, Turk's cap, wax mallow, white moho
Malvaviscus grandiflorus – aguate, bizil, chilmecate, malvavisco, manzanilla, manzanita, mazapán, monacillo, mosilado, tamanché – Mexican-apple, red mallow, Turk's cap, wax mallow
Malvaviscus popullifolius – chocho, manzanilla
Malvella lepidota – oreja de ratón, torinaca – alkali mallow
Mamillopsis = Mammillaria
Mammea americana – abrico de para, chacalhaaz, mamey, mamey amarillo, mamey de Santo Domingo, Santo Domingo, teoezquauitl, tetzonzapotl, zapote de niño, zapote de Santo Domingo, zapote domingo, zapote mamey, zapote niño – mammee apple, mamey
Mammillaria albicans – biznaguita, junco, viejito – pincushion cactus
Mammillaria brandegeei (dawsonii) – biznaguita, viejito
Mammillaria capensis – viejito
Mammillaria dioica – biznaga, biznaguita, llavina, viejito – fish-hook cactus, pincushion cactus
Mammillaria discolor – biznaguita de chilito
Mammillaria elongata – chilito – ladyfinger cactus

Mammillaria geminispina – metzolle, metzollin – twin-spined cactus, whitey
Mammillaria goodridgei – llavina
Mammillaria grahamii (microcarpa, milleri, oliviae) – cabeza de viejo, chollita – Arizona fish-hook cactus, tangled fishhook
Mammillaria haageana (collina) – biznaga
Mammillaria heyderi (gummifera) – biznaga de chilitos, chilito, witculíki – pancake pincushion
Mammillaria heyderi (gaumeri)* ssp *gaumeri – poltzacam – fishhook cactus, nipple cactus
Mammillaria karwinskiana (collinsii)* ssp *collinsii – yaga-bigi-lasi – royal cross
Mammillaria lasiacantha – chilito – golf-ball pincushion, lacyspine cactus
Mammillaria (Dolichothele) longimamma – biznaguita – finger cactus, nipple cactus
Mammillaria magnimamma (centricirrha) – biznaga, biznaga de chilitos, biznaguita de chilito, chilito – Mexican pincushion
Mammillaria (Neomammillaria) mazatlanensis – biznaga
Mammillaria pectinifera – conchilinque, peyote
Mammillaria peninsularis – pitayita, viejito
Mammillaria petrophila (gatesii) – corona del indio, pitayita, viejito – pincushion cactus
Mammillaria petrophila (baxteriana)* ssp *baxteriana – biznaguita, viejito – pincushion cactus
Mammillaria phitauiana – pitayita, viejito
Mammillaria pottsii – chilito – rattail cactus
Mammillaria prolifera (pusilla) – biznaga – Texas hair cactus
Mammillaria (Neomammillaria) rekoi – chilito
Mammillaria (Mamillopsis) senilis (diguetii) – biznaga, cabeza de viejo, hikora rosapara, hikuri, peyote de San Pedro, rosapara, viejo, wichuriki
Mammillaria sheldonii – cabeza de viejito – Sheldon's pincushion
Mammillaria solisioides – pitayita
Mammillaria sphaerica – biznaga
Mammillaria tetrancistra – chilito, viejito – fish-hook cactus, pincushion cactus, Yaqui fishhook
Mammillaria thornberi – cabeza de viejo – clustered fish-hook cactus
Mammillaria uncinata – biznaguita de chilito
Mammillaria wiesingeri (erectacantha) – biznaguita de chilito
Mammillaria zephyranthoides – biznaguita de chilito
Mammillaria = Cochemia, Coryphanta
Mandevilla foliosa – hierba de la cucaracha, hierba del piojo
Mandevilla laxa – jazmín de Chile – Chilean jasmine
Mandragora autumnalis (officinarum) – berenjenilla, mandrágora, uva de moro, vilanera – mandrake
Manfreda (Agave) brachystachya – amole, cal-paich
Manfreda guttata – amole, amole de raíz
Manfreda = Polianthes
Mangifera foetida – bachang – bachang mango
Mangifera indica – mango, mango criollo, mango de Manila, mangón – mango
Manihot angustiloba – pata de gallo, yuca del cerro – narrow-leaved cassava
Manihot caudata – pata de gallo, tayacua, teteque
Manihot chlorosticta – yuca
Manihot davisiae – pata de gallo – Arizona cassava
Manihot dulcis – guacamote, guacamote dulce, yuca, yuca dulce – manioc, sweet cassava
Manihot esculenta (utilissima) – guacamote, guacamotli, mandioca, manioc, pan de tierra caliente, quauhcámotl, quiscamote, tapioca, tariña, tsin, yuca, yuca amarga, yuca blanca, yuca brava, yuca dulce, yuca mansa – cassava, manioc, tapioca
Manihot glaziovii – caucho ceará – ceara rubber
Manihot isoloba – pata de gallo
Manihot palmata – mandioca dulce – sweet manioc
Manilkara = Achras
Marah (Echinocystis) fabaceus – chilicote, hierba del pescado – California man-root, wild cucumber
Marah (Echinocystis) lobata – estrella del norte – mock cucumber

Marah macrocarpus (Echinocystis macrocarpa) – chilicote – big-root, man-root, wild cucumber

Maranta arundinacea – arrurruz, arrurruz arundináceo, arrurruz común, arrurruz de las Bermudas, camotillo, chaak, maranta, perrito, platanillo, quentó, sagú, sagú cimarrón, sagú del monte, suco, viuxita, yuquilla – arrowroot

Maranta leuconeura – capuchina, malvito, maranta – maranta, prayer plant

Marasmius oreades – comasec, moxerno – fairy ring agaric

Marginatocereus = Pachycereus

Mariscus jamaicensis – jasam, serrucho – saw-grass

Marrubium vulgare – concha, juanrubio, marrubio, marrubio blanco, marrubio común, mastranzo, menta de burro, poleo, uitzacua – horehound

Marsdenia edulis – mabem, talayote, tonchi

Marsilea mexicana – trébol de cuatro hojas

Martynia = Proboscidea

Mascagnia (Callaeum) macroptera – bataneni, bejuco prieto, doncella amarilla, gallineta, gallinita, mantel, mantel rojo, matanene, matanene hembra, matanene macho, maxacotl, sana rogua – yellow orchid vine

Mastichodendron foetidissimum – tempisque, zapote de ave, zapotillo – false mastic

Matelea cordifolia – talayote – milkweed

Matelea parvifolia – talayote – spearleaf, talayote

Matricaria recutita (chamomilla) – camomila, camomila inglesa, manzanilla, manzanilla común, queza – English chamomile, wild chamomile

Matricaria = Chamomilla

Matteuccia struthiopteris – helecho avestruz – ostrich fern

Matthiola annua – alhelí blanco, alhelí encanto, alhelí encarnado

Matthiola incana – alhelí, alhelí blanco, alhelí cuarenteno, alhelí de diez semanas, alhelí encarnado, alhelí morado, matiola – gillyflower, stock

Matthiola longipetala* var *bicornis – alhelí de pétalos grandes, alhelí nocturno – evening scented stock

Maurandya = Asarina

Mauritia flexuosa – aguaje, moriche, palma mauritia – alta palm, moriche palm

Mauritia minor – moriche, palma moriche

Maxillaria cucullata – negrito

Maxillaria liliacea – azucena del monte

Maximiliana maripa – cucurita

Maximilianea = Cochlospermum

Maximowiczia insularis – melón de coyote

Maytenis boaria – boaria – mayten

Maytenis laevis – chucchu, chuchuhaso

Maytenus (Tricerma) phyllanthoides – aguabola, agua de bola, granadilla, limoncillo, mangle, mangle aguabola, mangle dulce, mangle rojo, palo blanco, pasillo – mangle dulce, sweet mangrove

Meconopsis betonicifolia (baileyi) – amapola azul, amapola azul del Himalaya, falsa amapola – Himalayan poppy

Meconopsis cambrica – amapola amarilla, amapola galesa, falsa amapola – Welsh poppy

Medicago arabica – carretón, trébol de carretilla – burclover, spotted burclover

Medicago lupulina – alfalfa lupulina, carretilla, lupulina, mielga negra – black medick, none-such, yellow trefoil

Medicago minima – alfalfa enana, alfalfa minima – burclover, burmedick

Medicago orbicularis – carretón, mielga de caracolillo, torteruelo

Medicago polymorpha (hispida) – alfalfa enhiesta, carretilla, carretón de amores, trébol carretón, trébol de carretilla – California burclover

Medicago praecox – alfalfa precoz – early medick

Medicago sativa – alfalfa, alfalfa del Perú, alfalfa sativa, alfalfa variagada, lucerna, mielga, mielga alcacer, trébol de carretilla – alfalfa, lucerne

Melaleuca alternifolia – árbol de té, cayeputi, corcho, melaleuca, niaulí – melaleuca, tea tree

Melaleuca armillaris – brazalete, melaleuca – drooping melaleuca

Melaleuca ericafolia – árbol de corteza de papel – heath melaleuca

Melaleuca leucodendron – cayeputi, melaleuca – paperbark tree

Melaleuca viridiflora – gomenol, niaulí

***Melia azedarach* (var *umbraculiformis*)** – agriaz, alhelí, árbol del paraíso, árbol de quitasol, árbol paraíso, baya china, canela, cinamomo, granillo, jacinto, lila, lila de China, lila de las Indias, lila de Persia, lilaila, margosa, melia, mirabobo, panjí, paraguas chino, paraíso, paraíso chino, paraíso morado, piocha, rosariera, sicomoro falso, violeta – China berry, Persian lilac, pride of India, Texas umbrella tree, umbrella tree

Melica montezumae – papelillo dulce – melic grass

Melica nitens – papelillo brillante – three-flower melic

Melicocca bijuga – mamón, mamoncillo – genip, Spanish lime

Melilotus alba – alfalfón, coronilla blanca, meliloto blanco, trébol blanco dulce, trébol blanco oloroso, trébol de olor blanco, trébol dulce, trébol hubam – white sweetclover

Melilotus indica – alfalfilla, alfombrilla, meliloto, meliloto de flores pequeñas, meliloto indio, trébol, trébol agrio, trébol amarillo, trébol de olor – sourclover, yellow sweet clover

Melilotus officinalis – alfalfón, alfombrilla, canacua, meliloto, meliloto amarillo, meliloto común, trébol almizclado, trébol amarillo oloroso, trébol de San Juan, trébol oloroso – yellow sweetclover

Melinis minutiflora – catinguero, gordura, hierba melada, melaza, pasto catinguero, pasto gordura, pasto melado, pasto melaza, pasto melinida, yaraguá, zacate gordura, zacate melado – molasses grass

Meliosma alba – ayón, palo blanco

Meliosma dentata – mato negro

Melissa officinalis – bálsamo de limón, cedrón, cidronela, cidronela abejera, hierba bergamota, hierba Luisa, limoncillo, limonero, melisa, melisa oficinal, toronjil, toronjil extranjero, toronjina – bee balm, garden balm, lemon balm, sweet balm

Melochia tomentosa – bretónica, malva, malva de los cerros, malva rosa, malvavisco, varita de San José, zacchichibi – broom-wood

Melothria pendula – meloncito – speckled gourd

Menodora helianthemoides – jazmincillo del monte – menodora

Mentha arvensis – hierbabuena, menta, menta campestre, menta silvestre, poleo, poleo del Paris, poleo grande, tabardillo – brook mint, corn mint, field mint

Mentha arvensis* var *piperascens – menta japonesa

Mentha canadensis – menta silvestre, poleo

***Mentha* x *gracilis* (x *gentilis*)** – menta romana – golden apple mint

***Mentha* x *piperita* (*citrata*)** – bete, hierbabuena, hierbabuena de olor, hortela, menta, menta montana, menta piperita, nocuana-piti, toronjil, xak'il-xiu – peppermint

Mentha pulegium – azotapulga, chichilticxoitl, hierba de garapata, menta polea, planta del mosquito, poleo, poleo chino, poleo común, poleo europeo, poleomenta, poleo negro – pennyroyal

Mentha requienii – menta de requien – jewel mint of Corsica

***Mentha spicata* (*longifolia*)** – hierbabuena, hierba sana, hierba santa, menta, menta crespa, menta de caballo, menta de Nuestra Señora, menta espídia, menta romana, menta verde – spearmint

***Mentha suaveolens* (*rotundifolia*)** – hierbabuena, hierbabuena de burro, hierba de la pulga, madrastra, mastranzo, menta, menta de burro, menta de hojas circulares, poleo, sacui, verbena, zacatlachichinoa – apple mint

Mentha sylvestris – hierbabuena, menta de hojas grandes, poleo – spearmint

Mentha viridis – alfabieso, hierbabuena, hierbabuena de comer, tlalatóchietl

Mentzelia adhaerens – pega pega

Mentzelia affinis – pega pega – triangle-seed blazing star

Mentzelia albicaulis – pega pega – whitestem blazing star, whitestem stickleaf

Mentzelia aspera – amor, pega pega, pegarropa, tzayuntzay

Mentzelia desertorum – pega pega – blazing star

Mentzelia hispida – amor, colo-tzitcaztli, pegajosa, pega pega, pegarropa, pegarropa amarillo, zazale, zazalic – blazing star

Mentzelia involucrata – pega pega, samija – blazing star, whitebract stickleaf
Mentzelia lindleyi – bartonia – blazing star
Mentzelia multiflora – buena mujer, hierba pegajosa, pegajosa, pega pega, pegarropa – blazing star, desert mentzelia, stickleaf
Mentzelia multiflora* ssp *longiloba – pega pega – blazing star
Mentzelia obscura – pega pega – blazing star
Mentzelia oligosperma – pegajosa – chicken thief, stickleaf
Mentzelia oreophila (leucophylla) – pega pega – blazing star
Mentzelia veatchiana – pega pega – blazing star
Menyanthes trifoliata – trébol de agua, trébol fibrino, trébol palustre – bogbean, buckbean
Mercurialis annua – malcoraje, mercurial, mercurial annual, ortiga muerta – mercury
Merremia aurea – yuca – yellow morning-glory
Merremia (Ipomoea) tuberosa – rosa de madera – Honolulu rose, pilikai, Spanish arbor vine, woodrose
Merremia umbellata – bolsa de gato
Merremia = Ipomoea
Mesembryanthemum criniflorum – margarita de Livingstone
Mesembryanthemum crystallinum – aguazul, aguazur, algazul, barilla, cortina, cristalillo, cristalino, escarchada, escarchosa, flor de sol, hielito, hierba de hielo, hierba escarchada, molina, rocío – crystalline iceplant, iceplant
Mesembryanthemum nodiflorum – algazul, escarchada sin tallo, flor de sol, molina – slender-leaved iceplant
Mesembryanthemum = Carpobrotus
Metasequoia glyptostroboides – metasecoya – dawn redwood
Metrosideros excelsa – árbol de hierro – New Zealand Christmas tree, pohutukawa
Metroxylon sagus – palma de sagú, palmera de sagú, sagú – sago palm
Miconia glaberrima – dos caras, tesguatillo – joint bush
Microcycas calocoma – palma corcho
Micromeria = Satureja
***Micronia* spp** – maya, sirin – chigger nuts
Microrhamnus ericoides – abrojo, comida de cuervo, garambullo, tecomblate – javelina-brush
Microsechium helleri – amole amargo, camote de cerro, chayotillo, chichicamole, neco, sanacoche
Milium effusum – mijo silvestre – wood millet
Milla biflora – azucena del campo, azucena silvestre, carcoma, estrella, estrella de San Nicolás, estrella mexicana, estrellita, flor de mayo, flor de San Nicolás, lilia cimarrón, San Nicolás – Mexican star, Mexican star of Bethlehem, wild lily
Milleria quinqueflora – chinquisque, cocolmeca, escobilla, xiutoloc
***Miltonia* spp.** – orquídea pensamiento – pansy orchid
Mimosa aculeaticarpa – garruño, gatuña, uña de gato – cat claw, wait-a-minute
Mimosa albida – chochohuisté, cuatantillo, dormilona, huihuitz-yo-cochis, sensitiva, tapa vergüenza, vergonzosa, zarza
Mimosa asperata – rama dormilera, rama dormilona
Mimosa (Mimosopsis) biuncifera – estrellita, garabatillo, garruño, gatuña, uña de gato – catclaw mimosa, wait-a-bit, wait-a-minute
Mimosa borealis – gatuña – fragrant mimosa, pink mimosa
Mimosa brandegeei – gatuña
Mimosa distachya (purpurascens) – cuca, cuca de árbol, cuilón, garbancillo, garabatillo, gatuña, iguano, uña de gato
Mimosa dysocarpa – gatuña, uña de gato – velvet-pod mimosa
Mimosa grahamii – gatuña – Graham mimosa
Mimosa laxiflora – garabatillo, gatuña, uña de gato – Arizona mimosa
Mimosa malacophylla – raspa huevos, raspilla
Mimosa margaritae – gatuña
Mimosa palmeri – chapo, chopó
Mimosa pudica – adormidera, dormidera, dormilona, mimosa púdica, mimosa vergonzosa, pinahuihuixtle, quecupatli, sensitiva, ten vergüenza, vergonzosa, vergüenza, vergüenza varonil, xmutz – mimosa, sensitive plant, shame plant, touch-me-not
Mimosa wrightii – gatuña
Mimosa xanti – celosa, uña de gato

Mimosopsis* = *Mimosa
Mimulus aurantiacus (puniceus) – perrito – monkeyflower
Mimulus cardinalis – mímulo – scarlet monkey flower
Mimulus guttatus – almizcle amarillo, berro, lama, llantén cimarrón, mímulo, tocasoiahui – monkeyflower
Mimulus moschatus – almizcle – musk monkeyflower
Mimulus primuloides – almizcleña – primrose monkeyflower
Mimulus verbenaceus – pico de pájaro
Mimusops balata – balata
Mimusops dariensis – níspero – bulletwood
Minuartia (Arenaria) rubella – álsine rugosa – mountain sandwort
Minuartia stricta – álsine rígida – bog sandwort
Mirabilis exserta – arrebolera – four-o'clock
Mirabilis jalapa – arrebolera, bella de noche, buenas tardes, Diego de noche, dondiego de noche, donjuan, flor de maravilla, linda tarde, maravilla, maravilla silvestre, pegajosa, tlalquilín, trompetilla, trompetilla de noche, tzutzuyxiu – garden four-o'clock, marvel of Peru
Mirabilis longiflora – acxoyatic, alzoyatic, atzoyatl, maravilla, maravilla de jardín, maravilla del cerro, pebete, suspiro – sweet four-o'clock
Mirabilis multiflora – maravilla – desert four-o'clock, marvel of Peru, wild four-o'clock
Mirabilis multiflora* var *pubescens (froebellii* var *glabrata) – hierba del indio
Mirabilis wrightiana – maravilla
Miscanthus sinensis – eulalia – eulalia grass, Japanese silver grass, zebra grass
Mitchella repens – eskuóvain, fruta de la perdiz, michela – red partridge berry, squaw vine, twinberry
Mitracarpus hirtus – zacate
Modiolastrum (Malvastrum) scoparium – alatle – false rose-mallow
Moenchia erecta – moenchia erecta – chickweed
Molinia caerulea – molinia azul – moor grass
Mollugo verticillata – anisillo, culantrillo, espuelita – carpetweed, Indian chickweed
Moluccella laevis – campanillas de Irlanda – bells-of-Ireland, molucca balm, shell flower
Mombus sapondia – jocote
Momordica balsamina – avellano de guía, balsamina, cundeamor, cundeamor chico – balsam apple, bitter apple, southern balsam pear
Momordica charantia – 1. balsamina, bálsamo, calabaza africana, catajera, cosquelite, cundeamor, cundeamor de Yucatán, cundeamor grande, estropajo, momordica, momordica amarga, pepino, pepino amargo, yacunahax – balsam pear, bitter melon; 2. balsamito, cohombrillo amargo – Indian bitter gourd
Momordica cochinchinensis – cundeamor, pepinillo del diablo, pepino amargo espinoso – Chinese bitter cucumber
Monanthochloë littoralis – zacate de la playa, zacate playero, zacate salado – shoregrass
Monarda austromontana – orégano, orégano de la sierra – bee balm
Monarda citriodora – bergamota, monarda escarlatina, orégano, té de Oswego – bee balm, bergamot, lemon horsemint, Oswego tea
Monarda didyma – bergamota, té de Pensilvania, toronja de abejas – bee balm, bergamot, horsemint, Oswego tea
Monarda fistulosa – bergamota silvestre, ortiga de los indios – bee balm, pink horsemint, wild bergamot
Monarda menthaefolia – monarda, orégano de la sierra – mintleaf beebalm, wild oregano
Monarda pectinata – orégano de la sierra, orégano del campo – horsemint, plains bee balm
Monardella macrantha – flor de roca
Monardella odoratissima – poleo chino – coyote mint, pennyroyal
Monardella villosa – poleo – coyote-mint
Moneses uniflora – pirola uniflora – woodnymph
Monolepis nuttalliana – patata, patota – poverty weed
Monotropa hypopitys – monótropa, monótropa amarilla, monótropa multiflora – pinesap
Monotropa (Hypopitys) uniflora – flor de tierra, pipa del indio – Indian pipe, pinesap

Monstera deliciosa (Philodendron pertusum) – acaxóchitl, amánchacua, árbol de pan mexicano, balazo, cerimán, costilla de Adán, filodendro, filodendro de hoja hendida, hojadillo, malanga ojal, mano de tigre, mimbre, monstera, paaktzatz, piñanona – ceriman, Mexican breadfruit, split-leaf philodendron, Swiss cheese plant

Montanoa hibiscifolia – telecate blanco

Montanoa patens – batayáqui, matayaqui, talacáo

Montanoa speciosa – acahual, vara blanca

Montanoa tomentosa – aclinán, cihuapatli, hierba de la parida, hierba de la primavera, margaritón, medicina de mujer, sihuapatl, sihuapatli, zihoapatle, zoapatle, zompantle – daisy tree

Montanoa tomentosa (rosei)* ssp *rosei – batayáqui, matayaqui, talacáo – tree daisy

Montbretia = Crocosmia

Montia fontana – hierba del manantial – blinks, water chickweed

Montia = Claytonia

Moquilea = Licania

Morangaya = Echinocereus

Morchella esculenta – cagarria, colmena, colmenilla, crespilla, elote, elotito, gallarda, guindo garrafal, hongo, mazorca, mazorquita, morilla, múrgola, pancita, rabassola – morel

Moringa oleifera – acacia, árbol de las perlas, ben, chinto borrego, flor de jacinto, jacinto, moringa, paraíso, paraíso blanco, paraíso de España, perla, perla de la India, perla de oriente, picante blanco, San Jacinto – drumstick tree, horseradish tree

Morisonia americana – árbol del diablo, chico, chico silvestre, chicozapote, mabolo, yoal, zapote blanco – wild sapodilla

Mormodes pardinum* ssp *unicolor – canela, orquídea china amarilla – leopard-like mormodes

Mortonia greggii – afinador

Mortonia utahensis – afinador

Morus alba – baya, moral, moral blanco, morera, morera blanca – silkworm mulberry, white mulberry

Morus celtidifolia – amacapulín, moral, moral negro, morera negra, palo moral, tzitzi – wild mulberry

Morus microphylla – apurí, baya, moral – littleleaf mulberry, Mexican mulberry, Texas mulberry

Morus nigra – mora de árbol, moral, moral negro, morera, morera negra – black mulberry, Persian mulberry, sycamine

Morus rubra – moral, moral rojo – red mulberry

Mospilus germanica – níspero – medlar

Mucuna deeringiana – frijol terciopelo de Florida – velvetbean

Mucuna pruriens – chiporro, cuecuexquic, gusano, picapica – cowitch, itchy bean

Mucuna pruriens* var *utilis – frijol terciopelo, guisante negro, haba terciopela, judía aterciopelada de Florida – Florida velvet bean

Mucuna sloanei – ojo de borrico, ojo de buey, ojo de venado, petama, petana – ox-eye bean

Muehlenbeckia platyclados – solitaria – wire vine

Muhlenbergia dubia – liendrilla del pinar – pine muhly

Muhlenbergia dumosa – carrizo, cola de zorra, liendrilla abierta, otate, otatillo, ototillo, totèkam – bamboo muhly

Muhlenbergia emersleyi – cola de ratón, cola de zorra – bullgrass

Muhlenbergia longiligula – liendrilla lengua larga, liendrilla ligulada

Muhlenbergia (Epicampes) macroura – gubaya, malinali, paja, raíz de zacatón, soromuta, zacate amacollado, zacate de escoba, zacate malinali, zacatón – broom root, whiskgrass muhly

Muhlenbergia microsperma – liendrilla chica, zacatón – littleseed muhly

Muhlenbergia montana (gracilis) – popote cambray, zacate aparejo, zacatón liso – mountain muhly

Muhlenbergia monticola – zacate liso – mesa muhly

Muhlenbergia parviglumis – liendrilla antenada

Muhlenbergia pauciflora – liendrilla de Nuevo México – few-flowered muhly, New Mexican muhly

Muhlenbergia porteri – liendrilla amacollada, telaraña, zacate aparejo – bush muhly

Muhlenbergia repens – liendrilla aparejo, salado – creeping muhly
Muhlenbergia (Epicampes) rigens – liendrilla de venado, zacatón – basket grass, deergrass
Muhlenbergia rigida – grama, liendrilla morada, zacatón – purple muhly
Muhlenbergia (Epicampes) utilis – aparejo, salado – aparejograss
Muntingia calabura – bersilana, cacanicua, calabura, capulín, capulín de mayo, cerecillo, guindo, huizlán, jonote, memiso, nigua, palmán, puan, pus – calabura, Jamaica cherry
Murraya koenigii – cari, curry – curry
Murraya paniculata (exotica) – boj de China, jazmín, limonaria, murraya – orange jessamine
Mus = Musa
Musa (Mus) acuminata – 1. plátano de China – Chinese banana; 2. guineo niño – lady finger banana; 3. colorado blanco, morado verde – Dacca banana; 4. banano enano, cachaco, cuatrofilos, popocho – dwarf Cavendish banana; 5. banano, guineo gigante, plátano gros Michel, plátano roatan – gros Michel banana; 6. banano color clarete, plátano colorado – red Dacca banana; 7. banano robusta, plátano poyo – poyo banana, robusta banana
Musa (Mus) acuminata* x *balbisiana – 1. plátano, plátano francés – green French plantain; 2. plátano macho – horn plantain; 3. plátano colorado – red plantain, wine plantain; 4. banano manzano, plátano manzano – apple banana; 5. largo, plátano burro – cooking banana
Musa (Mus) balbisiana – banano harinoso – balbis banana, starchy banana
Musa (Mus) cavendishii – plátano, plátano costeño, plátano de China, plátano de Costa Rica, plátano dominico, plátano enano – China banana
Musa (Mus) coccinea (uranoscopus) – plátano, plátano tuno – ornamental banana
Musa (Mus) errans* var *botoan – plátano butuhan – Butuhan banana
Musa (Mus) fehi – guineo, plátano – plantain
Musa (Mus) nana – banano enano, guineo enano, plátano enano – dwarf banana
Musa (Mus) nepalensis – banano de Nepal, banano nepalés – Nepal banana
Musa (Mus) paradisiaca – bananero, banano, boxhaas, guineo, plátano, plátano amarillo, plátano bárbaro, plátano costilla, plátano costillón, plátano de pájaro, plátano grande, plátano harton, plátano hembra, plátano largo, plátano macho, plátano vianda, roatán, sapane, zapalote – cooking banana, plantain
Musa (Mus) sapientum – bananero, banano, boxhaas, camburí, criollo, guineo, guineo de seda, guinia, haas, manzano, morado, morado amarillo, pantani, plátano, plátano camburi, plátano chico, plátano dominico, plátano enano, plátano fruta, plátano guineo, plátano morado, plátano roatán, plátano Tabasco, plátano zapalote, roatán, zachaax – banana
Musa (Mus) textilis – abacá, cáñamo de Manila, plátano textil – Manila hemp
Musa = Ensete
Muscari armeniacum – almizcleña, jacinto racemoso – grape hyacinth
Muscari botryoides – nazareno, nazareno azul, nazareno menor – grape hyacinth
Muscari comosum – almizcleña, nazareno, nazareno de penacho – fringe hyacinth, tassel hyacinth
Muscari moschatum – almizcleña – musk hyacinth
Muscari neglectum – nazareno descuidado – starch hyacinth
Myoporum laetum – mióporo
Myosotis discolor – raspilla versicolor – changing forget-me-not
Myosotis laxa – raspilla cespitosa – water forget-me-not
Myosotis micrantha – raspilla de flores pequeñas, raspilla micracanta – spring forget-me-not
Myosotis palustris – nomeolvides, nomeolvides de agua, raspilla de los pantanos, raspilla palustre – water forget-me-not
Myosotis scorpioides – nomeolvides – forget-me-not
Myosotis sylvatica (alpestris) – miosota, nomeolvides, nomeolvides alpino, nomeolvides de bosque, raspilla selvática – wood forget-me-not

Myosurus minimus – cola de ratón menor – mouse-tail
Myrciaria floribunda – mirto – rum berry
Myrica californica – árbol de la cera, arrayán – bayberry, California sweet bay, Pacific wax myrtle, wax myrtle
Myrica cerifera – árbol de la cera, arbolito de la cera, arrayán, arrayán cerífero, atocamay, cerero, guacanalá, huancanalá, mirica cerífera, mirto de cera, otocamay, vegetal – bayberry, southern bayberry, wax myrtle
Myrica gale – arrayán brabántico, arrayán de los pantanos, mirto de brabante – sweet gale
Myrica mexicana – árbol de la cera, chaclolo, chacolol, chilpanxohuitl, copalxihuitl, guacanalá, huancanalá
Myrica pensylvania (caroliniensis) – arrayán brabántico – bayberry
Myrica salicifolia – insulina vegetal, pedra hume caa
Myriophyllum verticillatum – filigrana menor – water-milfoil
Myristica fragrans – 1. mirística, mirística olorosa, moscadero, nuez de especia, nuez enemosa, nuez moscada – nutmeg; 2. macia, macis – mace
Myristica moschata – nuez moscada – nutmeg
Myroxylon (Toluifera) balsamum* var *pereirae – árbol de bálsamo, balsamito, bálsamo, bálsamo de las Indias, bálsamo de la unión, bálsamo del Perú, bálsamo de San Salvador, bálsamo de Tolú, bálsamo negro, cedro chino, chichipale, chuchupate, chucté, huitzilóxitl, nabá, nabal, palo de bálsamo, palo de trapiche, semilla del obispo, tache, yaga-guienite – balm tree, balsam of Peru, balsam of Tolu
Myroxylon flexuosum – coronilla, huichichilemetl, manzanillo
Myrrhis odorata – perifollo almizclado, perifollo oloroso – garden myrrh, sweet cicely
Myrtillocactus cochal – cochal – cochal
Myrtillocactus (Cereus) geometrizans – bastón, biti-xoba, garambullo, garbancillo, padrenuestro – blue candle, candelabra cactus, whortleberry cactus
Myrtillocactus schenckii – garambullo, vichishovo
Myrtus cerifera – cerero – wax myrtle
Myrtus communis – arrayán, arrayán blanco, arrayán común, mirto, mirto común, murta – common myrtle, myrtle, true myrtle
Myrtus* = *Luma

N

Najas flexilis – nayada flexuosa – slender water-nymph

Najas major – nayada mayor, zaragoza – water-nymph

Najas marina – nayada marina, sargazo – holly-leaved water-nymph, naiad-wort

Nama coulteri – hierba de la punzada, junco – purple mat

Nama demissum – morada – purple matseed

Nama hispidum – bentosidad, flor morada, morada, moradita, ventosidad – purple mat, sand bell

Nama undulatum – hierba de la pulga, hierba de la punzada, hierba de la ventosidad, matagusanos, ventosidad

Nannorrhops ritchiana – palma de mazari

Narcissus asturiensis – narciso asturiano, narciso de Asturias – Asturian daffodil

Narcissus bulbocodium – narciso de hojas tenues – hoop petticoat daffodil, trumpet daffodil

Narcissus cyclamineus – narciso cyclamineus, narciso de hojas de ciclaminos – cyclamen-leaved daffodil, cyclamineus daffodil

Narcissus jonquilla – junquillo, junquillo oloroso, narciso junquillo – jonquil, jonquilla daffodil, jonquilla hybrids

Narcissus poeticus – narciso, narciso blanco, narciso de los poetas, narciso poético – pheasant's eye, poet's narcissus

Narcissus pseudo-narcissus – 1. narciso, narciso de los prados, tragapán – daffodil; 2. narciso atrompetado, narciso trompeta, narciso trompón – trumpet daffodil; 3. narciso de tubo largo, narciso de corona grande – large-cupped daffodil; 4. narciso de tubo corto, narciso de corona pequeña – small-cupped daffodil; 5. narciso doble, narciso lleno – double daffodil; 6. narciso de roqueda, narciso enano, narciso miniatura – dwarf daffodil

Narcissus tazetta – narciso de manojo, narciso tazetta – multiflowering daffodil, rose of Sharon, tazetta daffodil

Narcissus triandrus – copa de los ángeles, narciso triandro, narciso triple – angel's tears, triandus hybrids

Nardostachys jatamansi – nardo – spikenard

Nasturtium = Rorippa

Nectandra ambigens – laurel, laurel de hojas grandes, laurelillo

Nectandra coriacea – laurel

Nectandra sanguinea – aguacatillo, laurel, laurel amarillo, piecito de paloma – timber sweet, white laurel

Nectandra sinuata – aguacate, aguacatillo, aguacatillo hediondo, palo de aguacate, palo de tejón – wild pear

Nelumbium = Nelumbo

Nelumbo lutea (Nelumbium luteum) – chaga, serrucho – American lotus

Nelumbo nucifera (Nelumbium nelumbo) – flor de loto, loto sagrado, rosa del Nilo, haba de Egipto – Chinese lotus, Indian lotus, sacred lotus

Nemophila menziesii (insignis) – azuelita, azuleja, azulejita, campánula de California, mariana, nemofila – baby blue-eyes

Neobuxbaumia (Cephalocereus) mezcalaensis – gigante

Neobuxbaumia (Pachycereus) tetetzo – cabeza de viejo, cardón, higo de tetetzo, órgano, tetetzo, tzetzetzo – column cactus

Neodypsis decaryi – palmera triangular

Neoevansia = Peniocereus

Neolloydia conoidea – biznaguita – Texas cone cactus, thimble cactus

Neomammillaria = Mammillaria

Neoregelia carolinae – nidularium

Neottia cinnabarina – cutzis

Neottia nidus-avis – nido de pájaro – bird's nest orchid

Nepeta cataria – calamento, gatera, hierba del gato, hierba gatera, menta de gato, nébeda, nébeda común, toronjil de China – catmint, catnip

Nepeta* x *faassenii (mussinii) – hierba gatera, nébeda – catmint

Nepeta grandiflora – nébeda de flores grandes – large-flowered catmint

Nephelea mexicana – cola de mono, pesma – tree fern

Nephelium lappaceum – lichi peludo, rambután – hairy lychee, rambutan

Nephelium longana – longán, longana – longan
Nephelium = Litchi
Nephrolepis biserrata – cola de pisote – callawalla, quash tail
Nephrolepis exaltata – guirnalda, helecho, helecho común, helecho de Boston, helecho madre, nefrolepis – Boston fern, sword fern
Nerine bowdenii – nerina
Nerium oleander – adelfa, adelfa olorosa, baladre, hojaranzo, laurel, laurel blanco, laurel colorado, laurel rosa, narciso, oleandro, rosa francesa, rosa laurel, trinitaria, yagaquiguece – oleander, rose bay
Neurolaena lobata – árnica, cola de faisán, lobata, rabo de faisán, tabaco cimarrón – jackass bitters
Nicandra physalodes (violacea) – belladona del país, manzano del Perú, nicandra, puchero de monja, tomate de burro, tomate de culebra – apple of Peru
Nicotiana alata (affinis) – nicociana, planta de tabaco, tabaco, tabaco alar, tabaco de adorno, tabaco ornamental – flowering tobacco, jasmine tobacco, nicotiana
Nicotiana attenuata – punche, tabaco de coyote – coyote tobacco
Nicotiana clevelandii – tabaco de coyote, tabaco de perro – desert tobacco
Nicotiana glauca – álamo loco, árbol de tabaco, buena moza, cornetón, cornetón del monte, donjuan, gandul, gigante, gretaña, hierba del gigante, hierba del zopilote, hoja de cera, Juan loco, lengua de buey, levántate Don Juan, levántate Juan, marijuana falsa, marquiana, mostaza montés, palo extranjero, palo loco, palo de San Juan, palo Virgen, punche, tabaco, tabaco amarillo, tabaco cimarrón, tabaco lampiño, tabaco moro, tabaco moruno, tabacón, tabaquillo, tacote, tepozán extranjero, toripipa, tronador, tronador de España, Virginio, wipaka, xiutecuitlanextli – mustard tree, tree tobacco
Nicotiana ipomopsiflora – tabaco de coyote, tabaco de papanta, tabaquillo
Nicotiana obtusifolia (trigonophylla) – tabaco cimarrón, tabaco de coyote, tabaco loco, tabaco papanta, tabaquillo, tabaquillo de coyote – desert tobacco
Nicotiana repanda – tabaco cimarrón – fiddle-leaf tobacco
Nicotiana rustica – andumuqua, macuche, noholcikutz, picietl, punche, punche mexicano, tabaco, tabaco del Brasil, tabaco de macuache, tabaco menor, tabaco pequeño, tabaco rústico, tabaquillo, tenapepe, teneshil, yetl – Aztec tobacco, Indian tobacco
Nicotiana sylvestris – tabaco de los bosques – wood tobacco
Nicotiana tabacum – cuauhyetl, cutz, guexa, hierba del diablo, huepaca, iyátl, kutz, macuchi, nicociana, pícietl, punche, quauhyetl, tabaco, tabaco cimarrón, tabaco mayor, tabaquera, uipa, yetl – tobacco
Nigella damascena – ajenuz, araña, arañuela, cabellos de Venus, estrella de mar, neguilla, neguilla de Damasco, nigela – bishop's wort, love-in-a-mist
Nigella sativa – comino negro, neguilla aromática – fitches, nutmeg flower
Nissolia fruticosa – bejuco de tripa, candelilla prieta, tripa de gallina
Nolina beldingii – palmilla, palmita, sotol – nolina
Nolina bigelovii – moho, palmita, sacahuista, sotol, zacate – beargrass, desert tree-beargrass, nolina, ribbon grass
Nolina cespitosa – palmilla
Nolina erumpens – sacahuista – basketgrass
Nolina longifolia – borracho – Mexican grass tree
Nolina matapensis – palmilla, palmita, sotol, tueya, tuyá – Sonoran tree-beargrass, tree beargrass
Nolina micrantha – sacahuiste – beargrass
Nolina microcarpa – palma, palmilla, sacahuiste, sotol chico, sotol chiquito, zacate, zacate cortador, zacate de aparejo, zacate de armazón – beargrass, saw-grass
Nolina palmeri – amole, palmita, sotol – nolina
Nolina parryi – sotol – beargrass, nolina, ribbon grass
Nolina (Beaucarnea) recurvata – palmita – bottle palm, ponytail palm
Nolina texana – sacahuiste – beargrass, Texas sacahuiste
Nopalea = Opuntia

Nopalxochia = Disocactus
Nothofagus antarctica – haya, haya antártica, haya austral, haya falsa, ñire, notofagus antártica – false beech, southern beech
Nothofagus dombeyi – coihue, haya chilena, notofagus de Dombey
Nothofagus obliqua – coyán, notofagus oblicua, roble, roble pelin – roble beech
Notholaena aurea – calaguala, palmilla
Notholaena lemmonii – helecho – cloak fern
Notholaena sinuata – calaguala, calaguala del indio, candelilla, doradilla, nacahuela – cloak fern
Nothoscordium bivalve – mayito – crow-poison
Nothoscordium fragrans – lágrimas de la Virgen
Nothoscordium striatum – cebolleta
Nuphar luteum – cubilete, lirio de agua, nenúfar, nenúfar amarillo – cow-lily, spatterdock, yellow pond lily, yellow water lily
Nyctanthes arbor-tristes – árbol triste – night-flowering jasmine
Nyctocereus = Peniocereus, Selenicereus
Nymphaea alba – albahaca de limón, apepe, cabeza de negro, flor de agua, lirio de agua, nelumbio blanco, nenúfar, nenúfar blanco, ninfea, pan de manteca, rosa del amor – white waterlily
Nymphaea ampla – flor de laguna, hoja de sol, lab, ninfea, sol de agua
Nymphaea caerulea – ninfea, quetzalaxochiacatl – blue waterlily
Nymphaea elegans – cabeza de negro, flor de capomo, nenúfar, ninfea azul – blue waterlily
Nymphaea flavovirens – apapatla, ninfea blanca – white waterlily
Nymphaea mexicana – apapatla, cabeza de negro, nenúfar, ninfea, ninfea amarilla, tambácuaro – banana waterlily, blue waterlily, yellow waterlily
Nymphaea odorata – cumbacuare, lampazo, nenúfar oloroso, nenúfar perfumado – fragrant waterlily, white waterlily
Nymphoides falax – ninfea
Nymphoides (Limnanthemum) humboldtianum – hoja de pescado, ninfea, nopalillo, pan de manteca – floating-heart
Nymphoides indica – lirio – floating heart
Nymphoides peltata – nenúfar peltado – water fringe
Nyssa sylvatica (multiflora) – cabo de luc, chisté, palo de papaxi, tetzicohuitl, tupelo, tupelo negro – pepperidge, sour gum, tupelo

O

Oatasetum maculatum – tzacouhxóchitl
Ochroma lagopus – árbol de algodón, balsa, ceibón lanero, lano, palo balsa, pepe balsa – balsa
Ochroma pyramidale – balsa, corcho, gonote real, jonote real, jopi, maho – balsa
Ocimum basilicum – albaca, albacar, albacarón, albahaca, albahaca blanca, albahaca dulce, alfábega, alhábega, basílico, hierba real – basil, common basil, sweet basil
Odontoglossum cervantesii – jazmincito, mariposa, tigrillo
Odontoglossum cordatum – Manuelito
Odontoglossum insleayii – terciopelo – Insleay's odontoglossum
Odontoglossum pendulum (citrosmum) – pozolillo
Odontoglossum pulchellum – tanalito
Odontoglossum rosii – Manuelito, orquídea payaso, tigrillo – tiger orchid
Odostemon = Berberis
Oenanthe aquatica – cicuta de agua, enante, enante de agua, felandrio, felandrio acuático, hinojo acuático – water dropwort, water fennel
Oenanthe crocata – cicuta acuática, nabo del diablo – water hemlock
Oenanthe pimpinelloides – enante falsa-pimpinela – water dropwort
Oenothera biennis – capa de San José, enotera, flor de Santa Rita, hierba del asno, onagra bienal, onagra común, primavera, primavera nocturna, prímula, prímula nocturna – common evening primrose
Oenothera californica – flor de arena
Oenothera deltoides (trichocalyx) – hierba salada – basket evening primrose, devil's lantern, lion-in-a-cage
Oenothera fruticosa – hierba del asno, onagra fruticosa – sundrops
Oenothera glazioviana – onagra de flores largas – large-flowered evening primrose
Oenothera hartweggi – amapola
Oenothera laciniata – amapola, onagra laciniada – cut-leaved evening primrose
Oenothera macrocarpa (missouriensis) – amapola, onagra – buttercup, flutter-mill, Missouri primrose, Ozark sundrops
Oenothera pubescens – tenalxíhuitl – evening primrose
Oenothera (Hartmannia) rosea – agua de azahar, amapola, hierba del golpe, onagra rosada, tarepeni – evening primrose
Oenothera (Hartmannia) speciosa – amapola, amapola del campo, onagra blanca, onagra elegante – Mexican evening primrose, Mexican primrose, pink evening primrose, pink ladies
Oenothera stricta – onagra de hojas estrechas – fragrant evening primrose
Oenothera (Hartmannia) tetraptera – hierba del golpe, linda tarde – evening primrose
Oenothera (Onagra) trichocalyx – flor de arena
Oenothera villosa (hookeri) – capa de San José, enotera, flor de Santa Rita, hierba del asno, onagra vellosa – evening primrose
Okenia hypogaea – cachaz
Olea europaea – aceituno, manzanilla, olivera, olivo – olive
Olea europaea* var *oleaster – acebuche, oleastro, olivastro, olivo silvestre – wild olive
Oliveranthus (Cotyledon) elegans – perico – Oliver's flower
Olneya tesota – árbol de hierro, comitín, hésen, palo de tinta, palo fierro, tésota, uña de gato – desert ironwood, ironwood
Olyra latifolia – carricillo, carrizo, carrizo verde, fisga de ratón – reed
Omphalea oleifera – aguacate de danta, chatet, corcho, piñoncillo
Omphalodes cappadocica – ombligo de Venus, oreja de monje
Omphalodes linifolia – aubliguera, ombligo de Venus, omfalodes de hojas de lino, oreja de monje – Venus' navelwort
Omphalodes verna – omfalodes primaveral – blue-eyed Mary
Onagra = Oenothera
Oncidium adscendens – ahoché, putsché
Oncidium carthaginensis – ahoché, chilillo, putsché, varita de San José
Oncidium cavendishianum – oreja de burro – the Duke of Devonshire's oncidium

Oncidium cebolleta – cebolleta, chilillo, putsché
Oncidium incurvum – angelito
Oncidium ornithorynchum – arañita
Oncidium tigrinum – flor de muerto, orquídea, parácata
Onobrychis viciifolia – esparceta, pipirigallo – sanfoin
Onopordum acanthium – alcachofa borriquera, cardo borriquero, cardo yesquero, manto de Judas – Scotch thistle, Scottish thistle
Onopordum bracteatum – alcachofa borriquera
Onopordum illyricum – cardo del demonio – Illyrian thistle
Operculina aurea – yuca
Ophrys apifera – abejera, flor de la abeja, hierba de la abeja – bee orchid
Ophrys speculum – espejo de Venus – mirror of Venus, mirror orchid
Opizia stolonifera – bramilla, pasto de conejo
Oplismenus burmannii – hayal-sitsuuk, zacate cadillo
Oplismenus setarius – pasto sombra, zacate barbón
Oplopanax horridus – club del diablo
Opopanax chironium – opopónace, pánace – Hercules' allheal
Opuntia amyclaea – nochtli, nopal, nopalli, tuna, tuna de alfajayuca
Opuntia auberi – lengua de vaca, nopal de lengüita
Opuntia azurea – coyotillo, nopal coyotillo, nopalillo
Opuntia basilaris – nopal – beavertail, beavertail cactus
Opuntia bravoana – nopal, tuna
Opuntia chaffeyi – organillo, organito, sacacil, viejo
Opuntia chavena – nopal cadillo, nopal cascarón, nopal chaveño
Opuntia chlorotica – nopal – pancake-pear, pancake prickly-pear, silver-dollar cactus
Opuntia (Nopalea) cochenillifera – nochez nopalli, nocheztli, nopal de cochinilla, nopal de San Gabriel, nopal nocheztli, pacam, palera, tlalnopal, tlatocnochtli – cochineal cactus
Opuntia decumbens – aprón, lengua de vaca, nopal de coyote, nopal de culebra, nopal de tortuga
Opuntia (Nopalea) dejecta – lengua de vaca, nopal chamacuero – spiny opuntia
Opuntia dillenii – chaparro, cuija, dildo, pacam, yaaxpakan – Eltham Indian fig, sweet prickly-pear
Opuntia durangensis – nopal mantequilla, nopal tapón
Opuntia engelmannii (cantabrigiensis) – abrojo, coyonoxtle, cuija, heel, joconostle, nopal, nopal cuixo, tuna, tuna cuija, vela de coyote – Engelmann prickly-pear
Opuntia engelmannii (lindheimeri)* var *lindheimeri – cuija, tuna – Texas prickly-pear
Opuntia ficus-indica (compressa, vulgaris) – chumbera, higuera chumba, nochtli, nopal, nopal castellano, nopal de Castilla, nopalli, nopal sin espinas, nopal tunero, noxtli, palera, tuna, tuna de alfajayuca, tuna de Castilla, tuna mansa, tuna pelona, tuna tempranilla, tuna verde – Indian-fig, mission cactus, tree prickly-pear
Opuntia fuliginosa – tuna
Opuntia guilanchi – nopal guilanchi
Opuntia hyptiacantha (matudae) – nopal cadillo, nopal cascarón, nopal chaveño, nopal memelo, tuna chaveña, tuna colorado con espinas, tuna corriente, tuna mansa, vela de coyote – prickly-pear
Opuntia inaperta (Nopalea escuintlensis, N. gaumeri) – tuna de playa, zacam, zacamsotz
Opuntia joconostle – joconostle, joconoxtle, tempranilla, tuna blanca, xoconoxtle
Opuntia (Nopalea) karwinskiana – jara, lengua de vaca, nopal, nopalillo, nopalillo de flor, tuna
Opuntia lagunae – tuna morada
Opuntia larreyi – nopal camueso
Opuntia lasiacantha (rzedowskii) – nopal, nopal de cerro, nopal de pedregal
Opuntia leucotricha – duraznillo, duraznillo blanco, duraznillo colorado, nopal blanco, nopal chaveño, nopal colorado, nopal duraznillo, tuna duraznillo, tuna duraznillo blanca, tuna duraznillo colorada – Aaron's beard prickly-pear
Opuntia macrocentra (violacea) – duraznillo,

nopal coyotillo – purple prickly-pear, violet prickly-pear
Opuntia megacantha – higo de mar, nochtli, nopal, nopal de Castilla, nopal de raíz, nopalli, nopal manso, nopal tempranillo, tuna, tuna brava, tuna colorada fina, tuna de alfajayuca, tuna de Castilla, tuna mansa – Indian-fig
Opuntia megarhiza – nopal de raíz, nopalillo, sacanochtli, soconochtli
Opuntia microdasys – cegador, ciega borrego, nopal cegador, nopalillo cegador, nopal real, tlatoc-nochtli – angel's wings, bunny-ear prickly-pear, bunny ears, polka dot cactus
Opuntia oricola – nopal – chaparral prickly-pear
Opuntia phaeacantha – abrojo, nopal, vela de coyote – Big Bend prickly-pear, Mojave prickly-pear
Opuntia pilifera – cocoche loco, crinado, nopal, nopal crinado, nopal de crinas
Opuntia polyacantha – nopal redondo – juniper prickly pear, plains prickly-pear
Opuntia polyacantha (arenaria)* var *arenaria – nopal – El Paso prickly-pear
Opuntia polyacantha (erinacea)* var *erinacea – nopal – porcupine prickly-pear
Opuntia puberula – nopal de culebra, nopal de tortuga
Opuntia pumila – bichi-bixui, cardo, nopal cardoso
Opuntia rastrera – cuija, nopal rastrero
Opuntia robusta (guerrana) – bartolona, camuesa, cochinera, nopal camueso, nopal tapón, nopal tapón blanco, rastrero, tuna camuesa, tuna morada, tuna tapona, tuna tapona cimarrona – prickly-pear
Opuntia rufida – nopal cegador – blind prickly-pear
Opuntia spinulifera (heliabravoana) – duraznillo blanco, nopal ardilla, nopal chaparro, nopal corriente, tetechal, xoconostle blanco
Opuntia stenopetala – arrastradillo, nopal colorado, nopal lasarón, nopal serrano
Opuntia streptacantha – cardón, cenizo, chaveño, nochtli, nopal, nopal cardón, nopal de tuna colorada, nopal hartón, nopalli, tuna, tuna cardona, tuna cascarona, tuna chaveña, tuna chica, tuna esferica, tuna hartona, tuna mansa – prickly-pear
Opuntia stricta – yaaxpakan – coastal prickly-pear, pest pear
Opuntia tapona – tapona, tuna tapona
Opuntia tomentosa (hernandezii, sarca) – nocheznopalli, nopal, nopal chamacuerito, nopal chamacuero, nopal común, nopal de San Gabriel, nopal nocheztli – velvet opuntia
Opuntia tuna – alquitira del país, culhua, flor de tuna blanca, nochtli, nopal de cochino, nopal huevo, tuna
Opuntia wilcoxii – navo, nopal, tuna – prickly-pear
Opuntia = Cylindropuntia, Grusonia
Orbignya (Attalea) cohune – chuncuy, coco de aceite, coco de Guadalajara, cocoyule corozo, cohune, coquillo, coquito, coquito de aceite, coquito de Colima, corozo, corozo de sábana, coyol, coyol de sábana, guacoyul, jamiltepec, juquilla, manaca, manzanillo, monaco, mop, palapa, palma, palma coyol, palma de coquito de aceite, palma real, quacoyul, tutz – cohune palm
Orbignya = Attalea
Orchidaceae – orquídeas – orchids
Oreodoxa oleracea – chaguaramo, palma real
Oreodoxa = Roystonea
Oreopanax capitatus – cabellera de palo, coletón, picón
Oreopanax peltatum (salvinii) – coleto, mano de danta, mano de león, palo de coleto, palo de danta, papayo cimarrón, tronador – pumpwood
Orestias agassiz – carachi negro
Orestias luteus – carachi amarillo
Origanum (Amaracus) dictamnus – orégano, orégano de Creta – Crete dittany, dittany of Crete, hop marjoram
Origanum majorana (Majorana hortensis) – almoraduj, almoradux, arnárico, mejorana, mejorana dulce, orégano hortense, orégano mayor, orégano mejorana, sampsuco, sarilla – knotted marjoram, marjoram, sweet marjoram
Origanum maru* var *aegyptiacum – hisopo – hyssop
Origanum vulgare – mejorana bastarda, mejorana silvestre, orégano, orégano común – oregano, wild marjoram

Ornithogalum arabicum – leche de gallina, leche de pájaro de Arabia – star of Bethlehem
Ornithogalum nutans – leche de gallina nutante – drooping star of Bethlehem
Ornithogalum umbellatum – dama de once horas, estrella de Belén, leche de gallina, matacandiles – dove's dung, star of Bethlehem
Ornithopus pinnatus – pie de pájaro ebracteado – orange birdsfoot
Ornithopus sativus – pie de pájaro, pie de pájaro cultivado, serradella – common serradella
Orobanche (Aphyllon) cooperi – acam, flor de tierra, nep-cha-ga, orobanca – cancer root, desert broom-rape, Indian asparagus
Orobanche crenata – orobanca crenada, topo de las habas – broomrape
Orobanche ramosa – hierba tora – broom-rape
Orthilia secunda – peralito, peralito unilateral – one-sided wintergreen
Orthocarpus = Castilleja
Orthosiphon aristatus – té de Java
Oryza sativa – arroz, hierba del ratón, jotei, xica – rice
Oryzopsis = Achnatherum, Piptatherum
Osmanthus fragrans – olivo oloroso – sweet olive
Osmorhiza brachypoda – cilantro – sweet cicely, sweet root
Osmorhiza longistylis – perifollo oloroso – sweet cicely
Osmunda regalis – helecho real – flowering fern, royal fern
Ostrya carpinifolia – carpe negro, falso carpe, ostria – European hop-hornbeam
Ostrya virginiana – guapaque, júcaro, tzutujté – American hop-hornbeam
Osyris alba – guardalobo, retama loca – poet's cassia
Otatea acuminata – otate, otate dulce – Mexican weeping bamboo
Otatea fimbriata – carrizo, mutu, otate de hojas anchas
Oxalis acetosella – acedera, acederilla, acetosilla, aleluya blanca, oxálida blanca, pan de cuchillo, vinagrillo – shamrock, wood sorrel
Oxalis adenophylla – acedera menor, acederilla
Oxalis albicans – agrito, xocoyol, xocoyole – oxalis, wood sorrel
Oxalis americana – quauhtlaxoxocoyolin, xocoyole de México
Oxalis corniculata – acedera, acederilla, agrito, agrito de maceta, aleluya amarilla, aleluya corniculada, pan de cuco, vinagrillo, xarimpuscua, xocoyol, xocoyole, xoxocoyolin – creeping wood sorrel, oxalis, sourgrass
Oxalis deppei – aleluya de Deppe, trébol de la suerte
Oxalis dillenii (stricta) – agrito, aleluya europea, chilito, oreja de ratón – yellow wood-sorrel
Oxalis divergens – agrito, cuiyul, xocoyole
Oxalis hernandesii – agrito
Oxalis incarnata – aleluya encarnada
Oxalis jacquiniana – agrito
Oxalis lanceolata – socoyole
Oxalis latifolia – acederilla, agrio, agrito, aleluya de hojas anchas, copalillo, elel, sosxocoyoli, yalaelel, zutskeyem – wood sorrel
Oxalis pes-caprae (cernua) – agrio, agrito, aleluya amarilla, xocoyole – Bermuda buttercup
Oxalis purpurea – aleluya de Bowie
Oxalis tuberosa – aleluya tuberosa, cuiba, huisisai, ibía, oca, papa inglesa – oca
Oxydendrum arboreum – acedero arbóreo – sorrel tree, sourwood
Oxyria digyna – oxiria de montaña – mountain sorrel
Oxytropis deflexa – astrágalo deflexo – blue pendent-pod oxytrope
Oxytropis lambertii – chachaquila, frijolillo, hierba loca, loco – Lambert's crazyweed, locoweed

P

Pachira aquatica – acamoyote, amapola, apompo, cabellos de ángel, castaño de agua, castaño de la Guayana, clavellina, clavellina blanca, clavellina roja, guacta, oje, pachira, palo de agua, palo de boya, pitón, pochote, tetón, tura, ture, xiloxóchitl, zapote bobo, zapote de agua, zapote reventador, zapote reventón, zapotón – Guiana chestnut, provision tree, saba nut, shaving-brush tree

Pachira insignis – castaño, ceibón

Pachira macrocarpa – apomo, cabeza de negro, guacta, pochote

Pachycereus (Lophocereus) gatesii – garambullo – old man cactus

Pachycereus gaumeri – kanzakán, kulul, zak-kulul

Pachycereus (Lemairocereus) hollianus – acompes, baboso, órgano

Pachycereus (Marginatocereus, Stenocereus) marginatus – cantueso, chilayo, jarrito, órgano, órgano de la costa, pitaya – organpipe cactus

Pachycereus pecten-aboriginum – candelabro, cardón, cardón-barbón, cardón de la costa, cardón espino, cardón espinoso, cardón hecho, cardón pelón, chik, chiquí, hecho, órgano, órgano de la costa, wichowaka – hairbrush cactus

Pachycereus pringlei (calvus) – cardón, cardón gigante, cardón hecho, cardón pelón, sagüera, sagüeso – elephant cactus

Pachycereus (Cereus, Lophocereus) schottii – barbón, cabeza de viejo, garambullo, hombre viejo, mochi, músaro, muse, pitahaya barbona, pitaya agria, pitaya barbona, pitayita, senita, sina, sina barbona, sinita, tuna barbona, viejo, zabila hecho – old man cactus, senita

Pachycereus (Lemaireocereus, Stenocereus) weberi – bitzuri, candebobe, candelabro, cardón, cardón chico, cardón de Tomellín, cardón espino, cardón espinoso, chico, órgano – candelabra

Pachycereus = Cephalocereu, Neobuxbaumia

Pachycormus discolor – árbol elefante, copalquín, torote, torote blanco – Baja elephant tree

Pachyrrhizus erosus – ajipa, catzotl, chata de agua, chicam, frijol de ñame, jícama, jícama de agua, judía batata, mechen, xícama, xicamatl, xiquima, yuco de bejuco – jicama, Mexican yam bean

Pachyrrhizus palmatilobus – jícama cimarrona, jícama de leche

Pachysandra terminalis – euforbio japonés – Japanese spurge

Pachystachys lutea – jacobinia

Packera candidissima – lechuguilla de la sierra

Paeonia broteroi – escaramón, peonía, peonía de brotero, rosa albardera

Paeonia brownii – peonía – peony

Paeonia californica – peonía de California – peony

Paeonia lactiflora – peonía china, peonía de junio – peony

Paeonia officinalis – albardera, hierba casta, hierba de Santa Rosa, peonía, peonía común, peonía de mayo, peonía oficinal, Reina Margarita, rosa de monte, rosa de Santa María – peony

Paeonia suffructicosa – peonía arbórea, peonía arborescente, peonía arbustiva, peonía leñosa, peonía sufruticosa – tree peony

Palaquium (Isonandra) gutta – gutapercha – gutta-percha

Paliurus spina-christi – espino – Christ thorn, crown of thorns, Jerusalem thorn

Panaeolus papilionaceus (sphinctrinus) – hongo, nanácate, nanácatl, ñui, teonanacatl, teunamacatlth, teyhuintli – psychedelic mushroom

Panax ginseng – ginseng, ginseng asiático, ginseng chino, ginseng coreano – Chinese ginseng, Korean ginseng, pannag

Panax notoginseng – ginseng tienchi – ginseng

Panax quinquefolium – ginseng, ginseng americano, ginseng silvestre americano – American ginseng

Pancratium illyricum – amormío estrellado

Pancratium maritimum – amormío, azucena de mar, nardo marítimo, nebulosa, pancracio, pancracio marítimo – sea-daffodil, sea lily

Pandanus tectorius – palma de cinta – hala, seashore screw pine, screw pine
Pandanus utilis – pah, pándano – screw pine
Panicum antidotale – pánico azul, panizo azul, panizo gigante, pasto azulado, zacate azulado, zacate panizo azul – blue panic grass
Panicum capillare – gallinita, matanena, mijo capilar, zacate lisonjero – old witch grass, witchgrass
Panicum coloratum – grama zaina, gramalote, guinea fino, guinea pajarito, hierba de búfalo, hierba de Guinea, hierba india, macaricari, murumbu, paja guinea carorena, pasto búfalo, pasto colorado, pasto indio, pasto klein, pasto sabaya, surbana, zacate berqueno, zacate colorado, zacate sabaya – colored Guinea grass, white buffalo grass
Panicum dichotomiflorum – mijo de flores dicotómicas – panicgrass
Panicum filipes – panizo, panizo corcolito
Panicum hallii – panizo aserrín, panizo rizado – Hall's panicum
Panicum havardii – zacate de los médanos
Panicum hirticaule – kanchin, panizo, panzio cauchín – millet, panicgrass
Panicum huachucae – huachuca
Panicum italicum – panizo – foxtail millet
Panicum maximum – camalote, colonial, gamelote, grama castilla, gramalote, graminea guinea, guineo, guineón, hierba de Guinea, hierba guinea, hoja fina, matalote, mijo de Guinea, mijo verde, morubú, paja de Guinea, panizo de Guinea, panizo verde, pasto castilla, pasto guinea, pasto guineo, pasto sabaya, privilegio, rabo de mula, zacate egipto, zacate guinea, zacate privilegio, zacatón, zaina – Guinea grass, millet
Panicum miliaceum – borona de Filipinas, maicillo, mijo, mijo común, mijo mayor, mijo miliar, proso – broomcorn millet, pannag, Turkish millet
Panicum molle – hierba de para, malojillo, pasto para, zacate corredor – para grass
Panicum obtusum – panizo, panizo mezquite, purga de paridas, zacate gramilla, zacate guía, zacate masarca – vine-mesquite
Panicum procerum – carricillo
Panicum repens – canota, grama de Castilla, grama portuguesa, mijo repente, pasto torpedo – torpedo grass, Victoria grass
Panicum schinzii – mijo de hojas lisas – smooth-leaved panic grass
Panicum sonorum – sauhui – panicgrass
Panicum trichoides – cuhuech, zacate carricillo
Panicum virgatum – mijo, pasto azulado, talquezal – switch grass
Panicum = Brachiaria, Dichanthelium
Papaver alpinum – adormidera alpina, amapola alpina, amapola de los Alpes – alpine poppy
Papaver argemone – amapola macho
Papaver atlanticum – amapola atlántica – Atlantic poppy
Papaver dubium – adormidera de hojas obtusas, amapola oblonga
Papaver glaucum – amapola tulipán
Papaver hybridum – amapola mestiza, amapola triste – rough poppy
Papaver nudicaule – adormidera de tallo nudo, amapola de Islandia – Iceland poppy
Papaver orientale – adormidera de Oriente, amapola oriental – oriental poppy
Papaver rhaeticum – amapola de los Alpes
Papaver rhoeas – ababol, abala, adormidera, amapola, amapola común, amapola silvestre de los campos – corn poppy, Flanders field poppy, red poppy, Shirley poppy
Papaver somniferum – adormidera, adormidera china, amapola, amapola azul, amapola blanca, amapola de la adormidera, amapola de opio, dormidera – breadbox poppy, gall, opium poppy
***Paphiopedilum* spp** – orquídea zapatilla, zapatillo de Venus – lady's slipper
Pappophorum bicolor – barbón bicolor – pappus grass
Pappophorum vaginatum – barbón puntiagudo
Parapholis incurva – vallico encurvado, vallico torcido – sickle grass
Parapholis strigosa – vallico estrigoso, vallico filiforme – sea hard-grass
Parentucella latifolia – algarabía hegajosa, eufrasia latifolia – broad-leaved bartsia, broad-leaved eyebright
Parentucella viscosa – algarabía viscosa, eufrasia viscosa – painted cup, yellow bartsia

Parietaria debilis – tumali
Parietaria judaica – cañarroya, parietaria ramiflora – pellitory-of-the-wall
Parietaria officinalis – cañarroya, caracolera, hierba del muro, hierba de San Pedro, parietaria, parietaria común, pelosilla – pellitory-of-the-wall
Parietaria pensylvanica – parietaria, tepanzozmahtli – Pennsylvania pellitory
Paris quadrifolia – hierba de París, uva de raposa, uva de zorro – herb Paris
Parkia speciosa – petai – nitta tree, pete bean
Parkinsonia aculeata – acacia de agüijote, bacaporo, bagoto, cacaporo, cahuinga, cinacina, espinillo, espino de Jerusalén, espino real de España, flor de rayo, guacaporo, guichibelle, junco, junco marino, mezquite extranjero, mezquite verde, palo de rayo, palo verde, parkinsonia, retama, retama china, retama de cerda – horse bean, Jerusalem thorn, Mexican palo verde
Parkinsonia = Cercidium
Parmelia molluscula – liquen de tierra
Parmelia tiliacea – liquen – rock lichen
Parmentiera aculeata – crucetillo, cuajilote, espiga, kaat, pepino, pepino de ardilla
Parmentiera cerifera – árbol de la cera – candle tree
Parmentiera edulis – chachi, chote, chotecuahuite, comate, cuachilote, cuajilote, cuauxílotl, guajilote, gueto xica, huajilote, jícaro, kaat, pepino, pepino de árbol, pepino de ardilla, quauhxilotl, turi – cow okra, food candle tree
Parnassia californica – parnasia de los pantanos – grass-of-Parnassus
Parnassia palustris – hepática blanca, parnasia de los pantanos – grass-of-Parnassus
Paronychia echinulata – nevadilla, paroniquia espinosa – prickly nailwort, whitlow-wort
Parosela = Dalea, Psorothamnus
Parrotia persica – árbol de hierro, parrotia de Persia – Persian parrotia
Parsonsia = Cuphea
Parthenice mollis – cotalulu, nendo de burro
Parthenium argentatum – afinador, copallín, guayule, hierba blanca, hierba ceniza, hierba del hule, hule, jihuite, tataniní, xihuite – Mexican rubber plant
Parthenium hysterophorus – altamisa, amargo, amargoso, arrocillo, chaile, cicutilla, confitillo, coriente, hauay, hierba amarga, hierba amargosa, hierba del burro, hierba del gusano, huachochole, jihuite amargo, zacate amargo – false ragweed, Santa Maria feverfew
Parthenium incanum – copalillo, guayule, guayule hembra, hembra de guayule, hierba blanca, hierba ceniza, mariola, tananiní – mariola
Parthenium stramonium – guasaraco, ocotillo
Parthenium tomentosum – guasaraco, huasaraco, ocotillo, otatillo
Parthenocissus quinquefolia – enredadera de Virginia, guaco, guau, labrusca quinquefolia, viña loca, viña virgen – five-finger ivy, Virginia creeper
Parthenocissus tricuspidata – hiedra de Boston, labrusca tricuspidada, viña loca, viña trepadora, viña virgen – Boston ivy
Pascopyrum (Agropyron) smithii – grama del colorado, planta de trigo – western wheatgrass
Paspalum compressum – pasto chato, zacate amargo – carpet grass, savannah grass
Paspalum conjugatum – amargo, camalote de antena, cambutera, cintillo, grama de antena, hierba agria, hierba criolla, hierba de la vaca sana, hierba verde, horqueta, jengibrillo, maleza hortequilla, paja de canela, papayote, papuano, pashma sacat, pasto amargo, pasto hilo, pasto horqueta, pasto horquilla, pata de gallo, torurco, turvará, zacate grama – sourgrass, West Indian sourgrass
Paspalum dilatatum – ajengibrillo dilatado, canara, dallis, grama alema, grama de agua, grama sanduva, gramilla blanca, hierba dallis, hierba de Australia, pasto dallis, pasto de Australia, pasto dulce, pasto meloso, sanduva, zacate dallis, zacate dulce, zacate meloso, zacate miel, zacate polaco – dallis grass
Paspalum distichum (paspalodes) – ajengibrillo dístico, camalote saladillo, grama, grama colorada, grama de agua, grama dulce, gramilla blanca, hierba de zanjas, pasto dulce, zacate grama – knot grass
Paspalum elephantipes – camalote

Paspalum fluitans – camalote, nudillo acuático, trenza acuática – water paspalum
Paspalum hartwegianum – baiburín, camalote lampiño, mal de ojo
Paspalum langei – camalote moreno
Paspalum lividum – camalote pálido – long-tom
Paspalum notatum – bahía, cañamazo, grama común, grama dulce, grama nativa, grama trenza, gramilla, gramilla blanca, hierba de bahía, hierba menuda, jengibrillo, pasto bahía, pasto horqueta, tejana, trencilla, zacate bahía, zacate horqueta – bahia grass
Paspalum plicatulum – caguazo, camalote, camalotillo, conejito, cortadera, gamalotillo, grama negra, hierba del caballo, paspalum de semilla parda, plicatulum, yáguna – brownseed paspalum
Paspalum pubiflorum – camalote velludo – hairy-seed paspalum
Paspalum racemosum – grama, nudillo – Peruvian paspalum
Paspalum scrobiculatum – mijo koda – kodo millet
Paspalum urvillei – camalote aglomerado, pasto vasey
Paspalum virgatum – camalote blanco, cortadera, grama de chosica, marciega, remolina, zacate cortador, zacate de burro – seashore paspalum, water couch
Passiflora arida – bolsita de la víbora, rosol de la pasión – passion flower
Passiflora ciliata – flor de la pasión, pasiflora, pasionaria
Passiflora caerulea – flor de la pasión, pasionaria – blue crown passion flower
Passiflora edulis – clavo del señor, flor de la pasión, granada, granada china, granadilla, granadilla de China, granadilla morada, granadita de China, murucuyá, parcha, parchita amarilla, pasiflora, pasiflorina, pasionaria, rosal de la pasión – passion flower, passion fruit, purple granadilla
Passiflora foetida – amapola, clavellina blanca, corona de Cristo, granadilla, jujito colorado, jujito peludo, ojo de venado, talayote – fetid passionflower
Passiflora incarnata – cohuanenepili, corona de Cristo, cuanenepili, flor de la pasión, granada china, granadilla, hierba de la paloma, murucuyá, parcha, pasiflora, pasionaria, pasión de Cristo – May-pop, passion flower, wild passion vine
Passiflora jorullensis – cuanenepili, granada, granada de gorro
Passiflora ligularis – granada de China, granadilla, granadilla dulce, granadita, granadita china, granadita de China, pelul – sweet granadilla
Passiflora manicata – flor de las cinco llagas
Passiflora mexicana – clavellina roja, itamo, itamo real, ojo de venado, pasiflora – Mexican passion flower
Passiflora mollissima – curuba, curuba de Castilla, tacso, tasco – tacso, tasco
Passiflora palmeri – bolsita de víbora, granadilla, granadito, sandía de la pasión, sandillita – passion flower
Passiflora quadrangularis – badea, fruta de la pasión, granadilla, granadilla real, murucuyá cascudo, parcha, tumbo – giant granadilla
Pastinaca sativa – cañota, china dulce, chirivía, pastinaca – parsnip
Paullinia = Cupania
Paulownia tomentosa (imperialis) – kiri, paulonia, paulonia imperial – empress tree
Paurotis wrightii – tasiste
Pausinystalia = Corynanthe
Pectis angustifolia – hierba de limón, limoncillo, mazcabmiz – lemonweed
Pectis coulteri – manzanilla de coyote – fetid marigold
Pectis cylindrica – hierba de chinche – cinchweed
Pectis filipes – limoncillo – fetid-marigold
Pectis imberbis – hierba del venado – fetid marigold
Pectis multiseta – limoncillo – cinchweed
Pectis palmeri – cominillo, comino, manzanilla del coyote – cinchweed
Pectis papposa – limoncillo, manzanilla del coyote – chinchweed
Pectis prostrata – limoncillo – fetid-marigold
Pectis stenophylla – cominillo, limoncillo, té limón
Pectis uniaristata – virtugosa
Pedicularis canadensis – gallarito, gorbiza – lousewort, wood betony

Pedilanthus macrocarpus – candelilla, gallito, zapato del diablo – lady's slipper, slipper plant, wax plant
Pedilanthus tithymaloides – candelilla, chapulxóchitl, comalpatli, ítamo real, pie de niño, pinipiniche, pono-pinito, saca espina, tenapalitl, vela de sebo, zapatito, zapatito de la Virgen – devil's backbone, Jacob's ladder
Pediocactus* = *Sclerocactus
Peganum harmala – alharma, harmaga, soma – harmal, Syrian rue
Peganum mexicanum – garbancillo, limoncillo, limoncillo del campo, romerillo
Pelargonium* x *domesticum – geranio, pelargonio – Lady Washington pellargonium, Martha Washington pellargonium, regal geranium, regal pelargonium
Pelargonium* x *fragrans – geranio malva – nutmeg geranium
Pelargonium graveolens – geranio de olor – rose geranium
Pelargonium* x *hortorum – geranio, malvón, novio – common geranium, garden geranium
Pelargonium odoratissimum – geranio, geranio aromático, geranio de olor, geranio rosa, malva bouquet, malva de geranio, malva de olor, malva Luisa, malva rosa, malva rubi – scented geranium
Pelargonium peltatum – geranio, geranio de enredadera, geranio de hiedra, geranio enredador, geranio trepador, gitanilla – ivy geranium
Pelecyphora aselliformis – peyote, peyote meco, peyotillo – hatchet cactus, little peyote, woodlouse cactus
Pelecyphora strobiliformis – peyote – pinecone cactus
Pellaea cordifolia – espárrago cimarrón, itamo real
Pellaea rotundifolia – helecho botón – button fern, roundleaf fern
Pellaea ternifolia – helecho, helecho peyote – cliff-brake
Pellaea truncata – calaguala – cliff brake
Peltophorum pterocarpum – San Francisco – copperpod tree, yellow flame tree, yellow poinciana
Penicillium notatum – penicilina – penicillin
Peniocereus cuixmalensis – pitahaya de aguas
Peniocereus (Cereus) greggii – huevo de venado, pitayita, reina de la noche, saramatraca – Arizona queen-of-the-night
Peniocereus greggii* var *transmontanus – huevo de venado, reina de la noche, saramatraca – desert night-blooming cereus, queen-of-the-night
Peniocereus johnstonii – pitayita, saramatraca
Peniocereus serpentinus (Nyctocereus castellanosii) – gigante, junco, junco blanco, junco espinoso, pitahaya, pitahayita, pitahayita colorada, pitaya, reina de la noche – Mexican night-blooming cereus, snake cactus
Peniocereus striatus (Neoevansia striata, Wilcoxia striata) – cardoncillo, pitahayita, pitayita, saramatraca, sina de la mar – dahlia-rooted cereus, gearstem cactus
Peniocereus viperinus (Wilcoxia papillosa) – cardoncillo, organito de víbora, siviri
Pennisetum americanum – mijo negro, mijo perla, panizo de Daimiel – bulrush millet, pearl millet
Pennisetum ciliare – buffel, zacate buffel – buffelgrass
Pennisetum clandestinum – hierba kikuyu, kikuyu, mijo perla, pasto africano, pasto de ciudad, pasto kikuyu, tapete, zacate africano, zacate alfombra, zacate kikuyu – kikuyu grass
Pennisetum glaucum – mijo, mijo candela, mijo negro, mijo perla, panizo negro – pearl millet
Pennisetum purpureum – elefante, falsa caña de azúcar, gigante, hierba elefante, hierba merker, marianga, merker, merkerón, napier, pasto candelaria, pasto elefante, pasto gigante, pasto merkerón, pasto napier, zacate elefante, zacate gigante – elephant grass, napier grass
Pennisetum setaceum – panizo setáceo, pasto fuerte – fountain grass
Pennisetum setosum – plumita, zacate gusano café – fountain grass
Pennisetum villosum – panizo velloso, zacate plumoso – Abyssinian feathertop
Penstemon barbatus – cebadilla falsa, chilpantlácotl, hierba de San Pedro, jarrito,

perrito, tarrito, vara de San José, varita de San José – southwestern penstemon
Penstemon campanulatus – cebadilla falsa, chilpantlácotl, jarrito, lengua de arista, tarrito – beardtongue
Penstemon centranthifolius – romerillo – scarlet bugler
Penstemon gentianoides – jarrito, tarrito
Penstemon* x *gloxinioides (hartwegii) – dedalera, lengua ursina – border penstemon, garden penstemon
Penstemon imberbis – cebadilla falsa, chilpantlácotl, jarrito, perrito silvestre
Penstemon labrosus – bastón de verano
Penstemon lanceolatus – pénstemon
Penstemon parryi – pajarito, San José del campo – Parry's penstemon
Penstemon spectabilis – pichel – royal beardtongue, tinted penstemon
Penstemon = Keckiella
Pentaclethra macrophylla – árbol de las suelas – owala oil tree
Peperomia campylotropa – ombligo de tierra, pimentero de tierra
Peperomia cordovana – oreja de león
Peperomia quadrifolia – retoño
Perdicium californianum – raíz barbuda
Pereskia aculeata – buganvilla blanca, camelia blanca, grosellero, grosellero americano, grosellero de Florida, guamacho, jazmín de uvas, palo de puerco espín, pitahayita de agua, ramo de novia, tsumya – Barbados gooseberry, Florida gooseberry, lemon vine, Spanish gooseberry, West Indian gooseberry
Pereskia grandifolia (tampicana) – grosellero, guamacho morado, ñajú de espinas, ora pro nobis, palo de puerco espín, patillón, pitahayita de agua, sabonete – lemon vine rose cactus, wax rose
Pereskia lychnidiflora (conzatii, Pereskiopsis pititache) – árbol del matrimonio, cruz del matrimonio, guichi-tache, manzanote, mateado, matial, palo de puerco espín, pata de león, patillón, pititache, tasajillo
Pereskiopsis aquosa – alfilerillo, chirrioncillo, pitaya de agua, tasajillo, tuna de agua
Pereskiopsis blakeana – joconoxtle, xoconoxtle
Pereskiopsis diguetii (velutina) – alfilerillo, cola del diablo, nopaleta, nopalito, patillón, tasajillo
Pereskiopsis porteri – alcahuésar, alcájer, joconoxtle, saramatraca, rosa amarilla, xoconoxtle
Pereskiopsis rotundifolia (chapistle) – chapistle, chapiztle, patillón, pititache, tzompahuiztle
Pereskiopsis = Pereskia
Perezia = Acourtia
Pericome caudata – hierba de chivato – pericome, taperleaf
Perilla frutescens – chiso, perilla – Japanese perilla, perilla, shiso
Perityle californica – manzanilla, manzanilla del coyote – laphamia, rock daisy
Perityle emoryi – manzanilla – rock daisy
Perityle microglossa – manzanilla, manzanilla bronca, manzanilla del burro, manzanilla del coyote, manzanilla del monte – rock daisy
Pernettya ciliata – arrayán, capirote, capulincillo, hierba loca, lengua de vaca, lolito, taglli, tecapulín, tesguate
Pernettya (Gaultheria) mucronata – brezo espinoso
Peronospora viticola – mildiu, tizón – downy mildew
Perovskia atriplicifolia – salvia rusa – Russian sage
Persea americana – abacate, aguacachile, aguacate, aguacate xinene, aguacatero, aguacatillo, ahuacaquahuitl, ahuácatl, cupanda, kuitum, kuitumkeip, laurelillo, on, pahua, pahuatl, palto, talzan, tonalahuate, tzits, um, xinene – avocado, Guatemalan avocado
Persea drymifolia – abacate, aguacate, aguacate anís, aguacate oloroso, ahuacaquahuitl, ahuácatl – avocado, Mexican avocado
Persea gratissima – aguacate, persea – avacado
Persea indica – viñátigo
Persea podadenia – amolillo, chihuana cara, laurel, laurel de la sierra, laurelón, sasafrás – avocado
Persea schiedeana – aguacate, aguacate chupón, aguacate xinene, aguacatillo, chinine,

chucpte, coyó, pahua, xinene, yas – avocado, coyo avocado
Persica = Prunus
Persicaria = Polygonum
Petasites japonicus – fárfara japonesa – fuki, Japanese coltsfoot
Peteria scoparia – camote de cerro, camote de monte
Petiveria alliacea – carricillo silvestre, cola de zorrillo, hierba de la gallina, jupachumi, paiché, rama de zorrillo, zorrillo, zorrillo silvestre, zorro – Guinea hen weed
Petrea arborea (volubilis) – adolfina, bejuco de caballo, bejuco de San Diego, capitán, capitán lila, corona de la reina, cuamecate, cuamecate azul, estrella azul, flor de Jesús, flor de Santa María, flor de Santa Rita, hierba del soltero, jazmín, jazmín azul, lengua de vaca, lila, opptzimin, pañuelo de Sarah, papel de lila, penitente, piocha viejo, raspa sombrero, Santa María, Santa Rita, solerillo, soltero, topostillo, tortilla tostado del caballo – blue bird vine, purple wreath, queen's wreath, sandpaper vine
Petrorhagia nanteuilii – clavellina de Nanteuil – childing pink
Petrorhagia prolifera – clavellina prolífera – proliferous pink
Petroselinum crispum – perejil – parsley
Petroselinum crispum* var *neapolitanum – perejil de hojas planas, perejil de hojas de apio, perejil italiano – flat-leaved parsley, Italian parsley
Petunia* x *hybrida – betunia, petunia – petunia
Petunia* x *hybrida* var *pendula – petunia de enredadera – cascade petunia, trailing petunia
Petunia parviflora – betunia, petunia, tabaquillo oloroso – midget petunia
Peucedanum officinale – oreoselino oficinal, servato – sea hog's fennel
Peucedanum oreoselinum – oreoselino, oreoselino montano, perejil de monte – mountain parsley
Peucedanum ostruthium – imperatorial, oreoselino de tallo lueco – great pellitory of Spain, masterwort
Peucephyllum schottii – romero – pygmy-cedar
Peumus boldus (Boldo fragrans) – boldino, boldo, boldo fragrante – boldo
Pfaffia paniculata – ginseng brasileño, suma, zanahoria brasileña – suma
Phacelia ambigua – heliotropo
Phacelia campanularia – phacelia – California desert bluebell
Phacelia congesta – tomasita – blue-curls, tight phacelia
Phacelia crenulata – rama de zorrillo – scalloped phacelia
Phacelia distans – vervenía – phacelia, wild-heliotrope
Phacelia pedicellata – rama de zorrillo – phacelia
Phacelia platycarpa – espuela, tlatomaxíhuitl
Phacelia scariosa – rama de zorrillo
Phacelia tanacetifolia – vervenía – phacelia
Phaedranthus = Distictis
Phalaris aquatica (tuberosa* var *stenoptera) – falaris perenne, rabillo de cordero, zacate de alpiste – Harding grass
Phalaris arundinacea – alpiste arundináceo, caña acintada, cinta, hierba cinta, pasto brasilero, pasto canario, pasto cinto, zacate de listón – gardener's garters, reed canary grass, ribbon grass
Phalaris brachystachys – alpiste de espigas cortas, triquera
Phalaris canariensis – alpiste, alpiste canario, alpistera – canary grass
Phalaris caroliniana – alpiste, alpiste silvestre, alpistillo – Carolina canary grass
Phalaris minor – alpiste, alpiste menor, alpiste silvestre, alpiste valillo, alpistillo, pasto romano, rabillo de cordero – little-seed canary grass
Phalaris paradoxa – alpiste, alpiste paradójico
Phalaris tuberosa – alpiste blanco, alpiste bulboso, alpiste perenne de Australia, falaris, mata dulce, pasto hardin, triguera – perennial canary grass
Phanerophlebia umbonata – helecho chino
Phaseolus acutifolius – ejotillo, escomite, frijol tepari, judía tepari, tepari – Mexican tepary bean, Texas bean
Phaseolus atropurpureus – chorequillo, ojo de zanate, oreja de ratón

Phaseolus calcaratus – frijol amarillo, frijol de arroz – rice bean
Phaseolus caracalla – caracól, caracolillo, flor de caracól, noua
Phaseolus coccineus (multiflorus) – alubia de España, ayacotli, ayécotl, ayocote, bótil, chamborote, chilipuca, címatl, frijol, frijol ayocote, frijol bótil, frijol chamborote, frijol de España, frijol del monte, frijolillo, haba coccinea, judía de España, judía encarnada, judía escarlata, judía pinta, judía roja, patol, poroto de España, rastrero escarlata, yegua – Mexican runner bean, scarlet runner bean
Phaseolus filiformis – frijol, frijolito, frijol silvestre – slender-stemmed bean, Wright's phaseolus
Phaseolus heterophyllus – chichi-mecapatli, jícama del cerro, trompillo
Phaseolus lunatus – 1. comba, frijol blanco, frijol comba, frijol de la peladilla, frijol de Lima, frijol de luna, frijol de medialuna, frijolillo, frijol lima, haba, haba blanca, haba lima, judía de la peladilla, judía de Lima, judía lima, judión, Lima, pallar, poroto de Lima – butter bean, Lima bean; 2. frijol de luna, furuna, sieva – Carolina bean; 3. faba, fava, poroto de manteca – fava bean
Phaseolus metcalfei – calcomeca, cocolmeca
Phaseolus sonorensis – güirote
Phaseolus vulgaris – 1. alubia, ayacotli, ayocote, bul, caraota, etl, frijol, haba, habichuela, judía, poroto, selaim, siquimuni, subaemunim, tatsin, tzama – bean; 2. bajoca, chaucha, ejote, faséolo, habichuela tierna, habichuela verde, judía común, judía verde, rodrigón, vainita – green bean, pole bean, stringbean; 3. ayécotl, ayocote, alubia chica, caraota blanca, haba blanca, habichuela blanca, pequeño frijol blanco común – haricot, navy bean; 4. frijol anasazi – Anasazi bean; 5. frijol blanco, habichuela del gran norte – great northern bean; 6. frijol blanco corto, habichuela blanca, panamito – small white bean; 7. alubia, frijol colorado, frijol colorado largo, frijol ordinario, frijol rojo, judía de campo, judía enana – kidney bean, red kidney bean; 8. frijol negro, espelól – black bean; 9. frijol pinto, habichuela pinta, judía pinta, pinto – pinto bean; 10. frijol peruano, bayo, canario, mayocoba – Peruvian bean; 11. frijol rojo, frijol rojo pequeño, pinquito – small red bean; 12. frijol rojo de seda – red bean; 13. frijol rosado – pink bean; 14. gallinita – jack bean; 15. judía blanca – butter bean; 16. judía blanca redonda – round-white-seeded French haricot
Phaseolus vulgaris* ssp *nunas – nuña, poroto, porotillo – popbeans
Phaseolus = Vigna
Phaulothamnus spinescens – mal de ojo, putia, tutuqui
Philadelphus coronarius – celinda, falso jazmín, jazmín mosqueta, jeringuilla, jeringuilla coronaria, mosqueta – sweet mock orange
Philadelphus mexicanus – acuilote, acuílotl, jazmín, jazmín de Hueyapán, jazmín del monte, jeringuilla, mosqueta – evergreen mock orange, syringa, wild jasmine
Philadelphus microphyllus – jeringuilla – littleleaf mock orange
Philodendron affine – huacalxóchitl – philodendron
Philodendron bipinnatifidum – gorro de león – upright philodendron
Philodendron radiatum – bejuco de casa, cola de faisán, macusey, paisaje, zingiblio – philodendron
Philodendron sagittifolium – conté
Philodendron scandens (oxycardium) – akalkumche, filodendro, filodendro trepador – philodendron
Philodendron = Monstera
Philoxerus vermicularis – hierba de sal, kuk'uk – beach carpet
Phlebodium (Polypodium) aureum – calaguala, licushi
Phleum alpinum – fleo alpino, timote negro – mountain timothy
Phleum pratense – cola de topo, fleo, fleo de los prados, fleo praderoso, pasto timoty, timote, zacate timoty – common timothy, cultivated timothy
Phlomis fruticosa – candilera fruticosa, matagallos, matulera, oreja de liebre, salvia amarilla, salvia de Jerusalén – Jerusalem sage
Phlomis italica – candilera itálica – Italian woolly sage

Phlomis purpurea – colorado, matagallos, matulera, mechera – purple woolly sage
Phlomis samia – candilera de Samos – Samos woolly sage
Phlomis tuberosa – candilera tuberosa – tuberous woolly sage
Phlox douglasii – flox alpino, flox enano – phlox
Phlox drummondii – alfombrilla, flox – annual phlox
Phlox nana – rosita morada – dwarf phlox, Santa Fe phlox
Phlox paniculata – flox, mil flores, flox paniculado – flame flowers, garden phlox, giant hardy phlox, summer phlox
Phlox subulata – flox, flox enano, flox musgoso – moss phlox, moss pink
Phoebe ehrenbergii – aguacatillo, bebelama
Phoebe mexicana – aguacate cimarrón, aguacatillo, quechol aguacate
Phoebe montana – laurel
Phoebe tampicensis – aguacatillo, laurel, palo de humo, tzenoiti – Mexican laurel
Phoenix canariensis – dátil, datilera, datilera canaria, datilera de las Canarias, palma canaria, palma de Canarias, palmera canaria – Canary Island date palm
Phoenix dactylifera – dátil, datilera, fenix, palma, palma común, palma datilera, palma de dátiles, palmera, palmera datilera, soyal capulín – date palm
Phoenix reclinata – palmera de Senegal – Senegal date palm
Phoenix roebelenii – datilera enana, palmera de Roebelen – pygmy date palm
Pholiota nameko – foliota nameko – nameko mushroom
Pholisma culiacanum – hongo
Pholisma (Ammobroma) sonorae – camote de los médanos, oyutch – sand food
Phoradendron brachystachyum – liga, muérdago, tepalcayo
Phoradendron californicum – chile de espino, guhoja, muérdago, toji – desert mistletoe, mesquite, mesquite mistletoe
Phoradendron commutatum – cabellera, toji
Phoradendron densum (bolleanum* var *densum) – injerto, muérdago, pasto de madroño, toji – dense mistletoe
Phoradendron diguetianum – toji
Phoradendron juniperinum – bellota de sabina, toji – juniper mistletoe
Phoradendron longifolium – liga
Phoradendron macrophyllum (flavescens* var *macrophyllum, tomentosum) – injerto, liga, muérdago americano, pasto de encino, silmo, toji, visco, visco cuercino – bigleaf mistletoe, greenleaf mistletoe
Phoradendron pauciflorum (bolleanum* var *pauciflorum) – injerto, pasto de madroño, toji – fir mistletoe
Phoradendron scaberrimum – toji
Phoradendron schumannii – injerto de encino, pasto de encino
Phoradendron tomentosum – cuashilaca, injerto, liga, seca palo, silmo
Phoradendron velutinum – barba, injerto, liga, muérdago, toji – mistletoe
Phoradendron villosum (flavescens* var *villosum) – muérdago, toji – oak mistletoe
Phormium tenax – cáñamo de Nueva Zelandia, formio, lino de Nueva Zelandia – New Zealand flax
Photinia serratifolia (serrulata) – fotinia – Chinese photinia
Photinia = Heteromeles
Phragmites australis (communis) – acatl, bixilla, caña de indio, cañota, carricillo, carrizo, carrizo común, halal, junco, picuáremu, zacate, zachalal – cane, carrizo grass, common reed, sweet flag
Phrygillanthus sonorae – chupón, injerto, toji – mistletoe
Phyla (Lippia) nodiflora – cedrón – lemon verbena
Phyla strigulosa* var *sericea – hembra, hierbabuena montés, turre – diamond leaf frog-fruit
Phyllanthus acidus – ciruelo costeño, cuatelolote, grosellero, manzano estrella, pimientillo, totolole – otaheite gooseberry
Phyllanthus acuminatus – ciruello, grosellero cimarrón, lavandero
Phyllanthus sellowianus – sarandí blanco
Phyllitis (Asplenium) scolopendrium – escolopendro, helecho lengua de ciervo, lengua cerval, lengua cervina, lengua de ciervo – hart's tongue fern

Phyllitis = Asplenium
***Phyllodoce* spp** – loiseleuria – mountain heather
Phyllostachys aurea – bambú, bambú amarillo
Phyllostylon brasiliensis – baitoa, boj de las antillas, cerón, jatía, membrillo, otélom
Phymosia = Sphaeralcea
Physalis acuminata – guajtomate, tomatillo – tomatillo
Physalis acutifolia (wrightii) – mala mujer, tomatillo, tomatillo cimarrón, tomatillo espinoso – Wright's ground-cherry
Physalis aeguata – miltomate, tomate
Physalis alkekengi (franchetii) – alquequenje, bolsa del pastor, carapulla, farolillo chino, halicácabo, solano vejigoso, tomate inglés, vejiga de perro, vejiguilla – Chinese lantern plant, strawberry ground-cherry, winter cherry
Physalis angulata – alquequenje, farolito, pacanil, pakmul, tlemoli, tomate, tomate de cáscara, tomate de culebra, tomatillo, tomatl – ground-cherry
Physalis coztomatl – coztomate, coztomate amarillo, coztomatl, tomate
Physalis crassifolia – tomate capotillo, tomate de culebra, tomate de víbora, tomatillo del desierto – desert ground-cherry
Physalis hederifolia – rama de amores – heartleaf ground-cherry
Physalis lancifolia – tomate, tomatillo – ground-cherry
Physalis leptophylla – tomate del monte, tulumisi
Physalis longifolia – tomatillo – ground-cherry
Physalis peruviana – alquequenje, capulín, miltomate, tomate, tomate de cáscara, uchuba – Cape gooseberry, strawberry tomato
Physalis philadelphica (ixocarpa) – coztomatl, fresadilla, miltomate, tomate, tomate de capote, tomate de cáscara, tomate de fresadilla, tomate milpero, tomate verde, tomatillo, tomatl, tulumisi – husk tomato, tomatillo
Physalis pubescens (neomexicana) – coztomate, coztomatl, farolito, guajtomate, miltomate, pacanil, tomate, tomate de cáscara, tomate del campo, tomate silvestre, tomate verde, tomatillo, tomatillo del campo, tomatl, vejiga de perro – ground-cherry, husk tomato
Physalis sordida – chaltoma
Physalis stapelioides – guajtomate, jaltomate, tomatillo
Physostegia virginiana – fisostegia, planta de bisagra, planta obediente – dragonhead, false dragonhead, lionheart, obedient plant
Physostigma venenosum – calabar
Phytelephas macrocarpa – marfil vegetal, nuez de marfil, palma de marfil, tagua, yarina – ivory nut, ivory palm
Phytolacca americana – bledo carbonero, fitolaca, grana encarnada, hierba carmín, hierba carmín decandra, mazorquilla – great pokeweed, pigeonberry, pokeberry, pokeroot, pokeweed, scoke
Phytolacca decandra – fitolaca, tintorera
Phytolacca dioica – bella sombra, fitolaca, hierba carmín dioica, malambo, ombú, tepezapote
Phytolacca icosandra – amolquilitl, carricillo, cónguera, mazorquilla, namole, quelite, telcocox
Phytolacca octandra – amole, amolxihuitl, barba china, bledo carbonero, cargamanta, cónguera, hierba china, jabonera, mazorquilla, ñamole, perla de China, quelite de toro, sicamole, tarasca, tarasca de negros, telkox, tonojoso, verbachina, yamole – pokeweed
Phytolacca rivinoides – jaboncillo – pokeberry
Phytophthora infestans – añublo, candelilla, clavo, hongo de la papa, podredumbre, tizón de la papa – potato blight
Picea abies (excelsa) – abeto del norte, abeto de Navidad, abeto de Noruega, abeto falso, abeto rojo, picea, picea común, picea de Noruega, picea europea, pinabete – Norway spruce
Picea alba – picea blanca
Picea breweriana – picea de Brewer – Brewer spruce, Brewer's weeping spruce
Picea chihuahuana – pinabete, pinabete de Chihuahua – Chihuahua spruce
Picea engelmannii – picea, picea de Engelmann, pino real – Engelmann spruce
Picea glauca – abeto del Canadá, picea blanca,

picea blanca del Canadá, picea glauca – white spruce
Picea mariana – picea mariana, picea negra, picea negra americana, picea negra norteamericana – black spruce
Picea mexicana – ciprés
Picea omorika – picea de Servia, picea servia
Picea pungens – abeto azul, abeto plateado, picea azul, picea de Colorado, pino real – blue spruce, Colorado spruce
Picea pungens* var *glauca – abeto azul glauco, picea azul de Colorado, picea de Parry – Colorado blue spruce
Picea rubens – picea roja del Canadá – red spruce
Picea sitchensis – abeto de Menzies, picea de Sitka, picea falcada, Sitka – Sitka spruce
Picea smithiana – picea de Afganistán, picea del Himalaya occidental, picea de Smith, picea morinda
Picramnia pistaciaefolia – cáscara amarga – bitter bush
Picris echioides – cardo perruno, rasparago – bristly ox-tongue
Pieris forrestii (formosa) – andrómeda – Chinese pieris
Pieris (Andromeda) japonica – andrómeda – lily-of-the-valley shrub
Pilea cadierei – gallineta, madreperla, pilea de aluminio, pilea plateada – aluminum plant, clearweed, creeping charley
Pilea involucrata – panamiga – pan-American friendship plant
Pilea microphylla – artillería, chisme, frescura, golondrina, hierba de la viruela, palo verde, pilea, yomha – clear weed, artillery plant
Pileus = Carica
Pilgerodendron uviferum – ciprés de la guaiteca
Pilocarpus jaborandi – arruda brava, cáñamo indígena, jaborandi, pernambuco, pilocarpo – Indian hemp
Pilosocereus (Cephalocereus) alensis – barba de viejo, matacachí, pitahaya barbona – woolly torch
Pilosocereus (Cephalocereus) chrysacanthus – pitaya vieja – golden old man
Pilosocereus (Cephalocereus) leucocephalus – cabeza de viejo, órgano, pitaya barbona, pitayón, tuna, viejo – old man cactus, old man of Mexico
Pilosocereus (Cephalocereus) purpusii – barba de viejo, viejo
Pilosocereus (Cephalocereus) royenii – dildo
Pimenta = Eugenia
Pimpinella anisum – anís, anís chico, anís común, anís verde, matalahúga, simiente dulce – anise, aniseed
Pimpinella major – pimpinela blanca, pimpinela mayor – burnet-saxifrage
Pinanga coronata – pinanga de corona
Pinaropappus roseus – escabiosa, escorzonera, ispile, ixpostli, ixpul, motita, motita morada – white rock-lettuce
Pinguicula caudata – violeta azteca, violeta cimarrona, violeta del campo, violeta de monte – butterwort
Pinguicula grandiflora – grasilla de flor grande
Pinguicula moranensis – violeta cimarrona, violeta de monte
Pinguicula vulgaris – grasilla, tiraña común – bog violet, butterwort
***Pinus* spp** – ococuahuitl, ocopino, ocote, ocotl, ocotle, pino – pine
Pinus arizonica – joco pino, pino arizónica, pino blanco, pino cenizo, pino fino, úkui – Arizona pine, western yellow pine
Pinus attenuata – chichonuda, pino, pino de cono pitonudo, pino de Eldorado, pino del Monte Shasta, pino de piñas chichonudas – knobcone pine
Pinus ayacahuite – acacalocahuite, acahuite, acalo cahuite, acalocote, acanita, acolote, ayacahuite, ayacahuite colorado, ayauhcuáhuitl, canite, ocote, ocote blanco, ocote gretado, pinabete, pino, pino acahuite, pino ayacahuite, pino barbón, pino blanco, pino blanco de México, pino cahuite, pino chimonque, pino de azúcar, pino gris mexicano, pino huiyoco, pino real, pino tabla, pino triste, sacalacahuite, salacahuite, wiyoko – Arizona white pine, Mexican white pine
Pinus balfouriana – pino de Balfour, pino de cola de zorro – foxtail pine
Pinus banksiana – pino banksiano, pino de Banks, pino de Labrador – jack pine

Pinus bungeana – pino chino de Bunges – lacebark pine
Pinus canariensis – pino canario, pino de las Canarias, pino de las Islas Canarias – Canary Island pine
Pinus caribaea – pino, pino blanco, pino colorado, pino macho – slash pine
Pinus caribaea* var *hondurensis – ocote, pino, pino amarillo, pino caribeño, pino costanero, pino costero, pino curtidor – Honduras Caribbean pine
Pinus cembra – cembro, pino del cembrio, pino de los Alpes – arolla pine, Swiss stone pine
Pinus cembroides – bishicuri, ocote, pino, piñón, piñonero, piñón prieto, pino piñón, pino piñonero, pino piñonero mexicano – Mexican pinyon pine
Pinus contorta* ssp *contorta – pino contorto, pino contorto de la costa, pino costero de Murray, pino de la playa del Pacífico – shore pine
Pinus contorta* ssp *murrayana – pino, pino ponderosa – lodgepole pine, tamarack
Pinus cooperi – albacarrote, pino albacarrote, pino amarillo
Pinus coulteri – pino, pino colorado, pino de Coulter, pino de piñas grandes – Coulter pine
Pinus culminicola – pino enano, piñón, piñón de Potosí, pino piñonero – Potosi pinyon
Pinus densiflora – akamatsu, pino rojo japonés – Japanese red pine
Pinus devoniana – mocochtaj, ocote, ocote escobetón, ocote gretado, pino, pino blanco, pino chantaj, pino escobetón, pino lacio, pino prieto, pucuri, tsihuírea – Michoacan pine
Pinus discolor – piñón, pino piñonero – border pinyon
Pinus douglasiana – ocote, ocote lacio, pinabete, pino, pino amarillo, pino avellano, pino blanco, pino canís, pino hallarín, pino real – Douglas pine
Pinus durangensis – ocote, pino, pino barbón, pino blanco, pino de Durango, pino de seis hojas, pino real, pino real de seis hojas – Durango pine
Pinus edulis – ocote, piñón, piñón colorado, piñonero, piñonero de dos hojas, pino piñonero, trementina – colorado pinyon, nut pine, piñon, pinyon pine
Pinus engelmanii (latifolia) – pino, pino prieto, pino real, pino real de barbas largas, pino real de tres hojas – Apache pine, Arizona longleaf pine
Pinus flexilis – ayacahuite, pino, pino enano, pino flexible – limber pine
Pinus greggii – ocote, pino, pino ocote, pino prieto – Gregg pine
Pinus halepensis – carrasco, pino carrasco, pino de Alepo, pino de Jerusalén – Aleppo pine
Pinus hartwegii – ocote, ocote blanco, ocote pardo, pino, pino amarillo, pino chino, pino colorado, pino de cumbres, pino de las alturas, pino de ocote – Donnell Smith pine
Pinus herrerae – ocote, pino chino, pino llanero – Herrera pine, shagbark pine
Pinus jaliscana – ocote, pino – Jalisco pine
Pinus jeffreyi – pino, pino de Jeffrey, pino negro, pino ponderosa, pino real, pino trusa – Jeffrey pine
Pinus lagunae – pino piñonero
Pinus lambertiana – ocote, pino, pino de azúcar, pino de Lambert, pino dulce, pino gigante de California – sugar pine
Pinus lawsonii – ocote, ocote chino, ortiguillo, pino, pino chino, pino ortiguillo – Lawson pine
Pinus leiophylla – ocote, ocote blanco, ocote chino, pino, pino chino, pino gretado, pino otomite, pino prieto, pino real, tlacocote, xalatlaco – five-needled Chihuahua pine
Pinus leiophylla (chihuahuana)* var *chihuahuana – ocote blanco, ocote chino, pino chamonque, pino chimonque, pino chino, pino otomite, pino prieto, pino real, pino saguaco, sahuaca – Chihuahua pine
Pinus longaeva (aristata) – pino de cola de zorro, pino de Colorado, pino de piñas aristadas de las rocosas – western bristlecone pine
Pinus lumholtzii – huiyoco, ocote, ocote dormido, pino, pino albacarrote, pino amarillo, pino barba caída, pino lumholtz, pino triste – Lumholtz pine, weeping pine
Pinus maximartinezii – ocote, pino, pino de

Martínez, piñón, piñón de Zacatecas – Martinez pine
Pinus michoacana – ocote escobetón, ocote gretado, pino blanco, pino lacio, tsihirén
Pinus monophylla – ocote, piñón, piñón de una hoja, piñonero, pino piñonero – singleleaf pinyon pine
Pinus montezumae – chalmaite blanco, ocote, ocote blanco, ocote hembra, ocote macho, ocotl, pino, pino blanco, pino de Moctezuma, pino mexicano, pino mexicano de corteza basta, pino real, yutnu-satnu – Montezuma pine
Pinus monticola – pino blanco americano, pino blanco de Idaho, pino blanco montano del oeste, pino estrobo del oeste, pino montícola – western white pine
Pinus mugo (montana) – barancio, negral, pino de montaña, pino montano, pino mugo, pino negro, pino rodeno – mugho pine, Swiss mountain pine
Pinus mugo* var *rostrata – aznacho – mountain pine
Pinus mugo (uncinata)* var *uncinata – pino moro, pino negro, pino negro de los Pirineos, pino uncinado – Swiss mountain pine
Pinus muricata – pino, pino de cono picudo, piñón, pino obispo, pino remorata – bishop pine
Pinus nelsoni – ocote, pino de Nelson, piñón, piñón colorado, piñonero, piñón prieto, pino prieto – Nelson pinyon
Pinus nigra (austriaca) – pino clusiano, pino laricio, pino laricio de Austria, pino negral, pino negral de Austria, pino negro de Austria, pino pudio, pino salgareño – Austrian black pine
Pinus nigra* ssp *salzmannii – pino negral español
Pinus oocarpa – itchai, ichtaj, ocote, ocote chino, ocote de carretillo, ocote macho, pino amarillo, pino avellano, pino chino, pino colorado, pino de Nicaragua, pino de ocote, pino prieto, pino real, pino resinoso – egg-cone pine, ocote pine, pitch pine
Pinus palustris – madera pino – longleaf pine
Pinus parviflora – pino blanco japonés, pino de Himekomatsu, pino japonés de flores pequeñas – Japanese white pine
Pinus patula – jelecote, ocote, ocote colorado, ocote macho, ocotl, pino colorado, pino lacio, pino llorón, pino mexicano, pino ocote de México, pino pátula, pino triste – jelecote pine
Pinus pinaster – pino gallego, pino marítimo, pino negral, pino resinero, pino rodeno, pino rubial – cluster pine, French turpentine pine, maritime pine
Pinus pinceana – ocote, pinete, pino, pino mexicano, piñón, pino piñón – Mexican pine
Pinus pinea – doncel, picea, pino albar, pino de Barcelona, pino doncel, pino manso, pino piñonero – Italian stone pine
Pinus ponderosa – pinabete, pino, pino blanco, pino ponderosa, pino real, pino real americano – Pacific ponderosa pine, ponderosa pine, western yellow pine, yellow pine
Pinus ponderosa* var *scopulorum – pino blanco, pino real – Rocky Mountain ponderosa pine, western yellow pine
Pinus praetermissa – pino, pino chino, pino prieto – smallleaf pine
Pinus pringlei – ocote, pino, pino rojo – Pringle pine
Pinus pseudostrobus – chalmaite, ocote, ocote blanco, pacingo, pino, pino blanco, pino chalmaite, pino chino, pino de cono chico, pino liso, pino ortiguillo, pino real – false white pine
Pinus quadrifolia – pino de cuatro hojas, piñón, piñón de la sierra de Juárez, piñonero de cuatro hojas, pino piñonero, piñón – Parry pinyon pine
Pinus radiata – ocote, pino, pino de Monterrey, pino de Santander, pino insigne, pino radiata – Monterey pine
Pinus reflexa – pino huiyoco, pino nayar
Pinus remota – piñón – remote pinyon, Texas pine
Pinus resinosa – pino resinoso, pino rojo americano, pino rojo del Canadá
Pinus rigida – pino bronco, pino rígido – northern pitch pine, southern yellow pine
Pinus roxburghii – pino asiático – chir pine, Indian longleaf pine
Pinus rzedowskii – ocote, pino, pino rzedowski – Rzedowski pine

Pinus sabiniana – piñón – digger pine, foothill pine, gray pine
Pinus strobiformis – acahuite, acamita, acanita, huiyoco, ocote, pinabete, pino acahuite, pino blanco, pino de azúcar, pino enano, pino hallarín, pino huiyoco, piñón, pino nayar, pino real, pino triste – Mexican white pine, southwestern white pine
Pinus strobus – acalocote, palo pique, pinabete, pinabete blanco, pino canadiense, pino de Lord Weymouth, pino de Weymouth, pino estrobo, pino estrobo del este, tonotzin – eastern white pine, northern white pine, white pine
Pinus strobus* var *chiapensis – acalocote, ayacahuite vidrioso, ocote, ocote blanco, palo pique, pinabete, pinabete blanco, pino ayahuite, pino blanco – Chiapas white pine
Pinus sylvestris – pino albar, pino del norte, pino de Noruega, pino de Valsain, pino rojo, pino serrano, pino silvestre, pino valsain – Scotch pine, Scots pine
Pinus taeda – pino americano, pino de incienso – loblolly pine
Pinus tecunumanii – ocote, ocote de carretilla, pino
Pinus tenuifolia (maximinoi) – cantaj, ocote, pino canís, pino corriente, tzit – Maximino pine
Pinus teocote – albacarrote amarillo, amarillo, jalocote, juncia, ocote, ocotl, ocotzotl, pino chino, pino colorado, pino escobetón, pino real, pino rosillo, teocote, teocotl, tosarza, tzatadi, uxitl, xalócotl – Aztec pine
Pinus thunbergii – kuromatsu, pino japonés de Thunberg, pino negro japonés – Japanese black pine
Pinus torreyana – pino torrey – Torrey pine
Pinus tropicalis – pino blanco, pino hembra – tropical pine
Pinus wallichiana (excelsa, griffithii) – pino alto, pino azul del Himalaya, pino de Bhután, pino del Himalaya, pino excelso, pino llorón del Himalaya – Himalayan white pine
Pinus yecorensis – pino colorado, pino rojo – yecora pine
Piper aduncum – cordoncillo, cordoncillo blanco, gusanillo – wild pepper
Piper amalago – cordoncillo, cordoncillo chico, yaxal – buttonwood, Spanish elder
Piper angustifolium – achiotlín, artanta, cordoncillo, mático, pimentero de hojas estrechas, platanillo de Cuba, rabo de zorro, saladilla, santilla montés, soldadillo
Piper auritum – acuyo, cordoncillo, corrimiento, coyoquelite, hierba santa, hoja de acuyo, hoja santa, jaco, maculán, mecaxóchitl, momo, obel, omoquelite, omoquilitl, Santa María, tlampa, tlanepaquilitl, tlanipa – bullhoof, cowfoot, pepper
Piper betle – betel, nuez de betel – betel, betel nut
Piper cubeda – cubeba, pimentero cubeba – cubeb berry
Piper methysticum – kava kava, pimentero ava – kava-kava
Piper nigrum – cordoncillo, pimentero, pimentero blanco, pimentero blanco de la India, pimentero negro – black pepper, white pepper
Piper palmeri – cordoncillo, hachogue, mático
Piper peltatum – chút-ik, San Diego, Santa María, Santiago – cowfoot
Piper sanctum – acuyo, akokojxiuit, cordoncillo, hierba santa, hoja de aján, hoja de anís, hoja santa, jaco, jeco, momo, palo de zanate, rama de queso, Santa María, santilla de comer, tapa cántaro, tlanepaquelite, tlanepaquilitl, tlanipa – pepper
Piptatherum (Oryzopsis) miliaceum – mijera, mijo negrillo – smilo grass
Piptochaetium fimbriatum – arrocillo, falso espartillo del pinar – pinyon ricegrass
Piqueria trinervia – altarreina, chichiquis, cuapopolchi, cuimic, empueshte, harta reina, hierba del perro, hierba del tabardillo, hierba del zopilote, hierba del zorrillo, hierba de San Nicolás, tabardillo, xonitlzal, xoxonitzal, yoloxiltic – stevia
Piscidia mollis – maxcuáhuitl, palo blanco, palo blanco duro, tzijol – fish-poison tree
Pisonia aculeata – coma de uña, cruceta espinuda, cruz espina, cumbro, espino y camote, garabato, garabato blanco, garabato prieto, guichigu, gumbio, huele de noche, huiscolote, huitzcocólotl, uña de gato, uña del

diablo, vainoro, vainoro prieto, zapallo prieto – cockspur
Pisonia capitata – baijuo, bainoro, bainoro prieto, caloca, garabato, garabato prieto, garambullo, susutí, uña de gato prieto, vainoro prieto – garabato
Pisonia flavescens – garabato, San Agustín
Pisoniella arborescens – jazmincillo
Pistacia atlantica – betoum – Mount Atlas pistache
Pistacia lentiscus – almácigo, árbol de la almáciga, charneca, lentisco, mata charneca – lentisc, mastic tree
Pistacia mexicana – achín, almácigo, lentisco, lentisco de Nuevo León, pistache cimarrón, ramón, yagaguieguei – pistachio
Pistacia texana – lentisco – American pistachio, Mexican pistachio, pistache
Pistacia terebinthus – albotín, cornicabra, terebinto – terebinth, turpentine tree
Pistacia vera – alfóncigo, lentisco, pistachero – pistachio
Pistia stratiotes – lechuga, lechuga de agua, lechuguilla, lechuguilla de agua, repollito de agua, verdolaga de agua – water-lettuce
Pisum sativum – 1. alberja, alberjón, alverja, alverjón, arveja, arvejita, chaucha, chícharo, chícharo inglés, guisante, guisante inglés, guisante sativo, pésol – garden pea, sugar pea; 2. guisante de grano liso, guisante liso – green pea; 3. guisante de grano rugoso, guisante de semilla rugosa – wrinkled pea; 4. frijol carita – black-eyed pea
Pisum sativum (arvense)* var *arvense – alverjón forrajero, arvejo, guisante de campo, guisante de huerta, guisante forragero – field pea
Pisum sativum* var *macrocarpon – arveja china, arveja cometodo, chícharo verde, guisante azucarado, guisante chino, guisante cometodo, guisante tirabeque, tirabeque – Chinese pea, edible-pod pea, snow pea
Pithecellobium acatlense – acinchete, barba de chivo, huisache de chintete, rabo de iguana
Pithecellobium brevifolium – carbonero, guajillo, huajillo, palo de tenaza, palo huácaro, tenacilla, tenaza, tijerilla
Pithecellobium confine – ejotón, guapinole, palo fierro
Pithecellobium dulce – bacochini, chucum blanco, civico, coacamachalli, cuamóchil, cuaumóchitl, guamoche, guamúchi, guamúchil, guamúchili, guamúchitl, huamúchil, huamúchil costeño, humo, humo cimarrón, humo costeño, macochín, muchite, pichejumo, pinzán, quamóchitl, quauhmóchitl, yaga-biguichi – Manila tamarind, monkey pod, white 'opiuma
Pithecellobium ebano – ébano
Pithecellobium echinatum – bejuco de huico, corneta, clarín, güico, hiedra bocina, lengua de vaca, mariposa, palomita, petaquilla
Pithecellobium flexicaule – ébano, guaypinole – ape's earring, Texas ebony
Pithecellobium leptophyllum – charrasquillo, huisache
Pithecellobium undulatum – ébano, palo ébano, palo fierro
Pithecellobium unguis-cati – coralito, guamuchilillo, pinzán, uña de gato – black bean
Pithecellobium = Chloroleucon, Havardia
Pittosporum crassifolium – pitosporo
Pittosporum phillyraeoides – pitosporo llorón – willow pittosporum
Pittosporum tenuifolium – pitosporo
Pittosporum tobira – azahar de la China, clavo, lila, pitosporo, pitosporo del Japón, pitosporo de la China – Japanese pittosporum, mock orange, pittosporum, tobira
Pittosporum undulatum – pitosporo de bayas anaranjadas – mock orange, Victorian box
Plagiobothrys nothofulvus – nievitas – popcornflower, redstain
Plantago coronopus – estrellania, hierba del costado, llantén cuerno de ciervo, pie de cuervo, zaragatona – buck's-horn plantain
Plantago fastigiata – trigo de los indios – Indian wheat
Plantago hirtella (australis) – llantén, plantago
Plantago indica – llantén arenario, zaragatona – sand plantain
Plantago lanceolata – llantén, llantén lanceolado, llantén menor, pastora, plantago inglés – buckhorn, English plantain, ribbon grass, ribgrass, ribwort

Plantago major – carmel, chía gorda, lengua de oveja, llantén, llantén grande, llantén mayor, plantago, vitsacua sipiati – broadleaf plantain, common plantain, English plantain
Plantago maritima – llantén marítima, zaragatona falsa – sea plantain
Plantago mexicana (linearis* var *mexicana) – acaxílotl, llantén, pastora, plantago
Plantago ovata – ispágula, llantén ovado, pastora, psilio, zaragatona – psyllium
Plantago patagonica – pastora – Patagonia plantain, woolly Indian wheat
Plantago psyllium – pulguera, zaragatona – psyllium
Plantago subnuda (hirtella* var *galeottiana) – llantén, plantago – plantain
Platanthera hyperborea – plantanera del norte – green-flowered bog-orchid
Platanus* x *acerifolia (hispanica) – plátano, plátano común, plátano de Londres, plátano de sombra, plátano híbrido común, plátano híbrido español – London planetree
Platanus lindeniana – álamo, canelo, chicolcohuite, haya – Jalapa planetree
Platanus mexicana – acuáhuitl, álamo, álamo blanco, chicolcohuite, guayabillo, haya – Mexican planetree
Platanus occidentalis – álamo, plátano de occidente, plátano de sombra, plátano de Virginia, plátano occidental, sicomoro – American sycamore, buttonwood
Platanus orientalis – plátano de oriente, plátano oriental, sicomoro – oriental plane
Platanus racemosa – álamo, aliso, ciclamor, sicomoro – California sycamore, planetree, western sycamore
Platanus wrightii (racemosa* var *wrightii) – álamo, álamo blanco, aliso, ciclamor – Arizona sycamore
Platonia insignis – babari, bacuri – bacury
Platycerium bifurcatum – cuerno de alce, cuerno de ciervo, platicerio – staghorn fern
Platycladus = Thuja
Platycodon grandiflorus – campanilla china, campánula china – balloon flower
Platymiscium dimorphandrum – hormigo, hormiguillo, hormiguillo colorado, marimba, palo de hormigo, palo de marimba, palo marimba – marimbawood
Platymiscium trifoliolatum – palo santo, tampicerán, tepezapote
Pleione bulbocodioides (formosana) – orquídea rocosa
Pleuraphis (Hilaria) belangeri (H. cenchroides) – grama, rizado, toboso menudo, zacate chino, zacate galleta – curly mesquite
Pleuraphis (Hilaria) cenchroides – espiga negra, grama, guishi, toboso, toboso menudo, zacate grama
Pleuraphis (Hilaria) jamesii – galleta – galleta
Pleuraphis (Hilaria) mutica – toboso, toboso común, zacate toboso – tobosa
Pleuraphis (Hilaria) rigida – galleta, toboso, zacate galleta gigante – big galleta
Pleuraphis (Hilaria) swalleni – toboso esbelto
Pleurotus cornucopiae – hongo de maguey – branched oyster mushroom
Pleurotus eryngii – boleto de espinacal, chirgola, presquilla, seta de cardo
Pleurotus eryngii* var *ferulae – seta de cañaheja
Pleurotus ostreatus* var *sativus – girbola de pollancre, orellana, seta de ostra – cultivated oyster mushroom
Pleurotus ostreatus* var *vulgaris – cazahuate, hongo de cazahuate, hongo de encino, hongo de maguey, oreja blanca, oreja de cazahuate, orellana, seta de alpaca, seta de cardo, seta de mimbre, seta de ostra – wild oyster mushroom
Pluchea odorata (purpurascens) – ahuapatli, alimanche, alinanché, bobche, canelo, canelón, chalché, cihuapatli, cipatle, clina, comalpatli, conolón, flor de ángel, flor de Guadalupe, hierba de Santa María, hoja de playa, jara, Santa María, tepozá – rosy cups, salt marsh fleabane
Pluchea salicifolia – batamote
Pluchea (Berthelotia, Tessaria) sericea – cachanilla, cochinilla – arrow weed, arrow-wood, marsh fleabane
Pluchea symphytifolia – canela, canelón
Plumbago auriculata (capensis) – azulino, bago, belesa, cañutillo, celestina, chapak, dentelaria, embeleso, hierba del alacrán, hierba del diablo, hierba del negro, jazmín azul, lagán de perro, nomeolvides, pililo,

plúmbago, tlalchichinolli, tlepatli, turicua, umbela – Cape plumbago, climbing plumbago, leadwort

Plumbago pulchella – chilillo, cola de iguana, cola de pescado, curicua, dominguilla, hierba del alacrán, hierba del diablo, hierba del negro, hierba del pescado, hierba lumbre, juricua, pañete, planta del negro, tianguis, tlachichinol, tlachichinole, tlepatli, turicua – leadwort

Plumbago scandens – cañutillo, chapak, dentelaria, embeleso, estrenina, hierba del alacrán, hierba del diablo, hierba del monte, hierba del negro, hierba del pájaro, hierba del pescado, jazmín azul, juricua, lagaña de perro, pañete, pegajosa, pitilla, plúmbago, plúmbago blanco, tlalchichinolli, tlepatli, turicua, umbela – blue jasmine, clammy plumbago, leadwort

Plumeria acutifolia (acuminata) – acalztatsim, ahaipuih, alejandría, alhelí, cacalosúchil, cacalosúchil blanco, cacaloxóchitl, campechana, campotonera, flor blanca, flor de cuervo, flor de mayo, flor de mayo blanca, franchipán, guiechachi, jacalosúchil, jacalosúchil blanco, lirio de la costa, quiechachi, rosal, súchil, suchilcahue, tizaxóchitl, tlapalito – frangipani, Mexican frangipani, temple flower

Plumeria mollis – cacalosúchil

Plumeria rubra – alejandría, amapola, atapaima, cacalosúchil, cacalosúchil rojo, cacaloxóchitl, campechana, chacnicté, chaczabacnicte, campotonera, cuauxúchil, cundá, ensalada, flor de cementerio, flor de corpus, flor de la cruz, flor de cuervo, flor de María, flor de mayo, flor de mayo roja, flor de templo, franchipán, frangipán, frangipani, guiechachi, jacalosúchil, jacalosúchil rojo, juchuch, lengua de toro, nicté, sabacuite, sabanicté, saugran, súchil, suchilete, turoco, zabacnicte – frangipani, nosegay frangipani, plumeria, temple tree

Poa annua – espiguilla, espiguilla anual, hierba de punta, pastillo de invierno, pastito de invierno, pasto azul anual, poa anual, zacate alpiste, zacate azul, zacate azul anual, zacate de ratón – annual bluegrass

Poa bigelovii – pasto azul precoz – bigelow bluegrass

Poa bulbosa – grama cebollera, poa bulbosa – bulbous bluegrass

Poa chaixii – poa de los bosques – Chaix's bluegrass

Poa compressa – poa chata, poa comprimida, zacate azul del Canadá – Canada bluegrass

Poa glauca – poa de color verdemar – timberline bluegrass

Poa hybrida – poa híbrida – hybrid meadow grass

Poa inferma – poa precoz, poa temprana – early meadow grass, weak bluegrass

Poa laxa – poa de flores dispuestas flojamente, poa laxiflora – tufted meadow grass

Poa nemoralis – poa de los bosques, poa espiguilla – wood bluegrass

Poa palustris – poa de los pantanos – fowl bluegrass

Poa pratensis – cañuela, espiguilla, grama de prados, pasto azul de Kentucky, pasto kentucky, pasto poa, poa común, poa de los prados, poa pratense, zacate azul de Kentucky, zacate azul de las praderas, zacate poa – Kentucky bluegrass

Poa pratensis* ssp *angustifolia – poa de hojas estrechas – narrow-leaf meadow grass

Poa trivialis – espiguilla, poa aspera, poa común – rough bluegrass, rough-stalked bluegrass

Poa violacea – poa violácea – violet meadow grass

Poaceae (Gramineae) – 1. grama, hierba, mijo, paja, pasto, xiuitl, zacate – grass; 2. cereales – grain

Podachaenium eminens – acasucut, calzadilla chica, hoja de manteca, tacote

Podocarpus andinus – leuque, lleuque, podocarpo andino – plum-fruited yew

Podocarpus guatemalensis – chusnito, ciprés de montaña, cipresillo, pinillo – mountain cypress

Podocarpus nubigenus – maniú de Patagonia, maniú hembra – Chilean totara

Podocarpus reichei – palmilla

Podocarpus totara – totara – totara

Podophyllum peltatum – mandrágora – mandrake, May apple

Poeticus narcissus – narcisco – poet's narcissus
Pogostemon cablin – pachulí – patchouli
Poinciana = Caesalpinia, Delonix
Poinsettia = Euphorbia
Polanisia uniglandulosa (dodecandra* var *uniglandulosa) – hierba del coyote, hierba del zorrillo, ortiga – clammy weed, large-flowered clammy weed
Polaskia (Lemairocereus) chende – chende, chinoa
Polaskia (Lemairocereus) chichipe – chichibe, chichipe, chichitun, chituna, dichitune
Polemonium caeruleum – escala de Jacob, polemonio, polemonio azul, valeriana azul, valeriana griega – Jacob's ladder
Polianthes (Manfreda) maculosa – amole – American aloe
Polianthes tuberosa – amiga de noche, amole, amolli, azucena, flor de hueso, guía-chilla, guiegana, jacinto occidental, margarita blanca, margarita olorosa, nardo, nardo común, nardo de jardín, nardo oloroso, omisúchil, omixóchitl, puchúngari, tuberosa, vara de Jesús, vara de nardo, vara de San José – tuberose
Poliomintha maderensis (longiflora) – orégano – Mexican oregano
Polycarpon tetraphyllum – policarpon de hojas en verticilos de cuatro – four-leaved allseed
Polygala apopetala – cordoncillo
Polygala chamaebuxus – polígala, polígala de hojas de boj – shrubby milkwort
Polygala myrtifolia – polígala de jardín
Polygala scoparia – flor de María, polígala, polígala de México, polígala medicinal, xochipitsahuac – milkwort
Polygala seneca – polígala de Virginia, serpentaria de Séneca – milkwort, Seneca snakeroot, senega
Polygonatum biflorum – sello de Salomón – Solomon's seal
Polygonatum multiflorum – sello de Salomón europeo
Polygonatum odoratum (officinale) – lágrimas de David, planta de los panadizos, poligonato, sello de Salomón, sello de Salomón oloroso, suelda blanca – Soloman's seal
Polygonum acre – chililllo, venenillo – knotweed, smartweed
Polygonum amphibium – bistorta, chilillo, persicaria anfibia, polígono anfibio, venenillo – water smartweed
Polygonum arenastrum – centinodia, correhuela de los ánimos, hierba rosa, lengua de pájaro, polígono de hojas pequeñas, sanguinaria mayor – common knotweed, doorweed
Polygonum argyrocoleon – alambrillo, huichuri – knotweed, smartweed
Polygonum aviculare – altabaquillo, centinodia, cien nudos, correhuela, lengua de pájaro, polígono heterófilo, sanguinaria mayor – knotgrass, prostrate knotweed, yard knotweed
Polygonum baldschuanicum – polígono de Aubert, polígono de Bokhara, vid rusa – Bokhara fleeceflower – knotgrass
Polygonum (Persicaria) bistorta – bistorta, hierba sanguinaria, nevadilla, polígono bistorto, romaza retorcida – bistort, knotweed, snake rod
Polygonum bistortoides – bistorta, hierba del pescado – western bistort
Polygonum (Fallopia) convolvulus – pimentilla trepadora, polígono correhuela – black bindweed
Polygonum cuspidatum – polígono cuspidado – Japanese knotweed
Polygonum hydropiper – a-chili, centinodia, chilillo, persicaria picante, pimienta acuática, pimienta de agua – marshpepper, smartweed, waterpepper
Polygonum hydropiperoides – camalote, chilillo – waterpepper
Polygonum (Persicaria) lapathifolium – ananash, chilillo, persicaria acederifolia, polígono pata de perdiz, venenillo, zarzaparrilla – willowleaf smartweed, willow weed
Polygonum orientale – disciplina de monja, polígono oriental – kiss-me-over-the-garden-gate
Polygonum patulum – polígono pátulo
Polygonum pensylvanicum – hierba del pescado, moco de cucuno, moco de guajolote – pinkweed

Polygonum persicaria – duraznillo, hierba pejiguera, moco de guajolote, pejiguera, persicaria, pimentilla – lady's thumb, redshank
Polygonum polystachyum – polígono de muchas espigas – Himalayan knotweed
Polygonum punctatum – ananash, chilillo, venenillo – water smartweed
Polygonum tinctorium – índigo chino – Chinese indigo
Polygonum = Fallopia
Polymnia edulis – aricoma, yacón – yacon strawberry
Polypodium californicum – lengua de ciervo – California polypody
Polypodium guttatum – calaguala
Polypodium hartwegianum – helecho polipodio
Polypodium lanceolatum – helecho macho, lengua de ciervo, mananepile – polypody
Polypodium polypodioides – calaguala
Polypodium = Dryopteris, Phlebodium
Polypogon elongatus – cola de zorra, polibarbado largo – beard grass
Polypogon interruptus – zacate natal – ditch beard grass
Polypogon maritimus – polipogon marítimo – Mediterranean beard grass
Polypogon monspeliensis – cola de zorra, colita de ratón, zacate cola de zorra – annual beard grass, rabbit-foot grass
Polyporus fomentarius – yesca – polypore
Polyporus officinalis – agárico blanco, agárico de las farmacias, agárico de los médicos, agárico purgante
Polyscias scutellaria (balfouriana) – aralia, balfour – dinner plate aralia
Polystichum = Dryopteris
Polytrichum juniperum – musgo – moss
Poncirus (Citrus) trifoliata – naranjo amargo espinoso, naranjo de tres folíolos, naranjo trébol, naranjo trifoliado – myrtle orange, trifoliate orange
Pontederia cordata – camalote, flor de agua, flor de laguna, lirio de laguna, pontederia, pontederia de hojas cordiformes, tule – pickerel weed
Pontederia sagittata – lirio de laguna, popalillo
Poponax = Acacia
Populus* x *acuminata – huérigo – lanceleaf cottonwood
Populus alba – álamo, álamo blanco, álamo común, chopo blanco – silver poplar, white poplar
Populus angustifolia – álamo, álamo saúco, jara – narrowleaf poplar, narrow-leaved cottonwood
Populus balsamifera (candicans) – álamo, álamo balsamero, álamo tacamahaca, balsámico, bálsamo de Gilead – balm-of-Gilead, balsam poplar
Populus balsamifera* ssp *trichocarpa – álamo, álamo balsámico de California, álamo balsámico del oeste, balsamero negro, chopo, tacansahac – black cottonwood
Populus brandegeei (monticola) – álamo, güérigo, huérigo, olmo
Populus* x *canadensis (angulata,* x *euramericana) – álamo de la Carolina, álamo del Canadá, carolina, chopo del Canadá, chopo euroamericano, chopo negro híbrido del Canadá – Carolina poplar
Populus* x *canescens – álamo canescente, álamo cano, álamo gris, álamo gris de picardía – gray poplar
Populus deltoides – álamo, chopo americano, chopo del Canadá, chopo de Virginia, chopo papelero – Carolina poplar, eastern cottonwood, eastern poplar
Populus dimorpha – abaso, álamo, chopo
Populus euphratica – álamo – poplar
Populus fremontii (arizonica, macdougalii) – álamo, álamo blanco, álamo chopo, álamo cimmarón, balsamillo, chopo, güérigo, itzohu, olmo, pepeyoca, pepeyócatl – alamo, Frémont cottonwood, western cottonwood
Populus grandidentata – álamo canadiense – bigtooth aspen
Populus mexicana – abaso, álamo, olmo, pepeyoca, volador – Mexican poplar
Populus nigra – álamo, álamo negro, chopo, chopo común, chopo mosquitero, chopo negro, olmo – black poplar
Populus nigra* var *italica – álamo, álamo de Italia, álamo negro, álamo negro de Italia, chopo, chopo de Italia, chopo de Lombardia,

chopo lombardo, chopo piramidal – Lombardy poplar
Populus sargentii – álamo – plains cottonwood
Populus simaroa – álamo, simaroa, tepalcoxtli
Populus tremula – álamo temblón, chopo temblón, lamparilla, temblón – European aspen
Populus tremuloides – alamillo, álamo, álamo blanco, álamo temblón, chopo temblón, huérigo – quaking aspen
Populus wislizenii – álamo, güérigo – Rio Grande cottonwood
Porlieria = Guaiacum
Porophyllum gracile – cuchu, hestej, hierba del venado, ojo de pescado, pápalo quelite, siendre – odora
Porophyllum macrocephalum – chaoacocopin, hierba del venado, liendrilla, matapiojos, pápalo quelite, tepelcasho
Porophyllum macrophyllum – hierba del venado, liendrilla, matapiojos
Porophyllum punctatum – hierba del venado, piojillo, piojo
Porophyllum scoparium – hierba del venado, jarilla, romerillo
Portulaca grandiflora – amor de un día, bella a las once, flor de seda, portulaca, verdolaga, verdolaga de flores grandes – garden portulaca, moss rose, portulaca, rose moss, sun-plant
Portulaca halimoides (mundula) – chisma, tsayochcan – desert portulaca, shaggy portulaca
Portulaca oleracea – acahuecaxaqua, itzmiquitl, portulaca, quauquilitl, tequitl, verdolaga, verdolaga amarilla, verdolaga silvestre, xenshe, xucul – common purslane, pigweed, purslane, pusley
Portulaca pilosa – arroz-xiu, hielito, mañanita – moss-rose
Portulaca umbraticola – verdolaga de la sierra
Potamogeton alpinus – espiga de agua alpina, espiga de agua rojiza – reddish pondweed
Potamogeton crispus – espiga de agua serrada – crispate-leaved pondweed
Potamogeton diversifolius – lama de agua – diverse-leaved pondweed
Potamogeton epihydrus – espiga de agua de Nuttall – nuttall's pondweed
Potamogeton filiformis – espiga de agua filiforme – slender-leaved pondweed
Potamogeton gramineus – espiga de agua heterofila – grass-leaved pondweed
Potamogeton illinoensis – putsurini tepari – shining pondweed
Potamogeton natans – acaxaxan, espiga de agua flotante – floating-leaved pondweed
Potamogeton nodosus – espiga de agua, espiga de agua nudosa – long-leaved pondweed
Potamogeton pectinatus – chira, espiga de agua, granza, hierba de agua, lino de zanja – fennel-leaf pondweed
Potamogeton praelongus – espiga de agua pedunculada – white-stemmed pondweed
Potamogeton pusillus – espiga de agua palermitana – small pondweed
Potamogeton zosteriformis – espiga de agua de hojas de zostera – eel-grass pondweed
Potentilla anserina – argentina, buen varón silvestre, buen varón silvestre de Jarava, hierba de la plata, plateada – cinquefoil, feather silverweed, silverweed
Potentilla atrorubens – hierba colorada
Potentilla atrosanguinea – potentilla roja – potentilla
Potentilla aurea – cincoenrama áurea, cincoenrama dorada – golden cinquefoil
Potentilla canadensis – cincoenrama, potentilla canadiense – cinquefoil
Potentilla candicans – atlanchana, coanenepili, consuelda, fresa, goñi, sínfito, sínfito falso, suelda, suelda consuelda, tormentilla – cinquefoil
Potentilla erecta – cincoenrama erecta, consuelda roja, sieteenrama, tormentilla – bloodroot, cinquefoil, tormentil
Potentilla fruticosa – cincoenrama, cincoenrama leñosa, potentilla arbustiva, quinquefolio arbustivo – bush cinquefoil, shrubby cinquefoil
Potentilla nitida – cincoenrama nítida, cincoenrama rosa – lustrous cinquefoil
Potentilla norvegica – cincoenrama noruega – Norwegian cinquefoil
Potentilla palustris – cincoenrama de los pantanos, cincoenrama palustre – marsh locks
Potentilla pensylvanica – cincoenrama de Pensilvania – Pennsylvanian cinquefoil

Potentilla recta – cincoenrama recta – sulphur cinquefoil
Potentilla rubra – sangre de grado, suelda consuelda, tormentilla
Potentilla tabernaemontanii (verna) – cincoenrama de primavera, cincoenrama rocosa, quinquefolio rocoso – spring cinquefoil
Potentilla thurberi – clameria, fresa cimarrona, hierba colorada – cinquefoil
Potentilla tormentilla – cinco dedos, tormentila – five finger grass
Poulsenia armata – carnero, guaimaro
Pouteria caimito – abiú, caimito
Pouteria (Lucuma) campechiana – atzápotl, caca de niño, canisté, canistel, costiczápotl, mamey ciruela, mamey de Campeche, tigrillo, zapote amarillo, zapote borracho, zapote de niño, zapote dormilón, zapote mante, zapote niño, zapotillo, zapotillo rojo – eggfruit, yellow sapote
Pouteria multiflora – jacana
Pouteria sapota – mamey, tetzonzapotl, zapote, zapote mamey – mamey apple
Pouteria viridis – injerto – green sapote
Primula auricula – auricula de hojas carnosas, auricula de los Alpes, oreja de oso, primavera, primavera rocosa – auricula
Primula denticulata – prímula denticulada – primrose
Primula obconica – primavera venenosa – poison primrose
Primula officinalis – flor de San José, hierba de San Pablo, primavera, prímula común, vellorita
Primula x _polyantha_ – primavera, prímula – English primrose, polyanthus primrose
Primula sinensis – primavera – Chinese primrose
Primula suffrutescens – primavera – Sierra primrose
Primula veris – flor de primavera, gayada, gordolobollo, hierba de San Pablo, primavera, prímula, vellorita, vellorita de oro – cowslip, primrose
Primula vulgaris (acaulis) – hierba de San Pablo, primavera, primavera cicaule, primavera común – English primrose, primrose
Prionosciadium thapsoides – comino rústico, jalocote, hierba del oso
Prionosciadium townsendii – saraviki, tápokoli
Pritchardia hillebrandii – palmera de Hillebrand – loulu palm
Pritchardia pacifica – palma de abanico, palma imperial, palmera de Fiji – Fiji fan palm
Pritchardia (Colpothrynax) wrightii – palma barrigona
Proboscidea (Martynia) althaeifolia – campanita, cuernito, cuernos del diablo, espuela del diablo, gato, torito, uña de gato, uña del diablo – desert unicorn plant, devil's claw
Proboscidea (Martynia) annua – aguaro, tancocochi, uña de gato – devil's claws
Proboscidea arenaria – cuernos del diablo, espuela del diablo – devil's claw
Proboscidea (Martynia) fragrans – aguaro, casa pulgas, catachío, cinco llagas, cuernito, cuernito de hormiguero, cuernos del toro, garambullo, gatuña, hierba del toro, hoja de la pulga, perrito, torito, toro, uña del diablo – devil's claw, unicorn plant
Proboscidea gracillima – espuela del diablo – devil's claw
Proboscidea louisianica – cuernito, torito – common unicorn plant, devil's claw
Proboscidea (Martynia) parviflora – aguaro, cuernero, cuernito, espuela del diablo, garambullo, torito, toroberico, uña de gato – red devil's claw, unicorn plant
Prochnyanthes viridescens – amole, amolillo, chupac
Prosopis articulata – mezquite, mezquite amargo – bitter mesquite
Prosopis chilensis – algarrobo chileno, hupala, mezquite, mezquite dulce, mizquitl, quetzalmizquitl – Chilean mesquite
Prosopis cinerascens – mezquite, tornillo
Prosopis glandulosa var _prostrata_ – mezquite rastrero – running mesquite
Prosopis glandulosa (juliflora) var _torreyana_ – algarrobo, chachaca, mezquite – honey mesquite, mesquite
Prosopis globosa var _mexicana_ – mezquite – dwarf mesquite

Prosopis juliflora – algarrobo, aromo, chachaca, chúcata, cují yaque, háas, huisache, huupa, jupala, katzimelk, mezquiquez, mezquite, mezquite amarillo, mezquite blanco, mezquite chachaca, mezquite chino, mezquite colorado, mezquite dulce, mezquitl, mixquicopalli, mizquitl, péchita, taji, tziritsecua, upala, valla honda – sweet mesquite

Prosopis laevigata – algarrobo, chachaca, chúcata, mezquite, mizquitl – mesquite

Prosopis palmeri – mezquite, palo fierro – mesquite

Prosopis pubescens – mezquite tornillo, tornillo – screw bean, tornillo

Prosopis velutina – algarrobo, chachaca, mezquite – Arizona mesquite, velvet mesquite

Protium copal – carana hediondo, copal, copalillo, incienso, latilla, pom – copa tree

Prunella grandiflora – brunela de hojas hastadas, brunela grande – large self-heal

Prunella vulgaris – brunela, brunela común, hierba de la muela, hierba de la herida, hierba del carpintero, sanícula – heal-all, self-heal

Prunus americana – ciruelo, ciruelo americano, ciruelo silvestre – American plum, goose plum, sloe, wild plum

Prunus armenica (Armeniaca vulgaris) – albaricoque, albaricoquero, amarrio, chabacano, damasco – apricot

Prunus avium (cerasus) – calabazón, cerezo, cerezo común, cerezo dulce, cerezo silvestre, guindo – cherry, sweet cherry

Prunus avium* var *cerasus – cerezo, cerezo ácido, cerezo común, guindo, guindo común – sour cherry

Prunus avium* var *duracina – cerezo costalero, cerezo durazno – Bigarreau cherry

Prunus avium* var *juliana – cerezo mollar – gean cherry

Prunus* x *blireiana – ciruelo-cerezo híbrido doble, ciruelo ornamental – flowering plum

Prunus brachybotrya – aguatero, aguacatillo, barranco, capulincillo, cerezo, cerezo montés, duraznillo, eucaz, pajarito, zarzafrás – shortclustered cherry

Prunus capuli – capulín, capulín blanco, capulín pequeño, capulxihuitl, cerezo, cerezo cimarrón, cusabi, detzse, jeco, pakshumk, tnundaya, uasiqui, xengua – Mexican chokecherry, wild cherry

Prunus cerasifera – cerezo mirobalán, ciruelo-cerezo, ciruelo chabacano, ciruelo de cereza, ciruelo mirobalán, guindo, marapa, mirobálano – cherry plum, myrobalan

Prunus davidiana – melocotonero chino, melocotonero ornamental – flowering peach

Prunus domestica – bruno, bruño, cirolero, ciruelo, ciruelo de España, ciruelo europeo – plum

Prunus domestica* ssp *insititia – 1. ciruelo quetche – German prune; 2. ciruelo damasceno, ciruelo de San Julián, enendrino grande, espino de injetar, niso – damson plum

Prunus domestica* ssp *syriaca – ciruelo mirabel, mirabella – mirabella, Syrian plum

Prunus dulcis (Amygdalus communis) – allozo, almendrero, almendro – almond

Prunus dulcis* var *amara – almendro amargo – bitter almond

Prunus dulcis* var *fragilis – almendro mollar – soft-shelled almond

Prunus emarginata – cerezo – bitter cherry

Prunus fremontii – damasquillo, duraznillo – desert apricot

Prunus fruticans – cerezo enano – dwarf cherry

Prunus fruticosa – cerezo enano – European dwarf cherry, ground cherry

Prunus gentryi – aigre, uasiqui

Prunus glandulosa – cerezo chinesco – Chinese bush cherry

Prunus ilicifolia – isláy, islaya – holly-leaved cherry, islay

Prunus ilicifolia* ssp *lyonii – almendro – Catalina cherry

Prunus insititia – ciruelo de San Julián – bullace plum tree

Prunus italica (domestica* ssp *italica) – ciruelo claudio, Reina Claudia – greengage plum, greengage tree

Prunus laurocerasus – laurel-cerezo, laurel de Trebisonda, laurel real, lauroceraso, loro – cherry laurel, English laurel, laurel cherry

Prunus lusitanica – azarero, azarezo, laurel de Portugal, laurel portugués, lauroceraso de

Portugal, lauroceraso portugués, loro, palo loro – Portugal laurel
Prunus maackii – cerezo de Manchuria – amur chokecherry
Prunus mahaleb – árbol de Santa Lucía, cerecino, cerezo de Santa Lucía, mahaleb, pudriero – mahleb cherry, St. Lucie cherry
Prunus mexicana – ciruelo, ciruelo silvestre – Mexican plum
Prunus mume – albaricoquero japonés – Japanese flowering apricot, Japanese flowering plum
Prunus padus – cerezo, cerezo-aliso, cerezo de racimo, cerezo ornamental, cerezo pado, cerezo silvestre – European bird cherry, maybush, mayday tree
Prunus pensylvanica – cerezo de Pensilvania – pin cherry
Prunus persica (Persica vulgaris) – albérchigo, alberchiguero, durazno, duraznero, melocotón, melocotonero, pash, pavía, pérsico duraznero, prisco, túrusi, ucansa – peach
Prunus persica* ssp *nectarina – abridero, abridor, blanquillo, ixi, nectarino, pelón – nectarine
Prunus persica* ssp *nucipersica – bruñón, griñón – smooth peach
Prunus pumila – cerezo del Canadá – sand cherry
Prunus rhamnoides – iza, mataiza
Prunus salicifolia – capulín – capulin cherry
Prunus salicina – cirolero japonés, ciruelo japonés – Japanese plum
Prunus sargentii – cerezo japonés de Sargent – flowering cherry
Prunus serotina – capulín, capulín grande, cerezo, cerezo americano, cerezo criollo, cerezo negro, cerezo negro americano, cerezo negro silvestre, cerezo serótino, cerezo silvestre, detze, estaquilla, usabi – capulin black cherry, capulin cherry, Mexican cherry, wild black cherry
Prunus serotina* ssp *virens – aguasique, capulín – southwestern chokecherry
Prunus serotina* var *rufula – capulín – gila chokecherry
Prunus serrula – cerezo tibetano – flowering cherry
Prunus serrulata – cerezo de adorno japonés, cerezo de hojas serradas, cerezo japonés, cerezo japonés de hojas serruladas, cerezo tibetano – flowering cherry, Japanese cherry
Prunus simonii – abricotina, albaricoquero de Simón – apricot-plum
Prunus spinosa – andrino, arañón, asarero, ciruelo silvestre, endrino, endrino azul, endrino bruño, espino negro – blackthorn, sloe
Prunus subhirtella – capullo de rosa, cerezo japonés – Japanese flowering cherry
Prunus syriaca – ciruelo amarillo pequeño – yellow mirabelle tree
Prunus tenella – almendro enano – dwarf almond tree
Prunus triloba – almendro con tres lobos, almendro florido, pruno trilobulado – flowering almond
Prunus virginiana – cerezo de Virginia, cerezo silvestre norteamericano, cerezo virginiano – chokecherry
Prunus virginiana* var *demissa – capulín, cerezo mexicano – western chokecherry
Prunus virginiana* var *melanocarpa – capulín – black western chokecherry
Prunus* x *yedoensis – cerezo híbrido japonés yoshino – yoshino flowering cherry
Prunus zinggii – jashuca, jeco, jeco guasiqui
Psacalium decompositum (Cacalia decomposita) – matarique, matarique blanco, matariqui, maturín – Indian plantain
Psaliota = Agaricus
Pseudobombax (Bombax) palmeri – clavellina, clavellina blanca, clavellina colorada, clavellina de la barranca, cuajilote, guajilote, taguapo – silk cotton tree
Pseudolarix amabilis – alerce chino, alerce de China, falso alerce dorado – golden larch
Pseudophoenix sargentii – cuca, kuka – cherry palm
Pseudoroegneria (Agropyron) spicata – agropiro azul
Pseudosasa japonica – bambú – arrow bamboo
Pseudotsuga flahaulti – guayamé colorado, hallarín
Pseudotsuga guinieri – pinabete

Pseudotsuga lindleyana – guayamé, hallarín, pinabete – Mexican Douglas-fir

Pseudotsuga macrocarpa – douglasia de California, douglasia de piñas grandes, falsa tsuga de California – bigcone Douglas-fir, bigcone spruce

Pseudotsuga menziesii (taxifolia) – abeto, abeto de Douglas, acahuite, cahuite, douglasia de Oregón, douglasia verde, falsa tsuga verde de las rocosas, guayamé, guayamé colorado, hallarín, jaboncillo, pinabete, pino de corcho, pino de Oregón, pino douglas, pino real, pino real colorado, romerillo – Douglas-fir

Pseudotsuga menziesii (macrolepis, mucronata)* var *glauca – abeto, acahuite, cahuite, guayamé, hallarín, hallarín colorado, jaboncillo, monje, pinabete, pino de corcho, pino real, romerillo, sabina – Rocky Mountain Douglas-fir

Pseudotsuga rehderi – guayamé, pinabete

Psidium cattleyanum – cas dulce, guayabo fresa, guayabo japonés, guayabo pequeño, guayabo rosado – strawberry guava

Psidium friedrich-stahlianum – guayabo agrio, guayabo montés – Costa Rican guava

Psidium guajava – chalxócotl, coloc, enandi, guáibasim, guayabero, guayabo, guayabo blanco, guayabo casero, guayabo colorado, guayabo común, guayabo de China, guayabo de las sabinas, guayabo de venado, guayabo dulce, guayabo manzana, guayabo peruano, guayabo perulero, guayabo rojo, guayaguasi, jalocote, pichí, pichiché, pooscuy, posh, poshkeip, xalxócotl, xoxocoenabitl – guava

Psidium guineense – guayabo sabanero, guayabo silvestre – Guinea guava

Psidium molle – guayabo, guayabo agrio, guisaro – guava

Psidium sartorianum – arrayán, choquey, guayabillo, pichiché, rayana – guava

Psilocybe aztecorum – apipiltzin, dormilón, niñito, niño de las agues

Psilocybe caerulescens – derrumbe, dichitekisho, razón bei

Psilocybe (Stropharia) cubensis – hongo de San Isidro, hongo kentesh, San Isidro, San Isidro Labrador, teotlaquilnanácatl – cubies, golden top

Psilocybe mexicana – amokid, angelito, dichitonizé, di-nizé, hongo enteógeno, nize, piule de churis, teonanácatl, teotlaquilnanácatl, teunamacatlth, teyhuintli – Mexican liberty cap

Psilocybe muliericula – cihuatsinsintle, mujercita, nano-catsintli, niño – Wasson's psilocybe

Psilocybe yungensis (acutissima) – derrumbe negro, pajarito de monte – divinatory mushroom, genius mushroom

Psilocybe zapotecorum – badao zoo, beya zoo, bi-neechi, derrumbe negro, piule de barda, razón guiol, razón viejo – zapos

Psittacanthus (Loranthus) calyculatus – batuu-chá, caballera, chac-xciu, chujquén, flor de palo, flor de papagayo, injerto, injerto de aguacate, liga, lirio, lirio parasito, mal de ojo, matapalo, muérdago, muérdago visco, planta quebradora, quauhtzictli, tapaojo, toji, visco, visco cuercino, yecapixtla – Mexican mistletoe, red parrot flower

Psittacanthus sonorae – chupón, injerto, muérdago, toji

Psophocarpus tetragonolobus – dólico de goa, frijol alado, judía careta, judía dragón, seguidilla – asparagus pea, winged-bean

Psoralea esculenta – contrahierba – breadroot, Indian breadroot

Psoralea pentaphylla – contrahierba, contrahierba aromatica, contrahierba blanca, contrahierba del interior, contrahierba de México, iztacpatli – blue pea

Psoralea rhombifolia – contrahierba – brown-flowered psoralea, wild-alfalfa

Psoralea = Bituminaria, Psoralidium

Psoralidium (Psoralea) lanceolatum – soralea – lemon scurf-pea, lemon weed, scurf-pea, scurvy-pea

Psorothamnus (Dalea) schottii – jiguata – indigo bush, mesa dalea

Psorothamnus (Dalea, Parosela) spinosus – corona de Cristo, mangle, palo blanco, palo cenizo, palo triste – indigo bush, smoke thorn, smoke tree

Psychotria emetica – raicilla – false ipecac

Psychotria nervosa – crucecillo, laurel amarillo, tepecajete blanco – wild coffee

Psylliostachys suworowii – espliego de mar

Ptelea angustifolia – cola de zorrillo, telea – common hop tree
Ptelea crenulata – cola de zorrillo, telea – hop tree
Ptelea trifoliata – cola de zorrillo, fresno, pinacatillo, telea, vara de zorro, zorrillo – hop tree, skunkbush, wafer-ash
Pteridaceae – helecho – fern
Pteridium aquilinum (Pteris aquilina) – calaguala, chipe, felguera común, helecho, helecho común, helecho hembra, ocopetate, pata de cuervo, pteris, tzin, yogo – bracken, bracken fern, brake
Pteris cretica – epteride, helecho, helecho bracken, polipodio – bracken fern, brake
Pteris = Pteridium
Pterocarpus officinalis – chabecté, drago, guayabillo, llora sangre, palo de sangre, sangre de drago, sangre de grado, sangre de toro, sangregado
Pterocarpus santalinus – sándalo – almug tree, red sandalwood, red sanders wood
Pterocarya fraxinifolia – nogal de alas común, nogal de Cáucaso, pterocaria de hojas de fresno, pterocaria del Cáucaso, terocaria – common wingnut
Pterocarya stenoptera – coi, nogal de alas chino, pterocaria china – Chinese wingnut
Pterostyrax hispida – árbol de la charretera – epaulette tree
Ptychopetalum ovata – raíz de macho – muirapuama
Ptychosperma elegans – palmera solitaria – Alexander palm, solitaire palm
Ptychosperma macarthuri – palmera de Macarthur – MacArthur palm
Puccinellia distans – gliceria de ramificación rezagada – European alkali grass
Puccinia graminis – roya de los cereales – black stem rust
Pueraria lobata – kudzu – kudzu, pueraria
Pueraria phaseoloides – kudzu, kudzu tropical, pasto kudzu, pueraria – kudzu
Pulicaria dysenterica – cunilago, hierba del gato, pulguera disentérica, pulicaria, zaragatona disentérica – common fleabane
Pulicaria paludosa – pulguera de los pantanos – marsh fleabane
Pulmonaria angustifolia – pulmonaria de hojas estrechas – cowslip lungwort
Pulmonaria longifolia – pulmonaria de hojas estrechas, pulmonaria de hojas largas – lungwort
Pulmonaria officinalis – pulmonaria, pulmonaria mayor – lungwort
Pulmonaria saccharata – pulmonaria italiana, salvia de Belén – Bethlehem sage
Pulsatilla patens – pulsatila patente – eastern pasque flower
Pulsatilla vulgaris (Anemone pulsatilla) – anémona, anémona pulsatila, flor del viento, flor de pascua, pulsatila, pulsatila vulgar – European pasque flower
Pulsatilla vulgaris* ssp *slavica – pulsatila de Eslovaquia
Punica granatum – balaustra, granado, granado agridulce, granado agrio, granado cordelina, granado de China, granado dulce, manglano, milgranos, raíz de granado, tzapyan, yagasachi, yanuco, yutnudidzi – pomegranate
Punica granatum* var *nana – granado enano – dwarf pomegranate
Purshia (Cowania) mexicana – chinacate, chivatillo, poño, romerillo cimarrón, romero cedro, romero cimarrón, rosa de la sierra, vara zarza, zarza vara – antelope bush, cliff rose, quinine bush
Puschkinia scilloides – escila rayada
Pycnanthemum muticum – poleo de monte – mountain mint
Pyracantha coccinea – arbusto ardiente, espino de fuego, piracanta – firethorn, pyracantha
Pyracantha koidzumii – piracanta – firethorn
Pyrethrum = Chrysanthemum, Tanacetum
Pyrola chlorantha – peralito virescente – green-flowered wintergreen
Pyrola minor – peralito menor – lesser wintergreen
Pyrostegia (Bignonia) venusta (ignea) – chiltote, chorro de oro, llamarada, lluvia de oro, lluvia dorada, San Carlos, triquitraque, trompeta, trompeta anaranjada – flame vine, golden shower, orange creeper, orange trumpet vine, sweetheart vine

Pyrrhopappus multicaulis – pata de león – Texas dandelion

Pyrus amygdaliformis – peral amigdaloide, peral de monte – almond-leaved pear

Pyrus communis – peral, peral común, peral de San Juan, peral gamboa, peral panochero, perón cristalino – pear

Pyrus nivalis – peral de las nieves – snow pear

Pyrus pyraster – peral silvestre, peruétano – wild pear

Pyrus salicifolia – peral de hojas de sauce – weeping willow-leaved pear

Pyrus = Malus

Q

Quamoclit coccinea – bandera española, chiqueo de monjas, clarín, clarincito, clarincito silvestre, cundeamor, estrella, hiedra, hiedra colorada, hiedra roja, quamoclit, trompillo – cardinal vine, ivy, star ipomoea

Quamoclit hederifolia – hiedra colorada

Quamoclit lobata – bandera española, estrella, palmira – exotic love, firecracker vine, Spanish flag

Quamoclit = Ipomoea

Quararibea cordata – zapote amarillo – South American sapote

Quararibea funebris – cacahuaxóchitl, cacaoxóchitl, canela, flor de cacao, madre de cacao, maha, molinillo, palo copado, rosa de cacao, tepecacao – swizzle-stick tree

Quassia amara – amargo, corteza de Jamaica, cuasia, cuasia amarga, hombre grande, quino de Cayena, quino de Surinam, tinco, tiquistiquis – bitterwood, quassia

***Quercus* spp** – 1. ahoaquahuitl, ahoatl, ahuaquahuitl, ahuatl, bellota, bellotero, charrasquillo, encino, encino blanco, roble – white oak; 2. ahuatetz, ahuatetzmolli, encino – live oak; 3. – encino negro – black oak

Quercus acutifolia – aguatle, ahuatl, ahuatle, encino blanco, encino chilillo, encino de asta, encino laurelillo, encino saucillo, encino teposcohuite, roble, teposcohuite

Quercus acutissima (serrata) – roble asiático kunigi – sawtooth oak

Quercus aegilops – roble – Valonia oak

Quercus affinis – encino, encino colorado, encino de asta, encino manzanillo, laurelillo

Quercus agrifolia – encino, encino de California, encino memelito, encino verde, roble costero – coast live oak, encina

Quercus alamoensis – encino

Quercus alba – roble blanco, roble blanco americano, roble blanco de América, roble claro – eastern white oak

Quercus albocincta – cusi, encino, encino blanco, encino negro, encino prieto, encino roble, hachucá, roble

Quercus aristata – encino

Quercus arizonica – encino, encino azul, encino blanco, encino manzano, roble, rojacasacame, toa – Arizona white oak, silverleaf oak

Quercus berberidifolia – encinillo, encinito chaparro – scrub oak

Quercus bicolor – roble bicolor americano, roble blanco americano de los pantanos – swamp white oak

Quercus brandegeei – encino, encino arroyero, encino bellotero, encino negro – black oak, Cape oak

Quercus canariensis – quejigo, roble andaluz, roble mediterráneo – Mediterranean oak

Quercus candicans – ahuahuaxl, ahuamextli, encino, encino blanco, encino cenizo, encino de asta, encino papatla, huilocualoni, roble, teposcohuite, tzacué blanco – whitish oak

Quercus castanea – bellota, capulincillo, encino, encino amarillo, encino blanco, encino chino, encino colorado, encino negro, encino roble, encino rojo, encino rosillo – chestnut oak

Quercus cedroensis – encinillo, encinito chaparro, encino – evergreen oak, live oak

Quercus cerrioides – roble cerrioide

Quercus cerris – encino cabelludo, roble cabelludo, roble cerris, roble de Turquía, roble musgoso, roble turco – Turkey oak

Quercus chartacea – encino amarillo

Quercus chihuahuensis – encino, encino blanco, encino chino, encino miscalme, encino roble, roja, sahuauó, sajaguo – Chihuahua oak

Quercus chrysolepis – encino roble – canyon live oak, canyon oak, goldencup oak, maul oak

Quercus coccifera – carrasca, chaparro, coscoja, matarrubia, roble calliprinos, roble coccifero – kermes oak, scarlet

Quercus coccinea – roble escarlata – scarlet oak

Quercus coccolobifolia – encino, encino manzano, encino negro, encino verde, gueja

Quercus coffeicolor – encino colorado, encino de asta

Quercus conspersa – chiquilín, encino, encino blanco, encino cascado, encino colorado, encino pipitillo, encino roble amarillo, encino

rojo, encino teposcohuite, roble, teposcohuite, yagareche – spreckled oak
Quercus corrugata – chícharro, encino rey
Quercus crassifolia – chanal, chicharrón, encino, encino blanco, encino carrocero, encino chicharrón, encino colorado, encino de hojas anchas, encino hojarasco, encino huaje, encino prieto, encino roble, encino tesmolillo, excalahuate, hoja ancha, roble – thickleaf oak
Quercus crassipes – blanco chino, chiquinib, encino, encino blanco, encino chilillo, encino colorado, encino laurel, encino manzanillo, encino negro, encino oreja de ratón, encino piptza
Quercus crenata – roble crenado – Spanish oak
Quercus crispipilis – chiquinib, encino
Quercus depressipes – encino
Quercus devia – encino colorado, encino negro – black oak, Cape oak
Quercus diversicolor – encino
Quercus dumosa – chaparro, encinillo, encino achaparrado – coastal scrub oak, Nuttall's scrub oak
Quercus durifolia – cusi, encino, encino colorado, encino laurelillo, roble prieto, saucillo, xipari
Quercus eduardii – encino colorado, encino manzano
Quercus elliptica – encino colorado, encino cucharita, encino laurel, encino nanche, encino tapahuite, tapahuile
Quercus emoryi – bellota, encino, encino bellotero, encino colorado, encino duraznillo, encino negro, encino prieto, roble negro, toa – black oak, blackjack oak, Emory oak
Quercus endlichiana – encino, encino hueja, popuixoli, roble, rocuró, rojaca
Quercus faginea (lusitanica) – quejigo, rebollo, roble carrasqueño – Portuguese oak
Quercus fendleri – encinillo, encino – Fendler's oak
Quercus frainetto – roble de Hungría, roble húngaro – Hungarian oak, Italian oak
Quercus fruticosa (humilis) – robledilla – shrubby oak
Quercus fulva – encino hueja
Quercus fusiformis – encino, encino negro, maculi, maquilihuatl, palo duro, roble, roble serrano, texmole – escarpment live oak, plateau oak
Quercus gambelii – encino, encino de hojas anchas – Gambel oak, Rocky Mountain white oak
Quercus gentryi – encino avellano cimarrón, encino cacachila, encino chilillo, encino colorado
Quercus glabrescens – encino, encino roble, roble
Quercus glaucescens – encino blanco, encino memelito, encino prieto, encino roble, encino roble amarillo
Quercus glaucoides (laceyi) – encino, encino tocuz – canyon oak, Lacey oak
Quercus gravesii – encino – Chisos red oak
Quercus greggii – encino – Gregg oak
Quercus grisea – encino, encino blanco, encino prieto – gray oak, mountain white oak
Quercus hypoleucoides (hypoleuca) – cusi, cusi prieto, encino, encino blanco, encino colorado, xipari – silverleaf oak
Quercus ilex – alsina, carrasca, chaparro, encino, matarrubia – holly oak, holm oak, Spanish holly oak
Quercus intricata – charrasquillo
Quercus jonesii – encino, encino roble, encino verde
Quercus laeta – encino prieto
Quercus laurina – ahualpitzahuac, capulincillo, encino, encino ahualpitzahuac, encino chino, encino colorado, encino de hojas chicas, encino de hojas de laurel, encino de hojas delgadas, encino delgado, encino jarilla, encino laurelillo, encino mamelito, encino nechilahue, encino roble, encino rojo, escobilla, laurelillo, pipitillo, teposcohuite, urina
Quercus lobata – jonjonata, roble, roble del valle – California white oak, roble, valley oak
Quercus macrocarpa – roble blanco americano de bellotas grandes, roble de bellotas grandes, roble macrocarpa – burr oak, mossycup oak
Quercus magnoliifolia – encino amarillo, encino napis
Quercus martinezii – corti tocuz, encinillo, encino chaparro, encino laurelillo, encino prieto, encino prieto sancón, encino roble, encino sancón

Quercus mcvaughii – cusi, encino roble, gueja
Quercus mexicana – cozahuatl, encino, encino blanco – evergreen oak, live oak, Mexican oak
Quercus microphylla – encino, encino capulincillo, encino chaparro, encino enano
Quercus mohriana – encino – Mohr's shin oak
Quercus muehlenbergii – aguacatillo, encino, encino amarillo, encino blanco, encino chaparro, encino negro, encino pipitillo, encino prieto, encino rojo, palo colorado, roble, roble de Muehlenberg, teposcohuite chino – chinkapin oak, chinquapin oak, yellow chestnut oak
Quercus myrsinifolia – encino asiático de hojas de bambú – Japanese live oak
Quercus oblongifolia – bellota, encino, encino azul, encino laurel – Mexican blue oak, white live oak
Quercus obtusata – encino, encino avellano, encino blanco, encino carrasco, encino negro, roble
Quercus oleoides – béek, encino, encino nanche, encino prieto, encino roble blanco, encino texmole, roble, roblecito, texmole, yagpsuy
Quercus oocarpa – cantulán colorado, encino roble
Quercus ovandensis – chícharro grande
Quercus palmeri – chaparro, encinillo, encino, encino achaparrado – Palmer oak
Quercus palustris – roble de los pantanos, roble palustre, roble palustre americano, roble-pino – pin oak, swamp oak
Quercus pennivenia – toché – hand-basin oak
Quercus perpallida – encino cacachila
Quercus petraea – cajiga, marojo, roble albar, roble de invierno, roble invernal – durmast oak
Quercus phellos – roble de hojas de sauce, roble rojo americano de hojas de sauce – willow oak
Quercus polymorpha – roble, xinil – net-leaf white oak
Quercus pungens – encino, encino chino – sandpaper oak, scrub oak
Quercus pyrenaica – melojo, roble borne, roble negral, roble negro, roble pirenaico, roble tozo – Pyrenean oak
Quercus resinosa – encino blanco, encino roble, roble
Quercus reticulata – ahoaquahuitl, ahoatl, chaparro, encino, encino blanco, encino de hojas anchas, encino de miel, encino memelito, encino prieto, encino quiebra hacha, mata parda, palo colorado, tnuya – chaparro oak, netleaf oak, white oak
Quercus robur – ahuatón, carvallo, roble, roble carvallo, roble común, roble fresnal, roble pedunculado, roble pedúnculo – English oak
Quercus rotundifolia – encino, roble de hojas redondeadas – round-leaved oak
Quercus rubra (borealis) – roble americano, roble boreal rojo americano, roble común, roble rojo, roble rojo americano, roble rojo del norte – northern red oak, red oak
Quercus rugosa (macrophylla) – ahoaquahuitl, ahuacocoztli, encino, encino avellano, encino blanco, encino blanco liso, encino cuero, encino de asta, encino de miel, encino negro, encino prieto, encino quiebra hacha, encino roble, encino tocuz, palo colorado, roble, roble rugoso, suchi, tulán – net-leaf oak, wrinkle-leaf oak
Quercus rugulosa – encino, encino cascalote, encino colorado, sipúraca
Quercus salicifolia – encino saucillo, encino chino, encino laurel, encino laurelillo, encino prieto, encino saucillo – willowleaf oak
Quercus sartorii – encino escobilla
Quercus scytophylla – encino blanco, encino prieto, encino rosillo
Quercus sideroxyla – encino, encino colorado, encino mamelito, encino prieto, encino tecomate – Santa Rosa oak
Quercus similis – roble rojo – delta oak
Quercus skinneri – chícharro, cololté, roble, tzajalchit
Quercus suber – alcornoque, alcornoque mediterráneo, sobrero – cork oak
Quercus subspathulata – encino de asta, encino borrego, encino cochi, encino mixcahue
Quercus tarahumara – encino cajete, gueja – handbasin oak
Quercus tomentella – encino – island oak

Quercus toumeyi – encino, encino chaparro – Toumey oak
Quercus tuberculata (idonea) – encino, encino amarillo, encino de miel, encino roble, roble, toché – golden oak
Quercus turbinella – chaparro, encinillo, encino, encino achaparrado, encinito chaparro – desert scrub oak, shrub live oak
Quercus undulata – encino, encino blanco – wavyleaf oak
Quercus urbanii – encino cucharillo, encino hueja, encino jumate, toché
Quercus vaseyana (pungens* var *vaseyana) – encino, encino chino – shin oak, Vasey oak
Quercus velutina – roble velutino – black oak, dyer's oak
Quercus viminea – bellota, cusi saucillo, encino sauce, machichare, saucillo – Madrean willow oak
Quercus virginiana – encino, roble de hojas anchas, roble de verdura perenne, roble serrano, texmole – coastal live oak, southern live oak
Quillaja saponaria – arbusto del jabón, pacón, palo de jabón, quillaya – soapbark tree
Quisqualis indica – barbudo, Santa Cecilia – Chinese honeysuckle, drunken sailors, Rangoon creeper

R

Raffia* = *Raphia
Rajania hastata – bejuco de ñame, ñame chato
Ramalina reticulata – orchilla, rocella – Spanish moss
Ramaria botrytis – clavito – red-tipped coral fungus
Ramaria flavia – escobeta, escobetilla, escobillón, patita de pájaro – golden coral fungus
Ramaria stricta – escobeta, pata de pájaro – upright coral fungus
Randia armata – canastilla, crucecito, cruceto, huele de noche, jazmín, maluco de montaña, palo de cruces, palo de la cruz, papache, zapotillo
Randia blepharophylla – papache
Randia echinocarpa – chacua, cirián chino, granjel, josocola, papache, papache borracho, papache picudo, tecolosapo – spiny-fruited randia
Randia laetevirens – capulín corona, crucero, crucero blanco
Randia laevigata – crucecillo de la sierra, guacabé, sapuchi
Randia megacarpa – papache
Randia mitis – crucecillo, crucecillo de la costa, crucecito, cruceto, crucillo, crucito, espino cruz, papachilla
Randia mollifolia – sapuchi de la sierra
Randia obcordata – cabijósocura, namuquia, papache, papache borracho, papachilla, pisi
Randia rosei – papache borracho, tecolotillo
Randia tetracantha – árbol de las cruces
Randia thurberi – papache, papache borracho, pusi
Randia watsoni – cruceto, huiscruceta, josocola, papache, papache de la sierra
Ranunculus acris – botón de oro, botón de oro amarillo, flor de grillo, hierba belida, ranúnculo – meadow buttercup
Ranunculus aquatilis – hierba lagunera, ranúnculo, ranúnculo acuático – water buttercup, water crowfoot
Ranunculus arvensis – ranúnculo arvense – corn buttercup
Ranunculus asiaticus – botón de oro, francesilla, marimoña, ranúnculo, ranúnculo asiático – Persian ranunculus, turban buttercup
Ranunculus bulbosus – hierba velluda, ranúnculo bulboso – bulbous buttercup
Ranunculus cymbalaria – oreja de ratón – desert buttercup
Ranunculus dichotomus – botón de oro, nextamalxóchitl, pata de león
Ranunculus fascicularis – botón de oro – pine buttercup
Ranunculus ficaria – cabeza de perro, celidonia, celidonia menor, hierba de almorranas – lesser celandine, pilewort
Ranunculus flammula – flámula – banewort, lesser spearwort
Ranunculus harveyi – botón de oro
Ranunculus hookeri – francesilla, francesita, pata de león
Ranunculus hydrocharis – ranúnculo – frog's-bit buttercup
Ranunculus montanus – botón de oro enano, ranúnculo de montaña – mountain buttercup
Ranunculus muricatus – ranúnculo espinoso, ranúnculo lleno de espinas, ranúnculo muricado – spinous buttercup
Ranunculus napellifolius – ranúnculo amarillo
Ranunculus parviflorus – ranúnculo con flores pequeñas – small-flowered buttercup
Ranunculus petiolaris – botón de oro, estamasúchil, hierba de la muela, nextamalxóchitl, pata de león, ranúnculo – buttercup, ranunculus
Ranunculus repens – botón de oro, hierba belida, ranúnculo repente – creeping buttercup
Ranunculus sardous – ranúnculo hirsuto, ranúnculo sardo – hairy buttercup
Ranunculus sceleratus – hierba belida, hierba de fuego, hierba sardonia, ranúnculo de pantano, revienta buey, sardonia – blister plant, celery-leaved crowfoot
Ranunculus tridentatus – oreja de ratón
Raphanus raphanistrum – erviana, flor de nabo, jaramago, jaramago blanco, labrestos, nabo, nabo blanco, nabo cimarrón, rabanillo, rabaniza, rábano cimarrón, rábano silvestre, raíz fuerte – jointed charlock, wild radish
Raphanus sativus – brusco esmuláceo,

rabanete, rabanillo, rabanito, rábano, rábano rojo – garden radish, radish

Raphanus sativus* var *longipinnatus – rábano daikon – daikon

Raphia (Raffia) ruffia – palma rafia, rafia – raffia, raphia palm

Raphiolepis = Rhaphiolepis

Rapistrum rugosum – rabanillo, rapistro anual, rapistro rugoso, tamarilla – wild turnip

Rathbunia = Stenocereus

Ratibida columnaris – conito, conito de campo, sombrero de zapata – Mexican hat

Ratibida mexicana – onova – coneflower

Ratibida tagetes – hierba de la tusa – short-rayed coneflower

Rauwolfia hirsuta (tetraphylla) – ajillo, chacmuc, chalchupa, chilillo, cinco negritos, cocotombo, coralillo, fruta de víbora, kabal-muk, pabilo, paulilla, sarna de perro, venenillo

Rauwolfia serpentina – boboró – Indian snakeroot

Ravenala madagascariensis – árbol de viajero, palma de viajero – traveler's palm, traveler's tree

Ravenea rivularis – palmera majestuosa

Reinhardtia gracilis – coquito, coyolillo, coyolito, jon-cháb, palmiche, tepejilote, xate

Reseda alba – reseda blanca – white mignonette

Reseda lutea – reseda amarilla – yellow mignonette

Reseda luteola – gabarro, gasparilla, gualda, luteolina, reseda – dyer's rocket

Reseda odorata (Lawsonia inermis) – alcana, alheña, cinamomo de manila, heno, hierba del amor, lausonia, miñoneta, reseda, reseda de jardín, reseda de olor, reseda francesa, reseda olorosa, resedón, rociado – camphire, henna, garden mignonette, mignonette

Retama raetam – retama – broom

Rhamnus alaternus – aladierno, carrasquilla, coscolina – Italian buckthorn, Mediterranean buckthorn

Rhamnus alpinus – aladierno alpino, pudio – alpine buckthorn

Rhamnus betulaefolia – ramno, salicieso – birch-leaf buckthorn, coffeeberry

Rhamnus californica – cacachila, capulincillo, cáscara sagrada, espino sagrado, hierba del oso – California buckthorn, California coffeeberry, coffee-berry, pigeon-berry

Rhamnus cathartica – aladierno, cambrón, cervispina, espino cerval, ramno catártico, uva cabruna – common buckthorn

Rhamnus frangula (Frangula alnus) – aladierno, arraclán, avellanillo, chopera, frángula, hediondo, sangüeño, sanguiñuelo – alder buckthorn

Rhamnus infectorius – aladierno fétido, arto, grana d'Avignon – Avignon berry

Rhamnus palestina – espino – Palestine buckthorn

Rhamnus purshiana – aladierno, burrobiote, cambrón, cáscara, cáscara sagrada, coscolina, espino cerval, tamujo – cascara, cascara sagrada, chittam bark, chittim

Rhamnus scopulorum – ojo de venado

Rhamnus serrata – ahuatl tepitón, capulincillo, capulín cimarrón, naranjillo, tlalcapolín – buckthorn

Rhamnus sharpii – duraznillo, palo liso

Rhamnus sphaerosperma – arracacho, cangica, duraznillo, mantequillo, yema de huevo – West Indian buckthorn

Rhaphiolepis (Raphiolepis) indica – rafiolepis – Indian hawthorn

Rhapis excelsa – palmera bambú – lady palm

Rhapis humilis – palmera china – rattan palm, slender lady palm

Rheedia edulis – chaparrón, limoncillo, naranjillo, toronjil, zapotillo

Rheum alexandrae – ruibarbo real

Rheum officinale – ruibardo, ruibardo chino, ruibardo común, ruibardo de la China, ruibardo oficinal, ruibardo medicinal – Chinese rhubarb, medicinal rhubarb

Rheum palmatum – ruibarbo palmeado – ornamental rhubarb

Rheum rhabarbarum – rabárbaro, ruibardo común

Rheum rhaponticum – rapóntico, ruibardo, ruibardo de comer, ruibardo de huerta, ruipóntico – garden rhubarb, rhubarb

Rhipidocladum pittieri – jimba

Rhipidocladum racemiflorum – cañutillo, chiquilla, chiquión, chiquita, otate, otate chino, otatillo

Rhipsalidopsis gaertneri – plumilla de Santa Teresa – easter cactus
Rhipsalis baccifera (cassutha) – bejuco de quebradura, cabellera, dedos de muerto, disciplinilla, nigüilla, pegapalo, tatache, tzalilte – mistletoe cactus
Rhizophora mangle – canaro, candelilla, candelón, mangle, mangle colorado, mangle dulce, mangle gateador, manglero, mangle rojo, mangle salado, mangle tinto, mangle zapatero, tabché – red mangrove
Rhododendron ferrugineum – rododendro, rododendro eruginoso, rododendro herrumbroso, rosa de los Alpes – rhododendron
Rhododendron indicum – azalea – azalea
Rhododendron japonicum – rododendro de Japón
Rhododendron luteum – rododendro amarillejo – pontic azalea
Rhododendron obtusum – azalea japonesa – azalea
Rhododendron ponticum – ojaranzo, rododendro del mar negro – rose-bay
Rhododendron simsii (Azalea indica) – azalea, azalea india
Rhodymenia palmata – dulse – dulse
Rhoeo = Tradescantia
Rhopalostylis sapida – palmera de nikau – nikau palm, shaving brush palm
Rhus aromatica – hierba del venado – aromatic sumac, fragrant sumac
Rhus (Schmaltzia) copallina – copal – flame-leaf sumac, shining sumac
Rhus coriaria – rus, zumaque, zumaque de los curtidores, zumaque de tenerías – Sicilian sumac, tanner's sumac
Rhus (Schmaltzia) glabra – zumaque, zumaque glabro, zumaque lampiño – Rocky Mountain sumac, scarlet sumac, smooth sumac
Rhus integrifolia – birringo, hiedra, lentisco, mangle – lemonadeberry
Rhus javanica – zumaque javanés
Rhus lentii – lentisco – sugar bush
Rhus microphylla – agrillo, agrito, correosa, lemita, limilla de la sierra, limita, limita de la sierra, pico de pájaro, salado – desert sumac
Rhus mollis – cimarrón, mangle, sumaco, sumaco cimarrón, tnu-ndé, vinagrillo, xoxoco, yucucaya, zumaque – sumac
Rhus ovata – lentisco, mangle – mountain lilac, sugar bush, sugar sumac
Rhus rydbergii – hiedra
Rhus schiedeana (tepetape) – agrín, lentisco, pajulul, palo agrín, palo pajulul, tepetape
Rhus sempervirens – ayumé, lambrisco, lentisco, zumaque – evergreen sumac
Rhus succedanea – árbol de la cera – Japanese wax tree, sumac
Rhus terebinthifolia – hierba del temascal, jaguay, sal de venado, temazcal, temazcalchihual, yagabiche, yaguay, zumaque cimarrón
Rhus (Schmaltzia) trilobata – acidito, agrillo, agrito, lambrisco, lemita, lentisco, pajul del norte – skunkbrush, skunkbush, squawbush
Rhus typhina – zumaque de astas de ciervo, zumaque de Virginia, zumaque mayor – staghorn sumac
Rhus verniciflua – árbol de la laca, laca – Japanese lacquer tree
Rhus virens (sempervirens) – capulín, lambrisco, lentisco, sumaco cimarrón, zumaque – evergreen sumac
Rhus = Cotinus, Malosma, Toxicodendron
Rhynchelytrum repens (Tricholaena rosea) – ilusión, pasto del Senegal, pasto de Natal, pasto senegalés, zacate aceitunillo, zacate de la loma, zacate de seda, zacate ilusión, zacate natal, zacate rosa, zacate rosado – Natal grass, Natal ruby grass
Rhynchosia macrocarpa – frijolín
Rhynchosia minima – arrocillo, ib-ché, mehenibbech, negrito, tlitliltzin – least snoutbean
Rhynchosia phaseoloides – negrito, ojito de pájaro, ojo de pajarito, piule, sinicuichi
Rhynchosia precatoria – atecuixtle, chanate pusi, cocolmeca, colorín chiquito, frijol – rosary bean
Rhynchosia pyramidalis – antipusi, atecuixtle, casa pulgas, chanate, chanate pusi, colorín chiquito, colorincito, coralito, frijol de chintlatlahua, frijolillo, guarecita, negrito, ojito de pájaro, ojito de pichón, ojo de cangrejo, ojo de chanate, ojo de culebra, ojo de pajarito, ojo de perico, ojo de zanate,

peonía, perico, pipilzíntli, pulguita, purensapichu, saltipuchi, saltipús, sasan, xenecuilche – rosary bean
Rhynchosia* = *Lablab
Rhynchostele cervantesii – mariposita blanca, tigrillo
Ribes affine – capulincillo, ciruelillo, garambullo
Ribes alpinum – calderilla, grosellero alpino, grosellero de los Alpes, grosellero de montaña – alpine currant, mountain currant
Ribes aureum – grosellero amarillo, grosellero dorado – golden currant
Ribes brandegeei – ciruelillo – currant
Ribes cereum – capulincillo, ciruelillo, grosellero – squaw currant, wax currant
Ribes cereum (inebrians)* var *inebrians – grosellero, manzanita – squaw currant
Ribes ceriferum – capulincillo
Ribes ciliatum – capulincillo, ciruelillo, saracuacho – gooseberry
Ribes grossularia – grosellero, grosellero espinosa, grosellero silvestre, uva espina – sweet gooseberry
Ribes microphylla – cocohuitztli
Ribes montigenum – grosellero espinoso, uva crespa, uva espina – gooseberry currant, mountain gooseberry
Ribes nigrum – casis, grosellero negro – black currant, squinancy berry
Ribes odoratum – grosellero ornamental – ornamental currant
Ribes petraeum – grosellero pétreo – rock currant
Ribes rubrum – grosellero, grosellero rojo – garnet berry, red currant
Ribes rugosum – capulincillo, ciruelillo
Ribes sanguineum – grosellero floreciente – pink winter currant, red flowering currant
Ribes uva-crispa – grosellero, grosellero espino, uva espina – gooseberry
Ribes vulgare – grosellero
Richardia scabra – alcatraz
Ricinus communis – axaxaxoxihuitl, castor, catapucia, cherva, coch, higuera, higuera del diablo, higuera del infierno, higuera infernal, higuereta, higuerilla, higuerilla pequeña, kevenia, koch, palmacristi, quebena, ricino, ricino común, tártago, tlapate, tlapatl, tulua, yagabilape, yagahigo, yorimaqui – castor bean
Rinorea guatemalensis – cafecillo, costarrica, frutillo, guayabillo, huesillo – wild coffee
Rivea* = *Turbina
Rivina humilis – baja tripa, colorín, coqueta, coral, coralillo, coralito, coralito cimarrón, coral xilacuaco, cuxubcan, hierba de la víbora, hierbamora, teihuisí, tomatillo, ucuquiro, xilacuaco – bloodberry rouge-plant, pigeonberry, pokeberry, rouge-plant, small pokeweed
Robinia neomexicana – garrobo, hojalito, uña de gato – desert locust, New Mexican locust
Robinia pseudoacacia – acacia, acacia bastarda, acacia blanca, acacia de flores blancas, acacia espinosa, acacia falsa, falsa acacia, loco, pan y quesillo, robinia, siete sayo – black locust
Robinia viscosa – acacia pegajosa – clammy locust
Roccella babingtonia – orchilla – lichen
Roccella fusiformis – toji
Roccella tinctoria – orchilla – lichen
Rollinia deliciosa – araticú, biribá, cachimán
Rollinia mucosa – anona, anona babosa, anona cimarrona, anonillo de monte, anonita, cachimán, guanábano cimarrón – wild soursop
Rollinia pulchrinervis – biribá, corosol – biriba
Romneya coulteri – amapola arbórea, amapola blanca, amapola del campo, cardo, chicalote, iztecauhticmixtl, matilija – Coulter's matilija poppy
Rorippa austriaca – berro austriaco – Austrian field-cress
Rorippa islandica – berro falso, berro islandés – marsh yellow cress
Rorippa (Radicula) nasturtium-aquaticum (Nasturtium officinale) – agrioris, berro, berro de agua, cresón, mastuerzo, mastuerzo acuático, mastuerzo de agua, mexixin – water cress
Rosa acicularis (sayi) – rosal – prickly rose
Rosa alba – rosal blanco, rosal de Cambray – white rose of York
Rosa arizonica – rosal – Arizona rose
Rosa arkansana – rosal – Arkansas rose

Rosa banksiae – lady Eubanksia – lady Banks' rose
Rosa bracteata – rosal – McCartney rose, wild white rose
Rosa californica – rosal de California, rosal del campo – California rose
Rosa canina – agavanzo, escaramojo, rosal canina, rosal de Jerico, rosal del diablo, rosal perruno, rosal silvestre, zarzarrosa – dog rose
Rosa centifolia – flor de Castilla, rosal, rosal común, rosal de Castilla, rosal de cien hojas, rosal de cien pétalos, rosal de Provenza – cabbage rose, Provence rose
Rosa chinensis (sinensis) – rosal de todos los meses – China rose
Rosa chinensis* var *minima – rosal enano – miniature rose
Rosa cinnamomea – rosal canela, rosal de mayo – cinnamon rose, May rose, Scotch rose
Rosa damascena – rosal de damasco, rosal de las cuatro estaciones, rosal de todos los meses – damask rose
Rosa damascena* var *semperflorens – damasco de otoño – autumn damask
Rosa eglanteria (rubiginosa) – eglantina, eglantina roja, escaramujo, mosqueta, rosal fétido – eglantine, stinking rose, sweet briar
Rosa foetida (lutea) – rosal de los capuchinos, rosal fétido – Austrian brier
Rosa fortuniana – fortuniana – double white banksia, fortuniana
Rosa gallica – rosal de Francia, rosal de Jerico, rosal del norte, rosal de Provenza, rosal francés, rosal castellano – French rose, Provence rose
Rosa glauca (rubrifolia) – rosal de hojas rojas – red-leaved rose
Rosa hibernica – rosal irlandés – Irish rose
Rosa hypoleuca – rosal – pale-leaf rose
Rosa indica – rosal de Bengala, rosal té – tea rose
Rosa laevigata – rosal té
Rosa micrantha – rosal balsamina menor – small-flowered rose
Rosa minutifolia – colguinero, rosal de hojas pequeñitas, rosal silvestre – desert rose, miniature rose, small-leaved rose
Rosa montezumae – agabanzo, cinorrodón, escaramujo, garambullo, rosal de Moctezuma, trompetillo, trompillo, uña de gato – wild rose
Rosa moschata – mosqueta, rosal mosqueta – elegantine rose, musk rose
Rosa multiflora – rosal enredadero, rosal multiflora – multiflora rose, shrub rose
Rosa muscosa – rosal musgoso – moss rose
Rosa neomexicana – rosal – New Mexico rose
Rosa nutkana (manca) – rosal – Nootka rose
Rosa parviflora – rosal miniaturo
Rosa pendulina – rosal alpino – alpine rose
Rosa pimpinellifolia (spinosissima) – rosal de hojas de pimpinela, rosal pimpinela – burnet rose, Scotch rose
Rosa polyantha – rosal multiflora – climbing rose
Rosa roxburghii – rosal chayote – chestnut rose
Rosa rugosa – rosal de Japón, rosal rugoso – ramanas rose, sea tomato
Rosa sempervirens – rosal de verdura perenne, rosal mosqueta común, trepadora – evergreen rose
Rosa serrulata – rosal mexicano
Rosa standleyi – rosal – Standley rose
Rosa stellata – rosal – desert rose, star rose
Rosa sulphurea – canaria, rosal amarillo
Rosa woodsii – champes, rosal – Woods' rose
Rosa woodsii (fendleri)* var *fendleri– rosal – Fendler rose, wild rose
Roseocactus = Ariocarpus
Roseodendron = Tabebuia
Rosmarinus officinalis – aroma de mar, el viejo, planta compás, romeo, romero, romero común, romero extranjero, rosmarino – rosemary
Rothrockia cordifolia – majoypiualá, talayote
Roystonea borinquena – palmera real de Puerto Rico
Roystonea oleracea – palmera imperial – cabbage palm, Caribbee royal palm
Roystonea (Oreodoxa) regia (elata) – coyol real, cubana, jagua, palmacte, palma de costa, palma de yaguas, palma real, palma real puertorriqueña, palmera de las Antillas, palmera real, palmiche, pumu, yagua – Cuban royal palm, royal palm
Rubia tinctorium – alizari, granza, royuela,

rubia, rubia de tintes, sangre lengua – alizarin, madder
Rubus arctius – zarza nórdica – arctic bramble, nagoonberry
Rubus caesius – zarza de los rastrojos, zarza terreña – dewberry
Rubus chamaemorus – frambueso amarillo – averin, cloudberry, knotberry
Rubus cockburnianus – frambueso encalado
Rubus coriifolius – citún, zarza
Rubus discolor – zarza – Himalayan blackberry
Rubus fruticosus – arto, rubo, zarza, zarza común, zarzaneda – blackberry bush, bramble
Rubus glaucus – moral de Castilla, moral fresa, moral piña – Andes berry
Rubus hispidus – zarza – bristly raspberry
Rubus idaeus – chardonera, churdón, frambueso, frambueso rojo, sangüeso, zarza, zarzaidea – European red raspberry, red raspberry
Rubus laciniatus – zarza de hojas laciniadas – cut-leaved blackberry
Rubus loganobaccus – frambueso americano, zarza de Logan, zarza frambuesa – loganberry
Rubus neomexicanus – frambueso – New Mexican raspberry, thimbleberry
Rubus occidentalis – frambueso negro – black-cap, black raspberry, trimbleberry
Rubus oligospermus – zarza
Rubus palmeri – güismora, huismora, moral, moral de la sierra
Rubus parviflorus – baya de dedal, zarza de flores pequeñas – thimbleberry
Rubus phoenicolasius – frambueso japonés, zarza japonesa – Japanese wineberry
Rubus rosaefolius – frambueso de África – Cape bramble, Mauritius raspberry
Rubus scolocaulon – zarza
Rubus spectabilis – frambueso espléndido – salmonberry
Rubus strigosus – coatlamitl, coatlanxocotl, frambueso, fresón, moral de Castilla, zarza, zarzaidea – American red raspberry
Rubus tridel – frambueso ornamental
Rubus trivialis – zarza silvestre – dewberry, wild blackberry
Rubus ulmifolius – moral de Santa Rosa, zarza, zarza de hojas de olmo – Santa Rosa blackberry
Rubus ursinus (vitifolius) – zarza – California blackberry
Rubus ursinus* var *loganobaccus – baya de Boysen – boysenberry
Rubus villosus – moral, moral negro, zarza – blackberry
Rudbeckia hirta – dormilón, rudbequia – black-eyed Susan, brown-eyed Susan, gloriosa daisy
Rudbeckia laciniata – dormilón, rudbeckia amarilla, rudbeckia laciniada – cutleaf cone-flower
Rudbeckia mexicana – honoba, jovinoba – cone-flower
Rudbeckia = Echinacea
Ruellia californica – flor del campo, rama parda, tronador – wild petunia
Ruellia intermedia – cuete, tronador
Ruellia nudiflora – cuetito – longneck ruel, wild petunia
Ruellia peninsularis – flor de campo, rama parda
Ruellia tuberosa – barreta, fulminante, hierba de chivo, hierba de la calentura, hierba del toro, maravilla del monte, pix-ulum, rama de toro
Rumex acetosa – acedera, acedera pratense, acedera silvestre, lengua de vaca, vinagrera, xoxoyolpapatla – garden sorrel
Rumex acetosella – acedera, acederilla, romaza acederilla, vinagrera, vinagrerita – sheep sorrel
Rumex alpinus – rapóntico, romaza alpina, ruipóntico – alpine dock, monk's rhubarb
Rumex conglomeratus – lengua de vaca, lengua de toro, romaza conglomerada – bull's tongue dock, whorled dock
Rumex crispus – acedera bendita, cañaigre, col agria, eviloriva, hierba colorada, lengua de vaca, pira de berraco, pira de puerco, raíz colorada, romaza amarilla, romaza crespa, vinagrillo – curly dock
Rumex dentatus – romaza dentada, romaza uliginosa – dock
Rumex flexicaulis – lengua de vaca cimarrona
Rumex hymenosepalus – cañaigre, hierba colorada, raíz colorada, raíz del indio –

canaigre, desert rhubarb, red dock, wild-rhubarb
Rumex inconspicuus – cañaigre, hierba colorado – dock
Rumex maritimus – lengua de vaca, romaza de mar – golden dock
Rumex mexicanus – amamastla, axixpatli cóztic, hierba colorada, lengua de vaca, lengua de vaca cimarrona – dock
Rumex obtusifolius – izcua, lengua de vaca, romaza de hojas grandes – bitter dock, mid patience
Rumex patientia – espinaca silvestre, hierba de la paciencia, romaza, romaza común, romaza hortense, romaza paciencia, ruibardo de frailes – herb patience
Rumex pulcher – romaza común – fiddle dock
Rumex salicifolius – romaza de hojas de sauce – Indian rhubarb, water dock, willow dock
Rumex scutatus – acedera romana, badolas, romaza escudada – French sorrel
Rumex stenophyllus – romaza de hojas estrechas – narrow-leaved dock
Rumex violascens – cañaigre, lengua de vaca – Mexican dock
Rumfordia connata – tacote cedroso
Ruppia cirrhosa – broza fina, ruppia cirrosa – Dutch-grass – spiral tasselweed
Ruppia maritima – lama, ruppia marina – Dutch-grass
Ruprechtia pallida – flor de corona, olmo
Ruscus aculeatus – amor oculto, arrayán, arrayán salvaje, brusco, brusco espinescente, galcerán, rusco – butcher's broom
Ruscus hypoglossum – bislingua, laurel alejandrino de hojas estrechas, uvularia – butcher's broom
Russelia equisetiformis – arete, arete de la cocinera, coral de la playa, coralillo, lágrimas de amor, lluvia de coral, lluvia de fuego, rubia – coral fountain, firecracker plant
Russelia retrorsa – cañutillo
Russelia sarmentosa – clavel, sapoyolilla, sonpante, tronador
Russula cyanoxantha – carbonero, chapiado, llora, santiaguero – blue and yellow russula, charcoal burner
Russula delica – hongo de venado, iztacnanácatl, oreja de puerco, quexque, taza blanca, totopixtle, trompa, trompa blanca, trompa de cochino, trompa de puerco, trompeta
Russula nigricans – borrego, trompa negra – black russula
Ruta chalepensis – acuitze varicua, ruda, ruda bracteosa – rue
Ruta graveolens – acuitze varicua, besasa, galega, ruda, ruda alemana, ruda común, ruda de jardín, ruda fétida – herb of grace, rue, wine rue
Ryncholaelia digbyana – monjita

S

Sabadilla officinalis – cebadilla, cebolleja, cunicho, quimichpatli – sabadilla
Sabal bermudana – sabal de Bermuda
Sabal blackburniana (umbraculifera) – sabal de Santo Domingo – Hispaniola palmetto
Sabal etonia – sabal achaparrado, sabal de Florida
Sabal (Inodes) japa – bayal, botán, guano, palma de guano, xan – cabbage palm, guano palm
Sabal mauritiiformis – botán – bayleaf palm
Sabal (Inodes) mexicana (texana) – apachitli, apatz, bayal, guano, guano redondo, palma apachite, palma de micharo, palma de petate, palma real, palma redonda, soyate – Mexican fan palm, Mexican hat fiber, Oaxaca palmetto
Sabal minor – sabal enano – palmetto
Sabal palmetto – sabal de Carolina – cabbage palm
Sabal parviflora – palma cana
Sabal pumos – palma real, pumos
Sabal (Inodes) rosei – palma de llano, viga de llano
Sabal uresana – babiso, coguegue, palma, palma blanca, palma real, palma verde, tachu – Sonoran palmetto
Sabal yapa – botán, cana, guano – bayleaf palm, thatch palm
Sabatia angularis – centaura americana – American centaury, squarestem rose gentian
Sabatia campestris – rosa de prado – meadow pink, rose-gentian, rose-pink, Texas-star
Saccharomyces cerevisiae – levadura, levadura de la cerveza – yeast
Saccharomyces ellipsodieus – levadura del vino – yeast
Saccharomyces minor – levadura del pan – bread yeast
Saccharum barberi – caña criolla – Indian sugarcane
Saccharum officinarum – caña, caña común, caña criolla, caña de azúcar, caña de Castilla, caña dulce, cañaduz, cañamiel, cha'ncat, dusho, gubaguihnashi, maca, pakab, pexa, sana, uhuatl – sugar cane
Saccharum ravennae – caña de ravenna, carricera – ravennagrass
Saccharum sinense – caña de azúcar forrajera, caña de uva, caña japonesa, pasto panamá, pennisetum híbrido – Chinese sugarcane, Japanese cane
Sagina apetala – sagina apétala – dwarf pearlwort
Sagina procumbens – sagina tendida – arctic pearlwort
Sagina saginoides – sagina linneana – alpine pearlwort
Sagina subulata (Arenaria verna) – arenaria de primavera, musgo irlandés, sagina subulada – heath pearlwort, Irish moss, Scotch moss, spring sandwort
Sagittaria latifolia – cola de pato, colomo delgado, sagitaria, sagitaria de hojas anchas – arrowhead, tule potato, wapato
Sagittaria macrophylla – apaclol, apaclilillo, apatlol, cucharilla, hierba de la flecha, papa de agua
Sagittaria sagittifolia – colomo, flecha de agua, hierba de la flecha, hoja flecha, saeta de agua, sagitaria, sagitaria sagitifolia – arrowhead
Saintpaulia ionantha – begonia africana, saintpaulia, santipaula, violeta africana, violeta de África – African violet
Saintpaulia jonantha – violeta africana – African violet
Salacca edulis – salak – salak palm
Salicornia europaea – hierba salada, lechuguina, planta barrillera, salicor europeo, vidrillo – glasswort, pickleweed
Salix alba – jara, salguera, salguero, sauce, sauce blanco, sauce desmochado, sauz – white willow
Salix alba* var *tristis (S. babylonica* var *aurea) – sauce llorón, sauce triste – golden weeping willow
Salix amygdaloides – sauce – peachleaf willow, southwestern peach willow
Salix babylonica – desmayo, llorón, sauce de Babilonia, sauce llorón – weeping willow
Salix bebbiana – sauce, sauz – Bebb willow, gray willow
Salix bonplandiana – aheujote, huata, huatosí, huecojote, huejote, quetzahuéyotl, sauce,

sauce macho, sauce ramagara, sauz – Bonpland willow, toumey willow, Xochimilco willow
Salix caprea – sauce blanco, sauce cabruno – French pussy willow, pink pussy willow
Salix chilensis (humboldtiana) – ahuejote, sauce, sauce blanco, sauce colorado, sauce de Humboldt, sauz
Salix chrysocoma – sauce llorón dorado
Salix cinerea – sauce ceniciento – gray sallow, gray willow
Salix daphnoides – lauréola, sauce dafnoide, sauce falsa-lauréola, sauce purpúreo, sauce violeta – violet willow
Salix exigua – jarita, taray – coyote willow, narrow-leaved willow, sandbar willow
Salix exilifolia – taraiz – silver willow
Salix fragilis – bardaguera, mimbrera, sauce – crack willow
Salix gooddingii – huata, huatasí, sauce, sauz, tewer – Gooding's black willow, Gooding willow
Salix jaliscana – jarilla
Salix laevigata – sauce, sauce rojo, sauz rojo – red willow
Salix lanata – sauce lanudo, sauce lanuginoso – woolly willow
Salix lasiolepis – ahuejote, awará, huecojote, huéxotl, quetzahuéyotl, quetzalhuexotl, sauce, sauz, tepehuexote – arroyo willow
Salix matsudana – sauce chino tortuoso, sauce sacacorchos, sauce tortuoso – hankow willow
Salix nigra – sauce, sauce negro, sauz, sauz serrano – black willow
Salix pentandra – quetzalhuexotl, sauce laurifolio – bay willow
Salix purpurea – mimbre púrpura, mimbrera púrpura, sargatillo, sauce colorado, sauce llorón americano, sauce rojo – Alaska blue willow, purple osier
Salix reticulata – sauce enano reticulado, sauce reticulado – netleaf willow, snow willow
Salix taxifolia – humpil, palo de agua, sauce, taraiz, taray, taray de río – yewleaf willow
Salix viminalis – mimbre, mimbrera, salguero, salguero blanco – common osier
Salpianthus arenarius – azozúcar, granadillo, guayabillo, jueoguo, pie de paloma, zazúcar
Salpianthus macrodontos – catarina, guayabillo, guayabito, quelite de fraile
Salpiglossis sinuata – salpiglosis, trompeta – painted tongue
Salsola soda – barilla, peralejo – soda saltwort
Salsola tragus (kali* var *tenuifolia) – barrilla, barilla de hojas delgadas, chamizo volador, maromero, peralejo, pincho, pinchosa, salicor borde, salicor de la mancha – Russian thistle, tumbleweed
Salsola vermiculata – barilla vermicular, caramillo – saltwort
Salvia aethiopis – etiópide villosa, oropesa – Mediterranean sage
Salvia apiana – salvia blanca, salvia orejona, salvia real – bee sage, California white sage, white sage
Salvia argentea – salvia blanca peluda – silver sage
Salvia arizonica – marrubio, salvia – Arizona sage
Salvia azurea – moradilla, salvia del monte – blue sage, pitcher sage, prairie sage
Salvia ballotaeflora – crespa, engordacabra, mejorana, mejorana del país, Santa Isabel – shrubby blue sage
Salvia brandegei – salvia – Brandegee's sage
Salvia californica – chica, salvia, ucasan – sage
Salvia chamaedryoides – mirto – germander sage, salvia
Salvia chian – chian
Salvia clevelandii – salvia – sage
Salvia coccinea – cardenal, chactzitz, mejorana, mirto, salvia cardenal, salvia roja, tsabtsits, tscixiu – cardinal sage, scarlet sage, tropical sage
Salvia columbariae – chía, salvia – chia
Salvia darcyi (schaffneri) – mirto grande – Darcy sage
Salvia divinorum – hembra, hierba de la pastora, hierba de la Virgen, hoja de la pastora, pipiltzintzintli
Salvia elegans – salvia roja – pineapple sage
Salvia eremostachya – salvia – desert sage, Santa Rosa sage
Salvia farinacea – salvia farinácea – blue sage, mealycup sage, mealy sage, Texas violet
Salvia fulgens (cardinalis) – cardenal,

huitzitzixóchitl, mirto, salvia cardinal, salvia roja – cardinal sage
Salvia galinsogifolia – cadenilla
Salvia gesneraeflora – mirto, salvia roja – cardinal salvia
Salvia goldmanii – salvia del monte
Salvia hispanica – chía, chía commercial, chía del campo, chía del monte, chian, chianpitzáhuac, chiantzozolli, romerillo, salvia, tepechian – chia
Salvia leucantha – hierba de la playa, pluma de Santa Teresa, salvia real, Santa María – Mexican bush sage, velvet sage
Salvia mellifera – estafiate – black sage
Salvia mexicana – flor de chuparrosa, mirto, tacote, tapachichi – Mexican sage
Salvia microphylla (grahamii) – bandera mexicana, mirto, pabellón mexicano, salvia del monte, toronjil – red bush salvia
Salvia miltiorrhiza – salvia roja asiática – Asian red sage
Salvia munzii – salvia – Munz's sage
Salvia nemorosa – salvia del bosque – wild sage
Salvia officinalis – salvia, salvia común, salvia de jardín, salvia de Moncayo, salvia de la pradera, salvia real, salvia virgen, selima, selima fina – common sage, culinary sage, garden sage
Salvia pachyphylla – salvia real de la sierra, salvia rosa – rose sage
Salvia patens – almoraduj, flor de gallito, gallito, mejorana del país, pájaro azul, quiquiriquí – gentian sage
Salvia platycheila – salvia – sage
Salvia polystachys – chía del campo, hisopo, romerillo, salvia de Guadalajara
Salvia pratensis – salvia del prado – meadow clary
Salvia retroflexa – chan, chía
Salvia sclarea – amaro, bácara, esclarea, maro, ormino, orvalle, salvia romana, tárrago – clary sage
Salvia sessei – ocotillo, sangre de toro, San Miguel – shrub salvia
Salvia setosa – hermano de sciatica
Salvia similis – salvia – sage
Salvia sonomensis – ramón – creeping sage, Sonoma sage
Salvia splendens – banderilla, güirirí, salvia roja – cardinal sage, scarlet sage
Salvia* x *superba – bellida, salvia soberbia – superb sage
Salvia sylvestris – salvia del bosque – wild sage
Salvia tiliaefolia – chía cimarrona, chiantzozolli, chulisi, quiebra plato
Salvia verbenacea – gallocresta, maro negro, verbenaca – wild clary
Salvia verticillata – salvia verticilada – whorled clary
Salvia viridis – gallocresta, ormino, ormino silvestre – annual clary, painted sage
Sambucus canadensis – saúco, saúco dulce – American elderberry
Sambucus glauca – saúco – blueberry elder
Sambucus mexicana (caerulea, glauca) – azumatl, bi-xhumí, capiro, capulín, capulín silvestre, coyapa, cumdemba, nttzirza, ocoquihui, saúco, saúco azul, saugua, tapiro, xiiksh, xómet, xúmetl, yutnucate – blue elderberry, desert elderberry, Mexican elder
Sambucus nigra – canillero, sabuco, saúco, saúco negro – black elderberry, European elder
Sambucus racemosa (callicarpa) – saúco, saúco de racimos, saúco rojo – red elderberry
Samolus parviflorus (ebracteatus) – pamplina, peonía – brookweed, water-pimpernel
Samuela = Yucca
Sanguinaria canadensis – orcaneta, litospermo, raíz de sangre, sanguinaria, sanguinaria del Canadá, tibinagua – bloodroot
Sanguisorba minor – hierba del cuchillo, pimpinella menor – garden burnet, salad burnet
Sanguisorba officinalis – pimpinela mayor – garden burnet, great burnet
Sanicula europaea – hierba de San Lorenzo, sanícula, sanícula europea – wood sanicle
Sanicula marilandica – sanícula – American sanicle, black snakeroot
Sansevieria guineensis – curarina, lengua de vaca, mapanare, sanseveria – bowstring hemp
Sansevieria roxburghiana – curarina
Sansevieria trifasciata (zeylanica) – cáñamo sansevieria de Ceilán, corbata, espada, espada japonesa, hoja de cebra, hoja pinta, lengua de

suegra, maguey pinto, mapanare, oreja de burro, sanseviera, sanseviera marginada, zeylanica – bowstring hemp, mother-in-law's tongue, snake plant
Santalum album – sándalo, sándalo blanco – sandalwood
Santolina chamaecyparisus – abrótano hembra, cipresillo, esparrillera, guardarropa, hierba piojera, santolina – lavender-cotton
Santolina rosmarinifolia – botonera, manzanilla borde, mataperros, tomillo perruno – rosemary-leaved lavender-cotton
Sanvitalia abertii – hierba fría – sanvitalia
Sanvitalia procumbens – ojo de gallo, ojo de gato, ojo de loro, ojo de negro, ojo de perico, ojo de pollo, sanguinaria, vaquita – creeping zinnia
Sapindus drummondii – amole, amole de bolita, amolillo, cirioni, jaboncillo, palo blanco, tehuistli – western soapberry
Sapindus saponaria – amole, amole de bolita, amolillo, árbol de las bolitas, árbol de las cuentas, arbolillo, bibí, boliche, cherioni, chirrión, cholulo, coyul, devanador, gualulo, guayul, hualule, huayul, iyamole, jaboncillo, jabonera, jutuhuí, matamuchachos, ojo de loro, pacón, palo blanco, palo de cuentas, palo de voladillo, parapara, pipal, pipe, saponaria, tehuitzli, tehuixtle, tupchi, tzatzupú, yamole, zubul – soapberry, western soapberry
Sapium appendiculatum – hierba de la flecha, hinchador, mincapatli, palo de la flecha, palo hinchador, tronador
Sapium biloculare – hierba de la flecha, hierba mala, mago, palo de la flecha – Mexican jumping bean
Sapium pedicellatum – hierba de la flecha, higuerilla brava, mataise
Sapium sebiferum – árbol del sebo – candle tree, Chinese tallow tree
Sapium stylare – chilamate, lechero, palo de leche, yos
Saponaria americana – amolli
Saponaria ocymoides – jabonera de roca, jabonera rocosa, saponaria albahaca, saponaria falsa-albahaca – red soapwort
Saponaria officinalis – clavellina, hierba de bataneros, hierba jabonera, jabonera, jabonera común, lanaria, saponaria, saponaria común, saponella – bouncing bet, soapwort
Saponaria pumilio – jabonera enana – dwarfish soapwort
Saracha jaltomata – ixpulul, jaltomate, pasaji, tomatillo, xaltómatl
Saracha (Jaltomata) procumbens – cachimbo, chicaltón, cinco negritos, jaltomate, pichecua, tonchichi, tulusisi
Sarcobatus vermiculatus – chico – black greasewood, greasewood
Sarcodon imbricatus – diente de venado – tiled hydnum
Sarcostemma crispum – hierba del indio – climbing milkweed
Sarcostemma cynanchoides – guichori, güirote, hierba lechosa, huichori, huirote – climbing milkweed, milkvine
Sargassum vulgare – sargazo, uva de mar – sea-of-Cortez kelp, seaweed
Sargentia greggii – chapote amarillo, limoncillo, naranjillo
Sarothamnus scoparius – escoba – broom
Sassafras albidum – árbol del sasafrás, sasafrás, sasafrás americano – sassafras
Sassafridium macrophyllum – bebelama, laurel, laurel blanco, laurelillo, laurelón
Satureja chandleri – poleo – San Miguel savory
Satureja douglasii (Micromeria chamissonis) – hierbabuena – California wilding, savory, yerba buena
Satureja hortensis – ajedrea, ajedrea de huerta, ajedrea de jardín, hierba del amor, saborija, satureja, tomillo real, tomillo salsero – summer savory
Satureja macrostema – cuencuentzpatli, hierba del borracho, isioche, menta, tabaquillo, tabaquillo grande, té del monte, té nurite, toronjil, tuché – savory
Satureja montana – ajedrea montana, hisopillo – winter savory
Saussurea lappa – costo – costus root
Saxifraga cespitosa – saxífraga cespitosa – tufted saxifrage
Saxifraga crassifolia – oreja de oso – saxifrage
Saxifraga decipiens – saxífraga rosácea – Irish saxifrage

Saxifraga longifolia – corona de rey
Saxifraga rivularis – saxífraga de los arroyos – brook saxifrage
Saxifraga stolonifera (sarmentosa) – saxífraga, saxífraga tricolor – strawberry geranium
Saxifraga umbrosa – orgullo de Londres, saxífraga de bosque – London pride
Scabiosa atropurpurea – ambar, ambarillo, ambarina, escabiosa, escabiosa morisca, escobilla, escobilla morisca, esmeralda, viuda – mourning bride, pincushion flower, pincushions, sweet scabious
Scabiosa caucasica – escabiosa – pincushion flower
Scabiosa columbaria – escabiosa menor – pincushion flower
Scabiosa fruticosa – tlahtlacotic – pincushion flower
Scabiosa ochroleuca – escabiosa blanca-amarilleante – yellowish scabious
Scabiosa stellata – botón de soldado, escabiosa estrellada, farolito, viuda – pincushion flower
Scadoxus multiflorus (Haemanthus katherinae) – oreja de asno – blood lily
Scaevola plumierii – bosborón, coralillo – inkberry, mauve clusters
Scandix pectin-veneris – aguja de pasta, peine de Venus – crow needles, Venus' needle
Schaefferia cuneifolia – capulín, panalero, sarompillón – desert yaupon
Schaefferia shrevei – piel de león
Scheelia costaricensis – palma real
Scheelia liebmannii – cocoyol, corozo, coyol, coyol real, mop, palma real – palm
Scheelia preussii – corozo, manaca, palma
Scheelia zonensis – palma real
Scheelia = Attalea
Schefflera (Brassaia) actinophylla – aralia arborícola, árbol paraguas, árbol pulpo, cheflera, pulpo – octopus tree, schefflera, Queensland umbrella tree
Schefflera (Dizygotheca) elegantissima – aralia elegantísima, dicigoteca, falsa aralia, planta araña – false aralia, finger aralia
Scheuchzeria palustris – scheuchzeria palustre – American scheuchzeria
Schinopsis quebracho-colorado (lorentzii) – quebracho, quebracho colorado – quebracho
Schinus molle – aguaribay, árbol de Júdas, árbol del Perú, árbol de pimienta, copalastle, copalquáhuitl, falso pimentero, lentisco del Perú, molle, peloncuáhuitl, Perú, pimentero, pimentero de América, pimentero falso, pimientillo, pirú, pirul, turbinto, tzactumi, tzantuni, xaza – California pepper tree, Peruvian pepper tree
Schinus terebinthifolius – turbinto – Brazilian pepper tree
Schizachyrium cirratum (Andropogon cirratus) – popotillo texano – Texas beard grass
Schizachyrium (Andropogon) hirtiflorum* var *feensi – pajón tallo azul, popotillo hirsuto, popotillo peludo
Schizachyrium sanguineum* var *brevipedicellatum – popotillo hirsuto
Schizachyrium scoparium – popotillo azul, popotillo cañuelo – little bluestem
Schizandra chinensis – esquizandra – schizandra
Schizanthus pinnatus – flor de mariposa, orquídea del pobre – butterfly flower, poor man's orchid
Schizocarpum palmeri – melón de coyote
Schizostylis coccinea – azucena de Kafiristán – crimson flag, Kafir lily
Schkuhria virgata – anisillo, anisillo acuatil, anisillo cimarrón, escobilla, hierba del tifo
Schlumbergera (Zygocactus) buckleyi (bridgesii) – cacto de Navidad – Christmas cactus
Schlumbergera truncata – cacto eslumbergera, nopalillo – crab cactus, thanksgiving cactus
Schmaltzia = Rhus
Schoenocaulon officinale – cebadilla, cunicho, etzemó, gusanillo, itzcuimpatli, quimichpatli – green-lily
Schoenocaulon pringlei – cebadilla, gusanillo, hierba del burro, quimichpatli
Schoenoplectus = Scirpus
Schoenus nigricans – juncia bastarda, juncia morisca, junco negro, junco nigricante – black sedge, bog-rush
Schoepfia californica – higuajil – graytwig
Schoepfia schreberi (parviflora) – palo cachorro, palo de hamaca, palo fierro,

pimentillo, tecolotillo, tecolotito, ximitzcuahuitl – white beefwood
Schoepfia shreveana – júchica, juia tóluco, palo verde
Schomburgkia = Laelia
Sciadopitys verticillata – koyamaki, pino parasol, pino sombrilla, pino sombrilla del Japón – umbrella pine
Scilla biflora – escila biflora – alpine squill
Scilla italica – escila de los bosques, escila itálica – Italian squill
Scilla peruviana – farolillo rojo de la floración, flor de corona, flor de la piña – Peruvian scilla
Scilla sibirica – escila primaveral, escila siberiana, escila sibérica – Siberian squill
Scilla = Hyacinthoides
Scirpus acutus – tule – tule
Scirpus americanus (olneyi) – junco espinoso, junco pungente, tule, tule esquinado – threesquare, tule
Scirpus californicus – tule, tule rollizo – California bulrush, tule
Scirpus cernuus – junco enano, junco incurvado, tule – drooping club-rush, fiber optics plant, tule
Scirpus holoschoenus – junco común europeo
Scirpus lacustris – castañuela, tule, tule bofo, tule grande
Scirpus maritimus – juncia marina, junco del mar – maritime clug-rush, sea club-rush
Scirpus mucronatus – junco mucronado – club-rush
Scirpus paradoxus – tusillo – bulrush
Scirpus pungens – junco espinoso, junco pungente – common threesquare
Scirpus robustus – tule
Scirpus setaceous – junco setáceo – bristly club-rush
Scirpus tabernaemontani (validus) (Schoenoplectus tabernaemontanus) – castañuela, halal, junco glauco, junco rugoso – bulrush, rough club-rush
Scleranthus annuus – escleranto annual – annual knawel
Scleria bracteata – navajuela, zacate navajuela – cutting-grass, saw-grass
Sclerocactus (Pediocactus) papyracanthus – toumeya – grama grass cactus, paperspine cactus
Sclerocactus whipplei – biznaguita – Whipple's claw cactus
Sclerocarpus spathulatus – gachupín, mirasol
Sclerochloa dura – esclerochloa dura – hardgrass
Scleropogon brevifolius – cola de zorra, zacate de burro – burro grass
Scolopendrium officinale – cervina, escolopendro, lengua de ciervo
Scolymus hispanicus – cardillo, cardillo bravillo, cardillo hispánico, cardo, cardón lechar, salsifí español – golden thistle, Spanish oyster plant
Scoparia dulcis – anisillo, culantrillo, escobilla, escobilla amarga, hierba del golpe
Scorzonera hispanica – churrimana, escorzonera, escorzonera hispánica, salsifí de España, salsifí negro – Spanish salsify, viper's grass
Scrophularia auriculata – escrofularia acuática, escrofularia auriculada, hierba del moro – water betony, water figwort
Scrophularia canina – escrofularia canina, hierba de Troya, ruda camera, ruda canina – French figwort
Scrophularia nodosa – escrofularia, escrofularia de los bosques, hierba de lamparones, hierba de San Pedro – figwort
Scutellaria alpina – tercianaria alpina – alpine skullcap
Scutellaria baicalensis – escutelaria baical – skullcap
Scutellaria galericulata – casida, hierba de la celada, tercianaria, tercianaria de los pantanos – marsh skullcap
Scutellaria gaumeri – agrimonia del monte, agrimonia silvestre, albacar, claudiosa amarga, claudiosa del monte
Scutellaria lateriflora – casquete, escutelaria, tercianaria – blue skullcap, mad-dog skullcap
Seaforthia = Archontophoenix
Sebastiania bilocularis – hierba de la flecha
Sebastiania cornuta – hierba de la flecha, jagüéy
Sebastiania pavoniana – brincadora, hierba de la flecha, jagüé, mincapatli, palo de la flecha, palo de leche, palo lechero, semilla

brincadora, tronador, túbucti – Mexican jumping bean
Sebastiania pringlei – brincadora
Secale cereale – centeno – rye
Sechium edule – apupo, camochayote, chayote, chayotera, chayotli, chinchayote, chocho, chouchoute, chuchu, cueza, guiquil, guisayote, huisquil, ichintla, mirlitón, pataste, pepino blanco, pipinella, pupa, quilxpachkuum, talote, tayota, uisquil, xuxu – chayote, chayote squash
Sedum acre – chisme, jaspalache, pampajarito, pan de cuco común, pimientra de muro, siempreviva menor, xalache – goldmoss sedum, stonecrop
Sedum album (brevifolium) – sedo, siempreviva menor, uva cana, uva de gato, vermicularia, vermicularia de Europa – white stonecrop
Sedum allantoides – dedito, dedos de Dios
Sedum anglicum – vermicularia inglesa – English stonecrop
Sedum brevifolium – vermicularia de hojas cortas – short-leaved stonecrop
Sedum caeruleum – vermicularia azul
Sedum chihuahuense – siempreviva
Sedum dasyphyllum – vermicularia de hojas espesas – thick-leaved stonecrop
Sedum dendroideum – sedo, siempreviva, siempreviva amarilla, tetzmitl, tetzmitic, texiote, texiotl, uñas, vermicularia dendroide – orpine, stonecrop
Sedum greggii – siempreviva
Sedum kamtschaticum – sedo de Kamtschatka – stonecrop
Sedum mexicanum – chisme, cola de borrego, cordoncillo – orpine, stonecrop
Sedum moranense – chisme, cordoncillo, espinaca, jaspalache, xalache
Sedum morganianum – cola de borrego – burro tail, donkey tail, sheep's tail
Sedum oxypetalum – siempreviva
Sedum praealtum – siempreviva, siempreviva amarilla, texiote – stonecrop
Sedum reflexum – vermicularia rupestre – reflexed stonecrop
Sedum roseum – itzmetl, itztamatl, sedo, siempreviva – king's crown, live-forever, stonecrop
Sedum sediforme – uña de gato, uva de pájaro – high stonecrop
Sedum spathulifolium – pan de cuco, siempreviva, siempreviva menor – stonecrop
Sedum spectabile – pan de cuco, telefio – stonecrop
Sedum spurium – telefio bastardo, vermicularia caucásica – Caucasian stonecrop
Sedum telephium – hierba callera – livelong, orpine, sedum
Selaginella eremophila – flor de piedra – desert spike-moss
Selaginella lepidophylla – doradilla, flor de peña, flor de piedra, helecho, inmortal, lama, muchcoc, siempreviva – resurrection plant
Selaginella novoleonensis – flor de piedra, teta
Selaginella pallescens – doradilla, flor de peña, tetacegua – resurrection plant
Selaginella pilifera – siempreviva – resurrection plant
Selaginella pringlei – flor de piedra
Selaginella rupestris – doradilla, flor de peña, flor de piedra, pata de rana, siempreviva
Selaginella rupincola – siempreviva
Selenicereus (Nyctocereus) chontalensis – lamish-que
Selenicereus donkelaari – choh-kan, pitahaya, pitaya, tzakamak, zacbacelcan
Selenicereus grandiflorus – gigante, organillo, reina de la noche, reina de las flores – queen-of-the-night
Selenicereus hamatus – organillo, reina de la noche – night-blooming cereus
Selenicereus pteranthus – pitaya real, princesa de la noche, reina de la noche
Selenicereus spinulosus – pitaya
Selloa (Gymnosperma) glutinosa – casal, cola de zorra, escobilla, hierba de tatalencho, hierba pegajosa, jarilla, jécota, jucee ndede, mariquita, mota, motita, pegajosa, tatalencho, tzitziu-ton, vara del burro, xonequílitl, xonequitl, yucundede, zazale
Sempervivum arachnoideum – siempreviva aracnoidea, siempreviva de telarañas – cobweb houseleek
Sempervivum tectorum – alcachofa de gato, consolva, hierba puntera, siempreviva común, siempreviva mayor, siempreviva mayor común – hen and chicks, houseleek

Senecio barba-johannis – barba de San Juan de Dios – groundsel
Senecio candidissimus – chucaca, hierba de fuego, lechuguilla – groundsel, ragwort
Senecio canicida – clarincillo, hierba de la puebla, hierba del perro, itzcuinpatli
Senecio cardiophyllus – cabeza de aura, cachana, caxane, peyote, peyotl xochimilcensis – groundsel
Senecio chenopodioides – flor de carnaval, flor de niño, llama trepadora – Mexican flame vine
Senecio cineraria – cineraria, senecio cinerario – cineraria, dusty miller
Senecio confusus – bejuco de sol, llama trepadora – Mexican flame vine
Senecio elegans – senecio elegante – purple ragwort
Senecio flaccidus – hierba cana, hierba de San Juan de Dios – groundsel, ragwort
Senecio (Brachyglottis) greyi – senecio común
Senecio hartwegii – hoja semita, peyote, peyote cimarrón, sopépari
Senecio* x *hybridus (cruentas) – cineraria – florist's cineraria
Senecio integrifolius – coniza, hierba pulguera, senecio de hojas integérrimas – fleawort
Senecio jacobaea – azuzón, hierba amargosa, hierba de Santiago, sacapelos, zuzón – ragweed, tansy ragwort
Senecio longilobus – gordolobo, hierba cana – threadleaf groundsel
Senecio multicapitus – hierba del caballo – groundsel, ragwort
Senecio praecox – candelero, hierba cana, palo bobo, palo bofo, palo loco, quantlapatziinzintli, texcapatli
Senecio salignus – achayotl, azumiatl, chilca, flor de dolores, iztacatzóyatl, jara, jara brava, jaral amarillo, jaralillo, jarilla – willow groundsel
Senecio sanguisorbae – rabanillo
Senecio sylvaticus – senecio de los bosques – wood groundsel
Senecio toluccanus – rabanillo
Senecio vulgaris – hierba cana, hierba de la quemadura, lechocino, senecio, senecio común, zuzón – common butterweed, common groundsel, common ragwort
Senecio vulneraria (bellidifolius) – calamápatl, calancapatle, calomopotl, gordolobo, huboquillo, iztacpalancapatli, lechuguilla, nanahuapatli, palancapatli – groundsel
Senecio = Delairea
Senna (Cassia) alata – árbol vela, candelero, capulco, ecapatli, flor del secreto, guajava, hierba de la fiebre, sen silvestre, tarantana, totoncaxíhuitl – candelabra bush, candle bush, candle-tree, emperor's cup, golden chain tree, ringworm cassia, ringworm senna
Senna (Cassia) angustifolia – casia, sen, té de sena – Indian senna
Senna atomaria (Cassia emarginata) – alcaparra, alcaparroso, barba de jolote, cañafistula, chile perro, flor de San José, hediondilla, jupachumi, juracosi, mora hedionda, palo de maya, palo hediondo, palo zorrillo, retama, vainilla, vara de San José, viche, xtuhabin, zorrillo
Senna (Cassia) bahuinoides – daisillo – two-leaved senna
Senna bicapsularis (Cassia emarginata) – alcaparrillo, biche, bicho, bricho, cachimbo, flor de San José, hediondilla, julahuasi, mora hedionda, palo hediondo, palo zorrillo, retama, vainilla, vara de San José – senna
Senna (Cassia) confinis – hojasén – cassia
Senna (Cassia) covesii – dais, daisillo, hojasén, rosa María – coues' cassia, desert senna
Senna italica (Cassia obovata) – hojasén – shower tree
Senna (Cassia) laevigata – bricho, cacaotl, cafecillo, café del país, cañafistula, chatalhuic, duerme de noche, guajillo, hierba del aire, retama, retama del país, sen del país, sen silvestre, yehcapaztzin – wild senna
Senna (Cassia) leptocarpa – biche, biche manso, hierba del piojo, nahuapoztli – senna
Senna multiglandulosa (Cassia tomentosa) – retama, retama de tierra caliente, retamalo – woolly senna
Senna (Cassia) occidentalis – biche prieto, bricho, ecapatl, ecapatli, ejotillo grande, frijolillo, guanina, habilla, habilla prieta,

hediondilla, hierba del aire, huashihua, mezquitillo, palo zorrillo, pico de pájaro, vainilla – coffee senna

Senna (Cassia) oxyphylla – candelilla, cañafistula, cuica, cuica de gato, frijol del monte, xihuihui joyo – four-leaf senna

Senna pallida (Cassia biflora) – abejón, ahumada, biche, biche silvestre, bicheviro, bicho, ejotillo, ejotillo del monte, flor de iguana, flor de San José, garbancillo, malva, ojotón, ronrón, rosa María – twin senna

Senna purpusii – ejotillo

Senna (Cassia) siamea – casia de Siam – Siamese senna

Senna (Cassia) spectabilis – canchín, candelilla, mucuteno, palo todos santos

Senna (Cassia) tora – biche manso, dormidera, ejotillo – senna

Senna (Cassia) uniflora – ejotillo chico

Senna (Cassia) wislizeni – hierba de Pinacate, palo prieto, Pinacate, pinacatillo – shrubby senna

Sequoia sempervirens – palo colorado, secoya, secoya de California, secoya roja – coast redwood, redwood

Sequoiadendron giganteum – secoya gigante, secoya wellingtoniana, wellingtonia – big tree, giant sequoia

Serenoa repens – palmera enana, palmito de juncia – saw palmetto

Serjania mexicana – barbasco, cuamecate, cuamecatl, culebra, diente de culebra, diente de víbora, güirote, güirote de culebra, ichuanitzo, quaumecatl, sierrilla

Serjania palmeri – camayia, güirote grueso, palo cachona

Serjania racemosa – bejuco costillón, contra ranilla de bestias, cuamecate, hierba del golpe, nueve hojas, temacate

Serratula tinctoria – costo, serratula de los tintoreros – sawwort

Sesamun indicum (orientale) – ajonjolí, ajonjolín, cuahuaxtli, mehensial, sésamo, zicilpuuz – sesame

Sesban = Sesbania

Sesbania grandiflora – agati sesbania, gallito, pico de flamenco

Sesbania herbacea – baiquillo, bequilla – coffee weed, Colorado River hemp

Sesbania (Sesban) macrocarpa – bequilla, curi, zacate de agua – Colorado River hemp

Sesuvium portulacastrum – cenicienta, cenicilla, chamis, tsaycan, verdolaga de la playa, vidrillo – sea-purslane

Sesuvium verrucosum – cenicienta, cenicilla – western sea-purslane

Setaria anceps – cola de perro, mijo silvestre, rabo de cachorro, setaria, setaria kazangula, setaria narok, zacate setaria – nandi grass

Setaria gracilis (geniculata) – gusanillo, gusano, mijillo, pajita, pajita cerdosa, remolín, setaria perenne, zacate amargo, zacate cedroso, zacate peludo – bristlegrass

Setaria grisebachii – cola de ardilla

Setaria italica – mijo, mijo de Italia, mijo de pájaros, mijo italiano, mijo menor, mijo pajarero, panizo de Italia, setaria italiana – foxtail millet

Setaria italica* var *moharica – moha, setaria húngara – European foxtail millet

Setaria leucopila – zacate temprano – white-haired bristlegrass

Setaria liebmannii – cola de ardilla, cola de zorra, lecualquimichi, ratoncillo

Setaria macrostachya – pajita tempranera, zacate tempranero – plains bristlegrass

Setaria pumila (lutescens) – almorejo, almorejo glauco, cola de gato, pajita amarilla, setaria glauca – cattail grass, yellow bristlegrass

Setaria scheelei – pajita abierta

Setaria sphacelata – fleo dorado, kazungula, nandi, narok, pasto nandi, zacate nandi – golden timothy

Setaria verticillata – almorejo verticilado, lagartera, setaria verticilada, zacate pegajoso – bristly foxtail

Setaria viridis – almorejo, almorejo verde, setaria verde – green foxtail

Setcreasea = Tradescantia

Sherardia arvensis – sherardia de los prados – field madder

Sibbaldia procumbens – sibaldia procumbente – Sibbald's potentilla

Sicana odorifera – calabaza melona, casabanana, cohombro de olor, melocotón

Sicyos angulatus – acarino, chayotillo – mapleleaf peamelon

Sicyos laciniatus – calabacilla, chayotillo – Arizona peamelon, one-seeded bur-cucumber
Sicyos parviflorus – calabacilla, chayotillo
Sida acuta – ajalache, calalagua, chichibe, chichipe, escoba, escoba fuerte, escobilla, malva, malva amarilla, malva colorada, malva de Castilla, malva de cochino, malva de platanillo, malvavisco – broomweed
Sida aggregata – malva blanca, rama lisita
Sida angustifolia – malva
Sida cordifolia – malva, zacmizbil
Sida (Disella, Malva) hederacea – meloncilla – alkali mallow
Sida procumbens – axocatzín, hauyxiu, hierba de la viejita, hierba del buen día, malva
Sida rhombifolia – axocatzín, chichipe, escobilla, güinare, huinare, malva, malva colorada, malvavisco, malvavisco escobilla, oreja de burro, popotalagua, sida – broom jute
Sida tragiifolia – mal de ojo
Sidalcea malvaeflora (malviflora) – malva, malva griega, tuchi – checker, checkerbloom, checker mallow
Sidalcea neomexicana – tuchi – checker, checkerbloom, prairie mallow
Sideroxylon angustifolium – tempisque, tempixtle
Sideroxylon capiri – capiri, huacux, tempisque, totozapotl, zapote de ave
Sideroxylon (Bumelia) laetevirens – abalo, bebelama, choboc, coma, cupia, huizilacate, tempisque, tempixquiztle, tempixtle, tilapo, tilzápotl
Sideroxylon (Bumelia) lanuginosum – bebelama, coma, mulcé, tempixtle – chittam wood, chittimwood, gum bumelia, shittimwood
Sideroxylon occidentale (Bumelia occidentalis) – bebelama, júchica
Sideroxylon (Bumelia) peninsularis – bebelama
Sideroxylon persimile – bebelama, cupia
Sideroxylon (Bumelia) rigida – huajuco
Sideroxylon (Bumelia) spiniflora – bagre, coma, coma resinera, jos – saffron plum
Sideroxylon tepicense – cozahuico, cuamirro, tempisque, tempixtle
Sigesbeckia jorullensis – flor de araña
Silene acaulis – cruz de Malta musgosa, silene acaule, silene de rocalla – cushion pink, moss campion, moss carchfly
Silene alpestris – silene alpestre – campion
Silene armeria – juliana falsa, silene armeria – sweet-William catchfly
Silene californica – hierba del indio – campion, catchfly, California Indian pink
Silene coeli-rosa – silene violácea – viscaria
Silene conoidea – silene conoidea – catchfly
Silene dichotoma – silene dicotomal – forked catchfly
Silene dioica – borbonesa, jabonera blanca – hare's eye, red campion
Silene gallica – silene francesa – small-flowered catchfly
Silene inflata – colleja – bladder campion
Silene laciniata – clavel del monte, flor del indio, mamota, metatera – catchfly, fire pink, Mexican campion, Mexican pink, Mexican silene
Silene latifolia – silene de hojas anchas – broad-leaved campion
Silene noctiflora – silene de flores nocturnas – night-flowering campion
Silene vulgaris (maritima, uniflora) – cruz de Malta marina, silene venosa – bladder campion, catchfly
Silphium perfoliatum – brújula, hierba de piloto – ragged cup
Silybum marianum – arzolla, cardo bendito, cardo de leche, cardo de María, cardo lechar, cardo lechero, cardo mariano, cardoncillo, cardo santo, lechero – holy thistle, milk thistle
Simaruba glauca – aceituno, negrito, pajulilté, pasaque, zapatero – paradise tree
Simaruba officinalis – simaruba
Simmondsia chinensis (californica) – jojoba, pnaocl – coffee bush, deer-nut, goat-nut, jojoba, pig-nut, quinine plant, wild hazel
Simsia amplexicaulis – acahual – bush sunflower
Sinapis alba (Brassica hirta) – jebana, mostaza, mostaza amarilla, mostaza blanca, mostaza inglesa – white mustard
Sinapis arvensis (Brassica kaber) – collejón, mostaza arvense, mostaza de los campos, mostaza silvestre – charlock, wild mustard

Sinningia (Gloxinia) speciosa – gloxínea – gloxinia
Sisymbrium altissimum – jaramago más alto – Jim Hill mustard, tall rocket, tumble mustard
Sisymbrium auriculatum – huachelai, mostaza – tumble mustard
Sisymbrium canescens – pamita
Sisymbrium erysimoides – jaramago falso-erisimo – treacle-mustard-like rocket
Sisymbrium irio – matacandiles, oruga leonia, pamita, pamitón, rabanillo amarillo – London rocket
Sisymbrium loeselii – jaramago de Loesel – false London rocket
Sisymbrium officinale – erismo, hierba del cantor, hierba de San Alberto, jaramago, mostacilla, sisimbrio – hedge mustard
Sisymbrium orientale – jaramago oriental – eastern rocket
Sisymbrium pinnatum – pamita – tansy mustard
Sisyrinchium angustifolium – moradilla
Sisyrinchium arizonicum – hierba de la muela
Sisyrinchium bellum – azulea, raíz de los cerdos, villela, zacate de muela, zacate de mula – blue-eyed-grass
Sitanion = Elymus
Sium latifolium – berrera, chirivía de hojas anchas – water parsnip
Sium sisarum – arsáfraga, chirivía de azúcar, escaravía – skirret
Sium suave – chirivía de agua – water-parsnip
Sium = Berula
Smilax aspera – uva de perro, zarza morisca, zarzaparrilla, zarzaparrilla de Europa, zarzaparrilla de país – sarsaparilla, thorny greenbriar
Smilax bona-nox – cabrastillo, cocolmeca, cozulmécatl, mecapatli, zarzaparilla – greenbriar
Smilax china – china, lampatán, zarzaparrilla china, zarzaparrilla de China – Chinaroot
Smilax medica (aristolochiaefolia) – mecapatli, nanahuapatli, purga de las ánimas, quauhmecapatli, tequammaytl, zarza, zarzaparrilla, zarzaparrilla de América, zarzaparrilla de la sierra, zarzaparrilla de Tulancingo – greenbriar, sarsaparilla, sasparilla
Smilax mexicana – bejuco de chiquihuite, bejuco de diente de perro, cocolmeca, cocomecan, cozolmecatl, diente de perro, lasha, mecapatli, xcocehak, zarza, zarzaparrilla – greenbriar
Smilax mollis – bejuco de chiquihuite, bejuco de diente de perro, pate – greenbriar
Smilax officinalis – cuculmeca, zarzaparrilla, zarzaparrilla carmelita – greenbriar, sarsaparilla, sasparilla
Smyrnium olustratum – apio caballar, cañareja, esmirnio común – alexanders, black lovage
Sobralia macrantha – babucha, lirio de San Antonio, tlazolteo-zácatl, zapatito
Solandra guerrerensis – hueipatl, kieli, tecomaxóchitl – chalicevine
Solandra maxima (guttata) – copa de oro, cútacua, floripondio del monte, floripondio del norte, perilla, tecomasúchil, tecomaxóchitl – cup-of-gold vine, trumpet plant
Solandra (Swartzia) nitida – bolsa de Judas, bonete, copa de oro, cútacua, floripondio del monte, gorro de Napoleón, hueypatli, hueytlaca, itzucuatziqui, mendieta, tecomasúchil, tecomaxóchitl, tetona, xochitecómatl – cup of gold, milkcup chalice vine
Solanum amazonium – berenjena silvestre, mala mujer, pacanul, palohuisi, pusera, sacamanteca, xkomyaxnic
Solanum americanum (nodiflorum) – bocano, cha-yúk, chichiquelite, hierbamora, hierbamora negra, mariola – white nightshade
Solanum bulbocastanum – papa cimarrona, papa coyote – wild potato
Solanum capsicastrum – farolito, tomatillo, tomatillo de Jerusalén – Christmas cherry, winter cherry
Solanum cardiophyllum* ssp *ehrenbergii – papa, patata – potato
Solanum cervantesii – chichimeca, hierba del perro, hierbamora, veneno de perro
Solanum demissum – papa, papa cimarrona, papita, patata – potato
Solanum diversifolium – berenjena, berenjena silvestre, lava plato, sacamanteca, salvadora
Solanum dulcamara – amargamiel, dulcamara,

dulciamarga, flor de gloria, gloria, guía de jazmincillo, jazmín morado, lazo de amor, matagallinas, morera trepadora, solano dulce, uva del diablo, uva de zorro, vid de Judea – bittersweet, nightshade, potato vine, purple jasmine

Solanum elaeagnifolium (hindsianum) – buena mujer, guajtomate, hierba de golpe, huitztomatzin, mala mujer, mariola, mariolona, ojo de liebre, pera, tomatillo, tomatillo del campo, tomatillo espinoso, tomatito, tomatito de buena mujer, tomatito pelón, trompillo – bull nettle, silverleaf nettle, silverleaf nightshade, trompillo, white horse-nettle

Solanum erianthum – berenjena, buey nini, cornetón del monte, huatauhui, lengua de buey, malabar, mambia, quita manteca, sacamanteca, salvadora – tree nightshade

Solanum gracile – chichiquelite, mambia, maniloche, solano grácil – elegant nightshade

Solanum grayi – churuni, hierba del gato, huevo de gato

Solanum jasminoides – dulcamara perenne, flor de San Diego, gloria, jazmín, jazmín morado – potato vine

Solanum lanceolatum – sosa – lanceleaf nightshade

Solanum macrantherum – gloria

Solanum madrense – berenjena, jenepene, sacamanteca, tlamatlantli

Solanum mammosum – berenjena, berenjena de gallina, berenjena de teta, berenjena peludita, berenjenita peluda, berenjenita peludita, chichigüita, cojón de gato, cuchito, tetilla – apple of Sodom

Solanum marginatum – bola de oro, cabalonga, mala mujer, sosa – white-margined nightshade

Solanum melongena* var *esculentum – berenjena, flor de huevo, nana – aubergine, eggplant

Solanum muricatum – manguena, pepino, pepino de maceta, pepino dulce, pera melón – melon pear, pepino

Solanum nigrescens – chichiquelite, hierbamora, mambia

Solanum nigrum – bahalcan, belladona, bi-taxe, chichiquelite, chuchilita, cuchilita, hierba de hechiceros, hierbamora, ixcapul, mambia, maniloche, mora, pakcan, pitoxe, solano, solano negro, thucupache xaqua, tojonechichi, tomatillo del diablo, tomatito, tomatito del diablo, tonchichi, tucupache xacua, tzopilotlacuatl, veneno de cuervo – black nightshade

Solanum nudum – hierba de barrer, huele de noche negro, lava pañal, sak-kol – Mayan soap, nightshade

Solanum pseudocapsicum – collar de la reina, coral, coral de los jardines, manzanita de amor, solano, solano seudopimienta – Jerusalem cherry

Solanum quitoense – lulo, naranjillo – naranjilla

Solanum rostratum – ayohuiztle, duraznillo, hierba del gato, hierba del sapo, huiztlecuatl, mala mujer, rabo de iguana, solano rostrado, tomatillo espinoso – buffalo berry, buffalo-bur

Solanum sarrachoides – mariola, solano verde – green nightshade

Solanum seaforthianum – bellisima, flor de gloria, gloria, guía de jazmincillo, lágrimas de la Virgen, lazo de amor, piocha – Brazilian nightshade

Solanum sisymbriifolium – jua, ocote mullaca – nightshade

Solanum sodomaeum – tomatillo del diablo – apple of Sodom, Dead Sea fruit, dust and ashes, vine of Sodom

Solanum tequilense – huevo de gato

Solanum topiro – cocona

Solanum tridynamum – mala mujer, ojo de liebre, paropus, poros pusi, sacamanteca

Solanum triflorum – solano de tres flores – three-flower nightshade

Solanum triquetrum – hierbamora – Texas nightshade, white nightshade

Solanum tuberosum – camotillo, papa, papa cimarrona, papa correlona, patata, patatera, pöpa, rerogüe, trumfera – potato

Solanum umbellatum – barba de chivato, berenjena, cazaniche, chivira, venenillo

Solanum verbascifolium – berenjena, berenjena cimarrona, berenjenilla, cornetón del monte, friega plato, galatea, gordolobo, guardalobo, hierba de San Pedro, hikuli, hoja de manteca, lengua de vaca, malabar, palo de chachalaca, palo de hojas de manteca, preadedera hedionda, sacamanteca, salvadora, solano, sosa, tabaco cimarrón, tompaap, trompillo, tunkux, tzom-pach-tzin, xaxox, xuxox – mullein nightshade

Solanum wendlandii – elisa, jazmín italiano, quixtán, solano – Costa Rican nightshade, divorce vine, giant potato creeper, grand potato vine, marriage vine

Solanum wrightii – árbol de papa, azulillo – potato tree

Solanum xanti – trompillo – purple nightshade

Soleirolia soleirolii – helxine de Soleirol – baby's tears

Solidago californica – oreja de liebre – California goldenrod

Solidago canadensis – mariquilla, palma de oro, plumero amarillo, vara de oro, vara de oro del Canadá, vara de San José – Canada goldenrod

Solidago gigantea – vara de oro gigantiflora – smooth goldenrod

Solidago hybrida – vara de oro – goldenrod

Solidago mexicana – calcanpatli, lechuguilla, nanahuapatli, palancapatli

Solidago velutina – calancapatle, escobilla, lechuguilla, nanahuapatli, palancapatli, perú cimarrón – goldenrod

Solidago virgaurea – vara de oro, vara de San José – Aaron's rod, goldenrod

Sonchus arvensis – cerraja arvense, lechuguilla – perennial sow thistle

Sonchus asper – cardimuelle, cerraja, cerraja común, cerraja espinosa, chinita, lechuguilla – prickly sow thistle, spiny sow thistle

Sonchus oleraceus – achicoria, achicoria dulce, cerraja, cerrajón, chichicaquilitl, chicoria, chinita, diente de león, envidia, falso diente de león, lechecino, lechuga silvestre, lechuguilla, lechuguilla silvestre, mitihuaxaqua, muela de caballo, nabukak, tlamatsalin – common sow thistle

Sonchus tenerrimus – cerraja delicada, cerraja de pared – delicate sow thistle

Sophia ochroleuca – pamita

Sophora japonica – acacia del Japón, árbol de la miel, árbol de las pagodas, falsa acacia del Japón, sofora – Chinese scholar tree, Japanese pagoda tree

Sophora secundiflora – colorín, coralillo, frigolito, frijol, frijolillo, frijolito, patol – coral bean, mescal bean, Texas mountain laurel

Sophora tomentosa – salche

Sorbus americana – serbal americano – American mountain ash

Sorbus aria – mojera, mostajo, mostellar, serbal, serbal blanco – whitebeam

Sorbus aucuparia – acerallero, acerolo, azarollo, serbal de los cazadores, serbal de los pájaros, serbal silvestre, serbo – European mountain ash, rowan

Sorbus domestica – acerolo, serbal, serbal común, serbal doméstico, serbo, sorbo – mountain ash, true service tree

Sorbus dumosa – serbal de los cazadores, serbo – Arizona mountain ash

Sorbus hupehensis – serbal chino de Hupeh – Hupeh rowan

Sorbus* x *hybrida – serbal finés – Finnish whitebeam

Sorbus intermedia – mojera sueca, serbal de Suecia, serbal sueco – mountain ash, Swedish whitebeam

Sorbus* x *thuringiaca – serbal híbrido de Turingia

Sorghastrum nutans – zacate indio – Indian grass

Sorghum almum – almum, garaví perenne, pasto colón, sorgo almum, sorgo negro, sorgo negro argentino, sorgo perenne, sudán perenne – black sorghum, Columbus grass

Sorghum bicolor – caña de espiga, carquexia, escoba maicera, hierba del Sudán, kafir, maíz de Guinea, maizmillo, maíz morocho, mijo de escoba, mijo maíz, mijo sorgo, molote, panizo moruno, pajón, panizo negro, pasto del Sudán, popote, sorgo, sorgo blanco, sorgo común, sorgo de escobas, sorgo de Sudán, sorgo forrajero, tasaui, zacate sudan – broomcorn, Guinea corn, milo, sorghum, Sudan grass

Sorghum (Holcus) halepense – akilzuuc, alpiste, cañutillo, grama china, hierba johnson, maicillo, pasto johnson, pasto ruso, sorgo de Alepo, sorgo selvático, sorguillo, tío del maíz, xkuku, zacate agrarista, zacate de Johnson, zacate egipto, zacate johnson, zacate milo, zacate parana, zacate ruso – Johnsongrass

Sorghum saccharatum (Andropogon sorghum) – caña, maíz de China, sorgo azucarado, sorgo sacarífero – sugar millet

Sorghum sudanense – hierba del Sudán, pasto del Sudán, sudán, zacate sudán – Sudangrass

Sorghum vulgare (Holcus sorghum) – caña, escoba, maicena, maíz de Guinea, maizmillo, mijo, pasto maicillo, sorgo, sorgo común, sorgo de grano, sorgo dulce, tasaui, zahina – sorghum

Sorghum vulgare* var *durra – caña, durra – dhura, hyssop, Jerusalem corn

Soya = Glycine

Sparaxis tricolor – arlequina – harlequin flower

Sparganium angustifolium – platanaria de hojas estrechas, platanaria natátil – narrow-leaved bur-reed

Sparganium emersum – platanaria emergente – unbranched bur-reed

Sparganium erectum – platanaria mayor, platanaria ramosa – greater bur-reed

Sparganium eurycarpum – palma roja – bur-reed

Sparganium natans – platanaria enana – small bur-reed

Sparmannia africana – esparmania – African hemp, African linden

Spartina alterniflora – espartillo de flores alternas – salt-water cord grass

Spartina densiflora – espartillo densifloro – dense-flowered cord grass

Spartina patens – espartillo de olor, retama – salt-meadow cord grass

Spartium junceum – cambroño, gayombo, genista, piorno, retama, retama de China, retama de escobas, retama de España, retama de flor, retama delgada, retama de olor, retama macho – Spanish broom

Spathiphyllum blandum – platanillo, tsabal – plantain grove

Spathiphyllum cochlearispathum – chile de gato, espatifilo, flor de gusano, gusano, hierba del manso, olotillo – spathe flower

Spathiphyllum wallisii – cala blanca, cuna de Moisés, espatifilo – spathe flower

Spathodea campanulata – árbol de fuente, árbol del fuego, caoba de Santo Domingo, cedro africano, espatodea, gallito, llama de bosque, llamarada de bosque, tulipán, tulipán africano, tulipán de Africa, tulipero del Gabón – African tulip tree, fire tree, flame of the forest, fountain tree

Specularia (Triodanis) speculum – espejo de Venus – Venus' looking glass

Spergula arvensis – cilantrillo, esparcilla – spurrey, starwort, stickwort

Spergularia bocconii – spergularia de boccone – Greek sea spurrey, sand-spurrey

Spergularia marina – rabaniza de los soseros, spergularia marina – salty sand-spurrey

Spergularia media – hierba de la golondrina, spergularia marginada – greater sea spurrey

Spergularia rubra – spergularia arenaria – sand-spurrey

Sphaeralcea ambigua – hierba del negro, mal de ojo, malva, planta muy mala – apricot mallow, desert globemallow, desert hollyhock, desert mallow

Sphaeralcea angustifolia (lobata) – hierba de la negrita, hierba del negrito, hierba del negro, mal de ojo, tlaltzacutli, tlilxihuitl – copper-mallow, narrowleaf globemallow

Sphaeralcea axillaris – mal de ojo, malva rosa

Sphaeralcea coccinea – hierba de la negrita, hierba del negro – scarlet globemallow

Sphaeralcea coulteri – mal de ojo – annual globemallow

Sphaeralcea emoryi – mal de ojo – Emory's globemallow

Sphaeralcea fendleri – hierba de la negrita – Fendler's globemallow

Sphaeralcea fulva – pintapán, quesito

Sphaeralcea incana – hierba del negro – globe mallow

Sphaeralcea (Phymosia) rosea – flor de San José, malvón

Sphaeropteris horrida – cola de chango, rabo de mico

***Sphagnum* spp** – musgo de pantano – peat moss

Sphenopholis obtusata – zacate de cuña – prairie wedgegrass, wedge-scale

Spigelia anthelmia – hierba lombricera, huambia, lombricero, tequitcua
Spigelia longiflora – hierba del burro, sangre de toro
Spilanthes acmella – berro de para, yambo – para cress
Spilanthes americana (oppositifolia) – botoncillo, botón de oro, tripa de gallo
Spinacia oloracea – espinaca – spinach
Spiraea arguta – corona nupcial
Spiraea douglasii – espírea, espírea de Douglas – western spiraea
Spiraea japonica – espírea del Japón, espírea japonesa, rosa japonesa – Japanese spiraea
Spiraea prunifolia – espírea de hojas de ciruelo – bridal-wort, bridal wreath spiraea, shoe button spiraea
Spiraea thunbergii – espírea de Thunberg – spiraea
Spiraea trilobata – espírea trilobulada – spiraea
Spiraea* x *vanhouttei – espírea de Vanhoutte – spiraea
Spiranthes aurantiaca – coyol cimarrón, cutzis, espirantes, orquídea del campo – ladies' tresses
Spiranthes cinnabarina – cutzis – cinnabar ladies' tresses, flame orchid
Spiranthes polyantha – cutzis
Spiranthes romanzoffiana – espirantes irlandesa
Spondias cytherea – ambarella, ciruelo dulce, jobo de la India – golden apple, otaheite apple
Spondias dulcis – jobo de la India – ambarella
Spondias mombin – abal, atoyaxócotl, biaxhi, capuatlcacao, chiabal, chupandilla, ciruelo, ciruelo agrio, ciruelo amarillo, ciruelo cochino, ciruelo colorado, ciruelo de jobo, ciruelo del país, ciruelo de México, ciruelo español, ciruelo mango, ciruelo obo, ciruelo rojo, ciruelo trópico, cítara, cozticxócotl, cupú, jobillo, jobo, jobo amarillo, jobo corronchoso, jobo de monte, jobo de zopilote, jobo espino, jobo roñoso, jocote, kanabal, kinin-jobo, mazaxócotl, mombin, pok, pompoqua, sahualiiotorosim, xocat, zuliabal – hog plum, Spanish plum, tropical plum, yellow mombin
Spondias purpurea – abal, atoyaxócotl, biaxhi, capuatlcacao, chiabal, ciriál, ciruelo, ciruelo agrio, ciruelo amarillo, ciruelo campechano, ciruelo chico, ciruelo cimmarón, ciruelo colorado, ciruelo coyote, ciruelo del país, ciruelo de México, ciruelo de San Juan, ciruelo rojo, cozticxócotl, cundaria, cupú, ixhouen, jobo, jobo de zopilote, jobo rojo, jocote, jocote de corona, jondura, kanabal, kinin-jobo, mazaxócotl, poon, quininhobon, tuxpana, xócotl, yoyoma – jocote, red mombin, Spanish plum
Sporobolus airoides – cresta de gallo, zacate de agua, zacatón, zacatón alcalino – alkali sacaton
Sporobolus buckleyi – zacatón pulguero
Sporobolus contractus – zacate alcalino espigado, zacate compacto – spike dropseed
Sporobolus cryptandrus – zacate encubierto, zacatón arenoso, zacatón desgranador, zacatón encubierto – sand dropseed
Sporobolus flexuosus – zacatón, zacatón de los médanos, zacatón flexible – mesa dropseed
Sporobolus giganteus – zacate gigante – giant dropseed
Sporobolus indicus – chilibsuuk, cola de ratón, espartillo, espartillo común, grama de la costa, guayacán, liendrecillo, liendrilla, liendrilla tropical, matojo, matojo de burro, pasto dulce, pitilla, saetilla, serillo, sinchiquigua, yunquillo, zacate de encinar, zacatón alcalino, zacatón de salitre, zacatón del desierto, zacatón indio – smutgrass
Sporobolus luchley – alpiste
Sporobolus pulvinatus – zacate de agua
Sporobolus pyramidatus – zacate pirámide, zacatón pyramidal – whorled dropseed
Sporobolus wrightii (airoides* var *wrightii) – cola de ratón, zacate alcalino, zacate de aparejo, zacatón – sacaton
Sprekelia formosissima – amacayo, atzcalxóchitl, capa de Santiago, chintule, cruz, flor de lis, flor de mayo, flor de Santiago, lirio jacobino, lirio de San Jaime, pata de gallo, tempranilla, venera de Santiago, zagalejo – Aztec lily, Jacobean lily, St. James lily

Stachys arvensis – hierba del gato, ortiga hedionda – field woundwort, hedge nettle
Stachys betonica – betónica – bishop's wort, lousewort, purple betony
Stachys byzantina (lanata, olympica) – oreja de cordero, ortiga lanosa, stachys lanosa – lamb's ears
Stachys coccinea – mirto – scarlet hedge nettle, scarlet sage
Stachys (Betonica) macrantha (grandiflora) – betónica mayor – big betony
Stachys (Betonica) officinalis – betónica, betónica oficinal – bishop's wort, wood betony
Stachys palustris – ortiga palustre, salvia del pantano – hedge nettle, marsh woundwort
Stachys sieboldii – alcachofa, crosne del Japón – Chinese artichoke
Stachys tuberifera – crosne del Japón – Chinese artichoke
Stanhopea devoniensis – cabeza de culebra, cabeza de tigre, cabeza de víbora, flor de cuerno, flor de serpiente, torito, yo-betz – tiger-like stanhopea
Stanhopea oculata – cozticcoatzontecomaxóchitl, torito
Stanhopea tigrina – cabeza de culebra, cabeza de víbora, coatzontecomaxóchitl, flor de cabeza, flor de torito, tecuanxóchitl, torito, toro
Stanhopea wardii – cozticcoatzontecomaxóchitl, gibguin
Stapelia gigantea – flor del lagarto – carrion flower
Staphylea bolanderi – estafilea – bladdernut
Statice = Limonium
Stegnosperma halimifolium – amole, bledo carbonero, chapacolor, hierba de la víbora, hierba del cuervo, ojo de zanate, tinta, vomitivo
Stellaria calycantha – álsine calicanta – starwort, stitchwort
Stellaria graminea – álsine gramínea – lesser stitchwort
Stellaria longifolia – álsine de hojas largas – long-leaved starwort
Stellaria media – álsine, hierba del pajarito, hierba del pollito, hierba riquera, oreja de ratón, pajarera, pamplina, pamplina de canarios, regojo – common chickweed
Stellaria nemorum – alahual, álsine de bosques, estrellita, limpia gallinas, matanza, morgelina – chichweed, starwort, wood chickweed, wood stitchwort
Stellaria pallida – álsine pálida – lesser chickweed
Stellaria prostrata – llovisna
Stemmadenia donnell-smithii – chapón, chapona, cojón de berraco, cojón de burro, cojón de danta, cojón de toro, cumicán, jazmín, tontzimin
Stemmadenia obovata – cojón de berraco, cojón de coche, cojón de puerco, lecherillo, tepechicle, palo bolero
Stemmadenia tomentosa (palmeri)* var *palmeri – berraco, cojón de puerco, huevo del toro, peychi, sicte, sictillo, tapaco, zapote
Stemodia durantifolia – tomosali
Stemodia pusilla – rosita amarilla
Stenanthium frigidum – cebadilla, cebolleja, cebolleta, cintul
Stenanthium occidentale – cebadilla
Stenocactus multicostatus – biznaguita – brain cactus
Stenocactus (Echinofossulocactus) obvallatus – biznaguita, tepenexcomitl
Stenocactus ochoterenanus (Echinofossulocactus heteracanthus) – biznaguita
Stenocactus (Echinofossulocactus) phyllacanthus – biznaguita
Stenocarpus sinuatus – árbol de ruedas de fuego – firewheel tree
Stenocereus (Rathbunia) alamosensis – nacido, sina, tasajo, tuna – octopus cactus
Stenocereus (Machaerocereus) eruca – casa de rata, chirinole – creeping devil
Stenocereus fimbriatus (Lemairocereus hystrix) – dildo español, sebucán español
Stenocereus fricii – pitaya de agua
Stenocereus (Lemaireocereus) griseus – pitahaya, pitaya de mayo, pitaya de Mitla
Stenocereus (Machaerocereus) gummosus – pitahaya, pitahaya agria, pitaya agria – galloping cactus, sour pitaya
Stenocereus (Rathbunia) kerberi – sina

Stenocereus (Lemaireocereus) martinezii – pitahaya, pitaya

Stenocereus (Lemaireocereus) montanus – pitahaya, pitaya colorada, sagüera, sahuí – mountain organ pipe

Stenocereus (Lemaireocereus) pruinosus – pitaya, pitaya de mayo, pitaya de octubre – gray ghost organ pipe

Stenocereus (Lemaireocereus) queretaroensis – cardón, pitahaya, pitaya, pitaya de Querétaro

Stenocereus (Lemaireocereus) quevedonis – pitahaya, pitaya, pitire

Stenocereus (Lemaireocereus) standleyi – pitahaya, pitaya marismeña

Stenocereus (Lemaireocereus) stellatus – joconostle, órgano, pitahaya, pitaya, pitaya roja, tuna, xoconóchitl, xoconochtli, xoconoxtle

Stenocereus (Lemaireocereus) thurberi – aaqui, marismeño, ool, órgano, órgano marismeño, pitahaya, pitahaya dulce, pitaya, pitaya dulce, tarrito – organ pipe cactus

Stenocereus (Lemaireocereus) treleasei – nutillo, tuna olorosa de ejutla, tunillo

Stenocereus = Isolatocereus, Pachycereus

Stenolobium = Tecoma

Stenorrhynchus cinnabarinus – cutzi

Stenorrhynchus lanceolatus – hulté-k'-aak, terciopelo

Stenorrhynchus orchioides – terciopelo

Stenotaphrum secundatum – carpeta de San Agustín, gramilla de San Agustín, gramillón, gramillón de San Agustín, grama dulce, pasto alfombra, pasto San Agustín, saladilla, San Agustín – Saint Augustine grass, St. Augustine grass

Stephanomeria pauciflora – pionilla – stick-weed, brownplume wire-lettuce, skeletonweed

Stephanotis floribunda – estefanote – Madagascar jasmine, stephanotis

Sterculia apetala – anacagüita, árbol del bellote, bellota, castaño, petaca, tepetaca – bottle tree, Panama tree

Sterculia carthaginensis – camaruca, Panamá

Sterculia foetida – anacagüita – Java olive

Sterculia mexicana – bellota, castaño, picapica

Sterculia urens – goma karaya

Sternbergia lutea – azucena amarilla, sternbergia amarilla – fall daffodil, lily of the field, winter daffodil

Stevia rebaudiana – estevia, hierba dulce del Paraguay

Stevia salicifolia – hierba del aire, hierba de la mula, hierba de San Nicolás, jarilla, zazale, zazale de olor

Stevia serrata – chile burro, cola de borrego, requesón, Santa Teresa

Stevia viscida – hierba de la pulga, matapulgas, pipitzáhuatl

Stewartia pseudocamellia – falsa camellia, falsa camellia decidua, stuartia – Japanese stewartia

Sticta pulmonacea – pulmonaria, pulmonaria de árbol – tree lichen

Stillingia acutifolia – pavil – queen's delight, queen's root

Stillingia paucidentata – pavil – queen's root

Stilozobium deeringianum – frijol terciopelo – velvet bean

Stipa eminens – agujilla grande, flechilla, flechilla grande

Stipa gigantea – banderilla, berceo, esteba pirenaica – Pyrenean feathergrass

Stipa ichu – ichú, paja de puna

Stipa leucotricha – flechilla bulbosa – Texas needlegrass

Stipa neomexicana – barba blanca, flechilla neomexicana – New Mexican feathergrass

Stipa pennata – espolín, esteba de hojas pinadas – woolly-stalked feathergrass

Stipa robusta – popote, popotón, zacatón – sleepy grass

Stipa (Macrochloa) tenacisima – esparto, esparto común, raigón – esparto grass, needlegrass

Stipa = Acantherum

Stizolobium deeringianum – frijol aterciopelado, frijol de terciopelo, frijol terciopelo de Florida, guisante mascate, haba terciopela, mucuna rayada, terciopelo – velvet bean

Stokesia laevis – aster de Stoke – Stoke's aster

Strelitzia reginae – ave del paraíso, estrelichia, estrelitzia, flor ave del paraíso, flor de pájaro, pájaro azul – bird-of-paradise, crane flower

Streptocarpus hybridus – estreptocarpo – Cape primrose
Streptopus amplexifolius – estreptopo amplexifolio – twisted-stalk
Strombocactus disciformis – peyote – top cactus
Stropharia = Psilocybe
Struthanthus haenkeanus – togue, toji
Struthanthus palmeri – chibchel, toji, toji de mezquite
Struthanthus quercicola – correhuela
Strychnos ignatii – cabalonga, caniran, haba de San Ignacio – Ignatius bean, St. Ignatius's bean
Strychnos nux-vomica – nuez vómica – nux vomica
Strychnos panamensis – cabalonga, caimitillo, cabalonga, pataste, pataste de caballo, pataste de mico
Strychnos tabascana – cabalonga de Tabasco, mataperros, veneno del diablo – logania
Strychnos tepicencis – cuero de vaca, mataperros
Strychnos toxifera – curare – curare
Stylosanthes guianensis – alfalfa tropical, estilosantes, tabardillero, trifolio brasileño – Brazilian lucerne, finestem stylo
Stylosanthes viscosa – hierba de la gallinera, hierba del pujo – pencil flower
Styrax argenteus – aguacatillo, capulín, chilacuate, chucamay, estoraque, hoja de jabón, levadura, ruín – snowdrop bush
Styrax benzoin – árbol de benjuí – benzoin tree
Styrax japonicus – estoraque japonés – Japanese snowbell, Japanese snowdrop tree
Styrax obassia – estoraque, estoraque japonés de hojas grandes – fragrant snowbell
Styrax officinalis – benjuí común, estoraque, estoraque común – snowdrop bush, stacte, storax
Suaeda calceoliformis (depressa, depressa var erecta) – quelite salado – horned sea-blite
Suaeda moquinii (fruticosa, torreyana) (Dondia fruticosa, D. ramosissima) – juaja, quelite salado, romerillo, romerito, romero, sosa, sosa fruticosa, vidrillo – bush seepweed, Mojave sea-blite
Suadea nigra – romerito
Suadea taxifolia (californica var taxifolia, Dondia brevifolia) – vidrillo – woolly sea-blite
Suadea = Dondia
Subularia aquatica – subularia acuática – awlwort
Suillus granulatus – bojín, boleto granulado – granulated bolete
Suillus luteus – pegajosa – butter mushroom, slippery Jack
Suriana maritima – pantzil – bay cedar
Swartzia = Solandra
Sweetia panamensis – bálsamo amarillo, bálsamo oloroso, cáscara amarga, chacté, corteza de Honduras, guayacán, huesillo – cascara amarga
Swertia perennis – swertia perenne – marsh felwort
Swertia radiata (Frasera speciosa) – cebadilla, cebadilla de la sierra, lengua de ciervo – deers ears, green gentian, monument plant
Swietenia humilis – caoba, caobilla, cirrhata, cóbano, flor de venadillo, gateado, palo de zopilote, tzopilotzontecómatl, venadillo, zapatón, zopilote – Mexican mahogany, Pacific coast mahogany
Swietenia macrophylla – caoba, cedro montés, cóbano, maro, mogno, mogno aguano, palo de zopilote, punab, rosadillo, tzopilocuáhuitl, tzopilotl, tzopilotzontecómatl, zontecomacuahuitl, zopilote, zopilozontecomacuahuitl – Brazilian mahogany, Honduras mahogany
Swietenia mahagoni – caoba, caoba de Santo Domingo – Dominican mahogany
Syagrus coronata – licuri – licuri palm
Syagrus (Arecastrum) romanzoffiana (Cocos plumosa) – coco de pluma, coco plumoso, palma real – feather palm, queen palm
Syagrus schizophylla – coquito de Brasil, palmera de la reina
Sympetaleia = Eucnide
Symphoricarpos albus (racemosus) – baya de nieve, bolita de nieve, perla, sinforina – common snowberry, snowberry
Symphoricarpos microphyllus – baya de nieve, huihuilán, perlilla, perlita – snowberry
Symphoricarpos rivularis – oreja de ratón – snowberry

Symphytum asperum – borraja silvestre, consuelda escabrosa – comfrey, wild borage
Symphytum officinale – conferva, consólida, consuelda, consuelda mayor, hierba de la cortadura, oreja de asno, sínfito, sínfito mayor, suelda, suelda consuelda – comfrey
Symplocarpus foetidus – col de pantano, col fétida – skunk cabbage
Synechanthus fibrosus – tu-kiak – monkey-tail palm
Syngonium podophyllum – carne de perro, conté siete dedos, contzontón, lengua de vaca, ochil, singonio, teliconté – African evergreen
Syringa vulgaris – lila – common lilac
Syzygium = Caryophyllus, Eugenia

T

Tabebuia (Tecoma) chrysantha – amapa, amapa amarilla, amapa prieta, amapa verde, araguaney, corteza amarilla, guayacán, lapacho, lombricillo, matilisguate, primavera, roble, roble amarillo, totcui, verdecillo – gold tree, yellow amapa

Tabebuia (Cybistax, Roseodendron) donnell-smithii – amapa, amapa amarilla, duranga, guayacán, palo blanco, primavera – gold tree, primavera, springbell, sunshine tree, white mahogany

Tabebuia guayacan – guayacán, palo blanco, primavera, roble serrano

Tabebuia impetiginosa (palmeri) – amapa, amapa colorada, amapa morada, amapa prieta, amapa rosa, amapa toubo, arbusto atrompetado, coyahual, maculiz, matilisguate, roble, tabuió, tanillo, toubo, tlamahual – pink amapa, pink trumpet tree, purple trumpet tree

Tabebuia (Tecoma) pentaphylla – amapa, amapa rosa, apamate, cachahua, coyolxóchitl, fresno, guayacán, hokab, icotl, lapacho, macuilixuatl, maculiz, maculiz de abajo, maculiz prieto, maquilicua, matilisguate, matilisguate rosado, mocoque, ocobo, palo blanco, palo de arco, palo de rosa, palo de yugo, primavera, roble, roble blanco,roble colorado, roble de San Luis, roble prieto, roble rosa morada, rosa morada, satanicua, tural, verdecillo, zapote – May flower, Mexican cherry, trumpet tree

Tabebuia rosea – amapa rosa, apamate, maculiz, mano de león, palo de rosa, roble, roble colorado – pink trumpet tree

Tabebuia = Cybistax

Tabernaemontana alba – chaklikin, chichihualcaxtli, cojón de gato, cojón de perro, cojón de toro, cojotón, huevo de caballo, huevo de chuchu, huevo de cucho, huevo de gato, huevo de toro, laurel blanco, lecherillo, ton-chá

Tagetes elongata – San Paulillo

Tagetes erecta – apátsicua, caléndula africana, caléndula azteca, cempasúchil, cempoal, cempoalxóchitl, clavel de las Indias, clavel de oro, clavelón, clavelón de la India, copete, damasquina, flor de muerto, flor de oro, jutús, maravilla africana, musajoyo, nulibé, picoa, rosa de muerto, tepecempoalxóchitl, tiringuini, xempoalxóchitl, xumpatsúchitl – African marigold, American marigold

Tagetes filifolia – anisillo, curucumín, flor de Santa María, hierbanís, manzanilla, pericón, periquillo, Santa María, teyatli, yauhtli – Irish lace

Tagetes florida – anisillo, curucumín, hierba de Santa María, hierbanís, pericón, Santa María, teyatli

Tagetes jaliscana – cempoal, San Paul

Tagetes lacera – cempasúchil

Tagetes lucida – anís, anisillo, coronilla, curucumín, falso hipericón, flor de Santa María, flor de tierradentro, guamusi, hierbanís, hierba de Santa María, pericón, periquillo, Santa María, yauhtli – Mexican marigold, Mexican tarragon, sweet-scented marigold

Tagetes lunulata – caléndula, cempasúchitl, cinco llagas, cocoyatón, cocozatona, flor de cinco llagas, flor de muerto, rosa de ángel, tagetes de hojas finas – fine-leaved marigold

Tagetes micrantha – anís, anís del campo, anisillo, temolillo, tlalanís, yeyésowa – marigold

Tagetes minuta – chinchilla enana – dwarf marigold

Tagetes patula – caléndula francesa, cempoal, cempoalxóchitl cimarrón, cepetillo, clavel de las Indias, clavel de moro, clemole, clemolito, copetillo, damasquina, expujuj, flor de los muertos, flor de muerto, flor de oro, iscoque, macehualpuhuk, maravilla francesa, molxóchitl, pastora, pastoral, pastorcilla, pastorcita, tepecempoalxóchitl, tlemole, tlemolito, xpuhuc – French marigold

Tagetes tenuifolia (signata) – cempasúchil, cempoal, chinchilla de hojas delgadas, clavel de China, clavel de las Indias, curucumín, flor de muerto, flor de Santa María, flor de tierradentro, hierba de Santa María, maravilla, Santa María, tagetes – signet marigold

Taiwania cryptomerioides – taiwania

Talauma mexicana – anonillo, chipahuac, chocoijoyo, cocté, flor del corazón, guielachi, haulhua, hierba de la matadura, jolmashté, laurel tulipán, magnolia, magnolia mexicana, súchil, yo-lachi, yollopatli, yolosúchil, yoloxóchitl, yucacaca – heartflower tree, Mexican magnolia

Talinum paniculatum – negrito, pionilla, rama de sapo, verdolaga – flameflower

Talinum triangulare – espinaca de Filipinas, espinaca de Surinam, hierba de cacolote, verdolaguilla – Surinam purslane

Tamarindus indica – hoaxin, huaje, pahxuhuc, tama, tamarindo – tamarind

Tamarix africana – atarfe, taraje, taray africano, taray hispánico – African tamarisk

Tamarix aphylla – pino, pino salado – athel, athel tree

Tamarix chinensis – tamarisco – salt cedar, tamarisk

Tamarix gallica – asul, atarfe, eshel, tamarisco, tamariz, taraje, taray, taray gálico – French tamarisk, manna bush, salt cedar

Tamarix parviflora (tetrandra) – pino salado, tamarisco, tamariz, taray de flores pequeñas – tamarisk

Tamarix ramosissima (pentandra) – atarfe, pino salado, salado, talaya, tamarisco, tamarisco común, tamariz, taray – salt cedar, tamarisk

Tamus communis – brionia negra, canduerca, espárrago de caña, nueza negra, uvilla de lagarto – black bryony

Tanacetum (Chrysanthemum, Pyrethrum) parthenium – altamisa, altamisa de Castilla, altamisa mexicana, bola de hilo, botón de plata, hierba de Santa María, magarza, magarzuela, margarita, matricaria, matricaria partenio, Santa María – feverfew

Tanacetum (Chrysanthemum) vulgare – altamisa, atanasia, hierba de Santiago, hierba lombriguera, hierba Santa María, ponso, tanaceto, tanaceto común, tansé – common tansy, tansy

Tanacetum = Chrysanthemum

Taraxacum officinale – achicoria amarga, amargón, cerraja, chicoria, consuelda, diente de león, moraja, pelosilla, taraxacón – common dandelion, dandelion

Tauschia arguta – panaripa

Taxodium distichum – calvo, ciprés, ciprés americano, ciprés calvo, ciprés de los pantanos, ciprés de Virginia, falso ciprés calvo, falso ciprés de los pantanos, taxodio, taxodio de pantano – bald cypress

Taxodium distichum (mucronatum)* var *mexicanum – ahuehuete, ahuéhuetl, bochil, cedro, ciprés, cipresco, ciprés de Moctezuma, ciprés mexicano, cipreso, jauolí, matéoco, naciño, pentamón, pentamú, sabina, sabina de Oaxaca, teponaxtle, teponaxtli, tnuyucú, xiña, yaga-guichi xiña, yucuchichicino, yucu-ndatura – Mexican cypress, Montezuma bald-cypress

Taxus baccata – tejo, tejo común, tejo negro – English yew

Taxus baccata* var *stricta (fastigata) – tejo irlandés – Irish yew

Taxus cuspidata – tejo de Japón, tejo japonés – Japenese yew

Tecoma (Tecomaria) capensis – bignonia – Cape honeysuckle

Tecoma incisa – caballito

Tecoma (Stenolobium) stans – batilini, bignonia, borla de San Pedro, botiline, caballito, campana amarilla, candelilla, candox, canlol, chilca, chirlobirlo, copete, corneta amarilla, esperanza, flor amarilla, flor de San Pedro, fluvia del oro, fresnillo, gloria, guarán amarillo, guie-biche, hierba de San Nicolás, hierba de San Pedro, hoja de baño, huachacata, ichculili, ixtontle, lluvia de oro, matilimi, mazorca, miñona, mixtontze, nextamalxóchitl, palo de arco, retama, roble amarillo, San Andrés, San Pedro, saúco amarillo, tache, talusúchil, trompeta, trompetilla, tronador, xochimitl – trumpet bush, trumpet flower, yellow bell, yellow elder, yellow trumpet

Tecoma = Campsis, Tabebuia

Tecomaria = Tecoma

Tectaria heracleifolia – cruxiv, morcegalo, pim

Tectona grandis – teca – indian oak, teak

Teesdalia coronopifolia – teesdalia de hojas de mastuerzo – lesser shepherd's cress

Teloxys = Chenopodium

Tephrosia cana – barbasco

Tephrosia nicaraguensis – barbasco, barbasco cimarrón, chilapate, hierba del piojo

Tephrosia tenella – añil

Tephrosia toxicaria – barbasco
Terminalia amazonia (obovata) – almendro, amarillo, cacho de toro, canolté, canxun, cochun, corteza amarilla, guayabo, guayabón, membrillo, sombrerete, tepesúchil, volador – Indian almond
Terminalia (Chuncoa) catappa – alcornoque, almendrillo, almendro, almendro de la India, almendro de Tehuántepec, almendrón, castaño, pardillo amarillo – Barbados almond, Indian almond, tropical almond tree
Terminalia chebula – chébula
Ternstroemia pringlei – tilo, tilo grande, trompillo
Ternstroemia sylvatica – hierba del cura, ixquefé, limoncillo de Meztitlán, ministro, palo agrio, tepezapote, tilo grande
Tessaria = Pluchea
Tetracera volubilis – bejuco colorado, bejuco de agua, bejuco de tachicón, bejuco de canasta, chaparillo, palo de tachicón, tachicón de sabana
Tetraclinis (Thuja) articulata – alerce africano, arar, ciprés de Cartagena, sabina de Cartagena, tuya articulada – arar tree, citronwood, sandarac, thyine wood
Tetragonia tetragonioides – espinaca de Nueva Zelandia, espinaca neozelandesa, tetrágona – New Zealand spinach
Tetranema mexicana – dedalera mexicana – Mexican foxglove
Tetraneuris = Hymenoxys
Teucrium cossonii – germandria pulverulenta
Teucrium cubense – cabeza de hormiga, epazotillo, gallina ciega, gallinita ciega, hierba de la gallina, hierba del negro, hierba del perro, quelite cuale
Teucrium fruticans – germandria fruticosa, olivilla, salvia amarga – bush germander
Teucrium* x *lucidrys (chamaedrys) – alzineta, camedrio, germandria, germandrina – germander, wall germander
Teucrium marum – hierba del papa, hierba fuerte, maro – cat thyme
Teucrium polium – hierba crispa, pollo de monte, teucrio, tomillo terrero, zamarilla – white-felted germander
Teucrium scordium – camedrio acuático, escordio, germandria de agua – water germander
Teucrium scorodonia – camedrio de los bosques, escordio bastardo, escorodonia – wood sage
Thalassia testudinum – pasto de tortuga, seibadal – turtlegrass
Thalia geniculata – banderilla, bojillo de popal, platanillo, popal, quentó, talia – arrowroot, swamp lily
Thalictrum alpinum – ruda alpina – alpine meadow-rue
Thalictrum aquilegifolium – pigamón de hojas de ancolia, ruda de hojas de aguileña – greater meadow-rue
Thalictrum dipterocarpum – ruda pratense – Chinese meadow-rue
Thalictrum fendleri – paloma consulta, ruda cimarrona, ruda de la sierra, tanito – meadow-rue
Thalictrum hernandezii – cocoztic, cozticpatli, cucoztli, paloma consulta – meadow-rue
Thamnosma montana – cordoncillo – turpentine-broom
Thapsia garganica – tapsia – death carrot
Thea = Camellia
Thelesperma gracile – cota – Hopi tea, Navajo tea
Thelesperma longipes – cota – long-stalk greenthread, Navajo tea
Thelesperma megapotamicum – cota, hierba de San Nicolás, té, té de cota, té de los Navajos, té silvestre – Hopi-tea greenthread, Indian tea, rayless thelesperma
Thelocactus (Hamatocactus) bicolor – biznaga – Texas pride
Thelocactus (Ferocactus, Hamatocactus) setispinus – biznaga de lima, biznaga de tuna – fishhook cactus, hedgehog, mesquite barrel cactus
Thelypteris puberula – helecho plumoso
Thelypteris puberula (Lastrea augescens)* var *sonorensis – hierba del golpe – Sonoran maiden fern
Thenardia floribunda – petatillo
Theobroma angustifolium – cacao, cacao de mono – cacao, sonusco cocoa
Theobroma bicolor – balam-té, cacao, cacao

blanco, cacao malacayo, patachtli, pataste, pataxtle – cacao, Peru cocoa

Theobroma cacao – bizoya, cacahuacuahuitl, cacahuatl, cacao, cacaocuáhuitl, cacaotero, cacaotlquahuitl, caceteiro, cajecua, chudechu, kako, tlapalcacauatl, trujillano, xau, yagabizoya – cacao, chocolate, cocoa

Thespesia populnea – cremón, majagua, majaguilla – cork tree, milo, Pacific rosewood, portia tree

Thevetia ovata – actiz, ayoyote, castaño de la India, chilamate, chilca, chilca silvestre, chilindrón, convulí, cumbili, cunduacán de Túxtepec, huevo de gato, manzano de burro, meriendita, naranjo amarillo, narciso amarillo, rejalgar, venenillo, yoyote, yoyotli – yellow oleander

Thevetia peruviana (neriifolia) – actiz, adelfa amarilla, ayoyote, cabalonga, cabalonga de la Huasteca, camín, campanilla, campanilla de oro, cascabel, cascabel de lacandón, chilca, chilindrón, chirca, codo de fraile, cojón de gato, copa de oro, flor trompeta, guayapol, guayocule, haba, jamín, naranjo amarillo, narciso amarillo, palo de San Antonio, retama, San Diego, San Nicolás, trompeta, venenillo, yambigo, yoyote, yoyotli, yucacaca – be-still tree, lucky nut, trumpet flower, yellow oleander

Thevetia thevetioides – actiz, ayoyote, cabrito, calaverita, chirca, codo de fraile, fraile, hueso de fraile, naranjo amarillo, narciso amarillo, petlacotl, rejalgar, tzinacanitlacual, yoyote, yoyotli, yucacaca – giant thevetia

Thlaspi arvense – carraspique, hierba del ochavo, telaspio de los campos – fan-weed, field penny-cress

Thlaspi montanum – telaspio montano – mountain penny-cress

Thouinia villosa – guasimilla roja

Thrinax radiata – chit, pelambrera de Florida – salt-water palmetto, thatch palm

Thrinax wendlandiana (parviflora) – bayaboon, bayal, chit, cul-tok, huano blanco

Thryallis = Galphimia

Thuja occidentalis – árbol de la vida, árbol de la vida occidental, cedro amarillo, ciprés palmita, tuya, tuya del Canadá, tuya occidental – American arborvitae, northern white cedar

Thuja (Platycladus) orientalis – árbol de la vida, árbol de la vida de oriente, árbol de la vida oriental, biota, tuya, tuya de la China, tuya oriental – Chinese arborvitae, oriental arborvitae

Thuja plicata – árbol de la vida gigante, árbol de la vida gigantesco, cedro rojo del Pacífico, tuya, tuya gigante – canoe cedar, western red cedar

Thuja = Tetraclinis

Thujopsis dolabrata – hiba, tuya japonesa – deerhorn cedar, false arborvitae, hiba cedar

Thunbergia alata – anteojo de poeta, colombiana, ojitos negros, ojo de pajarito, ojo de pájaro, ojo de poeta, ojo de Venus, principe Alberto – black-eyed Susan vine, clock vine

Thunbergia grandiflora – fausto, presidio de amor – Bengal clockvine, blue trumpet vine, sky flower

Thurberia = Gossypium

Thymophylla concinna – manzanilla del coyote

Thymophylla (Dyssodia) pentachaeta – limoncillo, parraleña – five-needle dogweed, golden dyssodia

Thymophylla = Dyssodia

Thymus camphorates – tomillo alcanferado – camphor thyme

Thymus capitatus – morquera, tomillo aceitunero, tomillo andaluz, tomillo arborescente – Spanish thyme

Thymus mastichina – almoradux, mejorana silvestre, sarilla, tomillo blanco – Spanish marjoram, white thyme

Thymus pulegioides – serpol, tomillo común – large thyme

Thymus serpyllum – hierba luna, madre de tomillo, salsa de pastor, serpillo, serpol, tomillo, tomillo de monte – creeping thyme, mother-of-thyme, wild thyme

Thymus vulgaris – carrasquilla, estremoncillo, salsero, senserina, tomillo, tomillo de jardín, tomillo vulgar, tremoncillo – common thyme, garden thyme, thyme

Tiarella cordifolia – flor de espuma – foamflower

Tibouchina spathulata – entrodelia, mistelajoyo – glorybush
Tibouchina urvilleana (semidecandra) – mosqueta silvestre, tibuchina – glory bush, princess flower
Tidestromia lanuginosa – espantavaqueros, hierba ceniza, hierba lanuda, mochis, verdolaga de cochi – honeysweet
Tigridia dugesii – cacomite, xahuique, xahuique de Tupátaro
Tigridia pavonia – atigrada, cabeza de culebra, cacomite, cacómitl, flor de concha, flor de maravilla, flor de tigre, flor de un día, flor manchada, hierba de la Trinidad, lirio azteca, ocelosúchil, oceloxóchitl, rodilla de Cristo, trinitaria, xahuique – Aztec lily, Mexican shell flower, tiger flower
Tigridia pringlei – chacual
Tilia americana – tilo americano – basswood, American linden
Tilia cordata – tilo de hojas pequeñas, tilo silvestre – little-leaf linden, small-leaved limetree
Tilia dasystyla – tilo de Cáucaso
Tilia* x *euchlora – tilo de Cáucaso, tilo ruso – Crimean linden
Tilia* x *europaea – arguma, tillón, tilo, tilo de Europa, tilo híbrido de Holanda, tilo intermediario – English lime, European basswood, European linden
Tilia mexicana – sirimo, tilo, tzirimu, yaco
Tilia occidentalis – árbol de lima, sirimo, tilo, yaco – basswood, linden
Tilia* x *petiolaris – tilo húngaro, tilo péndulo – weeping silver lime
Tilia platyphylla – tilo, tilo común, tilo de hojas grandes – broad-leaved limetree
Tilia tomentosa – tilo húngaro, tilo plateado, tilo tomentoso – silver linden
Tilia* x *vulgaris – tillera, tilo común, tilo híbrido de Holanda – common lime
Tillandsia exserta – mezcalito de huitlacochi
Tillandsia ferrisiana – gallito, heno pequeño – ball moss
Tillandsia flabellata – piña de palo, pino aéreo, trencilla – air pine
Tillandsia imperialis – guacamaya, pino aéreo, tecolote, tecolumate – air pine
Tillandsia inflata – mezcalito
Tillandsia limbata – flor de incienso, saripara
Tillandsia lindenii – flor de aire
Tillandsia prodigiosa – cadena, pie de gallo, tecolumate, tecomelé, tonjoyó – prodigious bromeliad
Tillandsia recurvata – agare palo, clavél del aire, curujey, gallito, heno, heno chino, heno pequeño, hichiconi, jaboncillo, magueyito, magueyito de árbol, magueyito parasito, mezcalito, paixtli, paxtle, paxtle de mezquite, salvajina macho, toji – ball moss
Tillandsia (Dendropogon) usneoides – barba de árbol, barba de palo, barba de viejo, barba española, cuampach, gallito, guajaca, heno, heno largo, mexnuxeb, musgo, musgo blanco, musgo español, paixtli, pañal de niño, patzueni, patzueni tacari, paxtle, salvajina, soscilchac, tacali – Louisiana moss, old man's beard, Spanish moss
Tilletia tritici – caries, tizoncillo – wheat smut
Tinantia erecta – felvira, hierba del pollo, matlalxóchitl
Tinantia modesta – felvira, hierba del pollo
Tipuana tipu – tipa, tipuana – tipu tree
Tiquilia (Coldenia) canescens – hierba de la Virgen, oreja de perro – dog's ear
Tiquilia (Coldenia) greggii – cenizo, hierba del cenizo
Tithonia calva* var *lancifolia – mirasol, tacón
Tithonia fruticosa – mirasol – tree sunflower
Tithonia rotundifolia (speciosa) – acahual, acaute, acaute de flor, andani, árnica, chaczuum, chiople, gigantón, girasol, girasol mexicano, mirasol, polocote, sunté, tacón, tanchichi, tzum, zuun – Mexican sunflower
Tithonia tubaeformis – acahual, acaute, andani, gigantón, girasol, palocote, sunté – bush sunflower
Tolpis barbata – tolpis umbelado
Toluifera = Myroxylon
Torenia fournieri – lazo de amor – Florida pansy, wishbone flower
Torilis arvensis – torilio de los campos – spreading bur parsley
Torilis nodosa – torilio nudoso – knotted bur parsley, knotted hedge parsley
Torreya californica – torreya de California – California nutmeg

Tournefortia capitata – tlachichinoa, tochimirtillo
Tournefortia hartwegiana – cola de alacrán, confite, confite coyote, confite negro, confiturilla negra, hierba del burro, hierba del negro, hierba del sapo, ortigüilla, tatachinole, tiña cuerda
Toxicodendron diversilobum (Rhus diversiloba) – hiedra, hiedra mala, hierba de la hiedra, hincha huevos, mala mujer, tztzil, zumaque venenoso – western poison oak
Toxicodendron (Rhus) radicans – árbol de guau, bembericua, chechén, dominguilla, fuego, guadalagua, guardalagua, guau, hiedra, hiedra mala, hiedra mala mujer, hiedra maligna, hiedra venenosa, hincha huevos, huembéreua, mala mujer, mexye, tetlatia, yagalache, zumaque – poison ivy
Trachelium caeruleum – alfiler, flor de viuda
Trachelospermum (Rhynchospermum) jasminoides – jazmín, trachelospermo – Confederate jasmine, star jasmine
Trachycarpus fortunei – palma excelsa, palmera china, palmera china de Chusán, palmera de fortuna, palmito de jardín, palmito elevado, palmito molino de viento – windmill palm
Trachycarpus martianus – palmera de Martius
Trachylobium verrucosum – copal de Zanzibar – Zanzibar copal
Trachypogon montufari – polmuc, quimec – crinkle-awn
Trachypogon secundus – alea-pajac, barba larga, zacate barba larga – crinkle-awn
Tradescantia (Setcreasea) brevifolia – pollo
Tradescantia crassifolia – coapatli, hierba del pollo
Tradescantia (Rhoeo) discolor – chactsán, maguey morado, yastsana, zebrina
Tradescantia fluminensis – hierba del pollo, mataliz, tradescantia manchada – spiderwort, wandering Jew
Tradescantia occidentalis – hierba del pollo – western spiderwort
Tradescantia (Setcreasea) pallida – maclale – purple heart, purple queen
Tradescantia peninsularis – hierba del pollo
Tradescantia (Rhoeo) spathacea – maguey silvestre – boat lily, Moses-in-the-boat, Moses-in-the-cradle, oyster plant
Tradescantia virginiana – lirio araña, pasajera, tradescantia – spiderwort
Tradescantia zebrina (Zebrina pendula) – cucarachita, hoja de plata, matal, matalí, moradilla, siempreviva, zebrina – wandering Jew
Tragia laciniata – mala mujer – noseburn
Tragia nepetaefolia – chichicastle, ortiga, ortiguilla, ortiguilla quemadora, salitrillo, tajehuali – noseburn
Tragopogon dubius – barba de cabra mayor – goat's beard, greater bock's beard
Tragopogon porrifolius – barba cabruna de hojas de puerro, salsifí, salsifí blanco – oyster plant, salsify, vegetable oyster
Tragopogon pratensis – barba cabruna, barba de cabra, barba de cabra de los prados, salsifí – goat's beard, salsify
Tragus berteronianus – abrojo espigado – spike burgrass
Trapa natans – abrojo de agua, castaña de agua, túbulo acuático – water chestnut, waternut
Trema micrantha – capulín, capulín cimarrón, equipal, guacimilla, ixpepe, kib, majagua blanca, majagua colorada, pellejo de viejo, pie de paloma, pixó, yaco de cuero, zonote – Mexican hackberry, wild bay cedar
Tremella fuciformis – gelatina de palo – silver ear, white jelley fungus
Trianthema portulacastrum – bledo, quelite, verdolaga blanca, verdolaga bronca, verdolaga de cochi – horse-purslane
Tribulus cistoides – abrojo, abrojo de tierra caliente, cabeza de arriero, cabeza de chivo, chanixnuc, hierba de la araña, pericón, punachi, raíz de abrojo, rosilla – Jamaica feverplant
Tribulus terrestris – abrojo, espigón, toboso, torito, tríbulo – bull-head, bur-nut, caltrop, goat-head, puncture vine
Tribulus = Kallstroemia
Tricerma = Maytens
Trichachne = Digitaria
Trichilia americana – bola colorada, piocha, síquiri
Trichilia hirta – asapescado, azuica, cabo de

hacha, cabra de hacha, cauache, cedrillo, choben-ché, chuminillo, cucharillo, cuju, culimziz, estribillo, garbancillo, garrapatilla, ixbahach, jumay, limoncillo, mapahuite, napahuite, palo amarillo, palo colorado chico, papelillo, piocha, piocha cimarrón, purbegan, síquiri tájcara, trompillo, xpukusikil, zorrillo – bastard lime, red cedar
Trichloris = Chloris
Trichocereus = Echinopsis
Tricholaena = Rhynchelytrum
Tricholoma acerbum – escarlata blanca – bitter agaric
Tricholoma columbetta – cogoma, culgro
Tricholoma equestre – calandria, calandrita, canario, nejo, tzenso, yema de huevo
Tricholoma flavovirens – canario, groguer, palomita, seta de los caballeros – firewood agaric, yellow night fungus
Tricholoma gambosum (Calocybe gambosa) – moixernó, seta de abril, seta de primavera, seta de San Jorge – St. George's mushroom
Tricholoma matsutake – champiñón de pino japonés – matsutake mushroom
Tricholoma sejunctum – pastel, yema de huevo
Tricholoma terreum – fredeluch, fredolic, negrilla, seta de estepa, tricoloma portentoso – gray agaric
Tricholoma vaccinum – corralito café – scaly agaric
Trichosanthes cucumerina* var *anguina – calabaza anguina, calabaza de culebra, patole, serpiente, serpiente vegetal – club gourd, Indian snake gourd, viper's gourd
Trichosanthes dioica – patole – pointed gourd
Trichosanthes kirilowii – pepino chino – Chinese cucumber
Trichostema lanatum – hierba del pescado, romero, romero de la sierra – romero, vinegar weed, woolly bluecurls
Trichostema parishii – romerillo del país, romero, romero del país – bluecurls
Tridens albescens – tridente agrio – white tridens
Tridens elongatus – tridente alargado – rough tridens
Tridens eragrostoides – tridente fino
Tridens muticus – tridente, tridente esbelto – slim tridens
Tridens texanus – tridente texano
Tridens = Erioneuron
Trientalis arctica – trientalis – arctic starflower, northern starflower
Trifolium alexandrinum – bersim, trébol alejandrino, trébol bersim, trébol de Alejandría, trébol verdadero – berseem clover, Egyptian clover
Trifolium angustifolium – trébol de hojas estrechas – narrow-leaved clover
Trifolium arvense – trébol, trébol arvense – rabbitfoot clover
Trifolium aureum – trébol dorado – hop clover
Trifolium campestre – fenarda, trébol amarillo, trébol campestre, trébol del campo – hop clover
Trifolium dubium – trébol amarillo, trébol dubio – little hop clover, shamrock
Trifolium fragiferum – fresa de burro, trébol fresa, trébol fresero – strawberry clover
Trifolium glomeratum – trébol de flores en glomérulo – clustered clover
Trifolium hirtum – trébol hirto, trébol roso – rose clover
Trifolium hybridum – trébol alsike, trébol bastardo, trébol de alsike, trébol de Suecia, trébol híbrido, trébol negro, trébol sueco – alsike clover
Trifolium incarnatum – trébol del rosellón, trébol encarnado, trébol morado, trébol rosado – crimson clover
Trifolium macrocephalum – trébol fresa – strawberry clover
Trifolium medium – trébol intermedio, trébol medio, trébol zigzag – mammoth clover
Trifolium microcephalum – cascabelito – maiden clover
Trifolium pannonicum – trébol de Hungría – Hungarian clover
Trifolium pratense – clavo rojo, melga, trébol, trébol común, trébol del prado, trébol morado, trébol rojo, trébol rosado, trébol violeto – red clover
Trifolium procumbens – trébol amarillo – hop clover
Trifolium repens – flor de primavera, ladino, trébol amargo, trébol blanco, trébol blanco

ladino, trébol de Holanda, trébol holandés, trébol ladigiano, trébol ladino, trébol rastrero – Dutch white clover, white clover

Trifolium repens **ssp** ***prostratum*** – trébol rampante

Trifolium resupinatum – trébol de Persia, trébol persa, trébol pérsico – Persian clover

Trifolium subterraneum – trébol rastrero, trébol subterráneo, trébol trepador – subterranean clover

Trifolium wormskioldii – trébol – clover

Triglochin maritima – triglochin marítimo – seaside arrow-grass

Triglochin palustris – triglochin de los pantanos – marsh arrow-grass

Trigonella foenum-graecum – alholva, fenogreco, heno griego – fenugreek

Trigonidium egertonianum – boca de dragón

Trillium grandiflorum – trilio, trilio blanco – trillium, wakerobin

Triodanis = Specularia

Triodia pulchella – zacate borrego, zacate pelillo, zacatito

Triphasia trifolia – limoncillo, limoncito – lime-berry

Triplaris americana – palo María, palo santo, triplaria – ant tree

Triplasis purpurea – arenoso purpúreo – purple sandgrass

Tripsacum dactyloides – maicillo, zacate maicero – eastern gamagrass

Tripsacum floridanum – zacate maicero floridano, zacate maizar – Florida gamagrass

Tripsacum lanceolatum – espadaña, maicero, mijo silvestre, milpa de venado, prodigioso, zacate tule, zacatón – Mexican gamagrass

Tripsacum laxum – caña brava, grama del este, grama verde, guatemale, hierba de Guatemala, hierba maravilla, hierba prodigio, malote, manillo, maravilla, nutriol, pal, pasto de Guatemala, prodigiosa, saladilla, zacate grama, zacate Guatemala, zacate prodigio – gamagrass, grama grass

Trisetum flavescens – avena rubia, triseto amarillo, triseto rubio – yellow oatgrass

Trisetum spicatum – tres cerdas espigado, triseto espigado – northern oatgrass

Trithrinax brasiliensis – palmera de Buruti

Trithrinax campestris – caranday

Triticum aestivum – 1. caztaz, trigo, trigo blando, trigo candela, trigo pelón – wheat; 2. trigo de invierno – winter wheat; 3. hierba de trigo – wheat grass

Triticum aestivum **var** ***spelta*** – espelta – spelt

Triticum durum – trigo duro, trigo moruno – hard wheat

Triticum secale – triticale – triticale

Triticum vulgare – grano, trigo, trigo tierno – wheat

Tritoma = Kniphofia

Tritonia crocata – avispa, tritonia – flame freesia

Tritonia = Crocosmia

Triumfetta galeottiana – pastora

Triumfetta goldmanii – chapote

Triumfetta oxyphylla – cajete

Triumfetta polyandra – cadillo, guachaporo de la sierra, jonote, pastora – paroquet-bur

Triumfetta semitriloba – abrojo, batel, cadillo, cadillo cimarrón, cadillo malva, cardillo, guachaporito, güizapol, güizapol de borrego, hierba de la hormiga, huizapolillo, majahuilla, mozote, mozote de caballo, muloch, ochmuul, pega pega, piojo de caballo – burr bush, linden

Trixis angustifolia – árnica, falsa árnica, hierba del aire, Moctezuma del campo, plumilla

Trixis californica – árnica, cachana, plumilla, Santa Lucía

Trixis radialis – falsa árnica, hierba del aire, plumilla, tabi, tokabán

Trixis wrightii – hierba del aire

Trollius europaeus – calderón, flor de San Pallari – globeflower

Tropaeolum majus – capuchina, capuchina mayor, cuitziquíendas, curutziti, espuela de gañán, flor de sangre, llagas de Cristo, marañuela, mastuerzo, mastuerzo de las Indias, nasturcio, pajarito, palomilla, pelón, pelonchili, pelonmexixquilitl – garden nasturtium

Trophis racemosa – campanilla, chacox, confitura, eldorado, granadilla, huanchal, jucite, leche María, papelillo, poxoncahuitl, pu'sí, ramón, ramón blanco, ramoncillo, ramón colorado, ramón de caballo, ramón de

Castilla, tulipán, yaxox – breadnut, ramoon, white ramon
Tsuga canadensis – tsuga, tsuga del Canadá, tsuga oriental – Canada hemlock, eastern hemlock
Tsuga caroliniana – tsuga de Carolina – Carolina hemlock
Tsuga heterophylla – tsuga, tsuga de Alberta, tsuga del Pacífico, tsuga heterofila – western hemlock
Tsuga mertensiana – tsuga de California, tsuga de la costa occidental, tsuga de montaña, tsuga de Patton, tsuga negra – mountain hemlock
Tuber aestivum – trufa de verano – red-grained black truffle, summer truffle
Tuber album – trufa blanca – white truffle
Tuber borchii – trufa blanca – whitish truffle
Tuber brumale – trufa de invierno, trufa violeta – violet truffle, winter truffle
Tuber cibarium – criadilla – black truffle
Tuber magnatum – trufa blanca, trufa de Piamonte, trufa gris – gray truffle, Italian truffle, Piedmont truffle
Tuber melanosporum – criadilla de tierra, trufa de invierno, trufa violeta – black truffle, French truffle
Tuber rufum – trufa roja – red truffle
Tuber uncinatum – trufa gris – burgundy truffle
Tuberaria guttata – hierba turmera, tuberaria manchada – spotted rock-rose
Tubiflora squamosa – anisillo, corazón de Jesús, cordoncillo, hierba del sagrado corazón de Jesús, pie de gallo
Tulipa australis – burilla, tulipán austral – southern tulip
Tulipa clusiana – tulipán de Clusius – candy tulip, lady tulip
Tulipa gesneriana – 1. tulipán, tulipán de los jardineros – garden tulip, tulip; 2. tulipán precoz, tulipán temprano sencillo – single early tulip; 3. tulipán temprano doble, tulipán temprano lleno – double early tulip; 4. tulipán fosteriana – emperor tulip; 5. tulipán triumph – triumph tulip; 6. tulipán darwin – Darwin tulip and Darwin hybrids; 7. tulipán de flor de azucena, tulipán de flor de lirio – lily-flowered tulip; 8. tulipán campestre – cottage tulip; 9. tulipán lefeber – mayflowering tulip, single late tulip; 10. tulipán tardio doble, tulipán tardio lleno – double late tulip, peony tulip; 11. tulipán rembrandt – Rembrandt tulip; 12. tulipán papagayo – parrot tulip; 13. tulipán verde – green tulip, viridiflora tulip
Tulipa saxatilis – tulipán saxátil – rock tulip
Tulipa sharonensis – rosa de Sarón – rose of Sharon, Sharon tulip
Tulipa sylvestris – burilla, tulipán silvestre – wild tulip
Turbina (Rivea) corymbosa – badoh, batatilla, campanilla, coatlxoxouqui, coaxihuitl, flor de la Virgen, flor de pascua, hierba de la Virgen, hueyytzontecon, loquetico, manto, ololiuquí, pascua, patito, piule, santa, semilla de la Virgen, señorita, tumba caballo, xtabentún – giant white morning-glory
Turnera aphrodisiaca – damiana – damiana
Turnera diffusa – damiana, damiana de California, damiana mexicana, escobilla, hierba de la pastora, hierba del pastor, hierba del venado, itamo real, misibcoc, pastorcita, peludillo, tabuquit – damiana
Turnera pumila – bruja, damiana
Turnera ulmifolia – amaranto, caléndula, clavel de oro, damiana, marilopez, peludillo, saráame – Bahama buttercup, large-leaf damiana, sage rose, West Indian holly, yellow alder
Turpinia occidentalis – manzanillo, manzanito
Tussilago farfara – fárfara, pata de asno, pata de mula, pata de toro, pie de caballo, tabaco británico, tusílago, uña de asno, uña de caballo – coltsfoot
Tynnanthus panurensis – clavohuasca
Typha angustifolia – anea, anea de hojas estrechas, bayón, espadaña, puh, tibinagua, tolpatlactli, tule – narrow-leaved cattail
Typha domingensis – cola de gato, cola pecho, espadaña, hueizacat, maza de agua, petalzimicua, puh, tule, vela de sabana, zacate ignea – southern cattail
Typha latifolia – aguapá, anea, anea de hojas anchas, bobordo, cola de gato, cola de topo, espada, espadaña, gladiolo, junco, maza de agua, ñisñil, plumilla, tolpatlacti, totora, tule, tule espadilla – broad-leaved cattail

U

Ulex europaea – árgoma, aulaga, escajo, gatuña, hiniesta espinosa, tojo – furze, gorse

Ulex parviflorus – árgoma de flores pequeñas, aulaga morisca

Ullucus – ulluco

Ulmus americana – olmedo, olmo, olmo americano, olmo blanco americano, olmo norteamericano – American elm

Ulmus campestris – álamo negro, negrillo, olmo, olmo de los campos

Ulmus crassifolia – olmo – cedar elm, Texas elm

Ulmus fulva – olmo norteamericano – slippery elm

Ulmus glabra – olmo de montaña, olmo de monte, olmo montano, sándalo falso – Scotch elm, wych elm

Ulmus hollandica – olmo holandés – Dutch elm

Ulmus laevis – olmo pedunculado, olmo temblón – European white elm

Ulmus (Chaetoptelea) mexicana – baqueta, cempoaléhuatl, chaperno, chuchum, cuerillo, cuero, ilite, mezcal, nuculpat, olmo, palo de baqueta, papalote, petatillo – Mexican elm, ramrod tree

Ulmus minor – negrillo, olmo, olmo común – English elm, smooth-leaved elm

Ulmus parvifolia – olmo chino, olmo de hojas pequeñas – Chinese elm, Chinese evergreen elm

Ulmus procera – olmo campestre, olmo común, olmo inglés, olmo negrillo – English elm

Ulmus pumila – olmo de Siberia, olmo siberiano – Siberian elm

Umbellularia californica – laurel, laurel de California, laurel de la sierra, laurel silvestre, mirto de Oregón – California bay, California laurel, Oregon myrtle, pepperwood

Uncaria tomentosa – uña de gato, uña de gavilán – cat's claw

Ungnadia speciosa – castaño de México, mona, monilla – Mexican buckeye, Spanish buckeye, Texas buckeye

Uniola latifolia – canastilla de hojas anchas – broadleaf uniola

Urena lobata – aramina, cadillo, guaxima, malva blanca, urena – aramina

Urera baccifera – chichicastle, cuahtzitzicaztli, ortiga, ortiga de caballo

Urera caracasana (Urtica chichicaztli) – carne de caballo, cuahtzitzicaztli, chichicastle, chichicazillo, laal, laltsimin, mala mujer, mal hombre, oncinapor, ortiga, ortiga de caballo, ortiga quemadora, ortiguilla, quemador, tachinole, tlachinole, tumalí, xiopatli – flameberry scratchbush, nettle

Urginea maritima – albarrana, cebolla albarrana, cebolla de grajo, escila, esquila – sea onion

Urginea scilla – cebolla marítima, escila, escila blanca, escila roja

Urospermum picroides – barba de viejo

Urtica dioica – chichicastle, dominguilla, guichibidu, hierba del ciego, mala mujer, ortiga, ortiga común, ortiga cuerda grande, ortiga mayor, ortiga punzante, solimán – spurge nettle, stinging nettle

Urtica dioica* ssp *gracilis – ortiguilla – American stinging nettle

Urtica serra (breweri) – ortiga

Urtica urens – chichicastle, ortiga, ortiga menor, ortiga pequeña – dwarf nettle

Urtica = Urera

Urticularia minor – lentibularia menor – bladderwort

Urticularia vulgaris – lama, lentibularia – common bladderwort

Usnea barbata – barba de árbol, barba de viejo, heno barbón, liquen – tree lichen

Ustilago carbo – carbón de la avena, carbón del pasto inglés – oat smut, rye smut

Ustilago hordei – carbón de la cebada – barley smut

Ustilago maydis – carbón del maíz, cuernecillo del maíz, cuerno, cuitlacoche, dentó, donta, hongo del maíz, huitlacoche, papiotl, tahchaac, tizón del maíz – corn smut

Ustilago tritici (segetum) – carbón del trigo, tizón del trigo – wheat smut

Utricularia gibba – lentibularia gibosa – bladderwort

Utricularia intermedia – lentibularia intermedia – bladderwort
Utricularia vulgaris – lama – common bladderwort, greater bladderwort

V

Vaccaria hispanica – hierba de la vaca, vaccaria – cow-herb
Vaccinium confertum – cerezo, congón, granjeno, madroñito, madroño chino – blueberry
Vaccinium corymbosum – arándano, arándano americano, arándano azul, arándano corimboso, madroñito – blueberry, highbush blueberry
Vaccinium macrocarpon – arándano, arándano agrio, arándano macrocarpo, arándano trepado, baya de grulla – cranberry
Vaccinium myrtillus – 1. ráspano – huckleberry; 2. arándano, mirtillo, rasponera, uva del bosque – bilberry
Vaccinium oxycoccus – arándano agrio, arándano europeo – cranberry
Vaccinium uliginosum – arándano de los pantanos, arándano negro, baya – alpine bilberry, European cranberry
Vaccinium vitis-idaea – arándano encarnado, arándano rojo – cowberry, foxberry, red whortleberry
Vaccinium vitis-idaea* var *minus – arándano encarnado – lingonberry
Valeriana acutiloba – raíz de valeriana – valerian
Valeriana ceratophylla – hierba del gato, mazatán, raíz del gato, raíz del oso, ucuares, valeriana
Valeriana clematitis – nube de campo, uña de gato
Valeriana edulis – chinonua, raíz del gato, valeriana
Valeriana mexicana – hierba del gato, raíz del gato, valeriana
Valeriana officinalis – hierba del gato, valeriana, valeriana medicinal, valeriana menor, valeriana oficinal – garden heliotrope, valerian
Valeriana sorbifolia – canónigo, hierba del gato, matazetes, mazatil, raíz del gato, raíz del oso, ucuares, valeriana, vauar – tobacco root, valerian, valeriana
Valeriana = Centranthus
Valerianella eriocarpa – canónigo de Italia, valerianela de Italia – Italian corn salad
Valerianella locusta (olitoria) – canónigo, hierba del canónigo, lechuga de campo, valerianela – corn salad, lamb's lettuce
Vallesia glabra – cacaragua, caracaline, citabaro, cristalillo, crucecillo, frutilla, huevito, huitatave, otatave, palo verde, sitavaro, uclatave
Vallesia laciniata – huitatave, jazmín de nacapule, otatave
Vanilla planifolia (fragrans) – canela de cuya, cuomecaxot, tlilxóchitl, vainilla, vainilla bastarda, vainilla cimarrona, vainilla conianona, vainilla de cochino, vainilla de mono, vainilla escarchada, vainilla fina, vainilla mansa, zizbic – vanilla
Vanilla pompona – plátano vainilla, senexante, vainilla, vainilla boba, vainilla francesa, vainilla plátano, vainilla pompona, vainilla senexante, vainillón – French vanilla, West Indian vanilla
Vanilla sylvestris – vainilla cimarrona
Vascellum pratense – bolita de tierra, cuesco de lobo
Vauquelinia californica – árbol prieto, guayule – Arizona rosewood
Vauquelinia corymbosa – árbol prieto, guayul, guayule, palo prieto, palo verde, ucas – slimleaf vauquelinia
Vauquelinia karwinskyi – palo de boda, vauquelinia – false rosewood
Veitchia merrillii – palma de Manila
Ventenata dubia – ventenata avena falsa – soft-bearded oatgrass
Veratrum album – ballestera, cebadilla, eléboro blanco, hierba ballestera, vedegambre, vedegambre blanco – false hellebore, white hellebore
Veratrum californicum – cebadilla
Veratrum frigidum – cebadilla, zohzoyatic
Veratrum officinale – itzcuimpatli, quimichpatli, veratro – corn-lily, false hellebore
Veratrum sabadillo – cebadilla de la sierra, eléboro negro
Veratrum viride – vedegambre verde – corn-lily, false hellebore
Verbascum blattaria – polillera – moth mullein

Verbascum bombyciferum – gordolobo – mullein
Verbascum chaixii – gordolobo austriaco, verbasco austriaco – Austrian mullein
Verbascum dumulosum – gordolobo rocoso – mullein
Verbascum phoeniceum – gordolobo púrpura, gordolobo purpúreo, verbasco purpúreo – purple mullein
Verbascum speciosum – gordolobo especioso, verbasco especioso – showy mullein
Verbascum thapsus – candela regia, candelaria, cirio de Nuestra Señora, gordolobo, guardalobo, hopo de zorra, punchón, tabaco cimarrón, tripa, verbasco – common mullein, great mullein, woolly mullein
Verbascum virgatum – gordolobo ramificado, verbasco ramificado – wand mullein
Verbena ambrosifolia – moradilla – Dakota vervain, pink vervain
Verbena carolina – hierba de San José, hierba de San Juan, verbena – verbena
Verbena ciliata – alfombrilla del campo, moradilla, verbena, verbena mexicana – Mexican vervain
Verbena corymbosa – alfombrilla, alfombrilla de jardín
Verbena ehrenbergiana – verbena
Verbena elegans* var *asperata – moradilla, verbena
Verbena halei – hierba del gusano – slender vervain, Texas vervain
Verbena hastata – verbena azul – blue vervain
Verbena* x *hybrida (*x *hortensis) – verbena – garden verbena
Verbena macdougali – verbena – blue vervain
Verbena menthifolia – bercul, ueijuli
Verbena neomexicana – verbena común
Verbena officinalis – dormilón, hierba de los males, hierba sagrada, hierba santa, verbena, verbena fina, verbena macho, verbena mayor – European vervain
Verbena orcuttiana – verbena azul
Verbena peruviana (chamaedrifolia) – alfombrilla
Verbena pumila – bartanucha, cahuillo – low verbena, pink vervain
Verbena wrightii – moradilla – desert vervain, Wright's verbena
Verbesina crocata – capitaneja, capitaneja anaranjada, nahuiliput
Verbesina encelioides – añil del muerto, capitaneja, girasolillo, hierba de la bruja, oreja de lobo, verbesina – cow pen daisy, golden crownbeard, goldweed
Verbesina erosa – capitaneja – crownbeard
Verbesina oncophora – apapatlaco
Verbesina pedunculosa – capitaneja, capitaneja de Tonila
Verbesina pustulata – tacote chino
Vernonia deppeana – cihuapatli, flor de cuaresma, hierba hermosa, malacate, siquinay, tachiste, tziquinay, tzitzit, xiquite – ironweed
Veronica alpina – verónica alpina – alpine speedwell
Veronica americana – verónica – American brooklime
Veronica anagallis-aquatica – anagálide acuática – water speedwell
Veronica arvensis – verónica arvense – speedwell
Veronica austriaca – verónica austriaca – large speedwell
Veronica beccabunga – becabunga, hierba azul, picante de la boca, verónica acuática – brooklime
Veronica catenata – verónica comosa – chain speedwell
Veronica chamaedrys – verónica común – germander speedwell
Veronica filiformis – verónica filiforme, verónica rocosa – speedwell
Veronica gentianoides – verónica – gentian speedwell
Veronica hederifolia – té de Europa, verónica de hojas de hiedra – ivy-leaved speedwell
Veronica officinalis – té de Europa, verónica, verónica macho – heath speedwell
Veronica peregrina – verónica, verónica peregrina – American speedwell, purslane speedwell
Veronica persica – verónica pérsica – Persian speedwell
Veronica prostrata – verónica prostrada – prostrate speedwell

Veronica repens – verónica repente – creeping speedwell
Veronica scutellata – verónica de los pantanos, verónica escutelada – marsh speedwell
Veronica serpyllifolia – verónica de hojas de serpillo – thyme-leaved speedwell
Veronica triphyllos – verónica trifoliada – fingered speedwell
Veronica spicata – nueva reforma, verónica, verónica espigada – speedwell
Veronica spicata* ssp *incana – veronica blanquecina – silver speedwell
Vetiveria zizanioides – pasto violeta, vetiver – vetiver
Viburnum awabukii (odoratissimum) – sardonia – sweet viburnum
Viburnum farreri (fragrans) – viburno – viburnum
Viburnum glabratum – achichil, carindapaz
Viburnum lantana – barbadejo, lantana, mentironera, morrionera, pierna, vitilaina – wayfaring tree
Viburnum opulus – bola de nieve, mundillo, rodela, rosa de Güeldres – European cranberry bush, guelder rose
Viburnum plicatum (tomentosum) – viburno lanudo – Japanese snowball
Viburnum prunifolium – viburno – black haw
Viburnum rhytidophyllum – viburno – leatherleaf viburnum
Viburnum tinus (laurustinus) – barbadejo, barbadija, bola de nieve, duraznillo, durillo, tino, viburno – laurustinus
Vicia articulata – algarroba, algarroba de Castilla, algarroba genuina, alverja, arveja articulada, garroba, lenteja de Aragón – one-flowered vetch
Vicia benghalensis – arveja atropurpúrea, veza purpúrea – purple vetch
Vicia cracca – arveja, arveja glabra, veza cracca, veza de pájaro, veza francesa – bird vetch
Vicia disperma – arveja de dos semillas – vetch
Vicia ervilia – alcarceña, alverja, ervilla, lenteja bastarda, tito, yero – bitter vetch
Vicia faba – alampepe, faba, frijol fava, haba, haba coba, haba común, haba de huerta, haba falsa, haba fava, haba loca, haba mayor, habichuela, veza – broad bean, fava bean
Vicia faba* var *equina – haba caballar, haba panosa – horsebean
Vicia faba* var *minuta – haba menor – tickbean
Vicia hirsuta – alverja erizada, alverjón, arveja hirsuta, veza hirsuta – hairy tare, tiny vetch
Vicia lathyroides – arveja de primavera – spring vetch
Vicia lutea – arveja amarilla, arvejón – yellow vetch
Vicia narbonensis – alverjón, arveja francesa, haba loca, haba silvestre – French vetch, Narbonne vetch
Vicia pannonica – arveja húngara, veza de Hungría, veza húngara, veza panonica, vicia de Hungría – Hungarian vetch
Vicia sativa – algarroba común, algarrobilla, alverja, arveja, arveja común, arveja sativa, ebo, garrobilla, janamargo, jara amarga, veza, veza común, veza forrajera – spring vetch
Vicia sativa* ssp *nigra – alverjilla, veza de hojas estrechas – common vetch, narrow-leaved vetch
Vicia sepium – arveja, arveja de los cercados, arveja de los setos, veza – bush vetch
Vicia tetrasperma – arveja de cuatro semillas, veza tetrasperma – sparrow vetch
Vicia villosa – arveja vellosa, arvejilla velluda, veza, veza de arena, veza vellosa, veza velluda, veza velluda de invierno – fodder vetch, hairy vetch, winter vetch
Victoria regia – irupé, maruru, Victoria regia – Queen Victoria waterlily, royal water-lily
Vigna (Phaseolus) angularis – frijol adzuki, judía adzuki, poroto arroz – adzuki bean
Vigna caracalla – caracolillo, caracól real – snail flower
Vigna radiata (Phaseolus aureus) – alubia mungo, azuquia radiata, frijol chino, frijol de oro, frijol mungo, judía de mungo, mungo, poroto mungo, semilla oriental – green gram, mung bean
Vigna sinensis – chícharo de vaca, frijol, frijol de campo, frijol de cuerno, frijol de maíz, frijol ojinegro, poroto, vigna – black-eye pea, cowpea
Vigna subterranea – bambarra, guandsú,

guisante de tierra, maní africano, maní de bambarra – earth pea, ground bean, hog peanut

Vigna umbellata – frijol de arroz, frijol rojo, judía arroz – rice bean

Vigna unguiculata – arveja de vaca, caragilate, caupí, chícharo de vaca, chícharo tropical, costeño, frijol, frijol de Castilla, frijol de cuerno, frijol de maíz, frijol de ojos negros, frijol de ombligo negro, frijol de pinta negra, frijol de vaca, frijol precioso, frijol tripa de tuza, guisante de carita, guisante de vaca, judía carilla, judía de careta, judía de vaca, rabiza – black-eyed pea, cowpea, crowder pea

Vigna unguiculata* ssp *cylindrica – judía catjang – Indian cowpea

Vigna unguiculata* ssp *sesquipedalis – colón rojo, dólico espárrago, dólico gigante, frijol castillo, frijol del Perú, frijolera, judía espárrago, poroto espárrago – asparagus bean, yard-long bean

Viguiera dentata – chamizo, girasol, jarilla, liga, raíz de manso, romerillo de la costa, tah, tlalpopolote – goldeneye

Viguiera laciniata – margarita, tacote, tecote, viguiera – San Diego county viguiera, San Diego sunflower

Viguiera linearis – romerillo

Viguiera montana – ariosa, huachomó

Viguiera parishii (deltoidea* var *parishii) – ariosa, tacote – desert goldenbush, golden-eye

Viguiera similis – tacote

Viguiera tomentosa – tecote – San Diego sunflower

Vinca major – cielo raso, hierba doncella, molinillo, pervinca, vincapervinca mayor – greater periwinkle

Vinca minor – brusela, cielo raso, flor de paragüito, hierba doncella, hierba doncella menor, pervinca, vinca, vincapervinca, violeta de la bruja – dwarf periwinkle, lesser periwinkle

Vinca = Catharanthus

Vincetoxicum caudatum – talayote

Vincetoxicum hastulatum – talayote, tlalayote – milkvine

Viola arvensis – pensamiento silvestre, violeta del campo – field pansy, heart's ease

Viola cornuta – violeta, violeta cornuda – horned violet, viola, tufted pansy

Viola flagelliformis – violeta amarilla, violeta de monte – yellow violet

Viola hederacea – violeta australiana – Australian violet

Viola labradorica – violeta rocosa

Viola odorata – hierba del aguardiente, viola, violeta, violeta común, violeta de jardín, violeta de olor, violeta española, violeta extranjera, violeta olorosa – English violet, sweet violet

Viola painteri – violeta de monte

Viola palustris – violeta de los pantanos, violeta palustre – marsh violet

Viola pedunculata – gallito, pensamiento, violeta – Johnny-jump-up, pansy, wild pansy

Viola riviniana – violeta de Rivini – common dog violet

Viola tricolor – pensamiento, pensamiento silvestre, suegras y nueras, trinitaria, violeta – Johnny-jump-up, viola, wild pansy

Viola* x *wittrockiana – flor de la trinidad, pensamiento, violeta de hortelano – garden pansy, garden violet, pansy

Viscainoa geniculata – garambullo, guayacán, viscainoa

Viscum album – almuérdago, arfueyo, liga, muérdago, muérdago europeo, vesque, visco blanco, visco cuercino – European mistletoe

Vitex agnus-castus – agnocasto, gatillo casto, pimiento loco, sauz gatillo – chaste tree

Vitex gaumeri – barrabás, carrete – fiddlewood

Vitex mollis – aguamalario, agüilolote, agüilote, ahuilote, angualamo, atuto, cajúguari, coyotomate, huhuhuali, igualamo, jari, júbare, juguari, jujuhuali, jupari, negro coyote, obalamo, tescalama, torete, uvalama – chaste tree

Vitex pyramidata – ahuilote, cajuari, jupari, nanche de perro, negrito, negrito coyote, negrito jupari, negro coyote, querendanicua, quereque, tescalama, uvalama, yascabté

Vitex trifolia – carnavalito – pigeon berry

Vitis arizonica – cerahui, envolver, parra, parra cimarrona, uva, uva cimarrona, vid – canyon grape

Vitis californica – parra, parra cimarrona, uva cimarrona – California wild grape

Vitis cinerea – parra silvestre, uva, uva cimarrona

Vitis coignetiae – vid ornamental, vid ornamental del Japón – crimson glory vine

Vitis labrusca – labrusca, parriza, parriza americana, uva, vid americana, vid silvestre – calawba grape, fox grape, grape

Vitis peninsularis – parra, uva cimarrona

Vitis rupestris – vid americana – bush grape

Vitis sylvestris – vid silvestre – woodland grape

Vitis tiliifolia – bejuco de agua, bejuco de cazadores, bicholi, brincadora, parra, parra brincadora, parra bronca, parra broncadora, parra silvestre, pichol, sanalotodo, temecate, temecatl, tripa de Judas, tripa de vaca, uva, uva cimarrona, uva del monte, uvilla, uvilla cimarrona, vid silvestre, xocomecatl – wild grape

Vitis vinifera – bicholi, parra, uva, vid, vid común, vid silvestre, yaga-bicholi – cultivated grape, grape vine, wine grape

Vitis vinifera (rotundifolia)* var *muscat – moscadel del grano menudo, uva de Muscadino – Muscat grape

Vitis vinifera* var *sativa – airén, garnacha, palomino, tempranilla – wine varietal grapes

Voandzeia subterranea – bambarra, guisante bambarra – bambarra groundnut, earth pea

Vulpia bromoides – vulpia bromo-falso – squirreltail fescue

Vulpia myuros – vulpia cola de rata – ratstail fescue

W

Waltheria americana – cuahuilotillo, cuaulotillo, hierba del ángel, hierba del cáncer, hierba del pasmo, hierba del soldado, malva, malva blanca, malva de llano, malva del monte, malva rastrera, tapacola, tzitziquilitl, zacmizib, zacxiu
Waltheria indica – hierba del pasmo
Warszewiczia coccinea – barba de gallo, crucero, guna, sangrenaria – chaconia, pride of Trinidad, wild poinsettia
Washingtonia filifera – palma, palma de abanico, palma de Castilla, palmera, palmera de California, palmito, palo amarillo, washingtonia – California fan palm
Washingtonia robusta (sonorae) – abanico, palma, palma blanca, palma colorada, palma de Castilla, palma negra, palmera, palmera mexicana – Mexican fan palm, skyduster
Wedelia incarnata – hierba de la hormiga
Wedelia = Zexmenia
Weigela florida – veigelia
Whipplea modesta – hierba de selva – modesty, yerba de selva, yerba selva
Wigandia caracasana – chichicastle, consuelda mayor, hoja de San Pablo, mala mujer, ortiga, ortiga de tierra caliente, ortiga grande, palo de San Pablo, quemador, San Pablo, sosa, tabaco cimarrón, tabacón, tabaquillo
Wigandia kunthii – chichicastle, mala mujer, ortiga, ortiga de tierra caliente, ortiga grande, ortiga prieta, ortiga real, pahpatlanuac, quemador, sosa, tabaco cimarrón, tabaquillo, tzitzicaxtli
Wigandia urens – ortiga, ortiga de tierra caliente, tabaco cimarrón, tabaquillo
Wilcoxia = Echinocereus, Peniocereus
Willardia = Lonchocarpus
Wimmeria confusa – acedilla, algodoncillo, cedilla, papelillo
Wimmeria mexicana – algodoncillo, chi'ini, machicari, papelillo
Wislizenia refracta – guaco – jackass clover
Wissadula amplissima – pintapan – velvet-leaf
Wisteria sinensis – glicina, vistaria – Chinese wisteria
Withania somnifera – ashwaganda, orovale
Wyethia helenioides – oreja de liebre – mules ears

X

Xanthium canadensis – abrojo, cadillo, piojo de burro – cocklebur

Xanthium spinosum – cachurrera menor, cepa caballo, pegote – spiny cocklebur

Xanthium strumarium – abrojo, arrancamoños, bardana menor, cachurra, cadillo, cadillo común, cardillo, wisapole – common cocklebur

Xanthocephalum = Gutierrezia

Xanthorrhiza simplicissima – raíz amarilla – yellowroot

Xanthosoma caracu – yautia, yautia horqueta

Xanthosoma nigrum – metstsonkilit

Xanthosoma robustum – arámicua, cacalacaxtli, capote, capote blanco grande, capote de monte, caramicua, colomo, exquiquílit, hoja elegante, lampazo, mafafa, malanga, pise, quequeste, rejalgar, tarabundí, yautia – elephant ear

Xanthosoma roseum – capote blanco, huacalxóchitl, poco

Xanthosoma sagittifolium – aramicua, caramicua, colomo anona, macal, mafafa, malanga, malanga amarilla, mangara, mangarito, manto de la reina, ocumo, pixi, qualuza, tarabundi, taya, tiquisque, uncucha, yuatia – elephant ear, new cocoyam, tannia

Xanthosoma violaceum – apish, capote de jardín, malanga, tequescamote

Xeranthemum annuum – inmortal, perpetua anual – everlasting, common immortelle

Ximenia americana – limoncillo, limón de mar, manzanillo, molcol, nanchicacao, untzincacá de montaña, ximenia – seaside plum, tallow wood, wild lime

Xylosma characantha – aguja de ara, aguja rabia, copalín, corona, puntezapat, tirazapato

Xylosma flexuosum – brujo, corona de santo, coronilla, huichichiltémel, malacate, manzanillo, palo de brujo – brush holly, sweetwood

Y

Yucca aloifolia – bayoneta, flor de pito, iczotl, iczotli, izote, yuca – aloe yucca, Spanish bayonet
Yucca arizonica – dátil, hamat, palma criolla – Arizona yucca, blue yucca, Spanish dagger
Yucca baccata – amole, dátil, palma, palma criolla, palmilla, palmilla ancha, yuca – banana yucca, blue yucca, datil, Spanish bayonet
Yucca brevifolia – árbol de Josué – Joshua tree
Yucca (Samuela) carnerosana – ixtle de palma, palma, palma barreta, palma de ixtle, palma de San José, palma de San Pedro, palma loca, palma samandoca, samandoca – giant Spanish dagger
Yucca decipiens – izote, palma, palma china, palma de San Pedro, palmilla, sotol
Yucca declinata – dátil
Yucca elata – amole, cortadillo, dátil, palma datillo, palmilla, palmito, soyate – soaptree yucca, soapweed
Yucca elephantipes (gigantea) – coyol, dátil, espadín, iczotl, izote, ocozote, palhmaj, palma, palmita, telicso, tuc, yuca – giant yucca, spineless yucca
Yucca enlichiana – pitilla
Yucca faxoniana – palma, palma de San Pedro – Eve's needle
Yucca filamentosa – palma china, yuca – Adam's needle
Yucca filifera (australis) – coyol, habilla, izote, palma, palma china, palma corriente, palma de aparejo, palma de San Pedro, palma de sudadero, palma grande, palma loca, palma samandoca, quiote, yuca – St. Peter's palm, tree yucca
Yucca gloriosa – yuca, yuca brillante – soft-tip yucca, Spanish dagger
Yucca grandiflora – dátil, sahuiliqui
Yucca guatemalensis – bayoneta, coyol, iczotl, itabo, izote, izotl, palma, palmita, tzamuc-cóyol, yuca – bluestem yucca, elephant yucca
Yucca macrocarpa – izote, palma, palma criolla
Yucca rigida – palma de San José, palmilla, palmita, socol – blue yucca
Yucca rostrata (thompsoniana) – amole, palmilla, palmita, soyate – beaked yucca
Yucca schidigera (mohavensis) – amole, dátil, datillo – Mojave yucca, Spanish dagger
Yucca schottii – socol, yuca – hoary yucca, mountain yucca, Schott's yucca
Yucca torreyi (macrocarpa) – palma, palma ceniza, palma china, palma criolla, palma de San Juan, palma loca, palma pita – Spanish bayonet, Spanish dagger, Torrey yucca
Yucca treculeana – coyol, izote, izotl, palma ceniza, palma de dátiles, palma de San Juan, palma loca, palma pita, pita, soyate – Spanish bayonet
Yucca valida – dátil cimarrón, dátil de monte, datilillo, datillo, láatiro – tree yucca
Yucca (Hesperoyucca) whipplei – lechuguilla, palmilla, quiote, quiotl, sotolillo – chaparral yucca, Our Lord's candle, quixote plant, Spanish bayonet

Z

Zaluzania augusta – caxtidán, cenicilla, limpiatunas
Zamia furfuracea – camotillo, palma fúnebre, palmita – cardboard palm
Zamia lindenii – palma tabaquillo
Zamia polymorpha – matarratón
Zamia pumila (integrifolia) – arrurruz de Florida – coontie, Florida arrowroot
Zannichellia palustris – zannichellia de los pantanos – horned-pondweed
Zantedeschia aethiopica – alcatraz, aro, cala, cala blanca, capote, cartucho, corneta, cucurucho, flor de embudo, lampazo, lirio de agua, lirio de Egipto, lirio de Etiopía, manto de la Virgen, tonalxóchitl – calla lily, common calla, white arum lily
Zantedeschia (Calla) palustris – aro acuático, cala – bog arum
Zanthoxylum americanum – árbol de dolor de muelas, fresno espino, fresno espinoso – prickly ash, toothache tree
Zanthoxylum arborescens – naranjillo, pipima, zorrillo – prickly ash
Zanthoxylum fagara – alacrán, colima, corriosa, espino, espino rubial, garabatillo, limoncillo, mata la sed, naranjillo, palo mulato, rabo de lagarto, tancazche, tenaza, tomequín, uña de gato, xik-ché – lime pricklyash, wild lime
Zapoteca (Calliandra) anomalia – cabellito, cabellitos de ángel, cabellitos de una vara, cabellos de ángel, cabeza de ángel, canela, clagot, coquito, hierba de canela, hierba del ángel, lele, pambotano, taguapillo, tentzonxóchitl, tepachera, tepemezquite, tepexiloxóchitl, texoxóchitl, timbrillo, tlacoxiloxóchitl, tlamacatzcatzotl, tzonxóchitl, xiloxochicuahuitl, xiloxóchitl – hair flower
Zapoteca formosa (Calliandra formosa, C. schottii) – ébano, ébano de Tabasco, huaje de caballo, tepeguaje, tosapolo – Schott caliandra
Zauschneria = Epilobium
Zea canina – maíz de coyote
Zea diploperennis – maíz chapule, milpilla
Zea mays – borona, cacahuazintle, capi, centli, cintli, cuatequil, guate, ixim, jinjillero, maíz, maíz de la India, malayo, mazorca, mijo turquesco, nahuo, nhal, panizo americano, panizo de Indias, siquilibachi, tlaoli, tlaolli, tlautlin, trigo de Turquía, tziri – corn, Indian corn, maize
Zea mays* var *amylacea – maíz blando, maíz harinoso, maíz tierno – flour corn, soft maize
Zea mays* var *ceratina – maíz ceroso – waxy corn
Zea mays* var *everia – izquitl, maíz de palomita, maíz de rosita, maíz muchito, maíz palomero, maíz reventador, maíz reventón, palomita – popcorn
Zea mays* var *gracillima – maíz enano – dwarf maize, slender corn
Zea mays* var *indentata – maíz dentado, maíz diente de caballo – dent corn
Zea mays* var *indurata – maíz córneo, maíz duro – flint corn
Zea mays* var *japonica – maíz japonés – Japanese corn
Zea mays* var *saccharata – choclo, elote, maíz azucarado, maíz de grano dulce, maíz dulce, maíz para choclo – corn-on-the-cob, sweet corn
Zea mays* var *tunicata – maíz tunicado – pod corn
Zea = Euchlaena
Zebrina = Tradescantia
Zelkova carpinifolia – azad, olmo del Cáucaso, zelkova del Cáucaso
Zelkova serrata – keyaki, zelkova japonesa, zelkova serrada – sawleaf zelkova
Zephyranthes candida – croco indio – fairy lily
Zephyranthes carinata – atzcalxóchitl, cebolla de mayo, cefirante, flor de mayo, lirio de lluvia, mañanita, mapoanita, mayito, mayo – fairy lily, zephyr flower
Zephyranthes citrina – mayitol – fairy lily
Zephyranthes longifolia – atzcalxóchitl, cebolla de mayo, cebollita, mayito, rayito – copper zephyr lily
Zephyranthes sessillis – mayito
Zephyranthes texana – atamasco – atamasco-lily, star-lily
Zexmenia aurea – atepocapatli, peonía

Zexmenia frutescens – malacate blanco de montaña, ponté, rodilla de vieja, zactah, zook
Zexmenia (Wedelia) hispida – botón de oro, clemolillo, sahumi, zexmenia
Zexmenia podocephala – peonía, pionilla, reyóchari
Zexmenia seemanni – airosa, uachomo
Zigadenus mexicanus – amole, cebadilla – death camas
Zingiber officinale – ajengibre, ancoas, jengibre, jengibre oficinal, labitz, siti – ginger, ginger root
Zinnia acerosa – hierba del burro, manzanilla del campo, zinia – dwarf zinnia, spinyleaf zinnia
Zinnia angustifolia – mal de ojo, malojo, Teresita – zinnia
Zinnia elegans – Carolina, mal de ojo, Miguel, peseta, rascamoño, Virginia, zinia – garden zinnia, youth-and-old-age, zinnia
Zinnia grandiflora – cinco llagas – plains zinnia
Zinnia haageana – rascamoño
Zinnia linearis – flor serrana, hierba del toro
Zinnia multiflora – mal de ojo, malojo, Teresita, tostón, tostón amarillo, Virginia – redstar zinnia
Zinnia pauciflora – hierba del gallo, malacatillo, malacaxóchitl, mal de ojo, malojo, viuda
Zinnia peruviana – Carolina, flor de julio, flor de junio, hierba de junio, hierba del gallo, junio del monte, mal de ojo, malojo, Teresita, Virginia – zinnia
Zinnia pumila – hierba del burro, manzanilla del campo, parraleña – desert zinnia
Zinnia tenuiflora – mal de ojo, malojo – red zinnia
Zinnia violacea – Carolina, Miguel, San Miguel, tacote
Zizania aquatica – arroz, cizaña – Canada rice, wild rice
Zizania palustris – arroz salvaja, arroz silvestre – wild rice
Ziziphus amole (sonorensis) – amole, amole dulce, bahi cápore, bicapora, brasililIo, cahuesquite, confite, corúngoro, frutillo, manzanita, manzanita de costoche, nanche de la costa, nanche saituna, naranjillo, naranjito, quesquite, saituna
Ziziphus jujuba – azufaifo, ciruelo gobernadora, jinjolero, jujube, pera cimarróna, ponsigue, quetembilla, yuyuba – jujube
Ziziphus mauritiana – aprín, azufaifo africano, ponsigue – Indian jujube
Ziziphus obtusifolia (Condalia lycoides) – abrojo, amole dulce, bachata, barchata, chaparral, chaparro prieto, ciruelo de monte, clepe, crucillo, crucillo blanco, garambullo, garrapata, jotora, palo blanco, tecomblate – graythorn, lotebush, southwestern condalia, white crucillo
Ziziphus (Condalia) parryi – crucillo – California lotebush, crucillo, lotebush
Ziziphus pedunculata – cholulo, manzanita de costoche
Ziziphus (Condalia) spathulata – abrojo, chamis, condalia, guichutilla, tecomblate – Mexican crucillo, squaw bush
Ziziphus spina-christi – azufaifo de la espina de Cristo – Christ's thorn, dhour tree
Zornia bracteata – viperina
Zornia diphylla – encarrugada, hierba de la víbora, raíz de víbora, viperina, zornia trencilla
Zornia reticulata – cascabelito, hierba de la víbora
Zornia thymifolia – hierba de la víbora, raíz de víbora
Zornia venosa – hierba de la víbora
Zostera marina – hierba de mar mayor, trigo de mar, zacate del mar – eel-grass
Zygadenus = Zigadenus
Zygophyllum fabago – morsana – bean-caper
Zygocactus = Schlumbergera

Spanish

aahuatin – *Karwinskia humboldtiana*
aaqui – *Stenocereus thurberi*
ababol – *Papaver rhoeas*
abacá – *Musa textilis*
abacachi – *Ananas sativus*
abacamiel – *Combretum farinosum*
abacate – *Persea americana, P. drymifolia*
abahuma – *Couroupita guianensis*
abal – *Spondias mombin, S. purpurea*
abala – *Papaver rhoeas*
abalo – *Sideroxylon laetevirens*
abanico – *Amaranthus caudatus, Celosia, Washingtonia robusta*
abaso – *Populus dimorpha, P. mexicana*
abedul – *Alnus acuminata, A. arguta, A. jorullensis, Betula*
abedul amarillo americano – *Betula alleghaniensis*
abedul amarillo canadiense – *Betula alleghaniensis*
abedul blanco – *Betula alba, B. verrucosa*
abedul-cerezo americano – *Betula lenta*
abedul común – *Betula pendula*
abedul común europeo – *Betula pendula*
abedul chino de corteza roja – *Betula albo-sinensis*
abedul de agua – *Betula nigra*
abedul de Europa – *Betula pendula*
abedul de hojas de carpe – *Betula lenta*
abedul de Jacquemont – *Betula jacquemontii*
abedul de la India – *Betula jacquemontii*
abedul del Himalaya – *Betula jacquemontii*
abedul del Himalaya de corteza blanca – *Betula jacquemontii*
abedul de los Alleghanys – *Betula alleghaniensis*
abedul de los cánoes – *Betula papyrifera*
abedul dulce americano – *Betula lenta*
abedul fluvial – *Betula nigra*
abedulillo – *Carpinus betulus*
abedul japonés de Maximowicz – *Betula maximowicziana*
abedul llorón plateado – *Betula pendula* var *youngii*
abedul negro – *Betula nigra*
abedul negro americano – *Betula nigra*
abedul papirífero – *Betula papyrifera*
abedul péndulo – *Betula pendula*
abedul plateado – *Betula verrucosa*
abedul pubescente – *Betula pubescens*
abedul verrugoso – *Betula pendula*
abejera – *Ophrys apifera*
abejón – *Senna pallida*
abelia – *Abelia floribunda, A.* x *grandiflora*
abelmosco – *Abelmoschus moschatus, Hibiscus esculentus*
abeto – *Abies, Liquidambar styraciflua, Pseudotsuga menziesii*
abeto amábilis – *Abies amabilis*
abeto azul – *Picea pungens*
abeto azul de Oregón – *Abies procera*
abeto azul glauco – *Picea pungens* var *glauca*
abeto balsámico – *Abies balsamea*
abeto blanco – *Abies pectinata*
abeto blanco americano – *Abies lasiocarpa*
abeto blanco de las rocosas – *Abies lasiocarpa*
abeto bracteado – *Abies bracteata*
abeto cedroso – *Abies bracteata*
abeto común – *Abies pectinata*
abeto chino – *Cunninghamia lanceolata*
abeto de Corea – *Abies koreana*
abeto de Douglas – *Pseudotsuga menziesii*
abeto de España – *Abies pinsapo*
abeto del Canadá – *Picea glauca*
abeto del Cáucaso – *Abies nordmanniana*
abeto del Colorado – *Abies concolor*
abeto del norte – *Picea abies*
abeto del Pacífico – *Abies amabilis*
abeto de Menzies – *Picea sitchensis*
abeto de Navidad – *Picea abies*
abeto de Nordmann – *Abies nordmanniana*
abeto de Noruega – *Picea abies*
abeto de Santa Lucía – *Abies bracteata*
abeto de Vancuver – *Abies grandis*
abeto español – *Abies pinsapo*
abeto falso – *Picea abies*
abeto gigante – *Abies grandis*
abeto grande americano – *Abies grandis*
abeto griego – *Abies cephalonica*
abeto guatemalteco – *Abies guatemalensis*
abeto lasiocarpa americano – *Abies lasiocarpa*
abeto magnífico – *Abies magnifica*
abeto noble – *Abies procera*
abeto plateado – *Abies pectinata, Picea pungens*
abeto prócer – *Abies procera*
abeto rojo – *Picea abies*
abeto rojo americano – *Abies amabilis*

abeto rojo shasta – *Abies magnifica* var *shastensis*
abeto ruso – *Abies siberica*
abeto shasta – *Abies magnifica* var *shastensis*
abeto siberiano – *Abies sibirica*
abiodo – *Cornus disciflora*
abiú – *Pouteria caimito*
ablano – *Corylus cornuta* var *californica*
abrasín del Japón – *Aleurites cordata*
abrecapalo – *Ficus tecolutensis*
abrepuño – *Centaurea solstitialis*
abrico de para – *Mammea americana*
abricotina – *Prunus simonii*
abridero – *Prunus persica* ssp *nectarina*
abridor – *Prunus persica* ssp *nectarina*
abrojo – *Adolphia infesta, Cenchrus incertus, Cirsium, Condalia, Cylindropuntia imbricata, C. tunicata, Koeberlinia spinosa* ssp *tenuispina, Microrhamnus ericoides, Opuntia engelmannii, O. phaeacantha, Tribulus cistoides, T. terrestris, Triumfetta semitriloba, Xanthium canadensis, X. strumarium, Ziziphus obtusifolia*
abrojo de agua – *Trapa natans*
abrojo de flor amarilla – *Kallstroemia grandiflora*
abrojo de tierra caliente – *Tribulus cistoides*
abrojo espigada – *Tragus berteronianus*
abrojo roseta – *Cenchrus incertus, C. longispinus*
abrótano – *Artemisia abrotanum*
abrótano hembra – *Santolina chamaecyparisus*
abrótano macho – *Artemisia abrotanum*
absintio – *Artemisia absinthium*
abutilon – *Abutilon vitifolium*
abutilon trepador – *Abutilon megapotamicum*
acacalocahuite – *Pinus ayacahuite*
acacia – *Acacia, Delonix regia, Leucaena leucophala, Moringa oleifera, Robinia pseudoacacia*
acacia amarilla – *Albizzia lebeck, Caesalpinia spinosa*
acacia bastarda – *Robinia pseudoacacia*
acacia bella rosa – *Leucaena leucocephala*
acacia blanca – *Robinia pseudoacacia*
acacia de agüijote – *Parkinsonia avuleata*
acacia de Constantinopla – *Albizzia julibrissin*
acacia de espinas – *Acacia paradoxa*
acacia de flores blancas – *Robinia pseudoacacia*
acacia de hojas azules – *Acacia saligna*
acacia de la costa – *Acacia cyclops*
acacia del Japón – *Sophora japonica*
acacia de madera negra – *Acacia melanoxylon*
acacia de Persia – *Albizzia julibrissin*
acacia de Rusia – *Caragana arborescens*
acacia de tres espinas – *Gleditsia triacanthos*
acacia espinosa – *Robinia pseudoacacia*
acacia falsa – *Acacia longifolia, Robinia pseudoacacia*
acacia frondosa – *Leucaena leucocephala*
acacia gomosa – *Acacia senegal*
acacia majorera – *Acacia cyclops*
acacia mimosa – *Acacia dealbata*
acacia negra – *Gleditsia triacanthos*
acacia pálida – *Leucaena glauca*
acacia pegajosa – *Robinia viscosa*
acacia plateada – *Acacia retinodes*
acaciste – *Cornus excelsa*
acahual – *Bidens, Encelia mexicana, Helianthus annuus, Heterotheca inuloides, Montanoa speciosa, Simsia amplexicaulis, Tithonia*
acahual blanco – *Bidens odorata, B. pilosa*
acahualillo – *Bidens aurea, B. grandiflora, B. tetragosa*
acahualillo de flor blanca – *Bidens pilosa*
acahualli – *Helianthus annuus*
acahuatl – *Heterotheca inuloides*
acahuecaxaqua – *Portulaca oleracea*
acahuite – *Pinus ayacahuite, Pinus strobiformis, Pseudotsuga menziesii*
acajú – *Anacardium occidentale*
acajuiba – *Anacardium occidentale*
acalifa – *Acalypha hispida, A. wilkesiana*
acalo cahuite – *Pinus ayacahuite*
acalocote – *Pinus ayacahuite, Pinus strobus*
acalztatsim – *Plumeria acutifolia*
acam – *Orobanche*
acamallotetl – *Lucuma salicifolia*
acamayo – *Chysis aurea, Lucuma salicifolia*
acamba – *Agave potatorum*
acamita – *Pinus strobiformis*
acamoyote – *Pachira aquatica*
acanguarica – *Bixa orellana*
acanita – *Pinus ayacahuite, Pinus strobiformis*
acanto – *Acanthus mollis, Cirsium*

acanto balcánico – *Acanthus balcanicus*
acanto de Cáucaso – *Heracleum mantegazzianum*
acanto espinoso – *Acanthus spinosus*
acanto mullido – *Acanthus mollis*
acapatle – *Iresine calea*
acarino – *Sicyos angulatus*
acasucut – *Podachaenium eminens*
acatl – *Arundo donax, Phragmites australis, Rosa*
acatlzauhtli – *Cranichis speciosa*
acaute – *Tithonia*
acaute de flor – *Tithonia rotundifolia*
acaxacahuitztli – *Cynodon dactylon*
acaxaxan – *Potamogeton natans*
acaxílotl – *Plantago mexicana*
acaxóchitl – *Lobelia cardinalis* var *pseudosplendens, L. laxiflora, Monstera deliciosa*
acayacoyotl – *Coix lacryma-jobi*
acazacahuiztli – *Elytrigia repens*
acebo – *Berberis fremontii, Ilex aquifolium, I. crenata, I. opaca*
acebo común – *Ilex aquifolium*
acebo chino de Perny – *Ilex pernyi*
acebo híbrido de Highclere – *Ilex* x *altaclerensis*
acebuche – *Celtis pallida, C. reticulata, Forestiera angustifolia, Olea europaea*
acecé – *Euchaena mexicana*
acecinte – *Acer negundo*
acedera – *Oxalis acetosella, O. corniculata, Rumex acetosa, R. acetosella*
acedera bendita – *Rumex crispus*
acedera de Guinea – *Hibiscus sabdariffa*
acedera menor – *Oxalis adenophylla*
acedera pratense – *Rumex acetosa*
acedera roja de Guinea – *Hibiscus sabdariffa*
acedera romana – *Rumex scutatus*
acedera silvestre – *Rumex acetosa*
acederilla – *Oxalis, Rumex acetosella*
acedero arbóreo – *Oxydendrum arboreum*
acedilla – *Wimmeria confusa*
aceite – *Calophyllum calaba*
aceitilla – *Aristida scabra, Bidens, Bouteloua aristidoides, Chiococca alba*
aceitilla blanca – *Bidens pilosa* var *radiata*
aceituna – *Celtis reticulata, Ilex tolucana*
aceituno – *Olea europaea, Simarouba glauca*
aceitunillo – *Cornus excelsa, Ilex tolucana*
acelga – *Beta vulgaris* ssp *cicla, Brassica rapa* var *chinensis*
acelga suiza – *Beta vulgaris* ssp *cicla*
acerallero – *Sorbus aucuparia*
acero blanco – *Acer saccharinum*
acero blanco americano – *Acer saccharinum*
acerolo – *Crataegus azarolus, Malpighia emarginata, M. glabra, Sorbus aucuparia, S. domestica*
acerolo rojo – *Crataegus mollis*
acetilla – *Cosmos parviflorus, Galinsoga parviflora*
acetosilla – *Oxalis acetosela*
acezintle – *Acer negundo* var *californicum*
aciano – *Centaurea cyanus*
aciano de montaña – *Centaurea montana*
acíbar – *Aloe socotrina, A. vera*
acidito – *Rhus trilobata*
acinchete – *Pithecellobium acatlense*
acitrón – *Echinocactus platyacanthus*
aclinán – *Montanoa tomentosa*
acmea – *Aechmea chantinii, A. fasciata, A. fulgens*
acocote – *Arracacia atropurpurea, A. brandegeei, Lagenaria siceraria*
acócotl – *Lagenaria siceraria*
acocotli – *Bidens pilosa, Dahlia coccinea*
acocotli tepecuacuilcense – *Arracacia atropurpurea*
acocoxóchitl – *Dahlia coccinea*
acolote – *Pinus ayacahuite*
acompes – *Pachycereus hollianus*
acónito – *Aconitum*
acónito americano – *Hydrastis canadensis*
acónito común – *Aconitum napellus*
acónito invernal – *Eranthis hyemalis*
ácoro – *Acorus calamus, A. gramineus*
ácoro bastardo – *Iris pseudacorus*
ácoro verdadero – *Acorus calamus*
acroclíneo – *Helipterum roseum*
actea – *Actaea spicata, Cimicifuga racemosa*
actimpatli – *Haplophyton cimicidum*
actiz – *Thevetia ovata, T. peruviana, T. thevetioides*
acuáhuitl – *Chilopsis linearis* ssp *arcuata, Platanus mexicana*
acuilote – *Philadelphus mexicanus*
acuilotl – *Philadelphus mexicanus*

acuitze varicua – *Ruta chalepensis, R. graveolens*
acuña – *Echinomastus erectocentrus*
acusá – *Erythroxylum mexicanum*
acuyo – *Piper auritum, P. sanctum*
acxoyatic – *Mirabilis longiflora*
acxoyatl – *Abies religiosa*
achayotl – *Senecio salignus*
achcuauit – *Jatropha curcas*
achicoria – *Cichorium intybus, Sonchus oleraceus*
achicoria amarga – *Cichorium intybus, Taraxacum officinale*
achicoria de Bruselas *Cichorium intybus*
achicoria de hojas blancas – *Cichorium intybus*
achicoria de hojas rojas – *Cichorium intybus*
achicoria de hojas verdes – *Cichorium intybus*
achicoria de raíz – *Cichorium intybus*
achicoria dulce – *Chondrilla juncea, Sonchus oleraceus*
achicoria juncal – *Chondrilla juncea*
achicoria silvestre – *Cichorium intybus*
achichil – *Viburnum*
achilea – *Achillea millefolium*
a-chili – *Polygonum hydropiper*
achin – *Pistacia mexicana*
achiote – *Bixa orellana*
achiotillo – *Bixa orellana, Caryocar amygdaliferum*
achiotl – *Bixa orellana*
achiotlín – *Piper angustifolium*
achira – *Canna coccinea, C. edulis, C. indica*
achita – *Amaranthus hybridus*
achojcha – *Cyclanthera pedata*
achuma – *Echinopsis pachanoi*
adelaida – *Fuchsia*
adelfa – *Nerium oleander*
adelfa amarilla – *Thevetia peruviana*
adelfa olorosa – *Nerium oleander*
adelfilla – *Bupleurum fruticosum, Daphne, Epilobium angustifolium*
adelfilla ciliada – *Epilobium ciliatum*
adelfilla de flores blancas – *Epilobium lactiflorum*
adelfilla de hojas de anagálida – *Epilobium anagallidifolium*
adelfilla de hojas estrechas – *Epilobium angustifolium*
adelfilla de los pantanos – *Epilobium palustre*
adelfilla glandulosa – *Epilobium ciliatum*
adelfilla hirsuta – *Epilobium hirsutum*
adelfilla pelosa – *Epilobium hirsutum*
adelia – *Forestiera pubescens*
adenófora olorosa – *Adenophora liliifolia*
adianto – *Adiantum capillus-veneris*
adianto del Canadá – *Adiantum aleuticum*
adiós – *Clarkia amoena*
adiós primavera – *Clarkia amoena*
adlay – *Coix lacryma-jobi*
admirable – *Brachiaria mutica, Kalanchoe pinnata*
adolfina – *Petrea arborea*
adonis – *Adonis vernalis, Hippeastrum*
adonis amarillo – *Adonis vernalis*
adonis blanco – *Duranta repens*
adonis de otoño – *Adonis annua*
adonis de primavera – *Adonis vernalis*
adonis de verano – *Adonis aestivalis*
adonis escarlato – *Adonis flamea*
adonis morada – *Duranta erecta, D. repens*
adonis vernal – *Adonis vernalis*
adormidera – *Mimosa pudica, Papaver rhoeas, P. somniferum*
adormidera alpina – *Papaver alpinum*
adormidera china – *Papaver somnifera*
adormidera de hojas obtusas – *Papaver dubium*
adormidera de Oriente – *Papaver orientale*
adormidera de tallo nudo – *Papaver nudicaule*
adormidera espinosa – *Argemone mexicana*
adormidera marina – *Glaucium flavum*
afaca – *Lathyrus aphaca*
afelandra – *Aphelandra squarrosa*
afelandra rayadita – *Aphelandra squarrosa*
aficoz – *Cucumis melo* var *flexuosus*
afinador – *Mortonia greggii, M. utahensis, Parthenium argentatum*
Africanas – *Freesia refracta*
agabanzo – *Rosa montezumae*
agapanto – *Agapanthus africanus*
agar-agar – *Gelidium cartilagineum*
agare palo – *Tillandsia recurvata*
agárico bianillado – *Agaricus bitorquis*
agárico blanco – *Polyporus officinalis*
agárico de las farmacias – *Polyporus officinalis*
agárico de los medicos – *Polyporus officinalis*
agárico purgante – *Polyporus officinalis*
agarito – *Berberis trifoliolata*
agati sesbania – *Sesbania grandiflora*

agavanzo – *Rosa canina*
agave – *Agave deserti*
agave azul – *Agave tequilana*
agave de la Reina Victoria – *Agave victoriae-reginae*
agérato – *Achillea ageratifolia, Ageratum houstonianum*
agí – *Echinocereus stoloniferus, E. subinermis*
agigandú – *Acacia brandegeana*
aglaonema – *Aglaonema modestum*
agnocasto – *Vitex agnus-castus*
agosteña – *Echinocereus stramineus*
agotope – *Inga spuria*
agracejina – *Berberis vulgaris*
agracejo – *Berberis fremontii, B. vulgaris*
agracejo de Darwin – *Berberis darwinii*
agracillo – *Berberis vulgaris*
agrarista – *Cynodon dactylon*
agrazón – *Berberis vulgaris*
agriaz – *Melia azederach* var *umbraculiformis*
agrifolio – *Ilex aquifolium*
agrillo – *Berberis trifoliata, Rhus microphylla, R. trilobata*
agrimonia – *Agrimonia gryposepala*
agrimonia común – *Agrimonia eupatoria*
agrimonia del monte – *Scutellaria*
agrimonia silvestre – *Scutellaria*
agrín – *Rhus schiedeana*
agrioris – *Rorippa nasturtium-aquaticum*
agrio – *Oxalis latifolia, O. pes-caprae*
agrios – *Citrus*
agripalma – *Leonurus cardiaca*
agrito – *Berberis trifoliata, Lippia ligustrina, Oxalis, Rhus microphylla, R. trilobata*
agrito de maceta – *Oxalis corniculata*
agropiro – *Elytrigia repens, E. trachycaulus*
agropiro azul – *Elytrigia arizonicus, Pseudoroegneria spicata*
agropiro delgado – *Elytrigia trachycaulus*
agropiro de Parish – *Elytrigia stebbinsii*
agróstide blanca – *Agrostis gigantean, A. stolonifera*
agróstide canina – *Agrostis canina*
agróstide común – *Agrostis idahoensis*
agróstide de las turberas – *Agrostis canina*
agróstide de los pantanos – *Agrostis palustris*
agróstide de perro – *Agrostis canina*
agróstide de renuevos – *Agrostis stolonifera*
agróstide escabra – *Agrostis scabra*
agróstide estolonifera – *Agrostis stolonifera*
agróstide mayor – *Agrostis gigantea*
agróstide rastrera – *Agrostis stolonifera*
agróstide palustre – *Agrostis palustris*
agua bendita – *Cestrum anagyris*
aguabola – *Maytenus phyllanthoides*
agua bonita – *Gilia tenuiflora*
aguacachile – *Persea americana*
aguacasle – *Enterolobium cyclocarpum*
aguacate – *Nectandra sinuata, Persea*
aguacate anís – *Persea drymifolia*
aguacate cimarrón – *Phoebe mexicana*
aguacate chupón – *Persea schiedeana*
aguacate de danta – *Omphalea oleifera*
aguacate de mono – *Calatola mollis*
aguacate oloroso – *Persea drymifolia*
aguacatero – *Persea americana, Prunus brachybotrya*
aguacate xinene – *Persea americana, P. schiedeana*
aguacatillo – *Clethra lanata, Nectandra, Persea americana, P. schiedeana, Phoebe ehrenbergii, P. mexicana, P. tampicensis, Prunus brachybotrya, Quercus muehlenbergii, Styrax*
aguacatillo hediondo – *Nectandra sinuata*
aguacolla – *Echinopsis pachanoi*
agua de azahar – *Oenothera rosea*
agua de bola – *Maytenus phyllanthoides*
aguaje – *Mauritia flexuosa*
aguajpo – *Heliocarpus tomentosus*
aguama – *Bromelia karatas, B. pinguin*
aguamalaria – *Vitex mollis*
agua maravilla – *Hamamelis virginiana*
aguapá – *Typha latifolia*
aguapepe – *Ardisia revoluta*
aguapié – *Hechtia texensis*
aguardientillo – *Cordia alliodora*
aguaribay – *Schinus molle*
aguaro – *Martynia annua, Proboscidea fragrans, P. parviflora*
aguasique – *Prunus serotina* ssp *virens*
aguate – *Malvaviscus grandiflorus*
aguatle – *Quercus acutifolia*
aguatosa – *Anoda cristata*
aguaturma – *Helianthus tuberosus*
aguava – *Bromelia karatas, B. pinguin*
aguavilla – *Arctostaphylos uva-ursi*
aguazul – *Mesembryanthemum crystallinum*

aguazur – *Mesembryanthemum crystallinum*
aguedita blanca – *Casearia sylvestris*
aguedita dulce – *Casearia sylvestris*
agui – *Echinocereus luteus*
aguileña – *Aquilegia*
aguileña alpina – *Aquilegia alpina*
aguileña azul – *Aquilegia caerulea*
aguileña común – *Aquilegia vulgaris*
agüilolote – *Vitex mollis*
agüilote – *Vitex mollis*
aguinaldo – *Campanula rapunculoides*
aguja – *Erodium moschatum*
aguja de ara – *Xylosma characantha*
aguja del pastor – *Erodium cicutarium, E. moschatum*
aguja de pasta – *Scandix pectin-veneris*
aguja rabia – *Xylosma characantha*
agujilla – *Cylindropuntia leptocaulis, C. molesta*
agujilla grande – *Stipa eminens*
agujita – *Erodium cicutarium, E. moschatum*
agusá – *Erythroxylum mexicanum*
ahaipuih – *Plumeria acutifolia*
ahate – *Annona purpurea, A. squamosa*
ahijado – *Coleus blumei*
ahoaquahuitl – *Quercus*
ahoatl – *Quercus*
ahoché – *Oncidium adscendens, O. carthaginensis*
ahogagatos – *Anthriscus caucalis*
ahogaviejas – *Anemone nemorosa*
ahuacachili – *Bocconia arborea*
ahuacaquahuitl – *Persea americana, P. drymifolia*
ahuácatl – *Persea americana, P. drymifolia*
ahuacocoztli – *Quercus rugosa*
ahuahuasúchil – *Disocactus speciosus*
ahuahuaxl – *Quercus candicans*
ahuahuaxóchitl – *Disocactus speciosus*
ahualxocotl – *Malpighia glabra*
ahuamextli – *Quercus candicans*
ahuapatli – *Calea zacatechichi, Pluchea odorata*
ahuaquahuitl – *Quercus*
ahuasúchil – *Laelia autumnalis*
ahuatetz – *Quercus*
ahuatetzmolli – *Quercus*
ahuatl – *Quercus*
ahuatle – *Quercus acutifolia*
ahuatl tepitón – *Rhamnus serrata*
ahuaxóchitl – *Disocactus speciosus, Laelia autumnalis*
ahucepitzahuac – *Quercus laurina*
ahuehuete – *Taxodium distichum* var *mexicanum*
ahuéhuetl – *Taxodium distichum* var *mexicanum*
ahuejote – *Astianthus viminalis, Salix bonplandiana, S. humboldtiana, S. lasiolepis*
ahuevado – *Amanita caesarea*
ahuilote – *Vitex mollis, V. pyramidata*
ahuiron – *Cichorium intybus*
ahuitzyó – *Lophophora williamsii*
ahumada – *Senna pallida*
ahuyama – *Cucurbita maxima, C. pepo*
aigre – *Prunus gentryi*
ailanto – *Ailanthus altissima*
aile – *Alnus, Arbutus glandulosa*
ailitl – *Alnus rubra*
aira de césped – *Deschampsia cespitosa*
aira flexuosa – *Deschampsia flexuosa*
airén – *Vitis vinifera* var *sativa*
airía – *Dodonaea viscosa*
airosa – *Zexmenia seemanni*
aja – *Brosimum alicastrum*
ajalache – *Sida acuta*
ajamete – *Asclepias subulata*
aje – *Dioscorea villosa*
ajedrea – *Satureja hortensis*
ajedrea de huerta – *Satureja hortensis*
ajedrea de jardín – *Satureja hortensis*
ajedrea montana – *Satureja montana*
ajenabo – *Brassica nigra*
ajengibre – *Zingiber officinale*
ajengibrillo dilatado – *Paspalum dilatatum*
ajengibrillo dístico – *Paspalum distichum*
ajenjo – *Artemisia*
ajenjo anual – *Artemisia annua*
ajenjo arbóreo – *Artemisia abrotanum*
ajenjo arborescente – *Artemisia arborescens*
ajenjo bisanuo – *Artemisia biennis*
ajenjo común – *Artemisia vulgaris*
ajenjo de las playas – *Artemisia stellerana*
ajenjo del Cáucaso – *Artemisia caucasica*
ajenjo de los campos – *Artemisia campestris*
ajenjo del país – *Artemisia ludoviciana, A. mexicana*
ajenjo dulce – *Artemisia dulce*

ajenjo extranjero – *Artemisia absinthium*
ajenjo lactescente – *Artemisia lactiflora*
ajenjo mayor – *Artemisia absinthium*
ajenjo moruno – *Artemisia abrotanum, A. arborescens*
ajenjo noruego – *Artemisia norvegica*
ajenjo póntico – *Artemisia pontica*
ajenuz – *Nigella damascena*
ají – *Capsicum annuum, C. annuum* var *grossum, C. frutescens, C. pubescens*
ají dátil – *Capsicum chinense*
ají de Cayena – *Capsicum fastigatum*
ají dulce – *Capsicum annuum* var *grossum*
ajillá – *Guazuma ulmifolia*
ajillo – *Rauwolfia hirsuta*
ajipa – *Pachyrhizus erosus*
ajito – *Capparis verrucosa*
ají turco – *Hibiscus esculentus*
ajo – *Allium oreophilim, A. sativum, Hesperocallis undulata*
ajo carinado – *Allium carinatum* ssp *pulchellum*
ajo cebolla – *Allium cernuum*
ajo cebollino – *Allium schoenoprasum*
ajocopaque – *Gaultheria acuminata*
ajo chalote – *Allium cepa* var *ascalonicum*
ajo de cabeza redonda – *Allium sphaerocephalum*
ajo de las viñas – *Allium vineale*
ajo del Turquestán – *Allium karataviense*
ajo dorado – *Allium moly*
ajo fistulosum – *Allium fistulosum*
ajo floreciente – *Allium christophii*
ajo francés – *Allium sativum* var *ophioscorodon*
ajo italiano – *Allium sativum* var *ophioscorodon*
ajo moruno – *Allium schoenoprasum*
ajo napolitano – *Allium neapolitanum*
ajonjera – *Chondrilla juncea*
ajonjolí – *Crotalaria incana, C. sagittalis, Sesamum indicum*
ajonjolí de monte – *Crotalaria sagittalis*
ajonjolín – *Sesamum indicum*
ajo paniculado – *Allium paniculatum*
ajopatla – *Gaultheria acuminata*
ajo porro – *Allium porrum*
ajo puerro – *Allium porrum*
ajo sativo – *Allium sativum*
ajo silvestre – *Allium moly, Hesperocallis undulata*
ajo triángulo – *Allium triquetrum*
ajté – *Casimiroa sapota*
ajuga – *Ajuga reptans*
akabyon – *Cestrum nocturnum*
akalkumche – *Philodendron scandens*
akamatsu – *Pinus densiflora*
akí – *Blighia sapida*
akilzuuc – *Sorghum halapense*
akokojxiuit – *Piper sanctum*
aktarcin – *Drimys winteri*
ak-té – *Astrocaryum mexicanum*
akund – *Calotropis procera*
akuum – *Cryosophila stauracantha*
ala – *Inula helenium*
alabarda – *Fouquieria campanulata, F. fasciculata*
alacate – *Lagenaria siceraria*
alacrán – *Heliotropium indicum, Zanthoxylum fagara*
alacrancillo – *Heliotropium curassavicum, H. indicum*
alacrancillo de playa – *Heliotropium curassavicum*
alachi – *Anoda cristata*
ala de ángel – *Begonia gracilis*
ala de cucaracha – *Bernoullia flammea*
ala de oro – *Amaranthus tricolor*
ala de perico – *Amaranthus tricolor*
aladierno – *Rhamnus*
aladierno alpino – *Rhamnus alpinus*
aladierno fétido – *Rhamnus infectorius*
alagüé – *Heliocarpus tomentosus*
alahual – *Stellaria*
alalatz – *Hibiscus bifurcatus*
alamanda – *Allamanda cathartica*
alambre – *Lygodium venustum*
alambrillo – *Polygonum argyrocoleon*
alamillo – *Alnus oblongifolia, Dodonaea viscosa, Populus tremuloides*
álamo – *Ficus, Platanus, Populus*
álamo balsámico – *Populus balsamifera*
álamo balsámico de California – *Populus balsamifera* ssp *trichocarpa*
álamo balsámico del oeste – *Populus balsamifera* ssp *trichocarpa*
álamo blanco – *Platanus mexicana, Platanus*

wrightii, Populus alba, P. fremontii, P. tremuloides
álamo canadiense – *Populus grandidentata*
álamo canescente – *Populus* x *canescens*
álamo cano – *Populus* x *canescens*
álamo cimarrón – *Populus fremontii*
álamo común – *Populus alba*
álamo cubano – *Ficus religiosa*
álamo chopo – *Populus fremontii*
álamo de Italia – *Populus nigra* var *italica*
álamo de la Carolina – *Populus canadensis*
álamo del Canadá – *Populus* x *canadensis*
álamo extranjero – *Ficus microcarpa*
álamo gris – *Populus canescens*
álamo gris de picardía – *Populus* x *canescens*
álamo loco – *Nicotiana glauca*
álamo negro – *Populus nigra, Ulmus campestris*
álamo negro de Italia – *Populus nigra* var *italica*
álamo saúco – *Populus angustifolia*
álamo tacamahaca – *Populus balsamifera*
álamo temblón – *Populus tremula, P. tremuloides*
alampepe – *Entada, Vicia faba*
alantana – *Lantana camara*
alarquez – *Buxus sempervirens*
alatle – *Eremalche scoparium, Modiolastrum scoparium*
alazor – *Carthamus tinctorius*
alazor bastardo – *Carthamus tinctorius*
albaca – *Ocimum basilicum*
albacar – *Ocimum basilicum, Scutellaria*
albacarón – *Ocimum basilicum*
albacarrote – *Pinus cooperi*
albacarrote amarillo – *Pinus teocote*
albahaca – *Ocimum basilicum*
albahaca blanca – *Ocimum basilicum*
albahaca cimarrona – *Lantana camara*
albahaca de caballo – *Lantana camara*
albahaca de limón – *Nymphaea alba*
albahaca dulce – *Ocimum basilicum*
albahaca larga – *Bassia hyssopifolia*
albarda – *Fouquieria campanulata, F. fasciculata, F. splendens*
albardera – *Paeonia officinalis*
albardín – *Lygeum spartum*
albaricoque – *Prunus armenica*
albaricoquero – *Prunus armenica*
albaricoquero de Simon – *Prunus simonii*
albaricoquero japonés – *Prunus mume*
albarines – *Anchusa officinalis*
albarrana – *Urginea maritima*
albarraz – *Delphinium staphisagria*
albérchigo – *Prunus persica*
alberchiguero – *Prunus persica*
alberja – *Pisum sativum*
alberjón – *Pisum sativum*
alboheza – *Malva sylvestris*
alborto – *Arbutus unedo*
albotín – *Pistacia terebinthus*
alcachofa – *Cynara scolymus, Stachys sieboldii*
alcachofa borriquera – *Onopordum acanthium, O. bracteatum*
alcachofa de gato – *Sempervivum tectorum*
alcachofa de Jerusalén – *Helianthus tuberosus*
alcachofero – *Cynara scolymus*
alcachola – *Beta vulgaris* ssp *cicla*
alcahuésar – *Pereskiopsis porteri*
alcaina – *Achillea millefolium*
alcájer – *Pereskiopsis porteri*
alcali zacate – *Panicum maximum*
alcana – *Alkanna tinctoria, Ancusa azurea, Reseda odorata*
alcáncer – *Cuphea aequipetala, C. jorullensis*
alcanfor – *Achillea millefolium, Cinnamomum camphora, Eucalyptus*
alcanforero – *Achillea millefolium, Cinnamomum camphora, Laurus camphora*
alcanforero amarillo – *Cinnamomum parthenoxylon*
alcaparra – *Capparis indica, C. spinosa, Erythrina, Ipomoea longifolia, Senna atomaria*
alcaparra común – *Capparis spinosa*
alcaparra parrera – *Capparis spinosa*
alcaparra parro – *Capparis spinosa*
alcaparrera – *Capparis spiniosa*
alcaparrillo – *Senna bicapsularis*
alcaparrón – *Capparis spinosa*
alcaparroso – *Senna atomaria*
alcaravea – *Carum carvi*
alcaravea de prado – *Carum carvi*
alcarceña – *Vicia ervilia*
alcatraz – *Aristolochia grandiflora, Richardia scabra, Zantedeschia aethiopica*
alcaucil – *Cynara carduncululs, C. scolymus*
alcaucil silvestre – *Cynara carduncululs*

alcaulera – *Cynara scolymus*
alcayota común – *Cucurbita ficifolia*
alcayuela – *Halimium ocymoides*
alcazuz – *Glycyrrhiza glabra*
alcea – *Alcea rosea*
alcocaz – *Colocasia esculenta*
alcornoque – *Quercus suber, Terminalia catappa*
alcornoque mediterráneo – *Quercus suber*
alcosoria – *Carum carvi*
alchicolia – *Dugesia mexicana*
aldiza – *Centaurea cyanus*
alea-pajac – *Trachypogon secundus*
alegría – *Amaranthus*
alegría de la casa – *Impatiens walleriana*
Alejandría – *Plumeria acutifolia, P. rubra*
aleluya amarilla – *Oxalis corniculata, O. pes-caprae*
aleluya blanca – *Oxalis acetosella*
aleluya corniculada – *Oxalis corniculata*
aleluya de Bowie – *Oxalis purpurea*
aleluya de Deppe – *Oxalis deppei*
aleluya de hojas anchas – *Oxalis latifolia*
aleluya encarnada – *Oxalis incarnata*
aleluya europea – *Oxalis dillenii*
aleluya tuberosa – *Oxalis tuberosa*
alemán – *Echinochloa polystachya*
alerce – *Fitzroya cupressoides, Larix decidua*
alerce africano – *Tetraclinis articulata*
alerce americano occidental – *Larix occidentalis*
alerce americano oriental – *Larix laricina*
alerce común – *Larix deciduas*
alerce chino – *Pseudolarix amabilis*
alerce dahuriano – *Larix gmelinii*
alerce deciduo – *Larix común*
alerce de China – *Pseudolarix amabilis*
alerce de Europa – *Larix decidua*
alerce de Gmelin – *Larix gmelinii*
alerce del Japón – *Larix kaempferi*
alerce europeo – *Larix decidua*
alerce japonés – *Larix kaempferi*
alerce melis – *Larix decidua*
alerce siberiano – *Larix sibirica*
alesia – *Halesia monticola*
alesta olorosa – *Anthoxanthum odoratum, Hierochloë odorata*
alesta aristosa – *Anthoxanthum aristatum*
aleurita – *Aleurites moluccana*
alfábega – *Ocimum basilicum*
alfabieso – *Mentha viridis*
alfalfa – *Medicago sativa*
alfalfa del campo – *Lotus repens*
alfalfa del Perú – *Medicago sativa*
alfalfa enana – *Medicago minima*
alfalfa enhiesta – *Medicago polymorpha*
alfalfa lupulina – *Medicago lupulina*
alfalfa minima – *Medicago minima*
alfalfa precoz – *Medicago praecox*
alfalfa sativa – *Medicago sativa*
alfalfa tropical – *Stylosanthes guianensis*
alfalfa variagada – *Medicago sativa*
alfalfilla – *Dicliptera resupinata, Melilotus indica*
alfalfón – *Melilotus alba, M. officinalis*
alfambrilla – *Abronia*
alfesira – *Bryonia dioica*
alficaz – *Cucurbita foetidissima*
alfiler – *Trachelium caeruleum*
alfilerillo – *Cylindropuntia leptocaulis, Erodium cicutarium, E. moschatum, E. texanum, Lopezia mexicana, L. racemosa, Pereskiopsis aquosa, P. diguetii*
alfilerillo de pastor – *Erodium cicutarium*
alfombra – *Axonopus compressus*
alfombrilla – *Abronia gracilis, A. maritima, Drymaria arenarioides, Melilotus indica, M. officinalis, Phlox drummondii, Verbena*
alfombrilla de jardín – *Verbena corymbosa*
alfombrilla del campo – *Verbena ciliata*
alfombrilla hedionda – *Lantana camara*
alfóncigo – *Pistacia vera*
alfóncigo de tierra – *Arachis hypogaea*
alforfón – *Fagopyrum esculentum*
alforfón común – *Fagopyrum esculentum*
algalia – *Abelmoschus moschatus, Hibiscus esculentus*
alga marina – *Fucus vesiculosus, Laminaria, Macrocystis pyrifera*
algarabía hegajosa – *Parentucella latifolia*
algarabía viscosa – *Parentucella viscosa*
algarroba – *Vicia articulata*
algarroba común – *Vicia sativa*
algarroba de Castilla – *Vicia articulata*
algarroba genuina – *Vicia articulata*
algarrobilla – *Vicia sativa*
algarrobo – *Acacia greggii, A. macracantha, A.*

pennatula, Ceratonia siliqua, Hymenaea courbaril, Prosopis
algarrobo chileno – *Prosopis chilensis*
algarrobo de España – *Ceratonia siliqua*
algarrobo de olor – *Albizzia occidentalis*
algarrobo espino – *Acacia pennatula*
algarrobo loco – *Cercis siliquastrum*
algazul – *Carpobrptus chilensis, Mesembryanthemum crystallinum, M. nodiflorum*
algerita – *Berberis fremontii, B. haematocarpa, B. trifoliata*
algodón – *Gossypium*
algodón bulboso – *Asclepias tuberosa*
algodoncillo – *Asclepias curassavica, A. linaria, A. syriaca, Croton guatemalensis, Gossypium herbaceum, G. thurberi, Wimmeria confusa, W. mexicana*
algodoncillo del campo – *Gossypium thurberi*
algodoncillo del monte – *Gossypium thurberi*
algodón cimarrón – *Gossypium davidsonii, G. harknessii*
algodón cornudo – *Asclepias syriaca*
algodón coyuche – *Gossypium hirsutum*
algodón de árbol – *Ceiba pentandra, Gossypium mexicanum*
algodón del país – *Gossypium barbadense*
algodón de seda – *Calotropis procera*
algodonero – *Gossypium*
algodonero americano – *Gossypium hirsutum*
algodonero herbáceo – *Gossypium herbaceum*
algodón rojo – *Asclepias incarnata*
algodón silvestre – *Cochlospermum vitifolium, Gossypium davidsonii, G. harknessii*
alguita – *Aframomum melegueta*
alhábega – *Ocimum basilicum*
alharma – *Peganum harmala*
alhelí – *Dianthus caryophyllus, Erysimum cheiri, Malcomia maritima, Matthiola incana, Melia azedarach, Plumeria acutifolia*
alhelí alpino – *Erysimum hieraciifolium*
alhelí amarillo – *Erysimum cheiri*
alhelí blanco – *Matthiola annua, M. incana*
alhelicillo – *Lobularia maritima*
alhelí cuarenteno – *Matthiola incana*
alhelí de diez semanas – *Matthiola incana*
alhelí de pétalos grandes – *Matthiola longipetala*
alhelí de Siberia – *Erysimum hieraciifolium*
alhelí de Virginia – *Malcolmia maritima*
alhelí encanto – *Matthiola annua*
alhelí encarnado – *Matthiola annua, M. incana*
alhelilla – *Gilia capitata*
alhelí morado – *Matthiola incana*
alhelí nocturno – *Matthiola longipetala* var *bicornis*
alheña – *Ligustrum vulgare, Reseda odorata*
alholva – *Trigonella foenum-graecum*
alhucema – *Lavandula angustifolia, L. lanata, L. latifolia*
alicoche – *Echinocereus*
aligonero – *Celtis australis*
aligustre – *Ligustrum japonica, L. vulgare*
aligustre de China – *Ligustrum lucidum*
aligustre japonés – *Ligustrum ovalifolium*
aligustre lustroso de la China – *Ligustrum lucidum*
aligustrón – *Ligustrum japonicum*
alimanche – *Pluchea odorata*
álimo – *Atriplex halimus*
alinanché – *Pluchea odorata*
alingaro – *Eleagnus philippensis*
álipo – *Globularia alypum*
alisma – *Alisma plantago-aquatica*
aliso – *Alnus, Platanus racemosa, P. wrightii*
aliso blanco – *Alnus incana, A. rhombifolia, Betula verrucosa*
aliso cano – *Alnus incana*
aliso común – *Alnus glutinosa*
aliso corso – *Alnus cordata*
aliso de Córcega – *Alnus cordata*
aliso de Italia – *Alnus cordata*
aliso de Oregón – *Alnus rubra*
aliso gris – *Alnus incana*
aliso italiano – *Alnus cordata*
aliso manchado – *Hamamelis virginiana*
alisón – *Aurinia saxatilis, Lobularia maritima*
aliso napolitano – *Alnus cordata*
aliso negro – *Alnus glutinosa*
aliso rojo – *Alnus rubra*
aliso rojo americano – *Alnus rubra*
aliso verde – *Alnus viridis*
alizari – *Rubia tinctorium*
aljófar – *Lithospermum officinale*
alkandal – *Citrullus colocynthis*
alma – *Desmodium cinerascens*
almácigo – *Bursera gummifera, B. simaruba, Pistacia lentiscus, P. mexicana*

almácigo amarillo – *Bursera gummifera*
almecino – *Celtis australis*
almendrero – *Prunus dulcis*
almendrillo – *Terminalia catappa*
almendro – *Prunus dulcis, Prunus ilicifolia* ssp *lyonii, Terminalia amazonia, T. catappa*
almendro amargo – *Prunus dulcis* var *amara*
almendro con tres lobos – *Prunus triloba*
almendro de Java – *Canarium commune*
almendro de la India – *Terminalia catappa*
almendro de Tehuántepec – *Terminalia catappa*
almendro enano – *Prunus tenella*
almendro florido – *Prunus triloba*
almendro mollar – *Prunus dulcis* var *fragilis*
almendrón – *Caryocar amygdaliferum, Terminalia catappa*
almez – *Celtis australis*
almez americano – *Celtis occidentalis*
almez del sur – *Celtis australis*
almez occidentalis – *Celtis occidentalis*
almirón – *Cichorium intybus*
almizcle – *Mimulus moschatus*
almizcle amarillo – *Mimulus guttatus*
almizcleña – *Erodium moschatum, Mimulus primuloides, Muscari armeniacum, M. comosum, M. moschatum*
almizclera – *Erodium moschatum*
almizclillo – *Abelmoschus moschatus, Brugmansia* x *candida*
almohada de culebra – *Byttneria aculeata*
almoraduj – *Origanum majorana, Salvia patens*
almoradux – *Origanum majorana, Thymus mastichina*
almorejo – *Setaria pumila, S. viridis*
almorejo glauco – *Setaria pumila*
almorejo verde – *Setaria viridis*
almorejo verticilado – *Setaria verticillata*
almorta – *Lathyrus sativus*
almorta del monte – *Lathyrus cicera*
almuérdago – *Viscum album*
almum – *Sorghum almum*
alno – *Alnus glutinosa*
áloe – *Aloe ferox, A. socotrina, A. vera, Aquilaria agallocha, Cordia sebestena*
áloe de Barbados – *Aloe vera*
áloe de Zanzíbar – *Aloe vera*
áloe vera – *Aloe vera*
aloloe amarillo – *Jacobinia umbrosa*
alopecuro – *Alopecurus pratensis*
alopecuro de los campos – *Alopecurus myosuroides*
alopecuro de los prados – *Alopecurus pratensis*
alopecuro geniculado – *Alopecurus geniculatus*
alopecuro pratense – *Alopecurus pratensis*
alopecuro rojizo – *Alopecurus aequalis*
alosna – *Artemisia absinthium*
alpina – *Alpina zerumbet*
alpiste – *Ichnanthus palleus, Phalaris, Sorghum halepense, Sporobolus luchley*
alpiste arundináceo – *Phalaris arundinacea*
alpiste blanco – *Phalaris tuberosa*
alpiste bulboso – *Phalaris tuberosa*
alpiste canario – *Phalaris canariensis*
alpiste de espigas cortas – *Phalaris brachystachys*
alpiste menor – *Phalaris minor*
alpiste paradójico – *Phalaris paradoxa*
alpiste perenne de Australia – *Phalaris tuberosa*
alpiste silvestre – *Phalaris caroliniana, P. minor*
alpistera – *Phalaris canariensis*
alpiste valillo – *Phalaris minor*
alpistillo – *Dactylis glomerata, Phalaris caroliniana, P. minor*
alquequenje – *Physalis angulata, P. francheti, P. peruviana*
alquimila – *Alchemilla mollis, A. vulgaris, Aphanes occidentalis, Geranium*
alquimila alpina – *Alchemilla alpina*
alquimila común – *Alchemilla vulgaris*
alquimila del país – *Geranium mexicanum*
alquimila híbrida – *Alchemilla glaucescens*
alquitira – *Astragalus gummifer, Lycium chinense*
alquitira del país – *Opuntia tuna*
alsina – *Quercus ilex*
álsine – *Stellaria*
álsine calicanta – *Stellaria calycantha*
álsine de bosques – *Stellaria nemorum*
álsine de hojas largas – *Stellaria longifolia*
álsine gramínea – *Stellaria graminea*
álsine pálida – *Stellaria pallida*
álsine rígida – *Minuartia stricta*
álsine rugosa – *Minuartia rubella*
alstroemeria – *Alstromeria pelegrina*
altabaquillo – *Polygonum aviculare*

altamisa – *Aloysia wrightii, Ambrosia artemisiifolia, A. psilostachya, Artemisia franserioides, A. mexicana, Parthenium hysteriopsis, P. hysterophorus, Tanacetum parthenium, T. vulgare*
altamisa de Castilla – *Tanacetum parthenium*
altamisa de la casa – *Artemisia ludoviciana*
altamisa de la sierra – *Artemisia franserioides, A. frigida*
altamisa del campo – *Ambrosia confertiflora*
altamisa mexicana – *Tanacetum parthenium*
altarreina – *Achillea ageratifolia, A. millefolium, Piqueria trinervia*
altea – *Alcea rosea, Althaea officinalis, Anoda cristata, Hibiscus trionum*
alternantera repente – *Alternanthera caracasana*
altramuz – *Lupinus*
altramuz amarillo – *Lupinus luteus*
altramuz arbóreo – *Lupinus arboreus*
altramuz arborescente – *Lupinus arboreus*
altramuz azul – *Lupinus angustifolius*
altramuz blanco – *Lupinus albus*
altramuz de hojas estrechas – *Lupinus angustifolius*
altramuz del diablo – *Anagyris foetida*
altramuz de Texas – *Lupinus texensis*
altramuz hediondo – *Anagyris foetida*
altramuz perenne – *Lupinus polyphyllus*
alubia – *Phaseolus vulgaris*
alubia chica – *Phaseolus vulgaris*
alubia de España – *Phaseolus coccineus*
alubia mungo – *Vigna radiata*
alva lupina – *Aconitum napellus*
alverja – *Lathyrus sativus, Pisum sativum, Vicia articulata, V. Ervilia, V. sativa*
alverja erizada – *Vicia hirsuta*
alverjana – *Lathyrus ochrus*
alverja silvestre – *Lathyrus aphaca*
alverjilla – *Vicia sativa* ssp *nigra*
alverjón – *Pisum sativum, Vicia hirsuta, V. narbonensis*
alverjón forrajero – *Pisum sativum* var *arvense*
alzineta – *Teucrium* x *lucidrys*
alzoyatic – *Mirabilis longiflora*
allozo – *Prunus dulcis*
am – *Encyclia cochleata*
amacahuitl – *Cordia boissieri*
amacapulín – *Morus celtidifolia*
amacayo – *Sprekelia formosissima*
amacoztic – *Ficus petiolaris*
amacuahuitl – *Ficus*
ama-flores – *Eschscholzia californica*
amamastla – *Rumex mexicanus*
amancay – *Alstromeria aurea, Hymenocallis amancaes*
amánchacua – *Monstera deliciosa*
amanda – *Allamanda cathartica*
amapa – *Cordia alliodora, Tabebuia*
amapa amarilla – *Tabebuia chrysantha, T. donnell-smithii*
amapa asta – *Cordia sebestena*
amapa blanca – *Cordia sonorae*
amapa boba – *Cordia alliodora, C. sebestena, C. sonorae*
amapa colorada – *Tabebuia impetiginosa*
amapa morada – *Tabebuia impetiginosa*
amapa prieta – *Cordia alliodora, Tabebuia chrysantha, T. impetiginosa*
amapa rosa – *Tabebuia impetiginosa, T. pentaphylla, T. rosea*
amapa verde – *Tabebuia chrysantha*
amapola – *Bombax ellipticum, Eschscholzia, Hunnemannia fumariifolia, Ipomoea fistulosa, Malvaviscus arboreus, Oenothera, Pachira aquatica, Papaver rhoeas, P. somniferum, Passiflora foetida, Plumeria rubra*
amapola alpinum – *Papaver alpinum*
amapola amarilla – *Argemone ochroleuca, Chelidonium majus, Eschscholzia californica, E. mexicana, Hunnemannia fumariifolia, Meconopsis cambrica*
amapola arbórea – *Romneya coulteri*
amapola atlántica – *Papaver atlanticum*
amapola azul – *Meconopsis betonicifolia, Papaver somniferum*
amapola azul del Himalaya – *Meconopsis betonicifolia*
amapola blanca – *Argemone mexicana, A. munita, Bombax ellipticum, Papaver somniferum, Romneya coulteri*
amapola colorada – *Bombax ellipticum*
amapola común – *Papaver rhoeas*
amapola de California – *Eschscholzia californica*
amapola de Islandia – *Papaver nudicaule*

amapola de la adormidera – *Papaver somniferum*
amapola del campo – *Eschscholzia mexicana, Oenothera speciosa, Romneya coulteri*
amapola de los Alpes – *Papaver alpinum, P. rhaeticum*
amapola de opio – *Papaver somniferum*
amapola de playa – *Glaucium flavum*
amapola galesa – *Meconopsis cambrica*
amapola grande – *Alcea rosea*
amapola macho – *Papaver argemone*
amapola mestiza – *Papaver hybridum*
amapola mexicana – *Argemone mexicana*
amapola oblonga – *Papaver dubium*
amapola orientale – *Papaver orientale*
amapola plumosa – *Macleaya cordata*
amapola silvestre – *Dahlia*
amapola silvestre de los campos – *Papaver rhoeas*
amapola triste – *Papaver hybridum*
amapola tulipán – *Papaver glaucum*
amapolita – *Anoda crenatiflora, A. cristata, Esenbeckia minutiflora*
amapolita del campo – *Anoda cristata*
amapolita morada – *Anoda cristata*
amarantina – *Gomphrena globosa*
amaranto – *Amaranthus, Turnera ulmifolia*
amaranto blanco – *Amaranthus albus*
amaranto cinco de mayo – *Amaranthus caudatus*
amaranto común – *Amaranthus retroflexus*
amaranto de espigas verdes – *Amaranthus hybridus*
amaranto espinoso – *Amaranthus spinosus*
amaranto pátulo – *Amaranthus cruentas*
amaranto plumoso – *Celosia plumosa*
amaranto prostrado – *Amaranthus deflexus*
amaranto rojo – *Amaranthus caudatus*
amarasillo – *Iresine calea*
amargamiel – *Solanum dulcamara*
amargo – *Artemisia judaica, Axonopus compressus, Hintonia latiflora, Parthenium hysterophorus, Paspalum conjugatum, Quassia amara*
amargón – *Cichorium intybus, Taraxacum officinale*
amargosa – *Ambrosia artemisiifolia, A. psilostachya, Castela, Helenium mexicanum, Helianthus ciliaris, Hymenoxys odorata, Iva ambrosiaefolia, Parthenium hysterophorus*
amargosillo – *Iresine*
amarguera – *Bupleurum fruticosum*
amarguera de hojas lanceoladas – *Bupleurum lancifolium*
amarguera fruticosa – *Bupleurum fruticosum*
amarilillo – *Hippeastrum vittatum*
amarilis – *Amaryllis belladonna, Hippeastrum*
amarilis de hojas largas – *Crinum bulbispermum*
amarilis del Perú – *Hippeastrum vittatum*
amarilis escarlata – *Hippeastrum equestre*
amarillo – *Cantharellus cibarius, Pinus teocote, Terminalia amazonia*
amaro – *Salvia sclarea*
amarrio – *Prunus armenica*
amashíhuitl – *Lupinus bilineatus*
amate – *Ficus*
amate amarillo – *Ficus petiolaris*
amate blanco – *Ficus glabrata*
amate prieto – *Ficus cotinifolia, F. glaucescens*
amatillo – *Ficus padifolia*
amatl – *Ficus*
amatúngula – *Carissa grandiflora*
amatzauhtli – *Epidendrum pastoris*
amaxíhuitl – *Lupinus aschenbornii, L. bilineatus*
amazaquitl – *Arbutus xalapensis*
ambar – *Scabiosa atropurpurea*
ambarella – *Spondias cytherea*
ambarillo – *Scabiosa atropurpurea*
ambarina – *Malva moschata, Scabiosa atropurpurea*
ambrosía de hojas de ajenjo – *Ambrosia artemisiifolia*
ambrosía de México – *Chenopodium ambrosioides*
amelanchier arbóreo – *Amelanchier laevis*
amenduai – *Ceratozamia mexicana, Dioon edule*
ameos común – *Ammi visnaga*
ameos mayor – *Ammi majus*
amezquite – *Ficus padifolia*
ami – *Ammi majus*
ami vulgare – *Ammi majus*
amiga de noche – *Polianthes tuberosa*
amistad – *Hibiscus mutabilis*
amistad del día – *Hibiscus mutabilis*

amizolillo – *Datura stramonium*
amoena – *Dieffenbachia amoena*
amokid – *Psilocybe mexicana*
amole – *Agave, Chlorogalum, Hymenocallis harrisiana, H. rotata, Manfreda, Nolina palmeri, Phytolacca octandra, Polianthes maculosa, P. tuberosa, Prochnyanthes viridescens, Sapindus drummondii, S. saponaria, Stegnosperma halimifolium, Yucca, Zigadenus, Ziziphus amole*
amole amargo – *Microsechium helleri*
amole de bolita – *Sapindus drummondii, S. saponaria*
amole de raíz – *Manfreda guttata*
amole dulce – *Ziziphus amole, Z. obtusifolia*
amolillo – *Agave ocahui, A. schottii, Colubrina glomerata, Glycyrrhiza lepidota, Persea podadenia, Prochnyanthes viridescens, Sapindus drummondii, S. saponaria*
amolquilitl – *Phytolacca isocandra*
amolxihuitl – *Phytolacca octandra*
amolli – *Chlorogalum, Hymenocallis rotata, Polianthes tuberosa, Saponaria americana*
amontonado – *Lyophyllum decastes*
amor – *Cosmos parviflorus, Eragrostis curvula, Mentzelia aspera, M. hispida*
amor al uso – *Hibiscus mutabilis*
amor de campo – *Desmodium canum*
amor de estos tiempos – *Hibiscus mutabilis*
amor de hortelano – *Galium*
amor de Pedro – *Clerodendrum buchananii* var *fallax*
amor de un día – *Portulaca grandiflora*
amormío – *Pancratium maritimum*
amormío estrellado – *Pancratium illyricum*
amor oculto – *Ruscus aculeatus*
amorseco – *Celosia argentea, Desmodium canum, Eragrostis cilianensis, E. palmeri, Gomphrena*
amorseco africano – *Eragrostis lehmanniana*
amorseco ciliado – *Eragrostis cilianensis*
amorseco curvado – *Eragrostis curvula*
amorseco llorón – *Eragrostis curvula, E. lugens*
amorseco norteño – *Eragrostis palmeri*
amorseco piloso – *Eragrostis pilosa*
amor tras de la puerta – *Kalanchoe pinnata*
amsinckia de Douglas – *Amsinckia douglasiana*
amsinckia intermedia – *Amsinckia menziesii* var *intermedia*
amsinckia teselada – *Amsinckia tessellata*
amul – *Agave deserti*
amule – *Ipomoea alba*
amush – *Chenopodium ambrosioides*
anaba – *Ficus*
anacagüita – *Sterculia apetala, S. foetida*
anacahuite – *Cordia boissieri, C. dodecandra, C. sebestena, Ehretia elliptica*
anacahuite del Istmo – *Cordia alliodora*
anacahuite de Tehuántepec – *Cordia eleagnoides*
anacardo – *Anacardium occidentale*
anacua – *Ehretia anacua, E. elliptica*
anacuáhuitl – *Cordia boissieri*
anagálide – *Anagallis arvensis*
anagálide acuática – *Veronica anagallis*
anaite – *Hymenocallis americana*
analkak – *Asclepias curassavica*
anama – *Ehretia anacuna*
ananás – *Ananas comosus, A. sativus*
ananash – *Polygonum lapathifolium, P. punctatum*
anapelo – *Aconitum*
anasillo – *Celtis trinervia*
anastasia – *Artemisia vulgaris*
ancoas – *Zingiber officinale*
ancolia – *Aquilegia hybrida*
ancusa – *Ancusa azurea, A. officinales*
ancho mulato – *Capsicum annuum*
ancho negro – *Capsicum annuum*
andani – *Tithonia rotundifolia, T. tubaeformis*
andrino – *Prunus spinosa*
andrómeda – *Andromeda polifolia, Pieris forrestii, P. japonica*
androsace alargada – *Androsace elongata*
androsace anual – *Androsace maxima*
androsace de los campos – *Androsace maxima*
androsace septentrional – *Androsace septentrionalis*
androsemo – *Hypericum androsaemum*
andumuqua – *Nicotiana rustica*
anea – *Typha angustifolia, T. latifolia*
anea de hojas anchas – *Typha latifolia*
anea de hojas estrechas – *Typha angustifolia*
anémona – *Anemone*
anémona coronaria – *Anemone coronaria*
anémona del Japón – *Anemone* x *hybrida*

anémona de los bosques – *Anemone nemorosa*
anémona japonesa – *Anemone* x *hybrida*
anémona ojo de pavo – *Anemone* x *fulgens*
anémona pulsatila – *Pulsatilla vulgaris*
anémona silvia – *Anemone nemorosa*
anéota – *Calamintha nepeta*
angarilla – *Combretum farinosum*
angélica – *Angelica archangelica*
angélica borde – *Anemone sylvestris*
angélica china – *Angelica sinensis*
angélica japonesa – *Aralia elata*
angelito – *Oncidium incurvum, Psilocybe mexicana*
angiolote – *Lennoa coerulea*
angú – *Hibiscus esculentus*
angualamo – *Vitex mollis*
anime – *Hymenaea courbaril*
anís – *Foeniculum vulgare, Pimpinella anisum, Tagetes lucida, T. micrantha*
anís común – *Pimpinella anisum*
anís chico – *Pimpinella anisum*
anís de la China – *Illicium anisatum*
anís del campo – *Tagetes micrantha*
anís estrella – *Illicium anisatum*
anís estrellado – *Illicium anisatum*
anís hinojo – *Foeniculum vulgare* var *azoricum*
anisillo – *Dyssodia chrysanthemoides, Elytraria squamosa, Hedyotis wrightii, Mollugo verticillata, Schkuhria virgata, Scoparia dulcis, Tagetes, Tubiflora squamosa*
anisillo acuatil – *Schkuhria virgata*
anisillo cimarrón – *Schkuhria virgata*
anisote – *Artemisia redolens*
anís verde – *Pimpinella anisum*
anona – *Annona cherimola, A. reticulata, A. squamosa, Rollinia mucosa*
anona amarilla – *Annona muricata*
anona babosa – *Rollina mucosa*
anona blanca – *Annona diversifolia, A. squamosa*
anona cimarrona – *Rollinia mucosa*
anona colorada – *Annona reticulata*
anona de broquel – *Annona muricata*
anona de tierra caliente – *Annona reticulata*
anona lisa – *Annona glabra*
anona morada – *Annona purpurea, A. reticulata*
anona tis – *Annona squamosa*
anonilla – *Annona diversifolia, A. reticulata*
anonillo – *Magnolia grandiflora, Talauma mexicana*
anonillo de monte – *Rollinia mucosa*
anonita – *Rollinia mucosa*
anorza – *Bryonia dioica*
anserina – *Chenopodium album, C. bonus-henricus*
anteojo de poeta – *Thunbergia alata*
antimaño – *Chrysanthemum coronarium*
antipusi – *Rhynchosia pyramidalis*
antirrino – *Antirrhinum majus*
antiviperina – *Abelmoschus moschatus*
anturio – *Anthurium*
añil – *Helianthus annus, Indigofera suffruticosa, Tephrosia tenella*
añil bastardo – *Amorpha fruticosa*
añil blanco – *Indigofera tinctoria*
añil colorado – *Indigofera suffruticosa*
añil del muerto – *Verbesina encelioides* ssp *exauriculata*
añililo – *Aphelandra deppeana*
añil indigo – *Indigofera suffruticosa*
añil jiquelite – *Indigo suffruticosa*
añil montés – *Indigofera fruticosa, I. suffruticosa*
añublo – *Phytophtora infestans*
aoatl – *Quercus*
aorí – *Juniperus deppeana*
apá – *Bebbia atriplicifolia, B. juncea*
apaclililo – *Sagittaria macrophylla*
apaclol – *Sagittaria macrophylla*
apachitli – *Sabal mexicana*
apagafuego – *Costus mexicanus*
apagallum – *Macrolepiota procera*
apamate – *Tabebuia pentaphylla, T. rosea*
apan – *Brickellia hastata*
apán – *Bebbia atriplicifolia*
apanchola – *Cuphea aequipetala*
apánico – *Cochlospermum vitifolium*
apapatla – *Nymphae flavovirens, N. mexicana*
apapatlaco – *Verbesina oncophora*
aparejo – *Muhlenbergia utilis*
apatlol – *Sagittaria macrophylla*
apátsicua – *Tagetes erecta*
apatz – *Sabal mexicana*
apepe – *Nymphaea alba*
apera de los campos – *Apera spica-venti*
apera interrumpida – *Apera interrupta*

apio – *Apium graveolens, Arracacia xanthorrhiza, Ciclospermum leptophyllum*
apio borde – *Apium graveolens*
apio caballar – *Smyrnium olustratum*
apio común – *Apium graveolens*
apio de hojas nudosas – *Apium nodiflorum*
apio de montaña – *Levisticum officinale*
apio de monte – *Levisticum officinale*
apio de perro – *Aethusa cynapium*
apio montano – *Levisticum officinale*
apio nabo – *Apium graveolens* var *rapaceum*
apipiltzin – *Psilocybe aztecorum*
apirrábano – *Apium graveolens* var *rapaceum*
apish – *Xanthosoma violaceum*
apocino – *Apocynum androsaemifolium*
apomo – *Brosimum alicastrum*
apompo – *Cochlospermum vitifolium, Pachira aquatica, P. macrocarpa*
aposhi – *Erythrina flabelliformis*
apoyamate – *Cyperus squarrosus*
apoyomatli – *Cyperus articulatus*
aprín – *Ziziphus mauritiana*
aprón – *Opuntia decumbens*
apupo – *Sechium edule*
apurí – *Morus microphylla*
aquiche – *Guazuma ulmifolia*
aquifolio – *Ilex aquifolium*
aquilea – *Achillea millefolium*
aquilea amarilla – *Achillea filipendulina*
aquilea de hojas de filipendula – *Achillea filipendulina*
aquilmacal – *Dioscorea alata*
arabeta – *Arabidopsis thaliana*
arabeta de Thalius – *Arabidopsis thaliana*
arábide – *Arabis blepharophylla*
arábide alpina – *Arabis alpina*
arábide caucásica – *Arabis caucasica*
arábide ciliada – *Arabis sparsiflora*
arábide de los Alpes – *Arabis alpina, A. caucasica*
arábide glabrescente – *Arabis glabra*
arábide hirsuta – *Arabis hirsuta*
arabota – *Juniperus communis*
araguaney – *Tabebuia chrysantha*
aral barilla – *Halogeton glomeratus*
aralia – *Aralia, Fatsia japonica, Polyscias scutellaria*
aralia arborícola – *Schefflera actinophylla*
aralia elegantísima – *Schefflera elegantissima*
arámicua – *Xanthosoma robustum, X. sagittifolium*
aramina – *Urena lobata*
arándano – *Vaccinium*
arándano agrio – *Vaccinium macrocarpon, V. oxycoccus*
arándano americano – *Vaccinium corymbosum*
arándano azul – *Caulophyllum thalictroides, Vaccinium corymbosum*
arándano corimboso – *Vaccinium corymbosum*
arándano de los pantanos – *Vaccinium uliginosum*
arándano encarnado – *Vaccinium vitis-idaea*
arándano europeo – *Vaccinium oxycoccus*
arándano macrocarpo – *Vaccinium macrocarpum*
arándano negro – *Vaccinium uliginosum*
arándano rojo – *Vaccinium vitis-idaea*
arándano trepado – *Vaccinium macrocarpum*
araña – *Hymenocallis harrisiana, H. rotata, Nigella damascena*
arañita – *Oncidium ornithorynchum*
arañón – *Prunus spinosa*
arañuela – *Adianthum capillus-veneris, Hymenocallis harrisiana, Nigella damascena*
aráquida – *Arachis hypogaea*
arar – *Tetraclinis articulata*
araticú – *Rollinia deliciosa*
araucaria – *Araucaria araucana, A. bidwillii, A. heterophylla*
araucaria de Bidwill – *Araucaria bidwillii*
araucaria de Chile – *Araucaria araucana*
araucaria de neuquén – *Araucaria araucana*
araucaria espinuda – *Araucaria araucana*
arbelcorán – *Hesperis matronalis*
árbol amargo – *Hintonia latiflora*
árbol atrompetado – *Cecropia peltata*
árbol balsámico – *Commiphora opobalsamum*
árbol candil – *Aleurites moluccana*
árbol contra la fiebre – *Eucalyptus globulus*
árbol chino del barniz – *Koelreuteria paniculata*
árbol chocolate – *Irvingia gabonensis*
árbol de aguila – *Conzattia multiflora*
árbol de algodón – *Ceiba pentandra, Ochroma lagopus*
árbol de bálsamo – *Myroxylon balsamum* var *pereirae*
árbol de barril – *Fouquieria fasciculata*

árbol de benjuí – *Styrax benzoin*
árbol de corteza de papel – *Melaleuca ericifolia*
árbol de cuerno – *Acacia spadicigera*
árbol de dolor de muelas – *Zanthoxylum americanum*
árbol de fuente – *Spathodea campanulata*
árbol de goma laca – *Ficus religiosa*
árbol de guau – *Toxicodendron radicans*
árbol de hierro – *Carpinus betulus, Metrosideros excelsa, Olneya tesota, Parrotia persica*
árbol de incienso – *Abies bracteata*
árbol de jack – *Artocarpus heterophyllus*
árbol de Josué – *Yucca brevifolia*
árbol de Judas – *Bocconia arborea, Cercidiphyllum japonicum, Cercis occidentalis, C. siliquastrum, Schinus molle*
árbol de Judea – *Cercidiphyllum japonicum, Cercis siliquastrum*
árbol de Júpiter – *Lagerstroemia indica*
árbol de katsura – *Cercidiphyllum japonicum*
árbol de la almáciga – *Pistacia lentiscus*
árbol de la brea – *Cercidium floridum*
árbol de la canela – *Cinnamomum zeylanicum*
árbol de la cera – *Aleurites moluccana, Myrica californica, M. cerifera, M. mexicana, Parmentiera cerifera, Rhus succedeana*
árbol de la charretera – *Pterostyrax hispida*
árbol de la felicidad – *Dracena fragrans*
árbol de la fiebre – *Cinchona officinalis, Eucalyptus fibrosa*
árbol de la fiebre australiano – *Eucalyptus globulus*
árbol de la gutapercha – *Eucommia ulmoides*
árbol de la hierba – *Ilex paraguayensis*
árbol de la laca – *Rhus verniciflua*
árbol del alcanfor – *Cinnamomum camphora*
árbol de la leche – *Brosimium utile*
árbol de la lengua de mujer – *Albizzia occidentalis*
árbol de la lluvia chino – *Koelreuteria bipinnata*
árbol de la manteca – *Butyrospermum parkii*
árbol del ambar – *Liquidambar styraciflua*
árbol de la miel – *Sophora japonica*
árbol del amor – *Cercis siliquastrum*
árbol de la muerte – *Hippomane mancinella, Ipomoea murucoides*
árbol de la nuez pili – *Canarium ovatum*
árbol de la quina – *Cinchona officinalis*
árbol de la sabiduría – *Betula alba*
árbol de las bolitas – *Sapindus saponaria*
árbol de las calabazas – *Crescentia cujete*
árbol de las campanillas – *Halesia monticola*
árbol de las cruces – *Randia*
árbol de las cuentas – *Sapindus saponaria*
árbol de las manitas – *Chiranthodendron pentadactylon*
árbol de las pagodas – *Sophora japonica*
árbol de las palomas – *Davidia involucrata*
árbol de las pelucas – *Cotinus coggygria*
árbol de las perlas – *Moringa oleifera*
árbol de las salchichas – *Kigelia africana*
árbol de las suelas – *Pentaclethra macrophylla*
árbol de las trompetas – *Catalpa bignonioides*
árbol de la vaca – *Brosimium utile*
árbol de la vida – *Thuja occidentalis, T. orientalis*
árbol de la vida de oriente – *Thuja orientalis*
árbol de la vida gigante – *Thuja plicata*
árbol de la vida gigantesco – *Thuja plicata*
árbol de la vida occidental – *Thuja occidentalis*
árbol de la vida oriental – *Thuja orientalis*
árbol del bellote – *Sterculia apetala*
árbol del bonete – *Leucopremna mexicana*
árbol del caucho – *Ficus elastica, Hevea brasiliensis*
árbol del cielo – *Ailanthus altissima*
árbol del clavo – *Caryophyllus aromaticus*
árbol del coral – *Erythrina crista-galli*
árbol del corcho – *Annona glabra*
árbol del diablo – *Hura polyandra, Morisonia americana*
árbol del fuego – *Brachychiton acerifolius, Delonix regia, Spathodea campanulata*
árbol del hule – *Castilla elastica, Ficus elastica*
árbol de lima – *Tilia occidentalis*
árbol de limón – *Citrus limon, C. limonia*
árbol del jebe – *Hevea brasiliensis*
árbol del maná – *Fresno ornus*
árbol del matrimonio – *Pereskia lycnidiflora*
árbol del neem – *Azadirachta indica*
árbol de los cuarenta escudos – *Ginkgo biloba*
árbol de los dioses – *Ailanthus altissima*
árbol de los farolitos – *Koelreuteria paniculata*
árbol de los pañuelos – *Davidia involucrata*

árbol del paraíso – *Eleagnus angustifolius, Melia azedarach*
árbol del Perú – *Schinus molle*
árbol del sebo – *Sapium sebiferum*
árbol del tomate – *Cyphomandra betacea*
árbol del tulipán – *Liriodendron tulipifera*
árbol del tung – *Aleurites fordii*
árbol del venado – *Ipomoea murucoides*
árbol de mamey – *Lucuma mammosum*
árbol de María – *Calophyllum calaba*
árbol de melón – *Carica papaya*
árbol de pan – *Artocarpus altilis, A. incisa, A. intergrifolia*
árbol de pan mexicano – *Monstera deliciosa*
árbol de papa – *Solanum wrightii*
árbol de pimienta – *Schinul molle*
árbol de quitasol – *Melia azedarach* var *umbraculiformis*
árbol de ruedas de fuego – *Stenocarpus sinuatus*
árbol de sal – *Avicennia germinans*
árbol de Santa Lucía – *Prunus mahaleb*
árbol de Santa María – *Ageratina*
árbol de sasafrás – *Sassafras albidum*
árbol de seda – *Albizzia julibrissin, Calotropis procera*
árbol de shea – *Butyrospermum parkii*
árbol de tabaco – *Nicotiana glauca*
árbol de té – *Camellia sinensis, Melaleuca alternifolia*
árbol de viajero – *Ravenala madagascariensis*
árbol elefante – *Bursera microphylla, Pachycormus discolor*
árbol huaje – *Leucaena esculenta*
arbolillo – *Sapindus saponaria*
arbolito – *Clavaria flava*
arbolito de la cera – *Myrica cerifera*
árbol lirio – *Magnolia soulangiana*
árbol María – *Ageratina*
árbol orquídea – *Bauhinia forficata, B. variegata*
árbol paraguas – *Schefflera actinophylla*
árbol paraíso – *Melia azedarach*
árbol pica-pica – *Lagunaria patersonii*
árbol prieto – *Vauquelinia californica, V. corymbosa* var *angustifolia*
árbol pulpo – *Schefflera actinophylla*
árbol que tulle – *Karwinskia humboldtiana*
árbol sagrada – *Clerodendron*
árbol santo – *Guaiacum coulteri*
árbol sombrero – *Brachychiton discolor*
árbol triste – *Nyctanthes arbor-tristis*
árbol vela – *Senna alata*
arbusto ardiente – *Pyracantha coccinea*
arbusto atrompetado – *Tabebuia impetiginosa*
arbusto de fuego chileno – *Embothrium coccineum*
arbusto de la llama – *Calliandra tweedii*
arbusto del Japón – *Quillaja saponaria*
arbusto escarlata – *Hamelia erecta*
arce – *Acer*
arce afelpado del Japón – *Acer japonicum*
arce afelpado japonés – *Acer japonicum*
arce azucarero – *Acer saccharum*
arce blanco – *Acer pseudoplatanus*
arce coreano – *Acer pseudosieboldianum*
arce chino del padre David – *Acer davidii*
arce chino gris – *Acer griseum*
arce de azúcar – *Acer saccharum*
arce de capadocia – *Acer cappadocicum*
arce de hoja ceniza – *Acer negundo*
arce de hojas de fresno – *Acer negundo* var *californicum*
arce de Japón – *Acer palmatum*
arce del Amur – *Acer tartaricum* var *ginnala*
arce de Manchuria – *Acer tartaricum* var *ginnala*
arce de Oregón – *Acer macrophyllum*
arce de Pensylvania – *Acer pensylvanicum*
arce japonés – *Acer japonicum, A. palmatum*
arce macrofilo – *Acer macrophyllum*
arce menor – *Acer campestre*
arce negundo – *Acer negundo*
arce noruego – *Acer platanoides*
arce palmado – *Acer palmatum*
arce palmado japonés – *Acer palmatum*
arce plátano falso – *Acer pseudoplatanus*
arce platanoide – *Acer platanoides*
arce plateado – *Acer saccharinum*
arce polimorfo – *Acer palmatum*
arce real – *Acer platanoides*
arce rojo americano – *Acer rubrum*
arce rojo japonés de corteza serpentina – *Acer capillipes*
arce ruso – *Acer tartaricum* var *ginnala*
arce sacarino – *Acer saccharinum, A. saccharum*
arce silvestre – *Acer campestre*

arce tártaro – *Acer tartaricum*
arce trífido – *Acer buergeranum*
arce trinerve – *Acer buergeranum*
arco iris – *Iris germanica*
arco negro – *Apoplanesia paniculata*
archenas – *Juniperus communis*
ardivieja – *Halimium umbellatum*
areca – *Areca catechu, Chrysalidocarpus lutescens*
areca catechu – *Areca catechu*
arejuela de ratón – *Dichondra argentea*
arellano – *Albizzia occidentalis, Caesalpinia platyloba*
arenaria – *Ammophila arenaria, Arenaria*
arenaria de hojas de tomillo – *Arenaria serpyllifolia*
arenaria de las Baleares – *Arenaria balearica*
arenaria de primavera – *Sagina subulata*
arenaria montana – *Arenaria montana*
arenoso purpureo – *Triplasis purpurea*
arestín – *Eryngium*
arete – *Bessera elegans, Capparis indica, Fuchsia, Russelia equisetiformis*
arete de India – *Bomarea hirtella, Clerodendrum thomsoniae*
arete de la cocinera – *Russelia equisetiformes*
aretillo – *Bessera elegans, Fuchsia, Lopezia racemosa*
aretillo del monte – *Fuchsia microphylla*
aretito – *Carlowrightia californica, Gaura coccinea, Lobelia laxiflora*
arfueyo – *Viscum album*
argentina – *Potentilla anserina*
árgoma – *Ulex europaea*
árgoma de flores pequeñas – *Ulex parviflorus*
argoña – *Erica cinerea*
arguma – *Tilia* x *europaea*
aricoma – *Polymnia edulis*
ariosa – *Viguiera montana, V. parishii*
aristoloquia – *Aristolochia clematitis*
aristoloquia clematítide – *Aristolochia clematitis*
aristoloquia de hojas grandes – *Aristolochia durior*
aristoloquia de Virginia – *Aristolochia serpentaria*
aristoloquia larga – *Aristolochia longa*
aristoloquia macrofila – *Aristolochia durior*
aristoloquia redonda – *Aristolochia rotunda*
Arizónica – *Cupressus arizonica*
arjorán – *Cercis siliquastrum*
arlequina – *Sparaxis tricolor*
arlo – *Berberis vulgaris*
armeria – *Armeria juniperifolia, A. maritima* ssp *californica*
armeria cespitosa – *Armeria juniperifolia*
armeria de jersey – *Armeria alliacea*
armeria falso junco – *Armeria girardii*
armeria maritima – *Armeria maritima*
armuelle – *Atriplex hortensis, A. patula, Beta vulgaris* ssp *cicla*
armuelle bueno – *Chenopodium bonus-henricus*
armuelle de huerta – *Atriplex hortensis*
armuelle glauco – *Atriplex halimus*
armuelle heterosperma – *Atriplex heterosperma*
armuelle molle – *Atriplex hortensis*
armuelle patente – *Atriplex patula*
armuelle pátulo – *Atriplex patula*
armuelle róseo – *Atriplex rosea*
arnárico – *Origanum majorana*
arnato – *Bixa orellana*
árnica – *Acourtia thurberi, Arnica, A. montana, Gaillardia nervosa, Heterotheca, Neurolaena lobata, Tithonia, Trixis angustifolia, T. californica*
árnica del país – *Helenium mexicanum, Heterotheca inuloides*
árnica flor – *Ericameria spinulosus, Grindelia oxylepis*
árnica mexicana – *Heterotheca psammophila*
árnica montana – *Arnica montana*
aro – *Arum italicum, A. maculatum, Zantedeschia aethiopica*
aro acuático – *Zantedeschia palustris*
aro de Levante – *Arum palestinum*
aro italiano – *Arum italicum*
aroma blanca – *Leucaena glauca*
aroma boba – *Leucaena leucocephala*
aroma de mar – *Rosmarinus officinalis*
aroma mansa – *Leucaena leucocephala*
aro manchado – *Arum maculatum*
aromo – *Acacia farnesiana, Prosopis juliflora*
aromo espinoso – *Acacia farnesiana*
aróracua – *Cattleya citrina*
aroyoguo – *Karwinskia humboldtiana*
arsáfraga – *Berula erecta, Sium sisarum*
artanta – *Piper angustifolium*

artemia – *Chrysanthemum indicum*
artemisa – *Ambrosia artemisiifolia, Artemisia*
artemisa bastarda – *Achillea millefolium*
artichaut – *Cynara scolymus*
artiguilla – *Bouteloua gracilis*
artillería – *Pilea microphylla*
arto – *Rhamnus infectorius, Rubus fruticosus*
artocárpeo – *Artocarpus heterophyllus*
arvejita – *Pisum sativum*
arveja – *Ceratonia siliqua, Pisum sativum, Vicia*
arveja amarilla – *Vicia lutea*
arveja articulada – *Vicia articulata*
arveja atropurpúrea – *Vicia benghalensis*
arveja cometodo – *Pisum sativum* var *macrocarpon*
arveja común – *Vicia sativa*
arveja china – *Pisum sativum* var *macrocarpon*
arveja de cuatro semillas – *Vicia tetrasperma*
arveja de dos semillas – *Vicia disperma*
arveja de los cercados – *Vicia sepium*
arveja de los setos – *Vicia sepium*
arveja de primavera – *Vicia lathyroides*
arveja de vaca – *Vigna unguiculata*
arveja francesa – *Vicia narbonensis*
arveja glabra – *Vicia cracca*
arveja hirsuta – *Vicia hirsuta*
arveja húngara – *Vicia pannonica*
arveja sativa – *Vicia sativa*
arveja tuberosa – *Lathyrus tuberosus*
arveja vellosa – *Vicia villosa*
arvejera – *Ceratonia siliqua*
arvejilla de olor – *Lathyrus odoratus*
arvejilla velluda – *Vicia villosa*
arvejo – *Lathyrus sativus, Pisum sativum* var *arvense*
arvejón – *Vicia lutea*
arzolla – *Silybum marianum*
arracacha – *Arracacia xanthorrhiza*
arracacho – *Rhamnus sphaerosperma*
arraclán – *Rhamnus frangula*
arrancamoños – *Xanthium strumarium*
arrastradillo – *Opuntia stenopetala*
arrayán – *Ardisia revoluta, Eugenia capuli, E. fragrans, Gaultheria acuminata, G. odorata, Myrica californica, M. cerifera, Myrtus communis, Pernettya ciliata, Psidium sartorianum, Ruscus aculeatus*
arrayán blanco – *Myrtus communis*
arrayán brabántico – *Myrica gale, M. pensylvania*
arrayán cerífero – *Myrica cerifera*
arrayancito – *Malpighia glabra*
arrayán común – *Myrtus communis*
arrayán de los pantanos – *Myrica gale*
arrayán prieto – *Eugenia fragrans*
arrayán salvaje – *Ruscus aculeatus*
arrebolera – *Mirabilis exserta, M. jalapa*
arrocera – *Echinochloa polystachya*
arrocillo – *Coix lacryma-jobi, Echinochloa colona, Leersia hexandra, Parthenium hysterophorus, Piptochaetium fimbriatum, Rhynchosia minima*
arrocillo áspero – *Leersia oryzoides*
arrocillo cimarrón – *Leersia monandra*
arroyuela – *Lythrum salicaria*
arroyuela de hojas de hisopo – *Lythrum hyssopifolium*
arroyuela de tres brácteas – *Lythrum tribracteatum*
arroyuela menor – *Lythrum virgatum*
arroyuela portulaca – *Lythrum portula*
arroz – *Oryza sativa, Zizania aquatica*
arroz bravo – *Leersia hexandra*
arroz del mar – *Distichlis palmeri*
arroz del monte – *Echinochloa colona*
arroz indio – *Achnatherum hymenoides*
arroz salvaja – *Zizania palustris*
arroz silvestre – *Zizania palustris*
arroz-xiu – *Portulaca pilosa*
arruda brava – *Pilocarpus jaborandi*
arrurruz – *Maranta arundinacea*
arrurruz arundináceo – *Maranta arundinacea*
arrurruz común – *Maranta arundinaceae*
arrurruz de Florida – *Zamia pumila*
arrurruz de las Indias Orientales – *Curcuma angustifolia*
arrurruz de las Islas Bermudas – *Maranta arundinacea*
asafétida – *Ferula asafoetida*
asapescado – *Trichilia hirta*
asarabácara – *Asarum europaeum*
asarero – *Prunus spinosa*
ásaro – *Asarum europaeum*
ásaro común – *Asarum europaeum*
ascalonia – *Allium cepa* var *ascalonicum*
asclepias – *Asclepias syriaca*
asenjo – *Artemisia absinthium*

aserina – *Asarum europaeum*
asfodelina – *Asphodeline lutea*
asfódelo – *Asphodelus fistulosus*
asfódelo blanco – *Asphodelus albus*
asfódelo fistuloso – *Asphodelus fistulosus*
ash – *Brosimum alicastrum*
ashecani – *Chenopodium album*
ashwaganda – *Withania somnifera*
asiento de la suegra – *Echinocactus grusoni, E. texensis*
asintla – *Cornus disciflora*
asisincle – *Cornus disciflora*
aspérgula – *Galium aparine*
asperilla dulce – *Galium odoratum*
asperilla olorosa – *Galium odoratum*
aspidistra – *Aspidistra elatior*
asplenio – *Asplenium nidus*
asplenio de roca – *Asplenium fontanum*
asplenio verde – *Asplenium trichomanes-ramosum*
asqui – *Cynodon dactylon*
asta – *Cordia sebestena, C. sonorae*
astabaca – *Inula helenium*
aster – *Aster, Callistephus chinensis*
aster alpino – *Aster alpinus*
aster brumal – *Aster novi-belgii*
aster de China – *Callistephus chinensis*
aster de la Nueva-Inglaterra – *Aster novae-angliae*
aster de los Pirineos – *Aster sibiricus*
aster de montaña – *Aster alpinus, A. amellus*
aster de otoño – *Aster amellus, A. dumosus*
aster de septiembre – *Callistephus chinensis*
aster de Stoke – *Stokesia laevis*
aster de verano – *Callistephus chinensis*
aster de virgilio – *Aster amellus*
asterisco del mar – *Asteriscus maritimus*
aster lanceolado – *Aster lanceolatus*
astilbe – *Astilbe arendsii*
astrágalo – *Astragalus*
astrágalo deflexo – *Oxytropis deflexa*
astrágalo de hojas de regaliz – *Astragalus glycyphyllos*
astrágalo español – *Astragalus boeticus*
astrancia – *Astrancia major*
astrancia mayor – *Astrantia major*
astras – *Callistephus chinensis*
astromelia – *Lagerstroemia indica, L. speciosa*
astromero – *Lagerstroemia indica, L. speciosa*
astronómica – *Lagerstroemia indica*
asul – *Tamarix gallica*
atamasco – *Zephyranthes texana*
atanasia – *Brickellia cavanillesi, Tanacetum vulgare*
atanasia amarga – *Brickellia cavanillesi, B. grandiflora*
atapaima – *Plumeria rubra*
atarfe – *Tamarix africana, T. gallica, T. ramosissima*
atecuixtle – *Dolicholus phaseoloides, Rhynchosia precatoria, R. pyramidalis*
atepocapatli – *Zexmenia aurea*
atepule – *Cephalanthus salicifolius*
atesúchil – *Fuchsia arborescens*
atexóchitl – *Fuchsia arborescens*
atigrada – *Tigridia pavonia*
atimorreal – *Aloe vera*
atizador al rojo vivo – *Kniphofia uvaria*
atlanchana – *Cuphea aequipetalia, Potentilla*
atlanchane – *Cuphea aequipetala, Lythrum*
atlatlancuaya – *Iresine calea*
atlinán – *Datura ceratocaula*
atmósfera – *Lagerstroemia indica*
atmosférica – *Lagerstroemia indica*
atocamay – *Myrica cerifera*
atoyaxocotl – *Spondias mombin, S. purpurea*
atractylis – *Atractylodes lancea*
atrapamoscas – *Dionaea muscipula, Drosera rotundifolia*
atropa – *Atropa belladona*
atuto – *Vitex mollis*
atzahutli – *Cranichis speciosa*
atzapotl – *Lucuma salicifolia*
atzapotlquahuitl – *Lucuma mammosa, L. salicifolia*
atzcalxóchitl – *Atamosco carinata, Sprekelia formosissima, Zephyranthes carinata, Z. longifolia*
atzitzintle – *Euchlaena mexicana*
atzoyatl – *Mirabilis longiflora*
auaquauitl – *Quercus*
auatetz – *Quercus*
auatetzmulli – *Quercus*
auatl – *Quercus*
auaton – *Quercus robur*
aubarell – *Boletus edulis, B. reticulatus*
aubliguera – *Omphalodes linifolia*
aubrecia – *Aubrieta deltoidea*

aucuba – *Aucuba japonica*
aulaga – *Genista anglica, G. hispanica, Ulex europaea*
aulaga morisca – *Ulex parviflorus*
aulaga occidental – *Genista hispanica*
auricula de hojas carnosas – *Primula auricula*
auricula de los Alpes – *Primula auricula*
aurora – *Ipomoea cairica, I. purpurea, I. scopulorum*
aurora común – *Hibiscus trionum*
ave del paraíso – *Strelitzia reginae*
avellanera – *Corylus avellana*
avellanillo – *Rhamnus frangula*
avellano – *Corylus*
avellano arbóreo – *Corylus colurna*
avellano común – *Corylus avellana*
avellano chasqueante – *Hamamelis virginiana*
avellano de bruja – *Hamamelis virginiana*
avellano de guía – *Momordica balsamina*
avellano de invierno – *Corylopsis spicata*
avellano de la hechicera – *Hamamelis virginiana*
avellano de Lambert – *Corylus maxima*
avellano de Levante – *Corylus colurna*
avellano mediterráneo – *Corylus colurna*
avellano purgante – *Jatropha curcas*
avellano turco – *Corylus colurna, C. maxima*
avena – *Avena sativa*
avena barbada – *Avena barbata*
avena cimarrona – *Avena fatua*
avena común – *Avena sativa*
avena de invierno – *Avena sterilis*
avena desarrollada – *Arrhenatherum elatius*
avena elevada – *Arrhenatherum elatius*
avena estéril – *Avena sterilis*
avena estrigosa – *Avena strigosa*
avena falsa alta – *Arrhenatherum elatius*
avena guacha – *Avena fatua*
avena loca – *Avena fatua*
avena mayor – *Avena sterilis*
avena perenne – *Helictotrichon sempervirens*
avena rubia – *Trisetum flavescens*
avena silvestre – *Avena fatua*
avenca – *Adiantum capillus-veneris*
avens – *Geum urbanum*
avens japonés – *Geum japonicum*
avispa – *Tritonia crocata*
awará – *Salix lasiolepis*
axalxóchitl – *Cosmos sulphureus*
axaxaxoxihuitl – *Ricinus communis*
axi – *Ipomoea batatas*
axixpatli cóztic – *Rumex mexicanus*
axocatzín – *Sida procumbens, S. rhombifolia*
axocopaque – *Gaultheria acuminata, G. odorata*
axóchitl – *Astianthus viminalis, Erblichia odorata*
axs'que – *Aechmea bracteata, A. mexicana*
axté – *Jatropha curcas*
axtlacapali – *Buddleja americana*
ayacahuite – *Pinus ayacahuite, P. flexilis*
ayacahuite colorado – *Pinus ayacahuite*
ayacahuite vidrioso – *Pinus strobus* var *chiapensis*
ayacotli – *Phaseolus coccineus, P. vulgaris*
ayahuasca – *Banisteriopsis caapi*
ayale – *Crescentia alata, C. cujete*
ayatito – *Calochortus*
ayauhcuahuitl – *Pinus ayacahuite*
ayécotl – *Phaseolus coccineus, P. vulgaris*
ayer ayer – *Lansium domesticum*
ayocote – *Phaseolus coccineus, P. vulgaris*
ayohuiztle – *Solanum rostratum*
ayón – *Meliosma alba*
ayote – *Cucurbita maxima, C. mixta, C. pepo*
ayotli – *Cucurbita*
ayoxochquílitl – *Cucurbita*
ayoyote – *Thevetia*
ayumé – *Rhus sempervirens*
azad – *Zelkova carpinifolia*
azafrán – *Buddleja marrubifolia, Carthamus tinctorius, Colchicum autumnale, Crocus sativus, C. speciosus, Ditaxis*
azafrán albo – *Crocus vernus*
azafrán amarillo – *Crocus chrysanthus, C. sativus*
azafrán bastardo – *Carthamus tinctorius, Colchicum autumnale*
azafrán bifloro – *Crocus biflorus*
azafrancillo – *Buddleja marrubifolia, Carthamus tinctorius, Cuscuta, Ditaxis*
azafrán cultivado – *Crocus sativus*
azafrán de bolita – *Ditaxis*
azafrán de la India – *Curcuma domestica*
azafrán de las Indias – *Curcuma domestica*
azafrán del campo – *Buddleja marrubifolia*
azafrán de los pobres – *Buddleja marrubifolia, Carthamnus tinctorius*

azafrán de los prados – *Colchicum speciosum*
azafrán dorado – *Crocus angustifolius*
azafranillo – *Buddleja marrubifolia, Carthamus tinctorius*
azafrán oficinal – *Crocus sativus*
azafrán otoñal – *Colchicum autumnale*
azafrán primaveral – *Crocus vernus*
azafrán rumi – *Carthamus tinctorius*
azafrán silvestre – *Colchicum autumnale, Crocus vernus*
azafrán verdadero – *Crocus sativus*
azafrán zacatlaxcale – *Cuscuta*
azahar – *Citrus aurantium, C. sinensis*
azahar de la China – *Pittosporum tobira*
azalea – *Rhododendron*
azalea de guía – *Bougainvillea spectabilis*
azalea de la barranca – *Exogonium bracteatum*
azalea india – *Rhododendron simsii*
azalea japonesa – *Rhododendron obtusum*
azamboero – *Citrus medica*
azarero – *Prunus lusitanica*
azarezo – *Prunus lusitanica*
azarollo – *Sorbus aucuparia*
azaya – *Helichrysum orientale*
azcapán – *Datura stramonium*
azcasúchil – *Atamosco carinata*
aznacho – *Pinus mugo* var *rostrata*
azota caballo – *Cordia brevispicata, Duranta repens*
azotacristos – *Carthamus lanatus*
azotalengua – *Asperugo procumbens, Galium aparine*
azotapulga – *Mentha pulegium*
azozúcar – *Salpianthus arenarius*
azúcar huayo – *Hymenaea courbaril*
azucena – *Hemerocallis fulva, Hippeastrum, Hymenocallis rotata, Lilium, Polianthes tuberosa*
azucena africana – *Agapanthus africanus*
azucena amarilla – *Cattleya citrina, Encyclia citrina, Govenia superba, Hemerocallis fulva, H. lilioasphodelus, Sternbergia lutea*
azucena anaranjada – *Lilium bulbiferum*
azucena atigrada – *Lilium lancifolium*
azucena áurea – *Lilium auratum*
azucena azafranada – *Hemerocallis fulva*
azucena blanca – *Lilium candidum*
azucena canadiense – *Lilium canadense*
azucena común – *Lilium candidum*
azucena de coral – *Lilium pumilum*
azucena de gorro de Turco – *Lilium martagon*
azucena de gorro de Turco dorada – *Lilium hansonii*
azucena de Henry – *Lilium henryi*
azucena de Kafiristán – *Schizostylis coccinea*
azucena de la sierra – *Lilium candidum*
azucena de la Virgen – *Lilium candidum*
azucena del campo – *Milla biflora*
azucena del monte – *Cattleya citrina, Echeandia brevifolia, Govenia liliacea, G. supurba, Maxillaria liliacea*
azucena de mar – *Pancratium maritimum*
azucena de México – *Hippeastrum*
azucena de pascua – *Lilium longiflorum*
azucena de rabo de zorro – *Eremurus robustus*
azucena de rayas doradas – *Lilium auratum*
azucena de trompeta – *Lilium longiflorum*
azucena dorada – *Lilium auratum*
azucena gigante – *Cardiocrinum giganteum*
azucena leopardo – *Lilium pardalinum*
azucena martagón – *Lilium martagon*
azucena peruana – *Alstroemeria aurantiaca*
azucena real – *Lilium regale*
azucena regia – *Lilium regale*
azucena roja – *Amaryllis reginae, Lilium bulbiferum*
azucena rosa – *Amaryllis belladonna*
azucena serrana – *Lilium candidum*
azucena silvestre – *Lilium martagon, Milla biflora*
azucena tabacal – *Hemerocallis fulva*
azucena tigrina – *Lilium lanciflorum*
azuche – *Erblichia odorata*
azúchil – *Astianthus viminalis*
azuica – *Trichilia hirta*
azuelita – *Nemophila menziesii*
azuelo – *Centaurea cyanus, Indigofera suffruticosa*
azufaifo – *Ziziphus jujuba*
azufaifo africano – *Ziziphus mauritiana*
azufaifo de la espina de Cristo – *Ziziphus spina-christi*
azufre vegetal – *Lycopodium clavatum*
azul – *Agave tequilana*
azulea – *Sisyrinchium bellum*
azuleja – *Nemophila menziesii*
azulejita – *Nemophila menziesii*

azulejo – *Centaurea cyanus, Clitoria terneata, Crusea calocephala*
azulete – *Centaurea cyanus*
azulillo – *Haematoxylon brasiletto, Solanum wrightii*
azulino – *Centaurea cyanus, Plumbago auriculata*
azul orchoro – *Dactylis glomerata*
azumate de Puebla – *Artemisia mexicana*
azumatl – *Sambucus mexicana*
azúmbar – *Alisma plantago-aquatica*
azumiatl – *Senecio salignus*
azuquia radiata – *Vigna radiata*
azuzón – *Senecio jacobaea*

babaco – *Carica pentagona*
babadai – *Eryngium hemsleyanum*
baba de buey – *Commelina dianthifolia*
babari – *Platonia insignis*
babasú – *Attalea speciosa*
babino – *Hyptis albida*
babiso – *Erythea brandegeei, E. nitida, Sabal uresana*
babiza – *Anemopsis californica*
babki – *Agave angustifolia*
babosito – *Armillaria mellea*
baboso – *Pachycereus hollianus*
babucha – *Sobralia macrantha*
baburia – *Ipomoea ancisa*
baca – *Arundo donax*
bacachari – *Euphorbia colletioides*
bacalché – *Bourreria huanita*
bacánoc – *Lophophora williamsii*
bacaporo – *Parkinsonia aculeata*
bácara – *Salvia sclarea*
bácaris – *Baccharis sarathroides*
bacatón – *Lippia umbellata*
bacáu – *Bambusa vulgaris*
bacochini – *Pithecellobium dulce*
baco muteca – *Cercidium floridum*
bacotomi – *Brickelia coulteri*
bactris garrocha – *Bactris mexicana*
bacuri – *Platonia insignis*
bachang – *Mangifera foetida*
bachata – *Chorizanthe fimbriata, Condalia globosa, Ziziphus obtusifolia*
bachomo – *Baccharis salicifolia*
badao zoo – *Psilocybe zapotecorum*
badea – *Passiflora quadrangularis*
badiana – *Illicium anisatum*
badoh – *Turbina corymbosa*
badoh negro – *Ipomoea tricolor, I. violacea*
badolas – *Rumex scutatus*
badungás – *Ipomoea tricolor, I. violacea*
bael – *Aegle marmelos*
bago – *Plumbago capensis*
bagoto – *Parkinsonia aculeata*
bagre – *Sideroxylon spiniflora*
baguela – *Bromelia pinguin*
bahalcan – *Solanum nigrum*
bahía – *Paspalum notatum*
bahi cápore – *Ziziphus amole*
baiburilla – *Dorstenia drakeana*
baiburín – *Kalanchoe pinnata, Kallstroemia grandiflora, K. parviflora, Paspalum hartwegianum*
baiburina – *Dorstenia drakeana*
baijuo – *Pisonia capitata*
bailador – *Bombax ellipticum*
bailarina – *Bombax ellipticum, Laelia nodosa*
baileya del desierto – *Baileya multiradiata*
bainoro – *Celtis iguanea, C. pallida, C. reticulata, Pisonia capitata*
bainoro prieto – *Pisonia capitata*
baiquillo – *Flaveria oppositifolia, Sesbania herbacea*
baitoa – *Phyllostylon bresiliensis*
baja tripa – *Rivina humilis*
bajoca – *Phaseolus vulgaris*
bakana – *Scirpus*
bakelus – *Ageratum conyzoides*
bala de cañón – *Couroupita guianensis*
baladre – *Nerium oleander*
balamte – *Theobroma bicolor*
balata – *Achras bidentata, A. zapota, Mimusops balata*
balaustra – *Punica granatum*
balazo – *Monstera deliciosa*
balche – *Lonchocarpus longistylus*
balfour – *Polyscias scutellaria*
balota – *Ballota nigra*
balota falso-tarraquillo – *Ballota pseudodictamnus*
balota negra – *Ballota nigra*
balsa – *Heliocarpus donnell-smithii, Ochroma lagopus, O. pyramidale*
balsamea – *Epilobium canum*
balsamero – *Populus balsamifera*

balsamero negro – *Populus balsamifera* ssp *trichocarpa*
balsamillo – *Balsamorhiza sagittata, Populus fremontii*
balsamina – *Impatiens balsamina, I. pallida, Momordica balsamina, M. charantia*
balsamina cultivada – *Impatiens balsamina*
balsamina del Cabo – *Impatiens capensis*
balsamina del monte – *Impatiens capensis*
balsamina glandulífera – *Impatiens glandulifera*
balsamina manchada silvestre – *Impatiens pallida*
balsamina nicaragua – *Impatiens walleriana*
balsamita – *Chrysanthemum balsamita*
balsamito – *Momordica charantia, Myroxylon balsamum* var *pereirae*
bálsamo – *Balanites aegyptiaca, Copaifera officinalis, Dalbergia, Epilobium canum, Hamelia erecta, Momordica charantia, Myroxylon balsamum* var *pereirae*
bálsamo amarillo – *Sweetia panamensis*
bálsamo copaiba – *Copaifera officinalis*
bálsamo copalme – *Liquidambar styraciflua*
bálsamo de copaifera – *Copaifera coriacea*
bálsamo de Gilead – *Commiphora opobalsamum, Populus balsamifera*
bálsamo de la Meca – *Commiphora opobalsamum*
bálsamo de las Indias – *Myroxylon balsamum* var *pereirae*
bálsamo de la unión – *Myroxylon balsamum* var *pereirae*
bálsamo del Canadá – *Abies balsamea*
bálsamo de limón – *Melissa officinalis*
bálsamo del Jesuita – *Copaifera officinalis*
bálsamo del Perú – *Myroxylon balsamum* var *pereirae*
bálsamo de María – *Calophyllum calaba*
bálsamo de San Salvador – *Myroxylon balsamum* var *pereirae*
bálsamo de Tolu – *Myroxylon balsamum* var *pereirae*
bálsamo negro – *Myroxylon balsamum* var *pereirae*
bálsamo oloroso – *Sweetia panamensis*
ballestera – *Veratrum album*
ballí – *Desmoncus chinantlensis, D. orthacanthos*
ballico – *Lolium perenne*
ballico anual – *Lolium multiflorum*
ballico doméstico – *Lolium multiflorum*
ballico inglés – *Lolium perenne*
ballico italiano – *Lolium multiflorum*
ballico perenne – *Lolium perenne*
ballico vivaz – *Lolium perenne*
ballueca – *Avena fatua*
bambarra – *Vigna subterranean, Voandzeia subterranea*
bambú – *Bambusa, Phyllostachys aurea, Pseudosasa japonica*
bambú amarillo – *Phyllostachys aurea*
bambú común – *Bambusa vulgaris*
bambú espinudo – *Guadua longifolia*
bambú gigante – *Dendrocalamus giganteus*
bambú japonés – *Bambusa arundinacea*
bambú patamba – *Bambusa vulgaris*
bananero – *Musa paradisiaca, M. sapientum*
banano – *Musa paradisiaca, M. sapientum*
banano color clarete – *Musa acuminata*
banano de Nepal – *Musa nepalensis*
banano enano – *Musa acuminata, M. nana*
banano harinoso – *Musa balbisiana*
banano manzano – *Musa acuminata* x *balbisiana*
banano nepalés – *Musa nepalensis*
banano robusta – *Musa acuminata*
bandarigu de juchitan – *Caesalpinia pulcherrima*
bandera – *Canna indica, Euphorbia pulcherrima*
bandera española – *Canna coccinea, C. indica, Kniphofia uvaria, Lantana camara, Quamoclit coccinea, Q. lobata*
bandera mexicana – *Salvia microphylla*
banderilla – *Bouteloua curtipendula, Salvia splendens, Stipa gigantean, Thalia geniculata*
banderilla simple – *Bouteloua uniflora*
bansu – *Abies religiosa*
banyan – *Ficus bengalensis, F. religiosa*
baobab – *Adansonia digitata*
baqueta – *Chaetoptelea mexicana*
baraima – *Bacopa monnieri*
barambuyan de nejapa – *Delonix regia*
barancio – *Pinus mugo*
baraña – *Alnus incana* ssp *tenuifolia*
barba – *Phoradendron velutinum*
barba blanca – *Stipa neomexicana*

barba cabruna – *Aruncus sylvester, Eryngium campestre, Tragopogon pratensis*
barba cabruna de hojas de puerro – *Tragopogon porrifolius*
barbachata – *Ziziphus obtusifolia*
barba china – *Phytolacca octandra*
barba de árbol – *Tillandsia usneoides, Usnea barbata*
barba de cabra – *Tragopogon pratensis*
barba de cabra de los prados – *Tragopogon pratensis*
barba de cabra mayor – *Tragopogon dubius*
barba de capuchino – *Cuscuta*
barba de cabrón – *Aruncus dioicus*
barba de chivato – *Calliandra houstoniana, Clematis, Eupatorium quadrangulare, Solanum umbellatum*
barba de chivo – *Acacia angustissima, Aruncus dioicus, Calliandra houstoniana, Clematis, Cymbalaria muralis, Pithecellobium acatlense*
barba de gallo – *Caesalpinia arenosa, Warszewiczia coccinea*
barba de gato – *Clematis, Hymenocallis harrisiana, H. rotata*
barba de indio – *Chloris virgata, Echinochloa frumentacea, Eragrostis pilosa*
barbadejo – *Viburnum lantana, V. tinus*
barba de jolote – *Senna atomaria*
barba de Júpiter – *Anthyllis barba-jovis*
barba del sol – *Caesalpinia pulcherrima*
barba de mantel – *Bauhinia divaricata*
barba de palo – *Tillandsia usneoides*
barba de San Juan de Dios – *Senecio barba-johannis*
barba de viejo – *Caesalpinia platyloba, Clematis, Laelia nodosa, Pilosocereus alansis, P. purpusii, Tillandsia usneoides, Urospermum picroidea, Usnea barbata*
barbadija – *Viburnum tinus*
barbadino – *Desmodium barbatum*
barba española – *Tillandsia usneoides*
barba larga – *Trachypogon secundus*
barba negra – *Heteropogon contortus*
barba negra dulce – *Heteropogon melanocarpus*
barbasco – *Croton californicus, C. texensis, Dioscorea composita, D. mexicana, Dorstenia drakeana, Entada phaseoloides, Jacquinia macrocarpa* ssp *pungens, Lonchocarpus utilis, Serjania mexicana, Tephrosia cana, T. nicaraguensis, T. toxicaria*
barbasco cimarrón – *Tephrosia nicaraguensis*
barbasco de camote – *Dioscorea composita*
barbasúchil – *Caesalpinia pulcherrima*
barbón – *Pachycereus schottii*
barbona – *Cleome guianensis*
barbona rosada – *Cleome hasslerana, Isomeris spinosa*
barbón bicolor – *Pappophorum bicolor*
barbón puntiagudo – *Pappophorum vaginatum*
barbudillo – *Dorstenia contrayerba, D. drakeana*
barbudo – *Quisqualis indica*
barco de caridad – *Heliconia*
barchata – *Condalia globosa, Lycium berlandieri, Ziziphus obtusifolia*
barda – *Fouquieria campanulata, F. fasciculata, F. splendens*
bardaguera – *Salix fragilis*
bardana – *Arctium lappa*
bardana común – *Arctium minus*
bardana mayor – *Arctium lappa*
bardana menor – *Arctium minus, Xanthium strumarium*
barí – *Calophyllum braziliense*
baría – *Cordia sebestena*
baricoca – *Cyrtocarpa procera*
barilla – *Halogeton glomeratus*
barillo – *Calophyllum braziliense*
barí oscuro – *Calophyllum braziliense*
barit – *Leersia hexandra*
barossa – *Maclura tinctoria*
bartanucha – *Verbena pumila*
bartolona – *Opuntia robusta*
bartonia – *Mentzelia lindleyi*
barú – *Arenga pinnata*
barrabás – *Vitex gaumeri*
barranco – *Prunus brachybotrya*
barrera – *Berula erectum*
barreta – *Ruellia tuberosa*
barreta china – *Fraxinus greggii*
barreta de cochino – *Fraxinus greggii*
barrilla – *Batis maritima, Mesembryanthemum crystallinum, Salsola soda, S. tragus*
barrilla de hojas delgadas – *Salsola tragus*
barrilla vermicular – *Salsola vermiculata*
barrillo – *Calophyllum calaba*

barriola – *Bassia hyssopifolia*
barrón – *Ammophila arenaria*
bas – *Desmoncus oxyacanthus*
basalaca – *Eleocharis atropurpurea*
basicuáare – *Bromus carinatus*
basílico – *Ocimum basilicum*
bassia de hojas de hisopo – *Bassia hyssopifolia*
bastón – *Myrtillocactus geometrizans*
bastón de Jacob – *Asphodeline lutea*
bastón de San Francisco – *Leonotis nepetaefolia*
bastón de verano – *Penstemon labrosus*
batácora – *Jatropha spathulata*
batachín – *Leucaena lanceolata*
batagá – *Coix lacryma-jobi*
batamote – *Baccharis salicifolia, B. sarathroides, Pluchea salicifolia*
batamote del monte – *Baccharis thesioides*
bataneni – *Mascagnia macroptera*
batata – *Ipomoea batatas*
batata acuática – *Ipomoea aquatica*
batata dulce – *Ipomoea batatas*
batata pavonada – *Ipomoea purpurea*
batatilla – *Turbina corymbosa*
batatilla de playa – *Ipomoea macrorrhiza*
batayáqui – *Lippia umbellata, Montanoa patens, M. tomentosa* ssp *rosei*
batel – *Triumfetta semitriloba*
batey – *Castilla elastica*
batilini – *Tecoma stans*
batillo – *Desmanthus covillei, Leucaena lanceolata*
batoba-xha – *Aloe variegata, A. vera*
batuuchá – *Psittacanthus calyculatus*
bavisa – *Cosmos crithmifolius, C. pringlei*
baya – *Morus alba, M. microphylla, Vaccinium uliginosum*
bayaba – *Psidium guajava*
bayaboon – *Thrinax wendlandiana*
baya China – *Melia azedarach*
baya de Boysen – *Rubus ursinis* var *loganobaccus*
baya de dedal – *Rubus parviflorus*
baya de grulla – *Vaccinium macrocarpon*
baya de la bruja – *Atropa belladonna*
baya de nieve – *Symphoricarpos albus, S. microphyllus*
bayal – *Desmoncus, Eragrostis mexicana, Sabal japa, S. mexicana, Thrinax wendlandiana*
bayal-suuk – *Eragrostis mexicana*
bayo – *Phaseolus vulgaris*
bayón – *Typha angustifolia*
bayoneta – *Epidendrum parkinsonianum, Yucca aloifolia, Y. guatemalensis*
bayrum – *Eugenia acris, E. racemosa*
beaumontia – *Beaumontia grandiflora*
bebelama – *Phoebe ehrenbergii, Sassafridium macrophyllum, Sideroxylon*
bebeta – *Euphorbia pulcherrima*
becabunga – *Veronica beccabunga*
becerra – *Antirrhinum majus, Croton lecheri*
becoquino – *Borago officinalis, Cerinthe major*
bedoll – *Betula pubescens*
beek – *Ehretia tinifolia*
béek – *Quercus oleoides*
begonia – *Begonia*
begonia africana – *Saintpaulia ionantha*
begonia bulbosa – *Begonia tuberosa*
begonia colgante – *Begonia solananthera*
begonia de hojas ornamentales – *Begonia rex*
begonia regia – *Begonia rex*
begonia rex – *Begonia rex*
begonia tuberosa – *Begonia tuberosa*
behek – *Bourreria huanita*
bejín – *Lycoperdon perlatum*
bejuco – *Calumus, Cissampelos pareira, Cryptostegia grandiflora, Desmoncus chinantlensis, Macfadyena unguis-cati*
bejuco angarilla – *Combretum farinosum*
bejuco blanco – *Exogonium bracteatum*
bejuco cachorro – *Macfadyena unguis-cati*
bejuco colorado – *Hippocratea, Tetracera volubilis*
bejuco costillón – *Serjania racemosa*
bejuco de agua – *Entada polystachya, Tetracera volubilis, Vitis tilifolia*
bejuco de alambre – *Lygodium venustum*
bejuco de amole – *Entada polystachya*
bejuco de biznaga – *Dioscorea laxiflora*
bejuco de caballo – *Heteropteris laurifolia, Petrea arborea*
bejuco de canasta – *Tetracera volubilis*
bejuco de canasto – *Desmoncus chinantlensis*
bejuco de casa – *Philodendron radiatum*
bejuco de cazadores – *Vitis tiliifolia*
bejuco de colación – *Antigonon leptopus*

bejuco de conchitas – *Clitoria terneata*
bejuco de coral – *Cissus cucurbitina*
bejuco de coraza – *Dioscorea*
bejuco de cuajar hule – *Ipomoea alba*
bejuco de culebra – *Dioscorea alata*
bejuco de chilillo – *Dioscorea laxiflora*
bejuco de chinaca – *Macfadyena unguis-cati*
bejuco de chiquihuite – *Smilax mexicana, S. mollis*
bejuco de diente de perro – *Smilax mexicana, S. mollis*
bejuco de estribo – *Entada polystachya*
bejuco de huico – *Pithecellobium echinatum*
bejuco de jaba – *Entada phaseoloides*
bejuco de la candelaria – *Exogonium bracteatum*
bejuco de leche – *Funastrum clausum*
bejuco de mondongo – *Entada polystachya*
bejuco de ñame – *Rajania hastata*
bejuco de panune – *Dalbergia glabra, Entada polystachya*
bejuco de peine – *Combretum farinosum*
bejuco de piedra – *Combretum farinosum*
bejuco de playa – *Ipomoea pes-caprae*
bejuco de puerco – *Ipomoea alba*
bejuco de purgación – *Commicarpus scandens*
bejuco de quebradura – *Rhipsalis baccifera*
bejuco de ratón – *Macfadyena unguis-cati*
bejuco de San Diego – *Petrea arborea*
bejuco de sol – *Senecio confuses*
bejuco de tacalote – *Entada phaseoloides*
bejuco de tachicón – *Tetracera volubilis*
bejuco de toro – *Combretum farinosum*
bejuco de tripa – *Nissolia fruticosa*
bejuco de vaca – *Ipomoea alba*
bejuco estrella – *Ipomoea quamoclit*
bejuco loco – *Cissus sicyoides*
bejuco loco rojo – *Cissus sicyoides*
bejuco prieto – *Entada polystachya, Mascagnia macroptera*
bejuco roñoso – *Hidalgoa ternata*
bejuquillo – *Equisetum robustum*
bela – *Aegle marmelos*
belamcanda – *Belamcanda chinensis*
belcho – *Ephedra sinica*
belehui – *Malus sylvestris*
Belén – *Cryptostegia grandiflora, Impatiens balsamina, I. walleriana*
beleño – *Hyoscyamus niger*
beleño negro – *Hyoscyamus niger*
belesa – *Plumbago auriculata*
beloperone – *Justicia californica*
bella a las once – *Portulaca grandiflora*
belladama – *Atropa belladonna*
bella de noche – *Mirabilis jalapa*
belladona – *Atropa belladona, Datura discolor, D. wrightii, Solanum nigrum*
belladona del país – *Nicandra physalodes*
bella Inés – *Castilleja arvensis, C. canescens, C. gracilis*
bella Luisa – *Chrysanthemum indicum*
bella sombra – *Phytolacca dioica*
bellida – *Salvia* x *superba*
bellísima – *Antigonon leptopus, Solanum seaforthianum*
bellorita – *Bellis perennis*
bellota – *Calliandra haematocephala, Cucurbita maxima, Quercus, Sterculia apetala, S. mexicana*
bellota de sabina – *Phoradendron juniperinum*
bellotero – *Quercus*
bembericua – *Toxicodendron radicans*
ben – *Moringa oleifera*
bendito – *Cnicus benedictus*
benjuí común – *Styrax officinalis*
bentosidad – *Nama hispidum*
bequilla – *Sesbania herbacea, S. macrocarpa*
berberís – *Berberis fremontii*
berberís común – *Berberis vulgaris*
berberís de hojas cortas – *Berberis* x *stenophylla*
berberís de hojas de palma – *Berberis buxifolia*
berberís japonés – *Berberis thunbergii*
bérbero – *Berberis fremontii, B. julianae, B. vulgaris*
berceo – *Stipa gigantea*
bercul – *Verbena menthaefolia*
berenjena – *Cyphomandra betacea, Solanum*
berenjena cimarrona – *Solanum verbascifolium*
berenjena de gallina – *Solanum mammosum*
berenjena del diablo – *Datura stramonium*
berenjena de teta – *Solanum mammosum*
berenjena peludita – *Solanum mammosum*
berenjena silvestre – *Solanum amazonium, S. diversifolium*
berenjenilla – *Mandragora autumnalis, Solanum verbascifolium*
berenjenita peluda – *Solanum mammosum*

berenjenita peludita – *Solanum mammosum*
bergamota – *Monarda citriodora, M. didyma*
bergamota silvestre – *Monarda fistulosa*
bergamoto – *Citrus bergamotia*
bergenia – *Bergenia cordifolia*
bermejo – *Agave tequilana, Hyparrhenia rufa*
Bermuda – *Cynodon dactylon*
Bermuda de la costa – *Cynodon dactylon*
bersilana – *Muntingia calabura*
bersim – *Trifolium alexandrinum*
berula acuática de hojas estrechas – *Berula erecta*
berza – *Brassica oleracea* var *acephala, B. o.* var *capitata, B. o.* var *viridis*
berza amarga – *Calystegia soldanella*
berza col – *Brassica oleracea* var *acephala, B. o.* var *sabellica*
berza común – *Brassica oleracea* var *acephala, Calystegia sepium*
berza de perro – *Cynanchum vincetoxicum*
berza de prado – *Cardamine pratensis*
berza de saboya – *Brassica oleracea* var *sabuada*
berza marina – *Calystegia soldanella*
berza perruna – *Cynanchum vincetoxicum*
berza roja – *Brassica oleracea* var *capitata*
berraco – *Aesculus californica, A. parryi, Cyrtocarpa procera, Stemmadenia tomentosa* var *palmeri*
berrera – *Sium latifolium*
berrillo – *Barbarea praecox*
berro – *Berula erecta, Coreopsis tinctoria, Mimulus guttatus, Rorippa nasturtium-aquaticum*
berro austriaco – *Rorippa austriaca*
berro de agua – *Rorippa nasturtium-aquaticum*
berro de ganso – *Arabis caucasica*
berro de jardín – *Barbarea praecox*
berro de para – *Spilanthes acmella*
berro de prado – *Cardamine pratensis*
berro de tierra – *Barbarea verna*
berro falso – *Rorippa palustris*
berro islandés – *Rorippa islandica*
berro terrestre – *Barbarea vulgaris*
besasa – *Ruta graveolens*
besico – *Campanula medium, Cardiospermum halicacabum*
betabel – *Beta vulgaris*
betarraga – *Beta vulgaris*
betarraga azucarera – *Beta vulgaris* ssp *saccharifera*
betarraga forrajera – *Beta vulgaris*
bete – *Mentha* x *piperita*
betel – *Piper betle*
betónica – *Stachys betonica, S. officinalis*
betónica mayor – *Stachys macrantha*
betónica oficinal – *Stachys officinalis*
betoum – *Pistacia atlantica*
bétula – *Betula alba*
bétula pubescente – *Betula pubescens*
bétula verrugosa – *Betula verrucosa*
betunia – *Petunia* x *hybrida, P. parviflora*
beyazoo – *Psilocybe zapotecorum*
bezinic-ché – *Alvaradoa amorphoides*
biahi – *Heliconia bihai*
biaqui – *Diospyros ebenaster*
biaxhi – *Spondias mombin, S. purpurea*
bibí – *Sapindus saponaria*
bicapora – *Ziziphus amole*
biche – *Caesalpinia mexicana, Senna*
biche manso – *Senna leptocarpa, S. tora*
biche prieto – *Senna occidentalis*
biche silvestre – *Senna pallida*
biche viro – *Senna pallida*
bichi-bixui – *Opuntia pumila*
bichicol – *Bromelia karatas*
bicho – *Senna bicapsularis, S. pallida*
bicholi – *Vitis tiliifolia, V. vinifera*
bidente de hojas desiguales – *Bidens aurea*
bidente pedunculado – *Bidens frondosa*
bidente penché – *Bidens cernua*
bidente piloso – *Bidens pilosa*
bidente tripartito – *Bidens tripartita*
bidueiro – *Betula pubescens*
biengranada – *Chenopodium botrys*
biércol – *Calluna vulgaris*
bignonia – *Bignonia capreolata, Campsis radicans, Macfadyena unguis-cati, Tecoma capensis, T. stans*
bigotito – *Heliotropium indicum*
biguero del infierno – *Datura stramonium*
bihagua – *Heliconia latispatha*
bihi – *Acacia farnesiana*
bija – *Bixa orellana*
bilimbi – *Averrhoa bilimbi*
billbergia – *Aechmea fasciata*
bindo – *Condalia mexicana*
bi-neechi – *Psilocybe zapotecorum*

biota – *Thuja orientalis*
biribá – *Rollinia deliciosa, R. pulchrinervis*
birreta de obispo – *Astrophytum myriostigma*
birringo – *Rhus integrifolia*
bisbirinda – *Castela texana, C. tortuosa*
biscoyol – *Bactris major*
bishicuri – *Pinus cembroides*
bislingua – *Ruscus hypoglossum*
bismalva – *Althaea officinalis*
bistorta – *Polygonium amphibium, P. bistorta, P. bistortoides*
bitaxe – *Solanum nigrum*
biti-xoba – *Myrtillocactus geometrizans*
bito – *Liquidambar styraciflua*
bitze – *Inga edulis, I. spuria*
bitzuri – *Pachycereus weberi*
bixhumí – *Sambucus mexicana*
bi-xilla – *Phragmites australis*
bizcolote – *Condalia mexicana*
bizil – *Malvaviscus arboreus, M. grandiflorus*
biznaga – *Ammi visnaga, Coryphantha elephantidens, C. pallida, Echinocactus, Ferocactus, Mammillaria, Thelocactus bicolor*
biznaga burra – *Echinocactus texensis*
biznaga caballona – *Ferocactus macrodiscus*
biznaga colorada – *Ferocactus gracilis, F. pilosus*
biznaga costillona – *Ferocactus hamatacanthus, F. histrix*
biznaga de acitrón – *Echinocactus platyacanthus*
biznaga de agua – *Ferocactus wislizeni*
biznaga de chilitos – *Echinocactus polycephalus, Ferocactus latispinus, Mammillaria heyderi, M. magnimamma*
biznaga de dulce – *Echinocactus horizonthalonius, E. platyacanthus, Ferocactus macrodiscus*
biznaga de estropajos – *Astrophytum capricorne*
biznaga de lana – *Echinocactus platyacanthus*
biznaga de lima – *Ferocactus pilosus, Thelocactus setispinus*
biznaga de piña – *Coryphanta longicornis*
biznaga de tuna – *Ferocactus hamatacanthus, Thelocactus setispinus*
biznaga espinosa – *Ferocactus hamatacanthus*
biznaga ganchuda – *Ferocactus hamatacanthus, F. latispinus*
biznaga gigante – *Echinocactus platyacanthus*
biznaga gigantesca – *Ferocactus wislizeni*
biznaga hembra – *Ferocactus wislizeni*
biznaga limilla – *Ferocactus hamatacanthus*
biznaga meloncillo – *Echinocactus horizonthalonius*
biznaguita – *Cochemia, Echinomastus erectocentrus, E. intertextus, Mammillaria, Neolloydia conoidea, Sclerocactus whipplei, Stenocactus*
biznaguita de chilito – *Mammillaria*
bizoya – *Theobroma cacao*
blanco chino – *Quercus crassipes*
blanco de las sierras – *Abies lasiocarpa*
blanquillo – *Prunus persica* ssp *nectarina*
bleda – *Beta vulgaris* ssp *cicla*
bledera – *Chenopodium foliosum*
bledo – *Amaranthus, Celosia, Chenopodium album, Trianthema portulacastrum*
bledo carbonero – *Phytolacca americana, P. octandra, Stegnosperma halimifolium*
bledo cimarrón – *Amaranthus spinosus*
bledo espinoso – *Amaranthus spinosus*
bledo macho – *Amaranthus spinosus*
bledo molle – *Atriplex hortensis*
bledo rojo – *Amaranthus spinosus*
boaria – *Maytenus boaria*
bobché – *Cocoloba barbadensis, Pluchea odorata*
bobordo – *Typha latifolia*
boboró – *Rauwolfia serpentina*
boca de dragón – *Antirrhinum majus, Croton lecheri, Etlingera elatior, Trigonidium egertonianum*
boca de dragón campestre – *Antirrhinum orontium*
boca de fuego – *Epidendrum ibaguense*
boca de león – *Acineta barkeri, Antirrhinum majus*
bocado de gallina – *Anagallis arvensis*
bocamelia – *Clerodendrum chinense*
bocano – *Solanum americanum*
bocina – *Distictis buccinatoria*
boconia – *Bocconia arborea, B. frutescens*
bocote – *Cordia elaeagnoides, C. sebestena*
boch – *Crescentia cujete*
bóchil – *Taxodium distichum* var *mexicanum*

Boer – *Eragrostis chloromelas*
boj – *Buxus sempervirens*
boja – *Artemisia abrotanum*
boj común – *Buxus sempervirens*
boj de China – *Murraya paniculata*
boj de las Antillas – *Phyllostylon brasiliensis*
bojillo de popal – *Thalia geniculata*
bojín – *Suillus granulatus*
bojón – *Cordia alliodora, C. sebestena*
bojón blanco – *Cordia alliodora*
bojón prieto – *Cordia alliodora*
bola – *Capsicum annuum, Lycoperdon perlatum*
bola colorada – *Trichilia americana*
bola de añil – *Echinops exaltatus*
bola de hilo – *Kallstroemia maxima, Lobularia maritima, Tanacetum parthenium*
bola de nieve – *Viburnum*
bola de oro – *Solanum marginatum*
boldino – *Peumus boldus*
boldo – *Peumus boldus*
boldo fragrante – *Peumus boldus*
boleto – *Boletus edulis*
boleto amarillento – *Boletus impolitus*
boleto de espinacal – *Pleurotis eryngii*
boleto granulado – *Suillus granulatus*
boliche – *Sapindus saponaria*
bolillo – *Albizzia occidentalis, Leucaena lanceolata*
bolita – *Capsicum annuum, Guazuma ulmifolia*
bolita de conejo – *Lycoperdon candidum, L. perlatum, L. umbrinum*
bolita de hongo – *Lycoperdon perlatum*
bolita de nieve – *Symphoricarpos albus*
bolita de San Juan – *Lycoperdon perlatum*
bolita de tierra – *Vascellum pratense*
bolita prieta – *Cordia brevispicata*
bolontobi – *Cissus trifoliate*
bolsa de clerigo – *Calceolaria mexicana*
bolsa de conejo – *Harfordia macroptetra*
bolsa de coyote – *Diphysa occidentalis*
bolsa de gato – *Merremia umbellata*
bolsa de Judas – *Solandra nitida*
bolsa del pastor – *Calceolaria mexicana, Capsella bursa-pastoris, Physalis alkekengi*
bolsilla – *Cardiospermum halicacabum*
bolsita – *Cypripedium*
bolsita de la víbora – *Passiflora arida*
bolsita de víbora – *Passiflora palmeri*
boludo – *Capsicum annuum*
bomba – *Brugmansia, Calotropis procera*
bomba reventadora – *Lycoperdon perlatum*
bombilla – *Bougainvillea spectabilis, Cardiospermum halicacabum*
bombonaje – *Carludovicia palmata*
bomcabalsáh – *Chamaedorea humilis*
bonete – *Aristolochia grandiflora, Astrophytum myriostigma, Carica mexicana, Jarilla caudata, Leucopremna mexicana, Solandra nitida*
bonete del diablo – *Aristolochia grandiflora*
bonetero – *Euonymus europaeus*
bonetero alato – *Euonymus alatus*
bonetero común – *Euonymus europaeus*
bonetero de Europa – *Euonymus europaeus*
bonetero radicante – *Euonymus fortunei*
bonetillo – *Cucurbita pepo*
boniato – *Ipomoea batatas*
boquillo – *Epithelantha bokei*
borbonesa – *Silene dioica*
bore – *Colocasia esculenta*
boria – *Celosia argentea*
borla de San Pedro – *Tecoma stans*
borla española – *Clematis dioica*
borona – *Zea mays*
borona de Filipinas – *Panicum miliaceum*
boroquim – *Jatropha cinerea*
borto – *Arbutus unedo*
borrachero – *Brugmansia, Datura stramonium*
borrachito – *Coryphanta macromeris*
borracho – *Nolina longifolia*
borrachuela – *Lolium temulentum*
borraja – *Borago officinalis*
borraja común – *Borago officinalis*
borraja fina – *Borago officinalis*
borraja silvestre – *Symphytum aspersum*
borrajina – *Borago officinalis*
borrajita – *Borago officinalis*
borrego – *Lactarius vellereus, Russula nigricans*
borrosa amarilla pálida – *Gnaphalium luteo-album*
borrubiete – *Arbutus unedo*
bosborón – *Scaevola plumierii*
boshi – *Baccharis pteronioides*
botán – *Sabal*
botavara – *Fraxinus velutina*
bote – *Bombax ellipticum*

botil – *Phaseolus coccineus*
botiline – *Tecoma stans*
botoncahui – *Conocarpus erecta*
botoncillo – *Casearia nitida, Centaurea cyanus, Conocarpus erecta, Spilanthes americana*
botoncillo azul – *Centaurea cyanus*
botón de mezcal – *Lophophora williamsii*
botón de oro – *Caltha palustris, Hydrastis canadensis, Ranunculus, Spilanthes americana, Zexmenia hispida*
botón de oro amarillo – *Ranunculus acris*
botón de oro enano – *Ranunculus montanus*
botón de plata – *Achillea ptarmica, Tanacetum parthenium*
botón de soldado – *Scabiosa stellata*
botón encarnado – *Cercis occidentalis*
botonera – *Santolina rosmarinifolia*
botón negro – *Atropa belladonna*
bototo – *Cochlospermum vitifolium*
boxhaas – *Musa paradisiacal, M. sapientum*
boxhuayalté – *Malpighia glabra*
boxosdá – *Isocoma menziesii*
bracillo – *Ceanothus cordulatus, C. cuneatus, C. leucodermis*
bramilla – *Opizia stolonifera*
branca ursina – *Acanthus mollis*
branca ursina falsa – *Heracleum mantegazzianum*
braquiquito blanco – *Brachychiton populneus*
braquiquito rojo – *Brachychiton acerifolius*
Brasil – *Caesalpinia crista, C. platyloba, Condalia hookeri, C. obovata, Haematoxylon brasiletto, H. campechianum*
Brasil azulillo – *Haematoxylon brasiletto*
brasilete – *Caesalpinia platyloba, Haematoxylum brasiletto*
brasilete colorado – *Caesalpinia crista*
brasilete de la India – *Caesalpinia sappan*
brasilillo – *Calliandra eriophylla, Haematoxylon brasiletto, Ziziphus amole*
brasilillo marismeño – *Haematoxylon brasiletto*
brazalete – *Melaleuca armillaris*
brea – *Cercidium floridum, C. praecox, C.* x *sonorae*
brecina – *Calluna vulgaris*
brecina común – *Calluna vulgaris*
brécol – *Brassica oleracea* var *italica*
brena – *Cynodon dactylon*
bretaña – *Basella alba*
bretónica – *Melochia tomentosa*
breva – *Ficus carica*
brezo – *Calluna vulgaris, Daboecia cantabrica, Erica*
brezo arbóreo – *Erica arborea*
brezo blanco – *Erica arborea*
brezo ciliado – *Erica ciliaris*
brezo colorado – *Erica australis*
brezo común – *Calluna vulgaris*
brezo de cornualles – *Erica vagans*
brezo de escobas – *Erica scoparia*
brezo de hojas en cruz – *Erica tetralix*
brezo de invierno – *Erica carnea*
brezo de Irlanda – *Daboecia cantabrica*
brezo de mar – *Frankenia laevis*
brezo de turbera – *Erica tetralix*
brezo encarnado – *Erica carnea*
brezo escocés – *Calluna vulgaris*
brezo español – *Erica australis*
brezo espinoso – *Pernettya mucronata*
brezo irlandés – *Daboecia cantabrica*
brezo mediterráneo – *Erica mediterranea*
brezo nazareno – *Erica cinerea*
brezo polifolio – *Daboecia cantabrica*
brezo portugués – *Erica lusitanica*
brezo vagabundo – *Erica vagans*
bricho – *Senna bicapsularis, S. laevigata, S. occidentalis*
brillo de sol – *Lasthenia californica*
brincadora – *Cylindropuntia fulgida, Sebastiania pavoniana, S. pringlei, Vitis tiliifolia*
brionia – *Bryonia alba, B. dioica*
brionia blanca – *Bryonia alba*
brionia negra – *Tamus communis*
briqueta – *Ilex nitida*
briza corta – *Briza minor*
briza menor – *Briza minor*
bróculi – *Brassica oleracea* var *italica*
bróculi chino – *Brassica oleracea* var *alboglabra*
bróculi de yemas – *Brassica oleracea* var *italica*
brocha de Indio – *Castilleja gracilis*
brocha india – *Castilleja indivisa, C. integre*
brocha tejana – *Castilleja indivisa*
brodiea – *Brodiaea*

bromelia – *Cryptanthus zonatus*
bromo – *Bromus stamineus, B. sterilis*
bromo blando – *Bromus hordeaceus*
bromo cebadilla – *Bromus catharticus, B. inermis*
bromo de California – *Bromus carinatus*
bromo de Hungría – *Bromus inermis*
bromo de los techos – *Bromus tectorum*
bromo de Madrid – *Bromus madritensis*
bromo dormilón – *Bromus anomalus*
bromo español – *Bromus madritensis*
bromo estamíneo – *Bromus stamineus*
bromo estéril – *Bromus sterilis*
bromo falso alopecuro – *Bromus alopecurus*
bromo falso centeno – *Bromus secalinus*
bromo frondoso – *Bromus anomalus*
bromo grande – *Bromus diandrus*
bromo inerme – *Bromus inermis*
bromo liso – *Bromus inermis*
bromo mediterráneo – *Bromus hordeaceus*
bromo mocho – *Bromus inermis*
bromo pátulo – *Bromus japonicus*
bromo peloso – *Bromus hordeaceus*
bromo perenne – *Bromus inermis*
bromo pratense – *Bromus japonicus*
bromo rojezo – *Bromus madritensis* ssp *rubens*
bromo rojo – *Bromus madritensis* ssp *rubens*
bromo suave – *Bromus inermis*
bromo suelto – *Bromus hordeaceus*
broqueleto – *Alyssum montanum*
broza fina – *Ruppia cirrhosa*
bruco – *Erica vagans*
bruja – *Kalanchoe pinnata, Turnera pumila*
brujo – *Juniperus communis, Xylosma flexuosum*
brújula – *Silphium perfoliatum*
brunela – *Prunella vulgaris*
brunela común – *Prunella vulgaris*
brunela de hojas hastadas – *Prunella grandiflora*
brunela grande – *Prunella grandiflora*
bruno – *Prunus domestica*
bruño – *Prunus domestica*
bruñón – *Prunus persica* ssp *nucipersica*
brusco – *Ruscus aculeatus*
brusco esmuláceo – *Raphanus sativus*
brusco espinescente – *Ruscus aculeatus*
brusela – *Vinca minor*
búcaro – *Lycaste virginalis*
buchón de agua – *Eichhornia crassipes*
buchú – *Barosma betulina*
budleya – *Buddleja davidii*
buena moza – *Nicotiana glauca*
buena mujer – *Chloracantha spinosa, Mentzelia multiflora, Solanum eleagnifolium*
buenas tardes – *Mirabilis jalapa*
buen Enrique – *Chenopodium bonus-henricus*
buen pan – *Artocarpus integrifolia*
buen varón silvestre – *Potentilla anserina*
buen varón silvestre de Jarava – *Potentilla anserina*
buey nini – *Solanum erianthum*
buffel – *Pennisetum ciliare*
buffel africano – *Cenchrus ciliaris*
buffel azul – *Cenchrus ciliaris*
buffel pretoria – *Cenchrus ciliaris*
buganvilla – *Bougainvillea spectabilis*
buganvilla blanca – *Pereskia aculeata*
buglosa – *Anchusa azurea*
buglosa común – *Anchusa officinales*
buglosa del Cabo – *Anchusa capensis*
buglosa de los campos – *Anchusa arvensis*
búgula menor – *Ajuga reptans*
búgula piramidal – *Ajuga pyramidalis*
búgula reptante – *Ajuga reptans*
buli – *Lagenaria siceraria*
bulines – *Guazuma ulmifolia*
bullonera – *Amelanchier ovalis, A. vulgaris*
bungya – *Araucaria bidwillii*
buniato – *Ipomoea batatas*
bupléurum – *Bupleurum chinense*
burilla – *Tulipa australis, T. sylvestris*
burladora – *Asclepias curassavica, Datura metel*
burrobiote – *Rhamnus purshiana*
bush – *Lagenaria siceraria*
butia lanosa – *Butia eriospatha*
butirospermo – *Butyrospermum parkii*
butua – *Cissampelos pareira*
buvardia – *Bouvardia longiflora*

caapi – *Banisteriopsis caapi*
cabalahauche – *Dorstenia contrayerba*
cabalonga – *Solanum marginatum, Strychnos ignatii, S. panamensis, Thevetia peruviana*
cabalonga de la Huasteca – *Thevetia peruviana*
cabalonga de Tabasco – *Strychnos tabascana*
cabalpixoy – *Guazuma ulmifolia*

cabalxaan – *Elytraria squamosa*
caballito – *Basistelma angustifolium, Tecoma incisa, T. stans*
caballo – *Equisetum*
cabellera – *Epidendrum vitellinum, Phoradendron commutatum, Psittacanthus calyculatus, Rhipsalis baccifera*
cabellera de la reina – *Carpobrotus chilensis*
cabellera de palo – *Oreopanax capitatus*
cabellito – *Basistelma angustifolium, Zapoteca anomala*
cabellitos de ángel – *Bombax ellipticum, Calliandra eriophylla, Ceiba pentandra, Zapoteca anomala*
cabellitos de una vara – *Zapoteca anomala*
cabellos de ángel – *Albizzia lebeck, Bombax ellipticum, Calliandra, Ceiba pentandra, Clematis, Ipomoea quamoclit, Pachira aquatica, Zapoteca anomala*
cabellos de Venus – *Adiantum capillus-veneris, Cuscuta, Nigella damascena*
cabelluda – *Eugenia tomentosa*
cabeza de ángel – *Calliandra eriophylla, Zapoteca anomala*
cabeza de arriero – *Tribulus cistoides*
cabeza de aura – *Senecio cardiophyllus*
cabeza de culebra – *Stanhopea devoniensis, S. tigrina, Tigridia pavonia*
cabeza de chamal – *Dioon edule*
cabeza de chivo – *Tribulus cistoides*
cabeza de dragón – *Antirrhinum majus*
cabeza de gorrión – *Castilleja arvensis*
cabeza de hormiga – *Teucrium cubense*
cabeza de ilama – *Annona purpurea*
cabeza de mico – *Hirtella americana, Licania platypus*
cabeza de negrito – *Eleocharis dulcis*
cabeza de negro – *Annona muricata, A. purpurea, Dioscorea alata, D. composita, D. mexicana, Echinacea angustifolia, Nymphaea, Pachira macrocarpa*
cabeza de perro – *Ranunculus ficaria*
cabeza de tigre – *Stanhopea devoniensis*
cabeza de Turco – *Cucurbita pepo*
cabeza de víbora – *Stanhopea devoniensis, S. tigrina*
cabeza de viejita – *Mammillaria sheldonii*
cabeza de viejo – *Cephalocereus sensilis, Clematis dioica, Echinocereus rigidissimus, Mammillaria, Neobuxbaumia tetetzo, Pachycereus schottii, Pilosocereus leucocephalus*
cabezona – *Gomphrena decumbens, Helenium autumnale, H. mexicanum*
cabezuela – *Centaurea cyanus, C. montana*
cabezuela purpúrea – *Echinacea purpurea*
cabijósocura – *Randia obcordata*
cabo de hacha – *Trichilia hirta*
cabo de luc – *Nyssa sylvatica*
cabonero – *Pithecellobium brevifolium*
cabra de hacha – *Trichilia hirta*
cabradora – *Lippia ligustrina*
cabra-higo – *Ficus padifolia*
cabrastillo – *Smilax bona-nox*
cabrito – *Thevetia thevetioides*
cabuya – *Agave americana*
cacachila – *Brongniartia alamosana, Citharexylum berlandieri, Karwinskia humboldtiana, Rhamnus californica*
cacachila blanca – *Chiococca petrina*
cacachila china – *Karwinskia humboldtiana*
cacachila silvestre – *Karwinskia humboldtiana*
caca de caballo – *Andira inermis*
caca de mono – *Lantana involucrata*
caca de niño – *Licania platypus, Pouteria campechiana*
cacahuacuahuitl – *Theobroma cacao*
cacahuananche – *Gliricidia sepium, Licania arborea, L. platypus*
cacahuanantzin – *Licania arborea*
cacahuano – *Gliricidia sepium*
cacahuapastle – *Hamelia erecta*
cacahuate – *Arachis hypogaea*
cacahuatl – *Theobroma cacao*
cacahuaxóchitl – *Hamelia erecta, Quararibea*
cacahuazintle – *Zea mays*
cacaiste – *Erythea edulis*
cacaíto – *Curatella Americana, Datura wrightii*
cacala – *Ehretia elliptica*
cacalacaxtli – *Xanthosoma robustum*
cacalaco – *Caesalpinia cacalaco, C. pulcherrima*
cacalaxtli – *Caesalpinia cacalaco*
cacalosúchil – *Plumeria acutifolia, P. mollis, P. rubra*
cacalosúchil blanco – *Plumeria acutifolia*
cacalosúchil rojo – *Plumeria rubra*
cacaloxóchitl – *Caesalpinia pulcherrima,*

Macrosiphonia brachysiphon, Plumeria acutifolia, P. rubra
cacanaguaste – *Apoplanesia paniculata*
cacanapazue – *Hamelia erecta*
cacanicua – *Muntingia calabura*
cacao – *Theobroma*
cacao blanco – *Theobroma bicolor*
cacaocuáhuitl – *Theobroma cacao*
cacao de mono – *Theobroma angustifolium*
cacao malacayo – *Theobroma bicolor*
cacaotero – *Theobroma cacao*
cacaotl – *Senna laevigata*
cacaotlquahuitl – *Theobroma cacao*
cacao volador – *Licania arborea*
cacaoxóchitl – *Quararibea funebris*
cacaporo – *Parkinsonia aculeata*
cacapuate – *Hamelia erecta*
cacaragua – *Vallesia glabra*
cacastlapa – *Ipomoea stans*
cacaya – *Agave kerchovei*
caceteiro – *Theobroma cacao*
caco – *Chrysobalanus icaco*
cacomite – *Dichelostemma capitatum, Tigridia dugesii, T. pavonia*
cacómitl – *Tigridia pavonia*
cacto – *Cactaceae*
cacto-agave – *Leuchtenbergia principis*
cacto arcoiris – *Echinocereus rigidissimus*
cacto aterciopelado – *Bergerocactus emoryi*
cacto cola de rata – *Disocactus flagelliformis*
cacto de Navidad – *Schlumbergera buckleyi*
cacto eslumbergera – *Schlumbergera truncata*
cacto fresa – *Echinocereus engelmannii*
cachaco – *Musa acuminata*
cachahua – *Tabebuia pentaphylla*
cachana – *Helianthella madrensis, Iostephane madrensis, Liatris punctata, L. spicata, Senecio cardiophyllus, Trixis californica*
cachanilla – *Pluchea sericea*
cachayumbo – *Gonolobus erianthus*
cachaz – *Okenia hypogaea*
cachete – *Heliconia latispatha*
cachete colorado – *Heliconia bihai*
cachimán – *Rollinia deliciosa, R. mucosa*
cachimbo – *Saracha procumbens, Senna bicapsularis*
cachisda – *Bocconia frutescens*
cachito – *Cajanus cajan*
cacho de toro – *Laelia tibicinis, Terminalia amazonia*
cacho de venado – *Eugenia capuli*
cachurra – *Xanthium strumarium*
cachurrera menor – *Xanthium spinosum*
cadena – *Tillandsia prodigiosa*
cadena del amor – *Antigonon leptopus*
cadenilla – *Drymaria gracilis, Salvia galinsogifolia*
cadenilla bronca – *Asclepias subulata*
cadillo – *Acaena elongata, Arctia lappa, Bouvardia ternifolia, Cenchrus, Dalea diffusa, Triumfetta, Urena lobata, Xanthium canadensis, X. strumarium*
cadillo aglomerado – *Cenchrus brownii*
cadillo austral – *Cenchrus echinatus, C. incertus*
cadillo cimarrón – *Triumfetta semitriloba*
cadillo común – *Xanthium strumarium*
cadillo de playa – *Cenchrus incertus*
cadillo malva – *Triumfetta semitriloba*
cadillo negro – *Bidens pilosa*
café – *Coffea arabica, C. liberica*
café arábica – *Coffea arabica*
cafecillo – *Rinorea, Senna laevigata*
café cimarrón – *Caesaria nitida*
café criollo – *Coffea arabica*
café de bugre – *Cordia salicifolia*
café de la China – *Astragalus boeticus*
café del duque – *Hibiscus esculentus*
café de Liberia – *Coffea liberica*
café de los bosques – *Cordia salicifolia*
café de los pobres – *Hibiscus esculentus*
café del país – *Hibiscus esculentus, Senna laevigata*
café de mato – *Cordia salicifolia*
café extranjero – *Hibiscus esculentus*
café liberia – *Coffea liberica*
café mareño – *Hibiscus esculentus*
café mexicano – *Astragalus boeticus*
café robusto – *Coffea canephora*
cafetillo – *Casearia nitida, C. sylvestris*
cafeto – *Coffea arabica, C. liberica*
cafeto arábica – *Coffea arabica*
cafeto de Kentucky – *Gymnocladus dioica*
cafeto de Liberia – *Coffea liberica*
cafeto del Perú – *Coffea racemosa*
cafeto robusto – *Coffea canephora*
cagalera – *Celtis iguanaea*

cagarria – *Morchella*
caguazo – *Paspalum plicatulum*
caguo rara – *Diospyros sonorae*
cahinca – *Chiococca alba*
cahualagua – *Heliocarpus tomentosus*
cahuapate – *Cephalanthus occidentalis*
cahuesquite – *Ziziphus amole*
cahuillo – *Verbena pumila*
cahuinga – *Parkinsonia aculeata*
cahuite – *Abies religiosa, Pseudotsuga menziesii*
cahum – *Agave fourcroydes*
caigua – *Cyclanthera pedata*
caimitillo – *Chrysophyllum bicolor, Strychnos panamensis*
caimitillo de perro – *Chrysophyllum oliviforme*
caimito – *Averrhoa carambola, Chrysophyllum cainito, C. mexicanum, Pouteria caimito*
caimito silvestre – *Chrysophyllum mexicanum*
caimito verde – *Chrysophyllum cainito*
cainia – *Chiococca alba*
cainito – *Chrysophyllum cainito*
cajecua – *Theobroma cacao*
cajeta – *Gilibertia arborea, Heliocarpus donnell-smithii*
cajete – *Triumfetta*
cajetillo – *Hirtella americana, H. racemosa*
cajiga – *Quercus petraea*
cajú – *Anacardium occidentale*
cajueiro – *Anacardium occidentale*
cajúguari – *Vitex mollis*
cajui – *Acacia pennatula*
cajuil – *Anacardium occidentale*
cajun – *Cajanus Cajun*
cala – *Zantedeschia aethiopica, Z. palustris*
calaba – *Calophyllum calaba*
calabacero – *Crescentia cujete, Lagenaria siceraria*
calabacilla – *Cucurbita, Ecballium elaterium, Sicyos laciniatus, S. parviflorus*
calabacilla amarga – *Cucurbita foetidissima, Ecballium elaterium*
calabacilla loca – *Cucurbita foetidissima*
calabacín – *Cucurbita pepo*
calabacita – *Cucurbita pepo*
calabacita italiana – *Cucurbita pepo*
calabacita verde – *Cucurbita pepo*
calabar – *Physostigma venenosum*
calabaza – *Cucurbita, Legenaria siceraria, L. vulgaris, Luffa acutangula*
calabaza africana – *Momordica charantia*
calabaza almisclada – *Cucurbita moschata*
calabaza amarga – *Cucurbita digitata, C. foetidissima, C. palmata, Ibervillea sonorae*
calabaza amarilla – *Cucurbita maxima, C. moschata*
calabaza anguina – *Trichosanthes cucumerina* var *anguina*
calabaza biche – *Cucurbita pepo*
calabaza blanca – *Benincasa hispida*
calabaza búfalo – *Cucurbita foetidissima*
calabaza común – *Cucurbita pepo*
calabaza confitera – *Cucurbita maxima*
calabaza chompa – *Cucurbis pepo*
calabaza de aristas – *Luffa acutangula, L. aegyptiaca*
calabaza de cuello torcido – *Cucurbita moschata*
calabaza de culebra – *Trichosanthes cucumerina* var *anguina*
calabaza de invierno – *Cucurbita maxima*
calabaza de peregrino – *Lagenaria siceraria*
calabaza de San Roque – *Lagenaria siceraria*
calabaza de verano – *Cucurbita pepo*
calabaza gigante – *Cucurbita maxima*
calabaza grande – *Cucurbita maxima*
calabaza india – *Cucurbita pepo*
calabaza melón – *Cucumis melo, Cucurbita moschata*
calabaza melona – *Sicana odorifera*
calabaza moscada – *Cucurbita moschata*
calabaza pamúe – *Cucumeropsis edulis*
calabaza silvestre – *Citrullus colocynthis*
calabaza tamala – *Cucurbita maxima*
calabaza tamalayota – *Cucurbita maxima*
calabaza turbante – *Cucurbita pepo*
calabaza vinatera – *Lagenaria siceraria*
calabazón – *Prunus avium*
cala blanca – *Spathiphyllum wallisii, Zantedeschia aethiopica*
calabón – *Cytisus purgans*
calabura – *Muntingia calabura*
caladio – *Caladium bicolor*
calafate – *Berberis buxifolia*
calaguala – *Asplenium monanthes, Notholaena aurea, N. sinuata, Pellaea truncata,*

Phlebodium aureum, Polypodium, Pteridium aquilinum
calaguala del indio – *Notholaena sinuata*
calalagua – *Sida acuta*
calalú – *Hibiscus esculentus*
calamagrostis rígida – *Calamagrostis stricta*
calamápatl – *Senecio vulneraria*
calamento – *Calamintha nepeta, Nepeta cataria*
calamento de flores grandes – *Calamintha grandiflora*
calamento de flores pequeñas – *Calamintha nepeta*
cálamo aromático – *Acorus calamus*
calamondín – *Citrus madurensis*
calanca – *Chrysactinia mexicana*
calancapatle – *Baileya multiradiata, Senecio vulneraria, Solidago velutina*
calancoe – *Kalanchoe blossfeldiana*
calandria – *Tricholoma equestre*
calandrita – *Tricholoma equestre*
cálanis – *Acorus calamus*
calarni – *Eugenia capuli*
calatea – *Calathea makoyana*
calatola – *Calatola mollis*
calatolazno – *Calatola mollis*
calavera – *Cuphea aequipetala*
calaverita – *Thevetia thevetioides*
calcanpatli – *Solidago mexicana*
calce de oro – *Eschscholzia californica*
calceolaria – *Calceolaria herbeohybrida*
calcitrapa – *Centaurea calcitrapa*
calcomeca – *Phaseolus metcalfei*
calderilla – *Ribes alpinum*
calderón – *Bocconia frutescens, Trollius europaeus*
calderona – *Krameria*
calderona amarilla – *Galphimia glauca*
caléndula – *Calendula officinalis, Tagetes lunulata, Turnera ulmifolia*
caléndula africana – *Tagetes erecta*
caléndula azteca – *Tagetes erecta*
caléndula campestre – *Calendula arvensis*
caléndula de jardín – *Calendula officinalis*
caléndula del Cabo – *Dimorphotheca sinuata*
caléndula francesa – *Tagetes patula*
caléndula oficinal – *Calendula officinalis*
caléndula silvestre – *Calendula arvensis*
calicanto – *Calycanthus occidentalis*
calicanto de Florida – *Calycanthus floridus*
calicanto del Japón – *Chimonanthus praecox*
calicanto de occidente – *Calycanthus occidentalis*
calicedra – *Cedrela mexicana, C. odorata*
calicó – *Kalmia latifolia*
caliche – *Bahia absinthifolia*
califa – *Acalypha hispida*
calipso bullosa – *Calypso bulbosa*
calisaya – *Cinchona calisaya*
calistemon – *Callistemon*
calmia – *Kalmia latifolia*
caloca – *Fucus vesiculosus, Pisonia capitata*
calomopotl – *Senecio vulneraria*
cálomo rota – *Calamus rotang*
cálomo roten – *Calamus rotang*
calotropis – *Calotropis procera*
cal-paich – *Manfreda brachystachya*
calta – *Caltha palustris*
calta palustre – *Caltha palustris*
calzadilla chica – *Podachaenium eminens*
calzoncillo – *Bauhinia divaricata, B. variegata*
calzon de indio – *Castilleja*
calliandra – *Calliandra haematocephala*
callistemo – *Callistemon viminalis*
camaba – *Lophophora williamsii*
camac – *Dalbergia glabra*
camaca – *Ardisia revoluta*
camajtli – *Ipomoea batatas*
camalote – *Brachiaria mutica, Digitaria insularis, Eichhornia crassipes, Fimbristylis thermalis, Panicum maximum, Paspalum, Polygonum hydropiperoides, Pontederia cordata*
camalote aglomerado – *Paspalum urvillei*
camalote blanco – *Paspalum virgatum*
camalote de antenna – *Paspalum conjugatum*
camalote lampiño – *Paspalum hartwegianum*
camalote moreno – *Paspalum langei*
camalote pálido – *Paspalum lividum*
camalote saladillo – *Paspalum distichum*
camalote velludo – *Paspalum pubiflorum*
camalotillo – *Paspalum plicatulum*
camándula – *Coix lacryma-jobi*
camará – *Lantana camara*
camarino negro – *Empetrum nigrum*
camarón – *Alvaradoa amorphoides, Caesalpinia pulcherrima, Calycophyllum candidissimum*

camaruca – *Sterculia carthaginensis*
camarroya – *Cichorium intybus*
camasa – *Lagenaria siceraria*
camasia – *Camassia quamash*
camayia – *Serjania palmeri*
cambanda – *Brugmansia arborea*
cambeba – *Ambrosia artemisiifolia*
cambo – *Guassia maya*
cambray – *Adenophyllum porophylloides*
cambrón – *Rhamnus*
cambronera – *Lycium chinense, L. pallidum* var *oligospermum*
cambronera de hojas de jarilla – *Lycium barbarum*
cambronera de la China – *Lycium chinense*
cambroño – *Spartium junceum*
cambuc – *Geonoma interrupta*
cambule – *Ipomoea pes-caprae*
camburi – *Musa sapientum*
cambustera – *Ipomoea quamoclit, Quamoclit pinnata*
cambutera – *Paspalum conjugatum*
camecíparis blanco americano – *Chamaecyparis thyoides*
camecíparis de Lawson – *Cupressus lawsoniana*
camecíparis de Nootka – *Cupressus nootkatensis*
camecíparis japonés hinoki – *Chamaecyparis obtusa*
camecíparis sawara – *Chamaecyparis psisfera*
camedor – *Chamaedora ernesti-augustii*
camedor de Cambray – *Chamaedorea elegans*
camedrio – *Teucrium chamaedrys*
camedrio acuático – *Teucrium scordium*
camedrio de los bosques – *Teucrium scorodonia*
cameleón blanco – *Atractylis gummifera*
camelia – *Camellia japonica*
camelia blanca – *Pereskia aculeata*
camelina – *Bougainvillea spectabilis, Camelina sativa*
camelina de fruto pequeño – *Camelina microcarpa*
camelina pilosa – *Camelina sativa*
camelote – *Cyperus articulatus*
camichín – *Ficus*
camín – *Thevetia peruviana*
camochayote – *Sechium edule*
camomila – *Chamaemelum nobile, Chamomilla suaveolens, Matricaria recutita*
camomila alemana – *Chamomilla suaveolens*
camomila de jardín – *Chamaemelum nobile*
camomila inglesa – *Matricaria recutita*
camomila romana – *Chamaemelum nobile*
camote – *Ipomoea batatas*
camote amarillo – *Ipomoea batatas*
camote blanco – *Dioscorea convolvulacea, Ipomoea batatas*
camote de cerro – *Dioscorea composita, Microsechium helleri, Peteria scoparia*
camote de los médanos – *Pholisma sonorae*
camote de monte – *Peteria scoparia*
camote de ratón – *Hoffmannseggia glauca*
camote de Tehuántepec – *Exogonium bracteatum*
camote gigante – *Dioscorea composita*
camote matapecado – *Dioscorea composita*
camote morado – *Ipomoea batatas*
camote real – *Dioscorea alata*
camotic – *Ipomoea batatas, I. purga*
camotillo – *Curcuma domestica, Maranta arundinacea, Solanum tuberosum, Zamia furfuracea*
camotli – *Ipomoea batatas*
campamocha – *Epilobium brachycarpum*
campana – *Allamanda cathartica, Brugmansia suaveolens*
campana amarilla – *Tecoma stans*
campanario – *Fritillaria biflora*
campanilla – *Brugmansia arborea, B. candida, B. suaveolens, Campanula isophylla, C. medium, Cobaea scandens, Convolvulus arvensis, C. tricolor, Coutarea latiflora, Dracaena Americana, Hintonia latiflora, Ipomoea tricolor, I. violacea, Jacquemontia abutiloides, Leucojum vernum, Thevetia peruviana, Trophis racemosa, Turbina corymbosa*
campanilla blanca – *Brugmansia* x *candida, B. suaveolens, Galanthus nivalis*
campanilla china – *Platycodon grandiflorus*
campanilla de coral – *Heuchera sanguinea*
campanilla de febrero – *Galanthus nivalis*
campanilla de hojas de melocotonero – *Campanula persicifolia*
campanilla de hojas gramíneas – *Leucojum trichophyllum*

campanilla de invierno – *Galanthus nivalis*
campanilla de invierno común – *Galanthus nivalis*
campanilla de invierno gigante – *Galanthus elwesii*
campanilla de oro – *Thevetia peruviana*
campanilla de otoño – *Leucojum autumnale*
campanilla de primavera – *Leucojum aestivum, L. verum*
campanilla de verano – *Leucojum aestivum, L. trichophyllum*
campanilla encarnada – *Brugmansia sanguinea*
campanilla menor – *Convolvulus arvensis*
campanilla primaveral – *Leucojum vernum*
campanilla rosa – *Convolvulus althaeoides*
campanilla silvestre – *Datura stramonium*
campanillas de Irlanda – *Moluccella laevis*
campanita – *Convolvulus, Ipomoea purpurea, Proboscidea althaefolia*
campanita amarilla – *Allamanda cathartica*
campanita china – *Forsythia suspensa*
campanita de suspiro – *Emmenanthe penduliflora*
campánula – *Campanula, Hyacinthoides non-scripta*
campánula azul – *Campanula rotundifolia*
campánula china – *Platycodon grandiflorus*
campánula de California – *Nemophila menziesii*
campánula de hojas circulares – *Campanula rotundifolia*
campánula de hojas de ortiga – *Campanula trachelium*
campánula de pared – *Campanula portenschlagiana*
campánula en ramilletes – *Campanula cochlearifolia*
campánula gigante – *Hyacinthoides hispanicus*
campánula media – *Campanula medium*
campechana – *Plumeria acutifolia, P. rubra*
campeche – *Haematoxylon brasiletto, H. campechianum, Leucaena glauca*
campotonera – *Plumeria acutifolia, P. rubra*
camuesa – *Opuntia robusta*
camxóchitl – *Chiranthodendron pentadactylon*
cana – *Sabal yapa*
canabal – *Spondias purpurea*
canaco – *Chiranthodendron pentadactylon*
canacua – *Melilotus officinalis*
canague – *Chiranthodendron pentadactylon*
canalete – *Cryptomeria japonica*
canan – *Hamelia erecta*
cananga – *Cananga odorata*
canara – *Paspalum dilatatum*
canari – *Canarium commune*
canaria – *Allamanda cathartica, Elytrigia repens, Rosa sulphurea*
canario – *Capsicum pubescens, Phaseolus vulgaris, Tricholoma equestre, T. flavovirens*
canaro – *Rhizophora mangle*
canastilla – *Abutilon notolophium, Aristolochia taliscana, Aurinia saxatilis, Randia*
canastilla de hojas anchas – *Uniola latifolia*
canavalia – *Canavalia ensiformis*
canavalia gigante – *Canavalia ensiformis*
cancachu – *Gossypium*
cáncer – *Cuphea aequipetala*
cancerillo – *Asclepias curassavica*
cancerina – *Asclepias curassavica*
canchacche – *Chiococca alba*
canchalagua – *Centaurium muehlenbergii, C. nudicaule, C. venustum*
canche – *Conocarpus erecta*
canchín – *Senna spectabilis*
canchoh – *Disocactus flagelliformis*
candebobe – *Pachycereus weberi*
candelabro – *Euphorbia lactea, Isolatocereus dumortieri, Pachycereus pecten-aboriginum, P. weberi*
candela regia – *Verbascum thapsus*
candelaria – *Agrostemma githago, Cattleya skinneri, Verbascum thapsis*
candelaria blanca – *Laelia glauca*
candelaria china – *Euphorbia plicata*
candelero – *Senna alata, Senecio praecox*
candelero de Salomón – *Lepidium campestre*
candelilla – *Asclepias subulata, Euphorbia antisiphilitica, E. ceroderma, E. colletioides, Notholaena sinuata, Pedilanthus macrocarpus, P. tithymaloides, Phytophthora infestans, Rhizophora mangle, Senna oxyphylla, S. spectabilis, Tecoma stans*
candelilla bronca – *Asclepias subulata, Euphorbia colletioides*
candelilla china – *Euphorbia plicata*
candelilla de palo – *Euphorbia plicata*
candelilla prieta – *Nissolia fruticosa*
candelón – *Rhizophora mangle*

cándida – *Candida albicans*
candil – *Amyris balsamifera*
candilejo – *Agrostemma githago*
candilera de Samos – *Phlomis samia*
candilera fruticosa – *Phlomis fruticosa*
candilera itálica – *Phlomis italica*
candilera tuberosa – *Phlomis tuberosa*
candox – *Tecoma stans*
canduerca – *Tamus communis*
canela – *Calycophyllum candidissimum, Chrysophyllum cainito, C. mexicanum, Cinnamomum zeylanicum, Cornus disciflora, C. excelsa, Drimys winteri, Lycaste aromatica, Melia azedarach, Mormodes pardinum* ssp *unicolor, Platanus lindeniana, Pluchea odorata, P. purpurascens, P. symphytifolia, Quararibea funebris, Zapoteca anomala*
canela de Ceilán – *Cinnamomum zeylanicum*
canela de cuya – *Vanilla planifolia*
canela de la China – *Cinnamomum cassia*
canela de paramo – *Drimys winteri*
canela montés – *Hamelia erecta*
canelero – *Cinnamomum zeylanicum*
canelilla – *Chamaedorea pochutlensis, Croton, Lippia berlandieri*
canelita – *Lycaste aromatica*
canelo – *Cornus disciflora, C. excelsa, Drimys winteri, Platanus lindeniana, Pluchea odorata*
canelo de invierno – *Drimys winteri*
canelo de paramo – *Drimys winteri*
canelón – *Pluchea odorata, P. symphytifolia*
canfor – *Cinnamomum camphora*
cangica – *Rhamnus sphaerosperma*
canibinche – *Malpighia glabra*
canica – *Chiococca alba*
canilla de venado – *Alibertia edulis*
canillero – *Sambucus nigra*
caninara – *Chiococca alba*
caniran – *Strychnos ignatii*
canisté – *Pouteria campechiana*
canistel – *Lucuma rivicoa, Pouteria campechiana*
canita agria – *Costus spicatus*
canite – *Pinus ayacahuite*
canlecay – *Cuscuta*
canlol – *Tecoma stans*
canlolxiu – *Flaveria*
canmul – *Bidens pilosa*
canolté – *Terminalia amazonia*
canónigo – *Valeriana sorbifolia, Valerianella locusta*
canónigo de Italia – *Valerianella eriocarpa*
canota – *Panicum repens*
cantaj – *Pinus tenuifolia*
cantala – *Agave cantala*
cantalupo – *Cucumis melo* var *reticulatus*
cantang – *Coleus rotundifolius*
cantarillo – *Androsace maxima*
cantemó – *Acacia angustifolia, A. angustissima*
cantiriz – *Acacia farnesiana*
cantua – *Cantua buxifolia*
cantueso – *Helichrysum orientale, Lavandula stoechas, Pachycereus marginatus*
cantueso dentado – *Lavandula dentata*
cantueso rizado – *Lavandula dentata*
cántula – *Agave angustifolia*
cantulán colorado – *Quercus oocarpa*
canxun – *Terminalia amazonia*
canzuuc – *Cynodon dactylon, Elytrigia repens*
caña – *Arundo donax, Canna indica, Saccharum officinarum, Sorghum saccharatum, S. vulgare*
caña acintada – *Phalaris arundinacea*
caña agria – *Costus mexicanus, C. spicatus*
caña amarga – *Costus spicatus, Gynerium sagittatum*
caña blanca – *Gynerium sagittatum*
caña brava – *Arthrostylidium racemiflora, Arundinaria longifolia, Bambusa aculeata, Bambusa arundinacea, Costus mexicanus, Guadua aculeata, G. longifolia, Gynerium sagittatum, Tripsacum laxum*
caña castilla – *Arundo donax*
caña comestible – *Canna edulis, C. indica*
caña común – *Arundo donax, Saccharum officinarum*
caña coro – *Canna indica*
caña criolla – *Saccharum barberi, S. officinarum*
caña de azúcar – *Saccharum officinarum*
caña de azúcar forrajera – *Saccharum sinense*
caña de Bengala – *Calamus rotang*
caña de Castilla – *Arundo donax, Gynerium sagittatum, Saccharum officinarum*
caña de Cristo – *Costus mexicanus, C. pulverulentus*

caña de cuentos – *Canna indica*
caña de espiga – *Sorghum bicolor*
caña de indio – *Cordyline fruticosa, Phragmites australis*
caña de jabalí – *Costus mexicanus, C. spicatus*
caña de las arenas – *Ammophila arenaria*
caña de las Indias – *Calamus draco, Canna indica*
caña de puerco – *Costus spicatus*
caña de Ravenna – *Saccharum ravennae*
caña de tigre – *Costus mexicanus*
caña de uva – *Saccharum sinense*
caña de venado – *Costus mexicanus*
caña dulce – *Axonopus compressus, Licania arborea, Saccharum officinarum*
cañaduz – *Saccharum officinarum*
caña espina – *Bambusa spinosa*
cañafistula – *Cassia grandis, Senna*
cañafistula grande – *Cassia grandis*
caña forrajera – *Saccharum officinarum*
caña gigante – *Arundo donax*
cañaheja – *Conium maculatum, Ferula communis*
caña hueca – *Arundo donax*
cañaigre – *Begonia gracilis, B. martiana, Rumex*
caña india – *Canna indica*
caña japonesa – *Saccharum sinense*
caña mansa – *Guadua amplexifolia*
cañamazo – *Paspalum notatum*
caña mexicana – *Costus speciosus*
cañamiel – *Saccharum officinarum*
cáñamo – *Cannabis sativa*
cáñamo acuático – *Bidens tripartita*
cáñamo común – *Cannabis sativa*
cáñamo cubano – *Furcraea hexapetala*
cáñamo cultivado – *Cannabis sativa*
cáñamo de Bengala – *Crotalaria juncea*
cáñamo de gombó – *Hibiscus cannabinus*
cáñamo de Guinea – *Hibiscus sabdariffa*
cáñamo de las Indias – *Corchorus olitorius*
cáñamo del Canadá – *Apocynum cannabinum*
cáñamo de Manila – *Musa textilis*
cáñamo de Mauricio – *Furcraea gigantea*
cáñamo de Nueva Zelandia – *Phormium tenax*
cáñamo indígena – *Pilocarpus jaborandi*
cáñamo sansevieria de Ceilán – *Sansevieria trifasciata*
cáñamo silvestre – *Cannabis ruderalis*
cáñamo sisal – *Agave sisalana*
caña muda – *Dieffenbachia amoena*
caña otate – *Guadua aculeate, G. amplexifolia*
cañareja – *Ferula communis, Smyrnium olustratum*
cañarroya – *Parietaria judaica, P. officinalis*
cañatilla – *Ephedra*
cañaveral – *Arundo donax*
cañerla – *Conium maculatum*
cañita agria – *Begonia nelumbiifolia*
cañizo – *Guadua longifolia*
cañota – *Pastinaca sativa, Phragmites australis, Sorghum halepense*
cañuela – *Carex arenaria, Costus mexicanus, Equisetum, Festuca pratensis, Poa pratensis*
cañuela alta – *Festuca arundinacea*
cañuela arizónica – *Festuca arizonica*
cañuela borreguerra – *Festuca ovina* var *brachyphylla*
cañuela común – *Festuca pratensis*
cañuela de los prados – *Festuca arundinacea, F. pratensis*
cañuela del Rhin – *Festuca arundinacea*
cañuela de oveja – *Festuca ovina*
cañuela descollada – *Festuca arundinacea*
cañuela durilla – *Festuca longifolia*
cañuela gigante – *Festuca arundinacea*
cañuela noble – *Festuca arundinacea*
cañuela roja – *Festuca rubra*
cañutillo – *Arundo donax, Axonopus affinis, Ephedra, Equisetum arvense, Hamelia erecta, Plumbago auriculata, P. scandens, Rhipidocladum racemiflorum, Russelia retrorsa, Sorghum halepense*
cañutillo del campo – *Ephedra trifurca*
cañutillo del llano – *Equisetum*
caoba – *Cercocarpus ledifolius, Khaya senegalensis, Swietenia humilis, S. macrophylla, S. mahagoni*
caoba de montaña – *Cercocarpus betuloides*
caoba de Santo Domingo – *Spathodea campanulata, Swietenia mahagoni*
caobilla – *Croton, Swietenia humilis*
capá – *Cordia sebestena*
capa de pobre – *Anthurium xanthosomifolium, Gunnera insignis, G. mexicana*
capa de rey – *Caladium bicolor, Jatropha podagrica*

capa de San José – *Oenothera biennis, O. villosa* ssp *strigosa*
capa de Santiago – *Sprekelia formosissima*
capa roja – *Acalypha wilkesiana*
caparossa – *Hulsea vestita* ssp *callicarpha*
capi – *Zea mays*
capiaj – *Jarilla chocola*
capicuerno – *Ligustrum vulgare*
capilaria – *Adianthum capillus-veneris*
capiri – *Sideroxylon capiri*
capiro – *Eleocharis montevidensis, Sambucus mexicana*
capirote – *Pernettya ciliata*
capitán – *Petrea arborea*
capitaneja – *Bidens aurea, Verbesina*
capitaneja anaranjada – *Verbesina crocata*
capitaneja de Tonila – *Verbesina pedunculosa*
capitán lila – *Petrea arborea*
capoca – *Ceiba pentandra*
capolín – *Cerasus capollin*
capollaxipehualli – *Chrysobalanus icaco*
capomo – *Brosimum alicastrum*
capoquero – *Ceiba pentandra*
capoquero blanco – *Cochlospermum religiosum*
capote – *Capparis spinosa, Xanthosoma robustum*
capote blanco – *Xanthosoma roseum*
capote blanco grande – *Xanthosoma robustum*
capote de jardín – *Xanthosoma violaceum*
capote de monte – *Xanthosoma robustum*
capote de payaso – *Caladium bicolor*
capotillo – *Caladium bicolor*
caprago – *Galega officinalis*
capuatlcacao – *Spondias mombin, S. purpurea*
capucha de monje – *Aconitum*
capuche – *Calyptrogyne ghiesbreghtiana*
capuchina – *Maranta leuconeura, Tropeaolum majus*
capuchina mayor – *Tropaeolum majus*
capulco – *Senna alata*
capulín – *Ardisia revoluta, Belotia mexicana, Bocconia arborea, Celtis pallida, Condalia hookeri, C. obovata, Conostegia xalapensis, Erythea dulcis, Eugenia capuli, Ficus cotinifolia, F. padifolia, Karwinskia humboldtiana, Malpighia glabra, Muntingia calabura, Physalis peruviana, Prunus capuli, P. salicifolia, P. serotina, P. virginiana, Rhus virens, Sambucus mexicana, Schaefferia cuneifolia, Styrax argenteus, Trema micrantha*
capulina – *Ficus cotinifolia*
capulín blanco – *Belotia mexicana, Prunus capuli*
capulincillo – *Ardisia revoluta, Belotia mexicana, Conostegia xalapensis, Karwinskia humboldtiana, Pernettya ciliata, Prunus brachybotrya, Quercus castanea, Q. laurina, Rhamnus, Ribes*
capulincillo cimarrón – *Karwinskia humboldtiana*
capulín cimarrón – *Ehretia tinifolia, Karwinskia humboldtiana, Rhamnus serrata, Trema micrantha*
capulín corona – *Randia laetevirens*
capulín de cotorra – *Conostegia xalapensis*
capulín de hueso – *Eugenia fragrans*
capulín de mayo – *Muntingia calabura*
capulín de zorra – *Karwinskia humboldtiana*
capulín grande – *Prunus serotina*
capulín manso – *Ardisia revoluta*
capúlín negro – *Condalia hookeri, C. obovata*
capulín pequeño – *Prunus capuli*
capulín silvestre – *Sambucus mexicana*
capulixtl – *Agave peacockii*
capulxihuitl – *Prunus capuli*
capullo de rosa – *Prunus subhirtella*
capúsari – *Crateva palmeri*
caqui – *Diospyros kaki, D. virginiana*
caqui americano – *Diospyros virginiana*
caqui de China – *Diospyros kaki*
caqui de Virginia – *Diospyros virginiana*
caquilero – *Diospyros kaki*
carabanz – *Canavalia gladiata*
carabullo – *Karwinskia humboldtiana*
caracaline – *Vallesia glabra*
cara-caro – *Enterolobium cyclocarpum*
caracól – *Phaseolus caracalla*
caracolera – *Parietaria officinalis*
caracolí – *Anacardium rhinocarpus*
caracolillo – *Kohleria deppeana, Phaseolus caracalla, Vigna caracalla*
caracolillo de olor – *Lathyrus odoratus*
caracól real – *Vigna caracalla*
carachi amarillo – *Orestias luteus*
carachi negro – *Orestias agassiz*
caragilate – *Vigna unguiculata*

caraguatá – *Bromelia lasiantha*
carambola – *Averrhoa carambola, Chrysophyllum cainito*
caramicua – *Colocasia esculenta, Xanthosoma robustum, X. sagittifolium*
caramillo – *Salsola vermiculata*
carana hediondo – *Protium copal*
caranday – *Copernicia australis, Trithrinax campestris*
caraota – *Phaseolus vulgaris*
caraota blanca – *Phaseolus vulgaris*
caraota grande – *Canavalia gladiata*
caraota soya – *Glycine max*
carape – *Combretum farinosum, C. fruticosum*
carapulla – *Physalis alkekengi*
carasu – *Crataegus mexicana*
carátacua – *Baccharis pteronioides*
caratero – *Bursera gummifera, B. simaruba*
carballo – *Quercus robur*
carbón de la avena – *Ustilago carbo*
carbón de la cebada – *Ustilago hordei*
carbón del centeno – *Claviceps purpurea*
carbón del maíz – *Ustilago maydis*
carbón del pasto inglés – *Ustilago carbo*
carbón del trigo – *Ustilago tritici*
carbonero – *Capparis odoratissima, Havardia pallens, Pithecellobium brevifolium, Russula cyanoxantha*
carcoma – *Milla biflora*
cardamomo – *Elettaria cardamomum, Laurus camphora*
cardenal – *Lobelia cardinalis, Salvia coccinea, S. fulgens*
cardenal de la laguna – *Lobelia cardinalis* var *pseudosplendens*
cardenal de maceta – *Lobelia cardinalis* var *pseudosplendens*
cardencha – *Cylindropuntia*
cardencha común – *Dipsacus fullonum*
cardencha cultivada – *Dipsacus sativus*
cardíaca – *Leonurus cardiaca*
cardiduca – *Echinops sphaerocephalis*
cardillo – *Acaena elongata, Cynoglossum, Scolymus hispanicus, Triumfetta semitriloba, Xanthium strumarium*
cardillo bravillo – *Scolymus hispanicus*
cardillo hispánico – *Scolymus hispanicus*
cardimea – *Echinops sphaerocephalus*
cardimuelle – *Sonchus asper*
cardito – *Carduus acanthoides*
cardo – *Argemone, Bromelia karatas, B. pinguin, Centaurea, Cirsium, Cylindropuntia rosea, Cynara, Opuntia pumila, Romneya coulteri, Scolymus hispanicus*
cardo ajonjero – *Arctium lappa, Carlina vulgaris*
cardo alcachofero – *Cynara scolymus*
cardo almizclero – *Carduus nutans*
cardo amarillo – *Argemone mexicana*
cardo bendito – *Cnicus benedictus, Silybum marianum*
cardo blanco – *Argemone mexicana*
cardo borriquero – *Carduus crispus, Onopordum acanthium*
cardo bronco – *Cirsium mexicanum*
cardo cabezudo – *Carduus pycnocephalus*
cardo cabrero – *Carthamus lanatus*
cardo cancelado – *Atractylis cancellata*
cardo colorado – *Cirsium ehrenbergii*
cardo comestible – *Cynara cardunculus*
cardo corredor – *Eryngium*
cardo corredor marino – *Eryngium maritimum*
cardo corriente – *Carduus cripus*
cardo cundidor – *Cirsium arvense*
cardo de arrecife – *Cynara cardunculus*
cardo de batanero – *Dipsacus fullonum*
cardo de burro – *Carduus acanthoides, C. crispus*
cardo de calvero – *Carduus pycnocephalus*
cardo de cardador – *Dipsacus fullonum*
cardo de cardar – *Dipsacus fullonum, D. sativus*
cardo de comer – *Cynara cardunculus*
cardo de huerta – *Cynara cardunculus*
cardo del demonio – *Onopordum illyricum*
cardo de leche – *Silybum marianum*
cardo de liga – *Atractylis gummifera*
cardo del valle – *Centaurea americana*
cardo de mar – *Eryngium maritimum*
cardo de María – *Silybum marianum*
cardo de montaña – *Cirsium ehrenbergii*
cardo emplumado – *Cirsium mexicanum*
cardo erizo – *Echinops exaltatus, E. sphaerocephalus*
cardo estrellado – *Centaurea calcitrapa*
cardo globoso – *Echinops exaltatus*
cardo lanceolado – *Cirsium vulgare*
cardo lechar – *Silybum marianum*

cardo lechero – *Silybum marianum*
cardo mariano – *Silybum marianum*
cardo marítimo – *Eryngium maritimum*
cardo menor – *Carduus tenuiflorus*
cardo multicolor – *Echinops exaltatus*
cardón – *Bromelia pinguin, Cereus hexagonus, C. peruvianus, Cylindropuntia imbricata, Euphorbia lactea, Neobuxbaumia tetetzo, Opuntia streptacantha, Pachycereus, Stenocereus queretaroensis*
cardón-barbón – *Pachycereus pecten-aboriginum*
cardón blanco – *Cephalocereus columna-trajani*
cardoncillo – *Cylindropuntia kleiniae, Echinocereus pentalophus, Peniocereus striatus, P. viperinus, Silybum marianum*
cardón chico – *Pachycereus weberi*
cardón de la costa – *Pachycereus pecten-aboriginum*
cardón de tomellin – *Pachycereus weberi*
cardo negro – *Cirsium arvense*
cardón espino – *Pachycereus pecten-aboriginum*
cardón espinoso – *Pachycereus weberi*
cardón gigante – *Pachycereus pringlei*
cardón-hecho – *Pachycereus pecten-aboriginum, P. pringlei*
cardón lechar – *Scolymus hispanicus*
cardón pelón – *Pachycereus pecten-aboriginum, P. pringlei*
cardo perruno – *Picris echioides*
cardo rojo – *Carduus nutans*
cardo santo – *Argemone, Centaurea americana, Cirsium, Cnicus benedictus, Silybum marianum*
cardo santo campestre – *Argemone mexicana*
cardo santo cimarrón – *Cirsium mexicanum*
cardo yesquero – *Echinops exaltatus, E. strigosus, Onopordum acanthium*
cargamanta – *Phytolacca octandra*
cari – *Murraya koenigii*
caria amarga – *Carya amara*
caria blanca – *Carya ovata*
cariaquito colorado – *Lantana camara*
caribe – *Cnidoscolus angustidens, C. maculates, Jatropha angustidens*
caries – *Tilletia tritici*
carindapaz – *Viburnum glabratum*
cariofilada – *Geum chiloense*
cariofilada de hojas grandes – *Geum macrophyllum*
cariofilada de montaña – *Dryas octopetala, Geum montanum*
cariofilada oficinal – *Geum urbanum*
cariofilada rígida – *Geum aleppicum*
cariopteris – *Caryopteris incana*
cariota – *Aiphanes caryotifolia*
carlina – *Carlina vulgaris*
carlina de monte – *Carlina vulgaris*
carlina silvestre – *Carlina vulgaris*
carmel – *Plantago major*
carmelita – *Lablab purpureus*
carnauba – *Copernicia cerifera*
carnavalito – *Vitex trifolia*
carne de caballo – *Urera caracasana*
carne de doncella – *Begonia gracilis*
carne de perro – *Cochlospermum vitifolium, Syngonium podophyllum*
carnero – *Cocoloba barbadensis, C. uvifera, Couepia polyandra, Licania arborea, Poulsenia armata*
carnero blanco – *Licania arborea*
carnero de la costa – *Cocoloba barbadensis*
carnerona – *Acanthus mollis*
carnestolenda – *Cochlospermum vitifolium, Exogonium bracteatum*
caro – *Enterolobium cyclocarpum*
caroba – *Ceratonia siliqua*
carolina – *Bombax ellipticum, Eichhornia crassipes, Populus canadensis, Zinnia elegans, Z. peruviana, Z. violacea*
carpe – *Carpinus betulus*
carpe blanco – *Carpinus betulus*
carpe común – *Carpinus betulus*
carpe europeo – *Carpinus betulus*
carpe japonés – *Carpinus japonica*
carpe negro – *Ostrya carpinifolia*
carpenteria – *Carpenteria californica*
carpeta – *Axonopus affinis, A. scoparius*
carpeta de San Agustín – *Stenotaphrum secundatum*
carpintero – *Justicia brandegeana*
carquexia – *Sorghum bicolor*
carquexia fina – *Genista sagittalis*
cártamo – *Carthamus tinctoriis*
cartucho – *Zantedeschia aethiopica*
carvallo – *Quercus robur*

carvi – *Carum carvi*
carrasca – *Quercus coccifera, Q, ilex*
carrasco – *Pinus halepensis*
carraspique – *Iberis sempervirens, I. umbellate, Thlaspi arvense*
carraspique amara – *Iberis amara*
carraspique de Gibraltar – *Iberis gibraltarica*
carraspique intermedia – *Iberis intermedia*
carraspique siempreverde – *Iberis sempervirens*
carraspique umbelada – *Iberis umbellata*
carrasquilla – *Amelanchier ovalis, Rhamnus alaternus, R. alpinus, Thymus vulgaris*
carrechuela – *Aspicarpa hirtella*
carretadero – *Acacia hindsii*
carrete – *Vitex gaumeri*
carretilla – *Medicago lupulina, M. polymorpha*
carretón – *Medicago arabica, M. orbicularis*
carretón de amores – *Medicago polymorpha*
carricera – *Saccharum ravennae*
carricera africana – *Cenchrus ciliaris*
carricillo – *Arundo donax, Capsicum annuum, Equisetum, Gouinia virgata, Iresine calea, Lasiacis divaricata, Olyra latifolia, Panicum procerum, Phragmites australis, Phytolacca icosandra*
carricillo silvestre – *Petiveria alliacea*
carrizo – *Arundo donax, Bambusa vulgaris, Calamagrostis, Carex buchananii, Chusquea, Cortaderia selloana, Guadua aculeata, Gynerium sagittatum, Lasiacis divaricata, Muhlenbergia dumosa, Olyra latifolia, Otatea fimbriata, Phragmites australis*
carrizo acuático – *Carex aquatilis*
carrizo arenario – *Carex arenaria*
carrizo aristado – *Carex atherodes*
carrizo canescente – *Carex canescens*
carrizo capitado – *Carex capitata*
carrizo común – *Phragmites australis*
carrizo de Buxbaum – *Carex buxbaumii*
carrizo de la selva – *Arundo donax*
carrizo de la sierra – *Arundo donax*
carrizo de las pampas – *Cortaderia selloana*
carrizo de ratón – *Lasiacis divaricata*
carrizo diándrico – *Carex diandra*
carrizo dispermo – *Carex disperma*
carrizo falso de zorro – *Carex vulpinoidea*
carrizo glauco – *Carex flacca*
carrizo hirsuto – *Carex echinata*
carrizo lasiocarpo – *Carex lasiocarpa*
carrizo limoso – *Carex limosa*
carrizo negro – *Carex nigra*
carrizo noruego – *Carex norvegica*
carrizo pie de liebre – *Carex ovalis*
carrizo reál – *Gynerium sagittatum*
carrizo rígido – *Carex elata*
carrizo rostrado – *Carex utriculata*
carrizo serótino – *Carex viridula*
carrizo verde – *Olyra latifolia*
carrizo vesicular – *Carex vesicaria*
carroncha – *Erica ciliaris, E. tetralix*
casabanana – *Sicana odorifera*
casa de indio – *Lotus scoparius*
casa de rata – *Echinocereus brandegeei, Stenocereus eruca*
casal – *Selloa glutinosa*
casa pulga – *Proboscidea fragrans, Rhynchosia pyramidalis*
casa rata – *Grusonia invicta*
cascabel – *Capsicum annuum, Crotalaria longirostrata, C. pumila, Enterolobium cyclocarpum, Thevetia peruviana*
cascabel de lacandón – *Thevetia peruviana*
cascabel de víbora – *Crotalaria longirostrata*
cascabelillo – *Astragalus, Crotalaria*
cascabelito – *Astragalus, Crotalaria, Trifolium microcephalum, Zornia reticulata*
cascabel sonaja – *Enterolobium cyclocarpum*
casca de anta – *Drimys winteri*
cascalote – *Acacia farnesiana, Caesalpinia cacalaco, C. coriaria, Colubrina ferruginosa*
cascamote – *Colocasia esculenta*
cáscara – *Rhamnus*
cáscara amarga – *Picramnia, Sweetia panamensis*
cáscara sagrada – *Rhamnus californica, R. purshiana*
cascarillo – *Cinchona calisaya, Colubrina ferruginosa, Croton guatemalensis, C. niveus*
cascarillo verde – *Cinchona officinalis*
cascarita – *Alibertia edulis*
cascarón – *Crateva palmeri*
casco de buey – *Bauhinia monandra*
casco de Júpiter – *Aconitum napellus*
cas dulce – *Psidium cattleyanum*
casia – *Cassia fistula, Cinnamomium cassia, Senna angustifolia*
casia acanelada – *Cinnamomum cassia*
casia amarilla – *Albizzia lebeck*

casia de Siam – *Senna siamea*
casia purgante – *Cassia fistula*
casia purgativa – *Cassia fistula*
casida – *Scutellaria galericulata*
casílago – *Hyoscyamus niger*
casis – *Ribes nigrum*
casquete – *Scutellaria lateriflora*
castaña de agua – *Eleocharis dulcis, Pachira aquatica, Trapa natans*
castaña de tierra – *Bunium bulbocastanum*
castaño – *Artocarpus altilis, A. integrifolia, Bertholletia excelsa, Castanea sativa, Pachira insignis, Sterculia apetala, S. mexicana, Terminalia catappa*
castaño americano – *Castanea dentata*
castaño caballuno – *Aesculus hippocastanum*
castaño común – *Castanea sativa*
castaño de Australia – *Castanospermum australe*
castaño de Brasil – *Bertholletia excelsa*
castaño de cajú – *Anacardium occidentale*
castaño de frutos lisos – *Aesculus pavia*
castaño de Indias – *Aesculus hippocastanum*
castaño de Indias común – *Aesculus hippocastanum*
castaño de Indias rojo – *Aesculus carnea*
castaño de la Guayana – *Pachira aquatica*
castaño de la India – *Aesculus hippocastanum, Thevetia ovata*
castaño de maranhoa – *Bertholletia excelsa*
castaño de México – *Ungnadia speciosa*
castaño de pará – *Bertholletia excelsa*
castaño falso – *Aesculus hippocastanum*
castaño japonés – *Aesculus urbinata*
castaño regoldo – *Castanea sativa*
castaño rubio – *Aesculus* x *carnea*
castañuela – *Scirpus lacustris, S. tabernaemontani*
castilla – *Hordeum vulgare*
castilleja – *Castilleja*
castillito – *Agrostis gigantea, A. viridis*
castillo de agua – *Agrostis viridis*
castlapa – *Ipomoea stans*
castor – *Ricinus communis*
casuarina – *Casuarina*
catachio – *Proboscidea fragrans*
catajera – *Momordica charantia*
catalina – *Euphorbia heterophylla, E. pulcherrima*
catalinaria – *Cylindropuntia leptocaulis*
catalpa – *Catalpa bignonioides*
catalpa común americana – *Catalpa bignonioides*
catalpa especiosa – *Catalpa speciosa*
catalpa híbrida americana – *Catalpa* x *erubescens*
catapucia – *Ricinus communis*
catapucia menor – *Euphorbia lathyris*
catarina – *Euphorbia heterophylla, E. pulcherrima, Salpianthus macrodontos*
caticón – *Curatella americana*
catinguero – *Melinis minutiflora*
catoxitzi – *Castilleja arvensis*
catuche – *Annona muricata*
catzotl – *Pachyrhizus erosus*
cauache – *Trichilia hirta*
cauchera – *Hevea brasiliensis*
caucho – *Achras zapota, Castilla elastica, Ficus elastica, Hevea brasiliensis*
caucho ceará – *Manihot glaziovii*
caucho de Castilla – *Castilla elastica*
caucho de la India – *Ficus elastica*
caucho macho – *Castilla fallax*
caujaro – *Cordia alba*
caumóchil – *Pithecellobium dulce*
caupi – *Vigna unguiculata*
cauro – *Guadua amplexifolia*
causamo – *Coursetia glandulosa*
caussú – *Calathea lutea*
cavicuerna – *Ligustrum vulgare*
caxane – *Senecio cardiophyllus*
caxtidán – *Zaluzania augusta*
cayábajo – *Albizzia sinaloensis*
cayena – *Hibiscus rosa-sinensis*
cayeputi – *Melaleuca alternifolia, M. leucodendron*
cayolizán – *Buddleja americana*
cayumito – *Chrysophyllum cainito*
cazahuate – *Armillaria mellea, Ipomoea arborescens, I. mucuroides, Pleurotis ostreatus*
cazahuate blanco – *Ipomoea arborescens*
cazahuate prieto – *Ipomoea murucoides*
cazahuatl – *Ipomoea murucoides*
cazaniche – *Solanum umbellatum*
caztaz – *Triticum aestivum*
cebada – *Hordeum*
cebada cimarrona – *Hordeum jubatum*

cebada común – *Hordeum vulgare*
cebada crinita – *Hordeum jubatum*
cebada cultivada – *Hordeum vulgare*
cebada de los muros – *Hordeum murinum*
cebada marítima – *Hordeum marinum*
cebada perla – *Hordeum vulgare*
cebadilla – *Bromus catharticus, Chloris virgata, Lasiacis divaricata, Sabadilla officinalis, Schoenocaulon officinale, S. pringlei, Stenanthium, Swertia radiata, Veratrum album, V. californicum, V. frigidum, Zigadenus mexicanus*
cebadilla criolla – *Bromus catharticus*
cebadilla de Hungría – *Bromus inermis*
cebadilla de la sierra – *Swertia radiata, Veratrum sabadillo*
cebadilla falsa – *Penstemon barbatus, P. campanulatus, P. imberbis*
cebadilla perenne – *Bromus inermis*
cebadilla silvestre – *Hordeum murinum* ssp *glaucum*
cebolillo – *Lolium multiflorum*
cebolino – *Allium schoenoprasum*
cebolla – *Allium cepa, A. schoenoprasum*
cebolla albarrana – *Hymenocallis harrisiana, H. rotata, Urginea maritima*
cebolla bulbífera – *Allium cepa* var *viviparum*
cebollada – *Globularia vulgaris*
cebolla de chucho – *Limnocharis flava*
cebolla de Egipto – *Allium cepa* var *viviparum*
cebolla de grajo – *Urginea maritima*
cebolla de hoja – *Allium fistulosum*
cebolla de invierno – *Allium fistulosum*
cebolla de mayo – *Atamosco carinata, Zephyranthes carinata, Z. longifolia*
cebolla de rabo – *Allium cepa*
cebolla de rama – *Allium schoenoprasum*
cebolla escalonia – *Allium cepa* var *ascalonicum*
cebollaja – *Allium fistulosum*
cebolla junca – *Allium fistulosum*
cebolla larga – *Allium cepa*
cebolla marítima – *Urginea scilla*
cebollana – *Allium schoenoprasum*
cebolla nativa – *Allium haematochiton*
cebolla patata – *Allium cepa* var *aggregatum*
cebolla subterránea – *Allium cepa* var *aggregatum*
cebolla verde – *Allium fistulosum*
cebolleja – *Allium schoenoprasum, Sabadilla officinalis, Stenanthium frigidum,*
cebolleta – *Allium, Asphodelus fistulosus, Cooperia drummondii, Ipomoea longifolia, Nothoscordium striatum, Oncidium cebolleta, Stenanthium frigidum*
cebolleta del campo – *Allium cernuum*
cebollete – *Allium fistulosum*
cebollín – *Allium fistulosum, A. scaposum, A. schoenoprasum, Cyperus esculentus, C. rotundus*
cebollina – *Allium scaposum, A. schoenoprasum, Cooperia drummondii*
cebollino – *Allium schoenoprasum*
cebollino común – *Allium ampeloprasum, A. schoenoprasum*
cebollino chino – *Allium tuberosum*
cebollino de verdeo – *Allium fistulosum*
cebollino francés – *Allium schoenoprasum*
cebollino inglés – *Allium fistulosum*
cebollita – *Cooperia drummondii, Zephyranthes longifolia*
cebollita de Cambray – *Allium cepa*
ceborrincha – *Allium haematochiton*
cecepatli – *Iresine calea*
cecetzin – *Iresine calea*
cedacillo – *Briza maxima*
cedilla – *Wimmeria confusa*
cedoaria – *Curcuma zedoaria*
cedrela americana – *Cedrela odorata*
cedrela de China – *Cedrela sinensis*
cedrillo – *Juniperus flaccida, Trichilia hirta*
cedro – *Cedrela, Cedrus, Cupressus, Juniperus, Taxodium distichum* var *mexicanum*
cedro africano – *Spathodea campanulata*
cedro amargo – *Cedrella mexicana*
cedro amarillo – *Thuja occidentalis*
cedro asiático – *Cedrus deodara*
cedro atlántico – *Cedrus atlantica*
cedro australiano – *Cedrela toona*
cedro azul – *Cedrus atlantica*
cedro blanco – *Cedrus lusitanica, Cupressus, Juniperus*
cedro blanco americano – *Chamaecyparis thyoides*
cedro canario – *Juniperus cedrus*
cedro cimarrón – *Calophyllum braziliense*
cedro colorado – *Cedrela mexicana, C.*

odorata, Juniperus flaccida, J. mexicana, J. monticola
cedro chino – *Cedrela mexicana, Juniperus deppeana, Myroxylon balsamum* var *pereirae*
cedro de busaco – *Cupressus lusitanica*
cedro de Canarias – *Juniperus cedrus*
cedro de China – *Cedrela sinensis*
cedro de chipre – *Cedrus brevifolia*
cedro de España – *Cedrela odorata, Juniperus thurifera*
cedro de hojas cortas – *Cedrus brevifolia*
cedro de incienso – *Calocedrus decurrens, Juniperus deppeana*
cedro de incienso de California – *Calocedrus decurrens*
cedro de la Habana – *Cedrela mexicana, C. odorata*
cedro de la India – *Cedrus deodara*
cedro de la sierra – *Cupressus arizonica, C. montana*
cedro del atlante – *Cedrus atlantica*
cedro del Atlas – *Cedrus atlantica*
cedro del Himalaya – *Cedrus deodara*
cedro del Japón – *Cryptomeria japonica*
cedro del Líbano – *Cedrus libani*
cedro de México – *Cedrela odorata*
cedro deodar – *Cedrus deodara*
cedro de Oregón – *Cupressus lawsoniana*
cedro de Puerto Oxford – *Cupressus lawsoniana*
cedro de Salomón – *Cedrus libani*
cedro de Virginia – *Juniperus virginiana*
cedro español – *Cedrela odorata*
cedro fino – *Cedrela mexicana*
cedro hembro – *Cedrela mexicana*
cedro índico – *Cedrela toona*
cedro japonés – *Cryptomeria japonica*
cedro limón – *Citrus medica*
cedro liso – *Cedrela mexicana, Juniperus flaccida*
cedro macho – *Cedrela mexicana*
cedro mexicano – *Cedrella odorata*
cedro montés – *Swietenia macrophylla*
cedrón – *Aloysia gratissima, A. triphylla, Casearia aculeata, Juniperus osteosperma, Melissa officinalis, Phyla nodiflora, Verbena*
cedrón de Castilla – *Aloysia triphylla*
cedro oloroso – *Cedrela mexicana, C. odorata*
cedro real – *Cedrela mexicana*
cedro rojo – *Cedrella mexicana, C. odorata, Juniperus scopulorum, J. virginiana*
cedro rojo del Pacífico – *Thuja plicata*
cedro tasco – *Juniperus flaccida*
cefalantera – *Cephalanthera austiniae*
cefalotaxo de fortune – *Cephalotaxus fortuni*
cefalotaxo drupáceo – *Cephalotaxus harringtonia*
cefalotejo de fortune – *Cephalotaxus fortuni*
cefalotejo drupáceo – *Cephanotaxus harringtonia*
cefalotejo oriental – *Cephalotaxus harringtonia*
cefirante – *Zephyranthes carinata*
cegador – *Opuntia microdasys*
ceiba – *Bombax ellipticum, Ceiba, Ficus cotinifolia*
ceiba del Brasil – *Ceiba speciosa*
ceibón – *Pachira insignis*
ceibón lanero – *Ochroma lagopus*
celastro – *Celastrus scandens*
celestina – *Plumbago auriculata*
celidonia – *Chamaesyce maculata, C. prostrata, Chelidonium majus, Ranunculus ficaria*
celidonia mayor – *Chelidonium majus*
celidonia menor – *Ranunculus ficaria*
celinda – *Philadelphus coronarius*
celosa – *Duranta erecta, Mimosa xanti*
celosa cimmarona – *Duranta erecta*
celosia – *Celosia*
cembro – *Pinus cembra*
cemita – *Boletus edulis, B. luridus, B. pinicola, B. reticulatus*
cempasúchil – *Tagetes*
cempasúchil chiquito – *Adenophyllum porophylloides*
cempoal – *Cosmos sulphureus, Tagetes*
cempoalehuatl – *Ulmus*
cempoalxóchitl – *Tagetes erecta, T. patula*
cempoalxóchitl cimarrón – *Tagetes patula*
cendrero – *Cereus hexagonus*
cenicienta – *Sesuvium portulacastrum, S. verrucosum*
cenicilla – *Helianthemum glomeratum, Leucophyllum frutescens, Sesuvium portulacastrum, S. verrucosum, Zaluzania augusta*
cenicilla del tomate – *Fusarium solani*
cenizo – *Agave colorata, A. shrevei, Atriplex,*

Chenopodium album, Leucophyllum, Opuntia streptacantha, Tiquilia greggii
cenizo blanco – *Chenopodium album*
cenizo botrioide – *Chenopodium chenopodioides*
cenizo de Berlandier – *Chenopodium berlandieri*
cenizo fétido – *Chenopodium foetidum*
cenizo folioso – *Chenopodium foliosum*
cenizo glauco – *Chenopodium glaucum*
cenizo leptofilo – *Chenopodium leptophyllum*
cenizo multifido – *Chenopodium multifidum*
cenizo mural – *Chenopodium murale*
cenizo rojo – *Chenopodium rubrum*
centaura alba – *Centaurea cineraria*
centaura americana – *Centaurea americana, Sabatia angularis*
centaura azufrada – *Centaurea sulphurea*
centaura de los prados – *Centaurea* x *pratensis*
centaura del solsticio – *Centaurea solstitialis*
centaura de Malta – *Centaurea melitensis*
centaura difusa – *Centaurea diffusa*
centaura fragil – *Centaurea diluta*
centaura macrocéfala – *Centaurea macrocephala*
centaura maculada – *Centaurea maculosa*
centaura mayor – *Centaurea scabiosa*
centaura menor – *Centaurium erythraea*
centaura montana – *Centaurea montana*
centaura negra – *Centaurea nigra*
centaura obscura – *Centaurea nigra*
centaura olor de almizcle – *Centaurea moschata*
centaurea – *Centaurea cyanus, C. montana, C. moschata*
centaurea bendita – *Cnicus benedictus*
centaurea real – *Anagallis monelli*
centaurea umbelada – *Centaurium erythraea*
centeno – *Secale cereale*
centeno italiano – *Lolium multiflorum*
centinodia – *Polygonum arenastrum, P. aviculare, P. hydropiper*
ceñidero – *Bryonia dioica*
ceñiglo – *Chenopodium album*
cepa – *Boletus edulis, B. pinicola, B. reticulatus*
cepa caballo – *Carlina vulgaris, Loranthus, Xanthium spinosum*
cepetillo – *Tagetes patula*
cepillito – *Lamarckia aurea*
cepillo – *Combretum farinosum, C. fruticosum, Entada polystachya*
cepillo del diablo – *Combretum farinosum*
cerafolio – *Anthriscus cerefolium*
cerafolio agreste – *Anthriscus sylvestris*
cerahui – *Vitis arizonica*
cerastio – *Cerastium tomentosum*
cerastio aglomerado – *Cerastium glomeratum*
cerastio de la fuente – *Cerastium fontanum*
cerastio de los campos – *Cerastium arvense*
cerastio tomentoso – *Cerastium tomentosum*
ceratófila común – *Ceratophyllum demersum*
cercis – *Cercis siliquastrum*
cereal – *Poaceae*
cerecillo – *Muntingia calabura*
cerecino – *Prunus mahaleb*
cerero – *Myrica cerifera, Myrtus cerifera*
cerezo – *Belotia mexicana, Malpighia glabra, Prunus, Vaccinium confertum*
cerezo ácido – *Prunus avium*
cerezo-aliso – *Prunus padus*
cerezo americano – *Prunus serotina*
cerezo cimarrón – *Prunus capuli*
cerezo común – *Prunus avium, P. cerasus*
cerezo costalero – *Prunus avium* var *duracina*
cerezo criollo – *Prunus serotina*
cerezo chinesco – *Prunus glandulosa*
cerezo de adorno japonés – *Prunus serrulata*
cerezo de Castilla – *Malpighia glabra*
cerezo de Cayena – *Eugenia uniflora*
cerezo de hojas serradas – *Prunus serrulata*
cerezo de Jamaica – *Malpighia glabra*
cerezo del Canadá – *Prunus pumila*
cerezo de Manchuria – *Prunus maackii*
cerezo de Natal – *Carissa grandiflora*
cerezo de Pensilvania – *Prunus pensylvanica*
cerezo de racimo – *Prunus padus*
cerezo de Santa Lucía – *Prunus mahaleb*
cerezo de Virginia – *Prunus virginiana* var *demissa*
cerezo dulce – *Prunus avium*
cerezo durazno – *Prunus avium* var *duracina*
cerezo enano – *Prunus fruticans, P. fruticosa*
cerezo híbrido japonés yoshino – *Prunus* x *yedoensis*
cerezo japonés – *Prunus serrulata, P. subhirtella*

cerezo japonés de hojas serruladas – *Prunus serrulata*
cerezo japonés de Sergeant – *Prunus sargentii*
cerezo mexicano – *Prunus virginiana* var *demissa*
cerezo mirobalán – *Prunus cerasifera*
cerezo mollar – *Prunus avium* var *juliana*
cerezo montés – *Prunus brachybotrya*
cerezo negro – *Prunus serotina*
cerezo negro americano – *Prunus serotina*
cerezo negro silvestre – *Prunus serotina*
cerezo ornamental – *Prunus padus*
cerezo pado – *Prunus padus*
cerezo serótino – *Prunus serotina*
cerezo silvestre – *Cornus mas, Prunus avium, P. padus, P. serotina*
cerezo silvestre norteamericano – *Prunus virginiana*
cerezo tibetano – *Prunus serrula, P. serrulata*
cerezo tropical – *Malpighia glabra*
cerezo virginiano – *Prunus virginiana*
cergazo – *Cistus*
ceriflor – *Cerinthe major*
cerimán – *Monstera deliciosa*
cerón – *Phyllostylon brasiliense*
cerote de coche – *Iresine calea*
cervina – *Scolopendrium officinale*
cervispina – *Rhamnus cathartica*
cerraja – *Sonchus asper, S. oleraceus, Taraxacum officinale*
cerraja arvense – *Sonchus arvensis*
cerraja común – *Sonchus asper*
cerraja delicada – *Sonchus tenerrimus*
cerraja de pared – *Sonchus tenerrimus*
cerraja espinosa – *Sonchus asper*
cerrajón – *Sonchus oleraceus*
cerrillo – *Elytrigia repens*
césped de Olimpo – *Armeria maritima*
césped espinoso – *Arenaria montana*
césped inglés – *Lolium perenne*
césped marino – *Armeria maritima*
cesta de oro – *Aurinia saxatilis*
cesta de plata – *Iberis sempervirens*
cestillo de oro – *Alyssum montanum*
ciali – *Avicennia germinans*
ciano – *Centaurea cyanus*
cianori – *Cochlospermum vitifolium*
cib – *Chamaedorea ibarrae*
cica – *Cycas revoluta*
cicércula – *Lathyrus cicera*
cicércula angulosa – *Lathyrus angulatus*
cicércula de frutos esféricos – *Lathyrus sphaericus*
cicércula de hojas anchas – *Lathyrus latifolius*
cicércula de hojas tenues – *Lathyrus clymenum*
cicércula de los pantanos – *Lathyrus palustris*
cicércula de Tánger – *Lathyrus tingitanus*
cicércula española – *Lathyrus clymenum*
cicércula hirsuta – *Lathyrus hirsutus*
cicércula latifolia – *Lathyrus latifolius*
cicércula marítima – *Lathyrus japonicus*
cicércula ocre – *Lathyrus ochrus*
cicércula olorosa – *Lathyrus odoratus*
cicércula primaveral – *Lathyrus vernus*
cicércula purpurina – *Lathyrus clymenum*
cicércula roja – *Lathyrus cicera*
cicércula silvestre – *Lathyrus sylvestris*
cicércula tuberosa – *Lathyrus tuberosus*
ciclamen – *Cyclamen*
ciclamen de Chile – *Cyclamen cilicum*
ciclamen de Nápoles – *Cyclamen neapolitanum*
ciclamino – *Cyclamen purpurascens*
ciclamino de Europa – *Cyclamen purpurascens*
ciclamino de hojas circulares – *Cyclamen coum*
ciclamino de hojas de hiedra – *Cyclamen hederifolium*
ciclamino de Persia – *Cyclamen persicum*
ciclamino europeano – *Cyclamen purpurascens*
ciclamino primaveral – *Cyclamen repandum*
ciclamor – *Cercis siliquastrum, Platanus recemosa, P. wrightii*
cico – *Albizzia occidentalis*
cicua de cueramo – *Cordia boissieri*
cicuta – *Conium maculatum*
cicuta acuática – *Oenanthe crocata*
cicuta de agua – *Oenanthe aquatica*
cicuta maculada – *Conium maculatum*
cicuta mayor – *Conium maculatum*
cicuta menor – *Aethusa cynapium*
cicuta roja – *Geranium robertianum*
cicutilla – *Parthenium hysterophorus*
cidra cayote – *Citrullus colocynthis, Cucurbita ficifolia*
cidrelo – *Citrus medica*
cidrera – *Citrus medica*
cidro – *Citrus medica*
cidro limón – *Citrus medica*
cidronela – *Melissa officinalis*

cidronela abejera – *Melissa officinalis*
ciega borrego – *Opuntia microdasys*
ciega vista – *Croton ciliato-glandulosas*
cielito – *Ageratum corymbosum*
cielo estrellado – *Aster novi-belgii*
cielo raso – *Vinca major, V. minor*
cien nudos – *Polygonum aviculare*
cientoenrama – *Achillea millefolium*
cigarrilla – *Cigarrilla mexicana*
cigró – *Boletus edulis, B. reticulatus*
cigüeña malva – *Erodium malacoides*
cihuapatli – *Ageratina, Montanoa tomentosa, Pluchea odorata, Vernonia deppeana*
cihuatsinsintle – *Psilocybe muliericula*
cilantrillo – *Spergula arvensis*
cilantro – *Coriandrum sativum, Eryngium foetidum, Osmorhiza brachypoda*
cilindrillo – *Lycium berlandieri, L, brevipes*
cimarrón – *Acacia greggii, Calophyllum braziliense, Eustoma exaltatum, Rhus mollis*
címatl – *Phaseolus coccineus*
cimbalaria – *Cymbalaria muralis*
cimbalaria de hojas de hiedra – *Cymbalaria muralis*
cimbalaria de tres lobos iguales – *Cymbalaria aequitriloba*
cimboa – *Citrus maxima*
cimbopogon mediterráneo – *Hyparrhenia hirta*
cimicífuga – *Cimicifuga racemosa*
cimicífuga negra – *Cimicifuga racemosa*
cimifuga azul – *Caulophyllum thalictroides*
cinacina – *Parkinsonia aculeata*
cinamomo – *Eleagnus pungens, Melia azederach* var *umbraculiformis*
cinamomo de Manila – *Reseda odorata*
cinamón – *Melia azedarach*
cinatoria – *Distictis buccinatoria*
cincahuite – *Laguncularia racemosa*
cinco dedos – *Potentilla tormentilla*
cinco de mayo – *Amaranthus caudatus, A. leucocarpus, Celosia*
cincoenrama – *Potentilla*
cincoenrama áurea – *Potentilla aurea*
cincoenrama de los pantanos – *Potentilla palustris*
cincoenrama de Pensilvania – *Potentilla pensylvanica*
cincoenrama de primavera – *Potentilla tabernaemontanii*
cincoenrama dorada – *Potentilla aurea*
cincoenrama erecta – *Potentilla erecta*
cincoenrama leñosa – *Potentilla fruticosa*
cincoenrama nítida – *Potentilla nitida*
cincoenrama noruega – *Potentilla norvegica*
cincoenrama palustre – *Potentilla palustris*
cincoenrama recta – *Potentilla recta*
cincoenrama rocosa – *Potentilla tabernaemontanii*
cincoenrama rosa – *Potentilla nitida*
cinco llagas – *Proboscidea fragrans, Tagetes lunulata, Zinnia grandiflora*
cinco negritos – *Conostegia xalapensis, Lantana camera, L. involucrata, Rauwolfia hirsute, Saracha procumbens*
cinchona – *Cinchona officinalis*
cineraria – *Senecio cineraria, S. hybridus*
cinoglosa – *Cynoglossum officinale*
cinorrodón – *Rosa montezumae*
cinosuro crestado – *Cynosurus cristatus*
cinosuro espinoso – *Cynosurus echinatus*
cinosuro gigantesco – *Cynosurus echinatus*
cinta – *Chlorophytum comosum, Phalaris arundinacea*
cintillo – *Codiaeum variegatum, Paspalum conjugatum*
cintli – *Zea mays*
cintoria – *Centaurium erythraea*
cintul – *Stenanthium frigidum*
cintura – *Cynosurus cristatus*
cipatle – *Pluchea odorata*
cípero – *Cyperus rotundus*
ciprés – *Calocedrus decurrens, Casuarina equisetifolia, Chamaecyparis, Cupressus, Juniperus californica, J. gamboana, J. monticola, Picea mexicana, Taxodium*
ciprés americano – *Cupressus macrocarpa, C. nootkatensis, Taxodium distichum*
ciprés calvo – *Taxodium distichum*
cipresco – *Taxodium distichum* var *mexicanum*
ciprés común – *Cupressus sempervirens*
ciprés de Arizona – *Cupressus arizonica*
ciprés de Arizona de corteza áspera – *Cupressus arizonica*
ciprés de California – *Cupressus goveniana, C. macrocarpa*
ciprés de Cartagena – *Tetraclinis articulata*
ciprés de frutos grandes – *Cupressus macrocarpa*

ciprés de Gowen – *Cupressus goveniana*
ciprés de guisantes – *Chamaecyparis pisifera*
ciprés de hondo – *Chamaecyparis pisifera*
ciprés de la cordillera – *Austrocedrus chilensis*
ciprés de la guaiteca – *Pilgerodendron uviferum*
ciprés de Lambert – *Cupressus macrocarpa*
ciprés de Lawson – *Cupressus lawsoniana*
ciprés de Leyland – *Cupressocyparis leylandii*
ciprés de los cementerios – *Cupressus sempervirens*
ciprés de los pantanos – *Taxodium*
ciprés de México – *Cupressus arizonica, C. lusitanica*
ciprés de Moctezuma – *Taxodium distichum* var *mexicanum*
ciprés de montaña – *Podocarpus guatemalensis*
ciprés de Monterrey – *Cupressus macrocarpa*
ciprés de Nootka – *Cupressus nootkatensis*
ciprés de Portugal – *Cupressus lusitanica*
ciprés de verano escobiforme – *Kochia scoparia*
ciprés de Virginia – *Taxodium distichum*
ciprés fúnebre – *Cupressus funebris*
ciprés glabro de Arizona – *Cupressus arizonica* ssp *glabra*
ciprés híbrido de Leyland – *Cupressocyparis leylandii*
cipresillo – *Podocarpus guatemalensis, Santolina chamaecyparisus*
ciprés italiano – *Cupressus sempervirens*
ciprés llorón chino – *Cupressus funebris*
ciprés mexicano – *Taxodium distichum* var *mexicanum*
ciprés palmita – *Thuja occidentalis*
ciprés pirámidal – *Cupressus sempervirens*
ciprés pisífero – *Chamaecyparis pisifera*
ciprés silvestre – *Juniperus mexicana*
cipripedio – *Cypripedium irapeanum*
cirial – *Spondias purpurea*
cirián – *Crescentia alata, C. cujete, Randia*
cirián chino – *Randia echinocarpa*
ciribe – *Cylindropuntia*
cirio – *Fouquieria columnaris, F. diguetii, F. splendens*
cirio de Chile – *Echinopsis chiloensis*
cirio de Nuestra Señora – *Verbascum thapsus*
cirioni – *Sapindus drummondii*
cirolero – *Prunus domestica*
cirolero japonés – *Prunus salicina*
ciruelillo – *Ribes*
ciruelo – *Cyrtocarpa edulis, Phyllanthus acuminatus, Prunus americana, P. domestica, Prunus mexicana, Spondias mombin, S. purpurea*
ciruelo agrio – *Spondias mombin, S. purpurea*
ciruelo amarillo – *Spondias mombin, S. purpurea*
ciruelo amarillo pequeño – *Prunus syriaca*
ciruelo americano – *Prunus americana*
ciruelo campechano – *Spondias purpurea*
ciruelo cimarrón – *Cyrtocarpa edulis*
ciruelo-cerezo – *Prunus cerasifera*
ciruelo-cerezo híbrido doble – *Prunus* x *blireiana*
ciruelo cimarrón – *Spondias purpurea*
ciruelo claudio – *Prunus italica*
ciruelo cochino – *Spondias mombin*
ciruelo colorado – *Spondias mombin, S. purpurea*
ciruelo costeño – *Phyllanthus acidus*
ciruelo coyote – *Spondias purpurea*
ciruelo chabacano – *Prunus cerasifera*
ciruelo damasceno – *Prunus domestica* ssp *insititia*
ciruelo de América – *Chrysobalanus icaco*
ciruelo de cereza – *Prunus cerasifera*
ciruelo de España – *Prunus domestica*
ciruelo de jobo – *Spondias mombin*
ciruelo del monte – *Cyrtocarpa edulis, Ziziphus obtusifolia*
ciruelo del país – *Spondias mombin, S. purpurea*
ciruelo de México – *Spondias mombin, S. purpurea*
ciruelo de paloma – *Chrysobalanus icaco*
ciruelo de San Juan – *Spondias purpurea*
ciruelo de San Julián – *Prunus insititia*
ciruelo dulce – *Spondias cytherea*
ciruelo español – *Spondias mombin*
ciruelo europeo – *Prunus domestica*
ciruelo gobernadoro – *Flacourtia indica, Ziziphus jujuba*
ciruelo japonés – *Prunus salicina*
ciruelo mango – *Spondias mombin*
ciruelo mirabel – *Prunus domestica* ssp *syriaca*
ciruelo mirobalán – *Prunus cerasifera*
ciruelo obo – *Spondias mombin*

ciruelo ornamental – *Prunus blireiana*
ciruelo quetche – *Prunus domestica* ssp *insititia*
ciruelo rojo – *Spondias mombin, S. purpurea*
ciruelo silvestre – *Prunus americana, P. mexicana, P. spinosa*
ciruelo trópico – *Spondias mombin*
cirrhata – *Swietenia humilis*
cisca – *Imperata cylindrica*
cisiquelite – *Bidens odorata*
cisto – *Cistus*
citabaro – *Vallesia glabra*
cítara – *Spondias mombin*
cítiso – *Cytisus scoparius, Laburnum anagyroides*
cítiso enano – *Chamaecytisus proliferus*
cítricos – *Citrus*
citrín – *Garcinia cambogia*
citrón – *Citrus medica*
citronela – *Cymbopogon citratus, C. nardus*
citún – *Rubus coriifolius*
ciunery – *Boletus edulis, B. reticulatus*
civico – *Pithecellobium dulce*
civil – *Malvaviscus arboreus*
cizaña – *Lolium perenne, L. temulentum, Zizania aquatica*
cizaña cilíndrica – *Hainardia cylindrica*
clacancauayo – *Iresine*
clacojiguite – *Bocconia frutescens*
cladrastis amarillo – *Cladrastis kentukea*
clagot – *Zapoteca anomalia*
claitonia – *Claytonia perfoliata*
clameria – *Cuphea jorullensis, Potentilla thurberi*
clarín – *Bougainvillea spectabilis, Distictis buccinatoria, Ipomoea quamoclit, Lathyrus odoratus, Pithecellobium echinatum, Quamoclit coccinea*
clarincillo – *Ipomoea quamoclit, Quamoclit pinnata, Senecio canicida*
clarincito – *Quamoclit coccinea*
clarincito silvestre – *Quamoclit coccinea*
clarkia – *Clarkia unguiculata*
claudiosa amarga – *Scutellaria*
claudiosa del monte – *Scutellaria*
clavel – *Dianthus caryophylla, D. chinensis, Gentiana superba, Russelia sarmentosa*
clavel alemán – *Cryptostegia grandiflora*
clavel alpino – *Dianthus alpinus*
clavel anual – *Dianthus chinensis*
clavel arenario – *Dianthus arenarius*
clavel barbado – *Dianthus barbatus*
clavel común – *Dianthus caryophylla*
clavel coronado – *Dianthus plumarius*
clavel de Cheddar – *Dianthus gratianopolitanus*
clavel de China – *Dianthus chinensis, Tagetes tenuifolia*
clavel de España – *Cryptostegia grandiflora*
clavel de jardín – *Dianthus caryophyllus*
clavel del aire – *Tillandsia recurvata*
clavel de las Indias – *Tagetes erecta, T. patula, T. tenuifolia*
clavel del monte – *Silene laciniata* ssp *major*
clavel de los Alpes – *Dianthus alpinus*
clavel de los prados – *Dianthus deltoides*
clavel deltoideo – *Dianthus deltoides*
clavel de moro – *Tagetes patula*
clavel de nobleza – *Dianthus barbatus*
clavel de oro – *Tagetes erecta, Turnera ulmifolia*
clavel de poeta – *Dianthus barbatus*
clavel de ramillete – *Dianthus barbatus*
clavel de San Isidro – *Dianthus barbatus*
clavel encarnado – *Dianthus chinensis*
clavel hortense – *Dianthus caryophyllus*
clavel imperial – *Dianthus chinensis*
clavelito – *Bouvardia multiflora, Malacothrix*
clavelito del monte – *Malacothrix xantii*
clavel japonés – *Hibiscus rosa-sinensis*
clavel minutisa – *Dianthus barbatus*
clavelón – *Tagetes erecta*
clavelón de la India – *Tagetes erecta*
clavel reventón – *Dianthus caryophyllus*
clavel rugoso – *Dianthus armeria*
clavel silvestre – *Armeria maritima*
clavellina – *Bombax ellipticum, Caesalpinia pulcherrima, Calliandra gentryi, Capparis flexuosa, Catharanthus roseus, Cylindropuntia alcahes, C. molesta, C. tunicata, Dianthus, Echinocereus pensilis, Grusonia schottii, Hibiscus cannabinus, Lychnis coronaria, Pachira aquatica, Pseudobombax palmeri, Saponaria officinalis*
clavellina blanca – *Pachira aquatica, Passiflora foetida, Pseudobombax palmeri*
clavellina colorada – *Caesalpinia pulcherrima, Pseudobombax palmeri*

clavellina de la barranca – *Bombax ellipticum, Pseudobombax palmeri*
clavellina de Nanteuil – *Petrorhagia nanteuilii*
clavellina prolífera – *Petrorhagia prolifera*
clavellina rocosa – *Dianthus deltoides*
clavellina roja – *Bombax palmeri, Pachira aquatica, Passiflora mexicana*
clavero – *Caryophyllus aromaticus*
clavero oloroso – *Caryophyllus aromaticus*
clavillo – *Choisya ternata, Dianthus caryophyllus, Ludwigia peploides*
clavillo blanco – *Ludwigia octovalvis*
clavito – *Ludwigia octovalvis, Lyophyllum decastes, Ramaria botrytis*
clavito grande – *Lyophyllum decastes*
clavo – *Armillaria mellea, Caryophyllus aromaticus, Ludwigia octovalvis, Phytophthora infestans, Pittosporum tobira, Taxodium distichum*
clavo del señor – *Passiflora edulis*
clavo de olor – *Caryophyllus aromaticus, Choisya ternata*
clavohuasca – *Tynnanthus panurensis*
clavo rojo – *Trifolium pratense*
clemátide – *Clematis*
clemátide alpina – *Clematis alpina*
clemátide de hojas enteras – *Clematis integrifolia*
clemátide de Virginia – *Clematis viticella*
clemátide flámula – *Clematis flammula*
clemátide italiana – *Clematis viticella*
clemátide jackmanni – *Clematis* x *jackmanii*
clemátide montana – *Clematis montana*
clemátide recta – *Clematis recta*
clemátide vitalba – *Clematis vitalba*
clematítide – *Aristolochia clematitis*
clementina – *Citrus clementine*
clemole – *Tagetes patula*
clemolillo – *Zexmenia hispida*
clemolito – *Tagetes patula*
clepe – *Ziziphus obtusifolia*
clérigo – *Aquilegia vulgaris*
clérigo boca abajo – *Aquilegia vulgaris*
clerodendro – *Clerodendrum*
clerodendro chino – *Clerodendrum bungei*
clerodendro japonés – *Clerodendrum trichotomum*
clina – *Pluchea odorata*
clinopodio – *Clinopodium vulgare*
clinopodio común – *Clinopodium vulgare*
clitoria – *Clitoria terneata*
clivia – *Clivia miniata*
clorique – *Clitoria terneata*
clorofito – *Chlorophytum comosum*
club del diablo – *Oplopanax horridus*
clusia – *Hintonia latiflora*
cneoro tricoco – *Cneorum tricoccon*
coacamachalli – *Pithecellobium dulce*
coacoyutillo – *Maba salicifolia*
coacuahuitl – *Bocconia frutescens*
coahuitzxóchit – *Lantana camera*
coamichin – *Ficus*
coanenepili – *Boerhavia coccinea, Potentilla candicans*
coanenepilli – *Dahlia*
coapatli – *Acourtia cordata, Commelina coelestis, C. tuberosa, Tradescantia crassifolia*
coate – *Eysenhardtia polystachya*
coatillo – *Eysenhardtia polystachya*
coatlamitl – *Rubus*
coatlanxocotl – *Rubus*
coatli – *Eysenhardtia polystachya, Jatropha spathulata*
coatlxoxouqui – *Turbina corymbosa*
coatzontecomaxóchitl – *Stanhopea tigrina*
coaxihuitl – *Turbina corymbosa*
coaxmuttza – *Casimiroa edulis*
cóbano – *Cedrela mexicana, Swietenia humilis, S. macrophylla*
cobea – *Cobaea scandens*
cobó – *Ficus cotinifolia*
cobresia falsa-cola de ratón – *Kobresia bellardii*
coca – *Erythroxylum coca, E. novagranatense, E. truxillense*
coca de Levante – *Anamirta paniculata*
coca de México – *Erythroxylum mexicanum*
cocaína – *Erythroxylum coca, E. novagranatense, E. truxillense*
cocaiste – *Erythea edulis*
cocax – *Annona reticulata*
coceh – *Smilax mexicana*
cocito – *Cochlospermum vitifolium*
coclearia común – *Cochlearia officinalis*
coclearia del país – *Lepidium latifolium, L. virginicum*
coco – *Cocos nucifera*

cocobolo – *Dalbergia granadillo, D. retusa*
cococha – *Erythrina leptorhiza*
cocoche loco – *Opuntia pilifera*
coco de aceite – *Orbignya cohune*
coco de agua – *Cocos nucifera*
coco de castillo – *Cocos nucifera*
coco de cerro – *Cyrtocarpa procera*
coco de Chile – *Jubaea spectabilis*
coco de Guadalajara – *Orbignya cohune*
coco de mono – *Couroupita guianensis, Lecythis ollaria*
coco de pluma – *Syagrus romanzoffiana*
cocohuite – *Gliricidium sepium*
cocohuitztli – *Ribes microphylla*
cocojegüite – *Bocconia frutescens*
cocolmeca – *Berlandiera lyrata, Dioscorea, Milleria quinqueflora, Phaseolus metcalfei, Rhynchosia precatoria, Smilax bona-nox, S. mexicana*
cocolmecatl – *Dioscorea*
cocoloba – *Coccoloba uvifera*
cocolobo – *Dalbergia retusa*
cocona – *Solanum topiro*
coco plumoso – *Syagrus romanzoffiana*
cocorit – *Capsicum baccatum*
cocotero – *Cocos nucifera*
cocotombo – *Rauwolfia hirsuta*
cocox – *Annona reticulata*
cocoxíhuitl – *Bocconia arborea, B. frutescens*
cocoyán – *Colocasia esculenta*
cocoyatón – *Tagetes lunulata*
cocoyol – *Acrocomia aculeata, A. mexicana, Scheelia liebmannii*
cocoyule corozo – *Orbignya cohune*
cocozatona – *Tagetes lunulata*
cocoztic – *Thalictrum hernandezii*
cocté – *Talauma mexicana*
coctzán – *Inga spuria*
cocuá – *Echinocereus cinerascens*
cocuí – *Agave cocui*
cocuite – *Gliricidia sepium*
cocuiza – *Furcraea humboldtiana*
cóculo – *Cocculus laurifolius*
cocurstle – *Bromelia*
coch – *Ricinus communis*
cochal – *Myrtillocactus cochal*
cochear – *Cuscuta*
cochil sapota – *Casimiroa edulis*
cochilla – *Goldmania foetida*
cochinera – *Opuntia robusta*
cochinilla – *Pluchea sericea*
cochinita – *Asclepias curassavica*
cochitzápotl – *Casimiroa edulis*
cochizquilitl – *Erythrina*
cocholloxin – *Asclepias curassavica*
cochun – *Terminalia amazonia*
codeso – *Laburnum anagyroides*
codeso alpino – *Laburnum alpinum*
codeso de los Alpes – *Laburnum alpinum*
codo de fraile – *Thevetia peruviana, T. thevetioides*
cógetelo todo – *Clerodendron*
cogoma – *Tricholoma columbetta*
cogón – *Imperata cylindrica*
cogorda – *Lagenaria siceraria*
coguegue – *Erythea brandegeei, Sabal uresana*
cogulla de fraile – *Aconitum napellus*
cohombrillo amargo – *Momordica charanita*
cohombro – *Cucumis anguria, C. sativus*
cohombro de olor – *Sicana odorifera*
cohombro espinoso – *Cucumis anguira*
cohombro largo – *Cucumis melo* var *flexuosus*
cohombro serpentino – *Cucumis melo* var *flexuosus*
cohuanenepili – *Dahlia variabilis, Dorstenia contrayerba, Passiflora incarnata*
cohuite – *Bursera simaruba*
cohune – *Orbignya cohune*
coi – *Pterocarya stenoptera*
coihue – *Nothofagus dombeyi*
coilotápalo – *Cecropia peltata*
cojón de berraco – *Stemmadenia donnell-smithii, S. obovata*
cojón de burro – *Stemmadenia donnell-smithii*
cojón de coche – *Stemmadenia obovata*
cojón de danta – *Stemmadenia donnell-smithii*
cojón de gato – *Asclepias curassavica, Caesalpinia crista, Solanum mammosum, Tabernaemontana alba, Thevetia peruviana*
cojón de oro – *Cochlospermum vitifolium*
cojón de perro – *Tabernaemontana alba*
cojón de puerco – *Stemmadenia obovata, S. tomentosa*
cojón de toro – *Cochlospermum vitifolium, Stemmadenia donnell-smithii, Tabernaemontana alba*
cojotón – *Tabernaemontana alba*

col – *Brassica oleracea* var *acephala, B. o.* var *capitata*
cola – *Kola acuminata*
colación – *Antigonon leptopus*
cola de alacrán – *Elytraria imbricata, Heliotropium angiospermum, H. curassavicum, H. indicum, Tournefortia hartwegiana*
cola de ardilla – *Agrostis viridis, Elymus longifolius, Jacobinia umbrosa, Justicia aurea, Setaria grisebachii, S. liebmannii*
cola de borrego – *Castilleja, Sedum mexicanum, S. morganianum, Stevia serrata*
cola de caballo – *Chloris, Conyza canadensis, Equisetum*
cola de caballo de invierno – *Equisetum hyemale*
cola de caballo de los campos – *Equisetum arvense*
cola de caballo de los pantanos – *Equisetum palustre*
cola de caballo mucho ramoso – *Equisetum ramosissimum*
cola de camarón – *Justica brandegeana*
cola de conejo – *Lagurus ovatus*
cola de chango – *Cyathea princeps, Sphaeropteris horrida*
cola del diablo – *Pereskiopsis diguetii*
cola de faisán – *Anthurium schlechtendalii, Neurolaena lobata, Philodendron radiatum*
cola de gallo – *Anisacanthus thurberi, Eleusine indica*
cola de gato – *Acalypha arvensis, A. hispida, Disocactus speciosus, Setaria pumila, Typha domingensis, T. latifolia*
cola de golondrina – *Damasonium alisma*
cola de iguana – *Disocactus flagelliformis, Plumbago pulchella*
cola de león – *Agave attenuata, Leonurus cardiaca*
cola de liebre – *Lagurus ovatus*
cola de macaco – *Araucaria araucaria*
cola de machín – *Cyathea princeps*
cola de mico – *Cyathea princeps, Heliotropium curassavicum*
cola de mono – *Cyathea mexicana, Nephelea mexicana*
cola de pato – *Sagittaria latifolia*
cola de perro – *Cynosurus cristatus, Setaria anceps*
cola de pescado – *Plumbago pulchella*
cola de pisote – *Nephrolepis biserrata*
cola de rata – *Capsicum annuum*
cola de rata de los campos – *Equisetum arvense*
cola de ratón – *Anthurium crassinervium, Muhlenbergia emersleyi, Sporobolus indicus, S. wrightii*
cola de ratón menor – *Myosurus minimus*
cola de topo – *Phleum pratense, Typha latifolia*
cola de zorra – *Acourtia adnata, A. thurberi, Alopecurus geniculatus, A. pratensis, Andropogon glomeratus, Aristida roemeriana, Cheilanthes, Leptochloa filiformis, Lobelia fenestralis, Lupinus elegans, Muhlenbergia dumosa, M. emersley, Polypogon elongatus, P. monspeliensis, Scleropogon brevifolius, Selloa glutinosa, Setaria liebmannii*
cola de zorrillo – *Agrostis viridis, Hordeum jubatum, Petiveria alliacea, Ptelea*
cola de zorro roja – *Acalypha hispida*
col agria – *Rumex crispus*
cola pecho – *Typha domingensis*
colatero – *Kola acuminata*
col caballar – *Brassica oleracea* var *acephala, B. o.* var *medullosa, B. o.* var *ramosa*
col caribe – *Colocasia antiquorum, C. esculenta*
col común – *Brassica oleracea* var *capitata*
colconab – *Acrocomia mexicana*
col crespa – *Brassica oleracea* var *laciniata, B. o.* var *sabellica*
col china – *Brassica rapa* var *chinensis, B. r.* var *pekinensis*
colcho – *Erythrina flabelliformis*
colchol – *Diphysa suberosa*
col de Bruselas – *Brassica oleracea* var *gemmifera*
col de China – *Brassica rapa* var *chinensis, B. r.* var *pekinensis*
col de hoja – *Brassica oleracea* var *acephala*
col de la Sibéria – *Brassica napus* var *pabulária*
col de mar – *Crambe maritima*
col de meollo – *Brassica oleracea* var *medullosa*

col de Milán – *Brassica oleracea* var *sabauda*
col de pantano – *Symplocarpus foetidus*
col de Saboya – *Brassica oleracea* var *sabuada*
coleo – *Coleus hybridus*
colera de novio – *Duranta repens*
coleto – *Oreopanax peltatum*
coletón – *Oreopanax capitatus*
coletuy – *Coronilla varia*
coletuy multicolor – *Coronilla varia*
col fétida – *Symplocarpus foetidus*
col forragera *Brassica oleracea* var *acephala, B. o.* var *medullosa, B. o.* var *viridis*
col frizada – *Brassica oleracea* var *laciniata*
col gallega – *Brassica oleracea* var *acephala, B. o.* var *tronchuda*
colguinero – *Rosa minutifolia*
coliflor – *Brassica oleracea* var *botrytis*
colima – *Zanthoxylum*
colinabo – *Brassica oleracea* var *gongylodes, B. napus* var *napobrassica, B. oleracea*
colirrábano – *Brassica oleracea* var *gongylodes*
colita – *Elyonurus tripsacoides*
colita de ratón – *Eriogonum fasciculatum, Polypogon monspeliensis*
col lombarda – *Brassica oleracea* var *capitata*
col marina – *Calystegia soldanella, Crambe maritima*
col medular – *Brassica oleracea* var *medullosa*
colmena – *Morchella esculenta*
colmenilla – *Morchella esculenta*
colmillo de víbora – *Lippia umbellata*
coloc – *Psidium guajava*
colocasia – *Colocasia antiquorum, C. esculenta*
cololté – *Quercus skinneri*
colombiana – *Thunbergia alata*
col meollosa – *Brassica oleracea* var *medullosa*
colomo – *Sagittaria sagittifolia, Xanthosoma robustum*
colomo anona – *Xanthosoma sagittifolium*
colomo delgado – *Sagittaria latifolia*
colonia – *Alpina zerumbet*
colonial – *Panicum maximum*
colón rojo – *Vigna unguiculata* ssp *sesquipedalis*
coloquíntida – *Citrullus colocynthis, Fructus colocyathidis*
colorado – *Calacophyllum candidissimum, Phlomis purpurea*
colorado blanco – *Musa acuminata*
colorín – *Capparis indica, Erythrina, Rivina humilis, Sophora secundiflora*
colorín cimmarón – *Fouquieria formosa, F. splendens*
colorincito – *Dolichos phaseoloides, Rhynchosia pyramidalis*
colorín chiquito – *Dolichos phaseoloides, Rhynchosia precatoria, R. pyramidalis*
colorín negro – *Erythrina leptorhiza*
colotahue – *Abutilon lignosum*
colo-tzitcaztli – *Mentzelia hispida*
col petsa – *Brassica rapa* var *pekinensis*
col picuda – *Brassica oleracea* var *conica*
col portuguesa – *Brassica oleracea* var *costata*
cólquico – *Colchicum autumnale, C. speciosum*
cólquico autumnal – *Colchicum autumnale*
cólquico de otoño – *Colchicum autumnale*
col rábano – *Brassica oleracea* var *gongylodes*
col repollo – *Brassica oleracea* var *capitata*
col rizada – *Brassica oleracea* var *acephala*
col roja – *Brassica oleracea* var *capitata*
colubrina – *Bryonia alba*
colza – *Brassica napus* var *arvensis, B. n.* var *oleifera*
collar de la reina – *Solanum pseudocapsicum*
collegón oriental – *Conringia orientalis*
colleja – *Silene inflata*
collejón – *Sinapis arvensis*
coma – *Sideroxylon laetevirens, S. lanuginosum, S. spiniflora*
coma de uña – *Pisonia*
comahuí – *Cordia sonorae*
comalpatli – *Pedilanthus tithymaloides, Pluchea odorata*
coma resinera – *Sideroxylon spiniflora*
comasec – *Marasmius oreades*
comate – *Parmentiera edulis*
comba – *Phaseolus lunatus*
combava – *Citrus hystrix*
comelina – *Commelina pallida*
come mano – *Cissus rhombifolia, C. sicyoides*
come mano de llano – *Cissus rhombifolia*
comida de cuervo – *Microrhamnus ericoides*
comida de pajaritos – *Capsella bursa-pastoris, Lepidium virginicum*
comida de víbora – *Ephedra compacta*

cominillo – *Lolium temulentum, Pectis palmeri, P. stenophylla*
comino – *Arracacia atropurpurea, Cuminum cyminum, Pectis palmeri*
comino cultivado – *Cuminum cyminum*
comino de los prados – *Carum carvi*
comino negro – *Nigella sativa*
comino rústico – *Arracacia atropurpurea, Prionosciadium thapsoides*
comitín – *Olneya tesota*
compio – *Combretum farinosum, Heteropteris palmeri*
compuesta – *Asteraceae*
conacaste – *Enterolobium cyclocarpum*
concha – *Marrubium vulgare*
conchilinque – *Mammillaria pectinifera*
conchita – *Centaurium nudicaule, Clitoria terneata, Cotyledon orbiculata* var *oblonga, Echeveria*
conchita azul – *Clitoria terneata*
condalia – *Condalia*
condori – *Adenanthera pavonina*
condrila – *Chondrilla juncea*
conduce – *Licania arborea*
coneja – *Castilleja tenuiflora, Impatiens balsamina*
conejillo – *Fumaria capreolata*
conejito – *Antirrhinum majus, Consolida ambigua, Delphinium consolida, Paspalum plicatulum*
conferva – *Symphytum officinale*
confesionario – *Laelia tibicinis*
confetillo – *Asplundia chiapensis*
confite – *Antigonon leptopus, Lantana camara, L. involucrate, Tournefortia hartwegiana, Ziziphus amole*
confite coyote – *Tournefortia hartwegiana*
confite negro – *Lantana camara, Tournefortia hartwegiana*
confitillo – *Gomphrena decumbens, Parthenium hysterophorus*
confitura – *Hyptis seemannii, H. suaveolens, Lantana camara, Trophis racemosa*
confitura blanca – *Lantana camara*
confiturilla – *Cordia brevispicata, Hyptis suaveolens, Lagascea decipiens, Lantana camara, L. horrida*
confiturilla amarilla – *Lagascea decipiens, Lantana glandulosissima*
confiturilla blanca – *Lantana camara, L. involucrata, L. velutina*
confiturilla grande – *Lagascea decipiens*
confiturilla negra – *Lantana horrida, Tournefortia hartwegiana*
congocopac – *Euphorbia dentata* var *cuphosperma*
congón – *Vaccinium confertum*
cónguera – *Phytolacca isocandra, P. octandra*
conibare – *Hyptis suaveolens*
conisaca – *Cyperus perennis, C. rotundus*
conito – *Ratibida columnaris*
conito de campo – *Ratibida columnaris*
coniza – *Senecio integrifolius*
coniza de Buenos Aires – *Conyza bonariensis*
coniza del Canadá – *Conyza canadensis*
coniza pálida – *Conyza floribunda*
conolón – *Pluchea odorata*
consentida – *Cannabis indica*
consólida – *Symphytum officinale*
consólida real – *Delphinium consolida*
consolva – *Sempervivum tectorum*
consuelda – *Delphinium consolida, Potentilla, Symphytum officinale, Taraxacum officinale*
consuelda ambigua – *Consolida ambigua*
consuelda escabrosa – *Symphytum asperum*
consuelda mayor – *Symphytum officinale, Wigandia caracasana*
consuelda media – *Ajuga reptans*
consuelda media rastrera – *Ajuga reptans*
consuelda roja – *Potentilla erecta*
consulita – *Galphimia glauca*
conté – *Philodendron sagittifolium*
conté cinco dedos – *Anthurium xanthosomifolium*
conté siete dedos – *Syngonium podophyllum*
contrahierba – *Bouvardia ternifolia, Dorstenia contrayerba, D. drakeana, Dyssodia acerosa, Euphorbia heterophylla, Flaveria, Kallstroemia, Psoralea esculenta, P. pentaphylla, P. rhombifolia*
contrahierba aromatica – *Psoralea pentaphylla*
contrahierba blanca – *Psoralea pentaphylla*
contrahierba colorada – *Bouvardia ternifolia*
contrahierba del interior – *Psoralea pentaphylla*
contrahierba de México – *Psoralea pentaphylla*
contra ranilla de bestias – *Serjania racemosa*
contribo – *Aristolochia grandiflora*

contzontón – *Syngonium podophyllum*
convalaria – *Convallaria majalis*
convulí – *Thevetia ovata*
convulvo campestre – *Convolvulus arvensis*
copa de Júpiter – *Helianthus annuus*
copa de los ángeles – *Narcissus triandrus*
copa de oro – *Allamanda cathartica, Erblichia odorata, Eschscholzia californica, E. glyptosperma, Solandra maxima, S. nitida, Thevetia peruviana*
copaiba – *Astronium graveolens, Copaifera officinalis*
copaiba de pará – *Copaifera reticulata*
copal – *Bursera, Copaifera copallifera, C. officinalis, Cyrtocarpa procera, Protium copal, Rhus copallina*
copalastle – *Schinus molle*
copal blanco – *Bursera jorullensis*
copalcocote – *Cyrtocarpa procera*
copalcoxócotl – *Cyrtocarpa procera*
copalcuáhuitl – *Bursera odorata*
copalchi – *Coutarea hexandra, C. pterosperma, Croton, Exostema caribaeum, Hintonia latiflora*
copalchi blanco – *Croton niveus*
copalchi de Colima – *Coutarea hexandra*
copalchi de Jojutla – *Exostema caribaeum, Hintonia latiflora*
copalchi de Michoacán – *Coutarea hexandra, Exostema caribaeum*
copal de Manila – *Agathis alba*
copal de penca – *Bursera jorullensis*
copal de santo – *Bursera jorullensis, B. penicillata*
copal de Veracruz – *Heliocarpus tomentosus*
copal de Zanzibar – *Trachylobium verrucosum*
copal grueso – *Bursera submoniliformis*
copali – *Bursera odorata*
copalillo – *Bursera penicillata, B. simaruba, B. submoniliformis, Casearia dolichophylla, Jatropha cordata, Liquidambar styraciflua, Oxalis latifolia, Parthenium incanum, Protium copal*
copalín – *Xylosma characantha*
copaljocote – *Cyrtocarpa procera*
copalme – *Liquidambar styraciflua*
copalquáhuitl – *Bursera jorullensis, Schinus molle*
copalquín – *Bursera cerasifolia, B. hindsiana, B. microphylla, B. odorata, Cinchona succirubra, Coutarea pterosperma, Croton californicus, Hintonia latiflora, Pachycormus discolor*
copalquín cortez amarilla – *Croton guatemalensis*
copal rojo – *Bursera hindsiana*
copalxihuitl – *Myrica mexicana*
copalxocotl – *Cyrtocarpa procera*
copalli – *Bursera jorullensis*
copallín – *Parthenium argentatum*
coparche – *Coutarea pterosperma*
copayero – *Copaifera officinalis*
copete – *Tagetes erecta, Tecoma stans*
copete de faisán – *Castilleja arvensis*
copetillo – *Tagetes erecta*
copey – *Clusia major, C. rosea*
copihue – *Lapageria rosea*
copito – *Cordia dodecandra*
coposo – *Ehretia elliptica*
copoy – *Ficus cotinifolia, F. mexicana*
coprino – *Coprinus*
copté – *Cordia dodecandra, C. sebestena*
coptis – *Coptis groenlandica*
coqueta – *Impatiens walleriana, Rivina humilis*
coquete – *Acacia pennatula*
coquillo – *Cyperus, Elytraria tridentata, Jatropha curcas, Orbignya cohune*
coquillo amarillo – *Cyperus esculentus*
coquillo azul – *Cyperus acuminatus*
coquillo morado – *Cyperus rotundus*
coquimbo – *Amblyopappus pusillus*
coquito – *Bombax ellipticum, Capparis verrucosa, Cochlospermum vitifolium, Cyperus laevigatus, C. rotundus, Hoffmannseggia glauca, Ipomoea capillacea, Jubaea spectabilis, Orbignya cohune, Reinhardtia gracilis, Zapoteca anomala*
coquito baboso – *Acrocomia mexicana*
coquito blanco – *Bombax ellipticum*
coquito de aceite – *Elaeis melanococca, Orbignya cohune*
coquito de Brasil – *Bertholletia excelsa, Syagrus schizophylla*
coquito de Colima – *Orbignya cohune*
coquito de Puerto Ángel – *Bombax ellipticum*
coquito de tierra – *Dichelostemma capitatum*
coracán – *Eleusine coracana*
coral – *Antigonon leptopus, Clavaria flava,*

Erythrina americana, Rivina humilis, Solanum pseudocapsicum
coral de la playa – *Russelia equisetiformes*
coral de los jardines – *Solanum pseudocapsicum*
coralillo – *Anagallis arvensis, Antigonon leptopus, Arctostaphylos nevadensis, A. uva-ursi, Hamelia erecta, Rauwolfia hirsuta, Rivina humilis, Russelia equisetiformes, Scaevola plumierii, Sophora secundiflora*
coralito – *Adenanthera pavonina, Antigonon leptopus, Bessera elegans, Pithecellobium unguis-cati, Rhynchosia pyramidalis, Rivina humilis*
coralito cimarrón – *Rivina humilis*
coraloriza trífida – *Corallorhiza trifida*
coral xilacuaco – *Rivina humilis*
corape – *Chamaesyce maculata*
corazón – *Annona reticulata, Dicentra formosa, D. spectabilis, Justica brandegeeana*
corazoncillo – *Hypericum perforatum*
corazón cimarrón – *Annona glabra*
corazoncito – *Hypericum coris*
corazón de ángel – *Begonia fuchsioides*
corazón de buey – *Annona reticulata*
corazón de cabrito – *Caladium bicolor*
corazón de Jesús – *Begonia fuchsioides, Caladium bicolor, Tubiflora squamosa*
corazón de María – *Dicentra formosa, D. spectabilis*
corazón de perro – *Ageratina*
corazón sangrante – *Dicentra formosa, D. spectabilis*
corazón sangriento – *Caladium bicolor*
corbata – *Sansevieria trifasciata*
corcalito – *Asclepias curassavica*
corcho – *Annona glabra, Diphysa suberosa, Erythrina, Heliocarpus donnell-smithii, Melaleuca alternifolia, Ochroma pyramidale, Omphalea oleifera*
corcho colorado – *Belotia mexicana*
cordiline – *Cordyline australis, C. fruticosa, C. indivisa*
cordiline crasa – *Cordyline stricta*
cordiline de Nueva Zelanda – *Cordyline australis*
cordillera cantábrica – *Geranium argenteum*
cordoncillo – *Elytraria imbricata, E. squamosa, Gomphrena globosa, Piper, Polygala, Sedum mexicanum, S. moranense, Thamnosma Montana, Tubiflora squamosa*
cordoncillo blanco – *Piper aduncum*
cordoncillo chico – *Piper amalago*
cordón de fraile – *Leonotis nepetaefolia*
cordón de obispo – *Amaranthus caudatus, Gomphrena decumbens, G. sonorae*
cordón de San Francisco – *Gomphrena nitida*
coreopsis – *Coreopsis grandiflora, C. tinctoria*
cori – *Hypericum perforatum*
coriandro – *Coriandrum sativum*
coriandro cultivado – *Coriandrum sativum*
corídalo amarillo – *Corydalis lutea*
corídalo digitado – *Corydalis solida*
corídalo hueco – *Corydalis cava*
corídalo primaveral – *Corydalis solida*
coriente – *Parthenium hysteriopsis*
cornejo – *Cornus*
cornejo albo – *Cornus alba*
cornejo americano – *Cornus florida*
cornejo blanco americano – *Cornus nuttallii*
cornejo de Bentham – *Cornus capitata*
cornejo de Florida – *Cornus florida*
cornejo del Japón – *Cornus controversa, C. kousa*
cornejo de Nuttall – *Cornus nuttallii*
cornejo estolonífero – *Cornus stolonifera*
cornejo floreciente – *Cornus florida*
cornejo hembra – *Cornus sanguinea*
cornejo macho – *Cornus mas*
cornejón variado – *Coronilla varia*
cornejo sanguino – *Cornus sanguinea*
corneta – *Clitocybe squamulosa, Pithecellobium echinatum, Zantedeschia aethiopica*
corneta amarilla – *Tecoma stans*
corneta prieta – *Craterellus cornucopioides*
cornetón – *Nicotiana glauca*
cornetón del monte – *Nicotiana glauca, Solanum erianthum, S. verbascifolium*
cornezuelo – *Acacia hindsii, A. spadicigera, Claviceps purpurea, Indigofera suffruticosa, Lotus corniculatus*
cornezuelo de la playa – *Acacia macracantha*
cornezuelo del centeno – *Claviceps purpurea*
cornicabra – *Lithospermum arvensis, Pistacia terebinthus*
cornijuelo – *Amelanchier alnifolia, A. ovalis*

cornillo – *Amelanchier alnifolia, A. vulgaris*
corocito – *Cyperus esculentus*
corojo – *Acrocomia crispa, A. mexicana, Ananas comosus*
corojo de Guinea – *Elaeis guineensis*
corona – *Antigonon leptopus, Xylosma characantha*
corona de Cristo – *Canotia holacantha, Castela emoryi, Euphorbia fulgens, E. milii, Koeberlinia spinosa* ssp *tenuispina, Passiflora, Psorothamnus spinosus*
corona de espinas – *Euphorbia milii*
corona de la reina – *Antigonon leptopus, Petrea arborea*
corona del indio – *Mammillaria petrophila*
corona del marques – *Crinum zeylanicum*
corona de Moctezuma – *Cornus excelsa*
corona de rey – *Chrysanthemum segetum, Lotus corniculatus, Saxifraga longifolia*
corona de San Juan – *Artemisia vulgaris*
corona de San Pedro – *Cornus excelsa, C. florida*
corona de santo – *Xylosma flexuosum*
corona de sol – *Lantana camara*
coronado – *Callistephus chinensis*
corona imperial – *Amorphophalus rivieri, Fritillaria imperialis*
corona nupcial – *Spiraea arguta*
corona real – *Helianthus annuus*
coronaria – *Anemone coronaria, Lychnis coronaria*
coronaria coriácea – *Lychnis coronaria*
coronaria de Júpiter – *Lychnis flos-jovis*
coronaria escarlata – *Lychnis chalcedonica*
coronaria viscosa – *Lychnis viscaria*
coronilla – *Antigonon leptopus, Berlandiera lyrata* var *macrophylla, Coronilla varia, Gaillardia pinnatifida, Myroxylon flexuosum, Tagetes lucida, Xylosoma flexuosum*
coronilla blanca – *Melilotus alba*
coronilla de fraile – *Globularia alypum*
coronilla de rey – *Globularia alypum*
coronilla real – *Globularia alypum*
coronita – *Centaurea cyanus*
coronita del sol – *Lantana camara*
corónopo dídimo – *Coronopus didymus*
corónopo procumbente – *Coronopus squamatus*
corosol – *Annona muricata, Rollinia pulchrinervis*
corosol espinoso – *Annona muricata*
corozo – *Acrocomia mexicana, Aiphanes caryophyllea, Attalea butyracea, Orbignya cohune, Scheelia liebmannii, S. preussii*
corozo del Orinoco – *Aiphanes caryophylea*
corozo de sábana – *Orbignya cohune*
corozo oleifera – *Elaeis melanococca*
corpionchi – *Galphimia glauca*
corpus – *Jacquinia macrocarpa* ssp *pungens, Laelia speciosa, Magnolia schiedeana*
corpus de cacique – *Magnolia schiedeana*
cortadera – *Cortaderia selloana, Geonoma congesta, Gynerium argenteum, Paspalum plicatulum, P. virgatum*
cortadillo – *Yucca elata*
cortejo – *Hibiscus mutabilis*
corteno – *Acacia cymbispina*
corteza amarilla – *Tabebuia chrysantha, Terminalia amazonia*
corteza de cinchona – *Cinchona officinalis*
corteza de Honduras – *Sweetia panamensis*
corteza de Jamaica – *Quassia amara*
corteza de Jojutla – *Hintonia latiflora*
corteza de pimienta – *Drimys winteri*
corteza de quina – *Cinchona officinalis*
corteza fuerte – *Desmodium canum*
corteza peruana – *Cinchona ledgeriana*
cortina – *Mesembryanthemum crystallinum*
cortinchi – *Galphimia glauca*
corti tocuz – *Quercus martinezii*
cortopico – *Drypetes gentryi*
corúngoro – *Zizyphus amole*
corralito – *Agaricus campestris, Boletus edulis*
corralito café – *Tricholoma vaccinum*
correa – *Correa speciosa*
corredor – *Eryngium campestre*
correhuela – *Cocculus diversifolius, Convolvulus arvensis, Gaudichaudia mucronata, Ipomoea purpurea, Polygonum aviculare, Struthanthus quercicola*
correhuela de la Riviera – *Convolvulus sabatius*
correhuela de las doce – *Ipomoea sinuata*
correhuela de los ánimos – *Polygonum arenastrum*
correhuela de los prados – *Convolvulus arvensis*

correhuela hembra – *Hippuris vulgaris*
correhuela itálica – *Convolvulus althaeoides*
correhuela lechosa – *Cynanchum acutum, C. utahense*
correhuela loca – *Convolvulus arvensis*
correhuela mayor – *Calystegia sepium*
correhuela tricolor – *Convolvulus tricolor*
correhuela turca – *Convolvulus cneorum*
corremiento – *Lygodium venustum*
correosa – *Rhus microphylla*
corrigia – *Digitalis obscura*
corrimiento – *Dioscorea, Piper auritum*
corriosa – *Zanthoxylum fagara*
cósahui – *Calliandra eriophylla, Krameria*
cósahui del norte – *Calliandra eriophylla*
cósahui del sur – *Krameria grayi, K. parvifolia* var *glandulosa*
coscoja – *Quercus coccifera*
coscolina – *Rhamnus alaternus, R. purshiana*
coshu – *Canna indica*
cosmos – *Cosmos bipinnatus*
cosquelite – *Momordica charantia*
costarrica – *Alibertia edulis, Rinorea guatemalensis*
costeño – *Acacia cochlicantha, Vigna unguiculata*
costeño amarillo – *Capsicum annuum*
costicoatenoxóchitl – *Cattleya*
costicoatxóchitl – *Cattleya*
costiczápotl – *Pouteria campechiana*
costilla de Adán – *Monstera deliciosa*
costilla de caballo – *Canavalia ensiformis*
costilla de león – *Ceratozamia latifolia, C. mexicana*
costilla de vaca – *Atriplex canescens*
costo – *Saussurea lappa, Serratula tinctoria*
costo hortense – *Chrysanthemum balsamita*
cota – *Thelesperma gracile, T. longipes, T. megapotamicum*
cotalulu – *Parthenice mollis*
cotasula – *Chenopodium album*
cotea – *Cottea pappophoroides*
cotinus – *Cotinus coggygria*
cotolahue – *Abutilon*
cotomate – *Belotia mexicana*
cotoneaster – *Cotoneaster horizontalis, C. salicifolius*
cotoneaster en espina de pescado – *Cotoneaster horizontalis*
cotoneaster de hojas pequeñas – *Cotoneaster microphyllus*
cotootaj – *Fouquieria columnaris*
cotorrera de playa – *Heliotropium curassavicum*
cotsnemax – *Heliotropium angiospermum*
cotufa – *Cyperus esculentus, Helianthus tuberosus*
cótula sudafricana – *Cotula coronopifolia*
couis – *Crescentia cujete*
courbaril – *Hymenaea courbaril*
cousano – *Coursetia glandulosa*
covena – *Dichelostemma capitatum*
coveria – *Dichelostemma capitatum* ssp *pauciflorum*
coyacal – *Carludovica palmata*
coyacate – *Goldmania foetida*
coyahual – *Tabebuia impetiginosa*
coyame-catl – *Costus spicatus*
coyamol – *Anthericum torreyi*
coyán – *Nothofagus obliqua*
coyapa – *Sambucus mexicana*
coyaye – *Gutierrezia sarothrae*
coyó – *Persea schiedeana*
coyol – *Acrocomia aculeata, A. mexicana, A. vinifera, Begonia gracilis, Canna indica, Orbignya cohune, Scheelia liebmannii, Yucca*
coyol baboso – *Acrocomia mexicana*
coyol cimarrón – *Spiranthes aurantiaca*
coyol de sábana – *Orbignya cohune*
coyolillo – *Bactris mexicana, Cyperus rotundus, Reinhardtia gracilis*
coyolito – *Beaucarnea inermis, Reinhardtia gracilis*
coyolizan – *Buddleja americana*
coyol real – *Attalea butyracea, Roystonea regia, Scheelia liebmannii*
coyol redondo – *Acrocomia mexicana*
coyolxóchitl – *Bomarea acutifolia, B. hirtella, Tabebuia pentaphylla*
coyonostle – *Cylindropuntia imbricata*
coyonostli – *Cylindropuntia imbricata*
coyonoxtle – *Cylindropuntia tunicata, Opuntia engelmannii*
coyoquelite – *Piper auritum*
coyor – *Aiphanes acanthophylla, A. minima*
coyotillo – *Belotia mexicana, Karwinskia humboldtiana, Opuntia azurea*
coyotomate – *Vitex mollis*

coyuacate – *Inga spuria*
coyul – *Sapindus saponaria*
coyur – *Annona glabra*
cozahuatl – *Quercus mexicana*
cozahuico – *Sideroxylon tepicense*
cozahuique – *Ficus padifolia*
cozcanantzi – *Exogonium bracteatum*
cozolmecatl – *Smilax mexicana*
cozquelite – *Erythrina*
cozticcoatzontecomaxóchitl – *Stanhopea oculata, S. wardii*
cozticoatzontecoxóchitl – *Cattleya*
cozticpatli – *Thalictrum hernandezii*
coztictepetlacuxóchitl – *Cattleya citrina*
cozticxócotl – *Spondias mombin, S. purpurea*
cozticzacatzacuxóchitl – *Govenia superba*
cozticzapotl – *Lucuma salicifolia*
coztomate – *Physalis coztomatl, P. pubescens*
coztomate amarillo – *Physalis coztomatl*
coztomatl – *Physalis coztomatl, P. philadelphica, P. pubescens*
cozulmécatl – *Smilax bona-nox*
crambe marítima – *Crambe maritima*
crameria – *Krameria*
crásula – *Crassula lycopodioides*
crásula acuática – *Crassula aquatica*
crásula musgosa – *Crassula tillaea*
cratego – *Crataegus*
cremón – *Thespesia populnea*
crepis capilar – *Crepis capillaris*
crepis setoso – *Crepis setosa*
crepis vesiculoso – *Crepis vesicaria*
cresón – *Lepidium sativum, Rorippa nasturtium-aquaticum*
crespa – *Salvia ballotaeflora*
crespilla – *Morchella esculenta*
crespón – *Lagerstroemia indica*
crespón de China – *Lagerstroemia indica*
cresta de gallo – *Anthurium andraeanum, Castilleja arvensis, Celosia, Cephaëlis tomentosa, Dorstenia contrayerba, Gladiolus hortulanus, Jacobinia umbrosa, Sporobolus airoides*
cresta de pavo – *Adenanthera pavonina*
cresta marina – *Crithmum maritimum*
criadilla – *Tuber cibarium*
criadilla de tierra – *Tuber melanosporum*
crinado – *Opuntia pilifera*
criollo – *Musa sapientum*
cripsis alopecuro – *Crypsis alopecuroides*
cripsis falso-junco – *Crypsis schoenoides*
criptomeria japonesa – *Cryptomeria japonica*
crisantemo – *Chrysanthemum*
crisantemo anual – *Chrysanthemum carinatum*
crisantemo carinado – *Chrysanthemum carinatum*
crisantemo coronado – *Chrysanthemum coronarium*
crisantemo de las mieses – *Chrysanthemum segetum*
crisantemo gigantesco – *Chrysanthemum maximum*
crisantemo máximo – *Chrysanthemum maximum*
crisantemo segetal – *Chrysanthemum segetum*
crisantemo silvestre – *Chrysanthemum coronarium*
cristalillo – *Mesembryanthemum crystallinum, Vallesia glabra*
cristalino – *Mesembryanthemum crystallinum*
croco – *Crocus sativus*
croco indio – *Zephyranthes candida*
crosne del Japón – *Stachys sieboldii, S. tubifera*
crotalaria – *Crotalaria juncea, C. mucronata*
croto – *Codiaeum variegatum*
crotón – *Codiaeum variegatum, Croton tiglium*
crucecillo – *Fraxinus trifoliata, Psychotria nervosa, Randia, Vallesia glabra*
crucecillo de la sierra – *Randia laevigata*
crucecito – *Randia*
crucerilla – *Condalia globosa* var *pubescens*
crucero – *Randia, Warszewiczia coccinea*
cruceta espinuda – *Pisonia aculeata*
crucetillo – *Eupatorium odoratum, Parmentiera aculeata*
cruceto – *Randia*
crucianella angustifolia – *Crucianella angustifolia*
crucianella de hojas estrechas – *Crucianella angustifolia*
cruciata – *Gentiana cruciata*
crucillo – *Condalia globosa* var *pubescens, Randia, Ziziphus obtusifolia, Z. parryi*
crucillo blanco – *Ziziphus obtusifloia*
crucito – *Ageratina, Randia*
crujía – *Digitalis obscura*
crupina común – *Crupina vulgaris*

cruxiv – *Tectaria heracleifolia*
cruz – *Sprekelia formosissima*
cruz de Jerusalén – *Lychnis chalcedonica*
cruz del matrimonia – *Pereskia lycnidiflora*
cruz de Malta – *Ixora coccinea, Lychnis chalcedonica*
cruz de Malta alpina – *Lychnis viscaria*
cruz de Malta marina – *Silene vulgaris*
cruz de Malta musgosa – *Silene acaulis*
cruz de pasto – *Eruca vesicaria* ssp *sativa*
cruz espina – *Pisonia aculeata*
cuaba blanca – *Amyris balsamifera*
cuaba de la costa – *Amyris balsamifera*
cuacicitlal – *Amanita muscaria*
cuachalalate – *Amphipterygium adstringens, A. glaucum, Dioscorea convolvulacea*
cuachalate – *Caesalpinia cacalaco*
cuachichile – *Cedrela poblensis*
cuachile – *Loeselia mexicana*
cuachilote – *Parmentiera edulis*
cuachipile – *Alvaradoa amorphoides*
cuaguayote – *Leucopremna mexicana*
cuahtzitzicaztli – *Urera baccifera, U. caracasana*
cuahuaxtli – *Sesamum indicum*
cuahuayote – *Carica mexicana*
cuahuilotillo – *Waltheria americana*
cuaicuastle – *Ceanothus coeruleus*
cuajachote – *Bixa orellana*
cuajaleche – *Galium, Ipomoea alba*
cuaje – *Cephalanthus salicifolius*
cuajilote – *Bombax ellipticum, Parmentiera aculeata, P. edulis, Pseudobombax palmeri*
cuajilotillo – *Aralia humilis*
cuajinicuil – *Inga spuria*
cuajinicuil silvestre – *Inga spuria*
cuajiote – *Bursera aptera, B. fagarioides*
cuajiote amarillo – *Bursera fagarioides, B. odorata*
cuajiote blanco – *Bursera fagarioides, B. microphylla, B. odorata*
cuajiote colorado – *Bursera fagarioides, B. lancifolia, B. microphylla*
cuajiote verde – *Bursera fagarioides, B. odorata*
cuamecate – *Antigonon leptopus, Petrea arborea, Serjania mexicana, S. racemosa*
cuamecate azul – *Petrea arborea*
cuamecatl – *Antigonon leptopus, Serjania mexicana*
cuamirro – *Sideroxylon tepicense*
cuampach – *Tillandsia usneoides*
cuamúchil – *Pithecellobium dulce*
cuanaxunaxe – *Croton ciliato-glandulosus*
cuanenepili – *Passiflora incarnata, P. jorullensis*
cuapetate – *Belotia mexicana*
cuapinol – *Hymenaea courbaril*
cuapinole – *Hymenaea courbaril*
cuapopolchi – *Piqueria trinervia*
cuashilaca – *Phoradendron tomentosum*
cuasia – *Quassia amara*
cuasia amarga – *Quassia amara*
cuasipinsé – *Hamelia erecta*
cuatantillo – *Mimosa albida*
cuatatachi – *Hura polyandra*
cuate – *Croton draco, Eysenhardtia polystachya*
cuatelolote – *Phyllanthus acidus*
cuatequil – *Zea mays*
cuatlatlatzin – *Hura polyandra*
cuatlatlaya – *Bocconia frutescens*
cuatle – *Eysenhardtia polystachya*
cuatolotlanenzi – *Iostephane heterophylla*
cuatomate – *Lycopersicon esculentum* var *cerasiforme*
cuatrofilos – *Musa acuminata*
cuatztli – *Chenopodium nuttaliae*
cuaucoyotli – *Acrocomia mexicana*
cuauchichic – *Garrya flavescens, G. laurifolia, G. ovata*
cuauchichil – *Loeselia mexicana*
cuauchichili – *Bocconia frutescens*
cuauchile – *Castilla elastica*
cuauhpinoli – *Hymenaea courbaril*
cuauhtzapotl – *Annona cherimola, A. reticulata, A. squamosa*
cuauhyetl – *Nicotiana tabacum*
cuauhzontli – *Chenopodium nuttalliae*
cuaulote – *Guazuma ulmifolia*
cuaulotillo – *Wlatheria americana*
cuaumecate – *Combretum farinosum*
cuaumecate chayote – *Cryptostegia grandiflora*
cuaumóchitl – *Pithecellobium dulce*
cuaunacaxtle – *Enterolobium cyclocarpum*
cuauolotl – *Guazyma ulmifolia*
cuautecomate – *Crescentia alata*

cuauteteco – *Arnica, Heterotheca inuloides*
cuautecomate – *Crescentia cujete*
cuautecomatl – *Crescentia alata*
cuautlahuizquílitl – *Cirsium ehrenbergii*
cuauxúchil – *Plumeria rubra*
cuaxilote – *Parmentiera edulis*
cuaxtololote – *Andira inermis*
cuayote – *Leucopremna mexicana*
cubana – *Roystonea elata*
cubanita – *Lobelia cardinalis* ssp *graminea*
cubata – *Acacia cymbispina*
cubé – *Lonchocarpus utilis*
cubeba – *Piper cubeda*
cubilete – *Nuphar luteum*
cuca – *Acacia farnesiana, Cyperus esculentus, Mimosa, Pseudophoenix sargentii*
cucabit – *Acacia pennatula*
cuca de árbol – *Mimosa distachya*
cucaracha – *Dodonea viscosa*
cucarachita – *Tradescantia zebrina*
cucarda – *Hibiscus rosa-sinensis*
cucoztli – *Thalictrum hernandezii*
cucuato – *Arctostaphylos polifolia*
cuculmeca – *Smilax officinalis*
cucurita – *Maximiliana maripa*
cucurucho – *Zantedeschia aethiopica*
cucut – *Allium cepa*
cucutchom – *Atamosco citrina*
cucharacho – *Cnidoscolus elasticus*
cuchareta – *Cochlearia officinalis*
cucharilla – *Acacia cymbispina, Dasylirion, Eichhornia crassipes, Sagittaria macrophylla, Trichilia hirta*
cucharita – *Acacia cymbispina, Dasylirion*
cucharo – *Acacia cymbispina*
cuchilita – *Solanum nigrum*
cuchilxiu – *Asclepias curassavica*
cuchipen – *Axonopus affinis*
cuchito – *Solanum mammosum*
cuchu – *Porophyllum gracile*
cuchucua – *Andira inermis*
cuchunuc – *Gliricidia sepium*
cuecuezquic – *Mucuna pruriens*
cuencuentz – *Clinopodium laevigatum*
cuencuentzpatli – *Satureja macrostema*
cuenta de oro – *Duranta erecta*
cuentas de Santa Elena – *Cyperus odoratus*
cueramo – *Cordia*
cuerda de violín – *Cuscuta americana*
cuerillo – *Ulmus mexicana*
cuernecillo – *Lotus corniculatus*
cuernecillo de centeno – *Claviceps purpurea*
cuernecillo del maíz – *Ustilago maydis*
cuernecillo de los pantanos – *Lotus uliginosus*
cuernecillo grande – *Lotus uliginosus*
cuernecillo pequeño – *Lotus angustissimus*
cuernero – *Proboscidea parviflora*
cuernito – *Acacia spadicigera, Proboscidea*
cuernito de hormiguero – *Proboscidea fragrans*
cuerno – *Cryptostegia grandiflora, Disocactus flagelliformis, Ustilago maydis*
cuerno de alce – *Platycerium bifurcatum*
cuerno de cabra – *Dodonaea viscosa*
cuerno de ciervo – *Platycerium*
cuernos del diablo – *Proboscidea althaeifolia, P. arenaria*
cuernos del toro – *Acacia spadicigera, Proboscidea fragrans*
cuero – *Ulmus mexicana*
cuero de sapo – *Ilex nitida*
cuero de vaca – *Strychnos tepicencis*
cuesa – *Bernoullia flammea*
cuesco de lobo – *Lycoperdon perlatum, L. pratense, Vascellum pratense*
cuesco grande de lobo – *Calvatia gigantea*
cuete – *Ruellia intermedia*
cuetito – *Ruellia nudiflora* var *glabrata*
cuetzé – *Alvaradoa amorphoides*
cueza – *Sechium edule*
cufea – *Cuphea ignea*
cuhrucumin – *Tagetes lucida*
cuhuech – *Panicum trichoides*
cuiba – *Oxalis tuberosa*
cuica – *Senna oxyphylla*
cuica de gato – *Senna oxyphylla*
cuiche – *Cedrela mexicana*
cuija – *Opuntia dillenii, O. engelmannii* var *lindheimeri, O. rastrera*
cuil de agua – *Inga edulis, I. spuria*
cuilimbuca – *Andira inermis*
cuilón – *Mimosa distachya*
cuilotillo – *Ageratina*
cuimic – *Piqueria trinervia*
cuipuy – *Jatropha curcas*
cuiro – *Crescentia alata*
cuitlacoche – *Ustilago maydis*
cuitlaxóchitl – *Euphorbia pulcherrima*
cuitziquiendas – *Tropaeolum majus*

cuiyul – *Oxalis divergens*
cujete – *Crescentia cujete*
cují negro – *Acacia macracantha*
cují yaque – *Prosopis juliflora*
cuju – *Trichilia hirta*
culantrillo – *Adiantum, Ginkgo biloba, Mollugo verticillata, Scoparia dulcis*
culantrillo bastardo – *Asplenium trichomanes*
culantrillo de México – *Adiantum concinnum, A. tenerum*
culantrillo de ojo de agua – *Adiantum capillus-veneris*
culantrillo de pozo – *Adiantum capillus-veneris, A. tenerum, A. trapeciforma*
culantrillo escamoso – *Ceterach officinarum*
culantro – *Coriandrum sativum, Eryngium foetidum*
culantro de zopilote – *Ciclospermum leptophyllum*
culash – *Brassica oleracea* var *tronchuda*
culche – *Cedrela mexicana*
culebra – *Serjania mexicana*
culebrera – *Dracunculus vulgaris*
culebrilla – *Arisaema dracontium, Dracunculus vulgaris*
culgro – *Tricholoma columbetta*
culhua – *Opuntia tuna*
culimziz – *Trichilia hirta*
culinché – *Astronium graveolens*
cul-tok – *Thrinax wendlandiana*
culumate – *Avicennia germinans*
culuncuahui – *Acacia millefolia*
cum – *Cucurbita moschata*
cúmaro – *Celtis reticulata*
cumarú – *Dipteryx odorata*
cumbacuare – *Nymphaea odorata*
cumbera – *Cylindropuntia bigelovii*
cumbili – *Thevetia ovata*
cumbro – *Celtis, Pisonia aculeata*
cúmero – *Celtis pallida*
cumicán – *Stemmadenia donnell-smithii*
cumquat – *Forestiera angustifolia*
cuna de Moisés – *Spathiphyllum wallisii*
cuna de Moisés cimarrona – *Aglaonema modestum*
cunda – *Plumeria rubra*
cundaria – *Spondias purpurea*
cundeamor – *Asarina barclaiana, Cobaea scandens, Ipomoea quamoclit, Momordica balsamina, M. cochinchinensis, M. charantia, Quamoclit coccinea*
cundeamor chico – *Momordica balsamina*
cundeamor de Yucatán – *Momordica charantia*
cundeamor grande – *Momordica charantia*
cundemba – *Sambucus mexicana*
cunduacán de Túxtepec – *Thevetia ovata*
cungo – *Celtis pallida*
cunicho – *Sabadilla officinalis, Schoenocaulon officinale*
cunilago – *Pulicaria dysenterica*
cuningamia – *Cunninghamia lanceolata*
cuningamia de la China – *Cunninghamia lanceolata*
cuña – *Clitoria terneata*
cuomecaxot – *Vanilla planifolia*
cupamu – *Bambusa vulgaris, Guadua aculeata*
cupana – *Cupania americana*
cupanda – *Persea americana*
cupandilla – *Cyrtocarpa procera*
cupape – *Cordia boissieri, C. dodecandra*
cupayaya – *Entada phaseoloides*
cupia – *Sideroxylon laetevirens, S. persimile*
cupú – *Spondias mombin*
cupuassú – *Guazuma grandiflora*
cuquet – *Acacia pennatula*
cúralo todo – *Lobelia*
curaqua – *Haematoxylon brasiletto*
curare – *Chondodendron tomentosium, Strychnos toxifera*
curarina – *Cissampelos pareira, Sansevieria guineensis, S. roxburghiana*
cúrcuma – *Curcuma domestica, C. zedoaria*
cúrcuma javanesa – *Curcuma xanthorrhiza*
cúrcuma larga – *Curcuma domestica*
curguatón – *Aralia humilis*
curi – *Sesbania macrocarpa*
curicua – *Plumbago pulchella*
curuba – *Passiflora mollissima*
curuba de Castilla – *Passiflora mollissima*
curucumín – *Tagetes*
curujey – *Tillandsia recurvata*
curutziti – *Tropaeolum majus*
curroñé – *Amelanchier ovalis, A. vulgaris*
curry – *Murraya koenigii*
cusa – *Goldmania foetida*
cusabi – *Prunus capuli*
cuscuta – *Cuscuta*
cuscuta de cáliz dorado – *Cuscuta approximata*

cuscuta del campo – *Cuscuta pentagona*
cuscuta glandulosa – *Cuscuta pentagona*
cusi – *Quercus albocincta, Q. durifolia, Q. hypoleucoides, Q. mcvaughii*
cusi prieto – *Quercus hypoleucoides*
cusi saucillo – *Quercus viminea*
cútacua – *Solandra maxima, S. nitida*
cuta guícori – *Colubrina triflora*
cutiperí – *Cordia sebestena*
cutz – *Nicotianum tabacum*
cutzi – *Stenorrhynchus cinnabarinus*
cutzis – *Neottia cinnabarina, Spiranthes aurantiaca, S. cinnabarinus, S. polyantha*
cuxub – *Bixa orellana*
cuxubcan – *Rivina humilis*
cuyabra – *Lagenaria siceraria*
cyperus – *Cyperus papyrus*

chaak – *Maranta arundinacea*
chabacano – *Prunus armenica*
chabecté – *Pterocarpus officinalis*
chaca – *Bursera simaruba*
chacalate – *Ficus radulina*
chacalhaaz – *Mammea americana*
chacalí – *Calacophyllum candidissimum*
chacam – *Hylocereus undatus*
chacamalxóchitl – *Caesalpinia pulcherrima*
chacámecua – *Cirsium mexicanum*
chacana – *Aechmea bracteata*
chacanalkak – *Asclepias curassavica*
chacara – *Cassia fistula*
chacasúchitl – *Caesalpinia pulcherrima*
chacate – *Eriogonum tenellum, Krameria*
chaccanahzihi – *Aechmea bracteata*
chacic – *Capsicum annuum*
chacilxiu – *Boerhavia coccinea*
chackuyche – *Bombax ellipticum*
chacloco – *Hamelia erecta*
chaclolo – *Myrica mexicana*
chaclomacal – *Antigonon leptopus*
chacmol – *Gomphrena globosa*
chacmolché – *Erythrina americana*
chacmuc – *Rauwolfia hirsuta*
chacnicté – *Plumeria rubra*
chaco – *Acanthocereus tetragonus*
chacolol – *Myrica mexicana*
chacoop – *Annona purpurea*
cha-cox – *Trophis racemosa*
chacsis – *Jacquinia aurantica*
chacte – *Caesalpinia platyloba*
chacté – *Sweetia panamensis*
chactsán – *Tradescantia discolor*
chactsinkin – *Caesalpinia pulcherrima*
chactzitz – *Salvia coccinea*
chactzotzmacal – *Alocasia cuprea, A. macrorrhiza*
chacua – *Randia echinocarpa*
chacual – *Tigridia pringlei*
chacxciu – *Psittacanthus calyculatus*
chaczabacnicte – *Plumeria rubra*
chaczuum – *Tithonia*
chacha – *Acanthocereus tetragonus*
chachaca – *Drimys winteri, Eriogonum rotundifolium, Prosopis*
chachahua – *Phoradendron*
chachalcahuite – *Caesalpinia pulcherrima*
chachalhaas – *Lucuma mammosa*
chachaquila – *Oxytropis lambertii*
chachi – *Parmentiera edulis*
chá de bugre – *Cordia salicifolia*
chaga – *Nelumbo lutea*
chaguar – *Bromelia lasiantha*
chaguaramo – *Oreodoxa oleracea*
chahuacho – *Cestrum dumentorum*
chahuiqui – *Agave multifilifera*
chaile – *Parthenium hysterophorus*
chajkal – *Leucaena esculenta, L. glauca*
chaklikin – *Tabernaemontana alba*
chak-suuk – *Aristida ternipes*
chakuyché – *Bombax ellipticum*
chala – *Ficus goldmanii*
chalahuite – *Inga edulis, I. spuria*
chalalá – *Caesalpinia cacalaco*
chalate – *Ficus*
chalate de burro – *Ficus maxima*
chalcahuite – *Casearia sylvestris*
chalcasúchil – *Caesalpinia pulcherrima*
chalché – *Pluchea odorata*
chalchuán – *Erigeron*
chalchupa – *Rauwolfia hirsuta*
chalmaite – *Pinus pseudostrobus*
chalmaite blanco – *Pinus montezumae*
chalote – *Allium cepa* var *aggregatum, A. c.* var *ascalonicum*
chaltoma – *Physalis sordida*
chalxócotl – *Psidium guajava*
chamadorea – *Chamaedorea*

chamadorea de Costa Rica – *Chamaedorea costaricana*
chamal – *Dioon edule*
chamalxóchitl – *Caesalpinia pulcherrima*
chambira – *Astrocaryum vulgare*
chamborote – *Phaseolus coccineus*
chamburo – *Erythrina glauca*
chamico – *Brugmansia suaveolens, Datura ferox, D. stramonium, D. wrightii*
chamis – *Batis maritima, Sesuvium portulacastrum, Ziziphus spathulata*
chamizo – *Adenostoma fasciculatum, A. sparsifolium, Ageratina, Allenrolfea occidentalis, Ambrosia deltoidea, A. dumosa, Artemisia tridentata, Atriplex, Baccharis salicifolia, Chrysothamnus nauseosus, Encelia farinosa, Erblichia odorata, Leucophyllum frutescens, Viguiera dentate*
chamizo amarillo – *Chrysothamnus nauseosus*
chamizo blanco – *Artemisia tridentata, Atriplex canescens, Chrysothamnus nauseosus*
chamizo cenizo – *Artemisia ludoviciana* ssp *albula, Atriplex*
chamizo colorado – *Adenostoma sparsifolium*
chamizo de vara colorada – *Adenostoma fasciculatum*
chamizo de vara prieta – *Adenostoma fasciculatum*
chamizo forrajero – *Ambrosia deltoidea*
chamizo ganado – *Artemesia tridentata*
chamizo hediondo – *Artemisia tridentata*
chamizo volador – *Salsola tragus*
chamol – *Caesalpinia pulcherrima, Colocasia esculenta*
chamolxóchitl – *Caesalpinia pulcherrima*
champes – *Rosa woodsii*
champiñón – *Agaricus bisporus* var *hortensis, A. campestris, Basidiomycetes*
champiñón anisado – *Agaricus silvicola*
champiñón chinesco – *Lentinus edodes*
champiñón de campo – *Agaricus campestris*
champiñón de Paris – *Agaricus bisporus* var *hortensis*
champiñón de pino japonés – *Tricholoma matsutake*
champiñón grande – *Agaricus arvensis, A. augustus, A. silvicola*
chan – *Hyptis suaveolens, Salvia retroflexa*
chanal – *Quercus crassifolia*
chanar – *Geoffroea decorticans*
chanate – *Ariocarpus retusus, Gliricidium sepium, Rhynchosia pyramidalis*
chanate pusi – *Rhynchosia precatoria, R. pyramidalis*
chancalá – *Canna indica*
chancaquilla – *Cenchrus longispinus, Lantana camara*
chancara xiutl – *Lantana camera*
chancarro – *Cecropia obtusifolia, C. peltata*
chancarro blanco – *Cecropia peltata*
cha'ncat – *Saccharum officinarum*
chancle – *Canna edulis, C. indica*
chanchanote – *Karwinskia humboldtiana*
changara – *Canna indica*
changuarica – *Bixa orellana*
changungo – *Byrsonima crassifolia*
chanita – *Malvaviscus arboreus*
chanixnuc – *Tribulus cistoides*
chantle – *Ariocarpus retusus*
chanuahko – *Aristolochia grandiflora*
chaoacocopin – *Porophyllum macrocephalum*
chapacolor – *Stegnosperma halimifolium*
chapai – *Chamaedorea graminifolia*
chapak – *Plumbago auriculata, P. scandens*
chapalhuacán – *Bocconia frutescens*
chapanillo – *Eysenhardtia spinosa*
chaparacoba – *Citharexylum scabrum*
chaparillo – *Tetracera volubilis*
chaparral – *Condalia obovata, Larrea tridentata, Ziziphus obtusifolia*
chaparral blanco – *Forestiera angustifolia*
chaparro – *Byrsonima coriacea, Castela texana, C. tortuosa, Curatella americana, Larrea divericata, Opuntia dillenii, Quercus*
chaparro amargoso – *Castela*
chaparro blanco – *Celtis iguana, Condalia hookeri*
chaparro de chinche – *Byrsonima crassifolia*
chaparro manteco – *Byrsonima crassifolia*
chaparrón – *Rheedia edulis*
chaparro prieto – *Acacia constricta, A. rigidula, Ziziphus obtusifolia*
chapaya – *Astrocaryum mexicanum, Bactris mexicana*
chaperla – *Astronium graveolens*
chaperno – *Ulmus mexicana*
chapiado – *Russula cyanoxantha*
chapizlillo – *Euphorbia fulgens*

chapiztle – *Pereskiopsis rotundifolia*
chapo – *Mimosa palmeri*
chapón – *Stemmadenia donnell-smithii*
chapona – *Stemmadenia donnell-smithii*
chapote – *Acacia farnesiana, Achras zapota, Casimiroa edulis, Diospyros californica, D. texana, Randia, Triumfetta goldmanii*
chapote amarillo – *Sargentia greggii*
chapote manzano – *Diospyros texana*
chapote prieto – *Diospyros ebenaster*
chapuiro – *Podocarpus*
chapul – *Lantana involucrata*
chapuliz – *Dodonaea viscosa*
chapuliztle – *Dodonaea viscosa*
chapulxóchitl – *Pedilanthus tithymaloides*
chapuz – *Helenium autumnale, H. mexicanum*
chaquira – *Ceanothus*
chaquirilla – *Ceanothus*
charagallo – *Alvaradoa amorphoides*
charahuesca – *Dahlia*
charaitzicua – *Justica spicigera*
charamusco – *Calliandra houstoniana*
chardonera – *Rubus idaeus*
charmilla – *Carpinus betulus*
charneca – *Pistacia lentiscus*
charresquillo – *Calliandra eriophylla, Pithecellobium leptophyllum, Quercus intricata*
chashi – *Leucaena glauca*
chata – *Catharanthus roseus*
chata de agua – *Pachyrrhizus erosus*
chatalhuic – *Senna laevigata*
chatet – *Omphalea oleifera*
chato – *Agave tequilana*
chaucha – *Phaseolus vulgaris, Pisum sativum*
chaucha turca – *Hibiscus esculentus*
chautle – *Ariocarpus fissuratus, A. retusus, Bletia campanulata, Cranichis speciosa*
chaveño – *Opuntia streptacantha*
chaya – *Bixa orellana, Cnidosculus aconitifolius, C. chayamansa, C. elasticus, C. urens*
chaya de Castilla – *Cnidoscolus chayamansa*
chaya mansa – *Cnidoscolus aconitifolius, C. chayamansa*
chaya pica – *Cnidoscolus aconitifolius*
chayote – *Sechium edule*
chayote de cerro – *Selaginella*
chayotera – *Sechium edule*
chayotillo – *Microsechium helleri, Sicyos angulatus, S. laciniatus, S. parviflorus*
chayotli – *Sechium edule*
chayte – *Crescentia cujete*
cha-yúk – *Solanum americanum*
chébula – *Terminalia chebula*
chechén – *Toxicodendron radicans*
cheflera – *Schefflera actinophylla*
chelele – *Inga spuria*
chelem – *Agave fourcroydes*
chem-chem – *Chamaedorea pinnatifrons*
chende – *Polaskia chende*
chengua – *Byrsonima crassifolia*
chepica – *Cynodon dactylon*
chepica alemana – *Agrostis capillaris*
cherahuí – *Acacia cymbispina, A. pringlei*
cheramusco – *Acacia angustissima*
cherémecua – *Cirsium mexicanum*
cherioni – *Sapindus saponaria*
cherva – *Ricinis communis*
chi – *Byrsonima crassifolia, Malpighia glabra*
chía – *Heteranthera inuloides, Hyptis emoryi, H. suaveolens, Malpighia glabra, Salvia*
chiabal – *Spondias mombin, S. purpurea*
chía cimarrona – *Salvia tiliaefolia*
chía commercial – *Salvia hispanica*
chía de Colima – *Hyptis suaveolens*
chía del campo – *Salvia hispanica, S. polystachya*
chía del monte – *Salvia hispanica*
chía gorda – *Hyptis suaveolens, Plantago major*
chía grande – *Hyptis suaveolens*
chian – *Salvia hispanica, S. chian*
chianpitzáhuac – *Salvia hispanica*
chiantzozolli – *Salvia hispanica, S. tiliaefolia*
chiat – *Chamaedorea graminifolia*
chibchel – *Struthanthus palmeri*
chib-kuúk – *Chamaedorea arenbergiana*
chica – *Salvia californica*
chicab – *Conostegia xalapensis*
chicalito de Túxtepec – *Astrocaryum mexicanum*
chicalote – *Argemone, Bocconia arborea, Romneya coulteri*
chicalote de árbol – *Bocconia arborea, C. frutescens*
chicalote grande – *Bocconia arborea*
chicalotl – *Argemone mexicana*

chicaltón – *Saracha procumbens*
chicam – *Pachyrhizus erosus*
chicata – *Flaveria*
chiceh – *Chrysophyllum mexicanum*
chicle – *Achras zapota*
chicle zapote – *Achras zapota*
chicnigui – *Capsicum microcarpum*
chico – *Achras zapota, Atriplex canescens, Lycium pallidum* var *oligospermum, Morisonia americana, Pachycereus weberi, Sarcobatus vermiculatus*
chicoche – *Goldmania foetida*
chicolcohuite – *Platanus lindeniana, P. mexicana*
chico muselo – *Hura polyandra*
chicopun – *Bursera grandifolia*
chicoria – *Acacia pringlei* ssp *california, Ambrosia ambrosioides, Bouvardia glaberrima, Cichorium intybus, Sonchus oleraceus, Taraxacum officinale*
chico silvestre – *Morisonia americana*
chicote – *Cryptostegia grandiflora*
chicote embarañada – *Lygodesmia juncea*
chicozapote – *Achras zapota, Morisonia americana*
chicozapote de hoja ancha – *Achras chicle*
chico zapotl – *Achras zapota*
chicuipo – *Gonolobus erianthus*
chicura – *Ambrosia*
chicura de la sierra – *Lepechinia hastata*
chicurilla – *Ambrosia ambrosioides, A. cordifolia, A. dumosa*
chícharo – *Cajanus cajun, Cicer arietinum, Lablab purpureus, Pisum sativum, Lathyrus*
chícharo de árbol – *Cajanus cajun*
chícharo de olor – *Lathyrus odoratus*
chícharo de vaca – *Vigna sinensis, V. unguiculata*
chícharo gandul – *Cajanus cajun*
chícharo inglés – *Pisum sativum*
chícharo tropical – *Vigna unguiculata*
chícharo verde – *Pisum sativum* var *macrocarpon*
chicharra de la hierba mate – *Fidicina mannifera*
chicharro – *Quercus corrugata, Q. skinneri*
chicharro grande – *Quercus ovadensis*
chichibe – *Corchorus siliquosus, Polaskia chichipe, Sida acuta*
chichibo – *Ambrosia confertiflora*
chichicahuite – *Garrya laurifolia*
chichicamole – *Cucurbita foetidissima, Microsechium helleri*
chichicaquilitl – *Sonchus oleraceus*
chichicastle – *Cnidoscolus multilobus, Lemna gibba, L. minor, Tragia nepetifolia, Urera baccifera, U. caracasana, Urtica dioica, U. urens, Wigandia caracasana*
chichicazillo – *Urera caracasana*
chichicoyote – *Cucurbita digitata, C. foetidissima, C. palmeri*
chichic-texcal-amatl – *Ficus petiolaris*
chichictzompotonic – *Gnaphalium conoideum*
chichicuáhuitl – *Garrya laurifolia, G. ovata*
chichichbe – *Malvastrum coromandelianum*
chichiguita – *Solanum mammosum*
chichihualcaxtli – *Tabernaemontana alba*
chichiltepetzacuxóchitl – *Laelia grandiflora*
chichilticxoitl – *Mentha pulegium*
chichillotl – *Argemone mexicana*
chichimán – *Amanita caesarea*
chichimeca – *Solanum cervantesii*
chichi-mecapatli – *Phaseolus heterophyllus*
chichioaquáhuitl – *Jatropha cardiophylla*
chichipale – *Myroxylon balsamum* var *pereirae*
chichipate – *Coutarea hexandra*
chichipe – *Polaskia chichipe, Sida acuta, S. rhombifolia*
chichipín – *Hamelia erecta*
chichiquelite – *Chenopodium ambrosioides, Lantana camara, Solanum*
chichiquis – *Piqueria trinervia*
chichita de burro – *Coryphantha pycnacantha*
chichitun – *Polaskia chichipe*
chichón – *Astrocaryum mexicanum*
chichona – *Citrus limetta*
chichón colorado – *Cupania americana*
chichonuda – *Pinus attenuata*
chichput – *Carica papaya*
chichuro – *Ambrosia confertiflora*
chidra – *Carludovicia palmata*
chifle de vaca – *Avicennia germinans*
chihuana cara – *Cinnamomum hartmanii, Persea podadenia*
chihue-chihue – *Bromelia pinguin*
chi'ini – *Wimmeria mexicana*
chi-it – *Acoelorraphe wrightii*
chijchijtli – *Maclura tinctoria*

chijol – *Erythrina*
chik – *Pachycereus pecten-aboriginum*
chikana – *Hechtia montana*
chiku – *Achras zapota*
chilaca seca – *Capsicum annuum*
chilacastle – *Lemna minor*
chilacayote – *Cucurbita ficifolia*
chilacayotl – *Cucurbita ficifolia*
chilacaxtli – *Azolla filiculoides*
chilacuate – *Styrax*
chilaile – *Capsicum annuum*
chilalaga – *Canna indica*
chilamate – *Ficus petiolaris, Sapium stylare, Thevetia ovata*
chilapate – *Tephrosia nicaraguensis*
chilayo – *Pachycereus marginatus*
chilca – *Baccharis salicifolia, Senecio salignus, Tecoma stans, Thevetia ovata, T. peruviana*
chilca silvestre – *Thevetia ovata*
chilcoztli – *Capsicum annuum*
chilcuán – *Erigeron*
chile – *Capsicum annuum*
chile achocolatado – *Capsicum annuum*
chile amash – *Capsicum microcarpum*
chile amomo – *Capsicum annuum* var *minimum*
chile ancho – *Capsicum annuum*
chile blanco – *Capsicum annuum*
chile burro – *Stevia serrata*
chile caballo – *Capsicum pubescens*
chile campana – *Capsicum annuum* var *grossum*
chile canario – *Capsicum pubescens*
chile carricillo – *Capsicum annuum*
chile cascabel – *Capsicum annuum*
chile cayena – *Capsicum fastigatum*
chile cera – *Capsicum annuum*
chile cimarrón – *Goldmania foetida*
chile colorado – *Capsicum annuum*
chile criollo – *Capsicum chinense*
chile cristal – *Capsicum annuum*
chile cuaresmeño – *Capsicum annuum*
chile cuernillo – *Capsicum annuum*
chile chilaca – *Capsicum annuum*
chile chilacate – *Capsicum annuum*
chile chilchote – *Capsicum annuum*
chile chipilquitín – *Capsicum annuum* var *minimum*
chile dátil – *Capsicum chinense*
chile de agua – *Capsicum annuum, C. annuum* var *grossum, Eclipta prostrata*
chile de árbol – *Capsicum annuum*
chile de Chiapas – *Capsicum annuum*
chile de espino – *Phoradendron*
chile de gato – *Spathiphyllum cochlearispathum*
chile del monte – *Capsicum annuum*
chile de pájaro – *Capsicum annuum* var *minimum*
chile de perro – *Capsicum annuum* var *minimum*
chile dulce – *Capsicum annuum* var *grossum*
chile enano – *Capsicum annuum* var *minimum*
chile gordo – *Capsicum annuum*
chile guajillo – *Capsicum annuum*
chile güero – *Capsicum annuum*
chile guindilla – *Capsicum frutescens*
chile habañero – *Capsicum chinense*
chile inayague – *Capsicum annuum*
chile jalapeño – *Capsicum annuum*
chile japonés – *Capsicum frutescens*
chile jarál – *Capsicum annuum*
chile joto – *Capsicum annuum*
chile jutiapa – *Capsicum pubescens*
chile largo – *Capsicum annuum*
chile manzano – *Capsicum pubescens*
chile max – *Capsicum annuum* var *minimum*
chile morrón – *Capsicum annuum* var *grossum*
chile mosquito – *Capsicum annuum*
chile mulato – *Capsicum annuum*
chile negro – *Capsicum annuum*
chile para deshebrar – *Capsicum annuum*
chile para rellenar – *Capsicum annuum*
chile pasilla – *Capsicum annuum*
chile perón – *Capsicum pubescens*
chile perro – *Senna atomaria*
chile piquín – *Capsicum annuun* var *minimum*
chile poblano – *Capsicum annuum*
chile puerco – *Amaranthus blitoides*
chile rocotillo – *Capsicum chinense*
chile roja – *Capsicum annuum*
chile rubio – *Capsicum annuum*
chile serrano – *Capsicum annuum*
chile silvestre – *Capsicum annuum* var *minimum*
chile tampiqueño – *Capsicum annuum*
chile travieso – *Capsicum annuum*

chile trompa – *Capsicum annuum*
chile verde – *Capsicum annuum*
chile verde del norte – *Capsicum annuum*
chile xcatic – *Capsicum annuum*
chilhuacle – *Capsicum annuum*
chilhuacle amarillo – *Capsicum annuum*
chilhuacle negro – *Capsicum annuum*
chilhuacle rojo – *Capsicum annuum*
chilibsuuk – *Sporobolus indicus*
chilicote – *Erythrina, Marah fabaceus, M. macrocarpus*
chilillo – *Agonandra racemosa, Asclepias curassavica, Casearia dolichophylla, C. nitida, Clematis, Drimys winteri, Ilex rubra, Malvaviscus arboreus, Oncidium carthaginensis, O. cebolleta, Plumbago pulchella, Polygonum, Rauwolfia hirsuta*
chilillo-ak – *Gaudichaudia mucronata*
chilillo del cerro – *Clematis dioica*
chilillo pequeño – *Capsicum annuum* var *minimum*
chilillosxiu – *Asclepias curassavica*
chilindrón – *Thevetia ovata, T. peruviana*
chilipiquín – *Capsicum annuun* var *minimum*
chilipuca – *Phaseolus coccineus*
chilito – *Coryphanta pycnacantha, C. retusa, Mammillaria, Oxalis dillenii*
chilito de biznaga – *Coryphanta potosiana*
chilmecate – *Malvaviscus grandiflorus*
chilmecatl – *Erigeron affinis*
chilocate – *Erythrina americana*
chilochuchi – *Bombax ellipticum*
chilpán – *Lactarius deliciosus*
chilpantlácotl – *Penstemon barbatus, P. campanulatus, P. imberbis*
chilpantlazolli – *Lobelia laxiflora*
chilpanxóchitl – *Lobelia laxiflora*
chilpanxohuitl – *Myrica mexicana*
chilte – *Cnidoscolus elasticus*
chilte colorado – *Cnidoscolus elasticus*
chiltepín – *Capsicum annuum, C. baccatum*
chiltepiquín – *Capsicum annuum* var *minimum, C. baccatum*
chilte rojo – *Cnidoscolus elasticus*
chiltote – *Pyrostegia venusta*
chi-lun-che – *Eugenia capuli*
chillazotl – *Argemone mexicana*
chilli – *Capsicum annuum*
chimaha – *Cymopterus fendleri, C. purpurascens*
chimalácatl – *Helianthus annuus*
chimalate – *Helianthus annuus*
chimay – *Acacia macracantha*
chimaya – *Cymopterus fendleri, C. purpurascens*
chimón – *Ficus glabrata*
chimu – *Cochlospermum vitifolium*
china – *Calendula officinalis, Smilax china*
chinaca – *Cirsium mexicanum, Eryngium longifolium*
chinacate – *Purshia mexicana*
china dulce – *Pastinaca sativa*
chincapín dorado – *Chrysolepis chrysophylla*
chincapino – *Castanea pumila*
chincuya – *Annona muricata, A. purpurea*
chinchayote – *Sechium edule*
chinchemalinche – *Caesalpinia pulcherrima*
chinchilla de hojas delgadas – *Tagetes tenuifolia*
chinchilla enana – *Tagetes minuta*
chinchin – *Lagenaria siceraria*
chinchín – *Astragalus*
chinchona – *Cinchona ledgeriana*
chindule – *Celtis pallida*
chinela – *Catasetum maculatum*
chiní – *Gossypium herbaceum*
chinine – *Persea schiedeana*
chinique – *Gossypium herbaceum*
chinita – *Dryopteris karwinskyana, Sonchus asper, S. oleraceus*
chinito – *Ficus padifolia, Impatiens walleriana*
chinito borrego – *Moringa oleifera*
chino – *Arctostaphylos polifolia, Coursetia glandulosa, Havardia mexicana, Impatiens balsamina, Randia*
chino azul – *Agave tequilana*
chino bermejo – *Agave tequilana*
chino grama – *Bouteloua breviseta*
chinola – *Polaskia chende*
chinonua – *Valeriana edulis*
chino pezmo – *Cyathea mexicana*
chinquisque – *Milleria quinqueflora*
chintule – *Cyperus articulatus, Eleocharis geniculata, Heliocarpus donnell-smithii, Sprekelia formosissima*
chintule grande – *Cyperus articulatus*
chiople – *Ageratina, Tithonia*

chiotilla – *Escontria chiotilla*
chipahuac – *Talauma mexicana*
chipe – *Pteridium aquilinum*
chipi – *Astrocaryum mexicanum*
chipil – *Crotalaria incana, C. longirostrata, C. pumila, Ficus cotinifolia*
chipile – *Coursetia glandulosa*
chipilillo – *Coursetia glandulosa*
chipilín – *Crotalaria longirostrata*
chipilín cimarrón – *Crotalaria incana, C. longirostrata*
chipil-max – *Crotalaria pumila*
chipiquelite – *Eruca vesicaria* ssp *sativa*
chiporro – *Mucuna pruriens*
chipuna – *Ambrosia psilostachya*
chiqueo – *Convolvulus*
chiqueo de monjas – *Quamoclit coccinea*
chiquete de embarañada – *Lygodesmia juncea*
chiquí – *Pachycereus pectin-aboriginum*
chiquilín – *Quercus conspersa*
chiquilla – *Rhipidocladum racemiflorum*
chiquinib – *Quercus crassipes, Q. crispipilis*
chiquiña – *Fouquieria campanulata, F. fasciculata*
chiquión – *Rhipidocladum racemiflorum*
chiquita – *Rhipidocladum racemiflorum*
chiquité – *Cnidoscolus elasticus*
chira – *Potamogeton pectinatus*
chírahui – *Acacia cymbispina, A. pennatula*
chirare – *Cordia sonorae*
chirca – *Croton ciliato-glandulosas, Thevetia peruviana, T. thevetioides*
chirgola – *Pleurotus eryngii*
chiribita – *Bellis perennis*
chiricote – *Cordia brevispicata, C. parvifolia*
chirigaita – *Cucurbita ficifolia*
chirimorriñón – *Annona cherimola*
chirimoyo – *Annona cherimola, A. reticulata, A. squamosa*
chirimoyo del Senegal – *Annona squamosa*
chirinole – *Stenocereus eruca*
chirivía – *Pastinaca sativa*
chirivía de agua – *Sium suave*
chirivía de azúcar – *Sium sisarum*
chirivía de hojas anchas – *Sium latifolium*
chirlobirlo – *Tecoma stans*
chironja – *Citrus sinensis* x *C. paradisi*
chiroui – *Acacia cymbispina*
chirrión – *Sapindus saponaria*
chirrioncillo – *Pereskiopsis aquosa*
chirrión del diablo – *Cryptostegia grandiflora*
chirrionera – *Cryptostegia grandiflora*
chischi – *Bactris mexicana*
chisis – *Chysis aurea*
chisma – *Portulaca mundula*
chisme – *Ipomoea quamoclit, Pilea microphylla, Quamoclit pinnata, Sedum acre, S. mexicanum, S. moranense*
chismito – *Ipomoea quamoclit*
chiso – *Perilla frutescens*
chisté – *Crataegus pubescens, Nyssa sylvatica*
chit – *Thrinax radiata, T. wendlandiana*
chit-cuuc – *Catasetum maculatum, C. punctatum*
chituna – *Polaskia chichipe*
chivatillo – *Purshia mexicana* var *stansburyana*
chivato grande – *Calliandra houstoniana*
chiverre – *Cucurbita ficifolia*
chivira – *Solanum umbellatum*
chixcanal – *Acacia spadicigera*
cho – *Ceiba pentandra*
choben-ché – *Trichilia hirta*
choboc – *Sideroxylon laetevirens*
choclo – *Zea mays* var *saccharata*
chocohuite – *Bursera simaruba*
chocoijoyo – *Talauma mexicana*
chocola – *Jarilla chocola*
chocolatillo – *Bourreria sonorae*
chocolín – *Erythrina*
chocon – *Astrocaryum mexicanum*
chochizapote – *Casimiroa edulis*
chocho – *Astrocaryum mexicanum, Hamelia erecta, Lupinus, Malvaviscus popullifolius, Sechium edule*
chocho blanco – *Geonoma interrupta*
chochohuisté – *Mimosa albida*
chochopali – *Cosmos sulphureus*
ch'oh – *Indigofera suffruticosa*
choh-kan – *Selenicereus donkelaari*
cholchonote – *Karwinskia humboldtiana*
cholulo – *Sapindus saponaria, Zizyphus pedunculata*
cholla – *Cylindropuntia*
cholla brincadora – *Cylindropuntia fulgida*
cholla del oso – *Cylindropuntia bigelovii*
cholla güera – *Cylindropuntia bigelovii*
cholla morada – *Cylindropuntia versicolor*

cholla pelona – *Cylindropuntia cholla*
chollita – *Cylindropuntia arbuscula, Mammillaria grahamii*
ch'om – *Bromelia karatas, B. pinguin*
chombrillo amargo – *Ecballium elaterium*
chonque – *Colocasia esculenta*
chonta – *Bactris gasipaes, Juania australis*
chontaduro – *Bactris gasipaes, Guilielma gasipaes*
chontalpa – *Asclepias curassavica*
chopera – *Rhamnus frangula*
chopo – *Mimosa palmeri, Populus*
chopo americano – *Populus deltoides*
chopo blanco – *Populus alba*
chopo común – *Populus nigra*
chopo de Italia – *Populus nigra* var *italica*
chopo del Canadá – *Populus* x *canadensis, P. deltoides*
chopo de Lombardia – *Populus nigra* var *italica*
chopo de Virginia – *Populus deltoides*
chopo euroamericano – *Populus* x *canadensis*
chopo lombardo – *Populus nigra* var *italica*
chopo mosquitero – *Populus nigra*
chopo negro – *Populus nigra*
chopo negro híbrido del Canadá – *Populus* x *canadensis*
chopo papelero – *Populus deltoides*
chopo piramidal – *Populus nigra* var *italica*
chopo temblón – *Populus tremula, P. tremuloides*
choquey – *Psidium sartorianum*
choreja – *Eichhornia crassipes*
choreque – *Lathyrus latifolius, L. odoratus*
chorequillo – *Phaseolus atropurpureus*
chorrito – *Cuitlauzinia pendula*
chorro – *Fuchsia arborescens*
chorro de oro – *Pyrostegia venusta*
chorro de sangre – *Croton draco*
chospo – *Bombax ellipticum*
chote – *Parmentiera edulis*
chotecuahuite – *Parmentiera edulis*
chouchoute – *Sechium edule*
choza de paja – *Imperata cylindrica*
chrimis-embra – *Eugenia capuli*
chual – *Boerhavia hirsuta, Chenopodium*
chual blanco – *Chenopodium album*
chual de cochi – *Chenopodium murale*
chual morado – *Chenopodium album, C. murale*
chub – *Gossypium*
chucaca – *Senecio candidissimus*
chucamay – *Styrax argenteus*
chúcata – *Prosopis juliflora, P. laevigata*
chucchu – *Maytenus laevis*
chucpte – *Persea schiedeana*
chucté – *Bombax ellipticum, Myroxylon balsamum* var *pereirae*
chucum blanco – *Pithecellobium dulce*
chuchaca – *Eriogonum tenellum*
chuchilitas – *Solanum nigrum*
chuchu – *Sechium edule*
chuchuhaso – *Maytenus laevis*
chuchum – *Ulmus mexicana*
chuchuni – *Ambrosia aptera*
chuchupate – *Arracacia brandegeei, Balsamorhiza sagittata, Lepechinia, Ligusticum grayi, L. porteri, Myroxylon balsamum* var *pereirae*
chudechu – *Theobroma cacao*
chufa – *Cyperus esculentus*
chuhucpakal – *Citrus sinensis*
chujquén – *Psittacanthus calyculatus*
chukuri to'oro – *Bursera microphylla*
chula – *Catharanthus roseus*
chulché – *Croton guatemalensis, C. niveus*
chulex-xiu – *Euphorbia heterophylla*
chulisi – *Salvia tiliaefolia*
chulté – *Bombax ellipticum*
chulul – *Apoplanesia paniculata, Lucuma viride*
chum – *Cochlospermum vitifolium*
chumari – *Fouquieria macdougalii*
chumbera – *Cylindropuntia arbuscula, Opuntia ficus-indica*
chumchintoc – *Guaiacum coulteri*
chumico – *Curatella americana*
chuminillo – *Trichilia hirta*
chuna – *Ficus*
chuncuy – *Orbignya cohune*
chupac – *Prochnyanthes viridescens*
chupachupa – *Combretum fruticosum*
chupamiel – *Anchusa azurea, Combretum farinosum, Cuphea llavea, Digitalis purpurea, Echium lycopsis*
chupamirto – *Malvaviscus arboreus*
chupandía – *Cyrtocarpa procera*

chupandilla – *Spondias mombin*
chupapaya – *Entada phaseoloides*
chuparrosa – *Anisacanthus thurberi, Bebbia juncea, Bouvardia glaberrima, Calliandra californica, Carlowrightia arizonica, Justicia, Loeselia mexicana*
chupirin – *Castilleja arvensis*
chupirin vitzaqua – *Castilleja canescens*
chupón – *Castilleja gracilis, Phoradendron, Psittacanthus sonorae*
chuqui – *Bromelia karatas*
churdón – *Rubus idaeus*
churumai – *Colubrina ferruginosa*
churuni – *Solanum grayi*
churutana – *Jacquinia aurantiaca*
churrimana – *Scorzonera hispanica*
chusnito – *Ficus padifolia, Podocarpus guatemalensis*
chuspatales – *Scirpus*
chuste – *Bombax ellipticum*
chutama – *Bursera*
chút-ik – *Piper peltatum*

dac kisin – *Laelia tibicinis*
dáctilo – *Dactylis glomerata*
dáctilo aglomerado – *Dactylis glomerata*
dáctilo apelotonado – *Dactylis glomerata*
dáctilo ramoso – *Dactylis glomerata*
dagame – *Calycophyllum candidissimum*
daí – *Desmanthus fruticosus*
dais – *Acacia villosa, Senna covesii*
daisillo – *Senna bahuinoides, S. covesii*
dalea – *Dalea hospes*
dalia – *Dahlia, Hidalgoa ternata*
dalia de árbol – *Dahlia excelsa, D. imperialis*
dalia del campo – *Dahlia coccinea*
dalia encarnada – *Dahlia coccinea*
dalia marina – *Coreopsis maritima*
dalia marítima – *Coreopsis maritima*
dalia morada – *Dahlia excelsa*
dalia plana – *Dahlia excelsa, D. imperialis*
dallis – *Paspalum dilatatum*
dama de noche – *Cestrum nocturnum, Epiphyllum oxypetalum, Ipomoea bona-nox*
dama de once horas – *Ornithogalium umbellatum*
damasco – *Prunus armenica*
damasco de otoño – *Rosa damascena* var *semperflorens*
damasquillo – *Prunus fremontii*
damasquina – *Tagetes erecta, T. patula*
damiana – *Chrysactinia mexicana, Dalea bicolor* var *orcuttiana, Isocoma menziesii, Turnera*
damiana de California – *Turnera diffusa*
damiana mexicana – *Turnera diffusa*
damianita – *Bigelovia venata, Ceanothus leucodermis, Chrysactinia mexicana, Turnera diffusa*
damzá – *Carya illinoensis*
dátil – *Phoenix canariensis, P. dactylifera, Yucca*
dátil cimarrón – *Yucca valida*
dátil de monte – *Yucca valida*
datilera – *Phoenix canariensis, P. dactylifera*
datilera canaria – *Phoenix canariensis*
datilera de las Canarias – *Phoenix canariensis*
datilera enana – *Phoenix roebelenii*
datilillo – *Agave datylio, Yucca valida*
datillo – *Agave datylio, Hechtia montana, Yucca schidigera, Y. valida*
datura estramonia – *Datura stramonium*
datura feroce – *Datura ferox*
datura inocua – *Datura wrightii*
dauco – *Aegopodium podagraria, Daucus carrota*
davidia – *Davidia involucrata*
day – *Acacia angustissima, Calliandra houstoniana*
dedalera – *Digitalis, Penstemon*
dedalera mexicana – *Tetranema mexicana*
dedito – *Batis maritima, Sedum allantoides*
dedo maloliente – *Hymenaea courbaril*
dedos de Dios – *Sedum*
dedos de muerto – *Rhipsalis baccifera*
dentabrón – *Dryopteris filix-mas*
dentaria – *Cardamine pratensis*
dentelaria – *Plumbago auriculata, P. scandens*
dentó – *Ustilago maydis*
deodar – *Cedrus deodara*
derrumbe – *Psilocybe caerulescens*
derrumbe negro – *Psilocybe yungensis, P. zapotecorum*
desmayo – *Salix babylonica*
desnudo – *Bursera simaruba*
desparramado dubiano – *Leptochloa dubia*
desparramadora – *Kickxia elatine*

desparramodo rojo – *Leptochloa filiformis, L. panicea* ssp *brachiata*
detze – *Prunus capuli, P. serotina*
devanador – *Sapindus saponaria*
diamela – *Jasminum sambac*
dicentra – *Dicentra formosa, D. spectabilis*
dicigoteca – *Schefflera elegantissima*
díctamo albo – *Dictamus albus*
díctamo blanco – *Dictamnus albus*
díctamo real – *Dictamnus albus*
dichitekisho – *Psilocybe caerulescens*
dichitonizé – *Psilocybe mexicana*
dichitune – *Polaskia chichipe*
Diego – *Laelia autumnalis*
Diego de noche – *Mirabilis jalapa*
diente de culebra – *Serjania mexicana*
diente de chucho – *Lobelia laxiflora*
diente de león – *Sonchus oleraceus, Taraxacum officinale*
diente de león hirto – *Leontodon taraxacoides*
diente de molino – *Karwinskia humboldtiana*
diente de perro – *Cynodon dactylon, Dioscorea esculenta, Erythronium dens-canis, Smilax mexicana*
diente de venado – *Sarcodon imbricatus*
diente de víbora – *Serjania mexicana*
difenbaquia – *Dieffenbachia*
difenbaquia manchada – *Dieffenbachia maculata*
digital – *Digitalis purpurea*
digital amarilla – *Digitalis lutea*
digital amarilla grande – *Digitalis grandiflora*
digital ferrogínea – *Digitalis ferruginea*
digital lanosa – *Digitalis lanata*
digital morada – *Digitalis purpurea*
digital negra – *Digitalis obscura*
digital purpúrea – *Digitalis purpurea*
digitaria ciliada – *Digitaria ciliaris*
digitaria común – *Digitaria sanguinalis*
digitaria glabra – *Digitaria ischaemum*
dildo – *Opuntia dillenii, Pilosocereus royenii*
dildo español – *Stenocereus fimbriatus*
dilenia – *Dillenia indica*
dimorfoteca – *Dimorphoteca sinuata*
di-nizé – *Psilocybe mexicana*
dionea – *Dionaea muscipula*
diosa – *Clematis ligusticifolia*
diospiro – *Diospyros virginiana*
diospiro loto – *Diospyros lotus*
dipsaco – *Dipsacus fullonum*
dipúa – *Cercidium floridum, C. microphyllum, C. peninsulare*
disciplina – *Bombax ellipticum*
disciplina de monja – *Polygonum orientale*
disciplinilla – *Rhipsalis baccifera*
dita – *Alstonia scholaris*
dividivi – *Caesalpinia cacalaco, C. coriaria*
dividivi de los Andes – *Caesalpinia spinosa*
dodecandra – *Cordia parvifolia*
dólico – *Lablab purpureus*
dólico de Goa – *Psophocarpus tetragonolobus*
dólico espárrago – *Vigna unguiculata* ssp *sesquipedalis*
dólico gigante – *Lablab purpureus, Vigna unguiculata* ssp *sesquipedalis*
dólicho – *Lablab purpureus*
dólicho rongai – *Lablab purpureus*
dominguilla – *Croton, Plumbago pulchella, Toxicodendron radicans, Urtica dioica*
donapé – *Krameria cytisoides*
doncel – *Pinus pinea*
doncella amarilla – *Mascagnia macroptera*
doncella de febrero – *Galanthus nivalis*
doncellita – *Bouvardia ternifolia*
dondiego de día – *Convolvulus tricolor, Fuchsia arborescens, Ipomoea purpurea, I. tricolor*
dondiego de día enano – *Convolvulus tricolor*
dondiego de noche – *Mirabilis jalapa*
dong quai – *Angelica sinensis*
donjuan – *Mirabilis jalapa, Nicotiana glauca*
donta – *Ustilago maydis*
doña Ana – *Coryphanta macromeris*
doña Juanita – *Cannabis sativa*
doradilla – *Cephaëlis tomentosa, Ceterach officinarum, Notholaena sinuata, Selaginella*
dormidera – *Eschscholzia californica, Mimosa pudica, Papaver somniferum, Senna tora*
dormilón – *Verbena, Psilocybe aztecorum, Rudbeckia*
dormilona – *Bellis perenis, Mimosa albida, M. pudica*
dorónico – *Doronicum plantagineum*
dos caras – *Miconia*
doucin – *Malus pumila* var *sylvestris*
douglasia de California – *Pseudotsuga macrocarpa*
douglasia de Oregón – *Pseudotsuga menziesii*

douglasia de piñas grandes – *Pseudotsuga menziesii*
douglasia verde – *Pseudotsuga menziesii*
draba – *Cardaria draba*
draba de los bosques – *Draba nemorosa*
draba de verdura perenne – *Draba aizoides*
draba primaveral – *Draba verna, Erophila verna*
dracena – *Cordyline australis, Dracaena deremensis*
dracena sander – *Dracaena sanderana*
dracena tricolor – *Dracaena marginata*
dracena verde – *Dracaena fragrans*
drago – *Croton draco, C. lecheri, Dracaena draco, Jatropha spathulata, Pterocarpus officinalis*
drago de Canarias – *Dracaena draco*
drago de Madagascar – *Dracaena marginata*
dragón – *Antirrhinum majus*
dragonal de Canarias – *Dracaena draco*
dragoncillo – *Antirrhinum majus, Artemisia dracunculus, Dracunculus vulgaris*
dragoneta – *Dracunculus vulgaris*
drago rojo – *Cordyline fruticosa*
dríada de ocho pétalos – *Dryas octopetala*
dríada octopétala – *Dryas octopetala*
drosera – *Drosera rotundifolia*
drosera de hojas circulares – *Drosera rotundifolia*
drosera mayor – *Drosera anglica*
duerme de noche – *Senna laevigata*
dulcamara – *Celastrus scandens, Solanum dulcamara*
dulcamara negra – *Hyoscyamus niger*
dulcamara perenne – *Solanum jasminoides*
dulciamarga – *Solanum dulcamara*
dulse – *Rhodymenia palmata*
duma – *Hyphaene thebaica*
duranga – *Tabebuia donnell-smithii*
duranta – *Duranta repens*
duraznero – *Prunus persica*
duraznillo – *Acalypha lindheimeri, Amelanchier, Calatola mollis, Cantharellus cibarius, Cercis canadensis, Croton magdalenae, Lantana involucrata, Opuntia leucotricha, O. macrocentra, Polygonum persicaria, Prunus brachybotrya, P. fremontii, Rhamnus sharpii, R. sphaerosperma, Solanum rostratum, Viburnum tinus*
duraznillo blanco – *Opuntia leucotricha, O. spinulifera*
duraznillo colorado – *Opuntia leucotricha*
durazno – *Prunus persica*
durazno de mono – *Calatola mollis*
durian – *Durio zibethinus*
durillo – *Viburnum tinus*
durillo encarnado – *Cornus sanguinea*
durio – *Durio zibethinus*
durión – *Durio zibethinus*
durra – *Sorghum vulgare* var *durra*
dusho – *Saccharum officinarum*

ébano – *Apoplanesia paniculata, Caesalpinia mexicana, C. sclerocarpa, Chloroleucon mangense* var *leucospermum, Diospyros ebenaster, D. ebenus, D. virginiana, Lysiloma acapulcensis, Maba, Pithecellobium ebano, P. flexicaule, P. undulatum, Zapoteca formosa*
ébano agrio – *Diospyros ebenaster*
ébano americano – *Brya ebenus*
ébano asiático – *Diospyros ebenum*
ébano del país – *Caesalpinia coriaria*
ébano de Mozambique – *Diospyros mespiliformis*
ébano de oriente – *Albizzia lebbeck*
ébano de Tabasco – *Zapoteca formosa*
ébano falso – *Laburnum anagyroides*
ébano loto – *Diospyros lotus*
ebo – *Vicia sativa*
ecapatli – *Senna alata, S. occidentalis*
echeveria – *Dudleya lanceolata, Echeveria agavoides*
edelweis – *Leontopodium alpinum*
edelweis alpino – *Leontopodium alpinum*
edichio – *Hedychium gardnerianum*
efedra – *Ephedra sinica*
efedra verde – *Ephedra viridis*
efedrina – *Ephedra sinica*
efémero – *Iris foetidissima*
egílope cilindrica – *Aegilops cylindrica*
egílope de tres uncias – *Aegilops triuncialis*
egílope geniculada – *Aegilops ovata*
Egipto – *Brachiaria mutica*
eglantina – *Rosa eglanteria*
eglantina roja – *Rosa eglanteria*
ehcaxtli – *Cosmos sulphureus*

ejote – *Phaseolus vulgaris*
ejotillo – *Isomeris arborea, Phaseolus acutifolius, Senna pallida, S. purpusii, S. tora*
ejotillo chico – *Senna uniflora*
ejotillo del monte – *Senna pallida*
ejotillo grande – *Senna occidentalis*
ejotón – *Pithecellobium confine*
ek – *Haematoxylon campechianum*
ek-kixilak – *Macfadyena unguis-cati*
elaterio – *Ecballium elaterium*
eldorado – *Trophis racemosa*
eleagno de Bohemia – *Eleagnus angustifolius*
eléboro – *Helleborus*
eléboro blanco – *Veratrum album*
eléboro de invierno – *Eranthis hyemalis*
eléboro fétido – *Helleborus foetidus*
eléboro negro – *Helleborus niger, Veratrum sabadillo*
eléboro pálido – *Helleborus lividus*
eléboro verde – *Helleborus viridis*
elefante – *Pennisetum purpureum*
elel – *Oxalis latifolia*
eleocharis acicular – *Eleocharis acicularis*
eleocharis enana – *Eleocharis parvula*
eleocharis roja subida – *Eleocharis atropurpurea*
eleusina – *Eleusine coracana*
eleuteria – *Croton eluteria*
elexóchitl – *Magnolia schiedeana*
elina de Bellard – *Kobresia bellardii*
elisa – *Solanum wendlandii*
elodea – *Elodea canadensis*
elodea del Canadá – *Elodea canadensis*
eloquiltic – *Bidens pilosa*
elosúchil – *Magnolia dealbata*
elote – *Morchella esculenta, Zea mays* var *saccharata*
elotillo – *Conopholis americana, C. mexicana*
elotito – *Conopholis americana, Morchella elata, M. esculenta*
embeleso – *Plumbago auriculata, P. scandens*
emborrachacabras – *Coriaria myrtifolia*
embotrio – *Embothrium coccineum*
embudo – *Cantharellus*
empanada – *Combretum farinosum, Exogonium bracteatum*
empanadilla – *Exogonium bracteatum*
empanadita – *Exogonium bracteatum*
empanochilla – *Exogonium bracteatum*
empueshte – *Piqueria trinervia*
enandi – *Psidium guajava*
enante – *Oenanthe aquatica*
enante de agua – *Oenanthe aquatica*
enante falsa-pimpinela – *Oenanthe pimpinelloides*
encaje – *Ammi majus, Drymaria glandulosa*
encarrugada – *Zornia diphylla*
encembla – *Croton dioicus*
encienso – *Artemisia absinthium*
encina – *Quercus*
encinilla – *Acacia cymbispina, Croton*
encinillo – *Quercus*
encinito chaparro – *Q. berberidifolia, Q. cedroensis, Q. turbinella*
encino – *Quercus*
encino achaparrado – *Quercus dumosa, Q. palmeri, Q. turbinella*
encino ahualpitzahual – *Quercus laurina*
encino amarillo – *Quercus*
encino arroyero – *Quercus brandegeei*
encino asiático de hojas de bamboo – *Quercus myrsinifolia*
encino avellano – *Quercus obtusata, Q. rugosa*
encino avellano cimarrón – *Quercus gentryi*
encino azul – *Quercus arizonica, Q. oblongifolia*
encino bellotero – *Quercus brandegeei, Q. emoryi*
encino blanco – *Quercus*
encino blanco liso – *Quercus rugosa*
encino borrego – *Quercus subspathulata*
encino cabelludo – *Quercus cerris*
encino cacachila – *Quercus gentryi, Q. perpallida*
encino cajete – *Quercus tarahumara*
encino capulincillo – *Quercus microphylla*
encino carocero – *Quercus crassifolia*
encino carrasco – *Quercus obtusata*
encino cascalote – *Quercus rugulosa*
encino cascata – *Quercus conspersa*
encino cenizo – *Quercus candicans*
encino cochi – *Quercus subspathulata*
encino colorado – *Quercus*
encino cucharillo – *Quercus urbanii*
encino cucharita – *Quercus elliptica*
encino cuero – *Quercus rugosa*
encino chaparro – *Quercus martinezii, Q. microphylla, Q. muehlenbergii, Q. toumeyi*

encino chicharrón – *Quercus crassifolia*
encino chilillo – *Quercus acutifolia, Q. crassipes, Q. gentryi*
encino chino – *Quercus*
encino de asta – *Quercus*
encino de California – *Quercus agrifolia*
encino de hoja ancha – *Quercus crassifolia, Q. gambelii, Q. reticulata*
encino de hoja chica – *Quercus laurina*
encino de hoja de laurel – *Quercus laurina*
encino de hoja delgada – *Quercus laurina*
encino delgado – *Quercus laurina*
encino de mar – *Fucus vesiculosus, Macrocystis pyrifera*
encino de miel – *Quercus reticulata, Q. rugosa, Q. tuberculata*
encino duraznillo – *Quercus emoryi*
encino enano – *Quercus microphylla*
encino escobilla – *Quercus sartorii*
encino haya – *Acer brachypterum*
encino hojarasco – *Quercus crassifolia*
encino huaje – *Quercus crassifolia*
encino hueja – *Quercus endlichiana, Q. fulva, Q. urbanii*
encino jarilla – *Quercus laurina*
encino jumate – *Quercus urbanii*
encino laurel – *Quercus crassipes, Q. elliptica, Q. oblongifolia, Q. salicifolia*
encino laurelillo – *Quercus*
encino manzanillo – *Quercus affinis, Q. crassipes*
encino manzano – *Quercus arizonica, Q. coccolobifolia, Q. eduardii*
encino marino – *Fucus vesiculosus*
encino memelito – *Quercus*
encino miscalme – *Quercus chihuahuensis*
encino mixcahue – *Quercus subspathulata*
encino nanche – *Quercus elliptica, Q. oleoides*
encino napis – *Quercus magnoliifolia*
encino nechilahue – *Quercus laurina*
encino negro – *Quercus*
encino oreja de ratón – *Quercus crassipes*
encino papatla – *Quercus candicans*
encino pipitillo – *Quercus conspersa, Q. muehlenbergii*
encino piptza – *Quercus crassipes*
encino prieto – *Quercus*
encino prieto sancón – *Quercus martinezii*
encino quiebra hacha – *Quercus reticulata, Q. rugosa*
encino rey – *Quercus corrugata*
encino roble – *Quercus*
encino roble amarillo – *Quercus conspersa, Q. glaucescens*
encino roble blanco – *Quercus oleoides*
encino rojo – *Quercus castanea, Q. conspersa, Q. laurina, Q. muehlenbergii*
encino rosillo – *Quercus castanea, Q. scytophylla*
encino sancón – *Quercus martinezii*
encino sauce – *Quercus viminea*
encino saucillo – *Quercus acutifolia, Q. salicifolia*
encino tapahuite – *Quercus elliptica*
encino tecomate – *Quercus sideroxyla*
encino teposcohuite – *Quercus conspersa*
encino tepozcohuite – *Quercus acutifolia*
encino tesmol – *Quercus oleoides*
encino tesmolillo – *Quercus crassifolia*
encino tocuz – *Quercus glaucoides, Q. rugosa*
encino verde – *Quercus agrifolia, Q. coccolobifolia, Q. jonesii*
enchilada – *Castilleja arvensis, C. canescens*
enchiladita – *Castilleja arvensis, C. canescens*
enchilado – *Cantharellus, Lactarius*
enchilado de ocote – *Lactarius deliciosus*
enchiladora – *Croton ciliato-glandulosas*
endibia – *Cichorium endiva, C. intybus*
endibia acogollada roja – *Cichorium intybus*
endibia de Belga – *Cichorium intybus*
endrino – *Prunus spinosa*
endrino azul – *Prunus spinosa*
endrino bruño – *Prunus spinosa*
endrino grande – *Prunus domestica* ssp *insititia*
enebro – *Juniperus*
enebro achaparrado – *Juniperus communis*
enebro albar – *Juniperus communis, J. oxycedrus*
enebro americano – *Juniperus virginiana*
enebro común – *Juniperus communis*
enebro de bayas rojas – *Juniperus oxycedrus*
enebro de Grecia – *Juniperus drupacea*
enebro de la China – *Juniperus chinensis*
enebro de la miera – *Juniperus oxycedrus*
enebro de Virginia – *Juniperus virginiana*
enebro enano – *Juniperus communis* ssp *nana, J. horizontalis*

enebro escamoso – *Juniperus squamata*
enebro español – *Juniperus sabina*
enebro espinoso – *Juniperus communis*
enebro fenicio – *Juniperus phoenicea*
enebro llorón – *Juniperus recurva*
enebro morisquillo – *Juniperus communis*
enebro rastrero – *Juniperus communis*
enebro real – *Juniperus communis*
enebro sabina – *Juniperus sabina*
enebro virginiano – *Juniperus virginiana*
eneldo – *Anethum graveolens, Foeniculum vulgare* var *azoricum*
engordacabra – *Dalea citriodora, D. seemannii, Salvia ballotaeflora*
enguambo – *Bocconia arborea, B. frutescens*
enotera – *Oenothera biennis, O. villosa* ssp *strigosa*
enovianto – *Enkianthus campanulatus*
enquianto – *Enkianthus campanulatus*
enredadera – *Convolvulus tricolor*
enredadera coralina – *Antigonon leptopus*
enredadera de campanilla – *Ipomoea*
enredadera de San Diego – *Antigonon leptopus*
enredadera de Virginia – *Parthenocissus quinquefolia*
ensalada – *Plumeria rubra*
ensangregado – *Jatropha cardiophylla, J. cinerea, J. platanifolia*
entraña – *Cylindropuntia imbricata*
entrodelia – *Tibouchina spathulata*
énula – *Inula helenium*
énula campana – *Inula helenium*
énula griega – *Inula helenium*
envidia – *Sonchus oleraceus*
envolver – *Vitis arizonica*
eonio de las Canarias – *Aeonium arboreum, A. canariense*
epazote – *Chenopodium ambrosioides, C. graveolens, C. nuttalliae, Lepidium lasiocarpum, Lippia berlandieri*
epazote cimarrón – *Chenopodium incisum*
epazote de comer – *Chenopodium ambrosioides*
epazote de coyote – *Chenopodium graveolens*
epazote de mula – *Amaranthus spinosus*
epazote de toro – *Chenopodium graveolens, C. incisum*
epazote de zorrillo – *Chenopodium ambrosioides, C. graveolens, C. incisum*
epazotillo – *Eclipta prostrata, Teucrium cubense*
epazotl – *Chenopodium ambrosioides, C. graveolens, Lippia berlandieri*
epimedio – *Epimedium grandiflorum*
epimedio alpino – *Epimedium alpinum*
epipactis común – *Epipactis helleborine*
epipógio sin hojas – *Epipogium aphyllum*
epteride – *Pteris cretica*
equimite – *Erythrina*
equinacia – *Echinacea angustifolia, E. purpurea*
equipal – *Trema micrantha*
equiseto – *Echinacea purpurea, Equisetum*
equiseto mayor – *Equisetum telmateia*
equiseto menor – *Equisetum arvense*
eragrostis – *Eragrostis curvula*
eragrostis de Barrelier – *Eragrostis barrelieri*
eragrostis llorona – *Eragrostis curvula*
eragrostis mayor – *Eragrostis cilianensis*
eragrostis mexicana – *Eragrostis mexicana*
eragrostis menor – *Eragrostis minor*
eragrostis peluda – *Eragrostis pilosa*
erguén – *Argania sideroxylon*
erigerón – *Erigeron modestus*
erigerón anual – *Erigeron annuus*
erigerón de color verdemar – *Erigeron glaucus*
erigerón glauco – *Erigeron glaucus*
erigerón mucronado – *Erigeron karvinskianus*
erigerón ramoso – *Erigeron strigosus*
eringe – *Eryngium*
eringe alpinum – *Eryngium alpinum*
eringe campestre – *Eryngium campestre*
eringe de color de amatista – *Eryngium amethystinum*
eringe de hojas aplanadas – *Eryngium planum*
eringe marfilado – *Eryngium giganteum*
eringe marítimo – *Eryngium maritimum*
eriojuelo – *Cymbopetalum penduliflorum*
erismo – *Sisymbrium officinale*
erismo común – *Erysimum cheiranthoides*
erismo de los campos – *Erysimum cheiranthoides*
erismo repandum – *Erisimo repando*
erofila temprana – *Erophila verna*
eruca – *Eruca vesicaria* ssp *sativa*
eruca roqueta – *Eruca vesicaria* ssp *sativa*
erviana – *Raphanus raphanistrum*
ervilla – *Vicia ervilia*

escabiosa – *Pinaropappus roseus, Scabiosa atropurpurea, S. caucasica*
escabiosa arvense – *Knautia arvensis*
escabiosa blanca – *Cephalaria leucantha*
escabiosa blanca-amarilleante – *Scabiosa ochroleuca*
escabiosa estrellada – *Scabiosa stellata*
escabiosa menor – *Scabiosa columbaria*
escabiosa morisca – *Scabiosa atropurpurea*
escabiosa velo de viuda – *Scabiosa atropurpurea*
escajo – *Ulex europaea*
escala de Jacob – *Polemonium caeruleum*
escalera de Jacob – *Convallaria majalis*
escalona – *Allium cepa* var *ascalonicum*
escalonia – *Allium cepa* var *ascalonicum, Escallonia illinita*
escaloña – *Allium cepa* var *aggregatum, A. c.* var *ascalonicum*
escamonea – *Convolvulus scammonia, Ipomoea orizabensis*
escamonea falsa – *Cynanchum acutum*
escaramón – *Paeonia broteroi*
escaramujo – *Rosa canina, R. eglanteria, R. montezumae*
escarapela – *Alcea rosea*
escaravía – *Sium sisarum*
escarchada – *Carpobrotus chilensis, Mesembryanthemum crystallinum*
escarchada sin tallo – *Mesembryanthemum nodiflorum*
escarchosa – *Mesembryanthemum crystallinum*
escarlata blanca – *Tricholoma acerbum*
escarola – *Cichorium endivia*
escarola crespa – *Cichorium endivia*
escarpela – *Gaillardia pulchella*
escila – *Urginea maritime, U. scilla*
escila bifolia – *Scilla bifolia*
escila blanca – *Urginea scilla*
escila de los bosques – *Scilla italica*
escila del país – *Hymenocallis rotata*
escila española – *Hyacinthoides hispanicus*
escila hispánica – *Hyacinthoides hispanicus*
escila itálica – *Scilla italica*
escila primaveral – *Scilla siberica*
escila rayada – *Pushkinia scilloides*
escila roja – *Urginea scilla*
escila siberiana – *Scilla siberica*
escila sibérica – *Scilla siberica*
esclarea – *Salvia sclarea*
escleranto anual – *Scleranthus annuus*
esclerochloa dura – *Sclerochloa dura*
esclerocio – *Claviceps purpurea*
escoba – *Abutilon incanum, Baccharis conferta, Cryosophila nana, C. stauracantha, Cytisus scoparius, Dalea diffusa, Desmanthus covillei, Heimia salicifolia, Sarothamnus scoparius, Sida acuta, Sorghum vulgare*
escoba amarga – *Baccharis sarathroides, B. sergiloides, Flaveria*
escoba blanca – *Cytisus multiflorus, Eremalche, Malpighia glabra, Malvastrum coromandelianum*
escoba colorada – *Dalea diffusa, Heimia salicifolia*
escoba de la víbora – *Gutierrezia sarothrae*
escoba del monte – *Baccharis conferta*
escoba fuerte – *Sida acuta*
escoba larga – *Dalea diffusa*
escoba maicera – *Sorghum bicolor*
escobeta – *Chlorgalum, Clavaria flava, Ramaria flava, R. stricta*
escobetilla – *Clavulina rugosa, Ramaria flava*
escobilla – *Aster pauciflorus, A. subulatus, Baccharis, Buddleja scordioides, Centaurea salmantica, Dipsacus fullonum, Eremalche, Erythroxylum mexicanum, Fraxinus greggii, Heteropteris laurifolia, Malpighia emarginata, M. glabra, Milleria quinqueflora, Quercus laurina, Scabiosa atropurpurea, Schkuhria virgata, Scoparia dulcis, Selloa glutinosa, Sida acuta, S. rhombifolia, Solidago velutina, Turnera diffusa*
escobilla amarga – *Scoparia dulcis*
escobilla de caléndula – *Bidens pilosa*
escobilla del carbonero – *Baccharis conferta*
escobilla del río – *Heimia salicifolia*
escobilla morisca – *Scabiosa atropurpurea*
escobilla parda – *Artemisia campestris*
escobilla savilla – *Buddleja*
escobillón – *Ramaria flava*
escobita – *Castilleja exserta, C. densiflora*
escobita de la Virgen – *Eriogonum ainsliei*
escobón – *Cytisus scoparius, Genista monspessulana*
escobón estriado – *Cytisus striatus*

escolopendro – *Phyllitis scolopendrium, Scolopendrium officinale*
escomite – *Phaseolus acutifolius*
escordio – *Teucrium scordium*
escordio bastardo – *Teucrium scordonia*
escordonia – *Teucrium scorodonia*
escorzonera – *Eryngium comosum, Iostephane heterophylla, Pinaropappus roseus, Scorzonera hispanica*
escorzonera hispánica – *Scorzonera hispanica*
escrofularia – *Scrophularia*
escrofularia acuática – *Scrophularia auriculata*
escrofularia auriculada – *Scrophularia auriculata*
escrofularia canina – *Scrophularia canina*
escrofularia de los bosques – *Scrophularia nodosa*
escuamaria común – *Lathraea squamaria*
escuradentis – *Ammi visnaga*
escutelaria – *Scutellaria lateriflora*
escutelaria baical – *Scutellaria baicalensis*
eshel – *Tamarix gallica*
eskuóvain – *Mitchella repens*
esmeralda – *Scabiosa atropurpurea*
esmilax – *Asparagus asparagoides*
esmirnio común – *Smyrnium olustratum*
espada – *Sansevieria trifasciata*
espada japonesa – *Sansevieria trifasciata*
espadaña – *Tripsacum lanceolatum, Typha*
espadaña fina – *Iris pseudacorus*
espadaña hedionda – *Iris foetidissima*
espadilla – *Gladiolus hortulanus*
espadillo – *Agave striata*
espadín – *Agave striata, Yucca*
espantachinches – *Cimifuga racemosa*
espantadiablos – *Hypericum perforatum*
espantalobos – *Colutea arborescens, Ipomoea stans*
espantalobos arborescente – *Colutea arborescens*
espantamuchachos – *Ecballium elaterium*
espantavaqueros – *Acourtia adnata, Ipomoea stans, Tidestromia lanuginosa*
esparceta – *Onobrychis viciifolia*
esparcilla – *Spergula arvensis*
esparmania – *Sparmannia africana*
espartillo – *Sporobolus indicus*
espartillo común – *Sporobolus indicus*
espartillo de flores alternas – *Spartina alterniflora*
espartillo densifloro – *Spartina densiflora*
espartillo de olor – *Spartina patens*
esparto *Fimbristylis thermalis, Lygeum spartum, Stipa tenacisima*
esparto basto – *Lygeum spartum*
esparto común – *Stipa tenacisima*
esparto chino – *Fimbristylis thermalis*
esparto mulato – *Fimbristylis thermalis*
espárrago – *Asparagus densiflorus* var *sprengeri, A. officinalis, A. setaceus*
espárrago cimarrón – *Pellaea cordifolia*
espárrago común – *Asparagus officinalis*
espárrago de caña – *Tamus communis*
espárrago densifloro – *Asparagus densiflorus* var *sprengeri*
espárrago de Sprenger – *Asparagus densiflorus* var *sprengeri*
espárrago falso – *Humulus lupulus*
espárrago plumoso – *Asparagus setaceus*
esparraguera – *Asparagus densiflorus* var *sprengeri, A. officinalis, A. setaceus*
esparraguera de adorno – *Asparagus setaceus*
esparrillera – *Santolina chamaecyparisus*
espatifilo – *Spathiphyllum*
espatodea – *Spathodea campanulata*
espátula fétida – *Iris foetidissima*
espejo de Venus – *Ophrys speculum, Specularia speculum*
espelón – *Phaseolus vulgaris*
espelta – *Triticum aestivum* var *spelta*
esperanza – *Tecoma stans*
espiga – *Parmentiera aculeata*
espiga de agua – *Potamogeton*
espiga de agua alpina – *Potamogeton alpinus*
espiga de agua de hojas de zostera – *Potamogeton zosteriformis*
espiga de agua de Nuttall – *Potamogeton epihydrus* ssp *nuttallii*
espiga de agua filiforme – *Potampgeton filiformis*
espiga de agua flotante – *Potamogeton natans*
espiga de agua heterofila – *Potamogeton gramineus*
espiga de agua nudosa – *Potamogeton nodosus*
espiga de agua palermitana – *Potamogeton pusillus*

espiga de agua pedunculada – *Potamogeton praelongus*
espiga de agua rojiza – *Potamogeton alpinus*
espiga de agua serrada – *Potamogeton crispus*
espigadilla – *Crucianella angustifolia*
espiga negra – *Pleuraphis cenchroides*
espigón – *Tribulus terrestris*
espiguilla – *Poa annua, P. pratensis, P. trivialis*
espiguilla anual – *Poa annua*
espinablo – *Crataegus monogyna*
espinaca – *Sedum moranense, Spinacia oleracea*
espinaca acuática – *Ipomoea aquatica*
espinaca blanca – *Basella alba*
espinaca de agua – *Ipomoea aquatica*
espinaca de Ceilán – *Basella alba*
espinaca de Filipinas – *Talinum triangulare*
espinaca de Nueva Zelandia – *Tetragonia tetragonioides*
espinaca de Surinam – *Talinum triangulare*
espinaca neozelandesa – *Tetragonia tetragonioides*
espinaca silvestre – *Rumex patientia*
espinillo – *Parkinsonia aculeata, Acacia farnesiana*
espino – *Acacia, Colubrina viridis, Crataegus, Genista anglica, Paliurus spina-christi, Rhamnus palestina, Zanthoxylum fagara*
espino albar – *Crataegus laevigata, C. monogyna*
espino amarillo – *Hippophae rhamnoides*
espino americano – *Crataegus crus-galli*
espino blanco – *Acacia farnesiana, A. pennatula, A. spadicigera, Crataegus laevigata, C. monogyna, Duranta repens*
espino cerval – *Rhamnus cathartica, R. purshiana*
espino corona – *Gleditsia amorphoides*
espino cruz – *Randia*
espino de canguro – *Acacia paradoxa*
espino de Cristo – *Euphorbia milii*
espino de dos estilos – *Crataegus laevigata*
espino de fuego – *Pyracantha coccinea*
espino de injertar – *Prunus domestica* ssp *insititia*
espino de Jerusalén – *Parkinsonia aculeata*
espino del aceite de castor – *Kalopanax septemlobus*
espino de paloma – *Duranta erecta*
espino de sol – *Cirsium mexicanum*
espino de tinta – *Acacia macracantha*
espino europeo – *Crataegus monogyna*
espino falso – *Hippophae rhamnoides*
espino híbrido – *Crataegus* x *lavallei*
espino majolar – *Crataegus laevigata*
espino majuelo – *Crataegus monogyna*
espino monógino – *Crataegus monogyna*
espino navarro – *Crataegus laevigata*
espino negro – *Duranta repens, Prunus spinosa*
espino pata de gallo – *Crataegus crus-galli*
espino real de España – *Parkinsonia aculeata*
espino rubial – *Zanthoxylum fagara*
espino sagrado – *Rhamnus californica*
espinosilla – *Acalypha arvensis, Karwinskia humboldtiana, Loeselia mexicana*
espino y camote – *Pisonia*
espirantes – *Spiranthes aurantiaca*
espirantes irlandesa – *Spiranthes romanzoffiana*
espírea – *Spiraea douglasii*
espírea de Douglas – *Spiraea douglasii*
espírea de hojas de ciruelo – *Spiraea prunifolia*
espírea del Japón – *Spiraea japonica*
espírea de Thunberg – *Spiraea thunbergii*
espírea de VanHoutte – *Spiraea* x *vanhouttei*
espírea japonesa – *Spiraea japonica*
espírea trilobulada – *Spiraea trilobata*
espíritu desilachado – *Eryngium rosei*
espliego – *Lavandula angustifolia, L. stoechas*
espliego común – *Lavandula angustifolia*
espliego de hojas divididas – *Lavandula multifida*
espliego de mar – *Limonium sinuatum, Psylliostachys suworowii*
espliego marino – *Limonium sinuatum*
espolín – *Stipa pennata*
espolón de gallo – *Crataegus crus-galli, C.* x *lavallei*
espondilio – *Heracleum mantegazzianum*
esponja – *Luffa aegyptiaca*
esponja vegetal – *Luffa acutangula, L. aegyptiaca*
esponjilla – *Luffa aegyptiaca*
espuela – *Delphinium, Phacelia platycarpa*
espuela de caballero – *Consolida ambigua, Delonix regia, Delphinium, Justicia candicans*

espuela de caballero nudicaule – *Delphinium nudicaule*
espuela de gañán – *Tropaeolum majus*
espuela del diablo – *Proboscidea*
espuela de Santiago – *Caesalpinia gilliesii*
espuelita – *Commelina erecta, Delphinium, Mollugo verticillata*
espuma de mar – *Ammi majus*
espumila – *Cryptantha grayi*
esquila – *Urginea maritima*
esquisúchil – *Bourreria huanita*
esquizandra – *Schizandra chinensis*
estacahuite – *Conocarpus erecta*
estafiate – *Ambrosia, Artemisia, Salvia mellifera*
estafiate del bosque – *Artemisia ludoviciana*
estafilea – *Staphylea bolanderi*
estafisagria – *Delphinium staphisagria*
estamasúchil – *Ranunculus*
estaquilla – *Prunus serotina*
estátice – *Armeria maritima* ssp *californica, Limonium sinuatum*
estaticia – *Limonium sinuatum*
estátil – *Limonium sinuatum*
esteba de hojas pinadas – *Stipa pennata*
esteba pirenaica – *Stipa gigantea*
estefanote – *Stephanotis floribunda*
estepa – *Cistus, Helianthemum nummularium*
estepa blanca – *Cistus albidus*
estepa de Creta – *Cistus creticus*
estepa de hojas de laurel – *Cistus laurifolius*
estepa de Montpellier – *Cistus monspeliensis*
estepa grisácea – *Cistus incanus*
estepa negra – *Cistus monspeliensis, C. salvifolius*
estepilla – *Digitalis obscura*
esterculia – *Firmiana simplex*
estevia – *Stevia rebaudiana*
estiércol de pájaro – *Capsicum annuum* var *minimum*
estilosanthes – *Stylosanthes guianensis*
estoque – *Gladiolus hortulanus*
estoquillo – *Agave striata*
estoraque – *Liquidambar orientalis, L. styraciflua, Styrax argenteus, S. officinalis, S. obassia*
estoraque común – *Styrax officinalis*
estoraque japonés – *Styrax japonicus*
estoraque japonés de hojas grandes – *Styrax obassia*
estornudadera – *Arnica montana*
estragón – *Artemisia dracunculus*
estramonio – *Brugmansia candida, Datura stramonium, D, wrightii*
estrara – *Cylindropuntia imbricata*
estrelichia – *Strelitzia reginae*
estrelitzia – *Strelitzia reginae*
estrella – *Callistephus chinensis, Ipomoea quamoclit, Milla biflora, Quamoclit coccinea, Q. lobata, Q. pinnata*
estrella africana – *Cynodon plectostachyus*
estrella azul – *Petrea arborea*
estrella de agua común – *Callitriche verna*
estrella de agua de los pantanos – *Callitriche verna*
estrella de agua de los terrenos paludosos – *Callitriche stagnalis*
estrella de agua de otoño – *Callitriche hermaphroditica*
estrella de Belén – *Campanula isophylla, Ornithogalum umbellatum*
estrella de Italia – *Campanula isophylla*
estrella de la tarde – *Escobaria vivipara*
estrella de la tierra – *Cryptanthus zonatus*
estrella del norte – *Asphodelus fistulosus, Marah lobata*
estrella del sol – *Ipomoea quamoclit*
estrella de mar – *Hymenocallis rotata, Nigella damascena*
estrella de San Nicolás – *Hymenocallis harrisiana, H. rotata, Milla biflora*
estrella gigante – *Cynodon dactylon*
estrella mexicana – *Milla biflora*
estrellania – *Plantago coronopus*
estrellita – *Aster subulatus, Galinsoga parviflora, Maba, Milla biflora, Mimosa biuncifera, Stellaria*
estremoncillo – *Thymus vulgaris*
estrenina – *Plumbago scandens*
estreptocarpo – *Streptocarpus hybridus*
estreptopo amplexifolio – *Streptopus amplexifolius*
estribillo – *Trichilia hirta*
estrilicia – *Strelitzia reginae*
estropajo – *Luffa acutangula, L. aegyptiaca, L. cylindrical, Momordica charantia*
estropajo silvestre – *Lagenaria siceraria*

etiónema – *Aethionema schistosum*
etiópide villosa – *Salvia aethiopis*
etl – *Phaseolus vulgaris*
etusa – *Aethusa cynapium*
etzcuáhuitl – *Croton draco*
etzemo – *Schoenocaulon officinale*
eucalipto – *Eucalyptus*
eucalipto azul – *Eucalyptus globulus*
eucalipto blanco – *Eucalyptus globulus*
eucalipto común – *Eucalyptus globulus*
eucalipto de Gunn – *Eucalyptus gunnii*
eucalipto de hojas de sauce – *Eucalyptus globulus*
eucalipto del alcanfor – *Eucalyptus robusta*
eucalipto de la sidra – *Eucalyptus gunnii*
eucalipto del maná – *Eucalyptus viminalis*
eucalipto de pantano – *Eucalyptus robusta*
eucalipto gigante – *Eucalyptus globulus*
eucalipto nival – *Eucalyptus pauciflora* ssp *niphophila*
eucalipto rojo – *Eucalyptus camaldulensis, E. ficifolia*
eucalipto salmón – *Eucalyptus salmonophloia*
eucaz – *Prunus brachybotrya*
euforbio – *Euphorbia epithymoides*
euforbio brillante – *Euphorbia fulgens, E. milii*
euforbio caracioide – *Euphorbia characias*
euforbio ciprés – *Euphorbia cyparissias*
euforbio de cabellos de tomillo – *Euphorbia epithymoides, E. polychroma*
euforbio de font quer – *Euphorbia myrsinites*
euforbio de hojas no numerosas – *Euphorbia esula*
euforbio del Himalaya – *Euphorbia griffithii*
euforbio de los pantanos – *Euphorbia palustris*
euforbio de Seguier – *Euphorbia seguierana*
euforbio dulce – *Euphorbia dulcis*
euforbio falsa-almendra – *Euphorbia amygdaloides*
euforbio fulgente – *Euphorbia fulgens*
euforbio helioscopo – *Euphorbia helioscopia*
euforbio japonés – *Pachysandra terminalis*
euforbio manchado – *Chamaesyce maculata*
euforbio menor – *Euphorbia peplus*
euforbio nutante – *Chamaesyce nutans*
euforbio policromo – *Euphorbia epithymoides, E. polychroma*
euforbio prostrado – *Chamaesyce prostrata*
euforbio rígido – *Euphorbia rigida*
euforbio serpenteado – *Chamaesyce serpens*
euforbio serrado – *Euphorbia serrata*
eufrasia – *Euphrasia officinalis*
eufrasia común – *Euphrasia officinalis*
eufrasia latifolia – *Parentucella latifolia*
eufrasia viscosa – *Parentucella viscosa*
eugenia aromatica – *Caryophyllus aromaticus*
eulalia – *Miscanthus sinensis*
eupatorio – *Eupatorium perfoliatum*
eupatorio adenóforo – *Ageratina adenophora*
eupatorio maculado – *Eupatorium maculatum*
eviloriva – *Rumex crispus*
evónimo – *Euonymus europaeus, E. japonica, E. occidentalis*
evónimo del Japón – *Euonymus japonica*
evónimo europeo – *Euonymus europaeus*
evónimo manchado – *Euonymus japonicus* var *aureo-variegatus*
excalahuate – *Quercus crassifolia*
exoxóchitl – *Helenium mexicanum*
expatli – *Bouvardia ternifolia*
expujuj – *Tagetes patula*
exquiquílit – *Xanthosoma robustum*
ezpatli – *Croton sanguifluus*

faba – *Phaseolus lunatus, Vicia faba*
faique – *Acacia macrantha*
faisán – *Leccinum*
faisán de la jara – *Leccinum corsicum*
falaris – *Phalaris tuberosa*
falaris perenne – *Phalaris aquatica*
falsa acacia – *Robinia pseudoacacia*
falsa acacia del Japón – *Sophora japonica*
falsa acacia de tres espinas – *Gleditsia triacanthos*
falsa alcaparra – *Capparis indica, Larrea tridentata*
falsa amapola – *Meconopsis betonicifolia, M. cambrica*
falsa aralia – *Schefflera elegantissima*
falsa árnica – *Heterotheca inuloides, Trixis angustifolia, T. radialis*
falsa barba de node – *Eragrostis curvula*
falsa begonia – *Heuchera mexicana*
falsa camelia – *Stewartia pseudocamellia*
falsa camelia decidua – *Stewartia pseudocamellia*
falsa caña de azúcar – *Pennisetum purpureum*
falsa cocolmeca – *Dioscorea*

falsa damiana – *Bigelovia venata, Chrysactinia mexicana, Isocoma menziesii*
falsa escamonea – *Cynanchum acutum*
falsa malva – *Malope trifida*
falsa oronja – *Amanita muscaria*
falsa palma de Nueva Zelanda – *Cordyline australis*
falsa spiraea – *Astilbe arendsii*
falsa quina – *Exostema caribaeum, Hintonia latiflora*
falsa tsuga de California – *Pseudotsuga macrocarpa*
falsa tsuga verde de las rocosas – *Pseudotsuga menziesii*
falso abeto – *Picea*
falso ácoro – *Iris pseudacorus*
falso alerce dorado – *Pseudolarix amabilis*
falso anís – *Anethum graveolens*
falso barniz – *Ailanthus altissima*
falso canelo de Sudamérica – *Drimys winteri*
falso carpe – *Ostrya carpinifolia*
falso castaño – *Aesculus hippocastanum*
falso castaño de flores rojas – *Aesculus pavia*
falso ciprés – *Chamaecyparis*
falso ciprés blanco americano – *Chamaecyparis thyoides*
falso ciprés calvo – *Taxodium distichum*
falso ciprés de Alaska – *Cupressus nootkatensis*
falso ciprés de Lawson – *Cupressus lawsoniana*
falso ciprés de los pantanos – *Taxodium distichum*
falso ciprés de Nootka – *Cupressus nootkatensis*
falso ciprés de Patagonia – *Fitzroya cupressoides*
falso ciprés japonés hinoki – *Chamaecyparis obtusa*
falso ciprés sawara – *Chamaecyparis pisifera*
falso diente de león – *Sonchus oleraceus*
falso ébano – *Laburnum anagyroides*
falso espartillo del pinar – *Piptochaetium fimbriatum*
falso hipericón – *Tagetes lucida*
falso jazmín – *Philadelphus coronarius*
falso peritre – *Erigeron*
falso pimentero – *Schinus molle*
falso plátano – *Acer pseudoplatanus*
falso plumbago – *Ceratostigma plumbaginoides*
falso ricino – *Fatsia japonica*
falso tridente borreguero – *Erioneuron pulchellum*
falso zumaque – *Ailanthus altissima*
fárfara – *Tussilago farfara*
fárfara japonesa – *Petasites japonicus*
farolillo – *Aquilegia vulgaris, Campanula medium*
farolillo blanco – *Campanula lactiflora*
farolillo chino – *Physalis alkekengi*
farolillo de los Carpatos – *Campanula carpatica*
farolillo rojo de la floración – *Scilla peruviana*
farolito – *Cardiospermum halicacabum, Physalis angulata, P. pubescens, Scabiosa stellata, Solanum capsicastrum*
farolito de enredadera – *Cardiospermum halicacabum*
farolito de jardín – *Cardiospermum halicacabum*
farolito de la Virgen – *Cardiospermum halicacabum*
faséolo – *Phaseolus vulgaris*
fatsia – *Fatsia japonica*
fausto – *Thunbergia grandiflora*
fava – *Phaseolus lunatus*
feijoa – *Feijoa sellowiana*
felandrio – *Oenanthe aquatica*
felandrio acuático – *Oenanthe aquatica*
felguera común – *Pteridium aquilinum*
felvira – *Tinantia erecta, T. modesta*
fenarda – *Trifolium campestre*
fenícula – *Foeniculum vulgare* var *azoricum*
fenix – *Phoenix dactylifera*
fenogreco – *Trigonella foenum-graecum*
fermina – *Janusia gracilis*
férula – *Ferula communis*
férula común – *Ferula communis*
ferrer – *Cynodon dactylon*
festuca alta – *Festuca arundinacea*
festuca arundinácea – *Festuca arundinacea*
festuca canosa – *Festuca arundinacea*
festuca de color amatista – *Festuca amethystina*
festuca de hojas ásperas – *Festuca trachyphylla*
festuca de hojas grandes – *Festuca longifolia*
festuca de los prados – *Festuca pratensis*

festuca de oveja – *Festuca ovina*
festuca descollada – *Festuca arundinacea*
festuca elevada – *Festuca arundinacea*
festuca galesa – *Festuca valesiaca*
festuca gigante – *Festuca arundinacea*
festuca glauca – *Festuca glauca*
festuca incurvada – *Festuca glauca*
festuca pratense – *Festuca pratensis*
festuca roja – *Festuca rubra*
ficus – *Ficus*
ficus llorón – *Ficus benjamina*
ficus rastrero – *Ficus pumila*
fideo – *Cuscuta*
figuera – *Ficus*
filago común – *Filago germanica*
filago de hojas tenues – *Filago gallica*
filago espatulado – *Filago pyramidata*
filago francés – *Filago gallica*
filigrana – *Lantana camara*
filigrana menor – *Myriophyllum verticillatum*
filipéndula – *Chrysanthemum coronarium, C. indicum, Filipendula vulgaris*
filipéndula de seis pétalos – *Filipendula vulgaris*
filipéndula ulmaria – *Filipendula ulmaria*
filodendro – *Monstera deliciosa, Philodendron*
filodendro de hoja hendida – *Monstera deliciosa*
filodendro trepador – *Philodendron scandens*
fiñisachi – *Acacia farnesiana*
fisga de ratón – *Olyra latifolia*
fisostegia – *Physostegia virginiana*
fistra – *Ammi visnaga*
fitolaca – *Phytolacca*
flamboyán – *Delonix regia*
flamboyán azul – *Jacaranda acutifolia*
flamboyán rojo – *Delonix regia*
flamenquilla – *Calendula officinalis*
flámula – *Ranunculus flammula*
flaveria – *Flaveria contrayerba*
flecha de agua – *Sagittaria sagittifolia*
flecha de Cupido – *Catananche caerulea*
flechilla – *Stipa eminens*
flechilla bulbosa – *Stipa leucotricha*
flechilla grande – *Stipa eminens*
flechilla neomexicana – *Stipa neomexicana*
fleo – *Phleum pratense*
fleo alpino – *Phleum alpinum*
fleo de los prados – *Phleum pratense*
fleo dorado – *Setaria sphacelata*
fleo praderoso – *Phleum pratense*
flor amarilla – *Tecoma stans*
flor ave del paraíso – *Strelitzia reginae*
florazul – *Jacaranda acutifolia*
flor blanca – *Euphorbia colletioides, Plumeria acutifolia*
flor blanca de la candelaria – *Laelia glauca*
flor de agua – *Astianthus viminalis, Nymphaea alba, Pontederia cordata*
flor de aire – *Kalanchoe pinnata, Tillandsia lindenii*
flor de ángel – *Pluchea odorata*
flor de araña – *Cleome hasslerana, Hymenocallis harrisiana, Sigesbeckia jorullensis*
flor de arena – *Oenothera californica, O. trichocalyx*
flor de arete – *Fuchsia*
flor de barbero – *Allamanda cathartica*
flor de bayoneta – *Epidendrum parkinsonianum*
flor de cabeza – *Stanhopea tigrina*
flor de cacao – *Quararibea funebris*
flor de cacho – *Laelia tibicinis*
flor de calavera – *Cypripedium irapeanum, Laelia autumnalis*
flor de calaverita – *Laelia digbyana*
flor de cáliz – *Hylocereus undatus*
flor de camarón – *Caesalpina pulcherrima, Delonix regia, Justica brandegeana, Ludwigia peploides*
flor de campana – *Brugmansia suaveolens, Cobaea scandens*
flor de campo – *Ruelia californica, R. peninsularis*
flor de cangrejo – *Canna indica*
flor de capomo – *Nymphaea elegans*
flor de caracól – *Phaseolus caracalla*
flor de carnaval – *Senecio chenopioides*
flor de Castilla – *Rosa*
flor de catarinas – *Laelia autumnalis*
flor de cementerio – *Plumeria rubra*
flor de cera – *Chysis bractescens, Hoya bella, H. carnosa*
flor de cesta – *Centaurea americana*
flor de cielo – *Duranta erecta*
flor de cinco llagas – *Tagetes lunulata*
flor de clavo – *Choisya ternata*

flor de coco – *Cattleya citrina*
flor de colibrí – *Loeselia mexicana*
flor de concha – *Encyclia cochleata, Tigridia pavonia*
flor de cono – *Echinacea angustifolia*
flor de corona – *Ruprechtia pallida, Scilla peruviana*
flor de corpus – *Cranichis speciosa, Laelia grandiflora, L. speciosa, Plumeria rubra*
flor de cuaresma – *Vernonia deppeana*
flor de cuclillo – *Lychnis coronaria*
flor de cuchillo – *Carpobrotus chilensis*
flor de cuerno – *Disocactus flagelliformis, Stanhopea devoniensis*
flor de cuervo – *Plumeria acutifolia, P. rubra*
flor de culebra – *Asclepias curassavica*
flor de China – *Impatiens balsamina*
flor de chinche – *Galphimia glauca*
flor de chupamirto – *Loeselia mexicana*
flor de chuparrosa – *Salvia mexicana*
flor de diciembre – *Galphimia glauca*
flor de dolores – *Senecio salignus*
flor de embudo – *Zantedeschia aethiopica*
flor de ensartar – *Jacquinia aurantica*
flor de espuma – *Tiarella cordifolia*
flor de estrella – *Callistephus chinensis, Hymenocallis harrisiana, H. rotata*
flor de fuego – *Euphorbia pulcherrima*
flor de gallito – *Salvia patens*
flor de gigantón – *Helianthus annuus*
flor de gloria – *Solanum dulcamara, S. seaforthianum*
flor de grillo – *Ranunculus*
flor de guacamaya – *Caesalpinia pulcherrima*
flor de Guadalupe – *Pluchea odorata*
flor de gusano – *Spathiphyllum cochlearispathum*
flor de hombre ahorcado – *Aceras anthropophorum*
flor de huachinango – *Eichhornia crassipes*
flor de hueso – *Poilanthes tuberosa*
flor de huevo – *Cattleya, Solanum melongena* var *esculentum*
flor de iguana – *Diphysa racemosa, Senna pallida*
flor de incienso – *Tillandsia limbata*
flor de invierno – *Dahlia coccinea, Hamamelis virginiana*
flor de jacinto – *Moringa oleifera*
flor de Jamaica – *Hibiscus sabdariffa*
flor de Jesús – *Petrea arborea*
flor de julio – *Zinnia peruviana*
flor de junio – *Zinnia peruviana*
flor de Júpiter – *Lychnis flos-jovis*
flor de la abeja – *Ophrys apifera*
flor de la alferecía – *Disocactus flagelliformis*
flor de la calentura – *Asclepias curassavica*
flor de la candelaria – *Cattleya skinneri* var *alba, Epidendrum atropurpureum, Exogonium bracteatum, Laelia superbiens*
flor de la candelaria de vara larga – *Laelia superbiens*
flor de la candelaria grande – *Laelia superbiens*
flor de la cruz – *Calliandra haematocephala, Plumeria rubra*
flor de laguna – *Nymphaea ampla, Pontederia cordata*
flor de la mañana – *Ipomoea fistulosa*
flor de la pasión – *Passiflora*
flor de la piña – *Scilla peruviana*
flor de la reina – *Lagerstroemia speciosa*
flor de las ánimas – *Laelia autumnalis*
flor de las cinco llagas – *Passiflora manicata*
flor de las manitas – *Chiranthodendron pentadactylon*
flor de la Trinidad – *Viola* x *wittrockiana*
flor de la vida – *Filago germanica*
flor de la Virgen – *Turbina corymbosa*
flor de lazo – *Lilium lancifolium*
flor del corazón – *Magnolia grandiflora, Talauma mexicana*
flor de leche – *Euphorbia colorata*
flor del indio – *Silene laciniata* ssp *major*
flor de lis – *Amaryllis formossisima, Iris florentina, I. germanica, Sprekelia formosissima*
flor del lagarto – *Stapelia gigantea*
flor del látigo – *Disocactus flagelliformis*
flor de los dioses – *Dodecatheon meadia*
flor de los hielos – *Gentiana adsurgens, G. spathacea*
flor de los muertos – *Lealia autumnalis, Oncidium tigrinum, Tagetes erecta, T. patula*
flor de los santos – *Laelia autumnalis*
flor de loto – *Nelumbo nucifera*
flor del secreto – *Senna alata*
flor del soldado – *Cestrum elegans*

flor de luna – *Ipomoea alba*
flor del viento – *Pulsatilla vulgaris*
flor de maravilla – *Mirabilis jalapa, Tigridia pavonia*
flor de María – *Aster moranensis, Calochortus barbatus, Lobelia gruina, Plumeria rubra, Polygala*
flor de mariposa – *Schizanthus pinnatus*
flor de mayo – *Atamosco carinata, Cattleya labiata, Cooperia pedunculata, Laelia grandiflora, L. speciosa, Milla biflora, Plumeria acutifolia, P. rubra, Sprekelia formosissima, Zephyranthes carinata*
flor de mayo blanca – *Plumeria acutifolia*
flor de mayo roja – *Plumeria rubra*
flor de mazorca – *Galphimia glauca*
flor de milpa – *Castilleja arvensis*
flor de mocasín – *Cypripedium irapeanum*
flor de muerto – *Bletia campanulata, B. reflexa, Calendula officinalis, Chrysanthemum coronarium, Datura stramonium, Dyssodia chrysanthemoides, D. papposa, Gonolobus erianthus, Laelia anceps, L. autumnalis, Oncidium tigrinum, Tagetes*
flor de nabo – *Brassica napus* var *napobrassica, B. rapa*
flor de nacimiento – *Jacquinia aurantiaca*
flor de Navidad – *Euphorbia heterophylla, E. pulcherrima*
flor de nieve – *Gentiana calyculata, Leontopodium alpinum*
flor de niño – *Acacia farnesiana, Jacquinia aurantica, Laelia superbiens, Senecio chenopioides*
flor de nochebuena – *Euphorbia pulcherrima*
flor de novia – *Eriogonum tenellum*
flor de ocote – *Arceuthobium globosum, A. vaginatum*
flor de oreja – *Cymbopetalum penduliflorum*
flor de oro – *Chrysanthemum coronarium, C. indicum, C. morifolium, Tagetes erecta, T. patula*
flor de otoño – *Colchicum autumnale*
flor de paisto – *Hibiscus bifurcatus*
flor de pájaro – *Brassica rapa, Strelitzia reginae*
flor de palma – *Yucca*
flor de palo – *Psittacanthus calyculatus*
flor de palomita – *Bourreria huanita*
flor de papagayo – *Psittacanthus calyculatus*
flor de papel – *Baileya multiradiata, Bougainvillea spectabilis*
flor de paragüito – *Vinca minor*
flor de pascua – *Euphorbia pulcherrima, Pusatilla vulgaris, Turbina corymbosa*
flor de pato – *Aristolochia grandiflora*
flor de pelícano – *Aristolochia grandiflora, Cypripedium*
flor de peña – *Drymaria leptophylla, Echeveria secunda, Selaginella*
flor de perrito – *Commelina dianthifolia*
flor de piedra – *Echeveria elegans, Eucnide rupestris, Selaginella*
flor de pito – *Yucca aloifolia*
flor de pluma – *Achillea millefolium*
flor de primavera – *Arum italicum, A. maculatum, Primula veris, Trifolium repens*
flor de príncipe – *Catharanthus roseus*
flor de pino – *Arceuthobium*
flor de piojo – *Alvaradoa amorphoides*
flor de rayo – *Parkinsonia aculeata*
flor de roca – *Monarda macrantha*
flor de San Cayetano – *Lantana camara*
flor de San Diego – *Antigonon leptopus, Laelia anceps, Solanum jasminoides*
flor de San Francisco – *Caesalpinia pulcherrima, Gomphrena decumbens, G. globosa, Heima salicifolia, Laelia albida, L. furfuracea*
flor de sangre – *Asclepias curassavica, Calliandra houstoniana, Tropaeolum majus*
flor de San José – *Alcea rosea, Gliricidia sepium, Primula officinalis, Senna atomaria, S. bicapsularis, S. pallida, Sphaeralcea rosea*
flor de San Juan – *Bouvardia longiflora, Macrosiphonia hypoleuca, M. macrosiphoni*
flor de San Miguel – *Antigonon leptopus, Laelia anceps*
flor de San Miguelito – *Antigonon leptopus*
flor de San Nicolás – *Milla biflora*
flor de San Pallari – *Trollius europaeus*
flor de San Pedro – *Cigarrilla mexicana, Tecoma stans*
flor de San Sebastián – *Cattleya deckeri, C. skinneri* var *alba, Epidendrum atropurpureum*

flor de Santa Catalina – *Euphorbia pulcherrima*
flor de Santa Catarina – *Euphorbia pulcherrima*
flor de Santa María – *Petrea arborea, Tagetes filifolia, T. lucida, T. tenuifolia*
flor de Santa Rita – *Castilleja linariifolia, Oenothera biennis, O. villosa, Petrea arborea*
flor de Santiago – *Sprekelia formosissima*
flor de Santo Domingo – *Gentiana calyculata*
flor de seda – *Amaranthus caudatus, Portulaca grandiflora*
flor de serpiente – *Stanhopea devoniensis*
flor de sol – *Carpobrotus chilensis, Helianthus annuus, Mesembryanthemum crystallinum, M. nodiflorum*
flor de sonaja – *Bomarea acutifolia*
flor de tecolón – *Aechmea bracteata*
flor de templo – *Plumeria rubra*
flor de tierra – *Lennoa coerulea, Monotropa uniflora, Orobanche cooperi*
flor de tierradentro – *Tagetes lucida, T. tenuifolia*
flor de tigre – *Tigridia pavonia*
flor de todos los meses – *Calendula officinalis*
flor de todos santos – *Laelia autumnalis*
flor de torito – *Stanhopea tigrina*
flor de trompeta – *Campsis radicans, Datura stramonium*
flor de tuna blanca – *Opuntia tuna*
flor de una hora – *Hibiscus trionum*
flor de un día – *Chysis aurea, Hemerocallis fulva, Hibiscus mutabilis, Tigridia pavonia*
flor de venadillo – *Swietenia humilis*
flor de verano – *Bougainvillea spectabilis, Ipomoea hederacea*
flor de viuda – *Trachelium caeruleum*
flor espinosa – *Disocactus speciosus*
floricuerno – *Disocactus flagelliformis*
floripondio – *Brugmansia*
floripondio blanco – *Brugmansia suaveolens*
floripondio del monte – *Solandra maxima, S. nitida*
floripondio del norte – *Solandra maxima*
floripondio encarnado – *Brugmansia sanguinea*
floripondio rojo – *Brugmansia sanguinea*
floripondio del norte – *Solandra maxima*
flor izquierda – *Cochlospermum vitifolium*
flor manchada – *Tigridia pavonia*
flor mechudo – *Calliandra houstoniana*
flor morada – *Nama hispidum* var *spathulatum*
flor porcelana – *Anthurium xanthosomifolium*
flor que pinta – *Cosmos sulphureus*
flor roja – *Cuphea llavea*
flor serrana – *Zinnia linearis*
flor trompeta – *Thevetia peruviana*
flox – *Phlox*
flox alpino – *Phlox douglasii* ssp *rigida*
flox enano – *Phlox douglasii* ssp *rigida, P. subulata*
flox musgoso – *Phlox subulata*
flox paniculado – *Phlox paniculata*
fluvia del oro – *Tecoma stans*
foliota nameko – *Pholiota nameko*
formio – *Phormium tenax*
forsitia – *Forsythia*
fortuniana – *Rosa fortuniana*
fotinia – *Photinia serratifolia*
fotuto – *Laelia tibicinis*
fraile – *Couepia polyandra, Lythrum, Thevetia thevetioides*
frailecillo – *Couepia polyandra, Jatropha gossypiifolia, Licania arborea, Lythrum salicaria*
frailejón – *Jatropha curcas*
framboyán – *Delonix regia*
framboyán azul – *Jacaranda acutifolia*
framboyán francés – *Caesalpinia pulcherrima*
framboyán rojo – *Delonix regia*
frambueso – *Rubus idaeus, R. neomexicanus, R. strigosus*
frambueso amarillo – *Rubus chamaemorus*
frambueso americano – *Rubus loganobaccus*
frambueso de África – *Rubus rosaefolius*
frambueso encalado – *Rubus cockburnianus*
frambueso espléndido – *Rubus spectabilis*
frambueso japonés – *Rubus phoenicolasius*
frambueso negro – *Rubus occidentalis*
frambueso ornamental – *Rubus tridel*
frambueso rojo – *Rubus idaesus*
francesilla – *Ranunculus*
francesita – *Ranunculus*
franchipán – *Plumeria acutifolia, P. rubra*
frangipán – *Plumeria rubra*
frangipani – *Plumeria rubra*
frángula – *Rhamnus frangula*
franseria – *Ambrosia deltoidea*
fraxinella – *Dictamnus albus*

fredeluch – *Tricholoma terreum*
fredolic – *Tricholoma terreum*
frejol – *Vigna unguiculata*
fremontia – *Fremontodendron californicum*
fresa – *Desmanthus, Fragaria, Potentilla*
fresa ananás – *Fragaria ananassa*
fresa cimarrona – *Potentilla thurberi*
fresa común – *Fragaria vesca*
fresa chilena – *Fragaria chiloensis*
fresa de burro – *Trifolium fragiferum*
fresa de Virginia – *Fragaria virginiana*
fresadilla – *Physalis philadelphica*
fresa silvestre – *Duchesnea indica, Fragaria mexicana, F. vesca* ssp *californica*
frescura – *Pilea microphylla*
fresera – *Fragaria ananassa, F. vesca*
fresia – *Freesia refracta*
fresia de exterior – *Freesia refracta*
fresnillo – *Dictamnus albus, Fraxinus gooddingii, Tecoma stans*
fresno – *Fraxinus, Ludwigia octovalvis, Ptelea trifoliate, Tabebuia pentaphylla*
fresno blanco – *Fraxinus americana*
fresno blanco americano – *Fraxinus americana*
fresno común – *Fraxinus excelsior*
fresno de flor – *Fraxinus ornus*
fresno de hojas estrechas – *Fraxinus angustifolia*
fresno de hojas pequeñas – *Fraxinus angustifolia*
fresno de la tierra – *Fraxinus angustifolia*
fresno del maná – *Fraxinus ornus*
fresno del sur – *Fraxinus angustifolia*
fresno de Pensilvania – *Fresno pennsylvanica*
fresno de viscaya – *Fraxinus excelsior*
fresno espino – *Zanthoxylum americanum*
fresno espinoso – *Zanthoxylum americanum*
fresno florido – *Fraxinus ornus*
fresno florifero – *Fraxinus ornus*
fresno grande – *Fraxinus excelsior*
fresno rojo americano – *Fraxinus pennsylvanica*
fresno terciopelo – *Fraxinus velutina*
fresno verde – *Fraxinus pennsylvanica*
fresón – *Fragaria chilensis, Rubus strigosus*
friega plato – *Solanum verbascifolium*
frigolito – *Sophora secundiflora*
frijol – *Phaseolus coccineus, P. filiformis, P. vulgaris, Rhynchosia precatoria, Sophora secundiflora, Vigna sinensis*
frijol adzuki – *Vigna angularis*
frijol alado – *Psophocarpus tetragonolobus*
frijol amarillo – *Phaseolus calcaratus*
frijol anasazi – *Phaseolus vulgaris*
frijol aterciopelado – *Stizolobium deeringianum*
frijol ayocote – *Phaseolus coccineus*
frijol blanco – *Canavalia ensiformis, Phaseolus lunatus, P. vulgaris*
frijol blanco corto – *Phaseolus vulgaris*
frijol botil – *Phaseolus coccineus*
frijol caballero – *Lablab purpureus*
frijol carita – *Pisum sativum*
frijol castillo – *Vigna unguiculata* ssp *sesquipedalis*
frijol colorado – *Phaseolus vulgaris*
frijol colorado largo – *Phaseolus vulgaris*
frijol comba – *Phaseolus lunatus*
frijol congo – *Cajanus cajun*
frijol chamborote – *Phaseolus coccineus*
frijol chino – *Vigna radiata*
frijol de adorno – *Lablab purpureus*
frijol de árbol – *Cajanus cajun*
frijol de arroz – *Phaseolus calcaratus, Vigna umbellata*
frijol de campo – *Vigna sinensis*
frijol de Castilla – *Vigna unguiculata*
frijol de cuerno – *Vigna sinensis, V. unguiculata*
frijol de chintlatlahua – *Dolichos phaseoloides, Rhynchosia pyramidalis*
frijol de enredadera – *Lablab purpureus*
frijol de España – *Phaseolus coccineus*
frijol de gordoniz – *Desmodium scorpiurus*
frijol de la India – *Cajanus cajan*
frijol de la peladilla – *Phaseolus lunatus*
frijol de Lima – *Phaseolus lunatus*
frijol del monte – *Cajanus cajun, Phaseolus coccineus, Senna oxyphylla*
frijol del Perú – *Vigna unguiculata* ssp *sesquipedalis*
frijol de luna – *Phaseolus lunatus*
frijol de machete – *Canavalia gladiata*
frijol de maíz – *Vigna sinensis, V. unguiculata*
frijol de medialuna – *Phaseolus lunatus*
frijol de ñame – *Pachyrrhizus erosus*
frijol de ojos negros – *Vigna unguiculata*

frijol de olor – *Lathyrus odoratus*
frijol de ombligo negro – *Vigna unguiculata*
frijol de oro – *Vigna radiata*
frijol de palo – *Cajanus cajun*
frijol de pinta negra – *Vigna unguiculata*
frijol de playa – *Canavalia ensiformis, C. maritima, C. villosa*
frijol de sable – *Canavalia ensiformis, C. gladiata*
frijol de terciopelo – *Stizolobium deeringianum*
frijol de vaca – *Vigna unguiculata*
frijolera – *Vigna unguiculata* ssp *sesquipedalis*
frijol espada – *Canavalia ensiformis, C. gladiata*
frijol fava – *Vicia faba*
frijol gandul – *Cajanus cajan*
frijol grande – *Canavalia gladiata*
frijolillo – *Acacia goldmanii, A. mcmurphyi, Canavalia villosa, , Cologania obovata, Desmanthus covillei, D. fruticosus, Dolichos phaseoloides, Gliricidia sepium, Oxytropis lambertii, Phaseolus coccineus, P. lunatus, Rhynchosia pyramidalis, Senna occidentalis, Sophora secundiflora*
frijolín – *Rhynchosia macrocarpa*
frijolito – *Canavalia villosa, Desmanthus fruticosus, Phaseolus filiformis, Sophora secundiflora*
frijol jacinto – *Lablab purpureus*
frijol lima – *Phaseolus lunatus*
frijol mungo – *Vigna radiata*
frijol negro – *Phaseolus vulgaris*
frijol ojinegro – *Vigna sinensis*
frijolón – *Canavalia ensiformis*
frijol ordinario – *Phaseolus vulgaris*
frijol peruano – *Phaseolus vulgaris*
frijol pinto – *Phaseolus vulgaris*
frijol precioso – *Vigna unguiculata*
frijol punzada – *Lablab purpureus*
frijol rojo – *Phaseolus vulgaris, Vigna umbellata*
frijol rojo de seda – *Phaseolus vulgaris*
frijol rojo pequeño – *Phaseolus vulgaris*
frijol rosado – *Phaseolus vulgaris*
frijol silvestre – *Phaseolus filiformis*
frijol soya – *Glycine max*
frijol tepari – *Phaseolus acutifolius*
frijol terciopelo – *Mucuna pruriens, Stilozobium deeringianum*
frijol terciopelo de Florida – *Mucuna deeringiana, Stizolobium deeringianum*
frijol trepador – *Lablab purpureus*
frijol tripa de tuza – *Vigna unguiculata*
fritilaria – *Fritillaria biflora*
fritilaria de cabeza de serpiente – *Fritillaria meleagris*
fritilaria imperial – *Fritillaria imperialis*
fritilaria pérsica – *Fritillaria persica*
fromental – *Arrhenatherum elatius*
fruta bomba – *Carica papaya*
fruta de condesa – *Annona squamosa*
fruta de iguana – *Duranta erecta*
fruta de la pasión – *Passiflora quadrangularis*
fruta de la perdiz – *Mitchella repens*
fruta del maní – *Arachis hypogaea*
fruta de pan – *Artocarpus altilis*
fruta de víbora – *Rauwolfia hirsuta*
fruta silvestre – *Rubus*
fruta sol – *Heliocarpus tomentosus*
fruta verada – *Citrus limettioides*
frutilla – *Canna indica, Conostegia xalapensis, Duchesnea indica, Fragaria, Lantana camara, Lycium, Vallesia glabra*
frutillo – *Desmanthus covillei, D. fruticosa, Ehretia tinifolia, Karwinskia humboldtiana, Rinorea guatemalensis, Ziziphus amole*
fuco – *Fucus vesiculosus*
fuco marino – *Fucus vesiculosus*
fucsia – *Fuchsia*
fuchila – *Cantharellus cibarius*
fuego – *Toxicodendron radicans*
fulgencia – *Evolvulus alsinoides*
fulmina – *Antigonon leptopus*
fulminante – *Ruellia tuberosa*
fumaria – *Corydalis, Fumaria capreolata, F. officinalis, F. parviflora*
fumaria amarilla – *Corydalis lutea*
fumaria común – *Fumaria officinalis*
fumaria de flores pequeñas – *Fumaria parviflora*
furuna – *Phaseolus lunatus*
fusique – *Heteromeles arbutifolia*
fustete – *Cotinus coggygria, Maclura tinctoria*
fustete amarillo – *Maclura tinctoria*

gabarro – *Reseda luteola*
gabat – *Gerando*

gachupín – *Helvella lacunosa, Sclerocarpus spathulatus*
gachupín negro – *Helvella lacunosa*
gaiaba – *Psidium guajava*
gaillardia – *Gaillardia aristata*
gaillardia anual – *Gaillardia pulchella*
galambo – *Boletus luridus*
galambo bueno – *Boletus luridus*
galán – *Cestrum dumentorum*
galán de noche – *Brugmansia, Brunfelsia pauciflora, Cestrum nocturnum, Datura*
galán de tarde – *Cestrum dumentorum, C. nocturnum*
galanga – *Alpinia officinarum, Cyperus diandrus, Laurus camphora*
galanga de la China – *Alpinia officinarum*
galanto de nieve – *Galanthus nivalis*
galatea – *Calathea, Dieffenbachia, Solanum verbascifolium*
galatea sapo – *Calathea makoyana*
gálbano – *Ferula galbaniflua*
galcerán – *Ruscus aculeatus*
galega – *Cladium californicum, Galega officinalis, Ruta graveolens*
galgana – *Lathyrus cicera*
galganor – *Lathyrus cicera*
galinsoga – *Galinsoga parviflora*
galinsoga ciliada – *Galinsoga quadriradiata*
galinsoga de flores pequeñas – *Galinsoga parviflora*
galio – *Galium odoratum G. verum*
galio boreal – *Galium boreale*
galio de las rocas – *Galium saxatile*
galio de los prados – *Galium mollugo*
galio de tres cuernos – *Galium tricornutum*
galio de tres flores – *Galium triflorum*
galio divaricado – *Galium divaricatum*
galio mural – *Galium murale*
galio oloroso – *Galium odoratum*
galio parisino – *Galium parisiense*
galio primaveral – *Galium verum*
galio trífido – *Galium trifidum*
galusa – *Dahlia*
gallarda – *Hibiscus rosa-sinensis, Morchella esculenta*
gallardete – *Hibiscus rosa-sinensis*
gallardía – *Gaillardia aristata, G. mexicana, G. pulchella*
gallarito – *Pedicularis canadensis*
galleta – *Pleuraphis jamesii, P. rigida*
gallibasco – *Aechmea lueddemanniana*
gallina ciega – *Teucrium cubense*
gallina con pollos – *Dudleya*
gallinazo blanco – *Lablab purpureus*
gallinazo morado – *Lablab purpureus*
gallineta – *Mascagnia macroptera, Pilea cadierei*
gallinita – *Canavalia villosa, Janusia gracilis, J. linearis, Lablab purpureus, Macrolepiota procera, Mascagnia macroptera, Panicum capillare, Phaseolus vulgaris*
gallinita ciega – *Teucrium cubense*
gallinita del cerro – *Exogonium bracteatum*
gallito – *Calochortus, Cosmos diversifolius, Cynodon dactylon, Pedilanthus macrocarpus, Salvia patens, Sesbania grandiflora, Spathodea campanulata, Tillandsia, Viola pedunculata*
gallo – *Dactylis glomerata*
gallocresta – *Bellardia trixago, Salvia verbenacea, S. viridis*
gamalotillo – *Paspalum plicatulum*
gamelote – *Panicum maximum*
gamón – *Asphodelus albus*
gamoncillo – *Asphodelus fistulosus*
gamuza – *Ipomoea alba*
gandul – *Cajanus cajan, Nicotiana glauca*
gandul verde – *Cajanus cajun*
garabatillo – *Caesaria dolichophylla, Mimosa biuncifera, M. distachya, M. laxiflora, Zanthoxylum fagara*
garabato – *Celtis iguanea, C. pallida, Pisonia*
garabato blanco – *Celtis iguanea, C. reticulata, Pisonia aculeata*
garabato prieto – *Pisonia aculeata, P. capitata*
garambullo – *Arctostaphylos, Celtis pallida, Lycium berlandieri, L. torreyi, Microrhamnus ericoides, Myrtillocactus geometrizans, M. schenckii, Pachycereus gatesii, P. schottii, Pisonia capitata, Proboscidea fragrans, P. parviflora, Ribes affine, Rosa montezumae, Viscainoa geniculata, Ziziphus obtusifolia*
garañona – *Castillejas, Chrysactinia mexicana, Heimia salicifolia*
garapalo – *Coriaria myrtifolia*
garaví perenne – *Sorghum almum*
garbancillo – *Astragalus, Bunchosia sonorensis, Crotalaria, Duranta repens,*

Galphimia glauca, Lupinus, Mimosa distachya, Myrtillocactus geometrizans, Peganum mexicanum, Senna pallida, Trichilia hirta
garbanzo – *Cicer arietinum*
garbanzo de coyote – *Bunchosia sonorensis*
garbanzo indio – *Lablab purpureus*
garcita – *Cleome hasslerana*
gardenia – *Gardenia florida, G. jasminoides*
garnacha – *Vitis vinifera* var *sativa*
garza real – *Epidendrum parkinsonianum*
garrachuelo – *Digitaria sanguinalis*
garrachuelo pequeño – *Digitaria ischaemum*
garrapata – *Ziziphus obtusifolia*
garrapatilla – *Trichilia hirta*
garroba – *Vicia articulata*
garrobilla – *Vicia sativa*
garrobo – *Acacia macrantha, A. pennatula, Robinia neomexicana*
garrofero – *Ceratonia siliqua*
garruño – *Mimosa aculeaticarpa, M. biuncifera*
garrya – *Garrya elliptica*
gasa – *Draba cuneifolia, Macfadyena unguis-cati*
gasparilla – *Reseda luteola*
gastridia ventricosa – *Gastridium ventricosum*
gateado – *Astronium graveolens, Swietenia humilis*
gateado galán – *Astronium graveolens*
gatera – *Nepeta cataria*
gatillo casto – *Vitex agnus-castus*
gatiña – *Genista anglica*
gatito – *Acalypha arvensis*
gato – *Acacia crinita, Havardia sonorae, Proboscidea altheaefolia*
gatuña – *Acacia greggii, Mimosa, Proboscidea fragrans, Ulex europaea*
gaulteria – *Gaultheria procumbens, G. shallon*
gavia – *Acacia, Fouquieria campanulata*
gayada – *Primula veris*
gayomba – *Spartium junceum*
gayuba – *Arctostaphylos uva-ursi*
gayuba alpina – *Arctostaphylos alpinus*
gayuba del país – *Arctostaphylos pungens*
gayuba de México – *Arctostaphylos pungens*
gazania – *Gazania nivea*
gazul – *Aizoon hispanicum*
gelatina de palo – *Tremella fuciformis*
gelatina vegetal – *Gelidium cartilagineum*
gelsemio – *Gelsemium sempervirens, Jasminum humile*
genciana – *Gentiana*
genciana acaule – *Gentiana acaulis, G. clusii*
genciana alpina – *Gentiana acaulis*
genciana amarilla – *Gentiana lutea*
genciana asclepiadea – *Gentiana asclepiadea*
genciana axilar – *Gentianella amarella*
genciana cruciata – *Gentiana cruciata*
genciana de Clausio – *Gentiana clusii*
genciana de flores grandes – *Gentiana clusii*
genciana de hojas de algodoncillo – *Gentiana asclepiadea*
genciana de primavera – *Gentiana verna*
genciana mayor – *Gentiana lutea*
genciana pedunculada – *Gentianella tenella*
genciana primaveral – *Gentiana verna*
genciana prostrada – *Gentiana prostrata*
gencianilla – *Gentiana clusii*
genipa – *Genipa americana*
genista – *Genista canariensis, Spartium junceum*
geonduna – *Kallstroemia*
geranio – *Geranium, Pelargonium*
geranio argénteo – *Geranium argenteum*
geranio aromático – *Pelargonium odoratissimum*
geranio cinéreo – *Geranium cinereum*
geranio de enredadera – *Pelargonium peltatum*
geranio de Endress – *Geranium endressii*
geranio de hiedra – *Pelargonium peltatum*
geranio de los caminos – *Geranium molle*
geranio de los prados – *Geranium pratense*
geranio de olor – *Pelargonium graveolens, P. odoratissimum*
geranio de raíces grandes – *Geranium macrorrhizum*
geranio disecado – *Geranium dissectum*
geranio enredador – *Pelargonium peltatum*
geranio malva – *Pelargonium* x *fragrans*
geranio menor – *Geranium pusillum*
geranio moreno – *Geranium phaeum*
geranio muelle – *Geranium molle*
geranio platipétalo – *Geranium platypetalum*
geranio rosa – *Pelargonium odoratissimum*
geranio sanguineo – *Geranium sanguineum*
geranio selvático – *Geranium sylvaticum*
geranio silvestre – *Geranium maculatum, G. mexicanum*

geranio trepador – *Pelargonium peltatum*
gerbera – *Gerbera jamesonii*
germandria – *Teucrium*
germandria de agua – *Teucrium scordium*
germandria fruticosa – *Teucrium fruticans*
germandria pulverulenta – *Teucrium cossonii*
germandrina – *Teucrium* x *lucidrys*
geum alpino – *Geum montanum*
geum de Chile – *Geum chiloense*
gibguin – *Stanhopea wardii*
gigante – *Eucalyptus fibrosa, E. globulus, Neobuxbaumia mezcalaensis, Nicotiana glauca, Peniocereus serpentines, Pennisetum purpureum, Selenicereus grandiflorus*
gigantillo – *Acacia constricta*
gigantón – *Dahlia excelsa, D. imperialis, Echinopsis pachanoi, Helianthus annuus, Tithonia*
giloria – *Digitalis purpurea, D. thapsi*
gimnoclado del Canadá – *Gymnocladus dioica*
gimnoclado dioico – *Gymnocladus dioica*
gimnoclado norteamericano – *Gymnocladus dioica*
ginebra – *Juniperus communis*
ginero – *Cortaderia selloana*
ginko – *Ginkgo biloba*
ginseng – *Panax ginseng, P. quinquefolium*
ginseng americano – *Panax quinquefolium*
ginseng asiático – *Panax ginseng*
ginseng brasileño – *Pfaffia paniculata*
ginseng coreano – *Panax ginseng*
ginseng chino – *Panax ginseng*
ginseng ruso – *Eleutherococcus senticosus*
ginseng siberiano – *Eleutherococcus senticosus*
ginseng silvestre americano – *Panax quinquefolium*
ginseng tienchi – *Panax notoginseng*
gipsófila – *Gypsophila elegans, G. paniculata*
gipsófila elegante – *Gypsophila elegans*
gipsófila paniculada – *Gypsophila paniculata*
gipsófila repente – *Gypsophila repens*
girasol – *Cosmos bipinnatus, Encelia mexicana, Helianthus, Tithonia, Viguiera dentata*
girasol amarillo – *Cosmos sulphureus*
girasol anual – *Helianthus annuus*
girasol común – *Helianthus annuus*
girasol de agua – *Bidens laevis*
girasolillo – *Verbesina encelioides* ssp *exauriculata*
girasol mexicano – *Tithonia rotundifolia*
girasol morado – *Cosmos bipinnatus*
girbola de bruch – *Lyophyllum aggregatum*
girbola de pollancre – *Pleurotus ostreatus* var *sativus*
gitanilla – *Pelargonium peltatum*
gitarán – *Dodonaea viscosa*
gladiolo – *Gladiolus colvillei, G. hortulanus, Typha latifolia*
gladiolo bizantino – *Gladiolus communis* ssp *byzantinum*
gladiolo de los campos – *Gladiolus italicus*
gladiolo menor – *Gladiolus* x *colvillei*
gladiolo oriental – *Gladiolus communis* ssp *byzantinum*
gladiolo primulinus – *Gladiolus dalenii*
glasto – *Isatis tinctoria*
glaucio – *Glaucium flavum*
glaucio amarillo – *Glaucium flavum*
glaucio roso – *Glaucium corniculatum*
glaux marítima – *Glaux maritima*
gleditsia de tres espinas – *Gleditsia triacanthos*
gliceria acuática – *Impatiens walleriana*
gliceria de ramificación rezagada – *Puccinellia distans*
gliceria estriada – *Glyceria striata*
glicina – *Wisteria sinensis*
globularia – *Globularia*
globularia común – *Globularia vulgaris*
globularia cordifoliada – *Globularia cordifolia*
globularia de hojas cordiformes – *Globularia cordifolia*
globularia mayor – *Globularia vulgaris*
globularia menor – *Globularia cordifolia*
gloria – *Solanum, Tecoma stans*
gloria de la mañana – *Ipomoea purpurea*
gloriosa – *Gloriosa superba*
gloxínea – *Sinningia speciosa*
gloxínea de intemperie – *Incarvillea delavayi*
gobernadora – *Brickellia veronicaefolia, Flacourtia indica, Larrea divaricata, L. tridentate, Zizyphus jujuba*
gobernadora de Puebla – *Brickellia cavanillesi, B. veronicaefolia*
godecia – *Clarkia amoena*
godiera – *Goodyera oblongifolia*
gogolo – *Brassica oleracea* var *capitata*

golondrilla – *Buddleja scordioides, Chamaesyce albomarginata*
golondrina – *Boerhavia erecta, Chamaesyce, Crusea calocephala, Dichondra argentea, Euphorbia, Kallstroemia californica, Pilea microphylla*
golondrina grande – *Euphorbia hirta*
golondrina serrana – *Euphorbia adenoptera*
golondrinera – *Chelidonium majus*
goma – *Castilla elastica*
goma adragante – *Astragalus gummifer*
goma arábiga – *Acacia senegal*
goma de guar – *Cyamopsis tetragonoloba*
goma elástica – *Ficus elastica*
gomaguta – *Garcinia lianburyi*
goma karaya – *Sterculia urens*
goma laca – *Ficus indica*
gombó – *Hibiscus esculentus*
gomea – *Coursetia glandulosa*
gomenol – *Melaleuca viridiflora*
gomero – *Ficus elastica*
gomilla – *Coursetia glandulosa*
gomo blanco – *Cordia alba*
gomosilla – *Bursera lancifolia*
gonote real – *Ochroma pyramidale*
goñi – *Potentilla candicans*
gorbiza – *Pedicularis canadensis*
gordolobo – *Anaphalis margaritacea, Bocconia frutescens, Eriogonum fasciculatum, Gnaphalium, Heterotheca psammophila, Senecio, Solanum verbascifolium, Verbascum bombyciferum, V. thapsus*
gordolobo austriaco – *Verbascum chaixii*
gordolobo del país – *Gnaphalium conoideum*
gordolobo especioso – *Verbascum speciosum*
gordolobollo – *Primula veris*
gordolobo púrpura – *Verbascum phoeniceum*
gordolobo purpúreo – *Verbascum phoeniceum*
gordolobo ramificado – *Verbascum virgatum*
gordolobo rocoso – *Verbascum dumulosum*
gordura – *Melinis minutiflora*
gorojo – *Juniperus communis*
gorrinchia – *Arctostaphylos uva-ursi*
gorro de cardenal – *Astrophytum myriostigma*
gorro de león – *Philodendron bipinnatifidum*
gorro de Napoleón – *Solandra nitida*
gorro de Turco – *Gloriosa superba*
gota de coral mexicano – *Bessera elegans*
gotu kola – *Hydrocotyle asiatica*
graciola – *Gratiola officinalis*
graciola común – *Gratiola officinalis*
grado – *Croton draco*
grama – *Bouteloua gracilis, B. hirsuta, Cathestecum erectum, Cynodon dactylon, Elytrigia repens, Muhlenbergia rigida, Paspalum distichum, P. racemosum, Pleuraphis belangeri, P. cenchroides*
grama alargada – *Elytrigia elongata*
grama alema – *Paspalum dilatatum*
grama argentina – *Axonopus compressus*
grama canina – *Elytrigia repens*
grama castilla – *Panicum maximum*
grama cebollera – *Poa bulbosa*
grama colorada – *Paspalum distichum*
grama común – *Cynodon dactylon, Paspalum notatum*
grama china – *Cathestecum erectum, C. prostratum, Sorghum halepense*
grama de agua – *Paspalum dilatatum, P. distichum*
grama de antena – *Paspalum conjugatum*
grama de Bermuda – *Cynodon dactylon*
grama de caballo – *Eleusine indica*
grama de camino – *Eleusine indica*
grama de Castilla – *Panicum repens*
grama de chosica – *Paspalum virgatum*
grama de la costa – *Cynodon dactylon, Sporobolus indicus*
grama de las dunas – *Ammophila arenaria*
grama de las playas – *Elytrigia juncea*
grama del colorado – *Pascopyrum smithii*
grama del este – *Tripsacum laxum*
grama del norte – *Elytrigia repens*
grama de los prados – *Anthoxanthum odoratum*
grama de marajó – *Cynodon dactylon*
grama de olor – *Anthoxanthum odoratum*
grama de pará – *Brachiaria mutica, Cynodon dactylon*
grama de prados – *Elytrigia repens, Poa pratensis*
grama de Rhodes – *Chloris gayana*
grama de seda – *Cynodon dactylon*
grama dorada – *Lamarckia aurea*
grama dulce – *Eleusine indica, Paspalum distichum, P. notatum, Stenotaphrum secundatum*
grama en jopillos – *Dactylis glomerata*
grama intermedia – *Elytrigia intermedia*

gramalote – *Axonopus scoparius, Panicum coloratum, P. maximum*
grama menor – *Cynodon dactylon*
grama morada – *Echinochloa crus-galli*
grama nativa – *Paspalum notatum*
grama negra – *Bouteloua eriopoda, Paspalum plicatulum*
grama oriental – *Cymbopogon citratus*
grama pata de gallina – *Eleusine indica*
grama pie de pollo – *Cynodon dactylon*
grama portuguesa – *Panicum repens*
grama repente – *Elytrigia repens*
gramari – *Arrhenatherum elatius*
grama roja – *Carex arenaria*
grama sanduva – *Paspalum dilatatum*
grama tapete – *Axonopus affinis*
grama trenza – *Axonopus compressus, Paspalum notatum*
grama verde – *Tripsacum laxum*
grama zaina – *Panicum coloratum*
grama zastrera – *Cynodon dactylon*
gramigna del Parnaso – *Maianthemum dilatatum*
gramilla – *Bouteloua curtipendula, Paspalum notatum*
gramilla blanca – *Paspalum dilatatum, P. distichum, P. notatum*
gramilla brava – *Cynodon dactylon*
gramilla colorada – *Cynodon dactylon*
gramilla de huerto – *Dactylis glomerata, Eragrostis mexicana*
gramilla de San Agustín – *Stenotaphrum secundatum*
gramilla forestal – *Cynodon dactylon*
gramilla italiana – *Cynodon dactylon*
gramilla perdiz – *Gouinia virgata*
gramilla zacate – *Digitaria sanguinalis*
gramillón – *Cynodon dactylon, Stenotaphrum secundatum*
gramillón de San Agustín – *Stenotaphrum secundatum*
graminea guinea – *Panicum maximum*
gramón – *Cynodon dactylon*
granada – *Passiflora edulis, P. jorullensis*
granada cimarrona – *Eugenia capuli*
granada china – *Passiflora edulis, P. incarnata*
granada de China – *Passiflora ligularis*
granada de gorro – *Passiflora jorullensis*
grana d'Avignon – *Rhamnus infectorius*
grana de paraíso – *Elettaria cardamomum*
granadilla – *Digitaria sanguinalis, Jarilla heterophylla, Malpighia emarginata, Maytenus phyllanthoides, Passiflora, Trophis racemosa*
granadilla de cuaresma – *Digitaria sanguinalis*
granadilla de China – *Passiflora edulis*
granadilla dulce – *Passiflora ligularis*
granadilla morada – *Passiflora edulis*
granadilla real – *Passiflora quadrangularis*
granadillo – *Couroupita guianensis, Dalbergia, Heimia salicifolia, Salpianthus arenarius*
granadillo blanco – *Chiococca petrina*
granadina – *Dodonaea viscosa*
granadita – *Bomarea acutifolia, B. hirtella, Colubrina viridis, Passiflora ligularis*
granadita china – *Passiflora ligularis*
granadita de China – *Passiflora edulis, P. ligularis*
granadito – *Passiflora palmeri*
granado – *Hibiscus trionum, Punicum granatum*
granado agridulce – *Punica granatum*
granado agrio – *Punica granatum*
granado blanco – *Hypericum calycinum*
granado cordelina – *Punica granatum*
granado de China – *Punica granatum*
granado dulce – *Punica granatum*
granado enano – *Punica granatum* var *nana*
grana encarnada – *Phytolacca americana*
granduque – *Jasminum*
grangene – *Ephedra aspera*
granillo – *Melia azedarach*
granjel – *Randia*
granjeno – *Celtis iguanea, C. pallida, Vaccinium confertum*
granjeno amarillo – *Celtis pallida*
granjeno huasteco – *Celtis pallida*
gran jicoria de corteza conchosa – *Carya laciniosa*
grano – *Triticum vulgare*
grano de amor – *Lithospermum diffusum, L. officinale*
grano del paraíso – *Aframomum melegueta*
grano de oro – *Hypericum formosum* var *scouleri*
granotino – *Leucaena glauca*
granza – *Potamogeton pectinatus, Rubia tinctorium*

grasilla – *Pinguicula vulgaris*
grasilla de flor grande – *Pinguicula grandiflora*
graviola – *Annona muricata*
gretado – *Cupressus*
gretado amarillo – *Cupressus lusitanica*
gretado galan – *Cupressus lusitanica*
gretaña – *Cordia eleagnoides, Nicotiana glauca*
grevillo – *Grevillea robusta*
grifa – *Cannabis sativa*
grindelia – *Grindelia aphanactis*
griñón – *Prunus persica* ssp *nucipersica*
grisiño – *Cordia eleagnoides, C. sebestena*
groguer – *Tricholoma flavovirens*
grosellero – *Cyphomandra betacea, Malpighia emarginata, Pereskia aculeata, P. grandifolia, Phyllanthus acidus, Ribes*
grosellero alpino – *Ribes alpinum*
grosellero amarillo – *Ribes aureum*
grosellero americano – *Pereskia aculeata*
grosellero cimarrón – *Phyllanthus acuminatus*
grosellero chino – *Actinidia deliciosa*
grosellero de Florida – *Pereskia aculeata*
grosellero de los Alpes – *Ribes alpinum*
grosellero de montaña – *Ribes alpinum*
grosellero dorado – *Ribes aureum*
grosellero espinoso – *Ribes*
grosellero floreciente – *Ribes sanguineum*
grosellero negro – *Ribes nigrum*
grosellero ornamental – *Ribes odoratum*
grosellero pétreo – *Ribes petraeum*
grosellero rojo – *Ribes rubrum*
grosellero silvestre – *Ribes*
grumichama – *Eugenia dombeyi*
guacabé – *Randia laevigata*
guacal – *Crescentia cujete*
guacamaya – *Caesalpinia gilliesii, C. pulcherrima, Delonix regia, Euphorbia pulcherrima, Heliconia latispatha, Tillandsia imperialis*
guacamote – *Manihot dulcis, M. esculenta*
guacamote dulce – *Manihot dulcis*
guacamotli – *Manihot esculenta*
guacanalá – *Myrica cerifera, M. mexicana*
guacapí – *Ceiba acuminata*
guacaporo – *Parkinsonia aculeata*
guacimilla – *Belotia mexicana, Colubrina glomerata, Trema micrantha*
guácimo – *Guazuma ulmifolia*
guácimo blanco – *Guazuma ulmifolia*
guácimo cimarrón – *Guazuma ulmifolia*
guácimo de ternero – *Guazuma ulmifolia*
guaco – *Aristolochia, Cleome serulata, Cleomella longipes, Mikania guaco, Parthenocissus quinquefolia, Wislizenia refracta* var *palmeri*
guaco del sur – *Aristolochia mexicana*
guaco de Merida – *Aristolochia pentandra*
guaco de San Cristobal – *Aristolochia pentandra*
guacolote – *Caesalpinia crista*
guacolote prieto – *Caesalpinia crista*
guacoyul – *Acrocomia mexicana, Bactris mexicana, Orbignya cohune*
guacta – *Pachira aquatica, P. macrocarpa*
guachalala – *Caesalpinia cacalaco*
guachapori – *Cenchrus*
guachaporito – *Triumfetta semitriloba*
guachaporo de la Sierra – *Triumfetta polyandra*
guachichi – *Garrya flavescens, G. laurifolia*
guachichil – *Bocconia frutescens*
guachichila – *Loeselia ciliata*
guachichile – *Bocconia frutescens*
guachichi ouitano – *Garrya laurifolia*
guachili – *Bocconia arborea, B. frutescens*
guachipil – *Alvaradoa amorphoides*
guachomó – *Dodonaea viscosa*
guadalagua – *Toxicodendron radicans*
Guadalupe – *Lobelia laxiflora*
guadua – *Guadua angustifolia, G. latifolia*
guaguayote – *Carica mexicana*
guáibasim – *Psidium guajava*
guaimaro – *Poulsenia armata*
guajaca – *Tillandsia usneoides*
guajava – *Psidium guajava, Senna alata*
guaje – *Alvaradoa amorphoides, Crescentia cujete, Lagenaria siceraria, Leucaena, Lycopersicon esculentum*
guaje cirial – *Cresentia alata, C. cujete*
guaje cirián – *Crescentia alata, C. cujete*
guaje de Castilla – *Leucaena glauca*
guaje de peregrino – *Lagenaria siceraria*
guajilote – *Bombax ellipticum, Parmentiera edulis, Pseudobombax palmeri*
guajillo – *Acacia angustissima, A. berlandieri, A. coulteri, Caesalpinia mexicana, Capsicum annuum, Havardia pallens, Leucaena*

lanceolata, Lycopersicon esculentum, Pithecellobium brevifolium, Senna laevigata
guajito – *Lagenaria siceraria*
guajtomate – *Physalis acuminata, P. pubescens, P. stapelioides, Solanum eleagnifolium*
gualanday – *Jacaranda acutifolia*
gualda – *Reseda luteola*
gualdaperra – *Digitalis purpurea*
gualeya – *Citrullus colocynthis*
gualulo – *Jatropha spathulata, Sapindus saponaria*
guamacho – *Pereskia aculeata*
guamacho morado – *Pereskia grandifolia*
guama de México – *Lonchocarpus longistylus*
guamara – *Bromelia pinguin*
guamis – *Larrea tridentata*
guamito – *Coccothrinax argentea*
guamo – *Inga edulis*
guamoche – *Pithecellobium dulce*
guamol – *Ipomoea alba*
guamú – *Goldmania foetida*
guamuchi – *Pithecellobium dulce*
guamúchil – *Acacia pringlei, Havardia mexicana, Lysiloma acapulcensis, Pithecellobium dulce*
guamúchili – *Pithecellobium dulce*
guamuchilillo – *Acacia pringlei* ssp *california, Pithecellobium unguis-cati*
guamúchitl – *Pithecellobium dulce*
guamusi – *Tagetes lucida*
guanaba – *Annona muricata*
guanábano – *Annona muricata, Cymbopetalum mayanum*
guanábano cimarrón – *Rollinia mucosa*
guanacaste – *Enterolobium cyclocarpum*
guananiquil – *Commicarpus brandegeei*
guandsú – *Vigna subterranea*
guandu – *Cajanus cajan*
guanina – *Senna occidentalis*
guanito talis – *Calyptrogyne ghiesbreghtiana*
guano – *Carludovica palmata, Sabal japa, S. mexicana*
guano de sombrero – *Erythea edulis*
guano redondo – *Sabal mexicana*
guante de Nuestra Señora – *Digitalis purpurea*
guantelete – *Digitalis purpurea*
guaorócochi – *Ficus cotinifolia*
guapaque – *Ostrya virginiana*
guapile – *Allionia incarnata*
guapilla – *Agave falcata, A. striata, Bromelia karatas, Hechtia podantha*
guapinole – *Pithecellobium confine*
guaposi – *Erythrina flabelliformis*
guar – *Cyamopsis tetragonoloba*
guara – *Cupania americana*
guarade – *Goldmania foetida*
guaraguao – *Celtis iguanea*
guaraná – *Cupania americana*
guarán amarillo – *Tecoma stans*
guaratara – *Axonopus scoparius, Eleusine indica*
guardalagua – *Toxicodendron radicans*
guardalobo – *Osyris alba, Solanum verbascifolium, Verbascum thapsus*
guardarropa – *Santolina chamaecyparisus*
guardia civil – *Clerodendron*
guarecita – *Rhynchosia pyramidalis*
guareque – *Ibervillea sonorae*
guariaca – *Ligusticum porteri*
guaria morada – *Cattleya skinneri*
guarín – *Boletus regius*
guarina – *Cecropia obtusifolia, C. peltata*
guaritoto – *Cnidoscolus urens*
guarumo – *Cecropia obtusifolia, C. peltata*
guarupa – *Jacaranda acutifolia*
guasalaco – *Karwinskia humboldtiana*
guasaraco – *Parthenium stramonium, P. tomentosum*
guasca – *Cyperus diandrus, Galinsoga parviflora*
guasimilla – *Aphananthe monoica*
guasimilla roja – *Thouinia villosa*
guatájori – *Cephalanthus salicifolius*
guatamote – *Baccharis salicifolia, B. sarathroides*
guate – *Cochlospermum vitifolium, Zea mays*
Guatemala – *Caesalpinia pulcherrima, Tripsacum laxum*
guatope – *Inga spuria*
guatoso – *Gustavia augusta*
guau – *Parthenocissus quinquefolia, Toxicodendron radicans*
guaute – *Amaranthus*
guaxi – *Leucaena esculenta*
guaxima – *Urena lobata*
guaya – *Calophyllum braziliense, Chamaedorea ernesti-augustii, C. tepejilote*

guayabero – *Psidium guajava*
guayabillo – *Caesalpinia mexicana, Casearia sylvestris, Cordia dodecandra, Dodonaea viscosa, Eriosema grandiflorum, Eugenia fragrans, Lopezia mexicana, L. racemosa, Platanus mexicana, Psidium sartorianum, Pterocarpus officinalis, Rinorea, Salpianthus arenarius, S. macrodontos*
guayabillo cimarrón – *Eugenia capuli*
guayabillo de tinta – *Couepia polyandra*
guayabito – *Alibertia edulis, Salpianthus macrodontos*
guayabito de tinta – *Couepia polyandra*
guayabo – *Psidium guajava, P. molle, Terminalia amazonia*
guayabo agrio – *Eugenia fragrans, Psidium friedrich-stahlianum, P. molle*
guayabo blanco – *Psidium guajava*
guayabo casero – *Psidium guajava*
guayabo colorado – *Psidium guajava*
guayabo común – *Psidium guajava*
guayabo de China – *Psidium guajava*
guayabo de las sabinas – *Psidium guajava*
guayabo del Brasil – *Feijoa sellowiana*
guayabo de monte – *Alibertia edulis*
guayabo de venado – *Alibertia edulis, Psidium guajava*
guayabo dulce – *Psidium guajava*
guayabo fresa – *Psidium cattleyanum*
guayabo japonés – *Psidium cattleyanum*
guayabo manzana – *Psidium guajava*
guayabo montés – *Psidium friedrich-stahlianum*
guayabón – *Terminalia amazonia*
guayabo pequeño – *Psidium cattleyanum*
guayabo peruano – *Psidium guajava*
guayabo perulero – *Psidium guajava*
guayabo pesgua – *Eugenia jambolana*
guayabo-piña – *Feijoa sellowiana*
guayabo rojo – *Psidium guajava*
guayabo rosado – *Psidium cattleyanum*
guayabo sabanero – *Psidium guineense*
guayabo silvestre – *Psidium guineense*
guayacán – *Guaiacum, Porlieria angustifolia, Sporobolus indicus, Sweetia panamensis, Tabebuia, Viscainoa*
guayacana – *Diospyros kaki*
guayacán amarillo – *Guaiacum officinale*
guayacán real – *Guaiacum sanctum*
guayaco – *Guaiacum officinale*
guaya de abajo – *Chamaedorea ernesti-augustii*
guaya de cerro – *Chamaedorea elatior*
guayaguasi – *Psidium guajava*
guayamé – *Abies durangensis, A. religiosa, Pseudotsuga lindleyana, P. menziesii, P. rehderi*
guayamé blanco – *Abies mexicana, A. vejari*
guayamé colorado – *Pseudotsuga flahaulti, P. menziesii*
guayapol – *Thevetia peruviana*
guayita – *Chamaedorea cataractarum, C. ernesti-augustii*
guayita de los arroyos – *Chamaedorea cataractarum*
guayo – *Couepia polyandra*
guayocule – *Thevetia peruviana*
guayparín – *Diospyros californica, D. sonorae, Maba latifolia*
guaypinole – *Pithecellobium flexicaule*
guayul – *Colubrina ferruginosa, Sapindus saponaria, Vauquelinia corymbosa*
guayule – *Diospyros sonorae, Parthenium argentatum, P. incanum, Vauquelinia californica, V. corymbosa* var *angustifolia*
guayule hembra – *Parthenium incanum*
guazoncles – *Chenopodium berlandieri, C. nuttaliae*
gubaguih – *Arundo donax*
gubaguihnashi – *Saccharum officinarum*
gubaya – *Muhlenbergia macroura*
güegüecho – *Aristolochia grandiflora*
gueguetzí – *Alvaradoa amorphoides*
gueja – *Quercus coccolobifolia, Q. mcvaughii, Q. tarahumara*
guendabixhu – *Annona muricata*
guendaxiña – *Achras zapota*
güerequi – *Ibervillea sonorae*
güérigo – *Populus fremontii, P. brandegeei, P. wislizenii*
güero – *Capsicum annuum*
guerra – *Cylindropuntia bigelovii*
gueto bichi – *Cucurbita palmata*
gueto bitoo – *Cucumis melo*
gueto guu – *Cucurbita maxima*
gueto pochi – *Caesalpinia caladenia*
gueto xica – *Parmentiera edulis*
gueto xiña – *Citrullus colocynthis*

guexa – *Nicotiana tabacum*
guhoja – *Phoradendron californicum*
guía-chilla – *Polianthes tuberosa*
guía de jazmincillo – *Solanum dulcamara, S. seaforthianum*
guía-lachi – *Magnolia dealbata*
güico – *Pithecellobium echinatum*
guicobo – *Diphysa occidentalis*
guichibelle – *Parkinsonia aculeata*
guichibezia – *Celtis pallida*
guichibidu – *Urtica dioica*
guichigu – *Pisonia aculeata*
güichila – *Laelia anceps*
guichitache – *Pereskia lychnidiflora*
guichori – *Sarcostemma cynanchoides* ssp *hartwegii*
guichutilla – *Ziziphus spathulata*
guiebiche – *Apoplanesia paniculata, Tecoma stans*
guiechachi – *Plumeria acutifolia, P. rubra*
guiegana – *Polianthes tuberosa*
guielachi – *Talauma mexicana*
guievada – *Asclepias glaucescens*
guiexoba – *Bourreria huanita*
guiezee – *Jacquinia aurantica*
güi-güi – *Bombax ellipticum*
guija – *Lathyrus sativus*
guija de hojas anchas – *Lathyrus latifolius*
guija de monte – *Lathyrus linifolius*
guija silvestre – *Lathyrus sylvestris*
guija velluda – *Lathyrus hirsutus*
guileña – *Aquilegia vulgaris*
guilmo – *Bromus catharticus*
guillomo – *Amelanchier ovalis*
guillomo arbóreo – *Amelanchier laevis*
guillomo común – *Amelanchier ovalis*
guillomo del Canadá – *Amelanchier canadensis*
guillomo grandifloro – *Amelanchier* x *grandiflora*
guinagua – *Eriogonum inflatum*
güinare – *Sida rhombifolia*
guindilla – *Capsicum chinense, C. annuum*
guindo – *Muntingia calabura, Prunus*
guindo común – *Prunus cerasus*
guindo de Indias – *Malpighia glabra*
guindo de la costa – *Atropa belladonna*
guindo garrafal – *Morchella esculenta*
guinea – *Panicum maximum*
guinea fino – *Panicum coloratum*
guinea pajarito – *Panicum coloratum*
guineillo – *Cymbopetalum penduliflorum*
guineillo prieto – *Cymbopetalum penduliflorum*
guineo – *Cymbopetalum penduliflorum, Ensete ventricosum, Musa, Panicum maximum*
guineo de seda – *Musa sapientum*
guineo enano – *Musa nana*
guineo gigante – *Musa acuminata*
guineón – *Panicum maximum*
guineo niño – *Musa acuminata*
guinia – *Musa sapientum*
güinote blanco – *Acacia macracantha*
guiquil – *Sechium edule*
güira – *Couroupita guianensis*
güirambo – *Macrosiphonia hypoleuca*
güirirí – *Salvia splendens*
guirnalda – *Gomphrena globosa, Jasminum grandiflorum, J. multiflorum, Nephrolepis exaltata*
güiro – *Crescentia alata, C. cujete, Legenaria siceraria, L. vulgaris*
güiro de Yucatán – *Crescentia cujete*
güirote – *Phaseolus sonorensis, Sarcostemma cynanchoides* ssp *hartwegii, Serjania mexicana*
güirote cachora – *Cardiospermum halicacabum*
güirote de culebra – *Serjania mexicana*
güirote de violín – *Gouania rosei*
güirote grueso – *Serjania palmeri*
güirote negro – *Arrabidaea*
güiro totumo – *Crescentia cujete*
guisante – *Pisum sativum*
guisante arbóreo – *Caragana arborescens*
guisante azucarado – *Pisum sativum* var *macrocarpon*
guisante bambarra – *Voandzeia subterranean*
guisante cometodo – *Pisum sativum* var *macrocarpon*
guisante chino – *Pisum sativum* var *macrocarpon*
guisante de Angola – *Cajanus cajun*
guisante de campo – *Pisum sativum* var *arvense*
guisante de carita – *Vigna unguiculata*
guisante de grano liso – *Pisum sativum*
guisante de grano rugoso – *Pisum sativum*
guisante de huerta – *Pisum sativum* var *arvense*
guisante del paraíso – *Clianthus formosus*

guisante de olor – *Lathyrus odoratus*
guisante de paloma – *Cajanus cajun*
guisante de semilla rugosa – *Pisum sativum*
guisante de tierra – *Vigna subterranea*
guisante de vaca – *Vigna unguiculata*
guisante forragero – *Pisum sativum* ssp *arvense*
guisante inglés – *Pisum sativum*
guisante liso – *Pisum sativum*
guisante mascate – *Stizolobium deeringianum*
guisante negro – *Mucuna pruriens* var *utilis*
guisante oloroso – *Lathyrus odoratus*
guisante pajarero – *Cajanus cajun*
guisante sable – *Canavalia ensiformis*
guisante sativo – *Pisum sativum*
guisante tangerino – *Lathyrus tingitanus*
guisante tirabeque – *Pisum sativum* var *macrocarpon*
guisaro – *Psidium molle*
guisaso – *Cenchrus echinatus*
guisayote – *Sechium edule*
güiscoyol – *Astrocaryum mexicanum*
guishi – *Pleuraphis cenchroides*
güismora – *Rubus palmeri*
güitzl – *Epidendrum erubescens*
gui-yaa – *Arthrostylidium racemiflora*
güizapol – *Triumfetta*
güizapol de borrego – *Triumfetta*
guizazo – *Cenchrus ciliaris*
gujavabaum – *Psidium guajava*
gulabere – *Cordia alba*
gullitalli – *Pereskia lychnidiflora*
gumbio – *Pisonia aculeata*
gumbo limbo blanco – *Bursera simaruba*
guna – *Warszewiczia coccinea*
gunabia prieto – *Caesalpinia mexicana*
guniró – *Ipomoea batatas*
guojtoli – *Ficus cotinifolia*
guonay – *Chamaedorea tuerckheimii*
gusanera – *Iresine*
gusanillo – *Iresine celosioides, Lobelia fenestralis, Piper adencum, Schoenocaulon officinale, S. pringlei, Setaria gracilis*
gusano – *Mucuna pruriens, Setaria gracilis, Spathiphyllum cochlearispathum*
gutapercha – *Palaquium gutta*
guteo xiga – *Parmentiera edulis*

haas – *Lucuma mammosa, Musa sapientum*
háas – *Prosopis juliflora*
ha'axat – *Larrea tridentata*
haba – *Entada phaseoloides, E. scandens, Hura polyandra, Lathyrus vernus, Phaseolus lunatus, Thevetia peruviana, Vicia faba*
haba blanca – *Canavalia ensiformis, Phaseolus lunatus, P. vulgaris*
haba caballar – *Vicia faba* var *equina*
haba coba – *Vicia faba*
haba coccinea – *Phaseolus coccineus*
haba común – *Vicia faba*
haba criolla – *Canavalia ensiformis*
haba de burro – *Canavalia ensiformis, C. gladiata*
haba de caballo – *Canavalia ensiformis*
haba de Egipto – *Nelumbo nucifera*
haba de Guatemala – *Hura polyandra*
haba de huerta – *Vicia faba*
haba de la costa – *Entada*
haba de las Indias – *Lathyrus odoratus*
haba del indio – *Hura polyandra*
haba del Yemen – *Coffea arabica*
haba de San Antonio – *Caesalpinia crista*
haba de San Ignacio – *Hura polyandra, Strychnos ignatii*
haba de soja – *Glycine max*
haba falsa – *Vicia faba*
haba fava – *Vicia faba*
haba grande – *Entada polystachya*
haba loca – *Vicia faba, V. narbonensis*
haba mayor – *Vicia faba*
haba menor – *Vicia faba* var *minuta*
habañero – *Capsicum chinense*
haba panosa – *Vicia faba* var *equina*
haba silvestre – *Vicia narbonensis*
haba terciopela – *Mucuna pruriens* var *utilis, Stizolobium deeringianum*
haba tonka – *Dipteryx odorata*
habeem – *Cordia sebestena*
habichuela – *Phaseolus vulgaris, Vicia faba*
habichuela azul – *Decaisnea fargesii*
habichuela blanca – *Phaseolus vulgaris*
habichuela del gran norte – *Phaseolus vulgaris*
habichuela pinta – *Phaseolus vulgaris*
habichuela soya – *Glycine max*
habichuela tierna – *Phaseolus vulgaris*
habichuela verde – *Phaseolus vulgaris*
habilla – *Hura polyandra, Senna occidentalis, Yucca filifera*
habilla de Guatemala – *Hura polyandra*

habilla de San Ignacio – *Hura polyandra*
habilla prieta – *Senna occidentalis*
haca – *Arundo donax*
hachís – *Cannabis sativa*
hachogue – *Piper palmeri*
hachucá – *Quercus albocinta*
hahuiya-xíhuitl – *Lippia berlandieri*
ha'kolel – *Heteranthera limosa*
halal – *Arundo donax, Phragmites australis, Scirpus tabernaemontani*
halanche – *Anoda cristata*
halicácabo – *Physalis alkekengi*
halu – *Cucumis anguria*
hallarín – *Abies vejari, Pseudotsuga flahaulti, P. lindleyana, P. menziesii*
hallarín colorado – *Pseudotsuga menziesii* var *glauca*
hamamelina – *Hamamelis virginiana*
hamamelis – *Hamamelis virginiana*
hamat – *Yucca arizonica*
hamélide de Virginia – *Hamamelis virginiana*
hamula – *Brickellia californica, B. grandiflora*
hannam – *Cylindropuntia acanthocarpa*
hapis-coil – *Cannabis sativa*
harmaga – *Peganum harmala*
harta reina – *Piqueria trinervia*
hauay – *Parthenium hysterophorus*
hauchinol – *Heima salicifolia*
hauchinoli – *Heima salicifolia*
haulhua – *Talauma mexicana*
hauya – *Fagus sylvatica*
hauyxiu – *Sida procumbens*
haxix – *Cannabis sativa*
haya – *Acer brachypterum, Caesalpinia mexicana, Cinnamomum hartmanii, Cucurbita palmeri, Erythroxylum coca, Fagus grandifolia, F. sylvatica, Nothofagus antarctica, Platanus lindeniana, P. mexicana*
haya americana – *Fagus grandifolia*
haya antártica – *Nothofagus antarctica*
haya austral – *Nothofagus antarctica*
haya cobriza – *Fagus sylvatica* var *purpurea*
haya común – *Fagus sylvatica*
haya chilena – *Nothofagus dombeyi*
haya falsa – *Nothofagus antarctica*
hayal-sitsuuk – *Oplismenus burmannii*
haya roja – *Fagus sylvatica*
hecho – *Pachycereus pecten-aboriginum, P. pringlei*
hedéota – *Calamintha nepeta*
hediondilla – *Bocconia arborea, Cestrum dumentorum, C. nocturnum, Chenopodium, Heliotropium curassavicum, Larrea tridentata, Leucaena glauca, Senna atomaria, S. bicapsularis, S. occidentalis*
hediondillón – *Garrya laurifolia*
hediondo – *Acacia macracantha, Larrea tridentata, Rhamnus frangula*
hedrón – *Ipomoea alba*
héel – *Opuntia engelmannii*
helecho – *Acrostichum aureum, Adiantum, Astrolepis cochisensis, Cheilanthes, Dicranopteris pectinata, Dryopteris, Nephrolepis exaltata, Notholaena lemmonii, Pellaea, Polypodium, Pteridaceae, Pteridium aquilinum, Pteris, Selaginella*
helecho arborescente – *Alsophila aculeate, A. salvinii, Cyathea delicatula*
helecho avestruz – *Matteuccia struthiopteris*
helecho botón – *Pellaea rotundifolia*
helecho bracken – *Pteris*
helecho cola de pavo – *Blechnum gibbum*
helecho común – *Nephrolepis exaltata, Pteridium aquilinum*
helecho chino – *Phanerophlebia umbonata*
helecho de agua – *Ceratopteris thalictroides*
helecho de Alabama – *Cheilanthes alabamensis*
helecho de árbol – *Alsophila salvinii, Cyathea delicatula*
helecho de Boston – *Nephrolepis exaltata*
helecho de culantrillo – *Adiantum capillus-veneris*
helecho de esparraguera – *Asparagus setaceus*
helecho de manglar – *Acrostichum aureum*
helecho hembra – *Asplenium filix-femina, Pteridium aquilinum* var *pubescens*
helecho lengua de ciervo – *Phyllitis scolopendrium*
helecho macho – *Dryopteris filix-mas, Polypodium lanceolatum*
helecho madre – *Nephrolepis exaltata*
helecho nido – *Asplenium nidus-avis*
helecho nido de pájaro – *Asplenium nidus-avis*
helecho peyote – *Pellaea ternifolia*
helecho plumoso – *Thelypteris puberula*
helecho polipodio – *Polypodium hartwegianum*

helecho real – *Cibotium regale, Osmunda regalis*
helecho sombrilla – *Llavea cordifolia*
helenio – *Helenium autumnale, H. hybridum, Inula helenium*
heliantemo – *Helianthemum*
heliantemo amarillo – *Helianthemum nummularium*
heliantemo de hojas de lavándula – *Helianthemum lavandulifolium*
heliantemo mayor – *Helianthemum nummularium*
heliotropo – *Heliotropium, Phacelia ambigua*
heliotropo cimarrón – *Heliotropium curassavicum*
heliotropo de curazao – *Heliotropium curassavicum*
heliotropo del cerro – *Heliotropium corymbosum*
heliotropo europeo – *Heliotropium europaeum*
heliotropo peruano – *Heliotropium arborescens*
helxine de Soleirol – *Soleirolia soleirolii*
hembra – *Phyla strigulosa* var *sericea, Salvia divinorum*
hembra de guayule – *Parthenium incanum*
hemerocálide – *Hemerocallis fulva*
henequén – *Agave*
henequén blanco – *Agave americana, A. fourcroydes*
henequén verde – *Agave sisalana*
heno – *Aira caryophyllea, Reseda odorata, Tillandsia recurvata, T. usneoides*
heno barbón – *Usnea barbata*
heno blanco – *Holcus lanatus*
heno chino – *Tillandsia recurvata*
heno elegante – *Aira elegantissima*
heno griego – *Trigonella foenum-graecum*
heno largo – *Tillandsia usneoides*
heno mayor – *Aira caryophyllea*
heno muelle – *Holcus mollis*
heno pequeño – *Tillandsia ferrisiana, T. recurvata*
heno precoz – *Aira praecox*
hepática – *Hepatica nobilis*
hepática blanca – *Parnassia palustris*
hepática de tres lóbulos – *Hepatica nobilis*
hepática estrellada – *Galium odoratum*
hepáticus – *Hepaticae*
hermana de albaca – *Delilia biflora*
hermano de sciatica – *Salvia setosa*
hermodátiles – *Colchicum autumnale*
hermosa – *Begonia semperflorens*
hermosa de día – *Hosta plantaginea*
hermosilla – *Campanula trachelium*
herniaria – *Herniaria glabra*
herniaria glabra – *Herniaria glabra*
herniaria hirsuta – *Herniaria hirsuta*
hésen – *Olneya tesota*
hestej – *Porophyllum gracile*
heteropogon de allioni – *Heteropogon contortus*
hiba – *Thujopsis dolabrata*
hibisco – *Hibiscus*
hibisco chino – *Hibiscus rosa-sinensis*
hibuero – *Bignonia capreolata*
hichiconi – *Hechtia montana, Tillandsia recurvata*
hidraste – *Hydrastis canadensis*
hidrastis – *Hydrastis canadensis*
hidrocotule común – *Nydrocotyle vulgaris*
hiedra – *Cobaea scandens, Convolvulus, Cryptostegia grandiflora, Distictis buccinatoria, Ficus pumila, Hedera helix, H. irlandesa, Macfadyena unguis-cati, Quamoclit coccinea, Rhus integrifolia, R. rydbergii, Toxicodendron diversilobum, T. radicans*
hiedra alemana – *Delairea odorata*
hiedra bocina – *Distictis buccinatoria, Pithecellobium echinatum*
hiedra colorada – *Quamoclit coccinea, Q. hederifolia*
hiedra común – *Hedera helix*
hiedra de Boston – *Parthenocissus tricuspidata*
hiedra de las Canarias – *Hedera canariensis*
hiedra del monte – *Echeveria simulans*
hiedra de Persia – *Hedera colchica*
hiedra de tierra – *Glechoma hederacea*
hiedra extranjera – *Hedera helix*
hiedra grande – *Echeveria simulans*
hiedra hélice – *Hedera helix*
hiedra inglesa – *Hedera helix*
hiedra mala – *Toxicodendron diversilobum, T. radicans*
hiedra mala mujer – *Toxicodendron radicans*
hiedra maligna – *Toxicodendron radicans*
hiedra morada – *Cobaea scandens, Ipomoea purpurea, I. scopulorum*

hiedra persa – *Hedera canariensis*
hiedra roja – *Quamoclit coccinea*
hiedra rosa – *Quamoclit pinnata*
hiedra sueca – *Glecoma hederacea*
hiedra terrestre – *Glecoma hederacea, Ipomoea violacea*
hiedra venenosa – *Toxicodendron radicans*
hiedra violeta – *Cobaea scandens*
hiel de tierra – *Centaurium erythraea*
hielera – *Gentiana spathacea*
hielito – *Mesembryanthemum crystallinum, Portulaca pilosa*
hierba – *Poaceae*
hierba ácida – *Begonia martiana*
hierba agria – *Paspalum conjugatum*
hierba amarga – *Chrysanthemum balsamita, Parthenium hysterophorus*
hierba amargosa – *Ambrosia ambrosioides, A. artemisiifolia, Parthenium hysterophorus, Senecio jacobaea*
hierba amarilla – *Baileya multiradiata, Leucanthemum vulgare*
hierba angélica – *Angelica archangelica*
hierba añil – *Justica spicigera*
hierba artemis – *Artemisia vulgaris*
hierba artética – *Ajuga chamaepitys*
hierba azul – *Justica spicigera, Veronica beccabunga*
hierba balsámica – *Chrysanthemum balsamita*
hierba ballestera – *Helleborus foetidus, Veratrum album*
hierba becerra – *Antirrhinum majus, A. orontium*
hierba belida – *Ranunculus acris, R. repens, R. scleratus*
hierba bergamota – *Melissa officinalis*
hierba bermeja – *Hyparrhenia rufa*
hierba blanca – *Boerhavia erecta, B. intermedia, Eriodictyon tomentosum, Parthenium argentatum, P. incanum*
hierba bruja – *Kalanchoe pinnata*
hierbabuena – *Eriodictyon angustifolium, Hedeoma, Lippia dulcis, Mentha, Satureja douglasii*
hierbabuena de burro – *Mentha suaveolens*
hierbabuena de comer – *Mentha viridis*
hierbabuena de olor – *Mentha* x *piperita*
hierbabuena montés – *Phyla strigulosa* var *sericea*
hierba cabezona – *Helenium mexicanum*
hierba cabruna – *Bituminaria bituminosa*
hierba callera – *Sedum telephium*
hierba cana – *Senecio*
hierba cañamera – *Althaea officinalis*
hierba caribe – *Eriochloa polystachya*
hierba carmín – *Phytolacca americana*
hierba carmín decandra – *Phytolacca americana*
hierba carmín dioica – *Phytolacca dioica*
hierba casta – *Paeonia officinalis*
hierba ceniza – *Encelia farinosa, Parthenium argentatum, P. incanum, Tidestromia lanuginosa*
hierba centella – *Caltha leptosepala, C. palustris*
hierba césped – *Axonopus compressus*
hierba cinta – *Phalaris arundinacea*
hierba colorada – *Eriogonum annuum, E. rupestre, Lythrum alatum, Potentilla atrorubens, P. thurberi, Rumex*
hierba contra los venenos – *Iostephane heterophylla*
hierba coral – *Anagallis arvensis*
hierba criolla – *Paspalum conjugatum*
hierba crispa – *Teucrium polium*
hierba cristalina – *Aptenia cordifolia*
hierba cristalina de hojas cordiformes – *Aptenia cordifolia*
hierba cupido – *Catananche caerulea*
hierba cupido azul – *Catananche caerulea*
hierba china – *Phytolacca octandra*
hierba dallis – *Paspalum dilatatum*
hierba de aguas – *Potamogeton pectinatus*
hierba de almorranas – *Ranunculus ficaria*
hierba de Alonso García – *Dalea formosa*
hierba de Apache – *Castilleja, Erysimum capitatum*
hierba de Aquiles – *Achillea millefolium*
hierba de arroz – *Leersia hexandra*
hierba de Australia – *Paspalum dilatatum*
hierba de bahía – *Paspalum notatum*
hierba de barrer – *Solanum nudum*
hierba de bataneros – *Saponaria officinalis*
hierba de Bermuda – *Cynodon dactylon*
hierba de búfalo – *Panicum coloratum*
hierba de cacolote – *Talinum triangulare*
hierba de caída – *Arnica montana*
hierba de canela – *Zapoteca anomala*

hierba de Cayena – *Echinochloa polystachya*
hierba de clavo – *Choisya ternata*
hierba de cochino – *Ligusticum porteri*
hierba de corpus – *Laelia grandiflora*
hierba de corso – *Echinochloa polystachya*
hierba de Cristo – *Lantana camara*
hierba de chinche – *Pectis cylindrica*
hierba de chivatito – *Chenopodium botrys, C. graveolens*
hierba de chivato – *Cardiospermum halicacabum, Clematis ligusticifolia, Pericome caudata*
hierba de chivo – *Clematis ligusticifolia, Leucothoe mexicana, Ruellia tuberosa*
hierba de chucho – *Ageratina, Eupatorium collinum*
hierba de chuparrosa – *Erodium cicutarium*
hierba de dallis – *Paspalum dilatatum*
hierba de fuego – *Ranunculus scleratus, Senecio candidissimus*
hierba de garrapata – *Mentha pulegium*
hierba de Guatemala – *Tripsacum laxum*
hierba de Guinea – *Panicum coloratum, P. maximum*
hierba de halcón lisa – *Hypochaeris glabra*
hierba de halcón radicada – *Hypochaeris radicata*
hierba de hechiceros – *Solanum nigrum*
hierba de hielo – *Mesembryanthemum crystallinum*
hierba de hilo – *Leptochloa dubia*
hierba de incienso – *Encelia farinosa*
hierba de jarazo – *Lepechinia, Digitalis purpurea*
hierba de Judas – *Cryptostegia grandiflora*
hierba de junio – *Zinnia peruviana*
hierba de la abeja – *Ophrys apifera*
hierba de la alferecía – *Disocactus falgelliformis*
hierba de la ánima – *Encelia farinosa, Helenium mexicanum*
hierba de la araña – *Euphorbia dentata, Tribulus cistoides*
hierba de la barranca – *Flaveria*
hierba de la bruja – *Verbesina encelioides*
hierba de la calentura – *Gratiola officinalis, Iresine, Ruellia tuberosa*
hierba de la celada – *Scutellaria galericulata*
hierba de la coralina – *Erodium cicutarium*
hierba de la cortadura – *Achillea millefolium, Symphytum officinale*
hierba de la coyuntura – *Ephedra trifurca*
hierba de la cruz – *Brickellia coulteri, Croton ciliato-glandulosas*
hierba de la cucaracha – *Cochlearia officinalis, Dodonaea viscosa, Haplophyton cimicidum, Macrosiphonia hypoleuca, Mandevilla foliosa*
hierba de la culebra – *Abelmoschus moschatus, Arisaema dracontium*
hierba de la China – *Abutilon avicennae*
hierba de la escoba – *Buddleja scordioides*
hierba de la fiebre – *Senna alata*
hierba de la flecha – *Sagittaria macrophylla, S. sagittifolia, Sapium appendiculatum, S. biloculare, S. pedicellatum, Sebastiana*
hierba de la gallina – *Halimium glomeratum, Helianthemum glomeratum, Petiveria alliacea, Teucrium cubense*
hierba de la gallinera – *Stylosanthes viscosa*
hierba de la gangrena – *Acalypha californica*
hierba de la gobernadora de Puebla – *Brickellia cavanillesi, B. veronicaefolia*
hierba de la golondrina – *Chamaesyce, Chelidonium majus, Euphorbia, Spergularia media*
hierba de la golondrina extranjera – *Chelidonium majus*
hierba de la gota – *Ammi visnaga, Drosera rotundifolia*
hierba del aguardiente – *Viola odorata*
hierba de la herida – *Hypericum perforatum, Prunella vulgaris*
hierba de la hiedra – *Toxicodendron diversilobium*
hierba de la hinchazón – *Iresine*
hierba de la hormiga – *Allionia incarnata, Galphimia glauca, Triumfetta semitriloba, Wedelia incarnata*
hierba del aire – *Baccharis thesioides, Gentiana adsurgens, , Senna laevigata, S. occidentalis, Stevia salicifolia, Trixis angustifolia, T. radialis, T. wrightii*
hierba de la junturas – *Ajuga chamaepitys*
hierba del ala – *Inula helenium*
hierba del alacrán – *Plumbago auriculata, P. pulchella, P. scandens*
hierba de la madre – *Lathraea squamaria*
hierba de la mariposa – *Asclepias tuberosa*

hierba de la matadura – *Talauma mexicana*
hierba de la moneda – *Lysimachia nummularia*
hierba del amor – *Arctium lappa, Reseda odorata, Satureja hortensis*
hierba de la mosca – *Allionia incarnata, Buddleja americana*
hierba de lamparones – *Scrophularia nodosa*
hierba de la muela – *Achillea millefolium, Ambrosia ambrosioides, Carphochaete pringlei, Eupatorium hyssopium, Prunella vulgaris, Ranunculus petiolaris, Sisyrinchium arizonicum*
hierba de la mula – *Croton boregensis, Heliotropium fruticosum, H. indicum, Lippia umbellata, Stevia salicifolia*
hierba de la negrita – *Sphaeralcea angustifolia, S. coccinea, S. fendleri*
hierba del ángel – *Ageratina, Ageratum corymbosum, Eupatorium collinum, E. deltoideum, Waltheria americana, Zapoteca anomala*
hierba de la paciencia – *Rumex patientia*
hierba de la paloma – *Passiflora incarnata*
hierba de la parida – *Montanoa tomentosa*
hierba de la partida – *Montanoa tomentosa*
hierba de la pastora – *Salvia divinorum, Turnera diffusa*
hierba de la peña – *Cheilanthes lindheimeri*
hierba de la perla – *Lithodora diffusa, Lithospermum*
hierba de la perlita – *Lithospermum calycosum*
hierba de la piedra – *Castilleja tenuiflora*
hierba de la plata – *Lunaria annua, L. biennis, Potentilla anserina*
hierba de la playa – *Salvia leucantha*
hierba de la princesa – *Aloysia triphylla, Lippia ligustrina*
hierba de la puebla – *Calea zacatechichi, Senecio canicida*
hierba de la pulga – *Dalea citriodora, Galium mexicanum, Mentha suaveolens, Nama undulatum, Stevia viscida*
hierba de la punzada – *Asclepias linaria, Hyoscyamus niger, Nama coulteri, N. undulatum*
hierba de la quemadora – *Senecio vulgaris*
hierba de la quintana – *Ericameria spinulosus*
hierba de la rabia – *Acleisanthes longiflora, Asperugo procumbens*
hierba de la raya – *Ipomoea pes-caprae*
hierba del arriero – *Dyssodia acerosa*
hierba de la Santa María – *Tagetes florida*
hierba de la sarna – *Ageratum houstonianum, Leucothoe mexicana*
hierba del asma – *Lobelia inflata*
hierba del asno – *Oenothera biennis, O. fruticosa, O. villosa* ssp *strigosa*
hierba de las pampas – *Cortaderia selloana*
hierba de la tarantuala – *Acourtia thurberi*
hierba de la tórtola – *Croton texensis*
hierba de la tos – *Chrysanthemum balsamita, Grindelia hirsuta*
hierba de la Trinidad – *Tigridia pavonia*
hierba de la tristeza – *Polygala*
hierba de la tusa – *Ratibida tagetes*
hierba de la vaca – *Brickellia californica, Vaccaria hispanica*
hierba de la vaca macho – *Digitaria sanguinalis*
hierba de la vaca sana – *Paspalum conjugatum*
hierba de la vainilla – *Heliotropium arborescens*
hierba del avaro – *Clematis drummondii*
hierba de la ventosidad – *Nama undulatum*
hierba de la verruga – *Heliotropium europaeum*
hierba de la víbora – *Aristolochia serpentaria, Berlandiera lyrata, Chamaesyce albomarginata, Cissampelos pareira, Cucurbita foetidissima, Daucus pusillus, Dyschoriste decumbens, Echium vulgare, Ericameria heterophyllus, E. sonorensis, E. spinulosus, Gutierrezia sarothrae, Helenium mexicanum, Rivina humilis, Stegnosperma halimifolium, Zornia*
hierba de la viejita – *Sida procumbens*
hierba de la Virgen – *Cephalanthus occidentalis* var *californicus, Cynodon dactylon, Gaura coccinea, Loeselia ciliata, L. mexicana, Salvia divinorum, Tiquilia canescens, Turbina corymbosa*
hierba de la viruela – *Pilea microphylla*
hierba del bazo – *Encelia farinosa*
hierba del becerro – *Brickellia cavanillesi, Hyptis emoryi*
hierba del borracho – *Clinopodium laevigatum, Satureja macrostema*
hierba del borrego – *Dalea bicolor* var *orcuttiana*

hierba del Brasil – *Hyparrhenia rufa*
hierba del buen día – *Sida procumbens*
hierba del buey – *Cissus sicyoides, C. trifoliate, Grindelia aphanactis, G. hirsutula*
hierba del burro – *Allenrolfea occidentalis, Ambrosia dumosa, Calliandra houstoniana, Distichlis spicata, Erigeron canadensis, Parthenium hysterophorus, Schoenocaulon officinale, S. pringlei, Spigelia longiflora, Tournefortia hartwegiana, Zinnia acerosa, Z. pumila*
hierba del caballo – *Lupinus, Paspalum plicatulum, Senecio multicapitus*
hierba del cáncer – *Acalypha, Anisacanthus thurberi, Ayenia pusilla, Castilleja, Cuphea aequipetalia, C. jorullensis, C. wrightii, Gomphrena globosa, Lythrum alatum, Waltheria americana*
hierba del cangrejo – *Digitaria sanguinalis*
hierba del canónigo – *Valerianella locusta*
hierba del cantor – *Sisymbrium officinale*
hierba del carbonero – *Archibaccharis serratifolia, Baccharis*
hierba del carpintero – *Achillea millefolium, Barbarea vulgaris, Prunella vulgaris*
hierba del cenizo – *Leucophyllum frutescens, Tiquilia greggii*
hierba del ciego – *Urtica dioica*
hierba del conejo – *Fumaria officinalis*
hierba del corazón – *Dalea pogonathera*
hierba del cortes – *Achillea millefolium*
hierba del costado – *Plantago coronopus*
hierba del coyote – *Euphorbia peplus, Galium, Leptodactylon, Polanisia uniglandulosa*
hierba del cuchillo – *Sanguisorba minor*
hierba del cuervo – *Asclepias erosa, A. linaria, Crotalaria, Stegnosperma halimifolium*
hierba del cura – *Ternstroemia sylvatica*
hierba del diablo – *Datura stramonium, Nicotiana tabacum, Plumbago auriculata, P. pulchella, P. scandens*
hierba del duende – *Eriosema grandiflorum*
hierba de leche – *Asclepias curassavica*
hierba del empacho – *Chorizanthe fimbriata*
hierba del empeine – *Leucothoe mexicana*
hierba del escorbuto – *Cochlearia officinalis*
hierba del espanto – *Lythrum hyssopifolium*
hierba del Espíritu Santo – *Angelica archangelica*
hierba del frío – *Iresine calea*
hierba del gallo – *Zinnia pauciflora, Z. peruviana*
hierba del gato – *Croton, Nepeta cataria, Pulicaria dysenterica, Solanum grayi, S. rostratum, Stachys arvensis, Valeriana*
hierba del gigante – *Nicotiana glauca*
hierba del golpe – *Ageratina, Allionia incarnata, Dryopteris arguta, Gaura coccinea, Lopezia mexicana, L. racemosa, Oenothera rosea, O. tetraptera, Scoparia dulcis, Serjania racemosa, Solanum eleagnifolium, Thelypteris puberula* var *sonorensis*
hierba del gusano – *Acalypha arvensis, Heliotropium procumbens, Parthenium hysterophorus, Verbena halei*
hierba del hígado – *Chimaphila dasystephana*
hierba del histerico – *Conyza lyrata*
hierba del huerto – *Dactylis glomerata*
hierba del hule – *Parthenium argentatum*
hierba de limón – *Cymbopogon citratus, C. nardus, Pectis angustifolia*
hierba del indio – *Aristolochia, Agdestis clematides, Bouvardia ternifolia, Mirabilis multiflora, Sarcostemma crispum, Silene californica*
hierba del jabali – *Costus mexicanus*
hierba del lobo – *Dugaldia hoopesii*
hierba del mal de ojos – *Commelina scabra*
hierba del maná – *Glyceria fluitans*
hierba del manantial – *Montia Fontana*
hierba del manso – *Anemopsis californica, Iostephane heterophylla, Spathiphyllum*
hierba del marrano – *Aster subulatus*
hierba del monte – *Lycium andersonii, Plumbago scandens*
hierba del moro – *Croton, Inula helenium, Scrophularia auriculata*
hierba del muro – *Parietaria officinalis*
hierba del naciente – *Erigeron pubescens*
hierba del negrito – *oSphaeralcea angustifolia*
hierba del negro – *Sphaeralcea, Plumbago auriculata, P. pulchella, P. scandens, Teucrium cubense, Tournefortia hartwegiana*
hierba del ochavo – *Thlaspi arvense*
hierba del oso – *Arracacia atropurpurea, A. edulis, Chamaebatia foliosa, Heracleum*

lanatum, Prionosciadium thapsoides, Rhamnus californica
hierba del paisano – *Eleusine indica*
hierba del pajarito – *Stellaria media*
hierba del pájaro – *Anagallis arvensis, Plumbago scandens*
hierba del papa – *Teucrium marum*
hierba del Paraguay – *Ilex paraguariensis*
hierba del parral – *Brachiaria mutica*
hierba del pasmo – *Adenostoma sparsifolium, Baccharis, Bigelovia diffusa, Bouvardia ternifolia, Dorstenia drakeana, Ericameria brachylepis, E. laricifolia, E. sonorensis, Machaeranthera carnosa, Waltheria americana, W. indica*
hierba del pastor – *Turnera diffusa*
hierba del peco – *Actaea arguta, A. rubra*
hierba del perro – *Brickellia veronicaefolia, Castela tortuosa, Cestrum fasciculatum, Chenopodium murale, Kickxia elatine, Piqueria trinervia, Senecio canicida, Solanum cervantesii, Teucrium cubense*
hierba del pescado – Aesculus californica, *Chenopodium neomexicanum, Croton, Eremocarpus setigerus, Lupinus elegans, Marah fabaceus, Plumbago pulchella, P. scandens, Polygonum bistortoides, P. pensylvanicum, Trichostema*
hierba del piojo – *Galphimia glauca, Haplophyton cimicidum, Mandevilla foliosa, Senna leptocarpa, Tephrosia nicaraguensis*
hierba del pobre – *Gratiola officinalis*
hierba del pobre hombre – *Filipendula vulgaris*
hierba del pollito – *Stellaria media*
hierba del pollo – *Commelina, Tinantia erecta, T. modesta, Tradescantia*
hierba del potro – *Caesalpinia mexicana*
hierba del puerco – *Dugesia mexicana*
hierba del pujo – *Stylosanthes viscosa*
hierba del ratón – *Oryza sativa*
hierba del rosario – *Caltha palustris, Canna indica*
hierba del sagrado corazón de Jesús – *Tubiflora squamosa*
hierba del Santo Espíritu – *Angelica archangelica*
hierba del sapo – *Ambrosia acanthicarpa, A. ambrosioides, A. confertifolia, Arabis glabra, Cirsium undulatum, Dorstenia contrayerba, Eryngium, Heliotropium curassavicum, Hydrocotyle vulgaris, Solanum rostratum, Tournefortia hartwegiana*
hierba del sol – *Gaillardia pinnatifida, Helianthus annuus*
hierba del soldado – *Achillea millefolium, Waltheria americana*
hierba del soltero – *Petrea arborea*
hierba del Sudán – *Sorghum bicolor, S. sudanense*
hierba del tabardillo – *Calliandra californica, Iresine calea, Piqueria trinervia*
hierba del temascal – *Rhus terebinthifolia*
hierba del tepozán – *Buddleja sessiliflora*
hierba del tifo – *Schkuhria virgata*
hierba del tiñoso – *Arctium lappa*
hierba del topo – *Datura stramonium*
hierba del toro – *Bonplandia geminiflora, Elytraria imbricata, Hamelia erecta, Lythrum hyssopifolium, Proboscidea fragrans, Ruellia tuberosa, Zinnia linearis*
hierba de Luisiana – *Axonopus compressus*
hierba del vaso – *Encelia farinosa*
hierba del venado – *Conium maculatum, Eremocarpus setigerus, Galphimia glauca, Pectis imberbis, Porophyllum, Rhus aromatica, Turnera diffusa*
hierba del zopilote – *Cestrum anagyris, Lobelia laxiflora, Nicotiana glauca, Piqueria trinervia*
hierba del zorrillo – *Bouvardia ternifolia, Chenopodium graveolens, C. incisum, Croton, Piqueria trinervia, Polanisia uniglandulosa*
hierba del zorrillo cimarrón – *Chenopodium botrys*
hierba del zorro – *Aspicarpa hirtella, Gaudichaudia mucronata*
hierba de mar mayor – *Zostera marina*
hierba de Nuestra Señora – *Chrysanthemum balsamita*
hierba de para – *Brachiaria mutica, Panicum molle*
hierba de Paris – *Paris quadrifolia*
hierba de petate – *Pteridium aquilinum*
hierba de piloto – *Silphium perfoliatum*
hierba de Pinacate – *Senna wislizeni*
hierba de pordiosero – *Clematis*
hierba de punta – *Poa annua*

hierba de Quirón – *Centaurium erythraea*
hierba de sal – *Philoxerus vermicularis*
hierba de San Alberto – *Sisymbrium officinale*
hierba de San Antonio – *Epilobium hirsutum, Loeselia mexicana*
hierba de San Benito – *Geum chiloense, G. urbanum*
hierba de San Cristóbal – *Actaea*
hierba de San Gerardo – *Aegopodium podagraria, Ammi visnaga*
hierba de sangre – *oBerberis aquifolium* var *repens, B. trifoliata*
hierba de San Guillermo – *Agrimonia eupatoria*
hierba de San Jorge – *Centranthus ruber*
hierba de San José – *Verbena carolina*
hierba de San Juan – *Artemisia vulgaris, Glechoma hederacea, Hypericum calycinum, H. perforatum, Lavandula stoechas, Macrosiphonia hypoleuca, M. macrosiphoni, Verbena carolina*
hierba de San Juan de Dios – *Senecio flaccidus*
hierba de San Lorenzo – *Sanicula europaea*
hierba de San Nicolás – *Chrysactinia mexicana, Dyssodia acerosa, Gutierrezia sarothrae, Ionidium polygalaefolium, Piqueria trinervia, Stevia salicifolia, Tecoma stans, Thelesperma megapotamicum*
hierba de San Pablo – *Primula officinalis, P. veris, P. vulgaris*
hierba de San Pedro – *Chrysanthemum balsamita, Cuphea llavea, Parietaria officinalis, Penstemon barbatus, Scrophularia nodosa, Solanum verbascifolium, Tecoma stans*
hierba de San Roberto – *Geranium robertianum*
hierba de San Simón – *Circaea lutetiana*
hierba de San Simón alpina – *Circaea alpina*
hierba de santa – *Chenopodium ambrosioides*
hierba de Santa Bárbara – *Barbarea verna, B. vulgaris*
hierba de Santa Bárbara común – *Barbarea vulgaris*
hierba de Santa Bárbara precoz – *Barbarea praecox*
hierba de Santa Bárbara primaveral – *Barbarea verna*
hierba de Santa Catalina – *Impatiens capensis, I. glandulifera*
hierba de Santa María – *Pluchea odorata, Tagetes, Tanacetum parthenium*
hierba de Santa Rosa – *Antigonon leptopus, Paeonia officinalis*
hierba de Santa Teresa – *Cortaderia selloana*
hierba de Santiago – *Senecio jacobaea, Tanacetum vulgare*
hierba de selva – *Whipplea modesta*
hierba desierto – *Fendlerella utahensis*
hierba de tajo – *Eclipta alba*
hierba de tatalencho – *Selloa glutinosa*
hierba de todos los males – *Verbena officinalis*
hierba de torojo – *Heliotropium curassavicum*
hierba de trigo – *Triticum aestivum*
hierba de troya – *Scrophularia canina*
hierba de zanjas – *Paspalum distichum*
hierba de zizotes – *Asclepias oenotheroides*
hierba doncella – *Catharanthus roseus, Vinca major, V. minor*
hierba doncella menor – *Vinca minor*
hierba dulce – *Aloysia gratissima, Eleusine indica, Hierochloë odorata, Lippia*
hierba dulce del Paraguay – *Stevia rebaudiana*
hierba elefante – *Pennisetum purpureum*
hierba escarchada – *Mesembryanthemum crystallinum*
hierba estañera – *Elytrigia repens*
hierba estonutatoria – *Achillea ptarmica*
hierba fina – *Agrostis capillaries*
hierba fría – *Sanvitalia abertii*
hierba fuerte – *Teucrium marum*
hierba gallina – *Cephaelis tomentosa*
hierba gatera – *Nepeta cataria, N. faassenii*
hierba gigante – *Acanthus mollis*
hierba golondrinera – *Chelidonium majus*
hierba guinea – *Panicum maximum*
hierba hedionda – *Cestrum nocturnum, Datura stramonium, Larrea tridentata*
hierba hermosa – *Vernonia deppeana*
hierba imperial – *Axonopus scoparius*
hierba india – *Panicum coloratum*
hierba jabonera – *Saponaria officinalis*
hierba johnson – *Sorghum halepense*
hierba jonequil – *Heima salicifolia*
hierba julia – *Achillea ageratifolia*
hierba kikuyo – *Pennisetum clandestinum*
hierba lagunera – *Ranunculus aquatilis*

hierba lanuda – *Tidestromia lanuginosa*
hierba lavamanos – *Calendula arvensis*
hierba lechera – *Asclepias vestita*
hierba lechosa – *Sarcostemma cynanchoides* ssp *hartwegii*
hierba liebrera – *Chrysothamnus nauseosus*
hierba limón – *Cymbopogon citratus*
hierba loca – *Argemone ochroleuca, Astragalus, Crotalaria lupulina, Croton fruticulosus, Datura stramonium, Hyoscyamus niger, Lupinus, Oxytropis lambertii, Pernettya*
hierba lombricera – *Spigelia anthelmia*
hierba lombriguera – *Artemisia abrotanum, Tanacetum vulgare*
hierba Luisa – *Aloysia triphylla, Cymbopogon citratus, Melissa officinalis*
hierba lumbre – *Plumbago pulchella*
hierba luna – *Thymus serpyllum*
hierba maestra – *Artemisia mexicana*
hierba mala – *Sapium biloculare*
hierba mansa – *Anemopsis californica*
hierba maravilla – *Tripsacum laxum*
hierba María – *Asclepias curassavica, Ipomoea violacea*
hierba mate – *Ilex paraguayensis*
hierba media – *Campanula medium*
hierba melada – *Melinus minutiflora*
hierba menuda – *Paspalum notatum*
hierba merker – *Pennisetum purpureum*
hierbamora – *Rivina humilis, Solanum*
hierba morada – *Cobaea scandens*
hierbamora negra – *Solanum americanum*
hierba musgo – *Erodium moschatum*
hierba negra de los prados – *Heteropogon contortus*
hierbanís – *Foeniculum vulgare, Tagetes filifolia, T. florida, T. lucida*
hierba niso – *Artemisia dracunculoides, A. dracunculus*
hierba nora – *Artemisia dracunculus*
hierba pajarera – *Anagallis arvensis*
hierba pangola – *Digitaria decumbens*
hierba parda – *Helianthus ciliaris*
hierba pastel – *Isatis tinctoria*
hierba pastel de los tintoreros – *Isatis tinctoria*
hierba pegajosa – *Mentzelia multiflora, Selloa glutinosa*
hierba pejiguera – *Polygonum persicaria*
hierba piojera – *Delphinium staphisagria, Santolina chamaecyparisus*
hierba plata – *Astragalus mollissimus*
hierba ponzoñosa – *Datura wrightii*
hierba prado – *Axonopus affinis*
hierba prieta – *Garrya salicifolia*
hierba prodigio – *Tripsacum laxum*
hierba pulguera – *Erigeron karvinskianus, E. speciosus, Senecio integrifolius*
hierba puntera – *Sempervivum tectorum*
hierba rastrera – *Digitaria, Elytriga repens*
hierba ratón – *Bahia absinthifolia*
hierba real – *Ocimum basilicum*
hierba reuma – *Frankenia palmeri, F. salina*
hierba rhodes – *Chloris gayana*
hierba riquera – *Stellaria media*
hierba rosa – *Polygonum arenastrum*
hierba sagrada – *Lantana brasilensis, Verbena officinalis*
hierba salada – *Gaura parviflora, Oenothera deltoids, Salicornia europaea*
hierba sana – *Helianthemum lavandulifolium, Mentha spicata*
hierba sanguinaria – *Polygonum bistorta*
hierba sanjuanera – *Galium verum*
hierba santa – *Acleisanthes longiflora, Artemisia absinthium, Dorstenia contrayerba, Eriodictyon, Mentha spicata, Piper auritum, P. sanctum, Verbena officinalis*
hierba Santa María – *Tanacetum vulgare*
hierba sardonia – *Ranunculus scleratus*
hierba sin raíz – *Cuscuta*
hierba sofia – *Descuriana sophia*
hierba té – *Corchorus siliquosus*
hierba tinta – *Justicia spicigera*
hierba topera – *Euphorbia lathyris*
hierba tora – *Orobanche ramosa*
hierba tostada – *Boerhavia hirsuta*
hierba turca – *Herniaria glabra*
hierba turmera – *Tuberaria guttata*
hierba turqueza – *Indigofera suffruticosa*
hierba velluda – *Caltha palustris, Ranunculus bulbosus*
hierba verde – *Paspalum conjugatum*
hierba verruguera – *Chelidonium majus*
hierba zapatera – *Coriaria myrtifolia*
higado – *Boletus luteus*
higado de buey – *Fistulina hepatica*
higo – *Ficus*

higo de Hotentote comestible – *Carpobrotus edulis*
higo del Cabo – *Caprobrotus edulis*
higo del mar – *Opuntia megacantha*
higo de tetetzo – *Cephalocereus columna-trajani, Neobuxbaumia tetetzo*
higoloxe grande – *Ficus glaucescens*
higo marino – *Caprobrotus edulis*
higuajil – *Schoepfia californica*
higuera – *Courcoupita guianensis, Ficus, Ricinus communis*
higuera abigarrada – *Ficus aspera*
higuera amate – *Ficus padifolia*
higuera americana – *Ficus cotinifolia*
higuera arqueada – *Ficus benghalensis*
higuera blanca – *Ficus mexicana*
higuera cauchera – *Ficus elastica*
higuera común – *Ficus carica*
higuera chumba – *Opuntia ficus-indica*
higuera de agua – *Ficus religiosa*
higuera de Bengala – *Ficus bengalensis*
higuera de las pagodas – *Ficus bengalensis*
higuera del caucho – *Ficus elastica*
higuera del diablo – *Ricinus communis*
higuera del hule – *Ficus elastica*
higuera del infierno – *Datura stramonium, Ricinis communis*
higuera del látex – *Ficus elastica*
higuera de Sumatra – *Ficus elastica*
higuera india – *Ficus benghalensis*
higuera infernal – *Ricinus communis*
higuera loca – *Datura stramonium*
higuera llorona – *Ficus benjamina*
higuera marítima – *Ficus superba*
higuera negra – *Ficus padifolia*
higuera serpenteante – *Ficus pumila*
higuera silvestre – *Ficus*
higuera trepadora – *Ficus pumila*
higuereta – *Ricinis communis*
higuerilla – *Ipomoea cristulata, Ricinus communis*
higuerilla brava – *Sapium pedicellatum*
higuerilla pequeña – *Ricinus communis*
higuero – *Crescentia cujete*
higuerón – *Ficus*
higuerote – *Ficus petiolaris*
higueruela – *Bituminaria bituminosa*
higuito – *Ficus padifolia*
hikora rosapara – *Mammillaria senilis*
hikora sunama – *Ariocarpus fissuratus*
hikora wanami – *Lophophora williamsii*
hikuli – *Lophophora williamsii, Solanum verbascifolium*
hikuri – *Echinocereus triglochidiatus, Lophophora williamsii, Mammillaria senilis*
hilite – *Alnus arguta*
hilo – *Eupatorium glabratum*
hilo de oro – *Cuscuta americana*
himenocalis – *Hymenocallis narcissiflora*
himoli – *Karwinskia humboldtiana*
hinchador – *Sapium appendiculatum*
hincha huevos – *Hippomane mancinella, Toxicodendron diversilobium, T. radicans*
hiniesta – *Cytisus scoparius*
hiniesta de escobas – *Cytisus scoparius*
hiniesta escocesa – *Cytisus scoparius*
hiniesta espinosa – *Ulex europaea*
hiniesta genista – *Cytisus multiflorus*
hinojo – *Foeniculum vulgare* var *azoricum*
hinojo acuático – *Oenanthe aquatico*
hinojo común – *Foeniculum vulgare*
hinojo de Castilla – *Foeniculum vulgare*
hinojo de prado – *Carum carvi*
hinojo fétido – *Anethum graveolens*
hinojo hediondo – *Anethum graveolens*
hinojo marino – *Crithmum maritimum*
hinojo marítimo – *Crithmum maritimum*
hinojo tuberoso – *Foeniculum vulgare*
hinseseli – *Loeselia mexicana*
hintimorreal – *Ephedra aspera*
hipechila – *Dalea diffusa*
hipérico – *Hypericum calycinum, H. perforatum*
hipérico androsemo – *Hypericum androsaemum*
hipérico arbustivo – *Hypericum androsaemum*
hipérico calicino – *Hypericum calycinum*
hipérico de hojas de conífera – *Hypericum coris*
hipericón – *Hypericum*
hipérico perforado – *Hypericum perforatum*
hirschfeldia incana – *Hirschfeldia incana*
hirundinaria – *Chelidonium majus*
hisopillo – *Satureja montana*
hisopo – *Hyssopus officinalis, Origanum maru* var *aegyptiacum, Salvia*
hisopo común – *Hyssopus officinalis*
hispidula – *Senna*

híxul – *Cylindropuntia leptocaulis*
hman – *Bromelia pinguin*
hoaxacan – *Guaiacum sanctum*
hoaxin – *Tamarindus indica*
hobon-k'aax – *Euphorbia heterophylla*
hocha de leche – *Caladium bicolor*
ho'i – *Condalia brandegeei*
hoitxóchitl – *Jacquinia aurantica*
hoitzcolotli – *Eryngium foetidum*
hoitzitziltentli – *Lobelia laxiflora*
hoitzmamazali – *Acacia spadicigera*
hoja ancha – *Quercus crassifolia*
hoja blanca – *Calathea*
hoja bruja – *Bryophyllum calycinum, Kalanchoe pinnata*
hoja de acuyo – *Piper auritum*
hoja de adorno – *Caladium bicolor*
hoja de aján – *Piper sanctum*
hoja de anís – *Piper sanctum*
hoja de baño – *Tecoma stans*
hoja de buitre – *Limnocharis flava*
hoja de cebra – *Sansevieria trifasciata*
hoja de cera – *Nicotiana glauca*
hoja de clavo – *Ludwigia peploides*
hoja de cobre – *Acalypha wilkesiana*
hoja de Columbia – *Bryophyllum calycinum*
hoja de cuero – *Anthurium crassinervium*
hoja de culebrón – *Arum maculatum*
hoja de chombo – *Calathea lutea*
hoja de jabón – *Styrax argenteus*
hoja del aire – *Kalanchoe pinnata*
hoja de la pastora – *Salvia divinorum*
hoja de la pulga – *Proboscidea fragrans*
hoja de maíz – *Aspidistra elatior*
hoja de manteca – *Podachaenium eminens, Solanum verbascifolium*
hoja de pescado – *Nymphoides humboldtianum*
hoja de petate – *Abies religiosa*
hoja de plata – *Tradescantia zebrina*
hoja de playa – *Pluchea odorata*
hoja de sal – *Calathea lutea*
hoja de salón – *Aspidistra elatior*
hoja de San Pablo – *Wigandia caracasana*
hoja de San Pedro – *Daphne mezereum*
hoja de sol – *Nymphaea ampla*
hoja de veneno – *Asclepias curassavica*
hoja de viento – *Anthurium crassinervium*
hoja de zope – *Calathea lancifolia*
hoja de zopilote – *Calathea*
hojadillo – *Monstera deliciosa*
hoja dura – *Cocoloba barbadensis*
hoja elegante – *Xanthosma robustum*
hoja fina – *Panicum maximum*
hoja flecha – *Sagittaria sagittifolia*
hoja fresca – *Gilibertia arborea, Kalanchoe pinnata*
hojalatillo blanco – *Conostegia xalapensis*
hojalito – *Robinia neomexicana*
hoja pinta – *Sansevieria trifasciata*
hojaranzo – *Carpinus betulus, Nerium oleander*
hojarasca – *Acourtia adnata*
hoja santa – *Piper auritum, P. sanctum*
hoja semita – *Senecio hartwegii*
hojasén – *Caesalpinia pulcherrima, Cassia fistula, Flourensia cernua, Senna*
hokab – *Tabebuia pentapylla*
holco lanoso – *Holcus lanatus*
holol – *Hibiscus tiliaceus*
holostea estrellada – *Holosteum umbellatum*
holostea umbelada – *Holosteum umbellatum*
holli – *Castilla elastica*
hombocoche – *Duranta repens*
hombrecillo – *Humulus lupulus*
hombre grande – *Quassia amara*
hombre viejo – *Pachycereus schottii*
homicin – *Laelia tibicinis*
hondapara – *Dillenia indica*
hongo – *Basidiomycetes, Agaricus campestris*
hongo amarillo – *Cantharellus cibarius*
hongo blanco – *Agaricus arvensis*
hongo común – *Agaricus campestris*
hongo corriente – *Agaricus campestris*
hongo de campo – *Agaricus campestris*
hongo de cazahuate – *Pleurotus ostreatus*
hongo de encino – *Pleurotus ostreatus*
hongo de huevo – *Lycoperdon mixtecorum*
hongo de la espiga del centeno – *Claviceps purpurea*
hongo de la papa – *Phytophthora infestans*
hongo del maíz – *Ustilago maydis*
hongo del oyamel – *Fomes pinicola*
hongo del pino – *Fomes pinicola*
hongo de maguey – *Pleurotis cornucopiae, P. ostreatus*
hongo de palo – *Fomes pinicola*
hongo de San Isidro – *Psilocybe cubensis*
hongo de San Juan – *Agaricus campestris*

hongo de sombrero – *Agaricus campestris*
hongo de venado – *Russula delica*
hongo kentesh – *Psilocybe cubensis*
hongo manzana – *Laccaria laccata*
hongorado – *Boletus regius*
hongo reishi – *Gandoderma lucidum*
hongo shiitake – *Lentinus edodes*
honoba – *Rudbeckia*
hopo de zorra – *Verbascum thapsus*
hordio – *Hordeum vulgare*
hormigo – *Platymiscium dimorphandrum*
hormiguero – *Cordia alliodora*
hormiguillo – *Cecropia obtusifolia, Croton magdalenae, Platymiscium dimorphandrum*
hormiguillo blanco – *Cordia alliodora*
hormiguillo colorado – *Platymiscium dimorphandrum*
hormino silvestre – *Salvia viridis*
horqueta – *Paspalum conjugatum*
hortela – *Mentha* x *piperata*
hortensia – *Hydrangea macrophylla*
hortensia arborea – *Hydrangea arborescens*
hortensia de guía – *Funastrum heterophyllum*
hortensia de hojas de roble – *Hydrangea quercifolia*
hortensia de invierno – *Bergenia crassifolia*
hortensia leñosa – *Hydrangea arborescens*
hortensia trepadora – *Hydrangea petiolaris*
howinowa – *Lepachys mexicana*
hoya – *Hoya carnosa*
hoya de cera – *Hoya carnosa*
huacacachu – *Brugmansia*
huacalillo – *Ceanothus coeruleus*
huacalxóchitl – *Philodendron affine, Xanthosoma roseum*
huacanalá – *Calathea lutea*
huacatay – *Chenopodium ambrosioides*
huacux – *Sideroxylon capiri*
huachacata – *Galphimia glauca, Tecoma stans, Thryallis glauca*
huachaco – *Haematoxylon brasiletto*
huachelai – *Sisymbrium auriculatum*
huachi blanco – *Leucaena esculenta*
huachochole – *Parthenium hysterophorus*
huachomó – *Viguiera montana*
huachuca – *Agave huachucensis, Panicum huachucae*
huachuma – *Echinopsis pachanoi*
huahuichó – *Havardia mexicana*
huaje – *Lagenaria siceraria, Leucaena esculenta, Tamarindus indica*
huaje de caballo – *Zapoteca formosa*
huaje de los niños – *Lagenaria siceraria*
huajilote – *Parmentiera edulis*
huajillo – *Acacia berlandieri, A. coulteri, Calliandra eriophylla, Havardia pallens, Leucaena lanceolata, Pithecellobium brevifolium*
huajillo prieto – *Calliandra houstoniana*
huajito – *Lagenaria siceraria*
huajito de niños – *Lagenaria siceraria*
huajuco – *Sideroxylon rigida*
hualcacho – *Echinochloa crus-galli*
hualeya – *Citrullus colocynthis*
hualule – *Sapindus saponaria*
huambia – *Spigelia anthelmia*
huamis – *Larrea tridentata*
huamúchil – *Bauhinia divaricata, Pithecellobium dulce*
huamúchil costeño – *Pithecellobium dulce*
huanacaxtle – *Enterolobium cyclocarpum*
huancanalá – *Myrica cerifera, M. mexicana*
huanchal – *Trophis racemosa*
huanita – *Bourreria huanita*
huano blanco – *Thrinax wendlandiana*
huanto – *Brugmansia*
huarumbo – *Cochlospermum vitifolium*
huas – *Crescentia cujete*
huasaraco – *Parthenium tomentosum*
huashihua – *Senna occidentalis*
huasi – *Leucaena esculenta, L. glauca*
huasteco – *Celtis pallida*
huata – *Juniperus californica, J. coahuilensis, J. deppeana, Salix bonplandiana, S. gooddingii, S. lasiolepis*
huatamote – *Baccharis salicifolia, B. sarathroides*
huatasí – *Salix gooddingii*
huatauhui – *Solanum erianthum, S. verbascifolium*
huatli – *Amaranthus*
huatosí – *Salix bonplandiana*
huatsin – *Leucaena glauca*
huauchinango – *Eichhornia crassipes*
huauquilitl – *Chenopodium berlandieri, C. nuttaliae*
huaute – *Amaranthus*
huautli – *Amaranthus*

huautzin – *Conzattia multiflora*
huauzontle – *Chenopodium berlandieri, C. nuttaliae*
huaxacote – *Malpighia glabra*
huaxaxán – *Guaiacum coulteri*
huaxcatol – *Euchlaena mexicana*
huaxin – *Leucaena esculenta, L. leucocephala*
huayul – *Sapindus saponaria*
huayunac – *Cardiospermum halicacabum*
huboquillo – *Senecio vulneraria*
huecojote – *Salix bonplandiana, S. lasiolepis*
huee'e – *Amaranthus blitoides*
hueilaqui – *Caesalpinia platyloba*
hueipatl – *Solandra guerrerensis*
hueizacat – *Typha domingensis*
huejote – *Salix bonplandiana*
huele de día – *Cestrum dumentorum*
huele de noche – *Bouvardia longiflora, Cestrum nocturnum, Cooperia drummondii, Lippia ligustrina, Pisonia, Randia*
huele de noche negro – *Solanum nudum*
huembéreua – *Toxicodendron radicans*
huepaca – *Nicotiana tabacum*
huérigo – *Populus* x *acuminata, P. brandegeei*
huesillo – *Exostema caribaeum, Rinorea guatemalensis, Sweetia panamensis*
huesito – *Dodonaea viscose, Guaiacum sanctum*
hueso del diablo – *Dioscorea villosa*
hueso de fraile – *Thevetia thevetioides*
hueso prieto – *Ilex nitida*
huetapoche – *Caesalpinia standleyi*
huevito – *Cardiospermum corundum, Vallesia glabra*
huevo de caballo – *Tabernaemontana alba*
huevo de coyote – *Coryphanta radians*
huevo de cucho – *Tabernaemontana alba*
huevo de chuchu – *Tabernaemontana alba*
huevo de gato – *Cardiospermum halicacabum, Solanum grayi, S. tequilense, Tabernaemontana alba, Thevetia ovata*
huevo de rana – *Evolvulus prostratus*
huevo de toro – *Cardiospermum halicacabum, Echinocereus polyacanthus, Stemmademia tomentosa* var *palmeri, Tabernaemontana alba*
huevo de venado – *Peniocereus greggii*
huexotl – *Salix lasiolepis*
hueynacaztli – *Cymbopetalum penduliflorum*
hueypali – *Drymaria gracilis*
hueypatli – *Solandra nitida*
hueytlaca – *Solandra nitida*
hueyytzontecon – *Turbina corymbosa*
huezequí – *Alvaradoa amorphoides*
huhuhuali – *Vitex mollis*
huiahuonahua – *Chiranthodendron pentadactylon*
huicicillo – *Loeselia mexicana*
huicume – *Licania platypus*
huicumo – *Lucuma salicifolia*
huichacame – *Cylindropuntia tunicata*
huichachago – *Haematoxylum brasiletto*
huichagorare – *Ceanothus huichagorae*
huichaxin – *Acacia farnesiana*
huichichil – *Loeselia mexicana*
huichichile – *Loeselia mexicana, L. scariosa*
huichichilemetl – *Myroxylon flexuosum*
huichichiltemel – *Xylosma flexuosum*
huichilaco – *Haematoxylum brasiletto*
huichipil – *Alvaradoa amorphoides*
huichocachacua – *Iostephane heterophylla*
huichori – *Sarcostemma cynanchoides* ssp *hartwegii*
huichuri – *Diphysa occidentalis, Polygonum argyrocoleon*
huihuilan – *Symphoricarpos microphyllus*
huihuitz-yo-cochis – *Mimosa albida*
huijol – *Andira inermis*
huilamole – *Costus mexicanus*
huilancha – *Opuntia stenopetala*
huilocochi – *Ficus insipida*
huilocualoni – *Quercus candicans*
huiloche – *Diphysa occidentalis, D racemosa*
huilote – *Croton guatemalensis*
huinare – *Eremalche scoparium, Sida rhombifolia*
huinare prieto – *Eremalche scoparium*
huingo – *Crescentia cujete*
huinol – *Acacia cymbispina*
huinole – *Acacia cymbispina*
huirote – *Ipomoea longifolia, Sarcostemma cynanchoides* ssp *hartwegii*
huisache – *Acacia, Caesalpinia cacalaco, C. gillesii, Goldmania foetida, Havardia pallens, Pithecellobium leptophyllum, Prosopis juliflora*
huisache bola – *Caesalpinia cacalaco, C. sclerocarpa*

huisache conchilla – *Goldmania foetida*
huisache costeño – *Acacia cymbispina, A. hindsii*
huisache chino – *Acacia schaffneri*
huisache de chintete – *Pithecellobium acatlense*
huisache de semilla – *Acacia farnesiana*
huisache marismeño – *Caesalpinia cacalaco*
huisache tepame – *Acacia cymbispina, A. pennatula*
huisache yóndiro – *Acacia farnesiana*
huiscole – *Agave marmorata*
huiscolote – *Pisonia aculeata*
huiscoyol – *Bactris major, B. mexicana*
huiscruceta – *Randia watsonii*
huisique – *Lepidium virginicum*
huisisai – *Oxalis tuberosa*
huismora – *Rubus palmeri*
huispancle – *Acacia berlandieri*
huisquil – *Sechium edule*
huitatave – *Vallesia glabra, V. laciniata*
huitlacoche – *Ustilago maydis*
huitlatexihuitl – *Galphimia glauca*
huitote – *Calliandra houstoniana*
huitz – *Inga edulis*
huitzapole – *Cenchrus echinatus, C. longispinus*
huitzaxin – *Acacia farnesiana*
huitzcocólotl – *Pisonia aculeata*
huitzilóxitl – *Myroxylon balsamum* var *pereirae*
huitzilzilxóchitl – *Loeselia mexicana*
huitzilzitzin – *Loeselia mexicana*
huitzitzixóchitl – *Salvia fulgens*
huitzmamaxalli – *Acacia spadicigera*
huitznáhuac – *Ferocactus viridescens*
huitzocuitlapilli – *Disocactus flagelliformis*
huitzquahuitl – *Caesalpinia crista, Condalia obovata, Haematoxylon brasiletto, H. campechianum*
huitzquilitl – *Argemone munita, Cirsium*
huitztomatzin – *Solanum eleagnifolium*
huixe – *Andira inermis*
huiyoco – *Pinus lumholtzii, P. strobiformis*
huizapol – *Ambrosia chenopodiifolia, A. dumosa, Cenchrus, Distichlis spicata*
huizapol gordo – *Cenchrus palmeri*
huizapolillo – *Triumfetta semitriloba*
huizapolito – *Ambrosia chenopodiifolia*
huizapori – *Cenchrus incertus, C. palmeri*
huizapotl – *Cenchrus echinatus, C. longispinus*
huizilacate – *Sideroxylon laetevirens*
huizlán – *Muntingia calabura*
huiztlecuatl – *Solanum rostratum*
huje – *Brosimum alicastrum*
hukup – *Bursera simaruba*
hule – *Castilla elastica, Cryptostegia grandiflora, Ficus elastica, Hevea brasiliensis, Parthenium argentatum*
hule cera – *Hoya carnosa*
hule de la India – *Ficus elastica*
hule macho – *Castilla fallax*
hulté-k'-aak – *Stenorrhynchus lanceolatus*
humero – *Alnus glutinosa*
humo – *Pithecellobium dulce*
humo cimarrón – *Pithecellobium dulce*
humo costeño – *Pithecellobium dulce*
humpil – *Salix taxifolia*
hunpetskinci – *Aloe vera*
húpachi – *Agastache micrantha*
hupala – *Prosopis chilensis*
hupari – *Begonia martiana*
husillo – *Chondrilla juncea*
hutatiyo – *Coutarea pterosperma*
huupa – *Prosopis juliflora*
hyuy – *Casimiroa edulis*

iati – *Gliricidia sepium*
ib-ché – *Rhynchosia minima*
ibía – *Oxalis tuberosa*
icaco – *Chrysobalanus icaco*
icahueyo – *Govenia liliacea*
icaquero – *Chrysobalanus icaco*
icaquillo – *Hirtella americana, H. racemosa*
icaquito – *Licania platypus*
icotl – *Tabebuia pentaphylla*
iczotl – *Yucca*
ichcatl – *Gossypium mexicanum*
ichculili – *Tecoma stans*
ichtaj – *Pinus oocarpa*
ichu – *Stipa ichu*
ichuanitzo – *Serjania mexicana*
igualamilla – *Eriosema grandiflorum*
igualamo – *Vitex mollis*
iguano – *Mimosa distachya*
ijinebre – *Juniperus communis*
ikilhaxiu – *Lantana camara*

ilama – *Annona diversifolia, A. purpurea, A. reticulata*
ilama de Tehuántepec – *Annona muricata, A. purpurea*
ilamatzapotl – *Annona diversifolia, A. muricata*
ilamaxóchitl – *Lupinus*
ilang-ilang – *Cananga odorata*
ilanilan – *Jasminum grandiflorum*
ilite – *Alnus acuminata, A. arguta, A. rubra, Ulmus mexicana*
ilite verde – *Alnus acuminata, A. jorullensis*
ilitl – *Alnus jorullensis*
illuale – *Genipa americana*
imbauha – *Cecropia peltata*
imperata – *Imperata cylindrica*
imperatoria – *Peucedanum ostruthium*
imperial – *Amorphophalus rivieri, Axonopus scoparius*
incarvillea – *Incarvillea delavayi*
incienso – *Boswellia carteri, B. thurifera, Bursera lancifolia, Encelia californica, E. farinosa, Protium copal*
incienso de la tierra – *Hymenaea courbaril*
incienso de petapa – *Hymenaea courbaril*
indianela – *Lablab purpureus*
índigo – *Indigofera gerardiana, I. suffruticosa*
índigo bastardo – *Amorpha fruticosa*
índigo chino – *Polygonum tinctorium*
índigo falso – *Galega officinalis*
indio – *Aristolochia californica, A. watsonii*
indio desnudo – *Bursera simaruba, Eugenia capuli*
indio peludo – *Bursera simaruba*
indita – *Bouvardia ternifolia, Catharanthus roseus*
Inglés – *Lolium perenne*
inguande – *Bocconia arborea, B. frutescens*
injerto – *Arceuthobium vaginatum, Lucuma viride, Phoradendron, Phrygillanthus sonorae, Pouteria viridis, Psittacanthus calyculatus, P. sonorae*
injerto de aguacate – *Psittacanthus calyculatus*
injerto de encino – *Phoradendron schumannii*
inmortal – *Asclepias asperula, A. lemmonii, A. linaria, A. tuberosa, Gomphrena decumbens, G. globosa, Helichrysum orientale, Selaginella lepidophylla, Xeranthemum annuum*
insulina vegetal – *Myrica salicifolia*
invierno – *Dahlia excelsa*
iñame – *Dioscorea*
ipeca – *Cephaëlis ipecacuanha*
ipecacuana – *Cephaëlis*
ipecacuana del país – *Ionidium polygalaefolium*
ipecacuana de México – *Ionidium polygalaefolium*
ipomea – *Ipomoea tricolor*
iraca – *Carludovicia palmata*
íride – *Iris foetidissima*
iris – *Iris* x *hollandica*
iris amarillo – *Acorus calamus*
irupé – *Victoria regia*
iscoque – *Tagetes patula*
ishcanal – *Acacia hindsii, A. spadicigera*
ishcanal negro – *Acacia gladiata*
isimac – *Cornus disciflora*
isioche – *Satureja macrostema*
isláy – *Prunus ilicifolia*
islaya – *Prunus ilicifolia*
isoca – *Ixoria coccinea*
isohuanquil – *Lepidium virginicum*
isopo rojizo – *Justicia canrea*
isoque – *Ixora coccinea*
ispágula – *Plantago ovata*
ispile – *Pinaropappus roseus*
istafiate – *Ambrosia psilostachya, Artemisia*
istaten – *Avicennia germinans*
itabo – *Yucca*
itamo – *Passiflora mexicana*
itamo real – *Aristolochia grandiflora, Ephedra, Passiflora mexicana, Pedilanthus tithymaloides, Pellaea cordifolia, Turnera diffusa*
itate – *Ipomoea arborescens*
itchai – *Pinus oocarpa*
itsi – *Citrullus colocynthis*
itzampi – *Licania platypus*
itza-zuuk – *Gouania virgata*
itzcuimpatli – *Schoenocaulon officinale, Veratrum officinale*
itzcuinpatli – *Senecio canicida*
itzil – *Karwinskia humboldtiana*
itzmal – *Bombax ellipticum*
itzmaqua – *Laelia grandiflora, L. speciosa*
itzmetl – *Sedum roseum* ssp *integrifolium*
itzmiquitl – *Portulaca oleracea*
itzohu – *Populus fremontii*

itztamatl – *Bombax ellipticum, Sedum roseum* ssp *integrifolium*
itzu – *Lagenaria siceraria*
itzucuatziqui – *Solandra nitida*
ixbahach – *Tritilia hirta*
ixcapatl – *Illicium floridanum*
ixcapul – *Solanum nigrum*
íxcatl – *Gossypium hirsutum, G. mexicanum*
ixcaxíhuitl – *Gossypium mexicanum*
ixcuicuil – *Anagallis arvensis*
ixhouen – *Spondias purpurea*
ixhuhuatl – *Erythea edulis*
ixi – *Prunus persica* ssp *nucipersica*
ixia – *Ixia maculata*
ixiconal – *Acacia hindsii*
ixim – *Zea mays*
iximche – *Andira inermis*
iximha – *Lemna minor*
ixora roja – *Ixoria coccinea*
ixpepe – *Trema micrantha*
ixpostli – *Pinaropappus roseus*
ixpul – *Pinaropappus roseus*
ixpulul – *Saracha jaltomata*
ixquefé – *Ternstroemia sylvatica*
ixtagua – *Cissampelos pareira*
ixtle – *Agave, Hesperaloe funifera, H. parviflora*
ixtle de jaumave – *Agave funkiana*
ixtle de jocuiztle – *Bromelia pinguin*
ixtle de palma – *Yucca carnerosana*
ixtle de Tampico – *Agave lechuguilla*
ixtle de Tula – *Agave heteracantha*
ixtontle – *Tecoma stans*
ix-tot – *Bromelia pinguin*
iyamole – *Sapindus saponaria*
iyátl – *Nicotiana tabacum*
iz – *Ipomoea batatas*
iza – *Prunus rhamnoides*
izcua – *Rumex obtusifolius*
izhuate – *Erythea edulis*
izote – *Beaucarnea gracilis, Yucca*
izote del monte – *Dracaena americana*
izotl – *Yucca guatemalensis, Y. treculeana*
izpaxíhuitl – *Croton dioicus*
izquitl – *Zea mays* var *everia*
izquixóchitl – *Bourreria huanita*
iztacatzóyatl – *Senecio salignus*
iztaccoanenepilli – *Cissampelos pareira*
iztac-cuahuitl – *Conocarpus erecta*
iztacnanácatl – *Russula delica*
iztacpalancapatli – *Senecio vulneraria*
iztacpatli – *Psoralea pentaphylla*
iztactepetzacuxóchitl – *Cissampelos pareira, Govenia liliacea*
iztactzápotl – *Casimiroa edulis*
iztauhyatl – *Artemisia ludoviciana, A. mexicana*
iztecauhticmixtl – *Romneya coulteri*
izuate – *Erythea edulis*

jabehui – *Erythea nitida*
jabillo – *Caesalpinia crista*
jabillo del mar – *Caesalpinia crista*
jabillo prieto – *Senna occidentalis*
jabino – *Juniperus communis*
jaboncillo – *Bursera stenophylla, Clethra mexicana, Fouquieria macdougalii, Lippia ligustrina, Phytolacca octandra, P. rivinoides, Pseudotsuga menziesii, Sapindus drummondii, S. saponaria, Tillandsia recurvata, Yucca*
jabonera – *Gypsophilia paniculata, Sapindus saponaria, Saponaria officinalis*
jabonera blanca – *Silene dioica*
jabonera común – *Saponaria officinalis*
jabonera de la China – *Koelreuteria paniculata*
jabonera de roca – *Saponaria ocymoides*
jabonera enana – *Saponaria pumilio*
jabonera rocosa – *Saponaria ocymoides*
jaborandi – *Pilocarpus jaborandi*
jaca – *Artocarpus heterophyllus, A. integrifolia, Lucuma viride*
jacalosúchil – *Plumeria acutifolia, P. rubra*
jacalosúchil blanco – *Plumeria acutifolia*
jacalosúchil rojo – *Plumeria rubra*
jacana – *Lucuma rivicoa, Pouteria multiflora*
jacaparí – *Crescentia cujete*
jacarandá – *Jacaranda acutifolia*
jacarandá de hojas de mimosa – *Jacaranda acutifolia*
jacia – *Hevea brasiliensis*
jacinto – *Eichhornia crassipes, Hyacinthus orientalis, Melia azedarach* var *umbraculiformis, Moringa oleifera*
jacinto acuático – *Eichhornia crassipes*
jacinto común – *Hyacinthus orientalis*
jacinto de agua – *Eichhornia crassipes*
jacinto del Cabo – *Galtonia candicans*

jacinto de Sudáfrica – *Galtonia candicans*
jacinto de verano – *Galtonia candicans*
jacinto estival – *Galtonia candicans*
jacinto holandés – *Hyacinthus orientalis*
jacinto legítimo – *Hyacinthus orientalis*
jacinto occidental – *Polianthes tuberosa*
jacinto racemoso – *Muscari armeniacum*
jacinto romano – *Hyacinthus orientalis* var *albulus*
jacinto silvestre – *Hyacinthoides non-scripta*
jaco – *Piper auritum, P. sanctum*
jacobina – *Dicliptera suberecta, Justicia carnea*
jacobinia – *Pachystachys lutea*
jacube – *Acanthocereus tetragonus, Hylocereus undatus*
jachuca – *Quercus albocinta*
jagua – *Roystonea regia*
jagua azul – *Genipa americana*
jagua blanca – *Genipa americana*
jaguarcillo – *Halimium umbellatum*
jaguarzo – *Cistus crispus, C. monspeliensis*
jaguarzo blanco – *Cistus albidus*
jaguarzo morisco – *Cistus salvifolius*
jaguarzo negro – *Cistus monspeliensis*
jaguarzo occidental – *Halimium lasianthum*
jaguarzo prieto – *Cistus crispus*
jaguay – *Rhus terebinthifolia*
jagüé – *Sebastiania cornuta, S. pavonia*
jahuacté – *Bactris major, B. mexicana*
jaiboli – *Agave jaiboli*
jalamata – *Ficus padifolia*
jalamate – *Ficus padifolia*
jalap – *Ipomoea purga*
jalapa – *Ipomoea purga, Jatropha macrorhiza*
jalapa de Veracruz – *Ipomoea purga*
jalapa limoncilla – *Ipomoea scopulorum*
jalapeño – *Capsicum annuum*
jal-au-té – *Euterpe precatoria*
jalocote – *Abies religiosa, Pinus teocote, Prionosciadium thapsoides, Psidium guajava*
jaltomate – *Physalis stapelioides, Saracha jaltomata, S. procumbens*
jamacha – *Cucurbita foetidissima*
jamaica – *Hibiscus rosa-sinensis, H. sabdariffa*
jamán – *Curatella americana*
jamatai – *Chlorogalum*
jambo – *Eugenia jambos, E. malaccensis*
jambolan – *Eugenia jambolana*
jambul – *Eugenia jambolana*
jamiltepec – *Orbignya cohune*
jamín – *Thevetia peruviana*
jamrosade – *Eugenia jambos*
janajuana – *Helianthemum glomeratum*
janamargo – *Vicia sativa*
jangó – *Dioon edule*
janiero – *Echinochloa polystachya*
jano – *Chilopsis linearis* ssp *arcuata*
japiró – *Lonchocarpus hermannii*
japito – *Chloris*
jaqueira – *Artocarpus integrifolia*
jara – *Arundo donax, Baccharis salicifolia, Bidens bigelovii, B. laevis, Cistus, Leonotis nepetaefolia, Opuntia karwinskiana, Pluchea odorata, Populus angustifolia, Salix alba, S. lasiolepis, Senecio salignus*
jara amarga – *Vicia sativa*
jara blanca – *Cistus albidus*
jara brava – *Senecio salignus*
jara de cádamo – *Cistus ladanifer*
jara de flores rosas – *Cistus incanus*
jara del Ládano – *Cistus ladanifer*
jara de pitayita – *Epiphyllum anguliger*
jara dulce – *Baccharis salicifolia*
jaraguá – *Hyparrhenia hirta, H. rufa*
jaraguá bermeja – *Hyparrhenia rufa*
jaraguá gris – *Eragrostis pilosa, Hyparrhenia hirta*
jaral – *Baccharis salicifolia*
jaral amarillo – *Senecio salignus*
jaral blanco – *Baccharis pteronioides*
jaralillo – *Senecio salignus*
jaramago – *Diplotaxis muralis, Eruca vesicaria* ssp *sativa, Raphanus raphanistrum, Sisymbrium officinale*
jaramago blanco – *Raphanus raphanistrum*
jaramago de Loesel – *Sisymbrium loeselii*
jaramago falso-erisimo – *Sisymbrium erysimoides*
jaramago francés – *Erucastrum gallicum*
jaramago más alto – *Sisymbrium altissimum*
jaramago oriental – *Sisymbrium orientale*
jarana de pitahayita – *Epiphyllum anguliger*
jara pringosa – *Cistus ladanifer*
jara rizada – *Cistus crispus*
jari – *Vitex mollis*
jarilla – *Baccharis conferta, B. neglecta, B. salicifolia, Cecropia obtusifolia, Dodonaea*

viscosa, Halimium umbellatum, Heima salicifolia, Helianthemum nummularium, Heterospermum pinnatum, Hybanthus mexicanus ssp *occidentalis, Jarilla caudata, Larrea tridentata, Porophyllum scoparium, Salix jaliscana, Selloa glutinosa, Senecio salignus, Stevia salicifolia, Viguiera dentata*
jarilla común – *Baccharis salicifolia*
jarilla chura – *Isocoma menziesii*
jarilla de la sierra – *Eriosema grandiflorum*
jarilla de loma – *Dodonaea viscosa*
jarilla del río – *Baccharis neglecta, B. salicifolia*
jarita – *Salix exigua, S. lasiolepis*
jaro – *Arum italicum, A. maculatum*
jarretadera – *Acacia hindsii*
jarrilla – *Jarilla caudata, J heterophylla, Larrea divericata*
jarrito – *Lamourouxia multifida, Lobelia laxiflora, Pachycereus marginatus, Penstemon*
jarro de oro – *Erblichia odorata*
jasam – *Mariscus jamaicensis*
jashis – *Cannabis sativa*
jashuca – *Prunus zinggii*
jaspalache – *Sedum acre, S. moranense*
jatía – *Phyllostylon brasiliensis*
jatoba – *Hymenaea courbaril*
jaumave – *Agave funkiana*
jaumave ixtle – *Agave funkiana*
jaumave lechuguilla – *Agave funkiana*
jauolí – *Taxodium distichum* var *mexicanum*
jauque – *Bocconia arborea*
jayacaste – *Crescentia alata*
jayacastle – *Crescentia alata*
jay sapo – *Jatropha cordata*
jazmín – *Cephalanthus occidentalis* var *californicus, Jasminum, Murraya paniculata, Petrea arborea, Philadelphus mexicanus, Randia, Solanum jasminoides, Stemmadenia donnell-smithii, Trachelospermum jasminoides*
jazmín amarillo – *Allamanda cathartica, Gelsemium sempervirens, Jasminum mesnyi*
jazmín azul – *Petrea arborea, Plumbago auriculata, P. scandens*
jazmín blanco – *Cephalanthus salicifolius, Jasminum grandiflorum*
jazmincillo – *Lippia ligustrina, Pisoniella arborescens*
jazmincillo del monte – *Menodora helianthemoides*
jazmín cimarrón – *Cornus excelsa*
jazmincito – *Odontoglossum cervantesii*
jazmín común – *Jasminum grandiflorum*
jazmín de Arabia – *Jasminum grandiflorum, J. sambac*
jazmín de Carolina – *Gelsemium sempervirens*
jazmín de Chile – *Mandevilla laxa*
jazmín de China – *Jasminum polyanthum*
jazmín de España – *Clerodendrum, Jasminum grandiflorum*
jazmín de flores grandes – *Jasminum grandiflorum*
jazmín de Hueyapán – *Philadelphus mexicanus*
jazmín de invierno – *Jasminum nudiflorum*
jazmín de Italia – *Clerodendrum*
jazmín de la India – *Bourreria huanita, Jasminum grandiflorum*
jazmín de la sierra – *Macrosiphonia hesperia*
jazmín del Cabo – *Gardenia augusta, G. florida*
jazmín del Istmo – *Bourreria huanita*
jazmín del monte – *Philadelphus mexicanus*
jazmín de nacapule – *Vallesia laciniata*
jazmín de Oaxaca – *Bourreria huanita*
jazmín de olor – *Jasminum grandiflorum*
jazmín de palo – *Bourreria huanita*
jazmín de San José – *Jasmium nudiflorum*
jazmín de San Juan – *Bouvardia longiflora*
jazmín de Tehuántepec – *Bourreria huanita*
jazmín de uvas – *Pereskia aculeata*
jazmín de Virginia – *Campsis radicans*
jazmín italiano – *Jasminum grandiflorum, Solanum wendlandii*
jazmín morado – *Solanum dulcamara, S. jasminoides*
jazmín mosqueta – *Philadelphus coronarius*
jazmín nudifloro – *Jasminum nudiflorum*
jazmín real – *Jasminum grandiflorum*
jazmín rocoso – *Androsace sarmentosa*
jazmín rosa – *Jasminum polyanthum*
jazmín silvestre – *Gelsemium sempervirens*
jazmín trompeta – *Campsis radicans*
jebana – *Sinapis alba*
jebe – *Hevea brasiliensis*
jeco – *Hymenoclea monogyra, Hyptis*

seemannii, Piper sanctum, Prunus capuli, P. zinggii
jeco guasiqui – *Prunus zinggii*
jécota – *Hymenoclea monogyra, H. salsola, Selloa glutinosa*
jegito – *Maba latifolia*
jejiri – *Lycium andersonii*
jelecote – *Pinus patula*
jenabe – *Brassica nigra*
jenepene – *Solanum madrense*
jenjibre – *Alpinia purpurata, Zingiber officinale*
jenjibre officinal – *Zingiber officinale*
jenjibre ornamental – *Alpinia purpurata*
jenjibre rojo – *Alpinia purpurata*
jenjibrillo – *Paspalum conjugatum, P. notatum*
jequirity – *Abrus precatorius*
jeringuilla – *Philadelphus coronarius, P. mexicanus, P. microphyllus*
jeringuilla coronaria – *Philadelphus coronarius*
jeroliz morado – *Distictis laxiflora*
Jerusalén – *Helianthus tuberosus*
jeseri – *Lysiloma divaricatum, L, microphylla*
jévero – *Erythrina flabelliformis*
jeyó – *Jatropha malacophylla*
jicaco – *Chrysobalanus icaco*
jícama – *Exogonium bracteatum, Pachyrhizus erosus*
jícama cimarrona – *Pachyrrhizus palmatilobus*
jícama de agua – *Pachyrhizus erosus*
jícama del cerro – *Phaseolus heterophyllus*
jícama del colera – *Dahlia coccinea, D. variabilis*
jícama de leche – *Pachyrhizus erosus*
jícama del monte – *Exogonium bracteatum*
jícama montés – *Bomarea hirtella*
jicamaxóchitl – *Dahlia variabilis*
jicamilla – *Jatropha macrorhiza*
jicamite – *Dahlia coccinea, D. variabilis*
jícamuchi – *Caesalpinia caladenia, C. palmeri*
jicarillo – *Clethra lanata*
jicarita – *Amanita caesarea, Crescentia alata*
jícaro – *Crescentia alata, C. cujete, Parmentiera edulis*
jícaro de Yucatán – *Crescentia cujete*
jicoria amarga – *Carya cordiformis*
jicoria glabra – *Carya glabra*
jicoria ovada – *Carya ovata*
jicoria pecán – *Carya illinoensis*
jicoria tomentosa – *Carya tomentosa*
jiguata – *Lotus scoparius, Psorothamnus schottii*
jihuite – *Parthenium argentatum*
jihuite amargo – *Parthenium hysterophorus*
jimba – *Bambusa guadua, Guadua, Rhipidocladum pittieri*
jimbilla – *Guadua longifolia*
jimuari – *Karwinskia humboldtiana*
jimuri – *Kallstroemia californica*
jinebro – *Juniperus communis*
jinicuile – *Inga spuria*
jinjillero – *Zea mays*
jinjolero – *Ziziphus jujuba*
jiogo – *Ambrosia ambrosioides*
jiote – *Bursera simaruba*
jiote blanco – *Bursera grandifolia*
jiote colorado – *Bursera simaruba*
jiotillo – *Bursera odorata, Escontria chiotilla, Jatropha cordata*
jipi – *Carludovica palmata*
jipijapa – *Carludovica palmata*
jipijapa de Panama – *Carludovica palmata*
jiquilite – *Indigofera suffruticosa*
jiquimite – *Erythrina*
jiquique – *Bombax ellipticum*
jira – *Iriartea*
jíride – *Iris foetidissima*
jirimú – *Dodonaea viscosa*
jito – *Capparis atamisquea, Forchammeria watsonii*
jitomate – *Lycopersicon esculentum*
jitomate guaje – *Lycopersicon esculentum*
jitomate guajito – *Lycopersicon esculentum* var *pyriforme*
jitomate redondo – *Lycopersicon esculentum*
jobillo – *Astronium graveolens, Spondias mombin*
jobo – *Bursera simaruba, Spondias mombin, S. purpurea*
jobo amarillo – *Spondias mombin*
jobo corronchoso – *Spondias mombin*
jobo de la India – *Spondias cytherea, S. dulcis*
jobo de monte – *Spondias mombin*
jobo de zopilote – *Spondias mombin, S. purpurea*
jobo espino – *Spondias mombin*
jobo rojo – *Spondias purpurea*
jobo roñoso – *Spondias mombin*

joco – *Hyptis seemannii*
jócono – *Esenbeckia hartmanii, Havardia sonorae*
joconostle – *Opuntia engelmannii, O. joconostle, Stenocereus stellatus*
joconoxtle – *Cylindropuntia imbricata, Opuntia joconostle, Pereskiopsis blakeana, P. porteri*
joco pino – *Pinus arizonica*
jocote – *Mombus sapondia, Spondias mombin, S. purpurea*
jocote de corona – *Spondias purpurea*
jocote marañón – *Anacardium occidentale*
jocotillo – *Astronium graveolens*
jocotillo de cerro – *Exostema caribaeum*
jocoyol color de rosa – *Begonia gracilis*
jocuiste – *Bromelia karatas*
jocuiste de majahua – *Bromelia karatas*
jocuistle – *Bromelia karatas, B. pinguin*
jojoba – *Simmondsia chinensis*
jolete – *Lyophyllum decastes*
jolmashté – *Talauma mexicana*
jolocín – *Heliocarpus donnell-smithii*
jololito – *Antigonon leptopus*
joltuluc – *Calathea*
jon-chab – *Reinhardtia gracilis*
jondura – *Spondias purpurea*
jonequil – *Heimia salicifolia*
jonjonata – *Quercus lobata*
jonote – *Heliocarpus donnell-smithii, H. tomentosus, Muntingia calabura, Triumfetta*
jonote blanco – *Heliocarpus donnell-smithii, H. tomentosus*
jonote coyotillo – *Belotia mexicana*
jonote real – *Ochroma pyramidale*
jopi – *Ochroma pyramidale*
jopo de las habas – *Orobanche*
jopoy – *Ficus mexicana*
jos – *Capparis flexuosa, Sideroxylon spiniflora*
José gris – *Croton magdalenae*
joso – *Albizzia sinaloensis, Conzattia multiflora, Havardia mexicana, Lycium berlandieri*
josocola – *Randia echinocarpa, R. watsonii*
joso de la sierra – *Conzattia multiflora*
jotei – *Oryza sativa*
jotora – *Ziziphus obtusifolia*
jovinoba – *Rudbeckia mexicana*
joyarí – *Drypetes gentryi*
joyo – *Lolium temulentum*
jua – *Solanum sisymbriifolium*
juaja – *Suaeda*
juájuas – *Guadua latifolia*
juajunesco – *Andira inermis*
Juana costa – *Enterolobium cyclocarpum*
Juana milpilla – *Gomphrena decumbens*
juanamipili – *Boerhavia*
Juan Diego – *Brosimum alicastrum*
Juan el conquistador – *Ipomoea purga*
Juanita – *Bourreria sonorae, Cardiospermum corindum, C. halicacabum, Halimium glomeratum, Helianthemum glomeratum*
Juanita mota – *Cannabis sativa*
Juan loco – *Nicotiana glauca*
Juan mecate – *Cissus gossypifolia*
Juanrubio – *Marrubium vulgare*
juantilipin – *Boerhavia coulteri, B. erecta*
jubaibena – *Atamisquea emarginata*
júbare – *Vitex mollis*
júcaro – *Alnus oblongifolia, Ostrya virginiana*
jucay – *Arbutus glandulosa*
jucee ndede – *Selloa glutinosa*
jucite – *Trophis racemosa*
júchica – *Schoepfia shreveana, Sideroxylon occidentale*
juchuch – *Plumeria rubra*
judía – *Phaseolus vulgaris*
judía adzuki – *Vigna angularis*
judía arroz – *Vigna umbellata*
judía aterciopelada de Florida – *Mucuna pruriens*
judía batata – *Pachyrhizus erosus*
judía blanca – *Phaseolus vulgaris*
judía blanca redonda – *Phaseolus vulgaris*
judía carilla – *Vigna unguiculata*
judía careta – *Psophocarpus tetragonolobus*
judía catjang – *Vigna unguiculata* ssp *cylindrica*
judía común – *Phaseolus vulgaris*
judía de caballo – *Canavalia ensiformis*
judía de campo – *Phaseolus vulgaris*
judía de careta – *Vigna unguiculata*
judía de Egipto – *Lablab purpureus*
judía de España – *Phaseolus coccineus*
judía de la peladilla – *Phaseolus lunatus*
judía de Lima – *Phaseolus lunatus*
judía de mungo – *Vigna radiata*

judía de puerco – *Canavalia ensiformis, C. villosa*
judía de vaca – *Vigna unguiculata*
judía dragón – *Psophocarpus tetragonolobus*
judía enana – *Phaseolus vulgaris*
judía encarnada – *Phaseolus coccineus*
judía escarlata – *Phaseolus coccineus*
judía espárrago – *Vigna unguiculata* ssp *sesquipedalis*
judía espía – *Canavalia ensiformis*
judía lima – *Phaseolus lunatus*
judía pinta – *Phaseolus coccineus, P. vulgaris*
judía roja – *Phaseolus coccineus*
judía sable – *Canavalia ensiformis, C. villosa*
judía tepari – *Phaseolus acutifolius*
judía tonquin – *Dipteryx odorata*
judía verde – *Phaseolus vulgaris*
judión – *Phaseolus lunatus*
jueoguo – *Salpianthus arenarius*
juguari – *Vitex mollis*
juia tóluco – *Schoepfia shreveana*
juilocuáhuitl – *Croton guatemalensis*
jujito colorado – *Passiflora foetida*
jujito peludo – *Passiflora foetida*
jujube – *Ziziphus jujuba*
jujuhuali – *Vitex mollis*
julahuasi – *Senna bicapsularis*
juliana matronal – *Hesperis matronalis*
juliana falsa – *Silene armeria*
jumallo – *Bombax ellipticum*
jumay – *Trichilia hirta*
jumete – *Asclepias albicans, A. subulata, Euphorbia colletioides, E. plicata, Heliotropium curassavicum*
jumetón – *Euphorbia californica, E. misera, E. xantii*
juncia – *Cyperus involucratus, Pinus teocote*
juncia avellanada – *Cyperus esculentus*
juncia bastarda – *Schoenus nigricans*
juncia comestible – *Cyperus eragrostis*
juncia común – *Cyperus longus*
juncia de flores pequeñas – *Cyperus difformis*
juncia de tallo circular – *Cyperus rotundus*
juncia diforme – *Cyperus difformis*
juncia distica – *Cyperus laevigatus*
juncia levigada – *Cyperus laevigatus*
juncia loca – *Cyperus longus*
juncia longus – *Cyperus longus*
juncia marina – *Scirpus maritimus*
juncia morena – *Cyperus fuscus*
juncia morisca – *Schoenus nigricans*
juncia olorosa – *Cyperus longus*
juncia redonda – *Cyperus esculentus, C. rotundus*
junco – *Adolphia californica, A. infesta, Bebbia juncea, Canotia holacantha, Carex morrowii* var *expallida, Ceanothus huichagorae, Cercidium microphyllum, Condalia brandegeei, Cyperus articulatus, Desmoncus chinantlensis, Disocactus, Eleocharis geniculata, Epiphyllum* x *Disocactus, Equisetum hyemale, Hylocereus undatus, Juncus, Koeberlinia spinosa* ssp *tenuispina, Mammillaria albicans, Nama coulteri, Parkinsonia aculeata, Peniocereus serpentinus, Phragmites australis, Typha latifolia*
junco agudo – *Juncus acutus*
junco articulado – *Juncus articulatus*
junco báltico – *Juncus balticus*
junco blanco – *Peniocereus serpentinus*
junco capitado – *Juncus capitatus*
junco común europeo – *Scirpus holoschoenus*
junco de bejuco – *Chamaedorea elatior*
junco de Dudley – *Juncus dudleyi*
junco de esteras – *Juncus effuses*
junco de hojas ensiformes – *Juncus ensifolius*
junco de Indias – *Canna indica*
junco de jaguay – *Desmoncus chinantlensis*
junco de las ranas – *Juncus ambiguus*
junco del mar – *Scirpus maritimus*
junco de los escuerzos – *Juncus bufonius*
junco enano – *Scirpus cernuus*
junco espinoso – *Peniocereus serpentines, Scirpus americanus, S. pungens*
junco florido – *Butomus umbellatus*
junco glauco – *Scirpus tabernaemontani*
junco incurvado – *Scirpus cernuus*
junco marino – *Parkinsonia aculeata*
junco mucronado – *Scirpus mucronatus*
junco negro – *Schoenus nigricans*
junco nigricante – *Schoenus nigricans*
junco oloroso – *Epiphyllum oxypetalum*
junco pungente – *Scirpus americanus, S. pungens*
junco rojo – *Disocactus speciosus*
junco rugoso – *Scirpus tabernaemontani*
junco serrano – *Ilex brandegeana*

junco setáceo – *Scirpus setaceous*
junco tapatío – *Hylocereus undatus*
junco tenue – *Juncus tenuis*
junio del monte – *Zinnia peruviana*
junípero – *Juniperus communis, J. monticola*
junípero turífero – *Juniperus thurifera*
junquera – *Juncus effuses*
junquillo – *Dasylirion longissimum, Disocactus flagelliformis, Narcissus jonquilla*
junquillo oloroso – *Narcissus jonquilla*
jupachomi – *Senna atomaria*
jupachumi – *Petiveria alliacea*
jupala – *Prosopis juliflora*
jupari – *Vitex mollis, V. pyramidata*
Jupiter – *Lagerstroemia indica*
juquilla – *Orbignya cohune*
juracosi – *Senna atomaria*
juralillo – *Calea zacatechichi*
juricua – *Plumbago pulchella, P. scandens*
jusiro – *Diphysa occidentalis*
jutitió – *Hintonia latiflora*
jutuhuí – *Sapindus saponaria*
jutuqui – *Condalia globosa* var *pubescens*
jutus – *Tagetes erecta*
juvavena – *Capparis atamisquea*
juve – *Bidens ferulaefolia*
juvia – *Bertholletia excelsa*
juyago – *Guaiacum coulteri*
juyapori – *Diphysa suberosa*

kaat – *Parmentiera aculeata, P. edulis*
kabal-muk – *Rauwolfia hirsuta*
kabak-xtes – *Alternanthera caracasana*
kafir – *Sorghum bicolor*
kakalché – *Bourreria huanita*
kako – *Theobroma cacao*
kalamoto – *Lycoperdon mixtecorum*
kamaba – *Lophophora williamsii*
kanabal – *Spondias mombin, S. purpurea*
kanchin – *Brachiara fasciculata, Panicum hirticaule*
kanibinche – *Malpighia glabra*
kankilizche – *Acacia farnesiana*
kanlal – *Argemone mexicana*
kanlolxiu – *Flaveria*
kanppocoche – *Duranta repens*
kantiriz – *Acacia farnesiana*
kanzakán – *Pachycereus gaumeri*
kanzinkin – *Caesalpinia pulcherrima*
kan-zuuk – *Aristida ternipes* var *gentilis*
kaquijá – *Amanita muscaria*
kasalá – *Commelina dianthifolia*
kat – *Catha edulis*
katsura – *Cercidiphyllum japonicum*
katzimelk – *Prosopis juliflora*
kaukpahk – *Lucuma mammosa*
kava kava – *Piper methysticum*
kazungula – *Setaria sphacelata*
kei – *Dovyalis caffra*
ken – *Guaiacum sanctum*
kenaf – *Hibiscus cannabinus*
kentia – *Howea belmoreana, H. forsteriana*
kevenia – *Ricinus communis*
keyaki – *Zelkova serrata*
khella – *Ammi majus*
khíb-shul – *Chamaedorea tepejilote*
kib – *Trema micrantha*
kiba – *Chamaedorea glaucifolia*
kiche – *Castilla elastica*
kickxia de hojas circulares – *Kickxia spuria*
kickxia elatina – *Kickxia elatine*
kieli – *Solandra guerrerensis*
kiiché – *Cocoloba uvifera*
kikí – *Laelia autumnalis*
kikuyu – *Pennisetum clandestinum*
kinim – *Ceiba pentandra*
kinin-jobo – *Spondias mombin, S. purpurea*
kiri – *Paulownia tomentosa*
kitembilla – *Dovyalis hebecarpa*
kiwano – *Cucumis metuliferus*
kiwi – *Actinidia deliciosa*
koch – *Ricinus communis*
koeleria grácil – *Koeleria macrantha*
koksigam – *Eysenhardtia orthocarpa*
koomene – *Justica californica*
koyamaki – *Sciadopitys verticillata*
kudzu – *Pueraria lobata*
kudzu tropical – *Pueraria phaseoloides*
kuitum – *Persea americana*
kuitumkeip – *Persea americana*
kuka – *Acacia farnesiana, Pseudophopenix sargentii*
kuk'uk – *Philoxerus vermicularis*
kulul – *Pachycereus gaumeri*
kunché – *Carica mexicana*
kunkuat – *Citrus japonica*
kuromatsu – *Pinus thunbergii*
kutz – *Nicotiana tabacum*

laaguia – *Lysiloma acapulcensis*
laal – *Urera caracasana*
láatrio – *Yucca valida*
lab – *Nymphaea ampla*
labitz – *Zingiber officinale*
lablab – *Lablab purpureus*
labrestos – *Raphanus raphanistrum*
labrusa – *Vitis labrusca*
labrusca – *Ampelopsis cordata*
labrusca quinquefolia – *Parthenocissus quinquefolia*
labrusca tricuspidada – *Parthenocissus tricuspidata*
laburno – *Laburnum anagyroides*
laburno común – *Laburnum anagyroides*
laburno de la India – *Cassia fistula*
laburno escocés – *Laburnum alpinum*
laburno híbrido de Voss – *Laburnum* x *watereri*
laca – *Rhus verniciflua*
lacayote – *Cucurbita ficifolia*
ladilla – *Cenchrus longispinus*
ladino – *Trifolium repens*
lady Eubanksia – *Rosa banksiae*
lagán de perro – *Plumbago auriculata*
lagaña de perro – *Plumbago scandens*
lagartera – *Setaria verticillata*
lágrimas – *Coix lacryma-jobi*
lágrimas de amor – *Russelia equisetiformes*
lágrimas de David – *Coix lacryma-jobi, Polygonatum odoratum*
lágrimas de Job – *Coix lacryma-jobi*
lágrimas de la Virgen – *Allium triquetrum, Ipomoea quamoclit, Lagurus ovatus, Nothoscordum fragrans, Solanum seaforthianum*
lágrimas de María – *Chiococca alba, Commelina coelestis*
lágrimas de Salomón – *Convallaria majalis*
lágrimas de San Diego – *Kohleria deppeana*
lágrimas de San Pedro – *Chiococca alba, Coix lacryma-jobi*
lagunaria – *Lagunaria patersonii*
lairén – *Calathea allouia*
laltsimin – *Urera caracasana*
lama – *Mimulus guttatus, Ruppia maritima, Selaginella, Utricularia vulgaris*
lama de agua – *Potamogeton diversifolius*
lamarckia dorada – *Lamarckia aurea*
lamatzapotl – *Annona cherimola*
lambedor – *Leersia hexandra*
lambrisco – *Rhus sempervirens, R. trilobata, R. virens*
lamio – *Lamium amplexicaule*
lamio maculado – *Lamium maculatum*
lamio púrpura – *Lamium purpureum*
lamish-que – *Selenicereus chontalensis*
lamparilla – *Hibiscus rosa-sinensis, Populus tremula*
lampatán – *Smilax china*
lampazo – *Arctium lappa, Eichhornia crassipes, Lemna minor, Nymphaea odorata, Xanthosoma robustum, Zantedeschia aethiopica*
lampazo menor – *Arctium minus*
lampote – *Helianthus annuus*
lampotillo – *Bidens aurea, Heterotheca inuloides*
lampranthus – *Lampranthus deltoides*
lámpsana – *Lapsana communis*
lámpsana común – *Lapsana communis*
lana – *Bombax ellipticum*
lanaria – *Saponaria officinalis*
lancetilla – *Astrocaryum mexicanum*
langosta antillana – *Hymenaea courbaril*
lano – *Ochroma lagopus*
lansa – *Lansium domesticum*
lantá – *Ceiba acuminata*
lantá de cerro – *Ceiba aesculifolia*
lantán – *Ixora coccinea*
lantana – *Lantana camara, Viburnum lantana*
lanzón – *Lansium domesticum*
lapa – *Arctium lappa, Galium aparine*
lapacho – *Tabebuia chrysantha, T. pentaphylla*
lapolla común – *Lappula squarrosa*
lárgalo – *Galium aparine*
largancillo – *Acacia constricta*
largo – *Musa acuminata* x *balbisiana*
largoncillo – *Acacia constricta*
lárice – *Larix decidua*
lasafras – *Laurus camphora*
lasha – *Smilax mexicana*
lastón anual – *Brachypodium distachyon*
lastón pinado – *Brachypodium pinnatum*
lata – *Gynerium sagittatum*
latania azul – *Latania loddigesii*
latania roja – *Latania lontaroides*
latilla – *Protium copal*

latonero – *Celtis australis, C. occidentalis*
laurel – *Ardisia revolute, Cordia alliodora, Ficus microcarpa, Laurus nobilis, Litsea glaucescens, Nectandra, Nerium oleander, Persea podadenia, Phoebe montana, P. tampicensis, Sassafridium macrophyllum, Umbellularia californica*
laurel alejandrino – *Danae racemosa*
laurel alejandrino de hojas estrechas – *Ruscus hypoglossum*
laurel amarillo – *Nectandra sanguinea, Psychotria nervosa*
laurel benjamín – *Ficus benjamina*
laurel blanco – *Cordia alliodora, Nerium oleander, Sassafridium macrophyllum, Tabernaemontana alba*
laurel-cerezo – *Laurus cerassus, Prunus laurocerasus*
laurel colorado – *Nerium oleander*
laurel de Apolo – *Laurus nobilis*
laurel de California – *Umbellularia californica*
laurel de hojas grandes – *Nectandra ambigens*
laurel de India – *Ficus benghalensis, F. microcarpa, F. religiosa*
laurel de la sierra – *Ardisia revoluta, Litsea glaucescens, Persea podadenia, Umbellularia californica*
laurel del país – *Litsea glaucescens*
laurel del poeta – *Laurus nobilis*
laurel de montaña – *Kalmia latifolia*
laurel de Nueva Zelanda – *Corynocarpus laevigata*
laurel de Portugal – *Prunus lusitanica*
laurel de San Antonio – *Epilobium angustifolium, E. hirsutum*
laurel de Trebisonda – *Prunus laurocerasus*
laurelillo – *Nectandra ambigens, Persea americana, Quercus affinis, Q. laurina, Sassafridium macrophyllum*
laurel manchado – *Aucuba japonica*
laurel noble – *Laurus nobilis*
laurelón – *Persea podadenia, Sassafridium macrophyllum*
laurel portugués – *Prunus lusitanica*
laurel real – *Prunus laurocerasus*
laurel rosa – *Nerium oleander*
laurel silvestre – *Umbellularia californica*
laurel tulipán – *Talauma mexicana*
lauréola – *Daphne laureola, D. mezereum*
lauréola hembra – *Daphne mezereum*
lauréola macho – *Daphne laureola*
lauro – *Laurus nobilis*
lauroceraso – *Prunus laurocerasus*
lauroceraso de Portugal – *Prunus lusitanica*
lauroceraso portugués – *Prunus lusitanica*
lausonia – *Reseda odorata*
lavanda – *Hyptis emoryi, Lavandula angustifolia, Salvia*
lavanda del mar – *Limonium californicum*
lavanda verdadera – *Lavandula angustifolia*
lavandero – *Phyllanthus acuminatus*
lavándula – *Lavandula angustifolia, L. latifolia*
lavándula de hojas anchas – *Lavandula latifolia*
lavándula francesa – *Lavandula stoechas*
lavándula hembra – *Lavandula angustifolia*
lavándula holandesa – *Lavandula angustifolia*
lavándula inglesa – *Lavandula angustifolia*
lavándula lanada – *Lavandula lanata*
lavándula multifida – *Lavandula multifida*
lavándula vera – *Lavandula angustifolia*
lavándula vividina – *Lavandula viridis*
lava pañal – *Solanum nudum*
lava plato – *Solanum diversifolium*
lavatera arbórea – *Lavatera arborea*
layolizán – *Buddleja americana*
lazo de amor – *Solanum dulcamara, S. seaforthianum, Torenia fournieri*
lebón – *Cercidium microphyllum*
lebón retama – *Cercidium microphyllum*
lecualquimichi – *Setaria liebmannii*
leche amarilla – *Calophyllum braziliense*
lechecino – *Sonchus oleraceus*
leche de gallina – *Ornithogalum arabicum, O. umbellatum*
leche de gallina nutante – *Ornithogalum nutans*
leche de pájaro de Arabia – *Ornithogalium arabicum*
leche María – *Calophyllum braziliense, Cordia alliodora, Trophis racemosa*
lecherillo – *Stemmadenia obovata, Tabernaemontana alba*
lecherina – *Euphorbia helioscopia*
lechero – *Asclepias speciosa, Sapium stylare, Silybum marianum*
lecheruela – *Euphorbia helioscopia*
lechetrenza – *Asclepias linaria, Euphorbia*
lechetrenza palustre – *Euphorbia palustris*

lechillo – *Carpinus betulus, C. caroliniana, Euphorbia dentata*
lechocino – *Senecio vulgaris*
lechón – *Asclepias latifolia, A. speciosa*
lechona – *Asclepias erosa*
lechoso – *Carica papaya*
lechuga – *Lactuca sativa, Pistia stratiotes*
lechuga acogollada – *Lactuca sativa* var *capitata*
lechuga arrepollada – *Lactuca sativa* var *capitata*
lechuga cos – *Lactuca sativa* var *longifolia*
lechuga de agua – *Eichhornia crassipes, Pistia stratiotes*
lechuga de bola – *Lactuca sativa* var *capitata*
lechuga de campo – *Valerianella locusta*
lechuga de hojas – *Lactuca sativa* var *longifolia*
lechuga de hojas de sauce – *Lactuca saligna*
lechuga de hojas sueltas – *Lactuca sativa* var *crispa*
lechuga de repollo abierto – *Lactuca sativa* var *crispa*
lechuga espárrago – *Lactuca sativa* var *angustana*
lechuga espinaca – *Lactuca serriola*
lechuga orejona – *Lactuca sativa* var *longifolia*
lechuga orejona común – *Lactuca sativa* var *capitata*
lechuga romana – *Lactuca sativa* var *longifolia*
lechuga romanita – *Lactuca sativa* var *longifolia*
lechuga silvestre – *Lactuca canadensis, L. serriola, L. virosa, Sonchus oleraceus*
lechuga tártara – *Lactuca tatarica*
lechuga venenosa – *Lactuca virosa*
lechuga virosa – *Lactuca virosa*
lechuguilla – *Abronia fragrans, Agave, Apocynum androsaemifolium, A. cannabinum, Chondrilla juncea, Hieracium, Lactuca serriola, Pistia stratiotes, Senecio candidissimus, S. vulneraria, Solidago mexicana, S. velutina, Sonchus arvensis, S. asper, S. oleraceus, Yucca whipplei*
lechuguilla agave – *Agave lechuguilla*
lechuguilla ceniza – *Agave shrevei*
lechuguilla de agua – *Pistia stratiotes*
lechuguilla de la sierra – *Packera candidissima*
lechuguilla mansa – *Agave filifera*
lechuguilla mezcal – *Agave aurea*
lechuguilla sabalí – *Agave bovicornuta*
lechuguilla silvestre – *Sonchus oleraceus*
lechuguilla verde – *Agave bovicornuta*
lechuguina – *Salicornia europaea*
lechuguino – *Lactuca sativa* var *longifolia*
lele – *Bombax ellipticum, Zapoteca anomala*
leleques – *Leucaena glauca*
lemilla – *Carlowrightia arizonica, C. lindauiana*
lemita – *Rhus microphylla, R. trilobata*
lengua cerval – *Phyllitis scolopendrium*
lengua cervina – *Phyllitis scolopendrium*
lengua de arista – *Penstemon campanulatus*
lengua de buey – *Anchusa azurea, A. officinalis, Buddleja crotonoides, B. sessiliflora, B. wrightii, Echium plantagineum, Fistulina hepatica, Nicotiana glauca, Solanum erianthum*
lengua de ciervo – *Phyllitis scolopendrium, Polypodium californicum, P. lanceolatum, Scolopendrium officinale, Swertia radiata*
lengua de dragón – *Canna edulis, C. generalis*
lengua del diablo – *Anthurium andraeanum, A. scherzerianum*
lengua de oveja – *Plantago major*
lengua de pájaro – *Caliandra ciliata, Polygonum arenastrum, P. aviculare*
lengua de perico – *Gliricidia sepium*
lengua de sapo – *Heliotropium indicum*
lengua de suegra – *Sansevieria trifasciata*
lengua de toro – *Plumeria rubra, Rumex conglomeratus*
lengua de vaca – *Buddleja sessiliflora, Opuntia, Pernettya ciliata, Petrea arborea, Pithecellobium echinatum, Rumex, Sansevieria guineensis, Solanum verbascifolium, Syngonium podophyllum*
lengua de vaca cimarrona – *Rumex flexicaulis, R. mexicanus*
lengua de víbora – *Erythronium americanum*
lengua ursina – *Penstemon gloxinioides*
lenguaza – *Anchusa azurea*
lenteja – *Lemna minor, Lens culinaris*
lenteja bastarda – *Vicia ervilia*
lenteja de agua – *Lemna minor*
lenteja de agua con gibas – *Lemna gibba*
lenteja de agua menor – *Lemna minor*
lenteja de agua minúscola – *Lemna minuscula*

lenteja de agua trisulca – *Lemna trisulca*
lenteja de Aragón – *Vicia articulata*
lentejilla – *Lemna gibba, L. minima, L. minor, Lepidium*
lentejilla de agua – *Lemna gibba, L. minor*
lentejuelilla – *Arabis mexicana*
lentibularia – *Urticularia vulgaris*
lentibularia gibosa – *Utricularia gibba*
lentibularia intermedia – *Utricularia intermedia*
lentibularia menor – *Urticularia minor*
lentil – *Malosma laurina*
lentisco – *Cercocarpus betuloides, C. montanus, Forestiera angustifolia, Malosma laurina, Pistacia lentiscus, P. texana, P. vera, Rhus*
lentisco del Perú – *Schinus molle*
lentisco de Nuevo León – *Pistacia mexicana*
leño colorado – *Arctostaphylos pungens*
leño gentil – *Daphne mezereum*
leño nefrítico – *Eysenhardtia polystachya*
leontopodio – *Leontopodium alpinum*
lepidio – *Descuriana pinnata, Lepidium*
lepidio campestre – *Lepidium campestre*
lepidio de hojas anchas – *Lepidium latifolium*
lepidio densifloro – *Lepidium densiflorum*
lepidio de ramificación compacta – *Lepidium ramosissimum*
lepidio de Virginia – *Lepidium virginicum*
lepidio draba – *Cardaria draba*
lepidio heterofilo – *Lepidium heterophyllum*
lepidio hortense – *Lepidium sativum*
lepidio perfoliado – *Lepidium perfoliatum*
lepra caballo – *Equisetum telmateia*
leque – *Lagenaria siceraria*
leucena – *Leucaena glauca, L. leucocephala*
leúque – *Podocarpus andinus*
levadura – *Lasiacis divaricata, Saccharomyces cerevisiae, Styrax argenteus*
levadura de la cerveza – *Saccharomyces cerevisae*
levadura del pan – *Saccharomyces minor*
levadura del vino – *Saccharomyces ellipsodieus*
levántate don Juan – *Nicotiana glauca*
levántate Juan – *Nicotiana glauca*
levístico – *Levisticum officinale, Ligustrum porteri*
lewisia – *Lewisia cotyledon*
liana amistad – *Cuscuta*
liátride – *Liatris spicata*
liátride en espiga – *Liatris spicata*
libocedro – *Calocedrus decurrens*
libocedro de California – *Calocedrus decurrens*
licio de China – *Lycium chinese*
licipio – *Lycopus americanus, L. asper*
licopodio – *Lycopodium clavatum*
licopodio chino – *Huperzia serrata*
licópside – *Anchusa arvensis*
licórice – *Glycyrrhiza glabra*
licóride – *Lycoris radiata*
licuala – *Licuala grandis*
licuala de manglar – *Licuala spinosa*
licuri – *Syagrus coronata*
licushi – *Phlebodium aureum*
lichi – *Litchi chinensis*
lichi peludo – *Nephelium lappaceum*
liebrecilla – *Centaurea cyanus, C. scabiosa*
liendrecillo – *Sporobolus indicus*
liendrilla – *Porophyllum macrocephalum, P. macrophyllum, Sporobolus indicus*
liendrilla abierta – *Muhlenbergia dumosa*
liendrilla amacollada – *Muhlenbergia porteri*
liendrilla antenada – *Muhlenbergia parviglumis*
liendrilla aparejo – *Muhlenbergia repens*
liendrilla chica – *Muhlenbergia microsperma*
liendrilla del pinar – *Muhlenbergia dubia*
liendrilla de Nuevo México – *Muhlenbergia pauciflora*
liendrilla de venado – *Muhlenbergia rigens*
liendrilla lengua larga – *Muhlenbergia longiligula*
liendrilla ligulada – *Muhlenbergia longiligula*
liendrilla morada – *Muhlenbergia rigida*
liendrilla tropical – *Sporobolus indicus*
liga – *Chamaesyce, Euphorbia, Iostephane heterophylla, Phoradendron, Psittacanthus calyculatus, Viguiera dentata, Viscum album*
lignum guaiaci – *Guaiacum sanctum*
lignum sanctum – *Guaiacum sanctum*
ligústico – *Levisticum officinale*
ligustro – *Ligustrum vulgare*
ligustro japonés – *Ligustrum japonicum*
lila – *Ceanothus, Melia azedarach* var *umbraculiformis, Petrea arborea, Pittosporum tobira, Syringa vulgaris*
lilac – *Ceanothus cordulatus, C. cuneatus, C. leucodermis*

lila de California – *Ceanothus thyrsiflorus*
lila de China – *Melia azedarach* var *umbraculiformis*
lila de las Indias – *Melia azedarach* var *umbraculiformis*
lila de Persia – *Melia azedarach*
lila de verano – *Buddleja davidii*
lilaila – *Melia azedarach*
lilia cimarrón – *Milla biflora*
lima – *Citrus aurantifolia, C. limetta, Phaseolus lunatus*
lima ácida – *Citrus aurantifolia*
lima agria – *Citrus aurantifolia, C. limetta*
lima común de Persia – *Citrus latifolia*
lima chica – *Citrus aurantifolia*
lima chichona – *Citrus aurantifolia, C. limetta*
lima del desierto – *Eremocitrus glauca*
lima dulce – *Citrus limetta, C. limettioides*
lima persa – *Citrus latifolia*
lima reina – *Citrus aurantifolia*
limero – *Citrus aurantifolia, C. limetta, C. limon*
limero de Tahiti – *Citrus latifolia*
limilla – *Ferocactus hamatacanthus*
limilla de la sierra – *Rhus microphylla*
limita – *Rhus microphylla*
limita de la sierra – *Rhus microphylla*
limnanta – *Limnanthes douglasii*
limón – *Citrus aurantifolia, C. limon*
limón agrio – *Citrus aurantifolia, C. limon*
limón amarillo – *Citrus limon*
limonaria – *Murraya paniculata*
limoncillo – *Agonandra racemosa, Amyris balsamifera, Capparis verrucosa, Casearia aculeata, Cymbopogon citratus, Dalea citriodora, D. lumholtzii, Hirtella americana, Ilex tolucana, Ipomoea purga, Jacquinia macrocarpa* ssp *pungens, Maytenus phyllanthoides, Melissa officinalis, Pectis, Peganum mexicanum, Rheedia edulis, Sargentia greggii, Thymophylla pentachaeta* var *belenidium, Trichilia hirta, Triphasia trifolia, Ximenia americana, Zanthoxylum fagara*
limoncillo de Cordóba – *Calophyllum braziliense*
limoncillo del campo – *Peganum mexicanum*
limoncillo de Meztitlán – *Ternstroemia sylvatica*
limoncito – *Encyclia citrina, Triphasia trifolia*
limón criollo – *Citrus aurantifolia*
limón chichón – *Citrus limon*
limón chiquito – *Citrus aurantifolia*
limón chivo – *Citrus medica*
limón de mar – *Ximenia americana*
limón dulce – *Citrus aurantifolia, C. limettioides, C. limetta, C. limon*
limonero – *Citrus aurantifolia, C. limon, Melissa officinalis*
limonero dulce – *Citrus limetta*
limón francés – *Citrus limon*
limonio – *Limonium sinuatum*
limonio sinuado – *Limonium sinuatum*
limón liso – *Citrus aurantifolia*
limón mexicano – *Citrus aurantifolia*
limón persa – *Citrus latifolia*
limón pérsico – *Citrus latifolia*
limón ponderosa – *Citrus limon* var *ponderosa*
limón real – *Citrus limetta, C. limon, C. limonia, C. medica*
limón rugoso – *Citrus jambhiri*
limón verde – *Citrus aurantifolia*
limosella acuática – *Limosella aquatica*
limpia gallinas – *Stellaria*
limpia platas – *Equisetum hyemale*
limpia tubos llorón – *Callistemon viminalis*
limpiatunas – *Baccharis conferta, Ipomoea stans, Kosteletzkya thurberi, Zaluzania augusta*
linaria – *Antirrhinum majus, Linaria maroccana, L. vulgaris*
linaria bipartida – *Linaria bipartita*
linaria cimbalaria – *Cymbalaria muralis*
linaria común – *Linaria vulgaris*
linaria dalmatica – *Linaria genistifolia* ssp *dalmatica*
linaria de hojas de retama – *Linaria genistifolia*
linaria marroquí – *Linaria maroccana*
linaria purpúrea – *Linaria purpurea*
linaza – *Linum lewisii, L. usitatissimum*
linda tarde – *Gaura coccinea, G. parviflora, Mirabilis jalapa, Oenothera tetraptera*
lindera – *Lindera benzoin*
lindernia dudosa – *Lindernia dubia*
linera – *Linum usitatissimum*
linnea boralis – *Linnaea borealis*
lino – *Linum grandiflorum, L. usitatissimum*
lino amarillo – *Linum flavum*

lino azul – *Linum narbonense*
lino bienal – *Linum bienne*
lino bravo – *Linum narbonense*
lino cultivado – *Linum usitatissimum*
lino de flores grandes – *Linum grandiflorum*
lino de Nueva Zelandia – *Phormium tenax*
lino de zanja – *Potamogeton pectinatus*
lino montesino – *Linaria vulgaris*
lino perenne – *Linum perenne*
liofilio agregado – *Lyophyllum decastes*
liquen – *Parmelia tiliacea, Usnea barbata*
liquen de Islandia – *Cetraria islandica*
liquen de tierra – *Parmelia molluscula*
liquidámbar – *Liquidambar styraciflua*
liquidámbar americano – *Liquidambar styraciflua*
liquidámbar chino – *Liquidambar formosana*
liquidámbar de Formosa – *Liquidambar formosana*
liquidámbar oriental – *Liquidambar orientalis*
lira de San Pedro – *Eustoma grandiflorum*
lirio – *Crinum zeylanicum, Eichhornia crassipes, Hymenocallis americana, Iris, Laelia, Lilium, Nymphoides indica, Psittacanthus calyculatus*
lirio acuático – *Eichhornia crassipes*
lirio africano – *Agapanthus africanus*
lirio alemán – *Iris germanica*
lirio amarillo – *Iris pseudacorus*
lirio araña – *Tradescantia virginiana*
lirio atrigado – *Lilium lancifolium*
lirio azteca – *Tigridia pavonia*
lirio azul – *Iris germanica, I. versicolor*
lirio blanco – *Hymenocallis rotata, Iris florentina, Laelia glauca, Lilium candidum*
lirio cárdeno – *Iris germanica*
lirio común – *Iris germanica*
lirio convalario – *Convallaria majalis*
lirio de agua – *Eichhornia crassipes, Nuphar luteum, Nymphaea alba, Zantedeschia aethiopica*
lirio de Alemania – *Iris germanica*
lirio de cinta – *Crinum zeylanicum*
lirio de Colón – *Alpinia zerumbet*
lirio de corpus – *Laelia speciosa*
lirio de Egipto – *Zantedeschia aethiopica*
lirio de España – *Iris xiphium*
lirio de Etiopía – *Zantedeschia aethiopica*
lirio de Florencia – *Iris florentina*
lirio de flores largas – *Lilium longiflorum*
lirio de Holanda – *Iris* x *hollandica*
lirio de Inglaterra – *Iris latifolia*
lirio de la costa – *Plumeria acutifolia*
lirio de laguna – *Pontederia cordata*
lirio del arroyo – *Hedychium coronarium*
lirio del Japón – *Lilium lancifolium*
lirio del monte – *Laelia anceps*
lirio del Perú – *Alstroemeria aurea*
lirio del valle – *Convallaria majalis*
lirio de lluvia – *Zephyranthes carinata*
lirio de Malabar – *Gloriosa superba*
lirio de mirto – *Acorus calamus*
liriodentro – *Liriodendron tulipifera*
lirio de pascua – *Lilium longiflorum*
lirio de San Antonio – *Lilium candidum, Sobralia macrantha*
lirio de San Francisco – *Laelia autumnalis*
lirio de San Jaime – *Amaryllis belladonna, Sprekelia formosissima*
lirio de San Juan – *Hemerocallis fulva, H. lilioasphodelus*
lirio de San Luis – *Lilium candidum*
lirio de San Sebastián – *Cattleya skinneri* var *alba*
lirio de tigre – *Lilium lancifolium*
lirio de todos santos – *Laelia anceps*
lirio dorado – *Lilium auratum*
lirio dulce – *Acorus calamus*
lirio enano – *Iris pumila*
lirio espadañal – *Iris pseudacorus*
lirio español – *Iris xiphium*
lirio espléndido – *Lilium speciosum*
lirio florentino – *Iris florentina*
lirio germánico – *Iris germanica*
lirio gigantesco – *Cardiocrinum giganteum*
lirio gloriosa – *Gloriosa superba*
lirio hediondo – *Iris foetidissima*
lirio holandés – *Iris* x *hollandica*
lirio iris – *Iris germanica*
lirio jacobino – *Sprekelia formosissima*
lirio japonés – *Eichhornia crassipes, Iris ensata*
lirio listado – *Amaryllis longifolia, Crinium zeylanicum*
lirio mariposa – *Calochortus flavus*
lirio morado – *Iris germanica*
lirio multicolor – *Iris versicolor*
lirio naranja – *Lilium bulbiferum*

lirio parasito – *Psittacanthus calyculatus*
lirio pascual – *Iris germanica*
lirio pata de gallo – *Epidendrum parkinsonianum*
lirio pata de paloma – *Epidendrum parkinsonianum*
liriope – *Liriope muscari*
lirio real – *Lilium regale*
lirio reticulado – *Iris reticulata*
lirio rocoso – *Iris innominata*
lirio rojo – *Amaryllis reginae*
lirio sego blanco – *Calochortus bruneaunis, C. panamintensis*
lirio siberiano – *Iris siberica*
lirio sibirica – *Iris sibirica*
lirio silvestre – *Lilium martagon*
lirio soberbio – *Lilium speciosum*
lirio tigrado – *Lilium lancifolium*
lirio unguiculado – *Iris unguicularis*
lisimaquia amarilla – *Lysimachia punctata*
lisimaquia ciliada – *Lysimachia ciliata*
lisimaquia china – *Lysimachia clethroides*
lisimaquia de hojas de clethra – *Lysimachia clethroides*
lisimaquia de los pantanos – *Lysimachia thyrsiflora*
lisimaquia numular – *Lysimachia nummularia*
lisimaquia punteada – *Lysimachia punctata*
lisimaquia purpúrea – *Lythrum salicaria*
lisimaquia rastrera – *Lysimachia nummularia*
lisimaquia roja – *Lysimachia ephemerum, L. salicaria, Lythrum salicaria*
listera de hojas cordiformes – *Listera cordata*
listoncillo – *Euphorbia pulcherrima*
litampa – *Chamaedorea oblongata*
litospermo – *Sanguinaria canadensis*
litro – *Lythrum salicaria*
livistona – *Livistona australis, L. chinensis*
lizarra – *Erica vagans*
lobata – *Neurolaena lobata*
lobelia – *Lobelia*
lobelia azulada – *Lobelia erinus*
lobelia cardinal – *Lobelia cardinalis*
lobelia del jardín – *Lobelia erinus*
lobelia glandulosa – *Lobelia siphilitica*
lobelia morada – *Lobelia fenestralis*
lobelia roja – *Lobelia cardinalis* var *pseudosplendens*
lobelia sifilítica – *Lobelia siphilitica*
lobito – *Gnaphalium bourgovii*
loco – *Astragalus mollissimus, A. nuttallianus, Oxytropis lambertii, Robinia pseudoacacia*
lofochloa crestada – *Koeleria phleoides*
loiseleuria – *Phyllodoce*
lolito – *Conostegia xalapensis, Pernettya ciliata*
lombarda – *Brassica oleracea* var *capitata*
lomboy – *Jatropha cinerea*
lomboy blanco – *Jatropha cinerea*
lomboy brilloso – *Jatropha vernicosa*
lomboy colorado – *Jatrophia vernicosa*
lomboy negro – *Jatropha vernicosa*
lomboy rojo – *Jatropha vernicosa*
lomboy vida – *Jatropha vellanowethii*
lombrecina – *Dorstenia contrayerba*
lombricero – *Andira inermis, Spigelia anthelmia*
lombricillo – *Tabebuia chrysantha*
longana – *Nephelium longana*
longán – *Nephelium longana*
loquetico – *Turbina corymbosa*
loro – *Prunus laurocerasus, P. lusitanica*
loto – *Lotus corniculatus*
loto corniculado – *Lotus corniculatus*
loto de cuernecillo – *Lotus corniculatus*
loto de hojas anchas – *Lotus corniculatus*
loto de los pantanos – *Lotus uliginosus*
loto de los prados – *Lotus corniculatus*
loto sagrado – *Nelumbo nucifera*
lovaja sichiana – *Ligusticum chuanxiong*
lovi-lovi – *Flacourtia inermis*
loyo – *Boletus edulis*
lucerillo – *Aster subulatus*
lucerna – *Medicago sativa*
lúcuma – *Lucuma rivicoa*
lucumxiu – *Chenopodium ambrosioides*
luch – *Crescentia alata, C. cujete*
ludwigia de los pantanos – *Ludwigia palustris*
lufa – *Luffa acutangula, L. aegyptiaca, L. cylindrica*
lulo – *Solanum quitoense*
lumbán – *Aleurites moluccana*
lunaria – *Botrychium lunaria, Lunaria annua*
lunaria anual – *Lunaria annua*
lupino – *Lupinus*
lupino amarillo – *Lupinus luteus*
lupino arbóreo – *Lupinus arboreus*
lupino azul – *Lupinus angustifolius*

lupino blanco – *Lupinus albus*
lupita – *Castilleja arvensis, Gomphrena decumbens*
lupo – *Humulus lupulus*
lupulina – *Medicago lupulina*
lúpulo – *Humulus lupulus*
lustiej – *Anisacanthus thurberi*
lustrosa – *Bougainvillea spectabilis, Anisacanthus thurberi*
luteolina – *Reseda luteola*
luya – *Bombax ellipticum*
lúzula de flores pequeñas – *Luzula parviflora*
lúzula espigada – *Luzula spicata*

llaba – *Andira inermis*
llagas de cristo – *Tropaeolum majus*
llama de bosque – *Spathodea campanulata*
llamarada – *Allamanda cathartica, Pyrostegia venusta*
llamarada de bosque – *Spathodea campanulata*
llama trepadora – *Senecio chenopodioides, S. confusus*
llanero – *Agaricus bitorquis, A. campestris*
llanta – *Brassica oleracea* var *acephala*
llantén – *Plantago*
llantén acuático – *Alisma plantago-aquatica*
llantén arenario – *Plantago indica*
llantén cimarrón – *Mimulus guttatus*
llantén cuerno de ciervo – *Plantago coronopus*
llantén de agua – *Alisma plantago-aquatica*
llantén de hojas estrechas – *Alisma gramineum*
llantén grande – *Plantago major*
llantén lanceolado – *Alisma lanceolatum, Plantago lanceolata*
llantén marítimo – *Plantago maritima*
llantén mayor – *Plantago major*
llantén menor – *Plantago lanceolata*
llantén ovado – *Plantago ovata*
llata – *Capsicum pubescens*
llaupangue – *Francoa sonchifolia*
llavina – *Mammillaria dioica, M. goodridgei*
lleuque – *Podocarpus andinus*
lleva del sapo – *Cirsium undulatum*
llora – *Russula cyanoxantha*
llora sangre – *Bocconia arborea, B. frutescens, Croton draco, Pterocarpus officinalis*
llorón – *Salix babylonica*
llorona – *Eragrostis curvula*
llovisna – *Stellaria prostrata*
llume – *Gaussia maya*
lluvia – *Cantua buxifolia, Duranta repens*
lluvia de coral – *Russelia equisetiformes*
lluvia de fuego – *Russelia equisetiformes*
lluvia de oro – *Cassia fistula, Cercidium floridium, Galphimia glauca, Laburnum anagyroides, Pyrostegia venusta, Tecoma stans*
lluvia dorada – *Cassia fistula, Pyrostegia venusta*

maaxcal – *Dioscorea*
mabem – *Marsdenia edulis*
mabolo – *Morisonia americana*
maca – *Lepidium, Saccharum officinarum*
maca colorada – *Andira inermis*
macadamia – *Macadamia integrifolia, M. ternifolia*
macahuite – *Cordia boissieri, Ficus*
macal – *Anthurium xanthosomifolium, Dioscorea, Xanthosoma sagittifolium*
macapura – *Heliotropium angiospermum*
macaricari – *Panicum coloratum*
macatlxóchitl – *Piper*
macayo – *Andira inermis*
macayo chico – *Andira inermis*
macazpipilloli – *Exogonium bracteatum*
macehualpuhuk – *Tagetes patula*
macetilla – *Dianthus barbatus*
macia – *Myristica fragrans*
macis – *Myristica fragrans*
maclale – *Tradescantia pallida*
maclura – *Maclura pomifera*
macochin – *Pithecellobium dulce*
macpalxochicuahuitl – *Chiranthodendron pentadactylon*
macpalxóchitl – *Chiranthodendron pentadactylon*
macuay – *Lippia umbellata*
macuche – *Nicotiana rustica*
macuchi – *Nicotiana tabacum*
macuilixuatl – *Tabebuia pentaphylla*
maculán – *Piper auritum*
maculi – *Quercus fusiformis*
maculiz – *Tabebuia impetiginosa, T. pentaphylla, T. rosea*
maculiz de abajo – *Tabebuia pentaphylla*
maculiz prieto – *Tabebuia pentaphylla*
macure – *Calathea allouia*

macús – *Calathea allouia*
macusey – *Philodendron radiatum*
macusi – *Cannabis sativa*
machacuana – *Ipomoea violacea*
machaquí – *Celtis reticulata, Lysiloma watsonii*
machicari – *Wimmeria mexicana*
machichare – *Quercus viminea*
macho – *Coleus pumilus*
machocote – *Cyrtocarpa procera*
madama – *Impatiens balsamina*
madera blanca – *Betula alba*
madera de arco – *Maclura pomifera*
madera de palisandro – *Dalbergia nigra*
madera de pasta – *Cochlospermum vitifolium*
madera negra – *Gliricidia sepium*
madera pino – *Pinus palustris*
maderista – *Eriogonum fasciculatum*
madia – *Madia sativa*
madrastra – *Mentha suaveolens*
madre alcaparra – *Erythrina glauca*
madre blanca – *Erythrina glauca*
madre brava – *Erythrina*
madre chontal – *Erythrina*
madre de agua – *Erythrina glauca, Fraxinus uhdei*
madre de cacao – *Erythrina, Gliricidia sepium, Quararibea funebris*
madre del maíz – *Dioscorea*
madre de sal – *Avicennia germinans, Conocarpus erecta*
madre de tomillo – *Thymus serpyllum*
madregosa – *Apocynum cannabinum*
madre mansa – *Erythrina glauca*
madreperla – *Pilea cadierei*
madre prieta – *Erythrina glauca*
madreselva – *Chiococca alba, Gelsemium sempervirens, Lonicera, Passiflora*
madreselva atlántica – *Lonicera periclymenum*
madreselva azul – *Lonicera caerulea*
madreselva de jardín – *Lonicera periclymenum*
madreselva etrusca – *Lonicera etrusca*
madreselva inglesa – *Lonicera periclymenum*
madreselva japonesa – *Lonicera japonica*
madreselva selvática – *Lonicera periclymenum*
madreselva siempreverde – *Lonicera sempervirens*
madreselva tártara – *Lonicera tatarica*
madronillo – *Amelanchier denticulata*
madroñero – *Arbutus unedo*
madroñito – *Vaccinium*
madroño – *Arbutus, Arctostaphylos, Clethra lanata, C. mexicana, Cupressus forbesii, Gomphrena globosa*
madroño blanco – *Arbutus xalapensis*
madroño borracho – *Arbutus xalapensis, Arctostaphylos*
madroño canario – *Arbutus canariensis*
madroño colorado – *Arbutus xalapensis*
madroño común – *Arbutus unedo*
madroño chino – *Arbutus tessellata, Arctostaphylos, Vaccinium confertum*
madroño de árbol – *Arctostaphylos tomentosa*
madroño de Grecia – *Arbutus andrachne*
madroño de las montañas rocosas – *Arbutus menziesii*
madroño del campo – *Gomphrena decumbens*
madroño negro – *Arbutus tessellata*
madroño oriental – *Arbutus andrachne*
madroño rastrero – *Arctostaphylos uva-ursi, Gomphrena decumbens*
madura plátano – *Hamelia erecta*
mafafa – *Xanthosoma robustum, X. sagittifolium*
mag – *Annona glabra*
magarza – *Chrysanthemum coronarium, Tanacetum parthenium*
magarzuela – *Anthemis cotula, Tanacetum parthenium*
magnolia – *Magnolia grandiflora, Talauma mexicana*
magnolia acuminada norteamericana – *Magnolia acuminata*
magnolia americana de hojas grandes – *Magnolia macrophylla*
magnolia común – *Magnolia grandiflora*
magnolia china – *Magnolia* x *soulangiana*
magnolia de Campbell – *Magnolia campbellii*
magnolia de flores pequeñas – *Magnolia sieboldii*
magnolia de Loebner – *Magnolia loebneri*
magnolia estrellada – *Magnolia stellata*
magnolia grandiflora – *Magnolia grandiflora*
magnolia híbrida de Veitch – *Magnolia* x *veitchii*
magnolia japonesa de hojas de sauce – *Magnolia salicifolia*

magnolia japonesa de hojas grandes – *Magnolia hypoleuca*
magnolia kobus del norte del Japón – *Magnolia kobus*
magnolia mexicana – *Talauma mexicana*
magnolia perenne de Delavay – *Magnolia delavayi*
magnolia perenne de la China – *Magnolia delavayi*
mago – *Hernandia sonora, Sapium biloculare*
maguey – *Agave*
maguey azul – *Agave tequilana*
maguey blanco – *Agave atrovirens* var *mirabilis*
maguey cenizo – *Agave asperrima, A. atrovirens* var *mirabilis, A. quiotifera*
magueycillo – *Hechtia montana*
maguey curandero – *Agave marmorata*
maguey chalqueño – *Agave salmiana*
maguey chato – *Agave guadalajarana*
maguey chichimeco – *Agave americana*
maguey de cerro – *Agave asperrima*
maguey de ixtle – *Agave applanata*
maguey de la casa – *Agave applanata*
maguey del cumbre – *Agave atrovirens*
maguey de maceta – *Agave filifera*
maguey de Manila – *Agave cantala*
maguey de mezcal – *Agave americana, A. angustifolia, A. tequilana*
maguey de pulque – *Agave americana, A. atrovirens, A. salmiana*
maguey de tequila – *Agave tequilana*
maguey de Yucatán – *Agave sisalana*
magueyito – *Agave hartmanii, Tillandsia recurvata*
magueyito de árbol – *Tillandsia recurvata*
magueyito parasito – *Tillandsia recurvata*
maguey javalin – *Agave atrovirens*
maguey lechuguilla – *Agave angustifolia, A. lechuguilla*
maguey manso – *Agave atrovirens, A. mapisaga, A. salmiana*
maguey manso fino – *Agave atrovirens*
maguey mapisaga – *Agave mapisaga*
maguey meco – *Agave americana*
maguey mezatillo – *Agave deweyana*
maguey morado – *Tradescantia discolor*
maguey pichomel – *Agave kerchovei*
maguey pinto – *Agave americana, Sansevieria trifasciata*
maguey pulquero – *Agave atrovirens*
maguey silvestre – *Tradescantia spathacea*
maguey tepestate – *Acacia marmorata*
maguey tequilero – *Agave tequilana*
maguey tobala – *Agave atrovirens*
maguey tuxtleco – *Agave sisalana*
maguey verde – *Agave macroculmis, A. salmiana*
maguey verde grande – *Agave atrovirens*
maguillo – *Malus sylvestris*
maguó – *Acacia coulteri*
maha – *Quararibea funebris*
mahaleb – *Prunus mahaleb*
maho – *Ochroma pyramidale*
mahonesa – *Malcolmia maritima*
mahonia – *Berberis*
ma-huang – *Ephedra sinica*
mahuitli – *Indigofera suffruticosa*
maicena – *Sorghum vulgare*
maicero – *Tripsacum lanceolatum*
maicillo – *Anthurium scandens, Euchlaena mexicana, Panicum miliaceum, Sorghum halepense, Tripsacum dactyloides*
mail – *Cucurbita ficifolia*
maíllo – *Malus sylvestris*
maíz – *Zea mays*
maíz azucarado – *Zea mays* var *saccharata*
maíz blando – *Zea mays* var *amylacea*
maíz ceroso – *Zea mays* var *ceratina*
maíz córneo – *Zea mays* var *indurata*
maíz correcaminos – *Euchlaena mexicana*
maíz chapule – *Zea diploperennis*
maíz de coyote – *Euchlaena mexicana, Zea canina*
maíz de China – *Sorghum saccharatum*
maíz de diente – *Euchlaena mexicana*
maíz de grano dulce – *Zea mays* var *saccharata*
maíz de Guinea – *Sorghum bicolor, S. vulgare*
maíz de la India – *Zea mays*
maíz del monte – *Anthurium scandens*
maíz dentado – *Zea mays* var *indentata*
maíz de palomita – *Zea mays* var *everia*
maíz de rosita – *Zea mays* var *everia*
maíz de teja – *Helianthus annuus*
maíz de Texas – *Helianthus annuus*

maíz diente de caballo – *Zea mays* var *indentata*
maíz dulce – *Zea mays* var *saccharata*
maíz duro – *Zea mays* var *indurata*
maíz enano – *Zea mays* var *gracillima*
maíz harinoso – *Zea mays* var *amylacea*
maíz japonés – *Zea mays* var *japonica*
maíz meco – *Helianthus annuus*
maizmillo – *Sorghum bicolor, S. vulgare*
maíz morocho – *Sorghum bicolor*
maíz muchito – *Zea mays* var *everia*
maíz palomero – *Zea mays* var *everia*
maíz para choclo – *Zea mays* var *saccharata*
maíz primitivo – *Euchlaena mexicana*
maíz reventador – *Zea mays* var *everia*
maíz reventón – *Zea mays* var *everia*
maíz silvestre – *Euchlaena mexicana*
maíz tierno – *Zea mays* var *amylacea*
maíz tunicado – *Zea mays* var *tunicata*
majagua – *Heliocarpus tomentosus, Hibiscus elatus, H. tiliaceus, Thespesia populnea*
majagua blanca – *Hibiscus tilliaceus, Trema micrantha*
majagua colorada – *Trema micrantha*
majagua de Cuba – *Hibiscus tilliaceus*
majagua de mar – *Hibiscus tilliaceus*
majagua de toro – *Guazuma ulmifolia*
majagua jolocín – *Heliocarpus donnell-smithii*
majahui – *Agave ocahui*
majahuilla – *Triumfetta semitriloba*
majorana – *Lantana macropoda*
majoypiualá – *Rothrockia cordifolia*
majuelo – *Crataegus*
majuelo navarro – *Crataegus laevigata*
makal – *Colocasia esculenta*
makopa – *Eugenia javanica*
malabar – *Solanum erianthum, S. verbascifolium*
malabarico – *Bombax ceiba*
malacate – *Vernonia deppeana, Xylosma flexuosum*
malacate blanco de montaña – *Zexmenia frutescens*
malacaxóchitl – *Zinnia*
malagueta – *Aframomum melegueta, Eugenia pimenta, E. racemosa*
mala madre – *Chlorophytum comosum*
malambo – *Phytolacca dioica*
mala mujer – *Chenopodium rubrum, Cirsium ehrenbergii, Cnidoscolus, Euphorbia cotinifolia, Hibiscus bifurcatus, Jatropha angustidens, J. gossypiifolia, Physalis acutifolia, Solanum, Toxicodendron diversilobium, T. radicans, Tragia laciniata, Urera caracasana, Urtica dioica, Wigandia caracasana, W. kunthii*
mala mujer china – *Jatropha angustidens*
mala mujer lisa – *Cnidoscolus multilobus*
malanga – *Colocasia antiquorum, C. esculenta, Xanthosoma*
malanga amarilla – *Xanthosoma sagittifolium*
malanga ojal – *Monstera deliciosa*
malaquito – *Alibertia edulis*
malayo – *Zea mays*
malcolmia africana – *Malcolmia africana*
malcolmia marítima – *Malcolmia maritima*
malcoraje – *Mercurialis annua*
mal de ojo – *Caesalpinia crista, Kallstroemia, Paspalum hartwegianum, Phaulothamnus spinescens, Psittacanthus calyculatus, Sphaeralcea, Sida tragiifolia, Sphaeralcea angustifolia, Zinnia*
maleagrida – *Fritillaria meleagris*
maleza de mariposa – *Asclepias curassavica*
maleza hortequilla – *Paspalum conjugatum*
mal hombre – *Urera caracasana*
malinali – *Muhlenbergia macroura*
malinche – *Ehretia tinifolia*
malmadurillo – *Ligustrum vulgare*
malojillo – *Brachiaria mutica, Eriochloa polystachya, Panicum molle*
malojo – *Zinnia*
malote – *Tripsacum laxum*
malque – *Cyathea mexicana*
maluco – *Genipa americana*
maluco de montaña – *Randia armata*
malva – *Abutilon incanum, A. lignosum, A. umbellatum, Alcea rosea, Anoda crenatiflora, Corchorus siliquosus, Eremalche scoparium, Herissantia crispa, Lavatera assurgentiflora, Malacothamnus fasciculatus, Malva, Malvastrum bicuspidatum, M. coromandelianum, M. sylvestrus, Malvaviscus arboreus, Melochia tomentosa, Senna pallida, Sida, Sidalcea malvaeflora, Sphaeralcea ambigua, Waltheria americana*
malva amarilla – *Sida acuta*
malva anual – *Lavatera trimestris*

malva arbórea – *Alcea rosea*
malva blanca – *Bastardiastrum cinctum, Horsfordia alata, Malva sylvestris, Sida aggregata, Urena lobata, Waltheria americana*
malva bouquet – *Pelargonium*
malva colorada – *Abutilon reventum, Sida acuta, S. rhombifolia*
malva común – *Malva sylvestris*
malva con espuelas – *Anoda cristata*
malva cubierta – *Corchorus siliquosus*
malva de Castilla – *Anoda cristata, Malva neglecta, M. parviflora, Sida acuta*
malva de cochino – *Sida acuta*
malva de Creta – *Lavatera cretica*
malva de flores grandes – *Lavatera trimestris*
malva de flores pequeñas – *Malva parviflora*
malva de geranio – *Pelargonium*
malva de hojas redondas – *Malva neglecta*
malva del campo – *Malva neglecta, M. parviflora*
malva del monte – *Waltheria americana*
malva de los cerros – *Melochia tomentosa*
malva de los Judios – *Kerria japonica*
malva de llano – *Waltheria americana*
malva de niza – *Malva nicaeensis*
malva de olor – *Pelargonium*
malva de olor de almizcle – *Malva moschata*
malva de platanillo – *Sida acuta*
malva de quesitos – *Malva neglecta, M. parviflora*
malva enana – *Malva neglecta*
malva grande – *Malvastrum sylvestrus*
malva griega – *Sidalcea malvaeflora*
malva índica – *Abutilon avicennae*
malva lanosa – *Croton borgensis*
malva loca – *Alcea rosea*
malva Luisa – *Pelargonium*
malva marítima – *Lavatera maritima*
malva medicinal – *Malva neglecta*
malva negra – *Herissantia crispa*
malva peluda – *Abutilon incanum*
malva rastrera – *Waltheria americana*
malva real – *Alcea rosea, Althaea officinalis, Lavatera assurgentiflora, L. trimestris, Malvastrum sylvestrus*
malva rosa – *Alcea rosea, Lavatera assurgentiflora, Melochia tomentosa, Pelargonium, Sphaeralcea axillaris*
malva rosa del monte – *Hibiscus bifurcatus, H. cardiophyllus*
malva rubi – *Pelargonium*
malva selvática – *Malva sylvestris*
malva té – *Corchorus siliquosus*
malvavisco – *Althea officinalis, Malvaviscus grandiflorus, Melochia tomentosa, Sida acuta, S. rhombifolia*
malvavisco de la India – *Abutilon avicrnnae*
malvavisco de las Indias – *Abutilon vitifolium*
malvavisco escobilla – *Sida rhombifolia*
malvavisco grande – *Corchorus siliquosus*
malvavisco marino – *Lavatera maritima*
malvavisco salvaje – *Malva alcea*
malvito – *Maranta leuconeura*
malvón – *Lavatera assurgentiflora, Pelargonium, Sphaeralcea rosea*
mamalacote – *Hydrocotyle ranunculoides*
mamalhuaztli – *Clethra lanata*
mambia – *Solanum erianthum, S. gracile, S. nigrescens, S. nigrum*
mamelique – *Krameria*
mamey – *Lucuma mammosa, Mammea americana, Pouteria sapota*
mamey amarillo – *Mammea americana*
mamey ciruela – *Pouteria campechiana*
mamey colorado – *Lucuma mammosa*
mamey de Campeche – *Pouteria campechiana*
mamey de Santo Domingo – *Mammea americana*
mameyito – *Clethra mexicana*
mameyito blanco – *Clethra lanata*
mameyito negro – *Clethra lanata*
mamey zapote – *Lucuma mammosa*
mamoa – *Erythroxylum mexicanum, Esenbeckia hartmanii*
mamojuaxtle – *Clethra mexicana*
mamón – *Annona reticulata, Melicocca bijuga*
mamoncillo – *Melicocca bijuga*
mamoncillo chino – *Litchi chinensis*
mamota – *Silene laciniata*
maná – *Alhagi pseudalhagi, Lecanora*
manaca – *Orbignya cohune, Scheelia preussii*
maná de Calabria – *Fraxinus ornus*
maná de Oriente – *Fraxinus ornus*
manalote – *Bocconia frutescens*
mananepile – *Polypodium lanceolatum*
manca – *Echinocactus texensis*

manca caballo – *Echinocactus horizonthalonius, E. texensis*
manca mula – *Echinocactus horizonthalonius*
mandarino – *Citrus deliciosa, C. nobilis, C. reticulata*
mandioca – *Manihot esculenta*
mandioca dulce – *Manihot palmata*
mandrágora – *Mandragora autumnalis, Podophyllum peltatum*
mangaba – *Landolphia florida*
mangabera – *Hancornia speciosa*
mangara – *Xanthosoma sagittifolium*
mangarito – *Xanthosoma sagittifolium*
manglano – *Punica gratatum*
mangle – *Avicennia germinans, Conocarpus erecta, Laguncularia racemosa, Maytenus phyllanthoides, Psorothamnus spinosus, Rhizophora mangle, R. mollis, Rhus integrifolia, R. ovata*
mangle aguabola – *Maytenus phyllanthoides*
mangle amarillo – *Laguncularia racemosa*
mangle blanco – *Avicennia germinans, Conocarpus erecta, Laguncularia racemosa*
mangle bobo – *Avicennia germinans, Laguncularia racemosa*
mangle botoncillo – *Conocarpus erectus*
mangle cenizo – *Conocarpus erecta*
manglecito – *Avicennia germinans*
mangle colorado – *Rhizophora mangle*
mangle chico – *Laguncularia racemosa*
mangle chino – *Laguncularia racemosa*
mangle de botón – *Conocarpus erecta*
mangle de la sierra – *Capparis indica*
mangle dulce – *Maytenus phyllanthoides, Rhizophora mangle*
mangle gateador – *Rhizophora mangle*
mangle negro – *Avicennia germinans, Conocarpus erecta*
mangle prieto – *Avicennia germinans, Conocarpus erecta, Laguncularia racemosa*
manglero – *Rhizophora mangle*
mangle rojo – *Avicennia marina* var *resinifera, Maytenus phyllanthoides, Rhizophora mangle*
mangle salado – *Avicennia germinans, Rhizophora mangle*
mangle tinto – *Rhizophora mangle*
mangle zapatero – *Rhizophora mangle*
mango – *Mangifera indica*
mango criollo – *Mangifera indica*
mango de Manila – *Mangifera indica*
mango de niña – *Cobaea scandens*
mangón – *Mangifera indica*
mangostán – *Garcinia mangostana*
manguena – *Solanum muricatum*
maní – *Arachis hypogaea*
maní africano – *Vigna subterranea*
maní de bambarra – *Vigna subterranea*
maní español – *Arachis hypogaea*
maniloche – *Solanum gracile, S. nigrum*
manillo – *Tripsacum laxum*
manioc – *Manihot esculenta*
manita – *Clavaria flava*
manita amarilla – *Clavaria flava*
manita de león – *Chiranthodendron pentadactylon*
maniú de Patagonia – *Podocarpus nubigenus*
maniú hembra – *Podocarpus nubigenus*
mano de cachorro – *Macfadyena unguis-cati*
mano de danta – *Gilibertia arborea, Oreopanax peltatum*
mano de dragón – *Chiranthodendron pentadactylon*
mano de león – *Begonia nelumbiifolia, Bocconiaarborea, B. frutescens, Celosia, Chiranthodendron pentadactylon, Geranium mexicanum, Gilibertia arborea, Oreopanax peltatum, Tabebuia rosea*
mano de muerto – *Castilleja*
mano de oso – *Gilibertia arborea*
mano de Santa María – *Leonurus cardiaca*
mano de sapo – *Gilibertia arborea*
mano de tigre – *Monstera deliciosa*
mano larga – *Agave tequilana*
manso – *Anemopsis californica*
manta – *Ipomoea batatas, Lysiloma microphylla*
manta de Judas – *Onopordum acanthium*
manta india – *Actinea gaillardia*
mantecoso – *Cercidium praecox*
mantel – *Coleus blumei, Mascagnia macroptera*
mantel de María – *Ipomoea pedicellaris*
mantel rojo – *Mascagnia macroptera*
mantequillo – *Rhamnus sphaerosperma*
manto – *Havardia mexicana, Ipomoea, Lysiloma divaricatum, Turbina corymbosa*
manto blanco – *Convolvulus arvensis*
manto colorado – *Lysiloma divaricatum*

manto de Jesús – *Bougainvillea spectabilis*
manto de la dama – *Alchemilla vulgaris*
manto de la reina – *Caladium bicolor, Xanthosoma sagittifolium*
manto de la Virgen – *Caladium bicolor, Cuscuta denticulata, Distictis laxiflora, Ipomoea, Zantedeschia aethiopica*
manto mexicano – *Ipomoea hirsutula*
manto real – *Aquilegia vulgaris*
manto regio – *Coleus blumeri*
Manuelito – *Epidendrum vitellinum, Odontoglossum cordatum, O. rosii*
manzanilla – *Anthemis cotula, Arctostaphylos polifolia, A. pungens, Chamaemelum nobile, Chamomilla courrantiana, C. suaveolens, Crataegus mexicana, C. pubescens, Ehretia elliptica, Fraxinus greggii, Gnaphalium bourgovii, G. leucocephalum, Helenium autumnale, Hippomane mancinella, Laccaria laccata, Malvaviscus grandiflorus, M. popullifolius, Matricaria recutita, Olea europaea, Perityle, Tagetes filifolia*
manzanilla alemana – *Chamomilla suaveolens*
manzanilla amarilla – *Anthemis tinctoria*
manzanilla bastarda – *Anthemis arvensis*
manzanilla borde – *Santolina rosmarinifolia*
manzanilla bronca – *Perityle microglossa*
manzanilla buena – *Chamaemelum nobile*
manzanilla cimarrona – *Erigeron pubescens*
manzanilla común – *Chamomilla suaveolens, Matricaria recutita*
manzanilla de Alemania – *Chamomilla suaveolens*
manzanilla de Aragón – *Chamomilla suaveolens*
manzanilla de Castilla – *Chamomilla courrantiana*
manzanilla de Indias – *Hippomane mancinella*
manzanilla del barrio – *Chamomilla suaveolens*
manzanilla del burro – *Perityle microglossa*
manzanilla del campo – *Gnaphalium leucocephalum, Zinnia acerosa, Z. pumila*
manzanilla del coyote – *Pectis coulteri, P. palmeri, P. papposa, Perityle californica, P. microglossa, Thymophylla concinna*
manzanilla del monte – *Perityle microglossa*
manzanilla de los tintoreros – *Anthemis tinctoria*
manzanilla del río – *Conyza sophiaefolia, Gnaphalium, Lithospermum cobrense*
manzanilla de playa – *Hippomane mancinella*
manzanilla de Yucatán – *Helenium autumnale*
manzanilla fétida – *Anthemis cotula*
manzanilla fina – *Chamaemelum nobile*
manzanilla hedionda – *Anthemis cotula*
manzanilla loca – *Anthemis tinctoria, Heliopsis helianthoides*
manzanilla montés – *Helenium mexicanum*
manzanilla noble – *Chameamelum nobile*
manzanilla olorosa – *Chamaemelum nobile*
manzanilla romana – *Chamaemelum nobile*
manzanilla verdadera – *Chamaemelum nobile*
manzanillo – *Myroxylon flexuosum, Orbignya cohune, Turnera occidentalis, Ximenia americana, Xylosma flexuosum*
manzanita – *Arbutus texana, A. xalapensis, Arctostaphylos, Colubrina ferruginosa, Crataegus mexicana, C. pubescens, Ehretia elliptica, E. tinifolia, Ilex brandegeana, Laccaria laccata, Lantana involucrata, Lycium andersonii, Malpighia diversifolia, M. glabra, Malvaviscus arboreus, M. grandiflorus, Ribes cereum* var *inebrians, Ziziphus obtusifolia*
manzanita de amor – *Solanum pseudocapsicum*
manzanita de cerro – *Heterotheca grandiflora*
manzanita de cogote – *Dyssodia concinna*
manzanita de costoche – *Ziziphus amole, Z. pedunculata*
manzanita de dama – *Crataegus azarolus*
manzanita del monte – *Colubrina triflora*
manzanito – *Eugenia jambos, Turnera occidentalis*
manzano – *Arctostaphylos pungens, Capsicum pubescens, Cocoloba uvifera, Ehretia tinifolia, Eugenia malaccensis, Malpighia diversifolia, Malus sylvestris, Musa sapientum*
manzano amargo – *Citrullus colocynthis*
manzano camueso – *Malus sylvestris*
manzano chato – *Malus sylvestris*
manzano de agua – *Eugenia malaccensis*
manzano de burro – *Thevetia ovata*
manzano del amor – *Lycopersicon esculentum*
manzano del diablo – *Datura stramonium*
manzano de Malaya – *Eugenia malaccensis*
manzano de oro – *Lycopersicon esculentum*

manzano del Perú – *Nicandra physalodes*
manzano de puya larga – *Crataegus erythropoda*
manzano espinoso – *Datura stramonium*
manzano estrella – *Phyllanthus acidus*
manzano floreciente – *Malus floribunda*
manzano floribundo del Japón – *Malus floribunda*
manzano híbrido purpúreo – *Malus purpurea*
manzano japonés – *Malus floribunda*
manzano multifloro – *Malus floribunda*
manzano rojo – *Malus purpurea*
manzano rosa – *Eugenia jambos, Malus floribunda*
manzano silvestre – *Malus sylvestris*
manzano silvestre chino de Hupeh – *Malus hupehensis*
manzano silvestre de la China – *Malus spectabilis*
manzano silvestre de Oregón – *Malus fusca*
manzano silvestre europeo – *Malus sylvestris*
manzanote – *Pereskia lycnidiflora*
mañanita – *Portulaca pilosa, Zephyranthes carinata*
mapá – *Hedeoma floribunda, H. patens*
mapahuite – *Trichilia hirta*
mapanare – *Sansevieria guineensis, S. trifasciata*
mapasúchil – *Chiranthodendron pentadactylon*
mapato – *Krameria trianda*
mapipitza – *Asclepias linaria*
maple – *Acer saccharinum*
mapoanita – *Zephyranthes carinata*
mapola – *Bernoullia flammea, Bombax ellipticum*
mapuey – *Dioscorea triphylla*
maque – *Ailanthus altissima*
maquilicua – *Tabebuia pentaphylla*
maquilihuatl – *Quercus fusiformis*
maraco – *Couroupita guianensis*
maranta – *Maranta arundinaceae, M. leuconeura*
marañón – *Anacardium occidentale*
marañón japonés– *Eugenia malaccensis*
marañuela – *Tropaeolum majus*
marapa – *Prunus cerasifera*
mararay – *Aiphanes caryotifolia*
marare – *Bursera simaruba*
maravilla – *Calendula officinalis, Callistephus chinensis, Convolvulus tricolor, Hamelia erecta, Helianthus annuus, Mirabilis, Tagetes tenuifolia, Tripsacum laxum*
maravilla africana – *Tagetes erecta*
maravilla común – *Calendula officinalis*
maravilla costeña – *Caesalpinia pulcherrima*
maravilla de España – *Catharanthus roseus*
maravilla de jardín – *Mirabilis longiflora*
maravilla del cerro – *Mirabilis longiflora*
maravilla del monte – *Ruellia tuberosa*
maravilla de los campos – *Calendula arvensis*
maravilla francesa – *Tagetes patula*
maravilla moreña – *Caesalpinia pulcherrima*
maravilla palustre – *Caltha palustris*
maravilla silvestre – *Calendula arvensis, Macrosiphonia hypoleuca, Mirabilis jalapa*
maravillosa – *Kalanchoe pinnata*
marceyalito – *Boletus reticulatus*
marciega – *Paspalum virgatum*
marfil vegetal – *Phytelephas macrocarpa*
margallón – *Chamaerops humilis*
margarita – *Agonandra racemosa, Aster novae-angliae, Bellis perennis, Bidens pilosa, Caesalpinia standleyi, Callistephus chinensis, Chrysanthemum, Karwinskia humboldtiana, Leucanthemum vulgare, Malpighia diversifolia, Tanacetum parthenium, Viguiera laciniata*
margarita africana – *Arctotis stoechadifolia*
margarita blanca – *Polianthes tuberosa*
margarita del Atlas – *Anacyclus depressus*
margarita del campo – *Lantana camara*
margarita del cerro – *Karwinskia humboldtiana*
margarita de Livingstone – *Carpobrotus chilensis, Mesembryanthemum criniflorum*
margarita de los prados – *Bellis perennis*
margarita de San Miguel – *Callistephus chinensis*
margarita enana – *Bellis perennis*
margarita gigante – *Chrysanthemum maximum*
margarita mayor – *Leucanthemum vulgare*
margarita menor – *Bellis perennis*
margarita michaelmas – *Aster novi-belgii*
margarita olorosa – *Polianthes tuberosa*
margarita pequeña – *Bellis perennis*
margarita reina – *Callistephus chinensis*
margarita roja – *Chrysanthemum coccineum*
margarita shasta – *Leucanthemum maximum*

margarita silvestre – *Baileya pleniradiata, Leucanthemum vulgare*
margaritón – *Leucanthemum vulgare, Montanoa tomentosa*
margaza – *Tanacetum parthenium*
margosa – *Melia azedarach* var *umbraculiformis*
María – *Calophyllum braziliense, C. calaba, Epidendrum mariae*
María Luisa – *Aloysia triphylla*
mariana – *Nemophila menziesii*
marianga – *Pennisetum purpureum*
marihuana – *Cannabis indica, C. sativa*
marihuanilla – *Leonurus cardiaca*
marihuela – *Cytisus multiflorus*
marijuana – *Cannabis indica, C. sativa*
marijuana falsa – *Nicotiana glauca*
marilopez – *Turnera ulmifolia*
marimba – *Platymiscium dimorphandrum*
marimoña – *Ranunculus asiaticus*
mariola – *Aloysia sonorensis, Chrysactinia mexicana, Dalea wislizeni, Horsfordia alata, Parthenium incanum, Solanum*
mariolona – *Solanum eleagnifolium*
maripenda – *Liquidambar styraciflua*
mariposa – *Calochortus, Hedychium coronarium, Odontoglossum cervantesii, Pithecellobium echinatum*
mariposa amarilla – *Lemboglossum maculatum*
mariposa blanca – *Rhynchostele cervantesii*
mariposa de David – *Buddleja davidii*
mariquilla – *Solidago canadensis* ssp *elongata*
mariquita – *Selloa glutinosa*
marismeño – *Haematoxylon brasiletto, Stenocereus thurberi*
marjoleto – *Crataegus laevigata, C. monogyna*
maro – *Salvia sclarea, Swietenia macrophylla, Teucrium marum*
marojo – *Quercus petraea*
maromero – *Ipomoea stans, Salsola tragus*
maro negro – *Salvia verbenacea*
mar pacífico – *Hibiscus rosa-sinensis*
marquesote – *Bernoullia flammea*
marquiana – *Nicotiana glauca*
marta – *Disocactus phyllanthoides*
marta boni – *Cymbopetalum mayanum*
martagón – *Lilium martagon*
maruru – *Victoria regia*
marrón – *Castanea sativa*
marrubio – *Marrubium vulgare, Salvia arizonica*
marrubio acuático – *Lycopus europaeus*
marrubio blanco – *Marrubium vulgare*
marrubio común – *Marrubium vulgare*
marrubio fétido – *Ballota nigra*
marrubio negro – *Ballota nigra*
mash – *Ipomoea alba*
masiega – *Cladium californicum*
masiola – *Halimium glomeratum*
masiyague – *Bromus carinatus*
masoco – *Dioscorea bulbifera*
mastranzo – *Hyptis suaveolens, Lippia callicarpaefolia, Marrubium vulgare, Mentha suaveolens*
mastuerzo – *Cardamine hirsute, Lepidium, Rorippa nasturtium-aquaticum, Tropaeolum majus*
mastuerzo acuático – *Rorippa nasturtium-aquaticum*
mastuerzo alpino – *Cardamine bellidifolia*
mastuerzo anual – *Alyssum alyssoides*
mastuerzo de agua – *Rorippa nasturtium-aquaticum*
mastuerzo de hojas de margaritilla – *Cardamine bellidifolia*
mastuerzo de jardín – *Barbarea verna*
mastuerzo de las Indias – *Tropaeolum majus*
mastuerzo de los campos – *Alyssum minus*
mastuerzo de prado – *Cardamine pratensis*
mastuerzo estrigoso – *Alyssum strigosum*
mastuerzo hirsuto – *Cardamine hirsuta*
mastuerzo hortense – *Lepidium sativum*
mastuerzo marítimo – *Lobularia maritima*
mastuerzo mayor – *Lepidium latifolium*
mastuerzo montano – *Alyssum montanum*
mastuerzo saxátil – *Aurinia saxatilis*
mastuerzo silvestre – *Coronopus squamatus*
mata – *Ageratina herbacea*
matacaballos – *Asclepias curassavica, Calatola mollis, Illicium floridanum, Lobelia urens*
matacachí – *Pilosocereus alensis*
matacán – *Cynanchum acutum, C. utahense*
matacandelillas – *Asclepias albicans, A. subulata*
matacandiles – *Ornithogalium umbellatum, Sisymbrium irio*
matacora – *Jatropha*
matacoyotes – *Asclepias curassavica*

mata chamaco – *Agonandra racemosa*
mata charneca – *Pistacia lentiscus*
matachín – *Caesalpinia pulcherrima*
mata de Dios – *Aeschynomene nivea*
mata de la seda – *Asclepias syriaca*
mata de pavo – *Boerhavia coccinea*
mata de San Juan – *Jacquinia aurantiaca*
mata dolor – *Dorstenia contrayerba*
mata dulce – *Phalaris tuberosa*
mata espinosa – *Rubus idaeus*
matagallinas – *Apoplanesia paniculata, Capparis incana, Euphorbia cotinifolia, Solanum dulcamara*
matagallos – *Phlomis fruticosa, P. purpurea*
matagusanos – *Acourtia thurberi, Cosmos pringlei, Croton cilatoglandulosus, Nama undulatum*
matahombres – *Ligustrum vulgare*
mataise – *Sapium pedicellatum*
mataiza – *Prunus rhamnoides*
matal – *Tradescantia zebrina*
matalahúga – *Pimpinella anisum*
mata la sed – *Zanthoxylum fagara*
matalí – *Hirtella racemosa, Tradescantia zebrina*
mataliste – *Commelina erecta, Ipomoea longifolia*
mataliz – *Commelina erecta, Tradescantia fluminensis*
matalobos – *Aconitum napellus*
matalobos de colonia – *Doronicum cordatum*
matalobos de flores compuestas – *Doronicum pardalianches*
matalobos de hojas cordiformes – *Doronicum cordatum*
matalobos de hojas de llantén – *Doronicum plantagineum*
matalote – *Panicum maximum*
mata mala – *Jatropha cordata*
matamatá – *Gustavia augusta*
matambilla – *Desmoncus chinantlensis, D. orthacanthos*
matamuchachos – *Jatropha cordata, J. curcas, Sapindus saponaria*
matanena – *Janusia gracilis, J. linearis, Panicum capillare*
matanene – *Mascagnia macroptera*
matanene de mar – *Aristolochia taliscana*
matanene hembra – *Erythroxylum mexicanum, Mascagnia macroptera*
matanene macho – *Mascagnia macroptera*
matanga – *Eriogonum fasciculatum*
matanza – *Stellaria nemorum*
matapalo – *Ficus, Psittacanthus calyculatus*
matapalo liso – *Ficus tecolutensis*
mata parda – *Quercus reticulata*
mataparen – *Boletus luridus*
mataperros – *Colchicum autumnale, Santolina rosmarinifolia, Strychnos tabascana, S. tepicencis*
matapiojos – *Heteropteris laurifolia, Porophyllum macrocephalum, P. macrophyllum*
matapulgas – *Stevia viscida*
matarique – *Psacalium decompositum*
matarique blanco – *Psacalium decompositum*
matarique morado – *Acourtia thurberi*
matariqui – *Eysenhardtia orthocarpa, Psacalium decompositum*
matarrata – *Gliricidia sepium*
matarratón – *Gliricidia sepium, Zamia polymorpha*
matarrubia – *Quercus coccifera, Q. ilex*
matasano – *Casimiroa edulis, C. sapota*
mata-torsalo – *Funastrum clausum*
matayaqui – *Montanoa patens, M. tomentosa* ssp *rosei*
matazetes – *Valeriana sorbifolia*
mate – *Crescentia cujete, Ilex paraguariensis*
mateado – *Pereskia lycnidiflora*
mate cicada – *Fidicina mannifera*
matéoco – *Taxodium distichum* var *mexicanum*
matial – *Pereskia lycnidiflora*
mático – *Piper angustifolium, P. palmeri*
matilija – *Romneya coulteri*
matilimi – *Tecoma stans*
matilisguate – *Tabebuia chrysantha, T. impetiginosa, T. pentaphylla*
matiola – *Matthiola incana*
matita – *Acacia farnesiana*
matizadilla – *Lantana camara*
matlalcuahuitl – *Guaiacum coulteri, G. officinale*
matlaliztic – *Commelina coelestis, C. pallida, C. tuberosa*
matlalxóchitl – *Tinantia erecta*
matojo – *Sporobolus indicus*

matojo de burro – *Sporobolus indicus*
matojo dulce – *Eleusine indica*
mato negro – *Meliosma dentata*
matorral – *Acacia berlandieri*
matricaria – *Chrysanthemum coccineum, Tanacetum parthenium*
matricaria partenio – *Tanacetum parthenium*
matruz – *Chenopodium ambrosioides*
matsatli – *Ananas sativus*
matulera – *Phlomis fruticosa, P. purpurea*
maturin – *Psacalium decomposita*
matzapotl – *Annona cherimola*
matze – *Lysiloma acapulcensis*
mauto – *Acacia millefolia, Lysiloma*
mauto colorado – *Lysiloma divaricata*
maxacocote – *Cyrtocarpa procera*
maxacotl – *Mascagnia macroptera*
maxak – *Capsicum frutescens*
maxcal – *Dioscorea*
maxcalzots – *Arum italicum*
maxcuáhuitl – *Piscidia mollis*
maxh – *Ipomoea muricata*
maxte – *Lantana camara*
maya – *Bellis perennis, Bromelia lasiantha, B. magdalenae, Callistephus chinensis, Leucanthemum vulgare, Miconia*
maya perenne – *Bellis perennis*
mayito – *Atamosco carinata, A. citrina, Cooperia drummondii, Nothoscordum bivalve, Zephranthes*
mayo – *Atamosco carinata, Lysiloma divaricatum, Zephyranthes carinata*
mayocoba – *Phaseolus vulgaris*
mayorana – *Lantana macropoda*
mayueta – *Fragaria vesca*
maza de agua – *Typha*
mazahua – *Hibiscus tiliaceus*
mazapán – *Malvaviscus arboreus, M. grandiflorus*
mazatán – *Valeriana ceratophylla*
mazatil – *Valeriana sorbifolia*
mazaxócotl – *Spondias mombin, S. purpurea*
mazayel – *Agaricus augustus, Boletus edulis, B. pinicola*
mazcabmiz – *Pectis angustifolia*
mazorca – *Morchella esculenta, Tecoma stans, Zea mays*
mazorca de cuervo – *Conopholis americana*
mazorca de culebra – *Helosis mexicana*
mazorquilla – *Conopholis americana, Phytolacca*
mazorquita – *Galphimia glauca, Morchella esculenta*
mazorra – *Arrhenatherum elatius*
mazuáa – *Iris sisyrinchium*
mecapatli – *Ipomoea purpurea, Smilax bona-nox, S. medica, S. mexicana*
mecaxóchitl – *Piper*
mecoztli – *Agave americana*
mechen – *Pachyrrhizus erosus*
mechera – *Phlomis purpurea*
mechuda – *Caesalpinia pulcherrima*
medesá – *Cercidium floridum, C. microphyllum*
medicina de mujer – *Montanoa tomentosa*
meex-mascual – *Chloris elegans*
mehenibbech – *Rhynchosia minima*
mehensial – *Sesamum indicum*
mehenxtohku – *Datura stramonium*
mehenzit – *Lasiacis divaricata*
mejorana – *Ageratum conyzoides, Brickellia veronicaefolia, Origanum majorana, Lantana macropoda, Salvia ballotaeflora, S. coccinea*
mejorana bastarda – *Origanum vulgare*
mejorana del país – *Salvia ballotaeflora, S. patens*
mejorana dulce – *Origanum majorana*
mejorana silvestre – *Origanum vulgare, Thymus mastichina*
melaleuca – *Melaleuca*
melampodo – *Helleborus niger*
melaza – *Melinus minutiflora*
melera – *Anchusa officinales*
melga – *Trifolium pratense*
melia – *Melia azederach* var *umbraculiformis*
meliloto – *Melilotus indica, M. officinalis*
meliloto amarillo – *Melilotus officinalis*
meliloto blanco – *Melilotus alba*
meliloto común – *Medicago officinalis*
meliloto de flores pequeñas – *Medicago indica*
meliloto indio – *Melilotus indica*
melisa – *Melissa officinalis*
melisa de Moldavia – *Lallemantia iberica*
melisa oficinal – *Melissa officinalis*
melocotón – *Cucurbita moschata, Prunus persica, Sicana odorifera*
melocotonero – *Prunus persica*
melocotonero chino – *Prunus davidiana*
melocotonero ornamental – *Prunus davidiana*

melojo – *Quercus pyrenaica*
melón – *Cucumis melo*
melón blanco – *Benincasa hispida, Cucumis melo*
melón borado – *Cucumis melo* var *reticulatus*
melón camote – *Carica papaya*
melón cantalupo – *Cucumis melo* var *reticulatus*
meloncillo – *Apodanthera undulata, Cucurbita digitata, C. palmate, Sida hederacea*
meloncillo de coyote – *Cucumis melo* var *dudaim*
melón chapote – *Carica papaya, C. pubescens*
melón chino – *Cucumis melo* var *cantalupensis, C. m.* var *indorus*
melón chito – *Cuicumis melo* var *chito*
melón de agua – *Citrullus colocynthis*
melón de cantalú – *Cucumis melo* var *cantalupensis*
melón de Castilla – *Cucumis melo* var *reticulatus*
melón de coyote – *Apodanthera undulata, Cucumis anguria, C. melo* var *dudaim, Cucurbita palmata, C. digitata, Ibervillea insularis, I. sonorae, Maximowiczia insularis, Schizocarpum palmeri*
melón de Indias – *Cucumis melo* var *indorus*
melón flexuoso – *Cucumis melo* var *flexuosus*
melón loco – *Apodanthera undulata*
melón mango – *Cucumis melo* var *chito*
melón manzana – *Cucumis melo* var *conomon*
melón pepino – *Cucumis melo* var *flexuosus*
melón serpentino – *Cucumis melo* var *flexuosus*
melón zapote – *Carica papaya, C. pubescens*
melosa – *Madia sativa*
membrillero – *Cydonia oblonga*
membrillero chino – *Chaenomeles sinensis*
membrillero del Japón – *Chaenomeles speciosa*
membrillero japonés – *Chaenomeles speciosa*
membrillero ornamental – *Chaenomeles speciosa*
membrillito – *Amelanchier*
membrillo – *Amelanchier, Cantharellus cibarius, Celtis reticulata, Cornus excelsa, Cydonia oblonga, Phyllostylon brasiliensis, Terminalia amazonia*
membrillo bronco – *Cinchona officinalis*
membrillo cimarrón – *Amelanchier, Condalia mexicana*
membrillo común – *Cydonia oblonga*
membrillo chino – *Chaenomeles sinensis*
membrillo de Bengala – *Aegle marmelos*
membrillo elegante – *Chaenomeles speciosa*
membrillo falso de Franchet – *Cotoneaster franchetii*
membrillo falso de hojas de sauce – *Cotoneaster salicifolius*
membrillo falso horizontal – *Cotoneaster horizontalis*
membrillo falso microfilo – *Cotoneaster microphyllus*
membrillo falso procumbente – *Cotoneaster adpressus*
membrillo japonés – *Chaenomeles speciosa*
memeya – *Chamaesyce maculata*
memiso – *Muntingia calabura*
mendieta – *Solandra nitida*
mendrasto – *Ageratum corymbosum*
menta – *Hedeoma patens, Mentha, Satureja macrostema*
menta campestre – *Mentha arvensis*
menta crespa – *Mentha spicata*
menta de burro – *Marrubium vulgare, Mentha suaveolens*
menta de caballo – *Mentha spicata*
menta de gato – *Nepeta cataria*
menta de hojas circulares – *Mentha suaveolens*
menta de hojas grandes – *Mentha sylvestris*
menta de lobo – *Lycopus americanus, L. asper, L. europaeus*
menta de Nuestra Señora – *Mentha spicata*
menta de requien – *Mentha requienii*
menta espídia – *Mentha spicata*
menta japonesa – *Mentha arvensis* var *piperascens*
menta montana – *Mentha* x *piperita*
menta piperita – *Mentha* x *piperita*
menta polea – *Mentha pulegium*
menta romana – *Mentha* x *gracilis, M. spicata*
menta silvestre – *Mentha arvensis, M. canadensis*
menta verde – *Mentha spicata*
mentironera – *Viburnum lantana*
menudo – *Lycopersicon esculentum* var *cerasiforme*
meple – *Acer*

mercadela – *Calendula officinalis*
mercurial – *Galinsoga parviflora, Mercurialis annua*
mercurial anual – *Mercurialis annua*
merey – *Anacardium occidentale*
meriendita – *Thevetia ovata*
merker – *Pennisetum purpureum*
merkeron – *Pennisetum purpureum*
mescal – see mezcal
mesembriantemo – *Carpobrotus chilensis*
mesquite – see mezquite
metálica – *Chamaedorea metallica*
metasecoya – *Metasequoia glyptostroboides*
metatera – *Silene laciniata*
metl – *Agave*
metlacasis – *Ipomoea purpurea*
metometl – *Agave lechuguilla*
metroceder – *Clerodendrum*
metstsonkilit – *Xanthosoma nigrum*
metzolle – *Mammillaria geminispina*
metzollin – *Mammillaria geminispina*
mexcalli – *Agave*
mexixin – *Rorippa nasturtium-aquaticum*
mexix-quílitl – *Lepidium virginicum*
mexnuxeb – *Tillandsia usneoides*
mexye – *Toxicodendron radicans*
mezcal – *Agave, Lophophora williamsii, Ulmus mexicana*
mezcal azul – *Agave tequilana*
mezcal blanco – *Agave pseudotequilana, A. shrevei* ssp *matapensis, A.tequilana*
mezcal ceniza – *Agave colorata, A. shrevei*
mezcal cucharo – *Agave pseudotequilana*
mezcal de bacanora – *Agave angustifolia*
mezcal de maguey – *Agave angustifolia, A. pseudotequilana*
mezcalillo – *Agave datylio*
mezcalito – *Agave felgeri, Hechtia montana, Tillandsia inflata, T. recurvata*
mezcalito de huitlacochi – *Tillandsia exserta*
mezcal lechuguilla – *Agave bovicornuta, A. lechuguilla, A. zebra*
mezcal pardo – *Agave sobria*
mezcal pelón – *Agave pelona*
mezereo – *Daphne mezereum*
mezéreon – *Daphne mezereum*
mezonzapote – *Licania platypus*
mezquiquez – *Prosopis juliflora*
mezquite – *Prosopis*
mezquite amargo – *Prosopis articulata, P. juliflora* var *articulata*
mezquite amarillo – *Prosopis juliflora*
mezquite blanco – *Prosopis juliflora*
mezquite colorado – *Prosopis juliflora*
mezquite chachaca – *Prosopis juliflora*
mezquite chino – *Prosopis juliflora*
mezquite dulce – *Prosopis chilensis, P. juliflora*
mezquite extranjero – *Parkinsonia aculeata*
mezquite prieto – *Lysiloma divaricata*
mezquite rastrero – *Prosopis glandulosa* var *prostrata*
mezquite sarnoso – *Condalia globosa* var *pubescens*
mezquite tornillo – *Prosopis pubescens*
mezquite verde – *Parkinsonia aculeata*
mezquitillo – *Acacia constricta, Calliandra eriophylla, Condalia obovata, Krameria, Senna occidentalis*
mezquitl – *Prosopis juliflora*
micacuáhuitl – *Cordia boissier, Ipomoea murucoidesi*
micay – *Axonopus scoparius*
michela – *Mitchella repens*
michiuauhtli – *Chenopodium nuttalliae*
miel de avispas – *Anchusa arvensis*
mielga – *Medicago sativa*
mielga alcacer – *Medicago sativa*
mielga de caracolillo – *Medicago orbicularis*
mielga negra – *Medicago lupulina*
mierda de loro – *Casearia sylvestris*
Miguel – *Zinnia elegans, Z. violacea*
Miguelito – *Epidendrum vitellinum, Jatropha cordata*
mijera – *Piptatherum miliaceum*
mijillo – *Setaria gracilis*
mijo – *Panicum miliaceum, P. virgatum, Pennisetum glaucum, Setaria italica, Sorghum vulgare*
mijo africano – *Eleusine coracana* ssp *africana*
mijo candela – *Pennisetum glaucum*
mijo capilar – *Panicum capilare*
mijo común – *Panicum miliaceum*
mijo coracano – *Eleusine coracana*
mijo de cuadra – *Echinochloa frumentacea*
mijo de escoba – *Sorghum bicolor*
mijo de flores dicotómicas – *Panicum dichotomiflorum*

mijo de Guinea – *Panicum maximum*
mijo de hojas lisas – *Panicum schinzii*
mijo de Italia – *Setaria italica*
mijo de los arrozales – *Echinochloa crus-galli, E. frumentacea, E. oryzoides*
mijo de pájaros – *Setaria italica*
mijo de sol – *Lithospermum officinale*
mijo de sol agreste – *Lithospermum arvensis*
mijo de sol común – *Lithospermum officinale*
mijo italiano – *Setaria italica*
mijo japonés – *Echinochloa frumentacea*
mijo koda – *Paspalum scrobiculatum*
mijo maíz – *Sorghum bicolor*
mijo mayor – *Panicum miliaceum*
mijo menor – *Setaria italica*
mijo miliar – *Panicum miliaceum*
mijo negrillo – *Piptatherum miliaceum*
mijo negro – *Pennisetum americanum, P. glaucum*
mijo pajarero – *Setaria italica*
mijo perla – *Pennisetum americanum, P. clandestinum, P. glaucum*
mijo repente – *Panicum repens*
mijo silvestre – *Milium effusum, Setaria anceps, Tripsacum lanceolatum*
mijo sorgo – *Sorghum bicolor*
mijo turqueso – *Zea mays*
mijo verde – *Panicum maximum*
milamores – *Centranthus ruber*
milamores roja – *Centranthus ruber*
milapa – *Fouquieria columnaris*
milchilli – *Capsicum annuum*
mildiu – *Peronospora viticola*
milefolio – *Achillea filipendulina, A. millefolium*
milefolio alpino – *Achillea tomentosa*
milena – *Eragrostis curvula*
milengrana – *Herniaria glabra*
milenrama – *Achillea filipendulina, A. millefolium*
milenrama alpina – *Achillea tomentosa*
milenrama amarilla – *Achillea tomentosa*
milenrama blanca – *Achillea clavennae*
mil flores – *Phlox*
milgranada – *Chenopodium botrys*
milgranos – *Herniaria glabra, Punica granatum*
milhojas – *Achillea millefolium*
milpa de venado – *Tripsacum lanceolatum*
milpilla – *Zea diploperennis*
miltomate – *Physalis aeguata, P. peruviana, P. philadelphica, P. pubescens*
mimbre – *Acacia berlandieri, Cephalanthus occidentalis, C. salicifolius, Chilopsis linearis, Cornus excelsa, Monstera deliciosa, Salix viminalis*
mimbre blanco – *Cornus excelsa*
mimbre del monte – *Capparis flexuosa*
mimbre pasilla – *Cornus disciflora*
mimbre púrpura – *Salix purpurea*
mimbrera – *Salix fragilis, S. viminalis*
mimbrera púrpura – *Salix purpurea*
mimosa – *Acacia, Leucaena leucocephala, Mimosa*
mimosa común – *Acacia decurrens*
mimosa dorada de Sidney – *Acacia longifolia*
mimosa plateada – *Acacia dealbata*
mimosa púdica – *Mimosa pudica*
mimosa vergonzosa – *Mimosa pudica*
mímulo – *Mimulus*
minari – *Cucumis melo*
mincapatli – *Sapium appendiculatum, Sebastiana pavoniana*
mini jitomate dulce – *Lycopersicon esculentum* var *cerasiforme*
ministro – *Ternstroemia sylvatica*
minutisa – *Dianthus barbatus*
miñona – *Tecoma stans*
miñoneta – *Reseda odorata*
miona – *Commicarpus scandens, Heterotheca grandiflora*
miona blanca – *Brickellia coulteri*
mioporo – *Myoporum laetum*
miosota – *Asperugo procumbens, Myosotis sylvatica*
mipil – *Asarina antirrhiniflora*
mirabel – *Bassia hyssopifolia, Chrysanthemum coronarium, Helianthus annuus, Kochia scoparia*
mirabella – *Prunus domestica* ssp *syriaca*
mirabobo – *Melia azederach*
miraguama – *Coccothrinax miraguama*
miramelindo – *Consolida ambigua, Impatiens balsamina, I. walleriana*
mírame-y-no-me-toques – *Impatiens pallida*
mirasol – *Capsicum annuum, Cochlospermum vitifolium, Cosmos bipinnatus, Cosmos sulphureus, Encelia palmeri, Helianthus*

annuus, Heliopsis rubra, Sclerocarpus spathulatus, Tithonia
mirasol amarillo – *Cosmos sulphureus*
mirasol anaranjado – *Helianthus annuus*
mirasol chiquito – *Dyssodia anomala*
mirasol xocoatole – *Cosmos bipinnatus*
mirica cerífera – *Myrica cerifera*
mirística – *Myristica fragrans*
mirística olorosa – *Myristica fragrans*
mirlitón – *Sechium edule*
mirobálano – *Prunus cerasifera*
mirtillo – *Vaccinium myrtillus*
mirto – *Bouvardia glaberrima, B. ternifolia, Loeselia mexicana, Myrciaria floribunda, Myrtus communis, Salvia, Stachys coccinea*
mirto cimarrón – *Castilleja canescens, C. gracilis*
mirto común – *Myrtus communis*
mirto de brabante – *Myrica gale*
mirto de cera – *Myrica cerifera*
mirto del campo – *Bouvardia ternifolia*
mirto del cerro – *Justica spicigera*
mirto de Oregón – *Umbellularia californica*
mirto grande – *Salvia darcyi*
mirto silvestre – *Loeselia mexicana*
mirra – *Commiphora abyssinica, C. myrrha*
misibcoc – *Turnera diffusa*
mispatle – *Buddleja*
mistelajoyo – *Tibouchina spathulata*
mitihuaxaqua – *Sonchus oleraceus*
mitlalxihuitl – *Lopezia mexicana, L. racemosa*
mitra – *Astrophytum myriostigma, A. ornatum, Cyclamen purpurascens*
mixquicopalli – *Prosopis juliflora*
mixtlacotl – *Ageratina*
mixtontze – *Tecoma stans*
mizbil – *Abutilon lignosum*
mízcalo – *Cantharellus, Lactarius deliciosus*
mizquitl – *Prosopis chilensis, P. juliflora, P. laevigata*
moca – *Andira inermis*
moca colorada – *Andira inermis*
mocaya – *Acrocomia sclerocarpa*
moco – *Celosia cristata*
mocochtaj – *Pinus devoniana*
moco de cucuno – *Polygonum pensylvanicum*
moco de guajolote – *Justica brandegeana, Polygonum pensylvanicum, P. persicaria*
moco de pavo – *Amaranthus caudatus, A. tricolor, Celosia argentea*
mocoque – *Bombax ellipticum, Tabebuia pentaphylla*
mocseté – *Lantana camara, L. involucrata*
mochi – *Boerhavia coccinea, B. erecta, Commicarpus scandens, Pachycereus schottii*
moenchia erecta – *Moenchia erecta*
mogno – *Swietenia macrophylla*
mogno aguano – *Swietenia macrophylla*
moha – *Setaria italica* var *moharica*
moho – *Nolina bigelovii*
mohuarí – *Aphananthe monoica*
mohuitle – *Justica spicigera*
mohuitli – *Jacobinia umbrosa*
moixernó – *Tricholoma gambosum*
mojagua – *Abutilon notolophium*
mojepe – *Carnegiea gigantea*
mojera – *Sorbus aria*
mojera sueca – *Sorbus intermedia*
mojito – *Brosimum alicastrum*
mojo – *Brosimum alicastrum*
molcol – *Ximenia americana*
molina – *Carpobrotus chilensis, Mesembryanthemum crystallinum, M. nodiflorum*
molinia azul – *Molinia caerulea*
molinillo – *Hura crepitans, Leonotis nepetaefolia, Quararibea funebris, Vinca major*
moloche – *Lyophyllum decastes*
molonqui – *Cissus sicyoides*
molote – *Sorghum bicolor*
molxóchitl – *Tagetes patula*
molle – *Schinus molle*
mombin – *Spondias mombin*
momo – *Piper auritum, P. sanctum*
mo'mo – *Jatropha cordata*
momordica – *Momordica charantia*
momordica amarga – *Momordica charanita*
mona – *Ungnadia speciosa*
monacillo – *Abutilon hypoleucum, Malvaviscus arboreus, M. grandiflorus*
monaco – *Orbignya cohune*
monapesto – *Campsis radicans*
monarda – *Monarda menthifolia*
monarda escarlatina – *Monarda citriodora*
moneda del papa – *Lunaria annua, L. biennis*
monedero – *Calceolaria herbeohybrida*

moniato – *Ipomoea batatas*
monilla – *Ungnadia speciosa*
monja blanca – *Lycaste virginalis*
monje – *Catasetum maculatum*
monjita – *Laelia, Lycaste virginalis, Ryncholaelia digbyana*
monótropa – *Monotropa hypopitys*
monótropa amarilla – *Monotropa hypopitys*
monótropa multiflora – *Monotropa hypopitys*
monstera – *Monstera deliciosa*
montbretia – *Crocosmia* x *crocosmiiflora*
monte de oro – *Jacobinia umbrosa, Justicia aurea*
Moctezuma del campo – *Trixis angustifolia*
moño – *Celosia argentea*
mooj – *Gossypium hirsutum*
mo'oso – *Chloracantha spinosa*
mop – *Orbignya cohune, Scheelia liebmannii*
mora – *Capsicum annuum, Conostegia xalapensis, Lantana camara, Solanum nigrum*
morada – *Lepista nuda, Nama demissum, N. hispidum*
mora de árbol – *Morus nigra*
mora del campo – *Malpighia emarginata*
moradilla – *Astragalus humboldtii, Cuphea ignea, Gomphrena decumbens, Maclura tinctoria, Salvia azurea, Sisyrinchium angustifolium, Tradescantia zebrina, Verbena*
moradita – *Nama hispidum* var *spathulatum*
morado – *Musa sapientum*
morado amarillo – *Musa sapientum*
morado verde – *Musa acuminata*
mora hedionda – *Senna atomaria, S. bicapsularis*
moraja – *Taraxacum officinale*
moral – *Maclura tinctoria, Morus, Rubus*
moral amarillo – *Maclura tinctoria*
moral blanco – *Morus alba*
moral de Castilla – *Rubus glaucus, R. strigosus*
moral de clavo – *Maclura tinctoria*
moral de la China – *Broussonetia papyrifera*
moral de la sierra – *Carpinus caroliniana, Rubus palmeri*
moral de papel – *Broussonetia papyrifera*
moral de Santa Rosa – *Rubus ulmifolius*
moral fresa – *Rubus glaucus*
moral liso – *Maclura tinctoria*
moral negro – *Morus celtidifolia, M. nigra, Rubus villosus*
moral piña – *Rubus glaucus*
moral rojo – *Morus rubra*
morcegalo – *Tectaria heracleifolia*
morello – *Prunus*
morenita – *Kochia scoparia*
morera – *Morus alba, M. nigra*
morera blanca – *Morus alba*
morera de papel – *Broussonetia papyrifera*
morera negra – *Morus celtidifolia, M. nigra*
morera trepadora – *Solanum dulcamara*
morga – *Anamirta paniculata*
morgelina – *Stellaria*
moriche – *Mauritia flexuosa, M. minor*
morilla – *Morchella esculenta*
moringa – *Moringa oleifera*
morita – *Capsicum annuum*
moronel – *Lonicera subspicata*
morquera – *Thymus capitatus*
morsana – *Zygophyllum fabago*
morubú – *Panicum maximum*
morrionera – *Viburnum lantana*
morro – *Crescentia alata, C. cujete*
morro guacalero – *Crescentia cujete*
mosca – *Linaria cymbalaria*
moscadel del grano menudo – *Vitis vinifera* var *muscat*
moscadero – *Myristica fragrans*
mosilado – *Malvaviscus arboreus, M. grandiflorus*
mosmote – *Ceiba acuminata, C. pentandra*
mosmote de cerro – *Ceiba aesculifolia*
mosquero – *Amanita muscaria*
mosqueta – *Philadelphus coronarius, P. mexicanus, Rosa eglanteria, R. moschata*
mosqueta silvestre – *Tibouchina urvilleana*
mosquita – *Eryngium carlinae, Lobelia erinus*
mostacilla – *Brassica rapa, Eruca vesicaria* ssp *sativa, Lepidium latifolium, Sisymbrium officinale*
mostafiete del campo – *Artemisia ludoviciana*
mostajo – *Sorbus aria*
mostaza – *Brassica nigra, B. rapa, B. tournefortii, Sinapis alba, Sisymbrium auriculatum*
mostaza amarilla – *Brassica rapa, Sinapis alba*
mostaza arvense – *Sinapis arvensis*
mostaza blanca – *Sinapis alba*

mostaza china – *Brassica juncea*
mostaza de Abisinia – *Brassica juncea*
mostaza de hojas – *Brassica juncea*
mostaza de la China – *Brassica juncea*
mostaza de la India – *Brassica juncea*
mostaza de los campos – *Sinapis arvensis*
mostaza de sarepta – *Brassica juncea* var *sareptana*
mostaza espinaca – *Brassica rapa* var *perviridis*
mostaza hindu – *Brassica juncea*
mostaza inglesa – *Sinapis alba*
mostaza japonesa – *Brassica japonica*
mostaza marrón – *Brassica nigra*
mostaza montés – *Nicotiana glauca*
mostaza negra – *Brassica nigra*
mostaza silvestre – *Lepidium campestre, Sinapis arvensis*
mostellar – *Diospyros lotus, Sorbus aria*
mosto – *Capparis flexuosa*
mota – *Cannabis indica, C. sativa, Dalea tomentosa, Selloa glutinosa*
mota morada – *Ageratum corymbosum*
moté – *Erythrina glauca*
motita – *Pinaropappus roseus, Selloa glutinosa*
motita morada – *Pinaropappus roseus*
motsoque – *Brosimum alicastrum*
moxerno – *Marasmius oreades*
moxete – *Lantana camara*
moya – *Chrysanthemum coronarium*
moyet – *Laguncularia racemosa*
moyote – *Justica spicigera*
moyotli – *Justica spicigera*
mozote – *Bidens pilosa, B. tetragosa, Cenchrus echinatus, Heliocarpus donnell-smithii, Triumfetia semitriloba*
mozote blanco – *Bidens pilosa*
mozote de caballo – *Triumfetia semitriloba*
muatzate – *Ananas comosus, A. sativus*
muc – *Celtis iguanea, Dalbergia*
muco – *Couroupita guianensis*
mucuna rayada – *Stizolobium deeringianum*
mucuteno – *Senna spectabilis*
muchcoc – *Selaginella*
muchite – *Pithecellobium dulce*
muchulcux – *Selaginella*
muela – *Lathyrus sativus*
muela de caballo – *Sonchus oleraceus*
muela de San Cristobál – *Aquilegia vulgaris*
muérdago – *Phoradendron, Psittacanthus calyculatus, P. sonorae, Viscum album*
muérdago americano – *Phoradendron macrophyllum*
muérdago europeo – *Viscum album*
muérdago visco – *Psittacanthus calyculatus*
muermera – *Clematis*
muguete – *Convallaria majalis*
muguete de los valles – *Convallaria majalis*
muicle – *Anisacanthus thurberi, A. quadrifidus* ssp *wrightii, Justica spicigera*
muitle – *Gliricidia sepium*
mujercita – *Psilocybe muliericula*
mujú – *Brosimum alicastrum*
mul – *Cenchrus echinatus*
mulato – *Bursera simaruba, Capsicum annuum, Coryphanta macromeris, Epithelantha micromeris*
mulce – *Sideroxylon lanuginosum*
mulito – *Bidens pilosa*
muloch – *Triumfetta semitriloba*
mundillo – *Viburnum opulus*
mundito – *Cardiospermum halicacabum, Dodonaea viscosa*
mungo – *Vigna radiata*
munisté – *Bourreria huanita*
muñeco – *Dracaena sanderana, Leccinum aurantiacum*
murajes – *Anagallis arvensis*
murajes arvenses – *Anagallis arvensis*
muraya – *Murraya paniculata*
mureo – *Fouquieria diguetii, F. macdougalii, F. splendens*
múrgola – *Morchella esculenta*
murgula – *Gyromitra esculenta*
murió – *Agonandra racemosa*
murta – *Myrus communis*
murucuyá – *Passiflora edulis, P. incarnata*
murucuyá cascudo – *Passiflora quadrangularis*
murumbu – *Panicum coloratum*
musá – *Gossypium hirsutum*
musajoyo – *Tagetes erecta*
músaro – *Pachycereus schottii*
muse – *Pachycereus schottii*
musgo – *Bryum, Polytrichum, Tillandsia usneoides*
musgo blanco – *Tillandsia usneoides*
musgo de Irlanda – *Chondrus crispus*
musgo de Islandia – *Cetraria islandica*

musgo de pantano – *Sphagnum*
musgo de perla – *Chondrus crispus*
musgo español – *Tillandsia usneoides*
musgo irlandés – *Sagina subulata*
musha – *Artemisia ludoviciana*
muste – *Clerodendron*
mutscoc – *Selaginella*
mutu – *Otatea fimbriata*
muul-suuk – *Cenchrus pilosus*
mux – *Lagenaria siceraria*
muyché – *Jacquinia aurantica*
muyozapot – *Achras zapota*

náaraso – *Citrus sinensis*
naba – *Brassica napus* var *oleifera, Brassica oleracea* var *gongylodes, B. rapa*
nabá – *Liquidambar styraciflua, Myroxylon balsamum* var *pereirae*
nabal – *Myroxylon balsamum* var *pereirae*
nabicol – *Brassica oleracea* var *gongylodes*
nabina – *Brassica rapa* var *oleifera*
nabiza – *Brassica rapa*
nabo – *Apium graveolens* var *rapaceum, Brassica napus* var *napobrassica, B. n.* var *oleifera, B. rapa, Eruca vesicaria* ssp *sativa, Raphanus raphanistrum*
nabo blanco – *Raphanus raphanistrum*
nabo cimarrón – *Raphanus raphanistrum*
nabo colorado de jarava – *Beta vulgaris*
nabo colza – *Brassica napus* var *oleifera*
nabo comestible – *Brassica napus* var *napobrassica*
nabo de invierno – *Brassica rapa* var *oleifera*
nabo del diablo – *Bryonia dioica, Oenanthe crocata*
nabo de verano – *Brassica rapa* var *campestris, B. r.* var *praecox*
nabo forrajero – *Brassica napus* var *napobrassica*
nabo hortelano – *Brassica rapa*
nabo sueco – *Brassica napus* var *napobrassica, Brassica rapa*
nabukak – *Sonchus oleraceus*
nacaco – *Licania arborea*
nacagua – *Cordia boissieri*
nacahuela – *Notholaena sinuata*
nacahuite – *Cordia boisieri*
nacahuitl – *Cordia boissieri*
nacapul – *Ficus cotinifolia, F. radulina*
nacapuli – *Ficus cotinifolia, F. glaucescens, F. padifolia*
nacare – *Lippia umbellata*
nacaste – *Enterolobium cyclocarpum, Leucaena glauca*
nacastillo – *Enterolobium cyclocarpum*
nacaxtle – *Enterolobium cyclocarpum*
nacazcólotl – *Caesalpinia cacalaco. C. coriaria*
nacazcul – *Caesalpinia cacalaco, C. coriaria, Datura stramonium, D. wrightii*
nacido – *Stenocereus alamosensis*
naciño – *Taxodium distichum* var *mexicanum*
nacuala – *Jatropha curcas*
nachácata – *Galphimia glauca*
nachachícoli – *Elytraria squamosa*
nafe – *Hibiscus esculentus*
nahuapoztli – *Senna leptocarpa*
nahuitiput – *Verbesina crocata*
nahuo – *Zea mays*
najcala – *Asclepias glaucescens*
nakaz – *Coccothrinax argentea*
namboca – *Juglans*
namole – *Phytolacca icosandra*
namuquia – *Randia obcordata*
nana – *Solanum melongena* var *esculentum*
nanácate – *Panaeolus sphinctrinus*
nanácatl – *Panaeolus sphinctrinus*
nanahuapatli – *Senecio vulneraria, Smilax medica, Solidago mexicana, S. velutina*
nancha – *Citrus sinensis*
nanche – *Byrsonima crassifolia*
nanche agrio – *Byrsonima crassifolia*
nanche amarillo – *Byrsonima crassifolia*
nanche de la costa – *Ziziphus amole*
nanche del monte – *Malpighia glabra*
nanche de perro – *Byrsonima crassifolia, Vitex pyramidata*
nanche dulce – *Byrsonima crassifolia*
nanche saituna – *Ziziphus amole*
nanchicacao – *Ximenia americana*
nandi – *Setaria sphacelata*
nandimbo – *Ehretia tinifolia*
nano-catsintli – *Psilocybe muliericula*
nantzinquáhuitl – *Byrsonima crassifolia*
nantzinxócotl – *Byrsonima crassifolia*
napacoma – *Carya*
napahuite – *Trichilia hirta*
napelo – *Aconitum*

napier – *Pennisetum purpureum*
napitoro – *Bursera grandifolia*
Napoleón – *Bougainvillea spectabilis*
naranjeto – *Citrus paradisi*
naranjillo – *Alibertia edulis, Capparis odoratissima, C. verrucosa, Casearia sylvestris, Ilex dugesii, Rhamnus serrata, Rheedia edulis, Sargentia greggii, Solanum quitoense, Zanthoxylum arborescens, Z. fagara, Ziziphus amole*
naranjillo cimarrón – *Celtis iguanea*
naranjito – *Fortunalla margarita, Ziziphus amole*
naranjito chino – *Citrus japonica*
naranjo – *Citrus aurantium, C. sinensis*
naranjo ácido – *Citrus aurantium*
naranjo agrio – *Citrus aurantium*
naranjo amargo – *Citrus aurantium*
naranjo amargo espinoso – *Poncirus trifoliata*
naranjo amarillo – *Thevetia ovata, T. peruviana, T. thevetioides*
naranjo amateco – *Citrus aurantium*
naranjo cajel – *Citrus aurantium*
naranjo chino – *Citrus sinensis, Fortunella margarita, Maclura pomifera*
naranjo de fruta agria – *Citrus aurantium*
naranjo de la China – *Citrus sinensis*
naranjo del diablo – *Citrullus colocynthis*
naranjo del Osage – *Maclura pomifera*
naranjo de tres folíolos – *Poncirus trifoliata*
naranjo dulce – *Citrus sinensis*
naranjo enano – *Citrus mitis*
naranjo hojas de mirto – *Citrus aurantium*
naranjo king – *Citrus nobilis*
naranjo lima – *Citrus aurantium*
naranjo mandarino – *Citrus deliciosa, C. reticulata*
naranjo mexicano – *Choisya ternata*
naranjo sanguino – *Citrus sinensis* var *melitense*
naranjo trébol – *Poncirus trifoliata*
naranjo trifoliata – *Poncirus trifoliata*
narciso – *Narcissus Poeticus, N. pseudo-narcissus, Nerium oleander, Poeticus*
narciso amarillo – *Thevetia ovata, T. peruviana, T. thevetioides*
narciso asturiano – *Narcissus asturiensis*
narciso atrompetado – *Cecropia peltata, Narcissus pseudo-narcissus*
narciso blanco – *Narcissus poeticus*
narciso cyclamineus – *Narcissus cyclamineus*
narciso de Asturias – *Narcissus asturiensis*
narciso de hojas de ciclaminos – *Narcissus cyclamineus*
narciso de hojas tenues – *Narcissus bulbocodium*
narciso de las nieves silvestre – *Galanthus nivalis*
narciso de los poetas – *Narcissus poeticus*
narciso de los prados – *Narcissus pseudo-narcissus*
narciso de manojo – *Narcissus tazetta*
narciso de otoño – *Colchicum autumnale*
narciso de roqueda – *Narcissus pseudo-narcissus*
narciso junquillo – *Narcissus jonquilla*
narciso lleno – *Narcissus pseudo-narcissus*
narciso miniatura – *Narcissus pseudo-narcissus*
narciso poético – *Narcissus poeticus*
narciso tazetta – *Narcissus tazetta*
narciso triandro – *Narcissus triandrus*
narciso triple – *Narcissus triandrus*
narciso trompeta – *Narcissus pseudo-narcissus*
nardo – *Argemone, Nardostachys jatamansi, Polianthes tuberosa*
nardo americano – *Aralia racemosa*
nardo común – *Polianthes tuberosa*
nardo de jardín – *Polianthes tuberosa*
nardo marítimo – *Pancratium maritimum*
nardo oloroso – *Polianthes tuberosa*
nariz – *Anacardium rhinocarpus*
narok – *Setaria sphacelata*
nasiva – *Leucaena lanceolata*
nasturcio – *Tropaeolum majus*
nata – *Ipomoea alba, I. muricata*
navajita – *Bouteloua*
navajita aguja – *Bouteloua aristidoides*
navajita anual – *Bouteloua barbata*
navajita azul – *Bouteloua gracilis*
navajita banderilla – *Bouteloua curtipendula*
navajita china – *Bouteloua breviseta, B. trifida*
navajita de agua – *Bouteloua barbata*
navajita liebrera – *Bouteloua barbata*
navajita morada – *Bouteloua chondrosioides, B. radicosa*
navajita negra – *Bouteloua eriopoda*
navajita peinada – *Bouteloua chondrosioides*
navajita pelillo – *Bouteloua repens*

navajita racimosa – *Bouteloua barbata*
navajita rastrera – *Bouteloua filiformis*
navajita roja – *Bouteloua trifida*
navajita Santa Rita – *Bouteloua eludens*
navajita simple – *Bouteloua simplex*
navajita velluda – *Bouteloua hirsuta*
navajuela – *Scleria bracteata*
navío – *Conzattia multiflora, C. sericea*
navo – *Opuntia wilcoxii*
nayada flexuosa – *Najas flexilis*
nayada marina – *Najas marina*
nayada mayor – *Najas major*
nazareno – *Brosimum alicastrum, Muscari botryoides, M. comosum*
nazareno azul – *Muscari botryoides*
nazareno de cuidado – *Muscari neglectum*
nazareno de penacho – *Muscari comosum*
nazareno menor – *Muscari botryoides*
ndexu – *Hordeum vulgare*
neanthe bella – *Chamaedorea elegans*
nébeda – *Nepeta cataria, N.* x *faassenii*
nébeda común – *Nepeta cataria*
nébeda de flores grandes – *Nepeta grandiflora*
nebrina – *Juniperus communis*
nebro – *Juniperus communis*
nebulosa – *Gypsophila paniculata, Pancratium maritimum*
neco – *Microsechium helleri*
nectarino – *Prunus persica* ssp *nectarina*
nefrolepis – *Nephrolepis exaltata*
negra – *Acacia wrightii*
negral – *Pinus montana*
negrilla – *Tricholoma terreum*
negrillo – *Ulmus campestris, U. minor*
negrito – *Ardisia revoluta, Asplenium monanthes, Citharexylum berlandieri, Dolicholus phaseoloides, Helvella lacunosa, Karwinskia humboldtiana, Maxillaria cucullata, Rhynchosia, Simarouba glauca, Talinum paniculatum, Vitex pyramidata*
negrito coyote – *Vitex pyramidata*
negrito jupuri – *Vitex pyramidata*
negro coyote – *Vitex mollis, V. pyramidata*
neguilla – *Agrostemma githago, Nigella damascena*
neguilla aromática – *Nigella sativa*
neguilla de damasco – *Nigella damascena*
neguillón – *Agrostemma githago*
negundo – *Acer negundo*
nejo – *Tricholoma equestre*
nela – *Euphorbia heterophylla*
neldo – *Anethum graveolens*
nelumbio blanco – *Nymphaea alba*
nemaax – *Heliotropium angiospermum, H. fruticosum, H. indicum*
nemaxtla – *Iresine calea*
nemiz – *Acalypha hispida*
nemofila – *Nemophila*
nemorosa – *Anemone nemorosa*
nendo de burro – *Parthenice mollis*
nenúfar – *Nuphar luteum, Nymphaea*
nenúfar amarillo – *Nuphar luteum*
nenúfar blanco – *Nymphaea alba*
nenúfar oloroso – *Nymphaea odorata*
nenúfar peltado – *Nymphoides peltata*
nenúfar perfumado – *Nymphaea odorata*
nep-cha-ga – *Orobanche cooperi*
nere – *Hymenaea courbaril*
nerina – *Nerine bowdenii*
nesco – *Lonchocarpus hermannii*
nesesehuilone – *Geranium carolinianum*
neuctixihuitl – *Lippia dulcis, L. graveolens*
neucxóchitl – *Jacquinia aurantica*
nevadilla – *Paronychia echinulata, Polygonum bistorta*
nevadilla de otoño – *Leucojum autumnale*
nevadilla de primavera – *Leucojum vernum*
nevadilla de verano – *Leucojum aestivum*
nexehuac – *Datura ceratocaula*
nexmitzi – *Artemisia mexicana*
nextamalxóchitl – *Ranunculus dichotomus, R. petiolaris, Tecoma stans*
nexticinatl – *Halimium glomeratum*
ngedni – *Bursera jorullensis*
nhal – *Zea mays*
niaulí – *Melaleuca alternifolia, N. viridiflora*
nicandra – *Nicandra physalodes*
nicaragua – *Impatiens balsamina*
nicociana – *Nicotiana alata, N. tabacum*
nicté – *Plumeria rubra*
nicua – *Ipomoea alba*
niche – *Cocoloba uvifera*
nido de ave – *Asplenium nidus*
nido de pájaro – *Neottia nidus*
nidularium – *Neoregelia carolinae*
nievitas – *Cryptantha, Plagiobothrys nothofulvus*
nigela – *Nigella damascena*

nigua – *Conostegia xalapensis, Muntingia calabura*
nigüilla – *Rhipsalis baccifera*
niin – *Jatropha curcas*
nilo – *Brachiaria mutica*
nim – *Azadirachta indica*
nimu – *Citrus limon*
ninfea – *Nymphaea, Nymphoides falax, N. humboldtianum*
ninfea amarilla – *Nymphea mexicana*
ninfea azul – *Nymphea elegans*
ninfea blanca – *Nymphea flavovirens*
niñito – *Psilocybe aztecorum*
niño – *Coleus blumei, Psilocybe muliericula*
niño de las aguas – *Psilocybe aztecorum*
niso – *Prunus domestica* ssp *insititia*
nisperero del Japón – *Eriobotrya japonica*
níspero – *Achras zapota, Eriobotrya japonica, Mimusops dariensis, Mospilus germanica*
níspero del Canadá – *Amelanchier canadensis*
níspero del Japón – *Eriobotrya japonica*
níspero japonés – *Eriobotrya japonica*
nite biito – *Liquidambar styraciflua*
nixtamalcuáhuitl – *Gilibertia arborea*
nixtamalillo – *Gilibertia arborea*
nize – *Psilocybe mexicana*
noa – *Agave parrasana, A. victoriae-reginae*
noai – *Ananas comosus, A. sativus*
nocuana-piti – *Mentha* x *piperita*
nochebuena – *Caesalpinia pulcherrima, Euphorbia pulcherrima*
nocheznopalli – *Opuntia cochenillifera, O. tomentosa*
nocheztli – *Opuntia cochenillifera*
nochi – *Lucuma salicifolia*
nochizo – *Corylus avellana*
nochtli – *Opuntia*
nogal – *Calatola mollis, Carya illinoensis, Juglans*
nogal americano – *Carya, Juglans nigra*
nogal blanco – *Juglans cinera*
nogal blanco americano – *Juglans cinera*
nogal ceniciento – *Juglans cinera*
nogal comùn – *Juglans regia*
nogal de alas común – *Pterocarya fraxinifolia*
nogal de alas chino – *Pterocarya stenoptera*
nogal de Australia – *Macadamia ternifolia*
nogal de Castilla – *Juglans regia*
nogal de Cáucaso – *Pterocarya fraxinifolia*
nogal de cerro – *Juglans major*
nogal de Cuba – *Juglans cinera*
nogal de Cuilapán – *Carya illinoensis, C. pecan*
nogal de Europa – *Juglans regia*
nogal de la brujería – *Hamamelis mollis*
nogal de la India – *Aleurites molucana*
nogal de las brujas – *Hamamelis virginiana*
nogal de las brujas chino – *Hamamelis mollis*
nogal de las brujas japonés – *Hamamelis japonica*
nogal de nuez chica – *Carya*
nogal de nuez lisa – *Carya illinoensis*
nogal encarcelado – *Carya illinoensis, Juglans mollis*
nogal español – *Juglans regia*
nogal europeo – *Juglans regia*
nogal gris americano – *Juglans cinera*
nogalillo – *Carya, Juglans microcarpa*
nogalito – *Juglans microcarpa*
nogal liso – *Carya illinoensis*
nogal morado – *Carya*
nogal motudo – *Carya*
nogal negro – *Juglans nigra*
nogal negro americano – *Juglans nigra*
nogal nuez meca – *Juglans mollis*
nogal pecanero – *Carya illinoensis*
nogal rayado – *Carya*
nogal satinado – *Liquidambar styraciflua*
nogal silvestre – *Juglans*
noholcikutz – *Nicotiana rustica*
nolí – *Elaeis melanococca*
nombó – *Jatropha cinerea*
nombre de Dios – *Agrostis stolonifera*
nomeolvides – *Cordia sebestena, Myosotis, Plumbago auriculata*
nomeolvides alpino – *Myosotis sylvatica*
nomeolvides de agua – *Myosotis palustris*
nomeolvides de bosque – *Myosotis sylvatica*
nomeolvides del Cáucaso – *Brunnera macrophylla*
nomeolvides perenne – *Brunnera macrophylla*
nopal – *Opuntia*
nopal ardilla – *Opuntia spinulifera*
nopal blanco – *Opuntia leucotricha*
nopal cadillo – *Opuntia chavena, O. hypticantha*
nopal camueso – *Opuntia larreyi, O. robusta*
nopal cardón – *Opuntia streptacantha*

nopal cardoso – *Opuntia pumila*
nopal cascarón – *Opuntia chavena, O. hypticantha*
nopal castellano – *Opuntia ficus-indica*
nopal cegador – *Opuntia microdasys, O. rufida*
nopal colorado – *Opuntia leucotricha*
nopal común – *Opuntia tomentosa*
nopal corriente – *Opuntia spinulifera*
nopal coyotillo – *Opuntia azurea, O. macrocentra*
nopal crinado – *Opuntia pilifera*
nopal cuixo – *Opuntia engelmannii*
nopal chamacuerito – *Opuntia tomentosa*
nopal chamacuero – *Opuntia dejecta, O. tomentosa*
nopal chaparro – *Opuntia spinulifera*
nopal chaveño – *Opuntia chavena, O. hyptiacantha, O. leucotricha*
nopal de Castilla – *Opuntia ficus-indica, O. megacantha*
nopal de cerro – *Opuntia lasiacantha*
nopal de cochinilla – *Opuntia cochenillifera*
nopal de cochino – *Opuntia tuna*
nopal de coyote – *Opuntia decumbens*
nopal de crines – *Opuntia pilifera*
nopal de culebra – *Opuntia decumbens, O. puberula*
nopal de lengüita – *Epiphyllum oxypetalum, Opuntia auberi*
nopal de pedregal – *Opuntia lasiacantha*
nopal de raíz – *Opuntia megacantha, O. megarhiza*
nopal de San Gabriel – *Opuntia cochenillifera, Opuntia tomentosa*
nopal de tortuga – *Opuntia decumbens, O. puberula*
nopal de tuna colorada – *Opuntia streptacantha*
nopal duraznillo – *Opuntia leucotricha*
nopaleta – *Pereskiopsis diguetii*
nopal guilanchi – *Opuntia guilanchi*
nopal hartón – *Opuntia streptacantha*
nopal huevo – *Opuntia tuna*
nopalillo – *Disocactus, Epiphyllum truncatum, Epiphyllum anguliger, E.* x *Disocactus, Nymphoides humboldtianum, Opuntia azurea, O. karwinskiana, O. megarhiza, Schlumbergera truncata*
nopalillo blanco – *Epiphyllum anguliger*
nopalillo cegador – *Opuntia microdasys*
nopalillo color de rosa – *Disocactus phyllanthoides*
nopalillo de flor – *Disocactus phyllanthoides, Opuntia karwinskiana*
nopalillo de olor – *Epiphyllum anguliger*
nopalito – *Pereskiopsis diguetii*
nopal lasarón – *Opuntia stenopetala*
nopal manso – *Opuntia megacantha*
nopal mantequilla – *Opuntia durangensis*
nopal memelo – *Opuntia hyptiacantha*
nopal nocheztli – *Opuntia cochenillifera, O. tomentosa*
nopal rastrero – *Opuntia rastrera*
nopal real – *Opuntia microdasys*
nopal redondo – *Opuntia polyacantha*
nopal serrano – *Opuntia stenopetala*
nopal sin espinas – *Opuntia ficus-indica*
nopal tapón – *Opuntia durangensis, O. robusta*
nopal tapón blanco – *Opuntia robusta*
nopal tempranillo – *Opuntia megacantha*
nopal tunero – *Opuntia ficus-indica*
nopalxochicuezaltic – *Disocactus phyllanthoides*
nopalli – *Opuntia*
nopo – *Cordia alliodora*
nopota peste – *Cordia alliodora*
noriba – *Agave bovicornuta, A. filifera*
noriten – *Cedronella mexicana*
notofagus antártica – *Nothofagus antarctica*
notofagus de Dombey – *Nothofagus dombeyi*
notofagus oblicua – *Nothofagus obliqua*
noua – *Phaseolus caracalla*
novio – *Pelargonium* x *hortorum*
noxtli – *Opuntia ficus-indica*
nttzirza – *Sambucus mexicana*
nube – *Gypsophila elegans, G. paniculata*
nube de campo – *Valeriana clematitis*
nube de novia – *Gypsophila paniculata*
nucalpat – *Cupressus lusitanica*
nuculpat – *Ulmus mexicana*
nudillo – *Paspalum racemosum*
nudillo acuático – *Paspalum fluitans*
nudita – *Acacia crinita*
nueva reforma – *Veronica spicata*
nueve hojas – *Serjania racemosa*
nuez – *Carya, Juglans*
nueza – *Bryonia alba, B. cretica, B. dioica*
nuez americana – *Juglans nigra*
nueza negra – *Tamus communis*

nuez chiquita – *Carya*
nuez de Australia – *Macadamia ternifolia*
nuez de betel – *Piper betle*
nuez de Brasil – *Bertholletia excelsa*
nuez de calatola – *Calatola mollis*
nuez de Castilla – *Carya, Juglans regia*
nuez de coco – *Cocos nucifera*
nuez de cola – *Kola acuminata*
nuez de Cuilapán – *Carya pecan*
nuez de especia – *Myristica fragrans*
nuez de gouro – *Kola acuminata*
nuez de la India – *Aleurites molucana, Anacardium occidentale*
nuez del paraíso – *Lecythis zabucajo*
nuez de macadamia – *Macadamia integrifolia*
nuez de marfil – *Phytelephas macrocarpa*
nuez de montaña – *Calatola mollis*
nuez de nogal – *Juglans*
nuez de pacana – *Carya*
nuez de pará – *Bertholletia excelsa*
nuez de Tahiti – *Inocarpus edulis*
nuez encarcelada – *Carya, Juglans*
nuez enemosa – *Myristica fragrans*
nuez grande – *Juglans*
nuez lisa – *Carya*
nuez meca – *Juglans*
nuez moscada – *Myristica fragrans, M. moschata*
nuez negra – *Juglans nigra*
nuez pacana – *Carya*
nuez pecan – *Carya illinoensis*
nuez pili – *Canarium ovatum*
nuez vómica – *Strychnos nux-vomica*
nulibé – *Tagetes erecta*
nuntzutzuy – *Acanthocereus tetragonus*
nuña – *Phaseolus vulgaris* ssp *nunas*
nurite – *Clinopodium laevigatum*
nusillo – *Chondrilla juncea*
nutillo – *Stenocereus treleasei*
nutriol – *Tripsacum laxum*
nuzu-ndu – *Arbutus glandulosa, A. texana, A. xalapensis*

ñajú – *Hibiscus esculentus*
ñajú de espinas – *Pereskia grandiflora*
ñame – *Colocasia antiquorum, C. esculenta, Dioscorea*
ñame blanco – *Dioscorea alata*
ñame común – *Dioscorea alata, D. sativa*
ñame chato – *Rajania hastata*
ñame de agua – *Dioscorea alata*
ñame de diez meses – *Dioscorea alata*
ñame grande – *Dioscorea alata*
ñame mexicano silvestre – *Dioscorea villosa*
ñame morado – *Dioscorea triphylla*
ñame silvestre – *Dioscorea villosa*
ñamole – *Phytolacca octandra*
ñampi – *Colocasia esculenta*
ñangate – *Dioscorea*
ñire – *Nothofagus antarctica*
ñisñil – *Typha latifolia*
ñongue – *Datura stramonium*
ñui – *Panaeolus sphinctrinus*

Oaxaqueña – *Calea zacatechichi*
obalamo – *Vitex mollis*
obatel – *Casearia nitida*
obel – *Piper auritum*
obelisco – *Hibiscus rosa-sinensis*
obelisco de la sierra – *Malvaviscus arboreus*
oca – *Oxalis tuberosa*
ocahui – *Agave ocahui*
ocalito – *Eucalyptus*
ocalo – *Eucalyptus*
ocatl – *Arundo donax*
ocelosúchil – *Tigridia pavonia*
oceloxóchitl – *Tigridia pavonia*
ocobo – *Tabebuia pentaphylla*
ococ – *Bombax ellipticum*
ococote – *Liquidambar styraciflua*
ococuahuitl – *Pinus*
ocopetate – *Alsophila salvinii, Cyathea delicatula, C. mexicana, Pteridium aquilinum*
ocopetla – *Abies hickeli, A. religiosa*
ocopiaxtli – *Eryngium comosum*
ocopino – *Pinus*
ocoquihui – *Sambucus mexicana*
ocosúchil – *Didymaea mexicana*
ocotate – *Guadua amplexifolia*
ocote – *Gochnatia arborescens, Pinus*
ocote blanco – *Pinus*
ocote colorado – *Pinus patula*
ocote chino – *Pinus lawsoniana, P. leiophylla, P. oocarpa*
ocote de carretillo – *Pinus oocarpa, P. tecunumanii*
ocote dormido – *Pinus lumholtzii*

ocote escobetón – *Pinus devoniana, P. michoacana*
ocote gretado – *Pinus ayacahuite, P. devoniana, P. michoacana*
ocote hembra – *Pinus montezumae*
ocotehue – *Dahlia imperialis*
ocote lacio – *Pinus douglasiana*
ocote macho – *Pinus montezumae, P. oocarpa, P. patula*
ocote mullaca – *Solanum sisymbrifolium*
ocote pardo – *Pinus hartwegii*
ocotillo – *Cardiospermum halicacabum, Cordia elaeagnoides, Croton, Dodonaea viscosa, Fouquieria, Parthenium stramonium, P. tomentosum, Salvia sessei*
ocotillo blanco – *Amyris balsamifera, Fouquieria burragei*
ocotillo de corral – *Fouquieria splendens*
ocotillo de flor blanca – *Fouquieria burragei*
ocotillo macho – *Fouquieria macdougallii*
ocotillo meco – *Cordia eleagnoides, C. sebestena*
ocotl – *Pinus*
ocotle – *Pinus*
ocotzoquiauitl – *Liquidambar styraciflua*
ocótzotl – *Liquidambar styraciflua, Pinus teocote*
ocoxóchitl – *Didymaea mexicana*
ocozote – *Liquidambar styraciflua, Yucca*
ocozotl – *Liquidambar styraciflua*
ocra – *Hibiscus esculentus*
octubre – *Chrysanthemum*
ocu – *Calophyllum braziliense*
ocueyanmona – *Croton*
ocuje – *Calophyllum calaba*
ocumo – *Xanthosoma sagittifolium*
ocumo chino – *Colocasia esculenta*
ochil – *Syngonium podophyllum*
ochmuul – *Triumfetta semitriloba*
oiticica – *Licania rigida*
ojaman – *Curatella americana*
ojaranzo – *Carpinus betulus, Rhododendron ponticum*
ojcome – *Agave wocomahi*
oje – *Pachira aquatica*
ojite – *Brosimum alicastrum*
ojito de pájaro – *Rhynchosia phaseoloides, R. pyramidalis*
ojito de pichón – *Dolicholus phaseoloides, Rhynchosia pyramidalis*
ojitos azules – *Evolvulus alsinoides*
ojitos negros – *Thunbergia alata*
ojob – *Dioscorea*
ojoche – *Brosimum alicastrum*
ojoche blanco – *Brosimum alicastrum*
ojochillo – *Brosimum alicastrum*
ojo de águila – *Caesalpinia crista*
ojo de arriera – *Cenchrus echinatus*
ojo de borrico – *Mucuna sloanei*
ojo de buey – *Heliopsis helianthoides, Mucuna*
ojo de caballo – *Inula helenium*
ojo de cangrejo – *Dolicholus phaseoloides, Rhynchosia pyramidalis*
ojo de culebra – *Rhynchosia pyramidalis*
ojo de chanate – *Dolicholus phaseoloides, Rhynchosia pyramidalis*
ojo de gallo – *Sanvitalia procumbens*
ojo de gato – *Sanvitalia procumbens*
ojo de hormiga – *Cenchrus echinatus*
ojo del diablo – *Hyoscyamus niger*
ojo de liebre – *Solanum eleagnifolium, S. tridynamum*
ojo de loro – *Sanvitalia procumbens, Sapindus saponaria*
ojo del sembrado – *Chrysanthemum segetum*
ojo de negro – *Sanvitalia procumbens*
ojo de pajarito – *Dolicholus phaseoloides, Rhynchosia phaseoloides, R. pyramidalis, Thunbergia alata*
ojo de pájaro – *Thunbergia alata*
ojo de perdiz – *Adonis flamea, A. vernalis*
ojo de perico – *Rhynchosia pyramidalis, Sanvitalia procumbens*
ojo de pescado – *Lantana camera, Porophyllum gracile*
ojo de poeta – *Thunbergia alata*
ojo de pollo – *Sanvitalia procumbens*
ojo de ratón – *Lantana hispida*
ojo de tecolote – *Clematis dioica*
ojo de venado – *Caesalpinia crista, Lycopersicon esculentum* var *cerasiforme, Mucuna, Passiflora foetida, P. mexicana, Rhamnus scopulorum*
ojo de Venus – *Thunbergia alata*
ojo de víbora – *Evolvulus alsinoides*
ojo de zanate – *Dolicholus phaseoloides,*

Phaseolus atropurpureus, Rhynchosia pyramidalis, Stegnosperma halimifolium
ojotate – *Chusquea nelsoni*
ojotón – *Senna pallida*
ojtotl – *Guadua aculeate*
olcuahuitl – *Castilla elastica*
oleandro – *Nerium oleander*
oleastro – *Olea europaea* var *oleaster*
olíbano – *Boswellia thurifera*
olicuahuitl – *Castilla elastica*
olivarda – *Inula helenium*
olivaseba – *Daphne mezereum*
olivastro – *Olea europaea* var *oleaster*
olivera – *Olea europaea*
olivilla – *Teucrium fruticans*
olivo – *Capparis odoratissima, Olea europaea*
olivo de Bohemia – *Eleagnus angustifolius*
olivo oloroso – *Osmanthus fragrans*
olivo silvestre – *Olea europaea* var *oleaster*
olmedo – *Ulmus americana*
olmo – *Chaetoptelea mexicana, Populus, Ruprechtia pallida, Ulmus*
olmo americano – *Ulmus americana*
olmo blanco americano – *Ulmus americana*
olmo campestre – *Ulmus procera*
olmo común – *Ulmus minor, U. procera*
olmo chino – *Ulmus parvifolia*
olmo de hojas pequeñas – *Ulmus parvifolia*
olmo del Cáucaso – *Zelkova carpinifolia*
olmo de los campos – *Ulmus campestris*
olmo del país – *Alnus arguta, A. rubra*
olmo de montaña – *Ulmus glabra*
olmo de monte – *Ulmus glabra*
olmo de Siberia – *Ulmus pumila*
olmo escocés – *Hamamelis virginiana*
olmo holandés – *Ulmus hollandica*
olmo inglés – *Ulmus procera*
olmo montano – *Ulmus glabra*
olmo negrillo – *Ulmus procera*
olmo norteamericano – *Ulmus americana, U. fulva*
olmo pedunculado – *Ulmus laevis*
olmo siberiano – *Ulmus pumila*
olmo temblón – *Ulmus laevis*
ololiuquí – *Ipomoea purpurea, Turbina corymbosa*
olóricua – *Encyclia citrina*
olosapo – *Couepia polyandra*
olotillo – *Dorstenia drakeana, Spathiphyllum*
olla de mono – *Lecythis zabucajo*
olli – *Castilla elastica*
ollin – *Castilla elastica*
omanche – *Calatola mollis*
ombligo de tierra – *Peperomia campylotropa*
ombligo de Venus – *Hydrocotyle, Omphalodes cappadocica, O. linifolia*
ombú – *Phytolacca dioica*
omfalodes primaveral – *Omphalodes verna*
omil – *Cirsium mexicanum*
omisúchil – *Polianthes tuberosa*
omixóchitl – *Polianthes tuberosa*
omoquelite – *Piper auritum*
omoquilitl – *Piper auritum*
omfalodes de hojas de lino – *Omphalodes linifolia*
on – *Persea americana*
onagra – *Oenothera*
onagra bienal – *Oenothera biennis*
onagra blanca – *Oenothera speciosa*
onagra común – *Oenothera biennis*
onagra de flores largas – *Oenothera glazioviana*
onagra de hojas estrechas – *Oenothera stricta*
onagra elegante – *Oenothera speciosa*
onagra fruticosa – *Oenothera fruticosa*
onagra laciniada – *Oenothera laciniata*
onagra rosada – *Oenothera rosea*
onagra vellosa – *Oenothera villosa*
ona jújugo – *Adelia virgata*
oncinapor – *Urera caracasana*
onmachinté – *Calatola mollis*
onoquiles – *Alkanna tinctoria*
onova – *Ratibida mexicana*
onza – *Capsicum chinense*
ool – *Stenocereus thurberi*
op – *Annona reticulata*
opay – *Bourreria huanita*
opobálsamo – *Commiphora opobalsamum*
opopónace – *Opopanax chironium*
opptzimin – *Petrea arborea*
oración – *Ipomoea alba*
ora pro nobis – *Pereskia grandifolia*
orcaneta – *Alkanna tinctoria, Sanguinaria canadensis*
orcaneta azul – *Alkanna tinctoria*
orchilla – *Ramalina reticulata, Roccella babingtonia, R. tinctoria*
orcho – *Dactylis glomerata*

oreganillo – *Aloysia gratissima, A. triphylla*
orégano – *Ageratina wrightii, Coleus ambolnicos, Hedeoma floribunda, H. patens, Hyptis albida, Lantana involucrata, Lippia, Monarda austromontana, M. citriodora, Origanum dictamnus, O. vulgare, Poliomintha longiflora, Salvia*
orégano brujo – *Coleus ambolnicos*
orégano castillo – *Lippia graveolens*
orégano cimarrón – *Dalea greggii, Hyptis suaveolens*
orégano común – *Origanum vulgare*
orégano de Creta – *Origanum dictamus*
orégano de España – *Coleus ambolnicos, C. blumei*
orégano de la sierra – *Monarda*
orégano del burro – *Lippia graveolens*
orégano del campo – *Brickellia veronicaefolia, Monarda pectinata*
orégano del cerro – *Brickellia veronicaefolia*
orégano del monte – *Brickellia veronicaefolia*
orégano hortense – *Origanum majorana*
orégano mayor – *Origanum majorana*
orégano mejorana – *Origanum majorana*
oreganón – *Lippia formosa*
oreja – *Auricularia polytricha, Clitocybe squamulosa*
oreja blanca – *Pleurotus ostreatus*
oreja de asno – *Scadoxus multiflorus, Symphytum officinale*
oreja de burro – *Aspidistra elatior, Cucurbita foetidissima, Echeveria gibbiflora, E. subrigida, Laelia nodosa, Oncidium cavendishianum, Sansevieria trifasciata, Sida rhombifolia*
oreja de cazahuate – *Pleurotus ostreatus*
oreja de cordero – *Stachys byzantina*
oreja de elefante – *Colocasia esculenta*
oreja de fraile – *Asarum europaeum*
oreja de gato – *Hieracium mexicanum, Lathyrus vernus*
oreja de Judas – *Auricularia auricula-judae, A. polytricha*
oreja de Judío – *Auricularia polytricha, Enterolobium cyclocarpum*
oreja de león – *Leonotis leonurus, Peperomia cordovana*
oreja de liebre – *Alisma plantago-aquatica, Cynoglossum officinale, Phlomis fruticosa, Solidago californica, Wyethia helenioides*
oreja de lobo – *Verbesina encelioides*
oreja de monje – *Omphalodes cappadocia, O. linifolia*
oreja de mono – *Enterolobium cyclocarpum*
oreja de mula – *Asclepias glaucescens*
oreja de mulato – *Enterolobium cyclocarpum*
oreja de oso – *Arctotis stoechadifolia, Primula auricula, Saxifraga crassifolia*
oreja de perro – *Tiquilia canescens*
oreja de puerco – *Lactarius, Russula delica*
oreja de ratón – *Bernardia myricifolia, Chiococca alba, Cissampelos pareira, Cordia brevispicata, Dichondra argentea, Echeveria secunda, Evolvulus alsinoides* var *angustifolia, Malvella lepidota, Oxalis dillenii, Phaseolus atropurpureus, Ranunculus, Stellaria media, Symphoricarpos albus*
oreja de vaina – *Enterolobium cyclocarpum*
oreja de venado – *Asclepias atropurpurea*
oreja retorcida – *Caesalpinia cacalaco*
orejera – *Enterolobium cyclocarpum*
orejita – *Echeveria pulverulenta*
orejón – *Buddleja perfoliata, Enterolobium cyclocarpum*
orejuela – *Citharexylum berlandieri, Cymbopetalum penduliflorum*
orejuela de ratón – *Dichondra argentea*
orejuela para chocolate – *Cymbopetalum penduliflorum*
orellana – *Pleurotus ostreatus*
oreoselino – *Peucedanum oreoselinum*
oreoselino de tallo lueco – *Peucedanum ostruthium*
oreoselino oficinal – *Peucedanum officinale*
oreoselino montano – *Peucedanum oreoselinum*
organillo – *Grusonia bradtiana, Hydrocotyle vulgaris, Selenicereus grandiflorus, S. hamatus*
organito – *Echinocereus cinerascens, Opuntia chaffeyi*
organito de víbora – *Peniocereus viperinus*
órgano – *Acanthocereus tetragonus, Disocactus speciosus, Echinocereus stramineus, Isolatocereus dumortieri, Neobuxbaumia tetetzo, Pachycereus,*

Pilosocereus leucocephalus, Stenocereus stellatus, S. thurberi
órgano de la costa – *Pachycereus marginatus, P. pecten-aboriginum*
órgano marismeño – *Stenocereus thurberi*
orgullo de la India – *Koelreuteria paniculata*
orgullo de Londres – *Saxifraga umbrosa*
orifela – *Cymbopetalum penduliflorum*
orilla de arroyo – *Ilex tolucana*
ormino – *Salvia sclarea, S. viridis*
ormino silvestre – *Salvia viridis*
orno – *Fraxinus ornus*
orobanca – *Orobanche*
orobanca crenada – *Orobanche crenata*
orobo vernal – *Lathyrus vernus*
oronja – *Amanita caesarea*
oropelo – *Cephaëlis tomentosa*
oropesa – *Salvia aethiopis*
orovale – *Withania somnifera*
orozuz – *Glycyrrhiza glabra, Lantana camara, Lippia dulcis*
orozuz americano – *Glycyrrhiza lepidota*
orozuz del país – *Lantana camara, Lippia graveolens*
orozuz falso – *Astragalus glycyphyllos*
orquídea – *Bauhinia variegata, Bletia stricta, Oncidium tigrinum, Orchidaceae*
orquídea blanca – *Chysis bractescens*
orquídea china amarilla – *Mormodes pardinum* ssp *unicolor*
orquídea del campo – *Bletia, Spiranthes aurantiaca*
orquídea del pobre – *Schizanthus pinnatus*
orquídea gigante – *Himantoglossum longibracteatum*
orquídea payaso – *Odontoglossum rosii*
orquídea pensamiento – *Miltonia*
orquídea rocosa – *Pleione bulbocodioides*
orquídea zapatilla – *Cypripedium irapeanum, Paphiopedilum*
ortiga – *Cnidoscolus angustidens, C. multilobus, C. urens, Jatropha urens, Polanisia uniglandulosa, Tragia nepetaefolia, Urera baccifera, U. caracasana, Urtica dioica, U. serra, Wigandia caracasana, W. kunthii, W. urens*
ortiga blanca – *Boehmeria nivea*
ortiga común – *Urtica dioica*
ortiga cuerda grande – *Urtica dioica*
ortiga de caballo – *Urera baccifera, U. caracasana*
ortiga de la China – *Boehmeria nivea*
ortiga de los Indios – *Monarda fistulosa*
ortiga de tierra caliente – *Wigandia*
ortiga española – *Bidens pilosa*
ortiga grande – *Wigandia caracasana, W. kunthii*
ortiga hedionda – *Stachys arvensis*
ortiga lanosa – *Stachys byzantina*
ortiga mayor – *Urtica dioica*
ortiga menor – *Urtica urens*
ortiga muerta – *Lamium amplexicaule, L. purpurea, Mercurialis annua*
ortiga muerta amarilla – *Lamium galeobdolon*
ortiga muerta maculada – *Lamium maculatum*
ortiga muerta menor – *Lamium galeobdolon*
ortiga muerta purpúrea – *Lamium purpureum*
ortiga palustre – *Stachys palustris*
ortiga pequeña – *Urtica urens*
ortiga prieta – *Wigandia kunthii*
ortiga punzate – *Urtica dioica*
ortiga quemadora – *Urera caracasana*
ortiga real – *Wigandia kunthii*
ortiguilla – *Cnidoscolus angustidens, C. palmeri, C. urens, Tournefortia hartwegiana, Tragia nepetaefolia, Urera caracasana, Urtica dioica* ssp *gracilis*
ortiguilla quemadora – *Tragia nepetaefolia*
ortiguillo – *Pinus lawsonii*
oruga – *Eruca vesicaria* ssp *sativa*
oruga común – *Eruca vesicaria* ssp *sativa*
oruga de los muros – *Diplotaxis muralis*
oruga de hojas tenues – *Diplotaxis tenuifolia*
oruga leonia – *Sisymbrium irio*
oruga marítima – *Cakile maritima*
oruga roqueta – *Eruca vesicaria* ssp *sativa*
oruga silvestre – *Diplotaxis erucoides, Dipsacus fullonum*
orvalle – *Salvia sclarea*
orzaga – *Atriplex halimus, A. hortensis*
orrea – *Juniperus communis*
oshá – *Ligusticum grayi, L. porteri, Lomatium dissectum*
oshá de la sierra – *Ligusticum porteri*
oshá del campo – *Angelica archangelica, A. arguta*
ostria – *Ostrya carpinifolia*
otatave – *Vallesia glabra, V. laciniata*

otate – *Arthrostylidium racemiflora, Arundinaria, Bambusa aculeata, B. arundinacea, Chusquea, Guadua, Muhlenbergia dumosa, Otatea acuminata, Rhipidocladum racemiflorum*
otate amargo – *Guadua paniculata*
otate chino – *Chusquea circinata, Rhipidocladum raceflorum*
otate de hojas anchas – *Otatea fimbriata*
otate dulce – *Otatea acuminata*
otatillo – *Gynerium sagittatum, Muhlenbergia dumosa, Parthenium tomentosum, Rhipidocladum racemiflorum*
otélom – *Phyllostylon brasiliensis*
otlatl – *Guada aculeata*
otocamay – *Myrica cerifera*
ototillo – *Muhlenbergia dumosa*
ovillo – *Hura polyandra*
ovitano – *Garrya laurifolia*
ox – *Brosimum alicastrum*
oxálida blanca – *Oxalis acetosella*
ox dulce – *Brosimum alicastrum*
oxiacanto – *Crataegus laevigata*
oxiria de montaña – *Oxyria digyna*
oxitl – *Brosimum alicastrum*
oxoak – *Abrus precatorius*
oxocoxóchitl – *Brosimum alicastrum*
oxotzin – *Brosimum alicastrum*
oyamel – *Abies durangensis, A. religiosa, A. vejari*
oyametl – *Abies religiosa*
oyutch – *Pholisma sonorae*
ozote – *Ipomoea arborescens, I. murucoides*

paaktzatz – *Monstera deliciosa*
pabellón mexicano – *Salvia microphylla*
pabilo – *Rauwolfia hirsuta*
pabola – *Cordyline terminalis*
pabona – *Hibiscus rosa-sinensis*
pacal – *Citrus sinensis*
pacam – *Opuntia cochenillifera, O. dillenii*
pacana – *Carya*
pacanero – *Carya*
pacanil – *Physalis angulata, P. pubescens*
pacanle – *Hydrocotyle verticillata*
pacanul – *Solanum amazonium*
pacay – *Andira inermis, Inga edulis*
pacaya – *Astrocaryum mexicanum, Chamaedorea, Geonoma interrupta*
pacingo – *Pinus pseudostrobus*
pacón – *Quillaja saponaria, Sapindus saponaria*
pacuy – *Hymenaea courbaril*
pachaba – *Brickellia californica, B. coulteri, B. desertorum*
pachiché – *Psidium guajava*
pachira – *Pachira aquatica*
pachulí – *Pogostemon cablin*
padillo – *Dasylirion longissimum*
padrenuestro – *Myrtillocactus geometrizans*
pagué – *Dyssodia papposa*
paguilla – *Clethra lanata, C. mexicana*
pah – *Pandanus utilis*
pahpatlanuac – *Wigandia kunthii*
pahtlaxoxoctic – *Buddleja sessiliflora*
pahtsab – *Costus mexicanus*
pahua – *Persea americana, P. schiedeana*
pahuatl – *Persea americana*
pahxuhuc – *Tamarindus indica*
paico – *Chenopodium ambrosioides*
paico mocho – *Chenopodium ambrosioides*
paiché – *Petiveria alliacea*
paisaje – *Philodendron radiatum*
paixtli – *Tillandsia recurvata, T. usneoides*
paja – *Chondrilla juncea, Muhlenbergia macroura*
paja americana – *Echinochloa colona*
paja bermuda – *Cynodon dactylon*
paja colchón – *Digitaria sanguinalis*
paja de canela – *Paspalum conjugatum*
paja de Guinea – *Panicum maximum*
paja de la Virgen – *Cynodon dactylon*
paja del camino – *Brachiara fasciculata*
paja de puna – *Stipa ichu*
paja guinea carorena – *Panicum coloratum*
pajarera – *Stellaria media*
pajarilla – *Aquilegia vulgaris, Fumaria parviflora*
pajarito – *Aquilegia vulgaris, Cordia alliodora, Linaria vulgaris, Loranthus, Penstemon parryi, Prunus brachybotrya, Tropaeolum majus*
pajarito del norte – *Lobelia cardinalis* ssp *graminea*
pajarito de monte – *Psilocybe yungensis*
pajarito prieto – *Cordia alliodora*
pájaro azul – *Salvia patens, Strelitzia reginae*

pájaro bobo – *Ipomoea arborescens, I. murucoides*
pajita – *Setaria gracilis*
pajita abierta – *Setaria scheelei*
pajita amarilla – *Setaria pumila*
pajita cerdosa – *Setaria gracilis*
pajita tempranera – *Setaria macrostachya*
pajito – *Chrysanthemum coronarium, Leucanthemum vulgare*
pajiza – *Hibiscus manihot*
pajón – *Sorghum bicolor*
pajón tallo azul – *Schizachyrium hirtiflorum* var *feensi*
pajul – *Anacardium occidentale*
pajul del norte – *Rhus trilobata*
pajulilté – *Simaruba glauca*
pajulul – *Rhus schiedeana*
pak – *Lycopersicon esculentum*
pakaab – *Arundo donax*
pakab – *Saccharum officinarum*
pakcan – *Solanum nigrum*
pakmul – *Physalis angulata*
pakshumk – *Prunus capuli*
pal – *Tripsacum laxum*
palabra de caballero – *Lantana camara*
palabra de mujer – *Lantana camara*
palagallina – *Kallstroemia peninsularis*
palancapatli – *Baileya multiradiata, Senecio vulneraria, Solidago mexicana, S. velutina*
palancapatli de Puebla – *Grindelia*
palapa – *Orbignya cohune*
palasán – *Calamus optimus*
palera – *Opuntia cochenillifera, O. ficus-indica*
paletaria – *Arenaria bourgaei*
palhmaj – *Yucca elephantipes*
palillo – *Colubrina glomerata, Croton ciliato-glandulosus, Erythroxylum mexicanum*
palisandro – *Centrolobium platinense, Jacaranda acutifolia*
palma – *Acoelorraphe wrightii, Acrocomia mexicana, Areaceae, Astrocaryum mexicanum, Ceratozamia mexicana, Chamaedorea graminifolia, Erythea, Gaussia maya, Lophosoria quadripinnata, Nolina microcarpa, Orbignya cohune, Phoenix dactylifera, Sabal uresana, Scheelia preussii, Washingtonia filifera, W. robusta, Yucca*
palma africana – *Elaeis guineensis*
palma alejandra – *Archontophoenix alexandrae*
palma apache – *Erythea edulis*
palma apachite – *Sabal mexicana*
palma areca – *Chrysalidocarpus lutescens*
palma avoira – *Eloeis guineensis*
palma azul – *Erythea armata*
palma barreta – *Yucca carnerosana*
palma barrigona – *Beaucarnea inermis, Pritchardia wrightii*
palma blanca – *Erythea aculeate, E. armata, Sabal uresana, Washingtonia robusta*
palma cambray – *Chamaedorea elegans*
palma camedora – *Chamaedora elegans, C. ernesti-augustii*
palma cana – *Sabal parviflora*
palma canaria – *Phoenix canariensis*
palma catechu – *Areca catechu*
palma ceniza – *Erythea armata, E. edulis, Yucca torreyi, Y. treculeana*
palma cimarrona – *Guassia maya*
palmacito – *Gaussia maya*
palma coca – *Acrocomia mexicana*
palma colorada – *Erythea brandegeei, Washingtonia robusta*
palma común – *Phoenix dactylifera*
palma corcho – *Microcycas calocoma*
palma coyol – *Acrocomia vinifera, Orbignya cohune*
palma criolla – *Yucca arizonica, Y. baccata, Y. macrocarpa, Y. torreyi*
palmacristi – *Ricinus communis*
palmacte – *Roystonea regia*
palma culona – *Beaucarnea inermis*
palma chapay – *Astrocaryum mexicanum*
palma china – *Yucca decipiens, Y. filamentosa, Y. filifera, Y. torreyi*
palma datilera – *Phoenix dactylifera*
palma datillo – *Yucca elata*
palma de abanico – *Cryosophila nana, Erythea edulis, Pritchardia pacifica, Washingtonia filifera*
palma de aceite – *Elaeis guineensis*
palma de agua – *Cocos nucifera*
palma de ardilla – *Catasetum maculatum*
palma de azúcar – *Arenga pinnata*
palma de Canarias – *Phoenix canariensis*
palma de Castilla – *Washingtonia filifera, W. robusta*
palma de cera – *Ceroxylon andicola*
palma de cinta – *Pandanus tectorius*

palma de coateco – *Cryosophila nana*
palma de cola de pescado – *Caryota mitis, C. urens*
palma de coquito de aceite – *Orbignya cohune*
palma de costa – *Roystonea elata*
palma de coyol – *Acrocomia mexicana*
palma de coyor – *Aiphanes acanthophylla*
palma de Cunningham – *Archontophoenix cunninghamiana*
palma de dátiles – *Phoenix dectylifera, Yucca treculeana*
palma de dolores – *Dioon edule*
palma de escoba – *Cryosophila nana*
palma de espina – *Bactris mexicana*
palma de Guadalupe – *Erythea edulis*
palma de guano – *Sabal japa*
palma de ixtle – *Yucca carnerosana*
palma de jango – *Dioon edule*
palma de jipijapa – *Carludovica palmata*
palma de la Virgen – *Dioon edule, D. purpusii*
palma del desierto – *Dracena draco*
palma de llano – *Sabal rosei*
palma de maceta – *Dioon edule*
palma de Madagascar – *Chrysalidocarpus lucubensis*
palma de manaca – *Calyptronoma rivalis*
palma de Manila – *Veitchia merrillii*
palma de marfil – *Phytelephas macrocarpa*
palma de matón – *Erythea edulis*
palma de mazari – *Nannorrhops ritchiana*
palma de micharo – *Sabal mexicana*
palma de oro – *Solidago canadensis* ssp *elongata*
palma de palmira – *Borassus flabellifer*
palma de petate – *Sabal mexicana*
palma de sagú – *Metroxylon sagus*
palma de sagú de Malabar – *Caryota urens*
palma de San José – *Yucca carnerosana, Y. rigida*
palma de San Juan – *Yucca torreyi, Y. treculeana*
palma de San Pedro – *Yucca carnerosana, Y. decipiens, Y. faxoniana*
palma de sombrero – *Carludovica palmata, Erythea edulis*
palma de taco – *Erythea brandegeei*
palma de tlaco – *Erythea brandegeei*
palma de viajero – *Ravenala madagascariensis*
palma de yaguas – *Roystonea elata*
palma dulce – *Erythea edulis*
palma enana – *Chamaerops humilis*
palma excelsa – *Trachycarpus fortunei*
palma fúnebre – *Zamia furfuracea*
palma gladiola – *Gladiolus*
palma helecho – *Cyathea mexicana*
palma hospita – *Copernicia hospita*
palma imperial – *Ceratozamia latifolia, C. mexicana, Pritchardia pacifica*
palma indiana – *Cocos nucifera*
palma jata – *Copernicia macroglossa*
palma kentia – *Howea forsteriana*
palma loca – *Yucca carnerosana, Y. filifera, Y. torreyi, Y. treculeana*
palma llanera – *Copernicia tectorum*
palma mauritia – *Mauritia flexuosa*
palma melocotonera – *Guilielma gasipaes*
palma moriche – *Mauritia minor*
palma murumuru – *Astrocaryum murumuru*
palman – *Muntingia calabura*
palma negra – *Erythea brandegeei, Washingtonia robusta*
palma palmilla – *Erythea brandegeei*
palma pita – *Yucca torreyi, Y. treculeana*
palma popotla – *Erythea edulis*
palma rafia – *Raphia ruffia*
palma real – *Attalea gomphococca, Orbignya cohune, Oreodoxa oleracea, Roystonea, Sabal mexicana, S. pumos, S. uresana, Scheelia costaricensis, S. liebmannii, S. zonensis, Syagrus romanzoffiana*
palma real puertorriqueña – *Roystonea regia*
palma redonda – *Acrocomia mexicana, Sabal mexicana*
palma roja – *Sparganium eurycarpum*
palma rosa – *Cymbopogon martini*
palma samandoca – *Yucca carnerosana, Y. filifera*
palma soyal – *Erythea edulis*
palma tabaquillo – *Zamia lindenii*
palma taliera – *Corypha taliera*
palma trepadora – *Chamaedorea elatior*
palma verde – *Sabal uresana*
palmera – *Arecaceae, Phoenix dactylifera, Washingtonia*
palmera amarilla – *Chrysalidocarpus lutescens*
palmera azucarera – *Arenga pinnata, A. saccharifera*
palmera azul – *Erythea armata*

palmera bambú – *Rhapis excelsa*
palmera botella – *Hyophorbe lagenicaulis*
palmera canaria – *Phoenix canariensis*
palmera cola de pescado – *Caryota urens*
palmera china – *Rhapis humilis, Trachycarpus fortunei*
palmera china de Chusán – *Trachycarpus fortunei*
palmera datilera – *Phoenix dactylifera*
palmera de abanico – *Chamaerops humilis*
palmera de abanico australiana – *Livistona australis*
palmera de abanico china – *Livistona chinensis*
palmera de abrigo – *Copernicia macroglossa*
palmera de aceite – *Elaeis guineensis*
palmera de betel – *Areca catechu*
palmera de Bismarck – *Bismarckia nobilis*
palmera de Buruti – *Trithrinax brasiliensis*
palmera de California – *Washingtonia filifera*
palmera de coco – *Cocos nucifera*
palmera de cola de pescado – *Caryota mitis*
palmera de Fiji – *Pritchardia pacifica*
palmera de fortune – *Trachycarpus fortunei*
palmera de Guadalupe – *Erythea edulis*
palmera de Hillebrand – *Pritchardia hillebrandii*
palmera de la jalea – *Butia capitata*
palmera de la reina – *Syagrus schizophylla*
palmera de las Antillas – *Roystonea regia*
palmera del vino – *Jubaea chilensis*
palmera de MacArthur – *Ptychosperma macarthuri*
palmera de Martius – *Trachycarpus martianus*
palmera de nikau – *Rhopalostylis sapida*
palmera de palmira – *Borassus flabellifer*
palmera de Roebelen – *Phoenix roebelenii*
palmera de sagú – *Cycas revoluta, Metroxylon sagus*
palmera de sagú de Malabar – *Caryota urens*
palmera de Senegal – *Phoenix reclinata*
palmera de sombrilla – *Corypha umbraculifera*
palmera enana – *Chamaerops humilis, Serenoa repens*
palmera imperial – *Roystonea oleracea*
palmera majestuosa – *Ravenea rivularis*
palmera mexicana – *Washingtonia robusta*
palmera plateada – *Coccothrinax argentea*
palmera real – *Cordyline australis, Roystonea regia*
palmera real de Puerto Rico – *Roystonea borinquena*
palmera solitaria – *Ptychosperma elegans*
palmera timbe – *Guilielma insignis*
palmera totai – *Acrocomia totai*
palmera triangular – *Neodypsis decaryi*
palmerita de Petrópolis – *Lytocaryum weddelianum*
palmetto – *Acoelorraphe wrightii*
palmiche – *Reinhardtia gracilis, Roystonea regia*
palmilla – *Bocconia arborea, Carludovicia palmata, Chamaedorea elegans, C. ernesti-augustii, C. microspadix, Cryosophila nana, Cyperus alternifolius, C. involucratus, Dasylirion acrotriche, Erythea aculeata, E. brandegeei, Nolina, Notholaena aurea, Podocarpus reichei, Yucca*
palmilla ancha – *Yucca baccata*
palmira – *Crocosmia* x *croccosmiiflora, Quamoclit lobata*
palmita – *Agave falcate, Chamaedorea elegans, Dioon purpusii, Nolina, Yucca, Zamia furfuracea*
palmita roja – *Cordyline fruticosa*
palmitera – *Chamaerops humilis*
palmito – *Bactris gasipaes, Chamaerops humilis, Erythea edulis, Washingtonia filifera*
palmito blandito – *Chamaerops humilis*
palmito de jardín – *Trachycarpus fortunei*
palmito de juncia – *Serenoa repens*
palmito elevado – *Trachycarpus fortunei*
palmito europeo – *Chamaerops humilis*
palmito molino de viento – *Trachycarpus fortunei*
palmitón – *Beaucarnea inermis*
palo agrín – *Rhus schiedeana*
palo agrio – *Ternstroemia sylvatica*
palo almizcle – *Croton niveus*
palo amargo – *Bocconia frutescens, Celtis trinervia, Coutarea pterosperma, Hintonia latiflora*
palo amargoso – *Castela tortuosa, Garrya longifolia*
palo amarillo – *Adenostoma sparsifolium, Berberis, Bocconia arborea, B. frutescens, Cochlospermum vitifolium, Esenbeckia flava, E. hartmanii, Ficus petiolaris, Maclura*

tinctoria, Trichilia hirta, Washingtonia filifera
palo apestoso – *Karwinskia humboldtiana*
palo atolito – *Haematoxylum campechianum*
palo azul – *Eysenhardtia polystachya, Haematoxylum campechianum*
palo balsa – *Ochroma lagopus*
palo barriga – *Cochlospermum vitifolium*
palo barril – *Cochlospermum vitifolium*
palo batea – *Clethra mexicana*
palo blanco – *Acacia coulteri, A. willardiana, Calycophyllum multiflorum, Casearia sylvestris, Celtis douglassii, C. reticulata, Conzattia multiflora, Croton guatemalensis, Encelia farinosa, Forestiera pubescens, Fraxinus trifoliata, Gilibertia arborea, Ipomoea arborescens, Lysiloma candidum, Maytenus phyllanthoides, Meliosma alba, Piscidia mollis, Psorothamnus spinosus, Sapindus drumondii, S. saponaria, Tabebuia donnell-smithii, T. guayacan, T. pentaphylla, Ziziphus obtusifolia*
palo blanco duro – *Piscidia mollis*
palo bobo – *Cochlospermum vitifolium, Ipomoea arborescens, I. murucoides, Senecio praecox*
palo bofo – *Senecio praecox*
palo bolero – *Haplophyton cimicidum, Stemmadenia obovata*
palo borracho – *Chorisia insignis*
palo brea – *Cercidium*
palo cabra – *Ipomoea arborescens*
palo cachona – *Serjania palmeri*
palo cachorro – *Colubrina triflora, Schoepfia schreberi*
palo calabaza – *Calacophyllum candidissimum*
palo canelo – *Cornus disciflora, C. excelsa*
palo cenizo – *Buddleja tomentella, Bunchosia sonorensis, Leucophyllum ambiguum, L. frutescens, Psorothamnus spinosus*
palo colorado – *Bursera microphylla, B. simaruba, Caesalpinia platyloba, Calacophyllum candidissimum, Ceanothus coeruleus, Colubrina californica, C. viridis, Quercus muehlenbergii, Q. reticulata, Q. rugosa, Sequoia sempervirens*
palo colorado chico – *Trichilia hirta*
palo copado – *Quararibea funebris*
palo copal – *Bursera stenophylla*
palocote – *Tithonia tubaeformis*
palo cuate – *Eysenhardtia polystachya*
palo cumbro – *Celtis reticulata*
palo chilamate – *Ficus petiolaris*
palo chino – *Acacia gregii, A. peninsularis, Bursera simaruba, Erythroxylum mexicanum, Havardia mexicana, Heliocarpus polyandrus*
palo de aceite – *Copaifera officinalis*
palo de Adán – *Fouquieria burragei, F. diguetii, F. splendens*
palo de agua – *Astianthus viminalis, Gilibertia arborea, Liabum glabrum, Pachira aquatica, Salix taxifolia*
palo de aguacate – *Nectandra sinuata*
palo de águila – *Alnus acuminata, A. arguta, Celtis pallida, Laelia superbiens*
palo de arco – *Acacia coulteri, A. roemeriana, Apoplanesia paniculata, Celtis iguana, Lysiloma divaricata, Tabebuia pentaphylla, Tecoma stans*
palo de arco negro – *Apoplanesia paniculata*
palo de asta – *Cordia sebestena, C. sonorae*
palo de azúcar – *Acer brachypterum, A. saccharum* ssp *grandidentatum*
palo de bálsamo – *Myroxylon balsamum* var *pereirae*
palo de baqueta – *Ulmus mexicana*
palo de barranco – *Carpinus caroliniana*
palo de berria – *Cercidium praecox*
palo de boda – *Vauquelinia karwinskyi*
palo de boya – *Pachira aquatica*
palo de Brasil – *Caesalpinia echinata, Dracaena fragrans, Haematoxylon brasiletto, H. campechana*
palo de brujo – *Xylosma flexuosum*
palo de buba – *Jacaranda acutifolia*
palo de burro – *Capparis flexuosa*
palo de cacique – *Magnolia grandiflora, M. schiedeana*
palo de Campeche – *Haematoxylon brasiletto, H. campechianum*
palo de canela – *Chrysophyllum cainito, C. mexicanum*
palo de carnero – *Cocoloba barbadensis, Licania arborea*
palo de cera – *Astronium graveolens*
palo de coco – *Ficus padifolia*
palo de coleto – *Oreopanax peltatum*
palo de corcho – *Annona glabra*

palo de corral – *Gliricidia sepium*
palo de cruces – *Randia*
palo de Cuba – *Maclura tinctoria*
palo de cuchara – *Bursera palmeri, Clethra mexicana, Cochlospermum vitifolium*
palo de cucharitas – *Acacia cymbispina*
palo de cuentas – *Sapindus saponaria*
palo de cuesa – *Bernoullia flammea*
palo de culebra – *Astronium graveolens*
palo de chachalaca – *Solanum verbascifolium*
palo de chicalpexte – *Hernandia sonorae*
palo de Chile – *Drimys winteri*
palo de chupandía – *Cyrtocarpa procera*
palo de danta – *Gilibertia arborea, Oreopanax peltatum*
palo de fraile – *Couepia polyandra, Licania arborea*
palo de gallina – *Malpighia glabra*
palo de gas – *Amyris balsamifera*
palo de goma – *Ficus elastica*
palo de granadillo morado – *Dalbergia granadillo*
palo de hamaca – *Schoepfia schreberi*
palo de hierro – see palo fierro
palo de hojas de manteca – *Solanum verbascifolium*
palo de hormiga – *Alvaradoa amorphoides, Cordia alliodora*
palo de hormigo – *Platymiscium dimorphandrum*
palo de huacal – *Crescentia cujete*
palo de hule – *Castilla elastica, Ficus elastica*
palo de humo – *Phoebe tampicensis*
palo de jabón – *Quillaja saponaria*
palo de jarro – *Alibertia edulis*
palo de Judas – *Bocconia arborea*
palo de la cruz – *Randia*
palo de la felicidad – *Dracaena fragrans*
palo de la flecha – *Sapium appendiculatum, S. biloculare, Sebastiana pavoniana*
palo de lana – *Bombax ellipticum*
palo de las ánimas – *Jacquinia aurantica*
palo del diablo – *Bocconia arborea*
palo de leche – *Sapium stylare, Sebastiania pavoniana*
palo del golpe – *Agonandra racemosa*
palo del muerto – *Cordia boissieri, Galphimia glauca, Ipomoea arborescens, I. murucoides*
palo de lumbre – *Malpighia glabra*
palo de maya – *Senna atomaria*
palo de mecate – *Chiranthodendron pentadactylon*
palo de membrillo – *Cornus excelsa*
palo de mora – *Maclura tinctoria*
palo de Oaxaca – *Dahlia excelsa, D. imperialis*
palo de orquídeas – *Bauhinia monandra, B. variegata*
palo de ozote – *Ipomoea murucoides*
palo de pajulul – *Rhus schiedeana*
palo de pan – *Artocarpus altilis, A. incisa*
palo de papaxi – *Nyssa sylvatica*
palo de patlaches – *Lonchocarpus longistylus*
palo de perdiz – *Bernoullia flammea*
palo de pingüica – *Arctostaphylos pungens*
palo de pua – *Cercidium floridum, C. penninsulare*
palo de puerco espín – *Pereskia aculeata, P. grandifolia, P. lycnidiflora*
palo de pulque – *Acacia angustissima*
palo de quina – *Coutarea hexandra*
palo de rayo – *Parkinsonia aculeata*
palo de rosa – *Amyris balsamifera, Cochlospermum vitifolium, Cordia alliodora, Guaiacum officinale, Tabebuia pentaphylla, T. rosea*
palo de rosa amarilla – *Cochlospermum vitifolium*
palo de sal – *Avicennia germinans*
palo de San Antonio – *Thevetia peruviana*
palo de sangre – *Haematoxylum campechianum, Pterocarpus officinalis*
palo de San Juan – *Forchammeria watsonii, Jacquinia macrocarpa* ssp *pungens, Nicotiana glauca*
palo de San Nicolás – *Astianthus viminalis, Liabum glabrum*
palo de San Pablo – *Wigandia caracasana*
palo de San Vicente – *Galphimia glauca*
palo de sombra – *Gliricidium sepium*
palo de tachicón – *Tetracera volubilis*
palo de taruga – *Bernardia myricifolia*
palo de tayuyo – *Chiranthodendron pentadactylon*
palo de tejón – *Nectandra sinuata*
palo de tenaza – *Pithecellobium brevifolium*
palo de tinta – *Calatola mollis, Haematoxylon brasiletto, H. campechianum, Olneya tesota*
palo de tomate – *Cyphomandra betacea*

palo de tortilla – *Bernoullia flammea*
palo de totole – *Conzattia multiflora*
palo de trapiche – *Myroxylon balsamum* var *pereirae*
palo de vaca – *Brosimum utile*
palo de violin – *Cecropia obtusifolia*
palo de voladillo – *Sapindus saponaria*
palo de yugo – *Tabebuia pentaphylla*
palo de zanate – *Piper sanctum*
palo de zopilote – *Swietenia humilis, S. macrophylla*
palo dulce – *Coursetia glandulosa, Eysenhardtia orthocarpa, E. polystachya, E. texana, Glycyrrhiza glabra, G. lepidota*
palo dulce blanco – *Eysenhardtia polystachya*
palo duro – *Celtis reticulata, Cercocarpus betuloides, C. montanus, Quercus fusiformis*
paloduz – *Glycyrrhiza glabra*
palo éban – *Chloroleucon mangense* var *leucospermum*
palo ébano – *Pithecellobium undulatum*
palo escopeta – *Albizzia occidentalis*
palo estaca – *Caesalpinia pannosa*
palo estribo – *Cercidium* x *sonorae*
palo extranjero – *Nicotiana glauca*
palo extraño – *Ilex tolucana*
palo fierro – *Albizzia occidentalis, Casuarina equisetifolia, Chloroleucon namgense* var *leucospermum, Olneya tesota, Pithecellobium confine, P. undulatum, Prosopis palmeri, Schoepfia schreberi*
palo fraile – *Couepia polyandra*
palo freno – *Caesalpinia sclerocarpa*
palo garrobo – *Acacia pennatula*
palo gato – *Havardia sonorae*
palo guayacán – *Guaiacum sanctum*
palo güemuchi – *Casearia dolichophylla*
palo hediondo – *Atamisquea emarginata, Capparis atamisquea, Cestrum dumentorum, Senna atomaria, S. bicapsularis*
palo hinchador – *Sapium appendiculatum*
palo huácaro – *Pithecellobium brevifolium*
palo huaco – *Cissus rhombifolia*
palo hueco – *Cissus rhombifolia*
palohuisi – *Asclepias glaucescens, Solanum amazonium*
palo jiote – *Bursera simaruba*
palo jito – *Forchammeria watsonii*
palo jócono – *Havardia sonorae*
palo joso – *Albizzia sinaloënsis, Conzattia multiflora, C. sericea*
palo lechero – *Sebastiania pavoniana*
palo levender – *Eucrypta chrysanthemifolia*
palo limón – *Amyris balsamifera*
palo liso – *Acacia willardiana, Carpinus caroliniana, Rhamnus sharpii*
palo loco – *Nicotiana glauca, Senecio praecox*
palo loro – *Prunus lusitanica*
palo ludo – *Cercocarpus betuloides*
paloma – *Aquilegia vulgaris, Cestrum anagyris*
paloma consulta – *Thalictrum fendleri, T. hernandezii*
paloma de mar – *Lavatera maritima*
palo mantecoso – *Cercidium praecox*
palo María – *Achras zapota, Calophyllum braziliense, C. calaba, Cordia alliodora, Ficus petiolaris, Triplaris americana*
palo marimba – *Platymiscium dimorphandrum*
palo matagallina – *Apoplanesia paniculata*
palomera – *Cerinthe major*
palomilto – *Dodonaea viscosa*
palomilla – *Alkanna tinctoria, Aquilegia vulgaris, Fumaria, Tropaeolum majus*
palomilla común – *Fumaria officinalis*
palomilla de tintes – *Alkanna tinctoria*
palomilla pintada – *Fumaria capreolata*
palomino – *Vitis vinifera* var *sativa*
palo mirto – *Cordia sonorae*
palomita – *Aquilegia, Distictis buccinatoria, Hedychium coronarium, Pithecellobium echinatum, Tricholoma flavovirens, Zea mays* var *everia*
palo moral – *Maclura tinctoria, Morus celtidifolia*
palo muela – *Coutarea pterosperma, Croton draco*
palo mulato – *Bursera, Celtis reticulata, Zanthoxylum fagara*
palo mulato de Chiapas – *Bursera simaruba*
palo nahuila – *Diphysa occidentalis*
palo negrito – *Karwinskia humboldtiana*
palo negro – *Adiantum tenerum, Garrya laurifolia*
palo nesco – *Lonchocarpus hermannii*
palo palomo – *Capparis incana*
palo picante – *Drimys winteri*
palo pinto – *Chloroleucon mangense* var *leucospermum*

palo piojo – *Alvaradoa amorphoides, Brongniartia alamosana, B. palmeri, Caesalpinia palmeri, Lonchocarpus hermannii*
palo piojo blanco – *Bauhinia pringlei, Caesalpinia caladenia*
palo pique – *Pinus strobus*
palo prieto – *Celaenodendron mexicanum, Cordia greggi, C. parvifolia, C. sonorae, Senna wislizeni, Vauquelinia corymbosa* var *angustifolia*
palo retinto – *Bursera simaruba*
palo rojo – *Haematoxylon brasiletto*
palo roñoso – *Fraxinus greggii*
palo santo – *Acer brachypterum, Aralia humilis, Bocconia frutescens, Colubrina viridis, Diospyros kaki, Diphysa suberosa, Fouquieria formosa, Gilibertia arborea, Guaiacum coulteri, G. officinale, G. sanctum, Ipomoea arborescens, I. murucoides, Platymiscium trifoliolatum, Triplaris americana*
palo santo amarillo – *Ipomoea arborescens*
palo santo de Castilla – *Ipomoea fistulosa*
palo sapán – *Caesalpinia sappan*
palo sapo – *Capparis indica, C. verrucosa*
palo silo – *Carpinus caroliniana*
palo sino – *Carpinus caroliniana*
palote negro – *Guazuma ulmifolia*
palo todos santos – *Senna spectabilis*
palo torcido – *Alvaradoa amorphoides*
palo triste – *Psorothamnus spinosus*
palo verde – *Agonandra racemosa, Canotia holacantha, Cercidium, Drypetes gentryi, Fouquieria macdougalii, Parkinsonia aculeate, Pilea microphylla, Schoepfia shreveana, Vallesia glabra, Vauquelinia corymbosa* var *angustifolia*
palo verde azul – *Cercidium floridum*
palo verde chino – *Cercidium floridum*
palo viga – *Cordia alliodora*
palo villa – *Hura polyandra*
palo virgen – *Nicotiana glauca*
palo zorrillo – *Atamisquea emarginata, Capparis atamisquea, Senna*
palpanichim – *Brugmansia arborea, B. candida*
palquis – *Cestrum elegans*
palto – *Persea americana*
palvadán – *Fouquieria diguetti*
palla – *Cyperus laevigatus*
pallar – *Phaseolus lunatus*
pambazo – *Boletus edulis, B. pinicola, B. reticulates, Leccinum aurantiacum*
pambotano – *Calliandra houstoniana, Zapoteca anomala*
pamita – *Brassica nigra, Descuriana pinnata, Sisymbrium, Sophia ochroleuca*
pamita cimarrón – *Lepidium eastwoodiae*
pamitón – *Descuriana pinnata, Sisymbrium irio*
pampajarito – *Sedum acre*
pamplemusa – *Citrus grandis, C. paradisi*
pamplina – *Samolus parviflorus, Stellaria media*
pamplina de canarios – *Stellaria media*
pamplina falsa espinosa – *Arenaria montana*
pamporcino – *Cyclamen purpurascens*
pana – *Artocarpus altilis*
pánace – *Opopanax chironium*
pana cimarrona – *Artocarpus altilis, A. integrifolia*
panaco – *Cochlospermum vitifolium*
pana de pepita – *Artocarpus altilis*
panadero – *Boletus reticulatus*
panadero de encino – *Boletus edulis*
panal – *Lobularia maritima*
panalero – *Forestiera angustifolia, Schaefferia cuneifolia*
panalillo – *Esenbeckia hartmanii, Lepidium, Lobularia maritima*
panalito – *Lobularia maritima*
panamá – *Sterculia carthaginensis*
panamiga – *Pilea involucrata*
panamito – *Phaseolus vulgaris*
panán-gsuni – *Arbutus glandulosa*
panapén – *Artocarpus altilis*
panaripa – *Tauschia arguta*
pan caliente – *Eragrostis hypnoides, E. reptans*
pancita – *Boletus edulis, Morchella esculenta*
pancita azul – *Boletus luridus*
pancita blanca – *Boletus edulis, B. pinicola*
pancracio – *Pancratium maritimum*
pancracio marítimo – *Pancratium maritimum*
pándano – *Pandanus utilis*
pan de cuco – *Oxalis corniculata, Sedum spathulifolium, S. spectabile*
pan de cuco común – *Sedum acre*

pan de cuchillo – *Oxalis acetosella*
pan de fruta – *Artocarpus integrifolia*
pan de manteca – *Nymphaea alba, Nymphoides humboldtianum*
pan de nuez – *Brosimum alicastrum*
pan de pepita – *Artocarpus altilis*
pan de puerco – *Cyclamen persicum, C. purpurascens*
pan de ranas – *Alisma plantago-aquatica*
pan de tierra caliente – *Manihot esculenta*
pandureca – *Ficus lyrata*
pandurita – *Ficus lyrata*
pangelín – *Andira inermis*
pangola – *Digitaria decumbens*
pangolote – *Cochlospermum vitifolium*
paniagua – *Capparis odoratissima*
pánico azul – *Panicum antidotale*
panigua – *Cochlospermum vitifolium*
panizo – *Panicum*
panizo americano – *Zea mays*
panizo aserrín – *Panicum hallii*
panizo azul – *Panicum antidotale*
panizo cauchín – *Panicum hirticaule*
panizo corcolito – *Panicum filipes*
panizo de Daimiel – *Pennisetum americanum*
panizo de Guinea – *Panicum maximum*
panizo de hojas anchas – *Dichanthelium oligosanthes* var *scribnerianum*
panizo de Indias – *Zea mays*
panizo de Italia – *Setaria italica*
panizo fasciculado – *Brachiaria fasciculata*
panizo gigante – *Panicum antidotale*
panizo mezquite – *Panicum obtusum*
panizo moruno – *Sorghum bicolor*
panizo negro – *Pennisetum glaucum, Sorghum bicolor*
panizo redondo – *Dichanthelium sphaerocarpon*
panizo rizado – *Panicum hallii*
panizo setáceo – *Pennisetum setaceum*
panizo texano – *Brachiaria texana*
panizo verde – *Panicum maximum*
panizo villosum – *Pennisetum villosum*
panjí – *Melia azedarach*
panochillo – *Citharexylum berlandieri*
pantani – *Musa sapientum*
pantzil – *Suriana maritima*
panuco – *Cochlospermum vitifolium*
panul – *Apium graveolens*
panxóchitl – *Lobelia laxiflora*
pan y quesillo – *Capsella bursa-pastoris, Robinia pseudoacacia*
panza – *Leccinum aurantiacum, Lycoperdon perlatum*
panza de culebra – *Dichaea muricata*
panza roja – *Boletus regius*
pañal de niño – *Tillandsia usneoides*
pañete – *Hamelia erecta, Plumbago pulchella, P. scandens*
paño de Holanda – *Euphorbia pulcherrima*
paño de oro – *Crocus angustifolius*
paño holandés – *Euphorbia pulcherrima*
pañuelo de Sarah – *Petrea volubilis*
papa – *Solanum tuberosum, S. demissum, S. cardiophyllum*
papa caribe – *Dioscorea bulbifera*
papa cimarrona – *Solanum bulbocastanum, S. demissum, S. tuberosum*
papacla – *Heliconia bihai*
papaconi – *Gnaphalium canescens*
papa correlona – *Solanum tuberosum*
papa coyote – *Solanum bulbocastanum*
papache – *Condalia globosa, Randia*
papache borracho – *Randia echinocarpa, R. obcordata, R. rosei*
papache de la sierra – *Randia watsoni*
papache picudo – *Randia echinocarpa*
papachilla – *Elytraria squamosa, Randia*
papada de gallo – *Exogonium bracteatum*
papa de agua – *Sagittaria macrophylla*
papa del aire – *Dioscorea bulbifera*
papa dulce – *Ipomoea batatas*
papagayo – *Amaranthus gangeticus, Caladium bicolor, Carica papaya, Euphorbia pulcherrima*
papagayo del Brasil – *Caladium bicolor*
papa inglesa – *Oxalis tuberosa*
papalacate – *Eichhornia crassipes*
papalocuáhuitl – *Bauhinia divaricata*
papalometl – *Agave marmorata, A. potatorum*
pápalo quelite – *Porophyllum gracile, P. macrocephalum*
papalote – *Ulmus*
papalotilla – *Guardiola platyphylla*
papamiel – *Combretum farinosum*
papanaca – *Annona purpurea*
papa real – *Helianthus tuberosus*
papatla – *Heliconia schiedeana*

papatlilla – *Heliconia schiedeana*
papaturro – *Cocoloba uvifera*
papausa – *Annona diversifolia*
papa voladora – *Dioscorea bulbifera*
papaya de árbol – *Carica papaya*
papayero – *Carica papaya*
papayito – *Carica cauliflora*
papayo – *Asimina triloba, Carica candamarcensis, C. papaya*
papayo cimarrón – *Oreopanax peltatum*
papayo del pájaro – *Carica papaya*
papayo montés – *Carica mexicana*
papayo oreja de mico – *Carica cauliflora*
papayo orejón – *Carica mexicana*
papayo real – *Carica papaya*
papayote – *Paspalum conjugatum*
papel de lija – *Petrea arborea*
papel de musica – *Calathea zebrina*
papelero – *Broussonetia papyrifera*
papelillo – *Bursera odorata, B. simaruba, Jatropha cordata, Trichilia hirta, Trophis racemosa, Wimmeria confusa, W. mexicana*
papelillo brillante – *Melica nitens*
papelillo dulce – *Melica montezumae*
papiotl – *Ustilago maydis*
papiro – *Cyperus papyrus, Epidendrum pastoris*
papiro de Egipto – *Cyperus papyrus*
papirus de Egipto – *Cyperus papyrus*
papita – *Solanum demissum*
papito – *Clitoria terneata*
papito de monte – *Malvaviscus arboreus*
papta – *Acoelorraphe wrightii*
papuano – *Paspalum conjugatum*
pará – *Hevea brasiliensis*
paracata – *Oncidium tigrinum*
paradís – *Malus pumila* var *paradisiacal*
paraguas – *Cyperus alternifolius*
paraguas chino – *Amorphophalus rivieri, Melia azedarach* var *umbraculiformis*
paraíso – *Eleagnus angustifolius, Melia azedarach* var *umbraculiformis, Moringa oleifera*
paraíso blanco – *Moringa oleifera*
paraíso chino – *Melia azedarach*
paraíso de España – *Moringa oleifera*
paraíso morado – *Melia azedarach* var *umbraculiformis*
paramo – *Drimys winteri*
paraná – *Brachiaria mutica*
parandesicua – *Guazuma ulmifolia*
parapara – *Sapindus saponaria*
parasol – *Macrolepiota procera*
parcha – *Passiflora edulis, P. incarnata, P. quadrangularis*
parchita amarilla – *Passiflora edulis*
pardal – *Aconitum napellus*
pardillo amarillo – *Terminalia catappa*
pardito – *Agave sobria*
pareira brava – *Cissampelos pareira*
parietaria – *Parietaria officinalis, P. pensylvanica*
parietaria común – *Parietaria officinalis*
parietaria ramiflora – *Parietaria judaica*
parkinsonia – *Parkinsonia aculeata*
parnasia de los pantanos – *Parnassia californica, P. palustrus*
paroniqua espinosa – *Paronychia echinulata*
paropus – *Solanum tridynamum*
paroqui – *Ericameria sonorensis*
parota – *Enterolobium cyclocarpum*
parote – *Enterolobium cyclocarpum*
parqui – *Cestrum nocturnum*
parsimonio – *Diospyros kaki*
parra – *Vitis arizonica, V. tiliifolia, V. vinifera*
parra brincadora – *Vitis tiliifolia*
parra bronca – *Vitis tiliifolia*
parra broncadora – *Vitis tiliifolia*
parra cimarrona – *Vitis arizonica, V. californica*
parra de monte – *Cissus sicyoides*
parraleña – *Thymophylla pentachaeta, Zinnia pumila*
parra silvestre – *Cissus sicyoides, Vitis cinerea, V. tiliifolia*
parriza – *Vitis labrusca*
parriza americana – *Vitis labrusca*
parrotia de Persia – *Parrotia persica*
pasajera – *Tradescantia virginiana*
pasaji – *Saracha jaltomata*
pasaque – *Simaruba glauca*
pascua – *Euphorbia pulcherrima, Turbina corymbosa*
paseo – *Avicennia germinans*
pash – *Prunus persica*
pashma sacat – *Paspalum conjugatum*
pasiflora – *Passiflora*
pasiflorina – *Passiflora*
pasilla – *Capsicum annuum*

pasilla blanca – *Cornus disciflora*
pasilla roja – *Capsicum annuum*
pasillo – *Maytenus phyllanthoides*
pasionaria – *Passiflora*
pasión de Cristro – *Passiflora*
pasita – *Conostegia xalapensis*
pasmore – *Ericameria palmeri*
paspalum de semilla parda – *Paspalum plicatulum*
pastel – *Tricholoma sejunctum*
pastillo de invierno – *Poa annua*
pastillo del pinar – *Blepharoneuron tricholepis*
pastinaca – *Pastinaca sativa*
pastito de agua – *Leersia hexandra*
pastito de invierno – *Poa annua*
pasto – *Poaceae*
pasto admirable – *Brachiaria mutica*
pasto africano – *Eragrostis lehmanniana, Pennisetum clandestinum*
pasto alemán – *Echinochloa crus-galli, E. polystachya*
pasto alfombra – *Axonopus compressus, Stenotaphrum secundatum*
pasto alicia – *Cynodon dactylon*
pasto amargo – *Paspalum conjugatum*
pasto avena – *Eragrostis superba*
pasto azulado – *Panicum antidotale, P. virgatum*
pasto azul anual – *Poa annua*
pasto azul de Kentucky – *Poa pratensis*
pasto azul precoz – *Poa bigelovii*
pasto bahia – *Paspalum notatum*
pasto banderilla – *Bouteloua curtipendula*
pasto blanco – *Bothriochloa saccharoides, Digitaria sanguinalis*
pasto boer – *Eragrostis curvula* var *conferta*
pasto brasilero – *Phalaris arundinacea*
pasto búfalo – *Panicum coloratum*
pasto búffel – *Cenchrus ciliaris*
pasto canario – *Phalaris arundinacea*
pasto candelaria – *Pennisetum purpureum*
pasto caribe – *Eriochloa polystachya*
pasto carpeta – *Axonopus scoparius*
pasto carpeta de hojas angostas – *Axonopus affinis, A. compressus*
pasto castilla – *Panicum maximum*
pasto catinguero – *Melinis multiflora*
pasto cinto – *Phalaris arundinacea*
pasto colón – *Sorghum almum*
pasto colorado – *Panicum coloratum*
pasto chato – *Axonopus compressus, A. scoparius, Paspalum compressum*
pasto chato común – *Axonopus affinis*
pasto chato de alfombra – *Axonopus compressus*
pasto chino – *Cathestecum erectum*
pasto dáctilo – *Dactyloctenium aegyptium*
pasto dallis – *Paspalum dilatatum*
pasto de agua – *Leersia hexandra*
pasto de Angola – *Echinochloa polystachya*
pasto de Australia – *Paspalum dilitatum*
pasto de Bermuda – *Cynodon dactylon*
pasto de cabra – *Bouteloua aristidoides*
pasto de ciudad – *Pennisetum clandestinum*
pasto de conejo – *Opizia stolonifera*
pasto de encino – *Phoradendron macrophyllum, P. schumannii*
pasto de gallina – *Digitaria ciliaris*
pasto de Guatemala – *Tripsacum laxum*
pasto de la China – *Boehmeria nivea*
pasto de laguna – *Brachiaria mutica*
pasto del arroz – *Echinochloa oryzoides*
pasto de limón – *Cymbopogon citratus*
pasto del Senegal – *Rhynchelytrum repens*
pasto del Sudán – *Sorghum bicolor, S. sudanense*
pasto de madroño – *Phoradendron densum, P. pauciflorum*
pasto de tortuga – *Thalassia testudinum*
pasto dulce – *Eleusine indica, Paspalum dilitatum, P. distichum, Sporobolus indicus*
pasto elefante – *Pennisetum purpureum*
pasto estrella – *Cynodon dactylon, C. plectostachyus*
pasto estrella gigante – *Cynodon plectostachyus*
pasto ferrer – *Cynodon dactylon*
pasto fuerte – *Pennisetum setaceum*
pasto gigante – *Leptochloa dubia, Pennisetum purpureum*
pasto gordura – *Melinis minutiflora*
pasto guinea – *Panicum maximum*
pasto guineo – *Panicum maximum*
pasto hardin – *Phalaris tuberosa*
pasto hediondo – *Eragrostis cilianensis*
pasto hilo – *Paspalum conjugatum*
pasto horqueta – *Paspalum conjugatum, P. notatum*

pasto horquilla – *Paspalum conjugatum*
pasto imperial – *Axonopus scoparius*
pasto indio – *Panicum coloratum*
pasto inglés – *Lolium perenne*
pasto italiano – *Lolium multiflorum*
pasto jaragua – *Hyparrhenia rufa*
pasto jigüite – *Eragrostis obtusiflora*
pasto johnson – *Sorghum halepense*
pasto kentucky – *Poa pratensis*
pasto kikuyo – *Pennisetum clandestinum*
pasto klein – *Panicum coloratum*
pasto kudzu – *Pueraria phaseoloides*
pasto limón – *Cymbopogon citratus*
pasto listón – *Chlorophytum comosum*
pasto llorón – *Eragrostis curvula*
pasto maicillo – *Sorghum vulgare*
pasto malojillo – *Brachiaria mutica*
pasto melado – *Melinus minutiflora*
pasto melaza – *Melinus minutiflora*
pasto melinida – *Melinus minutiflora*
pasto meloso – *Paspalum dilitatum*
pasto merkerón – *Pennisetum purpureum*
pasto miel – *Holcus lanatus*
pasto mijillo – *Echinochloa crus-galli*
pasto nandi – *Setaria sphacelata*
pasto napier – *Pennisetum purpureum*
pasto natural – *Axonopus compressus*
pasto olivo – *Dactylis glomerata*
pasto ovillo – *Dactylis glomerata*
pasto palojilla – *Eriochloa polystachya*
pasto panamá – *Saccharum sinense*
pasto pangola – *Digitaria decumbens*
pasto para – *Panicum molle*
pasto poa – *Poa pratensis*
pasto puna – *Acantherum brachychaetum*
pasto puntero – *Hyparrhenia rufa*
pastora – *Euphorbia pulcherrima, Plantago, Tagetes patula, Triumfetta*
pastoral – *Tagetes patula*
pastorcilla – *Tagetes patula*
pastorcillo – *Lobelia laxiflora*
pastorcita – *Tagetes patula, Turnera diffusa*
pasto rhodes – *Chloris gayana*
pasto romano – *Phalaris minor*
pasto ruso – *Sorghum halepense*
pasto sabaya – *Panicum coloratum, P. maximum*
pasto San Agustín – *Stenotaphrum secundatum*
pasto senegalés – *Rhynchelytrum repens*
pasto sombra – *Oplismenus setarius*
pasto timoty – *Phleum pratense*
pasto torpedo – *Panicum repens*
pasto vasey – *Paspalum urvillei*
pasto violeta – *Vetiveria zizanioides*
patabán – *Laguncularia racemosa*
pataca – *Helianthus tuberosus*
patachtli – *Theobroma bicolor*
pata de asno – *Tussilago farfara*
pata de cabra – *Bauhinia divaricata, B. variegata*
pata de cuervo – *Bouteloua barbata, Brassica rapa, Dactyloctenium aegyptium, Lepidium virginicum, Pteridium aquilinum*
pata de chivo – *Bauhinia divaricata*
pata de elefante – *Bursera microphylla*
pata de gallina – *Chloris, Digitaria sanguinalis, Eleusine indica*
pata de gallo – *Cynodon dactylon, Dactyloctenium aegyptium, Digitaria ciliaris, Dorstenia contrayerba, Echinochloa crus-galli, Eleusine indica, Epidendrum parkinsonianum, Manihot, Paspalum conjugatum, Sprekelia formosissima*
pata de gallo mexicano – *Chloris submutica*
pata de ganso – *Eleusine indica*
pata de grulla – *Dactyloctenium aegyptium*
pata de león – *Delonix regia, Geranium, Pereskia lycnidiflora, Pyrrhopappus multicaulis, Ranunculus*
pata de liebre – *Dactylis glomerata*
pata de mula – *Agave tequilana, Tussilago farfara*
pata de pájaro – *Erythroxylum mexicanum, Lotus corniculatus, L. scoparius, Ramaria stricta*
pata de paloma – *Epidendrum parkinsonianum*
pata de perdíz – *Cynodon dactylon*
pata de pollo – *Cynodon dactylon, Dactylis glomerata, Dactyloctenium aegyptium*
pata de rana – *Selaginella rupestris*
pata de res – *Bauhinia divaricata*
pata de toro – *Tussilago farfara*
pata de vaca – *Bauhinia divaricata, B. variegata, Cercis canadensis, Cleome hasslerana, Ipomoea pes-caprae*
pata de venado – *Ariocarpus kotschoubeyanus, Bauhinia forficata*
patalolote – *Eugenia pimenta*

patamba – *Bambusa vulgaris*
patamu – *Arundo donax*
patana – *Jessenia bataua*
patancán – *Ipomoea arborescens*
pataste – *Sechium edule, Strychnos panamensis, Theobroma bicolor*
pataste de caballo – *Strychnos panamensis*
pataste de mico – *Strychnos panamensis*
patata – *Solanum tuberosum, S. demissum, S. cardiophyllum, Monolepis nuttalliana*
patata de caña – *Helianthus tuberosus*
patata de los huasas – *Coleus rotundifolius*
patatera – *Solanum tuberosum*
pate – *Smilax mollis*
patilla – *Citrullus colocynthis*
patillón – *Pereskia grandifolia, P. lychnidiflora, Pereskiopsis diguetii, P. rotundifolia*
patita de burro – *Iris sisyrinchium*
patita de león – *Geranium caespitosum*
patita de pájaro – *Ramaria flava*
patito – *Acacia greggii, Achimenes grandiflora, Aristolochia taliscana, Astragalus, Carnavalia villosa, Heteranthera limosa*
patlahoac – *Bursera odorata*
patlahoachoixachin – *Caesalpinia cacalaco*
pato – *Aristolochia grandiflora*
patol – *Erythrina, Phaseolus coccineus, Sophora secundiflora*
patol colorín – *Erythrina*
patole – *Trichosanthes cucumerina* var *angoina, T. dioica*
patolote – *Eugenia pimenta*
patolli – *Erythrina*
patota – *Monolepis nuttalliana*
patusai – *Chenopodium ambrosioides*
patzueni – *Tillandsia usneoides*
patzueni tacari – *Tillandsia usneoides*
paulina – *Cupania americana*
paulilla – *Rauwolfia hirsuta*
paulonia – *Paulownia tomentosa*
paulonia imperial – *Paulownia tomentosa*
paulla – *Clerodendron thomsoniae*
pavía – *Prunus persica*
pavía roja – *Aesculus pavia*
pavil – *Stillingia*
pavito cimarrón – *Astragalus humboldtii*
pavona encendida – *Hibiscus cannabinus*
paxtle – *Tillandsia recurvata, T. usneoides*
paxtle de mezquite – *Tillandsia recurvata*
payasito – *Acalypha wilkesiana*
pebete – *Mirabilis longiflora*
pecanero – *Carya*
péchita – *Prosopis juliflora*
pedo de coyote – *Lycoperdon perlatum*
pedo de culebra – *Asclepias curassavica*
pedo del diablo – *Lycoperdon*
pedo de lobo – *Lycoperdon perlatum, L. umbrinum*
pedo de venado – *Calvatia gigantea*
pedra hume caa – *Myrica salicifolia*
peem – *Ceiba pentandra*
pegadillo – *Arctium lappa*
pega hueso – *Agonandra racemosa*
pegajosa – *Anulocaulis eriosolenus, Bonplandia geminiflora, Chiococea alba, Galium mexicanum, G. nuttallii, Ipomoea stans, Mentzelia, Mirabilis jalapa, Plumbago scandens, Selloa glutinosa, Suillus luteus*
pegamo – *Cissampelos pareira*
pegamosca – *Cuphea aequipetala*
pegapalo – *Rhipsalis baccifera*
pega pega – *Desmodium canum, D. styracifolium, Eucnide aurea, E. cordata, Galium nuttallii, Mentzelia, Triumfetia semitriloba*
pega pinto – *Boerhavia hirsuta*
pegarropa – *Acaean elongata, Desmodium canum, Encelia mexicana, Galium mexicanum, G. trifidum, Mentzelia aspera, M. hispida, M. multiflora*
pegarropa amarillo – *Mentzelia hispida*
pegote – *Xanthium spinosum*
pehuén – *Araucaria araucana*
peinecillo – *Agave striata, Agonandra racemosa, Alvaradoa amorphoides, Combretum farinosum, C. fruticosum*
peine de bruja – *Erodium cicutarium*
peine del diablo – *Combretum farinosum*
peine de mico – *Combretum farinosum*
peine de Venus – *Scandix pectin-veneris*
peineta – *Combretum farinosum, C. fruticosum*
peinetillo – *Combretum farinosum*
pejibaye – *Bactris gasipaes, Guilielma gasipaes*
pejiguera – *Polygonum persicaria*
pelagallina – *Kallstroemia peninsularis*
pelambrera de Florida – *Thrinax rediata*

pelampacate – *Ceanothus cuneatus*
pelargonio – *Pelargonium* x *domesticum*
pelícano – *Aquilegia vulgaris*
pelillo – *Bouteloua filiformis, Leersia hexandra*
pelitre – *Anacyclus depressus, Chrysanthemum cinerariifolium, Erigeron*
pelitre de Dalmacia – *Chrysanthemum coronarium*
pelitre del país – *Erigeron*
pelitre extranjero – *Anacyclus depressus*
pelma – *Alsophila salvinii, Cyathea delicatula, Lophosoria quadripinnata*
pelo de ángel – *Calliandra eriophylla*
pelo de gato – *Asclepias curassavica*
pelón – *Prunus persica* ssp *nucipersica, Tropaeolum majus*
peloncuáhuitl – *Schinus molle*
pelonchili – *Tropaeolum majus*
pelonmexixquilitl – *Tropaeolum majus*
pelo rojo – *Colubrina viridis*
pelosilla – *Parietaria officinalis, Taraxacum officinale*
pelotazo – *Abutilon*
pelotazo bronco – *Abutilon lignosum*
pelotazo chico – *Abutilon incanum*
pelotazo peludo – *Abutilon umbellatum*
peluda – *Cryptantha angustifolia, C. maritima*
peludillo – *Turnera diffusa, T. ulmifolia*
peludita – *Cheilanthes leucopoda, Chilopsis linearis* ssp *arcuata, Cryptantha maritima*
pelul – *Passiflora ligularis*
pelusilla – *Arabis alpina, Cottea pappophoroides*
pelusita – *Iresine calea*
pellejo de viejo – *Trema micrantha*
penacho – *Celosia argentea*
penadero de encino – *Boletus regius*
penadero de oyamel – *Boletus regius*
penicilina – *Penicillium notatum*
penitente – *Petrea arborea*
pennisetum híbrido – *Saccharum sinense*
pensamiento – *Viola*
pensamiento silvestre – *Viola arvensis, V. tricolor*
pénstemon – *Penstemon lanceolatus*
pentamón – *Taxodium distichum* var *mexicanum*
pentamú – *Taxodium distichum* var *mexicanum*
peonía – *Abrus precatorius, Acourtia runcinata, Berlandiera lyrata, Conostegia xalapensis, Cyperus esculentus, Dolicholus phaseoloides, Erythrina, Lantana involucrata, Lasianthaea podocephala, Paeonia, Rhynchosia pyramidalis, Samolus parviflorus, Zexmenia aurea, Z. podocephala*
peonía arbórea – *Paeonia suffructicosa*
peonía arborescente – *Paeonia suffructicosa*
peonía arbustiva – *Paeonia suffruticosa*
peonía colorada – *Lantana involucrata*
peonía común – *Paeonia officinalis*
peonía china – *Paeonia lactiflora*
peonía de botica – *Cyperus esculentus*
peonía de brotero – *Paeonia broteroi*
peonía de California – *Paeonia californica*
peonía de junio – *Paeonia lactiflora*
peonía del país – *Cyperus esculentus*
peonía de mayo – *Paeonia officinalis*
peonía de niños – *Dolicholus phaseoloides*
peonía leñosa – *Paeonia suffruticosa*
peonía negra – *Lantana camara*
peonía oficinal – *Paeonia officinalis*
peonía sufruticosa – *Paeonia suffruticosa*
pepebalsa – *Ochroma lagopus*
pepeyoca – *Populus fremontii, P. mexicana*
pepeyócatl – *Populus fremontii*
pepinillo – *Cucumis anguria, C. sativus*
pepinillo del diablo – *Ecballium elaterium, Momordica cochinchinensis*
pepinito – *Cucumis anguira*
pepino – *Cucumis anguira, C. sativus, Momordica charantia, Parmentiera aculeate, P. edulis, Solanum muricatum*
pepino amargo – *Momordica charantia*
pepino amargo espinoso – *Momordica cochinchinensis*
pepino blanco – *Sechium edule*
pepino chino – *Lagenaria siceraria, Trichosanthes kirilowii*
pepino de árbol – *Parmentiera edulis*
pepino de ardilla – *Parmentiera aculeata*
pepino de lagarto – *Ecballium elaterium*
pepino del diablo – *Cucurbita foetidissima*
pepino de maceta – *Solanum muricatum*
pepino de sabana – *Cucumis anguira*
pepino de San Ignacio – *Hura polyandra*
pepino dulce – *Solanum muricatum*
pepino espinosa – *Cucumis anguira*

pepino limón – *Cucumis melo* var *conomon*
pepino silvestre – *Cucumis anguria*
pepioita – *Arctostaphylos pungens*
pepita de San Ignacio – *Hura polyandra*
pepón – *Citrullus colocynthis*
pequeño flamboyant – *Caesalpinia pulcherrima*
pequeño frijol blanco común – *Phaseolus vulgaris*
pera – *Solanum eleagnifolium*
pera cimarrona – *Ziziphus jujuba*
pera de agua – *Eugenia malaccensis*
perado campo – *Eugenia klotzschiana*
peral – *Pyrus communis*
peral amigdaloide – *Pyrus amygdaliformis*
peral común – *Pyrus communis*
peral de hojas de sauce – *Pyrus salicifolia*
peral de las nieves – *Pyrus nivalis*
peral de monte – *Pyrus amygdaliformis*
peral de San Juan – *Pyrus communis*
peralejo – *Salsola soda, S. tragus*
peral gamboa – *Pyrus communis*
peralito – *Orthilia secunda*
peralito menor – *Pyrola minor*
peralito umbelado – *Chimaphila umbellata*
peralito unilateral – *Orthilia secunda*
peralito virescente – *Pyrola chlorantha*
peral panochero – *Pyrus communis*
peral silvestre – *Pyrus pyraster*
pera melón – *Solanum muricatum*
pera mexicana – *Casimiroa edulis*
percha – *Bursera simaruba*
peregrina de Lima – *Alstroemeria pelegrina*
perejil – *Coriandrum sativum, Petroselinum crispum*
perejil de hojas de apio – *Petroselinum crispum* var *neapolitanum*
perejil de hojas planas – *Petroselinum crispum* var *neapolitanum*
perejil del campo – *Ligusticum porteri*
perejil de lobo – *Conium maculatum*
perejil de mar – *Crithmum maritimum*
perejil de monte – *Peucedanum oreoselinum*
perejil de perro – *Aethusa cynapium*
perejilillo – *Eryngium nasturtiifolium*
perejil italiano – *Petroselinum crispum* var *neapolitanum*
perejil marino – *Crithmum maritimum*
perejil venenoso – *Conium maculatum*
perforada – *Hypericum perforatum*
perfume – *Astragalus tragacantha*
perico – *Oliveranthus elegans, Rhynchosia pyramidalis*
pericón – *Tagetes filifolia, T. florida, T. lucida, Tribulus cistoides*
perifollo – *Anthriscus cerefolium*
perifollo almizclado – *Myrrhis odorata*
perifollo de raíz de chirivía – *Chaerophyllium tuberosum*
perifollo oloroso – *Comptonia aspleniifolia, Myrrhis odorata, Osmorhiza longistylis*
perihuite – *Crateva palmeri*
perilla – *Perilla frutescens, Solandra maxima*
periquillo – *Tagetes filifolia, T. lucida*
perito – *Proboscidea fragrans*
perla – *Achillea millefolium, Moringa oleifera, Symphoricarpos albus*
perla de China – *Phytolacca octandra*
perla de la India – *Moringa oleifera*
perla de oriente – *Moringa oleifera*
perlilla – *Chiococca alba, Lopezia mexicana, L. racemosa, Symphoricarpos albus, S. microphyllus*
perlita – *Lithospermum, Symphoricarpos albus, S. microphyllus*
permanente – *Gomphrena globosa*
pernambuco – *Pilocarpus jaborandi*
perón – *Malus sylvestris*
perón cristalino – *Pyrus communis*
perota – *Enterolobium cyclocarpum*
perpetua – *Helichrysum bracteatum*
perpetua anual – *Xeranthemum annuum*
persea – *Persea gratissima*
persicaria – *Polygonum persicaria*
persicaria acederifolia – *Polygonum lapathifolium*
persicaria anfibia – *Polygonum amphibium*
persicaria picante – *Polygonum hydropiper*
pérsico – *Prunus persica*
pérsico duraznero – *Prunus persica*
persimo – *Diospyros kaki*
Perú – *Schinus molle*
Perú cimarrón – *Solidago velutina*
peruétano – *Achras zapota, Pyrus pyraster*
pervinca – *Catharanthus roseus, Vinca major, V. minor*
pervinca de Madagascar – *Catharanthus roseus*
pervinca rosada – *Catharanthus roseus*

perrito – *Antirrhinum majus, Grusonia schottii, Lathyrus odoratus, Maranta arundinacea, Mimulus aurantiacus, Penstemon*
perrito silvestre – *Penstemon imberbis*
peseta – *Zinnia elegans*
peseta del diablo – *Entada scandens*
pesjúa – *Gaultheria odorata*
pesma – *Alsophila salvinii, Cyathea delicatula, Lophosoria quadripinnata, Nephelea mexicana*
pésol – *Pisum sativum*
pesto – *Alsophila salvinii, Cyathea delicatula*
pestón – *Coleosanthus veronicaefolius, Hyptis seemannii*
petaca – *Sterculia apetala*
petai – *Parkia speciosa*
petalzimicua – *Typha domingensis*
petama – *Mucuna sloanei*
petana – *Mucuna sloanei*
petaquilla – *Funastrum clausum, Pithecellobium echinatum*
petatillo – *Thenardia floribunda, Ulmus mexicana*
petelkín – *Lantana camara*
peteltun – *Cissampelos pareira*
petlacotl – *Thevetia thevetioides*
petunia – *Petunia* x *hybrida, P. parviflora*
petunia de enredadera – *Petunia* x *hybrida* var *pendula*
petzacuaxóchitl – *Laelia autumnalis*
petzanxóchitl – *Govenia liliacea*
pexa – *Saccharum officinarum*
pextón – *Brickellia veronicaefolia*
peychi – *Stemmadenia tomentosa*
peyote – *Ariocarpus fissuratus, Astrophytum asterias, Echinocactus grusonii, Lophophora diffusa, L. williamsii, Mammillaria pectinifera, Pelecyphora aselliformis, P. strobiliformis, Senecio cardiophyllus, S. hartwegii, Strombocactus disciformis*
peyote cimarrón – *Ariocarpus fissuratus, A. retusus, Astrophytum myriostigma, Senecio haretwegii*
peyote de San Pedro – *Mammillaria senilis*
peyote liso – *Lophophora williamsii*
peyote meco – *Pelecyphora aselliformis*
peyote queso – *Astrophytum asterias*
peyote verde – *Echinocereus knippelianus*
peyotillo – *Pelecyphora aselliformis*
peyotl – *Ariocarpus fissuratus, Lophophora williamsii*
peyotl xochimilcensis – *Senecio cardiophyllus*
pezuña de venado – *Ariocarpus kotschoubeyanus*
phacelia – *Phacelia campanularia*
picacuello – *Androsace maxima*
picachlich – *Euphorbia dentata* var *cuphosperma, E. heterophylla*
picante blanco – *Moringa oleifera*
picante de la boca – *Veronica beccabunga*
picapica – *Mucuna pruriens, M. sloanei, Sterculia mexicana*
picea – *Picea, Pinus pinea*
picea azul – *Picea pungens*
picea azul de Colorado – *Picea pungens* var *glauca*
picea blanca – *Picea alba, P. glauca*
picea blanca del Canadá – *Picea glauca*
picea común – *Abies religiosa, Picea abies*
picea de Afganistán – *Picea smithiana*
picea de Brewer – *Picea breweriana*
picea de Colorado – *Picea pungens*
picea de Engelmann – *Picea engelmannii*
picea del Himalaya occidental – *Picea smithiana*
picea de Noruega – *Picea abies*
picea de Parry – *Picea pungens* var *glauca*
picea de Servia – *Picea omorika*
picea de Sitka – *Picea sitchensis*
picea de Smith – *Picea smithiana*
picea europea – *Picea abies*
picea falcada – *Picea sitchensis*
picea glauca – *Picea glauca*
picea mariana – *Picea mariana*
picea morinda – *Picea smithiana*
picea negra – *Picea mariana*
picea negra americana – *Picea mariana*
picea negra norteamericana – *Picea mariana*
picea roja del Canadá – *Picea rubens*
picea servia – *Picea omorika*
picietl – *Nicotiana rustica, N. tabacum*
picoa – *Tagetes erecta*
pico de ciqüeña – *Erodium cicutarium*
pico de flamenco – *Sesbania grandiflora*
pico de garza arracimado – *Erodium botrys*
pico de garza braquicarpo – *Erodium brachycarpum*
pico de garza de Reichard – *Erodium reichardii*

pico de gorrión – *Heliconia latispatha*
pico de grulla – *Geranium robertianum*
pico de pájaro – *Capsicum annuum, Forestiera angustifolia, Mimulus verbenaceus, Rhus microphylla, Senna occidentalis*
pico de paloma – *Capsicum annuum*
pico de pato – *Eichhornia crassipes*
picón – *Oreopanax capitatus*
picosa – *Croton ciliato-glandulosus*
picuaremu – *Phragmites australis*
pich – *Adiantum tenerum, Enterolobium cyclocarpum*
pichecua – *Saracha procumbens*
pichejumo – *Pithecellobium dulce*
pichel – *Penstemon spectabilis*
pichí – *Psidium guajava*
pichichagua – *Acourtia nana*
pichiché – *Psidium sartorianum*
pichilinge – *Fouquieria burragei*
pichoco – *Erythrina*
pichohuastle – *Cypripedium irapeanum*
pichol – *Vitis tiliifolia*
pichuco – *Cryptostegia grandiflora*
pichut – *Lasiacis divaricata*
pie azul – *Lepista saeva*
piecito de paloma – *Nectandra sanguinea*
pie de anade – *Chenopodium album*
pie de caballo – *Tussilago farfara*
pie de cabra – *Bauhinia divaricata*
pie de cuervo – *Plantago coronopus*
pie de gallina – *Echinochloa crus-galli, Elytrigia repens*
pie de gallo – *Alvaradoa amorphoides, Elytraria squamosa, Tillandsia prodigiosa, Tubiflora squamosa*
pie de ganzo – *Chenopodium murale*
pie de gato – *Anaphalis margaritacea, Antennaria*
pie de león – *Alchemilla mollis, A. vulgaris, Leontopodium alpinum*
pie de lobo – *Lycopus europaeus*
pie de mula – *Agave tequilana*
pie de niño – *Pedilanthus tithymaloides*
pie de oso – *Alchemilla vulgaris*
pie de pájaro – *Erythroxylum mexicanum, Hamelia erecta, Ornithopus sativus*
pie de pájaro cultivado – *Ornithopus sativus*
pie de pájaro ebracteado – *Ornithopus pinnatus*
pie de paloma – *Alkanna tinctoria, Iresine, Salpianthus arenarius, Trema micrantha*
pie de vaca – *Bauhinia divaricata*
pie de venado – *Bauhinia divaricata, Calliandra emarginata*
piedra del águila – *Caesalpinia crista*
piedra viva – *Lithops*
piel de león – *Schaefferia shrevei*
pierna – *Viburnum lantana*
pierna de gallo – *Echinochloa crus-galli, E. frumentacea*
pierna de gallo mayor – *Echinochloa oryzoides*
pierna de gallo meridional – *Echinochloa colona*
pie violeta – *Lepista nuda*
pigamón de hojas de ancolia – *Thalictrum aquilegifolium*
pilea – *Pilea microphylla*
pilea de aluminio – *Pilea cadierei*
pilea plateada – *Pilea cadierei*
pililllo – *Plumbago auriculata*
pilocarpo – *Pilocarpus jaborandi*
piltzintecouhxóchitl – *Lantana camera*
pim – *Tectaria heracleifolia*
pima – *Erythea edulis, Krameria erecta*
pimentero – *Eugenia pimenta, Piper nigrum, Schinus molle*
pimentero ava – *Piper methysticum*
pimentero blanco – *Piper nigrum*
pimentero blanco de la India – *Piper nigrum*
pimentero cubeda – *Piper cubeda*
pimentero de América – *Schinus molle*
pimentero de hojas estrechas – *Piper angustifolium*
pimentero de Jamaica – *Eugenia pimenta*
pimentero de Tabasco – *Eugenia pimenta*
pimentero de tierra – *Eugenia pimenta, Peperomia campylotropa*
pimentero dulce – *Eugenia pimenta*
pimentero falso – *Schinus molle*
pimentero gordo – *Eugenia pimenta*
pimentero inglés – *Eugenia pimenta*
pimentero negro – *Piper nigrum*
pimentilla – *Adelia virgata, Polygonum persicaria*
pimentilla trepadora – *Polygonum convolvulus*
pimentón – *Capsicum annuum, C. annuum* var *grossum, C. fastigatum, Eugenia pimenta*
pimentón paprika – *Capsicum annuum*

pimienta acuática – *Polygonum hydropiper*
pimienta de agua – *Polygonum hydropiper*
pimienta de muro – *Sedum acre*
pimientillo – *Ardisia revoluta, Eugenia fragrans, Karwinskia humboldtiana, Phyllanthus acidus, Schinus molle, Schoepfia schreberi*
pimiento – *Capsicum annuum*
pimiento alfilerillo – *Capsicum annuum*
pimiento bravo – *Capsicum pubescens*
pimiento caribe – *Capsicum pubescens*
pimiento cayena inglesa – *Capsicum fastigatum*
pimiento-che – *Colubrina ferruginosa*
pimiento de Cayena – *Capsicum fastigatum*
pimiento dulce – *Capsicum annuum* var *grossum*
pimiento loco – *Vitex agnus-castus*
pimiento morrón – *Capsicum annuum* var *grossum*
pimientón dulce – *Capsicum annuum*
pimiento picante – *Capsicum annuum, C. frutescens*
pimiento rojo – *Capsicum fastigatum*
pim-int – *Acoelorraphe wrightii*
pimpinela – *Lepista nuda*
pimpinela blanca – *Pimpinella major*
pimpinela mayor – *Pimpinella major, Sanguisorba officinalis*
pimpinela menor – *Sanguisorba minor* ssp *muricata*
pinabete – *Abies, Cupressus arizonica, Liquidambar styraciflua, Picea abies, P. chihuahuana, Pinus, Pseudotsuga*
pinabete blanco – *Pinus strobus*
pinabete de Chihuahua – *Picea chihuahuana*
pinacate – *Senna wislizeni*
pinacatillo – *Ptelea trifoliata, Senna wislizeni*
pinahuihuixtle – *Mimosa pudica*
pinanga de corona – *Pinanga coronata*
pincel – *Centuarea cyanus*
pincho – *Salsola tragus*
pinchorrero – *Juniperus communis*
pinchosa – *Salsola tragus*
pindo – *Butia capitata*
pinete – *Pinus pinceana, Podocarpus*
pingué – *Hymenoxys*
pingüica – *Arctostaphylos*
pingüico – *Ehretia tinifolia, Gilibertia arborea*
pinguin – *Bromelia pinguin*
pinicua – *Jacquinia macrocarpa* ssp *pungens*
pinillo – *Asclepias linaria, Lycopus americanus, L. asper, Podocarpus guatemalensis*
pinillo azul – *Ajuga genevensis*
pinillo búgula – *Ajuga reptans*
pinillo de ginebra – *Ajuga genevensis*
pinillo oloroso – *Ajuga chamaepitys*
pinillo rastrero – *Ajuga reptans*
pinipiniche – *Pedilanthus tithymaloides*
pinito de flor – *Iberis umbellata*
pino – *Abies religiosa, Casuarina, Juniperus deppeana, Pinus, Tamarix aphylla*
pino acahuite – *Pinus ayacahuite, P. strobiformis*
pino aéreo – *Tillandsia flabellata, T. imperialis*
pino albacarrote – *Pinus cooperi, P. lumholtzii*
pino albar – *Pinus pinea, P. sylvestris*
pino alto – *Pinus wallichiana*
pino amarillo – *Pinus*
pino americano – *Pinus taeda*
pino arizónica – *Pinus arizonica*
pino asiático – *Pinus roxburghii*
pino australiano – *Casuarina cunninghamiana, C. equisetifolia, Grevillea robusta*
pino avellano – *Pinus douglasiana, P. oocarpa*
pino ayacahuite – *Pinus ayacahuite*
pino ayahuite – *Pinus strobus* var *chiapensis*
pino azul del Himalaya – *Pinus wallichiana*
pino banksiano – *Pinus banksiana*
pino barba caida – *Pinus lumholtzii*
pino barbón – *Pinus ayacahuite, P. durangensis*
pino blanco – *Abies concolor, Pinus*
pino blanco americano – *Pinus monticola*
pino blanco de Idaho – *Pinus monticola*
pino blanco de México – *Pinus ayacahuite*
pino blanco japonés – *Pinus parviflora*
pino blanco montano del oeste – *Pinus monticola*
pino bronco – *Pinus rigida*
pino cahuite – *Pinus ayacahuite*
pino canadiense – *Pinus strobus*
pino canario – *Pinus canariensis*
pino canís – *Pinus douglasiana, P. tenuifolia*
pino caribeño – *Pinus caribaea* var *hondurensis*
pino carrasco – *Pinus halepensis*

pino cenizo – *Pinus arizonica*
pino clusiano – *Pinus nigra*
pino colorado – *Pinus*
pino contorto – *Pinus contorta* ssp *contorta*
pino contorto de la costa – *Pinus contorta* ssp *contorta*
pino corriente – *Pinus tenuifolia*
pino costanero – *Pinus caribaea* var *hondurensis*
pino costero – *Pinus caribaea* var *hondurensis*
pino costero de Murray – *Pinus contorta* ssp *contorta*
pino curtidor – *Pinus caribaea* var *hondurensis*
pino chalmaite – *Pinus pseudostrobus*
pino chamonque – *Pinus leiophylla*
pino chantaj – *Pinus devoniana*
pino chileno – *Araucaria araucana*
pino chimonque – *Pinus ayacahuite, P. leiophylla* var *chihuahuana*
pino chino – *Pinus*
pino chino de Bunges – *Pinus bungeana*
pino de Alepo – *Pinus halepensis*
pino de Australia – *Casuarina cunninghamiana, C. equisetifolia*
pino de azúcar – *Pinus ayacahuite, P. lambertiana, P. strobiformis*
pino de Balfour – *Pinus balfouriana*
pino de Banks – *Pinus banksiana*
pino de Barcelona – *Pinus pinea*
pino de Bhután – *Pinus wallichiana*
pino de cola de zorro – *Pinus balfouriana, P. longaeva*
pino de Colorado – *Pinus longaeva*
pino de cono chico – *Pinus pseudostrobus*
pino de cono picudo – *Pinus muricata*
pino de cono pitonudo – *Pinus attenuata*
pino de corcho – *Pseudotsuga menziesii*
pino de Coulter – *Pinus coulteri*
pino de cuatro hojas – *Pinus quadrifolia*
pino de cumbres – *Pinus hartwegii*
pino de Cunningham – *Cunninghamia lanceolata*
pino de Chile – *Araucaria araucana*
pino de Durango – *Pinus durangensis*
pino de Eldorado – *Pinus attenuata*
pino de Himekomatsu – *Pinus parviflora*
pino de incienso – *Pinus taeda*
pino de Jeffrey – *Pinus jeffreyi*
pino de Jerusalén – *Pinus halepensis*
pino de Labrador – *Pinus banksiana*
pino de Lambert – *Pinus lambertiana*
pino de la playa del Pacífico – *Pinus contorta* ssp *contorta*
pino de las alturas – *Pinus hartwegii*
pino de las Islas Canarias – *Pinus canariensis*
pino del cembrio – *Pinus cembra*
pino del Himalaya – *Pinus wallichiana*
pino del Monte Shasta – *Pinus attenuata*
pino del norte – *Pinus sylvestris*
pino de lord Weymouth – *Pinus strobus*
pino de los Alpes – *Pinus cembra*
pino de Martínez – *Pinus maximartinezii*
pino de Moctezuma – *Pinus montezumae*
pino de montaña – *Pinus mugo*
pino de Monterrey – *Pinus radiata*
pino de Nelson – *Pinus nelsoni*
pino de Nicaragua – *Pinus oocarpa*
pino de Norfolk – *Araucaria heterophylla*
pino de Noruega – *Pinus sylvestris*
pino de ocote – *Pinus hartwegii, P. oocarpa*
pino de Oregón – *Pseudotsuga menziesii*
pino de piñas aristadas de las rocosas – *Pinus longaeva*
pino de piñas chichonudas – *Pinus attenuata*
pino de piñas grandes – *Pinus coulteri*
pino de pisos – *Araucaria*
pino de santander – *Pinus radiata*
pino de seis hojas – *Pinus durangensis*
pino de Valsain – *Pinus sylvestris*
pino de Weymouth – *Pinus strobus*
pino doncel – *Pinus pinea*
pino douglas – *Pseudotsuga menziesii*
pino dulce – *Pinus lambertiana*
pino enano – *Pinus culminicola, P. flexilis, P. strobiformis*
pino escobetón – *Pinus devoniana, P. teocote*
pino estrobo – *Pinus strobus*
pino estrobo del este – *Pinus strobus*
pino estrobo del oeste – *Pinus monticola*
pino excelso – *Pinus wallichiana*
pino fino – *Pinus arizonica*
pino flexible – *Pinus flexilis*
pino gallego – *Pinus pinaster*
pino gigante de California – *Pinus lambertiana*
pino gretado – *Pinus leiophylla*
pino gris mexicano – *Pinus ayacahuite*
pino hallarín – *Pinus douglasiana, P. strobiformis*

pino hayuelo – *Podocarpus*
pino hembra – *Pinus tropicalis*
pino huiyoco – *Pinus ayacahuite, P. reflexa, P. strobiformis*
pino insigne – *Pinus radiata*
pino japonés de flores pequeñas – *Pinus parviflora*
pino japonés de Thunberg – *Pinus thunbergii*
pino lacio – *Pinus devoniana, P. michoacana, P. patula*
pino laricio – *Pinus nigra*
pino laricio de Austria – *Pinus nigra*
pino liso – *Pinus pseudostrobus*
pino lumholtz – *Pinus lumholtzii*
pino llanero – *Pinus herrerae*
pino llorón – *Pinus patula*
pino llorón del Himalaya – *Pinus wallichiana*
pino macho – *Juniperus sibirica, Pinus caribaea*
pino manso – *Pinus pinea*
pino marítimo – *Casuarina equisetifolia, Pinus pinaster*
pino mexicano – *Pinus montezumae, P. patula, P. pinceana*
pino mexicano de corteza basta – *Pinus montezumae*
pino montano – *Pinus mugo*
pino montícola – *Pinus monticola*
pino moro – *Pinus mugo* var *uncinata*
pino mugo – *Pinus mugo*
pino nayar – *Pinus reflexa, P. strobiformis*
pino negral – *Larix decidua, Pinus nigra, P. pinaster*
pino negral de Austria – *Pinus nigra*
pino negral español – *Pinus nigra* ssp *salzmannii*
pino negro – *Pinus jeffreyi, P. Montana, P. mugo* var *uncinata*
pino negro de Austria – *Pinus nigra*
pino negro de los Pirineos – *Pinus mugo* var *uncinata*
pino negro japonés – *Pinus thunbergii*
pino obispo – *Pinus muricata*
pino ocote – *Pinus greggii*
pino ocote de México – *Pinus patula*
pino ortiguillo – *Pinus lawsonii, P. pseudostrobus*
pino otomite – *Pinus leiophylla* var *chihuahuana*
pino oyamel – *Abies religiosa*
pino parasol – *Sciadopitys verticillata*
pino pátula – *Pinus patula*
pino piñón – *Pinus cembroides, P. pinceana*
pino piñonero – *Pinus*
pino piñonero mexicano – *Pinus cembroides*
pino ponderosa – *Pinus contorta* ssp *murrayana, P. jeffreyi, P. ponderosa*
pino prieto – *Pinus*
pino pudio – *Pinus nigra*
pino radiata – *Pinus radiata*
pino real – *Abies lasiocarpa, Picea engelmannii, Picea pungens, Pinus, Pseudotsuga menziesii*
pino real americano – *Pinus ponderosa*
pino real blanco – *Abies concolor, A. lasiocarpa*
pino real blanco de las sierras – *Abies lasiocarpa*
pino real colorado – *Pseudotsuga menziesii*
pino real de barbas largas – *Pinus engelmanii*
pino real de las sierras – *Abies lasiocarpa*
pino real de seis hojas – *Pinus durangensis*
pino real de tres hojas – *Pinus engelmannii*
pino remorata – *Pinus muricata*
pino resinero – *Pinus pinaster*
pino resinoso – *Pinus oocarpa, P. resinosa*
pino rígido – *Pinus rigida*
pino rodeno – *Pinus mugo, P. pinaster*
pino rojo – *Pinus pringlei, P. sylvestris, P. yecorensis*
pino rojo americano – *Pinus resinosa*
pino rojo del Canadá – *Pinus resinosa*
pino rojo japonés – *Pinus densiflora*
pino rosillo – *Pinus teocote*
pino rubial – *Pinus pinaster*
pino rzedowski – *Pinus rzedowski*
pino saguaco – *Pinus leiophylla*
pino salado – *Tamarix*
pino salgareño – *Pinus nigra*
pino serrano – *Pinus sylvestris*
pino silvestre – *Pinus sylvestris*
pino sombrilla – *Sciadopitys verticillata*
pino sombrilla del Japón – *Sciadopitys verticillata*
pino tabla – *Pinus ayacahuite*
pino torrey – *Pinus torreyana*
pino triste – *Pinus ayacahuite, P. patula, P. lumholtzii, P. strobiformis*

pino trusa – *Pinus jeffreyi*
pino uncinado – *Pinus mugo* var *uncinata*
pino valsain – *Pinus sylvestris*
pinquito – *Phaseolus vulgaris*
pinsapo – *Abies pinsapo*
pintapan – *Abutilon abutiloides, Anoda cristata, Heliotropium macrostachyum, Sphaeralcea fulva, Wissadula amplissima*
pintapan cimarrón – *Abutilon sonorae*
pintilla – *Agave victoriae-reginae*
pinto – *Phaseolus vulgaris*
pinto uus – *Chloroleucon mangense* var *leucospermum*
pinzán – *Pithecellobium dulce, P. unguis-cati*
pinzanguarimbo – *Goldmania foetida*
pinzanillo – *Heteropteris laurifolia*
piña – *Ananas comosus, A. sativus*
piña de América – *Ananas sativas*
piña de Indias – *Ananas comosus*
piña del monte – *Ceratozamia latifolia, C. mexicana*
piña de palo – *Tillandsia flabellata*
piñanona – *Monstera deliciosa*
piña ratón – *Bromelia pinguin*
piña silvestre – *Bromelia pinguin*
piñita – *Eryngium comosum, Ferocactus robustus*
piñón – *Jatropha curcas, Pinus*
piñoncillo – *Euphorbia cotinifolia, E. lathyris, Jatropha curcas, Omphalea oleifera*
piñón colorado – *Pinus edulis, P. nelsoni*
piñón de cerro – *Jatropha spathulata*
piñón de Indias – *Jatropha curcas*
piñón de la Sierra de Juárez – *Pinus quadrifolia*
piñón de Potosí – *Pinus culminicola*
piñón de Tlapa – *Datura stramonium*
piñón de una hoja – *Pinus monophylla*
piñón de Zacatecas – *Pinus maximartinezii*
piñonero – *Pinus cembroides, P. edulis, P. monophylla, P. nelsoni*
piñonero de cuatro hojas – *Pinus quadrifolia*
piñonero de dos hojas – *Pinus edulis*
piñón prieto – *Pinus cembroides, P. nelsoni*
piñón purgante – *Jatropha curcas*
piñón violento – *Gliricidia sepium*
piñuela – *Bromelia karatas, B. pinguin*
piñuelilla – *Bromelia pinguin*
pío – *Licania platypus*
piocha – *Bursera simaruba, Melia azedarach* var *umbraculiformis, Solanum seaforthianum, Trichilia americana, T. hirta*
piocha cimarrón – *Trichilia hirta*
piocha viejo – *Petrea arborea*
piojillo – *Karwinskia humboldtiana, Lupinus, Porophyllum punctatum*
piojillo para quixicoba – *Brachiaria mutica*
piojillo peludo – *Eriochloa punctata*
piojito – *Caesalpinia palmeri*
piojo – *Caesalpinia palmeri, Porophyllum punctatum*
piojo de burro – *Xanthium canadensis*
piojo de caballo – *Triumfetta semitriloba*
pionilla – *Dorstenia drakeana, Erythrina flabelliformis, Lasianthaea podocephala, Stephanomeria pauciflora, Talinum paniculatum, Zexmenia podocephala*
piorno – *Cytisus purgans, Spartinum junceum*
piorno serrano – *Cytisus purgans*
pipa del indio – *Monotropa uniflora*
pipa de los Holandeses – *Aristolochia durior*
pipa de mico – *Aristolochia californica*
pipal – *Erythrina americana, Sapindus saponaria*
pipe – *Sapindus saponaria*
piperisa – *Lepidium latifolium*
piperita de México – *Hedeoma piperita*
pipichowa – *Acourtia montana, A. thurberi*
pipilo – *Bromus carinatus*
pipilojihuite – *Cestrum nocturnum*
pipilosúchil – *Lobelia laxiflora*
pipiloxihuitl – *Cestrum nocturnum*
pipiloxóchitl – *Lobelia laxiflora*
pipiltzintzintli – *Salvia divinorum*
pipilzíntli – *Dolicholus phaseoloides, Rhynchosia pyramidalis*
pipima – *Zanthoxylum arborescens*
pipinance – *Maba*
pipinella – *Sechium edule*
pipirigallo – *Onobrychis viciifolia*
pipitillo – *Quercus laurina*
pipitzáhuac – *Acourtia adnata, A. cordata, A. wrightii*
pipitzáhuatl – *Stevia viscida*
pipixiuhatstli – *Cypripedium turgidum*
piquín – *Capsicum annuum* var *minimum*
pira brava – *Amaranthus spinosus*
piracanta – *Pyracantha*

pira de berraco – *Rumex crispus*
pira de puerco – *Rumex crispus*
piretro – *Chrysanthemum cinerariifolium, C. coccineum*
pirimu – *Dodonaea viscosa*
pirola – *Gaultheria procumbens*
pirola umbelada – *Chimaphila umbellata*
pirola uniflora – *Moneses uniflora*
pirú – *Schinus molle*
pirúl – *Schinus molle*
pisa – *Canarium album*
pise – *Xanthosoma robustum*
pishtin – *Ceiba pentandra*
pisi – *Randia obcordata*
pisilio – *Plantago*
pistache cimarrón – *Pistacia mexicana*
pistachero – *Pistacia vera*
piston-xiu – *Ayenia pusilla*
pita – *Aechmea magdalenae, Agave, Bromelia magdalenae, Furcraea gigantea, Opuntia, Yucca treculeana*
pita común – *Agave americana*
pita de corojo – *Ananas comosus*
pita floja – *Ananas magdalenae, Bromelia magdalenae*
pita grande – *Agave fourcroydes*
pitahaya – *Acanthocereus subinermis, A. tetragonus, Carnegia gigantea, Cereus fernambucensis, Echinocereus polyacanthus* ssp *pacificus, E. rosei, E. stramineus, Epiphyllum anguliger, Hylocereus ocamponis, H. trigonus, H. undatus, Peniocereus serpentinus, Selenicereus donkelaari, Stenocereus*
pitahaya agria – *Stenocereus gummosus*
pitahaya barbona – *Pachycereus schottii, Pilosocereus alensis*
pitahaya blanca – *Hylocereus undatus*
pitahaya de agosto – *Echinocereus stramineus*
pitahaya de aguas – *Peniocereus cuixmalensis*
pitahaya del cerro – *Disocactus speciosus, Epiphyllum* x *Disocactus*
pitahaya de roca – *Disocactus schrankii*
pitahaya de Yucatán – *Hylocereus undatus*
pitahaya dulce – *Stenocereus thurberi*
pitahaya marítima – *Acanthocereus tetragonus*
pitahaya morada – *Acanthocereus tetragonus*
pitahaya naranjada – *Acanthocereus tetragonus*
pitahaya orejona – *Hylocereus undatus*
pitahaya roja – *Hylocereus undatus*
pitahayita – *Echinocereus brandegeei, E. fasciculatus, Peniocereus serpentinus, P. striatus*
pitahayita colorada – *Peniocereus serpentinus*
pitahayita de agua – *Pereskia aculeata, P. grandifolia*
pitahayita del cerro – *Epiphyllum anguliger*
pitamo real – *Ephedra aspera, E. compacta*
pitanga – *Eugenia uniflora*
pita-puerco – *Bromelia pinguin*
pita samandoque – *Hesperaloe funifera*
pitaya – *Acanthocereus tetragonus, Carnegia gigantea, Echinocereus, Hylocereus ocamponis, H. undatus, Isolatocereus dumortieri, Pachycereus marginatus, Peniocereus serpentinus, Selenicereus, Stenocereus*
pitaya agria – *Pachycereus schottii, Stenocereus gummosus*
pitaya amarilla – *Echinocereus dasyacanthus*
pitaya barbona – *Pachycereus schottii, Pilocereus leucocephalis*
pitaya cimarrona – *Isolatocereus dumortieri*
pitaya colorada – *Stenocereus montanus*
pitaya de agua – *Disocactus speciosus, Pereskiopsis aquosa, Stenocereus fricii*
pitaya de mayo – *Stenocereus griseus, S. pruinosus*
pitaya de Mitla – *Stenocereus griseus*
pitaya de octubre – *Stenocereus pruinosus*
pitaya de Querétaro – *Stenocereus queretaroensis*
pitaya dulce – *Stenocereus thurberi*
pitaya marismeña – *Stenocereus standleyi*
pitaya orejona – *Hylocereus undatus*
pitaya real – *Selenicereus pteranthus*
pitaya roja – *Stenocereus stellatus*
pitaya verde – *Echinocereus viridiflorus*
pitaya vieja – *Pilosocereus chrysacanthus*
pitayita – *Echinocereus, Mammillaria phitauiana, M. solisioides, Pachycereus schottii, Peniocereus*
pitayita de agua – *Epiphyllum oxypetalum*
pitayita del cerro – *Epiphyllum anguliger*
pitayón – *Pilosocereus leucocephalus*
pitilla – *Plumbago scandens, Sporobolus indicus, Yucca enlichiana*
pitire – *Stenocereus quevedonis*

pititache – *Pereskia lycnidifolia, Pereskiopsis rotundifolia*
pito – *Erythrina*
pitón – *Ceiba pentandra, Pachira aquatica*
pitosporo – *Pittosporum crassifolium, P. tenuifolium, P. tobira*
pitosporo de bayas anaranjadas – *Pittosporum undulatum*
pitosporo de la China – *Pittosporum tobira*
pitosporo del Japón – *Pittosporum tobira*
pitosporo llorón – *Pittosporum phillyraeoides*
pitoxe – *Solanum nigrum*
pitre – *Furcraea hexapetala*
pitut – *Lasiacis divaricata*
pitzahuac – *Carica pubescens, Indigofera suffruticosa*
pitzometl – *Agave marmorata*
piule – *Ipomoea violacea, Rhynchosia phaseoloides, Turbina corymbosa*
piule de barda – *Psilocybe zapotecorum*
piule de churis – *Psilocybe mexicana*
pixi – *Xanthosoma sagittifolium*
pixó – *Trema micrantha*
pixoy – *Guazuma ulmifolia*
pixtonchich – *Ayenia pusilla*
pix-ulum – *Ruellia tuberosa*
pizle – *Lucuma mammosa*
placaminero – *Diospyros kaki, D. virginiana*
planta araña – *Schefflera elegantissima*
planta barrillera – *Salicornia europaea*
planta bistec – *Acalypha wilkesiana*
planta compás – *Rosmarinus officinalis*
planta de bisagra – *Physostegia virginiana*
planta de cera – *Hoya carnosa*
planta de la abeja – *Echinops sphaerocephalus*
planta del aceite de castor – *Fatsia japonica*
planta de la fiebre – *Eupatorium perfoliatum*
planta del caucho – *Ficus elastica*
planta del mosquito – *Mentha pulegium*
planta del negro – *Plumbago pulchella*
planta de los panadizos – *Polygonatum odoratum*
planta de San Juan – *Artemisia vulgaris*
planta de tabaco – *Nicotiana alata*
planta de trigo – *Pascopyrum smithii*
planta erizo – *Acaena novae-zelandiae*
planta flamingo – *Anthurium scherzerianum*
plantago – *Plantago*
plantago inglés – *Plantago lanceolata*
planta obediente – *Physostegia virginiana*
planta quebradora – *Psittacanthus calyculatus*
planta muy mala – *Sphaeralcea ambigua*
platanaria de hojas estrechas – *Sparganium angustifolium*
platanaria emergente – *Sparganium emersum*
platanaria enana – *Sparganium angustifolium*
platanaria mayor – *Sparganium erectum*
platanaria natátil – *Sparganium angustifolium*
platanaria ramosa – *Sparganium erectum*
platanillo – *Bernoullia flammea, Canna edulis, C. generalis, Heliconia, Maranta arundinacea, Spathiphyllum blandum, Thalia geniculata*
platanillo de Cuba – *Piper angustifolium*
platanillo enano – *Heliconia psitiacorum*
plátano – *Acer platanoides, Ensete ventricosum, Musa, Platanus hispanica*
plátano amarillo – *Musa paradisiaca*
plátano bárbaro – *Musa paradisiaca*
plátano burro – *Musa acuminata* x *balbisiana*
plátano butuhan – *Musa errans* var *botoan*
plátano camburi – *Musa sapientum*
plátano cimarrón – *Heliconia*
plátano colorado – *Musa acuminata, M. acuminata* x *balbisiana*
plátano común – *Acer pseudoplatanus, Platanus* x *acerifolia*
plátano costeño – *Musa cavendishii*
plátano costilla – *Musa paradisiaca*
plátano costillón – *Musa paradisiaca*
plátano chico – *Musa sapientum*
plátano de Abisinia – *Ensete ventricosum*
plátano de Costa Rica – *Musa cavendishii*
plátano de China – *Musa acuminata, M. cavendishii*
plátano de Londres – *Platanus* x *acerifolia*
plátano de occidente – *Platanus occidentalis*
plátano de oriente – *Platanus orientalis*
plátano de pájaro – *Musa paradisiaca*
plátano de sombra – *Platanus* x *acerifolia, P. occidentalis*
plátano de Virginia – *Platanus occidentalis*
plátano dominico – *Musa cavendishii, M. sapientum*
plátano enano – *Musa cavendishii, M. nana, M. sapientum*
plátano falso – *Acer pseudoplatanus*
plátano francés – *Musa acuminata* x *balbisiana*

plátano fruta – *Musa sapientum*
plátano gigante – *Ensete ventricosum*
plátano grande – *Musa paradisiaca*
plátano gros michel – *Musa acuminata*
plátano guineo – *Musa sapientum*
plátano hartón – *Musa paradisiaca*
plátano hembra – *Musa paradisiaca*
plátano híbrido común – *Platanus* x *acerifolia*
plátano híbrido español – *Platanus* x *acerifolia*
plátano largo – *Musa paradisiaca*
plátano macho – *Musa acuminata* x *balbisiana, M. paradisiaca*
plátano manzano – *Musa acuminata* x *balbisiana*
plátano morado – *Musa sapientum*
plátano occidental – *Platanus occidentalis*
plátano oriental – *Platanus orientalis*
plátano poyo – *Musa acuminata*
plátano roatán – *Musa acuminata, M. sapientum*
plátano silvestre – *Heliconia bihai*
plátano Tabasco – *Musa sapientum*
plátano textil – *Musa textilis*
plátano tuno – *Musa coccinea*
plátano vainilla – *Vanilla pompona*
plátano vianda – *Musa paradisiaca*
plátano zapalote – *Musa sapientum*
platantera del norte – *Platanthera hyperborea*
plateada – *Acleisanthes longiflora, Potentilla anserina*
platero – *Equisetum hyemale*
platicerio – *Platycerium bifurcatum*
plato y taza – *Asclepias curassavica*
pleomele – *Dracaena reflexa*
plicatulum – *Paspalum plicatulum*
pluma – *Gladiolus hortulanus*
pluma de ángel – *Lathyrus angulatus*
pluma de las pampas – *Cortaderia selloana*
pluma de oro – *Jacobinia umbrosa, Justicia aurea*
pluma de Santa Teresa – *Salvia leucantha*
plumajatzin – *Abies guatemalensis*
plumajillo – *Abies hickeli, Achillea millefolium, Alvaradoa amorphoides*
plumajillo de la sierra – *Artemisia frigida*
plumajillo de montaña – *Abies hickeli*
plumajo – *Achillea millefolium*
plúmbago – *Plumbago auriculata, P. scandens*
plúmbago blanco – *Plumbago scandens*
plumerillo – *Asclepias linaria*
plumerillo café – *Digitaria insularis*
plumero – *Digitaria, Fraxinus berlandieriana*
plumero amarillo – *Solidago canadensis*
plumero áspero – *Digitaria sanguinalis*
plumero blanco – *Digitaria californica*
plumilla – *Trixis angustifolia, T. californica, T. radialis, Typha latifolia*
plumilla de Santa Teresa – *Rhipsalidopsis gaertneri*
plumita – *Calliandra eriophylla, Pennisetum setosum*
pnaocl – *Simmondsia chinensis*
poa anual – *Poa annua*
poa aspera – *Poa trivialis*
poa bulbosa – *Poa bulbosa*
poa comprimida – *Poa compressa*
poa común – *Poa pratensis, P. trivialis*
poa chata – *Poa compressa*
poa de color verdemar – *Poa glauca* ssp *rupicola*
poa de flores dispuestas flojamente – *Poa laxa*
poa de hojas estrechas – *Poa pratensis* ssp *angustifolia*
poa de los bosques – *Poa chaixii, P. nemoralis*
poa de los patanos – *Poa palustris*
poa de los prados – *Poa pratensis*
poa espiguilla – *Poa nemoralis*
poa híbrida – *Poa hybrida*
poa laxiflora – *Poa laxa*
poa pratense – *Poa pratensis*
poa precoz – *Poa inferma*
poa temprana – *Poa inferma*
poa violácea – *Poa violacea*
poblano – *Capsicum annuum*
poco – *Xanthosoma roseum*
pocha – *Ferocactus latispinus*
pochote – *Bombax ellipticum, Ceiba acuminata, C. aesculifolia, C. pentandra, Cochlospermum vitifolium, Pachira aquatica, P. macrocarpa*
pochote de aguas – *Ceiba aesculifolia*
pochotillo – *Ceiba pentandra*
pochotl – *Bombax ellipticum*
pochotle – *Ceiba pentandra, Cochlospermum vitifolium*
podocarpo – *Podocarpus*
podocarpo andino – *Podocarpus andinus*

podredumbre – *Fusarium solani, Phytophthora infestans*
podsomatz – *Heliconia latispatha*
pog – *Crescentia cujete*
poinciana – *Caesalpinia gilliesii, C. pulcherrima*
poinciana amarilla – *Caesalpinia gilliesii*
poinciana enana – *Caesalpinia pulcherrima*
poinciana real – *Delonix regia*
poinsetia – *Euphorbia pulcherrima*
pojai – *Geonoma magnifica*
pojote – *Bernoullia flammea*
pok – *Spondias mombin*
polemonio – *Polemonium caeruleum*
polemonio azul – *Polemonium caeruleum*
poleo – *Clinopodium laevigatum, Hedeoma drummondii, H. nanum, H. pulegioides, Marrubium vulgare, Mentha arvensis, M. canadensis, M. pulegium, M. suaveolens, M. sylvestris, Monardella villosa, Satureja chandleri*
poleo americano – *Hedeoma pulegioides*
poleo común – *Mentha pulegium*
poleo chino – *Hedeoma nanum, H. oblongifolia, H. Piperita, H. pulegioides, Mentha pulegium, Monardella odoratissima*
poleo del Paris – *Mentha arvensis*
poleo de monte – *Pycnanthemum muticum*
poleo europeo – *Mentha pulegium*
poleo grande – *Mentha arvensis*
poleo menta – *Mentha pulegium*
poleo negro – *Mentha pulegium*
polibarbado largo – *Polypogon elongatus*
policarpon de hojas en verticilos de cuatro – *Polycarpon tetraphyllum*
polígala – *Polygala*
polígala de hojas de boj – *Polygala chamaebuxus*
polígala de jardín – *Polygala myrtifolia*
polígala de Virginia – *Polygala seneca*
poligonato – *Polygonatum odoratum*
polígono anfibio – *Polygonum amphibium*
polígono bistorto – *Polygonum bistorta*
polígono correhuela – *Polygonum convolvulus*
polígono cuspidado – *Polygonum cuspidatum*
polígono de Aubert – *Polygonum baldschuanicum*
polígono de Bokhara – *Polygonum baldschuanicum*
polígono de hojas pequeñas – *Polygonum arenastrum*
polígono de muchas espigas – *Polygonum polystachyum*
polígono heterófilo – *Polygonum aviculare*
polígono oriental – *Polygonum orientale*
polígono pata de perdiz – *Polygonum lapathifolium*
polígono pátulo – *Polygonum patulum*
polilla – *Caesalpinia palmeri*
polillera – *Verbascum blattaria*
polipodio – *Polypodium, Pteris*
polipogon marítimo – *Polypogon maritimus*
polkuch – *Asclepias curassavica*
polmuc – *Trachypogon montufari*
polocote – *Tithonia*
polvo de los Jesuitas – *Cinchona ledgeriana*
polvo de palo – *Lycopodium clavatum*
poltzacam – *Mammillaria heyderi* ssp *gaumeri*
polvox – *Annona muricata, A. purpurea*
pollita – *Amanita vaginatum*
pollo – *Tradescantia brevifolia*
pollo de monte – *Teucrium polium*
pom – *Protium copal*
pomagás – *Eugenia malaccensis*
pómajo – *Cordia sonorae*
pomalaca – *Eugenia malaccensis*
pomarrosa – *Cordia alliodora, Eugenia jambos*
pomarrosa americana – *Eugenia malaccensis*
pomarrosa del Brasil – *Eugenia malaccensis*
pomarrosa de Malaca – *Eugenia malaccensis*
pomelo – *Citrus grandis, C. paradisi*
pompilla – *Bougainvillea spectabilis*
pompoqua – *Cyrtocarpa edulis, Spondias mombin*
pomposúchitl – *Cochlospermum vitifolium*
poncidre – *Citrus medica*
poncil – *Citrus medica*
poncilero – *Citrus medica*
ponchilhuits – *Asclepias curassavica*
pongolote – *Bombax ellipticum, Gonolobus erianthus*
pono-pinito – *Pedilanthus tithymaloides*
ponsigue – *Ziziphus jujuba, Z. mauritiana*
ponso – *Tanacetum vulgare*
ponté – *Zexmenia frutescens*
pontederia – *Pontederia cordata*
pontederia de hojas cordiformes – *Pontederia cordata*

poñil – *Cercocarpus betuloides, Fallugia paradoxa*
poño – *Purshia mexicana* var *stansburyana*
poon – *Spondias purpurea*
pooscuy – *Psidium guajava*
pöpa – *Solanum tuberosum*
popal – *Calathea lutea, Thalia geniculata*
popalillo – *Pontederia cordata*
popatza – *Ficus goldmanii*
popo – *Dioscorea*
popoacua – *Cyrtocarpa procera*
popocotle – *Cordia alliodora*
popocho – *Musa acuminata*
poposito – *Boletus edulis*
popotalagua – *Sida rhombifolia*
popote – *Dalea grayi, D. pringlei* var *multijuga, Ephedra antisyphilitica, Sorghum bicolor, Stipa robusta*
popote cambray – *Muhlenbergia montana*
popote chiquito – *Dalea revoluta*
popotillo – *Aeschynomene nivea, Baccharis conferta, B. pteronioides, Bothriochloa saccharoides, Dalea parryi, D. pringlei* var *multijuga, Ephedra*
popotillo algodonoso – *Bothriochloa barbinodis*
popotillo arenoso – *Andropogon hallii*
popotillo azul – *Schizachyrium scoparium*
popotillo cañuelo – *Schizachyrium scoparium*
popotillo cola de coyote – *Bothriochloa barbinodis, B. wrightii*
popotillo cola de zorra – *Bothriochloa saccharoides*
popotillo del pinar – *Blepharoneuron tricholepis*
popotillo hirsuto – *Schizachyrium hirtiflorum, S. sanguineum* var *brevipedicellatum*
popotillo hoyuelo – *Bothriochloa wrightii*
popotillo matorralero – *Andropogon glomeratus*
popotillo pajón – *Andropogon virginicus*
popotillo peludo – *Schizachyrium hirtiflorum*
popotillo perforado – *Bothriochloa barbinodis*
popotillo plateado – *Bothriochloa barbinodis, B. saccharoides*
popotillo texano – *Schizachyrium cirratum*
popotón – *Stipa robusta*
popotzo – *Buddleja sessiliflora*
popu – *Conostegia xalapensis*
popuixoli – *Quercus endlichiana*
poro poro – *Cochlospermum vitifolium*
poros pusi – *Solanum tridynamum*
porotillo – *Phaseolus vulgaris* ssp *nunas*
poroto – *Phaseolus vulgaris, Vigna sinensis*
poroto arroz – *Vigna angularis*
poroto de España – *Phaseolus coccineus*
poroto de Lima – *Phaseolus lunatus*
poroto de manteca – *Phaseolus lunatus*
poroto espárrago – *Vigna unguiculata* ssp *sesquipedalis*
poroto gigante – *Canavalia plagiosperma*
poroto granado – *Cajanus cajun*
poroto mungo – *Vigna radiata*
poroto sable – *Canavalia ensiformis, C. gladiata*
poroto soya – *Glycine max*
portamoneda – *Calceolaria mexicana*
portulaca – *Portulaca grandiflora, P. oleracea*
porrilla – *Arrhenatherum elatius*
posh – *Psidium guajava*
poshkeip – *Psidium guajava*
potentilla – *Potentilla*
potentilla arbustiva – *Potentilla fruticosa*
potentilla canadiense – *Potentilla canadensis*
potentilla roja – *Potentilla atrosanguinea*
pothos – *Epipremnum pinnatum*
potija – *Cochlospermum vitifolium*
potkak – *Lucuma mammosa*
potonxihuite – *Cestrum dumentorum*
pox – *Annona cherimola, A. purpurea, A. reticulata, Hibiscus tilliaceus*
poxoncahuitl – *Trophis racemosa*
pozillo tosa – *Laguncularia racemosa*
pozolillo – *Cuitlauzinia pendula, Odontoglossum pendulum*
preadedera hedionda – *Solanum verbascifolium*
precioso – *Crocus imperati*
prementa – *Acoelorraphe wrightii*
presera – *Galium aparine, G. mexicanum, G. verum*
presidio de amor – *Thunbergia grandiflora*
presquilla – *Pleurotus eryngii*
prieta – *Capparis indica*
prim – *Ligustrum japonicum, L. vulgare*
primavera – *Oenothera biennis, Primula, Tabebuia*
primavera cicaule – *Primula vulgaris*
primavera común – *Primula vulgaris*

primavera nocturna – *Oenothera biennis*
primavera rocosa – *Primula auricula*
primavera venenosa – *Primula obconica*
primenta – *Acoelorraphe wrightii*
prímula – *Oenothera biennis, Primula*
prímula común – *Primula officinalis*
prímula del Cabo – *Lachenalia aloides*
prímula denticulada – *Primula denticulata*
prímula nocturna – *Oenothera biennis*
princesa de la noche – *Selenicereus pteranthus*
principe Alberto – *Thunbergia alata*
prisco – *Prunus persica*
privilegio – *Panicum maximum*
prodigiosa – *Brickellia californica, B. cavanillesi, B. grandiflora, Bryophyllum calycinum, Calea zacatechichi, Kalanchoe pinnata, Tripsacum laxum*
profeta – *Jacquinia aurantiaca*
proso – *Panicum miliaceum*
provisor – *Hyparrhenia rufa*
pruno trilobulado – *Prunus triloba*
psilio – *Plantago ovata*
psoralea bituminosa – *Bituminaria bituminosa*
ptármica – *Achillea ptarmica*
pteris – *Pteridium aquilinum*
pterocaria china – *Pterocarya stenoptera*
pterocaria de hojas de fresno – *Pterocarya fraxinifolia*
pterocaria del Cáucaso – *Pterocarya fraxinifolia*
puan – *Muntingia calabura*
pucuri – *Pinus devoniana*
puchero de monja – *Nicandra physalodes*
puchúngari – *Polianthes tuberosa*
pudio – *Rhamnus alpinus*
pudriero – *Prunus mahaleb*
pueraria – *Pueraria lobata, P. phaseoloides*
puerro – *Allium cepa, A. porrum*
puerro de piedra – *Allium fistulosum*
puerro gigante – *Allium giganteum*
puerro silvestre – *Allium ampeloprasum*
puh – *Typha angustifolia, T. domingensis*
puh-suuk – *Leptochloa filiformis*
pukil – *Duranta erecta*
pulguera – *Plantago psyllium*
pulguera de los pantanos – *Pulicaria paludosa*
pulguera disentérica – *Pulicaria dysenterica*
pulguita – *Dolicholus phaseoloides, Rhynchosia pyramidalis*
pulicaria – *Pulicaria dysenterica*
pulicua – *Calceolaria mexicana*
pulmonaria – *Pulmonaria officinalis, Sticta pulmonacea*
pulmonaria de árbol – *Sticta pulmonacea*
pulmonaria de hojas estrechas – *Pulmonaria angustifolia, P. longifolia*
pulmonaria de hojas largas – *Pulmonaria longifolia*
pulmonaria italiana – *Pulmonaria saccharata*
pulmonaria mayor – *Pulmonaria officinalis*
pulpito – *Encyclia cochleata*
pulpo – *Schefflera actinophylla*
pulsatila – *Anemone nemorosa, Pulsatilla vulgaris*
pulsatila de Eslovaquia – *Pulsatilla vulgaris* ssp *slavica*
pulsatila patente – *Pulsatilla patens*
pulsatila vulgar – *Pulsatilla vulgaris*
pumacua – *Bixa orellana*
pumacuchu – *Krameria trianda*
pumos – *Sabal pumos*
pumpo – *Lagenaria siceraria*
pumpunsuchi – *Cochlospermum vitifolium*
pumu – *Erythea edulis, Roystonea regia*
pumúk – *Calyptrogyne ghiesbreghtiana*
punab – *Swietenia macrophylla*
punachi – *Tribulus cistoides*
punche – *Nicotiana*
punche mexicano – *Nicotiana rustica*
punche ratón – *Gaura parviflora*
punchón – *Verbascum thapsus*
punchuhuiche – *Asclepias curassavica*
puntezapat – *Xylosma characantha*
pupa – *Sechium edule*
purbegan – *Trichilia hirta*
purenchecua – *Erythrina americana*
purensapichu – *Dolicholus phaseoloides, Rhynchosia pyramidalis*
pureque – *Erythrina*
purga – *Ipomoea purga*
purga de Jalapa – *Ipomoea purga*
purga de las ánimas – *Smilax medica*
purga de paridas – *Panicum obtusum*
purga macha – *Jatropha macrorhiza*
puru – *Cucurbita pepo*
pus – *Muntingia calabura*
pusera – *Solanum amazonium*
pusi – *Randia thurberi*

pu'sí – *Trophis racemosa*
put – *Carica papaya*
putcan – *Lepidium virginicum*
putia – *Phaulothamnus spinescens*
putsché – *Oncidium adscendens, O. cebolleta*
putsubche – *Oncidium cebolleta*
putsurini tepari – *Potamogeton illinoensis*
putxiu – *Lepidium virginicum*
puya – *Capsicum annuum*
puyeque – *Avicennia germinans*

quacoyul – *Orbignya cohune*
quahaychuachilli – *Jatropha curcas*
quaich – *Chenopodium foetidum*
qual – *Amaranthus hybridus*
qualuza – *Xanthosoma sagittifolium*
quamash – *Camassia quamash*
quamoclit – *Quamoclit coccinea*
quamóchitl – *Haematoxylon brasiletto, Pithecellobium dulce*
quanta hierba – *Euphorbia colorata*
quantlapatziinzintli – *Senecio praecox*
quaucacoatl – *Juglans regia*
quauhalagua – *Guazuma ulmifolia*
quauhayohuachtli – *Jatropha curcas*
quauhayohuatli – *Hura polyandra*
quauhcacoatl – *Carya*
quauhcámotl – *Manihot esculenta*
quauhcoyolli – *Bactris mexicana*
quauhchilli – *Capsicum annuum* var *minimum*
quauhizquitli – *Baccharis conferta*
quauhmecapatli – *Smilax medica*
quauhmóchitl – *Pithecellobium dulce*
quauhólotl – *Guazuma ulmifolia*
quauhpinoli – *Hymenaea courbaril*
quauhpinolli – *Hymenaea courbaril*
quauhtla huitzquilitl – *Cirsium mexicanum*
quauhtlatlatzin – *Hura polyandra*
quauhtlaxoxocoyolin – *Oxalis americana*
quauhtolotlanenci – *Iostephane heterophylla*
quauhtzictli – *Psittacanthus calyculatus*
quauhtzitli – *Psittacanthus calyculatus*
quauhuayo – *Cassia grandis*
quauhxilotl – *Parmentiera edulis*
quauhyetl – *Nicotiana tabacum*
quauhzahuatl – *Ipomoea arborescens*
quaumecatl – *Serjania mexicana*
quauquilitl – *Portulaca oleracea*
quautzontetl – *Chenopodium nuttaliae*
quebena – *Ricinus communis*
quebracho – *Aspidosperma quebracho, Caesalpinia platyloba, Schinopsis quebracho-colorado*
quebracho blanco – *Aspidosperma quebracho-blanco*
quebracho colorado – *Schinopsis quebracho-colorado*
quebrantapiedras – *Herniaria glabra*
quebra plata – *Ipomoea tricolor*
quecupatli – *Mimosa pudica*
quechesohui – *Janusia californica*
quechol aguacate – *Phoebe mexicana*
quejigo – *Quercus canariensis, Q. faginea*
quelite – *Amaranthus, Atriplex acanthocarpa, Brassica nigra, Chenopodium, Liabum glabrum, Malva sylvestris, Phytolacca isocandra, Rumex acetosella, Trianthema portulacastrum*
quelite apestoso – *Chenopodium graveolens*
quelite blanco – *Amaranthus hybridus*
quelite cenizo – *Chenopodium album, C. murale*
quelite colorado yus – *Amaranthus powellii*
quelite cuale – *Teucrium cubense*
quelite de agua – *Amaranthus blitoides*
quelite de cochino – *Amaranthus hybridus*
quelite de espiga – *Amaranthus hybridus*
quelite de fraile – *Salpianthus macrodontos*
quelite de puerco – *Atriplex linifolia, Calathea lutea*
quelite de toro – *Phytolacca octandra*
quelite de trigo – *Chenopodium album*
quelite de zorrillo – *Chenopodium graveolens, C. incisum*
quelite espinoso – *Amaranthus spinosus*
quelite hediondo – *Chenopodium graveolens*
quelite manso – *Amaranthus palmeri*
quelite morado – *Amaranthus cruentus, A. hybridus, A. paniculatus, Iresine herbstii*
quelite rodador – *Amaranthus blitoides*
quelite rojo – *Amaranthus palmeri*
quelite salado – *Chenopodium album, Suaeda*
quelite yus – *Amaranthus powellii*
quelitillo – *Amaranthus fimbriatus*
quelpo – *Fucus vesiculosus, Laminaria*
quema cosas – *Asclepias curassavica*
quemador – *Cnidoscolus angustidens, Jatropha*

angustidens, Urera caracasana, Wigandia caracasana, W. kunthii
quemite – *Erythrina americana*
quenopodio de los jardines – *Atriplex hortensis*
quentó – *Maranta arundinacea, Thalia geniculata*
quequeste – *Colocasia esculenta, Xanthosma robustum*
quequeste de puerco – *Calathea lancifolia*
querendanicua – *Vitex pyramidata*
quereque – *Vitex pyramidata*
quereque del monte – *Cissus trifoliata*
querihuela – *Calluna vulgaris*
querria – *Kerria japonica*
quesadilla – *Commelina*
quesadillita – *Commelina coelestis, C. tuberosa*
quesillo – *Malvaviscus arboreus*
quesito – *Anoda cristata, Lycoperdon, Malva parviflora, Malvaviscus arboreus, Sphaeralcea fulva*
quesquite – *Ziziphus amole*
quetchalalatl – *Amphipterygium adstringens*
quetembilla – *Ziziphus jujuba*
quetzahuéyotl – *Salix bonplandiana, S. lasiolepis*
quetzalaxochiacatl – *Nymphaea caerulea*
quetzalhuexotl – *Salix lasiolepis, S. pentandra*
quetzalmizquitl – *Prosopis chilensis*
quezque – *Russula delica*
queza – *Matricaria recutita*
queza-pijchi – *Berula erecta*
quiba – *Chamaedorea graminifolia*
quiche – *Castilla elastica*
quichi-nixe – *Funastrum clausum*
quiebracha – *Lysiloma divaricatum, L. microphylla*
quiebrahacha – *Lysiloma divaricatum*
quiebra muela – *Asclepias curassavica*
quiebra ojo – *Asclepias curassavica*
quiebra plato – *Cryptostegia grandiflora, Datura stramonium, Ipomoea purpurea, I. stans, I. tricolor, Salvia tiliaefolia*
quiechachi – *Plumeria acutifolia*
quie-quega – *Cochlospermum vitifolium*
quieriga – *Cochlospermum vitifolium*
quietzine – *Combretum farinosum*
quigua – *Amyris balsamifera*
quilalinacate – *Ehretia tinifolia*
quílitl – *Amaranthus, Chenopodium album, C. murale*
quilxpachkuum – *Sechium edule*
quillay – *Quillaja saponaria*
quimec – *Trachypogon montufari*
quimgombó – *Hibiscus esculentus*
quimichpatli – *Sabadilla officinalis, Schoenocaulon officinale, S. pringlei, Veratrum officinale*
quimixcuitlaxcoli – *Ionidium polygalaefolium*
quimixpatli – *Buddleja sessiliflora*
quinana – *Garrya fremontii*
quingombó – *Hibiscus esculentus*
quininhobon – *Spondias purpurea*
quino – *Cinchona, Colubrina glomerata, Coutarea hexandra, C. pterosperma, Croton, Hintonia latiflora*
quinoa – *Chenopodium quinoa*
quino amarillo – *Cinchona calisaya*
quino blanco – *Croton guatemalensis, C. niveus*
quino calisaya – *Cinchona calisaya*
quino de Cayena – *Quassia amara*
quino de Java – *Cinchona pitayensis*
quino de Michoacán – *Exostema caribaeum, Hintonia latiflora*
quino de Surinam – *Quassia amara*
quino gris – *Cinchona officinalis*
quino morado – *Croton eluteria*
quino negro – *Cinchona officinalis*
quino oficinal – *Cinchona officinalis*
quino rojo – *Cinchona officinalis, C. pitayensis, C. succirubra*
quinquefolio – *Potentilla*
quinquefolio arbustivo – *Potentilla fruticosa*
quinquefolio rocoso – *Potentilla tabernaemontanii*
quintonil – *Amaranthus*
quiote – *Agave quiotifera, Bursera simaruba, Yucca filifera, Y. whipplei*
quiotl – *Yucca whipplei*
quiquiriqui – *Salvia patens*
quirindol – *Licania arborea*
quirindol cacahuananche – *Licania arborea*
quiscamote – *Manihot esculenta*
quisco – *Echinopsis chiloensis*
quisnaniquitonale – *Helianthus annuus*
quita manteca – *Solanum erianthum*

quitameriendas – *Colchicum autumnale, C. speciosum*
quita pesar – *Lantana camara*
quixkanlol – *Argemone mexicana*
quixtán – *Solanum wendlandii*
quixxtez – *Amaranthus hybridus, A. spinosus*
quixzaclol – *Argemone mexicana*

rabanete – *Raphanus sativus*
rabanillo – *Raphanus raphanistrum, R. sativus, Rapistrum rugosum, Senecio sanguisorbae, S. toluccanus*
rabanillo amarillo – *Sisymbrium irio*
rabanito – *Raphanus sativus*
rabaniza – *Raphanus rapanistrum*
rabaniza de los soseros – *Spergularia marina*
rábano – *Dieffenbachia, Raphanus sativus*
rábano cimarrón – *Raphanus raphanistrum*
rábano daikon – *Raphanus sativus* var *longipinnatus*
rábano picante – *Armoracia rusticana*
rábano rojo – *Raphanus sativus*
rábano rusticano – *Armoracia rusticana*
rábano silvestre – *Armoracia rusticana, Raphanus rapanistrum*
rabárbaro – *Rheum rhabarbarum*
rabassola – *Morchella esculenta*
rabiacán – *Iris pseudacorus*
rabillo – *Lolium temulentum*
rabillo de cordero – *Phalaris aquatica, P. minor*
rabito cochino – *Anthurium scherzerianum*
rabiza – *Vigna unguiculata*
rabo de asno – *Equisetum telmateia, Heteropogon contortus*
rabo de cachorro – *Setaria anceps*
rabo de chango – *Cyathea princeps*
rabo de faisán – *Neurolaena lobata*
rabo de gato – *Acalypha hispida, Achyranthes aspera*
rabo de iguana – *Dioscorea, Fouquieria formosa, Pithecellobium acatlense, Solanum rostratum*
rabo de lagarto – *Zanthoxylum fagara*
rabo de león – *Agave kerchovei, A. stricta*
rabo de mico – *Alsophila salvinii, Cyathea delicatula, C. mexicana, C. princeps, Heliotropium angiospermum, H. curassavicum, H. indicum, Sphaeropteris horrida*
rabo de mono – *Alsophila salvinii, Cyathea delicatula, C. mexicana*
rabo de mula – *Panicum maximum*
rabo de ratón – *Gliricidia sepium*
rabo de zorro – *Alopecurus pratensis, Equisetum, Piper angustifolium*
rafia – *Raphia ruffia*
rafiolepis – *Rhaphiolepis indica*
raguño – *Leucophyllum ambiguum*
rah-li-ch'och – *Adiantum wilsonii*
raicilla – *Cephaelis ipecacuanha, Curcuma domestica, Psychotria emetica*
raigón – *Stipa tenacisima*
raigón del Canadá – *Gymnocladus dioica*
raíz amarilla – *Xanthorrhiza simpicissima*
raíz barbuda – *Perdicium californianum*
raíz brasileña – *Cephaëlis ipecacuanha*
raíz colorada – *Rumex crispus, R. hymenosepalus*
raíz de abrojo – *Tribulus cistoides*
raíz de ague – *Aletris farinosa*
raíz de cascabel – *Cimicifuga racemosa*
raíz de cuculillo – *Krameria*
raíz de culebra – *Cimicifuga racemosa*
raíz de fregar – *Cyclanthera dissecta*
raíz de fuego – *Anchusa azurea*
raíz de Jalapa – *Ipomoea purga*
raíz de la cucaracha – *Haplophyton cimicidum*
raíz del desierto – *Glycyrrhiza lepidota*
raíz del diablo – *Aconitum, Lophophora williamsii*
raíz del flato – *Aristolochia mexicana*
raíz del gato – *Valeriana*
raíz de liga – *Iostephane heterophylla*
raíz del indio – *Aristolochia watsonii, Rumex hymenosepalus*
raíz del lobo – *Dugaldia hoopesii*
raíz del manso – *Anemopsis californica, Iostephane heterophylla, Viguiera dentata*
raíz de los credos – *Sisyrinchium bellum*
raíz del oso – *Heracleum lanatum, Valeriana ceratophylla, V. sorbifolia*
raíz del sapo – *Eryngium*
raíz del suelo – *Dugesia mexicana*
raíz de macho – *Ptychopetalum ovata*
raíz de maguey – *Agave atrovirens*
raíz de oro – *Hydrastis canadensis*

raíz de pleuresia – *Asclepias tuberosa*
raíz de sangre – *Sanguinaria canadensis*
raíz de Sansón – *Gentiana villosa*
raíz de serpiente negra – *Cimicifuga racemosa*
raíz de valeriana – *Valeriana acutiloba*
raíz de víbora – *Zornia diphylla, Z. thymifolia*
raíz de zacatón – *Muhlenbergia macroura*
raíz diabolica – *Lophophora williamsii*
raíz fuerte – *Raphanus raphanistrum*
raíz granado – *Punica granatum*
raíz roja – *Ceanothus americanus*
rajador – *Lysiloma divaricata*
rama blanca – *Encelia farinosa*
rama de amores – *Physalis hederifolia*
rama de caballo – *Croton ciliato-glandulosus*
rama de chivato – *Havardia mexicana*
rama de queso – *Piper sanctum*
rama de sapo – *Talinum paniculatum*
rama de toro – *Carlowrightia californica, Dicliptera resupinata, Henrya insularis, Ruellia tuberosa*
rama de venado – *Justicia candicans*
rama de zorrillo – *Petiveria alliacea, Phacelia crenulata, P. pedicellata, P. scariosa*
rama dormilera – *Mimosa asperata*
rama dormilona – *Mimosa asperata*
rama hechicera – *Ericameria arborescens*
rama lisita – *Sida aggregata*
rama parda – *Ruellia californica, R. peninsularis*
rama pioja – *Bidens refracta*
rama prieta – *Garrya salicifolia, Indigofera nelsonii*
ramatilla – *Guizotia abyssinica*
rambután – *Nephelium lappaceum*
ramillete – *Aster laevis*
ramillete de Constantinopla – *Dianthus barbatus*
ramio – *Boehmeria nivea*
ramita colorada – *Dalea alamosana*
ramno – *Rhamnus betulaefolia*
ramno catártico – *Rhamnus cathartica*
ramo de novia – *Pereskia aculeata*
ramo de oro – *Galphimia glauca*
ramón – *Brosimum alicastrum, Cercocarpus betuloides, Pistacia mexicana, Salvia sonomensis, Trophis racemosa*
ramón blanco – *Brosimum alicastrum, Trophis racemosa*
ramoncillo – *Casearia aculeate, Trophis racemosa*
ramón colorado – *Brosimum alicastrum, Trophis racemosa*
ramón de caballo – *Trophis racemosa*
ramón de Castilla – *Trophis racemosa*
ramón de costa – *Celtis trinervia*
ramón de mico – *Brosimum alicastrum*
ramón rosa – *Brosimum alicastrum*
ramtil – *Guizotia abyssinica*
rancho viejo – *Epidendrum ibaguense*
rangpur – *Citrus limonia*
ranúnculo – *Ranunculus*
ranúnculo acuático – *Ranunculus aquatilis*
ranúnculo arvense – *Ranunculus arvensis*
ranúnculo asiático – *Ranunculus asiaticus*
ranúnculo bulboso – *Ranunculus bulbosus*
ranúnculo con flores pequeñas – *Ranunculus parviflorus*
ranúnculo de montaña – *Ranunculus montanus*
ranúnculo de pantano – *Ranunculus sceleratus*
ranúnculo espinoso – *Ranunculus muricatus*
ranúnculo hirsuto – *Ranunculus sardous*
ranúnculo lleno de espinas – *Ranunculus muricatus*
ranúnculo muricado – *Ranunculus muricatus*
ranúnculo repente – *Ranunculus repens*
ranúnculo sardo – *Ranunculus sardous*
rapistro anual – *Rapistrum rugosum*
rapistro rugoso – *Rapistrum rugosum*
rapo – *Brassica rapa*
rapónchigo – *Campanula rapunculoides, C. rapunculus*
rapóntico – *Rheum rhaponticum, Rumex alpinus*
rascalino – *Cuscuta*
rascamoño – *Zinnia*
rasca viejo – *Cordia boissieri, Curatella americana*
raspa huevos – *Mimosa malacophylla*
ráspano – *Vaccinium myrtillus*
rasparago – *Picris echioides*
raspasayo – *Dipsacus fullonum*
raspa sombrero – *Ehretia elliptica, Petrea arborea*
raspa viejo – *Curatella americana*
raspilla – *Mimosa malacophylla*
raspilla cespitosa – *Myosotis laxa*

raspilla de flores pequeñas – *Myosotis micrantha*
raspilla de los pantanos – *Myosotis palustris*
raspilla micracanta – *Myosotis micrantha*
raspilla palustre – *Myosotis palustris*
raspilla procumbente – *Asperugo procumbens*
raspilla selvática – *Myosotis sylvatica*
raspilla versicolor – *Myosotis discolor*
rasponera – *Vaccinium myrtillus*
rastrero – *Juniperus communis, Opuntia robusta*
rastrero escarlata – *Phaseolus coccineus*
ratán – *Calamus optimus*
ratania – *Krameria grayi, K. triandra*
ratania del Perú – *Krameria triandra*
ratania para los dientes – *Krameria triandra*
ratania peruana – *Krameria triandra*
ratón – *Dalea diffusa, Gliricidia sepium*
ratoncillo – *Setaria liebmannii*
ratowa – *Jatropha malacophylla*
raxtul – *Lucuma viride*
rayana – *Psidium sartorianum*
rayanillo – *Eugenia capuli*
raygrás – *Lolium multiflorum*
raygrás anual – *Lolium multiflorum*
raygrás criollo – *Lolium multiflorum*
raygrás inglés – *Lolium perenne*
raygrás italiano – *Lolium multiflorum*
rayito – *Zephyranthes longifolia*
rayo – *Cardiospermum halicacabum*
razón bei – *Psilocybe caerulescens*
razón guiol – *Psilocybe zapotecorum*
razón viejo – *Psilocybe zapotecorum*
real de oro – *Achillea lanulosa, A. millefolium*
rebollo – *Quercus faginea*
rebozuelo – *Cantharellus cibarius*
regadero – *Ipomoea quamoclit*
regaliera – *Glycyrrhiza glabra*
regaliz – *Glycyrrhiza glabra*
regaliz glabro – *Glycyrrhiza glabra*
regojo – *Stellaria media*
regoldo – *Castanea sativa*
reina – *Crinum zeylanicum, Eichhornia crassipes*
Reina Claudia – *Prunus domestica* ssp *italica*
reina de la noche – *Cereus hexagonus, Epiphyllum oxypetalum, Hylocereus purpusii, H. undatus, Peniocereus greggii, Selenicereus grandiflorus, S. hamatus, S. pteranthus*
reina de las flores – *Lagerstroemia indica, L. speciosa, Selenicereus grandiflorus*
reina de los bosques – *Galium odoratum*
reina de los prados – *Filipendula ulmaria*
reina del río – *Eichhornia crassipes*
Reina Luisa – *Aloysia triphylla*
Reina Margarita – *Callistephus, Paeonia officinalis*
reinita – *Calendula officinalis*
reishi – *Ganoderma lucidum*
rejalgar – *Asclepias curassavica, Thevetia ovata, T. thevetioides, Xanthosoma robustum*
relámpago – *Aulonemia fulgor*
remolacha – *Beta vulgaris*
remolacha azucarera – *Beta vulgaris* ssp *saccharifera*
remolacha comestible – *Beta vulgaris*
remolacha de mesa – *Beta vulgaris*
remolacha de huerto – *Beta vulgaris*
remolacha forrajera – *Beta vulgaris*
remolacha roja – *Beta vulgaris*
remolín – *Setaria gracilis*
remolina – *Paspalum virgatum*
renegada – *Chenopodium rubrum*
repollito de agua – *Pistia stratoides*
repollito de Bruselas – *Brassica oleracea* var *gemmifera*
repollo – *Brassica oleracea* var *capitata*
repollo colorado – *Brassica oleracea* var *capitata*
repollo chino – *Brassica rapa* var *pekinensis*
repollo de Bruselas – *Brassica oleracea* var *gemmifera*
repollo de China – *Brassica rapa* var *chinensis*
repollo rojo – *Brassica oleracea* var *capitata*
requesón – *Stevia serrata*
rerogüe – *Solanum tuberosum*
reseda – *Reseda luteola, R. odorata*
reseda amarilla – *Reseda lutea*
reseda blanca – *Reseda alba*
reseda de jardín – *Reseda odorata*
reseda de olor – *Reseda odorata*
reseda francesa – *Reseda odorata*
reseda olorosa – *Reseda odorata*
resedón – *Reseda odorata*
resina – *Commiphora africana*
retama – *Cercidium floridum, C. microphyllum, C. praecox, Cytisus scoparius, Genista anglica, G. lydia, Parkinsonia aculeata,*

Retama raetam, Senna, Spartium junceum, S. patens, Tecoma stans, Thevetia peruviana
retama china – *Cercidium texanum, Parkinsonia aculeata*
retama de Canarias – *Genista canariensis*
retama de cerda – *Parkinsonia aculeata*
retama de China – *Spartium junceum*
retama de escobas – *Cytisus scoparius, Spartium junceum*
retama de España – *Spartium junceum*
retama de flor – *Spartium junceum*
retama de hojas de lino – *Genista linifolia*
retama delgada – *Spartium junceum*
retama del Monte Etna – *Genista aetnensis*
retama de los montes – *Genista tinctoria*
retama del país – *Senna laevigata*
retama de olor – *Spartium junceum*
retama de tierra caliente – *Senna multiglandulosa*
retama de tintoreros – *Genista tinctoria*
retama escocesa – *Cytisus scoparius*
retama española – *Genista hispanica*
retama inglesa – *Genista anglica*
retamalo – *Senna multiglandulosa*
retama loca – *Osyris alba*
retama macho – *Spartium junceum*
retama negra – *Cytisus scoparius*
retama pelosa – *Genista pilosa*
retama real – *Ephedra compacta, E. pedunculata*
retama serrana – *Cytisus purgans*
retamilla – *Caesalpinia mexicana*
retoño – *Peperomia quadrifolia*
retoño de ocote – *Arceuthobium globosum*
retorcido moreno – *Heteropogon contortus, H. melanocarpus*
retorcido negro – *Heteropogon melanocarpus*
revienta – *Asclepias curassavica*
revienta buey – *Ranunculus scleratus*
revienta cabra – *Citharexylum berlandieri*
revienta muelas – *Asclepias curassavica*
reyóchari – *Zexmenia podocephala*
rhodes – *Chloris gayana*
rhodes masaba – *Chloris gayana*
rhodes pokot – *Chloris gayana*
ricino – *Ricinus communis*
ricino común – *Ricinus communis*
ricino de América – *Jatropha curcas*
riñón – *Annona squamosa*
riñonina – *Ipomoea pes-caprae*
risa de la Virgen – *Fagonia cretica*
rizado – *Pleuraphis belangeri*
roatán – *Musa paradisiacal, M. sapientum*
robaldo – *Croton dioicus*
robinia – *Robinia pseudoacacia*
roble – *Ehretia tinifolia, Nothofagus obliqua, Quercus, Tabebuia*
roble albar – *Quercus petraea*
roble amarillo – *Tabebuia chrysantha, Tecoma stans*
roble americano – *Quercus rubra*
roble andaluz – *Quercus canariensis*
roble asiático kunigi – *Quercus acutissima*
roble bicolor americano – *Quercus bicolor*
roble blanco – *Quercus alba, Tabebuia pentaphylla*
roble blanco americano – *Quercus alba*
roble blanco americano de bellotas grandes – *Quercus macrocarpa*
roble blanco americano de los pantanos – *Quercus bicolor*
roble blanco de América – *Quercus alba*
roble boreal rojo americano – *Quercus rubra*
roble borne – *Quercus pyrenaica*
roble cabelludo – *Quercus cerris*
roble calliprinos – *Quercus coccifera*
roble carvallo – *Quercus robur*
roble carrasqueño – *Quercus faginea*
roble cerrioide – *Quercus cerrioides*
roble cerris – *Quercus cerris*
roblecito – *Quercus oleoides*
roble claro – *Quercus alba*
roble coccifero – *Quercus coccifera*
roble colorado – *Tabebuia pentaphylla, T. rosea*
roble común – *Quercus robur, Q. rubra*
roble costero – *Quercus agrifolia*
roble crenado – *Quercus crenata*
roble de bellotas grandes – *Quercus macrocarpa*
roble de hojas anchas – *Quercus virgiliana*
roble de hojas de sauce – *Quercus phellos*
roble de hojas redondeadas – *Quercus rotundifolia*
roble de Hungría – *Quercus frainetto*
roble de invierno – *Quercus petraea*
roble de la costa – *Cocoloba barbadensis, C. uvifera*

roble de los pantanos – *Quercus palustris*
roble del valle – *Quercus lobata*
roble de Muehlenbergii – *Quercus muehlenbergii*
roble de San Luís – *Tabebuia pentaphylla*
roble de Turquía – *Quercus cerris*
roble de verdura perenne – *Quercus virgiliana*
robledilla – *Quercus fruticosa*
roble escarlata – *Quercus coccinea*
roble fresnal – *Quercus robur*
roble húngaro – *Quercus frainetto*
roble invernal – *Quercus petraea*
roble macrocarpa – *Quercus macrocarpa*
roble maguiligua – *Tabebuia pentaphylla*
roble mediterráneo – *Quercus canariensis*
roble musgoso – *Quercus cerris*
roble negral – *Quercus pyrenaica*
roble negro – *Quercus emoryi, Q. pyrenaica*
roble palustre – *Quercus palustris*
roble palustre americano – *Quercus palustris*
roble pedunculado – *Quercus robur*
roble pedúnculo – *Quercus robur*
roble pelin – *Nothofagus obliqua*
roble-pino – *Quercus palustris*
roble pirenaico – *Quercus pyrenaica*
roble prieto – *Quercus durifolia, Tabebuia pentaphylla*
roble rojo – *Quercus rubra, Q. similis*
roble rojo americano – *Quercus rubra*
roble rojo americano de hojas de sauce – *Quercus phellos*
roble rojo del norte – *Quercus rubra*
roble rosa morada – *Tabebuia pentaphylla*
roble rugoso – *Quercus rugosa*
roble serrano – *Quercus fusiformis, Q. virginiana, Tabebuia guayacan*
roble tozo – *Quercus pyrenaica*
roble turco – *Quercus cerris*
roble velutino – *Quercus velutina*
rocambola – *Allium sativum* var *ophioscorodon*
roca viviente – *Ariocarpus fissuratus*
rocella – *Ramalina reticulata*
rociado – *Reseda odorata*
rocío – *Mesembryanthemum crystallinum*
rocío del sol – *Drosera rotundifolia*
rocoto – *Capsicum pubescens*
rócot uchi – *Capsicum pubescens*
rocuro – *Quercus endlichiana*
rodela – *Viburnum opulus*
rodellón – *Boletus edulis, B. reticulatus*
rodilla de Cristo – *Tigridia pavonia*
rodilla de vieja – *Zexmenia frutescens*
rododendro – *Rhododendron*
rododendro amarillejo – *Rhododendron luteum*
rododendro de Japón – *Rhododendron japonicum*
rododendro del Mar Negro – *Rhododendron ponticum*
rododendro eruginoso – *Rhododendron ferrugineum*
rododendro herrumbroso – *Rhododendron ferrugineum*
rodrigón – *Phaseolus vulgaris*
roja – *Quercus chihuahuensis*
rojaca – *Quercus endlichiana*
rojaca sacamé – *Quercus arizonica*
rok-chik-kuán – *Adiantum tenerum*
roldón – *Coriaria myrtifolia*
romal – *Parietaria debilis*
romaza – *Rumex patientia*
romaza acederilla – *Rumex acetosella*
romaza alpina – *Rumex alpinus*
romaza amarilla – *Rumex crispus*
romaza común – *Rumex patientia, R. pulcher*
romaza conglomerada – *Rumex conglomeratus*
romaza crespa – *Rumex crispus*
romaza de hojas de sauce – *Rumex salicifolius*
romaza de hojas estrechas – *Rumex stenophyllus*
romaza de hojas grandes – *Rumex obtusifolius*
romaza de mar – *Rumex maritimus*
romaza dentada – *Rumex dentatus*
romaza escudada – *Rumex scutatus*
romaza hortense – *Rumex patientia*
romaza paciencia – *Rumex patientia*
romaza retorcida – *Polygonum bistorta*
romaza uliginosa – *Rumex dentatus*
romeo – *Rosmarinus officinalis*
romerillo – *Artemisia, Asclepias linaria, Atriplex linifolia, A. semibaccata, Baccharis sarathroides, Chrysactinia mexicana, Ericameria spinulosus, Helianthemum lavandulifolium, Hymenoclea monogyra, Keckiella antirrhinoides, Krascheninnikovia lanata, Peganum mexicanum, Penstemon centranthifolius, Porophyllum scoparium, Pseudotsuga menziesii, Salvia hispanica, S. polystachys, Suaeda, Viguiera linearis*

romerillo amargo – *Ericameria brachylepis, E. sonorensis*
romerillo cimarrón – *Purshia mexicana* var *stansburyana*
romerillo de la costa – *Viguiera dentata*
romerillo de la sierra – *Ipomoea ancisa*
romerillo del monte – *Lippia graveolens*
romerillo del país – *Trichostema*
romerillo dulce – *Hymenoclea monogyra*
romerillón – *Baccharis sarathroides*
romerito – *Artemisia californica, Dondia suffrutescens, Suaeda*
romero – *Peucephyllum schottii, Rosmarinus officinalis, Suaeda moquinii, Trichostema*
romero cedro – *Purshia mexicana* var *stansburyana*
romero cimarrón – *Purshia mexicana* var *stansburyana*
romero común – *Rosmarinus officinalis*
romero de la sierra – *Trichostema lanatum*
romero del país – *Trichostema*
romero de pantano – *Ledum palustre*
romero extranjero – *Rosmarinus officinalis*
rompecabeza de mono – *Araucaria araucana*
rompecapa – *Celtis pallida*
rompenieve – *Galanthus nivalis*
rompeplato – *Ipomoea tricolor*
ronrón – *Astronium graveolens, Senna pallida*
roqueta – *Eruca vesicaria* ssp *sativa*
roqueta común – *Eruca vesicaria* ssp *sativa*
roqueta marítima – *Cakile maritima*
rosa albardera – *Paeonia broteroi*
rosa amarilla – *Cochlospermum vitifolium, Pereskiopsis porteri*
rosabari – *Artemisia ludoviciana*
rosa china – *Cochlospermum vitifolium, Hibiscus roas-chinensis*
rosa de ángel – *Tagetes lunulata*
rosa de cacao – *Quararibea funebris*
rosa de Castilla – *Lippia callicarpaefolia*
rosa de Castilla cimarrona – *Guilleminea densa* var *aggregata, Lippia callicarpaefolia*
rosa de cuaresma – *Helleborus officinalis, H. orientalis*
rosa de China – *Hibiscus rosa-sinensis*
rosa de Güeldres – *Viburnum opulus*
rosa de Jamaica – *Hibiscus sabdariffa*
rosa del amor – *Nymphaea alba*
rosa de la sierra – *Purshia mexicana* var *stansburyana*
rosa de las nieves – *Cirsium ehrenbergii*
rosa del desierto – *Adenium obesum*
rosa del Nilo – *Nelumbo nucifera*
rosa de los Alpes – *Rhododendron ferrugineum*
rosa del pastor – *Euphorbia pulcherrima*
rosa de madera – *Ipomoea tuberosa*
rosa de mayo – *Antigonon leptopus*
rosa de mono – *Acalypha hispida*
rosa de montaña – *Antigonon leptopus, Brownea grandiceps*
rosa de monte – *Paeonia officinalis*
rosa de muerto – *Tagetes erecta*
rosa de navidad – *Helleborus niger*
rosa de prado – *Sabatia campestris*
rosa de San Juan – *Bouvardia longiflora, Cephalanthus occidentalis, Macrosiphonia hypoleuca*
rosa de San Pedro – *Macrosiphonia hypoleuca*
rosa de Santa María – *Paeonia officinalis*
rosa de Sarón – *Tulipa sharonensis*
rosa de Siria – *Hibiscus trionum, Hypericum calycinum*
rosadillo – *Cordia alliodora, Fouquieria formosa, Jacquinia aurantica, J. macrocarpa* ssp *pungens, Swietenia macrophylla*
rosafín – *Juniperus communis*
rosa francesa – *Nerium oleander*
rosa imperial – *Cochlospermum vitifolium*
rosa jamaica – *Hibiscus sabdariffa*
rosa japonesa – *Spiraea japonica*
rosákame – *Quercus arizonica*
rosal – *Plumeria acutifolia, Rosa*
rosal alpino – *Rosa pendulina*
rosal amarillo – *Rosa sulphurea*
rosa laurel – *Nerium oleander*
rosal balsamina menor – *Rosa micrantha*
rosal blanco – *Rosa alba*
rosal canela – *Rosa cinnamomea*
rosal canina – *Rosa canina*
rosal castellano – *Rosa gallica*
rosal común – *Rosa centifolia*
rosal chayote – *Rosa roxburghii*
rosal de Bengala – *Rosa indica*
rosal de California – *Rosa californica*
rosal de Cambray – *Rosa alba*
rosal de Castilla – *Rosa*
rosal de cien hojas – *Rosa centifolia*

rosal de cien pétalos – *Rosa centifolia*
rosal de damasco – *Rosa damascena*
rosal de Francia – *Rosa gallica*
rosal de hojas de pimpinela – *Rosa pimpinellifolia*
rosal de hojas pequeñitas – *Rosa minutifolia*
rosal de hojas rojas – *Rosa glauca*
rosal de Japón – *Rosa rugosa*
rosal de Jerico – *Rosa canina, R. gallica*
rosal de la pasión – *Passiflora edulis*
rosal de las cuatro estaciones – *Rosa damascena*
rosal del campo – *Rosa californica*
rosal del diablo – *Rosa canina*
rosal del norte – *Rosa gallica*
rosal de los capuchinos – *Rosa foetida*
rosal de mayo – *Rosa cinnamomea*
rosal de Moctezuma – *Rosa montezumae*
rosal de Provenza – *Rosa centifolia, R. gallica*
rosal de Siria – *Hypericum calycinum*
rosal de todos los meses – *Rosa chinensis, R. damascena*
rosal de verdura perenne – *Rosa sempervirens*
rosal enano – *Rosa chinensis* var *minima*
rosal enredadero – *Rosa multiflora*
rosal fétido – *Rosa eglanteria, R. foetida*
rosal francés – *Rosa gallica*
rosalía – *Jacquinia aurantiaca*
rosalillo – *Fouquieria formosa*
rosal irlandés – *Rosa hibernica*
rosal mexicano – *Rosa serrulata*
rosal miniaturo – *Rosa parviflora*
rosal mosqueta – *Rosa moschata*
rosal mosqueta común – *Rosa sempervirens*
rosal multiflora – *Rosa multiflora, R. polyantha*
rosal musgoso – *Rosa muscosa*
rosal perruno – *Rosa canina*
rosal pimpinela – *Rosa spinosissima*
rosal rugoso – *Rosa rugosa*
rosal silvestre – *Rosa canina, R. minutifolia*
rosal té – *Rosa indica, R. laevigata*
rosa maría – *Cannabis sativa, Senna covesii*
rosa morada – *Crusea calocephala, Tabebuia pentaphylla*
rosapara – *Mammillaria senilis*
rosariera – *Melia azedarach*
rosatilla – *Bidens odorata*
rosa trompetilla – *Bouvardia ternifolia*
rosáwari – *Artemisia ludoviciana*
rosea vieja – *Cordia boissieri*
rosella – *Hibiscus sabdariffa*
roseta – *Cenchrus incertus, C. longispinus*
rosetilla – *Ambrosia acanthicarpa, Bidens odorata, Cenchrus longispinus*
rosilla – *Bidens pilosa, Castilleja arvensis, Commelina, Eysenhardtia polystachya, Helenium mexicanum, H. puberulum, H. thurberi, Tribulus cistoides*
rosilla del valle – *Bidens pilosa*
rosilla de Puebla – *Helenium mexicanum*
rosita – *Centaurium calycosum, Gutierrezia sarothrae*
rosita amarilla – *Stemodia pusilla*
rosita morada – *Phlox nana*
rosmarino – *Rosmarinus officinalis*
rosol de la pasión – *Passiflora arida*
rosolí – *Drosera rotundifolia*
rossinyol – *Cantharellus cibarius*
rosval – *Croton dioicus*
rota – *Calamus*
rovelló – *Lactatius deliciosus*
roya – *Puccinia graminis*
royuela – *Rubia tinctorium*
rubaldo – *Croton dioicus*
rubellón – *Lactarius deliciosus*
rubia – *Rubia tinctorum, Russelia equisetiformis*
rubia de mar – *Crucianella maritima*
rubia de tintes – *Rubia tinctorium*
rubilla – *Galium odoratum*
rubo – *Rubus fruticosus*
rucamar – *Cakile maritima*
ruda – *Ruta chalepensis, R. graveolens*
ruda alemana – *Ruta graveolens*
ruda alpina – *Thalictrum alpinum*
ruda bracteosa – *Ruta chalepensis*
ruda cabruna – *Dalea citriodora, Galega officinalis*
ruda camera – *Scrophularia canina*
ruda canina – *Scrophularia canina*
ruda capraria *Dalea citriodora*
ruda cimarrona – *Alvaradoa amorphoides, Thalictrum fendleri*
ruda común – *Ruta graveolens*
ruda de hojas de guileña – *Thalictrum aquilegifolium*
ruda de jardín – *Ruta graveolens*
ruda de la sierra – *Thalictrum fendleri*

ruda fétida – *Ruta graveolens*
ruda inglesa – *Coronilla varia*
ruda pratense – *Thalictrum dipterocarpum*
rudbeckia – *Echinacea angustifolia*
rudbeckia amarilla – *Rudbeckia laciniata*
rudbeckia laciniada – *Rudbeckia laciniata*
rudbequia – *Rudbeckia*
rueca de Cleopatra – *Eremurus robustus*
rueda de la fortuna – *Echeveria pulverulenta*
rugón – *Erica vagans*
ruibardo – *Jatropha podagrica, Rheum officinale, R. rhaponticum*
ruibardo común – *Rheum officinale, R. rhabarbarum*
ruibardo chino – *Rheum palmatum*
ruibardo de comer – *Rheum rhaponticum*
ruibardo de frailes – *Rumex patientia*
ruibardo de huerta – *Rheum rhaponticum*
ruibardo de la China – *Rheum officinale*
ruibardo medicinal – *Rheum officinale*
ruibardo oficinal – *Rheum officinale*
ruibardo palmeado – *Rheum palmatum*
ruibardo real – *Rheum alexandrae*
ruín – *Styrax argenteus*
ruiponce – *Campanula rapunculus*
ruipóntico – *Rumex alpinus*
ruppia cirrosa – *Ruppia cirrhosa*
ruppia marina – *Ruppia maritima*
rus – *Rhus coriaria*
rusco – *Ruscus aculeatus*
rutabaga – *Brassica napus* var *napobrassica, B. rapa*

sabacuite – *Plumeria rubra*
sabakché – *Exostema caribaeum*
sabal achaparrado – *Sabal etonia*
sabal de Bermuda – *Sabal bermudana*
sabal de Carolina – *Sabal palmetto*
sabal de Santo Domingo – *Sabal blackburniana*
sabal enano – *Sabal minor*
sabal de Florida – *Sabal etonia*
sabali – *Ficus mexicana*
sabanicté – *Plumeria rubra*
sabina – *Cupressus arizonica, C. benthami, C. lusitanica, Juniperus, Taxodium*
sabina albar – *Juniperus thurifera*
sabina cimarrona – *Podocarpus*
sabina común – *Juniperus sabina*
sabina de arroyos – *Astianthus viminalis*
sabina de Cartagena – *Tetraclinis articulata*
sabina de la sierra – *Juniperus monosperma*
sabina de Oaxaca – *Taxodium distichum* var *mexicanum*
sabina de Virginia – *Juniperus virginiana*
sabina española – *Juniperus thurifera*
sabina macho – *Juniperus communis, J. sibirica*
sabina monje – *Pseudotsuga menziesii* var *glauca*
sabina morena – *Juniperus osteosperma*
sabina morisca – *Juniperus communis*
sabina negra – *Juniperus phoenicea*
sabina turífera – *Juniperus thurifera*
sabonete – *Pereskia aculeata*
sabor – *Maba latifolia*
saborija – *Satureja hortensis*
sabuco – *Sambucus nigra*
sacacil – *Echinocereus poselgeri, Opuntia chaffeyi*
saca espina – *Pedilanthus tithymaloides*
sacahuiste – *Nolina*
sacalacahuite – *Pinus ayacahuite*
sacamanteca – *Solanum*
saca miel – *Castilleja tenuiflora*
sacanochtli – *Opuntia megarhiza*
sacapelos – *Senecio jacobaea*
sacapeyote – *Acourtia microcephala*
saca teta – *Euphorbia hirta*
sacatinta – *Justicia carnea*
sacaú – *Bambusa vulgaris*
sacbayeck – *Bourreria huanita*
saccabah – *Ficus mexicana*
sac-chacah – *Bursera simaruba*
sacjavén – *Calathea lutea*
sacmumutz – *Lippia graveolens*
sacobari – *Citrullus colocynthis*
sacpet – *Crotalaria*
sacramento – *Jacquinia aurantiaca*
sac suuk – *Bothriochloa pertusa*
sacui – *Mentha suaveolens*
sacxiu – *Abutilon umbellatum*
sacha huasca blanca – *Lilium candidum*
saeta de agua – *Damasonium alisma, Sagittaria sagittifolia*
saetilla – *Bidens pilosa, Sporobolus indicus*
sadugua – *Chenopodium graveolens*
sagalejo – *Castilleja arvensis*

sagina apétala – *Sagina apetala*
sagina linneana – *Sagina saginoides*
sagina subulada – *Sagina subulata*
sagina tendida – *Sagina procumbens*
sagitaria – *Sagittaria latifolia, S. sagittifolia*
sagitaria de hojas anchas – *Sagittaria latifolia*
sagitaria sagitifolia – *Sagittaria sagittifolia*
sago – *Acrocomia mexicana*
sagú – *Maranta arundinacea, Metroxylon sagus*
saguaro – *Carnegia gigantea*
sagú cimarrón – *Maranta arundinacea*
sagú del monte – *Maranta arundinacea*
sagüera – *Pachycereus pringlei, Stenocereus montanus*
sagüeso – *Pachycereus pringlei*
saha – *Curatella americana*
sahmées – *Citrus sinensis*
sahuaca – *Pinus leiophylla* var *chihuahuana*
sahual – *Diospyros digyna*
sahualca – *Lagenaria siceraria*
sahualiiotorosim – *Spondias mombin*
sahuaro – *Carnegia gigantea*
sahuauó – *Quercus chihuahuensis*
sahuí – *Stenocereus montanus*
sahuiliqui – *Yucca grandiflora*
sahumi – *Zexmenia hispida*
sah-yak – *Laelia nodosa*
saibari – *Eryngium carlinae*
saintpaulia – *Saintpaulia ionantha*
saita – *Dichelostemma capitatum*
saituna – *Ziziphus amole*
sajaguo – *Quercus chihuahuensis*
sak-chum – *Dipholis salicifolia*
sak-kol – *Solanum nudum*
salacahuite – *Pinus ayacahuite*
salacasúchil – *Bauhinia variegata*
salacate – *Citharexylum berlandieri, C. scabrum*
saladilla – *Atriplex, Avicennia germinas, Batis marítima, Piper angustifolium, Stenotaphrum secundatum, Suaeda, Tripsacum laxum*
saladito – *Frankenia palmeri*
salado – *Atriplex halimus, Avicennia germinans, Muhlenbergia repens, M. utilis, Rhus microphylla, Tamarix ramosissima*
salak – *Salacca edulis*
salche – *Sophora tomentosa*
sal de venado – *Rhus terebinthifolia*
salgada – *Atriplex halimus*
salgadera – *Atriplex halimus*
salguero – *Salix alba, S. viminalis*
salguero blanco – *Salix viminalis*
salicaria – *Lythrum*
salicieso – *Lycium, Malva parviflora, Rhamnus betulaefolia*
salicor borde – *Salsola tragus*
salicor de la mancha – *Salsola tragus*
salicor europeo – *Salicornia europaea*
salitrillo – *Tragia nepetaefolia*
salpiglosis – *Salpiglossis sinuata*
salsa – *Leucaena leucocephala*
salsa de pastor – *Anthriscus cerefolium, Thymus serpyllum*
salsero – *Thymus vulgaris*
salsifí – *Tragopogon porrifolius, T. pratensis*
salsifí blanco – *Tragopogon porrifolius*
salsifí de España – *Scorzonera hispanica*
salsifí español – *Scolymus hispanicus*
salsifí negro – *Scorzonera hispanica*
saltipuchi – *Rhynchosia pyramidalis*
saltipús – *Rhynchosia pyramidalis*
salvadera – *Hura polyandra*
salvadora – *Solanum diversifolium, S. erianthum, S. verbascifolium*
salvajina – *Tillandsia usneoides*
salvajina macho – *Tillandsia recurvata*
salvia – *Buddleja, Hyptis, Iresine calea, Lepechinia caulescens, Salvia*
salvia amarga – *Teucrium fruticans*
salvia amarilla – *Phlomis fruticosa*
salvia aromatica – *Buddleja perfoliata*
salvia blanca – *Salvia apiana*
salvia blanca peluda – *Salvia argentea*
salvia cardenal – *Salvia coccinea*
salvia cardinal – *Salvia fulgens*
salvia común – *Salvia officinalis*
salvia de Belén – *Pulmonaria saccharata*
salvia de bolita – *Buddleja perfoliata*
salvia de Guadalajara – *Salvia polystachya*
salvia de jardín – *Salvia officinalis*
salvia de Jerusalén – *Phlomis fruticosa*
salvia de la pradera – *Salvia officinalis*
salvia del bosque – *Salvia nemorosa, S. sylvestris*
salvia del desierto – *Hyptis emoryi*
salvia del monte – *Salvia azurea, S. goldmanii, S. microphylla*

salvia del pantano – *Stachys palustris*
salvia del prado – *Salvia pratensis*
salvia de Moncayo – *Salvia officinalis*
salvia farinácea – *Salvia farinacea*
salvia india – *Buddleja perfoliata*
salvia orejona – *Salvia apiana*
salvia para la misa – *Hyptis collina*
salvia poblana – *Lippia callicarpaefolia, L. umbellata*
salvia real – *Buddleja, Hyptis, Lantana glandulosissima, Lippia callicarpaefolia, Salvia apiana, S. leucantha, S. officinalis*
salvia real de bolita – *Buddleja perfoliata*
salvia real de la sierra – *Salvia pachyphylla*
salvia real poblana – *Lippia callicarpaefolia*
salvia roja – *Salvia coccinea, S. elegans, S. gesneraeflora*
salvia roja asiática – *Salvia miltiorrhiza*
salvia romana – *Salvia sclarea*
salvia rosa – *Salvia pachyphylla*
salvia rusa – *Perovskia atriplicifolia*
salvia soberbia – *Salvia* x *superba*
salvia verticilada – *Salvia verticillata*
salvia virgen – *Salvia officinalis*
samán – *Albizzia saman, Coursetia glandulosa*
samandoca – *Yucca carnerosana*
samaritano – *Brosimum alicastrum*
samatito – *Ficus padifolia*
sambesarambe – *Boerhavia coccinea*
sambesarambe hembra – *Boerhavia spicata*
sambesarambe macho – *Boerhavia coccinea*
samija – *Mentzelia involucrata*
samo – *Coursetia glandulosa, Heliocarpus attenuatus, Lonchocarpus hermannii*
samo baboso – *Heliocarpus attenuatus*
samo blanco – *Heliocarpus attenuatus*
samo ochóname – *Heliocarpus palmeri*
samo prieto – *Coursetia glandulosa, Heliocarpus palmeri*
samota – *Coursetia glandulosa, Esenbeckia hartmanii*
sampaguita – *Jasminum sambac*
sampsuco – *Origanum majorana*
samuhú – *Chorisia insignis*
san – *Inga spuria*
sana – *Saccharum officinarum*
sanacoche – *Microsechium helleri*
San Agustín – *Pisonia flavescens, Stenotaphrum secundatum*
sanalotodo – *Cissus sicyoides, Kalanchoe pinnata, Vitis tiliifolia*
sanamunda – *Geum urbanum*
San Andrés – *Tecoma stans*
sanarí – *Heliocarpus attenuatus*
sana rogua – *Mascagnia macroptera*
San Bartolemé – *Cordia sebestina*
San Carlos – *Argemone ochroleuca, Pyrostegia venusta*
sándalo – *Juniperus excelsa, Pterocarpus santalinus, Santalum album*
sándalo blanco – *Santalum album*
sándalo falso – *Ulmus glabra*
sanderiana – *Dracaena sanderana*
sandía – *Citrullus colocynthis*
sandía chom – *Cucumis anguria*
sandía de la pasión – *Funastrum heterophyllum, Passiflora palmeri*
sandía del diablo – *Cucumis anguiria*
sandía de zopilote – *Cucumis anguria*
San Diego – *Antigonon leptopus, Laelia autumnalis, Piper peltatum, Thevetia peruviana*
San Dieguito – *Laelia anceps*
sandillita – *Passiflora palmeri*
sanduva – *Paspalum dilatatum*
San Francisco – *Peltophorum pterocarpum*
sangre de Cristo – *Berberis aquifolium* var *repens, Fumaria officinalis, Jatropha cardiophylla, J. curcas, Lychnis coronaria*
sangre de doncella – *Begonia gracilis*
sangre de drago – *Berberis aquifolium* var *repens, Croton draco, C. lecheri, C. sanguifluus, Jatropha, Pterocarpus officinalis*
sangre de dragón – *Daemonorops draco, Dracaena draco*
sangre de grado – *Croton draco, C. lecheri, Jatropha cinerea, J. spathulata, Potentilla rubra, Pterocarpus officinalis*
sangre de perro – *Croton draco*
sangre de toro – *Hamelia erecta, Pterocarpus officinalis, Salvia sessei, Spigelia longiflora*
sangre de venado – *Calamus draco, Daemonorops draco*
sangregado – *Croton draco, Euphorbia misera, Jatropha, Pterocarpus officinalis*
sangrenaria – *Warszewiczia coccinea*
sangre lengua – *Rubia tinctorium*
sangüeño – *Rhamnus frangula*

sangüeso – *Rubus idaeus*
sanguinaria – *Alternanthera sessilis, Ephedra aspera, E. compacta, E. pedunculata, Gomphrena decumbens, Halimium glomeratum, Helianthemum glomeratum, Hypericum pratense, Sanguinaria canadensis, Sanvitalia procumbens*
sanguinaria del Canadá – *Sanguinaria canadensis*
sanguinaria mayor – *Polygonum arenastrum, P. aviculare*
sanguinaria menor – *Draba cuneifolia*
sanguiñuelo – *Cornus sanguinea, Rhamnus frangula*
sanícula – *Prunella vulgaris, Prunus vulgaris, Sanicula europaea, S. marilandica*
sanícula europea – *Sanicula europaea*
sanícula hembra – *Astrancia major*
San Isidro – *Psilocybe cubensis*
San Isidro Labrador – *Psilocybe cubensis*
San Jacinto – *Moringa oleifera*
San José – *Allamanda cathartica*
San José del campo – *Penstemon parryi*
San Juan – *Jacquinia aurantica, Macrosiphonia hypoleuca*
San Juan del monte – *Brickellia greggii*
sanjuanero – *Echinocereus stramineus, Erblichia odorata*
San Juanito – *Cordia greggii, C. parvifolia, Jacquinia aurantiaca, J. macrocarpa* ssp *pungens, Microsiphonia hypoleuca*
San Miguel – *Antigonon leptopus, Cosmos sulphureus, Salvia sessei, Zinnia violacea*
San Miguelito – *Anoda thurberi, Antigonon leptopus, Ipomoea murucoides, Jacquemontia pringlei, Kallstroemia grandiflora*
San Nicolás – *Chrysactinia mexicana, Milla biflora, Thevetia peruviana*
sano – *Dasylirion wheeleri*
San Pablo – *Wigandia caracasana*
San Paul – *Tagetes jaliscana*
San Paulillo – *Tagetes elongata*
San Pedro – *Echinopsis pachanoi, Tecoma stans*
San Pedro de guía – *Macfadyena unguis-cati*
San Sebastián – *Cattleya skinneri*
sanseviera – *Sansevieria guineensis, S. trifasciata*
sanseviera marginada – *Sansevieria trifasciata*
san-shu – *Cunninghamia lanceolata*
Sansón negro – *Echinacea angustifolia*
santa – *Turbina corymbosa*
Santa Catarina – *Dahlia excelsa, Euphorbia pulcherrima*
Santa Cecilia – *Quisqualis indica*
Santa Isabel – *Salvia ballotaeflora*
Santa Lucía – *Trixis californica*
Santa María – *Adiantum wilsonii, Aristolochia brevipes, Calophyllum braziliense, Disocactus speciosus, Epiphyllum* x *Disocactus, Petrea arborea, Piper, Pluchea odorata, Salvia leucantha, Tagetes, Tanacetum parthenium*
Santa Marta – *Disocactus speciosus*
Santa Rita – *Epiphyllum hookeri, Ixora coccinea, Petrea arborea*
Santa Rosa – *Cannabis indica, C. sativa*
Santa Teresa – *Stevia serrata*
Santiago – *Piper peltatum*
santiaguero – *Russula cyanoxantha*
santilla de comer – *Piper sanctum*
santilla montés – *Piper angustifolium*
santimonia – *Chrysanthemum coronarium*
santipaula – *Saintpaulia ionantha*
Santo Domingo – *Cynodon oleufuensis, Hedeoma piperita, Mammea americana*
santolina – *Santolina chamaecyparisus*
santónico – *Artemisia pauciflora*
santónico americano – *Chenopodium ambrosioides*
San Vicente – *Galphimia glauca*
sapane – *Musa paradisiaca*
sapari – *Agave bovicornuta*
sape gigante – *Hyparrhenia rufa*
sapo – *Jatropha*
saponaria – *Anagallis arvensis, Gypsophilia paniculata, Sapindus saponaria, Saponaria officinalis*
saponaria albahaca – *Saponaria ocymoides*
saponaria común – *Saponaria officinalis*
saponaria falsa-albahaca – *Saponaria ocymoides*
saponella – *Saponaria officinalis*
sapoyolilla – *Russelia sarmentosa*
sapucaia – *Lecythis elliptica*
sapuchi – *Randia laevigata*
sapuchi de la sierra – *Randia mollifolia*
sapuli – *Agave bovicornuta*

saqui – *Agave fourcroydes*
saráame – *Turnera ulmifolia*
saracuacho – *Ribes*
saramago – *Eruca vesicaria* ssp *sativa*
saramatraca – *Peniocereus, Pereskiopsis porteri*
sarampión – *Condalia globosa* var *pubescens*
saramuyo – *Annona squamosa*
saranda – *Boerhavia erecta*
sarandí blanco – *Phyllanthus sellowianus*
sarapia – *Dipteryx odorata*
sarapico – *Dodecatheon clevelandii, D. meadia*
saraviki – *Prionosciadium townsendii*
sardonia – *Ranunculus sceleratus, Viburnum awabukii*
sargatillo – *Salix purpurea*
sargazo – *Najas marina, Sargassum herporhizum, S. vulgare*
sargazo gigante – *Macrocystis pyrifera*
sargazo vejigoso – *Fucus vesiculosus*
sarilla – *Origanum majorana, Thymus mastichina*
saripara – *Tillandsia limbata*
sarna de perro – *Rauwolfia hirsuta*
sarompillón – *Schaefferia cuneifolia*
sarumo – *Cecropia obtusifolia, C. peltata*
sarrapia – *Dipteryx odorata*
sasafrás – *Aletris farinosa, Laurus camphora, Persea podadenia, Sassafras albidum*
sasafrás americano – *Sassafras albidum*
sasan – *Rhynchosia pyramidalis*
sasaró – *Heliocarpus polyandrous*
saspuchi – *Bunchosia sonorensis*
satanicua – *Tabebuia pentaphylla*
satsuma – *Citrus reticulata*
satureja – *Satureja hortensis*
sauce – *Casuarina equisetifolia, Chilopsis linearis, Salix*
sauce blanco – *Salix alba, S. caprea, S. chilensis*
sauce cabruno – *Salix caprea*
sauce ceniciento – *Salix cinerea*
sauce colorado – *Salix chilensis, S. purpurea*
sauce chino tortuoso – *Salix matsudana*
sauce dafnoide – *Salix daphnoides*
sauce de Babilonia – *Salix babylonica*
sauce de Humboldt – *Salix chilensis*
sauce desmochado – *Salix alba*
sauce enano reticulado – *Salix reticulata*
sauce falsa-lauréola – *Salix daphnoides*
sauce lanudo – *Salix lanata*
sauce lanuginoso – *Salix lanata*
sauce laurifolio – *Salix pentandra*
sauce llorón – *Salix alba* var *tristis, S. babylonica*
sauce llorón americano – *Salix purpurea*
sauce llorón dorado – *Salix chrysocoma*
sauce macho – *Salix bonplandiana*
sauce negro – *Salix nigra*
sauce purpúreo – *Salix daphnoides*
sauce ramagara – *Salix bonplandiana*
sauce reticulado – *Salix reticulata*
sauce rojo – *Salix laevigata, S. purpurea*
sauce sacacorchos – *Salix matsudana*
sauce tortuoso – *Salix matsudana*
sauce triste – *Salix alba* var *tristis*
sauce violeta – *Salix daphnoides*
saucillo – *Quercus durifolia, Q. viminea*
saucle – *Cranichis speciosa, Epidendrum pastoris*
saúco – *Alnus rubra, Bocconia arborea, Ehretia tinifolia, Sambucus*
saúco amarillo – *Tecoma stans*
saúco azul – *Sambucus mexicana*
saúco de racimos – *Sambucus racemosa*
saúco dulce – *Sambucus canadensis*
saúco hediondo – *Citharexylum berlandieri*
saúco negro – *Sambucus nigra*
saúco rojo – *Sambucus racemosa*
saugran – *Plumeria rubra*
saugua – *Sambucus mexicana*
sauhui – *Panicum sonorum*
saumayate – *Castilleja arvensis*
sautle – *Salix nigra*
sauz – *Salix*
sauz gatillo – *Vitex agnus-castus*
sauz rojo – *Salix laevigata*
sauz serrano – *Salix nigra*
saxífraga – *Saxifraga*
saxífraga cespitosa – *Saxifraga cespitosa*
saxífraga de bosque – *Saxifraga umbrosa*
saxífraga de hojas grandes – *Bergenia cordifolia*
saxífraga de los arroyos – *Saxifraga rivularis*
saxífraga dorada – *Chrysosplenium glechomifolium*
saxífraga rosácea – *Saxifraga decipiens*
saxífraga tricolor – *Saxifraga stolonifera*

saya – *Amoreuxia gonzalezii, A. palmatifida*
saya mome – *Amoreuxia gonzalezii*
sayasaga – *Amoreuxia palmatifida*
sayolistle – *Ceanothus coeruleus*
scheuchzeria palustre – *Scheuchzeria palustris*
sciatica – *Bonplandia geminiflora, Dyschoriste hirsutissima*
scoba – *Chloracantha spinosa*
scocosacat – *Leptochloa filiformis*
seafortia – *Archontophoenix cunninghamiana*
sebestén – *Cordia myxa*
sebucán español – *Stenocereus fimbriatus*
secapalo – *Phoradendron*
secoya – *Sequoia sempervirens*
secoya de California – *Sequoia sempervirens*
secoya gigante – *Sequoiadendron giganteum*
secoya roja – *Sequoia sempervirens*
secoya wellingtoniana – *Sequoiadendron giganteum*
seda vegetal – *Boehmeria nivea*
sedita – *Conostegia xalapensis*
sedo – *Sedum*
sedo de Kamtschatka – *Sedum kamtschaticum*
sego – *Calochortus bruneaunis*
seguidilla – *Psophocarpus tetragonolobus*
sehualca – *Cucurbita moschata*
seibadal – *Thalassia testudinum*
seje – *Jessenia bataua*
selaim – *Phaseolus vulgaris*
selima – *Salvia officinalis*
selima fina – *Salvia officinalis*
sello de oro – *Hydrastis canadensis*
sello de Salomón – *Polygonatum biflorum, P. odoratum*
sello de Salomón europeo – *Polygonatum multiflorum*
sello de Salomón oloroso – *Polygonatum odoratum*
sello dorado – *Hydrastis canadensis*
semaluca – *Justica californica*
semencontra – *Artemisia pauciflora*
semeruco – *Malpighia emarginata*
semilla brincadora – *Sebastiana pavoniana*
semilla de culebra – *Abelmoschus moschatus, Abrus precatorius*
semilla del amor – *Lomatium foeniculaceum*
semilla de la Virgen – *Ipomoea violacea, Turbina corymbosa*
semilla del obispo – *Myroxylon balsamum* var *pereirae*
semilla oriental – *Vigna radiata*
semilla para los pájaros – *Brassica rapa*
semiramis – *Magnolia grandiflora*
sempiterna – *Anaphalis margaritacea, Gomphrena decumbens, G. globosa, Helichrysum orientale*
sen – *Kalopanax septemlobus, Senna*
sen del país – *Senna laevigata*
senecio – *Senecio vulgaris*
senecio cinerario – *Senecio cineraria*
senecio común – *Senecio greyi, S. vulgaris*
senecio de hojas integérrimas – *Senecio integrifolius*
senecio de los bosques – *Senecio sylvaticus*
senecio elegante – *Senecio elegans*
senecio escandente – *Delairea odorata*
senexante – *Vanilla pompona*
sen indigena – *Colutea arborescens*
senita – *Pachycereus schottii*
senna – *Cassia marilandica, Senna*
senserina – *Thymus vulgaris*
sen silvestre – *Senna alata, S. laevigata*
sensitiva – *Mimosa albida, M. pudica*
señif – *Lophophora williamsii*
señora de noche – *Clerodendron*
señorita – *Asclepias curassavica, A. glaucescens, A. incarnata, Clitocybe, Turbina corymbosa*
serbal – *Sorbus aria, S. domestica*
serbal americano – *Sorbus americana*
serbal blanco – *Sorbus aria*
serbal común – *Sorbus domestica*
serbal chino de Hupeh – *Sorbus hupehensis*
serbal de los cazadores – *Sorbus aucuparia, S. dumosa*
serbal de los pájaros – *Sorbus aucuparia*
serbal de Suecia – *Sorbus intermedia*
serbal doméstico – *Sorbus domestica*
serbal finés – *Sorbus* x *hybrida*
serbal híbrido de Turingia – *Sorbus* x *thuringiaca*
serbal silvestre – *Sorbus aucuparia*
serbal sueco – *Sorbus intermedia*
serbo – *Sorbus*
sereno – *Ageratum conyzoides*
serillo – *Sporobolus indicus*
seringa – *Hevea brasiliensis*

serita – *Conostegia xalapensis*
serpentaria – *Dracunculus vulgaris*
serpentaria de séneca – *Polygala seneca*
serpiente – *Trichosanthes cucumerina* var *anguina*
serpiente vegetal – *Trichosanthes cucumerina* var *anguina*
serpillo – *Thymus serpyllum*
serpol – *Thymus pulegioides, T. serpyllum*
servato – *Peucedanum officinale*
servilleta del pobre – *Luffa aegyptiaca*
serracenia – *Cecropia peltata*
serradella – *Ornithopus sativus*
serrallón – *Lactuca virosa*
serrano – *Capsicum annuum*
serrátula – *Liatris spicata*
serrátula de los tintoreros – *Serratula tinctoria*
serrucho – *Mariscus jamaicensis, Nelumbo lutea*
sésamo – *Sesamun indicum*
seso vegetál – *Blighia sapida*
seta – *Basidiomycetes, Agaricus campestris*
seta de abril – *Tricholoma gambosum*
seta de alpaca – *Pleurotus ostreatus* var *vulgaris*
seta de cañaheja – *Pleurotis eryngii* var *ferulae*
seta de cardo – *Pleurotus eryngii, P. ostreatus*
seta de estepa – *Tricholoma terreum*
seta del bonete rojo – *Amanita muscaria*
seta de los caballeros – *Tricholoma flavovirens*
seta de mimbre – *Pleurotus ostreatus* var *vulgaris*
seta de ostra – *Pleurotus ostreatus*
seta de primavera – *Tricholoma gambosum*
seta de San Jorge – *Tricholoma gambosum*
seta de vaca – *Lepista saeva*
seta japonesa – *Lentinus edodes*
seta mágica – *Amanita muscaria*
setaria – *Setaria anceps*
setaria glauca – *Setaria pumila*
setaria húngara – *Setaria italica* var *moharica*
setaria italiana – *Setaria italica*
setaria kazangula – *Setaria anceps*
setaria narok – *Setaria anceps*
setaria perenne – *Setaria gracilis*
setaria verde – *Setaria viridis*
setaria verticilada – *Setaria verticiliata*
seta sagrada – *Amanita muscaria*
setero – *Eryngium campestre*
sherardia de los prados – *Sherardia arvensis*
shiitake – *Lentinus edodes*
sibaldia procumbente – *Sibbaldia procumbens*
siberiana – *Brassica napus* var *pabularia*
sicaján – *Jacquinia aurantiaca*
sicamole – *Phytolacca octandra*
sicamor – *Cercis siliquastrum*
sicaté – *Jacquinia aurantica*
sicche – *Jacquinia aurantica*
sicjá – *Jacquinia aurantiaca*
sicoche – *Goldmania foetida*
sicomoro – *Acer pseudoplatanus, Azadirachta indica, Ficus vulgaris, Platanus occidentalis, P. orientalis, P. racemosa*
sicomoro falso – *Melia azedarach*
sicte – *Stemmadenia tomentosa*
sictillo – *Stemmadenia tomentosa*
siche – *Jacquinia aurantica*
sida – *Sida rhombifolia*
sidra – *Malpighia emarginata*
siempre me veras asi – *Iresine*
siempreviva – *Bryophyllum calycinum, Dudleya, Echeveria, Gomphrena decumbens, G. globosa, Helichrysum bracteatum, Kalanchoe pinnata, Limonium sinuatum, Sedum, Selaginella, Tradescantia zebrina*
siempreviva amarilla – *Sedum dendroideum, S. praealtum*
siempreviva aracnoidea – *Sempervivum arachnoideum*
siempreviva azul – *Limonium californicum, L. sinuatum*
siempreviva bracteada – *Helichrysum bracteatum*
siempreviva cimarrona – *Gomphrena sonorae*
siempreviva común – *Sempervivum tectorum*
siempreviva china – *Aglaonema modestum*
siempreviva de hojas estrechas – *Helichrysum italicum*
siempreviva de telarañas – *Sempervivum arachnoideum*
siempreviva itálica – *Helichrysum italicum*
siempreviva mayor – *Sempervivum tectorum*
siempreviva mayor común – *Sempervivum tectorum*
siempreviva menor – *Sedum acre, S. album, S. spathulifolium*
siempreviva silvestre – *Gomphrena decumbens*
siendre – *Porophyllum gracile*

sierrilla – *Serjania mexicana*
siete colores – *Lantana camara*
sieteenrama – *Potentilla erecta*
siete sayo – *Robinia pseudoacacia*
sieva – *Phaseolus lunatus*
sigreropo – *Lycium andersonii*
sihuapatl – *Montanoa tomentosa*
sihuapatli – *Montanoa tomentosa*
siitz – *Justica spicigera*
silene acaule – *Silene acaulis*
silene alpestre – *Silene alpestris*
silene armeria – *Silene armeria*
silene conoidea – *Silene conoidea*
silene de flores nocturnas – *Silene noctiflora*
silene de hojas anchas – *Silene latifolia*
silene de rocalla – *Silene acaulis*
silene dicotomal – *Silene dichotoma*
silene francesa – *Silene gallica*
silene venosa – *Silene vulgaris*
silene violácea – *Silene coeli-rosa*
silmo – *Phoradendron macrophyllum, P. tomentosum*
simaroa – *Populus simaroa*
simaruba – *Simaruba officinalis*
si me quieres, no me quieres – *Lasthenia californica*
simiente de papagayos – *Carthamnus tinctorius*
simiente dulce – *Pimpinella anisum*
simonillo – *Calea zacatechichi, Conyza filaginoides, Erigeron*
simpazúchil – *Tagetes erecta*
sina – *Pachycereus schottii, Stenocereus alamosensis, S. kerberi*
sina barbona – *Pachycereus schottii*
sina de la mar – *Peniocereus striatus*
sínala – *Acacia cymbispina*
sinapismo – *Descuriana pinnata*
sinchiquigua – *Sporobolus indicus*
sínfito – *Potentilla, Symphytum officinale*
sínfito mayor – *Symphytum officinale*
sinforina – *Symphoricarpos albus*
singonio – *Syngonium podophyllum*
sinicuichi – *Erythrina, Rhynchosia phaseoloides*
sinicuilche – *Heimia salicifolia*
sinita – *Pachycereus schottii*
sinita barbona – *Echinocereus engelmannii*
sintescochit – *Asclepias curassavica*
sinvergüenza – *Commelina nudiflora, Kalanchoe pinnata*
sinvergüenza silvestre – *Commelina nudiflora*
sipa – *Acrocomia aculeata*
sipariqui – *Eysenhardtia polystachya*
sipehui – *Euphorbia californica*
sipúraca – *Quercus rugulosa*
siqueté – *Jacquinia aurantiaca, J. macrocarpa, J. pungens*
siquilibachi – *Zea mays*
siquimuni – *Phaseolus vulgaris*
siquinay – *Vernonia deppeana*
síquiri – *Trichilia americana*
síquiri tájcara – *Trichilia hirta*
siranda – *Ficus*
sirasil – *Ardisia revoluta*
sire – *Malpighia emarginata*
siricote – *Cordia boissieri, C. dodecandra, C. sebestena*
siricote blanco – *Cordia sebestena*
sirimo – *Tilia mexicana, T. occidentalis*
sirin – *Miconia*
siringa – *Hevea brasiliensis*
sirundanicua – *Caesalpinia pulcherrima*
sirunda urueta – *Diospyros ebenaster*
sisal – *Agave sisalana, Ferocactus viridescens*
sisalxiu – *Kalanchoe pinnata*
sisallo – *Bassia hyssopifolia*
sisimbrio – *Sisymbrium officinale*
sit – *Lasiacis divaricata*
sitagape – *Haematoxylum brasiletto*
sitavaro – *Vallesia glabra*
siti – *Zingiber officinale*
sitiporo – *Desmanthus covillei*
sitka – *Picea sitchensis*
sit-suuk – *Hackelochloa granularis*
situni – *Rubus*
siviri – *Cylindropuntia arbuscula, C. thurberi, C. versicolor, Peniocereus viperinus*
sóbari – *Agave parviflora* ssp *flexiflora*
sobrepalo – *Aechmea bracteata*
sobrero – *Quercus suber*
socol – *Yucca rigida, Y. schottii*
socolume – *Agave applanata*
soconochtli – *Opuntia megarhiza*
sochi – *Amanita caesarea*
sofora – *Sophora japonica*
soguilla – *Eclipta prostrata, Elytraria squamosa, Heliotropium angiospermum*

soja – *Glycine max*
soja hispida – *Glycine max*
soja max – *Glycine max*
solano – *Solanum nigrum, S. pseudocapsicum, S. verbascifolium, S. wendlandii*
solano de tres flores – *Solanum triflorum*
solano dulce – *Solanum dulcamara*
solano furioso – *Atropa belladonna*
solano grácil – *Solanum gracile*
solano negro – *Solanum nigrum*
solano rostrado – *Solanum rostratum*
solano seudopimienta – *Solanum pseudocapsicum*
solano vejigoso – Physalis alkekengi
solano verde – Solanum sarrachoides
soldadiente – Geranium potentillifolium, G. potentilloides
soldadillo – Asclepias incarnata, Piper angustifolium
soldadito – Asclepias curassavica
soldanela mayor – Calystegia soldanella
sol de agua – Disocactus phyllanthoides, Nymphaea ampla
sol de las Indias – Helianthus annuus
solería – Cordia alliodora
solerillo – Petrea arborea
solimán – Asclepias linaria, Croton ciliato-glandulosus, Urtica dioica
solimán blanco – Croton ciliato-glandulosus
solimanché – Hura polyandra
solitaria – Muehlenbeckia platyclados
soltero – Petrea arborea, P. volubilis
soma – Peganum harmala
sombrerete – Terminalia amazonia
sombrerillo – Hydrocotyle bonariensis
sombrerillo de agua – Hydrocotyle mexicana, H. umbellata
sombrerito de agua – Hydrocotyle ranunculoides, H. vulgaris
sombrero de don Zope – Aristolochia grandiflora
sombrero de zapata – Ratibida columnaris
somerio – Liquidambar styraciflua
sonadora – Astragalus, Crotalaria pumila
sonaja – Crotalaria incana, C. sagittalis
soncoya – Annona purpurea
sonora – Lantana camara, L. involucrata
sonora roja – Lantana camara
sonorita – Commicarpus scandens
sonpante – Russelia sarmentosa
sopépari – Senecio hartwegii
sopitza – Armillaria mellea
soralea – Psoralidium lanceolatum
sorbo – Sorbus domestica
sorgo – Sorghum bicolor, S. vulgare
sorgo almum – Sorghum almum
sorgo azucarado – Sorghum saccharatum
sorgo blanco – Sorghum bicolor
sorgo común – Sorghum bicolor, S. vulgare
sorgo de Alepo – Sorghum halapense
sorgo de escobas – Sorghum bicolor
sorgo de grano – Sorghum vulgare
sorgo de Sudán – Sorghum bicolor
sorgo dulce – Sorghum vulgare
sorgo forrajero – Sorghum bicolor
sorgo negro – Sorghum almum
sorgo negro argentino – Sorghum almum
sorgo perenne – Sorghum almum
sorgo sacarífero – Sorghum saccaratum
sorgo selvático – Sorghum halepense
sorguillo – Sorghum halepense
soromuta – Muhlenbergia macroura
sosa – Solanum lanceolatum, S. marginatum, S. verbascifolium, Suaeda, Wigandia caracasana, W. kunthii
sosa fruticosa – Suaeda moquinii
soscilchac – Tillandsia usneoides
sospó – Bombax ellipticum
sosquil – Agave fourcroydes, A. sisalana
sosxocoyoli – Oxalis latifolia
sotol – Dasylirion, Dioon edule, Nolina, Yucca decipiens
sotol chico – Nolina microcarpa
sotol chiquito – Nolina microcarpa
sotolillo – Agave falcata, Yucca whipplei
sotolín – Beaucarnea gracilis
sotolito – Agave falcate, A. striata
soya – Glycine max
soyacal – Carludovica palmata
soyal capulín – Erythea edulis, Phoenix dactylifera
soyale – Erythea edulis
soyamiche – Cryosophila nana
soyate – Agave falcate, A. striata, Beaucarnea inermis, Erythea edulis, Yucca
soyatillo – Agave falcata
soyaviche – Cryosophila nana
spergularia arenaria – Spergularia rubra

spergularia de boccone – Spergularia bocconii
spergularia marginada – Spergularia media
spergularia marina – Spergularia marina
stachys lanosa – Stachys byzantina
sternbergia amarilla – Sternbergia lutea
stuartia – Stewartia pseudocamellia
suavoli – Descuriana pinnata
subaemunim – Phaseolus vulgaris
subinché – Acacia farnesiana
subularia acuática – Subularia aquatica
suco – Maranta arundinacea
suchete – Liquidambar styraciflua
suchi – Quercus rugosa
suchicahue – Cordia alliodora
súchil – Hibiscus rosa-sinensis, Plumeria acutifolia, P. rubra, Talauma mexicana
suchilcahue – Plumeria acutifolia
suchilete – Plumeria rubra
sudán – Sorghum sudanense
sudán perenne – Sorghum almum
suegras y nueras – Viola tricolor
suelda – Agonandra racemosa, Symphytum officinale, Potentilla
suelda blanca – Polygonatum odoratum
suelda consuelda – Agonandra racemosa, Chiococca alba, Erblichia odorata, Potentilla, Symphytum officinale
sueño – Cyperus laxus
suetsinic-che – Alvaradoa amorphoides
sufricago – Litsea glaucescens
sufricalla – Litsea glaucescens
sugi – Cryptomeria japonica
suma – Pfaffia paniculata
sumaco – Rhus mollis
sumaco cimarrón – Rhus mollis, R. virens
sumpual – Cosmos sulphureus
sunami – Ariocarpus fissuratus
sunté – Tithonia
surbana – Panicum coloratum
suruata – Gossypium hirsutum
suspinché – Hamelia erecta
suspiro – Mirabilis longiflora
susutí – Pisonia capitata
swertia perenne – Swertia perennis

tabaco – Lobelia, Nicotiana
tabaco alar – Nicotiana alata
tabaco amarillo – Nicotiana glauca
tabaco borde – Arnica montana, Atropa belladona
tabaco británico – Tussilago farfara
tabaco cimarrón – Neurolaena lobata, Nicotiana, Solanum verbascifolium, Verbascum thapsus, Wigandia caracasana, W. kunthii, W. urens
tabaco de adorno – Nicotiana alata
tabaco de coyote – Nicotiana
tabaco del Brasil – Nicotiana rustica
tabaco de los bosques – Nicotiana sylvestris
tabaco de macuache – Nicotiana rustica
tabaco de montaña – Arnica montana
tabaco de papanta – Nicotiana ipomopsiflora
tabaco de perro – Nicotiana clevelandii
tabaco indio – Lobelia inflata
tabaco lampiño – Nicotiana glauca
tabaco loco – Nicotiana obtusifolia
tabaco mayor – Nicotiana tabacum
tabaco menor – Nicotiana rustica
tabaco moro – Nicotiana glauca
tabaco moruno – Nicotiana glauca
tabacón – Nicotiana glauca, Wigandia caracasana
tabaco ornamental – Nicotiana alata
tabaco papanta – Nicotiana obtusifolia
tabaco pequeño – Nicotiana rustica
tabaco rústico – Nicotiana rustica
tabachín – Caesalpinia californica, C. crista, C. pulcherrima, Calliandra brandegei, Delonix regia, Jacaranda acutifolia
tabachín amarillo – Caesalpinia pulcherrima
tabachín chico – Hoffmannseggia multijuga
tabachín de la sierra – Caesalpinia pulcherrima
tabachín del monte – Caesalpinia mexicana
tabachín grande – Delonix regia
tabaniza blanca – Diplotaxis erucoides
tabaquera – Nicotiana tabacum
tabaquillo – Bouvardia longiflora, Clinopodium laevigatum, Hedeoma piperita, Lippia umbellata, Nicotiana, Satureja macrostema, Wigandia
tabaquillo chico – Hedeoma piperita
tabaquillo de coyote – Nicotiana obtusifolia
tabaquillo grande – Clinopodium laevigatum, Satureja macrostema
tabaquillo oloroso – Hedeoma piperita, Petunia parviflora

tabardillero – Stylosanthes guianensis
tabardillo – Ageratina herbacea, Calliandra, Eupatorium hyssopium, Mentha arvensis, Piqueria trinervia
tabardillo de la sierra – Calliandra peninsularis
tabarjeca – Capparis flexulosa
Tabasco – Capsicum frutescens
tabché – Conocarpus erecta, Rhizophora mangle
tabelojeca – Capparis flexuosa
tabi – Trixis radialis
tabla – Ficus padifolia
tablero de damas – Fritillaria meleagris
tablote – Guazuma ulmifolia
tabuió – Tabebuia impetiginosa
tabuquit – Turnera diffusa
tacali – Tillandsia usneoides
tacalote – Centaurea americana, Entada phaseoloides, E. polystachya, E. scandens
tacamahaca – Bursera gummifera, B. lancifolia
tacansahac – Populus balsamifera ssp trichocarpa
taciste – Acoelorraphe wrightii
tacón – Tithonia
tacopaste – Aristolochia californica
tacopate – Aristolochia californica, A. grandiflora
tacopatl – Aristolochia grandiflora
tacopaxtle – Aristolochia grandiflora, A. mexicana
tacora – Euphorbia misera
tacote – Galinsoga parviflora, Nicotiana glauca, Podachaenium eminens, Salvia mexicana, Viguiera, Zinnia violacea
tacote cedroso – Rumfordia connata
tacote chino – Verbesina pustulata
tacso – Passiflora mollissima
tacuara – Guadua latifolia
tache – Myroxylon balsamum var pereirae, Tecoma stans
tachicón – Curatella americana
tachicón de sabana – Tetracera volubilis
tachino – Argemone mexicana, A. ochroleuca
tachinole – Jatropha angustidens, Urera caracasana
tachiste – Vernonia deppeana
taeihcholi – Agave polianthiflora
tagetes – Tagetes tenuifolia
tagetes de hojas finas – Tagetes lunulata
taglli – Pernettya
tago-mani – Hordeum vulgare
tagua – Phytelephas macrocarpa
taguapillo – Zapoteca anomalia
taguapo – Pseudobombax palmeri
taguaro – Datura metel
tah – Viguiera dentata
tahcu – Sabal uresana
tahchaac – Ustilago maydis
taiché – Capparis indica
taiehcholi – Agave hartmanii
tainorej – Euphorbia colorata
tairago – Acourtia cordata
taiwania – Taiwania cryptomerioides
tajehuali – Tragia nepetaefolia
tají – Prosopus juliflora
takob – Annona muricata
tala – Celtis trinervia
talacáo – Lippia umbellata, Montanoa patens, M. tomentosa ssp rosei
talamapate – Gnaphalium leucophyllum
talampacate – Gnaphalium leucophyllum
talaya – Tamarix ramosissima
talayote – Gonolobus erianthus, Macroscepis obovata, Marsdenia edulis, Matelea cordifolia, M. parvifolia, Passiflora foetida, Rothrockia cordifolia, Vincetoxicum caudatum, V. hastulatum
talayote de costilla – Canavalia villosa
talcapache – Caesalpinia pulcherrima
talcoite – Brosimum alicastrum
talcote – Lobelia fenestralis
talia – Thalia geniculata
talife – Geonoma magnifica
talipot – Corypha umbraculifera
talipote – Corypha umbraculifera
talis – Geonoma magnifica
taliste – Lonchocarpus hermannii
talote – Sechium edule
talpakapache – Caesalpinia pulcherrima
talpica – Arnica montana
talquezal – Panicum virgatum
talusúchil – Tecoma stans
talzan – Persea americana
tallo azul de Kingranch – Bothriochloa ischaemum var songarica

tallo azul matorralero – *Andropogon glomeratus*
tama – *Tamarindus indica*
tamaloyota – *Cucurbita maxima*
taman – *Gossypium*
tamanché – *Malvaviscus grandiflorus*
tamanchich – *Malvaviscus arboreus*
tamarack – *Larix laricina*
tamarilla – *Cyphomandra betacea, Helianthemum nummularium, Rapistrum rugosum*
tamarindillo – *Aeschynomene americana, Leucaena glauca*
tamarindo – *Tamarindus indica*
tamarindo bastardo – *Leucaena leucocephala*
tamarindo culí – *Averrhoa carambola*
tamarindo chino – *Averrhoa carambola*
tamarindo malabar – *Garcinia cambogia*
tamarindo silvestre – *Leucaena leucocephala*
tamarisco – *Tamarix gallica, T. ramosissima*
tamariz – *Tamarix gallica, T. parviflora, T. ramosissima*
tamarrago – *Armoracia rusticana*
tambácuaro – *Nymphaea mexicana*
tambor – *Bombax ellipticum, Cordia alliodora*
tamborcillo – *Eichhornia crassipes*
tamborillo – *Combretum farinosum*
tamichil – *Krameria erecta*
tampala – *Amaranthus gangeticus*
tampicerán – *Platymiscium trifoliolatum*
tampico – *Agave lechuguilla, Clusia major*
tamujo – *Rhamnus purshiana*
tamulero – *Cocoloba barbadensis*
tanaceto – *Tanacetum vulgare*
tanaceto común – *Tanacetum vulgare*
tanal – *Laelia superbiens*
tanalito – *Odontoglossum pulchellum*
tananiní – *Parthenium argentatum, P. incanum*
tanay – *Heliconia bihai, H. latispatha*
tanay colorado – *Heliconia bihai*
tancazche – *Zanthoxylum fagara*
tancochi – *Martynia annua*
tanche – *Codiaeum variegatum, Croton fragilis*
tanchichi – *Tithonia*
tangelo – *Citrus reticulata x C. paradisi*
tangerina – *Citrus reticulata*
tangkong – *Ipomoea aquatica*
tangor – *Citrus reticulata x C. sinensis*
tango rojizo – *Justicia carnea*
tanibata – *Ipomoea stans*
tanillo – *Tabebuia impetiginosa*
tanito – *Thalictrum fendleri*
tansé – *Tanacetum vulgare*
tantsanacua – *Coriandrum sativum*
tapa cántaro – *Piper sanctum*
tapacapón – *Ceratozamia mexicana*
tapaco – *Stemmadenia tomentosa var palmeri*
tapacola – *Waltheria americana*
tapaculo – *Guazuma ulmifolia*
tapachichi – *Salvia mexicana*
tapachile – *Dodonaea viscosa*
tapahuile – *Quercus elliptica*
tapaojo – *Psittacanthus calyculatus*
taparera – *Capparis spinosa*
taparo – *Crescentia cujete*
taparo de chuco – *Couroupita guianensis*
taparón – *Couroupita guianensis*
tapate – *Datura stramonium*
tapa vergüenza – *Mimosa albida*
tapemete – *Agave vivipara*
tapete – *Pennisetum clandestinum*
tapichoqua – *Hintonia latiflora*
tapintzirán – *Dalbergia granadillo*
tapioca – *Manihot esculenta*
tapiro – *Buddleja wrightii, Sambucus mexicana*
tapisote – *Lathyrus ochrus*
tápokoli – *Prionosciadium townsendii*
tapona – *Lepidium virginicum, Opuntia tapona*
tapone – *Epithelantha micromeris*
tapsia – *Thapsia garganica*
tara – *Caesalpinia spinosa*
tarabundi – *Xanthosma robustum*
tarachique – *Dodonaea viscosa*
taraiz – *Salix exilifolia, S. taxifolia*
taraje – *Tamarix africana, T. gallica*
tarantacua – *Castilla elastica*
tarantana – *Senna alata*
tararaco – *Hippeastrum*
tarasca – *Phytolacca octandra*
tarasca de jardín – *Antirrhinum majus*
tarasca de negros – *Phytolacca octandra*
taraxacón – *Taraxacom officinale*
taray – *Caesalpinia crista, Eysenhardtia polystachya, Salix exigua, S. taxifolia, Tamarix gallica, T. ramosissima*
taray africano – *Tamarix africana*
taray de flores pequeñas – *Tamarix parviflora*
taray de México – *Eysenhardtia polystachya*

taray de río – Salix taxifolia
taray gálico – Tamarix gallica
taray hispánico – Tamarix africana
tarco – Jacaranda acutifolia
tarecua – Agaricus campestris
tarecua cahuica – Amanita muscaria
tarepe – Lantana camara, L. involucrata
tarepeni – Oenothera rosea
tariña – Manihot esculenta
taro – Colocasia antiquorum
tártago – Euphorbia lathyris, Ricinus communis
tártago mayor – Euphorbia characias
tárrago – Salvia sclarea
tarrito – Penstemon barbatus, P. campanulatus, P. gentianoides, Stenocereus thurberi
tarro – Arundo donax, Guadua
tasajillo – Cylindropuntia kleiniae, C. leptocaulis, C. ramosissima, Pereskia lycnidiflora, Pereskiopsis aquosa, P. diguetii
tasajillo rojo – Cylindropuntia kleiniae
tasajo – Acanthocereus occidentalis, Cylindropuntia, Hylocereus undatus, Stenocereus alamosensis
tasajo macho – Cylindropuntia imbricata
tasaui – Sorghum bicolor, S. vulgare
táscate – Cupressus arizonica, C. lusitanica, Juniperus
táscate blanco – Juniperus deppeana
táscate espinosa – Juniperus durangensis
tasco – Passiflora mollissima
tascuaya – Iresine calea
tásiro – Jacquinia macrocarpa ssp pungens
tasiste – Acoelorraphe wrightii, Paurotis wrightii
tasuchi – Bromelia karatas
tatache – Rhipsalis baccifera
tatachinole – Kohleria deppeana, Tournefortia hartwegiana
tatalencho – Gymnosperma glutinosium, Selloa glutinosa
tatio – Croton ciliato-glandulosus
tatsin – Phaseolus vulgaris
tatsingueni – Cattleya
tauchya – Diospyros ebenaster
tauta – Agave parviflora ssp flexiflora
tautilla – Agave parviflora ssp flexiflora
taxodio – Taxodium distichum
taxodio de pantano – Taxodium distichum
taya – Xanthosoma sagittifolium
tayacua – Manihot caudata
tayota – Sechium edule
taza blanca – Russula delica
té – Camellia assamica, C. sinensis, Eriogonum inflatum, Thelesperma megapotamicum
teatlale – Cupressus lusitanica
té borde – Chenopodium ambrosioides
té brigham –Ephedra nevadensis, E. viridis
teca – Tectona grandis
tecapulín – Conostegia xalapensis, Pernettya ciliata
tecats – Cucurbita maxima
tececeque – Alternanthera caracasana
té cedrón – Aloysia triphylla
teco – Bromelia lasiantha
tecointli – Euchlaena mexicana
tecolosapo – Randia echinocarpa
tecolote – Tillandsia imperialis
tecolotillo – Randia rosei, Schoepfia schreberi
tecolotito – Schoepfia schreberi
tecolumate – Tillandsia imperialis, T. prodigiosa
tecomasuche – Cochlospermum vitifolium
tecomasuchiamatl – Ficus involuta
tecomasúchil – Cochlospermum vitifolium, Solandra maxima, S. nitida
tecomate – Amanita caesarea, Cochlospermum vitifolium, Crescentia alata, C. cujete, Lagenaria siceraria
tecomate ayele – Crescentia alata
tecomate de sonajas – Crescentia alata
tecomatl – Crescentia alata, C. cujete
tecomaxóchitl – Cochlospermum vitifolium, Solandra guerrerensis, S. maxima, S. nitida
tecomblate – Baileya, Condalia, Microrhamnus ericoides, Ziziphus obtusifolia
tecomelé – Tillandsia prodigiosa
tecopalcuáhuitl – Bursera odorata, Copaifera officinalis
tecote – Viguiera lacinata, V. tomentosa
tecotéhue – Dahlia excelsa, D. imperialis
tecuanxóchitl – Stanhopea tigrina
tecuyani – Datura stramonium, D. wrightii
té de Arabia – Catha edulis
té de Castilla – Lippia alba
té de corral – Bidens aurea, B. pilosa

té de cota – Thelesperma megapotamicum
té de España – Chenopodium ambrosioides
té de Europa – Veronica hederifolia, V. officinalis
té de hiera – Corchorus siliquosus
té de Java – Orthosiphon aristatus
té de la tierra – Corchorus siliquosus
té del Canadá – Gaultheria odorata, G. procumbens
té de limón – Andropogon schaenanthus, Cymbopogon citratus
té del monte – Clinopodium laevigatum, Satureja macrostema
té de los Jesuitas – Chenopodium ambrosoides
té de los Navajos – Thelesperma megapotamicum
té del país – Lippia graveolens
té de malva – Corchorus siliquosus
té de México – Chenopodium ambrosioides
té de menta – Agastache micrantha
té de milpa – Bidens
té de milpa amarillo – Bidens aurea, B. tetragosa
té de milpa blanco – Bidens odorata, B. pilosa var radiata
té de Oswego – Monarda citriodora
té de Paraguay – Ilex paraguariensis
té de Pensilvania – Monarda didyma
té de rancheros – Hedeoma piperita
té de sena – Senna angustifolia
té de Suecia – Linnaea borealis
teesdalia de hojas de mastuerzo – Teesdalia coronopifolia
té español – Chenopodium ambrosioides
tehuitzli - Sapindis saponaria
tehuixtle – Sapindus saponaria
tehuizote – Dasylirion
teihuisí – Rivina humilis
teijoco de puerco – Canavalia ensiformis
teijoco de sable – Canavalia gladiata
tejamanil – Clitocybe squamulosa
tejamanilero – Clitocybe squamulosa, Laccaria laccata, Lyophyllum decastes
tejana – Paspalum notatum
tejedora – Cynodon dactylon
tejo – Taxus baccata
tejo común – Taxus baccata
tejocote – Crataegus mexicana, C. pubescens, C. wootoniana
tejo de fortune – Cephalotaxus fortunei
tejo de Hokkaido – Cephalotaxus harringtonia
tejo de Japón – Taxus cuspidata
tejo irlandés – Taxus baccata var fastigata
tejo japonés – Taxus cuspidata
tejón de pino – Phoradendron
tejo negro – Taxus baccata
tejoroso – Genipa americana
tekhalal – Arundo donax
telaraña – Muhlenbergia porteri
telaspio de los campos – Thlaspi arvense
telaspio montano – Thlaspi montanum
telcocox – Phytolacca icosandra
telea – Ptelea
telecate blanco – Montanoa hibiscifolia
telefio – Sedum spectabile
telefio bastardo – Sedum spurium
teléfono – Epipremnum pinnatum
teliconté – Syngonium podophyllum
telicso – Yucca elephantipes
té limón – Cymbopogon citratus, C. nardus, Dalea lumholtzii, Pectis stenophylla
telkox – Phytolacca octandra
telondillo – Jatropha spathulata
temacate – Serjania racemosa
témaqui – Amoreuxia palmatifida
temazcal – Rhus terebinthifolia
temazcalchihual – Rhus terebinthifolia
tembladera – Briza maxima
temblón – Populus tremula
temecate – Cissus sicyoides, Vitis tiliifolia
temecatl – Vitis tiliifolia
tememetlatl – Echeveria elegans, E. gibbiflora
temeshi – Agave jaiboli
té mexicano – Chenopodium ambrosoides
temitzo – Krameria grayi
témix-tapextli – Alternanthera caracasana
temolillo – Tagetes micrantha
té mormona – Ephedra
tempate – Jatropha curcas
tempisque – Mastichodendron foetidissimum, Sideroxylon
tempixquiztle – Sideroxylon laetevirens
tempixtle – Sideroxylon
tempranilla – Opuntia joconostle, Sprekelia formosissima, Vitis vinifera var sativa
temu – Luma apiculata
tenacilla – Pithecellobium brevifolium
tenalxíhuitl – Oenothera pubescens

tenapalitl – *Pedilanthus tithymaloides*
tenapepe – *Nicotiana rustica*
tenaza – *Havardia pallens, Pithecellobium brevifolium, Zanthoxylum fagara*
tenchalita – *Hypericum pratense*
tencholote – *Cylindropuntia tunicata*
tenegara – *Crotalaria incana*
té negro – *Camellia sinensis*
teneshil – *Nicotiana eustica*
tentzonxóchitl – *Zapoteca anomalia*
té nurite – *Clinopodium laevigatum, Satureja macrostema*
ten vergüenza – *Mimosa pudica*
teocomite – *Ferocactus alamosanus*
teocomitl – *Lophophora williamsii*
teocote – *Pinus teocote*
teocotillo – *Fouquieria campanulata, F. fasciculata*
teocotl – *Pinus teocote*
teocuahuitl – *Cedrela mexicana*
teoezquauitl – *Mammea americana*
teometl – *Agave americana, A. atrovirens*
teonanácatl – *Conocybe, Lophophora williamsii, Paneolus, Psilocybe*
teosinte – *Euchlaena mexicana*
teotlaquilnanácatl – *Psilocybe cubensis, P. mexicana*
teotlate – *Cupressus lusitanica*
teoyzquixóchitl – *Bourreria huanita*
teozintle – *Echinochloa crus-pavonis, Euchlaena mexicana*
tepachera – *Zapoteca anomala*
tepahuaje – *Leucaena esculenta*
tepalcahuite – *Cocoloba barbadensis*
tepalcayo – *Phoradendron brachystachyum*
tepalcoxtli – *Populus simaroa*
tepame – *Acacia greggii, A. pennatula, A. spadicigera*
tepanzozmahtli – *Parietaria pensylvanica*
tepari – *Phaseolus acutifolius*
tepeamate – *Ficus petiolaris*
tepeamatl – *Ficus petiolaris*
tepecacao – *Quararibea funebris*
tepecajete blanco – *Psychotria nervosa*
tepecamichín – *Ficus petiolaris*
tepecamote – *Dioscorea*
tepecempoalxóchitl – *Tagetes erecta, T. patula*
tepecentli – *Anthurium scandens*
tepecuilo – *Cornus excelsa*
tepecuilote – *Cornus excelsa*
tepecuilotl – *Cornus excelsa*
tepechayotli – *Selaginella*
tepechian – *Salvia hispanica*
tepechicle – *Cnidoscolus elasticus, Stemmadenia obovata*
tepechipile – *Coursetia glandulosa*
tepeguaje – *Acacia coulteri, Havardia mexicana, Lysiloma, Zapoteca formosa*
tepehoaxin – *Lysiloma acapulcensis*
tepehuaje – *Lysiloma acapulcensis*
tepehuexote – *Salix lasiolepis*
tepejilote – *Chamaedora, Reinhardtia gracilis*
tepejilotillo – *Chamaedorea pinnatifrons*
tepelcasho – *Porophyllum macrocephalum*
tepeme – *Agave atrovirens*
tepemecatl – *Cissus sicyoides*
tepemezquite – *Lysiloma divaricatum, Zapoteca anomalia*
tepemezquite blanco – *Acacia millefolia, Lysiloma divaricata*
tepemezquite colorado – *Lysiloma divaricatum*
tepemizquitl – *Lysiloma acapulcensis*
tepenexcomitl – *Stenocactus obvallatus*
teperomero – *Asclepias linaria*
tepesquisúchil – *Arctostaphylos pungens*
tepesúchil – *Cordia alliodora, Terminalia amazonia*
tepetaca – *Sterculia apetala*
tepetate – *Aralia humilis, Rhus schiedeana var tepetate*
tepetomate – *Arctostaphylos*
tepexilotl – *Chamaedorea tepejilote*
tepexiloxóchitl – *Zapoteca anomala*
tepezapote – *Casearia dolichophylla, Ilex tolucana, Phytolacca dioica, Platymiscium trifoliolatum, Ternstroemia sylvatica*
tepezquite – *Arctostaphylos pungens*
tepezquitl – *Arctostaphylos pungens*
tepín – *Capsicum annuum var minimum*
tepochilli – *Capsicum annuum*
teponaxtle – *Taxodium distichum var mexicanum*
teponaxtli – *Taxodium distichum var mexicanum*
tepopote – *Baccharis conferta, B. pteronioides, Ephedra*
tepopotl – *Baccharis conferta, B. pteronioides*
tepopotli – *Baccharis conferta*

teposcohuite – *Quercus acutifolia, Q. conspersa*
teposcohuite chino – *Quercus muehlenbergii*
tepoyán – *Ehretia latifolia*
tepozá – *Buddleja, Cornus excelsa, Pluchea odorata*
tepozán – *Buddleja, Cestrum dumentorum, Iresine, Lippia umbellata*
tepozán blanco – *Buddleja cordata*
tepozancillo – *Buddleja sessiliflora*
tepozán cimarrón – *Buddleja parviflora*
tepozán de cerro – *Buddleja parviflora*
tepozán extranjero – *Nicotiana glauca*
tepozán verde – *Buddleja sessiliflora*
tepozquehuite – *Quercus laurina*
tepum – *Cnidoscolus multilobus*
tepzotzo – *Bonplandia geminiflora*
tequammaytl – *Smilax medica*
tequescamote – *Xanthosoma violaceum*
tequila – *Agave tequilana*
tequitcua – *Spigelia anthelmia*
tequitl – *Portulaca oleracea*
tequixquiçacatl – *Distichlis spicata*
tercianaria – *Scutellaria galericulata, S. lateriflora*
tercianaria alpina – *Scutellaria alpina*
tercianaria de los pantanos – *Scutellaria galericulata*
terciopelillo – *Dalea citriodora, D. frutescens, D. lagopus*
terciopelo – *Celosia, Odontoglossum insleayii, Stenorrhynchus lanceolatus, S. orchioides, Stizolobium deeringianum*
terebinto – *Pistacia terebinthus*
teresita – *Zinnia angustifolia, Z. multiflora, Z. peruviana*
ternerita – *Lycoperdon perlatum, L. umbrinum*
ternerita de bosque – *Lycoperdon perlatum, L. rimulatum, L. umbrinum*
ternerita de llano – *Lycoperdon candidum*
ternerita de monte – *Lycoperdon perlatum, L. umbrinum*
terocaria – *Pterocarya fraxinifolia*
tescalama – *Ficus cotinifolia, F. palmeri, F. petiolaris, Hura polyandra, Vitex mollis, V. pyramidata*
tescalama lechosa – *Ficus petiolaris*
tescalamate – *Ficus petiolaris*
tesguate – *Pernettya ciliata*
tesguate colorado – *Conostegia xalapensis*
tesguatillo – *Miconia glaberrima*
té silvestre – *Thelesperma megapotamicum*
tesmoli – *Quercus fusiformis*
teso – *Acacia brandegeana, A. occidentalis, A. pringlei*
tésota – *Acacia greggii, A. occidentalis, Olneya tesota*
teta – *Selaginella novoleonensis*
tetacegua – *Selaginella pallescens*
tetechal – *Opuntia spinulifera*
teteque – *Manihot caudata*
tetetahuicha – *Heliotropium macrostachyum*
tetetzo – *Cephalocereus columna-trajani, Neobuxbaumia tetetzo*
tetezo – *Caesalpinia pulcherrima*
tetilla – *Solanum mammosum*
tetique – *Bombax ellipticum*
tetlatia – *Toxicodendron radicans*
tetón – *Pachira aquatica*
tetona – *Solandra nitida*
tetrágona – *Tetragonia tetragonioides*
tetzapotl – *Lucuma mammosa*
tetzicohuitl – *Nyssa sylvatica*
tetzmitic – *Sedum dendroideum*
tetzmitl – *Dudleya, Sedum dendroideum*
tetzonzapotl – *Mammea americana, Pouteria sapota*
tetzotzs – *Bonplandia geminiflora*
teucrio – *Teucrium*
teunacaztli – *Cymbopetalum penduliflorum*
teunamacatlth – *Conocybe, Paneolus, Psilocybe*
té verde – *Camellia sinensis*
tewer – *Salix gooddingii*
texaltzapotl – *Annona squamosa*
texcalamacoztli – *Ficus petiolaris*
texcalamatl – *Ficus petiolaris*
texcalamatlchichic – *Ficus petiolaris*
texcapatli – *Senecio praecox*
texcaxóchitl – *Disocactus schrankii*
texcocana – *Commelina coelestis, C. tuberosa*
texcomaxochiamatl – *Ficus petiolaris*
texiote – *Sedum dendroideum, S. praealtum*
texiotl – *Dudleya, Sedum dendroideum*
texmole – *Quercus fusiformis, Q. virginiana*
texócotl – *Crataegus mexicana, C. pubescens, C. wootiana*
texochilli – *Capsicum annuum*

texóchitl – Selaginella
texoxóchitl – Zapoteca anomala
teyacua – Chiranthodendron pentadactylon
teyatli – Tagetes filifolia, T. florida
teyhuintli – Conocybe, Paneolus, Psilocybe
tezonpatli – Asclepias linaria
tezonzapote – Lucuma mammosa, L. viride
tezonzapotl – Lucuma mammosa
thau – Liquidambar orientalis
thóbem – Acacia berlandieri
thucupache xaqua – Solanum nigrum
tianguis – Alternanthera caracasana, Euphorbia heterophylla, Plumbago pulchella
tianguispepetla – Alternanthera caracasana
tiati – Bombax ellipticum
tibet – Albizzia lebeck
tibinagua – Eriogonum inflatum, E. nudum, Sanguinaria Canadensis, Typha angustifolia
tibuchina – Tibouchina urvilleana
ticati – Cucurbita ficifolia
ticatsa – Cucurbita ficifolia
tigrillo – Odontoglossum cervantesii, O. rossi,i Pouteria campechiana, Rhynchostele cervantesii
tihuinahua – Hieracium fendleri
tijerilla – Pithecellobium brevifolium
tijerilla amarilla – Corydalis lutea
tijerilla bulbosa – Corydalis cava
tijerilla primaveral – Corydalis solida
tilapo – Sideroxylon laetevirens
tilihuete – Lantana involucrata
tilo – Ternstroemia pringlei, Tilia
tilo americano – Tilia americana
tilo común – Tilia platyphyllos, T. vulgaris
tilo de Cáucaso – Tilia dasystyla, T. x euchlora
tilo de Europa – Tilia x europaea
tilo de hojas grandes – Tilia platyphyllos
tilo de hojas pequeñas – Tilia cordata
tilo grande – Ternstroemia pringlei, T. sylvatica
tilo híbrido de Holanda – Tilia x europaea, T. x vulgaris
tilo húngaro – Tilia x petiolaris, T. tomentosa
tilo intermediario – Tilia x europaea
tilo péndulo – Tilia x petiolaris
tilo plateado – Tilia tomentosa
tilo ruso – Tilia x euchlora
tilo silvestre – Tilia cordata
tilo tomentoso – Tilia tomentosa
tiltikchalauij – Inga spuria
tilxóchitl – Vanilla planifolia
tilzápotl – Diospyros digyna, Sideroxylon laetevirens
tillera – Tilia x vulgaris
tillón – Tilia x europaea
timbiriche – Bromelia karatas, B. pinguin
timbó – Enterolobium cyclocarpum
timbre – Acacia angustifolia, A. angustissima, Inga spuria
timbrillo – Acacia angustissima, Zapoteca anomala
timote – Phleum pratense
timote negro – Phleum alpinum
tincl – Ambrosia ambrosioides
tinco – Andira inermis, Quassia amara
tindusa – Bombax ellipticum
tinguaraque – Lycopersicon esculentum var cerasiforme
tino – Viburnum tinus
tinta – Haematoxylum campechianum, Stegnosperma halimifolium
tinta roja – Ditaxis
tinte – Haematoxylum campechianum
tinto – Haematoxylon campechianum
tintorera – Phytolacca decandra
tiña – Loranthus
tiña cuerda – Tournefortia hartwegiana
tío de maíz – Sorghum halepense
tipa – Tipuana tipu
tipuana – Tipuana tipu
tiquisque – Xanthosoma sagittifolium
tiquistiquis – Quassia amara
tirabeque – Pisum sativum var macrocarpon
tirabuzón – Costus mexicanus, C. spicatus
tiraña común – Pinguicula vulgaris
tirazapato – Xylosma characantha
tiriguro – Averrhoa bilimbi
tiringuini – Tagetes erecta
tiripu – Cuscuta
tis – Acacia greggii
titarro – Lathyrus cicera
titilamatl – Bombax ellipticum
tito – Lathyrus sativus, Vicia ervilia
tizaxóchitl – Plumeria acutifolia
tizón – Peronospora viticola
tizoncillo – Tilletia tritici
tizón de la papa – Phytophthora infestans
tizón del maíz – Ustilago maydis

tizón del trigo – Ustilago tritici
tlacametl – Agave atrovirens, A. salmiana
tlaceoitli – Indigofera suffruticosa
tlacoamatl – Agave salmiana
tlacocote – Pinus leiophylla
tlacochichic – Gnaphalium canescens
tlacoguilla – Clethra mexicana
tlacopatl – Aristolochia grandiflora
tlacopatle – Aristolochia grandiflora, A. mexicana
tlacopatli – Aristolochia grandiflora, Helianthus annuus, Iostephane heterophylla
tlacopaxtle – Aristolochia grandiflora
tlaco-popotl – Flaveria angustifolia
tlacosúchil – Bouvardia ternifolia
tlacoxihuitl – Bocconia arborea, B. frutescens
tlacoxiloxóchitl – Zapoteca anomala
tlacoxóchitl – Bouvardia longiflora, B. ternifolia
tlachicón – Curatella americana
tlachichinoa – Tournefortia capitata
tlachichinol – Plumbago pulchella
tlachichinole – Kohleria deppeana, Plumbago pulchella
tlachinole – Urera caracasana
tlahtlacotic – Scabiosa fruticosa
tlahuilote – Lysiloma microphylla
tlailochtía – Aloe vera
tlalacxóyatl – Asclepias linaria, A. tuberosa
tlalahuacate – Ehretia elliptica
tlalámatl – Agastache mexicana
tlalanis – Tagetes micrantha
tlalatóchietl – Mentha viridis
tlalayote – Asclepias linaria, Gonolobus erianthus, Vincetoxicum hastulatum
tlalcacahuatl – Arachis hypogaea
tlalcapolín – Karwinskia humboldtiana, Rhamnus
tlalcipilli – Suaeda
tlalcoyote – Lophophora williamsii
tlalchichinolli – Plumbago auriculata, P. scandens
tlalixtli – Ceanothus coeruleus
tlalmemejan – Chamaesyce maculata
tlalnochtli – Asclepias linaria
tlalnopal – Opuntia cochenillifera
tlalocopetate – Coriaria ruscifolia
tlalometl – Agave peacockii
tlalpetate – Alternanthera caracasana
tlalpopolote – Iostephane heterophylla, Viguiera dentata
tlalquequetzal – Achillea millefolium, Adiantum trapeciforma
tlalquilín – Mirabilis jalapa
tlalsometl – Acaena elongata
tlaltecomaxóchitl – Asarina scandens
tlaltzacutli – Sphaeralcea angustifolia
tlamacatzcatzotl – Zapoteca anomalia
tlamahual – Tabebuia impetiginosa
tlamatlantli – Solanum madrense
tlamatsalin – Sonchus oleraceus
tlamexcaltzin – Argemone mexicana
tlamizquitl – Hoffmannseggia glauca
tlampa – Piper auritum
tlanchinol – Cestrum fasciculatum
tlanepaquelite – Piper sanctum
tlanepaquilitl – Piper auritum, P. sanctum
tlanipa – Piper auritum, P. sanctum
tlanoquiloni – Ipomoea purga
tlaoli – Zea mays
tlaolli – Zea mays
tlapa – Datura stramonium
tlapacón – Fouquieria formosa
tlapahuaxin – Leucaena esculenta
tlapahuaxpatli – Eysenhardtia polystachya
tlapalachiyotl – Bixa orellana
tlapalcacauatl – Theobroma cacao
tlapalezpatli – Eysenhardtia polystachya, Jatropha spathulata
tlapalito – Plumeria acutifolia
tlapalixquixóchitl – Bourreria huanita
tlapate – Ricinus communis
tlapatl – Datura, Ricinus communis
tlatlancuaye – Iresine calea
tlatlauhcapatli – Geranium carolinianum
tlatocnochtli – Opuntia cochenillifera, O. microdasys
tlatomaxíhuitl – Phacelia platycarpa
tlatón – Iresine
tlatzcan – Cupressus
tlautlin – Zea mays
tláxcal – Cupressus lusitanica, Juniperus
tlaxcalxíhuitl – Cissampelos pareira
tlaxcapán – Ipomoea stans
tlaxistle – Amelanchier denticulata, Ceanothus coeruleus
tlazol-patli – Datura stramonium
tlazolteo-zácatl – Sobralia macrantha

tlazoltomatl – *Cardiospermum halicacabum*
tlemole – *Tagetes patula*
tlemoli – *Physalis angulata*
tlemolito – *Tagetes patula*
tlepatli – *Plumbago auriculata, P. pulchella, P. scandens*
tlilamatl – *Ficus cotinifolia*
tlilchilli – *Capsicum annuum var minimum*
tliltzápotl – *Diospyros ebenaster*
tlilxihuitl – *Sphaeralcea angustifolia*
tlilxóchitl – *Vanilla planifolia*
tlinete – *Licania arborea*
tlipoton – *Eryngium foetidum*
tlitliltzin – *Ipomoea violacea, Rhynchosia minima*
tmuul – *Gomphrena globosa*
tnundaya – *Prunus capuli*
tnunde – *Rhus mollis*
tnundido – *Arctostaphylos pungens*
tnutinumi – *Malus sylvestris*
tnutque – *Arctostaphylos polifolia*
tnuya – *Quercus reticulata*
tnuyoocó – *Ceanothus coeruleus*
tnuyucú – *Ceanothus coeruleus, Taxodium distichum var mexicanum*
toa – *Quercus arizonica, Q. emoryi*
toaloatzin – *Datura*
toalxóchitl – *Epidendrum vitellinum*
tobala – *Agave atrovirens*
tobe – *Aechmea bracteata*
toboso – *Anagallis arvensis, Pleuraphis cenchroides, P. mutica, P. rigida, Tribulus terrestris*
toboso común – *Pleuraphis mutica*
toboso esbelto – *Pleuraphis swalleni*
toboso menudo – *Pleuraphis belangeri, P. cenchroides*
tocalote – *Centaurea melitensis*
tocasoiahui – *Mimulus guttatus*
tocte – *Juglans honorei, J. nigra*
toché – *Quercus pennivenia, Q. tuberculata, Q. urbanii*
tochimirtillo – *Tournefortia capitata*
tochimitillo – *Hamelia erecta, Kohleria deppeana*
tochite – *Bouteloua aristidoides*
tochomitl – *Kohleria deppeana*
todabuena – *Hypericum androsaemum*
todasana – *Hypericum androsaemum*
togue – *Strutanthus haenkeanus*
tohku – *Datura stramonium, D. wrightii*
tohza – *Lomatium dissectum*
toji – *Arceuthobium campylopodum, A. divaricatum, Phoradendron, Phrygillanthus sonorae, Psittacanthus calyculatus, P. sonorae, Roccella fusiformis, Struthanthus haenkeanus, S. palmeri, Tillandsia recurvata*
toji de mezquite – *Struthanthus palmeri*
tojo – *Ulex europaea*
tojonechichi – *Solanum nigrum*
tojtanyuc – *Amyris balsamifera*
tokabán – *Eupatorium, Trixis radialis*
tolicarochá – *Cardiospermum halicacabum*
toloache – *Datura*
toloache grande – *Datura wrightii*
tolohuaxihuitl – *Datura*
tololote – *Andira inermis*
tolompatl – *Ipomoea purga*
tolpatlacti – *Typha latifolia*
tolpatle – *Cyperus esculentus*
tolpatli – *Cyperus esculentus*
tolpis umbelado – *Tolpis barbata*
tomasita – *Phacelia congesta*
tomate – *Lycopersicon esculentum, Physalis*
tomate arbóreo – *Cyphomandra betacea*
tomate bola – *Lycopersicon esculentum*
tomate capotillo – *Physalis crassifolia*
tomate cereza – *Lycopersicon esculentum var cerasiforme*
tomate cimarrón – *Lycopersicon esculentum var cerasiforme, L. pimpinellifolium*
tomate chitalino – *Lycopersicon esculentum var cerasiforme*
tomate de árbol – *Cyphomandra betacea*
tomate de burro – *Nicandra physalodes*
tomate de capote – *Physalis philadelphica*
tomate de cáscara – *Physalis angulata, P. peruviana, P. philadelphica, P. pubescens*
tomate de culebra – *Nicandra physalodes, Physalis angulata, P. crassifolia*
tomate de fresadilla – *Physalis philadelphica*
tomate de la paz – *Cyphomandra betacea*
tomate del campo – *Physalis pubescens*
tomate del monte – *Lycopersicon esculentum var cerasiforme, Physalis leptophylla*
tomate de palo – *Cyphomandra betacea*
tomate de víbora – *Physalis crassifolia*
tomate inglés – *Physalis alkekengi*

tomate milpero – Physalis philadelphica
tomatera – Lycopersicon esculentum
tomatero – Cyphomandra betacea
tomate roma – Lycopersicon esculentum
tomate silvestre – Physalis pubescens
tomate verde – Physalis philadelphica, P. pubescens
tomatillo – Cardiospermum corindum, Lantana camara, Lycium, Malpighia glabra, Physalis, Rivina humilis, Saracha jaltomata, Solanum capsicastrum, S. eleagnifolium, S. nigrum
tomatillo cimarrón – Physalis acutifolia
tomatillo de Jerusalén – Solanum capsicastrum
tomatillo del campo – Physalis acutifolia, Solanum eleagnifolium
tomatillo del desierto – Physalis crassifolia
tomatillo del diablo – Solanum nigrum, S. sodomaeum
tomatillo espinoso – Physalis acutifolia, Solanum eleagnifolium, S. rostratum
tomatito – Solanum eleagnifolium
tomatito de buena mujer – Solanum eleagnifolium
tomatito del diablo – Solanum nigrum
tomatito pelón – Solanum eleagnifolium
tomatl – Lycopersicon esculentum, Physalis
tomazquitl – Arbutus unedo, Arctostaphylos pungens
tomequín – Zanthoxylum fagara
tomillo – Thymus serpyllum, T. vulgaris
tomillo aceitunero – Thymus capitatus
tomillo alcanferado – Thymus camphorates
tomillo andaluz – Thymus capitatus
tomillo arborescente – Thymus capitatus
tomillo blanco – Thymus mastichina
tomillo borriquero – Lavandula stoechas
tomillo común – Thymus pulegioides
tomillo de jardín – Thymus vulgaris
tomillo de monte – Thymus serpyllum
tomillo perruno – Santolina rosmarinifolia
tomillo real – Satureja hortensis
tomillo salsero – Satureja hortensis
tomillo terrero – Teucrium polium
tomillo vulgar – Thymus vulgaris
tomistlacati – Amelanchier denticulata
tomosali – Stemodia durantifolia
tompaap – Solanum verbascifolium
tonacaxóchitl – Distictis buccinatoria
tonalahuate – Persea americana
tonalcotl-xihuitl – Dodonaea viscosa
tonalchilli – Capsicum annuum
tonalxóchitl – Bletia coccinea, Zantedeschia aethiopica
tonati-shochit – Bouvardia ternifolia
toncnaya – Iresine celosioides
ton-chá – Tabernaemontana alba
tonchi – Marsdenia edulis
tonchichi – Saracha procumbens, Solanum nigrum
tongo – Dioscorea esculenta
tonjoyó – Tillandsia prodigiosa
tonka – Dipteryx odorata
tonojoso – Phytolacca octandra
tonotzin – Pinus strobus
tontzimin – Stemmadenia donnell-smithii
to'oro – Bursera fagarioides, B. microphylla, B. penicillata, Jatropha cordata
to'oro chucuri – Bursera laxiflora
to'oro chutama – Bursera lancifolia
to'oro mulato – Bursera grandifolia
to'oro ochónome – Bursera laxiflora
to'oro sajo – Bursera stenophylla
tootoco – Cylindropuntia fulgida
topinambur – Helianthus tuberosus
topocá – Bursera penicillata
topo de las habas – Orobanche crenata
toposcohuite – Quercus candicans
toposte – Licania arborea
topostle – Licania arborea
toquilla – Carludovica palmata
torbisco – Asclepias linaria
torescua – Datura stramonium
torete – Vitex mollis
torilio de los campos – Torilis arvensis
torilio nudoso – Torilis nodosa
torillo – Randia
torinaca – Malvella lepidota
toripipa – Nicotiana glauca
torito – Acacia spadicigera, Asclepias linaria, Proboscidea, Stanhopea devoniensis, S. oculata, S. tigrina, Tribulus terrestris
tormentilla – Potentilla
tornachile – Capsicum annuum
torna loco – Datura ceratocaula
tornasol – Helianthus annuus
tornillo – Prosopis cinerascens, P. pubescens
toro – Proboscidea fragrans, Stanhopea tigrina

toró – Bursera lancifolia, B. laxiflora, B. penicillata
toroberico – Proboscidea parviflora
tóroco – Bunchosia sonorensis
toró chukuri – Bursera microphylla
toró chutama – Bursera fagaroides
toronja – Eschscholtzia californica
toronjero – Citrus paradisi
toronjil – Agastache, Artemisia abrotanum, Brittonastrum mexicana, Cedronella mexicana, Mentha x piperita, Melissa officinalis, Rheedia edulis, Salvia microphylla, Satureja macrostema
toronjil de abejas – Monarda didyma
toronjil de China – Nepeta cataria
toronjil del país – Agastache mexicana, Cedronella mexicana
toronjil extranjero – Melissa officinalis
toronjil morado – Agastache mexicana
toronjillo – Cedronella mexicana
toronjina – Melissa officinalis
toronjo – Citrus bergamotia, C. grandis, C. medica, C. paradisi
toronjo agrio – Citrus paradisi
toronjo de semilla – Citrus nobilis
torosa – Eschscholzia californica
torote – Bursera, Fouquieria fasciculata, Jatropha, Pachycormus discolor
torote amarillo – Bursera fagaroides, B. lancifolia, Jatropha spathulata
torote blanco – Bursera, Jatropha cordata, Pachycormus discolor
torote camforado – Bursera stenophylla
torote colorado – Bursera
torote común – Jatropha cardiophylla
torote copal – Bursera inopinnata, B. penicillata, B. stenophylla
torote chutama – Bursera confusa
torote de incienso – Bursera lancifolia
torote de puntagruesa – Bursera penicillata
torote de vaca – Bursera fagarioides
torote de venado – Bursera fagarioides
torote espinoso – Fouquieria macdougalii
torote incienso – Bursera penicillata
torote jolopete – Bursera lancifolia
torote papelillo – Bursera, Jatropha cordata
torote prieto – Bursera, Jatropha cordata, J. spathulata
torote rojo – Bursera microphylla
torote verde – Fouquieria fasciculata, F. macdougalii
torotillo – Fouquieria macdougalii, Jatropha cinerea
torotito – Jatropha cinerea
tortero – Arrhenatherum elatius
torteruelo – Medicago orbicularis
tortilla de los sapos – Cissampelos pareira
tortilla de ratón – Desmodium canum
tortilla tostado del caballo – Petrea arborea
tortolita – Croton texensis
torurco – Paspalum conjugatum
torvisco – Daphne mezereum
torvisco de los Pirineos – Daphne cneorum
torvisco macho – Daphne laureola
torreya de California – Torreya californica
tosapolo – Calliandra gracilis, Hibiscus brasiliensis, Zapoteca formosa
tosarza – Pinus teocote
toshcura – Esenbeckia hartmanii
tostón – Cosmos parviflorus, Zinnia
tostón amarillo – Zinnia multiflora
totanera – Cucurbita maxima
totara – Podocarpus totara
totèkam – Muhlenbergia dumosa
totcui – Tabebuia chrysantha
totocuitlatl – Capsicum annuum var minimum
totolcal – Fagus americana
totolcuitlatzápotl – Diospyros digyna, D. ebenaster
totolole – Phyllanthus acidus
totoncaxihuitl – Senna alata
totoncaxoxocoyolin – Begonia gracilis
totopixtle – Russula delica
totopzots – Malvastrum coromandelianum
totora – Typha latifolia
totosá – Agave shrevei
totosali – Agave shrevei
totozapotl – Sideroxylon capiri
totumo – Couroupita guianensis, Crescentia cujete
toubo – Tabebuia impetiginosa
toumeya – Sclerocactus papyracanthus
toxcuitlapilxóchitl – Lobelia laxiflora
t'oyol – Cenchrus echinatus
toyon – Heteromeles arbutifolia
trachelospermo – Trachelospermum jasminoides
tradescantia – Tradescantia virginiana

tradescantia manchada – Tradescantia fluminensis
tragacanto – Astragalus gummifer
tragapán – Narcissus pseudo-narcissus
tragontina – Arum maculatum
trampa de perro – Gaultheria acuminata
trébol – Melilotus indica, Trifolium
trébol agrio – Melilotus indica
trébol alejandrino – Trifolium alexandrinum
trébol almizclado – Melilotus officinalis
trébol alsike – Trifolium hybridum
trébol amargo – Trifolium repens
trébol amarillo – Anthyllis vulneraria, Melilotus indica, Trifolium campestre, T. dubium, T. procumbens
trébol amarillo oloroso – Melilotus officinalis
trébol arvense – Trifolium arvense
trébol bastardo – Trifolium hybridum
trébol bersim – Trifolium alexandrinum
trébol blanco – Trifolium repens
trébol blanco dulce – Melilotus alba
trébol blanco ladino – Trifolium repens
trébol blanco oloroso – Melilotus alba
trébol campestre – Trifolium campestre
trébol carretón – Medicago polymorpha
trébol común – Trifolium pratense
trébol de agua – Menyanthes trifoliata
trébol de alejandria – Trifolium alexandrinum
trébol de alsike – Trifolium hybridum
trébol de carretilla – Medicago arabica, M. sativa, Trifolium polymorpha
trébol de cuatro hojas – Marsilea mexicana
trébol de cuernecillo – Lotus corniculatus
trébol de cuernos – Lotus corniculatus
trébol de flores en glomérulo – Trifolium glomeratum
trébol de hojas estrechas – Trifolium angustifolium
trébol de Holanda – Trifolium repens
trébol de Hungría – Trifolium pannonicum
trébol de la suerte – Oxalis deppei
trébol del campo – Trifolium campestre
trébol del Japón – Lespedeza striata
trébol del monte – Didymaea mexicana
trébol del prado – Trifolium pratense
trébol del rosellón – trifolium incarnatum
trébol de olor – Melilotus indica
trébol de olor blanco – Melilotus alba
trébol de Persia – Trifolium resupinatum
trébol de San Juan – Melilotus officinalis
trébol de Suecia – Trifolium hybridum
trébol dorado – Trifolium aureum
trébol dubio – Trifolium dubium
trébol dulce – Melilotus alba
trébol encarnado – Trifolium incarnatum
trébol español – Lotus corniculatus
trébol fibrino – Menyanthes trifoliata
trébol fresa – Trifolium fragiferum, T. macrocephalum
trébol fresero – Trifolium fragiferum
trébol hediondo – Bituminaria bituminosa
trébol híbrido – Trifolium hybridum
trébol hirto – Trifolium hirtum
trébol holandés – Trifolium repens
trébol hubam – Melilotus alba
trébol intermedio – Trifolium medium
trébol ladigiano – Trifolium repens
trébol ladino – Trifolium repens
trébol medio – Trifolium medium
trébol morado – Trifolium incarnatum, T. pratense
trébol negro – Trifolium hybridum
trébol oloroso – Melilotus officinalis
trébol palustre – Menyanthes trifoliata
trébol persa – Trifolium resupinatum
trébol pérsico – Trifolium resupinatum
trébol rampante – Trifolium repens ssp prostratum
trébol rastrero – Trifolium repens, T. subterraneum
trébol rojo – Trifolium pratense
trébol rosado – Trifolium incarnatum, T. pratense
trébol roso – Trifolium hirtum
trébol subterráneo – Trifolium subterraneum
trébol sueco – Trifolium hybridum
trébol trepador – Trifolium subterraneum
trébol verdadero – Trifolium alexandrinum
trébol violeto – Trifolium pratense
trébol zigzag – Trifolium medium
trementina – Liquidambar styraciflua, Pinus edulis
tremoncillo – Thymus vulgaris
trencilla – Axonopus compressus, Desmodium scorpiurus, Paspalum notatum, Tillandsia flabella
trenza acuática – Paspalum fluitans
trepacaballo – Centaurea calcitrapa

trepadora – Rosa
tres aristas abierto – Aristida divaricata
tres aristas arizónico – Aristida arizonica
tres aristas arqueado – Aristida ternipes var hamulosa
tres aristas barbado – Aristida barbata, A. divaricata
tres aristas de agua – Aristida adscensionis
tres aristas largo – Aristida purpurea var fendleriana
tres aristas mexicano – Aristida roemeriana
tres aristas morado – Aristida purpurea
tres aristas perenne – Aristida pansa
tres aristas rojo – Aristida purpurea var longiseta
tres barbas – Aristida adscensionis, Aristida purpurea var nealleyi
tres barbas abierto – Aristida divaricata
tres barbas anual – Aristida adscensionis
tres barbas arqueado – Aristida ternipes
tres barbas de California – Aristida californica var californica
tres barbas de semillas – Aristida adscensionis
tres barbas gentil – Aristida gentilis
tres barbas liso – Aristida purpurea var nealleyi
tres barbas perenne – Aristida purpurea var wrightii
tres barbas volador – Aristida orcuttiana
tres cerdas espigado – Trisetum spicatum
tres colores – Lantana camera
tres Marías – Bougainvillea spectabilis, Crateva palmeri, Forchammeria watsonii
tresmoles – Quercus oleoides
triatua – Jatropha gossypiifolia
tríbulo – Tribulus terrestris
tricoloma portentoso – Tricholoma terreum
tridente – Tridens muticus
tridente agrio – Tridens albescens
tridente alargado – Tridens elongatus
tridente esbelto – Tridens muticus
tridente fino – Tridens eragrostoides
tridente texano – Tridens texanus
trientalis – Trientalis arctica
trifolio brasileño – Stylosanthes guianensis
triglochin de los pantanos – Triglochin palustris
triglochin marítimo – Triglochin maritima
trigo – Triticum vulgare
trigo blando – Triticum aestivum
trigo candela – Triticum aestivum
trigo de invierno – Triticum aestivum
trigo de los indios – Plantago fastigiata
trigo de mar – Zostera marina
trigo de Turquía – Zea mays
trigo duro – Triticum durum
trigo mariso – Fagopyrum esculentum
trigo moruno – Triticum durum
trigo negro – Fagopyrum esculentum
trigo pelón – Triticum aestivum
trigo sarraceno – Fagopyrum esculentum
trigo tierno – Triticum vulgare
triguera – Phalaris tuberosa
triguerula – Bituminaria bituminosa
triguillo de Arizona – Elytrigia arizonicus
triguillo desértico – Elymus elymoides, E. longifolius
triguillo largo – Elytrigia trachycaulus
triguillo loco – Avena fatua
trilio – Trillium
trilio blanco – Trillium grandiflorum
trinitaria – Bougainvillea spectabilis, Nerium oleander, Tigridia pavonia, Viola tricolor
tripa – Verbascum thapsus
tripa baja – Agdestis clematides
tripa de aura – Ipomoea pes-caprae
tripa de gallina – Nissolia fruticosa
tripa de gallo – Spilanthes americana
tripa de Judas – Agdestis clematides, Cissus cucurbitina, C. sicyoides, Cuscuta americana, Vitis tiliifolia
tripa de pollo – Commelina coelestis, Kallstroemia maxima
tripa de ratón – Ionidium polygalaefolium
tripa de vaca – Cissus sicyoides, Vitis tiliifolia
tripa de zopilote – Cissus rhombifolia, C. sicyoides
tripal – Lysiloma acapulcensis
triplaria – Triplaris americana
triquera – Phalaris brachystachys
triquitraque – Pyrostegia venusta
triseto amarillo – Trisetum flavescens
triseto espigado – Trisetum spicatum
triseto rubio – Trisetum flavescens
triticale – Triticum secale
tritomo – Kniphofia uvaria
tritomo rojo – Kniphofia uvaria
tritonia – Tritonia crocata

triverdin de cuatro – Chloris pluriflora
triverdin de tres – Chloris crinita
trómbita – Brugmansia arborea, B. x candida
trompa – Lactarius piperatus, Russula
trompa blanca – Russula delica
trompa de cochino – Russula delica
trompa de puerco – Russula delica
trompa negra – Russula nigricans
trompeta – Brugmansia, Cecropia obtusifolia, C. peltata, Gomphus flocosus, Pyrostegia venusta, Russula delica, Salpiglossis sinuata, Tecoma stans, Thevetia peruviana
trompeta anaranjada – Pyrostegia venusta
trompeta de juicio – Brugmansia
trompeta de los muertos – Craterellus cornucopioides
trompeta de oro – Allamanda cathartica
trompeta trepadora – Campsis x tagliabuana
trompetilla – Acleisanthes longiflora, Campsis grandiflora, Cecropia obtusifolia, Datura stramonium, Hedyotis alexanderae, Justica spicigera, Mirabilis jalapa, Tecoma stans
trompetilla de noche – Mirabilis jalapa
trompetilla grande – Distictis buccinatoria
trompillo – Bouvardia ternifolia, Cordia boissieri, C. brevispicata, Epidendrum nemorale, Euphorbia cotinifolia, Ipomoea, Phaseolus heterophyllus, Quamoclit coccinea, Rosa montezumae, Solanum eleagnifolium, S. verbascifolium, S. xanti, Ternstroemia pringlei, Trichilia hirta
trompillo blanco – Ipomoea triloba
trompillo lila – Ipomoea triloba
trompillo morado – Ipomoea hederacea
trompo – Aesculus californica, A. parryi, Crateva palmeri
tronador – Abutilon incanum, Astianthus viminalis, Cannabis sativa, Cardiospermum corindum, C. halicacabum, Crotalaria, Hura polyandra, Nicotiana glauca, Oreopanax peltatum, Ruellia californica, R. intermedia, Russelia sarmentosa, Sapium appendiculatum, Sebastiania pavoniana, Tecoma stans
tronador de España – Nicotiana glauca
troncoso – Aloysia gratissima
truanilla – Ligustrum vulgare
trucha – Albizzia occidentalis, Croton
trueno – Ligustrum
trufa blanca – Tuber album, T. borchii, T. magnatum
trufa de invierno – Tuber brumale, T. melanosporum
trufa de Piamonte – Tuber magnatum
trufa de verano – Tuber aestivum
trufa gris – Tuber magnatum, T. uncinatum
trufa roja – Tuber rufum
trufa violeta – Tuber brumale, T. melanosporum
trujillano – Theobroma cacao
trumfera – Solanum tuberosum
tsabal – Spathiphyllum blandum
tsabtsits – Salvia coccinea
tsacixiu – Salvia coccinea
tsalbay – Bromelia pinguin
tsapassabani – Lucuma mammosa
tsaposh – Citrus limon
tsapquinc – Citrus sinensis
tsaycan – Sesuvium portulacastrum
tsayochcan – Portulaca halimoides
tsiim – Gnaphalium canescens
tsihirén – Pinus michoacana
tsihuírea – Pinus devoniana
tsin – Manihot esculenta
tsole – Cercidium praecox
tsomeltok – Bauhinia divaricata
tsopilotlsontecomatl – Swietenia humilis
tsotz – Brosimum alicastrum
tsuga de Alberta – Tsuga heterophylla
tsuga de California – Tsuga mertensiana
tsuga de Carolina – Tsuga caroliniana
tsuga de la costa occidental – Tsuga mertensiana
tsuga del Canadá – Tsuga canadensis
tsuga del Pacífico – Tsuga heterophylla
tsuga de montaña – Tsuga mertensiana
tsuga de Patton – Tsuga mertensiana
tsuga heterofila – Tsuga heterophylla
tsuga negra – Tsuga mertensiana
tsuga oriental – Tsuga canadensis
tsulipox – Annona cherimola, A. reticulata
tsulubppac – Lycopersicon esculentum
tsulubtok – Bauhinia divaricata
tsumya – Pereskia aculeata
tsuwíri – Ariocarpus retusus
tuah – Agave atrovirens
tubchi – Caesalpinia sclerocarpa
tuberaria manchada – Tuberaria guttata

tuberosa – Polianthes tuberosa
túbucti – Sebastiania pavoniana
túbulo acuático – Trapa natans
tuc – Yucca elephantipes
tucuchí – Ficus padifolia
tucumu – Abies religiosa
tucupache xacua – Solanum nigrum
tuché – Satureja macrostema
tuchi – Malvastrum bicuspidatum, Sidalcea malvaeflora, S. neomexicana
tuera – Citrullus colocynthis
tueya – Nolina matapensis
tuk – Acrocomia mexicana
tu-kiak – Synechanthus fibrosus
tuk'uch – Cyperus rotundus
tula istle – Agave lechuguilla
tulán – Quercus rugosa
tule – Cyperus, Eleocharis, Pontederia cordata, Scirpus, Typha
tule ahuapetla – Cyperus canus
tule bofo – Scirpus lacustris
tule chico – Cyperus articulatus
tule espadilla – Typha latifolia
tule esquinado – Scirpus americanus
tule grande – Scirpus lacustris
tule rollizo – Scirpus californicus
tulillo – Chloracantha spinosa, Cyperus, Eleocharis, Juncus, Scirpus
tulipán – Hibiscus coulteri, H. rosa-sinensis, Malvaviscus arboreus, Tulipa gesneriana, Spathodea campanulata, Trophis racemosa
tulipán africano – Spathodea campanulata
tulipán austral – Tulipa australis
tulipancillo – Malvaviscus arboreus
tulipán de Africa – Spathodea campanulata
tulipán de Clusius – Tulipa clusiana
tulipán de flor de lirio – Tulipa gesneriana
tulipán del monte – Hibiscus bifurcatus, H. cardiophyllus, Malvaviscus arboreus
tulipán de los jardineros – Tulipa gesneriana
tulipanero – Liriodendron tulipifera
tulipán fosteriana – Tulipa gesneriana
tulipán lefeber – Tulipa gesneriana
tulipán precoz – Tulipa gesneriana
tulipán rojo – Hibiscus rosa-sinensis
tulipán saxátil – Tulipa saxatilis
tulipán silvestre – Tulipa sylvestris
tulipán tardio lleno – Tulipa gesneriana
tulipán temprano lleno – Tulipa gesneriana
tulipán verde – Tulipa gesneriana
tulipero – Liriodendron tulipifera
tulipero del Gabón – Spathodea campanulata
tulipero de Virginia – Liriodendron tulipifera
tulipífero americano – Liriodendron tulipifera
tulipífero de Virginia – Liriodendron tulipifera
tulito – Juncus torreyi
tulua – Ricinus communis
tulumisi – Physalis leptophylla, P. philadelphica
tulusisi – Saracha procumbens
tullidor – Karwinskia humboldtiana
tumali – Parietaria debilis
tumalí – Urera caracasana
tumatle – Lycopersicon esculentum
tumba caballo – Turbina corymbosa
tumba vaquero – Cissus sicyoides, Ipomoea stans
tumbo – Passiflora quadrangularis
tumina – Alternanthera caracasana
tuna – Cylindropuntia, Opuntia, Pilosocereus leucocephalis, Stenocereus alamoensis, S. stellatus
tuna barbona – Pachycereus schottii
tuna blanca – Opuntia joconostle
tuna brava – Opuntia megacantha
tuna camuesa – Opuntia robusta
tuna cardona – Opuntia streptacantha
tuna cascarona – Opuntia streptacantha
tuna colorada fina – Opuntia megacantha
tuna corriente – Opuntia hyptiacantha
tuna cuija – Opuntia engelmannii
tuna chaveña – Opuntia hyptiacantha, O. streptacantha
tuna chica – Opuntia streptacantha
tuna de agua – Disocactus schrankii, Pereskiopsis aquosa
tuna de alfajayuca – Opuntia amyclaea, O. ficus-indica, O. megacantha
tuna de Castilla – Opuntia megacantha, Opuntia ficus-indica
tuna de playa – Opuntia inaperta
tuna de tierra – Lophophora williamsii
tuna duraznillo – Opuntia leucotricha
tuna duraznillo blanca – Opuntia leucotricha
tuna duraznillo colorada – Opuntia leucotricha
tuna esferica – Opuntia streptacantha
tuna hartona – Opuntia streptacantha
tuna huell – Cylindropuntia imbricata

tuna joconoxtli – Cylindropuntia imbricata
tuna mansa – Opuntia ficus-indica, O. hyptiacantha, O. megacantha, O. streptacantha
tuna morada – Opuntia lagunae, O. robusta
tuna olorosa de ejutla – Stenocereus treleasei
tuna pelona – Opuntia ficus-indica
tuna silvestre – Acanthocereus tetragonus
tuna tapona – Opuntia robusta, O. tapona
tuna tapona cimarrona – Opuntia robusta
tuna tempranilla – Opuntia ficus-indica
tuna verde – Disocactus schrankii, Opuntia ficus-indica
tungo – Aleurites fordii
tunillo – Stenocereus treleasei
tunkux – Solanum verbascifolium
tunuc – Antigonon leptopus
tupchi – Sapindus saponaria
tupelo – Nyssa sylvatica
tupelo negro – Nyssa sylvatica
tupinambo – Helianthus tuberosus
tura – Pachira aquatica
tural – Tabebuia pentaphylla
turasali – Dyssodia anomala
turbinto – Schinus molle, S. terebinthifolius
ture – Pachira aquatica
turi – Parmentiera edulis
turiara – Caladium bicolor
turicua – Plumbago auriculata, P. pulchella, P. scandens
turoco – Plumeria rubra
túrusi – Prunus persica
turvará – Paspalum conjugatum
turre – Phyla strigulosa var sericea
tush – Agdestis clematides
tusilago – Tussilago farfara
tusillo – Scirpus paradoxus
tutuqui – Phaulothamnus spinescens
tutz – Orbignya cohune
tuxpana – Spondias purpurea
tuxpata – Cyperus odoratus
tuya – Thuja occidentalis, T. orientalis, T. plicata
tuyá – Nolina matapensis
tuya articulada – Tetraclinis articulata
tuya del Canadá – Thuja occidentalis
tuya de la China – Thuja orientalis
tuya gigante – Thuja plicata
tuya japonesa – Thujopsis dolabrata
tuya occidental – Thuja occidentalis
tuya oriental – Thuja orientalis
tuzán – Citrus sinensis
tuzpatli – Dorstenia contrayerba, D. drakeana
tza – Cnidoscolus aconitifolius
tzacots – Licania platypus
tzacotxójol – Abutilon hypoleucum
tzacouhxóchitl – Epidendrum pastoris, Oatasetum maculatum
tzactumi – Schinus molle
tzacué blanco – Quercus candicans
tzacutli – Bletia campanulata, Epidendrum pastoris
tzacuxóchitl – Bletia campanulata, B. coccinea, Cranichis speciosa, Lealia autumnalis
tzajalchit – Quercus skinneri
tzajib – Guadua aculeata
tzakamak – Selenicereus donkelaari
tzalam – Lysiloma bahamensis
tzalilte – Rhipsalis baccifera
tzalmuy – Annona squamosa
tzalyaltzai – Anoda cristata
tzama – Phaseolus vulgaris
tzamal – Dioon edule
tzamuc-cóyol – Yucca guatemalensis
tzanalchilli – Capsicum annuum var minimum
tzantuni – Schinus molle
tzaposh – Citrus limon
tzápotl – Casimiroa edulis, Lucuma mammosa
tzaput – Achras zapota
tzapyan – Punica granatum
tzatadi – Pinus teocote
tzatzupú – Sapindus saponaria
tzautli – Bletia campanulata, Cranchis speciosa, C. tubularis, Laelia autumnalis
tzauxilotl – Arpophyllum spicatum
tzayuntzay – Mentzelia aspera
tzenoiti – Phoebe tampicensis
tzenso – Lyophyllum decastes, Tricholoma equestre
tzetzetzo – Neobuxbaumia tetetzo
tzicin – Calea zacatechichi
tzicxóchitl – Laelia albida, L. anceps
tzijol – Piscidia mollis
tzilacayotli – Cucurbita ficifolia
tzin – Pteridium aquilinum
tzinacancuáhuitl – Erythrina
tzinacanitlacual – Thevetia thevetioides

tzinquauhyo – Capsicum violaceum
tziquinay – Vernonia deppeana
tziri – Zea mays
tzirimu – Tilia mexicana
tziritsecua – Prosopis juliflora
tzit – Pinus tenuifolia
tzits – Persea americana
tzitzalxiu – Kalanchoe pinnata
tzitzi – Morus celtidifolia
tzitzicaitl – Cnidoscolus multilobus
tzitzicaxtli – Wigandia kunthii
tzitzil – Maclura tinctoria
tzitzim – Artemisia vulgaris
tzitzintlaptl – Datura stramonium
tzitziqulitl – Waltheria americana
tzitzit – Vernonia deppeana
tzitziu-ton – Selloa glutinosa
tzoalli – Chenopodium murale
tzol – Cucurbita pepo
tzompachtzin – Solanum verbascifolium
tzompahuiztle – Pereskiopsis rotundifolia
tzompancuáhuitl – Erythrina
tzompatli – Erythrina
tzompili-huitzpatli – Cedronella mexicana
tzompomitl – Erythrina americana
tzompotonic – Gnaphalium canescens
tzonayotli – Luffa aegyptiaca
tzonchilli – Capsicum annuum
tzontecomananácatl – Amanita muscaria
tzontecpatli – Jatropha cordata
tzonxóchitl – Calliandra grandiflora, Zapoteca anomala
tzopelchichilxóchitl – Malvaviscus arboreus
tzopelicxihuitl – Lippia dulcis
tzopilocuáhuitl – Swietenia macrophylla
tzopilotl – Swietenia macrophylla
tzopilotlacuatl – Solanum nigrum
tzopilotzontecómatl – Swietenia humilis, S. macrophylla
tzotzocoli – Calliandra houstoniana
tztzil – Toxicodendron diversilobum
tzucté – Lysiloma bahamensis
tzucui – Casimiroa sapota
tzum – Tithonia
tzuny – Guazuma ulmifolia
tzuta – Agave lechuguilla
tzutujté – Ostrya virginiana
tzutzuyxiu – Mirabilis jalapa
uachamo – Baccharis thesioides
uachomo – Zexmenia seemanni
uahko – Aristolochia grandiflora
uarasu – Ipomoea batatas
uasiqui – Prunus capuli, P. gentryi
uasiva – Acacia occidentalis
uaxim – Acacia angustissima
ubi – Gilia gloriosa, Loeselia tenuifolia
ucansa – Prunus persica
ucas – Vauquelinia corymbosa
ucasan – Salvia californica
uclatave – Vallesia glabra
ucuares – Valeriana ceratophylla, V. sorbifolia
ucuquiro – Rivina humilis
uchuba – Physalis peruviana
ud – Erica arborea
ueijuli – Verbena menthaefolia
uernia – Huernia schneideriana
ugedri – Bursera fagarioides
uhuatl – Saccharum officinarum
uili – Lagenaria siceraria
uipa – Nicotiana tabacum
uisquil – Sechium edule
uitzacua – Castilleja arvensis, Marrubium vulgare
ují – Arctostaphylos pungens, Brosimum alicastrum
úkui – Pinus arizonica
ulmaria – Filipendula ulmaria
última flor de primavera – Clarkia amoena
ulluco – Ullucus tuberosus
um – Persea americana
umbela – Plumbago auriculata, P. scandens
uncucha – Xanthosoma sagittifolium
ungurahui – Jessenia polycarpa
untzincacá – Ximenia americana
untzincacá de montaña – Ximenia americana
uña de águila – Ferocactus latispinus
uña de asno – Tussilago farfara
uña de caballo – Tussilago farfara
uña de cabra – Acacia farnesiana
uña de gato – Acacia greggii, A. occidentalis, A. wrightii, Celtis iguanea, Cynodon dactylon, Ferocactus latispinus, Havardia sonorae, Lantana camara, Martynia annua, Mimosa, Olneya tesota, Pisonia, Pithecellobium unguis-cati, Proboscidea, Robinia neomexicana, Rosa montezumae,

Sedum sediforme, Unicaria tomentosa, Valeriana clematitis, Zanthoxylum fagara
uña de gato negro – Acacia wrightii
uña de gato prieto – Pisonia capitata
uña de gavilán – Uncaria tomentosa
uña del diablo – Harpagophytum procumbens, Pisonia, Proboscidea
upala – Prosopis juliflora
urape – Bauhinia monandra, B. variegata
urape morado – Bauhinia variegata
urape orquídea– Bauhinia variegata
urani – Crescentia alata
urari – Chondodendron tomentosum
urce – Erica arborea
urena – Urena lobata
urina – Quercus laurina
urucu – Bixa orellana
uruata – Annona cherimola
urúbishi – Arbutus xalapensis
uruquenia – Croton ciliato-glandulosus
urza – Eysenhardtia polystachya
usabi – Prunus serotina
usimyuera – Agonandra racemosa
usipon – Astronium graveolens
uva – Cissus sicyoides, Cocoloba uvifera, Vitis
uva cabruna – Rhamnus cathartica
uva cana – Sedum album
uva cimarrona – Vitis
uva crespa – Ribes
uva de gato – Sedum album
uva de la costa – Cocoloba uvifera
uva de la playa – Cocoloba uvifera
uva del bosque – Vaccinium myrtillus
uva del diablo – Solanum dulcamara
uva del mar – Cocoloba uvifera
uva del monte – Vitis tiliifolia
uva del país – Dacryodes edulis
uva de mar – Sargassum vulgare
uva de moro – Mandragora autumnalis
uva de Muscadino – Vitis vinifera var muscat
uva de Oregón – Berberis aquifolium
uva de oso – Arctostaphylos uva-ursi
uva de pájaro – Sedum sediforme
uva de perro – Smilax aspera
uva de raposa – Paris quadrifolia
uva de zorro – Arctostaphylos uva-ursi, Paris quadrifolia, Solanum dulcamara
uvaduz – Arctostaphylos uva-ursi
uva espina – Ribes grossularia, R. montigenum, R. uva-crispa
uva gomosa – Cordia alba
uvalama – Lantana involucrata, Vitex mollis, V. pyramidata
uva silvestre – Cissus sicyoides
uva silvestre de Oregón – Berberis aquifolium var repens
uva trepadora – Cocculus diversifolius
uva tullidora – Cocculus diversifolius
uva verga – Aconitum napellus
úveda – Acacia macracantha
uveral – Cephalanthus occidentalis var californicus
uverillo – Cocoloba uvifera
uvero – Cephalanthus occidentalis var californicus, Cocoloba barbadensis, C. uvifera
uvero de la playa – Cocoloba uvifera
uvero del mar – Cocoloba uvifera
uvilla – Callicarpa americana, Vitis tiliifolia
uvilla cimarrona – Vitis tiliifolia
uvilla de lagarto – Tamus communis
uvita tullidora – Cocculus diversifolius
uvularia – Ruscus hypoglossum
uxe – Andira inermis
uxitl – Pinus teocote

vacabuey – Curatella americana
vaccaria – Vaccaria hispanica
vachácata – Galphimia glauca
vaga-biche – Astronium graveolens
vagiseo – Armoracia rusticana
vaina – Brassica rapa
vaina de espada – Eugenia capuli
vainilla – Inga spuria, Senna atomaria, S. bicapsularis, S. occidentalis, Vanilla planifolia, V. pompona
vainilla bastarda – Vanilla planifolia
vainilla boba – Vanilla pompona
vainilla cimarrona – Vanilla planifolia, V. sylvestris
vainilla conianona – Vanilla planifolia
vainilla de cochino – Vainilla planifolia
vainilla de mono – Vanilla planifolia
vainilla escarchada – Vanilla planifolia
vainilla fina – Vanilla planifolia
vainilla francesa – Vanilla pompona
vainilla mansa – Vanilla planifolia

vainilla plátano – Vanilla pompona
vainilla pompona – Vanilla pompona
vainilla senexante – Vanilla pompona
vainillón – Vanilla pompona
vainita – Phaseolus vulgaris
vainor – Celtis iguanea, C. reticulata
vainoro – Pisonia aculeata
vainoro prieto – Pisonia aculeata, P. capitata
valeriana – Eriogonum fasciculatum, Lippia alba, Valeriana
valeriana azul – Polemonium caeruleum
valeriana griega – Polemonium caeruleum
valeriana medicinal – Valeriana officinalis
valeriana menor – Valeriana officinalis
valeriana oficinal – Valeriana officinalis
valeriana roja – Centranthus ruber
valerianela – Valerianella locusta
valerianela de Italia – Valerianella eriocarpa
valla honda – Prosopis juliflora
valletilla – Hamelia erecta
vallico encurvado – Parapholis incurva
vallico estrigoso – Parapholis strigosa
vallico filiforme – Parapholis strigosa
vallico torcido – Parapholis incurva
vanita – Bourreria huanita, B. sonorae
vaquita – Sanvitalia procumbens
vara blanca – Acacia willardiana, Capparis incana, Croton, Leucaena involucrata, Lippia ligustrina, Montanoa speciosa
vara de cohete – Dasylirion longissimum
vara de cuete – Dasylirion longissimum
vara de Jesús – Polianthes tuberosa
vara del burro – Selloa glutinosa
vara de nardo – Polianthes tuberosa
vara de oro – Solidago
vara de oro del Canadá – Solidago canadensis
vara de oro gigantiflora – Solidago gigantea
vara de San Diego – Laelia anceps
vara de San José – Alcea rosea, Clarkia amoena, Jacobinia umbrosa, Laelia superbiens, Leonotis nepetaefolia, Penstemon barbatus, Polianthes tuberosa, Senna atomaria, S. bicapsularis, Solidago canadensis ssp elongata, S. virgaurea
vara de San Miguel – Arethusa grandiflora
vara de zorro – Ptelea trifoliata
vara dulce – Aloysia gratissima, A. wrightii, Baccharis salicifolia, Eysenhardtia polystachya, E. texana, Lippia ligustrina
varaduz – Eysenhardtia polystachya
varal – Dodonaea viscosa
vara mosqueda – Hamamelis virginiana
vara prieta – Acacia constricta, A. millefolia, Adenostoma fasciculatum, Aeschynomene nivea, Brongniartia alamosana, Caesalpinia californica, Capparis indica, C. verrucosa, Cordia greggi, C. parvifolia, Croton, Hamelia erecta, Lycium andersonii
vara zarza – Purshia mexicana var stansburyana
varejón – Fraxinus trifoliata
varirapeni – Amanita muscaria
varita de San José – Melochia tomentosa, Oncidium carthaginensis, Penstemon barbatus
vauar – Valeriana sorbifolia
vauquelinia – Vauquelinia karwinskyi
vaval – Cardiospermum halicacabum
vedegambre – Veratrum album
vedegambre azul – Aconitum napellus
vedegambre blanco – Veratrum album
vedegambre verde – Veratrum viride
vegetal – Myrica cerifera
veigelia – Weigela florida
veintiunilla – Asclepias curassavica
vejiga de perro – Physalis alkekengi, P. pubescens
vejiguilla – Physalis alkekengi
vela de coyote – Cylindropuntia, Opuntia engelmannii, O. hyptiacantha, O. phaeacantha
vela de sabana – Typha domingensis
vela de sebo – Pedilanthus tithymaloides
vela escamada – Adenophora liliifolia
velo de novia – Agdestis clematides, Duranta repens
velo de viuda – Duranta repens
vellorita – Bellis perennis, Primula officinalis, P. veris
vellorita de oro – Primula veris
vellosilla de color anaranjado – Hieracium aurantiacum
venadillo – Swietenia humilis
venado – Amanita vaginatum
vencetósigo – Cynanchum vincetoxicum
venenillo – Asclepias curassavica, A. linaria, Polygonum, Rauwolfia hirsute, Solanum umbellatum, Thevetia ovata, T. peruviana

veneno de cuervo – *Solanum nigrum*
veneno del diablo – *Strychnos tabascana*
veneno de perro – *Solanum cervantesii*
veneno rojo – *Asclepias curassavica*
venera de Santiago – *Sprekelia formosissima*
ventenata avena falsa – *Ventenata dubia*
ventosidad – *Nama hispidum, N. undulatum*
venturosa – *Lantana camara*
veranera – *Bougainvillea spectabilis*
veratro – *Veratrum officinale*
verbachina – *Phytolacca octandra*
verbasco – *Verbascum thapsus*
verbasco austriaco – *Verbascum chaixii*
verbasco especioso – *Verbascum speciosum*
verbasco purpúreo – *Verbascum phoeniceum*
verbasco ramificado – *Verbascum virgatum*
verbena – *Loeselia glandulosa, Mentha suaveolens, Verbena*
verbena azul – *Verbena hastate, V. orcuttiana*
verbenaca – *Salvia verbenacea*
verbena común – *Verbena neomexicana*
verbena fina – *Verbena officinalis*
verbena macho – *Verbena officinalis*
verbena mayor – *Verbena officinalis*
verbena mexicana – *Verbena ciliata*
verbesina – *Verbesina encelioides*
verdadera palma real – *Attalea gomphococca*
verdal – *Brassica oleracea var capitata*
verdecillo – *Tabebuia chrysantha, T. pentaphylla*
verdillo cacahuatoide – *Chloris chloridea*
verdillo de fleco – *Chloris ciliata*
verdillo esbelto – *Chloris andropogonoides*
verdillo norteño – *Chloris subdolichostachya*
verdillo papalote – *Chloris cucullata*
verdillo plumerito – *Chloris virgata*
verdolaga – *Lotus strigosus, Portulaca oleracea, P. grandiflora, Talinum paniculatum*
verdolaga acuática – *Ludwigia peploides*
verdolaga amarilla – *Portulaca oleracea*
verdolaga blanca – *Trianthema portulacastrum*
verdolaga bronca – *Trianthema portulacastrum*
verdolaga de agua – *Ludwigia peploides, L. repens, Pistia stratiotes*
verdolaga de cochi – *Tidestromia lanuginose, Trianthema portulacastrum*
verdolaga de Cuba – *Claytonia perfoliata*
verdolaga de flores grandes – *Portulaca grandiflora*
verdolaga de invierno – *Claytonia perfoliata*
verdolaga de la playa – *Sesuvium portulacastrum*
verdolaga de la sierra – *Portulaca umbraticola*
verdolaga de puerco – *Alternanthera achyrantha, A. caracasana*
verdolaga perfoliada – *Claytonia perfoliata*
verdolaga silvestre – *Portulaca oleracea*
verdolaguilla – *Talinum triangulare*
verduras de mostaza – *Brassica nigra*
verduras de nabo – *Brassica rapa*
vergonzosa – *Mimosa albida, M. pudica*
vergüenza – *Mimosa pudica*
vergüenza varonil – *Mimosa pudica*
verissima hierba – *Anemopsis californica*
vermicularia – *Sedum album*
vermicularia azul – *Sedum caeruleum*
vermicularia caucásica – *Sedum spurium*
vermicularia de Europa – *Sedum album*
vermicularia de hojas cortas – *Sedum brevifolium*
vermicularia de hojas espesas – *Sedum dasyphyllum*
vermicularia dendroide – *Sedum dendroideum*
vermicularia inglesa – *Sedum anglicum*
vermicularia rupestre – *Sedum reflexum*
verónica – *Veronica*
verónica acuática – *Veronica becabunga*
verónica alpina – *Veronica alpina*
verónica arbustiva – *Hebe buxifolia*
verónica arvense – *Veronica arvensis*
verónica austriaca – *Veronica austriaca*
verónica blanquecina – *Veronica spicata ssp incana*
verónica comosa – *Veronica catenata*
verónica común – *Veronica chamaedrys*
verónica de hojas de hiedra – *Veronica hederifolia*
verónica de hojas de serpillo – *Veronica serpyllifolia*
verónica de los pantanos – *Veronica scutellata*
verónica escutelada – *Veronica scutellata*
verónica espigada – *Veronica spicata*
verónica filiforme – *Veronica filiformis*
verónica macho – *Veronica officinalis*
verónica peregrina – *Veronica peregrina*
verónica pérsica – *Veronica persica*

verónica prostrada – Veronica prostrata
verónica repente – Veronica repens
verónica rocosa – Veronica filiformis
verónica trifoliada – Veronica triphyllos
vervenía – Phacelia distans, P. tanacetifolia
verrucaria – Heliotropium europaeum
vesque – Viscum album
vetiver – Vetiveria zizanioides
veylaqui – Caesalpinia platyloba
veza – Vicia
veza común – Vicia sativa
veza cracca – Vicia cracca
veza de arena – Vicia villosa
veza de hojas estrechas – Vicia sativa ssp nigra
veza de Hungría – Vicia pannonica
veza de pájaro – Vicia cracca
veza forrajera – Vicia sativa
veza francesa – Vicia cracca
veza hirsuta – Vicia hirsuta
veza húngara – Vicia pannonica
veza panonica – Vicia pannonica
veza purpúrea – Vicia benghalensis
veza tetrasperma – Vicia tetrasperma
veza vellosa – Vicia villosa
veza velluda – Vicia villosa
veza velluda de invierno – Vicia villosa
víbora – Asclepias curassavica
viborera – Echium vulgare
viborera común – Echium vulgare
viborona – Asclepias curassavica
viburno – Viburnum
viburno lanudo – Viburnum plicatum
vicaria – Catharanthus roseus
vico – Jatropha curcas
Victoria regia – Victoria regia
vichi – Crotalaria longirostrata
vichishovo – Myrtillocactus schenckii
vicia de Hungría – Vicia pannonica
vid – Vitis arizonica, V. vinifera
vidalba – Clematis vitalba
vid americana – Vitis labrusca, V. rupestris
vid común – Vitis vinifera
vid coral – Cissus cucurbitina
vid de hierro – Desmodium canum
vid de Judea – Solanum dulcamara
vid de Oregón – Berberis aquifolium
vid de taza y platillo – Cobaea scandens
vid de trompeta mexicana – Distictis buccinatorius
vid ornamental – Vitis coignetiae
vid ornamental del Japón – Vitis coignetiae
vidrillo – Batis maritima, Salicornia europaea, Sesuvium portulacastrum, Suadea
vidrioso – Croton guatemalensis, Gilibertia arborea
vid rusa – Polygonum baldschuanicum
vid silvestre – Cissus sicyoides, Vitis
viejito – Cephalocereus sensilis, Mammillaria
viejo – Cephalocereus sensilis, Grusonia bradtiana, Mammillaria senilis, Opuntia chaffeyi, Pachycereus schottii, Pilosocereus leucocephalus, P. purpusii, Rosmarinus officinalis
viejo arruga dulce – Capsicum annuum
viga de llano – Sabal rosei
vigna – Vigna sinensis
vilanera – Mandragora autumnalis
viluria – Digitalis purpurea
villela – Sisyrinchium bellum
vinagrera – Rumex acetosa, R. acetosella
vinagrerita – Rumex acetosella
vinagrillo – Oxalis acetosella, O. corniculata, Rhus crispus, R. mollis
vinca – Catharanthus roseus, Vinca minor
vincapervinca – Catharanthus roseus, Vinca minor
vincapervinca de Madagascar – Catharanthus roseus
vincapervinca mayor – Vinca major
vinca rosea – Catharanthus roseus
viniebla – Cynoglossum officinale
vinolo – Acacia cymbispina
vinolo blanco – Acacia macracantha
vinolo prieto – Acacia cymbispina
vinorama – Acacia
vinuela – Hibiscus sabariffa
viña loca – Parthenocissus quincuefolia, P. tricuspidata
viñátigo – Persea indica
viña trepadora – Parthenocissus tricuspidata
viña virgen – Parthenocissus quinquefolia, P. tricuspidata
viola – Viola odorata
violeta – Anoda acerifolia, A. cristata, Melia azedarach var umbraculiformis, Viola
violeta africana – Saintpaulia ionantha, S. jonantha
violeta amarilla – Viola flagelliformis

violeta australiana – Viola hederacea
violeta azteca – Pinguicula caudata
violeta cimarrona – Pinguicula caudate, P. moranensis
violeta común – Violeta odorata
violeta cornuta – Viola cornuta
violeta damasquina – Hesperis matronalis
violeta de África – Saintpaulia ionantha
violeta de agua – Eichhornia crassipes
violeta de diente de perro – Erythronium dens-canis
violeta de hortelano – Viola x wittrockiana
violeta de jardín – Viola odorata
violeta de la bruja – Vinca minor
violeta del campo – Anoda acerifolia, A. cristata, Pinguicula caudate, Viola arvensis
violeta de los Alpes – Cyclamen persicum, C. purpurascens
violeta de los pantanos – Viola palustris
violeta del país – Anoda cristata
violeta de monte – Pinguicula caudata, P. moranensis, Viola flagelliformis, V. painteri
violeta de olor – Viola odorata
violeta de rivini – Viola riviniana
violeta española – Viola odorata
violeta extranjera – Viola odorata
violeta olorosa – Viola odorata
violeta palustre – Viola palustris
violeta persa – Exacum affine
violeta rocosa – Viola labradorica
violincillo – Heliotropium macrostachyum
violo – Betula pubescens
viperina – Echium vulgare, Zornia bracteata, Z. diphylla
viperina común – Echium vulgare
viperina de Virginia – Aristolochia serpentaria
viperina falso-llantén – Echium lycopsis
Virginia – Zinnia elegans, Z. multiflora, Z. peruviana
Virginio – Nicotiana glauca
virtugosa – Pectis uniaristata
virreina – Calendula officinalis
virreinita – Calendula officinalis
viscainoa – Viscainoa geniculata
visco – Phoradendron, Pstiiacanthus calyculatus
visco blanco – Viscum album
visco cuercino – Phoradendron macrophyllum, Psittacanthus calyculatus, Viscum album
vistaria – Wisteria sinensis
vitilaina – Viburnum lantana
vito pijte – Liquidambar styraciflua
vitsacua sipiati – Plantago major
viuda – Abelmoschus manihot, Scabiosa atropurpurea, S. stellata, Zinnia pauciflora
viuda silvestre – Knautia arvensis
viuxita – Maranta arundinacea
volador – Aspidosperma quebracho-blanco, Dioscorea, Populus mexicana, Terminalia amazonia
volantín – Cleome hasslerana
volcameria – Clerodendrum chinense
vomitel – Cordia sebestena
vomitel colorado – Cordia sebestena
vomitivo – Stegnosperma halimifolium
vulneraria – Anthyllis vulneraria
vulneraria barba de Júpiter – Anthyllis barba-jovis
vulpia bromo-falso – Vulpia bromoides
vulpia cola de rata – Vulpia myuros
vulvaria – Chenopodium vulvaria

wame gobernadora – Larrea tridentata
wanamé – Ariocarpus fissuratus
washingtonia – Washingtonia filifera
wellingtonia – Sequoiadendron giganteum
wichowaka – Pachycereus pecten-aboriginum
wichuriki – Mammillaria senilis
wipaka – Nicotiana glauca
wisachito – Hoffmannseggia glauca
wisapole – Xanthium strumarium
witculíki – Mammillaria heyderi
wiyoko – Pinus ayacahuite
wocomahi – Agave wocomahi
wokowi – Lophophora williamsii

xaax – Acacia angustissima
xabalam – Codiaeum variegatum
xabxib – Acacia pennatula
xacil – Boerhavia erecta
xacilxacxiu – Boerhavia erecta
xahuique – Tigridia dugesii, T. pavonia
xahuique de Tupátaro – Tigridia dugesii
xakilche – Lippia graveolens
xakilxiu – Mentha x piperita
xalache – Sedum acre, S. moranense
xalama – Ficus
xalama limón – Ficus padifolia

xalamat – Ficus goldmanii
xalatlaco – Pinus leiophylla
xalócotl – Abies religiosa, Pinus teocote
xaltomatl – Saracha jaltomata
xalxócotl – Psidium guajava
xan – Sabal japa
xanabmucuy – Chamaesyce prostrata
xanacol – Bombax ellipticum
xanecuilche – Dolicholus phaseoloides
xantea – Citrullus colocynthis
xararo – Acourtia cordata
xaricamata – Cosmos bipinnatus, Helianthus annuus
xarimpuscua – Oxalis corniculata
xate – Argemone mexicana, Chamaedorea pinnatifrons, Reinhardtia gracilis
xate hembra – Chamaedorea elegans
xate macho – Chamaedorea oblongata
xau – Theobroma cacao
xaxachichitl – Calea zacatechichi
xaxaxacótic – Galphimia glauca
xaxox – Solanum verbascifolium
xaza – Schinus molle
xcacalche – Jatropha curcas
xcocehak – Smilax mexicana
xegua – Ipomoea arborescens
xeink – Achras zapota
xempoalxóchitl – Tagetes erecta
xenecuilche – Rhynchosia pyramidalis
xenelba – Brassica nigra
xengua – Prunus capuli
xenshe – Portulaca oleracea
xeshish – Fouquieria splendens
xhail – Ipomoea purpurea
xiazek – Lysiloma bahamensis
xica – Oryza sativa
xicaco – Chrysobalanus icaco
xícama – Pachyrrhizus erosus
xicamatl – Pachyrhizus erosus
xicólotl – Argemone mexicana
xicomini – Commicarpus scandens
xicotzapotl – Achras zapota
xictomatl – Lycopersicon esculentum
xidni – Citrus sinensis
xigaba – Cucurbita maxima, Lagenaria siceraria
xihuihui joyo – Senna oxyphylla
xihuite – Parthenium argentatum
xíhuitl – Chamaesyce maculata
xihuixcán – Bombax ellipticum
xiiksh – Sambucus mexicana
xik-ché – Zanthoxylum fagara
xilacuaco – Rivina humilis
xiloxochicuahuitl – Zapoteca anomala
xiloxóchitl – Bombax ellipticum, Caesalpinia pulcherrima, Ceiba pentandra, Pachira aquatica, Zapoteca anomala
ximenia – Ximenia americana
ximitzcuahuitl – Schoepfia schreberi
ximula – Cosmos sulphureus
xinene – Persea americana, P. schiedeana
xinil – Quercus polymorpha
xipari – Quercus durifolia, Q. hypoleucoides
xiña – Taxodium distichum var mexicanum
xiopatla – Gaultheria acuminata
xiopatli – Urera caracasana
xioquauitl – Bursera simaruba
xiquelite – Indigofera suffruticosa
xiquima – Pachyrhizus erosus
xiquite – Eupatorium, Vernonia deppeana
xisich – Acacia angustissima
xitotomatl – Lycopersicon esculentum
xiuh-elo-quílitl – Bidens aurea
xiuitl – Poaceae
xiuquílitl – Indigofera suffruticosa, I. tinctoria
xiutecuitlanextli – Nicotiana glauca
xiutoloc – Milleria quinqueflora
xix-el-bá – Lygodium venustum
xixinda – Lepidium virginicum
xixuega – Escontria chiotilla
xkomyaxnic – Solanum amazonium
xkuku – Sorghum halepense
xmutz – Mimosa pudica
xoalacatl – Disocactus speciosus, Epiphyllum x Disocactus
xoantepxilotl – Chamaedora ernesti-augustii
xobarroba – Lantana camara
xocat – Malpighia glabra, Spondias mombin
xocoac – Abrus precatorius
xocoatole – Cosmos bipinnatus
xococquauitl – Conostegia xalapensis
xocomecatl – Vitis tiliifolia
xoconóchitl – Cylindropuntia imbricata, Stenocereus stellatus
xoconochtli – Cylindropuntia imbricata, Stenocereus stellatus
xoconostle blanco – Opuntia spinulifera

xoconostle colorado con espinas – Opuntia hyptiacantha
xoconoxtle – Ocylindropuntia imbricata, puntia joconostle, Pereskiopsis blakeana, P. porteri, Stenocereus stellatus
xócotl – Spondias purpurea
xocoxóchitl – Eugenia pimenta
xocoyol – Acrocomia mexicana, Oxalis albicans, O. corniculata
xocoyol cimarrón – Begonia nelumbiifolia
xocoyole – Oxalis
xocoyole de México – Oxalis americana
xocoyotl – Begonia
xochicopalli – Bursera odorata
xochicotscuahuitl – Liquidambar styraciflua
xochicuahuitl – Liquidambar styraciflua
xochilcorona – Cornus disciflora
xochimitl – Tecoma stans
xochinacastle – Cymbopetalum penduliflorum
xochinanácatl – Amanita caesarea
xochiocotzocuahuitl – Liquidambar styraciflua
xochiocotzotl – Liquidambar styraciflua
xochipali – Cosmos sulphureus
xochipalli – Cosmos sulphureus
xochipatli – Jacquinia aurantica
xochipile – Lobelia laxiflora
xochipitsahuac – Polygala scoparia
xochitecómatl – Solandra nitida
xochtatl – Malpighia glabra
xohexnuc – Lantana camara
xokoklavoxochit – Citrus reticulata
xololco – Erythrina flabelliformis
xolotsin – Heliocarpus tomentosus
xolteco – Dalea diffusa
xómet – Sambucus mexicana
xonacatl – Allium cepa, A. scaposum
xonaxe – Croton ciliato-glandulosus
xonequitl – Selloa glutinosa
xong – Fouquieria splendens
xoniguilli – Cuphea llavea
xonitlzal – Piqueria trinervia
xooba-yati – Hordeum vulgare
xoop – Bursera microphylla
xoxoco – Rhus mollis
xoxocoenabitl – Psidium guajava
xoxocote – Arbutus xalapensis
xoxocoyolin – Oxalis corniculata
xoxonitzal – Piqueria trinervia
xoxouhca-pahtli – Ligusticum grayi
xoxoyolpapatla – Rumex acetosa
xoyo – Erythrina
xpolcútzil – Asclepias curassavica
xpukusikil – Trichilia hirta
xtabentún – Turbina corymbosa
xtuhabin – Senna atomaria
xucul – Portulaca oleracea
xucupara – Lycopersicon esculentum
xumetl – Sambucus mexicana
xumpatsúchitl – Tagetes erecta
xunalixase – Croton ciliato-glandulosas
xurata – Gossypium
xuxox – Solanum verbascifolium
xuxu – Sechium edule

ya – Achras zapota
ya'ax-kach – Bacopa monnieri
yaaxpakan – Opuntia dillenii, O. stricta
yabo – Andira inermis
yabona – Imperata cylindrical
yaco – Tilia mexicana, T. occidentalis
yaco de cuero – Trema micrantha
yaco de granadillo – Guazuma ulmifolia
yaco de la costa – Bombax ellipticum
yaco de venado – Belotia mexicana, Guazuma ulmifolia
yacón – Polymnia edulis
yacunahax – Momordica charantia
yagabecie – Celtis pallida
yagabicigui – Liquidambar styraciflua
yagabiche – Rhus terebinthifolia
yagabicholi – Vitis vinifera
yaga-bigi-lasi – Mammillaria karwinskiana ssp collinsii
yagabiguichi – Pithecellobium dulce
yagabilape – Ricinus communis
yagabito – Liquidambar styraciflua
yaga-bizie – Alnus acuminata
yagabizoya – Theobroma cacao
yagaguichixiña – Taxodium distichum var mexicanum
yagaguieguei – Pistacia mexicana
yagaguienite – Myroxylon balsamum var pereirae
yagaguiexoba – Bourreria huanita
yagahigo – Ricinus communis
yaga-huil – Maclura tinctoria
yagalache – Toxicodendron radicans
yagalan – Eugenia capuli

yagalán – Karwinskia humboldtiana
yagale – Hyparrhenia rufa
yagana – Guaiacum coulteri
yaganaraxo – Citrus sinensis
yagaquiguece – Nerium oleander
yagareche – Quercus conspersa
yagasachi – Punica granatum
yagaxiña – Erythea edulis
yagayace – Lysiloma acapulcensis
yaga-yaxo-Castilla – Ficus carica
yagpsuy – Quercus oleoides
yagrumo – Cecropia peltata
yagrumo hembra – Cecropia peltata
yagua – Roystonea regia
yaguaguyt – Gliricidia sepium
yaguay – Rhus terebinthifolia
yáguna – Paspalum plicatulum
yaha – Curatella americana
yaite – Gliricidia sepium
yaje – Banisteriopsis caapi
yalaelel – Oxalis latifolia
yamanqui – Selaginella
yambigo – Thevetia peruviana
yambo – Spilanthes acmella
yame de Canarias – Colocasia esculenta
yamete – Asclepias albicans, A. masonii, A. subulata
yamo baboso – Heliocarpus attenuatus
yamole – Phytolacca octandra, Sapindus saponaria
yampa – Perideridia gairdneri
yampi – Dioscorea
yana – Conocarpus erectus
yant – Agave utahensis
yanuco – Punica granatum
yara – Ehretia anacuna
yarabisco – Jacaranda acutifolia
yaraguá – Melinus minutiflora
yarcas – Madia sativa
yarina – Phytelephas macrocarpa
yaripas – Guadua latifolia
yas – Persea schiedeana
yascabté – Vitex pyramidata
yastsana – Tradescantia discolor
yatay – Butia yatay
yatolito – Haematoxylum campechianum
yatzí – Bomarea acutifolia
yauhtli – Tagetes filifolia
yaupan – Ilex deciduas, I. vomitoria
yautia – Colocasia esculenta, Xanthosoma caracu, X. robustum
yautia horqueta – Xanthosoma caracu
yaxal – Piper amalgo
yaxcelil – Ipomoea tricolor
yaxché – Ceiba pentandra
yaxha – Castilla elastica
yaxholche – Abutilon lignosum
yaxox – Trophis racemosa
yaxqui – Agave sisalana
yaxtabcanil – Cissus rhombifolia, C. sicyoides
yecapixtla – Psittacanthus calyculatus
yedra – see hiedra
yegua – Phaseolus coccineus
yehcapaztzin – Senna laevigata
yema – Amanita caesarea
yema ceniza – Amanita vaginatum
yema de huevo – Amanita caesarea, Rhamnus sphaerosperma, Tricholoma equestre, T. sejunctum
yemita – Amanita caesarea
yepouecha – Acacia pennatula
yerba – see hierba
yero – Vicia ervilia
yesca – Polyporus fomentarius
yetl – Nicotiana rustica, N. tabacum
yeyepaxtle – Chrysactinia mexicana
yeyésowa – Tagetes micrantha
yichcaan – Justica spicigera
yoal – Morisonia americana
yoale – Genipa americana
yo-betz – Stanhopea devoniensis
yocohuiro – Cissus sicyoides
yogo – Pteridium aquilinum var pubescens
yohimbe – Corynanthe yohimbe
yohimbina – Corynanthe yohimbe
yo-lachi – Talauma mexicana
yolochichitl – Eupatorium collinum, E. deltoideum
yolosúchil – Magnolia schiedeana, Talauma mexicana
yoloxiltic – Piqueria trinervia
yoloxóchitl – Talauma mexicana
yollopatli – Talauma mexicana
yomha – Pilea microphylla
yóndiro – Goldmania foetida
yopillo – Dactylis glomerata
yori – Arctostaphylos pungens
yorimaqui – Ricinus communis

yos – Sapium stylare
yotito – Agaricus campestris
yoyolxpchitl – Commelina tuberosa
yoyoma – Spondias purpurea
yoyote – Thevetia ovata, T. peruviana, T. thevetioides
yoyotli – Thevetia ovata, T. peruviana, T. thevetioides
yuatia – Xanthosoma sagittifolium
yuca – Beschorneria yuccoides, Hesperaloe parviflora, H. funifera, Manihot chlorosticta, M. dulcis, M. esculenta, Merremia aurea, Operculina aurea, Yucca
yuca amarga – Manihot esculenta
yuca blanca – Manihot esculenta
yuca brava – Manihot esculenta
yuca brillante – Yucca gloriosa
yucacaca – Talauma mexicana, Thevetia peruviana, T. thevetioides
yuca del cerro – Manihot angustiloba
yuca dulce – Manihot dulcis, M. esculenta
yuca mansa – Manihot esculenta
yucateco – Ficus nitida
yuco de bejuco – Pachyrhizus erosus
yucucaya – Rhus mollis
yucuchichicino – Taxodium distichum var mexicanum
yucundatura – Taxodium distichum var mexicanum
yucundede – Selloa glutinosa
yucuteyeye – Erythea edulis
yuchán – Chorisia insignis
yulan – Magnolia denudata
yumete – Asclepias albicans, A. erosa, A. subulata
yunquillo – Sporobolus indicus
yuquilla – Asclepias curassavica, Cana indica, Curcuma domestica, Maranta arundinacea
yuraguana – Copernicia vespertilionum
yute – Corchorus olitorius
yutnucate – Sambucus mexicana
yutnudidzi – Punica granatum
yutnuñun – Erythea edulis
yutnusatnu – Pinus montezumae
yutnutandaa – Guaiacum coulteri
yuy – Casimiroa edulis
yuyuba – Ziziphus jujuba

zabacnicte – Plumeria rubra
zabida – Aloe vera
zabila – Agave bovicornuta, Aloe socotrina, A. vera
zabila hecho – Pachycereus schottii
zabucajo – Lecythis zabucajo
zacacanzelxiu – Asclepias curassavica
zacachichi – Calea zacatechichi, Conyza filaginoides
zacachichic – Calea zacatechichi
zacahuistle – Distichlis spicata
zacam – Opuntia inaperta
zacamsotz – Opuntia inaperta
zacapal – Cuscuta
zacapoaxtla – Liquidambar styraciflua
zacate – Chloris elegans, C. virgata, Hackelochloa granularis, Mitracarpus hirtus, Nolina bigelovii, N. microcarpa, Phragmites australis, Poaceae
zacate aceite – Bothriochloa saccharoides
zacate aceitillo – Bouteloua aristidoides, Heteropogon contortus
zacate aceitunillo – Rhynchelytrum repens
zacate africano – Eragrostis lehmanniana, Pennisetum clandestinum
zacate agrarista – Cynodon dactylon, Sorghum halepense
zacate agrio – Digitaria sanguinalis
zacate agujilla – Cynodon dactylon
zacate alcalino – Sporobolus wrightii
zacate alcalino espigado – Sporobolus contractus
zacate alemán – Echinochloa polystachya
zacate alfombra – Axonopus affinis, Pennisetum clandestinum
zacate alicia – Cynodon dactylon
zacate alpiste – Poa annua
zacate amacollado – Muhlenbergia macroura
zacate amargo – Axonopus affinis, Calea zacatechichi, Parthenium hysterophorus, Paspalum compressum, Setaria gracilis
zacate angleton aristado – Dichanthium aristatum
zacate angleton pretoria – Dichanthium annulatum
zacate aparejo – Muhlenbergia montana, porteri
zacate apestoso – Eragrostis cilianensis
zacate araña – Aristida ternipes, Chloris dandyana, Leptochloa mucronata

zacate araña de tres barbas – Aristida ternipes var gentilis
zacate arrocillo – Echinochloa polystachya
zacate azul – Poa annua
zacate azulado – Panicum antidotale
zacate azul anual – Poa annua
zacate azul de Kentucky – Poa pratensis
zacate azul de las praderas – Poa pratensis
zacate azul del Canadá – Poa compressa
zacate bahía – Paspalum notatum
zacate barba larga – Trachypogon secundus
zacate barbón – Oplismenus setarius
zacate bermejo – Hyparrhenia rufa
zacate bermuda – Cynodon dactylon
zacate bermuda estrella – Cynodon plectostachys
zacate berqueno – Panicum coloratum
zacate bobo – Digitaria insularis
zacate borrego – Cynodon dactylon, Triodia pulchella
zacate borreguero – Cathestecum erectum, Erioneuron pulchellum
zacate búfalo – Buchloë dactyloides
zacate buffel – Cenchrus ciliaris, Pennisetum ciliare
zacate cabezón – Cyperus surinamensis
zacate cadillo – Oplismenus burmannii
zacate camalote – Brachiaria mutica
zacate camelote – Echinochloa crus-pavonis
zacate cangrejo piloso – Digitaria villosa
zacate cangrejo velludo – Digitaria ciliaris, D. sanguinalis
zacate caribe – Eriochloa polystachya
zacate carricillo – Panicum trichoides
zacate cebadilla – Elymus elymoides
zacate cedroso – Setaria gracilis
zacate cepillo – Bouteloua gracilis
zacate citronella – Cymbopogon citratus
zacate clavo – Imperata brasilensis
zacate cochinillo – Cenchrus incertus, C. longispinus
zacate cola de zorra – Aristida adscensionis, Chloris virgata, Enneapogon desvauxii, Lycurus pheoides, Polypogon monspeliensis
zacate colorado – Gouinia virgata, Heteropogon contortus, Panicum coloratum
zacate compacto – Sporobolus contractus
zacate cortador – Imperata brasiliensis, Nolina microcarpa, Paspalum virgatum
zacate corredor – Panicum molle
zacatechi – Calea zacatechichi
zacatechichi – Calea zacatechichi
zacatechichic – Conyza filaginoides
zacate chino – Buchloë dactyloides, Cynodon dactylon, Pleuraphis belangeri
zacate dallis – Paspalum dilatatum
zacate de agua – Aristida adscensionis, Echinochloa colona, E. crus-galli, E. frumentacea, Eleocharis atropurpurea, Eragrostis mexicana, Sesbania macrocarpa, Sporobolus airoides, S. pulvinatus
zacate de agua tres barbas – Aristida adscensionis
zacate de akalcheses – Imperata cylindrica
zacate de alpiste – Phalaris aquatica
zacate de amor – Eragrostis pilosa
zacate de Angola – Echinochloa polystachya
zacate de aparejo – Nolina microcarpa, Sporobolus wrightii
zacate de armazón – Nolina microcarpa
zacate de burro – Eleocharis atropurpurea, Scleropogon brevifolius
zacate de corral – Echinochloa crus-galli
zacate de cresta – Koeleria macrantha
zacate de cuentas – Coix lacryma-jobi
zacate de cuña – Sphenopholis obtusata
zacate de burro – Paspalum virgatum
zacate de encinar – Sporobolus indicus
zacate de escoba – Muhlenbergia macroura
zacate de gallina – Cynodon dactylon
zacate de ganso – Eleusine indica
zacate de Johnson – Sorghum halepense
zacate de la huerta – Dactylis glomerata
zacate de la loma – Rhynchelytrum repens
zacate de la playa – Monanthochloe littoralis
zacate del conejo – Cynodon dactylon
zacate de limón – Cymbopogon citratus, C. nardus
zacate de listón – Phalaris arundinacea
zacate del mar – Zostera marina
zacate de los médanos – Panicum antidotale
zacate de luna – Cynodon dactylon
zacate de manati – Echinochloa crus-galli
zacate de muela – Sisyrinchium bellum
zacate de mula – Sisyrinchium bellum
zacate de perro – Calea zacatechichi
zacate de piedras castillitos – Agrostis gigantea

zacate de piedras quillas – Agrostis palustris
zacate de ratón – Poa annua
zacate de rescate – Bromus catharticus
zacate de seda – Rhynchelytrum repens
zacate de semilla – Aristida adscensionis
zacate de toche – Cyperus sesleroides
zacate de tule – Cyperus
zacate de venado – Imperata brasilensis
zacate dulce – Paspalum dilitatum
zacate egipto – Brachiaria mutica, Panicum maximum, Sorghum halepense
zacate elefante – Pennisetum purpureum
zacate encubierto – Sporobolus cryptandrus
zacate escobilla – Leptoloma cognatum
zacate estrella – Cynodon plectostachyus
zacate ferrer – Cynodon dactylon
zacate galleta – Pleuraphis belangeri
zacate galleta gigante – Pleuraphis rigida
zacate gigante – Leptochloa dubia, L. filiformis, Pennisetum purpureum, Sporobolus giganteus
zacate gigante anual – Leptochloa fascicularis
zacate gordo – Amsinckia douglasiana, A. spectabilis
zacate gordura – Melinis minutiflora
zacate grama – Paspalum conjugatum, P. distichum, Pleuraphis cenchroides, Tripsacum laxum
zacate gramilla – Panicum obtusum
zacate guácima – Eleusine indica
zacate guatemala – Tripsacum laxum
zacate guía – Brachiaria arizonica, Panicum obtusum
zacate guinea – Panicum maximum
zacate gusano café – Pennisetum setosum
zacate hoja ancha – Dichanthelium oligosanthes
zacate horqueta – Paspalum notatum
zacate ignea – Typha domingensis
zacate ilusión – Rhynchelytrum repens
zacate imperial – Axonopus scoparius
zacate indio – Sorghastrum nutans
zacate inglés – Cynodon dactylon, Lolium perenne
zacate italiano – Lolium multiflorum
zacate jaragua – Hyparrhenia rufa
zacate jigüite – Eragrostis obtusiflora
zacate Johnson – Sorghum halepense
zacate kikuyu – Pennisetum clandestinum
zacate lagunero – Chloris virgata
zacate ladera – Enneapogon desvauxii
zacate liebre – Bouteloua barbata
zacate liebrero – Bouteloua barbata, Cathestecum erectum
zacate liendrilla – Leptochloa mucronata
zacate limón – Cymbopogon citratus, C. nardus
zacate liso – Muhlenbergia monticola
zacate lisonjero – Panicum capillare
zacate lobero – Enneapogon desvauxii, Lycurus phleoides
zacate llanero – Eragrostis intermedia
zacate llorón cimarrón – Eragrostis lugens
zacate maicero – Tripsacum dactyloides
zacate maicero floridano – Tripsacum floridanum
zacate maizar – Tripsacum floridanum
zacate malinali – Muhlenbergia macroura
zacate mano pangola – Digitaria pentzii var minor
zacate mano punta café – Digitaria insularis
zacate mano texano – Digiraria patens
zacate masarca – Panicum obtusum
zacate mateado – Bouteloua reflexa
zacate melado – Melinus minutiflora
zacate meloso – Paspalum dilatatum
zacate miel – Paspalum dilatatum
zacate milo – Sorghum halepense
zacate minero – Ageratina
zacate mota – Chloris virgata
zacate nandi – Setaria sphacelata
zacate navajuela – Scleria bracteata
zacate natal – Polypogon interruptus, Rhynchelytrum repens
zacate pangola – Digitaria decumbens
zacate panizo azul – Panicum antidotale
zacate panizo café – Brachiara fasciculata
zacate para – Panicum purpurascens
zacate parana – Sorghum halepense
zacate pata de cuervo – Dactyloctenium aegyptium
zacate pata de gallo – Chloris gayana
zacate pegajoso – Setaria verticillata
zacate pelillo – Cynodon dactylon, Triodia pulchella
zacate peludo – Setaria gracilis
zacate pinto – Echinochloa colona
zacate pirámide – Sporobolus pyramidatus

zacate playero – Monanthochloë littoralis
zacate plumoso – Pennisetum villosum
zacate poa – Poa pratensis
zacate polaco – Paspalum dilatatum
zacate popotillo – Bothriochloa barbinodis
zacate primavera – Anthoxanthum odoratum
zacate privilegio – Panicum maximum
zacate prodigio – Tripsacum laxum
zacate punta blanca – Digitaria californica
zacate puntero – Hyparrhenia rufa
zacate rayado – Echinochloa colona
zacate retorcido – Heteropogon contortus
zacate rhodes – Chloris gayana
zacate rosa – Rhynchelytrum repens
zacate rosado – Rhynchelytrum repens
zacate ruso – Sorghum halepense
zacate sabaya – Panicum coloratum
zacate salado – Distichlis palmeri, D. spicata, Eriochloa acuminata, Leptochloa mucronata, L. panicea ssp brachiata, Monanthochloë littoralis
zacate salado mexicano – Leptochloa fusca ssp uninervia
zacate setaria – Setaria anceps
zacate silvestre del Canadá – Elymus canadensis
zacate sudán – Sorghum bicolor, S. sudanense
zacate taiwan – Digitaria insularis
zacate taza – Eriochloa
zacate tempranero – Setaria macrostachya
zacate temprano – Setaria leucopila
zacate tigre – Echinochloa colona
zacate timoty – Phleum pratense
zacate toboso – Cenchrus echinatus, Pleuraphis mutica
zacate torcido – Agrostis tuberculosa
zacate tres barbas – Aristida adscensionis
zacate tule – Tripsacum lanceolatum
zacate volador – Aristida ternipes var gentilis
zacatillo – Heteropogon melanocarpus
zacatito – Triodia pulchella
zacatito del agua – Eleocharis atropurpurea
zacatlachichinoa – Mentha suaveolens
zacatlaxcale – Cuscuta
zacatlaxcalli – Cassytha filiformis
zacatón – Muhlenbergia, Panicum maximum, Sporobolus airoides, S. flexuosus, S. wrightii, Stipa robusta, Tripsacum lanceolatum
zacatón alcalino – Sporobolus airoides, S. indicus
zacatón arenoso – Sporobolus cryptandrus
zacatoncillo – Heteropogon melanocarpus
zacatón del desierto – Sporobolus indicus
zacatón de los médanos – Sporobolus flexuosus
zacatón de salitre – Sporobolus indicus
zacatón desgranador – Sporobolus cryptandrus
zacatón encubierto – Sporobolus cryptandrus
zacatón flexible – Sporobolus flexuosus
zacatón indio – Sporobolus indicus
zacatón liso – Muhlenbergia montana
zacatón pulguero – Sporobolus buckleyi
zacatón pyramidal – Sporobolus pyramidatus
zacbacelcan – Selenicereus donkelaari
zaccopté – Cordia dodecandra, C. sebestena
zacchichibi – Melochia tomentosa
zackuyche – Bombax ellipticum
zacmizbil – Sida cordifolia
zacmizib – Waltheria americana
zacolcom – Laguncularia racemosa
zacoub – Hylocereus undatus
zacoxóchitl – Echinocereus poselgeri
zacpah – Byrsonima crassifolia
zactah – Zexmenia frutescens
zacual – Crescentia cujete
zacualtipan – Ipomoea macorrhiza
zacxiu – Abutilon lignosum, Boerhavia erecta, Wlatheria americana
zacxiuthul – Boerhavia erecta
zachaax – Musa sapientum
zachalal – Phragmites australis
zagalejo – Castilleja arvensis, Sprekelia formosissima
zahina – Sorghum vulgare
zaina – Panicum maximum
zak-kulul – Pachycereus gaumeri
zalam – Lysiloma bahamensis
zalate – Ficus
zalate bronco – Ficus microchlamys
zamarilla – Teucrium polium
zamboa – Citrus aurantium
zanahoria – Daucus carota
zanahoria brasileña – Pfaffia paniculata
zanahoria silvestre – Daucus carota, D. pusillus
zancudo – Lopezia mexicana, L. racemosa
zannichellia de los pantanos – Zannichellia palustris

zapalote – Agave tequilana, Musa paradisiaca
zapalotl – Agave tequilana
zapallito – Cucurbita pepo
zapallo – Cucurbita pepo
zapallo prieto – Pisonia
zapatero – Gossypiospermum praecox, Simaruba glauca
zapatillo – Lantana camara
zapatillo de la reina – Clitoria terneata
zapatillo de mujer – Cypripedium
zapatillo de Venus – Cypripedium irapeanum, Paphiopedilum
zapatito – Aristolochia taliscana, Cypripedium, Pedilanthus tithymaloides, Sobralia macrantha
zapatito de dama – Cypripedium
zapatito de la Virgen – Pedilanthus tithymaloides
zapato del diablo – Pedilanthus macrocarpus
zapato de Venus – Cypripedium calceolus
zapote – Achras zapota, Casimiroa edulis, Diospyros californica, D. digyna, Lucuma mammosa, Pouteria sapota, Stemmadenia tomentosa var palmeri, Tabebuia pentaphylla
zapote amarillo – Couepia polyandra, Licania platypus, Lucuma salicifolia, Pouteria campechiana, Quararibea cordata
zapote blanco – Achras zapota, Casimiroa edulis, Morisonia americana
zapote bobo – Pachira aquatica
zapote borracho – Licania platypus, Lucuma salicifolia, Pouteria campechiana
zapote cabello – Licania platypus
zapote caimito – Chrysophyllum cainito, C. mexicanum
zapote colorado – Achras zapota, Lucuma mammosa
zapote chico – Achras zapota
zapote de abejas – Achras zapota
zapote de agua – Pachira aquatica
zapote de ave – Mastichodendron foetidissimum, Sideroxylon capiri
zapote de chicle – Achras zapota
zapote de mico – Licania platypus
zapote de mono – Calatola mollis, Couroupita guianensis, Licania platypus
zapote de niño – Mammea americana, Pouteria campechiana
zapote de perro – Capparis spinosa
zapote de rata – Casimiroa edulis
zapote de Santo Domingo – Mammea americana
zapote de viejas – Annona muricata
zapote domingo – Mammea americana
zapote dormilón – Casimiroa edulis, Pouteria campechiana
zapote dulce – Casimiroa edulis
zapote enano – Maba
zapote mamey – Lucuma mammosa, Mammea americana, Pouteria sapota
zapote mante – Pouteria campechiana
zapote montés – Achras zapota
zapote morado – Achras zapota
zapote negro – Diospyros digyna, D. ebenaster, D. texana
zapote niño – Mammea americana, Pouteria campechiana
zapote-níspero – Achras zapota
zapote prieto – Diospyros
zapote reventador – Pachira aquatica
zapote reventón – Pachira aquatica
zapote verde – Lucuma viride
zapotillo – Achras chicle, A. zapota, Calliandra californica, C. peninsularis, Couepia polyandra, Diospyros californica, D. intricata, Dipholis salicifolia, Erythroxylum mexicanum, Garrya laurifolia, Lantana camara, Lucuma rivicoa, Maba, Mastichodendron foetidissimum, Pouteria campechiana, Randia, Rheedia edulis
zapotillo amarillo – Crateva palmeri
zapotillo rojo – Pouteria campechiana
zapotón – Pachira aquatica, Swietenia humilis
zapupe – Agave angustifolia, A. deweyana, A. tequilana, A. zapupe
zapupe azul – Agave zapupe
zapupe de estopier – Agave zapupe
zapupe de Huastusco – Agave deweyana
zapupe de San Bernardo – Agave zapupe
zapupe de Tantoyuca – Agave deweyana
zapupe verde – Agave deweyana
zaracoche – Cucurbita foetidissima
zaragatona – Plantago
zaragatona disentérica – Pulicaria dysenterica
zaragatona falsa – Plantago maritima
zaragoza – Najas major
zaragutia – Dracunculus vulgaris

zarapinto – Iberis amara, I. sempervirens, I. umbellata
zarcilillo – Lobelia laxiflora
zarcillo – Bessera elegans, Lobelia laxiflora
zarza – Duranta erecta, Humulus lupulus, Mimosa albida, Rubus, Smilax medica, S. mexicana
zarza común – Rubus fruticosus
zarza de flores pequeñas – Rubus parviflorus
zarza de hojas de olmo – Rubus ulmifolius
zarza de hojas laciniadas – Rubus laciniatus
zarza del campo – Humulus lupulus
zarza de Logan – Rubus loganobaccus
zarza de los rastrojos – Rubus caesius
zarza frambuesa – Rubus loganobaccus
zarzafrás – Prunus brachybotrya
zarzaidea – Rubus idaeus, R. strigosus
zarza japonesa – Rubus phoenicolasius
zarzamora – Lantana camara
zarza morisca – Smilax aspera
zarzaneda – Rubus fruticosus
zarza nórdica – Rubus arctius
zarzaparrilla – Asparagus medeoloides, Eclipta alba, Humulus lupulus, Krameria secundiflora, Polygonum lapathifolium, Smilax
zarzaparrilla carmelita – Smilax officinalis
zarzaparrilla de Alemania – Carex arenaria
zarzaparrilla de América – Smilax medica
zarzaparrilla de China – Smilax china
zarzaparrilla de la sierra – Smilax medica
zarzaparrilla de país – Smilax aspera
zarzaparrilla de Tulancingo – Smilax medica
zarzarrosa – Rosa canina
zarza silvestre – Rubus trivialis
zarza terreña – Rubus caesius
zarza vara – Purshia mexicana var stansburyana
zarzilla – Erigeron nudiflorus
zarzuue – Cynodon dactylon
zate – Yucca elephantipes
zaya – Achras zapota
zayolizcán – Buddleja americana
zayolizco – Buddleja parviflora
zazacatzin – Karwinskia humboldtiana
zazale – Mentzelia hispida, Selloa glutinosa, Stevia salicifolia
zazale de olor – Stevia salicifolia
zazalic – Mentzelia hispida
zazalicpatli – Levisticum officinale
zazamil – Cordia alba
zazanaca – Acourtia cordata
zazúcar – Salpianthus arenarius
zebrina – Tradescantia discolor, T. zebrina
zeguín – Agave tequilana
zelkova del Cáucaso – Zelkova carpinifolia
zelkova japonesa – Zelkova serrata
zelkova serrada – Zelkova serrata
zempual – Cosmos sulphureus
zericote – Cordia dodecandra
zexmenia – Zexmenia hispida
Zeylanica – Sanseviera trifasciata
zicilhaxiu – Lantana involucrata
zicilpuuz – Sesamum indicum
zicilté – Jatropha curcas
zihoapatle – Montanoa tomentosa
zingiblio – Philodendron radiatum
zinia – Zinnia
zinkin – Caesalpinia pulcherrima
zipate – Hamelia erecta
ziquiltle – Jatropha curcas
zis coil – Citrullus colocynthis
zitlali – Lycopersicon esculentum var cerasiforme
zitsmuc – Celtis iguanea
zibic – Vanilla planifolia
zizuch – Litsea glaucescens
zoapatle – Montanoa tomentosa
zochipitzahoac – Polygala
zohzoyatic – Veratrum frigidum
zompantle – Buddleja americana, Erythrina, Montanoa tomentosa
zon – Guaiacum sanctum
zongolica – Astronium graveolens, Bursera simaruba
zonote – Trema micrantha
zonzapote – Licania platypus
zook – Zexmenia frutescens
zopilote – Agave tequilana, Conzattia multiflora, Swietenia humilis, S. macrophylla
zopilo zontecomacuahuitl – Swietenia macrophylla
zornia trencilla – Zornia diphylla
zorrillo – Choisya arizonica, C. dumosa, Crusea calocephala, Petiveria alliacea, Ptelea trifoliata, Senna atomaria, Trichilia hirta, Zanthoxylum arborescens
zorrillo blanco – Achyranthes aspera

zorrillo silvestre – Petiveria alliacea
zorro – Achyranthes aspera, Petiveria alliacea
zosa – Solanum verbascifolium
zoyalxóchitl – Commelina tuberosa
zubin – Acacia farnesiana, A. spadicigera
zubinché – Acacia farnesiana, A. spadicigera
zubul – Sapindus saponaria
zueco – Cypripedium calceolus
zuliabal – Spondias mombin
zulipox – Annona reticulata
zumaque – Rhus, Toxicodendron radicans
zumaque cimarrón – Rhus terebinthifolia
zumaque de astas de ciervo – Rhus typhina
zumaque de hojas de mirto – Coriaria myrtifolia
zumaque de los curtidores – Rhus coriaria
zumaque de tenerías – Rhus coriaria
zumaque de Virginia – Rhus typhina
zumaque glabro – Rhus glabra
zumaque javanés – Rhus javanica
zumaque lampiño – Rhus glabra
zumaque mayor – Rhus typhina
zumaque venenoso – Toxicodendron diversilobum
zunuiña – Cercocarpus betuloides
zurandanicua – Caesalpinia pulcherrima
zurrón – Chenopodium bonus-henricus
zurrón de pastor – Capsella bursa-pastoris
zutskeyem – Oxalis latifolia
zutspacal – Citrus aurantium
zutub – Ipomoea alba
zuun – Tithonia
zuzón – Senecio jacobaea, S. vulgaris

English

Aaron's beard – *Hypericum calycinum*
Aaron's beard prickly-pear – *Opuntia leucotrichia*
Aaron's rod – *Solidago virgaurea*
abelia, glossy – *Abelia* x *grandiflora*
abelia, Mexican – *Abelia floribunda*
absinth – *Artemisia absinthium, A. dracunculoides*
abutilon – *Abutilon*
abutilon, Rio Grande – *Abutilon hypoleucum*
abutilon, trailing – *Abutilon megapotamicum*
Abyssinian banana – *Ensete ventricosum*
Abyssinian feathertop – *Pennisetum villosum*
Abyssinian myrrh – *Commiphora abyssinica*
acacia – *Acacia*
acacia, blackbrush – *Acacia rigidula*
acacia, blackwood – *Acacia melanoxylon*
acacia, boat-thorn – *Acacia cymbispina*
acacia, borderpod – *Lysiloma divaricata*
acacia, bullhorn – *Acacia spadicigera*
acacia, catclaw – *Acacia greggii*
acacia, everblooming – *Acacia retinodes*
acacia, feather – *Acacia pennatula*
acacia, fern – *Acacia angustissima*
acacia, fernleaf – *Acacia pennatula*
acacia, flat-thorn – *Acacia gladiata*
acacia, hat-thorn – *Acacia hindsii*
acacia, Mescat – *Acacia constricta*
acacia, Persian – *Albizzia julibrissin*
acacia, prairie – *Acacia angustissima*
acacia, Roemer – *Acacia roemeriana*
acacia, Santa Rita – *Acacia millefolia*
acacia, Sierra Madre – *Acacia pennatula*
acacia, sweet – *Acacia farnesiana*
acacia, white-ball – *Acacia angustissima*
acacia, whitethorn – *Acacia constricta*
acacia, Willard's – *Acacia willardiana*
acanthus, Balkan – *Acanthus balcanicus*
acanthus, flame – *Anisacanthus quadrifidus*
acanthus, spiny – *Acanthus spinosus*
acerola cherry – *Malpighia glabra*
ackee – *Blighia sapida*
aconite, winter – *Eranthis hymenalis*
acorn squash – *Cucurbita maxima*
actinea, bitterweed – *Hymenoxys odorata*
acuña cactus – *Echinomastus erectocentrus*
Adam's apple – *Citrus limetta*
Adam's needle – *Yucca filamentosa*
Adam's tree – *Fouquieria diguetii*
adder's tongue – *Erythronium americanum*
adelia – *Forestiera neomexicana*
adolphia, California – *Adolphia californica*
adzuki bean – *Vigna angularis*
aerial yam – *Dioscorea bulbifera*
African corn-lily – *Ixia maculata*
African daisy – *Arctotis venusta, Dimorphotheca sinuata*
African daisy, blue-eyed – *Arctotis venusta*
African evergreen – *Syngonium podophyllum*
African grass – *Eragrostis lehmanniana*
African hemp – *Sparmannia africana*
African horned melon – *Cucumis metuliferus*
African linden – *Sparmannia africana*
African mahogany – *Khaya senegalensis*
African marigold – *Tagetes erecta*
African millet – *Eleusine coracana* ssp *africana*
African sausage tree – *Kigelia africana*
African stock – *Malcolmia africana*
African tamarisk – *Tamarix africana*
African tulip tree – *Spathodea campanulata*
African violet – *Saintpaulia ionantha, S. jonantha*
agapanthus – *Agapanthus africanus*
agaric, amethyst – *Lepista nuda*
agaric, bitter – *Tricholoma acerbum*
agaric, fairy ring – *Marasmius oreades*
agaric, firewood – *Tricholoma flavovirens*
agaric, fly – *Amanita muscaria*
agaric, gray – *Tricholoma terreum*
agaric, honey – *Armillaria mellea*
agaric, scaly – *Tricholoma vaccinum*
agarita – *Berberis trifoliata*
agave – *Agave*
agave, cabbage – *Agave perryi*
agave cactus – *Leuchtenbergia principis*
agave, coastal – *Agave shawii*
agave, cow horn – *Agave bovicornuta*
agave, desert – *Agave deserti*
agave, drooping – *Agave attenuata*
agave, false – *Hechtia montana*
agave, foxtail – *Agave attenuata*
agave, henequen – *Agave fourcroydes*
agave, Hohokam – *Agave murpheyi*
agave, mescal fiber – *Agave pseudotequilana*
agave, octopus – *Agave vilmoriniana*
agave, Palmer – *Agave palmeri*
agave, Parry – *Agave perryi*

agave, pulque – *Agave atrovirens, A. salmiana*
agave, Shaw's – *Agave shawii*
agave, sisal – *Agave sisalana*
agave, tequila – *Agave tequilana*
agave, Utah – *Agave utahensis*
agave, Weber blue – *Agave tequilana*
ageratum – *Ageratum*
agrimony, common – *Agrimonia gryposepala*
ague root – *Aletris farinosa*
air pine – *Tillandsia flabellata, T. imperialis*
airplant – *Kalanchoe pinnata*
air potato – *Dioscorea bulbifera*
ajamete – *Asclepias subulata*
akee – *Blighia sapida*
alamo – *Populus fremontii*
alamo vine – *Ipomoea sinuata*
Alaska blue willow – *Salix purpurea*
Alaska cedar – *Cupressus nootkatensis*
Aldenham crabapple – *Malus purpurea*
alder, Arizona – *Alnus oblongifolia*
alder, black – *Alnus glutinosa*
alder buckthorn – *Rhamnus frangula*
alder, Italian – *Alnus cordata*
alder-leaf mountain-mahogany – *Cercocarpus montanus*
alder, mountain – *Alnus incana* ssp *tenuifolia*
alder, New Mexican – *Alnus oblongifolia*
alder, red – *Alnus rubra*
alder, Sitka – *Alnus viridis*
alder, thinleaf – *Alnus tenuifolia, A. viridis*
alder, tropical – *Alnus acuminata*
alder, wax – *Alnus jorullensis*
alder, white – *Alnus rhombifolia*
alder, yellow – *Turnera ulmifolia*
Aleppo pine – *Pinus halepensis*
Alexander palm – *Ptychosperma elegans*
Alexanders – *Smyrnium olustratum*
Alexandra palm – *Archontophoenix alexandrae*
alfalfa – *Medicago sativa*
alfalfa, wild – *Psoralea rhombifolia*
Algerian ivy – *Hedera canariensis*
algerita, Texas – *Berberis trifoliata*
algum tree – *Juniperus excelsa*
alicoche – *Echinocereus papillosus*
alicoche, yellow – *Echinocereus papillosus*
Alison, field – *Alyssum minus*
Alison, golden – *Alyssum saxatile*
Alison, small – *Alyssum alyssoides*
alizarin – *Rubia tinctorium*
alkali aster, shrubby – *Machaeranthera carnosa*
alkali heath – *Frankenia salina*
alkali grass, European – *Puccinellia distans*
alkali lovegrass – *Eragrostis obtusiflora*
alkali mallow – *Malvella lepidota, Sida hederacea*
alkali sacaton – *Sporobolus airoides*
alkanet – *Alkanna tinctoria, Anchusa azurea, A. officinalis*
alkanet, dyer's – *Alkanna tinctoria*
Allegheny service-berry – *Amelanchier laevis*
all-good – *Chenopodium bonus-henricus*
allheal, Hercules' – *Opopanax chironium*
alligator apple – *Annona glabra*
alligator juniper – *Juniperus deppeana*
alligator weed – *Alternanthera caracasana*
allium, giant – *Allium giganteum*
allium, keeled – *Allium carinatum* ssp *pulchellum*
allium, ornamental – *Allium oreophilum*
allium, Turkestan – *Allium karataviense*
allium, yellow – *Allium moly*
all-scale – *Atriplex polycarpa*
allseed, four-leaved – *Polycarpon tetraphyllum*
allspice – *Eugenia pimenta*
allspice, Carolina – *Calycanthus floridus*
allthorn – *Koeberlinia spinosa* ssp *tenuispina*
allthorn castela – *Castela texana*
almond – *Prunus dulcis*
almond, Barbados – *Terminalia catappa*
almond, bitter – *Prunus dulcis* var *amara*
almond, dwarf – *Prunus tenella*
almond, earth – *Cyperus esculentus*
almond, flowering – *Prunus triloba*
almond, Indian – *Terminalia amazonia, T. catappa*
almond, Java – *Canarium commune*
almond-leaved pear – *Pyrus amygdaliformis*
almond, soft-shelled – *Prunus dulcis* var *fragilis*
almond tree, tropical – *Terminalia catappa*
almug tree – *Pterocarpus santalinus*
aloe – *Aloe socotrina*
aloe, American – *Polianthes maculosa*
aloe, Barbados – *Aloe vera*
aloe, bitter – *Aloe ferox*
aloe-leaved epidendrum – *Epidendrum parkinsonianum*

aloe, lign – *Aquilaria agallocha*
aloe, medicinal – *Aloe vera*
aloe, partridge-breast – *Aloe variegata*
aloes – *Aquilaria agallocha*
aloe, tiger – *Aloe variegata*
aloe vera – *Aloe vera*
aloe wood – *Cordia sebestena*
aloe yucca – *Yucca aloifolia*
aloe, Zanzibar – *Aloe vera*
alpine aster – *Aster alpinus*
alpine avens – *Geum montanum*
alpine barrenwort – *Epimedium alpinum*
alpine bear-berry – *Arctostaphylos alpinus*
alpine bilberry – *Vaccinium uliginosum*
alpine buckthorn – *Rhamnus alpinus*
alpine clematis – *Clematis alpina*
alpine columbine – *Aquilegia alpina*
alpine currant – *Ribes alpinum*
alpine dock – *Rumex alpinus*
alpine fescue – *Festuca ovina* var *brachyphylla*
alpine fir – *Abies lasiocarpa*
alpine meadow-rue – *Thalictrum alpinum*
alpine pearlwort – *Sagina saginoides*
alpine pink – *Dianthus alpinus*
alpine poppy – *Papaver alpinum*
alpine rose – *Rosa pendulina*
alpine sea holly – *Eryngium alpinum*
alpine skullcap – *Scutellaria alpina*
alpine speedwell – *Veronica alpina*
alpine squill – *Scilla biflora*
alpine strawberry – *Fragaria vesca*
alpine willow herb – *Epilobium anagallidifolium*
alsike clover – *Trifolium hybridum*
alta palm – *Mauritia flexuosa*
althea, shrub – *Hibiscus syriacus*
aluminum plant – *Pilea cadierei*
alum root – *Heuchera sanguinea*
alvaradoa, Mexican – *Alvaradoa amorphoides*
alyssum, sweet – *Lobularia maritima*
amapa, pink – *Tabebuia impetiginosa*
amapa, yellow – *Tabebuia chrysantha*
amaranth – *Amaranthus*
amaranth, fringed – *Amaranthus fimbriatus*
amaranth, globe – *Gomphrena globosa*
amaranth, green – *Amaranthus hybridus*
amaranth, prostrate – *Amaranthus blitoides, A. deflexus*
amaranth, spreading – *Amaranthus cruentas*
amaryllis – *Hippeastrum*
ambarella – *Spondias dulcis*
amber lily – *Anthericum torreyi*
ambrette – *Abelmoschus moschatus*
American aloe – *Polianthes maculosa*
American arborvitae – *Thuja occidentalis*
American beech – *Fagus grandifolia*
American bittersweet – *Celastrus scandens*
American broad-leaved bamboo – *Guadua latifolia*
American brooklime – *Veronica americana*
American centaury – *Sabatia angularis*
American chestnut tree – *Castanea dentata*
American elderberry – *Sambucus canadensis*
American elm – *Ulmus americana*
American galingale – *Cyperus eragrostis*
American ginseng – *Panax quinquefolium*
American holly – *Ilex opaca*
American hop-hornbeam – *Ostrya virginiana*
American hops – *Humulus lupulus* ssp *neomexicanus*
American hornbeam – *Carpinus caroliniana*
American jointvetch – *Aeschynomene americana*
American linden – *Tilia americana*
American long-leaved bamboo – *Guadua longifolia*
American lotus – *Nelumbo lutea*
American marigold – *Tagetes erecta*
American mountain ash – *Sorbus americana*
American narrow-leaved bamboo – *Guadua angustifolia*
American oil palm – *Elaeis melanococca*
American pawpaw – *Asimina triloba*
American pennyroyal – *Hedeoma pulegioides*
American persimmon – *Diospyros virginiana*
American pistachio – *Pistacia texana*
American plum – *Prunus americana*
American red raspberry – *Rubus strigosus*
American saffron – *Carthamnus tinctorius*
American sanicle – *Sanicula marilandica*
American scheuchzeria – *Scheuchzeria palustris*
American speedwell – *Veronica peregrina*
American spikenard – *Aralia racemosa*
American stinging nettle – *Urtica dioica* ssp *gracilis*
American sweet gum – *Liquidambar styraciflua*

American sycamore – *Platanus occidentalis*
American willow herb – *Epilobium ciliatum*
amethyst agaric – *Lepista nuda*
amethyst eryngium – *Eryngium amethystinum*
amole – *Chlorogalum*
Amur chokecherry – *Prunus maackii*
Amur maple – *Acer tartaricum* var *ginnala*
Anaheim chile – *Capsicum annuum*
Anasazi bean – *Phaseolus vulgaris*
ancho chile – *Capsicum annuum*
ancho pepper, chocolate – *Capsicum annuum*
Andalusian milk-vetch – *Astragalus boeticus*
Andes berry – *Rubus glaucus*
anemone – *Anemone*
anemone clematis – *Clematis montana*
anemone, Japanese – *Anemone hybrida*
anemone, poppy-flowered – *Anemone coronaria*
anemone, snowdrop – *Anemone sylvestris*
anemone, tree – *Carpenteria californica*
anemone, wood – *Anemone nemorosa*
angel herb – *Eupatorium collinum*
angelica – *Angelica*
angelica, Chinese – *Angelica sinensis*
angelica tree, Japanese – *Aralia elata*
angelin, cabbage – *Andira inermis*
angelin tree – *Andira inermis*
angel's tears – *Narcissus triandrus*
angel's trumpet – *Acleisanthes longiflora, Brugmansia*
angel's trumpet, mountain – *Brugmansia sanguinea*
angel's trumpet, yellow – *Brugmansia aurea*
angel's wings – *Opuntia microdasys*
angel trumpets – *Acleisanthes longiflora*
angel, yellow arch – *Lamium galeobdolon*
angled loofah – *Luffa acutangula*
angle-pod – *Gonolobus erianthus*
anil – *Indigofera fruticosa, I. nelsonii*
anise – *Pimpinella anisum*
anise, Chinese – *Illicium anisatum*
aniseed – *Pimpinella anisum*
anise, Japanese – *Illicium anisatum*
anise magnolia – *Magnolia salicifolia*
anise shrub – *Illicium anisatum*
anise, star – *Illicium anisatum*
anise tree, Florida – *Illicium floridanum*
annato – *Bixa orellana*
Annie, sweet – *Artemisia annua*
antelope bush – *Purshia mexicana*
antelope horns – *Asclepias asperula*
anthurium – *Anthurium*
ant tree – *Triplaris americana*
Apache pine – *Pinus engelmannii*
Apache plume – *Fallugia paradoxa*
aparejograss – *Muhlenbergia utilis*
ape's earring – *Havardia pallens, Pithecellobium flexicaule*
apple – *Malus sylvestris*
apple, Adam's – *Citrus limetta*
apple, alligator – *Annona glabra*
apple, Assyrian – *Citrus medica*
apple, balsam – *Momordica balsamina*
apple banana – *Mus acuminata* x *balbisiana*
apple, bitter – *Citrullus colocynthis, Momordica balsamina*
apple cactus, Peruvian – *Cereus repandus*
apple, Chinese crab – *Malus spectabilis*
apple, common – *Malus sylvestris*
apple, crab – *Malus sylvestris*
apple, custard – *Annona*
apple, elephant – *Dillenia indica*
apple, golden – *Spondias cytherea*
apple, Indian – *Datura wrightii*
apple, Java – *Eugenia javanica*
apple kei – *Dovyalis caffra*
apple, Malay – *Eugenia malaccensis*
apple, Malay rose – *Eugenia malaccensis*
apple, mamee – *Mammea americana*
apple, mamey – *Pouteria sapota*
apple, May – *Podophyllum peltatum*
apple, Mexican – *Casimiroa edulis, Malvaviscus grandiflorus*
apple mint – *Mentha suaveolens*
apple mint, golden – *Mentha* x *gracilis*
apple, monkey – *Licania platypus*
apple, mountain – *Eugenia malaccensis*
apple of Peru – *Nicandra physalodes*
apple of Sodom – *Solanum mammosum, S. sodomaeum*
apple, Oregon crab – *Malus fusca*
apple, otaheite – *Eugenia malaccensis, Spondias cytherea*
apple, paradise – *Malus pumila* var *paradisiaca*
apple, pond – *Annona glabra*
apple rootstock – *Malus pumila* var *sylvestris*
apple, rose – *Dillenia indica, Eugenia*

apple, Sodom – *Calotropis procera*
apple, star – *Averrhoa carambola, Chrysophyllum cainito*
apple, sugar – *Annona squamosa*
apple, tea crab – *Malus hupehensis*
apple, thorn – *Datura*
apple, wild star – *Chrysophyllum mexicanum*
apricot – *Prunus armenica*
apricot, desert – *Prunus fremontii*
apricot, Japanese – *Prunus mume*
apricot, Japanese flowering – *Prunus mume*
apricot mallow – *Sphaeralcea ambigua*
apricot-plum – *Prunus simonii*
Arabian jasmine – *Jasminum sambac*
Arabian tea – *Catha edulis*
arabica coffee – *Coffea arabica*
Arabic, gum – *Acacia senegal*
aralia, castor – *Kalopanax septemlobus*
aralia, dinner plate – *Polyscias scutellaria*
aralia, false – *Schefflera elegantissima*
aralia, finger – *Schefflera elegantissima*
aralia, Japanese – *Fatsia japonica*
aralis, tree – *Kalopanax septemlobus*
aramina – *Urena lobata*
arar tree – *Tetraclinis articulata*
arbor vine, Spanish – *Merremia tuberosa*
arborvitae, American – *Thuja occidentalis*
arborvitae, Chinese – *Thuja orientalis*
arborvitae, false – *Thujopsis dolabrata*
arborvitae, oriental – *Thuja orientalis*
archangel – *Angelica archangelica*
archangel, yellow – *Lamium galeobdolon*
arctic bramble – *Rubus arctius*
arctic pearlwort – *Sagina procumbens*
arctic starflower – *Trientalis arctica*
arecanut palm – *Areca catechu*
argan tree – *Argania sideroxylon*
Arizona alder – *Alnus oblongifolia*
Arizona ash – *Fraxinus velutina*
Arizona baccharis – *Baccharis thesioides*
Arizona barrel cactus – *Ferocactus wislizeni*
Arizona carlowrightia – *Carlowrightia arizonica*
Arizona cassava – *Manihot davisiae*
Arizona cypress – *Cupressus arizonica*
Arizona cypress, smooth – *Cupressus arizonica* ssp *glabra*
Arizona fescue – *Festuca arizonica*
Arizona fish-hook cactus – *Mammillaria grahamii*
Arizona honeysuckle – *Lonicera arizonica*
Arizona longleaf pine – *Pinus engelmanii*
Arizona lupine – *Lupinus arizonicus*
Arizona madrone – *Arbutus arizonica*
Arizona madroño – *Arbutus arizonica*
Arizona mesquite – *Prosopis velutina*
Arizona mimosa – *Mimosa laxiflora*
Arizona mountain ash – *Sorbus dumosa*
Arizona peamelon – *Sicyos laciniatus*
Arizona pine – *Pinus arizonica*
Arizona poplar – *Populus fremontii*
Arizona poppy – *Kallstroemia grandiflora*
Arizona queen-of-the-night – *Peniocereus greggii*
Arizona rainbow hedgehog cactus – *Echinocereus rigidissimus*
Arizona rose – *Rosa arizonica*
Arizona rosewood – *Vauquelinia californica*
Arizona sage – *Salvia arizonica*
Arizona spurge – *Chamaesyce arizonica*
Arizona star-leaf – *Choisya arizonica*
Arizona sycamore – *Platanus wrightii*
Arizona thistle – *Cirsium arizonicum*
Arizona three-awn – *Aristida arizonica*
Arizona walnut – *Juglans major*
Arizona wheatgrass – *Elytrigia arizonicus*
Arizona white oak – *Quercus arizonica*
Arizona white pine – *Pinus ayacahuite*
Arizona yellow show – *Amoreuxia palmatifida*
Arizona yucca – *Yucca arizonica*
Arkansas rose – *Rosa arkansana*
arnica – *Arnica*
arolla pine – *Pinus cembra*
aromatic holy grass – *Hierochloë odorata*
aromatic sumac – *Rhus aromatica*
arrow bamboo – *Pseudosasa japonica*
arrow-grass, marsh – *Triglochin palustris*
arrow-grass, seaside – *Triglochin maritima*
arrowhead – *Sagittaria latifolia, S. sagittifolia*
arrowroot – *Maranta arundinaceae, Thalia geniculata*
arrowroot, East Indian – *Curcuma angustifolia*
arrowroot, Florida – *Zamia pumila*
arrowroot, Queensland – *Canna edulis*
arrow weed – *Pluchea sericea*
arroyo lupine – *Lupinus succulentus*
arroyo willow – *Salix lasiolepis*

artichoke – *Cynara scolymus*
artichoke, Chinese – *Stachys sieboldii, S tuberifera*
artichoke, Jerusalem – *Helianthus tuberosus*
artichoke thistle – *Cynara cardunculus*
artillery plant – *Pilea microphylla*
arugula – *Eruca vesicaria* ssp *sativa*
arum – *Arum maculatum*
arum, bog – *Zantedeschia palustris*
arum, Italian – *Arum italicum*
arum lily, white – *Zantedeschia aethiopica*
asafoetida – *Ferula asafoetida*
asarabacca – *Asarum europaeum*
ash – *Fraxinus*
ash, Arizona – *Fraxinus velutina*
ash, Arizona mountain – *Sorbus dumosa*
ash, chaparral – *Fraxinus trifoliata*
ash, Chihuahua – *Fraxinus papillosa*
ash-coloured cranesbill – *Geranium cinereum*
Ashe juniper – *Juniperus ashei*
ashes, dust and – *Solanum sodomeum*
ash, European – *Fraxinus excelsior*
ash, European mountain – *Sorbus aucuparia*
ash, evergreen – *Fraxinus uhdei*
ash, flowering – *Fraxinus cuspidata, F. ornus*
ash, fragrant – *Fraxinus cuspidata* var *macropetala*
ash, Goodding – *Fraxinus gooddingii*
ash, green – *Fraxinus pennsylvanica*
ash, Gregg – *Fraxinus greggii*
ash, lime prickly – *Zanthoxylum fagara*
ash, littleleaf – *Fraxinus gooddingii, F. greggii*
ash, Mexican – *Fraxinus berlandieriana*
ash, mountain – *Sorbus*
ash, prickly – *Zanthoxylum americanum, Z. arborescens*
ash, red – *Fraxinus pennsylvanica*
ash, shamel – *Fraxinus uhdei*
ash, single-leaf – *Fraxinus anomala*
ash, velvet – *Fraxinus velutina*
ash, wafer – *Ptelea trifoliata*
ash, white – *Fraxinus americana*
ashy limberbush – *Jatropha cinerea*
Asian paddy weed – *Echinochloa oryzoides*
Asian red sage – *Salvia miltiorrhiza*
Asiatic redbud – *Cercis siliquastrum*
asparagus – *Asparagus officinalis*
asparagus bean – *Vigna unguiculata* ssp *sesquipedalis*
asparagus, fern – *Asparagus setaceus*
asparagus, Indian – *Orobanche cooperi*
asparagus, Japanese – *Asparagus densiflorus* var *sprengeri*
asparagus pea – *Psophocarpus tetragonolobus*
asparagus, smilax – *Asparagus asparagoides*
asparagus, Sprenger – *Asparagus densiflorus* var *sprengeri*
aspen, bigtooth – *Populus grandidentatum*
aspen, European – *Populus tremula*
aspen, quaking – *Populus tremuloides*
asphodel – *Asphodelus fistulosus*
asphodel, yellow – *Asphodeline lutea*
Assam rubber tree – *Ficus elastica*
Assam tea – *Camellia assamica*
Assyrian apple – *Citrus medica*
aster – *Aster, Chloracantha spinosa*
aster, alpine – *Aster alpinus*
aster, annual – *Aster subulatus*
aster, awlleaf – *Aster subulatus*
aster, China – *Callistephus chinensis*
aster, Geyer's – *Aster laevis*
asteriscus, marine – *Asteriscus maritimus*
aster, Italian – *Aster amellus*
aster, New England – *Aster novae-angliae*
aster, Pyrenean – *Aster sibiricus*
aster, shrubby alkali – *Machaeranthera carnosa*
aster, smooth – *Aster laevis*
aster, spiny – *Chloracantha spinosa, Ericameria spinulosus*
aster, Stoke's – *Stokesia laevis*
astragal, thorny – *Astragalus tragacantha*
Asturian daffodil – *Narcissus asturiensis*
atamasco-lily – *Zephranthes texana*
athel – *Tamarix aphylla*
athel tree – *Tamarix aphylla*
Atlantic poppy – *Papaver atlanticum*
Atlas cedar – *Cedrus atlantica*
aubergine – *Solanum melongena*
aubretia, common – *Aubrieta deltoidea*
aucuba, Japanese – *Aucuba japonica*
auricula – *Primula auricula*
Australian flame tree – *Brachychiton acerifolius*
Australian fuchsia – *Correa speciosa*
Australian pine – *Casuarina equisetifolia*
Australian saltbush – *Atriplex semibaccata*
Australian violet – *Viola hederacea*

Austrian black pine – *Pinus nigra*
Austrian brier – *Rosa foetida*
Austrian field-cress – *Rorippa austriaca*
Austrian mullein – *Verbascum chaixii*
autograph tree – *Clusia major*
autumn crocus – *Colchicum autumnale, C. speciosum*
autumn damask – *Rosa damascena* var *semperflorens*
autumn-flowering laelia – *Laelia autumnalis*
autumn gentian – *Gentianella amarella*
autumn snowflake – *Leucojum autumnale*
avens – *Geum*
avens, alpine – *Geum montanum*
avens, bigleaf – *Geum macrophyllum*
avens, mountain – *Dryas octopetala*
avens, wood – *Geum urbanum*
averin – *Rubus chamaemorus*
Avignon berry – *Rhamnus infectorius*
avocado – *Persea*
avocado, coyo – *Persea schiedeana*
avocado, Guatemalan – *Persea americana*
avocado, Mexican – *Persea drymifolia*
awlleaf aster – *Aster subulatus*
awlwort – *Subularia aquatica*
awn, crinkle – *Trachypogon montufari, T. secundus*
azalea – *Rhododendron indicum*
azalea, mock – *Adenium obesum*
azalea, pontic – *Rhododendron luteum*
azalea, western – *Rhododendron occidentale*
azarole tree – *Crataegus azarolus*
Aztec lily – *Sprekelia formosissima, Tigridia pavonia*
Aztec pine – *Pinus teocote*
Aztec tobacco – *Nicotiana rustica*

babaco – *Carica pentagona*
babassu palm – *Attalea speciosa*
baby blue-eyes – *Nemophila menziesii*
baby, cry – *Erythrina crista-galli*
baby gladiolus – *Gladiolus colvillei*
baby's breath – *Gypsophila elegans, G. paniculata*
baby's breath, false – *Galium mollugo*
baby snapdragon – *Linaria maroccana*
baby's tears – *Soleirolia soleirolii*
baby sun rose – *Aptenia cordifolia*
baccharis, Arizona – *Baccharis thesioides*
baccharis, broom – *Baccharis sarathroides*
baccharis, desert – *Baccharis sergiloides*
bachang mango – *Mangifera foetida*
bachelor's button – *Centaurea cyanus, Gomphrena globosa*
backbone, devil's – *Pedilanthus tithymaloides*
bacury – *Platonia insignis*
bael fruit – *Aegle marmelos*
bagflower – *Clerodendrum thomsoniae*
Bahama buttercup – *Turnera ulmifolia*
bahia grass – *Paspalum notatum*
bahia, hairy-seed – *Bahia absinthifolia*
Baja elephant tree – *Pachycormus discolor*
Baja fairyduster – *Calliandra californica*
balata – *Achras bidentata*
balbis banana – *Mus balbisiana*
bald-cypress – *Taxodium distichum*
bald-cypress, Mexican – *Taxodium distichum* var *mexicanum*
bald-cypress, Montezuma – *Taxodium distichum* var *mexicanum*
Balkan acanthus – *Acanthus balcanicus*
ball moss – *Tillandsia ferrisiana, T. recurvata*
balloon flower – *Platycodon grandiflorus*
balloon vine – *Cardiospermum halicacabum*
ball, starry – *Coryphantha elephantidens*
ball thistle, southern – *Echinops exaltatus*
balm – *Balanites aegyptiaca, Cedronella mexicana*
balm, bee – *Melissa officinalis, Monarda*
balm, garden – *Melissa officinalis*
balm, lemon – *Melissa officinalis*
balm, Mexican – *Brittonastrum mexicana*
balm, mintleaf bee – *Monarda menthaefolia*
balm, Molucca – *Moluccella laevis*
balm, mountain – *Eriodictyon angustifolium, E. trichocalyx*
balm-of-Gilead – *Commiphora opobalsamum, Populus balsamifera*
balm, plains bee – *Monarda pectinata*
balm, sweet – *Melissa officinalis*
balm tree – *Myroxylon balsamum* var *pereirae*
balsa – *Ochroma lagopus, O. pyramidale*
balsam – *Impatiens balsamina*
balsam apple – *Momordica balsamina*
balsam, copaiba – *Copaifera coriacea*
balsam fir – *Abies balsamea*
balsam, garden – *Impatiens balsamina*
balsam, Jesuit's – *Copaifera officinalis*

balsam of Peru – *Myroxylon balsamum* var *pereirae*
balsam of Tolu – *Myroxylon balsamum* var *pereirae*
balsam, orange – *Impatiens capensis*
balsam pear – *Momordica charantia*
balsam pear, southern – *Momordica balsamina*
balsam poplar – *Populus balsamifera*
balsam-root – *Balsamorhiza sagittata*
Baltic rush – *Juncus balticus*
bambarra groundnut – *Voandzeia subterranea*
bamboo – *Arundinaria, Bambusa arundinacea, B. guadua*
bamboo, American broad-leaved – *Guadua latifolia*
bamboo, American long-leaved – *Guadua longifolia*
bamboo, American narrow-leaved – *Guadua angustifolia*
bamboo, arrow – *Pseudosasa japonica*
bamboo, common – *Bambusa vulgaris*
bamboo, giant – *Dendrocalamus giganteus*
bamboo, Mexican weeping – *Otatea acuminata*
bamboo muhly – *Muhlenbergia dumosa*
bamboo, reed – *Bambusa arundinacea*
bamboo, spiny – *Bambusa spinosa*
banana – *Mus sapientum*
banana, Abyssinian – *Ensete ventricosum*
banana, apple – *Mus acuminata* x *balbisiana*
banana, balbis – *Mus balbisiana*
banana, butuhan – *Mus errans* var *botoan*
banana, China – *Mus cavendishii*
banana, Chinese – *Mus acuminata*
banana, cooking – *Mus acuminata* x *balbisiana, M. paradisiaca*
banana, Dacca – *Mus acuminata*
banana, dwarf – *Mus nana*
banana, dwarf Cavendish – *Mus acuminata*
banana, gros Michel – *Mus acuminata*
banana, lady finger – *Mus accuminata*
banana, Nepal – *Mus nepalensis*
banana, ornamental – *Mus coccinea*
banana, poyo – *Mus acuminata* x *balbisiana*
banana, red Dacca – *Mus acuminata*
banana, robusta – *Mus acuminata* x *balbisiana*
banana, starchy – *Mus balbisiana*
banana waterlily – *Nymphaea mexicana*
banana yucca – *Yucca baccata*
banano – *Mus acuminata*
baneberry – *Actaea*
bane, bug – *Cimicifuga racemosa*
bane, flea – *Erigeron speciosus, Pluchea*
bane, leopard's – *Doronicum*
bane, wolfs – *Aconitum napellus*
banewort – *Ranunculus flammula*
banksia, double white – *Rosa fortuniana*
Banks' rose, lady – *Rosa banksiae*
banyan – *Ficus microcarpa*
banyan, Chinese – *Ficus microcarpa*
banyan, Indian – *Ficus benghalensis*
banyan tree – *Ficus benghalensis*
banyan, weeping Chinese – *Ficus benjamina*
baobob – *Adansonia digitata*
Barbados almond – *Terminalia catappa*
Barbados aloe – *Aloe vera*
Barbados cherry – *Malpighia emarginata, M. glabra*
Barbados flower – *Caesalpinia pulcherrima*
Barbados gooseberry – *Pereskia aculeata*
Barbados nut – *Jatropha curcas*
Barbados pride – *Caesalpinia pulcherrima*
Barbados, pride of – *Caesalpinia pulcherrima*
Barbary nut – *Iris sisyrinchium*
barbed goatgrass – *Aegilops triuncialis*
barbed-wire cactus – *Acanthocereus tetragonus*
barberry – *Berberis*
barberry, creeping – *Berberis aquifolium* var *repens*
barberry, Darwin – *Berberis darwinii*
barberry, English – *Berberis vulgaris*
barberry, European – *Berberis vulgaris*
barberry, Japanese – *Berberis thunbergii*
barberry, Magellan – *Berberis buxifolia*
barberry, red – *Berberis haematocarpa*
barberry, rosemary – *Berberis* x *stenophylla*
barberry, western – *Berberis fendleri*
barberry, wintergreen – *Berberis julianae*
barberton daisy – *Gerbera jamesonii*
bark, cascarilla – *Croton eleuteria*
bark, chittam – *Rhamnus purshiana*
bark, crown – *Cinchona officinalis*
bark, ledger – *Cinchona ledgeriana*
bark, red iron – *Eucalyptus sideroxylon*
bark, shoemaker's – *Byrsonima coriacea*
bark, strong – *Desmodium canum*
bark, winter's – *Drimys winteri*
bark, yellow – *Cinchona calisaya*
barley – *Hordeum*

barley, foxtail – *Hordeum jubatum*
barley, Mediterranean – *Hordeum marinum*
barley smut – *Ustilago hordei*
barley, wild – *Hordeum murinum* ssp *glaucum*
barnyard grass – *Echinochloa crus-galli*
barrel cactus – *Echinocactus, Ferocactus*
barrel cactus, Arizona – *Ferocactus wislizeni*
barrel cactus, blue – *Echinocactus horizonthalonius*
barrel cactus, California – *Ferocactus cylindraceus*
barrel cactus, clustered – *Echinocactus polycephalis*
barrel cactus, coast – *Ferocactus viridescens*
barrel cactus, compass – *Ferocactus cylindraceus*
barrel cactus, fire – *Ferocactus gracilis*
barrel cactus, fishhook – *Ferocactus wislizenii*
barrel cactus, giant – *Echinocactus platyacanthus*
barrel cactus, golden – *Echinocactus grusonii*
barrel cactus, mesquite – *Thelocactus setispinus*
barrel cactus, red – *Ferocactus cylindraceus*
barrel cactus, viznaga – *Ferocactus hamatacanthus*
barrel cactus, whiskered – *Ferocactus hamatacanthus*
barrenwort, alpine – *Epimedium alpinum*
bartsia, broad-leaved – *Parentucella latifolia*
bartsia, yellow – *Parentucella viscosa*
basil – *Ocimum basilicum*
basil, common – *Ocimum basilicum*
basil, sweet – *Ocimum basilicum*
basil, wild – *Clinopodium laevigatum, C. vulgare*
basin sagebrush – *Artemisia tridentata*
basket evening primrose – *Oenothera deltoides*
basket flower – *Centaurea Americana, Hymenocallis narcissiflora*
basket grass – *Muhlenbergia rigens, Nolina erumpens*
basket-of-gold – *Aurinia saxatilis*
basket tie-tie – *Desmoncus orthacanthos*
basket whisk – *Desmoncus orthacanthos*
basswood – *Tilia americana, T. occidentalis*
basswood, European – *Tilia* x *europaea*
bastard cabbage tree – *Geoffroea decorticans*
bastard cedar – *Guazuma ulmifolia*
bastard cinnamon – *Cinnamomum cassia*
bastard lime – *Trichilia hirta*
bastard rocket, French – *Erucastrum gallicum*
bataua palm – *Jessenia bataua*
Batavian endive – *Cichorium endivia*
bat-faced cuphea – *Cuphea llavea*
bay – *Laurus nobilis*
bayberry – *Myrica cerifera, M. californica, M. pensylvanica*
bayberry, southern – *Myrica cerifera*
bay, bull – *Magnolia grandiflora*
bay, California – *Umbellularia californica*
bay, California sweet – *Myrica californica*
bay cedar – *Suriana maritima*
bay cedar, wild – *Trema micrantha*
bay laurel – *Laurus nobilis*
bay leaf, Mexican – *Litsea glaucescens*
bayleaf palm – *Sabal mauritiiformis, S. yapa*
bayonet, Spanish – *Yucca*
bay, poison – *Illicium floridanum*
bay, rose – *Nerium oleander, Rhododendron*
bay rum – *Eugenia acris*
bay, sweet – *Laurus nobilis*
bay tree – *Eugenia racemosa*
bay, West Indian – *Eugenia racemosa*
bay willow – *Salix pentandra*
bdellium – *Commiphora africana*
beach carpet – *Philoxerus vermicularis*
beachgrass, European – *Ammophila arenaria*
beach morning-glory – *Ipomoea pes-caprae*
beach strawberry – *Fragaria chiloensis*
beach wormwood – *Artemisia stellerana*
beachwort – *Batis maritima*
beadtree, sandal – *Adenanthera pavonina*
beak, crimson – *Krameria grayi*
beaked yucca – *Yucca rostrata*
beam, horn – *Caprinus*
beam, white – *Sorbus*
bean – *Phaseolus vulgaris*
bean, adzuki – *Vigna angularis*
bean, Anasazi – *Phaseolus vulgaris*
bean, asparagus – *Vigna unguiculata* ssp *sesquipedalis*
bean, black – *Phaseolus vulgaris, Pithecellobium unguis-cati*
bean, bog – *Menyanthes trifoliata*
bean, broad – *Vicia faba*
bean, buck – *Menyanthes trifoliata*
bean, butter – *Phaseolus lunatus, P. vulgaris*

bean-caper – *Zygophyllum fabago*
bean, Carolina – *Phaseolus lunatus*
bean, castor – *Ricinus communis*
bean, coral – *Sophora secundiflora*
bean, fava – *Phaseolus lunatus*
bean, Florida velvet – *Mucuna pruriens* var *utilis*
bean, giant – *Canavalia plagiosperma*
bean, great northern – *Phaseolus vulgaris*
bean, green – *Phaseolus vulgaris*
bean, ground – *Vigna subterranea*
bean, horse – *Parkinsonia aculeata, Vicia faba*
bean, hyacinth – *Lablab purpureus*
bean, Ignatius – *Strychnos ignatii*
bean, Indian – *Catalpa bignonioides, Erythrina flabelliformis*
bean, itchy – *Mucuna pruriens*
bean, jack – *Canavalia, Phaseolus vulgaris*
bean, jumpy – *Leucaena glauca*
bean, kidney – *Phaseolus vulgaris*
bean, Lima – *Phaseolus lunatus*
bean, match-box – *Entada scandens*
bean, mescal – *Sophora secundiflora*
bean, Mexican jumping – *Sapium biloculare, Sebastiana pavoniana*
bean, Mexican runner – *Phaseolus coccineus*
bean, Mexican tepary – *Phaseolus acutifolius*
bean, Mexican yam – *Pachyrhizus erosus*
bean, mung – *Vigna radiata*
bean, navy – *Phaseolus vulgaris*
bean, ox-eye – *Mucuna*
bean, Peruvian – *Phaseolus vulgaris*
bean, pete – *Parkia speciosa*
bean, pink – *Phaseolus vulgaris*
bean, pinto – *Phaseolus vulgaris*
bean, pole – *Phaseolus vulgaris*
bean, pop – *Phaseolus vulgaris*
bean, prayer – *Abrus precatorius*
bean, red – *Phaseolus vulgaris*
bean, red kidney – *Phaseolus vulgaris*
bean, rice – *Phaseolus calcaratus, Vigna umbellata*
bean, rosary – *Rhynchosia precatoria, R. pyramidalis*
bean, scarlet runner – *Phaseolus coccineus*
bean, sea – *Entada phaseoloides*
bean, slender-stemmed – *Phaseolus filiformis*
bean, small red – *Phaseolus vulgaris*
bean, small white – *Phaseolus vulgaris*
bean, snout – *Rhynchosia*
bean, snuffbox – *Entada polystachya*
bean, southwest coral – *Erythrina flabelliformis*
bean, soy – *Glycine max*
bean, St. Ignatius's – *Strychnos ignatii*
bean, string – *Phaseolus vulgaris*
bean, sword – *Cannavalia gladiata*
bean, tepary – *Phaseolus acutifolius*
bean, Texas – *Phaseolus acutifolius*
bean, tick – *Vicia faba* var *minuta*
bean, tonka – *Dipteryx odorata*
bean trefoil – *Anagyris foetida*
bean, velvet – *Mucuna deeringiana, Stizolobium*
bean, western coral – *Erythrina flabelliformis*
bean, winged – *Psophocarpus tetragonolobus*
bean, yam – *Pachyrrhizus erosus*
bean, yard-long – *Vigna unguiculata* ssp *sesquipedalis*
bearberry – *Arctostaphylos manzanita, A. uva-ursi*
bearberry, alpine – *Arctostaphylos alpinus*
bearberry honeysuckle – *Lonicera involucrata*
beard, blue – *Caryopteris incana*
beard, bock's – *Tragopogon dubius*
bearded creeper – *Crupina vulgaris*
bearded cupgrass – *Eriochloa aristata*
bearded dalea – *Dalea pogonathera*
bearded darnel – *Lolium temulentum*
bearded iris – *Iris barbata, I. germanica*
bearded sprangletop – *Leptochloa fascicularis*
beard, goat's – *Aruncus dioicus, Tragopogon dubius, T. pratensis*
beard grass – *Polypogon elongatus*
beard grass, annual – *Polypogon minspeliensis*
beard grass, ditch – *Polypogon interruptus*
beard grass, Mediterranean – *Polypogon maritimus*
beard grass, silver – *Bothriochloa saccharoides*
beard grass, Texas – *Schizachyrium cirratum*
beard, hawks – *Crepis*
beard, Jupiter's – *Anthyllis barba-jovis, Centranthus ruber*
beard, old man's – *Clematis dioica, C. vitalba, Tillandsia usneoides*
beardtongue – *Penstemon*
beardtongue, royal – *Penstemon spectabilis*
beargrass – *Nolina*

beargrass, desert tree – *Nolina bigelovii*
beargrass, Sonoran tree – *Nolina matapensis*
beargrass, tree – *Nolina matapensis*
bear's breech – *Acanthus mollis*
bear's clover – *Chamaebatia foliolosa*
bear's foot – *Helleborus viridis*
bear's-foot hellebore – *Helleborus foetidus*
beautyberry – *Callicarpa americana*
beavertail – *Opuntia basilaris*
beavertail cactus – *Opuntia basilaris*
Bebb willow – *Salix bebbiana*
bedding begonia – *Begonia semperflorens*
bedstraw – *Galium*
bedstraw, catchweed – *Galium aparine*
bedstraw, common – *Galium aparine*
bedstraw, heath – *Galium saxatile*
bedstraw, ladies' – *Galium verum*
bedstraw, lady's – *Galium vernum*
bedstraw, Lamarck's – *Galium divaricatum*
bedstraw, northern – *Galium boreale*
bedstraw, rough corn – *Galium tricornutum*
bedstraw, San Diego – *Galium nuttallii*
bedstraw, sweet-scented – *Galium triflorum*
bedstraw, tiny – *Galium murale*
bedstraw, wall – *Galium parisiense*
bedstraw, yellow – *Galium verum*
bee balm – *Melissa officinalis, Monarda*
bee balm, mintleaf – *Monarda menthaefolia*
bee balm, plains – *Monarda pectinata*
bee blossom – *Lippia ligustrina*
bee blossom, midget – *Gaura parviflora*
bee blossom, scarlet – *Gaura coccinea*
bee brush – *Aloysia gratissima, A. wrightii, Eysenhardtia texana, Lippia ligustrina*
beech – *Fagus*
beech, American – *Fagus grandifolia*
beech, copper – *Fagus sylvatica* var *purpurea*
beech, European – *Fagus sylvatica*
beech, false – *Nothofagus antarctica*
beech, roble – *Nothofagus obliqua*
beech, southern – *Nothofagus antarctica*
beefsteak fungus – *Fistulina hepatica*
beefsteak plant – *Acalypha wilkesiana*
beefwood – *Casuarina equisetifolia*
beefwood, white – *Schoepfia schreberi*
beehive cactus – *Echinomastus johnsonii, Escobaria vivipara*
bee orchid – *Ophrys apifera*
bee plant – *Cleome*
bee plant, Rocky Mountain – *Cleome serulata*
bee sage – *Hyptis emoryi, Salvia apiana*
beet – *Beta vulgaris*
beet, fodder – *Beta vulgaris*
beet, garden – *Beta vulgaris*
beetlesage cuphea – *Cuphea jorullensis*
beetle spurge – *Euphorbia eriantha*
beet, sugar – *Beta vulgaris* ssp *saccharifera*
beggar lice – *Desmodium styracifolium*
beggar's lice – *Galium aparine*
beggar-tick grass – *Aristida orcuttiana*
beggar-ticks, common – *Bidens pilosa*
begonia – *Begonia*
begonia, bedding – *Begonia semperflorens*
begonia, rex – *Begonia rex*
begonia, semperflorens – *Begonia semperflorens*
begonia, trailing – *Begonia solananthera*
begonia, tuberous – *Begonia tuberosa*
begonia, wax-leaf – *Begonia semperflorens*
Belgian endive – *Cichorium intybus*
belladonna – *Atropa belladona*
belladonna lily – *Amaryllis belladonna*
bell, blue – *Campanula rotundifolia, Eustoma grandiflorum, Hyacinthoides, Phacelia campanularia*
bell, Canterbury – *Campanula medium*
bell, cathedral – *Cobaea scandens*
bell, coral – *Heuchera sanguinea*
bellflower – *Campanula rapunculoides*
bellflower, Chilean – *Lapageria rosea*
bellflower, Dalmation – *Campanula portenschlagiana*
bellflower, Italian – *Campanula isophylla*
bellflower, tussock – *Campanula carpatica*
bell, golden – *Forsythia* x *intermedia*
bell, hare – *Campanula rotundifolia*
bell heather – *Erica cinera*
bell, lady – *Adenophora liliifolia*
bell, mission – *Fritillaria biflora*
bell, monkey dinner – *Hura polyandra*
bell, mountain silver – *Halesia monticola*
bell pepper – *Capsicum annuum* var *grossum*
bell, Persian – *Fritillaria persica*
bell, sand – *Nama hispidum*
bell, snow – *Styrax*
bells-of-Ireland – *Moluccella laevis*
bell, spring – *Tabebuia donnell-smithii*
bell, star – *Cordia dodecandra*

bell tree dahlia – *Dahlia imperialis*
bell, whispering – *Emmenanthe penduliflora*
bell, yellow – *Allamanda cathartica, Tecoma stans*
beloperone – *Justicia californica*
beloperone, California – *Justica californica*
Bengal clockvine – *Thunbergia grandiflora*
Benjamin fig – *Ficus benjamina*
Benjamin tree – *Ficus benjamina*
Bennet, herb – *Geum urbanum*
bent, brown – *Agrostis canina*
bent, colonial – *Agrostis capillaris*
bent, common – *Agrostis capillaris, A. idahoensis*
bent, creeping – *Agrostis palustris, A. stolonifera*
bent grass – *Agrostis*
bent grass, red-top – *Agrostis gigantea*
bent grass, water – *Agrostis viridis*
bent, marsh – *Agrostis palustris*
benzoin tree – *Styrax benzoin*
bergamot – *Citrus bergamotia, Monarda citriodora, M. didyma*
bergamot orange – *Citrus bergamotia*
bergamot, wild – *Monarda fistulosa*
bergenia, heartleaf – *Bergenia cordifolia*
bergenia, winter-blooming – *Bergenia crassifolia*
Bermuda buttercup – *Oxalis pes-caprae*
Bermuda grass – *Cynodon dactylon*
berry tree, India – *Anamirta paniculata*
berseem clover – *Trifolium alexandrinum*
be-still tree – *Thevetia peruviana*
betel – *Piper betle*
betelnut – *Areca catechu, Piper betle*
Bethlehem, Mexican star of – *Milla biflora*
Bethlehem sage – *Pulmonaria saccharata*
Bethlehem, star of – *Campanula isophylla, Ornithogalium*
betony, big – *Stachys macrantha*
betony, purple – *Stachys betonica*
betony, water – *Scrophularia auriculata*
betony, wood – *Pedicularis canadensis, Stachys officinalis*
biannual mugwort – *Artemisia biennis*
biddy-biddy – *Acaena novae-zelandiae*
biennial turnip rape – *Brassica rapa* var *oleifera*
bigarreau cherry – *Prunus avium* var *duracina*
Big Bend prickly-pear – *Opuntia phaeacantha*
Big Bend silverleaf – *Leucophyllum minus*
bigberry manzanita – *Arctostaphylos glauca*
big betony – *Stachys macrantha*
big blue lily turf – *Liriope muscari*
bigcone Douglas-fir – *Pseudotsuga macrocarpa*
bigcone spruce – *Pseudotsuga macrocarpa*
Bigelow bluegrass – *Poa bigelovii*
big galleta – *Pleuraphis rigida*
bigleaf avens – *Geum macrophyllum*
bigleaf hydrangea – *Hydrangea macrophylla*
bigleaf magnolia – *Magnolia macrophylla*
bigleaf maple – *Acer macrophyllum*
bigleaf mistletoe – *Phoradendron macrophyllum*
big needle cactus – *Coryphanta macromeris*
big-root – *Marah macrocarpus*
big sagebrush – *Artemisia tridentata*
big saltbush – *Atriplex lentiformis* ssp *lentiformis*
big shell-bark hickory – *Carya laciniosa*
bigtooth aspen – *Populus grandidentata*
bigtooth maple – *Acer saccharum* ssp *grandidentatum*
big tree – *Sequoiadendron giganteum*
bilberry – *Vaccinium myrtillus*
bilberry, alpine – *Vaccinium uliginosum*
bilimbi tree – *Averrhoa bilimbi*
billion dollar grass – *Echinochloa frumentacea*
bindweed – *Convolvulus*
bindweed, black – *Polygonum convolvulus*
bindweed, field – *Convolvulus arvensis*
bindweed, hedge – *Calystegia sepium*
bindweed, Italian – *Convolvulus althaeoides*
bindweed, sea – *Calystegia soldanella*
birch, canoe – *Betula papyrifera*
birch, cherry – *Betula lenta*
birch, common white – *Betula pubescens*
birch, European white – *Betula pendula*
birch, Himalayan – *Betula jacquemontii*
birch-leaf buckthorn – *Rhamnus betulaefolia*
birch-leaf mountain-mahogany – *Cercocarpus betuloides*
birch, monarch – *Betula maximowicziana*
birch, paper – *Betula papyrifera*
birch, red – *Betula nigra*
birch, river – *Betula nigra*
birch, silver – *Betula verrucosa*

birch, sweet – *Betula lenta*
birch, tropical – *Bursera simaruba*
birch, water – *Betula occidentalis*
birch, white – *Betula alba*
birch, Young's weeping – *Betula pendula* var *youngii*
birdbill dayflower – *Commelina dianthifolia*
bird bush, canary – *Crotalaria*
bird cherry, European – *Prunus padus*
birdfoot trefoil – *Lotus corniculatus*
bird-of-paradise – *Caesalpinia cacalaco, C. gilliesii, Strelitzia reginae*
bird-of-paradise, false – *Heliconia*
bird-of-paradise, Mexican – *Caesalpinia mexicana*
bird-of-paradise, red – *Caesalpinia pulcherrima*
bird-of-paradise, yellow – *Caesalpinia gilliesii*
bird-pepper – *Capsicum annuum* var *minimum, C. baccatum, C. frutescens*
bird rape – *Brassica rapa* var *campestris, B. r.* var *oleifera*
birdseed grass – *Brachiaria fasciculata*
bird's-foot, orange – *Ornithopus pinnatus*
bird's-foot trefoil – *Lotus corniculatus*
bird's-foot trefoil, large – *Lotus uliginosus*
bird's-foot trefoil, slender – *Lotus angustissimus*
bird's nest fern – *Asplenium nidus*
bird's nest orchid – *Neottia nidus-avis*
bird's nest, yellow – *Monotropa hypopitys*
birdsrape mustard – *Brassica rapa*
bird vetch – *Vicia cracca*
biriba – *Rollinia pulchrineruis*
birthwort – *Aristolochia*
birthwort, long – *Aristolochia longa*
birthwort, round-leaved – *Aristolochia rotunda*
Biscay heath – *Erica mediterranea*
biscuit root – *Cymopterus fendleri, C. purpurascens, Lomatium dissectum*
bishop pine – *Pinus muricata*
bishop's cap – *Astrophytum capricorne, A. myriostigma*
bishop's hat – *Astrophytum capricorne, Epimedium alpinum, E. grandiflorum*
bishop's weed – *Aegopodium podagraria, Ammi majus, A. visnaga*
bishop's weed, lesser – *Ammi visnaga*
bishop's wort – *Nigella damascena, Stachys betonica, S. officinalis*
bisnaga – *Ammi visnaga*
bistort – *Polygonum bistorta*
bistort, western – *Polygonium bistortoides*
bit, hawk – *Leontodon taraxacoides*
bit, hen – *Lamium*
bitter agaric – *Tricholoma acerbum*
bitter almond – *Prunus dulcis* var *amara*
bitter aloe – *Aloe ferox*
bitter apple – *Citrullus colocynthis, Momordica balsamina*
bitter bush – *Picramnia pistaciaefolia*
bitter cherry – *Prunus emarginata*
bitter condalia – *Condalia globosa* var *pubescens*
bitter-cress – *Cardamine bellidifolia, C. pratensis*
bitter-cress, hairy – *Cardamine hirsuta*
bitter cucumber, Chinese – *Momordica cochinchinensis*
bitter dock – *Rumex obtusifolius*
bitter dogbane – *Apocynum androsaemifolium*
bitter gourd – *Citrullus colocynthis*
bitter gourd, Indian – *Momordica charantia*
bitter herb – *Centaurium erythraea*
bitter lettuce – *Lactuca virosa*
bitter lime – *Citrus limetta*
bitter melon – *Momordica charantia*
bitter mesquite – *Prosopis articulata*
bitternut – *Carya amara, C. cordiformis*
bitter orange – *Citrus aurantium*
bitters – *Colubrina ferruginosa*
bitter's calea – *Calea zacatechichi*
bitters, jackass – *Neurolaena lobata*
bittersweet – *Solanum dulcamara*
bittersweet, American – *Celastrus scandens*
bitter vetch – *Vicia ervilia*
bitterweed actinea – *Hymenoxys odorata*
bitterweed, poison – *Hymenoxys odorata*
bitterweed, slender-stem – *Hymenoxys scaposa*
bitterwood – *Quassia amara*
black alder – *Alnus glutinosa*
black bean – *Phaseolus vulgaris, Pithecellobium unguis-cati*
blackberry – *Rubus*
blackberry bush – *Rubus fruticosus*
blackberry, California – *Rubus ursinus*
blackberry, cut-leaved – *Rubus laciniatus*

blackberry, Himalayan – *Rubus discolor*
blackberry lily – *Belamcanda chinensis*
blackberry, Santa Rosa – *Rubus ulmifolius*
blackberry, wild – *Rubus trivialis*
black bindweed – *Polygonum convolvulus*
blackbrush – *Flourensia cernua*
blackbrush acacia – *Acacia rigidula*
black bryony – *Tamus communis*
black calla – *Arum palestinum*
black-cap – *Rubus occidentalis*
black cherry – *Prunus*
black cherry, capulin – *Prunus serotina*
black cherry, wild – *Prunus serotina*
black chile – *Capsicum annuum*
black cohosh – *Cimicifuga racemosa*
black cottonwood – *Populus balsamifera* ssp *trichocarpa*
black crowberry – *Empetrum nigrum*
black currant – *Ribes nigrum*
black dalea – *Dalea frutescens*
black elderberry – *Sambucus nigra*
black-eyed pea – *Pisum sativum, Vigna sinensis, V. unguiculata*
black-eyed Susan – *Echinacea purpurea, Rudbeckia*
black-eyed Susan vine – *Thunbergia alata*
black grama – *Bouteloua eriopoda*
black greasewood – *Sarcobatus vermiculatus*
black haw – *Viburnum*
black henbane – *Hyoscyamus niger*
black horehound – *Ballota nigra*
blackjack oak – *Quercus emoryi*
black knapweed – *Centaurea nigra*
black locust – *Robinia pseudoacacia*
black lovage – *Smyrnium olustratum*
black mangrove – *Avicennia germinans*
black medick – *Medicago lupulina*
black mulberry – *Morus nigra*
black mustard – *Brassica nigra*
black nightshade – *Solanum nigrum*
black oak – *Quercus*
black oat – *Avena strigosa*
black pepper – *Piper nigrum*
black persimmon – *Diospyros texana*
black pine, Austrian – *Pinus nigra*
black pine, Japanese – *Pinus thunbergii*
black poplar – *Populus nigra*
black raspberry – *Rubus occidentalis*
black russula – *Russula nigricans*
black sage – *Salvia mellifera*
black Sampson – *Echinacea purpurea*
black sapote – *Diospyros digyna, D. ebenaster*
black sedge – *Schoenus nigricans*
black seeded pepper – *Capsicum pubescens*
black snakeroot – *Cimicifuga racemosa, Sanicula marilandica*
black sorghum – *Sorghum almum*
black spruce – *Picea mariana*
black stem rust – *Puccinia graminis*
black stick maidenhair – *Adiantum tenerum*
blackthorn – *Prunus spinosa*
black truffle – *Tuber cibarium, T. melanosporum*
black truffle, red-grained – *Tuber aestivum*
black twitch – *Alopecurus myosuroides*
black walnut – *Juglans nigra*
black walnut, California – *Juglans californica*
black willow – *Salix nigra*
black willow, Gooding's – *Salix gooddingii*
blackwood acacia – *Acacia melanoxylon*
bladder campion – *Silene inflata, S. vulgaris*
bladdernut – *Staphylea bolanderi*
bladderpod – *Isomeris arborea*
bladder sedge – *Carex vesicaria*
bladder senna – *Colutea arborescens*
bladder-stem – *Eriogonum inflatum*
bladderwort – *Urticularia gibba, U. lentibularia, U. minor*
bladderwort, common – *Urticularia vulgaris*
bladderwort, greater – *Urticularia vulgaris*
bladderwrack – *Fucus vesiculosus*
blanco juniper – *Juniperus blancoi*
blanket flower – *Gaillardia pinnatifida, G. pulchella*
blanket, Indian – *Gaillardia aristata, G. pulchella*
blazing star – *Liatris punctata, L. spicata, Mentzelia*
blazing star, triangle-seed – *Mentzelia affinis*
blazing star, whitestem – *Mentzelia albicaulis*
blechnum fern – *Blechnum gibbum*
bleeding heart – *Dicentra formosa*
bleeding heart, common – *Dicentra spectabilis*
bleeding heart glorybower – *Clerodendrum thomsoniae*
bleeding heart, western – *Dicentra formosa*
blessed thistle – *Cnicus benedictus*
bletia – *Bletia*

blewit, field – *Lepista saeva*
blewit, wood – *Lepista nuda*
blight, potato – *Phytophthora infestans*
blind prickly-pear – *Opuntia rufida*
blinks – *Montia fontana*
blister plant – *Ranunculus scleratus*
blite, sea – *Suadea*
blite, strawberry – *Chenopodium foliosum*
bloodberry rouge-plant – *Rivina humilis*
blood, dragon's – *Croton lechleri, Daemonorops draco, Jatropha cardiophylla*
blood flower – *Asclepias curassavica*
bloodleaf – *Iresine*
blood lily – *Scadoxus multiflorus*
blood orange – *Citrus sinensis* var *melitense*
blood-red spikerush, dark – *Eleocharis atropurpurea*
blood-red trumpet vine – *Distictis buccinatoria*
bloodroot – *Potentilla erecta, anguinaria canadensis*
blood, tears of – *Bocconia frutescens*
blood trumpet – *Distictis buccinatorius*
bloodtwig dogwood – *Cornus sanguinea*
bloodwood tree, Brazil – *Haematoxylon campechianum*
bloody cranesbill – *Geranium sanguineum*
bloom, checker – *Sidalcea malviflora*
blooming Sally – *Epilobium angustifolium*
blossom berry – *Eugenia capuli*
blossom, blue – *Ceanothus thyrsiflorus*
blue agave, Weber – *Agave tequilana*
blue and yellow russula – *Russula cyanoxantha*
blue barrel cactus – *Echinocactus horizonthalonius*
bluebeard, common – *Caryopteris incana*
bluebell – *Campanula rotundifolia, Eustoma grandiflorum*
bluebell, California desert – *Phacelia campanularia*
bluebell, English – *Hyacinthoides non-scripta*
bluebell, peach-leafed – *Campanula persicifolia*
bluebell, Spanish – *Hyacinthoides hispanicus*
blueberry – *Vaccinium*
blueberry elder – *Sambucus glauca*
blueberry, highbush – *Vaccinium corymbosum*
blueberry, tropical – *Conostegia xalapensis*
blueberry, western – *Vaccinium uliginosum*
blue bird vine – *Petrea arborea*
blue blossom – *Ceanothus thyrsiflorus*
bluebonnet – *Lupinus arizonicus, L. texensis*
bluebottle – *Centaurea cyanus*
blue bugle – *Ajuga genevensis*
blue buttons – *Knautia arvensis*
blue candle – *Myrtillocactus geometrizans*
blue cereus – *Cereus hexagonus*
blue cohosh – *Caulophyllum thalictroides*
blue crown passion flower – *Passiflora caerulea*
bluecurls – *Phacelia congesta, Trichostema*
bluecurls, woolly – *Trichostema lanatum*
blue dicks – *Dichelostemma capitatum*
blue dracena – *Cordyline indivisa*
blue elderberry – *Sambucus mexicana*
blue-eyed African daisy – *Arctotis venusta*
blue-eyed-grass – *Sisyrinchium bellum*
blue-eyed Mary – *Omphalodes verna*
blue-eyes, baby – *Nemophila menziesii*
blue fan palm – *Erythea armata*
blue fescue, common – *Festuca glauca*
blue flag – *Iris germanica, I. versicolor*
blue flag, western – *Iris missouriensis*
blue flax – *Linum lewisii, L. narbonense*
blue flax, perennial – *Linum perenne*
blue ginseng – *Caulophyllum thalictroides*
blue grama – *Bouteloua gracilis*
bluegrass, annual – *Poa annua*
bluegrass, bigelow – *Poa bigelovii*
bluegrass, bulbous – *Poa bulbosa*
bluegrass, Canadian – *Poa compressa*
bluegrass, Chaix's – *Poa chaixii*
bluegrass, fowl – *Poa palustris*
bluegrass, Kentucky – *Poa pratensis*
bluegrass, rough – *Poa trivialis*
bluegrass, rough-stalked – *Poa trivialis*
bluegrass, timberline – *Poa glauca*
bluegrass, weak – *Poa inferma*
bluegrass, wood – *Poa nemoralis*
blue-green lady's-mantle – *Alchemilla glaucescens*
blue gum – *Eucalyptis globulus*
blue honeysuckle – *Lonicera caerulea*
blue hyssop – *Hyssopus officinalis*
blue jasmine – *Plumbago scandens*
blue larkspur – *Delphinium carolinianum*
blue-leaf wattle – *Acacia saligna*
blue lily turf, big – *Liriope muscari*
blue lobelia – *Lobelia erinus*

blue lupine – *Lupinus angustifolius*
blue mahue – *Hibiscus elatus*
blue morning-glory – *Ipomoea lindheimeri*
blue oak, Mexican – *Quercus oblongifolia*
blue oatgrass – *Helictotrichon sempervirens*
blue palm – *Erythea armata*
blue palm, Mexican – *Erythea armata*
blue palo verde – *Cercidium floridum*
blue panicgrass – *Panicum antidotale*
blue pea – *Clitoria terneata, Psoralea pentaphylla*
blue pendent-pod oxytrope – *Oxytropis deflexa* var *sericea*
blue sage – *Salvia azurea, S. farnesiana*
blue sage, shrubby – *Salvia ballotaeflora*
blue sedge – *Carex flacca*
blues, heavenly – *Ipomoea violacea*
blue skullcap – *Scutellaria lateriflora*
blue spiraea – *Caryopteris incana*
blue spruce – *Picea pungens*
blue spruce, Colorado – *Picea pungens* var *glauca*
bluestem, broomsedge – *Andropogon virginicus*
bluestem, cane – *Bothriochloa barbinodis*
bluestem, little – *Schizachyrium scoparium*
bluestem, sand – *Andropogon hallii*
bluestem, southwestern bushy – *Andropogon glomeratus* var *scabriglumus*
bluestem yucca – *Yucca guatemalensis*
bluet – *Centaurea montana*
blue thimble flower – *Gilia capitata*
blue trumpet vine – *Thunbergia grandiflora*
bluets – *Hedyotis wrightii*
blue vervain – *Verbena hastata, V. macdougali*
blue waterlily – *Nymphaea caerulea, N. elegans*
blueweed – *Helianthus ciliaris*
blueweed, Texas – *Helianthus ciliaris*
blue willow, Alaska – *Salix purpurea*
bluewing – *Torenia fournieri*
bluewood – *Condalia hookeri*
bluewood condalia – *Condalia obovata*
bluewood condalia, Mexican – *Condalia mexicana*
blue yucca – *Agave arizonica, Yucca baccata, Y. rigida*
blushing epidendrum – *Epidendrum erubescens*
boat lily – *Tradescantia spathacea*
boat, Moses-in-the – *Tradescantia spathacea*
boat-thorn acacia – *Acacia cymbispina*
bock's beard, greater – *Tragopogon dubius*
Boer grass – *Eragrostis curvula* var *conferta*
Boer lovegrass – *Eragrostis chloromelas*
bog arum – *Zantedeschia palustris*
bogbean – *Menyanthes trifoliata*
bog-orchid, green-flowered – *Platanthera hyperborea*
bog rosemary – *Andromeda polifolia*
bog-rush – *Juncus effuses, Schoenus nigricans*
bog sandwort – *Minuartia stricta*
bog sedge – *Carex limosa*
bog violet – *Pinguicula vulgaris*
bok choy – *Brassica rapa* var *chinensis*
Boke's button cactus – *Epithelantha bokei*
bokhara fleeceflower – *Polygonum baldschuanicum*
boldo – *Peumus boldus*
bolete, edible – *Boletus edulis*
bolete, granulated – *Suillus granulatus*
bolete, lurid – *Boletus luridus*
bolete, summer king – *Boletus reticulatus*
bolete, yellow-brown – *Boletus luteus*
boneset – *Eupatorium perfoliatum*
bonnet, blue – *Lupinus arizonicus, L. texensis*
bonnet, Scotch – *Capsicum chinense*
Bonpland willow – *Salix bonplandiana*
boojum tree – *Fouquieria columnaris*
bootblack flowers – *Malvaviscus arboreus*
boquet, Cashmere – *Clerodendrum bungei*
borage – *Borago officinalis*
borage, country – *Coleus ambolnicos*
borage, wild – *Symphytum asperum*
border penstemon – *Penstemon* x *gloxinioides*
border pinyon – *Pinus discolor*
borderpod acacia – *Lysiloma divaricata*
boreal sagewort – *Artemisia norvegica* ssp *saxatilis*
borecole – *Brassica oleracea* var *laciniata*
Boston fern – *Nephrolepis exaltata*
Boston horsetail – *Equisetum ramosissimum*
Boston ivy – *Parthenocissus tricuspidata*
bo-tree – *Ficus religiosa*
bottle, blue – *Centaurea cyanus*
bottlebrush – *Callistemon*
bottlebrush, crimson – *Callistemon citrinus*
bottlebrush, lemon – *Callistemon citrinus*
bottlebrush, weeping – *Callistemon viminalis*

bottle gourd – *Lagenaria siceraria*
bottle palm – *Hyophorbe lagenicaulis, Nolina recurvata*
bottle sedge – *Carex utriculata*
bottle tree – *Brachychiton populneus, Sterculia apetala*
bougainvillea – *Bougainvillea spectabilis*
bouncing bet – *Saponaria officinalis*
boundary ephedra – *Ephedra aspera*
boundary goldenbush – *Ericameria brachylepis*
bouvardia, scarlet – *Bouvardia ternifolia*
bouvardia, white – *Bouvardia longiflora*
bower, vine – *Clematis viticella*
bower, virgin's – *Clematis*
bowstring hemp – *Sansevieria guineensis, S. trifasciata*
box elder – *Acer negundo* var *californicum*
box honeysuckle – *Lonicera nitida*
boxleaf hebe – *Hebe buxifolia*
box, rattle – *Crotalaria*
box thorn – *Lycium*
box thorn, Chinese – *Lycium chinense*
box, Victorian – *Pittosporum undulatum*
boxwood, common – *Buxus sempervirens*
boxwood, English – *Buxus sempervirens*
boxwood, Indian – *Caesaria nitida*
boxwood, West Indian – *Gossypiospermum praecox*
boysenberry – *Rubus ursinus* var *loganobaccus*
bracken – *Pteridium aquilinum*
bracken fern – *Pteridium aquilinum, Pteris*
bracted manzanita – *Arctostaphylos pringlei*
brain cactus – *Stenocactus multicostatus*
brain mushroom, poisonous – *Gyromitra esculenta*
brake – *Pteridium aquilinum* var *pubescens, Pteris*
brake, cliff – *Pellaea*
bramble – *Rubus*
bramble, arctic – *Rubus arctius*
bramble, Cape – *Rubus rosaefolius*
branched oyster mushroom – *Pleurotus cornucopiae*
Brandegee's sage – *Salvia brandegei*
brassavola, glaucous-stemmed – *Laelia glauca*
brass-buttons – *Cotula coronopifolia*
Brazil bloodwood tree – *Haematoxylon brasiletto*
Brazil cherry – *Eugenia uniflora*
Brazilian butterfly tree – *Bauhinia forficata*
Brazilian flamebush – *Calliandra tweedii*
Brazilian lucerne – *Stylosanthes guianensis*
Brazilian mahogany – *Swietenia macrophylla*
Brazilian nightshade – *Solanum seaforthianum*
Brazilian pepper tree – *Schinus terebinthifolius*
Brazilian plume flower – *Justicia carnea*
Brazilian rosewood – *Dalbergia nigra*
Brazil nut – *Bertholletia excelsa*
Brazilwood – *Caesalpinia echinata, Haematoxylum brasiletto*
breadbox poppy – *Papaver somniferum*
breadfruit – *Artocarpus altilis, A. incisa*
breadfruit, Mexican – *Monstera deliciosa*
breadnut – *Artocarpus altilis, Brosimum alicastrum, Trophis racemosa*
breadnut, ramon – *Brosimum alicastrum*
bread, poor man's – *Artocarpus heterophyllus*
breadroot – *Psoralea esculenta*
breadroot, Indian – *Psoralea esculentum*
bread, sow – *Cyclamen neapolitanum*
bread, St. John's – *Ceratona siliqua*
breath, baby's – *Gypsophila elegans, G. paniculata*
breath, false baby's – *Galium mollugo*
Brewer spruce – *Picea breweriana*
Brewer's weeping spruce – *Picea breweriana*
briar – *Rubus*
briar, green – *Smilax*
briar, sweet – *Rosa eglanteria*
brickellbush – *Brickellia*
brickellia – *Brickellia*
brickellia, desert – *Brickellia coulteri*
bridal-wort – *Spiraea prunifolia*
bridal wreath spiraea – *Spiraea prunifolia*
bride, mourning – *Scabiosa atropurpurea*
bride's tears – *Antigonon leptopus*
brier, Austrian – *Rosa foetida*
Brigham tea – *Ephedra torreyana*
bristlecone fir – *Abies bracteata*
bristlecone pine, western – *Pinus longaeva*
bristlegrass – *Setaria gracilis*
bristlegrass, plains – *Setaria macrostachya*
bristlegrass, white-haired – *Setaria leucopila*
bristlegrass, yellow – *Setaria pumila*
bristly club-rush – *Scirpus setaceous*
bristly fiddleneck – *Amsinckia tessallata*
bristly foxtail – *Setaria verticillata*
bristly ox-tongue – *Picris echioides*

bristly raspberry – *Rubus hispidus*
brittlebush – *Encelia farinosa*
brittlebush, white – *Encelia farinosa*
broad bean – *Vicia faba*
broccoli – *Brassica oleracea* var *italica*
broccoli, Calabrese – *Brassica oleracea* var *italica*
broccoli, Chinese – *Brassica oleracea* var *alboglabra*
brodiaea – *Brodiaea*
brodiaea, harvest – *Brodiaea coronaria*
broken heart – *Clerodendrum thomsoniae*
brome – *Bromus*
brome, California – *Bromus carinatus*
brome, downy – *Bromus tectorum*
brome, false – *Brachypodium distachyon*
brome, foxtail – *Bromus alopecurus*
brome, Japanese – *Bromus japonicus*
bromeliad, prodigious – *Tillandsia prodigiosa*
brome, mountain – *Bromus carinatus*
brome, poverty – *Bromus sterilis*
brome, red – *Bromus madritensis* ssp *rubens*
brome, rye – *Bromus secalinus*
brome, smooth – *Bromus inermis*
brome, soft – *Bromus hordeaceus*
brooklime – *Veronica beccabunga*
brooklime, American – *Veronica americana*
brook mint – *Mentha arvensis*
brook saxifrage – *Saxifraga rivularis*
brookweed – *Samolus parviflorus*
broom – *Genista canariensis, G. lydia, Retama raetam, Sarothamnus scoparius*
broom baccharis – *Baccharis sarathroides*
broom, butcher's – *Ruscus aculeatus, R. hypoglossum*
broom, California – *Lotus scoparius*
broom, Canary Island – *Genista canariensis*
broomcorn – *Sorghum bicolor*
broomcorn millet – *Panicum miliaceum*
broom, desert – *Baccharis sarathroides*
broom, dwarf – *Chamaecytisus proliferus*
broom, flax-leaved – *Genista linifolia*
broom heath – *Erica scoparia*
broom jute – *Sida rhombifolia*
broom, Mt. Etna – *Genista aetnensis*
broom, needle – *Genista anglica*
broom, Portuguese – *Cytisus multiflorus*
broom, Provence – *Cytisus purgans*
broom-rape – *Orobanche*
broom-rape, desert – *Orobanche cooperi*
broom root – *Muhlenbergia macroura*
broom, Scotch – *Cytisus scoparius*
broomsedge bluestem – *Andropogon virginicus*
broom, snake – *Gutierrezia sarothrae*
broom snakeweed – *Gutierrezia sarothrae*
broom, Spanish – *Cytisus multiflorus, Genista hispanica, Spartium junceum*
broom, turpentine – *Thamnosma montana*
broomweed – *Sida acuta*
broom-wood – *Melochia tomentosa*
brown bent – *Agrostis canina*
brown-eyed Susan – *Rudbeckia hirta*
brown-flowered psoralea – *Psoralea rhombifolia*
brownfoot – *Acourtia wrightii*
brown galingale – *Cyperus fuscus*
brownish sedge – *Carex buxbaumii*
brown mustard – *Brassica juncea*
brownplume wire-lettuce – *Stephanomeria pauciflora*
brownseed paspalum – *Paspalum plicatulum*
browntop – *Agrostis capillaris, A. idahoensis, Brachiaria fasciculata*
browse, deer – *Cercocarpus montanus*
brush, bee – *Aloysia gratissima, A. wrightii, Eysenhardtia texana, Lippia ligustrina*
brush, bottle – *Callistemon*
brush grass, haired – *Hyparrhenia hirta*
brush holly – *Xylosma flexuosum*
brush, lemon bottle – *Callistemon citrinus*
brush, weeping bottle – *Callistemon viminalis*
brush, white – *Aloysia gratissima, Lippia ligustrina*
Brussels sprouts – *Brassica oleracea* var *gemmifera*
bryony – *Bryonia alba*
bryony, black – *Tamus communis*
bryony, white – *Bryonia dioica*
buchu – *Barosma betulina*
buckbean – *Menyanthes trifoliata*
buck brush – *Ceanothus cuneatus*
buckeye – *Aesculus parryi*
buckeye, California – *Aesculus californica*
buckeye, Mexican – *Ungnadia speciosa*
buckeye, red – *Aesculus pavia*
buckeye, Spanish – *Ungnadia speciosa*
buckeye, Texas – *Ungnadia speciosa*
buckhorn – *Plantago lanceolata*

buckhorn cholla – *Cylindtropuntia acanthocarpa*
buck's-horn plantain – *Plantago coronopus*
buckthorn – *Karwinskia humboldtiana, Rhamnus*
buckthorn, alder – *Rhamnus frangula*
buckthorn, alpine – *Rhamnus alpinus*
buckthorn, birch-leaf – *Rhamnus betulaefolia*
buckthorn, California – *Rhamnus californica*
buckthorn, common – *Rhamnus cathartica*
buckthorn, Italian – *Rhamnus alaternus*
buckthorn, Mediterranean – *Rhamnus alaternus*
buckthorn, Palestine – *Rhamnus palestina*
buckthorn, sea – *Hippophae rhamnoides*
buckthorn, West Indian – *Rhamnus sphaerosperma*
buckwheat – *Eriogonum nudum, Fagopyrum esculentum*
buckwheat brush, flat-top – *Eriogonum fasciculatum*
buckwheat, California – *Eriogonum fasciculatum*
buckwheat, false – *Eriogonum*
buckwheat, wild – *Eriogonum*
Buddhist bauhinia – *Bauhinia variegata*
buddleja, rusty – *Buddleja cordata*
buffalo berry – *Solanum rostratum*
buffalo-bur – *Solanum rostratum*
buffalo gourd – *Cucurbita foetidissima*
buffalo grass – *Buchloë dactyloides*
buffalo grass, white – *Panicum coloratum*
buffelgrass – *Cenchrus ciliaris, Pennisetum ciliare*
bugbane – *Cimicifuga racemosa*
bugle, blue – *Ajuga genevensis*
bugle, carpet – *Ajuga pyramidalis, A. reptans*
bugleweed – *Lycopus americanus, L. asper, L. europaeus*
bugloss – *Anchusa arvensis, A. azurea, Echium lycopsis*
bugloss, common – *Borago officinalis*
bugloss, viper's – *Echium plantagineum, E. vulgare*
bulbous bluegrass – *Poa bulbosa*
bulbous buttercup – *Ranunculus bulbosus*
bulbous calypso – *Calypso bulbosa*
bulbous corydalis – *Corydalis cava*
bulbous fumitory – *Corydalis cava*
bullace plum tree – *Prunus insititia*
bull bay – *Magnolia grandiflora*
bulletwood – *Achras bidentata, Mimusops darienensis*
bullgrass – *Muhlenbergia emersleyi*
bull-head – *Tribulus terrestris*
bullhoof – *Bauhinia variegata, Piper auritum*
bullhorn acacia – *Acacia spadicigera*
bull mallow – *Malva nicaeensis*
bull-nettle – *Cnidoscolus texanus, Solanum elaeagnifolium*
bullock's-heart – *Annona reticulata*
bull's tongue dock – *Rumex conglomeratus*
bull thistle – *Cirsium vulgare*
bulrush – *Cyperus papyrus, Scirpus*
bulrush, California – *Scirpus californicus*
bulrush millet – *Pennisetum americanum*
bumelia, gum – *Sideroxylon lanuginosum*
bunching onion – *Allium fistulosum*
bundleflower – *Desmanthus fruticosus*
bunny-ear prickly-pear – *Opuntia microdasys*
bunny ears – *Opuntia microdasys*
bun, penny – *Boletus edulis*
bunya-bunya – *Araucaria bidwillii*
bupleurum – *Bupleurum chinense*
burbush – *Ambrosia chenopodiifolia*
bur-chervil – *Anthriscus caucalis*
burclover – *Medicago*
burclover, California – *Medicago polymorpha*
burclover, spotted – *Medicago arabica*
bur, cockle – *Arctium lappa, Xanthium*
bur-cucumber, one-seeded – *Sicyos laciniatus*
burdock – *Arctium lappa, A. minus*
burdock, lesser – *Arctium minus*
burgrass, spike – *Tragus berteronianus*
burgundy truffle – *Tuber uncinatum*
burhead, yellow – *Limnocharis flava*
bur-marigold – *Bidens grandiflora, B. laevis*
bur-marigold, nodding – *Bidens cernua*
burnet, garden – *Sanguisorba minor* ssp *muricata. S. officinalis*
burnet, great – *Sanguisorba officinalis*
burnet rose – *Rosa spinosissima*
burnet, salad – *Sanguisorba minor* ssp *muricata*
burnet-saxifrage – *Pimpinella major*
burning bush, western – *Euonymus occidentalis*
burning love – *Ixora coccinea*

bur-nut – *Tribulus terrestris*
bur, paroquet – *Triumfetta*
bur parsley, knotted – *Torilis nodosa*
bur parsley, spreading – *Torilis arvensis*
burr brush – *Triumfetta semitriloba*
bur-reed – *Sparganium eurycarpum*
bur-reed, greater – *Sparganium erectum*
bur-reed, narrow-leaved – *Sparganium angustifolium*
bur-reed, small – *Sparganium natans*
bur-reed, unbranched – *Sparganium emersum*
burr oak – *Quercus macrocarpa*
burrobrush – *Hymenoclea monogyra, H. salsola*
burrobush – *Ambrosia dumosa, Hymenoclea salsola*
burro fat – *Isomeris arborea*
burro grass – *Scleropogon brevifolius*
burro tail – *Sedum morganianum*
burro-weed – *Ambrosia dumosa*
bur-sage – *Ambrosia ambrosioides, A. confertiflora, A. cordifolia*
bur-sage, annual – *Ambrosia acanthicarpa*
bur-sage, San Diego – *Ambrosia chenopodiifolia*
bur-sage, triangle-leaf – *Ambrosia deltoidea*
bur-sage, white – *Ambrosia dumosa*
bur, sand – *Ambrosia acanthicarpa, Cenchrus ciliaris*
burstwort – *Herniaria glabra*
bur-weed – *Ambrosia acanthicarpa*
buryseed chloris – *Chloris chloridea*
bush butter tree – *Dacryodes edulis*
bush cherry, Chinese – *Prunus glandulosa*
bush cinquefoil – *Potentilla fruticosa*
bush-clover – *Lespedeza striata*
bush croton – *Croton fruticulosus*
bush germander – *Teucrium fruticans*
bush grape – *Vitis rupestris*
bush laurel – *Litsea glaucescens*
bush lupine, yellow – *Lupinus arboreus*
bushmint – *Hyptis albida*
bush morning-glory – *Convolvulus cneorum*
bush muhly – *Muhlenbergia porteri*
bush penstemon, yellow – *Keckiella antirrhinoides*
bush sage, Mexican – *Salvia leucantha*
bush salvia, red – *Salvia microphylla*
bush seepweed – *Suaeda moquinii*
bush sunflower – *Simsia amplexicaulis, Tithonia tubaeformis*
bush vetch – *Vicia sepium*
bushy bluestem, southwestern – *Andropogon glomeratus* var *scabriglumus*
bustic – *Dipholis salicifolia*
busy Lizzie – *Impatiens walleriana*
butcher's broom – *Ruscus aculeatus, R. hypoglossum*
butter-and-eggs – *Linaria vulgaris*
butter bean – *Phaseolus lunatus, P. vulgaris*
buttercup – *Allamanda cathartica, Oenothera macrocarpa, Ranunculus*
buttercup, Bahama – *Turnera ulmifolia*
buttercup, Bermuda – *Oxalis pes-caprae*
buttercup, bulbous – *Ranunculus bulbosus*
buttercup, corn – *Ranunculus arvensis*
buttercup, creeping – *Ranunculus repens*
buttercup, desert – *Ranunculus cymbalaria*
buttercup, frog's-bit – *Ranunculus hydrocharis*
buttercup, hairy – *Ranunculus sardous*
buttercup, meadow – *Ranunculus acris*
buttercup, mountain – *Ranunculus montanus*
buttercup, pine – *Ranunculus fascicularis*
buttercup, small-flowered – *Ranunculus parviflorus*
buttercup, spinous – *Ranunculus muricatus*
buttercup tree – *Cochlospermum vitifolium*
buttercup, water – *Ranunculus aquatilis*
butterfly bush – *Buddleja*
butterfly bush, woolly – *Buddleja marrubifolia*
butterfly flower – *Schizanthus pinnatus*
butterfly flower tree – *Bauhinia monandra*
butterfly gladiolus – *Gladiolus dalenii*
butterfly lily – *Hedychium coronarium*
butterfly palm – *Chrysalidocarpus lutescens*
butterfly pea – *Clitoria terneata*
butterfly plant – *Asclepias incarnata*
butterfly tree – *Erblichia odorata*
butterfly tree, Brazilian – *Bauhinia forficata*
butterfly-weed – *Asclepias curassavica, A. tuberosa*
butter lettuce – *Lactuca sativa* var *crispa*
butter mushroom – *Suillus luteus*
butternut – *Juglans cinera*
butter nut – *Caryocar amygdaliferum*
butternut squash – *Cucurbita moschata*
butter tree, bush – *Dacryodes edulis*
butterweed, common – *Senecio vulgaris*

butterwort – *Pinguicula caudate, P. vulgaris*
button, bachelor's – *Centaurea cyanus, Gomphrena globosa*
button bush – *Cephalanthus salicifolius*
button cactus – *Epithelantha micromeris*
button cactus, Boke's – *Epithelantha bokei*
button-celery – *Eryngium carlinac*
button fern – *Pellaea rotundifolia*
button, mescal – *Lophophora williamsii*
buttons, blue – *Knautia arvensis*
buttons, brass – *Cotula coronopifolia*
button snakeroot – *Eryngium lemmoni, Liatris punctata, L. spicata*
button weed – *Abutilon avicennae*
button willow, Californica – *Cephalanthus occidentalis* var *californicus*
buttonwood – *Piper amalago, Platanus occidentalis*
buttonwood mangrove – *Conocarpus erecta*
buttonwood, white – *Laguncularia racemosa*
Butuhan banana – *Mus errans* var *botoan*
Byzantine gladiolus – *Gladiolus communis* ssp *byzantinum*

cabbage – *Brassica oleracea* var *capitata*
cabbage agave – *Agave perryi*
cabbage angelin – *Andira inermis*
cabbage, Chinese – *Brassica rapa* var *pekinensis*
cabbage, Napa – *Brassica rapa* var *pekinensis*
cabbage, oxheart – *Brassica oleracea* var *conica*
cabbage palm – *Roystonea oleracea, Sabal japa, S. palmetto*
cabbage palm, mountain – *Euterpe precatoria*
cabbage, Portuguese – *Brassica oleracea* var *costata*
cabbage, red – *Brassica oleracea* var *capitata*
cabbage rose – *Rosa centifolia*
cabbage, Savoy – *Brassica oleracea* var *sabauda*
cabbage, skunk – *Symplocarpus foetidus*
cabbage tree, bastard – *Geoffroea decorticans*
cacao – *Theobroma*
cachibou – *Calathea lutea*
cachimant – *Annona reticulata*
cactus – *Cactaceae*
cactus, acuña – *Echinomastus erectocentrus*
cactus, agave – *Leuchtenbergia principis*
cactus, barbed-wire – *Acanthocereus tetragonus*
cactus, barrel – *Echinocactus, Ferocactus*
cactus, beavertail – *Opuntia basilaris*
cactus, beehive – *Echinomastus johnsonii, Escobaria vivipara*
cactus, big needle – *Coryphanta macromeris*
cactus, brain – *Stenocactus multicostatus*
cactus, button – *Epithelantha micromeris*
cactus, candelabra – *Isolatocereus dumortieri, Myrtillocactus geometrizans*
cactus, Christmas – *Cylindropuntia leptocaulis, Schlumbergera buckleyi*
cactus, claret-cup – *Echinocereus polyacanthus* ssp *pacificus*
cactus, cochineal – *Opuntia cochenillifera*
cactus, column – *Neobuxbaumia tetetzo*
cactus, compass – *Ferocactus cylindraceus*
cactus, cottontop – *Echinocactus polycephalus*
cactus, crab – *Schlumbergera truncata*
cactus, dahlia – *Echinocereus poselgeri*
cactus, devil – *Grusonia schottii*
cactus, Dutchman's pipe – *Epiphyllum oxypetalum*
cactus, eaglesclaw – *Echinocactus horizonthalonius*
cactus, easter – *Rhipsalidopsis gaertneri*
cactus, elephant – *Pachycereus pringlei*
cactus, finger – *Mammillaria longimamma*
cactus, fish-hook – *Mammillaria, Thelocactus setispinus*
cactus, galloping – *Stenocereus gummosus*
cactus, gearstem – *Peniocereus striatus*
cactus, grama grass – *Sclerocactus papyracanthus*
cactus, hairbrush – *Pachycereus pecten-aboriginum*
cactus, hanging – *Echinocereus pensilis*
cactus, hatchet – *Pelecyphora aselliformis*
cactus, hedgehog – *Echinocereus*
cactus, horse-crippler – *Echinocactus texensis*
cactus, jumping – *Cylindropuntia*
cactus, keg – *Ferocactus viridescens*
cactus, lacyspine – *Mammillaria lasiacantha*
cactus, lady finger – *Echinocereus pentalophus, Mammillaria elongata*
cactus, lead pencil – *Echinocereus poselgeri*
cactus, lily – *Echinocereus cinerascens*

cactus, living-rock – *Ariocarpus fissuratus, A. retusus*
cactus, Mexican lime – *Ferocactus pilosus*
cactus, mission – *Opuntia ficus-indica*
cactus, mistletoe – *Rhipsalis baccifera*
cactus, moon – *Epiphyllum anguliger*
cactus, nipple – *Mammillaria*
cactus, octopus – *Stenocereus alamosensis*
cactus, old man – *Cephalocereus sensilis, Pachycereus gatesii, P. schottii, Pilosocereus leucocephalus*
cactus, orchid – *Epiphyllum oxypetalum, E.* x *Disocactus*
cactus, organpipe – *Pachycereus marginatus, Stenocereus thurberi*
cactus, paperspine – *Sclerocactus papyracanthus*
cactus, pencil – *Cylindropuntia leptocaulis, C. ramosissima, Echinocereus poselgeri*
cactus, Peruvian apple – *Cereus repandus*
cactus, pickle – *Echinocereus berlandieri*
cactus, pincushion – *Coryphanta potosiana, C. radians, Mammillaria*
cactus, pineapple – *Echinomastus*
cactus, pinecone – *Pelecyphora strobiliformis*
cactus, pipestem – *Cylindropuntia leptocaulis*
cactus, polka dot – *Opuntia microdasys*
cactus, rainbow – *Echinocereus*
cactus, rain-stick – *Echinopsis chiloensis*
cactus, rat-tail – *Cylindropuntia leptocaulis, Disocactus flagelliformis, Mammillaria pottsii*
cactus, rose – *Pereskia grandifolia*
cactus, saguaro – *Carnegiea gigantea*
cactus, sand dollar – *Astrophytum asterias*
cactus, San Pedro – *Echinopsis pachanoi*
cactus, sea urchin – *Astrophytum asterias*
cactus, silver-dollar – *Opuntia chlorotica*
cactus, snake – *Echinocereus pensilis, Peniocereus serpentinus*
cactus, strawberry – *Echinocereus brandegeei, E. engelmannii, E. stramineus*
cactus, Texas cone – *Neolloydia conoidea*
cactus, Texas hair – *Mammillaria prolifera*
cactus, Thanksgiving – *Schlumbergera truncata*
cactus, thimble – *Neolloydia conoidea*
cactus, top – *Strombocactus disciformis*
cactus, Turk's head – *Echinocactus horizonthalonius*
cactus, twin-spined – *Mammillaria geminispina*
cactus, velvet – *Bergerocactus emoryi*
cactus, Whipple's claw – *Sclerocactus whipplei*
cactus, whisky – *Lophophora williamsii*
cactus, whortleberry – *Myrtillocactus geometrizans*
cactus, woodlouse – *Pelecyphora aselliformis*
cactus, woven-spine pineapple – *Echinomastus intertextus*
Caesar's mushroom – *Amanita caesarea*
cainit, wild – *Chrysophyllum bicolor*
Cairo morning-glory – *Ipomoea cairica*
calabash – *Legenaria vulgaris*
calabash tree – *Crescentia alata, C. cujete*
calabazilla – *Cucurbita foetidissima*
Calabrese broccoli – *Brassica oleracea* var *italica*
calabura – *Muntingia calabura*
caladium – *Caladium*
caladium, fancy-leaved – *Caladium bicolor*
calamint – *Calamintha grandiflora, C. nepeta*
calamint, lesser – *Calamintha nepeta*
calamondin – *Citrus madurensis*
calawba grape – *Vitis labrusca*
calea, bitter's – *Calea zacatechichi*
calendula – *Calendula officinalis*
caley pea – *Lathyrus hirsutus*
caliandra, Schott – *Zapoteca formosa*
calico bush – *Kalmia latifolia, Lantana horrida*
calico flower – *Aristolochia grandiflora*
California adolphia – *Adolphia californica*
California barrel cactus – *Ferocactus cylindraceus*
California bay – *Umbellularia californica*
California beloperone – *Justica californica*
California blackberry – *Rubus ursinus*
California black walnut – *Juglans californica*
California brome – *Bromus carinatus*
California broom – *Lotus scoparius*
California buckeye – *Aesculus californica*
California buckthorn – *Rhamnus californica*
California buckwheat – *Eriogonum fasciculatum*
California bulrush – *Scirpus californicus*
California burclover – *Medicago polymorpha*

California button willow – *Cephalanthus occidentalis* var *californicus*
California caltrop – *Kallstroemia californica*
California carpetweed – *Kallstroemia californica*
California cholla – *Cylindropuntia californica*
California coffeeberry – *Rhamnus californica*
California colubrina – *Colubrina californica*
California copper leaf – *Acalypha californica*
California cottontop – *Digitaria californica*
California desert bluebells – *Phacelia campanularia*
California Dutchman's pipe – *Aristolochia californica*
California fan palm – *Washingtonia filifera*
California fuchsia – *Epilobium canum*
California goldenrod – *Solidago californica*
California holly – *Heteromeles arbutifolia*
California Indian pink – *Silene californica*
California juniper – *Juniperus californica*
California laurel – *Umbellularia californica*
California-lilac – *Ceanothus*
California lotebush – *Ziziphus parryi*
California man-root – *Marah fabaceus*
California nutmeg – *Torreya californica*
California pepper tree – *Schinus molle*
California polypody – *Polypodium californicum*
California poppy – *Eschscholzia californica*
California, pride of – *Lathyrus splendens*
California privet – *Ligustrum ovalifolium*
California red fir – *Abies magnifica* var *shastensis*
California rock cress – *Arabis blepharophylla*
California rose – *Rosa californica*
California rose-bay – *Rhododendron macrophyllum*
California sagebrush – *Artemisia californica*
California saltbush – *Atriplex californica*
California slippery elm – *Fremontodendron californicum*
California sweet bay – *Myrica california*
California sycamore – *Platanus racemosa*
California three-awn – *Aristida californica* var *californica*
California white oak – *Quercus lobata*
California white sage – *Salvia apiana*
California wild grape – *Vitis californica*
California wilding – *Satureja douglasii*
California wood fern – *Dryopteris arguta*
calla, black – *Arum palestinum*
calla, common – *Zantedeschia aethiopica*
calla lily – *Zantedeshia aethiopica*
callawalla – *Nephrolepis biserrata*
calliopsis – *Coreopsis tinctoria*
caltrop – *Tribulus terrestris*
caltrop, California – *Kallstroemia californica*
caltrop, orange – *Kallstroemia grandiflora*
calypso, bulbous – *Calypso bulbosa*
camas – *Camassia quamash*
camas, death – *Zigadenus*
camass – *Camassia quamash*
camellia – *Camellia japonica*
camel thorn – *Alhagi pseudalhagi*
campeachy wood – *Haematoxylum campechianum*
camphire – *Reseda odorata*
camphor thyme – *Thymus camphoratus*
camphor tree – *Cinnamomum camphora*
camphor weed – *Heterotheca psammophila*
campion – *Silene*
campion, bladder – *Silene inflata, S. vulgaris*
campion, broad-leaved – *Silene latifolia*
campion, Jupiter's – *Lychnis flos-jovis*
campion, Mexican – *Silene laciniata*
campion, moss – *Silene acaulis*
campion, night-flowering – *Silene noctiflora*
campion, red – *Silene dioica*
campion, rose – *Lychnis coronaria*
campo pea – *Lathyrus splendens*
camwood – *Haematoxylum campechianum*
Canada goldenrod – *Solidago canadensis* ssp *elongata*
Canada hemlock – *Tsuga canadensis*
Canada rice – *Zizania aquatica*
Canada thistle – *Cirsium arvense*
Canadian bluegrass – *Poa compressa*
Canadian fleabane – *Conyza Canadensis*
Canadian wildrye – *Elymus canadensis*
canaigre – *Rumex hymenosepalus*
canary bird bush – *Crotalaria*
canary grass – *Phalaris canariensis*
canary grass, Carolina – *Phalaris caroliniana*
canary grass, littleseed – *Phalaris minor*
canary grass, perennial – *Phalaris tuberosa*
canary grass, reed – *Phalaris arundinacea*
Canary Island broom – *Genista canariensis*
Canary Island date palm – *Phoenix canariensis*

Canary Island pine – *Pinus canariensis*
Canary Island juniper – *Juniperus cedrus*
canary wood – *Centrolobium platinense*
cancer root – *Orobanche cooperi*
cancer-root, Mexican – *Conopholis mexicana*
canchalagua – *Centaurium venustum*
candelabra – *Pachycereus weberi*
candelabra bush – *Senna alata*
candelabra cactus – *Isolatocereus dumortieri, Myrtillocactus geometrizans*
candelilla – *Euphorbia antisyphilitica*
candida – *Candida albicans*
candle, blue – *Myrtillocactus geometrizans*
candle bush – *Senna alata*
candle delphinium – *Delphinium elatum*
candle, desert – *Dasylirion, Eremurus robustus*
candle larkspur – *Delphinium elatum*
candlenut tree – *Aleurites moluccana*
candle tree – *Parmentiera cerifera, Sapium sebiferum, Senna alata*
candle tree, food – *Parmentiera edulis*
candlewood – *Dracaena americana*
candytuft – *Iberis intermedia*
candytuft, evergreen – *Iberis sempervirens*
candytuft, Gibralter – *Iberis gibraltarica*
candytuft, globe – *Iberis umbellata*
candytuft, hyacinth-flowered – *Iberis amara*
candytuft, perennial – *Iberis sempervirens*
candytuft, rocket – *Iberis amara*
candy tulip – *Tulipa clusiana*
cane – *Arundinaria, Phragmites australis*
cane bluestem – *Bothriochloa barbinodis*
cane cholla – *Cylindropuntia californica, C. spinosior*
cane, dumb – *Dieffenbachia*
cane, Japanese – *Saccharum sinense*
cane, sour – *Costus spicatus*
cane, sugar – *Saccharum officinarum*
cane, wild – *Arthrostylidium racemiflora, Arundinaria longifolia, Gynerium sagittatum*
canistel – *Lucuma rivicoa*
cankerroot – *Coptis groenlandica*
canna – *Canna generalis, C. indica*
cannabis – *Cannabis sativa*
cannonball tree – *Couroupita guianensis, Hura polyandra*
canoe birch – *Betula papyrifera*
canoe cedar – *Thuja plicata*
canola – *Brassica napus* var *oleifera, Brassica rapa* var *campestris, B. r.* var *oleifera*
cantala maguey – *Agave cantala*
cantaloupe – *Cucumis melo* var *reticulatus*
Canterbury bell – *Campanula medium*
canyon grape – *Vitis arizonica*
canyon live oak – *Quercus chrysolepis*
canyon oak – *Quercus chrysolepis, Q. glaucoides*
canyon ragweed – *Ambrosia ambrosioides*
cap, bishop's – *Astrophytum capricorne, A. myriostigma*
cap, black – *Rubus occidentalis*
Cape bramble – *Rubus rosaefolius*
Cape cowslip – *Lachenalia aloides*
Cape forget-me-not – *Anchusa capensis*
Cape gooseberry – *Physalis peruviana*
Cape honeysuckle – *Tecoma capensis*
Cape ivy – *Delairea odorata*
Cape jasmine – *Gardenia augusta, G. florida*
Cape-marigold – *Dimorphotheca sinuata*
Cape oak – *Quercus brandegeei, Q. devia*
Cape plumbago – *Plumbago auriculata*
Cape primrose – *Streptocarpus hybridus*
caper – *Capparis spinosa*
caper, bean – *Zygophyllum fabago*
caper bush – *Capparis atamisquea, C. spinosa*
caper spurge – *Euphorbia lathyris*
caper tree – *Capparis flexuosa*
cap, ink – *Coprinus*
cap, Mexican liberty – *Psilocybe mexicana*
cap, skull – *Scutellaria*
cap, Turk's – *Malvaviscus arboreus, M. grandiflorus*
capulin black cherry – *Prunus serotina*
capulin cherry – *Prunus salicifolia, P. serotina*
carambola – *Averrhoa carambola*
caranday palm – *Copernicia australis*
caraway – *Carum carvi*
caraway, wild – *Perideridia gairdneri*
cardamom – *Elettaria cardamomum*
cardboard palm – *Zamia furfuracea*
cardinal climber – *Ipomoea quamoclit*
cardinal flower – *Lobelia cardinalis* var *pseudosplendens*
cardinal sage – *Salvia coccinea, S. fulgens, S. splendens*
cardinal salvia – *Salvia gesneriaeflora*
cardinal's hat – *Astrophytum myriostigma*

cardinal vine – *Quamoclit coccinea*
cardoon – *Cynara cardunculus*
careless weed – *Amaranthus palmeri, A. watsonii*
Caribbean copper plant – *Euphorbia cotinifolia*
Caribbean pine, Honduras – *Pinus caribaea* var *hondurensis*
Caribbee royal palm – *Roystonea oleracea*
carib grass – *Eriochloa polystachya*
carline thistle – *Carlina vulgaris*
carline thistle, gum-bearing – *Atractylis gummifera*
carlowrightia, Arizona – *Carlowrightia arizonica*
carnation – *Dianthus caryophyllus*
carob tree – *Ceratonia siliqua*
Carolina allspice – *Calycanthus floridus*
Carolina bean – *Phaseolus lunatus*
Carolina canary grass – *Phalaris caroliniana*
Carolina geranium – *Geranium carolinianum*
Carolina hemlock – *Tsuga caroliniana*
Carolina jessamine – *Gelsemium sempervirens*
Carolina poplar – *Populus canadensis, P. deltoides*
carpet, beach – *Philoxerus vermicularis*
carpet bugle – *Ajuga pyramidalis, A. reptans*
carpet grass – *Axonopus affinis, A. compressus, Paspalum compressum*
carpet, green – *Herniaria glabra*
carpetweed – *Mollugo verticillata*
carpetweed, California – *Kallstroemia californica*
carpetweed, warty – *Kallstroemia parviflora*
carrageen – *Chondrus crispus*
carriere hawthorn – *Crataegus* x *lavallei*
carrion flower – *Stapelia gigantea*
carrizo grass – *Phragmites australis*
carrot – *Daucus carota*
carrot, death – *Thapsia garganica*
carrot, garden – *Daucus carota* ssp *sativus*
carrot-leaf – *Lomatium dissectum*
carrot, Peruvian – *Arracacia xanthorrhiza*
carrot, wild – *Daucus pusillus, Lomatium dissectum*
casava – *Cucumis melo* var *indorus*
cascade fir – *Abies amabilis*
cascade palm – *Chamaedorea cataractarum*
cascade petunia – *Petunia* x *hybrida* var *pendula*
cascalote – *Caesalpinia cacalaco*
cascara – *Rhamnus purshiana*
cascara amarga – *Sweetia panamensis*
cascara sagrada – *Rhamnus purshiana*
cascarilla bark – *Croton eluteria*
cashew – *Anacardium occidentale*
Cashmere boquet – *Clerodendrum bungei*
cassava – *Manihot esculenta*
cassava, Arizona – *Manihot davisiae*
cassava, narrow-leaved – *Manihot angustiloba*
cassava, sweet – *Manihot dulcis*
casse – *Cassia fistula*
cassia – *Cassia, Cinnamomum cassia, Senna*
cassia, Coues' – *Senna covesii*
cassia, poet's – *Osyris alba*
cassia, purging – *Cassia fistula*
cassia tree, Chinese – *Cinnamomum cassia*
cassie – *Acacia farnesiana*
castela, allthorn – *Castela texana*
castilleja – *Castilleja*
castilloa rubber – *Castilla elastica*
cast-iron plant – *Aspidistra elatior*
castor aralia – *Kalopanax septemlobus*
castor bean – *Ricinus communis*
Catalina cherry – *Prunus ilicifolia* ssp *lyonii*
catalpa, common – *Catalpa bignonioides*
catalpa, desert – *Chilopsis linearis* ssp *arcuata*
catchfly – *Silene*
catchfly, forked – *Silene dichotoma*
catchfly gentian – *Eustoma exaltatum, E. grandiflorum*
catchfly, moss – *Silene acaulis*
catchfly, small-flowered – *Silene gallica*
catchfly, sticky – *Lychnis viscaria*
catchfly, sweet-William – *Silene armeria*
catchweed bedstraw – *Galium aparine*
catclaw – *Acacia rigidula, Mimosa aculeaticarpa*
catclaw acacia – *Acacia greggii*
catclaw mimosa – *Mimosa biuncifera*
catclaw, Texas – *Acacia wrightii*
catclaw, thornless – *Acacia berlandieri*
catclaw, tree – *Acacia occidentalis*
caterpillar, orange – *Combretum farinosum*
cathedral bells – *Cobaea scandens*
catmint – *Nepeta cataria, N.* x *faassenii*
catmint, large-flowered – *Nepeta grandiflora*
catnip – *Nepeta cataria*
cat palm – *Chamaedorea cataractarum*

cat's claw – *Macfadyena unguis-cati, Uncaria tomentosa*
cat's-ear, rough – *Hypochaeris radicata*
cat's-ear, smooth – *Hypochaeris glabra*
cat's-tail, red – *Acalypha hispida*
cattail – *Acalypha arvensis, Typha*
cattail, broad-leaved – *Typha latifolia*
cattail grass – *Setaria pumila*
cattail, narrow-leaved – *Typha angustifolia*
cattail, southern – *Typha domingensis*
cat thyme – *Teucrium marum*
cattle spinach – *Atriplex polycarpa*
cattleya, Skinner's – *Cattleya skinneri*
Caucasian stonecrop – *Sedum spurium*
cauliflower – *Brassica oleracea* var *botrytis*
Cavendish banana, dwarf – *Mus acuminata*
cayenne – *Capsicum fastigatum*
Cayenne cherry – *Eugenia uniflora*
ceanothus, desert – *Ceanothus greggii*
ceanothus, greenbark – *Ceanothus spinosus*
ceanothus, wart-stemmed – *Ceanothus verrucosus*
ceara rubber – *Manihot glaziovii*
cedar – *Cupressus*
cedar, Alaska – *Cupressus nootkatensis*
cedar, Atlantic – *Cedrus atlantica*
cedar, bastard – *Guazuma ulmifolia*
cedar, bay – *Suriana maritima*
cedar, Chilean – *Austrocedrus chilensis*
cedar, canoe – *Thuja plicata*
cedar, Cyprian – *Cedrus brevifolia*
cedar, Cyprus – *Cedrus brevifolia*
cedar, deerhorn – *Thujopsis dolabrata*
cedar, deodar – *Cedrus deodara*
cedar, eastern red – *Juniperus virginiana*
cedar elm – *Ulmus crassifolia*
cedar, hiba – *Thujopsis dolabrata*
cedar, incense – *Calocedrus decurrens*
cedar, Lawson – *Cupressus lawsoniana*
cedar, northern white – *Thuja occidentalis*
cedar of Lebanon – *Cedrus libani*
cedar, pygmy – *Peucephyllum schottii*
cedar, Port Orford – *Cupressus lawsoniana*
cedar, Portuguese – *Cedrus lusitanica*
cedar, pygmy – *Peucephyllum schottii*
cedar, red – *Cedrela odorata, C. toona, Trichilia hirta*
cedar, royal – *Cedrela mexicana*
cedar, salt – *Tamarix*
cedar, Spanish – *Cedrela odorata*
cedar, western red – *Thuja plicata*
cedar, white – *Chamaecyparis thyoides*
cedar, wild bay – *Trema micrantha*
cedrat – *Citrus medica*
cedrella, Mexican – *Cedrela mexicana*
celandine – *Chelidonium majus*
celandine, lesser – *Ranunculus ficaria*
celandine, tree – *Bocconia arborea*
celeriac – *Apium graveolens*
celery – *Apium graveolens*
celery, button – *Eryngium carlinae*
celery-leaved crowfoot – *Ranunculus sceleratus*
cenizo – *Leucophyllum frutescens, L. minus, L. violaceum*
centaury – *Centaurium*
centaury, American – *Sabatia angularis*
centaury, marsh – *Centaurium calycosum*
century plant – *Agave americana, A. asperrima, A. palmeri*
cèpe – *Boletus edulis*
cereus, blue – *Cereus hexagonus*
cereus, dahlia-rooted – *Peniocereus striatus*
cereus, desert night-blooming – *Peniocereus greggii* var *transmontanus*
cereus, golden-spined – *Bergerocactus emoryi*
cereus, Mexican night-blooming – *Peniocereus serpentinus*
cereus, night-blooming – *Hylocereus trigonus, H. undatus, Selenicereus hamatus*
cereus, sun – *Disocactus speciosus*
ceriman – *Monstera deliciosa*
cerro hawthorn – *Crataegus erythropoda*
cespitose globe daisy – *Globularia alypium*
cestrum, early – *Cestrum fasciculatum*
cestrum, red – *Cestrum elegans*
Ceylon gooseberry – *Dovyalis hebecarpa*
Ceylon spinach – *Basella alba*
chaconia – *Warszewiczia coccinea*
chaff-flower – *Alternanthera caracasana*
chaff-flower, prickly – *Achyranthes aspera*
chain-fruit cholla – *Cylindropuntia fulgida*
chain-link cholla – *Cylindropuntia cholla*
chain of love – *Antigonon leptopus*
chain speedwell – *Veronica catenata*
chain tree, golden – *Laburnum anagyroides, Senna alata*
Chaix's bluegrass – *Poa chaixii*

chalicevine – *Solandra guerrerensis*
chalk dudleya – *Dudleya pulverulenta*
chalk-lettuce – *Dudleya pulverulenta*
chamaedor palm – *Chamaedora ernesti*
chamisa – *Chrysothamnus nauseosus*
chamise – *Adenostoma fasciculatum*
chamomile – *Chamaemelum nobile, Anthemis cotula*
chamomile, corn – *Anthemis arvensis*
chamomile, dyer's – *Anthemis tinctoria*
chamomile, English – *Matricaria recutita*
chamomile, false – *Chamomilla courrantiana*
chamomile, German – *Chamomilla suaveolens*
chamomile, rayless – *Chamomilla suaveolens*
chamomile, Roman – *Chamaemelum nobile*
chamomile, Russian – *Chamaemelum nobile*
chamomile, wild – *Matricaria recutita*
changing forget-me-not – *Myosotis discolor*
chanterelle – *Cantharellus, C. cibarius*
chaparral – *Larrea divericata*
chaparral ash – *Fraxinus trifoliata*
chaparral mallow – *Malacothamnus fasciculatus*
chaparral prickly-pear – *Opuntia oricola*
chaparral whitethorn – *Ceanothus leucodermis*
chaparral yucca – *Yucca whipplei*
chaparro oak – *Quercus reticulata*
charcoal burner – *Russula cyanoxantha*
charcoal shrub – *Baccharis conferta*
chard – *Beta vulgaris* ssp *cicla*
chard, Swiss – *Beta vulgaris* ssp *cicla*
charlock – *Brassica kaber, Sinapis arvensis*
charlock, jointed – *Raphanus raphanistrum*
chaste tree – *Vitex agnus-castus, V. mollis*
chate of Egypt – *Cucumis melo* var *chito*
chayote – *Sechium edule*
cheat grass – *Bromus tectorum*
checker – *Sidalcea malvaeflora, S. neomexicana*
checkerbark madrone – *Arbutus tessellata*
checkerberry – *Gaultheria procumbens*
checkerbloom – *Sidalcea malviflora, S. neomexicana*
checkered lily – *Fritillaria meleagris*
checker mallow – *Sidalcea malviflora*
cheddar pink – *Dianthus gratianopolitanus*
cheeses – *Malva neglecta*
cheeseweed – *Malva parviflora*
chenille plant – *Acalypha hispida*
cherimoya – *Annona cherimola*
cherimoya-of-the-lowlands – *Annona diversifolia*
cherry – *Prunus avium*
cherry, acerola – *Malpighia glabra*
cherry, Amur choke – *Prunus maackii*
cherry, Barbados – *Malpighia emarginata, M. glabra*
cherry, Bigarreau – *Prunus avium* var *duracina*
cherry birch – *Betula lenta*
cherry, bitter – *Prunus emarginata*
cherry, black – *Prunus*
cherry, Brazil – *Eugenia uniflora*
cherry, capulin – *Prunus salicifolia, P. serotina*
cherry, capulin black – *Prunus serotina*
cherry, Catalina – *Prunus ilicifolia* ssp *lyonii*
cherry, Cayenne – *Eugenia uniflora*
cherry, Chinese bush – *Prunus glandulosa*
cherry, choke – *Prunus virginiana* var *demisa*
cherry, Christmas – *Solanum capsicastrum*
cherry, Cornelian – *Cornus mas*
cherry, desert ground – *Physalis crassifolia* var *versicolor*
cherry, dwarf – *Prunus fruticans, P. fruticosa*
cherry, European bird – *Prunus padus*
cherry, European dwarf – *Prunus fruticosa*
cherry, Florida – *Eugenia uniflora*
cherry, flowering – *Prunus sargentii, P. serrula*
cherry, gean – *Prunus avium* var *juliana*
cherry, ground – *Physalis*
cherry, holly-leaved – *Prunus ilicifolia*
cherry, Jamaica – *Muntingia calabura*
cherry, Japanese – *Prunus serrulata*
cherry, Japanese flowering – *Prunus subhirtella*
cherry, Jerusalem – *Solanum pseudocapsicum*
cherry, laurel – *Prunus laurocerasus*
cherry, mahleb – *Prunus mahaleb*
cherry, Mexican – *Prunus serotina, Tabebuia pentaphylla*
cherry, Mexican choke – *Prunus capuli*
cherry palm – *Pseudophoenix sargentii*
cherry pie – *Lantana camara*
cherry, pin – *Prunus pensylvanica*
cherry plum – *Prunus cerasifera*
cherry, shortclustered – *Prunus brachybotrya*
cherry, sand – *Prunus pumila*
cherry, sour – *Prunus*

cherry, southwestern choke – *Prunus serotina* ssp *virens*
cherry, St. Lucie – *Prunus mahleb*
cherry, strawberry ground – *Physalis alkekengi*
cherry, Surinam – *Eugenia uniflora*
cherry, sweet – *Prunus avium*
cherry tomato – *Lycopersicon esculentum* var *cerasiforme*
cherry, western choke – *Prunus virginiana* var *demissa*
cherry, West Indian – *Malpighia emarginata, M. glabra*
cherry, wild – *Prunus capuli*
cherry, wild black – *Prunus serotina*
cherry, winter – *Physalis alkekengi, Solanum capsicastrum*
cherry, yoshino flowering – *Prunus* x *yedoensis*
chervil – *Anthriscus cerefolium*
chervil, bur – *Anthriscus caucalis*
chervil root – *Chaerophyllium tuberosum*
chess, foxtail – *Bromus madritensis*
chestnut – *Castanea sativa*
chestnut dion – *Dioon edule*
chestnut, dwarf – *Castanea pumila*
chestnut, Guiana – *Pachira aquatica*
chestnut, horse – *Aesculus hippocastanum*
chestnut oak – *Quercus castanea*
chestnut oak, yellow – *Quercus muehlenbergii*
chestnut, Polynesian – *Inocarpus edulis*
chestnut, red horse – *Aesculus carnea*
chestnut rose – *Rosa roxburghii*
chestnut, Spanish – *Castanea sativa*
chestnut tree, American – *Castanea dentata*
chestnut tree, Japanese horse – *Aesculus urbinata*
chestnut tree, Moreton Bay – *Castanospermum australe*
chestnut, water – *Eleocharis dulcis, Trapa natans*
chestnut, water, of the West Indies – *Calathea lutea*
chia – *Salvia columbariae, S. hispanica*
Chiapas white pine – *Pinus strobus* var *chiapensis*
chicalote – *Argemone munita*
chickabiddy – *Asarina barclaiana*
chicken thief – *Mentzelia oligosperma*
chick-pea – *Cicer arietinum, Lathyrus sativus*
chickweed – *Cerastium, Moenchia erecta, Stellaria*
chickweed, common – *Stellaria media*
chickweed, field – *Cerastium arvense*
chickweed, Indian – *Mollugo verticillata*
chickweed, jagged – *Holosteum unbellatum*
chickweed, lesser – *Stellaria pallida*
chickweed, mouse-ear – *Cerastium fontanum, C. glomeratum*
chickweed, water – *Montia fontana*
chickweed, wood – *Stellaria nemorum*
chicle – *Achras zapota*
chicory – *Cichorium intybus*
chigger nuts – *Miconia*
Chihuahua ash – *Fraxinus papillosa*
Chihuahua oak – *Quercus chihuahuensis*
Chihuahua pine – *Pinus leiophylla* var *chihuahuana*
Chihuahua pine, five-needled – *Pinus leiophylla*
Chihuahua spruce – *Picea chihuahuana*
childing pink – *Petrorhagia nanteuilii*
chile – *Capsicum annuum*
chile, Anaheim – *Capsicum annuum*
Chilean bellflower – *Lapageria rosea*
Chilean cedar – *Austrocedrus chilensis*
chile, ancho – *Capsicum annuum*
Chilean cyclamen – *Cyclamen cilicum*
Chilean fire bush – *Embothrium coccineum*
Chilean jasmine – *Mandevilla laxa*
Chilean mesquite – *Prosopis chilensis*
Chilean tarweed – *Madia sativa*
Chilean totara – *Podocarpus nubigenus*
Chilean wine palm – *Jubaea chilensis*
chile, dried – *Capsicum annuum*
chile, fresno – *Capsicum annuum*
chile, habanero – *Capsicum chinense*
chile, jalapeno – *Capsicum annuum*
chile, New Mexico – *Capsicum annuum*
chile, poblano – *Capsicum annuum*
chile, serrano – *Capsicum annuum*
chile, tree – *Capsicum annuum*
chile, water – *Capsicum annuum*
chile, wax – *Capsicum annuum*
Chile wine palm – *Jubaea spectabilis*
chilte rubber – *Cnidoscolus elasticus*
China aster – *Callistephus chinensis*
China banana – *Mus cavendishii*

China berry – *Melia azedarach* var *umbraculiformis*
China fir – *Cunninghamia lanceolata*
Chinaroot – *Smilax china*
China rose – *Rosa chinensis*
chinchweed – *Pectis papposa* var *papposa*
Chinese angelica – *Angelica sinensis*
Chinese anise – *Illicium anisatum*
Chinese arbor-vitae – *Thuja orientalis*
Chinese artichoke – *Stachys sieboldii, S. tuberifera*
Chinese banana – *Mus acuminata*
Chinese banyan – *Ficus microcarpa*
Chinese banyan, weeping – *Ficus benjamina*
Chinese bitter cucumber – *Momordica cochinchinensis*
Chinese box thorn – *Lycium chinense*
Chinese broccoli – *Brassica oleracea* var *alboglabra*
Chinese bush cherry – *Prunus glandulosa*
Chinese cabbage – *Brassica rapa* var *pekinensis*
Chinese cassia tree – *Cinnamomum cassia*
Chinese chives – *Allium tuberosum*
Chinese crabapple – *Malus spectabilis*
Chinese cucumber – *Trichosanthes kirilowii*
Chinese elm – *Ulmus parvifolia*
Chinese ephedra – *Ephedra sinica*
Chinese evergreen – *Aglaonema modestum*
Chinese evergreen elm – *Ulmus parvifolia*
Chinese fan palm – *Livistona chinensis*
Chinese flame tree – *Koelreuteria bipinnata*
Chinese fountain palm – *Livistona chinensis*
Chinese ginseng – *Panax ginseng*
Chinese hibiscus – *Hibiscus rosa-sinensis*
Chinese honeysuckle – *Quisqualis indica*
Chinese indigo – *Polygonum tinctorium*
Chinese kale – *Brassica oleracea* var *alboglabra*
Chinese lantern – *Abutilon*
Chinese lantern plant – *Physalis alkekengi*
Chinese lemon – *Citrus limonia*
Chinese lotus – *Nelumbo nucifera*
Chinese meadow-rue – *Thalictrum dipterocarpum*
Chinese olive – *Canarium album*
Chinese parasol tree – *Firmiana simplex*
Chinese parsley – *Coriandrum sativum*
Chinese pea – *Pisum sativum* var *macrocarpon*
Chinese photinia – *Photinia serratifolia*
Chinese pieris – *Pieris forrestii*
Chinese pink – *Dianthus chinensis*
Chinese primrose – *Primula sinensis*
Chinese pusley – *Heliotropium curassavicum*
Chinese quince – *Chaenomeles sinensis*
Chinese rhubarb – *Rheum officinale*
Chinese rose – *Hibiscus rosa-sinensis*
Chinese scholar tree – *Sophora japonica*
Chinese sugarcane – *Saccharum sinense*
Chinese sweet gum – *Liquidambar formosana*
Chinese tallow tree – *Sapium sebiferum*
Chinese trumpet creeper – *Campsis grandiflora*
Chinese water spinach – *Ipomoea aquatica*
Chinese wingnut – *Pterocarya stenoptera*
Chinese wisteria – *Wisteria sinensis*
Chinese witch hazel – *Hamamelis mollis*
Chinese woolflower – *Celosia argentea*
chinkapin oak – *Quercus muehlenbergii*
chinquapin – *Castanea pumila*
chinquapin, giant – *Chrysolepis chrysophylla*
chinquapin oak – *Quercus muehlenbergii*
chironja – *Citrus sinensis* x *C. paradisi*
chir pine – *Pinus roxburghii*
Chisos red oak – *Quercus gravesii*
chittam bark – *Rhamnus purshiana*
chittam wood – *Sideroxylon lanuginosum*
chittim – *Rhamnus purshiana*
chittimwood – *Sideroxylon lanuginosum*
chive – *Allium schoenoprasum*
chives – *Allium schoenoprasum*
chives, Chinese – *Allium tuberosum*
chives, garlic – *Allium tuberosum*
chloris, buryseed – *Chloris chloridea*
chocolate – *Theobroma cacao*
chocolate ancho pepper – *Capsicum annuum*
chocolate flower – *Berlandiera lyrata*
chocolate lily – *Fritillaria biflora*
chokecherry – *Prunus virginiana* var *demissa*
chokecherry, Amur – *Prunus maackii*
chokecherry, gila – *Prunus serotina* var *rufula*
chokecherry, Mexican – *Prunus capuli*
chokecherry, southwestern – *Prunus serotina* ssp *virens*
chokecherry, western – *Prunus virginiana* var *demissa*
cholla, buckhorn – *Cylindropuntia acanthocarpa*
cholla, California – *Cylindropuntia californica*

cholla, cane – *Cylindropuntia californica, C. spinosior*
cholla, chain-fruit – *Cylindropuntia fulgida*
cholla, chain-link – *Cylindropuntia cholla*
cholla, dagger – *Grusonia invicta*
cholla, deerhorn – *Cylindropuntia versicolor*
cholla, diamond – *Cylindropuntia ramosissima*
cholla, dog – *Grusonia schottii*
cholla, handlegrip – *Cylindropuntia spinosior*
cholla, jumping – *Cylindropuntia bigelovii, C. fulgida*
cholla, Klein's – *Cylindropuntia kleiniae*
cholla, Klein's pencil – *Cylindropuntia kleiniae*
cholla, pencil – *Cylindropuntia arbuscula, C. tesajo*
cholla, Schott's dwarf – *Grusonia schottii*
cholla, sheathed – *Cylindropuntia tunicata*
cholla, snake – *Cylindropuntia californica*
cholla, Sonoran jumping – *Cylindropuntia fulgida* var *mamillata*
cholla, staghorn – *Cylindropuntia versicolor*
cholla, teddy-bear – *Cylindropuntia bigelovii*
cholla, thistle – *Cylindropuntia tunicata*
cholla, tree – *Cylindropuntia imbricata*
cholla, walkingstick – *Cylindropuntia imbricata*
chop suey greens – *Chrysanthemum coronarium*
Christmas berry – *Heteromeles arbutifolia*
Christmas cactus – *Schlumbergera buckleyi*
Christmas cactus, desert – *Cylindropuntia leptocaulis*
Christmas cherry – *Solanum capsicastrum*
Christmas holly – *Ilex aquifolium*
Christmas rose – *Helleborus niger*
Christmas tree, New Zealand – *Metrosideros excelsa*
Christopher, herb – *Actaea spicata*
Christ's thorn – *Gleditsia triacanthos, Ziziphus spina-christi*
Christ thorn – *Euphorbia milii, Paliurus spina-christi*
chrysanthemum – *Chrysanthemum*
chrysanthemum, corn – *Chrysanthemum segetum*
chrysanthemum, florist's – *Chrysanthemum morifolium*
chrysanthemum, summer – *Chrysanthemum carinatum, C. coronarium*
chrysanthemum, tricolor – *Chrysanthemum carinatum*
chuckwalla's delight – *Bebbia juncea*
chufa – *Cyperus esculentus*
chuparosa – *Anisacanthus thurberi, Justica californica*
chysis, smooth-lipped – *Chysis bractescens*
cicely, sweet – *Myrrhis odorata, Osmorhiza brachypoda, O. longistylis*
cider gum – *Eucalyptus gunnii*
cigarbox tree – *Cedrela mexicana*
cigarbox wood – *Cedrela odorata*
cigar flower – *Cuphea ignea*
cigar plant – *Cuphea ignea*
cilantro – *Coriandrum sativum*
cinchona – *Cinchona ledgeriana*
cinchona, red – *Cinchona pitayensis*
cinchona, redbark – *Cinchona succirubra*
cinchweed – *Dyssodia papposa, Pectis*
cineraria – *Senecio cineraria*
cineraria, florist's – *Senecio hybridus*
cinnabar ladies' tresses – *Spiranthes cinnabarina*
cinnamon – *Cinnamomum zeylanicum*
cinnamon rose – *Rosa cinnamomea*
cinquefoil – *Potentilla*
cinquefoil, bush – *Potentilla fruticosa*
cinquefoil, golden – *Potentilla aurea*
cinquefoil, lustrous – *Potentilla nitida*
cinquefoil, Norwegian – *Potentilla norvegica*
cinquefoil, Pennsylvanian – *Potentilla pensylvanica*
cinquefoil, shrubby – *Potentilla fruticosa*
cinquefoil, spring – *Potentilla tabernaemontanii*
cinquefoil, sulphur – *Potentilla recta*
cistus, gum – *Cistus ladanifer*
citron – *Citrus medica*
citronella grass – *Cymbopogon citratus, C. nardus*
citronelle – *Citrus jambhiri*
citronwood – *Tetraclinis articulata*
citrus fruit – *Citrus*
clammy locust – *Robinia viscosa*
clammy plumbago – *Plumbago scandens*
clammy weed – *Polanisia uniglandulosa*
clammy weed, large-flowered – *Polanisia uniglandulosa*
clap-weed – *Ephedra antisyphilitica*

claret cup – *Echinocereus triglochidiatus*
claret-cup cactus – *Echinocereus polyacanthus* ssp *pacificus*
clarkia – *Clarkia*
clary, annual – *Salvia viridis*
clary, meadow – *Salvia pratensis*
clary sage – *Salvia sclarea*
clary, whorled – *Salvia verticillata*
clary, wild – *Salvia verbenacea*
claw cactus, Whipple's – *Sclerocactus whipplei*
claw, cat – *Mimosa aculeaticarpa*
claw, cat's – *Macfadyena unguis-cati, Uncaria tomentosa*
claw, devil's – *Acacia greggii, Harpagophytum procumbens, Proboscidea*
claw, lobster – *Heliconia*
claws, devil's – *Martynia annua*
claws, owl – *Dugaldia hoopesii*
clearweed – *Pilea cadierei, P. microphylla*
cleavers – *Galium aparine*
cleaverwort – *Galium aparine*
clematis, alpine – *Clematis alpina*
clematis, anemone – *Clematis montana*
clematis, evergreen – *Clematis armandii*
clematis, fragrant – *Clematis flammula*
clematis, upright – *Clematis recta*
clementine – *Citrus reticulata*
clementine tree – *Citrus clementine*
cliff-brake – *Pellaea*
cliff fig – *Ficus palmeri, F. petiolaris*
cliff maids – *Lewisia cotyledon*
cliff rose – *Purshia mexicana*
cliff spurge – *Euphorbia californica, E. misera, E. xantii*
climbing dahlia – *Hidalgoa ternata*
climbing hydrangea – *Hydrangea anomala* ssp *petiolaris*
climbing lily – *Gloriosa superba*
climbing milkweed – *Funastrum heterophyllum, Sarcostemma cynanchoides* ssp *hartwegii*
climbing plumbago – *Plumbago auriculata*
climbing rose – *Rosa multiflora*
climbing snapdragon – *Asarina barclaiana*
cloak fern – *Notholaena lemmonii, N. sinuata*
cloak fern, scaly – *Astrolepis cochisensis*
clockvine – *Thunbergia alata*
clockvine, Bengal – *Thunbergia grandiflora*
cloth of gold crocus – *Crocus angustifolius*
cloudberry – *Rubus chamaemorus*
cloud ear fungus – *Auricularia auricula-judae*
clove – *Caryophyllus aromaticus*
clove pink – *Dianthus caryophyllus*
clover – *Trifolium*
clover, alsike – *Trifolium hybridum*
clover, bear's – *Chamaebatia foliolosa*
clover, berseem – *Trifolium alexandrinum*
clover, bur – *Medicago*
clover, bush – *Lespedeza striata*
clover, clustered – *Trifolium glomeratum*
clover, crimson – *Trifolium incarnatum*
clover, Dutch white – *Trifolium repens*
clover, Egyptian – *Trifolium alexandrinum*
clover, hop – *Trifolium aureum, T. campestre, T. dubium, T. procumbens*
clover, Hungarian – *Trifolium pannonicum*
clover, jackass – *Wislizenia refracta*
clover, maiden – *Trifolium microcephalum*
clover, mammoth – *Trifolium medium*
clover, musk – *Erodium moschatum*
clover, narrow-leaved – *Trifolium angustifolium*
cloveroot – *Geum urbanum*
clover, owl's – *Castilleja*
clover, Persian – *Trifolium resupitanum*
clover, pin – *Erodium cicutarium*
clover, prairie – *Dalea tomentosa*
clover, red – *Trifolium pratense*
clover, rose – *Trifolium hirtum*
clover, sour – *Melilotus indica*
clover, subterranean – *Trifolium subterraneum*
clover, strawberry – *Trifolium fragiferum, T. macrocephalum*
clover, sweet – *Melilotus*
clover, tick – *Desmodium cinerascens*
clover, white – *Trifolium repens*
clove, wild – *Ludwigia octovalvis*
club gourd – *Trichosanthes cucumerina* var *anguina*
club-moss – *Lycopodium clavatum*
club palm, spiny – *Bactris mexicana*
club-rush, bristly – *Scirpus setaceous*
club-rush, drooping – *Scirpus cernuus*
club-rush, maritime – *Scirpus maritimus*
club-rush, rough – *Scirpus tabernaemontani*
club-rush, sea – *Scirpus maritimus*
club sedge – *Carex buxbaumii*
clusia – *Hintonia latiflora*

clustered clover – *Trifolium glomeratum*
clustered fish-hook cactus – *Mammillaria thornberi*
clustered fishtail palm – *Caryota mitis*
clustered goldflower – *Hymenoxys scaposa*
clustered tricholoma – *Lyophyllum aggregatum*
cluster pine – *Pinus pinaster*
clusters, mauve – *Scaevola plumierii*
coach-whip – *Fouquieria splendens*
coastal agave – *Agave shawii*
coastal live oak – *Quercus virginiana*
coastal scrub oak – *Quercus dumosa*
coastal wood fern – *Dryopteris agruta*
coast barrel cactus – *Ferocactus viridescens*
coast desert thorn – *Lycium californicum*
coast hibiscus – *Hibiscus tilliaceus*
coast live oak – *Quercus agrifolia*
coast redwood – *Sequoia sempervirens*
coast rock cress – *Arabis blepharophylla*
coast saltbush – *Atriplex barclayana*
coast sandbur – *Cenchrus incertus*
coast silk tassel – *Garrya eliptica*
coast silk tassel bush – *Garrya eliptica*
coast tarweed – *Madia sativa*
coat, Jacob's – *Acalypha wilkesiana*
coat, Joseph's – *Amaranthus gangeticus, A. tricolor*
cobweb houseleek – *Sempervivum arachnoideum*
cocaine – *Erythroxylum coca, E. novagranatense, E. truxillense*
cocaine tree, huanuco – *Erythroxylum coca*
cocaine tree, Peruvian – *Erythroxylum novagranatense*
cocaine tree, Trujillo – *Erythroxylum truxillense*
coca plant – *Erythroxylum coca, E. novagranatense, E. truxillense*
cochal – *Myrtillocactus cochal*
cochineal cactus – *Opuntia cochenillifera*
cocklebur – *Arctium lappa, Xanthium canadensis, X. strumarium*
cocklebur, spiny – *Xanthium spinosum*
cockle, corn – *Agrostemma githago*
cockroach plant – *Haplophyton cimicidum*
cockscomb – *Celosia*
cockscomb, crested – *Celosia cristata*
cockscomb hedgehog – *Echinocereus fasciculatus*
cockscomb, plumed – *Celosia plumosa*
cocksfoot – *Dactylis glomerata*
cockspur – *Pisonia aculeata*
cockspur coral tree – *Erythrina crista-galli*
cockspur thorn – *Crataegus crus-galli*
cocoa – *Theobroma cacao*
cocoa, Peru – *Theobroma bicolor*
cocoa-plum – *Chrysobalanus icaco*
cocoa, sonusco – *Theobroma angustifolium*
cocobolo – *Dalbergia granadillo, D. retusa*
coconut palm – *Cocos nucifera*
coco-plum, wild – *Hirtella americana*
cocoyam – *Colocasia antiquorum, C. esculenta*
cocoyam, new – *Xanthosoma sagittifolium*
cocuswood, ebony – *Brya ebenus*
coffee – *Coffea arabica*
coffee, arabica – *Coffea arabica*
coffeeberry – *Karwinskia humboldtiana, Rhamnus*
coffeeberry, California – *Rhamnus californica*
coffee bush – *Simmondsia chinensis*
coffee, inhambane – *Coffee racemosa*
coffee, liberica – *Coffee liberica*
coffee, robusta – *Coffee canephora*
coffee senna – *Senna occidentalis*
coffee, Swedish – *Astragalus boeticus*
coffee tree, Congo – *Coffea canephora*
coffee tree, Kentucky – *Gymnocladus dioica*
coffee tree, Liberian – *Coffea liberica*
coffee weed – *Sesbania herbacea*
coffee, wild – *Casearia sylvestris, Psychotria nervosa, Rinorea*
cohosh, black – *Cimicifuga racemosa*
cohosh, blue – *Caulophyllum thalictroides*
cohune – *Astrocaryum mexicanum*
cohune palm – *Orbignya cohune*
cohune, warree – *Astrocaryum mexicanum*
coleus – *Coleus*
coliseum maple – *Acer cappadocicum*
collard greens – *Brassica oleracea* var *acephala*
collards – *Brassica oleracea* var *viridis*
colocynth – *Citrullus colocynthis* var *lanatus*
colonial bent – *Agrostis capillaris*
Colorado blue spruce – *Picea pungens* var *glauca*
Colorado River hemp – *Sesbania herbacea, S. macrocarpa*

Colorado rubber-weed – *Hymenoxys richardsoni*
Colorado spruce – *Picea pungens*
colored Guinea grass – *Panicum coloratum*
coltsfoot – *Tussilago farfara*
coltsfoot, Japanese – *Petasites japonicus*
colubrina, Californica – *Colubrina californica*
colubrina, Las Animas – *Colubrina californica*
columbine – *Aquilegia*
columbine, alpine – *Aquilegia alpina*
columbine, European – *Aquilegia vulgaris*
columbine, Rocky Mountain – *Aquilegia caerulea*
Columbus grass – *Sorghum almum*
column cactus – *Neobuxbaumia tetetzo*
colza – *Brassica napus* var *arvensis*
comfrey – *Symphytum asperum, S. officinale*
comitan juniper – *Juniperus comitana*
comp – *Asteraceae*
compass barrel cactus – *Ferocactus cylindraceus*
compass cactus – *Ferocactus cylindraceus*
composite – *Asteraceae*
condalia, bitter – *Condalia globosa* var *pubescens*
condalia, bluewood – *Condalia obovata*
condalia, Mexican bluewood – *Condalia mexicana*
condalia, southwestern – *Ziziphus obtusifolia*
cone cactus, Texas – *Neolloydia conoidea*
coneflower – *Ratibida mexicana, Rudbeckia*
coneflower, cutleaf – *Rudbeckia laciniata*
coneflower, handleleaf – *Echinacea angustifolia*
coneflower, purple – *Echinacea angustifolia, E. purpurea*
coneflower, short-rayed – *Ratibida tagetes*
Confederate jasmine – *Trachelospermum jasminoides*
Confederate rose – *Hibiscus mutabilis*
Confederate vine – *Antigonon leptopus*
Congo coffee tree – *Coffea canephora*
cooking banana – *Mus acuminata* x *balbisiana, M. paradisiaca*
coontie – *Zamia pumila*
copaiba – *Copaifera officinalis*
copaiba balsam – *Copaifera coriacea*
copaiba, para – *Copaifera reticulata*
copal, gum – *Copaifera copallifera*
copal, Zanzibar – *Trachylobium verrucosum*
copa tree – *Protium copal*
copper beech – *Fagus sylvatica* var *purpurea*
copper leaf – *Acalypha*
copper leaf, California – *Acalypha californica*
copper-mallow – *Sphaeralcea angustifolia*
copper plant, Caribbean – *Euphorbia cotinifolia*
copperpod tree – *Peltophorum pterocarpum*
copper zephyrlily – *Zephyranthes longifolia*
coracan – *Eleusine corocana*
coral bean – *Sophora secundiflora*
coral bean, southwest – *Erythrina flabelliformis*
coral bean, western – *Erythrina flabelliformis*
coral bells – *Heuchera sanguinea*
coral drops, Mexican – *Bessera elegans*
coral fountain – *Russelia equisetiformes*
coral fungus – *Clavaria, Ramaria flavia*
coral fungus, golden – *Ramaria flava*
coral fungus, red-tipped – *Ramaria botrytis*
coral fungus, upright – *Ramaria stricta*
coral honeysuckle – *Lonicera sempervirens*
coral iris – *Iris foetidissima*
coral lily – *Lilium pumilum*
coralroot, northern – *Corallorhiza trifida*
coral shower – *Cassia grandis*
coral tree – *Erythrina*
coral tree, cockspur – *Erythrina cristi-galli*
coral tree, common – *Erythrina crista-galli*
coral tree, naked – *Erythrina coralloides*
coral-vine – *Antigonon leptopus, Cissus cucurbitina*
cord grass, dense-flowered – *Spartina densiflora*
cord grass, salt-meadow – *Spartina patens*
cord grass, salt-water – *Spartina alterniflora*
cordia – *Cordia*
cordia, littleleaf – *Cordia parvifolia*
cordia, onion – *Cordia alliodora*
cordia, scarlet – *Cordia sebestena*
cordofan pea – *Clitoria terneata*
coreopsis – *Coreopsis grandiflora*
coreopsis, annual – *Coreopsis tinctoria*
coriander – *Coriandrum sativum*
coriander, wild – *Eryngium foetidum*
coriaria, Mediterranean – *Coriaria myrtifolia*
cork oak – *Quercus suber*
cork tree – *Thespesia populnea*

corkwood tree – *Heliocarpus donnell-smithii*
corn – *Zea mays*
corn bedstraw, rough – *Galium tricornutum*
corn, broom – *Sorghum bicolor*
corn buttercup – *Ranunculus arvensis*
corn chamomile – *Anthemis arvensis*
corn chrysanthemum – *Chrysanthemum segetum*
corn cleavers – *Galium tricornutum*
corn-cockle – *Agrostemma githago*
corn, dent – *Zea mays* var *indentata*
Cornelian cherry – *Cornus mas*
cornel, red – *Cornus sanguinea*
corn flag – *Gladiolus italicus*
corn, flint – *Zea mays* var *indurata*
corn, flour – *Zea mays* var *amylacea*
cornflower – *Centaurea cyanus*
cornflower, perennial – *Centaurea montana*
corn, Guinea – *Sorghum bicolor*
corn, Indian – *Zea mays*
Cornish heath – *Erica vagans*
corn, Japanese – *Zea mays* var *japonica*
corn, Jerusalem – *Sorghum vulgare* var *durra*
corn-lily – *Veratrum officinale, V. viride*
corn-lily, African – *Ixia maculata*
corn mint – *Mentha arvensis*
corn-on-the-cob – *Zea mays* var *saccharata*
cornplant – *Dracaena fragrans*
corn, pod – *Zea mays* var *tunicata*
corn, pop – *Zea mays* var *everia*
corn poppy – *Papaver rhoeas*
corn root, sweet – *Calathea allouia*
corn salad – *Valerianella locusta*
corn salad, Italian – *Valerianella eriocarpa*
corn, slender – *Zea mays* var *gracillima*
corn smut – *Ustilago maydis*
corn, sweet – *Zea mays* var *saccharata*
cornucopia – *Craterellus cornucopioides*
corn, waxy – *Zea mays* var *ceratina*
corojo palm – *Acrocomia crispa*
corpse flower – *Amorphophalus rivieri*
corpus christi flower – *Laelia speciosa*
Corsica, jewel mint of – *Mentha requienii*
Corsican sandwort – *Arenaria balearica*
corydalis – *Corydalis*
corydalis, bulbous – *Corydalis cava*
corydalis, yellow – *Corydalis lutea*
cosmos, garden – *Cosmos bipinnatus*
cosmos, yellow – *Cosmos sulphureus*
Costa Rican guava – *Psidium friedrich*
Costa Rican nightshade – *Solanum wendlandii*
costmary – *Chrysanthemum balsamita*
costus root – *Saussurea lappa*
cotoneaster – *Cotoneaster*
cotoneaster, rock – *Cotoneaster horizontalis*
cotoneaster, rockspray – *Cotoneaster microphyllus*
cotoneaster, willowleaf – *Cotoneaster salicifolius*
cottage pink – *Dianthus plumarius*
cotta grass – *Cottea pappophoroides*
cotton – *Gossypium*
cotton, desert – *Gossypium thurberi*
cotton, Egyptian – *Gossypium barbadense*
cotton grass – *Digitaria insularis*
cotton, lavender – *Santolina*
cotton, Pima – *Gossypium barbadense*
cotton plant, white silk – *Cochlospermum religiosum*
cotton rose – *Filago germanica*
cotton, sea island – *Gossypium barbadense*
cotton, silk – *Ceiba pentandra*
cottontop cactus – *Echinocactus polycephalus*
cottontop, California – *Digitaria californica*
cotton tree – *Bombax ellipticum*
cotton tree, silk – *Pseudobombax palmeri*
cotton tree, wild – *Cochlospermum vitifolium, Hibiscus tilliaceus*
cotton, upland – *Gossypium hirsutum*
cottonwood – *Populus macdougalii*
cottonwood, black – *Populus balsamifera* ssp *trichocarpa*
cottonwood, eastern – *Populus deltoides*
cottonwood, Frémont – *Populus fremontii*
cottonwood, lanceleaf – *Populus* x *acuminata*
cottonwood, narrow-leaved – *Populus angustifolia*
cottonwood, plains – *Populus sargentii*
cottonwood, Rio Grande – *Populus wislizenii*
cottonwood, western – *Populus fremontii*
couch-grass, sand – *Elytrigia juncea*
couch, water – *Paspalum virgatum*
Coues' cassia – *Senna covesii*
Coulter pine – *Pinus coulteri*
Coulter's matilija poppy – *Romneya coulteri*
country borage – *Coleus ambolnicos*
courbaril – *Hymenaea courbaril*
courgette – *Cucurbita pepo*

cowberry – *Vaccinium vitis-idaea*
cowboy's fried egg – *Argemone gracilenta*
cowfoot – *Piper auritum, P. peltatum*
cow-herb – *Vaccaria hispanica*
cow horn agave – *Agave bovicornuta*
cowitch – *Mucuna pruriens*
cow itch tree – *Lagunaria patersonii*
cow-lily – *Nuphar luteum*
cow okra – *Parmentiera edulis*
cow parsley – *Anthriscus sylvestris*
cow parsnip – *Heracleum lanatum*
cow pea – *Vigna sinensis, V. unguiculata*
cowpea, Indian – *Vigna unguiculata* ssp *cylindrica*
cow pen daisy – *Verbesina encelioides*
cowslip – *Primula veris*
cowslip, Cape – *Lachenalia aloides*
cowslip lungwort – *Pulmonaria angustifolia*
cow tamarind – *Albizzia saman*
cow tree – *Brosimum utile*
coyal palm – *Acrocomia mexicana*
coyo avocado – *Persea schiedeana*
coyor ruffle-palm – *Aiphanes acanthophylla*
coyote melon – *Cucurbita foetidissima, C. palmata, Ibervillea insularis, I. sonorae*
coyote-mint – *Monarda odoratissima, Monardella villosa*
coyote tobacco – *Nicotiana attenuata*
coyote willow – *Salix exigua*
coyotillo, Humboldt – *Karwinskia humboldtiana*
crabapple – *Malus sylvestris*
crabapple, Aldenham – *Malus purpurea*
crabapple, Chinese – *Malus spectabilis*
crabapple, Japanese flowering – *Malus floribunda*
crabapple, Oregon – *Malus fusca*
crabapple, tea – *Malus hupehensis*
crab cactus – *Schlumbergera truncata*
crabgrass – *Digitaria*
crabgrass, fringed – *Digitaria ciliaris*
crab, purple – *Malus purpurea*
crack star – *Ariocarpus fissuratus*
crack willow – *Salix fragilis*
cradle, Moses-in-the – *Tradescantia spathacea*
crag-lily, Torrey's – *Anthericum torreyi*
cranberry – *Vaccinium macrocarpon, V. oxycoccus*
cranberry bush, European – *Viburnum opulus*
cranberry, European – *Vaccinium uliginosum*
cranberry, tropical – *Hibiscus sabdariffa*
crane flower – *Strelitzia reginae*
cranesbill – *Erodium reichardii, Geranium*
cranesbill, ash-coloured – *Geranium cinereum*
cranesbill, bloody – *Geranium sanguineum*
cranesbill, cut-leaved – *Geranium dissectum*
cranesbill, dove's foot – *Geranium molle*
cranesbill, dusky – *Geranium phaeum*
cranesbill, French – *Geranium endressii*
cranesbill, large-flowered – *Geranium platypetalum*
cranesbill, meadow – *Geranium pratense*
cranesbill, rock – *Geranium macrorrhizum*
cranesbill, silver – *Geranium argenteum*
cranesbill, small-flowered – *Geranium pusillum*
cranesbill, spotted – *Geranium maculatum*
cranesbill, wild – *Geranium maculatum*
cranesbill, wood – *Geranium sylvaticum*
crape ginger – *Costus speciosus*
crape myrtle – *Lagerstroemia indica*
crape myrtle, giant – *Lagerstroemia speciosa*
crazyweed, Lambert's – *Oxytropis lambertii*
creeping barberry – *Berberis aquifolium* var *repens*
creeping bent – *Agrostis palustris, A. stolonifera*
creeping buttercup – *Ranunculus repens*
creeping Charley – *Pilea cadierei*
creeping devil – *Stenocereus eruca*
creeping fig – *Ficus pumila*
creeping gypsophila – *Gypsophila repens*
creeping heliotrope – *Heliotropium procumbens*
creeping Jenny – *Lysimachia nummularia*
creeping mahonia – *Berberis aquifolium*
creeping muhly – *Muhlenbergia repens*
creeping sage – *Salvia sonomensis*
creeping speedwell – *Veronica repens*
creeping St. John's wort – *Hypericum calycinum*
creeping thistle – *Cirsium arvense*
creeping thyme – *Thymus serpyllum*
creeping velvet grass – *Holcus mollis*
creeping wood sorrel – *Oxalis corniculata*
creeping zinnia – *Sanvitalia procumbens*
creole pepper – *Capsicum chinense*
creosote bush – *Larrea tridentata*

crepe myrtle – *Lagerstroemia speciosa*
crepe myrtle, queen's – *Lagerstroemia*
cress, bitter – *Cardamine*
cress, California rock – *Arabis blepharophylla*
cress, coast rock – *Arabis blepharophylla*
cress, field – *Rorippa austriaca*
cress, garden – *Lepidium sativum*
cress, heart-podded hoary – *Cardaria draba*
cress, marsh yellow – *Rorippa palustris*
cress, mouse-ear – *Arabidopsis thaliana*
cress, para – *Spilanthes acmella*
cress, penny – *Thlaspi*
cress, rock – *Arabis, Aubrieta deltoidea*
cress, rose – *Arabis blepharophylla*
cress, shepherd's – *Teesdalia coronopifolia*
cress, stone – *Aethionema schistosum*
cress, swine – *Coronopus didymus, C. squamatus*
cress, thale – *Arabidopsis thaliana*
cress, wart – *Coronopus squamatus*
cress, water – *Apium nodiflorum, Rorippa nasturtium-aquaticum*
cress, winter – *Barbarea*
crested cockscomb – *Celosia cristata*
crested dogtail – *Cynosurus cristatus*
crested hairgrass – *Koeleria phleoides*
Cretan sun-rose – *Cistus creticus*
Crete dittany – *Origanum dictamnus*
Crete, dittany of – *Origanum dictamnus*
Crimean linden – *Tilia* x *euchlora*
crimson beak – *Krameria grayi*
crimson bottlebrush – *Callistemon citrinus*
crimson clover – *Trifolium incarnatum*
crimson flag – *Schizostylis coccinea*
crimson glory vine – *Vitis coignetiae*
crimson-spot rock-rose – *Cistus ladanifer*
crinkle-awn – *Trachypogon montufari, T. secundus*
crinkled hairgrass – *Deschampsia flexuosa*
crispate-leaved pondweed – *Potamogeton crispus*
crisp fleabane – *Conyza bonariensis*
crocus, autumn – *Colchicum autumnale, C. speciosum*
crocus, cloth of gold – *Crocus angustifolius*
crocus, Dutch – *Crocus vernus*
crocus, garden – *Crocus chrysanthus*
crocus, Scottish – *Crocus biflorus*
cross gentian – *Gentiana cruciata*
cross-leaved heath – *Erica tetralix*
cross, Maltese – *Lychnis chalcedonica*
cross, royal – *Mammillaria karwinskiana* ssp *collinsii*
cross-vine – *Anisostichus capreolatus*
cross-wort – *Crucianella angustifolia*
crotalaria, striped – *Crotalaria mucronata*
croton – *Codiaeum variegatum, Croton californicus, C. tiglium*
croton, bush – *Croton fruticulosus*
croton, New Mexico – *Croton dioicus*
croton, purging – *Croton tiglium*
croton, scented – *Croton ciliato-glandulosus*
croton, Sonora – *Croton sonorae*
croton, Texas – *Croton texensis*
crowberry, black – *Empetrum nigrum*
Crowder pea – *Vigna unguiculata*
crowfoot, celery-leaved – *Ranunculus sceleratus*
crowfoot grass – *Dactyloctenium aegyptium*
crowfoot grass, Durban – *Dactyloctenium aegyptium*
crowfoot, water – *Ranunculus aquatilis*
crow garlic – *Allium vineale*
crown bark – *Cinchona officinalis*
crownbeard – *Verbesina*
crownbeard, golden – *Verbesina encelioides* ssp *exauriculata*
crown daisy – *Chrysanthemum coronarium*
crown gum – *Achras chicle*
crown imperial – *Fritillaria imperialis*
crown, king's – *Sedum roseum*
crown-of-gold tree – *Cassia fistula*
crown-of-thorns – *Euphorbia milii, E. splendens, Koeberlinia spinosa* ssp *tenuispina, Paliurus spina-christi*
crown-pink – *Lychnis coronaria*
crown vetch – *Coronilla varia*
crow needles – *Scandix pectin-veneris*
crow-poison – *Nothoscordium bivalve*
crucifixion thorn – *Canotia holacantha, Castela emoryi, Koberlinia spinosa*
crucillo – *Ziziphus parryi*
crucillo, Mexican – *Ziziphus spathulata*
crucillo, white – *Ziziphus obtusifolia*
cry baby – *Erythrina crista-galli*
cryptantha, ribbed – *Cryptantha costata*
cryptantha, winged – *Cryptantha holoptera*
cryptomeria, Japanese – *Cryptomeria japonica*

crystalline iceplant – *Mesembryanthemum crystallinum*
Cuban royal palm – *Roystonea regia*
cubeb-berry – *Piper cubeda*
cubies – *Psilocybe cubensis*
cucumber – *Cucumis sativus*
cucumber, Chinese – *Trichosanthes kirilowii*
cucumber, Chinese bitter – *Momordica cochinchinensis*
cucumber, globe – *Citrullus colocynthis*
cucumber melon – *Cucumis melo* var *chito*
cucumber, mock – *Marah lobata*
cucumber, one-seeded bur – *Sicyos laciniatus*
cucumber, pickling – *Cucumis sativus*
cucumber, prophet's – *Citrullus colocynthis*
cucumber, slicing – *Cucumis sativus*
cucumber, snake – *Cucumis melo* var *flexuosus*
cucumber, squirting – *Ecballium elaterium*
cucumber tree – *Magnolia acuminata*
cucumber, wild – *Marah fabaceus, M. macrocarpus*
cudjoe wood – *Jacquinia macrocarpa* ssp *pungens*
cudweed – *Filago, Gnaphalium*
cudweed, broad-leaved – *Filago pyramidata*
cudweed, common – *Filago germanica*
cudweed, Jersey – *Gnaphalium luteo-album*
cudweed, narrow – *Filago gallica*
cudweed, white – *Gnaphalium leucocephalum*
culantro – *Eryngium foetidum*
culinary sage – *Salvia officinalis*
cultivated flax – *Linum usitatissimum*
cultivated grape – *Vitis vinifera*
cultivated mushroom – *Agaricus bisporus* var *hortensis*
cultivated oat – *Avena sativa*
cultivated oyster mushroom – *Pleurotus ostreatus* var *sativus*
cultivated timothy – *Phleum pratense*
cumin – *Cuminum cyminum*
cup-and-saucer – *Campanula medium*
cup-and-saucer vine – *Cobaea scandens*
cup, butter – *Ranunculus*
cup, claret – *Echinocereus triglochidiatus*
cup, emperor's – *Senna alata*
cupgrass, bearded – *Eriochloa aristata*
cupgrass, prairie – *Eriochloa contracta*
cupgrass, southwestern – *Eriochloa acuminata*
cuphea, bat-faced – *Cuphea llavea*
cuphea, beetlesage – *Cuphea jorullensis*
Cupid flower – *Ipomoea quamoclit*
Cupid's dart – *Catananche caerulea*
cup, monkey – *Couepia polyandra*
cup-of-gold – *Solandra nitida*
cup-of-gold vine – *Solandra maxima*
cup, painted – *Parentucella viscosa*
cup, ragged – *Silphium perfoliatum*
cup, rosy – *Pluchea odorata*
Curaçao milkweed – *Asclepias curassavica*
curare – *Chondodendron tomentosium, Strychnos toxifera*
curls, blue – *Phacelia congesta, Trichostemma*
curly dock – *Rumex crispus*
curly kale – *Brassica oleracea* var *sabellica*
curly mesquite – *Pleuraphis belangeri*
curly welted thistle – *Carduus crispus*
currant – *Ribes*
currant, alpine – *Ribes alpinum*
currant, black – *Ribes nigrum*
currant, golden – *Ribes aureum*
currant, gooseberry – *Ribes montigenum*
currant, mountain – *Ribes alpinum*
currant, ornamental – *Ribes odoratum*
currant, pink winter – *Ribes sanguineum*
currant, red – *Ribes rubrum*
currant, red flowering – *Ribes sanguineum*
currant, rock – *Ribes petraeum*
currant, squaw – *Ribes cereum*
currant tomato – *Lycopersicon pimpinellifolium*
currant, wax – *Ribes cereum*
curry – *Murraya koenigii*
curry plant – *Helichrysum italicum*
cushaw – *Cucurbita moschata*
cushaw, green-striped – *Cucurbita mixta*
cushion pink – *Silene acaulis*
custard apple – *Annona*
cutgrass – *Leersia hexandra*
cutgrass, rice – *Leersia oryzoides*
cutleaf cone-flower – *Rudbeckia laciniata*
cutleaf water-parsnip – *Berula erecta*
cut-leaved blackberry – *Rubus laciniatus*
cut-leaved cranesbill – *Geranium dissectum*
cut-leaved evening primrose – *Oenothera laciniata*
cutting-grass – *Scleria bracteata*
Cuyamaca cypress – *Cupressus arizonica*
cycad – *Ceratozamia mexicana*
cyad palm – *Cycas revoluta*

cyclamen – *Cyclamen*
cyclamen, Chilean – *Cyclamen cilicum*
cyclamen, florist's – *Cyclamen persicum*
cyclamen, hardy – *Cyclamen hederifolium*
cyclamen-leaved daffodil – *Narcissus cyclamineus*
cyclamineus daffodil – *Narcissus cyclamineus*
cynanchum, Utah – *Cynanchum utahense*
cypress – *Cupressus, Taxodium*
cypress, Arizona – *Cupressus arizonica*
cypress, bald – *Taxodium distichum*
cypress, Cuyamaca – *Cupressus arizonica*
cypress, false – *Chamaecyparis*
cypress, Gowen – *Cupressus goveniana*
cypress, Guadalupe – *Cupressus guadalupensis*
cypress, Hinoki false – *Chamaecyparis obtusa*
cypress, Italian – *Cupressus sempervirens*
cypress, Lawson – *Cupressus lawsoniana*
cypress, Mexican – *Cupressus lusitanica, Taxodium mucronatum*
cypress, mock – *Bassia hyssopifolia, Kochia scoparia*
cypress, Monterey – *Cupressus macrocarpa*
cypress, Montezuma bald – *Taxodium mucronatum*
cypress, mountain – *Podocarpus guatemalensis*
cypress, Nootka – *Cupressus nootkatensis*
cypress, Patagonian – *Fitzroya cupressoides*
cypress, Portuguese – *Cupressus lusitanica*
cypress, Sargent – *Cupressus sargentii*
cypress, sawara false – *Chamaecyparis pisifera*
cypress, smooth Arizona – *Cupressus arizonica* ssp *glabra*
cypress spurge – *Euphorbia cyparissias*
cypress, summer – *Kochia scoparia*
cypress, Tecate – *Cupressus forbesii*
cypress vine – *Ipomoea quamoclit, Quamoclit pinnata*
cypress, yellow – *Cupressus nootkatensis*
Cyprian cedar – *Cedrus brevifolia*
Cyprus cedar – *Cedrus brevifolia*
Cyprus vetch – *Lathyrus ochrus*

Dacca banana – *Mus acuminata*
Dacca banana, red – *Mus acuminata*
daffodil – *Narcissus pseudo-narcissus*
daffodil, Asturian – *Narcissus asturiensis*
daffodil, cyclamen-leaved – *Narcissus cyclamineus*
daffodil, cyclamineus – *Narcissus cyclamineus*
daffodil, fall – *Sternbergia lutea*
daffodil, hoop petticoat – *Narcissus bulbocodium*
daffodil, jonquilla – *Narcissus jonquilla*
daffodil, multiflowering – *Narcissus tazetta*
daffodil, Peruvian – *Hymenocallis amancaes, H. narcissiflora*
daffodil, sea – *Pancratium maritimum*
daffodil, tazetta – *Narcissus tazetta*
daffodil, trumpet – *Narcissus bulbocodium*
daffodil, winter – *Sternbergia lutea*
dagger cholla – *Grusonia invicta*
dagger, giant Spanish – *Yucca carnerosana*
dagger, Spanish – *Yucca*
dahlia – *Dahlia*
dahlia, bell tree – *Dahlia imperialis*
dahlia cactus – *Echinocereus poselgeri*
dahlia, climbing – *Hidalgoa ternata*
dahlia, flat tree – *Dahlia excelsa*
dahlia-rooted cereus – *Peniocereus striatus*
dahlia, sea – *Coreopsis maritima*
dahlia, tree – *Dahlia imperialis*
dahlia, wild – *Dahlia rosea*
Dahurian larch – *Larix gmelinii*
daikon – *Raphanus sativus* var *longipinnatus*
daisy – *Chrysanthemum leucanthemum*
daisy, African – *Arctotis venusta, Dimorphotheca sinuata*
daisy, barberton – *Gerbera jamesonii*
daisy, blue-eyed African – *Arctotis venusta*
daisy, cow pen – *Verbesina encelioides*
daisy, crown – *Chrysanthemum coronarium*
daisy, English – *Bellis perennis*
daisy, English lawn – *Bellis perennis*
daisy, false – *Eclipta prostrata*
daisy, fleabane – *Erigeron*
daisy, garland – *Chrysanthemum coronarium*
daisy, globe – *Globularia*
daisy, gloriosa – *Rudbeckia hirta*
daisy, Mexican – *Erigeron karvinskianus*
daisy, Michaelmas – *Aster novi-belgii*
daisy, Mount Atlas – *Anacyclus depressus*
daisy, ox-eye – *Leucanthemum vulgare*
daisy, painted – *Chrysanthemum coccineum*
daisy, paper – *Baileya multiradiata*
daisy, Paris – *Chrysanthemum frutescens*
daisy, pink paper – *Helipterum roseum*
daisy, rock – *Perityle*

daisy, Santa Barbara – *Erigeron karvinskianus*
daisy, seaside – *Erigeron glaucus*
daisy, Shasta – *Chrysanthemum maximum*
daisy, southern – *Erigeron speciosus*
daisy, Transvaal – *Gerbera jamesonii*
daisy tree – *Montanoa tomentosa*
daisy, yellow – *Hymenoxys scaposa*
Dakota vervain – *Verbena ambrosifolia*
dalea, bearded – *Dalea pogonathera*
dalea, black – *Dalea frutescens*
dalea, mesa – *Psorothamnus schottii*
dalea, shrubby – *Dalea frutescens*
dallis grass – *Paspalum dilatatum*
Dalmation bellflower – *Campanula portenschlagiana*
Dalmatian toadflax – *Linaria genistifolia*
damask, autumn – *Rosa damascena* var *semperflorens*
damask rose – *Rosa damascena*
dame's rocket – *Hesperis matronalis*
damiana – *Turnera aphrodisiaca, T. diffusa*
damiana, large-leaf – *Turnera ulmifolia*
damianita – *Chrysactinia mexicana*
dammar pine, white – *Agathis alba*
damson plum – *Prunus domestica* ssp *insititia*
dandelion – *Taraxicum officinale*
dandelion, common – *Taraxacum officinale*
dandelion, Texas – *Pyrrhopappus multicaulis*
daphne – *Daphne*
daphne, February – *Daphne mezereum*
daphne, garland – *Daphne cneorum*
Darcy sage – *Salvia darcyi*
dark blood-red spikerush – *Eleocharis atropurpurea*
dark foxglove – *Digitalis obscura*
darnel – *Lolium temulentum*
darnel, bearded – *Lolium temulentum*
dart, Cupid's – *Catananche caerulea*
Darwin barberry – *Berberis darwinii*
dasheen – *Colocasia esculenta*
date – *Phoenix dactylifera*
date palm – *Phoenix dactylifera*
date palm, Canary Island – *Phoenix canariensis*
date palm, pygmy – *Phoenix roebelenii*
date palm, Senegal – *Phoenix reclinata*
date-plum – *Diospyros lotus*
datil – *Yucca baccata*
datil pepper – *Capsicum chinense*
datura, sacred – *Datura metel, D. wrightii*
datura, tree – *Brugmansia*
David's maple – *Acer davidii*
David's root – *Chiococca alba*
dawn redwood – *Metasequoia glyptostroboides*
dayflower – *Commelina*
dayflower, birdbill – *Commelina dianthifolia*
dayflower, whitemouth – *Commelina erecta*
daylily – *Hemerocallis fulva*
daylily, common orange – *Hemerocallis fulva*
daylily, lemon – *Hemerocallis lilioasphodelus*
daylily, tawny – *Hemerocallis fulva*
deadly nightshade – *Atropa belladona*
dead nettle – *Lamium*
Dead Sea fruit – *Solanum sodomeum*
death camas – *Zigadenus*
death carrot – *Thapsia garganica*
deceiver, the – *Laccaria laccata*
deer browse – *Cercocarpus montanus*
deergrass – *Muhlenbergia rigens*
deerhorn cedar – *Thujopsis dolabrata*
deerhorn cholla – *Cylindropuntia versicolor*
deer nut – *Simmondsia chinensis*
deers ears – *Swertia radiata*
deer vetch – *Lotus repens*
deerweed – *Lotus scoparius*
delicate sow thistle – *Sonchus tenerrimus*
delicious milk mushroom – *Lactarius deliciosus*
delight, heart's – *Abronia fragrans*
delight, queen's – *Stillingia*
Del Norte county iris – *Iris innominata*
delphinium, annual – *Consolida ambigua*
delphinium, candle – *Delphinium elatum*
delta oak – *Quercus similis*
dense mistletoe – *Phoradendron densum*
dent corn – *Zea mays* var *indentata*
deodar cedar – *Cedrus deodara*
Deptford pink – *Dianthus armeria*
desert agave – *Agave deserti*
desert apricot – *Prunus fremontii*
desert baccharis – *Baccharis sergiloides*
desert bluebells, California – *Phacelia campanularia*
desert brickellia – *Brickellia coulteri*
desert broom – *Baccharis sarathroides*
desert broom-rape – *Orobanche cooperi*
desert buttercup – *Ranunculus cymbalaria*
desert-candle – *Dasylirion, Eremurus robustus*

desert catalpa – *Chilopsis linearis* ssp *arcuata*
desert ceanothus – *Ceanothus greggii*
desert Christmas cactus – *Cylindropuntia leptocaulis*
desert cotton – *Gossypium thurberi*
desert elderberry – *Sambucus mexicana*
desert four-o'clock – *Mirabilis multiflora*
desert goldenbush – *Viguiera parishii*
desert globemallow – *Sphaeralcea ambigua*
desert ground-cherry – *Physalis crassifolia* var *versicolor*
desert ground-cover – *Abronia villosa*
desert hackberry – *Celtis pallida*
desert hibiscus – *Hibiscus coulteri*
desert holly – *Acourtia nana, A. wrightii*
desert hollyhock – *Sphaeralcea ambigua*
desert honeysuckle – *Anisacanthus quadrifidus* ssp *wrightii, A. thurberi*
desert hyacinth – *Dichelostemma capitatum*
desert Indian paintbrush – *Castilleja angustifolia*
desert ironwood – *Olneya tesota*
desert lantana – *Lantana macropoda*
desert-lavender – *Hyptis emoryi, H. laniflora*
desert lily – *Hesperocallis undulata*
desert locust – *Robinia neomexicana*
desert mahonia – *Berberis fremontii*
desert mallow – *Sphaeralcea ambigua*
desert-marigold – *Baileya multiradiata*
desert marigold, few-flowered – *Baileya pauciradiata*
desert marigold, woolly – *Baileya pleniradiata*
desert mariposa – *Calochortus kennedyi*
desert mentzelia – *Mentzelia multiflora*
desert milkweed – *Asclepias subulata, A. erosa*
desert mistletoe – *Phoradendron californicum*
desert night-blooming cereus – *Peniocereus greggii* var *transmontanus*
desert olive – *Forestiera neomexicana*
desert poppy – *Kallstroemia grandiflora*
desert portulaca – *Portulaca halimoides*
desert rhubarb – *Rumex hymenosepalus*
desert rock fig – *Ficus palmeri*
desert rose – *Adenium obesum, Rosa minutifolia, R. stellata*
desert rose mallow – *Hibiscus coulteri*
desert sage – *Salvia eremostachya*
desert saltbush – *Atriplex polycarpa*
desert sand verbena – *Abronia villosa*
desert scrub oak – *Quercus turbinella*
desert senna – *Senna covesii*
desert spike-moss – *Selaginella eremophila*
desert spoon – *Dasylirion wheeleri, Eriogonum inflatum*
desert sumac – *Rhus microphylla*
desert tea – *Ephedra californica*
desert-thorn – *Lycium*
desert thorn-apple – *Datura discolor*
desert-thorn, coast – *Lycium californicum*
desert-thorn, Parish's – *Lycium parishii*
desert-thorn, Santa Catalina Island – *Lycium brevipes* var *hassei*
desert tobacco – *Nicotiana clevelandii, N. obtusifolia*
desert tree-beargrass – *Nolina bigelovii*
desert trumpet – *Eriogonum inflatum*
desert unicorn plant – *Proboscidea althaeifolia*
desert vervain – *Verbena wrightii*
desert-willow – *Chilopsis linearis* ssp *arcuata*
desert wolfberry – *Lycium andersonii, L. fremontii*
desert yaupon – *Schaefferia cuneifolia*
desert zinnia – *Zinnia pumila*
desire – *Capparis spinosa*
devil cactus – *Grusonia schottii*
devil, creeping – *Stenocereus eruca*
devil's backbone – *Pedilanthus tithymaloides*
devil's claw – *Acacia greggii, Harpagophytum procumbens, Proboscidea*
devil's claw, red – *Proboscidea parviflora*
devil's claws – *Martynia annua*
devil's fig – *Argemone mexicana*
devil's head – *Echinocactus horizonthalonius, E. texensis*
devil's ivy – *Epipremnum pinnatum*
devil's lantern – *Oenothera deltoides*
devil's lettuce – *Amsinckia tessellata*
devil's pincushion – *Echinocactus texensis*
devilweed, Mexican – *Chloracantha spinosa*
dewberry – *Rubus caesius, R. trivialis*
dewdrop, golden – *Duranta repens*
dhour tree – *Ziziphus spina-christi*
dhura – *Sorghum vulgare* var *durra*
diamond cholla – *Cylindropuntia ramosissima*
diamond leaf frog-fruit – *Phyla strigulosa* var *sericea*
dichondria – *Dichondra argentea*
diffuse knapweed – *Centaurea diffusa*

digger pine – *Pinus sabiniana*
digitalis – *Digitalis purpurea*
dika – *Irvingia gabonensis*
dill – *Anethum graveolens*
dillenia, Indian – *Dillenia indica*
dillweed – *Anethum graveolens*
dinner bell, monkey – *Hura polyandra*
dinner plate aralia – *Polyscias scutellaria*
dion, chestnut – *Dioon edule*
distaff thistle – *Atractylis cancellata*
distaff thistle, woolly – *Carthamus lanatus*
dita tree – *Alstonia scholaris*
ditch beard grass – *Polypogon interruptus*
dittander – *Lepidium latifolium*
dittany – *Dictamnus albus, Origanum dictamus*
dittany, Crete – *Origanum dictamnus*
dittany, false – *Ballota pseudodictamnus*
dittany of Crete – *Origanum dictamnus*
diverse-leaved pondweed – *Potamogeton diversifolius*
divi-divi – *Caesalpinia coriaria*
divinatory mushroom – *Psilocybe yungensis*
divorce vine – *Solanum wendlandii*
dock – *Rumex*
dock, alpine – *Rumex alpinus*
dock, bitter – *Rumex obtusifolius*
dock, bull's tongue – *Rumex conglomeratus*
dock, curly – *Rumex crispus*
dock, elf – *Inula helenium*
dock, fiddle – *Rumex pulcher*
dock, golden – *Rumex maritimus*
dock, Mexican – *Rumex violascens*
dock, narrow-leaved – *Rumex stenophyllus*
dock, red – *Rumex hymenosepalus*
dock, spatter – *Nuphar luteum* ssp *polysepalum*
dock, water – *Rumex salicifolius*
dock, whorled – *Rumex conglomeratus*
dock, willow – *Rumex salicifolius*
dodder – *Cuscuta*
dogbane, bitter – *Apocynum androsaemifolium*
dog cholla – *Grusonia schottii*
dog-fennel – *Anthemis cotula*
dog rose – *Rosa canina*
dog's ear – *Tiquilia canescens*
dogtail, crested – *Cynosurus cristatus*
dogtail, hedgehog – *Cynosurus echinatus*
dog-tooth violet – *Erythronium dens-canis*
dog violet, common – *Viola riviniana*
dogweed, five-needled – *Thymophylla pentachaeta* var *belenidium*
dogwood – *Cornus*
dogwood, bloodtwig – *Cornus sanguinea*
dogwood, eastern – *Cornus florida*
dogwood, evergreen – *Cornus capitata*
dogwood, flowering – *Cornus florida*
dogwood, giant – *Cornus controversa*
dogwood, Japanese – *Cornus kousa*
dogwood, kousa – *Cornus kousa*
dogwood, mountain – *Cornus nuttallii*
dogwood, Pacific – *Cornus nuttallii*
dogwood, red-osier – *Cornus stolonifera*
dogwood, redtwig – *Cornus stolonifera*
dogwood, Tartarian – *Cornus alba*
dogwood, western – *Cornus nuttallii*
Dominican mahogany – *Swietenia mahagoni*
donard – *Eryngium alpinum*
dong quai – *Angelica sinensis*
donkey tail – *Sedum morganianum*
Donnell Smith pine – *Pinus hartwegii*
doorian – *Durio zibethinus*
doorweed – *Polygonum arenastrum*
Dorset heath – *Erica ciliaris*
dot and dash plant – *Karwinskia humboldtiana*
dotted loosestrife – *Lysimachia punctata*
double white banksia – *Rosa fortuniana*
Douglas-fir – *Pseudotsuga menziesii*
Douglas-fir, bigcone – *Pseudotsuga macrocarpa*
Douglas-fir, Mexican – *Pseudotsuga lindleyana*
Douglas hackberry – *Celtis douglassii*
Douglas pine – *Pinus douglasiana*
doum plum – *Hyphaene thebaica*
dove's dung – *Ornithogalum umbellatum*
dove's foot cranesbill – *Geranium molle*
dove tree – *Davidia involucrata*
dove weed – *Croton texensis, Eremocarpus setigerus*
downy hawthorn – *Crataegus mollis*
downy brome – *Bromus tectorum*
downy jasmine – *Jasminum multiflorum*
downy mildew – *Peronospora viticola*
dracena – *Cordyline*
dracena, blue – *Cordyline indivisa*
dragon, fire – *Acalypha wilkesiana*
dragon, green – *Arisaema dracontium, Dracunculus vulgaris*

dragonhead – *Physostegia virginiana*
dragonhead, false – *Physostegia virginiana*
dragon's blood – *Croton lechleri, Daemonorops draco, Jatropha cardiophylla*
dragon tree – *Dracaena draco*
dragon tree, flaming – *Cordyline fruticosa*
dragon tree, Madagascar – *Dracaena marginata*
dragonwort – *Dracunculus vulgaris*
dried chile – *Capsicum annuum*
drooping agave – *Agave attenuata*
drooping club-rush – *Scirpus cernuus*
drooping juniper, Mexican – *Juniperus flaccida*
drooping marigold – *Baileya pauciradiata*
drooping melaleuca – *Melaleuca armillaris*
drooping spurge – *Chamaesyce nutans*
drooping star of Bethlehem – *Ornithogalum nutans*
dropseed, giant – *Sporobolus giganteus*
dropseed, hairy – *Blepharoneuron tricholepis*
dropseed, mesa – *Sporobolus flexuosus*
dropseed, pine – *Blepharoneuron tricholepis*
dropseed, sand – *Sporobolus cryptandrus*
dropseed, spike – *Sporobolus contractus*
dropseed, whorled – *Sporobolus pyramidatus*
dropwort – *Filipendula vulgaris*
dropwort, water – *Oenanthe aquatica, O. pimpinelloides*
drug fustic tree – *Maclura tinctoria*
drug Osageorange – *Maclura tinctoria*
Drummond's rain lily – *Cooperia drummondii*
drumsticks – *Allium sphaerocephalum*
drumstick tree – *Moringa oleifera*
drunken sailors – *Quisqualis indica*
dry whisky – *Lophophora williamsii*
duckweed – *Lemna*
duckweed, common – *Lemna minor*
duckweed, fat – *Lemna gibba*
duckweed, ivy-leaved – *Lemna trisulca*
duckweed, lesser – *Lemna minuscula*
dudaim melon – *Cucumis melo* var *dudaim*
dudleya – *Dudleya lanceolata*
dudleya, chalk – *Dudleya pulverulenta*
dudleya, Orcutt's – *Dudleya attenuata* ssp *orcuttii*
Dudley's rush – *Juncus dudleyi*
Duke of Argyll's tea plant – *Lycium barbarum*
Duke of Argyll's tea-tree – *Lycium chinense*
Duke of Devonshire's oncidium, the – *Oncidium cavendishianum*
dulse – *Rhodymenia palmata*
dumb cane – *Dieffenbachia*
dune sunflower – *Helianthus niveus*
dung, dove's – *Ornithogalum umbellatum*
Durango fir – *Abies durangensis*
Durango juniper – *Juniperus durangensis*
Durango pine – *Pinus durangensis*
Durban crowfoot grass – *Dactyloctenium aegyptium*
durian fruit – *Durio zibethinus*
durmast oak – *Quercus petraea*
dusky cranesbill – *Geranium phaeum*
dust and ashes – *Solanum sodomeum*
dusty miller – *Artemisia stellerana, Centaurea cineraria, Cerastium tomentosum, Lychnis coronaria, Senecio cineraria*
Dutch crocus – *Crocus vernus*
Dutch elm – *Ulmus hollandica*
Dutch-grass – *Ruppia cirrhosa, R. maritima*
Dutch iris – *Iris* x *hollandica*
Dutchman's pipe – *Aristolochia*
Dutchman's pipe cactus *Epiphyllum oxypetalum*
Dutchman's pipe, Californica – *Aristolochia californica*
Dutchman's pipe, Watson's – *Aristolochia watsonii*
Dutch white clover – *Trifolium repens*
dwarf almond tree – *Prunus tenella*
dwarf banana – *Mus nana*
dwarf broom – *Chamaecytisus proliferus*
dwarf Cavendish banana – *Mus acuminata*
dwarf cherry – *Prunus fruticans, P. fruticosa*
dwarf cherry, European – *Prunus fruticosa*
dwarf chestnut – *Castanea pumila*
dwarf cholla, Schott's – *Grusonia schottii*
dwarf iris – *Iris pumila*
dwarfish soapwort – *Saponaria pumilio*
dwarf juniper – *Juniperus communis* ssp *nana, J. sibirica*
dwarf maize – *Zea mays* var *gracillima*
dwarf marigold – *Tagetes minuta*
dwarf mesquite – *Prosopis globosa* var *mexicana*
dwarf mistletoe – *Arceuthobium*
dwarf mistletoe, pinyon – *Arceuthobium divaricatum*

dwarf mistletoe, western – *Arceuthobium campylopodum*
dwarf morning-glory – *Convolvulus tricolor, Evolvulus alsinoides*
dwarf nettle – *Urtica urens*
dwarf orange – *Citrus mitis*
dwarf pearlwort – *Sagina apetala*
dwarf pennyroyal – *Hedeoma nanum*
dwarf periwinkle – *Vinca minor*
dwarf phlox – *Phlox nana*
dwarf plumbago – *Ceratostigma plumbaginoides*
dwarf poinciana – *Caesalpinia pulcherrima*
dwarf pomegranate – *Punica granatum* var *nana*
dwarf powder puff – *Calliandra emarginata*
dwarf tree fern – *Blechnum gibbum*
dwarf zinnia – *Zinnia acerosa*
dyer's alkanet – *Alkanna tinctoria*
dyer's chamomile – *Anthemis tinctoria*
dyer's greenweed – *Genista tinctoria*
dyer's mulberry – *Maclura tinctoria*
dyer's oak – *Quercus velutina*
dyer's rocket – *Reseda luteola*
dyer's woad – *Isatis tinctoria*
dyssodia, golden – *Thymophylla pentachaeta* var *belenidium*

eaglesclaw cactus – *Echinocactus horizonthalonius*
eagle wood – *Aquilaria agallocha*
ear, cat's – *Hypochaeris*
ear, dog's – *Tiquilia canescens*
eardrop tree – *Enterolobium cyclocarpum*
ear, elephant – *Xanthosoma robustum, X. Sagittifolium*
ear, elephant's – *Alocasia cuprea, A. macrorrhiza, Colocasia esculenta, Enterolobium cyclocarpum*
ear fungus, cloud – *Auricularia auricula-judae*
ear fungus, wood – *Auricularia polytricha*
ear, hare's – *Bupleurum fruticosum, Conringia orientalis*
early cestrum – *Cestrum fasciculatum*
early hairgrass – *Aira praecox*
early meadow grass – *Poa inferma*
early medick – *Medicago praecox*
early winter cress – *Barbarea verna*
earring, ape's – *Havardia pallens, Pithecellobium flexicaule*
ears, bunny – *Opuntia microdasys*
ears, deers – *Swertia radiata*
ears, elephant – *Colocassia esculenta*
ear, shrubby hare's – *Bupleurum fruticosum*
ear, silver – *Tremella fuciformis*
ears, lamb's – *Stachys byzantina*
ears, mules – *Wyethia helenioides*
earth almond – *Cyperus esculentus*
earth nut – *Bunium bulbocastanum*
earthnut pea – *Lathyrus tuberosus*
earth pea – *Vigna subterranean, Voandzeia subterranea*
easter cactus – *Rhipsalidopsis gaertneri*
easter lily – *Lilium candidum, L. longiflorum*
easter lily vine – *Beaumontia grandiflora*
eastern cottonwood – *Populus deltoides*
eastern dogwood – *Cornus florida*
eastern gamagrass – *Tripsacum dactyloides*
eastern hemlock – *Tsuga canadensis*
eastern larch – *Larix laricina*
eastern pasque flower – *Pulsatilla patens*
eastern poplar – *Populus deltoides*
eastern redbud – *Cercis canadensis*
eastern red cedar – *Juniperus virginiana*
eastern rocket – *Sisymbrium orientale*
eastern shooting star – *Dodecatheon meadia*
eastern white oak – *Quercus alba*
eastern white pine – *Pinus strobus*
East Indian arrowroot – *Curcuma angustifolia*
East Indian fig – *Ficus benghalensis*
East Indian walnut – *Albizzia lebbeck*
ebony – *Diospyros ebenus*
ebony cocuswood – *Brya ebenus*
ebony, macassar – *Diospyros ebenum*
ebony, Mexican *Caesalpinia sclerocarpa, Havardia mexicana*
ebony, mountain – *Bauhinia variegata*
ebony persimmon, Indian – *Diospyros ebenaster*
ebony, Sonoran – *Havardia sonorae*
ebony, Texas – *Pithecellobium flexicaule*
ebony, west African – *Diospyros mespiliformis*
echinacea – *Echinacea angustifolia, E. purpurea*
Ecuador walnut – *Juglans honorei*
edelweiss – *Leontopodium alpinum*
edible bolete – *Boletus edulis*

edible fig – *Ficus carica*
edible galingale – *Cyperus eragrostis*
edible-pod pea – *Pisum sativum* var *macrocarpon*
eel-grass – *Zostera marina*
eel-grass pondweed – *Potamogeton zosteriformis*
egg-cone pine – *Pinus oocarpa*
egg, cowboy's fried – *Argemone gracilenta*
eggfruit – *Pouteria campechiana*
eggplant – *Solanum melongena* var *esculentum*
eggs, butter-and – *Linaria vulgaris*
eggs, golden – *Cochlospermum vitifolium*
eggs, poached – *Limnanthes douglasii*
eglantine – *Rosa eglanteria*
egusi melon – *Cucumeropsis edulis*
Egypt, chate of – *Cucumis melo* var *chito*
Egyptian clover – *Trifolium alexandrinum*
Egyptian cotton – *Gossypium barbadense*
Egyptian paper plant – *Cyperus papyrus*
elderberry, American – *Sambucus canadensis*
elderberry, black – *Sambucus nigra*
elderberry, blue – *Sambucus mexicana*
elderberry, desert – *Sambucus mexicana*
elderberry, red – *Sambucus racemosa*
elder, blueberry – *Sambucus glauca*
elder, box – *Acer negundo* var *californicum*
elder, European – *Sambucus nigra*
elder, ground – *Aegopodium podagraria*
elder, horse – *Inula helenium*
elder, marsh – *Iva ambrosiaefolia*
elder, Mexican – *Sambucus mexicana*
elder, Spanish – *Piper amalago*
elder, yellow – *Tecoma stans*
elecampane – *Inula helenium*
elegant European hairgrass – *Aira elegantissima*
elegantine rose – *Rosa moschata*
elegant nightshade – *Solanum gracile*
elemi, gum – *Bursera simaruba*
elemi, West Indian – *Bursera gummifera*
elephant apple – *Dillenia indica*
elephant cactus – *Pachycereus pringlei*
elephant ear – *Xanthosoma robustum, X. sagittifolium*
elephant ears – *Colocasia esculenta*
elephant grass – *Pennisetum purpureum*
elephant's ear – *Alocasia cuprea, A. macrorrhiza, Colocasia esculenta, Enterolobium cyclocarpum*
elephant's tooth – *Coryphantha elephantidens*
elephant tree – *Bursera*
elephant tree, Baja – *Pachycormus discolor*
elephant tree, fragrant – *Bursera fagarioides*
elephant tree, red – *Bursera hindsiana*
elephant yucca – *Yucca guatemalensis*
elf dock – *Inula helenium*
elm – *Ulmus*
elm, American – *Ulmus americana*
elm, California slippery – *Fremontodendron californicum*
elm, cedar – *Ulmus crassifolia*
elm, Chinese – *Ulmus parvifolia*
elm, Chinese evergreen – *Ulmus parvifolia*
elm, Dutch – *Ulmus hollandica*
elm, English – *Ulms minor, U. procera*
elm, European white – *Ulmus laevis*
elm, Mexican – *Ulmus mexicana*
elm, Scotch – *Ulmus glabra*
elm, Siberian – *Ulmus pumila*
elm, slippery – *Ulmus fulva*
elm, smooth-leaved – *Ulmus minor*
elm, Spanish – *Cordia alliodora*
elm, Texas – *Ulmus crassifolia*
elm, wych – *Ulmus glabra*
El Paso prickly-pear – *Opuntia polyacantha* var *arenaria*
Eltham Indian fig – *Opuntia dellenii*
Emory oak – *Quercus emoryi*
emperor's cup – *Senna alata*
emperor tulip – *Tulipa gesneriana*
empress tree – *Paulownia tomentosa*
enchanter's nightshade – *Circaea alpina, C. lutetiana*
encina – *Quercus agrifolia*
endive – *Cichorium endivia*
endive, Batavian – *Cichorium endivia*
endive, Belgian – *Cichorium intybus*
Engelmann prickly-pear – *Opuntia engelmannii*
Engelmann spruce – *Picea engelmannii*
English barberry – *Berberis vulgaris*
English bluebell – *Hyacinthoides non-scripta*
English boxwood – *Buxus sempervirens*
English chamomile – *Matricaria recutita*
English daisy – *Bellis perennis*
English elm – *Ulmus minor, U. procera*

English hawthorn – *Crataegus laevigata*
English holly – *Ilex aquifolium*
English iris – *Iris latifolia*
English ivy – *Hedera helix*
English laurel – *Prunus laurocerasus*
English lavender – *Lavandula angustifolia*
English lawn daisy – *Bellis perennis*
English lime – *Tilia* x *europaea*
English oak – *Quercus robur*
English plantain – *Plantago lanceolata, P. major*
English primrose – *Primula* x *polyantha, P. vulgaris*
English stonecrop – *Sedum anglicum*
English sundew – *Drosera anglica*
English violet – *Viola odorata*
English wallflower – *Erysimum cheiri*
English walnut – *Juglans regia*
English yew – *Taxus baccata*
epaulette tree – *Pterostyrax hispida*
epazote – *Chenopodium ambrosioides*
ephedra, boundary – *Ephedra aspera*
ephedra, Chinese – *Ephedra sinica*
ephedra, green – *Ephedra viridis*
epidendrum, aloe-leaved – *Epidendrum parkinsonianum*
epidendrum, blushing – *Epidendrum erubescens*
ergot – *Claviceps purpurea*
eryngo – *Eryngium foetidum, E. lemmoni*
eryngo, field – *Eryngium campestre*
escallonia, white – *Escallonia illinita*
escarole – *Cichorium endivia*
escarpment live oak – *Quercus fusiformis*
esparto grass – *Lygeum spartum, Stipa tenacisima*
ethrog – *Citrus medica*
Etruscan honeysuckle – *Lonicera etrusca*
eucalyptus – *Eucalyptus*
eulalia grass – *Miscanthus sinensis*
euonymus, evergreen – *Euonymus japonicus*
euonymus, variegated – *Euonymus japonicus* var *aureo-variegatus*
euonymus, winged – *Euonymus alata*
European alkali grass – *Puccinellia distans*
European ash – *Fraxinus excelsior*
European aspen – *Populus tremula*
European barberry – *Berberis vulgaris*
European basswood – *Tilia* x *europaea*
European beachgrass – *Ammophila arenaria*
European beech – *Fagus sylvatica*
European bird cherry – *Prunus padus*
European columbine – *Aquilegia vulgaris*
European cranberry – *Vaccinium uliginosum*
European cranberry bush – *Viburnum opulus*
European dwarf cherry – *Prunus fruticosa*
European elder – *Sambucus nigra*
European filbert – *Corylus avellana*
European hackberry – *Celtis australis*
European hairgrass, elegant – *Aira elegantissima*
European hairgrass, silver – *Aira caryophylea*
European hop – *Humulus lupulus*
European hornbeam – *Carpinus betulus, Ostrya carpinifolia*
European larch – *Larix decidua*
European linden – *Tilia europaea*
European mistletoe – *Viscum album*
European mountain ash – *Sorbus aucuparia*
European pasque flower – *Pulsatilla vulgaris*
European red raspberry – *Rubus idaeus*
European silver fir – *Abies pectinata*
European vervain – *Verbena officinalis*
European white birch – *Betula pendula*
European white elm – *Ulmus laevis*
evening primrose – *Oenothera*
evening primrose, basket – *Oenothera deltoides*
evening primrose, common – *Oenothera biennis*
evening primrose, cut-leaved – *Oenothera laciniata*
evening primrose, fragrant – *Oenothera stricta*
evening primrose, large-flowered – *Oenothera glazioviana*
evening primrose, Mexican – *Oenotheria speciosa*
evening primrose, pink – *Oenothera speciosa*
evening scented stock – *Matthiola longipetala*
evening star rain lily – *Cooperia drummondii*
everblooming acacia – *Acacia retinodes*
Everglades palm – *Acoelorraphe wrightii*
evergreen, African – *Syngonium podophyllum*
evergreen ash – *Fraxinus uhdei*
evergreen candytuft – *Iberis sempervirens*
evergreen, Chinese – *Aglaonema modestum*
evergreen clematis – *Clematis armandii*
evergreen dogwood – *Cornus capitata*

evergreen elm, Chinese – *Ulmus parvifolia*
evergreen euonymus – *Euonymus japonicus*
evergreen holly – *Ilex vomitoria*
evergreen mock orange – *Philadelphus mexicanus*
evergreen oak – *Quercus mexicana, Q. cedroensis, Q. virgiliana*
evergreen rose – *Rosa sempervirens*
evergreen sumac – *Rhus sempervirens, R. virens*
everlasting – *Gnaphalium, Helipterum roseum, Xeranthemum annuum*
everlasting flower, yellow – *Helichrysum orientale*
everlasting pea – *Lathyrus latifolius*
everlasting, pearly – *Anaphalis margaritacea*
everlasting, pink and white – *Helipterum roseum*
Eve's needle – *Yucca faxoniana*
exotic love – *Quamoclit lobata*
eyebright – *Euphrasia officinalis*
eye, golden – *Viguiera dentata, V. parishii*
eye, hare's – *Silene dioica*
eye, ox – *Heliopsis helianthoides, H. rubra*
eye, pheasant's – *Adonis, Narcissus poeticus*
eyes, baby blue – *Nemophila menziesii*
eyes, lyreleaf green – *Berlandiera lyrata*

face, stone – *Lithops*
fairyduster – *Calliandra*
fairyduster, Baja – *Calliandra californica*
fairy lily – *Zephyranthes candida, Z. carinata, Z. citrina*
fairy ring agaric – *Marasmius oreades*
fairy slipper – *Calypso bulbosa*
fairy thimbles – *Campanula cochlearifolia*
fall daffodil – *Sternbergia lutea*
fall witchgrass – *Leptoloma cognatum*
false-agave – *Hechtia montana*
false aralia – *Schefflera elegantissima*
false arborvitae – *Thujopsis dolabrata*
false baby's breath – *Galium mollugo*
false beech – *Nothofagus antarctica*
false bird-of-paradise – *Heliconia*
false-brome, annual – *Brachypodium distachyon*
false-brome, heath – *Brachypodium pinnatum*
false buckwheat – *Eriogonum rotundifolium*
false chamomile – *Chamomilla courrantiana*
false cypress – *Chamaecyparis*
false cypress, hinoki – *Chamaecyparis obtusa*
false cypress, sawara – *Chamaecyparis pisifera*
false daisy – *Eclipta prostrata*
false dittany – *Ballota pseudodictamnus*
false dragonhead – *Physostegia virginiana*
false flax – *Camelina microcarpa*
false hellebore – *Veratrum album, V. officinale, V. viride*
false hop – *Justicia brandegeana*
false indigo – *Amorpha fruticosa*
false ipecac – *Psychotria emetica*
false lily-of-the-valley – *Maianthemum dilatatum*
false London rocket – *Sisymbrium loeselii*
false loosestrife – *Ludwigia peploides*
false mallow – *Eremalche*
false mastic – *Mastichodendron foetidissimum*
false mesquite – *Calliandra eriophylla*
false mint – *Dicliptera resupinata*
false morel – *Gyromitra esculenta*
false nettle – *Boehmeria nivea*
false peyote – *Ariocarpus fissuratus*
false pimpernel – *Lindernia dubia*
false poinsettia – *Euphorbia heterophylla*
false ragweed – *Parthenium hysterophorus*
false rose-mallow – *Modiolastrum scoparium*
false rosewood – *Vauquelinia karwinskyi*
false saffron – *Carthamnus tinctorius*
false sea holly – *Eryngium planum*
false spiraea – *Astilbe arendsii*
false tarragon – *Artemisia dracunculoides*
false white pine – *Pinus pseudostrobus*
fancy-leaved caladium – *Caladium bicolor*
fan palm – *Licuala grandis*
fan palm, blue – *Erythea armata*
fan palm, California – *Washingtonia filifera*
fan palm, Chinese – *Licuala grandis*
fan palm, Fiji – *Pritchardia pacifica*
fan palm, Mediterranean – *Chamaerops humilis*
fan palm, Mexican – *Sabal mexicana, Washingtonia robusta*
fan-weed – *Thlaspi arvense*
farewell-to-spring – *Clarkia amoena*
fat duckweed – *Lemna gibba*
fat hen – *Chenopodium album*
fat, winter – *Krascheninnikovia lanata*
fava bean – *Phaseolus lunatus*

feather acacia – *Acacia pennatula*
feather bush – *Lysiloma microphylla, L. divaricatum*
feather finger grass – *Chloris virgata*
feather, gay – *Liatris punctata, L. spicata*
feather geranium – *Chenopodium botrys*
feathergrass, New Mexican – *Stipa neomexicana*
feathergrass, Pyrenean – *Stipa gigantea*
feathergrass, woolly-stalked – *Stipa pennata*
feather palm – *Acrocomia mexicana, Syagrus romanzoffiana*
feather plume – *Dalea formosa*
feather, prince's – *Amaranthus hybridus*
feather silverweed – *Potentilla anserina*
feathertop, Abyssinian – *Pennisetum villosum*
feather tree – *Lysiloma watsonii*
February daphne – *Daphne mezereum*
feijoa – *Feijoa sellowiana*
felwort, marsh – *Swertia perennis*
Fendler rose – *Rosa woodsii* var *fendleri*
Fendler's globemallow – *Sphaeralcea fendleri*
Fendler's hawkweed – *Hieracium fendleri*
Fendler's oak – *Quercus fendleri*
Fendler three-awn – *Aristida purpurea* var *fendleriana*
fennel – *Foeniculum vulgare* var *azoricum*
fennel, dog – *Anthemis cotula*
fennel, giant – *Ferula communis*
fennel-leaf pondweed – *Potamogeton pectinatus*
fennel, sea hog's – *Peucedanum officinale*
fennel, water – *Oenanthe aquatica*
fenugreek – *Trigonella foenum-graecum*
fern – *Nephelea mexicana, Pteridaceae*
fern acacia – *Acacia angustissima*
fern asparagus – *Asparagus setaceus*
fern, bird's nest – *Asplenium nidus*
fern, blechnum – *Blechnum gibbum*
fern, Boston – *Nephrolepis exaltata*
fern, bracken – *Pteridium aquilinum, Pteris*
fern, button – *Pellaea rotundifolia*
fern, California wood – *Dryopteris arguta*
fern, cloak – *Notholaena lemmonii, N. sinuata*
fern, coastal wood – *Dryopteris arguta*
fern, five-finger – *Adiantum pedatum*
fern, flowering – *Osmunda regalis*
fern, hart's tongue – *Phyllitis scolopendrium*
fern, lace – *Cheilanthes*
fern, lady – *Asplenium filix-femina*
fernleaf acacia – *Acacia pennatula*
fernleaf lavender – *Lavandula multifida*
fernleaf yarrow – *Achillea filipendulina*
fern, lip – *Cheilanthes lindheimeri*
fern, maidenhair – *Adiantum*
fern, male – *Dryopteris filix-mas*
fern, mangrove – *Acrostichum aureum*
fern, mosquito – *Azolla filiculoides*
fern-of-the-desert – *Lysiloma microphylla*
fern, ostrich – *Matteuccia struthiopteris*
fern, roundleaf – *Pellaea rotundifolia*
fern, royal – *Osmunda regalis*
fern, scale – *Ceterach officinarum*
fern, scaly cloak – *Astrolepis cochisensis*
fern, shield – *Dryopteris filix-mas*
fern, Sonoran maiden – *Thelypteris puberula* var *sonorensis*
fern, staghorn – *Platycerium*
fern, sweet – *Comptonia aspleniifolia*
fern, sword – *Nephrolepis exaltata*
fern, tree – *Alsophila aculeata, A. salvinii, Blechnum gibbum, Cibotium regale, Cyathea delicatula, C. mexicana, Nephelium mexicana*
fern, water – *Ceratopteris thalictroides*
fern, wood – *Dryopteris*
fescue, alpine – *Festuca ovina* var *brachyphylla*
fescue, Arizona – *Festuca arizonica*
fescue, blue – *Festuca glauca*
fescue, hard – *Festuca longifolia, F. trachyphylla*
fescue, meadow – *Festuca pratensis*
fescue, ratstail – *Vulpia myuros*
fescue, red – *Festuca rubra*
fescue, sheep – *Festuca ovina, F. trachyphylla*
fescue, squirreltail – *Vulpia bromoides*
fescue, tall – *Festuca arundinacea*
fescue, tufted – *Festuca amethystina*
fescue, Valais – *Festuca valesiaca*
fescue, Wallis – *Festuca valesiaca*
fetid marigold – *Dyssodia papposa, Pectis*
fetid passion flower – *Passiflora foetida*
feverfew – *Tanacetum parthenium*
feverfew, Santa Maria – *Parthenium hysterophorus*
feverplant, Jamaica – *Tribulus cistoides*
fever tree – *Cinchona officinalis*

few-flowered desert marigold – *Baileya pauciradiata*
few-flowered muhly – *Muhlenbergia pauciflora*
fiber, Mexican hat – *Sabal mexicana*
fiber optics plant – *Scirpus cernuus*
fiber, Tampico – *Hesperaloe parviflora*
fiddle dock – *Rumex pulcher*
fiddleleaf fig – *Ficus lyrata*
fiddleneck – *Amsinckia douglasiana, A. spectabilis*
fiddleneck, bristly – *Amsinckia tessallata*
fiddlewood – *Citharexylum scabrum, Vitex gaumeri*
field alison – *Alyssum minus*
field bindweed – *Convolvulus arvensis*
field blewit – *Lepista saeva*
field chickweed – *Cerastium arvense*
field-cress, Austrian – *Rorippa austriaca*
field eryngo – *Eryngium campestre*
field madder – *Sherardia arvensis*
field-marigold – *Calendula arvensis*
field mint – *Mentha arvensis*
field mushroom – *Agaricus campestris*
field mustard – *Brassica rapa*
field pansy – *Viola arvensis*
field parasol – *Macrolepiota procera*
field pea – *Pisum sativum* var *arvense*
field penny-cress – *Thlaspi arvense*
field pepperwort – *Lepidium campestre*
field poppy, Flander's – *Papaver rhoeas*
fields, gold – *Lasthenia californica*
field southernwood – *Artemisia campestris*
field woundwort – *Stachys arvensis*
fiery reed orchid – *Epidendrum ibaguense*
fig – *Ficus*
fig, Benjamin – *Ficus benjamina*
fig, cliff – *Ficus palmeri, F. petiolaris*
fig, creeping – *Ficus pumila*
fig, desert rock – *Ficus palmeri*
fig, devil's – *Argemone mexicana*
fig, East Indian – *Ficus benghalensis*
fig, edible – *Ficus carica*
fig, Eltham Indian – *Opuntia dillenii*
fig, fiddleleaf – *Ficus lyrata*
fig, goat – *Ficus vulgaris*
fig, Hottentot – *Carpobrotus edulis*
fig, Indian – *Opuntia megacantha*
fig, Indian laurel – *Ficus microcarpa*
fig-leaved gourd – *Cucurbita ficifolia*
fig-marigold – *Carpobrotus chilensis, C. edulis*
fig, rock – *Ficus padifolia, F. petiolaris*
fig, sacred – *Ficus religiosa*
fig, strangler – *Ficus cotinifolia, F. padifolia*
fig, sycomore – *Ficus sycomorus*
fig, weeping – *Ficus benjamina*
fig, wild – *Ficus*
figwort – *Scrophularia*
figwort, French – *Scrophularia canina*
figwort, water – *Scrophularia auriculata*
Fiji fan palm – *Pritchardia pacifica*
filaree – *Erodium*
filaree, Texas – *Erodium texanum*
filbert – *Corylus cornuta* var *californica, C. maxima*
filbert, European – *Corylus avellana*
filbert, giant – *Corylus maxima*
fimbristylis, hot-springs – *Fimbristylis thermalis*
fine-leaved marigold – *Tagetes lunulata*
finestem stylo – *Stylosanthes guianensis*
finger aralia – *Schefflera elegantissima*
finger cactus – *Mammillaria longimamma*
fingered speedwell – *Veronica triphyllos*
finger grass – *Chloris virgata, Digitaria*
finger grass, feather – *Chloris virgata*
finger grass, smooth – *Digitaria ischaemum*
finger-leaved gourd – *Cucurbita digitata*
finger millet – *Eleusine coracana*
Finnish whitebeam – *Sorbus* x *hybrida*
finocchio – *Foeniculum vulgare* var *azoricum*
fir – *Abies*
fir, alpine – *Abies lasiocarpa*
fir, balsam – *Abies balsamea*
fir, bristlecone – *Abies bracteata*
fir, California red – *Abies magnifica*
fir, cascade – *Abies amabilis*
fir, China – *Cunninghamia lanceolata*
fir, Douglas – *Pseudotsuga menziesii*
fir, Durango – *Abies durangensis*
fire barrel cactus – *Ferocactus gracilis*
fire bush, Chilean – *Embothrium coccineum*
firecracker plant – *Russelia equisetiformis*
firecracker vine – *Quamoclit lobata*
fire dragon – *Acalypha wilkesiana*
fireman's helmet – *Erythrina crista-galli*
fire pink – *Silene laciniata*
fire plant, Mexican – *Euphorbia heterophylla*

firethorn – *Pyracantha coccinea, P. koidzumii*
fire tree – *Spathodea campanulata*
fir, European silver – *Abies pectinata*
fireweed – *Celosia argentea, Epilobium*
fireweed, parched – *Epilobium brachycarpum*
fireweed, rancher's – *Amsinckia menziesii* var *intermedia*
fireweed, red – *Epilobium angustifolium*
firewheel – *Gaillardia pulchella*
firewheel tree – *Stenocarpus sinuatus*
firewood agaric – *Tricholoma flavovirens*
fir, grand – *Abies grandis*
fir, Guatemalan – *Abies guatemalensis*
fir, joint – *Ephedra nevadensis, E. trifurca*
fir, Korean – *Abies koreana*
fir, Mexican Douglas – *Pseudotsuga lindleyana*
fir mistletoe – *Phoradendron pauciflorum*
fir, noble – *Abies procera*
fir, Nordmann – *Abies nordmanniana*
fir, Pacific silver – *Abies amabilis*
fir, Rocky Mountain – *Abies lasiocarpa*
fir, sacred – *Abies religiosa*
fir, Santa Lucia – *Abies bracteata*
fir, Shasta red – *Abies magnifica* var *shastensis*
fir, Siberian – *Abies sibirica*
fir, silver – *Abies amabilis*
fir, Spanish – *Abies pinsapo*
fir, subalpine – *Abies lasiocarpa*
fir, vejar – *Abies vejarii*
fir, white – *Abies concolor*
fishhook barrel cactus – *Ferocactus wislizenii*
fish-hook cactus – *Mammillaria, Thelocactus setispinus*
fish-hook cactus, Arizona – *Mammillaria grahamii*
fish-hook cactus, clustered – *Mammillaria thornberi*
fishhook, tangled – *Mammillaria microcarpa*
fishhook, Yaqui – *Mammillaria tetrancistra*
fish-poison tree – *Piscidia mollis*
fishtail palm, clustered – *Caryota mitis*
fishtail wine palm – *Caryota urens*
fitches – *Nigella sativa*
five-finger fern – *Adiantum pedatum*
five-finger grass – *Potentilla tormentilla*
five-finger ivy – *Parthenocissus quinquefolia*
five-leaf morning-glory – *Ipomoea cairica*
five-needled Chihuahua pine – *Pinus leiophylla*
five-needled dogweed – *Thymophylla pentachaeta* var *belenidium*
flag, blue – *Iris germanica, I. versicolor*
flag, corn – *Gladiolus italicus*
flag, crimson – *Schizostylis coccinea*
flag iris, yellow – *Iris pseudacorus*
flag, Japanese sweet – *Acorus gramineus*
flagroot – *Acorus calamus*
flag, Spanish – *Quamoclit lobata*
flag, sweet – *Acorus calamus, Phragmites australis*
flag, western blue – *Iris missouriensis*
flag, white – *Iris germanica*
flag, yellow – *Iris pseudoacorus*
flamboyant tree – *Delonix regia*
flame acanthus – *Anisacanthus quadrifidus*
flameberry scratchbush – *Urera caracasana*
flamebush, Brazilian – *Calliandra tweedii*
flamebush, Trinidad – *Calliandra tweedii*
flameflower – *Phlox paniculata, Talinum paniculatum*
flame freesia – *Tritonia crocata*
flame, heart of – *Bromelia balansae*
flame, jungle – *Ixora coccinea*
flame-leaf sumac – *Rhus copallina*
flame lily – *Gloriosa superba*
flame of the forest – *Spathodea campanulata*
flame of the woods – *Ixora coccinea*
flame orchid – *Spiranthes cinnabarina*
flame tree – *Brachychiton acerifolius, Delonix regia*
flame tree, Australian – *Brachychiton acerifolius*
flame tree, Chinese – *Koelreuteria bipinnata*
flame tree, pink – *Brachychiton discolor*
flame tree, yellow – *Peltophorum pterocarpum*
flame vine – *Pyrostegia venusta*
flame vine, Mexican – *Senecio chenopodioides, S. confusus*
flaming dragon tree – *Cordyline fruticosa*
flaming Katy – *Kalanchoe blossfeldiana*
flamingo flower – *Anthurium scherzerianum*
flamingo lily – *Anthurium andraeanum*
flamingo plant – *Justicia carnea*
Flander's field poppy – *Papaver rhoeas*
flannelbush – *Fremontodendron californicum*
flatbud prickly poppy – *Argemone munita*
flat-leaved parsley – *Petroselinum crispum* var *neapolitanum*

flat peavine – *Lathyrus sylvestris*
flatpod peavine – *Lathyrus cicera*
flat sedge – *Cyperus hermaphroditus, C. laevigatus*
flat-thorn acacia – *Acacia gladiata*
flat-top buckwheat brush – *Eriogonum fasciculatum*
flat tree dahlia – *Dahlia excelsa*
flax – *Linum*
flax, blue – *Linum lewisii, L. narbonense*
flax, common – *Linum usitatissimum*
flax, cultivated – *Linum usitatissimum*
flax, false – *Camelina microcarpa*
flax, flowering – *Linum grandiflorum*
flax, golden – *Linum flavum*
flax-leaved broom – *Genista linifolia*
flax, New Zealand – *Phormium tenax*
flax, pale – *Linum bienne*
flax, perennial blue – *Linum perenne*
flax, prairie – *Linum lewisii*
flax, scarlet – *Linum grandiflorum*
flax, toad – *Cymbalaria, Linaria*
fleabane – *Erigeron speciosus*
fleabane, Canadian – *Conyza Canadensis*
fleabane, common – *Pulicaria dysenterica*
fleabane, crisp – *Conyza bonariensis*
fleabane daisy – *Erigeron*
fleabane, marsh – *Pluchea sericea, Pulicaria paludosa*
fleabane, naked stem – *Erigeron nudiflorus*
fleabane, plains – *Erigeron modestus*
fleabane, prairie – *Erigeron modestus*
fleabane, running – *Erigeron nudiflorus*
fleabane, salt marsh – *Pluchea odorata*
fleawort – *Senecio integrifolius*
fleeceflower, bokhara – *Polygonum baldschuanicum*
fleur-de-lis – *Iris germanica, I. pseudacorus*
flint corn – *Zea mays* var *indurata*
flixweed – *Descuriana sophia*
floating-heart – *Nymphoides humboldtianum, N. indica*
floating-leaved pondweed – *Potamogeton natans*
floating pennywort – *Hydrocotyle ranunculoides*
Florentine iris – *Iris florentina*
Florida anise tree – *Illicium floridanum*
Florida arrowroot – *Zamia pumila*
Florida cherry – *Eugenia uniflora*
Florida gamagrass – *Tripsacum floridanum*
Florida gooseberry – *Pereskia aculeata*
Florida pansy – *Torenia fournieri*
Florida velvet bean – *Mucuna pruriens* var *utilis*
florist's chrysanthemum – *Chrysanthemum morifolium*
florist's cineraria – *Senecio hybridus*
florist's cyclamen – *Cyclamen persicum*
florist's smilax – *Asparagus medeoloides*
floss flower – *Ageratum houstonianum*
floss silk tree – *Ceiba speciosa*
flour corn – *Zea mays* var *amylacea*
flowering almond – *Prunus triloba*
flowering apricot, Japanese – *Prunus mume*
flowering ash – *Fraxinus cuspidata, F. ornus*
flowering cherry – *Prunus sargentii, P. serrula*
flowering cherry, Japanese – *Prunus subhirtella*
flowering cherry, yoshino – *Prunus* x *yedoensis*
flowering crabapple, Japanese – *Malus floribunda*
flowering dogwood – *Cornus florida*
flowering fern – *Osmunda regalis*
flowering flax – *Linum grandiflorum*
flowering maple – *Abutilon*
flowering peach – *Prunus davidiana*
flowering plum – *Prunus* x *blireiana*
flowering plum, Japanese – *Prunus mume*
flowering quince – *Chaenomeles speciosa*
flowering rush, sawah – *Limnocharis flava*
flowering tobacco – *Nicotiana alata*
flowering-willow – *Chilopsis linearis*
flower-of-an-hour – *Hibiscus trionum*
fluellin – *Kickxia elatine, K. spuria*
fluff grass – *Erioneuron pulchellum*
flutter-mill – *Oenothera macrocarpa*
fly agaric – *Amanita muscaria*
foamflower – *Tiarella cordifolia*
fodder beet – *Beta vulgaris*
fodder vetch – *Vicia villosa*
fog, Yorkshire – *Holcus lanatus*
food candle tree – *Parmentiera edulis*
food inga – *Inga edulis*
food, sand – *Pholisma sonorae*
fool's parsley – *Aethusa cynapium*
fool's watercress – *Apium nodiflorum*
foot, bear's – *Helleborus viridis*

foot, cocks – *Dactylis glomerata*
foot, colts – *Petasites, Tussilago farfara*
foot, cow – *Piper auritum, P. peltatum*
foot, crow – *Ranunculus*
foothill palo verde – *Cercidium microphyllum*
foothill pine – *Pinus sabiniana*
foot, owl's – *Hymenoxys richardsonii*
foot, turkey – *Andropogon hallii*
forestiera, narrow-leaf – *Forestiera angustifolia*
forget-me-not – *Anchusa capensis, Myosotis*
forget-me-not, Cape – *Anchusa capensis*
forget-me-not, changing – *Myosotis discolor*
forget-me-not, spring – *Myosotis micrantha*
forget-me-not, summer – *Anchusa capensis*
forget-me-not tree – *Duranta erecta*
forget-me-not, water – *Myosotis laxa, M. palustris*
forget-me-not, wood – *Myosotis sylvatica*
forked catchfly – *Silene dichotoma*
forking larkspur – *Delphinium consolida*
forsythia – *Forsythia suspensa*
forsythia, weeping – *Forsythia suspensa*
fortuniana – *Rosa fortuniana*
fountain grass – *Pennisetum setaceum*
fountain palm, Chinese – *Livistona chinensis*
fountain tree – *Spathodea campanulata*
four-leaf senna – *Senna oxyphylla*
four-leaved allseed – *Polycarpon tetraphyllum*
four-o'clock – *Mirabilis*
four-o'clock, desert – *Mirabilis multiflora*
four-o'clock, garden – *Mirabilis jalapa*
four-o'clock, trailing – *Allionia incarnata*
four-o'clock, sweet – *Mirabilis longiflora*
four-o'clock, wild – *Mirabilis multiflora*
fourwing saltbush – *Atriplex canescens, A. linearis*
fowl bluegrass – *Poa palustris*
foxberry – *Vaccinium vitis-idaea*
foxglove – *Digitalis purpurea, D. thapsi*
foxglove, common – *Digitalis purpurea*
foxglove, dark – *Digitalis obscura*
foxglove, Grecian – *Digitalis lanata*
foxglove, Mexican – *Tetranema mexicana*
foxglove, narrow-leaf – *Digitalis obscura*
foxglove, rusty – *Digitalis ferruginea*
foxglove, small yellow – *Digitalis lutea*
foxglove, yellow – *Digitalis grandiflora*
fox grape – *Vitis labrusca*
fox, red – *Celosia argentea*
fox sedge – *Carex vulpinoidea*
foxtail – *Alopecurus*
foxtail agave – *Agave attenuata*
foxtail barley – *Hordeum jubatum*
foxtail, bristly – *Setaria verticillata*
foxtail-brome – *Bromus alopecurus*
foxtail chess – *Bromus madritensis*
foxtail, green – *Setaria viridis*
foxtail lily – *Eremurus robustus*
foxtail, meadow – *Alopecurus pratensis*
foxtail millet – *Panicum italicum, Setaria italica*
foxtail millet, European – *Setaria italica* var *moharica*
foxtail pine – *Pinus balfouriana*
foxtail, short-awn – *Alopecurus aequalis*
foxtail, water – *Alopecurus geniculatus*
fragrant ash – *Fraxinus cuspidata* var *macropetala*
fragrant clematis – *Clematis flammula*
fragrant elephant tree – *Bursera fagarioides*
fragrant evening primrose – *Oenothera stricta*
fragrant mimosa – *Mimosa borealis*
fragrant oriental lily – *Lilium auratum*
fragrant plantain lily – *Hosta plantaginea*
fragrant snowbell – *Styrax obassia*
fragrant sumac – *Rhus aromatica*
fragrant waterlily – *Nymphaea odorata*
fraise du bois – *Fragaria vesca*
francoa, pink – *Francoa sonchifolia*
frangipani – *Plumeria acutifolia, P. rubra*
frangipani, Japanese – *Adenium obesum*
frangipani, Mexican – *Plumeria acutifolia*
frangipani, nosegay – *Plumeria rubra*
frankenia, Palmer's – *Frankenia palmeri*
frankincense – *Boswellia carteri, B. thurifera*
fraxinella – *Dictamnus albus*
freesia – *Freesia*
freesia, flame – *Tritonia crocata*
Frémont cottonwood – *Populus fremontii*
fremontia – *Fremontodendron californicum*
French bastard rocket – *Erucastrum gallicum*
French cranesbill – *Geranium endressii*
French figwort – *Scrophularia canina*
French haricot, round-white-seeded – *Phaseolus vulgaris*
French lavender – *Lavandula dentata, L. stoechas*

French marigold – *Tagetes patula*
French plantain, green – *Mus acuminata* x *balbisiana*
French pussy willow – *Salix caprea*
French Roman hyacinth – *Hyacinthus orientalis* var *albulus*
French rose – *Rosa gallica*
French sorrel – *Rumex scutatus*
French tamarisk – *Tamarix gallica*
French tarragon – *Artemisia dracunculus*
French truffle – *Tuber melanosporum*
French turpentine pine – *Pinus pinaster*
French vanilla – *Vanilla pompona*
French vetch – *Vicia narbonensis*
Fresno chile – *Capsicum annuum*
fried egg, cowboy's – *Argemone gracilenta*
friendship plant, pan-American – *Pilea involucrata*
fringed amaranth – *Amaranthus fimbriatus*
fringed crabgrass – *Digitaria ciliaris*
fringed spineflower – *Chorizanthe fimbriata*
fringed spurge – *Chamaesyce setiloba*
fringed wormwood – *Artemisia frigida*
fringe hyacinth – *Muscari comosum*
fringe-leaved loosestrife – *Lysimachia ciliata*
fringe, water – *Nymphoides peltata*
fritillary, Persian – *Fritillaria persica*
frog-fruit – *Lippia callicarpaefolia*
frog-fruit, diamond leaf – *Phyla strigulosa* var *sericea*
frog rush – *Juncus ambiguous*
frog's-bit buttercup – *Ranunculus hydrocharis*
fuchsia – *Fuchsia*
fuchsia, Australian – *Correa speciosa*
fuchsia, California – *Epilobium canum*
fuchsia, hybrid – *Fuchsia* x *hybrida*
fuchsia, mountain – *Fuchsia microphylla*
fuki – *Petasites japonicus*
Fuller's teasel – *Dipsacus fullonum, D. sativus*
fullmoon maple – *Acer japonicum*
fumitory – *Fumaria officinalis, F. parviflora*
fumitory, bulbous – *Corydalis cava*
fumitory, ramping – *Fumaria capreolata*
fumitory, spring – *Corydalis solida*
fungus, beefsteak – *Fistulina hepatica*
fungus, cloud ear – *Auricularia auricula-judae*
fungus, coral – *Clavaria, Ramaria flavia*
fungus, golden coral – *Ramaria flava*
fungus, red belt – *Fomes pinicola*
fungus, red-tipped coral – *Ramaria botrytis*
fungus, upright coral – *Ramaria stricta*
fungus, white jelly – *Tremella fuciformis*
fungus, wood ear – *Auricularia polytricha*
fungus, yellow night – *Tricholoma flavovirens*
furze – *Ulex europaea*
fustic – *Maclura tinctoria*
fustic tree, drug – *Maclura tinctoria*

Gairdner's yampah – *Perideridia gairdneri*
galangal, lesser – *Alpinia officinarum*
galbanum – *Ferula galbaniflua*
gale of the east – *Centaurium erythraea*
gale, sweet – *Myrica gale*
galingale – *Alpinia officinarum, Cyperus*
galingale, American – *Cyperus eragrostis*
galingale, brown – *Cyperus fuscus*
galingale, edible – *Cyperus eragrostis*
galingale, purple – *Cyperus rotundus*
galingale, small-flowered – *Cyperus difformis*
gall – *Papaver somniferum*
gallant soldier – *Galinsoga parviflora*
galleta – *Pleuraphis jamesii*
galleta, big – *Pleuraphis rigida*
galloping cactus – *Stenocereus gummosus*
gamagrass – *Tripsacum laxum*
gamagrass, eastern – *Tripsacum dactyloides*
gamagrass, Florida – *Tripsacum floridanum*
gamagrass, Mexican – *Tripsacum lanceolatum*
Gambell oak – *Quercus gambellii*
gamboa juniper – *Juniperus gamboana*
gamboge – *Garcinia hanburyi*
garabato – *Pisonia capitata*
garbanzo – *Cicer arietinum*
garden balm – *Melissa officinalis*
garden balsam – *Impatiens balsamina*
garden beet – *Beta vulgaris*
garden burnet – *Sanguisorba minor* ssp *muricata, S. officinalis*
garden carrot – *Daucus carota* ssp *sativus*
garden cosmos – *Cosmos bipinnatus*
garden cress – *Lepidium sativum*
garden crocus – *Crocus chrysanthus*
gardener's garters – *Phalaris arundinacea*
garden four-o'clock – *Mirabilis jalapa*
garden geranium – *Pelargonium* x *hortorum*
garden heliotrope – *Heliotropium arborescens, Valeriana officinalis*
garden hydrangea – *Hydrangea macrophylla*

gardenia – *Gardenia jasminoides*
gardenia, common – *Gardenia augusta*
garden iris – *Iris germanica*
garden lupin – *Lupinus polyphyllus*
garden mignonette – *Reseda odorata*
garden monkshood – *Aconitum napellus*
garden myrrh – *Myrrhis odorata*
garden nasturtium – *Tropaeolum majus*
garden orache – *Atriplex hortensis*
garden oracle – *Atriplex hortensis*
garden pansy – *Viola* x *wittrockiana*
garden pea – *Pisum sativum*
garden penstemon – *Penstemon* x *gloxinioides*
garden phlox – *Phlox paniculata*
garden pink – *Dianthus plumarius*
garden portulaca – *Portulaca grandiflora*
garden radish – *Raphanus sativus*
garden rhubarb – *Rheum rhaponticum*
garden-rocket – *Eruca vesicaria* ssp *sativa*
garden sage – *Salvia officinalis*
garden sorrel – *Rumex acetosa*
garden strawberry – *Fragaria ananassa*
garden thyme – *Thymus vulgaris*
garden tulip – *Tulipa gesneriana*
garden verbena – *Verbena* x *hybrida*
garden violet – *Viola* x *wittrockiana*
garden zinnia – *Zinnia elegans*
garland daisy – *Chrysanthemum coronarium*
garland daphne – *Daphne cneorum*
garland flower – *Hedychium coronarium*
garland, mountain – *Clarkia unguiculata*
garlic – *Allium sativum*
garlic chives – *Allium tuberosum*
garlic, crow – *Allium vineale*
garlic, golden – *Allium moly*
garlic, Levant – *Allium ampeloprasum*
garlic, Naples – *Allium neapolitanum*
garlic pear – *Crateva palmeri*
garlic, round-headed – *Allium sphaerocephalum*
garnet berry – *Ribes rubrum*
garters, gardener's – *Phalaris arundinacea*
gayfeather – *Liatris punctata, L. spicata*
gazania – *Gazania nivea*
gean cherry – *Prunus avium* var *juliana*
gearstem cactus – *Peniocereus striatus*
geiger tree – *Cordia dodecandra, C. sebestena*
genip – *Melicocca bijuga*
genipap – *Genipa americana*
genius mushroom – *Psilocybe yungensis*
gentian – *Gentiana*
gentian, autumn – *Gentianella amarella*
gentian, catchfly – *Eustoma exaltatum, E. grandiflorum*
gentian, cross – *Gentiana cruciata*
gentian, green – *Swertia radiata*
gentian, large-flowered – *Gentiana clusii*
gentian, pygmy – *Gentiana prostrata*
gentian, prairie – *Eustoma grandiflorum*
gentian, rose – *Sabatia*
gentian sage – *Salvia patens*
gentian, slender – *Gentianella tenella*
gentian speedwell – *Veronica gentianoides*
gentian, spring – *Gentiana verna*
gentian, squarestem rose – *Sabatia angularis*
gentian, stemless – *Gentiana acaulis*
gentian, willow – *Gentiana asclepiadea*
gentian, yellow – *Gentiana lutea*
geranium – *Geranium*
geranium, Carolina – *Geranium carolinianum*
geranium, common – *Pelargonium* x *hortorum*
geranium, feather – *Chenopodium botrys*
geranium, garden – *Pelargonium* x *hortorum*
geranium, ivy – *Pelargonium peltatum*
geranium, Mexican – *Geranium mexicanum*
geranium, nutmeg – *Pelargonium* x *fragrans*
geranium, purple – *Geranium caespitosum*
geranium, regal – *Pelargonium* x *domesticum*
geranium, rose – *Pelargonium graveolens*
geranium, scented – *Pelargonium odoratissimum*
geranium, strawberry – *Saxifraga stolonifera*
German chamomile – *Chamomilla suaveolens*
germander – *Teuerium*
germander, bush – *Teucrium fruticans*
germander sage – *Salvia chamaedryoides*
germander speedwell – *Veronica chamaedrys*
germander, wall – *Teucrium* x *lucidrys*
germander, water – *Teucrium scordium*
germander, white-felted – *Teucrium polium*
German iris – *Iris germanica*
German ivy – *Delairea odorata*
German prune – *Prunus domestica* ssp *insititia*
Geyer's aster – *Aster laevis*
gherkin, West Indian – *Cucumis anguira*
ghost, Miss Willmott's – *Eryngium giganteum*
ghost orchid – *Epipogium aphyllum*
giant allium – *Allium giganteum*

giant bamboo – *Dendrocalamus giganteus*
giant barrel cactus – *Echinocactus platyacanthus*
giant bean – *Canavalia plagiosperma*
giant chinquapin – *Chrysolepis chrysophylla*
giant crape myrtle – *Lagerstroemia speciosa*
giant dogwood – *Cornus controversa*
giant dropseed – *Sporobolus giganteus*
giant fennel – *Ferula communis*
giant filbert – *Corylus maxima*
giant granadilla – *Passiflora quadrangularis*
giant hardy phlox – *Phlox paniculata*
giant hogweed – *Heracleum mantegazzianum*
giant horsetail – *Equisetum telmateia*
giant hyssop – *Agastache micrantha*
giant Mexican lemon hyssop – *Agastache mexicana*
giant orange lobelia – *Lobelia laxiflora*
giant orchid – *Himantoglossum longibracteatum*
giant potato creeper – *Solanum wendlandii*
giant puffball – *Calvatia gigantea*
giant reed – *Arundo donax*
giant sandbur – *Cenchrus palmeri*
giant scabious, white-headed – *Cephalaria leucantha*
giant sequoia – *Sequoiadendron giganteum*
giant snowdrop – *Galanthus elwesii*
giant thevetia – *Thevetia thevetioides*
giant white morning-glory – *Turbina corymbosa*
giant yucca – *Yucca elephantipes*
gibnut fruit – *Alibertia edulis*
Gibraltar candytuft – *Iberis gibraltarica*
Gila chokecherry – *Prunus serotina* var *rufula*
Gilead, balm-of – *Commiphora opobalsamum, Populus balsamifera*
gilia – *Gilia*
gillyflower – *Matthiola incana*
ginger – *Zingiber officinale*
ginger, crape – *Costus speciosus*
ginger, kahili – *Hedychium gardnerianum*
ginger lily, white – *Hedychium coronarium*
ginger, Malay – *Costus speciosus*
ginger, red – *Alpinia purpurata*
ginger root – *Zingiber officinale*
ginger, shell – *Alpina zerumbet*
ginger, spiral – *Costus*
ginger, torch – *Etlingera elatior*
ginger, white – *Hedychium coronarium*
ginger, wild – *Asarum europaeum*
ginseng, American – *Panax quinquefolium*
ginseng, blue – *Caulophyllum thalictroides*
ginseng, Chinese – *Panax ginseng*
ginseng, Korean – *Panax ginseng*
ginseng, Russian – *Eleutherococcus senticosus*
ginseng, Siberian – *Eleutherococcus senticosus*
gipsywort – *Lycopus europaeus*
give-and-take palm – *Cryosophila stauracantha*
gladiola – *Gladiolus hortulanus*
gladiolus, baby – *Gladiolus colvillei*
gladiolus, butterfly – *Gladiolus dalenii*
gladiolus, Byzantine – *Gladiolus communis* ssp *byzantinum*
Gladwin iris – *Iris foetidissima*
glass, Venus' looking – *Specularia speculum*
glasswort – *Salicornia europaea*
glaucous-stemmed brassavola – *Laelia glauca*
globe amaranth – *Gomphrena globosa*
globe daisy – *Globularia*
globe daisy, cespitose – *Globularia alypium*
globe daisy, heart-leaved – *Globularia cordifolia*
globe candytuft – *Iberis umbellata*
globe cucumber – *Citrullus colocynthis*
globeflower – *Trollius europaeus*
globemallow – *Sphaeralcea*
globemallow, annual – *Sphaeralcea coulteri*
globemallow, desert – *Sphaeralcea ambigua*
globemallow, Fendler's – *Sphaeralcea fendleri*
globemallow, narrowleaf – *Sphaeralcea angustifolia*
globemallow, scarlet – *Sphaeralcea coccinea*
globe thistle – *Echinops exaltatus, E. sphaerocephalus*
gloriosa daisy – *Rudbeckia hirta*
gloriosa lily – *Gloriosa superba*
glorybower – *Clerodendrum*
glorybower, bleeding heart – *Clerodendrum thomsoniae*
glorybower, harlequin – *Clerodendrum trichotomum*
glorybower, Java – *Clerodendrum buchananii*
glorybush – *Tibouchina spathulata, T. urvilleana*
glory lily – *Gloriosa superba*

glory, morning – *Calystegia, Convolvulus, Ipomoea, Jacquemontia abutiloides*
glory pea – *Clianthus formosus*
glory, star – *Ipomoea quamoclit*
glory vine, crimson – *Vitis coignetiae*
glossy abelia – *Abelia* x *grandiflora*
glossy privet – *Ligustrum lucidum*
glove, fox – *Digitalis*
gloxinia – *Sinningia speciosa*
gloxinia, tree – *Kohleria deppeana*
goat fig – *Ficus vulgaris*
goat-foot creeper – *Ipomoea pes-caprae*
goatgrass – *Aegilops ovata*
goatgrass, barbed – *Aegilops triuncialis*
goatgrass, jointed – *Aegilops cylindrica*
goat head – *Tribulus terrestris*
goat-nut – *Simmondsia chinensis*
goat's beard – *Aruncus dioicus, Tragopogon dubius, T. pratensis*
goat's horn – *Astrophytum capricorne*
goat's rue – *Galega officinalis*
godetia – *Clarkia amoena*
gold-ban lily – *Lilium auratum*
gold, basket-of – *Aurinia saxatilis*
golden alison – *Alyssum saxatile*
golden apple – *Spondias cytherea*
golden apple mint – *Mentha* x *gracilis*
goldenaster – *Arnica, Heterotheca*
golden barrel cactus – *Echinocactus grusonii*
golden bells – *Forsythia* x *intermedia*
goldenbush – *Ericameria, Isocoma menziesii*
goldenbush, boundary – *Ericameria brachylepis*
goldenbush, desert – *Viguiera parishii*
goldenbush, Palmer's – *Ericameria palmeri*
goldenchain, common – *Laburnum anagyroides*
golden chain tree – *Laburnum anagyroides, Senna alata*
golden cinquefoil – *Potentilla aurea*
golden coral fungus – *Ramaria flava*
golden crownbeard – *Verbesina encelioides* ssp *exauriculata*
golden cup – *Hunnemannia fumariifolia*
goldencup oak – *Quercus chrysolepis*
golden currant – *Ribes aureum*
golden dewdrop – *Duranta repens*
golden dock – *Rumex maritimus*
golden dyssodia – *Thymophylla pentachaeta* var *belenidium*
golden eggs – *Cochlospermum vitifolium*
golden-eye – *Viguiera dentata, V. parishii*
golden flax – *Linum flavum*
golden-fleece – *Ericameria arborescens*
golden garlic – *Allium moly*
goldenglow, wild – *Bidens laevis*
golden heliconia – *Heliconia latispatha*
golden larch – *Pseudolarix amabilis*
golden marguerite – *Anthemis tinctoria*
golden oak – *Quercus tuberculata*
golden old man – *Pilosocereus chrysacanthus*
goldenrain tree – *Koelreuteria paniculata*
golden-rayed lily – *Lilium auratum*
goldenrod – *Solidago*
goldenrod, Californica – *Solidago californica*
goldenrod, Canada – *Solidago canadensis* ssp *elongata*
goldenrod, rayless – *Ericameria heterophyllus*
goldenrod, smooth – *Solidago gigantea*
golden saxifrage – *Chrysosplenium glechomifolium*
golden seal – *Hydrastis canadensis*
golden shower – *Cassia fistula, Pyrostegia venusta*
golden-spined cereus – *Bergerocactus emoryi*
golden spoon – *Byrsonima crassifolia*
golden thistle – *Scolymus hispanicus*
golden timothy – *Setaria sphacelata*
goldentop – *Lactuca virosa, Psilocybe cubensis*
golden trumpet – *Allamanda cathartica*
golden wattle – *Acacia longifolia*
golden wattle, Sydney – *Acacia longifolia*
golden-wave – *Coreopsis grandiflora*
goldenwaves – *Coreopsis tinctoria*
goldenweed, larchleaf – *Ericameria laricifolia*
golden weeping willow – *Salix alba* var *tristis*
goldfields – *Lasthenia californica*
goldflower, clustered – *Hymenoxys scaposa*
goldmoss sedum – *Sedum acre*
gold of pleasure – *Camelina sativa*
goldpoppy – *Eschscholzia glyptosperma*
gold, rain of – *Galphimia glauca*
goldthread – *Coptis groenlandica*
gold torch – *Heliconia psittacorum*
gold tree – *Tabebuia chrysantha, T. donnell-smithii*
goldweed – *Verbesina encelioides*

golf-ball pincushion – *Mammillaria pottsii*
gongo pea – *Cajanus cajun*
Goodding ash – *Fraxinus gooddingii*
Gooding's black willow – *Salix gooddingii*
Gooding willow – *Salix gooddingii*
good King Henry – *Chenopodium bonus-henricus*
good luck plant – *Cordyline fruticosa*
goodly fruit – *Citrus medica*
gooseberry – *Ribes*
gooseberry, Barbados – *Pereskia aculeata*
gooseberry, Cape – *Physalis peruviana*
gooseberry, Ceylon – *Dovyalis hebecarpa*
gooseberry currant – *Ribes montigenum*
gooseberry, Florida – *Pereskia aculeata*
gooseberry, mountain – *Ribes montigenum*
gooseberry, otaheite – *Phyllanthus acidus*
gooseberry, Spanish – *Pereskia aculeata*
gooseberry, sweet – *Ribes grossularia*
gooseberry, West Indian – *Pereskia aculeata*
goosefoot – *Chenopodium*
goosefoot, narrow-leaved – *Chenopodium leptophyllum*
goosefoot, nettle-leaf – *Chenopodium murale*
goosefoot, oak-leaved – *Chenopodium glaucum*
goosefoot, pitseed – *Chenopodium berlandieri*
goosefoot, red – *Chenopodium rubrum*
goosefoot, small red – *Chenopodium chenopodioides*
goose grass – *Eleusine indica, Galium aparine*
gooseneck loosestrife – *Lysimachia clethroides*
goose plum – *Prunus americana*
gopher plant – *Euphorbia lathyris*
gorgeous laelia – *Laelia superbiens*
gorse – *Ulex europaea*
gotu kola – *Hydrocotyle asiatica*
gourd – *Cucurbita maxima*
gourd, bitter – *Citrullus colocynthis*
gourd, bottle – *Lagenaria siceraria*
gourd, buffalo – *Cucurbita foetidissima*
gourd, club – *Trichosanthes cucumerina* var *anguina*
gourd, fig-leaved – *Cucurbita ficifolia*
gourd, finger-leaved – *Cucurbita digitata*
gourd, Indian bitter – *Momordica charantia*
gourd, Indian snake – *Trichosanthes cucumerina* var *anguina*
gourd, Malabar – *Cucurbita ficifolia*
gourd, pointed – *Trichosanthes dioica*
gourd, speckled – *Melothria pendula*
gourd tree – *Crescentia alata*
gourd, viper's – *Trichosanthes cucumerina* var *anguina*
gourd, wax – *Benincasa hispida*
gourd, white-flowered – *Lagenaria siceraria*
gourd, wild – *Cucurbita foetidissima*
gout plant – *Jatropha podagrica*
gout weed – *Aegopodium podagraria*
governor's plum – *Flacourtia indica, F. inermis*
Gowen cypress – *Cupressus goveniana*
grace, herb of – *Ruta graveolens*
Graham mimosa – *Mimosa grahamii*
grain – *Poaceae*
grains of paradise – *Aframomum melegueta*
grama – *Bouteloua hirsuta, B. reflexa*
grama, black – *Bouteloua eriopoda*
grama, blue – *Bouteloua gracilis*
grama grass – *Tripsacum laxum*
grama grass cactus – *Sclerocactus papyracanthus*
grama, hairy – *Bouteloua hirsuta*
grama, mat – *Bouteloua simplex*
grama, needle – *Bouteloua aristidoides*
grama, purple – *Bouteloua radicosa*
grama, red – *Bouteloua trifida*
grama, side-oats – *Bouteloua curtipendula*
grama, six-weeks – *Bouteloua barbata*
grama, slender – *Bouteloua filiformis, B. repens*
gram, green – *Vigna radiata*
granadilla, giant – *Passiflora quadrangularis*
granadilla, purple – *Passiflora edulis*
granadilla, sweet – *Passiflora ligularis*
grand fir – *Abies grandis*
grand potato vine – *Solanum wendlandii*
granjeno – *Celtis pallida*
granulated bolete – *Suillus granulatus*
grape – *Vitis*
grape, bush – *Vitis rupestris*
grape, calawba – *Vitis labrusca*
grape, California wild – *Vitis californica*
grape, canyon – *Vitis arizonica*
grape, cultivated – *Vitis vinifera*
grape flower – *Lagerstroemia indica*
grape, fox – *Vitis labrusca*
grapefruit – *Citrus paradisi*

grape, holly – *Berberis aquifolium*
grape hyacinth – *Muscari armeniacum, M. botryoides*
grape ivy – *Cissus rhombifolia, C. trifoliata*
grape, Muscat – *Vitis vinifera* var *muscat*
grape, Oregon – *Berberis*
grape, possum – *Cissus cucurbitina*
grape, sea – *Coccoloba uvifera, Ipomoea pes-caprae*
grape vine – *Vitis vinifera*
grape, wild – *Vitis tiliifolia*
grape, wine – *Vitis vinifera*
grape, wine varietal – *Vitis vinifera* var *sativa*
grape, woodland – *Vitis sylvestris*
grapple tree – *Harpagophytum procumbens*
grass – *Poaceae, Cannabis sativa*
grass, African – *Eragrostis lehmanniana*
grass, alkali – *Puccinellia distans*
grass, aparejo – *Muhlenbergia utilis*
grass, arrow – *Triglochin*
grass, bahia – *Paspalum notatum*
grass, barnyard – *Echinochloa crus-galli*
grass, basket – *Muhlenbergia rigens, Nolina erumpens*
grass, beach – *Ammophila*
grass, bear – *Nolina*
grass, beard – *Bothriochloa saccharoides, Polypogon, Schizachyrium cirratum*
grass, beggar-tick – *Aristida orcuttiana*
grass, bent – *Agrostis*
grass, Bermuda – *Cynodon dactylon*
grass, billion dollar – *Echinochloa frumentacea*
grass, birdseed – *Brachiaria fasciculata*
grass, blue – *Poa*
grass, blue-eyed – *Sisyrinchium bellum*
grass, Boer – *Eragrostis curvula* var *conferta*
grass, bristle – *Setaria*
grass, brush – *Hyparrhenia hirta*
grass, buffalo – *Buchloë dactyloides, Panicum coloratum*
grass, buffel – *Cenchrus ciliaris, Pennisetum ciliare*
grass, bull – *Muhlenbergia emersleyi*
grass, bur – *Tragus berteronianus*
grass, burro – *Scleropogon brevifolius*
grass, canary – *Phalaris*
grass, carib – *Eriochloa polystachya*
grass, carpet – *Axonopus compressus, Paspalum compressum*
grass, carrizo – *Phragmites australis*
grass, cattail – *Setaria pumila*
grass, cheat – *Bromus tectorum*
grass, citronella – *Cymbopogon citratus, C. nardus*
grass, Colombus – *Sorghum almum*
grass, cord – *Spartina*
grass, cotta – *Cottea pappophoroides*
grass, cotton – *Digitaria insularis*
grass, couch – *Elytrigia juncea, Paspalum virgatum*
grass, crab – *Digitaria*
grass, crowfoot – *Dactyloctenium*
grass, cup – *Eriochloa*
grass, cut – *Leersia hexandra, L. oryzoides*
grass, cutting – *Scleria bracteata*
grass, dallis – *Paspalum dilatatum*
grass, deer – *Muhlenbergia*
grass, ditch beard – *Polypogon interruptus*
grass, Dutch – *Ruppia cirrhosa, R. maritima*
grass, eel – *Zostera marina*
grass, elephant – *Pennisetum purpureum*
grass, esparto – *Lygeum spartum, Stipa tenacisima*
grass, eulalia – *Miscanthus sinensis*
grass, feather – *Stipa*
grass, finger – *Chloris virgata, Digitaria*
grass, five-finger – *Potentilla tormentilla*
grass, fluff – *Erioneuron pulchellum*
grass, fountain – *Pennisetum setaceum*
grass, gama – *Tripsacum*
grass, goat – *Aegilops*
grass, goose – *Eleusine indica, Galium aparine*
grass, grama – *Tripsacum laxum*
grass, Guinea – *Panicum*
grass, hair – *Aira, Deschampsia, Koeleria phleoides*
grass, hard – *Hainardia cylindrica, Parapholis strigosa, Sclerochloa dura*
grass, Harding – *Phalaris aquatica*
grass, holy – *Hierochloë odorata*
grass, imperial – *Axonopus scoparius*
grass, Indian – *Sorghastrum nutans*
grass, jaragua – *Hyparrhenia rufa*
grass, Johnson – *Sorgum halepense*
grass, June – *Koeleria macrantha*
grass, kikuyu – *Pennisetum clandestinum*
grass, knot – *Paspalum distichum, P. paspalodes, Polygonum*

grass-leaved pondweed – *Potamogeton gramineus*
grass, lemon – *Andropogon schaenanthus, Cymbopogon citratus*
grass, love – *Eragrostis*
grass, manna – *Glyceria fluitans, G. striata*
grass, marram – *Ammophila arenaria*
grass, meadow – *Poa*
grass, melic – *Melica montezumae*
grass, molasses – *Melinis minutiflora*
grass, Moor – *Molinia caerulea*
grass, mosquito – *Bouteloua gracilis*
grass, nandi – *Setaria anceps*
grass, napier – *Pennisetum purpureum*
grass, nard – *Cymbopogon nardus*
grass, natal – *Rhynchelytrum repens*
grass, needle – *Stipa*
grass, nit – *Gastridium ventricosum*
grass, nut – *Cyperus*
grass nuts – *Dichelostemma capitatum*
grass, oat – *Arrhenatherum elatius, Helictotrichon sempervirens, Trisetum flavescens, T. spicatum, Ventenata dubia*
grass-of-Parnassus – *Parnassia californica, P. palustris*
grass, orchard – *Dactylis glomerata*
grass, Palmer – *Distichlis palmeri*
grass, Pampas – *Cortaderia selloana*
grass, panic – *Panicum*
grass, pangola – *Digitaria decumbens*
grass, pappus – *Enneapogon desvauxii, Pappophorum bicolor*
grass, para – *Brachiaria mutica, Panicum molle*
grass, pepper – *Lepidium*
grass pink – *Dianthus armeria, D. plumarius*
grass poly – *Lythrum hyssopifolium*
grass, prickle – *Crypsis alopecuroides*
grass, quack – *Elytrigia repens*
grass, quail – *Celosia argentea*
grass, quaking – *Briza*
grass, rabbit-foot – *Polypogon monspeliensis*
grass, Ravenna – *Saccharum ravennae*
grass, reed – *Calamagrostis*
grass, rescue – *Bromus catharticus*
grass, Rhodes – *Chloris gayana*
grass, rib – *Plantago lanceolata*
grass, ribbon – *Nolina bigelovii, N. parryi, Phalaris arundinacea, Plantago lanceolata*
grass, rice – *Achnatherum hymenoides, Echinochloa colona, Leersia, Piptochaetium fimbriatum*
grass, ripgut – *Bromus diandrus*
grass, rye – *Lolium*
grass, Saint Augustine – *Stenotaphrum secundatum*
grass, salt – *Distichlis spicata*
grass, sand – *Triplasis purpurea*
grass, savannah – *Axonopus compressus, Paspalum compressum*
grass, saw – *Cladium californicum, Mariscus jamaicensis, Nolina microcarpa, Scleria bracteata*
grass, scurvy – *Barbarea verna, Cochlearia officinalis*
grass, shore – *Monanthochloë littoralis*
grass, sickle – *Parapholis incurva*
grass, silk – *Bromelia magdalenae*
grass, silver – *Miscanthus*
grass, sleepy – *Stipa robusta*
grass, smilo – *Piptatherum miliaceum*
grass, smut – *Sporobolus indicus*
grass, sour – *Digitaria insularis, Oxalis corniculata, Paspalum*
grass, spear – *Acantherum brachychaetum, Imperata cylindrica*
grass, spider – *Aristida ternipes*
grass, star – *Aletris farinosa, Cynodon plectostachyus*
grass, St. Augustine – *Stenotaphrum secundatum*
grass, stink – *Eragrostis cilianensis*
grass, Sudan – *Sorgum bicolor, S. sudanense*
grass, swamp – *Crypsis schoenoides*
grass, sweet – *Puccinellia distans*
grass, switch – *Panicum virgatum*
grass, torpedo – *Panicum repens*
grass tree, Mexican – *Dasylirion longissimum, Nolina longifolia*
grass tree, yellow – *Astianthus viminalis*
grass, turtle – *Thalassia testudinum*
grass, uva – *Gynerium sagittatum*
grass, vanilla – *Hierochloë odorata*
grass, velvet – *Holcus*
grass, vernal – *Anthoxanthum*
grass, Victoria – *Panicum repens*
grass, viper's – *Scorzonera hispanica*
grass, wedge – *Sphenopholis obtusata*

grass, wheat – *Elytrigia, Pascopyrum smithii, Triticum aestivum*
grass, whitlow – *Draba, Erophila verna*
grass, wire – *Juncus balticus, J. effuses*
grass, witch – *Leptoloma cognatum, Panicum capillare*
grass, yard – *Eleusine indica*
grass, zebra – *Miscanthus sinensis*
gray agaric – *Tricholoma terreum*
gray agave – *Agave fourcroydes*
gray ghost organ pipe – *Stenocereus pruinosus*
gray oak – *Quercus grisea*
gray pine – *Pinus sabiniana*
gray poplar – *Populus* x *canescens*
gray sallow – *Salix cinerea*
gray sun-rose – *Cistus incanus*
graythorn – *Ziziphus obtusifolia* var *canescens*
gray truffle – *Tuber magnatum*
graytwig – *Schoepfia californica*
gray willow – *Salix bebbiana, S. cinera*
greasewood – *Larrea tridentata, Sarcobatus vermiculatus*
greasewood, black – *Sarcobatus vermiculatus*
great northern bean – *Phaseolus vulgaris*
great pellitory of Spain – *Peucedanum ostruthium*
Grecian foxglove – *Digitalis lanata*
Grecian juniper – *Juniperus excelsa*
Grecian laurel – *Laurus nobilis*
Greek milkweed – *Asclepias speciosa*
Greek sea spurrey – *Spergularia bocconii*
Greek yarrow – *Achillea ageratifolia*
green amaranth – *Amaranthus hybridus*
green ash – *Fraxinus pennsylvanica*
greenbark ceanothus – *Ceanothus spinosus*
green bean – *Phaseolus vulgaris*
greenbriar – *Smilax*
greenbriar, thorny – *Smilax aspera*
green carpet – *Herniaria glabra*
green dragon – *Arisaema dracontium, Dracunculus vulgaris*
green ephedra – *Ephedra viridis*
green eyes, lyreleaf – *Berlandiera lyrata*
green-flowered bog-orchid – *Platanthera hyperborea*
green-flowered wintergreen – *Pyrola chlorantha*
green foxtail – *Setaria viridis*
green French plantain – *Mus acuminata* x *balbisiana*
greengage plum – *Prunus domestica* ssp *italica*
green gentian – *Swertia radiata*
green gram – *Vigna radiata*
green heather – *Erica scoparia*
green hedgehog – *Echinocereus viridiflorus*
green hellebore – *Helleborus viridis*
Greenland's cochlearia – *Cochlearia officinalis* var *arctica*
green lavender – *Lavandula viridis*
greenleaf manzanita – *Arctostaphylos patula*
greenleaf mistletoe – *Phoradendron macrophyllum*
green, lesser winter – *Pyrola minor*
green-lily – *Schoenocaulon officinale*
green milkweed – *Asclepias oenotheroides*
green Mormon tea – *Ephedra viridis*
green nightshade – *Solanum sarrachoides*
green, one-sided winter – *Orthilia secunda*
green onion – *Allium cepa*
green pea – *Pisum sativum*
green poinsettia – *Euphorbia dentata*
green sapote – *Lucuma viride, Pouteria viridis*
greens, chop suey – *Chrysanthemum coronarium*
greens, mustard – *Brassica nigra*
green spangletop – *Leptochloa dubia*
green spleenwort – *Asplenium trichomanes*
green-striped cushaw – *Cucurbita mixta*
greens, turnip – *Brassica rapa*
greenthread, Hopi-tea – *Thelesperma megapotamicum*
greenthread, long-stalk – *Thelesperma longipes*
green tulip – *Tulipa gesneriana*
greenweed, dyer's – *Genista tinctoria*
greenweed, hairy – *Genista pilosa*
green, winter – *Gaultheria acuminata, G. procumbens, Pyrola chlorantha*
Gregg ash – *Fraxinus greggii*
Gregg oak – *Quercus greggii*
Gregg pine – *Pinus greggii*
gromwell – *Lithospermum arvensis, L. officinale*
gromwell, mountain – *Lithospermum cobrense*
gros Michel banana – *Mus acuminata*
ground bean – *Vigna subterranea*
ground-cherry – *Physalis*

ground-cherry, desert – *Physalis crassifolia* var *versicolor*
ground-cherry, heartleaf – *Physalis hederifolia*
ground-cherry, strawberry – *Physalis alkekengi*
ground-cherry, Wright's – *Physalis acutifolia*
ground-cover, desert – *Abronia villosa*
ground elder – *Aegopodium podagraria*
ground ivy – *Glecoma hederacea*
ground morning-glory – *Convolvulus sabatius*
groundnut, bambarra – *Voandzeia subterranea*
ground pine – *Ajuga chamaepitys*
groundsel – *Senecio*
groundsel, common – *Senecio vulgaris*
groundsel, threadleaf – *Senecio longilobus*
groundsel, willow – *Senecio salignus*
groundsel, wood – *Senecio sylvaticus*
grove, plantain – *Spathiphyllum blandum*
Guadalupe cypress – *Cupressus guadalupensis*
Guadalupe Island palm – *Erythea edulis*
guajillo – *Acacia berlandieri*
guano palm – *Sabal japa*
guarana – *Cupania americana*
guar gum – *Cyamopsis tetragonoloba*
Guatemalan fir – *Abies guatemalensis*
Guatemala rhubarb – *Jatropha podagrica*
guava – *Psidium*
guava, Costa Rican – *Psidium friedrich*
guava, Guinea – *Psidium guineense*
guava, pineapple – *Feijoa sellowiana*
guava, strawberry – *Psidium cattleyanum*
Guelder rose – *Viburnum opulus*
Guiana chestnut – *Pachira aquatica*
Guinea corn – *Sorghum bicolor*
Guinea grass – *Panicum maxicum*
Guinea grass, colored – *Panicum coloratum*
Guinea guava – *Psidium guineense*
Guinea hen flowers – *Fritillaria meleagris*
Guinea hen weed – *Petiveria alliacea*
gum, American sweet – *Liquidambar styraciflua*
gum Arabic – *Acacia senegal*
gum-bearing carline thistle – *Atractylis gummifera*
gum, blue – *Eucalyptis globulus*
gumbo – *Hibiscus esculentus*
gumbolimbo – *Bursera simaruba*
gumbolimbo, white – *Gilibertia arborea*
gum bumelia – *Sideroxylon lanuginosum*
gum, Chinese sweet – *Liquidambar formosana*
gum, cider – *Eucalyptus gunnii*
gum cistus – *Cistus ladanifer*
gum copal – *Copaifera copallifera*
gum, crown – *Achras chicle*
gum elemi – *Bursera simaruba*
gum, guar – *Cyamopsis tetragonoloba*
gum, manna – *Eucalyptus viminalis*
gum, oriental sweet – *Liquidambar orientalis*
gumplant – *Grindelia*
gumplant, mountain – *Grindelia aphanactis*
gum, red – *Eucalyptus camaldulensis, E. rostratus*
gum, river red – *Eucalyptus camaldulensis*
gum, salmon – *Eucalyptus salmonophloia*
gum, snow – *Eucalyptus pauciflora* ssp *niphophila*
gum, sour – *Nyssa sylvatica*
gum, sweet – *Liquidambar styraciflua*
gumtree – *Castilla elastica*
gum vine – *Landolphia florida*
gumweed – *Grindelia aphanactis*
gunnera – *Gunnera*
gutta-percha – *Palaquium gutta*
gypsophila, creeping – *Gypsophila repens*

habanero chile – *Capsicum chinense*
hackberry – *Celtis*
hackberry, common – *Celtis occidentalis*
hackberry, desert – *Celtis pallida*
hackberry, Douglas – *Celtis douglassii*
hackberry, European – *Celtis australis*
hackberry, Mexican – *Trema micrantha*
hackberry, netleaf – *Celtis reticulata*
hackberry, western – *Celtis reticulata*
hackmatack – *Larix laricina*
hairbrush cactus – *Pachycereus pecten-aboriginum*
hairbrush vine – *Combretum farinosum*
hair cactus, Texas – *Mammillaria prolifera*
haired brush grass – *Hyparrhenia hirta*
hair flower – *Zapoteca anomalia*
hairgrass, annual – *Koeleria phleoides*
hairgrass, crested – *Koeleria phleoides*
hairgrass, crinkled – *Deschampsia flexuosa*
hairgrass, early – *Aira praecox*
hairgrass, elegant European – *Aira elegantissima*
hairgrass, silver European – *Aira caryophylea*
hairgrass, tufted – *Deschampsia cespitosa*

hair, witches – *Cuscuta denticulata*
hairy bitter-cress – *Cardamine hirsuta*
hairy buttercup – *Ranunculus sardous*
hairy dropseed – *Blepharoneuron tricholepis*
hairy grama – *Bouteloua hirsuta*
hairy greenweed – *Genista pilosa*
hairy lychee – *Nephelium lappaceum*
hairy peppergrass – *Lepidium lasiocarpum*
hairy prairie clover – *Dalea tomentosa*
hairy rock cress – *Arabis hirsuta*
hairy-seed bahia – *Bahia absinthifolia*
hairy-seed paspalum – *Paspalum pubiflorum*
hairy tare – *Vicia hirsuta*
hairy Tom palmetto – *Acoelorraphe wrightii*
hairy vetch – *Vicia villosa*
hairy wood rush – *Luzula parviflora, L. spicata*
hala – *Pandanus tectorius*
Hall's panicum – *Panicum hallii*
halogeton – *Halogeton glomeratus*
Hampshire purslane – *Ludwigia palustris*
handbasin oak – *Quercus pennivenia, Q. tarahumara*
hand-flower tree – *Chiranthodendron pentadactylon*
handlegrip cholla – *Cylindropuntia spinosior*
handleleaf coneflower – *Echinacea angustifolia*
hanging cactus – *Echinocereus pensilis*
hankow willow – *Salix matsudana*
hard fescue – *Festuca longifolia, F. trachyphylla*
hard-grass – *Hainardia cylindrical, Sclerochloa dura*
hard-grass, sea – *Parapholis strigosa*
hardheads – *Centaurea nigra*
Harding grass – *Phalaris aquatica*
hardtack – *Cercocarpus betuloides*
hard wheat – *Triticum duram*
hardy cyclamen – *Cyclamen hederifolium*
hardy phlox, giant – *Phlox paniculata*
hardy rubber tree – *Eucommia ulmoides*
harebell – *Campanula rotundifolia*
hare's ear – *Conringia orientalis*
hare's ear, shrubby – *Bupleurum fruticosum*
hare's eye – *Silene dioica*
hare's foot sedge – *Carex ovalis*
hare's tail – *Lagurus ovatus*
haricot – *Phaseolus vulgaris*
haricot, round-white-seeded French – *Phaseolus vulgaris*
harlequin flower – *Sparaxis tricolor*
harlequin glorybower – *Clerodendrum trichotomum*
harmal – *Peganum harmala*
hart's tongue fern – *Phyllitis scolopendrium*
harvest brodiaea – *Brodiaea coronaria*
hat, bishop's – *Astrophytum capricorne, Epimedium alpinum, E. grandiflorum*
hat, cardinal's – *Astrophytum myriostigma*
hatchet cactus – *Pelecyphora aselliformis*
hat fiber, Mexican – *Sabal mexicana*
hat, Mexican – *Ratibida columnaris*
hat-thorn acacia – *Acacia hindsii*
hausa potato – *Coleus rotundifolius*
havard three-awn – *Aristida barbata*
haw, black – *Viburnum*
hawkbit – *Leontodon taraxacoides*
hawksbeard – *Crepis capillaris, C. setosa, C. vesicaria*
hawkweed – *Hieracium*
hawkweed, Fendler's – *Hieracium fendleri*
hawkweed, orange – *Hieracium aurantiacum*
haw, possum – *Ilex decidua*
hawthorn – *Crataegus*
hawthorn, carriere – *Crataegus* x *lavallei*
hawthorn, cerro – *Crataegus erythropoda*
hawthorn, downy – *Crataegus mollis*
hawthorn, English – *Crataegus laevigata*
hawthorn, Indian – *Rhaphiolepsis indica*
hawthorn, Mexican – *Crataegus mexicana*
hawthorn, midland – *Crataegus laevigata*
hawthorn, Wooton – *Crataegus wootoniana*
hazel, Chinese witch – *Hamamelis mollis*
hazel, Japanese witch – *Hamamelis japonica*
hazelnut – *Corylus cornuta* var *californica, C. maxima*
hazel, spike winter – *Corylopsis spicata*
hazel tree – *Corylus avellana*
hazel, Turkish – *Corylus colurna*
hazel, wild – *Simmondsia chinensis*
hazel, witch – *Hamamelis virginiana*
head, bull – *Tribulus terrestris*
head, devil's – *Echinocactus horizonthalonius, E. texensis*
head, dragon – *Physostegia virginiana*
head, goat – *Tribulus terrestris*
head lettuce – *Lactuca sativa* var *capitata*

head, Turk's – *Ferocactus hamatacanthus*
heal-all – *Prunella vulgaris*
heal, self – *Prunella vulgaris*
heart, bleeding – *Dicentra formosa*
heart, broken – *Clerodendrum thomsoniae*
heart, bullock's – *Annona reticulata*
heart, common bleeding – *Dicentra spectabilis*
heart, floating – *Nymphoides humboldtianum, N. indica*
heartflower tree – *Talauma mexicana*
heartleaf bergenia – *Bergenia cordifolia*
heartleaf ground-cherry – *Physalis hederifolia*
heartleaf hibiscus – *Hibiscus cardiophyllus*
heart-leaved globe daisy – *Globularia cordifolia*
heart-leaved twayblade – *Listera cordata*
heart, lion – *Physostegia virginiana*
heart of flame – *Bromelia balansae*
heart of Jesus – *Caladium bicolor*
heart-podded hoary cress – *Cardaria draba*
heart, purple – *Tradescantia pallida*
heart's-delight – *Abronia fragrans*
heart's ease – *Viola arvensis*
heart, western bleeding – *Dicentra formosa*
heath – *Erica*
heath, alkali – *Frankenia salina*
heath bedstraw – *Galium saxatile*
heath, Biscay – *Erica mediterranea*
heath, broom – *Erica scoparia*
heath, Cornish – *Erica vagans*
heath, cross-leaved – *Erica tetralix*
heath, Dorset – *Erica ciliaris*
heather – *Erica*
heather, bell – *Erica cinera*
heather, green – *Erica scoparia*
heather, mountain – *Phyllodoce*
heather, purple – *Krameria erecta*
heather, Scotch – *Calluna vulgaris*
heather, tree – *Erica arborea*
heath false-brome – *Brachypodium pinnatum*
heath, Irish – *Daboecia cantabrica*
heath lobelia – *Lobelia urens*
heath melaleuca – *Melaleuca ericafolia*
heath pearlwort – *Sagina subulata*
heath, Portuguese – *Erica lusitanica*
heath, Scotch – *Erica cinera*
heath, southern – *Erica australis*
heath, Spanish – *Erica australis*
heath speedwell – *Veronica officinalis*
heath, spring – *Erica carnea*
heath, tree – *Erica arborea*
heath, twisted – *Erica cinerea*
heavenly-blue morning-glory – *Ipomoea tricolor*
heavenly blues – *Ipomoea violacea*
heaven, tree of – *Ailanthus altissima*
hebe, boxleaf – *Hebe buxifolia*
hedge bindweed – *Calystegia sepium*
hedgehog – *Thelocactus setispinus*
hedgehog cactus – *Echinocereus*
hedgehog cactus, Arizona rainbow – *Echinocereus rigidissimus*
hedgehog cactus, Lindsay's – *Echinocereus ferreirianus*
hedgehog cactus, nylon – *Echinocereus viridiflorus*
hedgehog cactus, strawberry – *Echinocereus enneacanthus*
hedgehog, cockscomb – *Echinocereus fasciculatus*
hedgehog dogtail – *Cynosurus echinatus*
hedgehog, green – *Echinocereus viridiflorus*
hedgehog, rainbow – *Echinocereus pectinatus*
hedgehog, spineless – *Echinocereus triglochidiatus*
hedgehog, spiny – *Echinocereus polyacanthus* ssp *pacificus*
hedgehog, spiny strawberry – *Echinocereus stramineus*
hedgehog, strawberry – *Echinocereus fasciculatus*
hedgehog, Texas rainbow – *Echinocereus dasyacanthus*
hedge-hyssop – *Gratiola officinalis*
hedge maple – *Acer campestre*
hedge mustard – *Sisymbrium officinale*
hedge nettle – *Stachys arvensis, S. coccinea, S. palustris*
hedge nettle, scarlet – *Stachys coccinea*
hedge parsley, knotted – *Torilis nodosa*
heliconia – *Heliconia*
heliconia, golden – *Heliconia latispatha*
heliconia, parrot – *Heliconia psittacorum*
heliotrope – *Heliotropium*
heliotrope, common – *Heliotropium arborescens*
heliotrope, creeping – *Heliotropium procumbens*

heliotrope, garden – *Heliotropium arborescens, Valeriana officinalis*
heliotrope, salt – *Heliotropium curassavicum*
heliotrope, wild – *Phacelia distans*
hellebore – *Helleborus viridis*
hellebore, bear's-foot – *Helleborus foetidus*
hellebore, false – *Veratrum album, V. officinale, V. viride*
hellebore, green – *Helleborus viridis*
hellebore, pale – *Helleborus lividus*
hellebore, white – *Veratrum album*
helleborine, broad-leaved – *Epipactis helleborine*
helmet, fireman's – *Erythrina crista-galli*
hemlock, Canada – *Tsuga canadensis*
hemlock, Carolina – *Tsuga caroliniana*
hemlock, eastern – *Tsuga canadensis*
hemlock, mountain – *Tsuga mertensiana*
hemlock, poison – *Conium maculatum*
hemlock, water – *Oenanthe crocata*
hemlock, western – *Tsuga heterophylla*
hemp – *Cannabis sativa*
hemp, African – *Sparmannia africana*
hemp, bowstring – *Sansevieria guineensis, S. trifasciata*
hemp, Colorado River – *Sesbania herbacea, S. macrocarpa*
hemp, Indian – *Apocynum cannabinum, Pilocarpus jaborandi*
hemp, Manila – *Mus textilis*
hemp, Mauritius – *Furcraea gigantea*
hemp, sunn – *Crotalaria juncea*
hemp, wild – *Cannabis ruderalis*
hen-and-chicks – *Echeveria elegans, E. secunda, Sempervivum tectorum*
henbane, black – *Hyoscyamus niger*
henbit – *Lamium amplexicaule*
henbit, red – *Lamium purpureum*
henequen agave – *Agave fourcroydes*
hen, fat – *Chenopodium album*
henna – *Lawsonia inermis*
herald's trumpet – *Beaumontia grandiflora*
herba impia – *Filago gallica, F. germanica, F. pyramidata*
herb Bennet – *Geum urbanum*
herb, bitter – *Centaurium erythraea*
herb Christopher – *Actaea spicata*
herb, cow – *Vaccaria hispanica*
herb Margaret – *Bellis perennis*
herb of grace – *Ruta graveolens*
herb Paris – *Paris quadrifolia*
herb patience – *Rumex patientia*
herb Robert – *Geranium robertianum*
herb royal – *Artemisia abrotanum*
herb, snake – *Dyschoriste decumbens*
herb, willow – *Epilobium, Lythrum*
Hercules' allheal – *Opopanax chironium*
heronbill – *Erodium texanum*
Herrera pine – *Pinus herrerae*
hesper palm, San José – *Erythea brandegeei*
hesper palm, Sonoran – *Erythea aculeata*
hiba cedar – *Thujopsis dolabrata*
hibiscus, Chinese – *Hibiscus rosa-sinensis*
hibiscus, coast – *Hibiscus tilliaceus*
hibiscus, desert – *Hibiscus coulteri*
hibiscus, heartleaf – *Hibiscus cardiophyllus*
hibiscus, sleeping – *Malvaviscus arboreus*
hibiscus, tree – *Hibiscus tilliaceus*
hibiscus, tropical – *Hibiscus rosa-sinensis*
hickory – *Carya*
hickory, big shell-bark – *Carya laciniosa*
hickory, pignut – *Carya glabra*
hickory, shagbark – *Carya ovata*
hierba santa – *Eriodictyon*
highbush blueberry – *Vaccinium corymbosum*
highclere holly – *Ilex* x *altaclerensis*
high mallow – *Malva sylvestris*
high stonecrop – *Sedum sediforme*
Himalayan birch – *Betula jacquemontii*
Himalayan blackberry – *Rubus discolor*
Himalayan knotweed – *Polygonum polystachyum*
Himalayan poppy – *Meconopsis betonicifolia*
Himalayan white pine – *Pinus wallichiana*
Hinoki false cypress – *Chamaecyparis obtusa*
Hispaniola palmetto – *Sabal blackburniana*
hoary cress, heart-podded – *Cardaria draba*
hoary mustard – *Hirschfeldia incana*
hoary yucca – *Yucca schottii*
hog peanut – *Vigna subterranea*
hog plum – *Spondias mombin*
hog potato – *Hoffmannseggia glauca*
hog weed – *Boerhavia hirsuta*
hogweed, giant – *Heracleum mantegazzianum*
Hohokam agave – *Agave murpheyi*
holly – *Ilex*
holly, American – *Ilex opaca*
holly, brush – *Xylosma flexuosum*

holly, California – *Heteromeles arbutifolia*
holly, Christmas – *Ilex aquifolium*
holly, desert – *Acourtia nana, A. wrightii*
holly, English – *Ilex aquifolium*
holly, evergreen – *Ilex vomitoria*
holly-grape – *Berberis aquifolium*
holly, highclere – *Ilex* x *altaclerensis*
hollyhock – *Alcea rosea*
hollyhock, desert – *Sphaeralcea ambigua*
holly, Japanese – *Ilex crenata*
holly-leaved cherry – *Prunus ilicifolia*
holly-leaved water-nymph – *Najas marina*
holly oak – *Quercus ilex*
holly oak, Spanish – *Quercus ilex*
holly, sea – *Eryngium*
holly, West Indian – *Turnera ulmifolia*
holly, Wilson – *Ilex* x *altaclerensis*
hollywood – *Heteromeles arbutifolia*
hollywood lignum vitae – *Guaiacum sanctum*
Holm oak – *Quercus ilex*
holy grass, aromatic – *Hierochloë odorata*
holy thistle – *Silybum marianum*
holywood lignum vitae – *Guaiacum sanctum*
Honduras Caribbean pine – *Pinus caribaea* var *hondurensis*
honesty – *Lunaria annua, L. biennis*
honey agaric – *Armillaria mellea*
honeylocust – *Gleditsia amorphoides*
honey locust – *Gleditsia triacanthos*
honey mangrove – *Avicennia germinans*
honey mesquite – *Prosopis glandulosa* var *torreyana*
honey mushroom – *Armillaria mellea*
honey plant – *Hoya carnosa*
honeysuckle – *Justicia california, Lonicera*
honeysuckle, Arizona – *Lonicera arizonica*
honeysuckle, bear-berry – *Lonicera involucrata*
honeysuckle, blue – *Lonicera caerulea*
honeysuckle, box – *Lonicera nitida*
honeysuckle, Cape – *Tecoma capensis*
honeysuckle, Chinese – *Quisqualis indica*
honeysuckle, coral – *Lonicera sempervirens*
honeysuckle, desert – *Anisacanthus quadrifidus* ssp *wrightii, A. thurberi*
honeysuckle, Etruscan – *Lonicera etrusca*
honeysuckle, Japanese – *Lonicera japonica*
honeysuckle, Mexican – *Justica spicigera*
honeysuckle, Tartarian – *Lonicera tatarica*
honeysuckle, trumpet – *Lonicera sempervirens*
honeysuckle, wild – *Guara coccinea, G. parviflora*
honeysweet – *Tidestromia lanuginosa*
honeywort – *Cerinthe major*
Honolulu creeper – *Antigonon leptopus*
Honolulu rose – *Merremia tuberosa*
hood, monk's – *Astrophytum ornatum*
hoof, bull – *Bauhinia variegata, Piper auritum*
hook three-awn – *Aristida ternipes* var *hamulosa*
hoop petticoat daffodil – *Narcissus bulbocodium*
hopbush – *Dodonaea viscosa*
hop clover – *Trifolium aureum, T. campestre, T. procumbens*
hop clover, little – *Trifolium dubium*
hop, common – *Humulus lupulus*
hop, European – *Humulus lupulus*
hop, false – *Justicia brandegeana*
hop-hornbeam, American – *Ostrya virginiana*
hop-hornbeam, European – *Ostrya carpinifolia*
Hopi tea – *Thelesperma gracile*
Hopi-tea greenthread – *Thelesperma megapotamicum*
hop marjoram – *Origanum dictamnus*
hops – *Humulus lupulus*
hops, American – *Humulus lupulus* ssp *neomexicanus*
hopseed bush – *Dodonea viscosa*
hop tree – *Ptelea crenulata, P. trifoliata*
hop tree, common – *Ptelea angustifolia*
hop, wild – *Bryonia alba*
horehound – *Marrubium vulgare*
horehound, black – *Ballota nigra*
horehound, water – *Lycopus americanus, L. europaeus*
hornbeam, American – *Carpinus caroliniana*
hornbeam, American hop – *Ostrya virginiana*
hornbeam, European – *Carpinus betulus*
horn, buck – *Plantago lanceolata*
horncone, Mexican – *Ceratozamia latifolia, C. mexicana*
horned poppy, red – *Glaucium corniculatum*
horned sea-blite – *Suaeda calceoliformis*
horned violet – *Viola cornuta*
horn, goat's – *Astrophytum capricorne*
horn-of-plenty – *Craterellus cornucopioides*
horn plantain – *Mus acuminata* x *balbasiana*
hornwort – *Ceratophyllum demersum*

horse bean – *Parkinsonia aculeata, Vicia faba*
horse chestnut – *Aesculus hippocastanum*
horse chestnut, Japanese – *Aesculus urbinata*
horse chestnut, red – *Aesculus carnea*
horse-crippler cactus – *Echinocactus texensis*
horse-elder – *Inula helenium*
horsemint – *Agastache micrantha, A. urticifolia, Monarda didyma, M. pectinata*
horsemint, lemon – *Monarda citriodora*
horsemint, pink – *Monarda fistulosa*
horse mushroom – *Agaricus arvensis*
horse-nettle, white – *Solanum eleagnifolium*
horse-purslane – *Trianthema portulacastrum*
horseradish – *Armoracia rusticana*
horseradish tree – *Moringa oleifera*
horsetail – *Equisetum*
horsetail, Boston – *Equisetum ramosissimum*
horsetail, common – *Equisetum arvense*
horsetail, giant – *Equisetum telmateia*
horsetail, marsh – *Equisetum palustre*
horsetail tree – *Casuarina equisetifolia*
horseweed – *Conyza canadensis, Erigeron canadensis*
hospita palm – *Copernicia hospita*
hot pepper – *Capsicum annuum*
hot-springs fimbristylis – *Fimbristylis thermalis*
Hottentot fig – *Carpobrotus edulis*
hound's tongue – *Cynoglossum*
houseleek – *Sempervivum tectorum*
houseleek, cobweb – *Sempervivum arachnoideum*
houseplant – *Cuphea jorullensis*
huanita tree – *Bourreria huanita*
huanuco cocaine tree – *Erythroxylum coca*
huckleberry – *Vaccinium myrtillus*
huisache – *Acacia farnesiana*
Humboldt coyotillo – *Karwinskia humboldtiana*
hummingbird bush – *Anisacanthus thurberi*
hummingbird flower – *Justica californica, J. purpusii, Loeselia mexicana*
Hungarian clover – *Trifolium pannonicum*
Hungarian oak – *Quercus frainetto*
Hungarian vetch – *Vicia pannonica*
hupeh rowan – *Sorbus hupehensis*
husk tomato – *Physalis philadelphica, P. pubescens*
hyacinth – *Hyacinthus orientalis*
hyacinth bean – *Lablab purpureus*
hyacinth, common – *Hyacinthus orientalis*
hyacinth, desert – *Dichelostemma capitatum*
hyacinth-flowered candytuft – *Iberis amara*
hyacinth, French Roman – *Hyacinthus orientalis* var *albulus*
hyacinth, fringe – *Muscari comosum*
hyacinth, grape – *Muscari armeniacum, M. botryoides*
hyacinth, Indian – *Camassia quamash*
hyacinth, musk – *Muscari moschatum*
hyacinth, Roman – *Hyacinthus orientalis* var *albulus*
hyacinth, starch – *Muscari neglectum*
hyacinth, summer – *Galtonia candicans*
hyacinth, tassel – *Muscari comosum*
hyacinth, water – *Eichhornia crassipes*
hyacinth, wood – *Hyacinthoides hispanicus, H. non-scripta*
hybrid fuchsia – *Fuchsia* x *hybrida*
hybrid meadow grass – *Poa hybrida*
hydnum, tiled – *Sarcodon imbricatus*
hydrangea – *Hydrangea macrophylla*
hydrangea, bigleaf – *Hydrangea macrophylla*
hydrangea, climbing – *Hydrangea anomala* ssp *petiolaris*
hydrangea, garden – *Hydrangea macrophylla*
hydrangea, oakleaf – *Hydrangea quercifolia*
hydrangea, smooth – *Hydrangea arborescens*
hyssop – *Agastache urticifolia, Hyssopus officinalis, Origanum maru* var *aegyptiacum, Sorghum vulgare* var *durra*
hyssop, blue – *Hyssopus officinalis*
hyssop, giant – *Agastache micrantha*
hyssop, giant Mexican lemon – *Agastache mexicana*
hyssop, hedge – *Gratiola officinalis*
hyssop loosestrife – *Lythrum hyssopifolium*
hyssop, water – *Bacopa monnieri*

iceberg lettuce – *Lactuca sativa* var *capitata*
Iceland moss – *Cetraria islandica*
Iceland poppy – *Papaver nudicaule*
ice plant – *Carpobrotus chilensis, Lampranthus deltoides, Mesembryanthemum*
iceplant, crystaline – *Mesembryanthemum crystallinum*
iceplant, slender-leaved – *Mesembryanthemum nodiflorum*

Ignatius bean – *Strychnos ignatii*
ilama – *Annona diversifolia*
Illyrian thistle – *Onopordum illyricum*
immortelle, common – *Xeranthemum annuum*
impala lily – *Adenium obesum*
impatiens – *Impatiens balsamina*
imperial, crown – *Fritillaria imperialis*
imperial grass – *Axonopus scoparius*
imperial, rose – *Cochlospermum vitifolium*
incarvillea – *Incarvillea delavayi*
Incas, sacred flower of the – *Cantua buxifolia*
incense cedar – *Calocedrus decurrens*
incienso – *Encelia farinosa*
India – *Ziziphus mauritiana*
India berry tree – *Anamirta paniculata*
India lovegrass – *Eragrostis pilosa*
Indian almond – *Terminalia amazonia, T. catappa*
Indian apple – *Datura wrightii*
Indian asparagus – *Orobanche cooperi*
Indian banyan – *Ficus benghalensis*
Indian bean – *Catalpa bignonioides, Erythrina flabelliformis*
Indian bitter gourd – *Momordica charantia*
Indian-blanket – *Gaillardia aristata, G. pulchella*
Indian boxwood – *Caesaria nitida*
Indian breadroot – *Psoralea esculenta*
Indian chickweed – *Mollugo verticillata*
Indian corn – *Zea mays*
Indian cowpea – *Vigna unguiculata* ssp *cylindrica*
Indian dillenia – *Dillenia indica*
Indian ebony persimmon – *Diospyros ebenaster*
Indian fig – *Opuntia ficus-indica, O. megacantha*
Indian fig, Eltham – *Opuntia dillenii*
Indian grass – *Sorghastrum nutans*
Indian hawthorn – *Rhaphiolepis indica*
Indian hemp – *Apocynum cannabinum, Pilocarpus jaborandi*
Indian hyacinth – *Camassia quamash*
Indian jujube – *Ziziphus mauritiana*
Indian laburnum – *Cassia fistula*
Indian laurel – *Ficus microcarpa*
Indian laurel fig – *Ficus microcarpa*
Indian lettuce – *Claytonia perfoliata*
Indian longleaf pine – *Pinus roxburghii*
Indian lotus – *Nelumbo nucifera*
Indian mango – *Licania*
Indian mallow – *Abutilon*
Indian mock strawberry – *Duchesnea indica*
Indian mustard – *Brassica juncea*
Indian oak – *Tectona grandis*
Indian paintbrush – *Castilleja, C. gracilis*
Indian paintbrush, desert – *Castilleja angustifolia*
Indian parsley – *Ligusticum grayi*
Indian pink – *Ipomoea quamoclit, Silene californica*
Indian pink, California – *Silene californica*
Indian pipe – *Monotropa uniflora*
Indian pipeweed – *Eriogonum inflatum*
Indian plantain – *Psacalium decompositum*
Indian rhubarb – *Rumex salicifolius*
Indian ricegrass – *Achnatherum hymenoides*
Indian root – *Aristolochia brevipes, A. watsonii, Asclepias curassavica, Lomatium dissectum*
Indian roseapple – *Dillenia indica*
Indian rubber plant – *Ficus elastica*
Indian rush pea – *Hoffmannseggia glauca*
Indian senna – *Senna angustifolia*
Indian shot – *Canna coccinea, C. indica*
Indian snake gourd – *Trichosanthes cucumerina* var *anguina*
Indian snakeroot – *Rauwolfia serpentina*
Indian sugarcane – *Saccharum barberi*
Indian sweet lime – *Citrus limettioides*
Indian tea – *Ephedra viridis, Thelesperma megapotamicum*
Indian tobacco – *Lobelia fenestralis, L. inflata, Nicotiana rustica*
Indian wheat – *Plantago fastigiata*
Indian wheat, woolly – *Plantago patagonica*
India, pride of – *Koelreuteria paniculata, Lagerstroemia indica, Melia azedarach*
India, song of – *Dracaena reflexa*
indigo – *Indigofera gerardiana, I. suffruticosa*
indigo bush – *Dalea bicolor* var *orcuttiana, D. citriodora, D. pringlei, Psorothamnus schottii, P. spinosus*
indigo bush, trailing – *Dalea greggii*
indigo, Chinese – *Polygonum tinctorium*
indigo, false – *Amorpha fruticosa*
inga – *Inga spuria*
inga, food – *Inga edulis*

inhambane coffee – *Coffea racemosa*
inkberry – *Lonicera involucrata, Scaevola plumierii*
ink-cap – *Coprinus*
Insleay's odontoglossum – *Odontoglossum insleayii*
intermediate wheatgrass – *Elytrigia intermedia*
iodine bush – *Allenrolfea occidentalis*
ipecac – *Cephaelis*
ipecac, false – *Psychotria emetica*
ipomoea, star – *Quamoclit coccinea*
Ireland, bells-of – *Moluccella laevis*
iris – *Iris germanica*
iris, bearded – *Iris barbata, I. germanica*
iris, coral – *Iris foetidissima*
iris, Del Norte county – *Iris innominata*
iris, Dutch – *Iris* x *hollandica*
iris, dwarf – *Iris pumila*
iris, English – *Iris latifolia*
iris, Florentine – *Iris florentina*
iris, garden – *Iris germanica*
iris, German – *Iris germanica*
iris, Gladwin – *Iris foetidissima*
Irish heath – *Daboecia cantabrica*
Irish lace – *Tagetes filifolia*
Irish moss – *Sagina subulata*
Irish rose – *Rosa hibernica*
Irish saxifrage – *Saxifraga decipiens*
Irish yew – *Taxus baccata* var *fastigiata*
iris, Japanese – *Iris ensata*
iris, reticulata – *Iris reticulata*
iris, Rocky Mountain – *Iris missouriensis*
iris, Siberian – *Iris sibirica*
iris, Spanish – *Iris xiphium*
iris, winter – *Iris unguicularis*
iris, yellow flag – *Iris pseudacorus*
iron bark, red – *Eucalyptus sideroxylon*
iron plant, cast – *Aspidistra elatior*
iron vine – *Desmodium canum*
iron weed – *Centaurea nigra, Vernonia deppeana*
ironwood – *Olneya tesota*
ironwood, desert – *Olneya tesota*
ironwood, mountain – *Cercocarpus betuloides*
island mallow – *Lavatera assurgentiflora*
island oak – *Quercus tomentilla*
island poppy – *Eschscholzia ramosa*
islay – *Prunus ilicifolia*
istle, tula – *Agave heteracantha*
Italian alder – *Alnus cordata*
Italian arum – *Arum italicum*
Italian aster – *Aster amellus*
Italian bellflower – *Campanula isophylla*
Italian bindweed – *Convolvulus althaeoides*
Italian buckthorn – *Rhamnus alaternus*
Italian corn salad – *Valerianella eriocarpa*
Italian cypress – *Cupressus sempervirens*
Italian jasmine – *Jasminum humile*
Italian oak – *Quercus frainetto*
Italian parsley – *Petroselinum crispum*
Italian ryegrass – *Lolium multiflorum*
Italian squill – *Scilla italica*
Italian stone pine – *Pinus pinea*
Italian thistle – *Carduus pycnocephalus*
Italian truffle – *Tuber magnatum*
Italian woolly sage – *Phlomis italica*
itchy bean – *Mucuna pruriens*
ivory nut – *Phytelephas macrocarpa*
ivory palm – *Phytelephas macrocarpa*
ivy – *Quamoclit coccinea*
ivy, Algerian – *Hedera canariensis*
ivy, Boston – *Parthenocissus tricuspidata*
ivy, Cape – *Delairea odorata*
ivy, devil's – *Epipremnum pinnatum*
ivy, English – *Hedera helix*
ivy, five-finger – *Parthenocissus quinquefolia*
ivy geranium – *Pelargonium peltatum*
ivy, German – *Delairea odorata*
ivy, grape – *Cissus rhombifolia, C. trifoliata*
ivy, ground – *Glecoma hederacea*
ivy, Kenilworth – *Cymbalaria muralis*
ivyleaf morning-glory – *Ipomoea hederacea*
ivy-leaved duckweed – *Lemna trisulca*
ivy-leaved speedwell – *Veronica hederifolia*
ivy-leaved toadflax – *Cymbalaria muralis*
ivy, Mexican – *Cobaea scandens*
ivy morning-glory – *Ipomoea nil*
ivy, Persian – *Hedera colchica*
ivy, poison – *Toxicodendron radicans*
ivy vine – *Ampelopsis cordata*
ivy, violet – *Cobaea scandens*
ixia – *Ixia maculata*
ixora – *Ixora coccinea*

jacaranda – *Jacaranda acutifolia*
jackass bitters – *Neurolaena lobata*
jackass clover – *Wislizenia refracta*
jackbean – *Canavalia, Phaseolus vulgaris*

jacket, water – *Lycium andersonii*
jackfruit – *Artocarpus heterophyllus, A. integrifolia*
jack-in-the-pulpit – *Arisaema dracontium*
jack pine – *Pinus banksiana*
jack, slippery – *Suillus luteus*
jacktree – *Artocarpus integrifolia*
Jacobean lily – *Sprekelia formosissima*
Jacob's coat – *Acalypha wilkesiana*
Jacob's ladder – *Pedilanthus tithymaloides, Polemonium caeruleum*
jagged chickweed – *Holosteum umbellatum*
Jalapa planetree – *Platanus lindeniana*
jalapeno pepper – *Capsicum annuum*
Jalisco pine – *Pinus jaliscana*
Jamaica cherry – *Muntingia calabura*
Jamaica feverplant – *Tribulus cistoides*
Jamaica flower – *Hibiscus sabdariffa*
Jamaica sorrel – *Hibiscus sabdariffa*
James lily, St. – *Sprekelia formosissima*
Japanese anemone – *Anemone hybrida*
Japanese angelica tree – *Aralia elata*
Japanese anise – *Illicium anisatum*
Japanese apricot – *Prunus mume*
Japanese aralia – *Fatsia japonica*
Japanese asparagus – *Asparagus densiflorus* var *sprengeri*
Japanese aucuba – *Aucuba japonica*
Japanese barberry – *Berberis thunbergii*
Japanese black pine – *Pinus thunbergii*
Japanese brome – *Bromus japonicus*
Japanese cane – *Saccharum sinense*
Japanese cherry – *Prunus serrulata*
Japanese coltsfoot – *Petasites japonicus*
Japanese corn – *Zea mays* var *japonica*
Japanese cryptomeria – *Cryptomeria japonica*
Japanese dogwood – *Cornus kousa*
Japanese flowering apricot – *Prunus mume*
Japanese flowering cherry – *Prunus subhirtella*
Japanese flowering crabapple – *Malus floribunda*
Japanese flowering plum – *Prunus mume*
Japanese frangipani – *Adenium obesum*
Japanese holly – *Ilex crenata*
Japanese honeysuckle – *Lonicera japonica*
Japanese horse chestnut tree – *Aesculus urbinata*
Japanese iris – *Iris ensata*
Japanese knotweed – *Polygonum cuspidatum*
Japanese lacquer tree – *Rhus verniciflua*
Japanese larch – *Larix kaempferi*
Japanese live oak – *Quercus myrsinifolia*
Japanese maple – *Acer palmatum*
Japanese millet – *Echinochloa frumentacea*
Japanese pagoda tree – *Sophora japonica*
Japanese perilla – *Perilla frutescens*
Japanese persimmon – *Diospyros kaki*
Japanese pittosporum – *Pittosporum tobira*
Japanese plum – *Prunus salicina*
Japanese privet – *Ligustrum japonicum*
Japanese red pine – *Pinus densiflora*
Japanese sedge, variegated – *Carex morrowii* var *expallida*
Japanese silver grass – *Miscanthus sinensis*
Japanese snowball – *Viburnum plicatum*
Japanese snowbell – *Styrax japonicus*
Japanese snowdrop tree – *Styrax japonicus*
Japanese spiraea – *Spiraea japonica*
Japanese spurge – *Pachysandra terminalis*
Japanese stewartia – *Stewartia pseudocamellia*
Japanese sweet flag – *Acorus gramineus*
Japanese tung – *Aleurites cordata*
Japanese wax tree – *Rhus succedeana*
Japanese white pine – *Pinus parviflora*
Japanese wineberry – *Rubus phoenicolasius*
Japanese witch hazel – *Hamamelis japonica*
Japanese yew – *Taxus cuspidata*
jaragua grass – *Hyparrhenia rufa*
jasmine – *Jasminum*
jasmine, Arabian – *Jasminum sambac*
jasmine, blue – *Plumbago scandens*
jasmine, Cape – *Gardenia augusta, G. florida*
jasmine, Chilean – *Mandevilla laxa*
jasmine, common white – *Jasminum grandiflorum*
jasmine, Confederate – *Trachelospermum jasminoides*
jasmine, downy – *Jasminum multiflorum*
jasmine, Italian – *Jasminum humile*
jasmine, Madagascar – *Stephanotis floribunda*
jasmine, night-flowering – *Nyctanthes arbor-tristis*
jasmine, poet's – *Jasminum grandiflorum*
jasmine, primrose – *Jasminum mesnyi*
jasmine, purple – *Solanum dulcamara*
jasmine, red – *Ipomoea quamoclit*
jasmine, rock – *Androsace*
jasmine, Spanish – *Jasminum grandiflora*

jasmine tobacco – *Nicotiana alata*
jasmine, wild – *Philadelphus mexicanus*
jasmine, winter – *Jasminium nudiflorum*
jasmine, yellow – *Jasminum humile*
jasmine, wild – *Philadelphus mexicanus*
jatropha, sycamore-leaf – *Jatropha malacophylla*
Java almond – *Canarium commune*
Java apple – *Eugenia javanica*
Java glorybower – *Clerodendrum buchananii*
Java olive – *Sterculia foetida*
Java plum – *Eugenia jambolana*
javelina-brush – *Microrhamnus ericoides*
javelina bush – *Condalia ericoides*
Jeffrey pine – *Pinus jeffreyi*
jelecote pine – *Pinus patula*
jelly fungus, white – *Tremella fuciformis*
jelly palm – *Butia capitata*
Jersey cudweed – *Gnaphalium luteo-album*
Jersey thrift – *Armeria alliacea*
Jerusalem artichoke – *Helianthus tuberosus*
Jerusalem cherry – *Solanum pseudocapsicum*
Jerusalem corn – *Sorghum vulgare* var *durra*
Jerusalem oak – *Chenopodium botrys*
Jerusalem sage – *Phlomis fruticosa*
Jerusalem thorn – *Paliurus spina-christi, Parkinsonia aculeata*
Jerusalem willow – *Eleagnus angustifolia*
jessamine, Carolina – *Gelsemium sempervirens*
jessamine, night – *Cestrum nocturnum*
jessamine, orange – *Murraya paniculata*
Jesuit's balsam – *Copaifera officinalis*
Jesus, heart of – *Caladium bicolor*
jewel mint of Corsica – *Mentha requienii*
jewelweed – *Impatiens capensis*
Jew's mallow – *Kerria japonica*
Jew, wandering – *Tradescantia fluminensis, T. zebrina*
jicama – *Pachyrhizus erosus*
Jim Hill mustard – *Sisymbrium altissimum*
jimmyweed – *Ericameria heterophyllus*
jimson weed – *Datura*
Job's tears – *Coix lacryma*
jocote – *Spondias purpurea*
Joe Pye weed – *Eupatorium maculatum*
Johnny-jump-up – *Viola pedunculata, V. tricolor*
John's bread, St. – *Ceratona siliqua*
Johnsongrass – *Sorgum halepense*
Johnson's pineapple cactus – *Echinomastus johnsonii*
John's wort, creeping St. – *Hypericum calycinum*
joint bush – *Miconia*
jointed charlock – *Raphanus raphanistrum*
jointed goatgrass – *Aegilops cylindrica*
jointed rush – *Juncus articulatus*
joint fir – *Ephedra nevadensis, E. trifurca*
jointfir, Nevada – *Ephedra nevadensis*
jointvetch – *Aeschynomene*
jointvetch, American – *Aeschynomene americana*
jojoba – *Simmondsia chinensis*
jonquil – *Narcissus jonquilla*
jonquilla daffodil – *Narcissus jonquilla*
Joseph's coat – *Amaranthus gangeticus, A. tricolor*
Joshua tree – *Yucca brevifolia*
joy, traveler's – *Clematis vitalba*
Judas tree – *Cercis occidentalis, C. siliquastrum*
jujube – *Ziziphus jujuba*
jujube, Indian – *Ziziphus mauritiana*
jujube tree – *Cordia myxa*
jumping bean, Mexican – *Sapium biloculare, Sebastiana pavoniana*
jumping cactus – *Cylindropuntia*
jumping cholla – *Cylindropuntia bigelovii, C. fulgida*
jumping cholla, Sonoran – *Cylindropuntia fulgida* var *mamillata*
jumpy bean – *Leucaena glauca*
June berry – *Amelanchier canadensis*
June berry, large-flowered – *Amelanchier* x *grandiflora*
Junegrass – *Koeleria macrantha*
June rose – *Lagerstroemia indica*
jungle flame – *Ixora coccinea*
jungle plum – *Dipholis salicifolia*
jungle ricegrass – *Echinochloa colona*
juniper – *Juniperus*
juniper, alligator – *Juniperus deppeana*
juniper, Ashe – *Juniperus ashei*
juniper, blanco – *Juniperus blancoi*
juniper, California – *Juniperus californica*
juniper, Canary Island – *Juniperus cedrus*
juniper, comitan – *Juniperus comitana*
juniper, common – *Juniperus communis*

juniper, Durango – *Juniperus durangensis*
juniper, dwarf – *Juniperus communis* ssp *nana, J. sibirica*
juniper, gamboa – *Juniperus gamboana*
juniper, Grecian – *Juniperus excelsa*
juniper, Mexican – *Juniperus mexicana*
juniper, Mexican drooping – *Juniperus flaccida*
juniper mistletoe – *Phoradendron juniperinum*
juniper, oneseed – *Juniperus monosperma*
juniper, prickly – *Juniperus oxycedrus*
juniper prickly-pear – *Opuntia polyacantha*
juniper, redberry – *Juniperus pinchotii*
juniper, Rocky Mountain – *Juniperus scopulorum*
juniper, rose-fruited – *Juniperus coahuilensis*
juniper, savin – *Juniperus sabina*
juniper, Syrian – *Juniperus drupacea*
juniper, Utah – *Juniperus osteosperma*
juniper, weeping – *Juniperus flaccida*
juniper, western – *Juniperus osteosperma*
Jupiter's beard – *Anthyllis barba-jovis, Centranthus ruber*
Jupiter's campion – *Lychnis flos-jovis*
jute – *Corchorus olitorius*
jute, broom – *Sida rhombifolia*
jute, Thai – *Hibiscus cannabinus*

Kaffir lily – *Clivia miniata, Schizostylis coccinea*
kahili ginger – *Hedychium gardnerianum*
kalallo bush – *Corchorus siliquosus*
kalanchoe – *Kalanchoe blossfeldiana*
kale – *Brassica oleracea* var *acephala*
kale, Chinese – *Brassica oleracea* var *alboglabra*
kale, curly – *Brassica oleracea* var *sabellica*
kale, marrow-stemmed – *Brassica oleracea* var *medullosa*
kale, Portuguese – *Brassica oleracea* var *tronchuda*
kale, sea – *Crambe maritima*
kale, Siberian – *Brassica napus* var *pabulária, B. oleracea* var *ramosa*
kangaroo thorn – *Acacia paradoxa*
kapok – *Ceiba acuminata*
kapok tree – *Ceiba pentandra*
katsura tree – *Cercidiphyllum japonicum*
Katy, flaming – *Kalanchoe blossfeldiana*
kava kava – *Piper methysticum*
keeled allium – *Allium carinatum* ssp *pulchellum*
keg cactus – *Ferocactus viridescens*
kei apple – *Dovyalis caffra*
kelp – *Fucus vesiculosus, Laminaria, Macrocystis pyrifera*
kelp, Sea-of-Cortez – *Sargassum*
kenaf – *Hibiscus cannabinus*
Kenilworth ivy – *Cymbalaria muralis*
Kentia palm – *Howea forsteriana*
Kentucky bluegrass – *Poa pratensis*
Kentucky coffee tree – *Gymnocladus dioica*
kermes oak – *Quercus coccifera*
Key lime – *Citrus aurantifolia*
khat – *Catha edulis*
khella – *Ammi visnaga*
kidney bean, red – *Phaseolus vulgaris*
kidney vetch – *Anthyllis vulneraria*
kidneywood – *Eysenhardtia orthocarpa, E. polystachya*
kidneywood, Texas – *Eysenhardtia texana*
kikuyu grass – *Pennisetum clandestinum*
killer, mouse – *Gliricidia sepium*
king bolete, summer – *Boletus reticulatus*
King Henry, good – *Chenopodium bonus-henricus*
king orange – *Citrus nobilis*
king palm – *Archontophoenix cunninghamiana*
king's crown – *Sedum roseum*
king's spear – *Asphodeline lutea, Asphodelus albus*
kings, tree of – *Cordyline fruticosa*
kinnikinnik – *Arctostaphylos uva-ursi*
kiss-me-over-the-garden-gate – *Polygonum orientale*
kiwi – *Actinidia deliciosa*
Klamathweed – *Hypericum perforatum*
Klein's cholla – *Cylindropuntia kleiniae*
Klein's pencil cholla – *Cylindropuntia kleiniae*
knackaway – *Ehretia anacuna*
knapweed – *Centaurea*
knapweed, black – *Centaurea nigra*
knapweed, diffuse – *Centaurea diffusa*
knapweed, greater – *Centaurea scabiosa*
knapweed, meadow – *Centaurea* x *pratensis*
knapweed, mountain – *Centaurea montana*
knapweed, spotted – *Centaurea maculosa*
knapweed, weak – *Centaurea diluta*
knapweed, white – *Centaurea cineraria*

knapweed, wide-spreading – *Centaurea diffusa*
knawel, annual – *Scleranthus annuus*
knobcone pine – *Pinus attenuata*
knotberry – *Rubus chamaemorus*
knot grass – *Paspalum distichum, P. paspalodes, Polygonum*
knotted bur parsley – *Torilis nodosa*
knotted hedge parsley – *Torilis nodosa*
knotted marjoram – *Origanum majorana*
knotweed – *Polygonum*
knotweed, common – *Polygonum arenastrum*
knotweed, Himalayan – *Polygonum polystachyum*
knotweed, Japanese – *Polygonum cuspidatum*
knotweed, prostrate – *Polygonum aviculare*
knotweed, yard – *Polygonum aviculare*
kobus magnolia – *Magnolia kobus*
kodo millet – *Paspalum scrobiculatum*
kohlrabi – *Brassica oleracea* var *gongylodes*
kola, gotu – *Hydrocotyle asiatica*
kola nut – *Kola acuminata*
kola tree – *Kola acuminata*
Korean fir – *Abies koreana*
Korean ginseng – *Panax ginseng*
kousa dogwood – *Cornus kousa*
kudzu – *Pueraria lobata, P. phaseoloides*
kukui – *Aleurites moluccana*
kumquat – *Fortunella margarita*
kumquat, nagami – *Fortunella margarita*

laburnum, Indian – *Cassia fistula*
laburnum, Scotch – *Laburnum alpinum*
lacebark pine – *Pinus bungeana*
lacebark, Queensland – *Brachychiton discolor*
lace fern – *Cheilanthes*
lace, Irish – *Tagetes filifolia*
lace vine – *Fallopia baldschuanica*
lace vine, silver – *Fallopia baldschuanica*
lacey oak – *Quercus glaucoides*
lacquer tree, Japanese – *Rhus verniciflua*
lacyspine cactus – *Mammillaria lasiacantha*
ladder, Jacob's – *Pedilanthus tithymaloides, Polemonium caeruleum*
ladies' bedstraw – *Galium verum*
ladies, pink – *Oenothera speciosa*
ladies' tresses – *Spiranthes aurantiaca*
ladies' tresses, cinnabar – *Spiranthes cinnabarina*
Lady Banks' rose – *Rosa banksiae*
lady bells – *Adenophora liliifolia*
lady fern – *Asplenium filix-femina*
lady finger banana – *Mus acuminata*
lady finger cactus – *Echinocereus pentalophus, Mammillarea elongata*
lady palm – *Rhapis excelsa*
lady palm, slender – *Rhapis humilis*
lady's bedstraw – *Galium vernum*
lady's-mantle – *Alchemilla*
lady's-mantle, blue-green – *Alchemilla glaucescens*
lady's-slipper – *Cypripedium, Paphiopedilum, Pedilanthus macrocarpus*
lady's smock – *Cardamine pratensis*
lady's thumb – *Polygonum persicaria*
lady tulip – *Tulipa clusiana*
Lady Washington pelargonium – *Pelargonium* x *domesticum*
laelia, autumn-flowering – *Laelia autumnalis*
laelia, gorgeous – *Laelia superbiens*
laelia, May-flowering – *Laelia speciosa*
lalang – *Imperata cylindrica*
Lamarck's bedstraw – *Galium divaricatum*
Lambert's crazyweed – *Oxytropis lambertii*
lamb's ears – *Stachys byzantina*
lamb's lettuce – *Valerianella locusta*
lamb's quarters – *Chenopodium album*
lanceleaf cottonwood – *Populus* x *acuminata*
lanceleaf nightshade – *Solanum lanceolatum*
lance pod – *Lonchocarpus longistylus*
langsat – *Lansium domesticum*
lantana – *Lantana camara*
lantana, desert – *Lantana macropoda*
lantana, Texas – *Lantana horrida*
lantern, Chinese – *Abutilon*
lantern, devil's – *Oenothera deltoides*
lantern plant, Chinese – *Physalis alkekengi*
laphamia – *Perityle*
larch, eastern – *Larix laricina*
larch, Dahurian – *Larix gmelinii*
larch, European – *Larix decidua*
larch, golden – *Pseudolarix amabilis*
larch, Japanese – *Larix kaempferi*
larchleaf goldenweed – *Ericameria laricifolia*
larch, Siberian – *Larix sibirica*
larch, western – *Larix occidentalis*
larkspur – *Consolida ambigua, Delphinium*
larkspur, blue – *Delphinium carolinianum*
larkspur, candle – *Delphinium elatum*

larkspur, forking – *Delphinium consolida*
larkspur, orange – *Delphinium nudicaule*
larkspur, red – *Delphinium nudicaule*
Las Animas colubrina – *Colubrina californica*
Lasater's pride – *Abronia fragrans*
late-flowering sedge – *Carex viridula*
laurel – *Kalmia latifolia, Laurus nobilis*
laurel, bay – *Laurus nobilis*
laurel, bush – *Litsea glaucescens*
laurel, California – *Umbellularia californica*
laurel cherry – *Prunus laurocerasus*
laurel, English – *Prunus laurocerasus*
laurel fig, Indian – *Ficus microcarpa*
laurel, Grecian – *Laurus nobilis*
laurel, Indian – *Ficus microcarpa*
laurel-leaved sun-rose – *Cistus laurifolius*
laurel, Mexican – *Phoebe tampicensis*
laurel, mountain – *Kalmia latifolia, Rhus ovata*
laurel, New Zealand – *Corynocarpus laevigata*
laurel, Portugal – *Prunus lusitanica*
laurel, spotted – *Aucuba japonica*
laurel, spurge – *Daphne laureola*
laurel sumac – *Malosma laurina*
laurel, sweet – *Illicium floridanum*
laurel, Texas mountain – *Sophora secundiflora*
laurel, variegated – *Codiaeum variegatum*
laurel, white – *Nectandra*
laurustinus – *Viburnum tinus*
lavender – *Lavandula spica*
lavender, broadleaf – *Lavandula latifolia*
lavender-cotton – *Santolina chamaecyparisus*
lavender-cotton, rosemary-leaved – *Santolina rosmarinifolia*
lavender, desert – *Hyptis emoryi, H. laniflora*
lavender, English – *Lavandula angustifolia*
lavender, fernleaf – *Lavandula multifida*
lavender, French – *Lavandula dentata, L. stoechas*
lavender, green – *Lavandula viridis*
lavender, sea – *Limonium californicum, L. sinuatum*
lavender, Spanish – *Lavandula stoechas*
lavender, spike – *Lavandula latifolia*
lavender, toothed – *Lavandula dentata*
lavender, woolly – *Lavandula lanata*
lavender, yellow – *Lavandula viridis*
lawn daisy, English – *Bellis perennis*
Lawson cedar – *Cupressus lawsoniana*
Lawson cypress – *Cupressus lawsoniana*
Lawson pine – *Pinus lawsonii*
lead pencil cactus – *Echinocereus poselgeri*
lead-tree, white – *Leucaena leucocephala*
lead-tree, white popinac – *Leucaena glauca*
leadwort – *Plumbago auriculata, P. pulchella, P. scandens*
leaf lettuce – *Lactuca sativa* var *longifolia*
leafy spurge – *Euphorbia esula*
least lettuce – *Lactuca saligna*
least snoutbean – *Rhynchosia minima*
leatherleaf viburnum – *Viburnum rhytidophyllum*
leatherplant – *Jatropha cordata, J. spathulata*
Lebanon, cedar of – *Cedrus libani*
ledger bark – *Cinchona ledgeriana*
leek – *Allium porrum*
leek, house – *Sempervivum*
leek, wild – *Allium ampeloprasum*
Lehmann lovegrass – *Eragrostis lehmanniana*
Lemmon's milkweed – *Asclepias lemmonii*
lemon – *Citrus limon*
lemonade berry – *Rhus integrifolia*
lemon balm – *Melissa officinalis*
lemon bottlebrush – *Callistemon citrinus*
lemon, Chinese – *Citrus limonia*
lemon daylily – *Hemerocallis lilioasphodelus*
lemon grass – *Andropogon schaenanthus, Cymbopogon citratus*
lemon horsemint – *Monarda citriodora*
lemon hyssop, giant Mexican – *Agastache mexicana*
lemon, Mediterranean sweet – *Citrus limetta*
lemon, ponderosa – *Citrus limon* var *ponderosa*
lemon, rough – *Citrus jambhiri*
lemon scurf-pea – *Psoralea lanceolata*
lemon, sweet – *Citrus limetta*
lemon verbena – *Aloysia triphylla, Lippia triphylla, Phyla nodiflora*
lemon vine – *Pereskia aculeate, P. grandifolia*
lemonweed – *Pectis angustifolia, Psoralea lanceolata*
lemon wood – *Calycophyllum candidissimum*
lens scale – *Atriplex lentiformis*
lenten rose – *Helleborus orientalis*
lentil – *Lens culinaris*
lentisc – *Pistacia lentiscus*
leopard-like mormodes – *Mormodes pardinum* ssp *unicolor*

leopard lily – *Lilium pardalinum*
leopard's bane – *Doronicum cordatum, D. pardalianches*
leopard's bane, plantain – *Doronicum plantagineum*
lespedeza, common – *Lespedeza striata*
lettuce – *Lactuca canadensis, L. sativa*
lettuce, bitter – *Lactuca virosa*
lettuce, butter – *Lactuca sativa* var *crispa*
lettuce, chalk – *Dudleya pulverulenta*
lettuce, devil's – *Amsinckia tessellata*
lettuce, head – *Lactuca sativa* var *capitata*
lettuce, iceberg – *Lactuca sativa* var *capitata*
lettuce, Indian – *Claytonia perfoliata*
lettuce, lamb's – *Valerianella locusta*
lettuce, leaf – *Lactuca sativa* var *longifolia*
lettuce, least – *Lactuca saligna*
lettuce, miner's – *Claytonia perfoliata*
lettuce, prickly – *Lactuca serriola*
lettuce, rock – *Pinaropappus roseus*
lettuce, romaine – *Lactuca sativa* var *longifolia*
lettuce, sawah – *Limnocharis flava*
lettuce, Tartaric – *Lactuca tatarica*
lettuce, water – *Pistia stratiotes*
lettuce, wire – *Stephanomeria pauciflora*
Levant garlic – *Allium ampeloprasum*
Liberian coffee tree – *Coffea liberica*
Liberica coffee – *Coffea liberica*
liberty cap, Mexican – *Psilocybe mexicana*
lice, beggar – *Desmodium styracifolium*
lice, beggar's – *Galium aparine*
lichen – *Roccella*
lichen, rock – *Parmelia tiliacea*
lichen, tree – *Sticta pulmonacea, Usnea barbata*
licorice – *Glycyrrhiza glabra*
licorice root – *Glycyrrhiza glabra*
licorice, wild – *Astragalus glycyphyllos, Glycyrrhiza lepidota*
licuri palm – *Syagrus coronata*
life plant – *Kalanchoe pinnata*
lightningweed – *Drymaria arenarioides*
lign aloe – *Aquilaria agallocha*
lignum vitae – *Guaiacum coulteri, G. officinale*
lignum vitae, hollywood – *Guaiacum sanctum*
lignum vitae, holywood – *Guaiacum sanctum*
ligonberry – *Vaccinium vitis-idaea* var *minus*
lilac, California – *Ceanothus*
lilac, common – *Syringa vulgaris*
lilac, Persian – *Melia azedarach* var *umbraculiformis*
lilac, summer – *Buddleja davidii*
lilac, wild – *Leucophyllum frutescens*
lilies of the field – *Anemone coronaria*
lily – *Lilium*
lily, African corn – *Ixia maculata*
lily, amber – *Anthericum torreyi*
lily, atamasco – *Zephranthes texana*
lily, Aztec – *Sprekelia formosissima, Tigridia pavonia*
lily, belladonna – *Amaryllis belladonna*
lily, blackberry – *Belamcanda chinensis*
lily, blood – *Scadoxus multiflorus*
lily, boat – *Tradescantia spathacea*
lily, butterfly – *Hedychium coronarium*
lily cactus – *Echinocereus cinerascens*
lily, calla – *Zantedeschia aethiopica*
lily, checkered – *Fritillaria meleagris*
lily, chocolate – *Fritillaria biflora*
lily, climbing – *Gloriosa superba*
lily, copper zephyr – *Zephyranthes longifolia*
lily, coral – *Lilium pumilum*
lily, corn – *Veratrum officinale, V. viride*
lily, cow – *Nuphar luteum*
lily, day – *Hemerocallis*
lily, desert – *Hesperocallis undulata*
lily, Drummond's rain – *Cooperia drummondii*
lily, easter – *Lilium candidum, L. longiflorum*
lily, fairy – *Zephyranthes candida, Z. carinata*
lily, flame – *Gloriosa superba*
lily, flamingo – *Anthurium andraeanum*
lily, foxtail – *Eremurus robustus*
lily, fragrant plantain – *Hosta plantaginea*
lily, gloriosa – *Gloriosa superba*
lily, glory – *Gloriosa superba*
lily, gold-ban – *Lilium auratum*
lily, golden-rayed – *Lilium auratum*
lily, green – *Schoenocaulon officinale*
lily, impala – *Adenium obesum*
lily, Jacobean – *Sprekelia formosissima*
lily, Kaffir – *Clivia miniata, Schizostylis coccinea*
lily, leopard – *Lilium pardalinum*
lily, madonna – *Lilium candidum*
lily, magic – *Lycoris radiata*
lily, Malabar – *Gloriosa superba*
lily, mariposa – *Calochortus*
lily of the field – *Sternbergia lutea*

lily-of-the-Nile – *Agapanthus africanus*
lily-of-the-valley – *Convallaria majalis*
lily-of-the-valley, false – *Maianthemum dilatatum*
lily-of-the-valley shrub – *Pieris japonica*
lily, orange – *Lilium bulbiferum*
lily, oriental – *Lilium auratum*
lily, Panamint mariposa – *Calochortus panamintensis*
lily, panther – *Lilium pardalinum*
lily, Peruvian – *Alstroemeria aurantiaca, A. pelegrina*
lily, porcelain – *Alpinia zerumbet*
lily, rain – *Cooperia*
lily, regal – *Lilium regale*
lily, royal water – *Victoria regia*
lily, sea – *Pancratium maritimum*
lily, sego – *Calochortus bruneaunis*
lily, spider – *Crinum zeylanicum, Hymenocallis americana, H. rotata, Lycoris radiata*
lily, splendid – *Lilium speciosum*
lily, star – *Zephyranthes texana*
lily, St. James – *Sprekelia formosissima*
lily, superb – *Gloriosa superba*
lily, swamp – *Thalia geniculata*
lily, tiger – *Lilium lancifolium*
lily, torch – *Kniphofia uvaria*
lily, Torrey's crag – *Anthericum torreyi*
lily turf, big blue – *Liriope muscari*
lily, Turk's cap – *Lilium henryi, L. martagon*
lily, white arum – *Zantedeschia aethiopica*
lily, white ginger – *Hedychium coronarium*
lily, wild – *Milla biflora*
lily, yellow pond – *Nuphar luteum* ssp *polysepalum*
lily, yellow water – *Nuphar luteum, Nymphaea mexicana*
Lima bean – *Phaseolus lunatus*
limberbush – *Jatropha cardiophylla, J. cuneata*
limberbush, ashy – *Jatropha cinerea*
limberbush, Sonoran tree – *Jatropha cordata*
limber pine – *Pinus flexilis*
lime – *Citrus aurantifolia*
lime, American brook – *Veronica americana*
lime, bastard – *Trichilia hirta*
lime-berry – *Triphasia trifolia*
lime, bitter – *Citrus limetta*
lime, brook – *Veronica beccabunga*
lime cactus, Mexican – *Ferocactus pilosus*
lime, common – *Tilia* x *vulgaris*
lime, English – *Tilia* x *europaea*
lime, Indian sweet – *Citrus limettioides*
lime, Key – *Citrus aurantifolia*
lime, Mandarin – *Citrus limonia*
lime, Mexican – *Citrus aurantifolia*
lime, Persian – *Citrus latifolia*
lime pricklyash – *Zanthoxylum fagara*
lime, Rangpur – *Citrus limonia*
lime, Spanish – *Melicocca bijuga*
lime, sweet, of India – *Citrus limettioides*
lime, Tahiti – *Citrus latifolia*
limetree, broad-leaved – *Tilia platyphyllos*
limetree, small-leaved – *Tilia cordata*
limetta – *Citrus limetta*
lime, weeping silver – *Tilia* x *petiolaris*
lime, West Indian – *Citrus aurantifolia*
lime, wild – *Casearia aculeata, Randia, Ximenia americana, Zanthoxylum fagara*
linda tarde – *Gaura coccinea*
linden – *Tilia europaea, T. occidentalis, Triumfetta semitriloba*
linden, African – *Sparmannia africana*
linden, American – *Tilia americana*
linden, Crimean – *Tilia* x *euchlora*
linden, European – *Tilia europaea*
linden, littleleaf – *Tilia cordata*
linden, silver – *Tilia tomentosa*
Lindsay's hedgehog cactus – *Echinocereus ferreirianus*
ling chih – *Ganoderma lucidum*
linseed – *Linum usitatissimum*
lionheart – *Physostegia virginiana*
lion-in-a-cage – *Oenothera deltoides*
lion's tail – *Leonotis leonurus*
lip fern – *Cheilanthes lindheimeri*
lipstick tree – *Bixa orellana*
liquidambar – *Liquidambar styraciflua*
liquorice – *Glycyrrhiza glabra*
litchi nut – *Litchi chinensis*
live-forever – *Dudleya, Sedum*
livelong – *Sedum telephium*
live oak – *Quercus*
live oak, canyon – *Quercus chrysolepis*
live oak, coast – *Quercus agrifolia*
live oak, coastal – *Quercus virginiana*
live oak, escarpment – *Quercus fusiformis*
live oak, Japanese – *Quercus myrsinifolia*
live oak, shrub – *Quercus turbinella*

live oak, southern – *Quercus virginiana*
live oak, white – *Quercus oblongifolia*
liverleaf – *Hepatica nobilis*
liverwort – *Hepaticae*
living rock – *Ariocarpus kotschoubeyanus*
living-rock cactus – *Ariocarpus fissuratus, A. retusus*
living stone – *Lithops*
lizard's tail – *Anemopsis californica*
lizard-tail – *Gaura parviflora*
Lizzy, busy – *Impatiens walleriana*
lobelia – *Lobelia*
lobelia, blue – *Lobelia erinus*
lobelia, giant orange – *Lobelia laxiflora*
lobelia, heath – *Lobelia urens*
loblolly pine – *Pinus taeda*
lobster-claw – *Heliconia*
lobster plant – *Heliconia latispatha*
locks, marsh – *Potentilla palustris*
locoweed – *Astragalus, Oxytropis lambertii*
locoweed, Nuttall's – *Astragalus nuttallianus*
locoweed, woolly – *Astragalus mollissimus*
locust – *Ceratonia siliqua*
locust, black – *Robinia pseudoacacia*
locust, clammy – *Robinia viscosa*
locust, desert – *Robinia neomexicana*
locust, honey – *Gleditsia amorphoides, G. triacanthos*
locust, New Mexican – *Robinia neomexicana*
locust tree – *Bernoullia flammea*
locust, West Indian – *Hymenaea courbaril*
lodgepole pine – *Pinus contorta* ssp *murrayana*
loganberry – *Rubus loganobaccus*
logania – *Gelsemium sempervirens, Strychnos tabascana*
logwood – *Condalia obovata, Haematoxylon campechianum*
lollipop tree – *Forchhammeria watsonii*
lomatium – *Lomatium foeniculaceum*
Lombardy poplar – *Populus nigra* var *italica*
lomboy – *Jatropha cinerea*
London planetree – *Platanus* x *acerifolia*
London pride – *Saxifraga umbrosa*
London rocket – *Sisymbrium irio*
London rocket, false – *Sisymbrium loeselii*
longan – *Nephelium longana*
long birthwort – *Aristolochia longa*
longleaf pine – *Pinus palustris*
longleaf pine, Arizona – *Pinus engelmanii*
longleaf pine, Indian – *Pinus roxburghii*
longneck ruel – *Ruellia nudiflora*
longspur epimedium – *Epimedium grandiflorum*
long-tom – *Paspalum lividum*
lontar palm – *Borassus flabellifer*
loofah – *Luffa aegyptiaca*
loofah, angled – *Luffa acutangula*
looking glass, Venus' – *Specularia speculum*
loosestrife – *Lysimachia punctata, Lythrum*
loosestrife, dotted – *Lysimachia punctata*
loosestrife, false – *Ludwigia peploides*
loosestrife, fringe-leaved – *Lysimachia ciliata*
loosestrife, gooseneck – *Lysimachia clethroides*
loosestrife, hyssop – *Lythrum hyssopifolium*
loosestrife, purple – *Lysimachia salicaria, Lythrum virgatum*
loosestrife, tufted – *Lysimachia thyrsiflora*
loquat – *Eriobotrya japonica*
lotebush – *Condia, Ziziphus obtusifolia* var *canescens, Z. parryi*
lotebush, California – *Ziziphus parryi*
lot plant – *Cistus creticus*
lotus – *Nelumbo nucifera*
lotus, American – *Nelumbo lutea*
lotus, Chinese – *Nelumbo nucifera*
lotus, Indian – *Nelumbo nucifera*
lotus, sacred – *Nelumbo nucifera*
Louisiana moss – *Tillandsia usneoides*
loulu palm – *Pritchardia hillebrandii*
lousewort – *Pedicularis canadensis, Stachys betonica*
lovage – *Levisticum officinale, Ligusticum porteri*
lovage, black – *Smyrnium olustratum*
love, burning – *Ixora coccinea*
love, chain of – *Antigonon leptopus*
love, exotic – *Quamoclit lobata*
lovegrass – *Eragrostis*
lovegrass, alkali – *Eragrostis obtusiflora*
lovegrass, Boer – *Eragrostis chloromelas*
lovegrass, India – *Eragrostis pilosa*
lovegrass, Lehmann – *Eragrostis lehmanniana*
lovegrass, lesser – *Eragrostis minor*
lovegrass, Mexican – *Eragrostis mexicana*
lovegrass, plains – *Eragrostis intermedia*
lovegrass, sawtooth – *Eragrostis superba*
lovegrass, weeping – *Eragrostis curvula*
love-in-a-mist – *Nigella damascena*

love-in-a-puff – *Cardiospermum halicacabum*
love-lies-bleeding – *Amaranthus caudatus*
low ragweed – *Ambrosia artemisiifolia*
low rattlebox – *Crotalaria pumila*
low verbena – *Verbena pumila*
loxa – *Cinchona officinalis*
lucerne – *Medicago sativa*
lucerne, Brazilian – *Stylosanthes guianensis*
luck plant, good – *Cordyline fruticosa*
lucky nut – *Thevetia peruviana*
lucky plant – *Echeveria gibbiflora*
Lumholtz pine – *Pinus lumholtzii*
lungwort – *Pulmonaria longifolia, P. officinalis*
lungwort, cowslip – *Pulmonaria angustifolia*
lupine – *Lupinus*
lupine, annual – *Lupinus angustifolius*
lupine, Arizona – *Lupinus arizonicus*
lupine, arroyo – *Lupinus succulentus*
lupine, blue – *Lupinus angustifolius*
lupine, garden – *Lupinus polyphyllus*
lupine, Russell hybrid – *Lupinus russeli*
lupine, sweet – *Lupinus luteus*
lupine, tree – *Lupinus arboreus*
lupine, white – *Lupinus albus*
lupine, yellow – *Lupinus luteus*
lupine, yellow bush – *Lupinus arboreus*
lurid bolete – *Boletus luridus*
lustrous cinquefoil – *Potentilla nitida*
lychee – *Litchi chinensis*
lychee, hairy – *Nephelium lappaceum*
lyreleaf green eyes – *Berlandiera lyrata*

macadamia nut – *Macadamia integrifolia, M. ternifolia*
macadamia, smooth-shell – *Macadamia integrifolia*
MacArthur palm – *Ptychosperma macarthuri*
macassar ebony – *Diospyros ebenum*
macaw palm – *Acrocomia sclerocarpa*
mace – *Myristica fragrans*
Madagascar dragon tree – *Dracaena marginata*
Madagascar jasmine – *Stephanotis floribunda*
Madagascar palm – *Chrysalidocarpus lutescens*
Madagascar periwinkle – *Catharanthus roseus*
Madagascar plant – *Dracaena marginata*
madder – *Rubia tinctorium*
madder, field – *Sherardia arvensis*
mad-dog skullcap – *Scutellaria lateriflora*
madia oil plant – *Madia sativa*
madonna lily – *Lilium candidum*
Madrean willow oak – *Quercus viminea*
madrona – *Arbutus menziesii*
madrone – *Arbutus*
madrone, Arizona – *Arbutus arizonica*
madrone, checkerbark – *Arbutus tessellata*
madrone, Mexican – *Arbutus xalapensis*
madrone, Pacific – *Arbutus menziesii*
madrone, smooth – *Arbutus xalapensis*
madrone, Texas – *Arbutus texana*
madroño, Arizona – *Arbutus arizonica*
madwort – *Asperugo procumbens*
Magellan barberry – *Berberis buxifolia*
magic flower – *Cantua buxifolia*
magic lily – *Lycoris radiata*
magnolia, anise – *Magnolia salicifolia*
magnolia, bigleaf – *Magnolia macrophylla*
magnolia, kobus – *Magnolia kobus*
magnolia, Mexican – *Talauma mexicana*
magnolia, oyama – *Magnolia sieboldii*
magnolia, saucer – *Magnolia soulangiana*
magnolia, southern – *Magnolia grandiflora*
magnolia, star – *Magnolia stellata*
magnolia, veitch – *Magnolia* x *veitchii*
magnolia, yulan – *Magnolia denudata*
maguey, cantala – *Agave cantala*
maguey, Manila – *Agave cantala*
mahleb cherry – *Prunus mahleb*
mahoe – *Hibiscus tiliaceus*
mahogany – *Swietenia macrophylla*
mahogany, African – *Khaya senegalensis*
mahogany, alder-leaf mountain – *Cercocarpus montanus*
mahogany, birch-leaf mountain – *Cercocarpus betuloides*
mahogany, Brazilian – *Swietenia macrophylla*
mahogany, Dominican – *Swietenia mahagoni*
mahogany, Honduras – *Swietenia macrophylla*
mahogany, Mexican – *Swietenia humilis*
mahogany, mountain – *Cercocarpus montanus*
mahogany, Pacific coast – *Swietenia humilis*
mahogany, swamp – *Eucalyptus robusta*
mahogany, white – *Tabebuia donnell-smithii*
mahonia, creeping – *Berberis aquifolium*
mahonia, desert – *Berberis fremontii*
mahonia, red – *Berberis haematocarpa*
mahuang – *Ephedra sinica*
mahue, blue – *Hibiscus elatus*

maiden clover – *Trifolium microcephalum*
maiden fern, Sonoran – *Thelypteris puberula* var *sonorensis*
maidenhair, black stick – *Adiantum tenerum*
maidenhair fern – *Adiantum*
maidenhair, rosy – *Adiantum hispidulum*
maidenhair, southern – *Adiantum capillus-veneris*
maidenhair spleenwort – *Asplenium trichomanes*
maidenhair tree – *Ginkgo biloba*
maiden pink – *Dianthus deltoides*
maid, old – *Catharanthus roseus*
maids, cliff – *Lewisia cotyledon*
maids, red – *Calandrinia ciliata*
maize – *Zea mays*
maize, dwarf – *Zea mays* var *gracillima*
maize, soft – *Zea mays* var *amylacea*
Malabar gourd – *Cucurbita ficifolia*
Malabar lily – *Gloriosa superba*
malady – *Aspidosperma quebracho-blanco*
Malaguetta pepper – *Aframomum melegueta*
Malay apple – *Eugenia malaccensis*
Malay ginger – *Costus speciosus*
Malay rose-apple – *Eugenia malaccensis*
male fern – *Dryopteris filix-mas*
mallow – *Malva, Malvastrum sylvestrus*
mallow, alkali – *Malvella lepidota, Sida hederacea*
mallow, annual – *Lavatera trimestris*
mallow, annual globe – *Sphaeralcea coulteri*
mallow, apricot – *Sphaeralcea ambigua*
mallow, bull – *Malva nicaeensis*
mallow, chaparral – *Malacothamnus fasciculatus*
mallow, checker – *Sidalcea malviflora*
mallow, common – *Malva neglecta*
mallow, copper – *Sphaeralcea angustifolia*
mallow, desert – *Sphaeralcea ambigua*
mallow, desert globe – *Sphaeralcea ambigua*
mallow, desert rose – *Hibiscus coulteri*
mallow, false – *Eremalche*
mallow, globe – *Sphaeralcea*
mallow, high – *Malva sylvestris*
mallow, Indian – *Abutilon*
mallow, island – *Lavatera assurgentiflora*
mallow, Jew's – *Kerria japonica*
mallow, marsh – *Althea officinalis*
mallow, mountain rose – *Hibiscus bifurcatus, H. cardiophyllus*
mallow, musk – *Abelmoschus moschatus, Malva moschata*
mallow, prairie – *Sidalcea neomexicana*
mallow, red – *Malvaviscus grandiflorus*
mallow, rose – *Hibiscus rosa-sinensis*
mallows – *Atriplex halimus*
mallow, scarlet globe – *Sphaeralcea coccinea*
mallow, sea – *Lavatera maritima*
mallow, Texas – *Malvaviscus arboreus*
mallow, tree – *Lavatera arborea, L. cretica, L. maritima*
mallow, umbrella – *Malva neglecta*
mallow, vervain's – *Malva alcea*
mallow, wax – *Malvaviscus arboreus*
Malta star-thistle – *Centaurea melitensis*
Maltese cross – *Lychnis chalcedonica*
malva rosa – *Lavatera assurgentiflora*
mamee apple – *Mammea americana*
mamee sapote – *Lucuma mammosa*
mamey – *Mammea americana*
mamey apple – *Pouteria sapota*
mamey sapote – *Lucuma mammosa*
mammoth clover – *Trifolium medium*
manchineel – *Hippomane mancinella*
Mandarine tree – *Citrus deliciosa*
Mandarin lime – *Citrus limonia*
Mandarin orange – *Citrus nobilis, C. reticulata*
mandrake – *Mandragora autumnalis, Podophyllum peltatum*
mangabeira – *Hancornia speciosa*
mangle dulce – *Maytenus phyllanthoides*
mango – *Mangifera indica*
mango, bachang – *Mangifera foetida*
mango, Indian – *Licania*
man, golden old – *Pilosocereus chrysacanthus*
mango melon – *Cucumis melo* var *chito*
mangosteen – *Garcinia mangostana*
mango, wild – *Irvingia gabonensis*
mangrove, black – *Avicennia germinans*
mangrove, buttonwood – *Conocarpus erecta*
mangrove fern – *Acrostichum aureum*
mangrove, honey – *Avicennia germinans*
mangrove, red – *Avicennia marina* var *resinifera, Rhizophora mangle*
mangrove, sweet – *Maytenus phyllanthoides*
mangrove, white – *Laguncularia racemosa*
Manila hemp – *Mus textilis*

Manila maguey – *Agave cantala*
Manila tamarind – *Pithecellobium dulce*
manioc – *Manihot dulcis, M. esculenta*
manioc, sweet – *Manihot palmata*
manna – *Alhagi pseudalhagi, Lecanora*
manna bush – *Tamarix gallica*
mannagrass – *Glyceria fluitans, G. striata*
manna gum – *Eucalyptus viminalis*
manna hebraica – *Alhagi pseudalhagi*
manna, Sinai – *Alhagi pseudalhagi*
man, old – *Artemisia abrotanum*
man orchid – *Aceras anthropophorum*
man-root, California – *Marah fabaceus, M. macrocarpus*
mantle flower – *Asplundia chiapensis*
mantle, lady's – *Alchemilla*
manzanita – *Arctostaphylos*
manzanita, bigberry – *Arctostaphylos glauca*
manzanita, bracted – *Arctostaphylos pringlei*
manzanita, common – *Arctostaphylos manzanita*
manzanita, greenleaf – *Arctostaphylos patula*
manzanita, Mexican – *Arctostaphylos pungens*
manzanita, Parry – *Arctostaphylos manzanita*
manzanita, pinemat – *Arctostaphylos nevadensis*
manzanita, pink-bracted – *Arctostaphylos pringlei* ssp *drupacea*
manzanita, point-leaf – *Arctostaphylos pungens*
manzanita, shaggy-barked – *Arctostaphylos tomentosa*
manzanita, whiteleaf – *Arctostaphylos viscida*
maple – *Acer*
maple, Amur – *Acer tartaricum* var *ginnala*
maple, bigleaf – *Acer macrophyllum*
maple, bigtooth – *Acer saccharum* ssp *grandidentatum*
maple, coliseum – *Acer cappadocicum*
maple, David's – *Acer davidii*
maple, flowering – *Abutilon*
maple, fullmoon – *Acer japonicum*
maple, hedge – *Acer campestre*
maple, Japanese – *Acer palmatum*
mapleleaf peamelon – *Sicyos angulatus*
maple, Norway – *Acer platanoides*
maple, paperbark – *Acer griseum*
maple, red – *Acer rubrum*
maple, Rocky Mountain sugar – *Acer saccharum* ssp *grandidentatum*
maple, scarlet – *Acer rubrum*
maple, silver – *Acer saccharinum*
maple, sugar – *Acer saccharum*
maple, sycamore – *Acer pseudoplatanus*
maple, Tartarian – *Acer tartaricum*
maple, trident – *Acer buergeranum*
maranta – *Calathea, Maranta leuconeura*
mare's-tail – *Hippuris vulgaris*
marganita – *Arctostaphylos tomentosa*
Margaret, herb – *Bellis perennis*
marguerite – *Chrysanthemum frutescens*
marguerite, golden – *Anthemis tinctoria*
marigold, African – *Tagetes erecta*
marigold, American – *Tagetes erecta*
marigold, bur – *Bidens cernua, B. grandiflora, B. laevis*
marigold, Cape – *Dimorphotheca sinuata*
marigold, desert – *Baileya multiradiata*
marigold, drooping – *Baileya pauciradiata*
marigold, dwarf – *Tagetes minuta*
marigold, fetid – *Dyssodia papposa, Pectis*
marigold, few-flowered desert – *Baileya pauciradiata*
marigold, field – *Calendula arvensis*
marigold, fig – *Carpobrotus chilensis, C. edulis*
marigold, fine-leaved – *Tagetes lunulata*
marigold, French – *Tagetes patula*
marigold, marsh – *Caltha leptosepala, C. palustris*
marigold, Mexican – *Tagetes lucida*
marigold, pot – *Calendula officinalis*
marigold, signet – *Tagetes tenuifolia*
marigold, sweet-scented – *Tagetes lucida*
marigold, woolly – *Baileya pleniradiata*
marigold, woolly desert – *Baileya pleniradiata*
marijuana – *Cannabis sativa*
marimbawood – *Platymiscium dimorphandrum*
marine asteriscus – *Asteriscus maritimus*
mariola – *Parthenium incanum*
mariposa, desert – *Calochortus kennedyi*
mariposa lily – *Calochortus*
mariposa lily, Panamint – *Calochortus panamintensis*
maritime club-rush – *Scirpus maritimus*
maritime pine – *Pinus pinaster*
maritime vetchling – *Lathyrus japonicus*
marjoram – *Origanum majorana*
marjoram, hop – *Origanum dictamnus*

marjoram, knotted – *Origanum majorana*
marjoram, Mexican – *Lantana macropoda*
marjoram, Spanish – *Thymus mastichina*
marjoram, sweet – *Origanum majorana*
marjoram, wild – *Origanum vulgare*
marlberry – *Ardisia revolute*
marmalade fruit – *Lucuma mammosa*
marram grass – *Ammophila arenaria*
marriage vine – *Solanum wendlandii*
marrow-stemmed kale – *Brassica oleracea* var *medullosa*
marsh arrow-grass – *Triglochin palustris*
marsh bent – *Agrostis palustris*
marsh centaury – *Centaurium calycosum*
marsh elder – *Iva ambrosiaefolia*
marsh felwort – *Swertia perennis*
marsh fleabane – *Pluchea sericea, Pulicaria paludosa*
marsh fleabane, salt – *Pluchea odorata*
marsh horsetail – *Equisetum palustre*
marsh locks – *Potentilla palustris*
marshmallow – *Althea officinalis*
marsh marigold – *Caltha leptosepala, C. palustris*
marsh parsley – *Ciclospermum leptophyllum*
marsh pea – *Lathyrus palustris*
marsh pennywort – *Hydrocotyle umbellata*
marshpepper – *Polygonum hydropiper*
marsh-rosemary, western – *Limonium californicum*
marsh sedge – *Cyperus squarrosus*
marsh skullcap – *Scutellaria galericulata*
marsh speedwell – *Veronica scutellata*
marsh spurge – *Euphorbia palustris*
marsh violet – *Viola palustris*
marsh willow herb – *Epilobium palustre*
marsh wort – *Apium nodiflorum*
marsh woundwort – *Stachys palustris*
marsh yellow cress – *Rorippa palustris*
martagon, scarlet – *Lilium chalcedonicum*
Martha Washington pelargonium – *Pelargonium* x *domesticum*
Martinez pine – *Pinus maximartinezii*
marvel of Peru – *Mirabilis jalapa, M. multiflora*
Mary, blue-eyed – *Omphalodes verna*
masterwort – *Astrantia major, Peucedanum ostruthium*
mastic – *Pistacia lentiscus*
mastic, false – *Mastichodendron foetidissimum*
mastic tree – *Bursera simaruba*
match-box bean – *Entada scandens*
match-me-if-you-can – *Acalypha wilkesiana*
matchweed – *Grindelia oxylepis, Gutierrezia sarothrae*
maté – *Ilex paraguariensis*
mat grama – *Bouteloua simplex*
matilija poppy, Coulter's – *Romneya coulteri*
mat, purple – *Nama coulteri, N. hispidum*
matrimony vine – *Lycium barbarum, L. chinense*
mat sandbur – *Cenchrus longispinus*
matseed, purple – *Nama demissum*
matsutake mushroom – *Tricholoma matsutake*
matte – *Ilex paraguariensis*
maul oak – *Quercus chrysolepis*
Mauritius hemp – *Furcraea gigantea*
Mauritius papeda – *Citrus hystrix*
Mauritius raspberry – *Rubus rosaefolius*
mauve clusters – *Scaevola plumierii*
maxillaria, Skinner's – *Lycaste virginalis*
Maximino pine – *Pinus tenuifolia*
Mayan soap – *Solanum nudum*
Maya palm – *Gaussia maya*
May apple – *Podophyllum peltatum*
Maybush – *Prunus padus*
Mayday tree – *Prunus padus*
Mayflower – *Crataegus laevigata*
May flower – *Tabebuia pentaphylla*
May-flowering laelia – *Laelia speciosa*
Mayflowering tulip – *Tulipa gesneriana*
May-pop – *Passiflora incarnata*
May rose – *Rosa cinnamomea*
mayten – *Maytenus boaria*
Mayweed – *Anthemis cotula*
McCartney rose – *Rosa bracteata*
meadow buttercup – *Ranunculus acris*
meadow clary – *Salvia pratensis*
meadow cord grass, salt – *Spartina patens*
meadow cranesbill – *Geranium pratense*
meadow fescue – *Festuca pratensis*
meadowfoam – *Limnanthes douglasii*
meadow foxtail – *Alopecurus pratensis*
meadow grass, early – *Poa inferma*
meadow grass, hybrid – *Poa hybrida*
meadow grass, narrow-leaf – *Poa pratensis* ssp *angustifolia*
meadow grass, tufted – *Poa laxa*

meadow grass, violet – *Poa violacea*
meadow knapweed – *Centaurea* x *pratensis*
meadow mushroom – *Agaricus campestris*
meadow pink – *Dianthus deltoids, Sabatia campestris*
meadow-rue – *Thalictrum fendleri, T. hernandezii*
meadow-rue, alpine – *Thalictrum alpinum*
meadow-rue, Chinese – *Thalictrum dipterocarpum*
meadow-rue, greater – *Thalictrum aquilegifolium*
meadow saffron – *Colchicum autumnale, C. speciosum*
meadow sweet – *Astilbe arendsii, Filipendula ulmaria*
mealycup sage – *Salvia farinacea*
mealy sage – *Salvia farinacea*
Mecca myrrh – *Commiphora opobalsamum*
medicinal aloe – *Aloe vera*
medicinal rhubarb – *Rheum officinale*
medick, black – *Medicago lupulina*
medick, early – *Medicago praecox*
Mediterranean barley – *Hordeum marinum*
Mediterranean beard grass – *Polypogon maritimus*
Mediterranean buckthorn – *Rhamnus alaternus*
Mediterranean coriaria – *Coriaria myrtifolia*
Mediterranean fan palm – *Chamaerops humilis*
Mediterranean oak – *Quercus canariensis*
Mediterranean sage – *Salvia aethiopis*
Mediterranean saltbush – *Atriplex halimus*
Mediterranean sweet lemon – *Citrus limetta*
medlar – *Mospilus germanica*
medlar tree, Neapolitan – *Crataegus azarolus*
melaleuca, drooping – *Melaleuca armillaris*
melaleuca, heath – *Melaleuca ericafolia*
melic grass – *Melica montezumae*
melic, three-flower – *Melica nitens*
melon – *Cucumis melo*
melon, African horned – *Cucumis metuliferus*
melon, bitter – *Momordica charantia*
melon, coyote – *Cucurbita foetidissima, C. palmata, Ibervillea insularis, I. sonorae*
melon, cucumber – *Cucumis melo* var *chito*
melon, dudaim – *Cucumis melo* var *dudaim*
melon, egusi – *Cucumeropsis edulis*
melon-loco – *Apodanthera undulata*
melon, mango – *Cucumis melo* var *chito*
melon pear – *Solanum muricatum*
melon, Persian – *Cucumis melo* var *reticulatus*
melon, pickling – *Cucumis melo* var *conomon*
melon, snake – *Cucumis melo* var *flexuosus*
menodora – *Menodora helianthemoides*
mentzelia, desert – *Mentzelia multiflora*
mercury – *Mercurialis annua*
mercury, three-seeded – *Acalypha lindheimeri*
mesa dalea – *Psorothamnus schottii*
mesa dropseed – *Sporobolus flexuosus*
mesa muhly – *Muhlenbergia monticola*
mescalbean – *Sophora secundiflora*
mescal buttons – *Lophophora williamsii*
mescal fiber agave – *Agave pseudotequilana*
Mescat acacia – *Acacia constricta*
mespilus, snowy – *Amelanchier ovalis*
mesquite – *Phoradendron californicum, Prosopis*
mesquite, Arizona – *Prosopis velutina*
mesquite barrel cactus – *Thelocactus setispinus*
mesquite, bitter – *Prosopis articulata*
mesquite, Chilean – *Prosopis chilensis*
mesquite, curly – *Pleuraphis belangeri*
mesquite, dwarf – *Prosopis globosa* var *mexicana*
mesquite, false – *Calliandra eriophylla*
mesquite, honey – *Prosopis glandulosa* var *torreyana*
mesquite mistletoe – *Phoradendron californicum*
mesquite, running – *Prosopis glandulosa* var *prostrata*
mesquite, sweet – *Prosopis juliflora*
mesquite, velvet – *Prosopis velutina*
mesquite, vine – *Panicum obtusum*
Mexican abelia – *Abelia floribunda*
Mexican alvaradoa – *Alvaradoa amorphoides*
Mexican apple – *Casimiroa edulis, Malvaviscus grandiflorus*
Mexican ash – *Fraxinus berlandieriana*
Mexican bald-cypress – *Taxodium distichum* var *mexicanum*
Mexican balm – *Brittonastrum mexicana*
Mexican bay leaf – *Listea glaucescens*
Mexican bird-of-paradise – *Caesalpinia mexicana*
Mexican blue oak – *Quercus oblongifolia*
Mexican blue palm – *Erythea armata*

Mexican bluewood condalia – *Condalia mexicana*
Mexican breadfruit – *Monstera deliciosa*
Mexican buckeye – *Ungnadia speciosa*
Mexican bush sage – *Salvia leucantha*
Mexican campion – *Silene laciniata*
Mexican cancer-root – *Conopholis mexicana*
Mexican cedrella – *Cedrela mexicana*
Mexican cherry – *Prunus serotina, Tabebuia pentaphylla*
Mexican chokecherry – *Prunus capuli*
Mexican coral drops – *Bessera elegans*
Mexican creeper – *Antigonon leptopus*
Mexican crucillo – *Ziziphus spathulata*
Mexican cypress – *Cupressus lusitanica, Taxodium distichum* var *mexicanum*
Mexican daisy – *Erigeron karvinskianus*
Mexican devilweed – *Chloracantha spinosa*
Mexican dock – *Rumex violascens*
Mexican Douglas-fir – *Pseudotsuga lindleyana*
Mexican drooping juniper – *Juniperus flaccida*
Mexican ebony – *Caesalpinia sclerocarpa, Havardia mexicana*
Mexican elder – *Sambucus mexicana*
Mexican elm – *Ulmus mexicana*
Mexican evening primrose – *Oenothera speciosa*
Mexican fan palm – *Sabal mexicana, Washingtonia robusta*
Mexican fire plant – *Euphorbia heterophylla*
Mexican flame vine – *Senecio chenopodioides, C. confusus*
Mexican foxglove – *Tetranema mexicana*
Mexican frangipani – *Plumeria acutifolia*
Mexican gamagrass – *Tripsacum lanceolatum*
Mexican geranium – *Geranium mexicanum*
Mexican grass tree – *Dasylirion longissimum, Nolina longifolia*
Mexican hackberry – *Trema micrantha*
Mexican hat – *Ratibida columnaris*
Mexican hat fiber – *Sabal mexicana*
Mexican hawthorn – *Crataegus mexicana*
Mexican honeysuckle – *Justica spicigera*
Mexican horncone – *Ceratozamia latifolia, C. mexicana*
Mexican ivy – *Cobaea scandens*
Mexican jumping bean – *Sapium biloculare, Sebastiana pavoniana*
Mexican juniper – *Juniperus mexicana*
Mexican laurel – *Phoebe tampicensis*
Mexican lemon hyssop, giant – *Agastache mexicana*
Mexican liberty cap – *Psilocybe mexicana*
Mexican lime – *Citrus aurantifolia*
Mexican lime cactus – *Ferocactus pilosus*
Mexican lovegrass – *Eragrostis cilianensis*
Mexican madrone – *Arbutus xalapensis*
Mexican magnolia – *Talauma mexicana*
Mexican mahogany – *Swietenia humilis*
Mexican manzanita – *Arctostaphylos pungens*
Mexican marigold – *Tagetes lucida*
Mexican marjoram – *Lantana macropoda*
Mexican mistletoe – *Psittacanthus calyculatus*
Mexican mulberry – *Morus microphylla*
Mexican night-blooming cereus – *Peniocereus serpentinus*
Mexican oak – *Quercus mexicana*
Mexican orange – *Choisya*
Mexican oregano – *Aloysia wrightii, Lippia, Poliomintha maderensis*
Mexican palo verde – *Parkinsonia aculeata*
Mexican passion flower – *Passiflora mexicana*
Mexican persimmon – *Diospyros texana*
Mexican pincushion – *Mammillaria magnimamma*
Mexican pine – *Pinus pinceana*
Mexican pink – *Silene laciniata*
Mexican pinyon pine – *Pinus cembroides*
Mexican pistachio – *Pistacia texana*
Mexican planetree – *Platanus mexicana*
Mexican plum – *Prunus mexicana*
Mexican poplar – *Populus mexicana*
Mexican poppy – *Argemone mexicana, Eschscholzia mexicana, Kallstroemia*
Mexican primrose – *Oenothera speciosa*
Mexican rubber plant – *Parthenium argentatum*
Mexican rubber tree – *Castilla elastica*
Mexican runner bean – *Phaseolus coccineus*
Mexican rush – *Juncus mexicanus*
Mexican sage – *Salvia mexicana*
Mexican scammony – *Ipomoea orizabensis*
Mexican shell flower – *Tigridia pavonia*
Mexican silene – *Silene laciniata* ssp *major*
Mexican sisal – *Agave fourcroydes*
Mexican spangletop – *Leptochloa fusca* ssp *uninervia*
Mexican star – *Milla biflora*

Mexican star of Bethlehem – *Milla biflora*
Mexican sunflower – *Tithonia*
Mexican tarragon – *Tagetes lucida*
Mexican tea – *Chenopodium ambrosioides, Ephedra torreyana, E. trifurca*
Mexican tepary bean – *Phaseolus acutifolius*
Mexican thistle – *Eryngium heterophyllum*
Mexican tree ocotillo – *Fouquieria macdougalii*
Mexican trumpet vine – *Distictis buccinatoria*
Mexican tulip poppy – *Hunnemannia fumariifolia*
Mexican vervain – *Verbena ciliata*
Mexican walnut – *Juglans microcarpa*
Mexican weeping bamboo – *Otatea acuminata*
Mexican white pine – *Pinus ayacahuite, P. strobiformis*
Mexican yam bean – *Pachyrhizus erosus*
Mexican yam, wild – *Dioscorea villosa*
mezcal – *Agave tequilana*
mezereon – *Daphne mezereum*
Michaelmas daisy – *Aster novi-belgii*
Michoacan pine – *Pinus devoniana*
midget bee blossom – *Gaura parviflora*
midget petunia – *Petunia parviflora*
midland hawthorn – *Crataegus laevigata*
mid patience – *Rumex obtusifolius*
mignonette – *Reseda odorata*
mignonette, garden – *Reseda odorata*
mignonette, white – *Reseda alba*
mignonette, yellow – *Reseda lutea*
mildew, downy – *Peronospora viticola*
mile-a-minute – *Ipomoea cairica*
milfoil – *Achillea millefolium*
milfoil, water – *Myriophyllum verticillatum*
milkberry – *Chiococca alba*
milk cap, peppery – *Lactarius piperatus*
milk cap, saffron – *Lactarius deliciosus*
milk cap, velvet – *Lactarius vellereus*
milkcup chalice vine – *Solandra nitida*
milk mushroom, delicious – *Lactarius deliciosus*
milk thistle – *Silybum marianum*
milk tree – *Brosimum utile*
milk-vetch – *Astragalus nuttallianus*
milk-vetch, Andalusian – *Astragalus boeticus*
milkvine – *Sarcostemma cynanchoides, Vincetoxicum hastulatum*
milkweed – *Asclepias, Matelea cordifolia*
milkweed, broad-leaved – *Asclepias latifolia*
milkweed, climbing – *Funastrum heterophyllum, Sarcostemma cynanchoides* ssp *hartwegii*
milkweed, Curaçao – *Asclepias curassavica*
milkweed, desert – *Asclepias erosa, A. subulata*
milkweed, Greek – *Asclepias speciosa*
milkweed, green – *Asclepias oenotheroides*
milkweed, Lemmon's – *Asclepias lemmonii*
milkweed, orange – *Asclepias curasavica, A. tuberosa*
milkweed, pine-needle – *Asclepias linaria*
milkweed, rush – *Asclepias subulata*
milkweed, showy – *Asclepias speciosa*
milkweed, spider – *Asclepias asperula*
milkweed, swamp – *Asclepias incarnata*
milkweed, vine – *Cynanchum acutum, C. utahense*
milkweed, wax – *Asclepias albicans*
milkweed, white-stemmed – *Asclepias albicans*
milkweed, woolly – *Asclepias vestita*
milkwort – *Polygala*
milkwort, sea – *Glaux maritima*
milkwort, shrubby – *Polygala chamaebuxus*
miller, dusty – *Artemisia stellerana, Centaurea cineraria, Cerastium tomentosum, Lychnis coronaria, Senecio cineraria*
millet – *Panicum, Pennisetum*
millet, African – *Eleusine corcacana* ssp *africana*
millet, broomcorn – *Panicum miliaceum*
millet, bulrush – *Pennisetum americanum*
millet, European foxtail – *Setaria italica* var *moharica*
millet, finger – *Eleusine coracana*
millet, foxtail – *Panicum italicum, Setaria italica*
millet, Japanese – *Echinochloa frumentacea*
millet, kodo – *Paspalum scrobiculatum*
millet, pearl – *Pennisetum americanum, P. glaucum*
millet, sugar – *Sorghum saccharatum*
millet, Texas – *Brachiaria texana*
millet, Turkish – *Panicum miliaceum*
millet, wood – *Milium effusum*
mill, flutter – *Oenothera macrocarpa*
millsweet – *Citrus limetta*
milo – *Sorgum bicolor, Thespesia populnea*

mimosa – *Albizzia julibrissin, Mimosa*
mimosa, Arizona – *Mimosa laxiflora*
mimosa bush – *Havardia pallens*
mimosa, catclaw – *Mimosa biuncifera*
mimosa, fragrant – *Mimosa borealis*
mimosa, Graham – *Mimosa grahamii*
mimosa, pink – *Mimosa borealis*
mimosa, velvet-pod – *Mimosa dysocarpa*
miner's lettuce – *Claytonia perfoliata*
miniature rose – *Rosa chinensis* var *minima, R. minutifolia*
mint – *Mentha*
mint, apple – *Mentha* x *gracilis, M. suaveolens*
mint, brook – *Mentha arvensis*
mint, bush – *Hyptis albida*
mint, cat – *Nepeta cataria, N.* x *faassenii*
mint, corn – *Mentha arvensis*
mint, coyote – *Monarda odoratissima, Monardella villosa*
mint, false – *Dicliptera resupinata*
mint, field – *Mentha arvensis*
mint, horse – *Agastache micrantha, A. urticifolia, Monarda didyma, M. pectinata*
mintleaf bee balm – *Monarda menthaefolia*
mint, lemon horse – *Monarda citriodora*
mint, mountain – *Pycnanthemum muticum*
mint of Corsica, jewel – *Mentha requienii*
mint, pepper – *Mentha* x *piperita*
mint, pink horse – *Monarda fistulosa*
mint, spear – *Mentha spicata, M. sylvestris*
mirabella – *Prunus syriaca*
mirabelle tree, yellow – *Prunus syriaca*
miraguama palm – *Coccothrinax miraguama*
mirasol pepper – *Capsicum annuum*
mirror of Venus – *Ophrys speculum*
mirror orchid – *Ophrys speculum*
mission bells – *Fritillaria biflora*
mission cactus – *Opuntia ficus-indica*
Missouri primrose – *Oenothera macrocarpa*
Miss Willmott's ghost – *Eryngium giganteum*
mistletoe – *Phoradendron*
mistletoe, bigleaf – *Phoradendron macrophyllum*
mistletoe cactus – *Rhipsalis baccifera*
mistletoe, dense – *Phoradendron densum*
mistletoe, desert – *Phoradendron californicum*
mistletoe, dwarf – *Arceuthobium*
mistletoe, European – *Viscum album*
mistletoe, fir – *Phoradendron pauciflorum*
mistletoe, greenleaf – *Phoradendron macrophyllum*
mistletoe, juniper – *Phoradendron juniperinum*
mistletoe, mesquite – *Phoradendron californicum*
mistletoe, Mexican – *Psittacanthus calyculatus*
mistletoe, oak – *Phoradendron villosum*
mistletoe, pine – *Arceuthobium*
mistletoe, pinyon dwarf – *Arceuthobium divaricatum*
mistletoe, western dwarf – *Arceuthobium campylopodum*
mock azalea – *Adenium obesum*
mock cucumber – *Marah lobata*
mock cypress – *Bassia hyssopifolia, Kochia scoparia*
mockernut – *Carya tomentosa*
mock orange – *Maclura pomifera, Philadelphus mexicanus, Pittosporum tobira, P. undulatum*
mock orange, evergreen – *Philadelphus mexicanus*
mock orange, littleleaf – *Philadelphus microphyllus*
mock orange, sweet – *Philadelphus coronarius*
mock pennyroyal – *Hedeoma*
mock strawberry – *Duchesnea indica*
mock strawberry, Indian – *Duchesnea indica*
modesty – *Whipplea modesta*
Mohave spurge – *Euphorbia incisa*
moho, white – *Malvaviscus arboreus*
Mohr's shin oak – *Quercus mohriana*
Mojave prickly-pear – *Opuntia phaeacantha*
Mojave prickly poppy – *Argemone corymbosa*
Mojave sea-blite – *Suaeda moquinii*
Mojave thistle – *Cirsium mohavense*
Mojave yucca – *Yucca schidigera*
molasses grass – *Melinis minutiflora*
mole plant – *Euphorbia lathyris*
molucca balm – *Moluccella laevis*
moly – *Allium moly*
mombin, red – *Spondias purpurea*
mombin, yellow – *Spondias mombin*
monarch birch – *Betula maximowicziana*
money plant – *Lunaria annua*
moneywort – *Lysimachia nummularia*
monkey apple – *Licania platypus*
monkey cup – *Couepia polyandra*
monkey dinner bell – *Hura polyandra*

monkeyflower – *Mimulus*
monkeyflower, musk – *Mimulus moschatus*
monkeyflower, primrose – *Mimulus primuloides*
monkeyflower, scarlet – *Mimulus cardinalis*
monkey pod – *Pithecellobium dulce*
monkey pod tree – *Albizzia saman*
monkey pot – *Lecythis zabucajo*
monkey puzzle tree – *Araucaria araucana*
monkey-tail palm – *Chamaedorea pinnatifrons, Geonomia interrupta, Synechanthus fibrosus*
monkshood – *Aconitum*
monk's hood – *Astrophytum ornatum*
monk's rhubarb – *Rumex alpinus*
montbretia – *Crocosmia crocosmiiflora*
Monterey cypress – *Cupressus macrocarpa*
Monterey pine – *Pinus radiata*
Montezuma bald-cypress – *Taxodium distichum* var *mexicanum*
Montezuma pine – *Pinus montezumae*
Montpellier sun-rose – *Cistus monspeliensis*
monument plant – *Swertia radiata*
moon cactus – *Epiphyllum anguliger*
moonflower – *Ipomoea alba, I. bona-nox*
moonwort – *Botrychium lunaria, Lunaria annua*
moor grass – *Molinia caerulea*
morel – *Morchella*
morel, false – *Gyromitra esculenta*
Moreton Bay chestnut tree – *Castanospermum australe*
moriche palm – *Mauritia flexuosa*
mormodes, leopard-like – *Mormodes pardinum* ssp *unicolor*
Mormon tea – *Ephedra nevadensis, E. torreyana*
Mormon tea, green – *Ephedra viridis*
morning-glory – *Calystegia, Convolvulus, Ipomoea, Jacquemontia abutiloides*
morning-glory, beach – *Ipomoea pes-caprae*
morning-glory, blue – *Ipomoea lindheimeri*
morning-glory, bush – *Convolvulus cneorum*
morning-glory, Cairo – *Ipomoea cairica*
morning-glory, common – *Ipomoea purpurea*
morning-glory, dwarf – *Convolvulus tricolor, Evolvulus alsinoides*
morning-glory, five-leaf – *Ipomoea cairica*
morning-glory, giant white – *Turbina corymbosa*
morning-glory, ground – *Convolvulus sabatius*
morning-glory, heavenly-blue – *Ipomoea tricolor*
morning-glory, ivy – *Ipomoea nil*
morning-glory, ivyleaf – *Ipomoea hederacea*
morning-glory, longleaf – *Ipomoea longifolia*
morning-glory, orchard – *Convolvulus arvensis*
morning-glory, plummer's – *Ipomoea plummerae*
morning-glory tree – *Ipomoea arborescens, I. murucoides*
morning-glory, tree – *Ipomoea arborescens*
morning-glory, yellow – *Merremia aurea*
Moses-in-the-boat – *Tradescantia spathacea*
Moses-in-the-cradle – *Tradescantia spathacea*
mosquito fern – *Azolla filiculoides*
mosquito grass – *Bouteloua gracilis*
moss – *Bryum musgo, Polytrichum*
moss, ball – *Tillandsia ferrisiana, T. recurvata*
moss campion – *Silene acaulis*
moss catchfly – *Silene acaulis*
moss, club – *Lycopodium clavatum*
moss, desert spike – *Selaginella eremophila*
moss, Iceland – *Cetraria islandica*
moss, Irish – *Sagina subulata*
moss, Louisiana – *Tillandsia usneoides*
moss, peat – *Sphagnum*
moss phlox – *Phlox subulata*
moss pink – *Phlox subulata*
moss rose – *Portulaca grandiflora, P. pilosa, Rosa muscosa*
moss, rose – *Portulaca grandiflora*
moss, Scottish – *Sagina subulata*
moss, Spanish – *Ramalina reticulata, Tillandsia usneoides*
moss, spike – *Selaginella*
mossycup oak – *Quercus macrocarpa*
mossy stonecrop – *Crassula tillaea*
mother-in-law's tongue – *Sansevieria trifasciata*
mother-of-thyme – *Thymus serpyllum*
motherwort – *Leonurus cardiaca*
moth mullein – *Verbascum blattaria*
motie, water – *Baccharis salicifolia*
mountain alder – *Alnus incana* ssp *tenuifolia*
mountain angel's trumpet – *Brugmansia sanguinea*
mountain apple – *Eugenia malaccensis*
mountain ash – *Sorbus*

mountain ash, American – *Sorbus americana*
mountain ash, Arizona – *Sorbus dumosa*
mountain ash, European – *Sorbus aucuparia*
mountain avens – *Dryas octopetala*
mountain balm – *Eriodictyon angustifolium, E. trichocalyx*
mountain brome – *Bromus carinatus*
mountain buttercup – *Ranunculus montanus*
mountain cabbage palm – *Euterpe precatoria*
mountain currant – *Ribes alpinum*
mountain cypress – *Podocarpus guatemalensis*
mountain dogwood – *Cornus nuttallii*
mountain ebony – *Bauhinia variegata*
mountain fuchsia – *Fuchsia microphylla*
mountain garland – *Clarkia unguiculata*
mountain gooseberry – *Ribes montigenum*
mountain gromwell – *Lithospermum cobrense*
mountain gumplant – *Grindelia aphanactis*
mountain heather – *Phyllodoce*
mountain hemlock – *Tsuga mertensiana*
mountain ironwood – *Cercocarpus betuloides*
mountain knapweed – *Centaurea montana*
mountain laurel – *Kalmia latifolia, Rhus ovata*
mountain laurel, Texas – *Sophora secundiflora*
mountain-mahogany – *Cercocarpus montanus*
mountain-mahogany, alder-leaf – *Cercocarpus montanus*
mountain-mahogany, birch-leaf – *Cercocarpus betuloides*
mountain-mahogany, curl-leaf – *Cercocarpus ledifolius*
mountain mint – *Pycnanthemum muticum*
mountain misery – *Chamaebatia foliolosa*
mountain mugwort – *Artemisia franserioides*
mountain muhly – *Muhlenbergia montana*
mountain organ pipe – *Stenocereus montanus*
mountain parsley – *Peucedanum oreoselinum*
mountain pawpaw – *Carica candamarcensis*
mountain penny-cress – *Thlaspi montanum*
mountain pine – *Pinus mugo* var *rostrata*
mountain pine, swiss – *Pinus mugo*
mountain rock cress – *Arabis alpina*
mountain rose-mallow – *Hibiscus bifurcatus, H. cardiophyllus*
mountain sagewort – *Artemisia ludoviciana* ssp *incompta*
mountain sandwort – *Arenaria montana, Minuartia rubella*
mountain silver bell – *Halesia monticola*
mountain sorrel – *Oxyria digyna*
mountain strawberry – *Fragaria virginiana*
mountain timothy – *Phleum alpinum*
mountain white oak – *Quercus grisea*
mountain whitethorn – *Ceanothus cordulatus*
mountain yucca – *Yucca schottii*
Mount Atlas daisy – *Anacyclus depressus*
Mount Atlas pistache – *Pistacia atlantica*
mourning bride – *Scabiosa atropurpurea*
mourning widow – *Geranium phaeum*
mouse-ear chickweed – *Cerastium fontanum, C. glomeratum*
mouse-ear cress – *Arabidopsis thaliana*
mouse killer – *Gliricidia sepium*
mouse-tail – *Myosurus minimus*
Mt. Etna broom – *Genista aetnensis*
mud plantain – *Heteranthera limosa*
mud sedge – *Carex limosa*
mudwort – *Limosella aquatica*
mugho pine – *Pinus mugo*
mugweed, western – *Artemisia ludoviciana* ssp *albula*
mugwort – *Artemisia douglasiana, A. vulgaris*
mugwort, biannual – *Artemisia biennis*
mugwort, mountain – *Artemisia franserioides*
mugwort, tree-like – *Artemisia arborescens*
mugwort, western – *Artemisia ludoviciana*
mugwort, white – *Artemisia lactiflora*
muhly, bamboo – *Muhlenbergia dumosa*
muhly, bush – *Muhlenbergia porteri*
muhly, creeping – *Muhlenbergia repens*
muhly, few-flowered – *Muhlenbergia pauciflora*
muhly, littleseed – *Muhlenbergia microsperma*
muhly, mesa – *Muhlenbergia monticola*
muhly, mountain – *Muhlengerbia montana*
muhly, New Mexican – *Muhlenbergia pauciflora*
muhly, pine – *Muhlenbergia dubia*
muhly, purple – *Muhlenbergia rigida*
muhly, whiskgrass – *Muhlenbergia macroura*
muirapuama – *Ptychopetalum ovata*
mulato pepper – *Capsicum annuum*
mulberry – *Morus celtidifolia*
mulberry, black – *Morus nigra*
mulberry, dyer's – *Maclura tinctoria*
mulberry, littleleaf – *Morus microphylla*
mulberry, Mexican – *Morus microphylla*
mulberry, paper – *Broussonetia papyrifera*

mulberry, Persian – *Morus nigra*
mulberry, red – *Morus rubra*
mulberry, silkworm – *Morus alba*
mulberry, Texas – *Morus microphylla*
mulberry, white – *Morus alba*
mulberry, wild – *Morus celtidifolia*
mule fat – *Baccharis salicifolia*
mules ears – *Wyethia helenioides*
mullein – *Verbascum*
mullein, Austrian – *Verbascum chaixii*
mullein, common – *Verbascum thapsus*
mullein, great – *Verbascum thapsus*
mullein, moth – *Verbascum blattaria*
mullein nightshade – *Solanum verbascifolium*
mullein pink – *Lychnis coronaria*
mullein, purple – *Verbascum phoeniceum*
mullein, showy – *Verbascum speciosum*
mullein, turkey – *Eremocarpus setigerus*
mullein, wand – *Verbascum virgatum*
mullein, woolly – *Verbascum thapsus*
multiflora rose – *Rosa multiflora*
multiplier onion – *Allium cepa* var *aggregatum*
mung bean – *Vigna radiata*
Munz's sage – *Salvia munzii*
murumuru palm – *Astrocaryum murumuru*
Muscat grape – *Vitis vinifera* var *muscat*
mushroom – *Basidiomycetes*
mushroom, branched oyster – *Pleurotus cornucopiae*
mushroom, butter – *Suillus luteus*
mushroom, Caesar's – *Amanita caesarea*
mushroom, cultivated – *Agaricus bisporus* var *hortensis*
mushroom, cultivated oyster – *Pleurotus ostreatus* var *sativus*
mushroom, delicious milk – *Lactarius deliciosus*
mushroom, divinatory – *Psilocybe yungensis*
mushroom, field – *Agaricus campestris*
mushroom, genius – *Psilocybe yungensis*
mushroom, horse – *Agaricus arvensis*
mushroom, matsutake – *Tricholoma matsutake*
mushroom, meadow – *Agaricus campestris*
mushroom, nameko – *Pholiota nameko*
mushroom, oyster – *Pleurotus ostreatus*
mushroom, parasol – *Lepiota procera*
mushroom, poisonous brain – *Gyromitra esculenta*
mushroom, psychedelic – *Psilocybe, Paneolus, Conocybe, Stropharia cubensis*
mushroom, reishi – *Gandoderma lucidum*
mushroom, shiitake – *Lentinus edodes*
mushroom, St. George's – *Calocybe gambosa*
mushroom, white – *Agaricus bitorquis*
mushroom, wild oyster – *Pleurotus ostreatus* var *vulgaris*
mushroom, wood – *Agaricus silvicola*
musk clover – *Erodium moschatum*
musk hyacinth – *Muscari moschatum*
musk mallow – *Abelmoschus moschatus, Malva moschata*
musk monkeyflower – *Mimulus moschatus*
musk rose – *Rosa moschata*
musk thistle – *Carduus nutans*
mustard – *Brassica*
mustard, birdsrape – *Brassica rapa*
mustard, black – *Brassica nigra*
mustard, brown – *Brassica juncea*
mustard, field – *Brassica rapa*
mustard greens – *Brassica japonica, B. nigra*
mustard, hedge – *Sisymbrium officinale*
mustard, hoary – *Hirschfeldia incana*
mustard, Indian – *Brassica juncea*
mustard, Jim Hill – *Sisymbrium altissimum*
mustard, sahara – *Brassica tournefortii*
mustard, sareptia – *Brassica juncea* var *sareptana*
mustard, spinach – *Brassica rapa* var *perviridis*
mustard, tansy – *Descuriana pinnata, Sisymbrium*
mustard, tower – *Arabis glabra*
mustard tree – *Nicotiana glauca*
mustard, tumble – *Sisymbrium altissimum, S. auriculatum*
mustard, white – *Sinapis alba*
mustard, wild – *Brassica juncea, Sinapis arvensis*
mustard, wormseed – *Erysimum cheiranthoides*
my lady – *Aspidosperma quebracho-blanco*
myrobalan – *Prunus cerasifera*
myrrh – *Cistus creticus, Commiphora*
myrrh, Abyssinian – *Commiphora abyssinica*
myrrh, common – *Commiphora myrrha*
myrrh, garden – *Myrrhis odorata*
myrrh, Mecca – *Commiphora opobalsamum*
myrtle – *Myrtus communis*

myrtle, common – *Myrtus communis*
myrtle, crape – *Lagerstroemia indica*
myrtle, crepe – *Lagerstroemia speciosa*
myrtle, giant crape – *Lagerstroemia speciosa*
myrtle orange – *Poncirus trifoliata*
myrtle, Oregon – *Umbellularia californica*
myrtle, Pacific wax – *Myrica californica*
myrtle, true – *Myrtus communis*
myrtle, wax – *Myrica californica, M. cerifera, Myrtus cerifera*

nagami kumquat – *Fortunella margarita*
nagoonberry – *Rubus arctius*
naiad – *Najas marina*
naiad-wort – *Najas marina*
nailwort – *Parnassia palustris*
nailwort, prickly – *Paronychia echinulata*
naked coral tree – *Erythrina coralloides*
naked lady – *Amaryllis belladonna*
naked stem fleabane – *Erigeron nudiflorus*
naked, walk – *Eugenia capuli*
nakedwood – *Colubrina viridis*
nameko mushroom – *Pholiota nameko*
nandi grass – *Setaria anceps*
Napa cabbage – *Brassica rapa* var *pekinensis*
napier grass – *Pennisetum purpureum*
Naples garlic – *Allium neapolitanum*
Naples, onion of – *Allium neapolitanum*
naranjilla – *Solanum quitoense*
Narbonne vetch – *Vicia narbonensis*
narcissus, poet's – *Narcissus poeticus*
nard grass – *Cymbopogon nardus*
nard, spike – *Aralia, Nardostachys jatamansi*
narrow cudweed – *Filago gallica*
narrow thorow wax – *Bupleurum lancifolium*
narrow-wing saltbush – *Atriplex linearis*
naseberry – *Achras zapote*
nasturtium, garden – *Tropaeolum majus*
Natal grass – *Rhynchelytrum repens*
Natal plum – *Carissa grandiflora*
Natal ruby grass – *Rhynchelytrum repens*
Navajo tea – *Thelesperma gracile, T. longipes*
navelwort, Venus' – *Omphalodes linifolia*
navy bean – *Phaseolus vulgaris*
Nealley three-awn – *Aristida purpurea* var *nealleyi*
Neapolitan medlar tree – *Crataegus azarolus*
nectarine – *Prunus persica* ssp *nectarina, P. p.* ssp *nucipersica*
needle, Adam's – *Yucca filamentosa*
needle broom – *Genista anglica*
needle, Eve's – *Yucca faxoniana*
needle grama – *Bouteloua aristidoides*
needlegrass – *Stipa tenacisima*
needlegrass, Texas – *Stipa leucotricha*
needle-leaved St. John's wort – *Hypericum coris*
needles, crow – *Scandix pectin-veneris*
needles, Spanish – *Bidens*
needle, Venus' – *Scandix pectin-veneris*
needle whim – *Genista anglica*
neem – *Azadirachta indica*
neem tree – *Azadirachta indica*
Nelson pinyon – *Pinus nelsoni*
Nepal banana – *Mus nepalensis*
netleaf hackberry – *Celtis reticulata*
netleaf oak – *Quercus reticulata, Q. rugosa*
netleaf white oak – *Quercus polymorpha*
netleaf willow – *Salix reticulata*
nettle – *Cnidoscolus, Urera, Urtica*
nettle, American stinging – *Urtica dioica* ssp *gracilis*
nettle, bull – *Cnidoscolus texanus, Solanum elaeagnifolium*
nettle, dead – *Lamium*
nettle, dwarf – *Urtica urens*
nettle, false – *Boehmeria nivea*
nettle, hedge – *Stachys*
nettle-leaf goosefoot – *Chenopodium murale*
nettle, rock – *Eucnide rupestris*
nettle, scarlet hedge – *Stachys coccinea*
nettle, silverleaf – *Solanum eleagnifolium*
nettle, spotted – *Lamium maculatum*
nettle, spurge – *Cnidoscolus urens, Urtica dioica*
nettle, stinging – *Jatropha urens, Urtica dioica*
nettle-tree – *Celtis australis*
nettle, white horse – *Solanum eleagnifolium*
Nevada jointfir – *Ephedra nevadensis*
new cocoyam – *Xanthosoma sagittifolium*
New England aster – *Aster novae-angliae*
New Mexican alder – *Alnus oblongifolia*
New Mexican feathergrass – *Stipa neomexicana*
New Mexican locust – *Robinia neomexicana*
New Mexican muhly – *Muhlenbergia pauciflora*
New Mexican privet – *Forestiera neomexicana*

New Mexican raspberry – *Rubus neomexicanus*
New Mexico chile – *Capsicum annuum*
New Mexico croton – *Croton dioicus*
New Mexico rainbow cactus – *Echinocereus viridiflorus*
New Mexico rose – *Rosa neomexicana*
New York aster – *Aster novi-belgii*
New Zealand christmas tree – *Metrosideros excelsa*
New Zealand flax – *Phormium tenax*
New Zealand laurel – *Corynocarpus laevigata*
New Zealand spinach – *Tetragonia tetragonioides*
nicker nut – *Caesalpinia crista*
nicotiana – *Nicotiana alata*
night-blooming cereus – *Hylocereus trigonus, H. undatus, Selenicereus hamatus*
night-blooming cereus, deseret – *Peniocereus greggii* var *transmontanus*
night-blooming cereus, Mexican – *Peniocereus serpentinus*
night-flowering campion – *Silene noctiflora*
night-flowering jasmine – *Nyctanthes arbor-tristis*
night fungus, yellow – *Tricholoma flavovirens*
night jessamine – *Cestrum nocturnum*
night, queen-of-the – *Cereus hexagonus, Epiphyllum anguliger, Hylocereus undulatus, Peniocereus greggii, Seleniocereus grandiflorus*
nightshade – *Solanum*
nightshade, black – *Solanum nigrum*
nightshade, Brazilian – *Solanum seaforthianum*
nightshade, Costa Rican – *Solanum wendlandii*
nightshade, deadly – *Atropa belladona*
nightshade, elegant – *Solanum gracile*
nightshade, enchanter's – *Circaea alpina, C. lutetiana*
nightshade, green – *Solanum sarrachoides*
nightshade, lanceleaf – *Solanum lanceolatum*
nightshade, mullein – *Solanum verbascifolium*
nightshade, purple – *Solanum xanti*
nightshade, silverleaf – *Solanum eleagnifolium*
nightshade, Texas – *Solanum triquetrum*
nightshade, three-flower – *Solanum triflorum*
nightshade, tree – *Solanum erianthum*
nightshade, white – *Solanum americanum, S. triquetrum*
nightshade, white-margined – *Solanum marginatum*
nikau palm – *Rhopalostylis sapida*
nine-awned pappus grass – *Enneapogon desvauxii*
nipple cactus – *Mammillaria heyderi, M. longimamma*
nipplewort – *Lapsana communis*
nit grass – *Gastridium ventricosum*
nitta tree – *Parkia speciosa*
noble fir – *Abies procera*
nodding bur-marigold – *Bidens cernua*
nodding onion – *Allium cernuum*
nogal – *Juglans major*
nolina – *Nolina*
none-such – *Medicago lupulina*
noon flower – *Ipomoea sinuata*
Nootka cypress – *Cupressus nootkatensis*
Nootka rose – *Rosa nutkana*
nopal, spiny – *Opuntia dejecta*
Nordmann fir – *Abies nordmanniana*
Norfolk Island pine – *Araucaria heterophylla*
northern bedstraw – *Galium boreale*
northern coralroot – *Corallorhiza trifida*
northern oatgrass – *Trisetum spicatum*
northern pitch pine – *Pinus rigida*
northern red oak – *Quercus rubra*
northern starflower – *Trientalis arctica*
northern white cedar – *Thuja occidentalis*
northern white pine – *Pinus strobus*
Norway maple – *Acer platanoides*
Norway spruce – *Picea abies*
Norwegian cinquefoil – *Potentilla norvegica*
noseburn – *Tragia nepetaefolia*
nosegay frangipani – *Plumeria rubra*
nut, Barbados – *Jatropha curcas*
nut, Barbary – *Iris sisyrinchium*
nut, betel – *Areca catechu, Piper betle*
nut, bitter – *Carya amara, C. cordiformis*
nut, bladder – *Staphylea bolanderi*
nut, bread – *Artocarpus altilis, Brosimum alicastrum, Trophis racemosa*
nut, bur – *Tribulus terrestris*
nut, butter – *Caryocar amygdaliferum, Juglans cinera*
nut, chigger – *Miconia*
nut, Chinese wing – *Pterocarya stenoptera*
nut, common wing – *Pterocarya fraxinifolia*
nut, deer – *Simmondsia chinensis*

nut, earth – *Bunium bulbocastanum*
nut-grass – *Cyperus rotundus*
nut-grass, purple – *Cyperus rotundus*
nut-grass, yellow – *Cyperus esculentus*
nut, hazel – *Corylus*
nut, ivory – *Phytelephas macrocarpa*
nut, kola – *Cola acuminata*
nut, lucky – *Thevetia peruviana*
nut, macadamia – *Macadamia*
nutmeg – *Myristica fragrans, M. moschata*
nutmeg, California – *Torreya californica*
nutmeg flower – *Nigella sativa*
nutmeg geranium – *Pelargonium* x *fragrans*
nut, mocker – *Carya tomentosa*
nut, nicker – *Caesalpinia crista*
nut, paradise – *Lecythis zabucajo*
nut, physic – *Jatropha curcas*
nut, pig – *Hoffmannseggia glauca, Simmondsia chinensis*
nut, pili – *Canarium ovatum*
nut pine – *Pinus edulis*
nut, prickle – *Guazuma ulmifolia*
nut, ramon bread – *Brosimum alicastrum*
nut, rush – *Cyperus esculentus*
nut, sapucaia – *Lecythis elliptica, L. zabucajo*
nutsedge – *Cyperus*
nutsedge, purple – *Cyperus rotundus*
nutsedge, umbrella – *Cyperus eragrostis*
nutsedge, yellow – *Cyperus esculentus*
nuts, grass – *Dichelostemma capitatum*
nut, shea – *Butyrospermum parkii*
nut, souari – *Caryocar amygdaliferum*
nut, suari – *Caryocar amygdaliferum*
Nuttall's locoweed – *Astragalus nuttallianus*
Nuttall's pondweed – *Potamogeton epihydrus*
Nuttall's scrub oak – *Quercus dumosa*
nut, water – *Trapa natans*
nux vomica – *Strychnos nux-vomica*
nylon hedgehog cactus – *Echinocereus viridiflorus*
nymph, holly-leaved water – *Najas marina*
nymph, water – *Najas*
nymph, wood – *Moneses uniflora*

oak, Arizona white – *Quercus arizonica*
oak, black – *Quercus*
oak, blackjack – *Quercus emoryi*
oak, burr – *Quercus macrocarpa*
oak, California white – *Quercus lobata*
oak, canyon – *Quercus chrysolepis, Q. glaucoides*
oak, canyon live – *Quercus chrysolepsis*
oak, Cape – *Quercus brandegeei, Q. devia*
oak, chaparro – *Quercus reticulata*
oak, chestnut – *Quercus castanea*
oak, Chihuahua – *Quercus chihuahuensis*
oak, chinkapin – *Quercus muehlenbergii*
oak, chinquapin – *Quercus muehlenbergii*
oak, Chisos red – *Quercus gravesii*
oak, coastal live – *Quercus virginiana*
oak, coastal scrub – *Quercus dumosa*
oak, coast live – *Quercus agrifolia*
oak, cork – *Quercus suber*
oak, delta – *Quercus similis*
oak, desert scrub – *Quercus turbinella*
oak, durmast – *Quercus petraea*
oak, dyer's – *Quercus velutina*
oak, eastern white – *Quercus alba*
oak, Emory – *Quercus emoryi*
oak, English – *Quercus robur*
oak, escarpment live – *Quercus fusiformis*
oak, evergreen – *Quercus cedroensis, Q. mexicana, Q. virgiliana*
oak, Fendler's – *Quercus fendleri*
oak fungus – *Armillaria mellea*
oak, Gambell – *Quercus gambellii*
oak, golden – *Quercus tuberculata*
oak, goldencup – *Quercus chrysolepis*
oak, gray – *Quercus grisea*
oak, Gregg – *Quercus greggii*
oak, handbasin – *Quercus pennivenia, Q. tarahumara*
oak, holly – *Quercus ilex*
oak, holm – *Quercus ilex*
oak, Hungarian – *Quercus frainetto*
oak, Indian – *Tectona grandis*
oak, island – *Quercus tomentella*
oak, Italian – *Quercus frainetto*
oak, Japanese live – *Quercus myrsinifolia*
oak, Jerusalem – *Chenopodium botrys*
oak, kermes – *Quercus coccifera*
oak, lacey – *Quercus glaucoides*
oakleaf hydrangea – *Hydrangea quercifolia*
oak-leaved goosefoot – *Chenopodium glaucum*
oak, live – *Quercus*
oak, Madrean willow – *Quercus viminea*
oak, maul – *Quercus chrysolepis*
oak, Mediterranean – *Quercus canariensis*

oak, Mexican – *Quercus mexicana*
oak, Mexican blue – *Quercus oblongifolia*
oak mistletoe – *Phoradendron villosum*
oak, Mohr's shin oak – *Quercus mohriana*
oak, mossycup – *Quercus macrocarpa*
oak, mountain white – *Quercus grisea*
oak, netleaf – *Quercus reticulata, Q. rugosa*
oak, netleaf white – *Quercus polymorpha*
oak, northern red – *Quercus rubra*
oak, Nuttall's scrub – *Quercus dumosa*
oak, Palmer's – *Quercus palmeri*
oak, pin – *Quercus palustris*
oak, plateau – *Quercus fusiformis*
oak, Portuguese – *Quercus faginea*
oak, Pyrenean – *Quercus pyrenaica*
oak, red – *Quercus rubra*
oak, river she – *Casuarina cunninghamiana*
oak, Rocky Mountain white – *Quercus gambellii*
oak, round-leaved – *Quercus rotundifolia*
oak, sandpaper – *Quercus pungens*
oak, Santa Rosa – *Quercus sideroxyla*
oak, sawtooth – *Quercus acutissima*
oak, scarlet – *Quercus coccinea*
oak, scrub – *Quercus berberidifolia, Q. pungens*
oak, shin – *Quercus vaseyana*
oak, shrubby – *Quercus fruticosa*
oak, shrub live – *Quercus turbinella*
oak, silk – *Grevillea robusta*
oak, silverleaf – *Quercus arizonica, Q. hypoleucoides*
oak, southern live – *Quercus virginiana*
oak, Spanish – *Quercus crenata*
oak, Spanish holly – *Quercus ilex*
oak, speckled – *Quercus conspersa*
oak, swamp – *Quercus palustris*
oak, swamp white – *Quercus bicolor*
oak, thickleaf – *Quercus crassifolia*
oak, Toumey – *Quercus toumeyi*
oak, turkey – *Quercus cerris*
oak, valley – *Quercus lobata*
oak, Valonia – *Quercus aegilops*
oak, Vasey – *Quercus vaseyana*
oak, wavyleaf – *Quercus undulata*
oak, western poison – *Toxicodendron diversilobum*
oak, white – *Quercus*
oak, white live – *Quercus oblongifolia*
oak, whitish – *Quercus candicans*
oak, willow – *Quercus phellos*
oak, willowleaf – *Quercus salicifolia*
oak, wrinkle-leaf – *Quercus rugosa*
oak, yellow chestnut – *Quercus muehlenbergii*
oat, black – *Avena strigosa*
oat, cultivated – *Avena sativa*
oatgrass, blue – *Helictotrichon sempervirens*
oatgrass, northern – *Trisetum spicatum*
oatgrass, soft-bearded – *Ventenata dubia*
oatgrass, tall – *Arrhenatherum elatius*
oatgrass, yellow – *Trisetum flavescens*
oat smut – *Ustilago carbo*
oat, wild – *Avena barbata, A. fatua, A. sterilis*
Oaxaca palmetto – *Sabal mexicana*
obedient plant – *Physostegia virginiana*
oca – *Oxalis tuberosa*
ocote pine – *Pinus oocarpa*
ocotillo – *Fouquieria splendens, F. formosa*
ocotillo, Mexican tree – *Fouquieria macdougalii*
ocotillo, tree – *Fouquieria macdougalii*
octopus agave – *Agave vilmoriniana*
octopus cactus – *Stenocereus alamosensis*
octopus tree – *Schefflera actinophylla*
odontoglossum, Insleay's – *Odontoglossum insleayii*
odora – *Porophyllum gracile*
oil palm – *Elaeis guineensis*
oil palm, American – *Elaeis melanococca*
oil plant, madia – *Madia sativa*
oilseed rape – *Brassica napus* var *oleifera*
oil tree – *Eleagnus angustifolia*
oil tree, owala – *Pentaclethra macrophylla*
okra – *Hibiscus esculentus*
okra, cow – *Parmentiera edulis*
old maid – *Catharanthus roseus*
old man – *Artemisia abrotanum*
old man cactus – *Cephalocereus sensilis, Pachycereus gatesii, P. schottii, Pilosocereus leucocephalus*
old man, golden – *Pilosocereus chrysacanthus*
old man of Mexico – *Pilosocereus leucocephalus*
old man's beard – *Clematis dioica, C. vitalba, Tillandsia usneoides*
old witch grass – *Panicum capillare*
old woman – *Artemisia stellerana*
oleander – *Nerium oleander*

oleander, yellow – *Thevetia ovata, T. peruviana*
oleaster – *Eleagnus angustifolius*
olive – *Olea europaea*
olive, Chinese – *Canarium album*
olive, desert – *Forestiera neomexicana*
olive, Java – *Sterculia foetida*
Oliver's flower – *Oliveranthus elegans*
olive, Russian – *Eleagnus angustifolius*
olive, spurge – *Cneorum tricoccon*
olive, sweet – *Osmanthus fragrans*
olive, Texas – *Cordia boissieri*
olive, wild – *Olea europaea*
oncidium, the Duke of Devonshire's – *Oncidium cavendishianum*
one-flowered vetch – *Vicia articulata*
one-seeded bur-cucumber – *Sicyos laciniatus*
one-seed juniper – *Juniperus monosperma*
one-sided wintergreen – *Orthilia secunda*
onion – *Allium cepa*
onion, bunching – *Allium fistulosum*
onion cordia – *Cordia alliodora*
onion, green – *Allium cepa*
onion, multiplier – *Allium cepa* var *aggregatum*
onion, nodding – *Allium cernuum*
onion of Naples – *Allium neapolitanum*
onion, peasant – *Allium cepa*
onion, potato – *Allium cepa* var *aggragatum*
onion, sea – *Urginea maritima*
onion, spring – *Allium fistulosum*
onion, tree – *Allium cepa* var *viviparum*
onion, Welsh – *Allium fistulosum*
onion, wild – *Allium*
onycha – *Cistus ladanifer*
opiuma, white – *Pithecellobium dulce*
opium poppy – *Papaver somniferum*
opuntia, velvet – *Opuntia tomentosa*
orache, common – *Atriplex patula*
orache, garden – *Atriplex hortensis*
oracle, garden – *Atriplex hortensis*
oracle, spear – *Atriplex patula*
oracle, tumbling – *Atriplex rosea*
orange – *Citrus sinensis*
orange balsam – *Impatiens capensis*
orange, bergamot – *Citrus bergamotia*
orange bird's-foot – *Ornithopus pinnatus*
orange, bitter – *Citrus aurantium*
orange, blood – *Citrus sinensis* var *melitense*
orange caltrop – *Kallstroemia grandiflora*
orange caterpillar – *Combretum farinosum*
orange creeper – *Pyrostegia venusta*
orange daylily, common – *Hemerocallis fulva*
orange, drug Osage – *Maclura tinctoria*
orange, dwarf – *Citrus mitis*
orange, evergreen mock – *Philadelphus mexicanus*
orange hawkweed – *Hieracium aurantiacum*
orange jessamine – *Murraya paniculata*
orange, king – *Citrus nobilis*
orange larkspur – *Delphinium nudicaule*
orange lily – *Lilium bulbiferum*
orange, littleleaf mock – *Philadelphus microphyllus*
orange lobelia, giant – *Lobelia laxiflora*
orange, Mandarin – *Citrus nobilis, C. reticulata*
orange, Mexican – *Choisya*
orange milkweed – *Asclepias curassavica, A. tuberosa*
orange, mock – *Maclura pomifera, Philadelphus mexicanus, Pittosporum tobira, P. undulatum*
orange, myrtle – *Poncirus trifoliata*
orange, Osage – *Maclura pomifera*
orange, Seville – *Citrus aurantiacum* var *bergamia*
orange sneezeweed – *Dugaldia hoopesii*
orange, sour – *Citrus aurantiacum* var *bergamia*
orange, sweet mock – *Philadelphus coronarius*
orange, trifoliate – *Poncirus trifoliata*
orange trumpet vine – *Pyrostegia venusta*
orange, wild – *Esenbeckia hartmanii*
orchard grass – *Dactylis glomerata*
orchard morning-glory – *Convolvulus arvensis*
orchid – *Orchidaceae*
orchid, bee – *Ophrys apifera*
orchid, bird's nest – *Neottia nidus-avis*
orchid, bog – *Platanthera hyperborea*
orchid cactus – *Epiphyllum oxypetalum, E.* x *Disocactus*
orchid, flame – *Spiranthes cinnabarina*
orchid, ghost – *Epipogium aphyllum*
orchid, giant – *Himantoglossum longibracteatum*
orchid, man – *Aceras anthropophorum*
orchid, mirror – *Ophrys speculum*

orchid, pansy – *Miltonia*
orchid, phantom – *Cephalanthera austiniae*
orchid, poor man's – *Schizanthus pinnatus*
orchid, reed – *Epidendrum ibaguense*
orchid, tiger – *Odontoglossum rosii*
orchid tree – *Bauhinia divaricata, B. monandra*
orchid tree, purple – *Bauhinia variegata*
orchid vine, yellow – *Mascagnia macroptera*
Orcutt's dudleya – *Dudleya attenuata* ssp *orcuttii*
oreganillo – *Aloysia wrightii*
oregano – *Origanum vulgare*
oregano, Mexican – *Aloysia wrightii, Lippia, Poliomintha maderensis*
oregano, wild – *Monarda menthaefolia*
Oregon crabapple – *Malus fusca*
Oregon-grape – *Berberis*
Oregon myrtle – *Umbellulatia californica*
organpipe cactus – *Pachycereus marginatus, Stenocereus thurberi*
organ pipe, gray ghost – *Stenocereus pruinosus*
organ pipe, mountain – *Stenocereus montanus*
oriental arborvitae – *Thuja orientalis*
oriental lily, fragrant – *Lilium auratum*
oriental plane – *Platanus orientalis*
oriental poppy – *Papaver orientale*
oriental sweet gum – *Liquidambar orientalis*
ornamental allium – *Allium oreophilum*
ornamental currant – *Ribes odoratum*
ornamental banana – *Mus coccinea*
ornamental kale – *Brassica oleracea* var *acephala*
ornamental rhubarb – *Rheum palmatum*
orpine – *Sedum*
Osage orange – *Maclura pomifera*
Osageorange, drug – *Maclura tinctoria*
osier, common – *Salix viminalis*
osier, purple – *Salix purpurea*
ostrich fern – *Matteuccia struthiopteris*
Oswego tea – *Monarda citriodora, M. didyma*
otaheite apple – *Eugenia malaccensis, Spondias cytherea*
otaheite gooseberry – *Phyllanthus acidus*
Our Lord's candle – *Yucca whipplei*
owala oil tree – *Pentaclethra macrophylla*
owl claws – *Dugaldia hoopesii*
owl's-clover – *Castilleja*
owl's-clover, purple – *Castilleja exserta*
owl's foot – *Hymenoxys richardsonii*
oxalis – *Oxalis*
ox-eye – *Heliopsis helianthoides, H. rubra*
ox-eye bean – *Mucuna*
ox-eye daisy – *Leucanthemum vulgare*
oxheart cabbage – *Brassica oleracea* var *conica*
ox hoof – *Bauhinia monandra*
ox-tongue, bristly – *Picris echioides*
oxytrope, blue pendent-pod – *Oxytropis deflexa* var *sericea*
oyama magnolia – *Magnolia sieboldii*
oyster mushroom – *Pleurotus ostreatus*
oyster mushroom, branched – *Pleurotus cornucopiae*
oyster mushroom, cultivated – *Pleurotus ostreatus* var *sativus*
oyster mushroom, wild – *Pleurotus ostreatus* var *vulgaris*
oyster plant – *Tradescantia spathacea, Tragopogon porrifolius*
oyster plant, Spanish – *Scolymus hispanicus*
oyster, vegetable – *Tragopogon porrifolius*
Ozark sundrops – *Oenothera macrocarpa*

pacaya – *Chamaedora tepejilote*
Pacific coast mahogany – *Swietenia humilis*
Pacific dogwood – *Cornus nuttallii*
Pacific madrone – *Arbutus menziesii*
Pacific ponderosa pine – *Pinus ponderosa*
Pacific rosewood – *Thespesia populnea*
Pacific silver fir – *Abies amabilis*
Pacific wax myrtle – *Myrica californica*
paddy weed, Asian – *Echinochloa oryzoides*
pagoda flower – *Clerodendrum buchananii* var *fallax*
pagoda tree, Japanese – *Sophora japonica*
paintbrush, Indian – *Castilleja, C. gracilis*
paintbrush, scarlet – *Castilleja indivisa*
paintbrush, southwestern – *Castillja integra*
paintbrush, Wyoming – *Castilleja linariifolia*
painted cup – *Parentucella viscosa*
painted daisy – *Chrysanthemum coccineum*
painted sage – *Salvia viridis*
painted spurge – *Euphorbia heterophylla*
painted tongue – *Salpiglossis sinuata*
pale flax – *Linum bienne*
pale hellebore – *Helleborus lividus*
pale-leaf rose – *Rosa hypoleuca*
Palestine buckthorn – *Rhamnus palestina*

pale wolfberry – *Lycium pallidum*
palm – *Arecaceae*
palm, Alexander – *Ptychosperma elegans*
palm, Alexandra – *Archontophoenix alexandrae*
palm, alta – *Mauritia flexuosa*
palm, American oil – *Elaeis melanococca*
palm, arecanut – *Areca catechu*
palmarosa – *Cymbopogon martini*
palm, babassu – *Attalea speciosa*
palm, bataua – *Jessenia bataua*
palm, bayleaf – *Sabal mauritiiformis, S. yapa*
palm, blue – *Erythea armata*
palm, blue fan – *Erythea armata*
palm, bottle – *Hyophorbe lagenicaulis, Nolina recurvata*
palm, butterfly – *Chrysalidocarpus lutescens*
palm, cabbage – *Roystonea oleracea, Sabal japa, S. palmetto*
palm, California fan – *Washingtonia filifera*
palm, Canary Island date – *Phoenix canariensis*
palm, caranday – *Copernicia australis*
palm, cardboard – *Zamia furfuracea*
palm, Caribbee royal – *Roystonea oleracea*
palm, cascade – *Chamaedorea cataractarum*
palm, cat – *Chamaedorea cataractarum*
palm, chamaedor – *Chamaedora ernesti*
palm, cherry – *Pseudophoenix sargentii*
palm, Chilean wine – *Jubaea chilensis*
palm, Chile wine – *Jubaea spectabilis*
palm, Chinese fan – *Livistona chinensis*
palm, Chinese fountain – *Livistona chinensis*
palm, clustered fishtail – *Caryota mitis*
palm, coconut – *Cocos nucifera*
palm, cohune – *Orbignya cohune*
palm, corojo – *Acrocomia crispa*
palm, coyal – *Acrocomia mexicana*
palm, coyor ruffle – *Aiphanes acanthophylla*
palm, Cuban royal – *Roystonea regia*
palm, cycad – *Cycas revoluta*
palm, date – *Phoenix dactylifera*
Palmer agave – *Agave palmeri*
Palmer grass – *Distichlis palmeri*
Palmer's frankenia – *Frankenia palmeri*
Palmer's goldenbush – *Ericameria palmeri*
Palmer's oak – *Quercus palmeri*
palmetto – *Sabal minor*
palmetto, hairy Tom – *Acoelorraphe wrightii*
palmetto, Hispaniola – *Sabal blackburniana*
palmetto, Oaxaca – *Sabal mexicana*
palmetto, salt-water – *Thrinax radiata*
palmetto, saw – *Serenoa repens*
palmetto, Sonoran – *Sabal uresana*
palm, Everglades – *Acoelorraphe wrightii*
palm, fan – *Licuala grandis*
palm, feather – *Acromia mexicana, Syagrus romanzoffiania*
palm, Fiji fan – *Pritchardia pacifica*
palm, fishtail wine – *Caryota urens*
palm, give-and-take – *Cryosophila stauracantha*
palm, Guadalupe Island – *Erythea edulis*
palm, guano – *Sabal japa*
palm, hospita – *Copernicia hospita*
palm, ivory – *Phytelephas macrocarpa*
palm, jelly – *Butia capitata*
palm, Kentia – *Howea forsteriana*
palm, king – *Archontophoenix cunninghamiana*
palm, lady – *Rhapis excelsa*
palm, licuri – *Syagrus coronata*
palm, lontar – *Borassus flabellifer*
palm, loulu – *Pritchardia hillebrandii*
palm, MacArthur – *Ptychosperma macarthuri*
palm, macaw – *Acrocomia sclerocarpa*
palm, Madagascar – *Chrysalidocarpus lutescens*
palm, Maya – *Gaussia maya*
palm, Mediterranean fan – *Chamaerops humilis*
palm, Mexican blue – *Erythea armata*
palm, Mexican fan – *Sabal mexicana, Washingtonia robusta*
palm, miraguamo – *Coccothrinax miraguama*
palm, monkey-tail – *Chamaedorea pinnatifrons, Geonoma interrupta, Synechanthus fibrosus*
palm, moriche – *Mauritia flexuosa*
palm, mountain cabbage – *Euterpe precatoria*
palm, murumuru – *Astrocaryum murumuru*
palm, nikau – *Rhopalostylis sapida*
palm, oil – *Elaeis guineensis*
palm, palmyra – *Borassus flabellifer*
palm, Panama hat – *Carludovica palmata*
palm, paradise – *Howea forsteriana*
palm, parlor – *Chamaedorea elegans*
palm, paurotis – *Acoelorraphe wrightii*

palm, peach – *Bactris gasipaes, Guilielma gasipaes*
palm, petticoat – *Copernicia macroglossa*
palm, pimento – *Acoelorraphe wrightii*
palm, pindo – *Butia capitata*
palm, ponytail – *Nolina recurvata*
palm, pygmy date – *Phoenix roebelenii*
palm, queen – *Syagrus romanzoffiana*
palm, raphia – *Raphia ruffia*
palm, rattan – *Calamus draco, C. optimus, C. rotang, Rhapis humilis*
palm, rotang – *Calamus rotang*
palm, royal – *Roystonea regia*
palm, sago – *Cycas revoluta, Metroxylon*
palm, salak – *Salacca edulis*
palm, San José hesper – *Erythea brandegeei*
palm, San Miguel – *Calyptrogyne ghiesbreghtiana*
palm, Senegal date – *Phoenix reclinata*
palm, sentry – *Howea belmoreana*
palm, shaving brush – *Rhopalostylis sapida*
palm, silver thatch – *Coccothrinax argentata*
palm, slender lady – *Rhapis humilis*
palm, solitaire – *Ptychosperma elegans*
palm, Sonoran hesper – *Erythea aculeata*
palm, spiny club – *Bactris mexicana*
palm, squirrel – *Chamaedorea arenbergiana*
palm, stilt – *Iriartea*
palm, St. Peter's – *Yucca filifera*
palm, sugar – *Arenga pinnata, A. saccharifera*
palm, suppa – *Acrocomia aculeata*
palm, talipot – *Corypha umbraculifera*
palm, thatch – *Sabal yapa, Thrinax radiata*
palm, toddy – *Borassus flabellifer, Caryota urens*
palm, totai – *Acrocomia totai*
palm, traveler's – *Ravenala madagascariensis*
palm, tucuma – *Astrocaryum vulgare*
palm, wax – *Copernicia cerifera*
palm, windmill – *Trachycarpus fortunei*
palm, wine – *Borassus flabellifer*
palm, yellow – *Chrysalidocarpus lutescens*
palmyra palm – *Borassus flabellifer*
palo blanco – *Acacia willardiana*
palo brea – *Cercidium praecox*
palo verde – *Cercidium praecox*
palo verde, blue – *Cercidium floridum*
palo verde, foothill – *Cercidium microphyllum*
palo verde, littleleaf – *Cercidium microphyllum*
palo verde, Mexican – *Parkinsonia aculeata*
palo verde, Sonoran – *Cercidium praecox*
palo verde, Texas – *Cercidium texanum*
pampas grass – *Cortaderia selloana*
Panama hat palm – *Carludovica palmata*
Panama rubber – *Castilla elastica, C. fallax*
Panama tree – *Sterculia apetala*
pan-American friendship plant – *Pilea involucrata*
pancake pincushion – *Mammillaria heyderi*
pancake-pear – *Opuntia chlorotica*
pancake prickly-pear – *Opuntia chlorotica*
panicgrass – *Panicum*
panicgrass, blue – *Panicum antidotale*
panicgrass, smooth-leaved – *Panicum schinzii*
panicum, Hall's – *Panicum hallii*
pangola grass – *Digitaria decumbens*
pannag – *Panax ginseng, Panicum miliaceum*
pansy – *Viola pedunculata, V.* x *wittrockiana*
pansy, field – *Viola arvensis*
pansy, Florida – *Torenia fournieri*
pansy, garden – *Viola* x *wittrockiana*
pansy orchid – *Miltonia*
pansy, tufted – *Viola cornuta*
pansy, wild – *Viola pedunculata, V. tricolor*
panther lily – *Lilium pardalinum*
papaw – *Carica papaya*
papaya – *Carica pubescens, C. papaya*
papaya, wild – *Carica cauliflora, C. mexicana*
papedo, Mauritius – *Citrus hystrix*
paperbark maple – *Acer griseum*
paperbark tree – *Melaleuca leucadendron*
paper birch – *Betula papyrifera*
paper daisy – *Baileya multiradiata*
paper daisy, pink – *Helipterum roseum*
paper mulberry – *Broussonetia papyrifera*
paper plant, Egyptian – *Cyperus papyrus*
paperspine cactus – *Sclerocactus papyracanthus*
pappus grass – *Pappophorum bicolor*
pappus grass, nine-awned – *Enneapogon desvauxii*
paprika – *Capsicum annuum*
papyrus – *Cyperus papyrus*
para copaiba – *Copaifera reticulata*
para cress – *Spilanthes acmella*
paradise apple – *Malus pumila* var *paradisiaca*
paradise, grains of – *Aframomum melegueta*
paradise nut – *Lecythis zabucajo*

paradise palm – *Howea forsteriana*
paradise tree – *Simarouba glauca*
paradise wood – *Aquilaria agallocha*
para grass – *Brachiaria mutica, Panicum molle*
para rubber tree – *Hevea brasiliensis*
parasol, field – *Macrolepiota procera*
parasol mushroom – *Lepiota procera*
parasol tree, Chinese – *Firmiana simplex*
parched fireweed – *Epilobium brachycarpum*
Paris daisy – *Chrysanthemum frutescens*
Paris, herb – *Paris quadrifolia*
Parish's desert-thorn – *Lycium parishii*
parlor palm – *Chamaedorea elegans*
Parnassus, grass of – *Parnassia californica, P. palustris*
paroquet-bur – *Triumfetta*
parrot flower, red – *Psittacanthus calyculatus*
parrot heliconia – *Heliconia psittacorum*
parrotia, Persian – *Parrotia persica*
Parry agave – *Agave perryi*
Parry manzanita – *Arctostaphylos manzanita*
Parry's penstemon – *Penstemon parryi*
parsley – *Petroselinum crispum*
parsley, bur – *Torilis*
parsley, cow – *Anthriscus sylvestris*
parsley, flat-leaved – *Petroselinum crispum* var *neapolitanum*
parsley, fool's – *Aethusa cynapium*
parsley, hedge – *Torilis*
parsley, Indian – *Ligusticum grayi*
parsley, Italian – *Petroselinum crispum*
parsley, marsh – *Ciclospermum leptophyllum*
parsley, mountain – *Peucedanum oreoselinum*
parsnip – *Pastinaca sativa*
parsnip, cow – *Heracleum lanatum*
parsnip, wafer – *Cymopterus purpurascens*
parsnip, water – *Berula erecta, Sium*
partridge berry, red – *Mitchella repens*
partridge-breast aloe – *Aloe variegata*
partridge wood – *Andira inermis*
paspalum, brownseed – *Paspalum plicatulum*
paspalum, hairy-seed – *Paspalum pubiflorum*
paspalum, Peruvian – *Paspalum racemosum*
paspalum, seashore – *Paspalum virgatum*
paspalum, water – *Paspalum fluitans*
pasque flower, eastern – *Pulsatilla patens*
pasque flower, European – *Pulsatilla vulgaris*
passion flower – *Passiflora*
passion flower, blue crown – *Passiflora caerulea*
passion flower, fetid – *Passiflora foetida*
passion flower, Mexican – *Passiflower mexicana*
passion fruit – *Passiflora*
passion vine, wild – *Passiflora incarnata*
pasture thistle – *Cirsium undulatum*
Patagonian cypress – *Fitzroya cupressoides*
Patagonia plantain – *Plantago patagonica*
patchouli – *Pogostemon cablin*
patience, herb – *Rumex patientia*
patience, mid – *Rumex obtusifolius*
paurotis palm – *Acoelorraphe wrightii*
pawpaw – *Carica candamarcensis, C. papaya*
pawpaw, American – *Asimina triloba*
pawpaw, mountain – *Carica candamarcensis*
pea, asparagus – *Psophocarpus tetragonolobus*
pea, black-eyed – *Pisum sativum, Vigna sinensis, V. unguiculata*
pea, blue – *Clitoria terneata, Psoralea pentaphylla*
pea bush – *Dalea lumholtzii*
pea butterfly – *Clitoria terneata*
pea, caley – *Lathyrus hirsutus*
pea, campo – *Lathyrus splendens*
peach – *Prunus persica*
peach, flowering – *Prunus davidiana*
pea, chick – *Cicer arietinum, Lathyrus sativus*
pea, Chinese – *Pisum sativum* var *macrocarpon*
peachleaf willow – *Salix amygdaloides*
peach palm – *Bactris gasipeas, Guilielma gasipaes*
peach, smooth – *Prunus persica* ssp *nucipersica*
peach willow, southwestern – *Salix amygdaloides*
peacock flower – *Caesalpinia pulcherrima*
peacock plant – *Calathea makoyana*
pea, cordofan – *Clitoria terneata*
pea, cow – *Vigna sinensis, V. unguiculata*
pea, Crowder – *Vigna unguiculata*
pea, earth – *Vigna subterranean, Voandzeia subterranea*
pea, earthnut – *Lathyrus tuberosus*
pea, edible-pod – *Pisum sativum* var *macrocarpon*
pea, everlasting – *Lathyrus latifolius*

pea, field – *Pisum sativum* var *arvense*
pea, garden – *Pisum sativum*
pea, glory – *Clianthus formosus*
pea, gongo – *Cajanus cajun*
pea, green – *Pisum sativum*
pea, marsh – *Lathyrus palustris*
peamelon, Arizona – *Sicyos laciniatus*
peamelon, mapleleaf – *Sicyos angulatus*
peanut – *Arachis hypogaea*
peanut, hog – *Vigna subterranea*
pea, pigeon – *Cajanus cajun*
pear – *Pyrus communis*
pear, almond-leaved – *Pyrus amygdaliformis*
pear, balsam – *Momordica charantia*
pear, garlic – *Crateva palmeri*
pear, melon – *Solanum muricatum*
pearl millet – *Pennisetum americanum, P. glaucum*
pearlwort, alpine – *Sagina saginoides*
pearlwort, arctic – *Sagina procumbens*
pearlwort, dwarf – *Sagina apetala*
pearlwort, heath – *Sagina subulata*
pearly everlasting – *Anaphalis margaritacea*
pearly gates – *Ipomoea violacea*
pea, rosary – *Abrus precatorius*
pear, pancake – *Opuntia chlorotica*
pear, pest – *Opuntia stricta*
pear, prickly – *Opuntia*
pear, snow – *Pyrus nivalis*
pea, rush – *Hoffmannseggia glauca*
pear, weeping willow-leaved – *Pyrus salicifolia*
pear, wild – *Nectandra, Pyrus pyraster*
pear, willow-leaved – *Pyrus salicifolia*
peasant onion – *Allium cepa*
pea, scurf – *Psoralea lanceolata, Psoralidium lanceolatum*
pea, scurvy – *Psoralidium lanceolatum*
pea, sea – *Lathyrus japonicus*
peashrub, Siberian – *Caragana arborescens*
pea, singletary – *Lathyrus hirsutus*
pea, snow – *Pisum sativum* var *macrocarpon*
pea, spring – *Lathyrus vernus*
pea, sugar – *Pisum sativum*
pea, sweet – *Lathyrus latifolius, L. odoratus, L. sativus*
pea, Tangier – *Lathyrus tingitanus*
peat moss – *Sphagnum*
pea tree, Siberian – *Caragana arborescens*
pea, turkey – *Astragalus nuttallianus*
peavine, flat – *Lathyrus sylvestris*
peavine, flatpod – *Lathyrus cicera*
pea, wild – *Lathyrus sativus*
pea, wood – *Lathyrus linifolius*
pea, wrinkled – *Pisum sativum*
pebble plant – *Lithops*
pecan – *Calathea mollis, Carya*
peepul – *Ficus religiosa*
pelargonium, Lady Washington – *Pelargonium* x *domesticum*
pelargonium, Martha Washington – *Pelargonium* x *domesticum*
pelargonium, regal – *Pelargonium* x *domesticum*
pelican flower, large-flowered – *Aristolochia grandiflora*
pellitory of Spain, great – *Peucedanum ostruthium*
pellitory-of-the-wall – *Parietaria judaica, P. officinalis*
pellitory, Pennsylvania – *Parietaria pensylvanica*
pencil cactus – *Cylindropuntia leptocaulis, C. ramosissima, Echinocereus poselgeri*
pencil cholla – *Cylindropuntia arbuscula, C. tesajo*
pencil cholla, Klein's – *Cylindropuntia kleiniae*
pencil flower – *Stylosanthes viscosa*
pendent-pod oxytrope, blue – *Oxytropis deflexa* var *sericea*
penicillin – *Penicillium notatum*
Pennsylvanian cinquefoil – *Potentilla pennsylvanica*
Pennsylvania pellitory – *Parietaria pensylvanica*
penny bun – *Boletus edulis*
penny-cress, field – *Thlaspi arvense*
penny-cress, mountain – *Thlaspi montanum*
pennyroyal – *Hedeoma oblongifolia, H. pulegioides, Mentha pulegium, Monardella odoratissima*
pennyroyal, American – *Hedeoma pulegioides*
pennyroyal, dwarf – *Hedeoma nanum*
pennyroyal, mock – *Hedeoma*
pennywort – *Hydrocotyle*
pennywort, floating – *Hydrocotyle ranunculoides*
pennywort, marsh – *Hydrocotyle umbellata*

pennywort, salt – *Hydrocotyle bonariensis*
pennywort, water – *Hydrocotyle verticillata*
penstemon – *Penstemon*
penstemon, border – *Penstemon* x *gloxinioides*
penstemon, garden – *Penstemon* x *gloxinioides*
penstemon, Parry's – *Penstemon parryi*
penstemon, southwestern – *Penstemon barbatus*
penstemon, tinted – *Penstemon spectabilis*
penstemon, yellow bush – *Keckiella antirrhinoides*
peony – *Paeonia*
peony, tree – *Paeonia suffructicosa*
peony tulip – *Tulipa gesneriana*
pepino – *Solanum muricatum*
pepper – *Piper auritum, P. sanctum*
pepper, bell – *Capsicum annuum* var *grossum*
pepper, bird – *Capsicum annuum* var *minimum, C. baccatum, C. frutescens*
pepper, black – *Piper nigrum*
pepper, black chile – *Capsicum annuum*
pepper, black seeded – *Capsicum pubescens*
pepperbush – *Clethra mexicana*
pepperbush, sweet – *Clethra mexicana*
pepper, chile – *Capsicum annuum*
pepper, chocolate ancho – *Capsicum annuum*
pepper, creole – *Capsicum chinense*
pepper, datil – *Capsicum chinense*
peppergrass – *Lepidium*
peppergrass, hairy – *Lepidium lasiocarpum*
peppergrass, Virginia – *Lepidium virginicum*
pepper, hot – *Capsicum annuum*
pepperidge – *Nyssa sylvatica*
pepper, jalapeno – *Capsicum annuum*
pepper, malaguetta – *Aframomum melegueta*
pepper, marsh – *Polygonum hydropiper*
peppermint – *Mentha* x *piperita*
pepper, mirasol – *Capsicum annuum*
pepper, mulato – *Capsicum annuum*
pepper, rocotillo – *Capsicum chinense*
pepper, rocoto – *Capsicum pubescens*
pepper, tree – *Capsicum annuum*
pepper tree, Brazilian – *Schinus terebinthifolius*
pepper tree, California – *Schinus molle*
pepper tree, Peruvian – *Schinus molle*
pepper, Turkish – *Capsicum annuum*
pepper, water – *Polygonum hydropiper, P. hydropiperoides*
pepper, wax – *Capsicum annuum*
pepper, white – *Piper nigrum*
pepper, wild – *Capsicum annuum* var *minimum, Piper aduncum*
pepperwood – *Umbellularia californica*
pepperwort – *Lepidium*
pepperwort, broad-leaved – *Lepidium latifolium*
pepperwort, field – *Lepidium campestre*
pepperwort, poor man's – *Lepidium virginicum*
pepperwort, Smith's – *Lepidium heterophyllum*
pepper, yellow-fruited rocoto – *Capsicum pubescens*
peppery milk cap – *Lactarius piperatus*
perennial blue flax – *Linum perenne*
perennial canary grass – *Phalaris tuberosa*
perennial candytuft – *Iberis sempervirens*
perennial cornflower – *Centaurea montana*
perennial ryegrass – *Lolium perenne*
perennial sow thistle – *Sonchus arvensis*
perennial sweet pea – *Lathyrus latifolius*
perezia – *Acourtia wrightii*
pericome – *Pericome caudata*
perilla – *Perilla frutescens*
perilla, Japanese – *Perilla frutescens*
periwinkle – *Catharanthus roseus*
periwinkle, dwarf – *Vinca minor*
periwinkle, greater – *Vinca major*
periwinkle, lesser – *Vinca minor*
periwinkle, Madagascar – *Catharanthus roseus*
periwinkle, rosy – *Catharanthus roseus*
Persian acacia – *Albizzia julibrissin*
Persian bells – *Fritillaria persica*
Persian clover – *Trifolium resupitanum*
Persian fritillary – *Fritillaria persica*
Persian ivy – *Hedera colchica*
Persian lilac – *Melia azedarach* var *umbraculiformis*
Persian lime – *Citrus latifolia*
Persian melon – *Cucumis melo* var *reticulatus*
Persian mulberry – *Morus nigra*
Persian parrotia – *Parrotia persica*
Persian ranunculus – *Ranunculus asiaticus*
Persian speedwell – *Veronica persica*
Persian violet – *Exacum affine*
Persian walnut – *Juglans regia*
Persia, star of – *Allium christophii*
persimmon, American – *Diospyros virginiana*
persimmon, black – *Diospyros texana*

persimmon, Indian ebony – *Diospyros ebenaster*
persimmon, Japanese – *Diospyros kaki*
persimmon, Mexican – *Diospyros texana*
persimmon, Sonoran – *Diospyros sonorae*
persimmon, Texas – *Diospyros texana*
persimmon, wild – *Diospyros californica*
Peru, apple of – *Nicandra physalodes*
Peru cocoa – *Theobroma bicolor*
Peru, marvel of – *Mirabilis jalapa, M. multiflora*
Peruvian apple cactus – *Cereus repandus*
Peruvian bean – *Phaseolus vulgaris*
Peruvian carrot – *Arracacia xanthorrhiza*
Peruvian cocaine tree – *Erythroxylum novagranatense*
Peruvian daffodil – *Hymenocallis amancaes, H. narcissiflora*
Peruvian lily – *Alstroemeria aurantiaca, A. pelegrina*
Peruvian paspalum – *Paspalum racemosum*
Peruvian pepper tree – *Schinus molle*
Peruvian rhatany – *Krameria trianda*
Peruvian scilla – *Scilla peruviana*
pest pear – *Opuntia stricta*
pete bean – *Parkia speciosa*
petticoat daffodil, hoop – *Narcissus bulbocodium*
petticoat palm – *Copernicia macroglossa*
petty spurge – *Euphorbia peplus*
petunia – *Petunia* x *hybrida*
petunia, cascade – *Petunia* x *hybrida* var *pendula*
petunia, midget – *Petunia parviflora*
petunia, trailing – *Petunia* x *hybrida* var *pendula*
petunia, wild – *Ruellia californica, R. nudiflora*
peyote – *Lophophora williamsii*
peyote, false – *Ariocarpus fissuratus*
peyote, little – *Pelecyphora aselliformis*
phacelia – *Phacelia*
phacelia, scalloped – *Phacelia crenulata*
phacelia, tight – *Phacelia congesta*
phantom orchid – *Cephalanthera austiniae*
phaseolus, Wright's – *Phaseolus filiformis*
pheasant's eye – *Adonis annua, A. vernalis, Narcissus poeticus*
pheasant's eye, large – *Adonis flamea*
pheasant's eye, summer – *Adonis aestivalis*
pheasant's eye, yellow – *Adonis vernalis*
pheasant's tail – *Anthurium schlechtendalii*
Philippine wax flower – *Etlingera elatior*
philodendron – *Philodendron*
philodendron, split-leaf – *Monstera deliciosa*
philodendron, upright – *Philodendron bipinnatifidum*
phlox – *Phlox*
phlox, annual – *Phlox drummondii*
phlox, dwarf – *Phlox nana*
phlox, garden – *Phlox paniculata*
phlox, giant hardy – *Phlox paniculata*
phlox, moss – *Phlox subulata*
phlox, prickly – *Leptodactylon*
phlox, Santa Fe – *Phlox nana*
phlox, summer – *Phlox paniculata*
photinia, Chinese – *Photinia serratifolia*
physic-nut – *Jatropha curcas*
pickle cactus – *Echinocereus berlandieri*
pickle tree – *Byrsonima crassifolia*
pickleweed – *Allenrolfea occidentalis, Salicornia europaea*
pickling melon – *Cucumis melo* var *conomon*
pie, cherry – *Lantana camara*
Piedmont truffle – *Tuber magnatum*
pieris, Chinese – *Pieris forrestii*
piert – *Aphanes occidentalis*
pigeonberry – *Duranta repens, Phytolacca americana, Rhamnus californica, Rivina humilis, Vitex trifolia*
pigeon bush – *Lonicera involucrata*
pigeon pea – *Cajanus cajun*
pigeon-plum – *Chrysobalanus icaco, Hirtella americana, H. racemosa*
pig-nut – *Hoffmannseggia glauca, Simmondsia chinensis*
pignut hickory – *Carya glabra*
pigtail – *Anthurium scherzerianum*
pigweed – *Amaranthus, Boerhavia hirsuta, Chenopodium, Portulaca oleracea*
pikake – *Jasminum sambac*
pilewort – *Ranunculus ficaria*
pilikai – *Merremia tuberosa*
pili nut tree – *Canarium ovatum*
Pima cotton – *Gossypium barbadense*
Pima rhatany – *Krameria erecta*
pimento – *Capsicum annuum, Eugenia pimenta*
pimento palm – *Acoelorraphe wrightii*

pimiento – *Capsicum annuum*
pimpernel – *Anagallis monelli*
pimpernel, false – *Lindernia dubia*
pimpernel, scarlet – *Anagallis arvensis*
pimpernel, water – *Samolus parviflorus*
pimpernel willow herb – *Epilobium anagallidifolium*
pin cherry – *Prunus pensylvanica*
pin-clover – *Erodium cicutarium*
pincushion cactus – *Coryphanta potosiana, C. radiansMammillaria*
pincushion, devil's – *Echinocactus texensis*
pincushion flower – *Scabiosa*
pincushion, golf-ball – *Mammillaria lasiacantha*
pincushion, Mexican – *Mammillaria magnimamma*
pincushion, pancake – *Mammillaria heyderi*
pincushions – *Scabiosa atropurpurea*
pincushion, Sheldon's – *Mammillaria sheldonii*
pindo palm – *Butia capitata*
pine – *Pinus*
pine, air – *Tillandsia flabellata, T. imperialis*
pine, Aleppo – *Pinus halepensis*
pine, Apache – *Pinus engelmannii*
pineapple – *Ananas sativus, Coryphanta macromeris*
pineapple cactus – *Echinomastus erectocentrus*
pineapple cactus, Johnson's – *Echinomastus johnsonii*
pineapple cactus, red – *Echinomastus erectocentrus*
pineapple cactus, woven-spine – *Echinomastus intertextus*
pineapple, corojo – *Ananas comosus*
pineapple guava – *Feijoa sellowiana*
pineapple sage – *Salvia elegans*
pineapple weed – *Chamomilla suaveolens*
pineapple, wild – *Bromelia pinguin*
pine, Arizona – *Pinus arizonica*
pine, Arizona longleaf – *Pinus engelmanii*
pine, Arizona white – *Pinus ayacahuite*
pine, arolla – *Pinus cembra*
pine, Australian – *Casuarina equisetifolia*
pine, Austrian black – *Pinus nigra*
pine, Aztec – *Pinus teocote*
pine, bishop – *Pinus muricata*
pine buttercup – *Ranunculus fascicularis*
pine, Canary Island – *Pinus canariensis*
pine, Chiapas white – *Pinus strobus* var *chiapensis*
pine, Chihuahua – *Pinus leiophylla* var *chihuahuana*
pine, chir – *Pinus roxburghii*
pine, cluster – *Pinus pinaster*
pinecone cactus – *Pelecyphora strobiliformis*
pine, Coulter – *Pinus coulteri*
pine, digger – *Pinus sabiniana*
pine, Donnell Smith – *Pinus hartwegii*
pine, Douglas – *Pinus douglasiana*
pine dropseed – *Blepharoneuron tricholepis*
pine, Durango – *Pinus durangensis*
pine, eastern white – *Pinus strobus*
pine, egg-cone – *Pinus oocarpa*
pine, false white – *Pinus pseudostrobus*
pine, five-needled Chihuahua – *Pinus leiophylla*
pine, foothill – *Pinus sabiniana*
pine, foxtail – *Pinus balfouriana*
pine, French turpentine – *Pinus pinaster*
pine, gray – *Pinus sabiniana*
pine, Gregg – *Pinus greggii*
pine, ground – *Ajuga chamaepitys*
pine, Herrera – *Pinus herrerae*
pine, Himalayan white – *Pinus wallichiana*
pine, Honduras Caribbean – *Pinus caribaea* var *hondurensis*
pine, Indian longleaf – *Pinus roxburghii*
pine, Italian stone – *Pinus pinea*
pine, jack – *Pinus banksiana*
pine, Jalisco – *Pinus jaliscana*
pine, Japanese black – *Pinus thunbergii*
pine, Japanese red – *Pinus densiflora*
pine, Japanese white – *Pinus parviflora*
pine, Jeffrey – *Pinus jeffreyi*
pine, jelecote – *Pinus patula*
pine, knobcone – *Pinus attenuata*
pine, lacebark – *Pinus bungeana*
pine, Lawson – *Pinus lawsonii*
pine, limber – *Pinus flexilis*
pine, loblolly – *Pinus taeda*
pine, lodgepole – *Pinus contorta* ssp *murrayana*
pine, longleaf – *Pinus palustris*
pine, Lumholtz – *Pinus lumholtzii*
pine, maritime – *Pinus pinaster*
pine, Martinez – *Pinus maximartinezii*

pinemat manzanita – *Arctostaphylos nevadensis*
pine, Maximino – *Pinus tenuifolia*
pine, Mexican – *Pinus pinceana*
pine, Mexican pinyon – *Pinus cembroides*
pine, Mexican white – *Pinus ayacahuite*
pine, Michoacan – *Pinus devoniana*
pine mistletoe – *Arceuthobium*
pine, Monterey – *Pinus radiata*
pine, Montezuma – *Pinus montezumae*
pine, mountain – *Pinus mugo* var *rostrata*
pine, mugho – *Pinus mugo*
pine muhly – *Muhlenbergia dubia*
pine-needle milkweed – *Asclepias linaria*
pine, Norfolk Island – *Araucaria heterophylla*
pine, northern pitch – *Pinus rigida*
pine, northern white – *Pinus strobus*
pine, nut – *Pinus edulis*
pine, ocote – *Pinus oocarpa*
pine, Pacific ponderosa – *Pinus ponderosa*
pine, Parry pinyon – *Pinus quadrifolia*
pine, pinyon – *Pinus edulis*
pine, pitch – *Pinus oocarpa*
pine, ponderosa – *Pinus ponderosa*
pine, prince's – *Chimaphila dasystephana, C. umbellata*
pine, Pringle – *Pinus pringlei*
pine, Rocky Mountain ponderosa – *Pinus ponderosa* var *scopulorum*
pine, running – *Lycopodium clavatum*
pine, Rzedowski – *Pinus rzedowski*
pinesap – *Monotropa hypopitys, M. uniflora*
pine, Scotch – *Pinus sylvestris*
pine, Scots – *Pinus sylvestris*
pine, screw – *Pandanus utilis*
pine, shagbark – *Pinus herrerae*
pine, shore – *Pinus contorta* ssp *contorta*
pine, singleleaf pinyon – *Pinus monophylla*
pine, slash – *Pinus caribaea*
pine, smallleaf – *Pinus praetermissa*
pine, southern yellow – *Pinus rigida*
pine, sugar – *Pinus lambertiana*
pine, Swiss mountain – *Pinus mugo*
pine, Swiss stone – *Pinus cembra*
pine, Texas – *Pinus remota*
pine, Torrey – *Pinus torreyana*
pine, tropical – *Pinus tropicalis*
pine, umbrella – *Sciadopitys verticillata*
pine, weeping – *Pinus lumholtzii*
pine, western bristlecone – *Pinus longaeva*
pine, western white – *Pinus monticola*
pine, western yellow – *Pinus arizonica, P. ponderosa*
pine, white – *Pinus strobus*
pine, white dammar – *Agathis alba*
pine, wild – *Aechmea lueddemanniana, Bromelia pinguin*
pine, yecora – *Pinus yecorensis*
pine, yellow – *Pinus ponderosa, P. tenuifolia*
pingwing – *Aechmea magdalenae, Hymenoxys richardsoni*
pink, alpine – *Dianthus alpinus*
pink amapa – *Tabebuia impetiginosa*
pink and white everlasting – *Helipterum roseum*
pink bean – *Phaseolus vulgaris*
pink-bracted manzanita – *Arctostaphylos pringlei* ssp *drupacea*
pink, California Indian – *Silene californica*
pink, cheddar – *Dianthus gratianopolitanus*
pink, childing – *Petrorhagia nanteuilii*
pink, Chinese – *Dianthus chinensis*
pink, clove – *Dianthus caryophyllus*
pink, cottage – *Dianthus plumarius*
pink, crown – *Lychnis coronaria*
pink, cushion – *Silene acaulis*
pink, Deptford – *Dianthus armeria*
pink evening primrose – *Oenothera speciosa*
pink, fire – *Silene laciniata*
pink flame tree – *Brachychiton discolor*
pink francoa – *Francoa sonchifolia*
pink, garden – *Dianthus plumarius*
pink, grass – *Dianthus armeria, D. plumarius*
pink horsemint – *Monarda fistulosa*
pink, Indian – *Ipomoea quamoclit, Silene californica*
pink ladies – *Oenothera speciosa*
pink, maiden – *Dianthus deltoides*
pink, meadow – *Dianthus deltoids, Sabatia campestris*
pink, Mexican – *Silene laciniata*
pink mimosa – *Mimosa borealis*
pink, moss – *Phlox subulata*
pink, mullein – *Lychnis coronaria*
pink paper daisy – *Helipterum roseum*
pink porcelain lily – *Alpinia zerumbet*
pink powder puff – *Calliandra haematocephala*

pink, proliferous – *Petrorhagia prolifera*
pink pussy willow – *Salix caprea*
pink, rainbow – *Dianthus chinensis*
pink, rose – *Sabatia campestris*
pink, sand – *Dianthus arenarius*
pink, sea – *Armeria juniperifolia, A. maritima* ssp *californica*
pink shower – *Cassia grandis*
pink trumpet tree – *Tabebuia impetiginosa, T. rosea*
pink vervain – *Verbena ambrosifolia, V. pumila*
pink vine – *Antigonon leptopus*
pink viznagita – *Echinomastus johnsonii*
pinkweed – *Polygonum pensylvanicum*
pinkwing – *Alvaradoa amorphoides*
pink winter currant – *Ribes sanguineum*
pink, Wisley – *Helianthemum apenninum* var *roseum*
pin oak – *Quercus palustris*
piñon – *Pinus edulis*
pinto bean – *Phaseolus vulgaris*
pinyon, border – *Pinus discolor*
pinyon dwarf mistletoe – *Arceuthobium divaricatum*
pinyon, Nelson – *Pinus nelsoni*
pinyon pine – *Pinus edulis*
pinyon pine, Mexican – *Pinus cembroides*
pinyon pine, Parry – *Pinus quadrifolia*
pinyon pine, singleleaf – *Pinus monophylla*
pinyon, Potosi – *Pinus culminicola*
pinyon, remote – *Pinus remota*
pinyon ricegrass – *Piptochaetium fimbriatum*
pipe, California Dutchman's – *Aristolochia californica*
pipe, Dutchman's – *Aristolochia california, A. durior*
pipe, Indian – *Monotropa uniflora*
pipestem cactus – *Cylindropuntia leptocaulis*
pipestems – *Clematis lasiantha*
pipevine – *Aristolochia californica*
pipevine, southwestern – *Aristolochia watsonii*
pipeweed, Indian – *Eriogonum inflatum*
pistache – *Pistacia texana*
pistache, Mount Atlas – *Pistacia atlantica*
pistachio – *Pistacia atlantica, P. vera*
pistachio, American – *Pistacia texana*
pistachio, Mexican – *Pistacia texana*
pitanga – *Eugenia uniflora*
pitaya, sour – *Stenocereus gummosus*
pitaya, yellow-flowered – *Echinocereus dasyacanthus*
pitcher sage – *Lepechinia hastate, Salvia azurea*
pitch pine – *Pinus oocarpa*
pitch pine, northern – *Pinus rigida*
pitre – *Furcraea hexapetala*
pitseed goosefoot – *Chenopodium berlandieri*
pittosporum – *Pittosporum tobira*
pittosporum, Japanese – *Pittosporum tobira*
pittosporum, willow – *Pittosporum phillyraeoides*
plains bee balm – *Monarda pectinata*
plains bristlegrass – *Setaria macrostachya*
plains cottonwood – *Populus sargentii*
plains fleabane – *Erigeron modestus*
plains lovegrass – *Eragrostis intermedia*
plains prickly-pear – *Opuntia polyacantha*
plains zinnia – *Zinnia grandiflora*
plane, oriental – *Platanus orientalis*
planetree – *Platanus racemosa*
planetree – *Platanus* x *acerifolia*
planetree, Jalapa – *Platanus lindeniana*
planetree, London – *Platanus* x *acerifolia*
planetree, Mexican – *Platanus mexicana*
plantain – *Ensete ventricosum, Mus fehi, M. paradisiaca, Plantago*
plantain, broadleaf – *Plantago major*
plantain, buck's-horn – *Plantago coronopus*
plantain, common – *Plantago major*
plantain, English – *Plantago lanceolata, P. major*
plantain, green French – *Mus acuminata* x *balbasiana*
plantain grove – *Spathiphyllum blandum*
plantain, horn – *Mus acuminata* x *balbisiana*
plantain, Indian – *Psacalium decompositum*
plantain leopard's bane – *Doronicum plantagineum*
plantain lily, fragrant – *Hosta plantaginea*
plantain, mud – *Heteranthera limosa*
plantain, Patagonia – *Plantago patagonica*
plantain, rattlesnake – *Goodyera oblongifolia*
plantain, red – *Mus acuminata* x *balbisiana*
plantain, sand – *Plantago indica*
plantain, sea – *Plantago maritima*
plantain, water – *Alisma gramineum, A. lanceolatum, A. plantago-aquatica*
plantain, wild – *Heliconia bihai, H. caribaea*

plantain, wine – *Mus acuminata* x *balbisiana*
plateau oak – *Quercus fusiformis*
pleasure, gold of – *Camelina sativa*
pleurisy root – *Asclepias tuberosa*
plum – *Prunus domestica*
plum, American – *Prunus americana*
plum, apricot – *Prunus simonii*
plumbago, dwarf – *Ceratostigma plumbaginoides*
plumbago, Cape – *Plumbago auriculata*
plumbago, clammy – *Plumbago scandens*
plumbago, climbing – *Plumbago auriculata*
plum, bullace – *Prunus insititia*
plum, cherry – *Prunus cerasifera*
plum, cocoa – *Chrysobalanus icaco*
plum, damson – *Prunus domestica* ssp *insititia*
plum, date – *Diospyros lotus*
plum, doum – *Hyphaene thebaica*
plume cockscomb – *Celosia plumosa*
plumed thistle – *Cirsium mexicanum*
plume, feather – *Dalea formosa*
plume flower, Brazilian – *Justicia carnea*
plumeless thistle – *Carduus acanthoides*
plume poppy – *Macleaya cordata*
plumeria – *Plumeria rubra*
plume, scarlet – *Euphorbia fulgens*
plum, flowering – *Prunus* x *blireiana*
plum-fruited yew – *Podocarpus andinus*
plum, goose – *Prunus americana*
plum, governor's – *Flacourtia indica, F. inermis*
plum, greengage – *Prunus domestica* ssp *italica*
plum, hog – *Spondias mombin*
plum, Japanese – *Prunus salicina*
plum, Japanese flowering – *Prunus mume*
plum, Java – *Eugenia jambolana*
plum, jungle – *Dipholis salicifolia*
plumleaf fig – *Ficus padifolia*
plummer's morning-glory – *Ipomoea plummerae*
plum, Mexican – *Prunus mexicana*
plum, Natal – *Carissa grandiflora*
plum, pigeon – *Chrysobalanus icaco, Hirtella americana, H. racemosa*
plum rose – *Eugenia jambos*
plum, saffron – *Sideroxylon spiniflora*
plum, seaside – *Ximenia americana*
plum, sebesten – *Cordia myxa*
plum, Spanish – *Spondias mombin, S. purpurea*
plum, Syrian – *Prunus domestica* ssp *syriaca*
plum, tropical – *Spondias mombin*
plum, wild – *Cyrtocarpa edulis, Prunus americana*
plum, wild coco – *Hirtella americana*
plum yew – *Cephalotaxus harringtonia*
poached eggs – *Limnanthes douglasii*
poblano chile – *Capsicum annuum*
pod corn – *Zea mays* var *tunicata*
poet's cassia – *Osyris alba*
poet's jasmine – *Jasminum grandiflorum*
poet's narcissus – *Narcissus poeticus*
poinciana – *Caesalpinia palmeri, C. platyloba, Delonix regia*
poinciana, dwarf – *Caesalpinia pulcherrima*
poinciana, royal – *Delonix regia*
poinciana, yellow – *Peltophorum pterocarpum*
poinsettia – *Euphorbia pulcherrima*
poinsettia, flase – *Euphorbia heterophylla*
poinsettia, green – *Euphorbia dentata*
poinsettia, wild – *Warszewiczia coccinea*
pointed gourd – *Trichosanthes dioica*
point-leaf manzanita – *Arctostaphylos pungens*
pointsettia – *Euphorbia pulcherrima*
poison bay – *Illicium floridanum*
poison bitterweed – *Hymenoxys odorata*
poison, crow – *Nothoscordium bivalve*
poison hemlock – *Conium maculatum*
poison ivy – *Toxicodendron radicans*
poison oak, western – *Toxicodendron diversilobum*
poisonous brain mushroom – *Gyromitra esculenta*
poison primrose – *Primula obconica*
pokeberry – *Phytolacca americana, Phytolacca rivinoides, Rivina humilis*
pokeroot – *Phytolacca americana*
poker, red-hot – *Kniphofia uvaria*
pokeweed – *Phytolacca americana, P. octandra*
pokeweed, great – *Phytolacca americana*
pokeweed, small – *Rivina humilis*
pole bean – *Phaseolus vulgaris*
polka dot cactus – *Opuntia microdasys*
polyanthus primrose – *Primula* x *polyantha*
polychrome spurge – *Euphorbia polychroma*
poly, grass – *Lythrum hyssopifolium*
Polynesian chestnut – *Inocarpus edulis*

polypody – *Polypodium*
polypody, California – *Polypodium californicum*
polypore – *Polyporus*
pomegranate – *Punica granatum*
pomegranate, dwarf – *Punica granatum* var *nana*
pomerac – *Eugenia malaccensis*
pond apple – *Annona glabra*
ponderosa lemon – *Citrus limon* var *ponderosa*
ponderosa pine – *Pinus ponderosa*
ponderosa pine, Pacific – *Pinus ponderosa*
ponderosa pine, Rocky Mountain – *Pinus ponderosa* var *scopulorum*
pond-lily, yellow – *Nuphar luteum* ssp *polysepalum*
pondweed, crispate-leaved – *Potamogeton crispus*
pondweed, diverse-leaved – *Potamogeton diversifolius*
pondweed, eel-grass – *Potamogeton zosteriformis*
pondweed, fennel-leaf – *Potamogeton pectinatus*
pondweed, floating-leaved – *Potamogeton natans*
pondweed, grass-leaved – *Potamogeton gramineus*
pondweed, long-leaved – *Potamogeton nodosus*
pondweed, Nuttall's – *Potamogeton epihydrus* ssp *nuttallii*
pondweed, reddish – *Potamogeton alpinus*
pondweed, shining – *Potamogeton illinoensis*
pondweed, slender-leaved – *Potamogeton filiformis*
pondweed, small – *Potamogeton pusillus*
pondweed, white-stemmed – *Potamogeton praelongus*
ponil – *Fallugia paradoxa*
pontic azalea – *Rhododendron luteum*
ponytail palm – *Nolina recurvata*
poor man's bread – *Artocarpus heterophyllus*
poor man's orchid – *Schizanthus pinnatus*
poor man's pepperwort – *Lepidium virginicum*
poor man's weatherglass – *Anagallis arvensis*
popbeans – *Phaseolus vulgaris* ssp *nanas*
popcorn – *Zea mays* var *everia*
popcornflower – *Cryptantha, Plagiobothrys nothofulvus*
poplar – *Populus*
poplar, Arizona – *Populus fremontii*
poplar, balsam – *Populus balsamifera*
poplar, black – *Populus nigra*
poplar, Carolina – *Populus canadensis, P. deltoides*
poplar, eastern – *Populus deltoides*
poplar, gray – *Populus canescens*
poplar, Lombardy – *Populus nigra* var *italica*
poplar, Mexican – *Populus mexicana*
poplar, narrowleaf – *Populus angustifolia*
poplar, silver – *Populus alba*
poplar, white – *Populus alba*
poppy – *Eschscholzia*
poppy, alpine – *Papaver alpinum*
poppy, Arizona – *Kallstroemia grandiflora*
poppy, Atlantic – *Papaver atlanticum*
poppy, breadbox – *Papaver somniferum*
poppy, California – *Eschscholzia californica*
poppy, corn – *Papaver rhoeas*
poppy, Coulter's matilija – *Romneya coulteri*
poppy, desert – *Kallstroemia grandiflora*
poppy, Flander's field – *Papaver rhoeas*
poppy-flowered anemone – *Anemone coronaria*
poppy, gold – *Eschscholzia glyptosperma*
poppy, Himalayan – *Meconopsis betonicifolia*
poppy, horned – *Glaucium corniculatum*
poppy, Iceland – *Papaver nudicaule*
poppy, island – *Eschscholzia ramosa*
poppy, Mexican – *Argemone mexicana, Eschscholzia mexicana, Kallstroemia*
poppy, Mexican tulip – *Hunnemannia fumariifolia*
poppy, opium – *Papaver somniferum*
poppy, oriental – *Papaver orientale*
poppy, plume – *Macleaya cordata*
poppy, prickle – *Argemone intermedia, A. platyceras*
poppy, prickly – *Argemone mexicana, A. munita*
poppy, pygmy – *Eschscholzia minutiflora*
poppy, red – *Papaver rhoeas*
poppy, rough – *Papaver hybridum*
poppy, sea – *Glaucium flavum*
poppy, Shirley – *Papaver rhoeas*
poppy, summer – *Kallstroemia grandiflora*

poppy, Welsh – *Meconopsis cambrica*
poppy, white-flowered prickly – *Argemone gracilenta*
poppy, yellow prickly – *Argemone ochroleuca*
porcelain flower – *Hoya carnosa*
porcelain lily, pink – *Alpinia zerumbet*
porcupine prickly-pear – *Opuntia polyacantha* var *erinacea*
portia tree – *Thespesia populnea*
Port Orford cedar – *Cupressus lawsoniana*
Portugal laurel – *Prunus lusitanica*
Portuguese broom – *Cytisus multiflorus*
Portuguese cabbage – *Brassica oleracea* var *costata*
Portuguese cedar – *Cedrus lusitanica*
Portuguese cypress – *Cupressus lusitanica*
Portuguese heath – *Erica lusitanica*
Portuguese kale – *Brassica oleracea* var *tronchuda*
Portuguese oak – *Quercus faginea*
portulaca – *Portulaca grandiflora*
portulaca, desert – *Portulaca halimoides*
portulaca, garden – *Portulaca grandiflora*
portulaca, shaggy – *Portulaca halimoides*
possum grape – *Cissus cucurbitina*
possum haw – *Ilex decidua*
possumwood – *Hura polyandra*
pot – *Cannabis sativa*
potato – *Solanum cardiophyllum, S. demissum, S. tuberosum*
potato, air – *Dioscorea bulbifera*
potato blight – *Phytophthora infestans*
potato creeper, giant – *Solanum wendlandii*
potato, hausa – *Coleus rotundifolius*
potato, hog – *Hoffmannseggia glauca*
potato onion – *Allium cepa* var *aggregatum*
potato, sweet – *Ipomoea batatas*
potato tree – *Solanum wrightii*
potato, tule – *Sagittaria latifolia*
potato vine – *Solanum dulcamara, S. jasminoides*
potato vine, grand – *Solanum wendlandii*
potato, wild – *Solanum bulbocastanum*
potentilla – *Potentilla atrosanguinea*
potentilla, Sibbald's – *Sibbaldia procumbens*
pothos – *Epipremnum pinnatum*
pot-marigold – *Calendula officinalis*
pot, monkey – *Lecythis zabucajo*
Potosi pinyon – *Pinus culminicola*
poverty brome – *Bromus sterilis*
poverty three-awn – *Aristida divaricata, A. ternipes* var *gentilis*
poverty weed – *Baccharis neglecta, Iva ambrosiaefolia*
powder puff, dwarf – *Calliandra emarginata*
powder puff, pink – *Calliandra haematocephala*
powder puff, red – *Calliandra haematocephala*
poyo banana – *Mus acuminata*
prairie acacia – *Acacia angustissima*
prairie clover, hairy – *Dalea tomentosa*
prairie cupgrass – *Eriochloa contracta*
prairie flax – *Linum lewisii*
prairie fleabane – *Erigeron modestus*
prairie gentian, showy – *Eustoma grandiflorum*
prairie mallow – *Sidalcea neomexicana*
prairie sage – *Salvia azurea*
prairie wedgegrass – *Sphenopholis obtusata*
prayer bean – *Abrus precatorius*
prayer plant – *Maranta leuconeura*
prickle grass – *Crypsis alopecuroides*
pricklenut – *Guazuma ulmifolia*
prickle poppy – *Argemone intermedia, A. platyceras*
prickly ash – *Zanthoxylum americanum, Z. arborescens*
pricklyash, lime – *Zanthoxylum fagara*
prickly chaff-flower – *Achyranthes aspera*
prickly juniper – *Juniperus oxycedrus*
prickly lettuce – *Lactuca serriola*
prickly nailwort – *Paronychia echinulata*
prickly-pear – *Opuntia*
prickly-pear, Aaron's beard – *Opuntia leucotricha*
prickly-pear, Big Bend – *Opuntia phaeacantha*
prickly-pear, blind – *Opuntia rufida*
prickly-pear, bunny-ear – *Opuntia microdasys*
prickly-pear, chaparral – *Opuntia oricola*
prickly-pear, El Paso – *Opuntia polyacantha* var *arenaria*
prickly-pear, Engelmann – *Opuntia engelmannii*
prickly-pear, juniper – *Opuntia polyacantha*
prickly-pear, Mojave – *Opuntia phaeacantha*
prickly-pear, pancake – *Opuntia chlorotica*
prickly-pear, plains – *Opuntia polyacantha*
prickly-pear, porcupine – *Opuntia polyacantha* var *erinacea*

prickly-pear, purple – *Opuntia macrocentra*
prickly-pear, sweet – *Opuntia dillenii*
prickly-pear, Texas – *Opuntia engelmannii* var *lindheimeri*
prickly-pear, tree – *Opuntia ficus-indica*
prickly-pear, violet – *Opuntia macrocentra*
prickly phlox – *Leptodactylon*
prickly poppy – *Argemone*
prickly poppy, flatbud – *Argemone munita*
prickly poppy, Mojave – *Argemone corymbosa*
prickly poppy, white-flowered – *Argemone gracilenta*
prickly poppy, yellow – *Argemone ochroleuca*
prickly rose – *Rosa acicularis*
prickly sow thistle – *Sonchus asper* ssp *asper*
pride, Barbados – *Caesalpinia pulcherrima*
pride, Lasater's – *Abronia fragrans*
pride, London – *Saxifraga umbrosa*
pride of Barbados – *Caesalpinia pulcherrima*
pride of California – *Lathyrus splendens*
pride of India – *Koelreuteria paniculata, Lagerstroemia indica, Melia azedarach*
pride of Trinidad – *Warszewiczia coccinea*
pride, Texas – *Thelocactus bicolor*
primavera – *Tabebuia donnell-smithii*
primrose – *Primula*
primrose, basket evening – *Oenothera deltoides*
primrose, Cape – *Streptocarpus hybridus*
primrose, Chinese – *Primula sinensis*
primrose, English – *Primula* x *polyantha, P. vulgaris*
primrose, evening – *Oenothera*
primrose jasmine – *Jasminum mesnyi*
primrose, Mexican – *Oenothera speciosa*
primrose, Missouri – *Oenothera macrocarpa*
primrose monkeyflower – *Mimulus primuloides*
primrose, poison – *Primula obconica*
primrose, polyanthus – *Primula* x *polyantha*
primrose, Sierra – *Primula suffrutescens*
primrose tree – *Lagunaria patersonii*
primrose, water – *Ludwigia octovalvis, L. peploides*
prince's feather – *Amaranthus hybridus*
prince's pine – *Chimaphila dasystephana, C. umbellata*
princess flower – *Tibouchina urvilleana*
princewood – *Exostema caribaeum*
pringle pine – *Pinus pringlei*
privet, California – *Ligustrum ovalifolium*
privet, common – *Ligustrum vulgare*
privet, glossy – *Ligustrum lucidum*
privet, Japanese – *Ligustrum japonicum*
privet, New Mexican – *Forestiera neomexicana*
privet, wax-leaf – *Ligustrum japonicum*
prodigious bromeliad – *Tillandsia prodigiosa*
proliferous pink – *Petrorhagia prolifera*
prophet's cucumber – *Citrullus colocynthis*
prostrate amaranth – *Amaranthus blitoides, A. deflexus*
prostrate knotweed – *Polygonum aviculare*
prostrate speedwell – *Veronica prostrata*
prostrate spurge – *Chamaesyce*
Provence broom – *Cytisus purgans*
Provence rose – *Rosa centifolia, R. gallica*
provision tree – *Pachira aquatica*
prune, German – *Prunus domestica* ssp *italica*
psilocybe, Wasson's – *Psilocybe muliericula*
psoralea, brown-flowered – *Psoralea rhombifolia*
psychedelic mushroom – *Psilocybe, Panaeolus, Conocybe, Stropharia cubensis*
psyllium – *Plantago ovata, P. psyllium*
puccoon – *Lithospermum cobrense*
puffball – *Lycoperdon*
puffball, common – *Lycoperdon perlatum*
puffball, giant – *Calvatia gigantea*
puff, powder – *Calliandra*
pulque agave – *Agave atrovirens, A. salmiana*
pummelo – *Citrus grandis*
pumpkin – *Cucurbita maxima, C. moscada, C. pepo*
pumpwood – *Cecropia obtusifolia, C. peltata, Oreopanax peltatum*
puncture vine – *Tribulus terrestris*
purging cassia – *Cassia fistula*
purging croton – *Croton tiglium*
purple betony – *Stachys betonica*
purple coneflower – *Echinacea angustifolia, E. purpurea*
purple crab – *Malus purpurea*
purple galingale – *Cyperus rotundus*
purple geranium – *Geranium caespitosum*
purple grama – *Bouteloua radicosa*
purple granadilla – *Passiflora edulis*
purple heart – *Tradescantia pallida*

purple heather – *Krameria erecta*
purple jasmine – *Solanum dulcamara*
purple loosestrife – *Lysimachia salicaria, Lythrum virgatum*
purple mat – *Nama coulteri, N. hispidum*
purple matseed – *Nama demissum*
purple muhly – *Muhlenbergia rigida*
purple mullein – *Verbascum phoeniceum*
purple nightshade – *Solanum xanti*
purple nut-grass – *Cyperus rotundus*
purple nutsedge – *Cyperus rotundus*
purple orchid tree – *Bauhinia variegata*
purple osier – *Salix purpurea*
purple owl's-clover – *Castilleja exserta*
purple prickly-pear – *Opuntia macrocentra*
purple queen – *Tradescantia pallida*
purple ragwort – *Senecio elegans*
purple sage – *Leucophyllum frutescens*
purple sandgrass – *Triplasis purpurea*
purple-spined viznagita – *Echinomastus erectocentrus*
purple star-thistle – *Centaurea calcitrapa*
purple three-awn – *Aristida purpurea*
purple toadflax – *Linaria purpurea*
purple trumpet tree – *Tabebuia impetiginosa*
purple vetch – *Vicia benghalensis*
purple woolly sage – *Phlomis purpurea*
purple wreath – *Petrea arborea, P. volubilis*
purse, rabbit's – *Harfordia macroptera*
purse, shepherd's – *Capsella bursa-pastoris*
purslane – *Portulaca oleracea*
purslane, common – *Portulaca oleracea*
purslane, Hampshire – *Ludwigia palustris*
purslane, horse – *Trianthema portulacastrum*
purslane, sea – *Atriplex halimus, Sesuvium portulacastrum*
purslane speedwell – *Veronica peregrina* ssp *xalapensis*
purslane, Surinam – *Talinum triangulare*
purslane, water – *Lythrum portula*
purslane, western sea – *Sesuvium verrucosum*
pusley – *Portulaca oleracea*
pusley, Chinese – *Heliotropium curassavicum*
pussy-toes – *Antennaria dioica*
pussy willow, French – *Salix caprea*
pussy willow, pink – *Salix caprea*
pygmy cedar – *Peucephyllum schottii*
pygmy date palm – *Phoenix roebelenii*
pygmy gentian – *Gentiana prostrata*
pygmy poppy – *Eschscholzia minutiflora*
pyracantha – *Pyracantha*
Pyrenean aster – *Aster sibiricus*
Pyrenean feathergrass – *Stipa gigantea*
Pyrenean oak – *Quercus pyrenaica*
pyrethrum – *Chrysanthemum cinerariifolium, C. coccineum*

quackgrass – *Elytrigia repens*
quail brush – *Atriplex lentiformis*
quail bush – *Atriplex lentiformis*
quailgrass – *Celosia argentea*
quail plant – *Heliotropium curassavicum*
quaking aspen – *Populus tremuloides*
quaking grass – *Briza maxima, B. minor*
quash tail – *Nephrolepis biserrata*
quassia – *Quassia amara*
quebracho – *Schinopsis quebracho-colorado*
quebracho, white – *Aspidosperma quebracho-blanco*
Queen Anne's lace – *Daucus carota*
queen flower – *Lagerstroemia speciosa*
queen of flowers – *Lagerstroemia speciosa*
queen-of-the-night – *Cereus hexagonus, Epiphyllum anguliger, Hylocereus undatus, Peniocereus greggii, Selenicereus grandiflorus*
queen-of-the-night, Arizona – *Peniocereus greggii*
queen palm – *Syagrus romanzoffiana*
queen, purple – *Tradescantia pallida*
queen's crepe myrtle – *Lagerstroemia speciosa*
queen's delight – *Stillingia*
Queensland arrowroot – *Canna edulis*
Queensland lacebark – *Brachychiton discolor*
Queensland umbrella tree – *Schefflera actinophylla*
queen's root – *Stillingia*
queen's wreath – *Antigonon leptopus, Petrea volubilis*
Queen Victoria waterlily – *Victoria regia*
quickweed – *Galinsoga parviflora*
quilete – *Amaranthus caudatus*
quince – *Cydonia oblonga*
quince, Chinese – *Chaenomeles sinensis*
quince, flowering – *Chaenomeles speciosa*
quinine – *Cinchona*
quinine bush – *Garrya fremontii, Purshia mexicana* var *stansburyana*

quinine plant – *Simmondsia chinensis*
quinoa – *Chenopodium quinoa*
Quixote plant – *Yucca whipplei*

rabbitbrush, rubber – *Chrysothamnus nauseosus*
rabbit-foot grass – *Polypogon monspeliensis*
rabbit's purse – *Harfordia macroptera*
radicchio – *Cichorium intybus*
radish – *Raphanus sativus*
radish, garden – *Raphanus sativus*
radish, wild – *Raphanus raphanistrum*
raffia – *Raphia ruffia*
ragged cup – *Silphium perfoliatum*
ragged robin – *Lychnis flos-cuculi*
ragweed – *Ambrosia, Senecio jacobaea*
ragweed, canyon – *Ambrosia ambrosioides*
ragweed, false – *Parthenium hysterophorus*
ragweed, low – *Ambrosia artemisiifolia*
ragweed, winged – *Hymenoclea monogyra*
ragwort – *Senecio*
ragwort, common – *Senecio vulgaris*
ragwort, purple – *Senecio elegans*
ragwort, tansy – *Senecio jacobaea*
railroad vine – *Ipomoea pes-caprae*
railway creeper – *Ipomoea cairica*
rainbow cactus – *Echinocereus pectinatus*
rainbow cactus, New Mexico – *Echinocereus viridiflorus*
rainbow cactus, Texas – *Echinocereus dasyacanthus*
rainbow hedgehog – *Echinocereus pectinatus*
rainbow hedgehog cactus, Arizona – *Echinocereus rigidissimus*
rainbow hedgehog, Texas – *Echinocereus dasyacanthus*
rainbow pink – *Dianthus chinensis*
rain lily – *Cooperia*
rain lily, Drummond's – *Cooperia drummondii*
rain lily, evening-star – *Cooperia drummondii*
rain of gold – *Galphimia glauca*
rain-stick cactus – *Echinopsis chiloensis*
rain tree – *Albizzia saman*
ramanas rose – *Rosa rugosa*
rambutan – *Nephelium lappaceum*
ramie plant – *Boehmeria nivea*
ramon breadnut – *Brosimum alicastrum*
ramon, white – *Trophis racemosa*
ramoon – *Trophis racemosa*
ramping fumitory – *Fumaria capreolata*
rampion – *Campanula rapunculus*
ramrod tree – *Chaetoptelea mexicana*
ramtilla – *Guizotia abyssinica*
rancher's fireweed – *Amsinckia menziesii* var *intermedia*
randia, spiny-fruited – *Randia echinocarpa*
range rhatany – *Krameria erecta, K. lanceolata*
ranger, Texas – *Leucophyllum frutescens*
Rangoon creeper – *Quisqualis indica*
rangpur lime – *Citrus limonia*
ranunculus – *Ranunculus*
ranunculus, Persian – *Ranunculus asiaticus*
ranunculus, turban – *Ranunculus asiaticus*
rape – *Brassica napus* var *oleifera*
rape, annual turnip – *Brassica rapa* var *campestris*
rape, biennial turnip – *Brassica rapa* var *oleifera*
rape, bird – *Brassica rapa* var *campestris, B. r.* var *oleifera*
rape, broom – *Orobanche*
rape, oilseed – *Brassica napus* var *oleifera*
rapeseed – *Brassica napus* var *napobrassica*
rape, summer turnip – *Brassica rapa* var *praecox*
rape, Swede – *Brassica napus* var *napobrassica*
raphia palm – *Raphia ruffia*
raspberry – *Rubus strigosus*
raspberry, American red – *Rubus strigosus*
raspberry, black – *Rubus occidentalis*
raspberry, bristly – *Rubus hispidus*
raspberry, Mauritius – *Rubus rosaefolius*
raspberry, New Mexican – *Rubus neomexicanus*
raspberry, red – *Rubus idaesus, R. strigosus*
ratstail fescue – *Vulpia myuros*
rat-tail cactus – *Cylindropuntia leptocaulis, Disocactus flagelliformis, Mammillaria pottsii*
rattan – *Calamus*
rattan palm – *Calamus draco, C. optimus, C. rotang, Rhapis humilis*
rattlebox – *Crotalaria sagittalis*
rattlebox, low – *Crotalaria pumila*
rattlepod – *Crotalaria longirostrata*
rattlesnake plantain – *Goodyera oblongifolia*
rattlesnake weed – *Chamaesyce*

albomarginata, C. polycarpa, Daucus pusillus
rattleweed – *Astragalus*
Ravennagrass – *Saccharum ravennae*
rayless chamomile – *Chamomilla suaveolens*
rayless goldenrod – *Ericameria heterophyllus*
rayless thelesperma – *Thelesperma megapotamicum*
red alder – *Alnus rubra*
red ash – *Fraxinus pennsylvanica*
red barberry – *Berberis haematocarpa*
redbark cinchona – *Cinchona succirubra*
red barrel cactus – *Ferocactus cylindraceus*
red bean – *Phaseolus vulgaris*
red bean, small – *Phaseolus vulgaris*
red belt fungus – *Fomes pinicola*
redberry – *Rhamnus*
redberry juniper – *Juniperus pinchotii*
red birch – *Betula nigra*
red bird-of-paradise – *Caesalpinia pulcherrima*
red brome – *Bromus madritensis* ssp *rubens*
red buckeye – *Aesculus pavia*
redbud, Asiatic – *Cercis siliquastrum*
redbud, eastern – *Cercis canadensis*
redbud, western – *Cercis occidentalis*
red bush salvia – *Salvia microphylla*
red cabbage – *Brassica oleracea* var *capitata*
red campion – *Silene dioica*
red cat's-tail – *Acalypha hispida*
red cedar – *Cedrela odorata, C. toona, Trichilia hirta*
red cedar, eastern – *Juniperus virginiana*
red cedar, western – *Juniperus scopulorum, Thuja plicata*
red cestrum – *Cestrum elegans*
red cinchona – *Cinchona pitayensis*
red clover – *Trifolium pratense*
red cornel – *Cornus sanguinea*
red currant – *Ribes rubrum*
red Dacca banana – *Mus acuminata*
red devil's claw – *Proboscidea parviflora*
reddish pondweed – *Potamogeton alpinus*
red dock – *Rumex hymenosepalus*
red elderberry – *Sambucus racemosa*
red elephant tree – *Bursera hindsiana*
red fescue – *Festuca rubra*
red fir, California – *Abies magnifica*
red fireweed – *Epilobium angustifolium*
red fir, Shasta – *Abies magnifica* var *shastensis*
red flowering currant – *Ribes sanguineum*
red fox – *Celosia argentea*
red ginger – *Alpinia purpurata*
red goosefoot – *Chenopodium rubrum*
red goosefoot, small – *Chenopodium chenopodioides*
red-grained black truffle – *Tuber aestivum*
red grama – *Bouteloua trifida*
red gum – *Eucalyptus camaldulensis, E. rostratus*
red gum, river – *Eucalyptus camaldulensis*
redhead – *Adenanthera pavonina, Euphorbia heterophylla*
red henbit – *Lamium purpureum*
red horned poppy – *Glaucium corniculatum*
red horse chestnut – *Aesculus carnea*
red-hot poker – *Kniphofia uvaria*
red iron bark – *Eucalyptus sideroxylon*
red jasmine – *Ipomoea quamoclit*
red kidney bean – *Phaseolus vulgaris*
red larkspur – *Delphinium nudicaule*
red-leaved rose – *Rosa glauca*
red mahonia – *Berberis haematocarpa*
red maids – *Calandrinia ciliata*
red mallow – *Malvaviscus grandiflorus*
red mangrove – *Avicennia marina* var *resinifera, Rhizophora mangle*
red maple – *Acer rubrum*
red mombin – *Spondias purpurea*
red mulberry – *Morus rubra*
red oak – *Quercus rubra*
red oak, Chisos – *Quercus gravesii*
red oak, northern – *Quercus rubra*
red-osier dogwood – *Cornus stolonifera*
red parrot flower – *Psittacanthus calyculatus*
red partridge berry – *Mitchella repens*
red pineapple cactus – *Echinomastus erectocentrus*
red pine, Japanese – *Pinus densiflora*
red plantain – *Mus acuminata* x *balbisiana*
red poppy – *Papaver rhoeas*
red powder puff – *Calliandra haematocephala*
red raspberry – *Rubus idaesus*
red raspberry, American – *Rubus strigosus*
red raspberry, European – *Rubus idaeus*
redroot – *Amaranthus palmeri, Ceanothus americanus*
red sage, Asian – *Salvia miltiorrhiza*
red sandalwood – *Pterocarpus santalinus*

red sanders wood – *Pterocarpus santalinus*
red shank – *Adenostoma sparsifolium, Polygonum persicaria*
red silk cotton tree – *Bombax ceiba*
red soapwort – *Saponaria ocymoides*
red sorrel – *Hibiscus sabdariffa*
red spangletop – *Leptochloa filiformis, L. panicea* ssp *brachiata*
red spider lily – *Lycoris radiata*
red spruce – *Picea rubens*
redstain – *Plagiobothrys nothofulvus*
redstar zinnia – *Zinnia multiflora*
red strawberry spinach – *Chenopodium foliosum*
red three-awn – *Aristida purpurea* var *longiseta*
red-tipped coral fungus – *Ramaria botrytis*
red-top bent grass – *Agrostis gigantea*
red truffle – *Tuber rufum*
redtwig dogwood – *Cornus stolonifera*
red twincup – *Cephaelis tomentosa*
red valerian – *Centranthus ruber*
red vetchling – *Lathyrus cicera*
red whortleberry – *Vaccinium vitis-idaea*
red willow – *Salix laevigata*
redwood – *Sequoia sempervirens*
redwood, coast – *Sequoia sempervirens*
redwood, dawn – *Metasequoia glyptostroboides*
red yucca – *Hesperaloe parviflora*
red zinnia – *Zinnia tenuiflora*
reed – *Olyra latifolia*
reed bamboo – *Bambusa arundinacea*
reed, bur – *Sparganium*
reed canary grass – *Phalaris arundinacea*
reed, common – *Phragmites australis*
reed, giant – *Arundo donax*
reed grass – *Calamagrostis*
reed orchid, fiery – *Epidendrum ibaguense*
reed stem epidendrum – *Epidendrum ibaguense*
regal gerenanium – *Pelargonium* x *domesticum*
regal pelargonium – *Pelargonium* x *domesticum*
regal lily – *Lilium regale*
reishi mushroom – *Gandoderma lucidum*
remote pinyon – *Pinus remota*
rescue grass – *Bromus catharticus*
resurrection plant – *Selaginella*
reticulata iris – *Iris reticulata*
rex begonia – *Begonia rex*
rhatany – *Krameria*
rhatany, Peruvian – *Krameria trianda*
rhatany, Pima – *Krameria erecta*
rhatany, range – *Krameria erecta, K. lanceolata*
rhatany, white – *Krameria grayi*
Rhodes grass – *Chloris gayana*
rhododendron – *Rhododendron*
rhubarb – *Rheum rhaponticum*
rhubarb, Chinese – *Rheum officinale*
rhubarb, desert – *Rumex hymenosepalus*
rhubarb, garden – *Rheum rhaponticum*
rhubarb, Guatemala – *Jatropha podagrica*
rhubarb, Indian – *Rumex salicifolius*
rhubarb, medicinal – *Rheum officinale*
rhubarb, monk's – *Rumex alpinus*
rhubarb, ornamental – *Rheum palmatum*
rhubarb, wild – *Rumex hymenosepalus*
ribbed cryptantha – *Cryptantha costata*
ribbon grass – *Nolina bigelovii, N. parryi, Phalaris arundinacea, Plantago lanceolata*
ribbon plant – *Dracaena sanderana*
ribgrass – *Plantago lanceolata*
ribwort – *Plantago lanceolatum*
rice – *Oryza sativa*
rice bean – *Phaseolus calcaratus, Vigna umbellata*
rice, Canada – *Zizania aquatica*
rice cutgrass – *Leersia oryzoides*
ricegrass, Indian – *Achnatherum hymenoides*
ricegrass, pinyon – *Piptochaetium fimbriatum*
ricegrass, swamp – *Leersia hexandra*
rice, wild – *Zizania aquatica, Z. palustris* var *interior*
ringflower – *Heliopsis helianthoides*
ringstem – *Anulocaulis eriosolenus*
ringworm cassia – *Senna alata*
ringworm senna – *Senna alata*
Rio Grande abutilon – *Abutilon hypoleucum*
Rio Grande cottonwood – *Populus wislizenii*
ripgut grass – *Bromus diandrus*
river birch – *Betula nigra*
river red gum – *Eucalyptus camaldulensis*
river she-oak – *Casuarina cunninghamiana*
roach plant, cock – *Haplophyton cimicidum*
Robert, herb – *Geranium robertianum*
robin, ragged – *Lychnis flos-cuculi*

robin-run-in-the-hedge – *Glecoma hederacea*
robin, wake – *Trillium*
roble – *Quercus lobata*
roble beech – *Nothofagus obliqua*
robusta banana – *Mus acuminata*
robusta coffee – *Coffea canephora*
rocambole – *Allium sativum* var *ophioscorodon*
rock cotoneaster – *Cotoneaster horizontalis*
rock cranesbill – *Geranium macrorrhizum*
rock cress – *Arabis, Aubrieta deltoidea*
rock cress, coast – *Arabis blepharophylla*
rock cress, hairy – *Arabis hirsuta*
rock cress, mountain – *Arabis alpina*
rock cress, wall – *Arabis caucasia*
rock currant – *Ribes petraeum*
rock daisy – *Perityle*
rocket candytuft – *Iberis amara*
rocket, dame's – *Hesperis matronalis*
rocket, dyer's – *Reseda luteola*
rocket, eastern – *Sisymbrium orientale*
rocket, false London – *Sisymbrium loeselii*
rocket, French bastard – *Erucastrum gallicum*
rocket, garden – *Eruca vesicaria* ssp *sativa*
rocket, London – *Sisymbrium irio*
rocket salad – *Eruca vesicaria*
rocket, salad – *Eruca vesicaria* ssp *sativa*
rocket, sea – *Cakile maritima*
rocket, sweet – *Hesperis matronalis*
rocket, tall – *Sisymbrium altissimum*
rocket, treacle-mustard-like – *Sisymbrium erysimoides*
rocket, wall – *Diplotaxis erucoides, D. muralis, D. tenuifolia*
rocket, yellow – *Barbarea vulgaris*
rock fig – *Ficus padifolia, F. petiolaris*
rock fig, desert – *Ficus palmeri*
rock-jasmine – *Androsace*
rock-lettuce, white – *Pinaropappus roseus*
rock lichen – *Parmelia tiliacea*
rock nettle – *Eucnide rupestris*
rock-rose – *Cistus, Helianthemum glomeratum, H. nummularium*
rock-rose, crimson-spot – *Cistus ladanifer*
rock-rose, sageleaf – *Cistus salvifolius*
rock-rose, spotted – *Tuberaria guttata*
rock-rose, white-leaved – *Cistus albidus*
rock spleenwort – *Asplenium fontanum*
rockspray cotoneaster – *Cotoneaster microphyllus*
rocktrumpet – *Macrosiphonia macrosiphoni*
rock tulip – *Tulipa saxatilis*
Rocky Mountain bee plant – *Cleome serulata*
Rocky Mountain columbine – *Aquilegia caerulea*
Rocky Mountain fir – *Abies lasiocarpa*
Rocky Mountain iris – *Iris missouriensis*
Rocky Mountain juniper – *Juniperus scopulorum*
Rocky Mountain ponderosa pine – *Pinus ponderosa* var *scopulorum*
Rocky Mountain sugar maple – *Acer saccharum* ssp *grandidentatum*
Rocky Mountain sumac – *Rhus glabra*
Rocky Mountain white oak – *Quercus gambellii*
rocotillo pepper – *Capsicum chinense*
rocoto – *Capsicum pubescens*
rocoto pepper – *Capsicum pubescens*
rocoto pepper, yellow-fruited – *Capsicum pubescens*
rod, Aaron's – *Solidago virgaurea*
rod, golden – *Solidago*
rod, snake – *Polygonum bistorta*
Roemer acacia – *Acacia roemeriana*
romaine lettuce – *Lactuca sativa* var *longifolia*
Roman chamomile – *Chamaemelum nobile*
Roman hyacinth – *Hyacinthus orientalis* var *albulus*
Roman hyacinth, French – *Hyacinthus orientalis* var *albulus*
Roman wormwood – *Artemisia pontica*
roma tomato – *Lycopersicon esculentum*
romero – *Trichostema lanatum*
Roosevelt weed – *Baccharis neglecta*
rooster tree – *Calotropis procera*
root, ague – *Aletris farinosa*
root, alum – *Heuchera sanguinea*
root, arrow – *Maranta arundinaceae, Thalia geniculata*
root, balsam – *Balsamorhiza sagittata*
root, big – *Marah macrocarpus*
root, biscuit – *Cymopterus fendleri, C. purpurascens, Lomatium dissectum*
root, blood – *Potentilla erecta, Sanguinaria canadensis*
root, bread – *Psoralea esculenta*
root, broom – *Muhlenbergia macroura*

root, cancer – *Conopholis mexicana, Orobanche cooperi*
root, canker – *Coptis groenlandica*
root, chervil – *Chaerophyllium tuberosum*
root, China – *Smilax china*
root, costus – *Saussurea lappa*
root, David's – *Chiococca alba*
root, ginger – *Zingiber officinale*
root, Indian – *Aristolochia brevipes, A. watsonii, Asclepias curassavica, Lomatium dissectum*
root, licorice – *Glycyrrhiza glabra*
root, pleurisy – *Asclepias tuberosa*
root, poke – *Phytolacca americana*
root, queen's – *Stillingia*
root, red – *Amaranthus palmeri, Ceanothus americanus*
root, snake – *Eryngium lemmoni, Liatris punctata, L. spicata*
root, soap – *Chlorogalium*
root, squaw – *Conopholis americana*
root, swamp – *Anemopsis californica*
root, sweet – *Osmorhiza brachypoda*
root, sweet corn – *Calathea allouia*
root, tobacco – *Valeriana sorbifolia*
root, yellow – *Xanthorrhiza simplicissima*
roquette – *Eruca vesicaria*
rosa de montana – *Antigonon leptopus*
rosary bean – *Rhynchosia precatoria, R. pyramidalis*
rosary pea – *Abrus precatorius*
rose – *Rosa*
rose, alpine – *Rosa pendulina*
roseapple – *Eugenia jambos*
roseapple, Indian – *Dillenia indica*
roseapple, Malay – *Eugenia malaccensis*
rose, Arizona – *Rosa arizonica*
rose, Arkansas – *Rosa arkansana*
rose-bay – *Nerium oleander, Rhododendron ponticum*
rose-bay, California – *Rhododendron macrophyllum*
rose, burnet – *Rosa spinosissima*
rose, cabbage – *Rosa centifolia*
rose cactus – *Pereskia grandifolia*
rose campion – *Lychnis coronaria*
rose, chestnut – *Rosa roxburghii*
rose, China – *Rosa chinensis*
rose, Chinese – *Hibiscus rosa-sinensis*
rose, Christmas – *Helleborus niger*
rose, cinnamon – *Rosa cinnamomea*
rose, cliff – *Purshia mexicana*
rose, climbing – *Rosa multiflora*
rose clover – *Trifolium hirtum*
rose, Confederate – *Hibiscus mutabilis*
rose, cotton – *Filago germanica*
rose cress – *Arabis blepharophylla*
rose, damask – *Rosa damascena*
rose, desert – *Adenium obesum, Rosa minutifolia, R. stellata*
rose, dog – *Rosa canina*
rose, elegantine – *Rosa moschata*
rose, evergreen – *Rosa sempervirens*
rose, Fendler – *Rosa woodsii* var *fendleri*
rose, French – *Rosa gallica*
rose-fruited juniper – *Juniperus coahuilensis*
rose gentian – *Sabatia campestris*
rose gentian, squarestem – *Sabatia angularis*
rose geranium – *Pelargonium graveolens*
rose, Guelder – *Viburnum opulus*
rose, Honolulu – *Merremia tuberosa*
rose imperial – *Cochlospermum vitifolium*
rose, Irish – *Rosa hibernica*
rose, June – *Lagerstroemia indica*
rose, Lady Banks' – *Rosa banksiae*
rose, lenten – *Helleborus orientalis*
roselle – *Hibiscus sabdariffa*
rose-mallow – *Hibiscus rosa-sinensis*
rose mallow, desert – *Hibiscus coulteri*
rose-mallow, false – *Modiolastrum scoparium*
rose-mallow, mountain – *Hibiscus bifurcatus, H. cardiophyllus*
rosemary – *Rosmarinus officinalis*
rosemary barberry – *Berberis* x *stenophylla*
rosemary, bog – *Andromeda polifolia*
rosemary-leaved lavender-cotton – *Santolina rosmarinifolia*
rosemary, western marsh – *Limonium californicum*
rose, May – *Rosa cinnamomea*
rose, McCartney – *Rosa bracteata*
rose, miniature – *Rosa chinensis* var *minima, R. minutifolia*
rose moss – *Portulaca grandiflora*
rose, moss – *Portulaca grandiflora, P. pilosa, Rosa muscosa*
rose, multiflora – *Rosa multiflora*
rose, musk – *Rosa moschata*

rose, New Mexico – *Rosa neomexicana*
rose, Nootka – *Rosa nutkana*
rose of Sharon – *Hibiscus syriacus, Hypericum calycinum, Narcissus tazetta, Tulipa sharonensis*
rose of Venezuela – *Brownea grandiceps*
rose of York, white – *Rosa alba*
rose, pale-leaf – *Rosa hypoleuca*
rose-pink – *Sabatia campestris*
rose, plum – *Eugenia jambos*
rose, prickly – *Rosa acicularis*
rose, Provence – *Rosa centifolia, R. gallica*
rose, ramanas – *Rosa rugosa*
rose, red-leaved – *Rosa glauca*
rose, rock – *Cistus, Helianthemum glomeratum, H. nummularium, Tuberaria guttata*
rose, rush – *Helianthemum glomeratum*
rose sage – *Salvia pachyphylla*
rose, sage – *Turnera ulmifolia*
rose, Scotch – *Rosa cinnamomea, R. spinosissima*
rose, shrub – *Rosa polyantha*
rose, small-flowered – *Rosa micrantha*
rose, small-leaved – *Rosa minutifolia*
rose, Standley – *Rosa standleyi*
rose, star – *Rosa stellata*
rose, stinking – *Rosa eglanteria*
rose, sun – *Aptenia cordifolia, Cistus, Helianthemum*
rose, tea – *Rosa indica*
rose, wax – *Pereskia grandiflora*
rose, wild – *Rosa fendleri*
rose, wild white – *Rosa bracteata*
rosewood – *Dalbergia*
rose, wood – *Merremia tuberosa*
rosewood, Arizona – *Vauquelinia californica*
rosewood, Brazilian – *Dalbergia nigra*
rosewood, false – *Vauquelinia karwinskyi*
rosewood, Pacific – *Thespesia populnea*
rose, woods' – *Rosa woodsii*
rosin brush – *Baccharis sarothroides*
rosy cups – *Pluchea odorata*
rosy maidenhair – *Adiantum hispidulum*
rosy periwinkle – *Catharanthus roseus*
rotang palm – *Calamus rotang*
rouge-plant – *Rivina humilis*
rouge-plant, bloodberry – *Rivina humilis*
rough bluegrass – *Poa trivialis*
rough cat's-ear – *Hypochaeris radicata*
rough club-rush – *Scirpus tabernaemontani*
rough corn bedstraw – *Galium tricornutum*
rough lemon – *Citrus jambhiri*
rough poppy – *Papaver hybridum*
rough-stalked bluegrass – *Poa trivialis*
rough tridens – *Tridens elongatus*
round-headed garlic – *Allium sphaerocephalum*
round-white-seeded French haricot – *Phaseolus vulgaris*
rowan – *Sorbus aucuparia*
rowan, hupeh – *Sorbus hupehensis*
royal beardtongue – *Penstemon spectabilis*
royal cedar – *Cedrela mexicana*
royal cross – *Mammillaria karwinskiana* ssp *collinsii*
royal fern – *Osmunda regalis*
royal, herb – *Artemisia abrotanum*
royal palm – *Roystonea regia*
royal palm, Caribbee – *Roystonea oleracea*
royal palm, Cuban – *Roystonea regia*
royal poinciana – *Delonix regia*
royal water-lily – *Victoria regia*
rubber bush – *Hymenoxys richardsonii*
rubber, castilloa – *Castilla elastica*
rubber, ceara – *Manihot glaziovii*
rubber, chilte – *Cnidoscolus elasticus*
rubber, Panama – *Castilla elastica, C. fallax*
rubber plant – *Jatropha spathulata*
rubber plant, Indian – *Ficus elastica*
rubber plant, Mexican – *Parthenium argentatum*
rubber rabbitbrush – *Chrysothamnus nauseosus*
rubber tree – *Castilla elastica, Ficus elastica*
rubber tree, Assam – *Ficus elastica*
rubber tree, hardy – *Eucommia ulmoides*
rubber tree, Mexican – *Castilla elastica*
rubber tree, para – *Hevea brasiliensis*
rubber-weed, Colorado – *Hymenoxys richardsoni*
ruby grass, Natal – *Rhynchelytrum repens*
rue – *Ruta chalepensis, R. graveolens*
rue, Chinese meadow – *Thalictrum dipterocarpum*
rue, goat's – *Galega officinalis*
ruel, longneck – *Ruellia nudiflora*
rue, meadow – *Thalictrum*
rue, Syrian – *Peganum harmala*
rue, wine – *Ruta graveolens*

ruffle-palm, coyor – *Aiphanes acanthophylla*
rum berry – *Myrciaria floribunda*
runner bean, Mexican – *Phaseolus coccineus*
runner bean, scarlet – *Phaseolus coccineus*
running fleabane – *Erigeron nudiflorus*
running mesquite – *Prosopis glandulosa* var *prostrata*
running pine – *Lycopodium clavatum*
rupture wort – *Herniaria glabra, H. hirsuta*
rush – *Juncus*
rush, Baltic – *Juncus balticus*
rush, bog – *Juncus effuses, Schoenus nigricans*
rush, bul – *Cyperus papyrus, Scirpus*
rush, club – *Scirpus*
rush, Dudley's – *Juncus dudleyi*
rush, flowering – *Limnocharis flava*
rush, frog – *Juncus ambiguus*
rush, jointed – *Juncus articulatus*
rush-like thrift – *Armeria girardii*
rush, Mexican – *Juncus mexicanus*
rush milkweed – *Asclepias subulata*
rush-nut – *Cyperus esculentus*
rush pea, Indian – *Hoffmannseggia glauca*
rush-rose – *Helianthemum glomeratum*
rush, scouring – *Equisetum*
rush, slender – *Juncus tenuis*
rush, spike – *Eleocharis*
rush, spiny – *Juncus acutus*
rush, sword-leaved – *Juncus ensifolius*
rush, toad – *Juncus bufonius*
rush, wood – *Luzula*
Russell hybrid lupine – *Lupinus russeli*
Russian ginseng – *Eleutherococcus senticosus*
Russian olive – *Eleagnus angustifolius*
Russian sage – *Perovskia atriplicifolia*
Russian thistle – *Salsola tragus*
russula, black – *Russula nigricans*
russula, blue and yellow – *Russula cyanoxantha*
rust, black stem – *Puccinia graminis*
rusty buddleja – *Buddleja cordata*
rusty foxglove – *Digitalis ferruginea*
rutabaga – *Brassica napus* var *napobrassica, B. rapa*
rye – *Secale cereale*
rye-brome – *Bromus secalinus*
rye, Canadian wild – *Elymus canadensis*
ryegrass, Italian – *Lolium multiflorum*
ryegrass, perennial – *Lolium perenne*
rye smut – *Ustilago carbo*
Rzedowski pine – *Pinus rzedowski*

sabadilla – *Sabadilla officinalis*
saba nut – *Pachira aquatica*
sabi star – *Adenium obesum*
sacahuiste, Texas – *Nolina texana*
sacaton – *Sporobolus wrightii*
sacaton, alkali – *Sporobolus airoides*
sacred datura – *Datura metel, D. wrightii*
sacred fig – *Ficus religiosa*
sacred fir – *Abies religiosa*
sacred flower of the Incas – *Cantua buxifolia*
sacred lotus – *Nelumbo nucifera*
safflower – *Carthamus tinctorius*
saffron – *Crocus sativus*
saffron, American – *Carthamnus tinctorius*
saffron, false – *Carthamnus tinctorius*
saffron, meadow – *Colchicum autumnale, C. speciosum*
saffron milk cap – *Lactarius deliciosus*
saffron plum – *Sideroxylon spiniflora*
saffron-spike – *Aphelandra squarrosa*
sage – *Ambrosia, Artemisia, Atriplex, Salvia*
sage, annual bur – *Ambrosia acanthicarpa*
sage, Arizona – *Salvia arizonica*
sage, Asian red – *Salvia miltiorrhiza*
sage, bee – *Hyptis emoryi, Salvia apiana*
sage, Bethlehem – *Pulmonaria saccharata*
sage, black – *Salvia mellifera*
sage, blue – *Salvia azurea, S. ballotaeflora, S. farinacea*
sage, Brandegee's – *Salvia brandegei*
sagebrush – *Atriplex canescens, A. polycarpa*
sagebrush, basin – *Artemisia tridentata*
sagebrush, California – *Artemisia californica*
sagebrush, big – *Artemisia tridentata*
sagebrush, sand – *Artemisia filifolia*
sage, bur – *Ambrosia ambrosioides*
sage, California white – *Salvia apiana*
sage, cardinal – *Salvia coccinea, S. fulgens, S. splendens*
sage, clary – *Salvia sclarea*
sage, common – *Salvia officinalis*
sage, creeping – *Salvia sonomensis*
sage, culinary – *Salvia officinalis*
sage, darcy – *Salvia darcyi*
sage, desert – *Salvia eremostachya*
sage, garden – *Salvia officinalis*

sage, gentian – *Salvia patens*
sage, germander – *Salvia chamaedryoides*
sage, Jerusalem – *Phlomis fruticosa*
sageleaf rock-rose – *Cistus salvifolius*
sage, mealy – *Salvia farinacea*
sage, mealycup – *Salvia farinacea*
sage, Mediterranean – *Salvia aethiopis*
sage, Mexican – *Salvia mexicana*
sage, Mexican bush – *Salvia leucantha*
sage, Munz's – *Salvia munzii*
sage, painted – *Salvia viridis*
sage, pineapple – *Salvia elegans*
sage, pitcher – *Lepechinia hastate, Salvia azurea*
sage, prairie – *Salvia azurea*
sage, purple – *Leucophyllum frutescens*
sage rose – *Turnera ulmifolia*
sage, rose – *Salvia pachyphylla*
sage, Russian – *Perovskia atriplicifolia*
sage, sand – *Artemisia filifolia*
sage, San Diego bur – *Ambrosia chenopodifolia*
sage, Santa Rosa – *Salvia eremostachya*
sage, scarlet – *Salvia coccinea, S. splendens, Stachys coccinea*
sage, silver – *Artemisia frigida, Salvia argentea*
sage, Sonoma – *Salvia sonomensis*
sage, superb – *Salvia* x *supurba*
sage, Texas – *Leucophyllum frutescens*
sage, tropical – *Salvia coccinea*
sage, velvet – *Salvia leucantha*
sage, white – *Artemisia ludoviciana* ssp *albula, Salvia apiana*
sage, white bur – *Ambrosia dumosa*
sage, wild – *Lantana camara, Salvia nemorosa, S. sylvestris*
sage, wood – *Teucrium scorodonia*
sage, woolly – *Phlomis*
sagewort, boreal – *Artemisia norvegica* ssp *saxatilis*
sagewort, mountain – *Artemisia ludoviciana* ssp *incompta*
sago palm – *Cycas revoluta, Metroxylon*
saguaro – *Carnegiea gigantea*
saguaro cactus – *Carnegiea gigantea*
Sahara mustard – *Brassica tournefortii*
sailors, drunken – *Quisqualis indica*
Saint Augustine grass – *Stenotaphrum secundatum*
salad burnet – *Sanguisorba minor* ssp *muricata*
salad, corn – *Valerianella*
salad-rocket – *Eruca vesicaria* ssp *sativa*
salad, rocket – *Eruca vesicaria*
salak palm – *Salacca edulis*
salal – *Gaultheria shallon*
sallow, gray – *Salix cinerea*
Sally, blooming – *Epilobium angustifolium*
salmonberry – *Rubus spectabilis*
salmon gum – *Eucalyptus salmonophloia*
salsify – *Tragopogon porrifolius, T. pratensis*
salsify, Spanish – *Scorzonera hispanica*
saltbush – *Atriplex*
saltbush, Australian – *Atriplex semibaccata*
saltbush, big – *Atriplex lentiformis* ssp *lentiformis*
saltbush, California – *Atriplex californica*
saltbush, coast – *Atriplex barclayana*
saltbush, desert – *Atriplex polycarpa*
saltbush, fourwing – *Atriplex canescens, A. linearis*
saltbush, Mediterranean – *Atriplex halimus*
saltbush, narrow-wing – *Atriplex linearis*
salt cedar – *Tamarix*
saltgrass – *Distichlis spicata*
salt heliotrope – *Heliotropium curassavicum*
salt marsh fleabane – *Pluchea odorata*
salt-meadow cord grass – *Spartina patens*
salt pennywort – *Hydrocotyle bonariensis*
salt-water cord grass – *Spartina alterniflora*
salt-water palmetto – *Thrinax radiata*
saltwort – *Batis maritima, Salsola vermiculata*
saltwort, soda – *Salsola soda*
salty sand-spurrey – *Spergularia bocconii*
salvia – *Salvia*
salvia, cardinal – *Salvia gesneriaeflora*
salvia, red bush – *Salvia microphylla*
salvia, shrub – *Salvia sessei*
Samos woolly sage – *Phlomis samia*
samphire – *Crithmum maritimum*
Sampson, black – *Echinacea purpurea*
samwood – *Cordia alliodora*
sandal beadtree – *Adenanthera pavonina*
sandalwood – *Santalum album*
sandalwood, red – *Pterocarpus santalinus*
sandarac – *Tetraclinis articulata*
sandbar willow – *Salix exigua*

sand bell – *Nama hispidum*
sand bluestem – *Andropogon hallii*
sandbox tree – *Hura polyandra*
sandbur – *Ambrosia acanthicarpa, Cenchrus ciliatus*
sandbur, coast – *Cenchrus incertus*
sandbur, giant – *Cenchrus palmeri*
sandbur, mat – *Cenchrus longispinus*
sandbur, southern – *Cenchrus echinatus*
sand cherry – *Prunus pumila*
sand couch-grass – *Elytrigia juncea*
sand dollar cactus – *Astrophytum asterias*
sand dropseed – *Sporobolus cryptandrus*
sanders wood, red – *Pterocarpus santalinus*
sand food – *Pholisma sonorae*
sandgrass, purple – *Triplasis purpurea*
San Diego bedstraw – *Galium nuttallii*
San Diego bur-sage – *Ambrosia chenopodiifolia*
San Diego county viguiera – *Viguiera laciniata*
San Diego sunflower – *Viguiera laciniata, V. tomentosa*
sandmat, whitemargin – *Chamaesyce albomarginata*
sandpaper oak – *Quercus pungens*
sandpaper tree – *Curatella americana*
sandpaper vine – *Petrea volubilis*
sand pink – *Dianthus arenarius*
sand plantain – *Plantago indica*
sand sage – *Artemisia filifolia*
sand sagebrush – *Artemisia filifolia*
sand sedge – *Carex arenaria*
sand-spurrey – *Spergularia bocconii, S. marina, S. rubra*
sand-spurrey, salty – *Spergularia marina*
sand strawberry – *Fragaria chiloensis*
sand verbena – *Abronia*
sand verbena, desert – *Abronia villosa*
sand verbena, sweet – *Abronia fragrans*
sandwort – *Arenaria*
sandwort, bog – *Minuartia stricta*
sandwort, Corsican – *Arenaria balearica*
sandwort, mountain – *Arenaria montana, Minuartia rubella*
sandwort, spring – *Sagina subulata*
sanfoin – *Onobrychis viciifolia*
sanicle, American – *Sanicula marilandica*
sanicle, wood – *Sanicula europaea*
San Jose hesper palm – *Erythea brandegeei*
San Miguel palm – *Calyptrogyne ghiesbreghtiana*
San Miguel savory – *Satureja chandleri*
San Pedro cactus – *Echinopsis pachanoi*
Santa Barbara daisy – *Erigeron karvinskianus*
Santa Catalina Island desert-thorn – *Lycium brevipes* var *hassei*
Santa Fe phlox – *Phlox nana*
Santa Lucia fir – *Abies bracteata*
Santa Maria feverfew – *Parthenium hysterophorus*
Santa Rita acacia – *Acacia millefolia*
Santa Rosa blackberry – *Rubus ulmifolius*
Santa Rosa oak – *Quercus sideroxyla*
Santa Rosa sage – *Salvia eremostachya*
sanvitalia – *Sanvitalia abertii*
sanvitalia, Abert's – *Sanvitalia abertii*
sapodilla – *Achras zapota*
sapodilla, wild – *Morisonia americana*
sapote – *Lucuma mammosa*
sapote, black – *Diospyros digyna, D. ebenaster*
sapote, green – *Lucuma viride, Pouteria viridis*
sapote, mamee – *Lucuma mammosa*
sapote, mamey – *Lucuma mammosa*
sapote, South American – *Quararibea cordata*
sapote, white – *Casimiroa edulis*
sapote, yellow – *Lucuma salicifolia, Pouteria campechiana*
sappan wood tree – *Caesalpinia sappan*
sap, pine – *Monotropa hypopitys, M. uniflora*
sapucaia nut – *Lecythis elliptica, L. zabucajo*
sarepta mustard – *Brassica juncea* var *sareptana*
sargassum – *Sargassum herporhizum*
Sargent cypress – *Cupressus sargentii*
sarsaparilla – *Smilax*
sasparilla – *Smilax medica, S. officinalis*
sassafras – *Sassafras albidum*
satin leaf – *Chrysophyllum oliviforme*
saucer magnolia – *Magnolia soulangiana*
sausage tree – *Hymenaea courbaril*
sausage tree, African – *Kigelia africana*
savannah grass – *Axonopus compressus, Paspalum compressum*
savin – *Juniperus excelsa*
savin juniper – *Juniperus sabina*
savory – *Satureja*
savory, San Miguel – *Satureja chandleri*
savory, summer – *Satureja hortensis*

savory, winter – *Satureja montana*
savoy cabbage – *Brassica oleracea* var *sabauda*
sawah flowering rush – *Limnocharis flava*
sawah lettuce – *Limnocharis flava*
sawara false cypress – *Chamaecyparis pisifera*
saw-grass – *Cladium californicum, Mariscus jamaicensis, Nolina microcarpa, Scleria bracteata*
sawleaf zelkova – *Zelkova serrata*
saw palmetto – *Serenoa repens*
saw-toothed spurge – *Euphorbia serrata*
sawtooth lovegrass – *Eragrostis superba*
sawtooth oak – *Quercus acutissima*
sawwort – *Serratula tinctoria*
saxifrage – *Saxifraga crassifolia*
saxifrage, brook – *Saxifraga rivularis*
saxifrage, burnet – *Pimpinella major*
saxifrage, golden – *Chrysosplenium glechomifolium*
saxifrage, Irish – *Saxifraga decipiens*
saxifrage, tufted – *Saxifraga cespitosa*
saya – *Amoreuxia palmatifida*
scabious, sweet – *Scabiosa atropurpurea*
scabious, white-headed giant – *Cephalaria leucantha*
scabious, yellowish – *Scabiosa ochroleuca*
scale fern – *Ceterach officinarum*
scale, lens – *Atriplex lentiformis*
scale, wedge – *Sphenopholis obtusata*
scale, wing – *Atriplex canescens*
scallion – *Allium cepa*
scallion, Texas – *Allium cepa*
scalloped phacelia – *Phacelia crenulata*
scaly agaric – *Tricholoma vaccinum*
scaly cloak fern – *Astrolepis cochisensis*
scammony – *Convolvulus scammonia*
scammony, Mexican – *Ipomoea orizabensis*
Scandanavian sedge – *Carex norvegica*
scarlet – *Quercus coccifera*
scarlet bee blossom – *Gaura coccinea*
scarlet bouvardia – *Bouvardia ternifolia*
scarlet bugler – *Penstemon centranthifolius*
scarlet-bush – *Hamelia erecta*
scarlet cordia – *Cordia sebestena*
scarlet flax – *Linum grandiflorum*
scarlet globemallow – *Sphaeralcea coccinea*
scarlet hedge nettle – *Stachys coccinea*
scarlet lobelia – *Lobelia cardinalis* var *pseudosplendens*
scarlet maple – *Acer rubrum*
scarlet martagon – *Lilium chalcedonicum*
scarlet monkey flower – *Mimulus cardinalis*
scarlet oak – *Quercus coccinea*
scarlet paintbrush – *Castilleja indivisa*
scarlet pimpernel – *Anagallis arvensis*
scarlet plume – *Euphorbia fulgens*
scarlet runner bean – *Phaseolus coccineus*
scarlet sage – *Salvia coccinea, S. splendens, Stachys coccinea*
scarlet spiderling – *Boerhavia coccinea*
scarlet sumac – *Rhus glabra*
scarlet windflower – *Anemone* x *fulgens*
scented croton – *Croton ciliato-glandulosus*
scented geranium – *Pelargonium odoratissimum*
scented stock, evening – *Matthiola longipetala*
schefflera – *Schefflera actinophylla*
scheuchzeria, American – *Scheuchzeria palustris*
schizandra – *Schizandra chinensis*
scholar tree, Chinese – *Sophora japonica*
schomburgkia, trumpeter's – *Lealia tibicinis*
Schott caliandra – *Zapoteca formosa*
Schott's dwarf cholla – *Grusonia schottii*
Schott's yucca – *Yucca schottii*
scoke – *Phytolacca americana*
scorpion tail – *Heliotropium angiospermum*
Scotch bonnet – *Capsicum chinense*
Scotch broom – *Cytisus scoparius*
Scotch elm – *Ulmus glabra*
Scotch heath – *Erica cinera*
Scotch heather – *Calluna vulgaris*
Scotch laburnum – *Laburnum alpinum*
Scotch pine – *Pinus sylvestris*
Scotch rose – *Rosa cinnamomea, R. spinosissima*
Scotch thistle – *Onopordum acanthium*
Scots pine – *Pinus sylvestris*
Scottish crocus – *Crocus biflorus*
Scottish moss – *Sagina subulata*
Scottish thistle – *Onopordum acanthium*
scouring rush – *Equisetum*
scouring rush, common – *Equisetum hyemale* ssp *affine*
scouring rush, smooth – *Equisetum laevigatum*
scratchbush, flameberry – *Urera caracasana*

screw bean – *Prosopis pubescens*
screw pine – *Pandanus tectorius, P. utilis*
screw pine, seashore – *Pandanus tectorius*
scrub oak – *Quercus berberidifolia, Q. pungens*
scrub oak, coastal – *Quercus dumosa*
scrub oak, desert – *Quercus turbinella*
scrub oak, Nuttall's – *Quercus dumosa*
scurf-pea – *Psoralidium lanceolatum*
scurf-pea, lemon – *Psoralea lanceolata*
scurvy grass – *Barbarea verna, Cochlearia officinalis*
scurvy pea – *Psoralidium lanceolatum*
sea-bean – *Entada phaseoloides*
sea bindweed – *Calystegia soldanella*
sea-blite – *Suaeda*
sea-blite, horned – *Suaeda calceoliformis*
sea-blite, Mojave – *Suaeda moquinii*
sea-blite, woolly – *Suadea taxifolia*
sea buckthorn – *Hippophae rhamnoides*
sea club-rush – *Scirpus maritimus*
sea-daffodil – *Pancratium maritimum*
sea-dahlia – *Coreopsis maritima*
sea-grape – *Ipomoea pes-caprae, Coccoloba uvifera*
sea hard-grass – *Parapholis strigosa*
sea hog's fennel – *Peucedanum officinale*
sea holly – *Eryngium amethystinum, E. maritimum*
sea holly, alpine – *Eryngium alpinum*
sea holly, false – *Eryngium planum*
cotton, sea island – *Gossypium barbadense*
sea kale – *Crambe maritima*
Sea-of-Cortez kelp – *Sargassum*
sea-lavender – *Limonium californicum, L. sinuatum*
seal, golden – *Hydrastis canadensis*
sea lily – *Pancratium maritimum*
seal, Solomon's – *Polygonatum biflorum, P. odoratum*
sea mallow – *Lavatera maritima*
sea milkwort – *Glaux maritima*
sea onion – *Urginea maritima*
sea pea – *Lathyrus japonicus*
sea-pink – *Armeria juniperifolia, A. maritima* ssp *californica*
sea plantain – *Plantago maritima*
sea poppy – *Glaucium flavum*
sea-purslane – *Atriplex halimus, Sesuvium portulacastrum*
sea-purslane, western – *Sesuvium verrucosum*
sea rocket – *Cakile maritima*
seashore paspalum – *Paspalum virgatum*
seaside arrow-grass – *Triglochin maritima*
seaside daisy – *Erigeron glaucus*
seaside plum – *Ximenia americana*
sea spurrey, greater – *Spergularia media*
sea spurrey, Greek – *Spergularia bocconii*
sea tomato – *Rosa rugosa*
sea urchin cactus – *Astrophytum asterias*
seaweed – *Fucus vesiculosus, Laminaria, Macrocystis pyrifera, Sargassum*
sebesten plum – *Cordia myxa*
sedge – *Carex*
sedge, black – *Schoenus nigricans*
sedge, bladder – *Carex vesicaria*
sedge, blue – *Carex flacca*
sedge, bluestem broom – *Andropogon virginicus*
sedge, bog – *Carex limosa*
sedge, bottle – *Carex utriculata*
sedge, brownish – *Carex buxbaumii*
sedge, club – *Carex buxbaumii*
sedge, flat – *Cyperus hermaphroditus, C. laevigatus*
sedge, fox – *Carex vulpinoidea*
sedge, hare's foot – *Carex ovalis*
sedge, Japanese variegated – *Carex morrowii* var *expallida*
sedge, late-flowering – *Carex viridula*
sedge, leather leaf – *Carex buchananii*
sedge, marsh – *Cyperus squarrosus*
sedge, mud – *Carex limosa*
sedge, nut – *Cyperus*
sedge, purple nut – *Cyperus rotundus*
sedge, sand – *Carex arenaria*
sedge, Scandanavian – *Carex norvegica*
sedge, slender – *Carex lasiocarpa*
sedge, tufted – *Carex elata*
sedge, tussock – *Carex diandra*
sedge, water – *Carex aquatilis*
sedge whitish – *Carex canescens*
sedum, goldmoss – *Sedum acre*
seed-ticks – *Daucus pusillus*
seepweed – *Suaeda*
seepweed, bush – *Suaeda moquinii*
seep-willow – *Baccharis salicifolia*
sego lily – *Calochortus bruneaunis*
self-heal – *Prunella vulgaris*

self-heal, large – *Prunella grandiflora*
semperflorens begonia – *Begonia semperflorens*
Seneca snakeroot – *Polygala seneca*
senega – *Polygala seneca*
Senegal date palm – *Phoenix reclinata*
senita – *Pachycereus schottii*
senna – *Cassia, Senna*
senna, bladder – *Colutea arborescens*
senna, coffee – *Senna occidentalis*
senna, desert – *Senna covesii*
senna, four-leaf – *Senna oxyphylla*
senna, Indian – *Senna angustifolia*
senna, ringworm – *Senna alata*
senna, shrubby – *Senna wislizeni*
senna, Siamese – *Senna siamea*
senna, twin – *Senna pallida*
senna, two-leaved – *Senna bahuinoides*
senna, wild – *Senna laevigata*
senna, woolly – *Senna multiglandulosa*
sensitive plant – *Mimosa pudica*
sentry palm – *Howea belmoreana*
sequoia, giant – *Sequoiadendron giganteum*
serradella, common – *Ornithopus sativus*
serrano chile – *Capsicum annuum*
service-berry – *Amelanchier*
service-berry, Allegheny – *Amelanchier laevis*
service tree, true – *Sorbus domestica*
sesame – *Sesamum indicum*
seven stars – *Ariocarpus retusus*
Seville orange – *Citrus aurantiacum* var *bergamia*
shad bush – *Amelanchier alnifolia, A. canadensis*
shaddock – *Citrus grandis*
shad-scale – *Atriplex canescens*
shagbark hickory – *Carya ovata*
shagbark pine – *Pinus herrerae*
shaggy-barked manzanita – *Arctostaphylos tomentosa*
shaggy portulaca – *Portulaca halimoides*
shaggy soldier – *Galinsoga quadriradiata*
shallot – *Allium cepa* var *ascalonicum*
shamel ash – *Fraxinus uhdei*
shame plant – *Mimosa pudica*
shame weed – *Acacia angustissima*
shamrock – *Oxalis acetosella, Trifolium dubium*
shank, red – *Adenostoma sparsifolium, Polygonum persicaria*
Sharon, rose of – *Hibiscus syriacus, Hypericum calycinum, Narcissus tazetta, Tulipa sharonensis*
Sharon tulip – *Tulipa sharonensis*
Shasta daisy – *Chrysanthemum maximum*
Shasta red fir – *Abies magnifica* var *shastensis*
shavegrass – *Equisetum*
shaving brush palm – *Rhopalostylis sapida*
shaving brush tree – *Bombax ellipticum, Pachira aquatica, Pseudobombax palmeri*
Shaw's agave – *Agave shawii*
shea-nut tree – *Butyrospermum parkii*
sheathed cholla – *Cylindropuntia tunicata*
sheep fescue – *Festuca ovina, F. trachyphylla*
sheep sorrel – *Rumex acetosella*
sheep's tail – *Sedum morganianum*
Sheldon's pincushion – *Mammillaria sheldonii*
shell-bark hickory, big – *Carya laciniosa*
shell flower – *Alpina zerumbet, Moluccella laevis*
shell flower, Mexican – *Tigridia pavonia*
shell ginger – *Alpina zerumbet*
shell seed – *Cochlospermum vitifolium*
she-oak, river – *Casuarina cunninghamiana*
shepherd's cress, lesser – *Teesdalia coronopifolia*
shepherd's purse – *Capsella bursa-pastoris*
shield fern – *Dryopteris filix-mas*
shiitake mushroom – *Lentinus edodes*
shindagger – *Agave schottii*
shining pondweed – *Potamogeton illinoensis*
shining sumac – *Rhus copallina*
shin oak – *Quercus vaseyana*
shin oak, Mohr's – *Quercus mohriana*
Shirley poppy – *Papaver rhoeas*
shiso – *Perilla frutescens*
shittah tree – *Acacia seyal*
shittimwood – *Sideroxylon lanuginosum*
shoe button spiraea – *Spiraea prunifolia*
shoemaker's bark – *Byrsonima coriacea*
shooting star – *Dodecatheon clevelandii*
shooting star, eastern – *Dodecatheon meadia*
shoregrass – *Monanthochloë littoralis*
shore pine – *Pinus contorta* ssp *contorta*
shot, Indian – *Canna coccinea, C. indica*
show, Arizona yellow – *Amoreuxia palmatifida*
shower, coral – *Cassia grandis*

shower, golden – *Cassia fistula, Pyrostegia venusta*
shower, pink – *Cassia grandis*
shower tree – *Senna italica*
shower, yellow – *Cassia fistula*
showy milkweed – *Asclepias speciosa*
showy mullein – *Verbascum speciosum*
showy prairie gentian – *Eustoma grandiflorum*
shrimp plant – *Justica brandegeana*
shrub althea – *Hibiscus syriacus*
shrubby alkali aster – *Machaeranthera carnosa*
shrubby blue sage – *Salvia ballotaeflora*
shrubby cinquefoil – *Potentilla fruticosa*
shrubby dalea – *Dalea frutescens*
shrubby hare's ear – *Bupleurum fruticosum*
shrubby milkwort – *Polygala chamaebuxus*
shrubby oak – *Quercus fruticosa*
shrubby senna – *Senna wislizeni*
shrub live oak – *Quercus turbinella*
shrub rose – *Rosa polyantha*
shrub salvia – *Salvia sessei*
shrub verbena – *Lantana camara*
Siamese senna – *Senna siamea*
Sibbald's potentilla – *Sibbaldia procumbens*
Siberian elm – *Ulmus pumila*
Siberian fir – *Abies sibirica*
Siberian ginseng – *Eleutherococcus senticosus*
Siberian iris – *Iris sibirica*
Siberian kale – *Brassica napus* var *pabularia, B. oleracea* var *ramosa*
Siberian larch – *Larix sibirica*
Siberian peashrub – *Caragana arborescens*
Siberian pea tree – *Caragana arborescens*
Siberian squill – *Scilla sibirica*
Siberian wallflower – *Erysimum hieraciifolium*
Sicilian star-thistle – *Centaurea sulphurea*
Sicilian sumac – *Rhus coriaria*
sickle grass – *Parapholis incurva*
side-oats grama – *Bouteloua curtipendula*
Sierra Madre acacia – *Acacia pennatula*
Sierra primrose – *Primula suffrutescens*
signet marigold – *Tagetes tenuifolia*
silene, Mexican – *Silene laciniata* ssp *major*
silk cotton plant, white – *Cochlospermum religiosum*
silk cotton tree – *Pseudobombax palmeri, Ceiba pentandra*
silk cotton tree, red – *Bombax ceiba*
silk flower – *Abelmoschus moschatus*
silk grass – *Bromelia magdalenae*
silk oak – *Grevillea robusta*
silk tassel bush – *Garrya*
silk tassel bush, coast – *Garrya eliptica*
silk tassel, coast – *Garrya eliptica*
silk tassel, longleaf – *Garrya longifolia*
silk tassel, yellow – *Garrya flavescens*
silk tree – *Albizzia julibrissin*
silk tree, floss – *Ceiba speciosa*
silkworm mulberry – *Morus alba*
silver beard grass – *Bothriochloa saccharoides*
silver bell, mountain – *Halesia monticola*
silverberry – *Eleagnus pungens*
silver birch – *Betula verrucosa*
silver cranesbill – *Geranium argenteum*
silver-dollar cactus – *Opuntia chlorotica*
silver ear – *Tremella fuciformis*
silver European hairgrass – *Aira caryophylea*
silver fir – *Abies amabilis*
silver fir, European – *Abies pectinata*
silver fir, Pacific – *Abies amabilis*
silver grass, Japanese – *Miscanthus sinensis*
silver king artemisia – *Artemisia ludoviciana*
silver lace vine – *Fallopia baldschuanica*
silverleaf, Big Bend – *Leucophyllum minus*
silverleaf nettle – *Solanum eleagnifolium*
silverleaf nightshade – *Solanum eleagnifolium*
silverleaf oak – *Quercus arizonica, Q. hypoleucoides*
silverleaf, spreading – *Dyschoriste decumbens*
silverleaf, Texas – *Leucophyllum frutescens*
silver lime, weeping – *Tilia* x *petiolaris*
silver linden – *Tilia tomentosa*
silver maple – *Acer saccharinum*
silver poplar – *Populus alba*
silver sage – *Artemisia frigida, Salvia argentea*
silver speedwell – *Veronica spicata* ssp *incana*
silver spreader – *Artemisia caucasica*
silver thatch palm – *Coccothrinax argentata*
silver vase – *Aechmea fasciata*
silver wattle – *Acacia dealbata*
silverweed – *Potentilla anserina*
silverweed, feather – *Potentilla anserina*
silver willow – *Salix exilifolia*
silver wormwood – *Artemisia ludoviciana*
silvery yarrow – *Achillea clavennae*
Sinai manna – *Alhagi pseudalhagi*
single late tulip – *Tulipa gesneriana*
single-leaf ash – *Fraxinus anomala*

singletary pea – *Lathyrus hirsutus*
siris tree – *Albizzia lebbeck*
sisal – *Agave fourcroydes, A. sisalana*
sisal agave – *Agave sisalana*
sisal hemp – *Agave sisalana*
sisal, Mexican – *Agave fourcroydes*
Sitka alder – *Alnus viridis*
Sitka spruce – *Picea sitchensis*
six-weeks grama – *Bouteloua barbata*
six-weeks three-awn – *Aristida adscensionis*
skeleton plant – *Lygodesmia juncea*
skeleton weed – *Chondrilla juncea, Stephanomeria pauciflora*
Skinner's cattleya – *Cattleya skinneri*
Skinner's maxillaria – *Lycaste virginalis*
skirret – *Sium sisarum*
skullcap – *Scutellaria*
skullcap, alpine – *Scutellaria alpina*
skullcap, blue – *Scutellaria lateriflora*
skullcap, mad-dog – *Scutellaria lateriflora*
skullcap, marsh – *Scutellaria galericulata*
skunkbrush – *Rhus trilobata*
skunkbush – *Ptelea trifoliata, Rhus trilobata*
skunk cabbage – *Symplocarpus foetidus*
skyduster – *Washingtonia robusta*
sky flower – *Commelina coelestis, Duranta erecta, Thunbergia grandiflora*
sky flower, Yucatan – *Duranta erecta*
slash pine – *Pinus caribaea*
sleeping hibiscus – *Malvaviscus arboreus*
sleepy grass – *Stipa robusta*
slimwood – *Fouquieria splendens*
slipper, fairy – *Calypso bulbosa*
slipper, lady's – *Cypripedium, Paphiopedilum, Pedilanthus macrocarpus*
slipper plant – *Pedilanthus macrocarpus*
slippery elm – *Ulmus fulva*
slippery elm, California – *Fremontodendron californicum*
slippery jack – *Suillus luteus*
sloe – *Prunus americana, P. spinosa*
smartweed – *Polygonum*
smartweed, water – *Polygonum amphibium, P. punctatum*
smartweed, willowleaf – *Polygonum lapathifolium*
smilax asparagus – *Asparagus asparagoides*
smilax, florist's – *Asparagus medeoloides*
smilo grass – *Piptatherum miliaceum*
Smith's pepperwort – *Lepidium heterophyllum*
smock, lady's – *Cardamine pratensis*
smoke thorn – *Psorothamnus spinosus*
smoke tree – *Cotinus coggygria, Psorothamnus spinosus*
smooth Arizona cypress – *Cupressus arizonica* ssp *glabra*
smooth aster – *Aster laevis*
smooth brome – *Bromus inermis*
smooth cat's-ear – *Hypochaeris glabra*
smooth finger grass – *Digitaria ischaemum*
smooth goldenrod – *Solidago gigantea*
smooth hydrangea – *Hydrangea arborescens*
smooth-leaved elm – *Ulmus minor*
smooth-leaved panicgrass – *Panicum schinzii*
smooth-lipped chysis – *Chysis bractescens*
smooth madrone – *Arbutus xalapensis*
smooth peach – *Prunus persica* ssp *nucipersica*
smooth scouring rush – *Equisetum laevigatum*
smooth-shell macadamia – *Macadamia integrifolia*
smooth sumac – *Rhus glabra*
smother weed – *Bassia hyssopifolia*
smut, barley – *Ustilago hordei*
smut, corn – *Ustilago maydis*
smutgrass – *Sporobolus indicus*
smut, oat – *Ustilago carbo*
smut, rye – *Ustilago carbo*
smut, wheat – *Tilletia tritici, Ustilago tritici*
snail flower – *Vigna caracalla*
snailseed vine – *Coculus diversifolius*
snake broom – *Gutierrezia sarothrae*
snake cactus – *Echinocereus pensilis, Peniocereus serpentinus*
snake cholla – *Cylindropuntia californica*
snake cucumber – *Cucumis melo* var *flexuosus*
snake gourd, Indian – *Trichosanthes cucumerina* var *anguina*
snake herb – *Dyschoriste decumbens*
snake melon – *Cucumis melo* var *flexuosus*
snake plant – *Sansevieria trifasciata*
snake rod – *Polygonum bistorta*
snakeroot, black – *Cimicifuga racemosa, Sanicula marilandica*
snakeroot, button – *Eryngium lemmoni, Liatris punctata, L. spicata*
snakeroot, Indian – *Rauwolfia serpentina*
snakeroot, Seneca – *Polygala seneca*
snakeroot, Virginia – *Aristolochia serpentaria*

snakeshead – *Fritillaria meleagris*
snakeweed, broom – *Gutierrezia sarothrae*
snakewood – *Colubrina californica, C. ferruginosa*
snapdragon – *Antirrhinum majus*
snapdragon, baby – *Linaria maroccana*
snapdragon, climbing – *Asarina barclaiana*
snapdragon vine – *Asarina antirrhiniflora*
sneezeweed – *Dugaldia hoopesii, Helenium*
sneezeweed, orange – *Dugaldia hoopesii*
sneezewort – *Achillea ptarmica*
snoutbean, least – *Rhynchosia minima*
snout, weasel's – *Antirrhinum orontium*
snowball bush – *Viburnum*
snowball, Japanese – *Viburnum plicatum*
snowbell, fragrant – *Styrax obassia*
snowbell, Japanese – *Styrax japonicus*
snowberry – *Chiococca alba, Symphoricarpos*
snowberry, common – *Symphoricarpos albus*
snowdrop anemone – *Anemone sylvestris*
snowdrop bush – *Styrax*
snowdrop, common – *Galanthus nivalis*
snowdrop, giant – *Galanthus elwesii*
snowdrops, white – *Galanthus nivalis*
snowdrop tree, Japanese – *Styrax japonicus*
snowflake, autumn – *Leucojum autumnale*
snowflake, spring – *Leucojum vernum*
snowflake, summer – *Leucojum aestivum*
snow gum – *Eucalyptus pauciflora* ssp *niphophila*
snow-in-summer – *Cerastium tomentosum*
snow pea – *Pisum sativum* var *macrocarpon*
snow pear – *Pyrus nivalis*
snow willow – *Salix reticulata* var *nivalis*
snowy mespilus – *Amelanchier ovalis*
snuffbox bean – *Entada polystachya*
soapbark tree – *Quillaja saponaria*
soapberry – *Sapindus saponaria*
soapberry, western – *Sapindus drummondii, S. saponaria*
soapbush – *Guaiacum angustifolium*
soap, Mayan – *Solanum nudum*
soap plant – *Chlorogalum*
soaproot – *Chlorogalum*
soaptree yucca – *Yucca elata*
soapweed – *Yucca elata*
soapwort – *Saponaria officinalis*
soapwort, dwarfish – *Saponaria pumilio*
soapwort, red – *Saponaria ocymoides*
soda saltwort – *Salsola soda*
Sodom apple – *Calotropis procera*
Sodom, apple of – *Solanum mammosum, S. sodomaeum*
Sodom, vine of – *Solanum sodomeum*
soft-bearded oatgrass – *Ventenata dubia*
soft brome – *Bromus hordeaceus*
softly, tread – *Cnidoscolus angustidens, C. urens*
soft maize – *Zea mays* var *amylacea*
soft-shelled almond – *Prunus dulcis* var *fragilis*
soft storksbill – *Erodium malacoides*
soft-tip yucca – *Yucca gloriosa*
soldier, gallant – *Galinsoga parviflora*
soldier, shaggy – *Galinsoga quadriradiata*
solitaire palm – *Ptychosperma elegans*
Solomon's seal – *Polygonatum biflorum, P. odoratum*
song of India – *Dracaena reflexa*
Sonoma sage – *Salvia sonomensis*
Sonora croton – *Croton sonorae*
Sonoran ebony – *Havardia sonorae*
Sonoran hesper palm – *Erythea aculeata*
Sonoran jumping cholla – *Cylindropuntia fulgida* var *mamillata*
Sonoran maiden fern – *Thelypteris puberula* var *sonorensis*
Sonoran palmetto – *Sabal uresana*
Sonoran palo verde – *Cercidium praecox*
Sonoran persimmon – *Diospyros sonorae*
Sonoran tree-beargrass – *Nolina matapensis*
Sonoran tree-limberbush – *Jatropha cordata*
sonusco cocoa – *Theobroma angustifolium*
sorghum – *Sorgum bicolor, S. vulgare*
sorghum, black – *Sorghum almum*
sorrel, creeping wood – *Oxalis corniculata*
sorrel, French – *Rumex scutatus*
sorrel, garden – *Rumex acetosa*
sorrel, Jamaica – *Hibiscus sabdariffa*
sorrel, mountain – *Oxyria digyna*
sorrel, red – *Hibiscus sabdariffa*
sorrel, sheep – *Rumex acetosella*
sorrel, switch – *Dodonaea viscosa*
sorrel tree – *Oxydendrum arboreum*
sorrel, wood – *Oxalis acetosella, O. albicans, O. latifolia*
sorrel, yellow wood – *Oxalis dillenii*
sotol – *Dasylirion wheeleri*

souari nut – *Caryocar amygdaliferum*
sour cane – *Costus spicatus*
sour cherry – *Prunus*
sourclover – *Melilotus indica*
sourgrass – *Digitaria insularis, Oxalis corniculata, Paspalum conjugatum*
sourgrass, West Indian – *Paspalum conjugatum*
sour gum – *Nyssa sylvatica*
sour orange – *Citrus aurantiacum* var *bergamia*
sour pitaya – *Stenocereus gummosus*
soursop – *Annona muricata, A. reticulata*
soursop, wild – *Rollinia mucosa*
sourwood – *Oxydendrum arboreum*
South American sapote – *Quararibea cordata*
southern ball thistle – *Echinops exaltatus*
southern balsam pear – *Momordica balsamina*
southern bayberry – *Myrica cerifera*
southern beech – *Nothofagus antarctica*
southern cattail – *Typha domingensis*
southern daisy – *Erigeron speciosus*
southern heath – *Erica australis*
southern live oak – *Quercus virginiana*
southern maidenhair – *Adiantum capillus-veneris*
southern magnolia – *Magnolia grandiflora*
southern sandbur – *Cenchrus echinatus*
southern tulip – *Tulipa australis*
southernwood – *Artemisia abrotanum*
southernwood, field – *Artemisia campestris*
southern yellow pine – *Pinus rigida*
Southwest coral bean – *Erythrina flabelliformis*
Southwestern bushy bluestem – *Andropogon glomeratus* var *scabriglumus*
Southwestern chokecherry – *Prunus serotina* ssp *virens*
Southwestern condalia – *Ziziphus obtusifolia*
Southwestern cupgrass – *Eriochloa acuminata*
Southwestern paintbrush – *Castilleja integra*
Southwestern peach willow – *Salix amygdaloides*
Southwestern penstemon – *Penstemon barbatus*
Southwestern pipevine – *Aristolochia watsonii*
Southwestern white pine – *Pinus strobiformis*
sowbread – *Cyclamen neapolitanum*
sow thistle, common – *Sonchus oleraceus*
sow thistle, delicate – *Sonchus tenerrimus*
sow thistle, perennial – *Sonchus arvensis*
sow thistle, prickly – *Sonchus asper* ssp *asper*
sow thistle, spiny – *Sonchus asper*
soyate – *Beaucarnea inermis*
soy bean – *Glycine max*
Spain, great pellitory of – *Peucedanum ostruthium*
spangletop, green – *Leptochloa dubia*
spangletop, Mexican – *Leptochloa fusca* ssp *uninervia*
spangletop, red – *Leptochloa filiformis, L. panicea* ssp *brachiata*
Spanish arbor vine – *Merremia tuberosa*
Spanish bayonet – *Yucca*
Spanish bluebells – *Hyacinthoides hispanicus*
Spanish broom – *Cytisus multiflorus, Genista hispanica, Spartium junceum*
Spanish buckeye – *Ungnadia speciosa*
Spanish cedar – *Cedrela odorata*
Spanish chestnut – *Castanea sativa*
Spanish dagger – *Yucca*
Spanish dagger, giant – *Yucca carnerosana*
Spanish elder – *Piper amalago*
Spanish elm – *Cordia alliodora*
Spanish fir – *Abies pinsapo*
Spanish flag – *Quamoclit lobata*
Spanish gooseberry – *Pereskia aculeata*
Spanish heath – *Erica australis*
Spanish holly oak – *Quercus ilex*
Spanish iris – *Iris xiphium*
Spanish jasmine – *Jasminum grandiflorum*
Spanish lavender – *Lavandula stoechas*
Spanish lime – *Melicocca bijuga*
Spanish marjoram – *Thymus mastichina*
Spanish moss – *Tillandsia usneoides, Ramalina reticulata*
Spanish-needles – *Bidens*
Spanish oak – *Quercus crenata*
Spanish oyster plant – *Scolymus hispanicus*
Spanish plum – *Spondias mombin, purpurea*
Spanish salsify – *Scorzonera hispanica*
Spanish tea – *Chenopodium ambrosioides*
Spanish thyme – *Coleus ambolnicos, Lippia graveolens, Thymus capitatus*
Spanish vetchling – *Lathyrus clymenum*
sparrow vetch – *Vicia tetrasperma*
spathe flower – *Anthurium, Spathiphyllum cochlearispathum, S. wallisii*
spatterdock – *Nuphar luteum* ssp *polysepalum*

spear grass – *Acantherum brachychaetum, Imperata cylindrica*
spear, king's – *Asphodeline lutea, Asphodelus albus*
spearleaf – *Matelea parvifolia*
spearmint – *Mentha spicata, M. sylvestris*
spear oracle – *Atriplex patula*
spearwort, lesser – *Ranunculus flammula*
speckled gourd – *Melothria pendula*
speckled oak – *Quercus conspersa*
speedwell – *Veronica*
speedwell, alpine – *Veronica alpina*
speedwell, American – *Veronica peregrina*
speedwell, chain – *Veronica catenata*
speedwell, creeping – *Veronica repens*
speedwell, fingered – *Veronica triphyllos*
speedwell, gentian – *Veronica gentianoides*
speedwell, germander – *Veronica chamaedrys*
speedwell, heath – *Veronica officinalis*
speedwell, ivy-leaved – *Veronica hederifolia*
speedwell, large – *Veronica austriaca*
speedwell, marsh – *Veronica scutellata*
speedwell, Persian – *Veronica persica*
speedwell, prostrate – *Veronica prostrata*
speedwell, purslane – *Veronica peregrina* ssp *xalapensis*
speedwell, silver – *Veronica spicata* ssp *incana*
speedwell, thyme-leaved – *Veronica serpyllifolia*
speedwell, water – *Veronica anagallis*
spelt – *Triticum aestivum* var *spelta*
spicebush – *Calycanthus occidentalis, Lindera benzoin*
spider flower – *Cleome hasslerana*
spider flower, common – *Isomeris spinosa*
spidergrass – *Aristida ternipes*
spider lily – *Crinum zeylanicum, Hymenocallis americana, H. rotata*
spider lily, red – *Lycoris radiata*
spiderling – *Boerhavia*
spiderling, scarlet – *Boerhavia coccinea*
spider milkweed – *Asclepias asperula*
spider plant – *Chlorophytum comosum*
spiderwort – *Tradescantia*
spiderwort, western – *Tradescantia occidentalis*
spike burgrass – *Tragus berteronianus*
spike dropseed – *Sporobolus contractus*
spike lavender – *Lavandula latifolia*
spike-moss – *Selaginella*
spike-moss, desert – *Selaginella eremophila*
spikenard – *Aralia humilis, Nardostachys jatamansi*
spikenard, American – *Aralia racemosa*
spikerush – *Eleocharis*
spikerush, dark blood-red – *Eleocharis atropurpurea*
spikerush, small – *Eleocharis parvula*
spikerush, traveling – *Eleocharis rostellata*
spike winter hazel – *Corylopsis spicata*
spinach – *Spinacia oloracea*
spinach, cattle – *Atriplex polycarpa*
spinach, Ceylon – *Basella alba*
spinach, Chinese water – *Ipomoea aquatica*
spinach mustard – *Brassica rapa* var *perviridis*
spinach, New Zealand – *Tetragonia tetragonioides*
spinach, red strawberry – *Chenopodium foliosum*
spindle tree – *Euonymus europaeus, E. fortunei*
spinebush – *Adolphia californica*
spineflower, fringed – *Chorizanthe fimbriata*
spineless hedgehog – *Echinocereus triglochidiatus*
spineless yucca – *Yucca elephantipes*
spineshrub – *Adolphia canifornica*
spinous buttercup – *Ranunculus muricatus*
spiny acanthus – *Acanthus spinosus*
spiny aster – *Chloracantha spinosa, Ericameria spinulosus*
spiny bamboo – *Bambusa spinosa*
spiny club palm – *Bactris mexicana*
spiny cocklebur – *Xanthium spinosum*
spiny-fruited randia – *Randia echinocarpa*
spiny hedgehog – *Echinocereus polyacanthus* ssp *pacificus*
spinyleaf zinnia – *Zinnia acerosa*
spiny nopal – *Opuntia dejecta*
spiny rush – *Juncus acutus*
spiny sow thistle – *Sonchus asper*
spiny strawberry hedgehog – *Echinocereus stramineus*
spiraea – *Spiraea*
spiraea, blue – *Caryopteris incana*
spiraea, bridal wreath – *Spiraea prunifolia*
spiraea, false – *Astilbe arendsii*
spiraea, Japanese – *Spiraea japonica*
spiraea, shoe button – *Spiraea prunifolia*

spiraea, western – *Spiraea douglasii*
spiral ginger – *Costus*
spiral tasselweed – *Ruppia cirrhosa*
spleenwort – *Asplenium monanthes*
spleenwort, green – *Asplenium trichomanes*
spleenwort, maidenhair – *Asplenium trichomanes*
spleenwort, rock – *Asplenium fontanum*
splendid lily – *Lilium speciosum*
split-leaf philodendron – *Monstera deliciosa*
spoon, desert – *Dasylirion wheeleri, Eriogonum inflatum*
spoon, golden – *Byrsonima crassifolia*
spoonplant – *Dasylirion leiophyllum*
spotted burclover – *Medicago arabica*
spotted cranesbill – *Geranium maculatum*
spotted knapweed – *Centaurea maculosa*
spotted laurel – *Aucuba japonica*
spotted nettle – *Lamium maculatum*
spotted rock-rose – *Tuberaria guttata*
spotted spurge – *Chamaesyce maculata*
spotted touch-me-not – *Impatiens pallida*
sprangletop, bearded – *Leptochloa fascicularis*
spreader, silver – *Artemisia caucasica*
spreading amaranth – *Amaranthus cruentas*
spreading bur parsley – *Torilis arvensis*
spreading silverleaf – *Dyschoriste decumbens*
Sprenger asparagus – *Asparagus densiflorus* var *sprengeri*
springbells – *Tabebuia donnell-smithii*
spring cinquefoil – *Potentilla tabernaemontanii*
spring forget-me-not – *Myosotis micrantha*
spring fumitory – *Corydalis solida*
spring gentian – *Gentiana verna*
spring heath – *Erica carnea*
spring onion – *Allium fistulosum*
spring pea – *Lathyrus vernus*
spring sandwort – *Sagina subulata*
spring snowflake – *Leucojum vernum*
spring vetch – *Vicia lathyroides, V. sativa*
spring whitlow-grass – *Draba verna*
spruce – *Picea*
spruce, bigcone – *Pseudotsuga macrocarpa*
spruce, black – *Picea mariana*
spruce, blue – *Picea pungens*
spruce, Brewer – *Picea breweriana*
spruce, Brewer's weeping – *Picea breweriana*
spruce, Chihuahua – *Picea chihuahuana*
spruce, Colorado – *Picea pungens*
spruce, Colorado blue – *Picea pungens* var *glauca*
spruce, Engelmann – *Picea engelmannii*
spruce, Norway – *Picea abies*
spruce, red – *Picea rubens*
spruce, Sitka – *Picea sitchensis*
spruce, white – *Picea glauca*
spur, cock – *Pisonia aculeata*
spurge – *Chamaesyce, Euphorbia*
spurge, Arizona – *Chamaesyce arizonica*
spurge, beetle – *Euphorbia eriantha*
spurge, caper – *Euphorbia lathyris*
spurge, cliff – *Euphorbia californica, E. misera, E. xantii*
spurge, cypress – *Euphorbia cyparissias*
spurge, drooping – *Chamaesyce nutans*
spurge, fringed – *Chamaesyce setiloba*
spurge, Japanese – *Pachysandra terminalis*
spurge laurel – *Daphne laureola*
spurge, leafy – *Euphorbia esula*
spurge, marsh – *Euphorbia palustris*
spurge, Mohave – *Euphorbia incisa*
spurge nettle – *Cnidoscolus urens, Urtica dioica*
spurge olive – *Cneorum tricoccon*
spurge, painted – *Euphorbia heterophylla*
spurge, petty – *Euphorbia peplus*
spurge, polychrome – *Euphorbia polychroma*
spurge, prostrate – *Chamaesyce*
spurge, saw-toothed – *Euphorbia serrata*
spurge, spotted – *Chameasyce maculata*
spurge, sun – *Euphorbia helioscopia*
spurge, sweet – *Euphobria dulcis*
spurge, threaded – *Euphorbia eriantha*
spurge, thyme-leaved – *Chamaesyce serpyllifolia*
spurge, toothed – *Euphorbia dentata*
spurge, twisting – *Chamaesyce serpens*
spurge, valley – *Chamaesyce ocellata*
spurge, wood – *Euphorbia amygdaloides*
spurrey – *Spergula arvensis*
spurrey, sand – *Spergularia bocconii, S. marina, S. rubra*
spurrey, sea – *Spergularia bocconii, S. media*
squarestem rose gentian – *Sabatia angularis*
square, three – *Scirpus americanus*
squash, acorn – *Cucurbita maxima*
squash, butternut – *Cucurbita moschata*
squash, crookneck – *Cucurbita moschata*

squash, summer – *Cucurbita pepo*
squash, winter – *Cucurbita maxima, C. moschata*
squawbush – *Rhus trilobata, Ziziphus spathulata*
squaw currant – *Ribes cereum*
squaw root – *Conopholis americana*
squaw vine – *Mitchella repens*
squaw waterweed – *Baccharis sergiloides*
squill, alpine – *Scilla biflora*
squill, Italian – *Scilla italica*
squill, Siberian – *Scilla sibirica*
squinancy berry – *Ribes nigrum*
squirrel palm – *Chamaedorea arenbergiana*
squirreltail – *Elymus elymoides, Hordeum jubatum*
squirreltail fescue – *Vulpia bromoides*
squirreltail, longleaf – *Elymus longifolius*
squirting cucumber – *Ecballium elaterium*
stacte – *Styrax officinalis*
staghorn cholla – *Cylindropuntia versicolor*
staghorn fern – *Platycerium*
staghorn sumac – *Rhus typhina*
Standley rose – *Rosa standleyi*
stanhopea, tiger-like – *Stanhopea devoniensis*
star anis – *Illicium anisatum*
star apple – *Averrhoa carambola, Chrysophyllum cainito*
star apple, wild – *Chrysophyllum mexicanum*
starbell – *Cordia dodecandra*
star, blazing – *Liatris punctata, L. spicata, Mentzelia*
starch hyacinth – *Muscari neglectum*
starchy banana – *Mus balbisiana*
star, crack – *Ariocarpus fissuratus*
star, eastern shooting – *Dodecatheon meadia*
starflower, arctic – *Trientalis arctica*
starflower, northern – *Trientalis arctica*
starfruit – *Averrhoa carambola, Damasonium alisma*
star glory – *Ipomoea quamoclit*
star grass – *Aletris farinosa, Cynodon plectostachyus*
star ipomoea – *Quamoclit coccinea*
star-leaf – *Choisya arizonica, C. dumosa*
star-leaf, Arizona – *Choisya arizonica*
star-lily – *Zephranthes texana*
star magnolia – *Magnolia stellata*
star, Mexican – *Milla biflora*
star of Bethlehem – *Campanula isophylla, Ornithogalum arabicum, O. umbellatum*
star of Bethlehem, drooping – *Ornithogalum nutans*
star of Bethlehem, Mexican – *Milla biflora*
star of Persia – *Allium christophii*
star rose – *Rosa stellata*
starry ball – *Coryphantha elephantidens*
star, sabi – *Adenium obesum*
star, shooting – *Dodecatheon*
stars, seven – *Ariocarpus retusus*
star, Texas – *Sabatia campestris*
star-thistle – *Centaurea*
star-thistle, Malta – *Centaurea melitensis*
star-thistle, purple – *Centaurea calcitrapa*
star-thistle, Sicilian – *Centaurea sulphurea*
star-thistle, yellow – *Centaurea solstitialis*
star tree – *Inga spuria*
star-tulip – *Calochortus*
starwort – *Spergula arvensis, Stellaria*
starwort, long-leaved – *Stellaria longifolia*
starwort, water – *Callitriche hermaphroditica, C. stagnalis, C. verna*
statice – *Limonium californicum*
St. Augustine grass – *Stenotaphrum secumdatum*
stavesacre – *Delphinium staphisagria*
stemless gentian – *Gentiana acaulis*
stem rust, black – *Puccinia graminis*
stephanotis – *Stephanotis floribunda*
stevia – *Piqueria trinervia*
stewartia, Japanese – *Stewartia pseudocamellia*
St. George's mushroom – *Tricholoma gambosa*
stickleaf – *Mentzelia*
stickleaf, white bract – *Mentzelia involucrata*
stickleaf, whitestem – *Mentzelia albicaulis*
stickseed – *Lappula squarrosa*
sticktight – *Bidens frondosa*
stick-weed – *Stephanomeria pauciflora*
stickwort – *Spergula arvensis*
sticky catchfly – *Lychnis viscaria*
St. Ignatius's bean – *Strychnos ignatii*
stilt palm – *Iriartea*
stinging nettle – *Jatropha urens, Urtica dioica*
stinging nettle, American – *Urtica dioica* ssp *gracilis*
stinkgrass – *Eragrostis cilianensis*
stinking rose – *Rosa eglanteria*
stinking toe – *Hymenaea courbaril*

stinkweed – *Anthemis cotula*
stitchwort – *Stellaria calycantha*
stitchwort, lesser – *Stellaria graminea*
stitchwort, wood – *Stellaria nemorum*
St. James lily – *Sprekelia formosissima*
St. John's bread – *Ceratonia siliqua*
St. John's flower – *Bouvardia longiflora*
St. John's wort – *Hypericum*
St. John's wort, needle-leaved – *Hypericum coris*
St. Lucie cherry – *Prunus mahleb*
stock – *Matthiola incana*
stock, African – *Malcolmia africana*
stock, evening scented – *Matthiola longipetala*
stock, Virginian – *Malcolmia maritima*
Stoke's aster – *Stokesia laevis*
stonecress – *Aethionema schistosum*
stonecrop – *Dudleya, Sedum*
stonecrop, Caucasian – *Sedum spurium*
stonecrop, English – *Sedum anglicum*
stonecrop, high – *Sedum sediforme*
stonecrop, mossy – *Crassula tillaea*
stonecrop, reflexed – *Sedum reflexum*
stonecrop, short-leaved – *Sedum brevifolium*
stonecrop, thick-leaved – *Sedum dasyphyllum*
stonecrop, white – *Sedum album*
stoneface – *Lithops*
stone, living – *Lithops*
stone pine, Italian – *Pinus pinea*
stone pine, Swiss – *Pinus cembra*
stoneseed – *Lithodora, Lithospermum*
storax – *Styrax officinalis*
storksbill – *Erodium, Geranium caespitosum*
storksbill, soft – *Erodium malacoides*
storksbill, Texas – *Erodium texanum*
St. Peter's palm – *Yucca filifera*
strangler fig – *Ficus cotinifolia, F. padifolia*
stranglewort – *Cynanchum acutum*
straw, bed – *Galium*
strawberry – *Fragaria*
strawberry, alpine – *Fragaria vesca*
strawberry, beach – *Fragaria chiloensis*
strawberry blite – *Chenopodium foliosum*
strawberry cactus – *Echinocereus brandegeei, E. engelmannii, E. stramineus*
strawberry clover – *Trifolium fragiferum, T. macrocephalum*
strawberry, garden – *Fragaria ananassa*
strawberry geranium – *Saxifraga stolonifera*
strawberry ground-cherry – *Physalis alkekengi*
strawberry guava – *Psidium cattleyanum*
strawberry hedgehog – *Echinocereus fasciculatus, E. stramineus*
strawberry hedgehog cactus – *Echinocereus enneacanthus*
strawberry, Indian mock – *Duchesnea indica*
strawberry, mock – *Duchesnea indica*
strawberry, mountain – *Fragaria virginiana*
strawberry, sand – *Fragaria chiloensis*
strawberry spinach, red – *Chenopodium foliosum*
strawberry-tomato – *Physalis peruviana*
strawberry tree – *Arbutus unedo*
strawberry, wood – *Fragaria vesca*
strawberry, yacon – *Polymnia edulis*
strawflower – *Helichrysum bracteatum*
straw, windle – *Apera spica-venti*
stringbean – *Phaseolus vulgaris*
striped crotalaria – *Crotalaria mucronata*
strong bark – *Desmodium canum*
stylo, finestem – *Stylosanthes guianensis*
suari nut – *Caryocar amygdaliferum*
subalpine fir – *Abies lasiocarpa*
subterranean clover – *Trifolium subterraneum*
succulent – *Aeonium canariense*
Sudan grass – *Sorgum bicolor, S. sudanense*
sugar-apple – *Annona squamosa*
sugar beet – *Beta vulgaris* ssp *saccharifera*
sugarberry – *Celtis reticulata, Ehretia anacuna, E. elliptica*
sugar bush – *Rhus lentii, R. ovata*
sugar cane – *Saccharum officinarum*
sugarcane, Indian – *Saccharum barberi*
sugarcane, Chinese – *Saccharum sinense*
sugar maple – *Acer saccharum*
sugar maple, Rocky Mountain – *Acer saccharum* ssp *grandidentatum*
sugar millet – *Sorghum saccharatum*
sugar palm – *Arenga pinnata, A. saccharifera*
sugar pea – *Pisum sativum*
sugar pine – *Pinus lambertiana*
sugar sumac – *Rhus ovata*
sulphur cinquefoil – *Potentilla recta*
sultan, sweet – *Centaurea moschata, Impatiens walleriana*
suma – *Pfaffia paniculata*
sumac – *Rhus*
sumac, aromatic – *Rhus aromatica*

sumac, desert – *Rhus microphylla*
sumac, evergreen – *Rhus sempervirens, R. virens*
sumac, flame-leaf – *Rhus copallina*
sumac, fragrant – *Rhus aromatica*
sumac, laurel – *Malosma laurina*
sumac, Rocky Mountain – *Rhus glabra*
sumac, scarlet – *Rhus glabra*
sumac, shining – *Rhus copallina*
sumac, Sicilian – *Rhus coriaria*
sumac, smooth – *Rhus glabra*
sumac, staghorn – *Rhus typhina*
sumac, sugar – *Rhus ovata*
sumac, tanner's – *Rhus coriaria*
sumac, Venetian – *Cotinus coggygria*
summer chrysanthemum – *Chrysanthemum carinatum, C. coronarium*
summer cypress – *Kochia scoparia*
summer forget-me-not – *Anchusa capensis*
summer hyacinth – *Galtonia candicans*
summer king bolete – *Boletus reticulatus*
summer lilac – *Buddleja davidii*
summer pheasant's eye – *Adonis aestivalis*
summer phlox – *Phlox paniculata*
summer poppy – *Kallstroemia grandiflora*
summer savory – *Satureja hortensis*
summer snowflake – *Leucojum aestivum*
summer, snow-in – *Cerastium tomentosum*
summer squash – *Cucurbita pepo*
summer truffle – *Tuber aestivum*
summer turnip rape – *Brassica rapa* var *praecox*
sun cereus – *Disocactus speciosus*
sunchoke – *Helianthus tuberosus*
sundew, English – *Drosera anglica*
sundew, long-leaved – *Drosera anglica*
sundew, round-leaved – *Drosera rotundifolia*
sundrops – *Oenothera fruticosa*
sundrops, Ozark – *Oenothera macrocarpa*
sunflower – *Helianthus*
sunflower, bush – *Simsia amplexicaulis, Tithonia tubaeformis*
sunflower, common – *Helianthus annuus*
sunflower, dune – *Helianthus niveus*
sunflower, Mexican – *Tithonia*
sunflower, San Diego – *Viguiera laciniata, V. tomentosa*
sunflower, tree – *Tithonia fruticosum*
sunfruit – *Heliocarpus attenuatus, H. tomentosus*
sunn hemp – *Crotalaria juncea*
sun-plant – *Portulaca grandiflora*
sunray – *Helipterum roseum*
sun-rose – *Helianthemum*
sun rose, baby – *Aptenia cordifolia*
sun-rose, Cretan – *Cistus creticus*
sun-rose, gray – *Cistus incanus*
sun-rose, laurel-leaved – *Cistus laurifolius*
sun-rose, Montpellier – *Cistus monspeliensis*
sunshine tree – *Tabebuia donnell-smithii*
sun spurge – *Euphorbia helioscopia*
superb lily – *Gloriosa superba*
superb sage – *Salvia sylvestris*
suppa palm – *Acrocomia aculeata*
Surinam cherry – *Eugenia uniflora*
Surinam purslane – *Talinum triangulare*
swallow herb – *Chamaesyce maculata*
swallow-wort – *Asclepias syriaca*
swallow-wort, white – *Cynanchum vincetoxicum*
swamp grass – *Crypsis schoenoides*
swamp lily – *Thalia geniculata*
swamp mahogany – *Eucalyptus robusta*
swamp milkweed – *Asclepias incarnata*
swamp oak – *Quercus palustris*
swamp ricegrass – *Leersia hexandra*
swamproot – *Anemopsis californica*
swamp white oak – *Quercus bicolor*
Swede – *Brassica napus* var *napobrassica*
Swede rape – *Brassica napus* var *napobrassica*
Swedish coffee – *Astragalus boeticus*
Swedish whitebeam – *Sorbus intermedia*
sweet acacia – *Acacia farnesiana*
sweet alyssum – *Lobularia maritima*
sweet-and-vile vine – *Agdestis clematides*
sweet Annie – *Artemisia annua*
sweet balm – *Melissa officinalis*
sweet basil – *Ocimum basilicum*
sweet bay – *Laurus nobilis*
sweet bay, California – *Myrica californica*
sweet birch – *Betula lenta*
sweet, bitter – *Solanum dulcamara*
sweet briar – *Rosa eglanteria*
sweet brush – *Cercocarpus betuloides*
sweetbush – *Bebbia juncea*
sweet cassava – *Manihot dulcis*
sweet cherry – *Prunus avium*

sweet cicely – *Myrrhis odorata, Osmorhiza brachypoda, O. longistylis*
sweetclover – *Melilotus officinalis*
sweetclover, white – *Melilotus alba*
sweetclover, yellow – *Melilotus indica, M. officinalis*
sweet corn – *Zea mays* var *saccharata*
sweet corn root – *Calathea allouia*
sweet fern – *Comptonia aspleniifolia*
sweet flag – *Acorus calamus, Phragmites australis*
sweet flag, Japanese – *Acorus gramineus*
sweet four-o'clock – *Mirabilis longiflora*
sweet gale – *Myrica gale*
sweet gooseberry – *Ribes grossularia*
sweet granadilla – *Passiflora ligularis*
sweet-grass, wide-spread – *Puccinellia distans*
sweet gum – *Liquidambar styraciflua*
sweet gum, American – *Liquidambar styraciflua*
sweet gum, Chinese – *Liquidambar formosana*
sweet gum, oriental – *Liquidambar orientalis*
sweetheart vine – *Pyrostegia venusta*
sweet, honey – *Tidestromia lanuginosa*
sweet laurel – *Illicium floridanum*
sweet lemon – *Citrus limetta*
sweet lemon, Mediterranean – *Citrus limetta*
sweet lime, Indian – *Citrus limettioides*
sweet lime of India – *Citrus limettioides*
sweet lupine – *Lupinus luteus*
sweet mangrove – *Maytenus phyllanthoides*
sweet manioc – *Manihot palmata*
sweet marjoram – *Origanum majorana*
sweet, meadow – *Astilbe arendsii, Filipendula ulmaria*
sweet mesquite – *Prosopis juliflora*
sweet, mill – *Citrus limetta*
sweet mock orange – *Philadelphus coronarius*
sweet olive – *Osmanthus fragrans*
sweet pea – *Lathyrus odoratus*
sweet pea, common – *Lathyrus odoratus*
sweet pea, perennial – *Lathyrus latifolius*
sweet pepperbush – *Clethra mexicana*
sweet potato – *Ipomoea batatas*
sweet prickly-pear – *Opuntia dillenii*
sweet rocket – *Hesperis matronalis*
sweet root – *Osmorhiza brachypoda*
sweet sand verbena – *Abronia fragrans*
sweet scabious – *Scabiosa atropurpurea*
sweet-scented bedstraw – *Galium triflorum*
sweet-scented marigold – *Tagetes lucida*
sweet-shrub – *Calycanthus occidentalis*
sweetsop – *Annona squamosa*
sweet spurge – *Euphorbia dulcis*
sweet sultan – *Centaurea moschata, Impatiens walleriana*
sweet tanglehead – *Heteropogon melanocarpus*
sweet, timber – *Nectandra*
sweet vernal grass – *Anthoxanthum odoratum*
sweet viburnum – *Viburnum awabukii*
sweet violet – *Viola odorata*
sweet William – *Dianthus barbatus*
sweet-William catchfly – *Silene armeria*
sweetwood – *Xylosma flexuosum*
sweet woodruff – *Galium odoratum*
sweet wormwood – *Artemisia annua*
swine cress – *Coronopus didymus, C. squamatus*
Swiss chard – *Beta vulgaris* ssp *cicla*
Swiss cheese plant – *Monstera deliciosa*
Swiss mountain pine – *Pinus mugo*
Swiss stone pine – *Pinus cembra*
switch grass – *Panicum virgatum*
switch sorrel – *Dodonaea viscosa*
swizzle-stick tree – *Quararibea funebris*
swordbean – *Canavalia gladiata*
sword fern – *Nephrolepis exaltata*
sword-leaved rush – *Juncus ensifolius*
sycamine – *Ficus sycomorus, Morus nigra*
sycamore, American – *Platanus occidentalis*
sycamore, Arizona – *Platanus wrightii*
sycamore, California – *Platanus racemosa*
sycamore-fig – *Ficus sycomorus*
sycamore-leaf jatropha – *Jatropha malacophylla*
sycamore maple – *Acer pseudoplatanus*
sycamore, western – *Platanus racemosa*
Sydney golden wattle – *Acacia longifolia*
Syrian juniper – *Juniperus drupacea*
Syrian plum – *Prunus domestica* ssp *syriaca*
Syrian rue – *Peganum harmala*
syringa – *Philadelphus mexicanus*

tabasco – *Capsicum frutescens*
tacso – *Passiflora mollissima*
Tahiti lime – *Citrus latifolia*
tail, beaver – *Opuntia basilaris*
tail, burro – *Sedum morganianum*

tail, cat – *Acalypha, Typha*
tail, dog – *Cynosurus*
tail, donkey – *Sedum morganianum*
tail flower – *Anthurium andraeanum*
tail, fox – *Alopecurus, Setaria*
tail, hare's – *Lagurus ovatus*
tail, horse – *Equisetum*
tail, mare's – *Hippuris vulgaris*
tail, mouse – *Myosurus minimus*
tail, pheasant's – *Anthurium schlechtendalii*
tail, pig – *Anthurium scherzerianum*
tail, quash – *Nephrolepis biserrata*
tail, scorpion – *Heliotropium angiospermum*
tail, sheep's – *Sedum morganianum*
tail, squirrel – *Elymus elymoides, Hordeum jubatum*
talipot palm – *Corypha umbraculifera*
tall fescue – *Festuca arundinacea*
tall oatgrass – *Arrhenatherum elatius*
tallow tree, Chinese – *Sapium sebiferum*
tallow wood – *Ximenia americana*
tall rocket – *Sisymbrium altissimum*
tall wheatgrass – *Elytrigia elongata*
tamarack – *Larix laricima, L. occidentalis, Pinus contorta* ssp *murrayana*
tamarillo – *Cyphomandra betacea*
tamarind – *Tamarindus indica*
tamarind, cow – *Albizzia saman*
tamarind, Manila – *Pithecellobium dulce*
tamarind, wild – *Leucaena glauca, Lysiloma bahamensis*
tamarisk – *Tamarix ramosissima*
tamarisk, African – *Tamarix africana*
tamarisk, French – *Tamarix gallica*
Tampico fiber – *Hesperaloe parviflora*
tangelo – *Citrus reticulata* x *C. paradisi*
tangerine – *Citrus reticulata*
Tangier pea – *Lathyrus tingitanus*
tangled fishhook – *Mammillaria microcarpa*
tanglehead – *Heteropogon contortus, H. melanocarpus*
tanglehead, sweet – *Heteropogon melanocarpus*
tangor – *Citrus reticulata* x *C. sinensis*
tanner's sumac – *Rhus coriaria*
tannia – *Xanthosoma sagittifolium*
tansy – *Tanacetum vulgare*
tansy, common – *Tanacetum vulgare*
tansy mustard – *Descuriana pinnata, Sisymbrium*
tansy ragwort – *Senecio jacobaea*
taper leaf – *Pericome caudata*
tapioca – *Manihot esculenta*
tara – *Caesalpinia spinosa*
tarbush – *Flourensia cernua*
tare, hairy – *Vicia hirsuta*
tares – *Lolium temulentum*
taro – *Colocasia antiquorum, C. esculenta*
tarragon – *Artemisia dracunculus*
tarragon, false – *Artemisia dracunculoides*
tarragon, French – *Artemisia dracunculus*
tarragon, Mexican – *Tagetes lucida*
tartago – *Jatropha podagrica*
Tartarian dogwood – *Cornus alba*
Tartarian honeysuckle – *Lonicera tatarica*
Tartarian maple – *Acer tartaricum*
Tartaric lettuce – *Lactuca tatarica*
tarweed, Chilean – *Madia sativa*
tarweed, coast – *Madia sativa*
tasco – *Passiflora mollissima*
tassel bush, silk – *Garrya*
tassel flower – *Amaranthus caudatus, Brickellia grandiflora*
tassel hyacinth – *Muscari comosum*
tasselweed, spiral – *Ruppia cirrhosa*
tassel, yellow silk – *Garrya flavescens*
tawny daylily – *Hemerocallis fulva*
tazetta daffodil – *Narcissus tazetta*
tea – *Camellia sinensis*
tea, Arabian – *Catha edulis*
tea, Assam – *Camellia assamica*
teaberry – *Gaultheria procumbens*
tea, Brigham – *Ephedra torreyana*
tea crabapple – *Malus hupehensis*
tea, desert – *Ephedra californica*
tea, Hopi – *Thelesperma gracile*
tea, Indian – *Ephedra viridis, Thelesperma megapotamicum*
teak – *Tectona grandis*
tea, Mexican – *Chenopodium ambrosioides, Ephedra torreyana, E. trifurca*
tea, Mormon – *Ephedra*
tea, Navajo – *Thelesperma gracile, T. longipes*
tea, Oswego – *Monarda citriodora, M. didyma*
tea plant, Duke of Argyll's – *Lycium barbarum*
tea rose – *Rosa indica*
tears, angel's – *Narcissus triandrus*

tears, baby's – *Soleirolia soleirolii*
tears, bride's – *Antigonon leptopus*
tears, Job's – *Coix lacryma-jobi*
tears of blood – *Bocconia frutescens*
tears, widow's – *Commelina erecta*
teasel, Fuller's – *Dipsacus fullonum, D. sativus*
teasel, wild – *Dipsacus fullonum*
tea, Spanish – *Chenopodium ambrosioides*
tea tree – *Melaleuca alternifolia*
tea-tree, Duke of Argyll's – *Lycium chinense*
Tecate cypress – *Cupressus forbesii*
teddy-bear cholla – *Cylindropuntia bigelovii*
telegraph plant – *Heterotheca psammophila*
telegraph weed – *Arnica, Heterotheca*
temple flower – *Plumeria acutifolia*
temple tree – *Plumeria rubra*
temu – *Luma apiculata*
tenaza – *Havardia pallens*
teosinte – *Euchaena mexicana*
tepary bean – *Phaseolus acutifolius*
tepary bean, Mexican – *Phaseolus acutifolius*
tequila agave – *Agave tequilana*
terebinth – *Pistacia terebinthus*
Texas algerita – *Berberis trifoliata*
Texas bean – *Phaseolus acutifolius*
Texas beard grass – *Schizachyrium cirratum*
Texas blueweed – *Helianthus ciliaris*
Texas buckeye – *Ungnadia speciosa*
Texas catclaw – *Acacia wrightii*
Texas cone cactus – *Neolloydia conoidea*
Texas croton – *Croton texensis*
Texas dandelion – *Pyrrhopappus multicaulis*
Texas ebony – *Pithecellobium flexicaule*
Texas elm – *Ulmus crassifolia*
Texas filaree – *Erodium texanum*
Texas hair cactus – *Mammillaria prolifera*
Texas kidneywood – *Eysenhardtia texana*
Texas lantana – *Lantana horrida*
Texas madrone – *Arbutus texana*
Texas mallow – *Malvaviscus arboreus*
Texas millet – *Brachiaria texana*
Texas mountain laurel – *Sophora secundiflora*
Texas mulberry – *Morus microphylla*
Texas needlegrass – *Stipa leucotricha*
Texas nightshade – *Solanum triquetrum*
Texas olive – *Cordia boissieri*
Texas paloverde – *Cercidium texanum*
Texas persimmon – *Diospyros texana*
Texas pine – *Pinus remota*
Texas prickly-pear – *Opuntia engelmannii* var *lindheimeri*
Texas pride – *Thelocactus bicolor*
Texas rainbow cactus – *Echinocereus dasyacanthus*
Texas rainbow hedgehog – *Echinocereus dasyacanthus*
Texas ranger – *Leucophyllum frutescens*
Texas sacahuiste – *Nolina texana*
Texas sage – *Leucophyllum frutescens*
Texas scallion – *Allium cepa*
Texas silverleaf – *Leucophyllum frutescens*
Texas-star – *Sabatia campestris*
Texas storksbill – *Erodium texanum*
Texas thistle – *Cirsium texanum*
Texas timothy – *Lycurus phleoides*
Texas umbrella tree – *Melia azedarach* var *umbraculiformis*
Texas vervain – *Verbena halei*
Texas violet – *Salvia farinacea*
Texas virgin's bower – *Clematis drummondii*
Thai jute – *Hibiscus cannabinus*
Thale cress – *Arabidopsis thaliana*
Thanksgiving cactus – *Schlumbergera truncata*
thatch palm – *Sabal yapa, Thrinax radiata*
thatch palm, silver – *Coccothrinax argentata*
the deceiver – *Laccaria laccata*
the Duke of Devonshire's oncidium – *Oncidium cavendishianum*
thelesperma, rayless – *Thelesperma megapotamicum*
thevetia, giant – *Thevetia thevetioides*
thickleaf oak – *Quercus crassifolia*
thimbleberry – *Rubus neomexicanus, R. parviflorus*
thimble cactus – *Neolloydia conoidea*
thimble flower, blue – *Gilia capitata*
thimbles, fairy – *Campanula cochlearifolia*
thinleaf alder – *Alnus tenuifolia, A. viridis*
thistle – *Cirsium*
thistle, Arizona – *Cirsium arizonicum*
thistle, artichoke – *Cynara cardunculus*
thistle, blessed – *Cnicus benedictus*
thistle, bull – *Cirsium vulgare*
thistle, Canada – *Cirsium arvense*
thistle, carline – *Carlina vulgaris*
thistle cholla – *Cylindropuntia tunicata*
thistle, creeping – *Cirsium arvense*
thistle, curly welted – *Carduus crispus*

thistle, distaff – *Atractylis cancellata*
thistle, globe – *Echinops exaltatus, E. sphaerocephalus*
thistle, golden – *Scolymus hispanicus*
thistle, gum-bearing carline – *Atractylis gummifera*
thistle, holy – *Silybum marianum*
thistle, Illyrian – *Onopordum illyricum*
thistle, Italian – *Carduus pycnocephalus*
thistle, Mexican – *Eryngium heterophyllum*
thistle, milk – *Silybum marianum*
thistle, Mojave – *Cirsium mohavense*
thistle, musk – *Carduus nutans*
thistle, pasture – *Cirsium undulatum*
thistle, plumed – *Cirsium mexicanum*
thistle, plumeless – *Carduus acanthoides*
thistle, Russian – *Salsola tragus*
thistle, Scotch – *Onopordum acanthium*
thistle, Scottish – *Onopordum acanthium*
thistle, slender – *Carduus tenuiflorus*
thistle, southern ball – *Echinops exaltatus*
thistle, sow – *Sonchus*
thistle, star – *Centaurea*
thistle, Texas – *Cirsium texanum*
thistle, thornless – *Centaurea americana*
thistle, wavyleaf – *Cirsium undulatum*
thistle, white – *Atriplex lentiformis*
thistle, woolly distaff – *Carthamus lanatus*
thorn-apple – *Datura*
thorn-apple, desert – *Datura discolor*
thorn, box – *Lycium*
thorn, camel – *Alhagi pseudalhagi*
thorn, Christ – *Euphorbia milii, Paliurus spina-christi*
thorn, Christ's – *Gleditsia triacanthos, Ziziphus spina-christi*
thorn, coast desert – *Lycium californicum*
thorn, cockspur – *Crataegus crus-galli*
thorn, crucifixion – *Canotia holacantha, Castela emoryi, Koberlinia spinosa*
thorn, fire – *Pyracantha*
thorn, Jerusalem – *Paliurus spina-christi, Parkinsonia aculeata*
thornless catclaw – *Acacia berlandieri*
thornless thistle – *Centaurea americana*
thorn, Parish's desert – *Lycium parishii*
thorns, crown of – *Euphorbia milii, E. splendens, Koeberlinia spinosa* ssp *tenuispina, Paliurus spina-christi*
thorn, smoke – *Psorothamnus spinosus*
thorny astragal – *Astragalus tragacantha*
thorny greenbriar – *Smilax aspera*
thoroughwort – *Eupatorium hyssopium*
thoroughwort, western – *Ageratina herbacea*
thoroughwort, Wright – *Ageratina wrightii*
thorow wax, narrow – *Bupleurum lancifolium*
threaded spurge – *Euphorbia eriantha*
thread, gold – *Coptis groenlandica*
threadleaf groundsel – *Senecio longilobus*
three-awn, Arizona – *Aristida arizonica*
three-awn, California – *Aristida californica* var *californica*
three-awn, Fendler – *Aristida purpurea* var *fendleriana*
three-awn, havard – *Aristida barbata*
three-awn, hook – *Aristida ternipes* var *hamulosa*
three-awn, Nealley – *Aristida purpurea* var *nealleyi*
three-awn, poverty – *Aristida divaricata, A. ternipes* var *gentilis*
three-awn, purple – *Aristida purpurea*
three-awn, red – *Aristida purpurea* var *longiseta*
three-awn, six-weeks – *Aristida adscensionis*
three-awn, wooton – *Aristida pansa*
three-flower melic – *Melica nitens*
three-flower nightshade – *Solanum triflorum*
three-seeded mercury – *Acalypha lindheimeri*
threesquare – *Scirpus americanus*
threesquare, common – *Scirpus pungens*
thrift – *Armeria juniperifolia, A. maritima* ssp *californica*
thrift, Jersey – *Armeria alliacea*
thrift, rush-like – *Armeria girardii*
throatwort – *Campanula trachelium*
thrumwort – *Damasonium alisma*
thumb, lady's – *Polygonum persicaria*
thyine wood – *Tetraclinis articulata*
thyme – *Thymus vulgaris*
thyme, camphor – *Thymus camphoratus*
thyme, cat – *Teucrium marum*
thyme, common – *Thymus vulgaris*
thyme, creeping – *Thymus serpyllum*
thyme, garden – *Thymus vulgaris*
thyme, large – *Thymus pulegioides*
thyme leaved speedwell – *Veronica serpyllifolia*

thyme-leaved spurge – *Chamaesyce serpyllifolia*
thyme, mother-of – *Thymus serpyllum*
thyme, Spanish – *Coleus ambolnicos, Lippia graveolens, Thymus capitatus*
thyme, white – *Thymus mastichina*
thyme, wild – *Thymus serpyllum*
tibisee – *Lasiacis divaricata*
tickbean – *Vicia faba* var *minuta*
tick clover – *Desmodium cinerascens*
ticks, beggar – *Bidens pilosa*
tickseed – *Coreopsis grandiflora*
ticks, seed – *Daucus pusillus*
tie-tie, basket – *Desmoncus orthacanthos*
tiger aloe – *Aloe variegata*
tiger bush – *Acrostichum aureum*
tiger flower – *Tigridia pavonia*
tiger-like stanhopea – *Stanhopea devoniensis*
tiger lily – *Lilium lancifolium*
tiger orchid – *Odontoglossum rosii*
tight phacelia – *Phacelia congesta*
tiled hydnum – *Sarcodon imbricatus*
tillaea, water – *Crassula aquatica*
timberline bluegrass – *Poa glauca*
timber sweet – *Nectandra*
timothy, common – *Phleum pratense*
timothy, cultivated – *Phleum pratense*
timothy, golden – *Setaria sphacelata*
timothy, mountain – *Phleum alpinum*
timothy, Texas – *Lycurus phleoides*
tinted penstemon – *Penstemon spectabilis*
tiny bedstraw – *Galium murale*
tiny vetch – *Vicia hirsuta*
ti plant – *Cordyline terminalis*
tipu tree – *Tipuana tipu*
toadflax – *Cymbalaria aequitriloba, C. muralis, Linaria*
toadflax, Dalmatian – *Linaria genistifolia*
toadflax, ivy-leaved – *Cymbalaria muralis*
toadflax, purple – *Linaria purpurea*
toad rush – *Juncus bufonius*
tobacco – *Nicotiana*
tobacco, Aztec – *Nicotiana rustica*
tobacco, coyote – *Nicotiana attenuata*
tobacco, desert – *Nicotiana clevelandii, N. obtusifolia*
tobacco, fiddle-leaf – *Nicotiana repanda*
tobacco, flowering – *Nicotiana alata*
tobacco, Indian – *Lobelia fenestralis, L. inflata, Nicotiana rustica*
tobacco, jasmine – *Nicotiana alata*
tobacco root – *Valeriana sorbifolia*
tobacco, tree – *Nicotiana glauca*
tobacco, wood – *Nicotiana sylvestris*
tobira – *Pittosporum tobira*
tobosa – *Pleuraphis mutica*
tocalote – *Centaurea melitensis*
toddy palm – *Borassus flabellifer, Caryota urens*
toes, pussy – *Antennaria dioica*
toe, stinking – *Hymenaea courbaril*
tomatillo – *Physalis philadelphica, P. acuminata, P. francheti*
tomato – *Lycopersicon*
tomato berry – *Lycium andersonii*
tomato, cherry – *Lycopersicon esculentum* var *cerasiforme*
tomato, currant – *Lycopersicon pimpinellifolium*
tomato, husk – *Physalis philadelphica, P. pubescens*
tomato, roma – *Lycopersicon esculentum*
tomato, sea – *Rosa rugosa*
tomato, strawberry – *Physalis peruviana*
tomato, tree – *Cyphomandra betacea*
tomato wilt – *Fusarium solani*
tongue, adder's – *Erythronium americanum*
tongue, beard – *Penstemon*
tongue, hound's – *Cynoglossum*
tongue, ox – *Picris echioides*
tongue, painted – *Salpiglossis sinuata*
tongue, royal beard – *Penstemon spectabilis*
tonka bean – *Dipteryx odorata*
toothache tree – *Zanthoxylum americanum*
toothed lavender – *Lavandula dentata*
toothed spurge – *Euphorbia dentata*
tooth, elephant's – *Coryphantha elephantidens*
toothwort – *Cardamine pratensis, Lathraea squamaria*
top cactus – *Strombocactus disciformis*
topee tambu – *Calathea allouia*
torch ginger – *Etlingera elatior*
torch, gold – *Heliconia psittacorum*
torch-lily – *Kniphofia uvaria*
torchwood – *Amyris balsamifera*
torch, woolly – *Pilosocereus alensis*
tormentil – *Potentilla erecta*

tornillo – *Prosopis pubescens*
torpedo grass – *Panicum repens*
Torrey pine – *Pinus torreyana*
Torrey's crag-lily – *Anthericum torreyi*
Torrey yucca – *Yucca torreyi*
torus herb – *Dorstenia contrayerba*
totai palm – *Acrocomia totai*
totara – *Podocarpus totara*
totara, Chilean – *Podocarpus nubigenus*
touch-me-not – *Mimosa pudica*
touch-me-not, spotted – *Impatiens pallida*
Toumey oak – *Quercus toumeyi*
Toumey willow – *Salix bonplandiana*
tower mustard – *Arabis glabra*
toyon – *Heteromeles arbutifolia*
tragacanth – *Astragalus gummifer*
tragacanth, wild – *Astragalus tragacantha*
trailing abutilon – *Abutilon megapotamicum*
trailing begonia – *Begonia solananthera*
trailing four-o'clock – *Allionia incarnata*
trailing indigo bush – *Dalea greggii*
trailing petunia – *Petunia* x *hybrida* var *pendula*
Transvaal daisy – *Gerbera jamesonii*
traveler's joy – *Clematis vitalba*
traveler's palm – *Ravenala madagascariensis*
traveler's tree – *Ravenala madagascariensis*
treacle-mustard-like rocket – *Sisymbrium erysimoides*
treadsoftly – *Cnidoscolus*
treasure vine – *Hidalgoa ternata*
tree-anemone – *Carpenteria californica*
tree aralia – *Kalopanax septemlobus*
tree bear grass – *Nolina matapensis*
tree-beargrass, desert – *Nolina bigelovii*
tree-beargrass, Sonoran – *Nolina matapensis*
treebine – *Cissus sicyoides*
tree catclaw – *Acacia occidentalis*
tree celandine – *Bocconia arborea*
tree chile – *Capsicum annuum*
tree cholla – *Cylindropuntia imbricata*
tree dahlia – *Dahlia imperialis*
tree daisy – *Montanoa tomentosa*
tree datura – *Brugmansia*
tree fern – *Alsophila aculeata, A. salvinii, Cibotium regale, Cyathea delicatula, C. mexicana, Nephelium mexicana*
tree fern, dwarf – *Blechnum gibbum*
tree gloxinia – *Kohleria deppeana*
tree heath – *Erica arborea*
tree heather – *Erica arborea*
tree hibiscus – *Hibiscus tilliaceus*
tree lichen – *Sticta pulmonacea, Usnea barbata*
tree-like mugwort – *Artemisia arborescens*
tree limberbush, Sonoran – *Jatropha cordata*
tree lupin – *Lupinus arboreus*
tree mallow – *Lavatera arborea, L. maritima*
tree mallow, smaller – *Lavatera cretica*
tree morning-glory – *Ipomoea arborescens*
tree nightshade – *Solanum erianthum*
tree ocotillo – *Fouquieria macdougalii*
tree ocotillo, Mexican – *Fouquieria macdougalii*
tree of heaven – *Ailanthus altissima*
tree-of-kings – *Cordyline fruticosa*
tree onion – *Allium cepa* var *viviparum*
tree peony – *Paeonia suffructicosa*
tree pepper – *Capsicum annuum*
tree prickly-pear – *Opuntia ficus-indica*
tree sunflower – *Tithonia fruticosum*
tree tobacco – *Nicotiana glauca*
tree tomato – *Cyphomandra betacea*
tree yucca – *Yucca filifera, Y. valida*
trefoil, bean – *Anagyris foetida*
trefoil, birdfoot – *Lotus corniculatus*
trefoil, bird's-foot – *Lotus corniculatus*
trefoil, slender bird's-foot – *Lotus angustissimus*
trefoil, yellow – *Medicago lupulina*
tresses, ladies – *Spiranthes*
triandrus hybrid – *Narcissus triandrus*
triangle-leaf bur-sage – *Ambrosia deltoidea*
triangle-seed blazing star – *Mentzelia affinis*
tricholoma, clustered – *Lyophyllum aggregatum*
tricolor chrysanthemum – *Chrysanthemum carinatum*
tridens, rough – *Tridens elongatus*
tridens, slim – *Tridens muticus*
tridens, white – *Tridens albescens*
trident maple – *Acer buergeranum*
trifoliate orange – *Poncirus trifoliata*
trimbleberry – *Rubus occidentalis*
Trinidad falmebush – *Calliandra tweedii*
Trinidad, pride of – *Warszewiczia coccinea*
triticale – *Triticum secale*
trompetillo – *Bouvardia ternifolia*

trompillo – *Solanum eleagnifolium*
tropical alder – *Alnus acuminata*
tropical almond tree – *Terminalia catappa*
tropical birch – *Bursera simaruba*
tropical blueberry – *Conostegia xalapensis*
tropical cranberry – *Hibiscus sabdariffa*
tropical hibiscus – *Hibiscus rosa-sinensis*
tropical pine – *Pinus tropicalis*
tropical plum – *Spondias mombin*
tropical sage – *Salvia coccinea*
true myrtle – *Myrtus communis*
true service tree – *Sorbus domestica*
truffle – *Tuber*
truffle, black – *Tuber cibarium, T. melanosporum*
truffle, Burgundy – *Tuber uncinatum*
truffle, French – *Tuber melanosporum*
truffle, gray – *Tuber magnatum*
truffle, Italian – *Tuber magnatum*
truffle, Piedmont – *Tuber magnatum*
truffle, red – *Tuber rufum*
truffle, red-grained black – *Tuber aestivum*
truffle, summer – *Tuber aestivum*
truffle, violet – *Tuber brumale*
truffle, white – *Tuber album*
truffle, whitish – *Tuber borchii*
truffle, winter – *Tuber brumale*
Trujillo cocaine tree – *Erythroxylum truxillense*
trumpet, angel's – *Acleisanthes longiflora, Brugmansia*
trumpet, blood – *Distictis buccinatorius*
trumpet bush – *Tecoma stans*
trumpet creeper, Chinese – *Campsis grandiflora*
trumpet creeper, common – *Campsis radicans*
trumpet daffodil – *Narcissus bulbocodium*
trumpet, desert – *Eriogonum inflatum*
trumpeter's schomburgkia – *Laelia tibicinis*
trumpet flower – *Datura discolor, Incarvillea delavayi, Tecoma stans, Thevetia peruviana*
trumpet, golden – *Allamanda cathartica*
trumpet, herald's – *Beaumontia grandiflora*
trumpet honeysuckle – *Lonicera sempervirens*
trumpet, mountain angel's – *Brugmansia sanguinea*
trumpet plant – *Solandra maxima*
trumpet, rock – *Macrosiphonia macrosiphoni*
trumpets, angel – *Acleisanthes longiflora*
trumpet tree – *Cecropia obtusifolia, C. peltata, Tabebuia*
trumpet tree, pink – *Tabebuia impetiginosa, T. rosea*
trumpet tree, purple – *Tabebuia impetiginosa*
trumpet vine, blood-red – *Distictis buccinatoria*
trumpet vine, blue – *Thunbergia grandiflora*
trumpet vine, Mexican – *Distictis buccinatoria*
trumpet vine, orange – *Pyrostegia venusta*
trumpet vine, vanilla – *Distictis laxiflora*
trumpet vine, yellow – *Macfadyena unguis-cati*
trumpet, yellow – *Tecoma stans*
tuberose – *Polianthes tuberosa*
tuberous begonia – *Begonia gracilis, B. tuberosa*
tuberous vetch – *Lathyrus tuberosus*
tuberous woolly sage – *Phlomis tuberosa*
tucuma palm – *Astrocaryum vulgare*
tuft, candy – *Iberis*
tufted fescue – *Festuca amethystina*
tufted hairgrass – *Deschampsia cespitosa*
tufted loosestrife – *Lysimachia thyrsiflora*
tufted pansy – *Viola cornuta*
tufted saxifrage – *Saxifraga cespitosa*
tufted sedge – *Carex elata*
tula istle – *Agave heteracantha*
tule – *Scirpus*
tule potato – *Sagitaria latifolia*
tulip – *Tulipa*
tulip, candy – *Tulipa clusiana*
tulip, garden – *Tulipa gesneriana*
tulip, lady – *Tulipa clusiana*
tulip poppy, Mexican – *Hunnemannia fumariifolia*
tulip, rock – *Tulipa saxatilis*
tulip, Sharon – *Tulipa sharonensis*
tulip, southern – *Tulipa australis*
tulip, star – *Calochortus*
tulip tree – *Liriodendron tulipifera, Magnolia* x *soulangeana*
tulip tree, African – *Spathodea campanulata*
tulip, wild – *Tulipa sylvestris*
tumble mustard – *Sisymbrium altissimum, S. auriculatum*
tumbleweed – *Amaranthus albus, Salsola tragus*
tumbling oracle – *Atriplex rosea*
tung – *Aleurites fordii*

tung, Japanese – *Aleurites cordata*
tungoil tree – *Aleurites molucana*
tupelo – *Nyssa sylvatica*
turban ranunculus – *Ranunculus asiaticus*
turf, big blue lily – *Liriope muscari*
Turkestan allium – *Allium karataviense*
turkey foot – *Andropogon hallii*
turkey mullein – *Eremocarpus setigerus*
turkey oak – *Quercus cerris*
turkey pea – *Astragalus nuttallianus*
Turkish hazel – *Corylus colurna*
Turkish millet – *Panicum miliaceum*
Turkish pepper – *Capsicum annuum*
Turk's cap – *Malvaviscus arboreus, M. grandiflorus*
Turk's cap lily – *Lilium henryi, L. martagon*
Turk's head – *Ferocactus hamatacanthus*
Turk's head cactus – *Echinocactus horizonthalonius*
turmeric – *Curcuma domestica*
turnip – *Brassica rapa*
turnip greens – *Brassica rapa*
turnip rape, annual – *Brassica rapa* var *campestris*
turnip rape, biennial – *Brassica rapa* var *oleifera*
turnip rape, summer – *Brassica rapa* var *praecox*
turnip, wild – *Brassica tournefortii, Rapistrum rugosum*
turnsole – *Heliotropium indicum*
turpentine-broom – *Thamnosma montana*
turpentine-brush – *Ericameria laricifolia*
turpentine bush – *Ericameria laricifolia*
turpentine pine, French – *Pinus pinaster*
turpentine tree – *Pistacia terebinthus*
turtlegrass – *Thalassia testudinum*
tussock bellflower – *Campanula carpatica*
tussock sedge, lesser – *Carex diandra*
tutsan – *Hypericum androsaemum*
twayblade, heart-leaved – *Listera cordata*
twinberry – *Lonicera involucrata, Mitchella repens*
twincup, red – *Cephaelis tomentosa*
twin flower – *Linnaea borealis*
twin senna – *Senna pallida*
twin-spined cactus – *Mammillaria geminispina*
twisted heath – *Erica cinerea*
twisted-stalk – *Streptopus amplexifolius*
twisting spurge – *Chamaesyce serpens*
twitch, black – *Alopecurus myosuroides*
two-leaved senna – *Senna bahuinoides*

umbrella mallow – *Malva neglecta*
umbrella nutsedge – *Cyperus eragrostis*
umbrella pine – *Sciadopitys verticillata*
umbrella plant – *Cyperus alternifolius, C. involucratus*
umbrella tree – *Melia azedarach* var *umbraculiformis*
umbrella tree, Queensland – *Schefflera actinophylla*
umbrella tree, Texas – *Melia azedarach* var *umbraculiformis*
umbrella-wort – *Allionia incarnata*
unicorn plant – *Proboscidea fragrans*
unicorn plant, common – *Proboscidea louisianica*
unicorn plant, desert – *Proboscidea althaeifolia, P. parviflora*
uniola, broadleaf – *Uniola latifolia*
upland cotton – *Gossypium hirsutum*
upright coral fungus – *Ramaria stricta*
upright philodendron – *Philodendron bipinnatifidum*
urn plant – *Aechmea fasciata*
Utah agave – *Agave utahensis*
Utah cynanchum – *Cynanchum utahense*
Utah juniper – *Juniperus osteosperma*
uva grass – *Gynerium sagittatum*

Valais fescue – *Festuca valesiaca*
valerian – *Valeriana*
valerian, red – *Centranthus ruber*
valley oak – *Quercus lobata*
valley spurge – *Chamaesyce ocellata*
Valonia oak – *Quercus aegilops*
vanilla – *Vanilla planifolia*
vanilla, French – *Vanilla pompona*
vanilla grass – *Hierochloë odorata*
vanilla trumpet vine – *Distictis laxiflora*
vanilla, West Indian – *Vanilla pompona*
variegated euonymus – *Euonymus japonicus* var *aureo-variegatus*
variegated Japanese sedge – *Carex morrowii* var *expallida*
variegated laurel – *Codiaeum variegatum*
varnishbrush – *Flourensia cernua*

vase, silver – *Aechmea fasciata*
Vasey oak – *Quercus vaseyana*
vauquinia, slimleaf – *Vauquinia corymbosa*
vegetable oyster – *Tragopogon porrifolius*
veitch magnolia – *Magnolia* x *veitchii*
vejar fir – *Abies vejarii*
velvet ash – *Fraxinus velutina*
velvet bean – *Mucuna deeringiana, Stizolobium*
velvet bean, Florida – *Mucuna pruriens* var *utilis*
velvet cactus – *Bergerocactus emoryi*
velvet grass, common – *Holcus lanatus*
velvet grass, creeping – *Holcus mollis*
velvet leaf – *Wissadula amplissima*
velvetleaf, yellow – *Limnocharis flava*
velvet mesquite – *Prosopis velutina*
velvet milk cap – *Lactarius vellereus*
velvet opuntia – *Opuntia tomentosa*
velvet-pod mimosa – *Mimosa dysocarpa*
velvet sage – *Salvia leucantha*
velvet weed – *Gaura parviflora*
Venetian sumac – *Cotinus coggygria*
Venezuela, rose of – *Brownea grandiceps*
Venus fly trap – *Dionoea muscipula*
Venus' looking glass – *Specularia speculum*
Venus, mirror of – *Ophrys speculum*
Venus' navelwort – *Omphalodes linifolia*
Venus' needle – *Scandix pectin-veneris*
Venus tree – *Lonchocarpus hermannii*
verbena – *Verbena*
verbena, desert sand – *Abronia villosa*
verbena, garden – *Verbena* x *hybrida*
verbena, lemon – *Aloysia triphylla, Lippia triphylla, Phyla nodiflora*
verbena, low – *Verbena pumila*
verbena, sand – *Abronia*
verbena, shrub – *Lantana camara*
verbena, Wright's – *Verbena wrightii*
vernal grass, annual – *Anthoxanthum aristatum*
vernal grass, sweet – *Anthoxanthum odoratum*
vervain – *Verbena*
vervain, blue – *Verbena hastata, V. macdougali*
vervain, Dakota – *Verbena ambrosifolia*
vervain, desert – *Verbena wrightii*
vervain, European – *Verbena officinalis*
vervain, Mexican – *Verbena ciliata*
vervain, pink – *Verbena ambrosifolia, V. pumila*
vervain, slender – *Verbena halei*
vervain's mallow – *Malva alcea*
vervain, Texas – *Verbena halei*
vetch – *Vicia*
vetch, bird – *Vicia cracca*
vetch, bitter – *Vicia ervilia*
vetch, bush – *Vicia sepium*
vetch, common – *Vicia sativa* ssp *nigra*
vetch, crown – *Coronilla varia*
vetch, Cyprus – *Lathyrus ochrus*
vetch, deer – *Lotus repens*
vetch, fodder – *Vicia villosa*
vetch, French – *Vicia narbonensis*
vetch, hairy – *Vicia villosa*
vetch, Hungarian – *Vicia pannonica*
vetch, joint – *Aeschynomene*
vetch, kidney – *Anthyllis vulneraria*
vetchling – *Lathyrus vernus*
vetchling, maritime – *Lathyrus japonicus*
vetchling, red – *Lathyrus cicera*
vetchling, Spanish – *Lathyrus clymenum*
vetchling, yellow – *Lathyrus aphaca*
vetch, milk – *Astragalus*
vetch, Narbonne – *Vicia narbonensis*
vetch, narrow-leaved – *Vicia sativa* ssp *nigra*
vetch, one-flowered – *Vicia articulata*
vetch, purple – *Vicia benghalensis*
vetch, sparrow – *Vicia tetrasperma*
vetch, spring – *Vicia lathyroides, V. sativa*
vetch, tiny – *Vicia hirsuta*
vetch, tuberous – *Lathyrus tuberosus*
vetch, winter – *Vicia villosa*
vetch, yellow – *Vicia lutea*
vetiver – *Vetiveria zizanioides*
viburnum – *Viburnum farreri*
viburnum, leatherleaf – *Viburnum rhytidophyllum*
viburnum, sweet – *Viburnum awabukii*
Victoria grass – *Panicum repens*
Victorian box – *Pittosporum undulatum*
Victoria waterlily, Queen – *Victoria regia*
viguiera, San Diego county – *Viguiera laciniata*
vine bower – *Clematis viticella*
vinegar weed – *Trichostema*
vine, grape – *Vitis vinifera*
vine-mesquite – *Panicum obtusum*
vine milkweed – *Cynanchum acutum, C. utahense*

vine of Sodom – *Solanum sodomeum*
viola – *Viola cornuta, V. tricolor*
violet – *Viola arvensis*
violet, African – *Saintpaulia ionantha, S. jonantha*
violet, Australian – *Viola hederacea*
violet, bog – *Pinguicula vulgaris*
violet, dog – *Viola riviniana*
violet, dog-tooth – *Erythronium dens-canis*
violet, English – *Viola odorata*
violet, garden – *Viola* x *wittrockiana*
violet, horned – *Viola cornuta*
violet ivy – *Cobaea scandens*
violet, marsh – *Viola palustre*
violet meadow grass – *Poa violacea*
violet, Persian – *Exacum affine*
violet prickly-pear – *Opuntia macrocentra*
violet, sweet – *Viola odorata*
violettas – *Anoda cristata*
violet, Texas – *Salvia farinacea*
violet truffle – *Tuber brumale*
violet willow – *Salix daphnoides*
violet, yellow – *Viola flagelliformis*
viper's bugloss – *Echium plantagineum, E. vulgare*
viper's gourd – *Trichosanthes cucumerina* var *anguina*
viper's grass – *Scorzonera hispanica*
Virginia creeper – *Parthenocissus quinquefolia*
Virginian stock – *Malcolmia maritima*
Virginia peppergrass – *Lepidium virginicum*
Virginia snakeroot – *Aristolochia serpentaria*
virgin's bower – *Clematis*
virgin's bower, Texas – *Clematis drummondii*
virgin's bower, western – *Clematis ligusticifolia*
viridiflora tulip – *Tulipa gesneriana*
viscaria – *Silene coeli-rosa*
viznaga barrel cactus – *Ferocactus hamatacanthus*
viznagita, pink – *Echinomastus johnsonii*
viznagita, purple-spined – *Echinomastus erectocentrus*
viznagita, white-flowered – *Echinomastus intertextus*

wafer ash – *Ptelea trifoliata*
wafer parsnip – *Cymopterus purpurascens*
wait-a-bit – *Mimosa biuncifera*
wait-a-minute – *Mimosa aculeaticarpa, M. biuncifera*
wakerobin – *Trillium*
walkingstick cholla – *Cylindropuntia imbricata*
walk naked – *Eugenia capuli*
wall bedstraw – *Galium parisiense*
wallflower – *Erysimum cheiri, E. repandum*
wallflower, English – *Erysimum cheiri*
wallflower, Siberian – *Erysimum hieraciifolium*
wallflower, western – *Erysimum capitatum*
wall germander – *Teucrium* x *lucidrys*
Wallis fescue – *Festuca valesiaca*
wall rock cress – *Arabis caucasica*
wall rocket – *Diplotaxis erucoides, D. muralis, D. tenuifolia*
walnut – *Juglans*
walnut, Arizona – *Juglans major*
walnut, black – *Juglans nigra*
walnut, California black – *Juglans californica*
walnut, East Indian – *Albizzia lebbeck*
walnut, Ecuador – *Juglans honorei*
walnut, English – *Juglans regia*
walnut, Mexican – *Juglans microcarpa*
walnut, Persian – *Juglans regia*
wandering Jew – *Tradescantia fluminensis, T. zebrina*
wand mullein – *Verbascum virgatum*
wapato – *Sagittaria latifolia*
warree cohune – *Astrocaryum mexicanum*
wart-cress – *Coronopus squamatus*
wart-stemmed ceanothus – *Ceanothus verrucosus*
wartweed – *Euphorbia helioscopia*
warty carpetweed – *Kallstroemia parviflora*
Washington pelargonium, Lady – *Pelargonium* x *domesticum*
Washington pelargonium, Martha – *Pelargonium* x *domesticum*
Wasson's psilocybe – *Psilocybe muliericula*
water bent grass – *Agrostis viridis*
water betony – *Scrophularia auriculata*
water birch – *Betula occidentalis*
water blossom – *Astianthus viminalis*
water buttercup – *Ranunculus aquatilis*
water chile – *Capsicum annuum*
water chestnut – *Eleocharis dulcis, Trapa natans*
water chestnut of the West Indies – *Calathea lutea*

waxy corn – *Zea mays* var *ceratina*
wayfaring tree – *Viburnum lantana*
weak bluegrass – *Poa inferma*
weak knapweed – *Centaurea diluta*
weasel's snout – *Antirrhinum orontium*
weatherglass, poor man's – *Anagallis arvensis*
Weber blue agave – *Agave tequilana*
wedgegrass, prairie – *Sphenopholis obtusata*
wedge-scale – *Sphenopholis obtusata*
weed, green – *Genista*
weeping bamboo, Mexican – *Otatea acuminata*
weeping birch, Young's – *Betula pendula* var *youngii*
weeping bottlebrush – *Callistemom viminalis*
weeping Chinese banyan – *Ficus benjamina*
weeping fig – *Ficus benjamina*
weeping forsythia – *Forsythia suspensa*
weeping juniper – *Juniperus flaccida*
weeping lovegrass – *Eragrostis curvula*
weeping pine – *Pinus lumholtzii*
weeping silver lime – *Tilia* x *petiolaris*
weeping spruce, Brewer's – *Picea breweriana*
weeping willow – *Salix babylonica*
weeping willow, golden – *Salix alba* var *tristis*
weeping willow-leaved pear – *Pyrus salicifolia*
Welsh onion – *Allium fistulosum*
Welsh poppy – *Meconopsis cambrica*
West African ebony – *Diospyros mespiliformis*
western azalea – *Rhododendron occidentale*
western barberry – *Berberis fendleri*
western bistort – *Polygonium bistortoides*
western bleeding heart – *Dicentra formosa*
western blueberry – *Vaccinium uliginosum*
western blue flag – *Iris missouriensis*
western bristlecone pine – *Pinus longaeva*
western burning bush – *Euonymus occidentalis*
western chokecherry – *Prunus virginiana* var *demissa*
western coral bean – *Erythrina flabelliformis*
western cottonwood – *Populus fremontii*
western dogwood – *Cornus nuttallii*
western dwarf mistletoe – *Arceuthobium campylopodum*
western hackberry – *Celtis reticulata*
western hemlock – *Tsuga heterophylla*
western juniper – *Juniperus osteosperma*
western larch – *Larix occidentalis*
western marsh-rosemary – *Limonium californicum*
western mugweed – *Artemisia ludoviciana* ssp *albula*
western mugwort – *Artemisia ludoviciana*
western poison oak – *Toxicodendron diversilobum*
western redbud – *Cercis occidentalis*
western red cedar – *Juniperus scopulorum, Thuja plicata*
western sea-purslane – *Sesuvium verrucosum*
western soapberry – *Sapindus drummondii, S. saponaria*
western spiderwort – *Tradescantia occidentalis*
western spiraea – *Spiraea douglasii*
western sycamore – *Platanus racemosa*
western thoroughwort – *Ageratina herbacea*
western virgin's bower – *Clematis ligusticifolia*
western wallflower – *Erysimum capitatum*
western wheatgrass – *Pascopyrum smithii*
western white pine – *Pinus monticola*
western yarrow – *Achillea lanulosa*
western yellow pine – *Pinus arizonica, P. ponderosa*
West Indian bay – *Eugenia racemosa*
West Indian boxwood – *Gossypiospermum praecox*
West Indian buckthorn – *Rhamnus sphaerosperma*
West Indian cherry – *Malpighia emarginata, M. glabra*
West Indian elemi – *Bursera gummifera*
West Indian gherkin – *Cucumis anguria*
West Indian gooseberry – *Pereskia aculeata*
West Indian holly – *Turnera ulmifolia*
West Indian lime – *Citrus aurantifolia*
West Indian locust – *Hymenaea courbaril*
West Indian sourgrass – *Paspalum conjugatum*
West Indian vanilla – *Vanilla pompona*
West Indies, water chestnut of the – *Calathea lutea*
wheat – *Triticum aestivum, T. vulgare*
wheat grass – *Elytrigia, Pascopyrum smithii, Triticum aestivum*
wheatgrass, Arizona – *Elytrigia arizonicus*
wheatgrass, intermediate – *Elytrigia intermedia*
wheatgrass, slender – *Elytrigia trachycaulus*
wheatgrass, tall – *Elytrigia elongata*
wheatgrass, western – *Elytrigia smithii*
wheat, hard – *Triticum durum*
wheat, Indian – *Plantago*

wheat smut – *Tilletia tritici, Ustilago tritici*
wheat, winter – *Triticum aestivum*
wheel, fire – *Gaillardia pulchella*
wheelscale – *Atriplex elegans*
whim, needle – *Genista anglica*
whip, coach – *Fouquieria splendens*
Whipple's claw cactus – *Sclerocactus whipplei*
whisk, basket – *Desmoncus orthacanthos*
whiskered barrel cactus – *Ferocactus hamatacanthus*
whiskgrass muhly – *Muhlenbergia macroura*
whisk, wire – *Lygodium venustum*
whisky cactus – *Lophophora williamsii*
whisky, dry – *Lophophora williamsii*
whispering bells – *Emmenanthe penduliflora*
white alder – *Alnus rhombifolia*
white arum lily – *Zantedeschia aethiopica*
white ash – *Fraxinus americana*
white-ball acacia – *Acacia angustissima*
white banksia, double – *Rosa fortuniana*
whitebeam – *Sorbus*
whitebeam, Finnish – *Sorbus* x *hybrida*
whitebeam, Swedish – *Sorbus intermedia*
white bean, small – *Phaseolus vulgaris*
white beefwood – *Schoepfia schreberi*
white birch – *Betula alba*
white birch, common – *Betula pubescens*
white birch, European – *Betula pendula*
white bouvardia – *Bouvardia longiflora*
white bract stickleaf – *Mentzelia involucrata*
white brittlebush – *Encelia farinosa*
white brush – *Aloysia gratissima, A. wrightii, Lippia ligustrina*
white bryony – *Bryonia dioica*
white buffalo grass – *Panicum coloratum*
white bur-sage – *Ambrosia dumosa*
white buttonwood – *Laguncularia racemosa*
white cedar – *Chamaecyparis thyoides*
white cedar, northern – *Thuja occidentalis*
white clover – *Trifolium repens*
white clover, Dutch – *Trifolium repens*
white costus – *Costus speciosus*
white crucillo – *Ziziphus obtusifolia*
white cudweed – *Gnaphalium leucocephalum*
white dammar pine – *Agathis alba*
white elm, European – *Ulmus laevis*
white escallonia – *Escallonia illinita*
white everlasting, pink and – *Helipterum roseum*
white-felted germander – *Teucrium polium*
white fir – *Abies concolor*
white flag – *Iris germanica*
white-flowered gourd – *Lagenaria siceraria*
white-flowered prickly poppy – *Argemone gracilenta*
white-flowered viznagita – *Echinomastus intertextus*
white-flowered willow herb – *Epilobium lactiflorum*
white ginger – *Hedychium coronarium*
white ginger lily – *Hedychium coronarium*
white gumbolimbo – *Gilibertia arborea*
white-haired bristlegrass – *Setaria leucopila*
white-headed giant scabious – *Cephalaria leucantha*
white hellebore – *Veratrum album*
white horse-nettle – *Solanum eleagnifolium*
white jasmine, common – *Jasminum grandiflorum*
white jelly fungus – *Tremella fuciformis*
white knapweed – *Centaurea cineraria*
white laurel – *Nectandra*
white lead-tree – *Leucaena leucocephala*
whiteleaf manzanita – *Arctostaphylos viscida*
white-leaved rock-rose – *Cistus albidus*
white live oak – *Quercus oblongifolia*
white lupine – *Lupinus albus*
white mahogany – *Tabebuia donnell-smithii*
white mangrove – *Laguncularia racemosa*
white-margined nightshade – *Solanum marginatum*
whitemargin sandmat – *Chamaesyce albomarginata*
white mignonette – *Reseda alba*
white moho – *Malvaviscus arboreus*
white morning-glory, giant – *Turbina corymbosa*
whitemouth dayflower – *Commelina erecta*
white mugwort – *Artemisia lactiflora*
white mulberry – *Morus alba*
white mushroom – *Agaricus bitorquis*
white mustard – *Sinapis alba*
white nightshade – *Solanum americanum, S. triquetrum*
white oak – *Quercus*
white oak, Arizona – *Quercus arizonica*
white oak, California – *Quercus lobata*
white oak, eastern – *Quercus alba*

white oak, mountain – *Quercus grisea*
white oak, net-leaf – *Quercus polymorpha*
white oak, Rocky Mountain – *Quercus gambellii*
white oak, swamp – *Quercus bicolor*
white 'opiuma – *Pithecellobium dulce*
white pepper – *Piper nigrum*
white pine – *Pinus strobus*
white pine, Arizona – *Pinus ayacahuite*
white pine, Chiapas – *Pinus strobus* var *chiapensis*
white pine, eastern – *Pinus strobus*
white pine, false – *Pinus pseudostrobus*
white pine, Himalayan – *Pinus wallichiana*
white pine, Japanese – *Pinus parviflora*
white pine, Mexican – *Pinus ayacahuite, P. strobiformis*
white pine, northern – *Pinus strobus*
white pine, Southwestern – *Pinus strobiformis*
white pine, western – *Pinus monticola*
white popinac lead-tree – *Leucaena glauca*
white poplar – *Populus alba*
white quebracho – *Aspidosperma quebracho-blanco*
white ramon – *Trophis racemosa*
white rhatany – *Krameria grayi*
white rock-lettuce – *Pinaropappus roseus*
white rose of York – *Rosa alba*
white rose, wild – *Rosa bracteata*
white sage – *Artemisia ludoviciana* ssp *albula, Salvia apiana*
white sage, California – *Salvia apiana*
white sapote – *Casimiroa edulis*
white silk cotton plant – *Cochlospermum religiosum*
white snowdrops – *Galanthus nivalis*
white spruce – *Picea glauca*
whitestem blazing star – *Mentzelia albicaulis*
white-stemmed milkweed – *Asclepias albicans*
white-stemmed pondweed – *Potamogeton praelongus*
whitestem stickleaf – *Mentzelia albicaulis*
white swallow-wort – *Cynanchum vincetoxicum*
white sweetclover – *Melilotus alba*
white thistle – *Atriplex lentiformis*
whitethorn – *Acacia constricta*
whitethorn acacia – *Acacia constricta*
whitethorn, chaparral – *Ceanothus leucodermis*
whitethorn, mountain – *Ceanothus cordulatus*
white thyme – *Thymus mastichina*
whitetop – *Cardaria draba*
white tridens – *Tridens albescens*
white truffle – *Tuber album*
white waterlily – *Nymphaea alba, N. flavovirens, N. odorata*
white willow – *Salix alba*
whitey – *Mammillaria geminispina*
whitish oak – *Quercus candicans*
whitish sedge – *Carex canescens*
whitish truffle – *Tuber borchii*
whitlow grass – *Draba cuneifolia, Erophila verna*
whitlow-grass, spring – *Draba verna*
whitlow-grass, wood – *Draba nemorosa*
whitlow-grass, yellow – *Draba aizoides*
whitlow-wort – *Paronychia echinulata*
whorled clary – *Salvia verticillata*
whorled dock – *Rumex conglomeratus*
whorled dropseed – *Sporobolus pyramidatus*
whortleberry cactus – *Myrtillocactus geometrizans*
whortleberry, red – *Vaccinium vitis-idaea*
wide-spreading knapweed – *Centaurea diffusa*
wide-spread sweet-grass – *Puccinellia distans*
widow, mourning – *Geranium phaeum*
widow's tears – *Commelina erecta*
wild alfalfa – *Psoralea rhombifolia*
wild barley – *Hordeum murinum* ssp *glaucum*
wild-basil – *Clinopodium laevigatum, C. vulgare*
wild bay cedar – *Trema micrantha*
wild bergamot – *Monarda fistulosa*
wild blackberry – *Rubus trivialis*
wild black cherry – *Prunus serotina*
wild borage – *Symphytum asperum*
wild buckwheat – *Eriogonum*
wild buckwheat, roundleaf – *Eriogonum rotundifolium*
wild cainit – *Chrysophyllum bicolor*
wild cane – *Arthrostylidium racemiflora, Arundinaria longifolia, Gynerium sagittatum*
wild caraway – *Perideridia gairdneri*
wild carrot – *Daucus pusillus, Lomatium dissectum*
wild chamomile – *Matricaria recutita*
wild cherry – *Prunus capuli*
wild clary – *Salvia verbenacea*

wild clove – *Ludwigia octovalvis*
wild coco-plum – *Hirtella americana*
wild coffee – *Casearia sylvestris, Psychotria nervosa, Rinorea*
wild coriander – *Eryngium foetidum*
wild cotton tree – *Cochlospermum vitifolium, Hibiscus tilliaceus*
wild cranesbill – *Geranium maculatum*
wild cucumber – *Marah fabaceus, M. macrocarpus*
wild dahlia – *Dahlia rosea*
wild fig – *Ficus*
wild four-o'clock – *Mirabilis multiflora*
wild ginger – *Asarum europaeum*
wild goldenglow – *Bidens laevis*
wild gourd – *Cucurbita foetidissima*
wild grape – *Vitis tiliifolia*
wild grape, California – *Vitis californica*
wild hazel – *Simmondsia chinensis*
wild-heliotrope – *Phacelia distans*
wild hemp – *Cannabis ruderalis*
wild honeysuckle – *Guara coccinea, G. parviflora*
wild hop – *Bryonia alba*
wilding, California – *Satureja douglasii*
wild jasmine – *Philadelphus mexicanus*
wild leek – *Allium ampeloprasum*
wild licorice – *Astragalus glycyphyllos, Glycyrrhiza lepidota*
wild lilac – *Leucophyllum frutescens*
wild lily – *Milla biflora*
wild lime – *Casearia aculeata, Randia, Ximenia americana, Zanthoxylum fagara*
wild mango – *Irvingia gabonensis*
wild marjoram – *Origanum vulgare*
wild Mexican yam – *Dioscorea villosa*
wild mulberry – *Morus celtidifolia*
wild mushroom – *Basidiomycetes*
wild mustard – *Brassica juncea, Sinapis arvensis*
wild oat – *Avena fatua*
wild oat, slender – *Avena barbata*
wild oat, winter – *Avena sterilis*
wild olive – *Olea europaea*
wild onion – *Allium*
wild orange – *Esenbeckia hartmanii*
wild oregano – *Monarda menthaefolia*
wild oyster mushroom – *Pleurotus ostreatus* var *vulgaris*
wild pansy – *Viola pedunculata*
wild papaya – *Carica cauliflora, C. mexicana*
wild passion vine – *Passiflora incarnata*
wild pea – *Lathyrus*
wild pear – *Nectandra, Pyrus pyraster*
wild pepper – *Capsicum annuum* var *minimum, Piper aduncum*
wild persimmon – *Diospyros californica*
wild petunia – *Ruellia californica, R. nudiflora*
wild pine – *Aechmea lueddemanniana, Bromelia pinguin*
wild pineapple – *Bromelia pinguin*
wild plantain – *Heliconia bihai, H. caribaea*
wild plum – *Cyrtocarpa edulis, Prunus americana*
wild poinsettia – *Warszewiczia coccinea*
wild potato – *Solanum bulbocastanum*
wild radish – *Raphanus raphanistrum*
wild rhubarb – *Rumex hymenosepalus*
wild rice – *Zizania aquatica, Z. palustris* var *interior*
wild rose – *Rosa fendleri*
wildrye, Canadian – *Elymus canadensis*
wild sage – *Lantana camara, Salvia nemorosa, S. sylvestris*
wild sapodilla – *Morisonia americana*
wild senna – *Senna laevigata*
wild soursop – *Rollinia mucosa*
wild star apple – *Chrysophyllum mexicanum*
wild tamarind – *Leucaena glauca, Lysiloma bahamensis*
wild teasel – *Dipsacus fullonum*
wild thyme – *Thymus serpyllum*
wild tragacanth – *Astragalus tragacantha*
wild tulip – *Tulipa sylvestris*
wild turnip – *Brassica tournefortii, Rapistrum rugosum*
wild white rose – *Rosa bracteata*
Willard's acacia – *Acacia willardiana*
William, sweet – *Dianthus barbatus*
Willmott's ghost, Miss – *Eryngium giganteum*
willow – *Salix*
willow, Alaska blue – *Salix purpurea*
willow, arroyo – *Salix lasiolepis*
willow, bay – *Salix pentandra*
willow, Bebb – *Salix bebbiana*
willow, black – *Salix nigra*
willow, Bonpland – *Salix bonplandiana*

willow, California button – *Cephalanthus occidentalis* var *californicus*
willow, coyote – *Salix exigua*
willow, crack – *Salix fragilis*
willow, desert – *Chilopsis linearis* ssp *arcuata*
willow dock – *Rumex salicifolius*
willow, flowering – *Chilopsis linearis*
willow, French pussy – *Salix caprea*
willow gentian – *Gentiana asclepiadea*
willow, golden weeping – *Salix alba* var *tristis*
willow, Gooding – *Salix gooddingii*
willow, Gooding's black – *Salix gooddingii*
willow, gray – *Salix bebbiana, S. cinerea*
willow groundsel – *Senecio salignus*
willow, hankow – *Salix matsudana*
willow herb – *Lythrum tribracteatum, L. virgatum*
willow herb, alpine – *Epilobium anagallidifolium*
willow herb, American – *Epilobium ciliatum*
willow herb, great – *Epilobium hirsutum*
willow herb, marsh – *Epilobium palustre*
willow herb, pimpernel – *Epilobium anagallidifolium*
willow herb, white-flowered – *Epilobium lactiflorum*
willow, Jerusalem – *Eleagnus angustifolia*
willow-leaved catalpa – *Chilopsis linearis*
willow, narrow-leaved – *Salix exigua*
willow, net-leaf – *Salix reticulata*
willow oak – *Quercus phellos*
willow oak, Madrean – *Quercus viminea*
willow, peachleaf – *Salix amygdaloides*
willow, pink pussy – *Salix caprea*
willow pittosporum – *Pittosporum phillyraeoides*
willow, red – *Salix laevigata*
willow, sandbar – *Salix exigua*
willow, seep – *Baccharis salicifolia*
willow, silver – *Salix exilifolia*
willow, snow – *Salix reticulata*
willow, Southwestern peach – *Salix amygdaloides*
willow, toumey – *Salix bonplandiana*
willow, violet – *Salix daphnoides*
willow, water – *Baccharis salicifolia, Justicia spicigera*
willow weed – *Polygonum lapathifolium*
willow, weeping – *Salix babylonica*
willow, white – *Salix alba*
willow, woolly – *Salix lanata*
willow, Xochimilco – *Salix bonplandiana*
willow, yewleaf – *Salix taxifolia*
Wilson holly – *Ilex* x *altaclerensis*
wilt, tomato – *Fusarium solani*
windflower – *Anemone blanda*
windflower, scarlet – *Anemone* x *fulgens*
windlestraw – *Apera spica-venti*
windmill palm – *Trachycarpus fortunei*
windmills – *Allionia incarnata*
wineberry, Japanese – *Rubus phoenicolasius*
wine flower – *Boerhavia coccinea*
wine grape – *Vitis vinifera*
wine palm – *Borassus flabellifer*
wine palm, Chile – *Jubaea spectabilis*
wine palm, Chilean – *Jubaea chilensis*
wine palm, fishtail – *Caryota urens*
wine plantain – *Mus acuminata* x *balbisiana*
wine rue – *Ruta graveolens*
wine varietal grape – *Vitis vinifera* var *sativa*
wing, blue – *Torenia fournieri*
winged-bean – *Psophocarpus tetregonolobus*
winged cryptantha – *Cryptantha holoptera*
winged euonymus – *Euonymus alatus*
winged ragweed – *Hymenoclea monogyra*
winged yam – *Dioscorea alata*
wingnut, Chinese – *Pterocarya stenoptera*
wingnut, common – *Pterocarya fraxinifolia*
wing, ping – *Aechmea magdalenae, Hymenoxys richardsoni*
wing, pink – *Alvaradoa amorphoides*
wings, angel's – *Opuntia microdasys*
wingscale – *Atriplex canescens*
winter aconite – *Eranthis hyemalis*
winter-blooming bergenia – *Bergenia crassifolia*
winter cherry – *Physalis alkekengi, Solanum capsicastrum*
winter cress – *Barbarea praecox*
winter cress, common – *Barbarea vulgaris*
winter cress, early – *Barbarea verna*
winter currant, pink – *Ribes sanguineum*
winter daffodil – *Sternbergia lutea*
winter fat – *Krascheninnikovia lanata*
wintergreen – *Gaultheria acuminata, G. procumbens*
wintergreen barberry – *Berberis julianae*

wintergreen, green-flowered – *Pyrola chlorantha*
wintergreen, lesser – *Pyrola minor*
wintergreen, one-sided – *Orthilia secunda*
winter hazel, spike – *Corylopsis spicata*
winter iris – *Iris unguicularis*
winter jasmine – *Jasminum nudiflorum*
winter savory – *Satureja montana*
winter's bark – *Drimys winteri*
winter squash – *Cucurbita maxima, C. Moschata*
wintersweet – *Chimonanthus praecox*
winter truffle – *Tuber brumale*
winter vetch – *Vicia villosa*
winter wheat – *Triticum aestivum*
winter wild oat – *Avena sterilis*
wire-grass – *Juncus balticus, J. effuses*
wire lettuce, brownplume – *Stephanomeria pauciflora*
wire vine – *Muehlenbeckia platyclados*
wire whisk – *Lygodium venustum*
wishbone flower – *Torenia fournieri*
Wisley pink – *Helianthemum apenninum* var *roseum*
wisteria, Chinese – *Wisteria sinensis*
witchgrass – *Panicum capillare*
witchgrass, fall – *Leptoloma cognatum*
witch grass, old – *Panicum capillare*
witch hazel, Chinese – *Hamamelis mollis*
witch hazel, common – *Hamamelis virginiana*
witch hazel, Japanese – *Hamamelis japonica*
witch's hair – *Cuscuta denticulata*
withe, water – *Cissus sicyoides*
witloof – *Cichorium intybus*
woad – *Isatis tinctoria*
woad, dyer's – *Isatis tinctoria*
woadwaxen – *Genista tinctoria*
wolfberry – *Lycium berlandieri, L. chinese*
wolfberry, desert – *Lycium andersonii, L. fremontii*
wolfberry, pale – *Lycium pallidum*
wolfsbane – *Aconitum napellus*
wolftail – *Lycurus phleoides*
woman, old – *Artemisia stellerana*
wood, aloe – *Cordia sebestena*
wood anemone – *Anemone nemorosa*
wood avens – *Geum urbanum*
wood betony – *Pedicularis canadensis, Stachys officinalis*
woodbine – *Lonicera periclymenum*
wood, bitter – *Quassia amara*
wood blewit – *Lepista nuda*
wood, blue – *Condalia hookeri*
wood bluegrass – *Poa memoralis*
wood, Brazil – *Caesalpinia echinata, Haematoxylum brasiletto*
wood, bullet – *Achras bidentata, Mimusops darienensis*
wood, cam – *Haematoxylum campechianum*
wood, campeachy – *Haematoxylum campechianum*
wood, canary – *Centrolobium platinense*
wood chickweed – *Stellaria nemorum*
wood, cigarbox – *Cedrela odorata*
wood cranesbill – *Geranium sylvaticum*
wood, dog – *Cornus*
wood, eagle – *Aquilaria agallocha*
wood fern – *Dryopteris*
wood fern, California – *Dryopteris arguta*
wood fern, coastal – *Dryopteris arguta*
wood, fiddle – *Citharexylum scabrum, Vitex gaumeri*
wood forget-me-not – *Myosotis sylvatica*
wood, grease – *Larrea tridentata, Sarcobatus vermiculatus*
wood groundsel – *Senecio sylvaticus*
wood hyacinth – *Hyacinthoides hispanicus, H. non-scripta*
woodland grape – *Vitis sylvestris*
wood, lemon – *Calycophyllum candidissimum*
woodlouse cactus – *Pelecyphora aselliformis*
wood millet – *Milium effusum*
wood mushroom – *Agaricus silvicola*
wood, naked – *Colubrina viridis*
woodnymph – *Moneses uniflora*
wood, paradise – *Aquilaria agallocha*
wood, partridge – *Andira inermis*
wood pea – *Lathyrus linifolius*
wood, possum – *Hura polyandra*
wood, prince – *Exostema caribaeum*
wood, pump – *Cecropia obtusifolia, C. peltata, Oreopanax peltatum*
wood, red – *Sequoia sempervirens*
wood rose – *Merremia tuberosa*
woodruff, sweet – *Galium odoratum*
wood rush, hairy – *Luzula parviflora, L. spicata*
wood sage – *Teucrium scorodonia*

wood, sandal – *Pterocarpus santalinus, Santalum album*
wood sanicle – *Sanicula europaea*
woods, flame of the – *Ixora coccinea*
wood, slim – *Fouquieria splendens*
wood, snake – *Colubrina californica, C. ferruginosa*
wood sorrel – *Oxalis acetosella, O. albicans, O. latifolia*
wood sorrel, creeping – *Oxalis corniculata*
wood sorrel, yellow – *Oxalis dillenii*
wood, sour – *Oxydendrum arboreum*
wood, southern – *Artemisia abrotanum, A. campestris*
wood spurge – *Euphorbia amygdaloides*
woods rose – *Rosa woodsii*
wood stitchwort – *Stellaria nemorum*
wood strawberry – *Fragaria vesca*
wood, sweet – *Xylosma flexuosum*
wood, tallow – *Ximenia americana*
wood tobacco – *Nicotiana sylvestris*
wood, torch – *Amyris balsamifera*
wood whitlow-grass – *Draba nemorosa*
wood, worm – *Artemisia, Iva ambrosiaefolia*
wood, yellow – *Cladrastis kentukea*
woolflower, Chinese – *Celosia argentea*
woolly bluecurls – *Trichostema lanatum*
woolly butterfly bush – *Buddleja marrubifolia*
woolly desert marigold – *Baileya pleniradiata*
woolly distaff thistle – *Carthamus lanatus*
woolly Indian wheat – *Plantago patagonica*
woolly lavender – *Lavandula lanata*
woolly locoweed – *Astragalus mollissimus*
woolly marigold – *Baileya pleniradiata*
woolly milkweed – *Asclepias vestita*
woolly mullein – *Verbascum thapsus*
woolly sage, Italian – *Phlomis italica*
woolly sage, purple – *Phlomis purpurea*
woolly sage, Samos – *Phlomis samia*
woolly sage, tuberous – *Phlomis tuberosa*
woolly sea-blite – *Suadea taxifolia*
woolly senna – *Senna multiglandulosa*
woolly-stalked feathergrass – *Stipa pennata*
woolly torch – *Pilosocereus alensis*
woolly willow – *Salix lanata*
woolly yarrow – *Achillea tomentosa*
wool tree – *Bombax ellipticum*
Wooton hawthorn – *Crataegus wootoniana*
Wooton three-awn – *Aristida pansa*
wormseed – *Chenopodium ambrosioides*
wormseed mustard – *Erysimum cheiranthoides*
wormwood – *Artemisia, Iva ambrosiaefolia*
wormwood, beach – *Artemisia stellerana*
wormwood, common – *Artemisia absinthium*
wormwood, fringed – *Artemisia frigida*
wormwood, Roman – *Artemisia pontica*
wormwood, silver – *Artemisia ludoviciana*
wormwood, sweet – *Artemisia annua*
wort, awl – *Subularia aquatica*
wort, bane – *Ranunculus flammula*
wort, barren – *Epimedium alpinum*
wort, beach – *Batis maritima*
wort, birth – *Aristolochia*
wort, bishop's – *Nigella damascena, Stachys betonica, S. officinalis*
wort, bladder – *Urticularia*
wort, bridal – *Spiraea prunifolia*
wort, burst – *Herniaria glabra*
wort, butter – *Pinguicula caudate, P. vulgaris*
wort, cleaver – *Galium aparine*
wort, cross – *Crucianella angustifolia*
wort, dragon – *Dracunculus vulgaris*
wort, drop – *Filipendula vulgaris, Oenanthe*
wort, fel – *Swertia perennis*
wort, fig – *Scrophularia*
wort, flea – *Senecio integrifolius*
wort, gipsy – *Lycopus europaeus*
wort, glass – *Salicornia europaea*
wort, honey – *Cerinthe major*
wort, horn – *Ceratophyllum demersum*
wort, lead – *Plumbago auriculata, P. pulchella, P. scandens*
wort, liver – *Hepaticae*
wort, louse – *Pedicularis canadensis, Stachys betonica*
wort, lung – *Pulmonaria*
wort, mad – *Asperugo procumbens*
wort, marsh – *Apium nodiflorum*
wort, master – *Astrantia major, Peucedanum ostruthium*
wort, milk – *Glaux maritima, Polygala*
wort, moon – *Botrychium lunaria, Lunaria annua*
wort, mother – *Leonurus cardiaca*
wort, mud – *Limosella aquatica*
wort, mug – *Artemisia*
wort, naiad – *Najas marina*

wort, nail – *Parnassia palustris, Paronychia echinulata*
wort, navel – *Omphalodes linifolia*
wort, nipple – *Lapsana communis*
wort, pearl – *Sagina*
wort, penny – *Hydrocotyle*
wort, pepper – *Lepidium*
wort, pile – *Ranunculus ficaria*
wort, rag – *Senecio*
wort, rib – *Plantago lanceolatum*
wort, rupture – *Herniaria glabra, H. hirsuta*
wort, sage – *Artemisia norvegica*
wort, salt – *Batis maritima, Salsola*
wort, sand – *Arenaria, Minuartia, Sagina*
wort, saw – *Serratula tinctoria*
wort, sneeze – *Achillea ptarmica*
wort, soap – *Saponaria*
wort, spear – *Ranunculus flammula*
wort, spider – *Tradescantia*
wort, spleen – *Asplenium*
wort, star – *Callitriche, Spergula arvensis, Stellaria*
wort, stick – *Spergula arvensis*
wort, stitch – *Stellaria*
wort, St. John's – *Hypericum*
wort, strangle – *Cynanchum acutum*
wort, swallow – *Asclepias syriaca*
wort, thorough – *Ageratina, Eupatorium hyssopium*
wort, throat – *Campanula trachelium*
wort, thrum – *Damasonium alisma*
wort, tooth – *Cardamine pratensis, Lathraea squamaria*
wort, umbrella – *Allionia incarnata*
wort, white swallow – *Cynanchum vincetoxicum*
wort, whitlow – *Paronychia echinulata*
wort, wound – *Stachys*
woundwort, field – *Stachys arvensis*
woundwort, marsh – *Stachys palustris*
woven-spine pineapple cactus – *Echinomastus intertextus*
wreath, purple – *Petrea arborea, P. volubilis*
wreath, queen's – *Antigonon leptopus, Petrea volubilis*
wrinkled pea – *Pisum sativum*
wrinkle-leaf oak – *Quercus rugosa*
wych elm – *Ulmus glabra*
Wyoming paintbrush – *Castilleja linariifolia*

Xochimilco willow – *Salix bonplandiana*

yacon strawberry – *Polymnia edulis*
yam – *Colocasia esculenta, Dioscorea, Ipomoea batatus*
yam, aerial – *Dioscorea bulbifera*
yam bean – *Pachyrrhizus erosus*
yam bean, Mexican – *Pachyrhizus erosus*
yam, coco – *Colocasia antiquorum, C. esculenta*
yam, lesser – *Dioscorea esculenta*
yampah, Gairdner's – *Perideridia gairdneri*
yam, wild Mexican – *Dioscorea villosa*
yam, winged – *Dioscorea alata*
Yaqui fishhook – *Mammillaria tetrancistra*
yard grass – *Eleusine indica*
yard knotweed – *Polygonum aviculare*
yard-long bean – *Vigna unguiculata* ssp *sesquipedalis*
yarrow – *Achillea*
yarrow, fernleaf – *Achillea filipendulina*
yarrow, Greek – *Achillea ageratifolia*
yarrow, silvery – *Achillea clavennae*
yarrow, western – *Achillea lanulosa*
yarrow, woolly – *Achillea tomentosa*
yaupon – *Ilex vomitoria*
yaupon, desert – *Schaefferia cuneifolia*
yeast – *Saccharomyces*
yeast, bread – *Saccharomyces minor*
yecora pine – *Pinus yecorensis*
yellow alicoche – *Echinocereus papillosus*
yellow alder – *Turnera ulmifolia*
yellow allamanda – *Allamanda cathartica*
yellow allium – *Allium moly*
yellow amapa – *Tabebuia chrysantha*
yellow angel's trumpet – *Brugmansia aurea*
yellow archangel – *Lamium galeobdolon*
yellow asphodel – *Asphodeline lutea*
yellow bark – *Cinchona calisaya*
yellow bartsia – *Parentucella viscosa*
yellow bedstraw – *Galium verum*
yellow bell – *Allamanda cathartica*
yellow bells – *Tecoma stans*
yellow bird-of-paradise – *Caesalpinia gilliesii*
yellow bird's nest – *Monotropa hypopitys*
yellow bristlegrass – *Setaria pumila*
yellow-brown bolete – *Boletus luteus*
yellow bush lupine – *Lupinus arboreus*

yellow bush penstemon – *Keckiella antirrhinoides*
yellow burhead – *Limnocharis flava*
yellow chestnut oak – *Quercus muehlenbergii*
yellow corydalis – *Corydalis lutea*
yellow cosmos – *Cosmos sulphureus*
yellow cress, marsh – *Rorippa palustris*
yellow cypress – *Cupressus nootkatensis*
yellow daisy – *Hymenoxys scaposa*
yellow elder – *Tecoma stans*
yellow everlasting flower – *Helichrysum orientale*
yellow flag – *Iris pseudacorus*
yellowflag iris – *Iris pseudacorus*
yellow flame tree – *Peltophorum pterocarpum*
yellow-flowered pitaya – *Echinocereus dasyacanthus*
yellow foxglove – *Digitalis grandiflora*
yellow foxglove, small – *Digitalis lutea*
yellow-fruited rocoto pepper – *Capsicum pubescens*
yellow gentian – *Gentiana lutea*
yellow grass tree – *Astianthus viminalis*
yellowish scabious – *Scabiosa ochroleuca*
yellow jasmine – *Jasminum humile*
yellow lavender – *Lavandula viridis*
yellow lupine – *Lupinus luteus*
yellow mignonette – *Reseda lutea*
yellow mirabelle tree – *Prunus syriaca*
yellow mombin – *Spondias mombin*
yellow morning-glory – *Merremia aurea*
yellow night fungus – *Tricholoma flavovirens*
yellow nut-grass – *Cyperus esculentus*
yellow nutsedge – *Cyperus esculentus*
yellow oatgrass – *Trisetum flavescens*
yellow oleander – *Thevetia ovata, T. peruviana*
yellow orchid vine – *Mascagnia macroptera*
yellow palm – *Chrysalidocarpus lutescens*
yellow pheasant's eye – *Adonis vernalis*
yellow pine – *Pinus ponderosa, P. tenuifolia*
yellow pine, southern – *Pinus rigida*
yellow pine, western – *Pinus arizonica, P. ponderosa*
yellow poinciana – *Peltophorum pterocarpum*
yellow pond-lily – *Nuphar luteum* ssp *polysepalum*
yellow prickly poppy – *Argemone ochroleuca*
yellow rocket – *Barbarea vulgaris*
yellowroot – *Xanthorrhiza simplicissima*
yellow sapote – *Lucuma salicifolia, Pouteria campechiana*
yellow show, Arizona – *Amoreuxia palmatifida*
yellow shower – *Cassia fistula*
yellow silk tassel – *Garrya flavescens*
yellow star-thistle – *Centaurea solstitialis*
yellow sweet clover – *Melilotus indica, M officinalis*
yellow trefoil – *Medicago lupulina*
yellow trumpet – *Tecoma stans*
yellow trumpet vine – *Macfadyena unguis-cati*
yellow velvetleaf – *Limnocharis flava*
yellow vetch – *Vicia lutea*
yellow vetchling – *Lathyrus aphaca*
yellow violet – *Viola flagelliformis*
yellow waterlily – *Nuphar luteum, Nymphaea mexicana*
yellow waterweed – *Ludwigia peploides*
yellow whitlow-grass – *Draba aizoides*
yellow wood – *Cladrastis kentukea*
yellow wood sorrel – *Oxalis dillenii*
yerba buena – *Satureja douglasii*
yerba de chiva – *Clematis ligusticifolia*
yerba de selva – *Whipplea modesta*
yerba desierto – *Fendlerella utahensis*
yerba mansa – *Anemopsis californica*
yerba santa – *Eriodictyon*
yerba selva – *Whipplea modesta*
yesterday, today and tomorrow – *Brunfelsia pauciflora*
yew, English – *Taxus baccata*
yew, Irish – *Taxus baccata* var *fastigiata*
yew, Japanese – *Taxus cuspidata*
yewleaf willow – *Salix taxifolia*
yew, plum – *Cephalotaxus harringtonia*
yew, plum-fruited – *Podocarpus andinus*
ylang-ylang – *Cananga odorata*
yohimbe – *Corynanthe yohimbe*
yohimbine – *Corynanthe yohimbe*
Yorkshire fog – *Holcus lanatus*
yoshino flowering cherry – *Prunus* x *yedoensis*
Young's weeping birch – *Betula pendula* var *youngii*
youth-and-old-age – *Zinia elegans*
Yucatan sky flower – *Duranta erecta*
yucca – *Hesperaloe funifera, Yucca*
yucca, aloe – *Yucca aloifolia*
yucca, Arizona – *Yucca arizonica*
yucca, banana – *Yucca baccata*

yucca, beaked – *Yucca rostrata*
yucca, blue – *Agave arizonica, Yucca baccata, Y. rigida*
yucca, bluestem – *Yucca guatemalensis*
yucca, chaparral – *Yucca whipplei*
yucca, elephant – *Yucca guatemalensis*
yucca, giant – *Yucca elephantipes*
yucca, hoary – *Yucca schottii*
yucca, Mohave – *Yucca schidigera*
yucca, mountain – *Yucca schottii*
yucca, red – *Hesperaloe parviflora*
yucca, Schott's – *Yucca schottii*
yucca, soaptree – *Yucca elata*
yucca, soft-tip – *Yucca gloriosa*
yucca, spineless – *Yucca elephantipes*
yucca, Torrey – *Yucca torreyi*
yucca, tree – *Yucca filifera, Y. valida*
yulan magnolia – *Magnolia denudata*

Zanzibar aloe – *Aloe vera*
Zanzibar copal – *Trachylobium verrucosum*
zauschneria – *Epilobium canum*
zebra grass – *Miscanthus sinensis*
zebra plant – *Aphelandra squarrosa, Calathea zebrina*
zedoary – *Curcuma zedoaria*
zelkova, sawleaf – *Zelkova serrata*
zephyr flower – *Zephyranthes carinata*
zephyrlily, copper – *Zephyranthes longifolia*
zerecote – *Cordia sebestena*
zinnia – *Zinnia*
zinnia, desert – *Zinnia pumila*
zinnia, creeping – *Sanvitalia procumbems*
zinnia, dwarf – *Zinnia acerosa*
zinnia, garden – *Zinnia elegans*
zinnia, plains – *Zinnia grandiflora*
zinnia, red – *Zinnia tenuiflora*
zinnia, redstar – *Zinnia multiflora*
zinnia, spinyleaf – *Zinnia acerosa*
zorillo – *Choisya arizonica*
zucchini – *Cucurbita pepo*